Sets and Extensions in the Twentieth Century

Handbook of the History of Logic

Volume 6

Handbook of the History of Logic

Volume 6

Sets and Extensions in the Twentieth Century

Edited by

Dov M. Gabbay
Department of Computer Science
King's College London
Strand, London, WC2R 2LS, UK

Akihiro Kanamori
Department of Mathematics
Boston University
Boston, Massachusetts, USA

John Woods
Department of Philosophy
University of British Columbia
1866 Main Mall
Vancouver BC
Canada V6T 1Z1

ELSEVIER

AMSTERDAM • BOSTON • HEIDELBERG • LONDON • NEW YORK • OXFORD
PARIS • SAN DIEGO • SAN FRANCISCO • SINGAPORE • SYDNEY • TOKYO
North-Holland is an imprint of Elsevier

N·H

North-Holland is an imprint of Elsevier
Radarweg 29, PO Box 211, 1000 AE Amsterdam, The Netherlands
The Boulevard, Langford Lane, Kidlington, Oxford OX5 1GB, UK

First edition 2012

British Library Cataloguing in Publication Data
A catalogue record for this book is available from the British Library

Library of Congress Cataloging-in-Publication Data
A catalog record for this book is available from the Library of Congress

ISBN: 978-0-444-51621-3

For information on all North-Holland publications
visit our website at elsevierdirect.com

Transferred to Digital Printing in 2012

CONTENTS

PREFACE

Most of the chapters of this volume present the historical development of aspects of modern, ZFC set theory as a field of mathematics, and several chapters at the end, of categorical logic and type theory. Be that as it may, as part of the *Handbook of the History of Logic* the volume can be construed as the one focusing on *extensions vis-à-vis* the intension vs. extension distinction. That distinction, traditionally attributed to Antoine Arnauld in the 1662 *Logique de Port Royal*, is exemplified by "featherless biped", a conceptualization, and the corresponding extension, the collection of all *homo sapiens sapiens*. However this distinction was historically developed and worked in logic, it was in the mathematical development of set theory that extensions became explicit through the mathematization of infinitary concepts and the further development of mathematics, itself increasingly based on informal concepts of set.

How many points are there on the line? This would seem to be a fundamental, even primordial, question. However, to cast it as a *mathematical* question, underlying concepts would have to be invested with mathematical sense and a way of mathematical thinking provided that makes an answer possible, if not informative. First, the real numbers as representing points on the linear continuum would have to be described precisely and extensionally. A coherent concept of cardinality and cardinal number would have to be developed for infinite mathematical collections. And, the real numbers would have to be enumerated in such a way so as to accommodate this concept of cardinality. Georg Cantor made all of these moves as part of the seminal advances that have led to modern set theory. His Continuum Hypothesis would propose a specific, structured resolution about the size of the continuum in terms of his transfinite numbers, and the continuum problem, whether the Continuum Hypothesis holds or not, would henceforth stimulate the development of set theory as its major outstanding problem.

From Cantor's work would come forth new developments drawing out the intension vs. extension distinction, for analyzing his famous diagonal argument Bertrand Russell came in 1901 to his well-known paradox. Thus full "comprehension" was ruled out, as self-applicability precluded that every intension corresponds to a viable extension. Exercised by the implications for any theory of extensions, Russell the artful dodger eventually developed his theory of types, a complicated formal system of logic for mathematics featuring type distinctions for functional application as well as a further ramification into orders. He thus enshrined non-self-applicability, but in his aspiring for universality he had to build in encumbrances like the Axiom of Reducibility. Russell's theory stood as a remarkable achievement of mathematical logic into the 1920s, when David Hilbert and Paul Bernays focused attention on the restricted, first-order logic as already the provenance of significant issues and Frank Ramsey directed criticism at the intensional aspects of the theory as alien to mathematics.

During this period, set theory was crucially advanced in coordination with emerging mathematical practice. Ernst Zermelo effected a first transmutation of the concept of set with his first, 1908 axiomatization, which made explicit principles like the existence of an infinite set, the power set of any set, and the Axiom of Separation, a natural, restricted form of comprehension. That axiom together with the Russell paradox argument established Zermelo's first theorem, that the universe is not a set, so that "is a set" has no extension. In the 1920s John von Neumann incorporated the transfinite numbers as sets, his ordinals, drawing in the Axiom of Replacement, and promoted the use of proper classes, like the universe, as intrinsic to a coherent axiomatic framework. In Zermelo's final, 1930 axiomatization the now-standard set theory ZFC is recognizable.

It was Kurt Gödel's 1930s work on the constructible universe L that launched set theory on an independent course as a distinctive field of mathematics. With L he established the relative consistency of the Continuum Hypothesis, synthesizing what had come before in axiomatic set theory with first-order logic through the infusion of metamathematical methods. Through definability, intensionality became woven into a fully extensionalized context. Gödel himself regarded his work on L as a transfinite extension of Russell's types and a rectification of the ramified theory, but in any case set theory would henceforth proceed with a further transmutation of the concept of set as stratified into a cumulative hierarchy.

Having become a theory of extensions *par excellence* in coordination with the mathematization of logic, set theory began its transformation into a sophisticated, autonomous field of mathematics with Paul Cohen's 1963 creation of forcing, with which he established the independence of the Continuum Hypothesis. Forcing quickly became a general method; a myriad of propositions and models were investigated; and lasting structural incentives were put into place.

The first chapter of this volume describes the historical development of set theory up to this point. Its length and detail, as well as those of the other set-theoretic chapters, speaks to the extent and richness of modern set theory. The chapter on the continuum reaches back to Cantor's original incursions and engagingly proceeds forward to how the arithmetical view of the continuum as a collection of points has led to the investigation of combinatorial structures and cardinalities. The chapter on infinite combinatorics describes, from its early beginnings, the direct investigation of the Cantorian transfinite as extension of number, mainly ordering and partition properties. Bringing the subject up through the 20th Century, the chapter draws in the wide-ranging enrichments resulting from the infusion of modern methods and initiatives. Its range and reach speaks both to the author's enthusiasm and expertise and to the remarkable and myriad ways in which set theory has expanded and transmuted.

The next several chapters deal with those aspects of modern set theory that came into prominence largely after the great expansion of the subject since 1963. The chapter on large cardinals with forcing brings to the fore the mainstream investigation of "strong axioms of infinity" and their intimate relationship with strong propositions through relative consistency, developed through the complementary methods of forcing and inner models. The articulation and analysis of inner models of set theory has itself become an abiding, expansive subject through the infusion of the "fine structure" definability method and pivotal "covering theorems". The next chapter, written by one of the principals involved, proves an informative account of the development of the subject through its self-fueling initiatives.

The investigation of the "determinacy" of infinite games is the most distinctive and intriguing development of modern set theory, and the deep correlations achieved with large cardinals and inner models the most remarkable and synthetic. The chapter on determinacy provides a detailed history and compendium of results that chronicles the emergence of the subject as a central theme of modern set theory.

The chapter on singular cardinals describes the most recent major initiative in modern set theory, singular cardinal combinatorics. A critical point was reached in set theory when singular cardinals, owing to their fast approachability from below, provide the setting and motivation for fresh, combinatorial arguments and sophisticated relative consistency results that have articulated the transfinite landscape in new and distinctive ways.

Are there alternatives to the new-standard ZFC set theory? None has led to a comparable, sustained development, but many have arisen owing to a variety of motivations, and their comparative study have generated interesting mathematical analysis and problems, as described in the chapter on alternative set theories.

The alternative to set theory as such for developing extensions had been, from the time of Russell, to have explicit types. But Russell's original theory was complicated and artificial as a matter of logic, and set-theoretic, extensional thinking became basic to mathematics. Types would come to the fore again however, as new ingredients within mathematics and as part of system building, with the development of category theory and the emergence of computer science. The objects of a category can be taken to be types and the arrows, mappings between the corresponding types. The chapter on sets, types, and categories provides a sweeping, integrative view through to the working out of the "propositions-as-types" doctrine in Martin-Löf's constructive dependent type theory. The chapter on categorical logic provides a detailed, tightly-knit account of the work of the "Montreal school" of F. William Lawvere, André Joyal, and others. These categorical developments have secured theories of extensions separate from set theory and more directly embeddable into the motivational frameworks of mathematical logic. Finally, going full circle back to Russell, the last chapter of this volume discusses uses of Russell's orders, his original ramification of types, in modern logic and computer science.

For support and encouragement we warmly thank the following persons: Professor Paul Bartha, Head of the Department of Philosophy, and Professor Nancy Gallini, Dean of Arts, at the University of British Columbia; Professor Michael Stingl, Chair of the Department of Philosophy, and Professor Christopher Nicol, Dean of Arts and Science, at the University of Lethbridge; Jane Spurr, Publications Administrator in London; Carol Woods, Production Associate in Vancouver; and our colleagues at Elsevier, Lauren Schultz and Derek Coleman.

As a final note, Aki Kanamori would like to express a special gratitude to the Lichtenberg-Kolleg at Göttingen. Awarded an inaugural 2009-2010 fellowship there, he was able to carry out editorial work in a particularly supportive environment at the Gauss Sternwarte.

The Editors

CONTRIBUTORS

John L. Bell
University of Western Ontario, Canada.
jbell@uwo.ca
Robert Constable
Cornell University, USA.
rc@cs.cornell.edu
Thomas Forster
University of Cambridge, UK.
tf@dpmms.cam.ac.uk
M. Randall Holmes
Boise State University, USA.
holmes@math.boisestate.edu
Akihiro Kanamori
Boston University, USA.
aki@math.bu.edu
Fairouz Kamareddine
Heriot-Watt University, Scotland, UK.
fairouz@macs.hw.ac.uk
Menachem Kojman
Ben Gurion University, Israel.
kojman@cs.bgu.ac.il
Twan Laan
Switzerland.
twan.laan@gmail.com
Jean A. Larson
University of Florida, USA.
jal@math.ufl.edu
Paul B. Larson
Miami University, USA.
larsonpb@muohio.edu
Thierry Libert
Université Libre de Bruxelles, Belgium.
tlibert@ulb.ac.be

Jean-Pierre Marquis
Université de Montréal, Canada.
jean-pierre.marquis@umontreal.ca
William J. Mitchell
University of Florida, USA.
wjm@math.ufl.edu
Gonzalo E. Reyes
Université de Montréal, Canada.
reyes@dms.umontreal.ca
Juris Steprāns
York University, Canada.
steprans@yorku.ca

SET THEORY FROM CANTOR TO COHEN

Akihiro Kanamori

Set theory is an autonomous and sophisticated field of mathematics, enormously successful not only at its continuing development of its historical heritage but also at analyzing mathematical propositions and gauging their consistency strength. But set theory is also distinguished by having begun intertwined with pronounced metaphysical attitudes, and these have even been regarded as crucial by some of its great developers. This has encouraged the exaggeration of crises in foundations and of metaphysical doctrines in general. However, set theory has proceeded in the opposite direction, from a web of intensions to a theory of extension *par excellence*, and like other fields of mathematics its vitality and progress have depended on a steadily growing core of mathematical proofs and methods, problems and results. There is also the stronger contention that from the beginning set theory actually developed through a progression of *mathematical* moves, whatever and sometimes in spite of what has been claimed on its behalf.

What follows is an account of the development of set theory from its beginnings through the creation of forcing based on these contentions, with an avowedly Whiggish emphasis on the heritage that has been retained and developed by the current theory. The whole transfinite landscape can be viewed as having been articulated by Cantor in significant part to solve the Continuum Problem. Zermelo's axioms can be construed as clarifying the set existence commitments of a single proof, of his Well-Ordering Theorem. Set theory is a particular case of a field of mathematics in which seminal proofs and pivotal problems actually shaped the basic concepts and forged axiomatizations, these transmuting the very notion of set. There were two main junctures, the first being when Zermelo through his axiomatization shifted the notion of set from Cantor's range of inherently structured sets to sets solely structured by membership and governed and generated by axioms. The second juncture was when the Replacement and Foundation Axioms were adjoined and a first-order setting was established; thus transfinite recursion was incorporated and results about all sets could established through these means, including results about definability and inner models. With the emergence of the cumulative hierarchy picture, set theory can be regarded as becoming a theory of well-foundedness, later to expand to a study of consistency strength. Throughout, the subject has not only been sustained by the axiomatic tradition through Gödel and Cohen but also fueled by Cantor's two legacies, the extension of number into the transfinite as transmuted into the theory of large cardinals and the investigation of definable sets of reals as transmuted into descriptive set theory. All this

Handbook of the History of Logic: Sets and Extensions in the Twentieth Century.
Volume editor: Akihiro Kanamori. General editors: Dov M. Gabbay, Paul Thagard and John Woods.

can be regarded as having a historical and mathematical logic internal to set theory, one that is often misrepresented at critical junctures in textbooks (as will be pointed out). This view, from inside set theory and about itself, serves to shift the focus to those tensions and strategies familiar to mathematicians as well as to those moves, often made without much fanfare and sometimes merely linguistic, that have led to the crucial advances.

1 CANTOR

1.1 Real numbers and countability

Set theory had its beginnings in the great 19th-Century transformation of mathematics, a transformation beginning in analysis. Since the creation of the calculus by Newton and Leibniz the function concept had been steadily extended from analytic expressions toward arbitrary correspondences. The first major expansion had been inspired by the explorations of Euler in the 18th Century and featured the infusion of infinite series methods and the analysis of physical phenomena, like the vibrating string. In the 19th-Century the stress brought on by the unbridled use of series of functions led first Cauchy and then Weierstrass to articulate convergence and continuity. With infinitesimals replaced by the limit concept and that cast in the ϵ-δ language, a level of deductive rigor was incorporated into mathematics that had been absent for two millenia. Sense for the new functions given in terms of infinite series could only be developed through carefully specified deductive procedures, and proof reemerged as an extension of algebraic calculation and became basic to mathematics in general, promoting new abstractions and generalizations.

Working out of this tradition Georg Cantor[1](1845–1918) in 1870 established a basic uniqueness theorem for trigonometric series: If such a series converges to zero everywhere, then all of its coefficients are zero. To generalize Cantor [1872] started to allow points at which convergence fails, getting to the following formulation: For a collection P of real numbers, let P' be the collection of limit points of P, and $P^{(n)}$ the result of n iterations of this operation. *If a trigonometric series converges to zero everywhere except on a P where $P^{(n)}$ is empty for some n, then all of its coefficients are zero.*[2]

It was in [1872] that Cantor provided his formulation of the real numbers in terms of fundamental sequences of rational numbers, and significantly, this was for the specific purpose of articulating his proof. With the new results of analysis to be secured by proof and proof in turn to be based on prior principles the regress led in the early 1870s to the appearance of several independent formulations of the real numbers in terms of the rational numbers. It is at first quite striking that the real numbers came to be developed so late, but this can be viewed as part of

[1] Dauben [1979], Meschkowski [1983], and Purkert-Ilgauds [1987] are mathematical biographies of Cantor.

[2] See Kechris-Louveau [1987] for recent developments in the Cantorian spirit about uniqueness for trigonometric series converging on definable sets of reals.

the expansion of the function concept which shifted the emphasis from the continuum taken as a whole to its extensional construal as a collection of objects. In mathematics objects have been traditionally introduced only with reluctance, but a more arithmetical rather than geometrical approach to the continuum became necessary for the articulation of proofs.

The other well-known formulation of the real numbers is due to Richard Dedekind [1872], through his cuts. Cantor and Dedekind maintained a fruitful correspondence, especially during the 1870s, in which Cantor aired many of his results and speculations.[3] The formulations of the real numbers advanced three important predispositions for set theory: the consideration of infinite collections, their construal as unitary objects, and the encompassing of arbitrary such possibilities. Dedekind [1871] had in fact made these moves in his creation of ideals, infinite collections of algebraic numbers,[4] and there is an evident similarity between ideals and cuts in the creation of new numbers out of old.[5] The algebraic numbers would soon be the focus of a major breakthrough by Cantor. Although both Cantor and Dedekind carried out an arithmetical reduction of the continuum, they each accommodated its antecedent geometric sense by asserting that each of their real numbers actually corresponds to a point on the line. Neither theft nor honest toil sufficed; Cantor [1872: 128] and Dedekind [1872: III] recognized the need for an *axiom* to this effect, a sort of Church's Thesis of adequacy for the new construal of the continuum as a collection of objects.

Cantor recalled[6] that around this time he was already considering infinite iterations of his P' operation using "symbols of infinity":

$$P^{(\infty)} = \bigcap_{n}^{\infty} P^{(n)}, P^{(\infty+1)} = P^{(\infty)\prime}, P^{(\infty+2)}, \ldots P^{(\infty\cdot 2)}, \ldots P^{(\infty^2)}, \ldots P^{(\infty^\infty)}, \ldots$$

In a crucial conceptual move he began to investigate infinite collections of real numbers and infinitary enumerations for their own sake, and this led first to a basic articulation of size for the continuum and then to a new, encompassing theory of counting. Set theory was born on that December 1873 day when Cantor established

[3]The most complete edition of Cantor's correspondence is Meschkowski-Nilson [1991]. Excerpts from the Cantor-Dedekind correspondence from 1872 through 1882 were published in Noether-Cavaillès [1937], and excerpts from the 1899 correspondence were published by Zermelo in the collected works of Cantor [1932]. English translations of the Noether-Cavaillès excerpts were published in Ewald [1996: 843ff.]. An English translation of a Zermelo excerpt (retaining his several errors of transcription) appeared in van Heijenoort [1967: 113ff.]. English translations of Cantor's 1899 correspondence with both Dedekind and Hilbert were published in Ewald [1996: 926ff.].

[4]The algebraic numbers are those real numbers that are the roots of polynomials with integer coefficients.

[5]Dedekind [1872] dated his conception of cuts to 1858, and antecedents to ideals in his work were also entertained around then. For Dedekind and the foundation of mathematics see Dugac [1976] and Ferreirós [2007], who both accord him a crucial role in the development of the framework of set theory.

[6]See his [1880: 358].

that *the real numbers are uncountable.*[7] In the next decades the subject was to blossom through the prodigious progress made by him in the theory of transfinite and cardinal numbers.

The uncountability of the reals was established, of course, via *reductio ad absurdum* as with the irrationality of $\sqrt{2}$. Both impossibility results epitomize how a *reductio* can compel a larger mathematical context allowing for the deniability of hitherto implicit properties. Be that as it may, Cantor the mathematician addressed a specific *problem*, embedded in the mathematics of the time, in his seminal [1874] entitled "On a property of the totality of all real algebraic numbers". After first establishing this property, the countability of the algebraic numbers, Cantor then established: *For any (countable) sequence of reals, every interval contains a real not in the sequence.* Cantor appealed to the order completeness of the reals:

Suppose that s is a sequence of reals and I an interval. Let $a < b$ be the first two reals of s, if any, in I. Then let $a' < b'$ be the first two reals of s, if any, in the open interval (a, b); $a'' < b''$ the first two reals of s, if any, in (a', b'); and so forth. Then however long this process continues, the (non-empty) intersection of these nested intervals cannot contain any member of s.

By this means Cantor provided a new proof of Joseph Liouville's result [1844, 1851] that there are transcendental numbers (real non-algebraic numbers) and only afterward did Cantor point out the uncountability of the reals altogether. This presentation is suggestive of Cantor's natural caution in overstepping mathematical sense at the time.[8]

Accounts of Cantor's work have mostly reversed the order for deducing the existence of transcendental numbers, establishing first the uncountability of the reals and only then drawing the existence conclusion from the countability of the algebraic numbers.[9] In textbooks the inversion may be inevitable, but this has promoted the misconception that Cantor's arguments are non-constructive.[10] It depends how one takes a proof, and Cantor's arguments have been implemented

[7]A set is *countable* if there is a bijective correspondence between it and the natural numbers $\{0, 1, 2, \ldots\}$. The exact date of birth can be ascertained as December 7. Cantor first gave a proof of the uncountability of the reals in a letter to Dedekind of 7 December 1873 (Ewald [1996: 845ff]), professing that ". . . only today do I believe myself to have finished with the thing . . .".

[8]Dauben [1979: 68ff] suggests that the title and presentation of Cantor [1874] were deliberately chosen to avoid censure by Kronecker, one of the journal editors.

[9]Indeed, this is where Wittgenstein [1956: I,Appendix II, 1-3] located what he took to be the problematic aspects of the talk of uncountability.

[10]A non-constructive proof typically deduces the existence of a mathematical object without providing a means for specifying it. Kac-Ulam [1968: 13] wrote: "The contrast between the methods of Liouville and Cantor is striking, and these methods provide excellent illustrations of two vastly different approaches toward proving the *existence* of mathematical objects. Liouville's is purely *constructive*; Cantor's is purely *existential*." See also Moore [1982: 39]. One exception to the misleading trend is Fraenkel [1930: 237][1953: 75], who from the beginning emphasized the constructive aspect of diagonalization.

The first non-constructive proof widely acknowledged as such was Hilbert's [1890] of his basis theorem. Earlier, Dedekind [1888: §159] had established the equivalence of two notions of being finite with a non-constructive proof that made an implicit use of the Axiom of Choice.

as algorithms to generate the successive digits of new reals.[11]

1.2 Continuum Hypothesis and transfinite numbers

By his next publication [1878] Cantor had shifted the weight to getting bijective correspondences, stipulating that two sets have the same *power* [Mächtigkeit] *iff* there is such a correspondence between them, and established that the reals $\mathbb{R}$ and the n-dimensional spaces $\mathbb{R}^n$ all have the same power. Having made the initial breach in [1874] with a negative result about the *lack* of a bijective correspondence, Cantor secured the new ground with a positive investigation of the *possibilities* for having such correspondences.[12] With "sequence" tied traditionally to countability through the indexing, Cantor used "correspondence [Beziehung]". Just as the discovery of the irrational numbers had led to one of the great achievements of Greek mathematics, Eudoxus's theory of geometrical proportions presented in Book V of Euclid's *Elements* and thematically antecedent to Dedekind's [1872] cuts, Cantor began his move toward a full-blown mathematical theory of the infinite.

Although holding the promise of a rewarding investigation Cantor did not come to any powers for infinite sets other than the two as set out in his [1874] proof. Cantor claimed at the end of [1878: 257]:

> Every infinite set of reals either is countable or has the power of the continuum.

This was the *Continuum Hypothesis* (CH) in the nascent context. The conjecture viewed as a primordial question would stimulate Cantor not only to approach the reals *qua* extensionalized continuum in an increasingly arithmetical fashion but also to grapple with fundamental questions of set existence. His triumphs across a new mathematical context would be like a brilliant light to entice others into

---

[11] Gray [1994] shows that Cantor's original [1874] argument can be implemented by an algorithm that generates the first n digits of a transcendental number with time complexity $O(2^{n^{1/3}})$, and his later diagonal argument, with a tractable algorithm of complexity $O(n^2 \log^2 n \log\log n)$. The original Liouville argument depended on a simple observation about fast convergence, and the digits of the Liouville numbers can be generated much faster. In terms of 2.3 below, the later Baire Category Theorem can be viewed as a direct generalization of Cantor's [1874] result, and the collection of Liouville numbers provides an explicit example of a co-meager yet measure zero set of reals (see Oxtoby [1971: §2]). On the other hand, Gray [1994] shows that every transcendental real is the result of diagonalization applied to *some* enumeration of the algebraic reals.

[12] Cantor developed a bijective correspondence between $\mathbb{R}^2$ and $\mathbb{R}$ by essentially interweaving the decimal expansions of a pair of reals to define the associated real, taking care of the countably many exceptional points like $.100\ldots = .099\ldots$ by an *ad hoc* shuffling procedure. Such an argument now seems straightforward, but to have bijectively identified the plane with the line was a stunning accomplishment at the time. In a letter to Dedekind of 29 June 1877 Cantor (Ewald [1996: 860]) wrote, in French in the text, "I see it, but I don't believe it."

Cantor's work inspired a push to establish the "invariance of dimension", that there can be no *continuous* bijection of any $\mathbb{R}^n$ onto $\mathbb{R}^m$ for $m < n$, with Cantor [1879] himself providing an argument. As topology developed, the stress brought on by the lack of firm ground led Brouwer [1911] to definitively establish the invariance of dimension in a seminal paper for algebraic topology.

the study of the infinite, but his inability to establish CH would also cast a long shadow. Set theory had its beginnings not as some abstract foundation for mathematics but rather as a setting for the articulation and solution of the *Continuum Problem*: to determine whether there are more than two powers embedded in the continuum.

In his magisterial *Grundlagen* [1883] Cantor developed the *transfinite numbers* [Anzahlen] and the key concept of *well-ordering*. A *well-ordering* of a set is a linear ordering of it according to which every non-empty subset has a least element. No longer was the infinitary indexing of his trigonometric series investigations mere contrivance. The "symbols of infinity" became autonomous and extended as the transfinite numbers, the emergence signified by the notational switch from the ∞ of potentiality to the ω of completion as the last letter of the Greek alphabet. With this the progression of transfinite numbers could be depicted:

$$0, 1, 2, \ldots \omega, \omega + 1, \omega + 2, \ldots, \omega + \omega (= \omega \cdot 2), \ldots, \omega^2, \ldots, \omega^\omega, \ldots, \omega^{\omega^\omega}, \ldots$$

A corresponding transition from subsets of $\mathbb{R}^n$ to a broader concept of set was signaled by the shift in terminology from "point-manifold [Punktmannigfaltigkeit]" to "set [Menge]". In this new setting well-orderings conveyed the sense of sequential counting and transfinite numbers served as standards for gauging well-orderings.

As Cantor pointed out, every linear ordering of a finite set is already a well-ordering and all such orderings are isomorphic, so that the general sense is only brought out by infinite sets, for which there are non-isomorphic well-orderings. Cantor called the set of natural numbers the first number class (I) and the set of numbers whose predecessors are countable the second number class (II). Cantor conceived of (II) as being bounded above according to a limitation principle and showed that (II) itself is not countable. Proceeding upward, Cantor called the set of numbers whose predecessors are in bijective correspondence with (II) the third number class (III), and so forth. Cantor took a set to be of a *higher power* than another if they are not of the same power yet the latter is of the same power as a subset of the former. Cantor thus conceived of ever higher powers as represented by number classes and moreover took every power to be so represented. With this "free creation" of numbers, Cantor [1883: 550] propounded a basic principle that was to drive the analysis of sets:

> "It is always possible to bring any *well-defined* set into the form of a *well-ordered* set."

He regarded this as a "an especially remarkable law of thought which through its general validity is fundamental and rich in consequences." Sets are to be well-ordered, and thus they and their powers are to be gauged via the transfinite numbers of his structured conception of the infinite.

The well-ordering principle was consistent with Cantor's basic view in the *Grundlagen* that the finite and the transfinite are all of a piece and uniformly comprehendable in mathematics,[13] a view bolstered by his systematic develop-

[13]This is emphasized by Hallett [1984] as Cantor's "finitism".

ment of the arithmetic of transfinite numbers seamlessly encompassing the finite numbers. Cantor also devoted several sections of the *Grundlagen* to a justificatory philosophy of the infinite, and while this metaphysics can be separated from the mathematical development, one concept was to suggest ultimate delimitations for set theory: Beyond the transfinite was the "Absolute", which Cantor eventually associated mathematically with the collection of all ordinal numbers and metaphysically with the transcendence of God.[14]

The Continuum Problem was never far from this development and could in fact be seen as an underlying motivation. The transfinite numbers were to provide the framework for Cantor's two approaches to the problem, the approach through power and the more direct approach through definable sets of reals, these each to initiate vast research programs.

As for the approach through power, Cantor in the *Grundlagen* established that the second number class (II) is uncountable, yet *any infinite subset of* (II) *is either countable or has the same power as* (II). Hence, (II) has exactly the property that Cantor sought for the reals, and he had reduced CH to the positive assertion that the reals and (II) have the same power. The following in brief is Cantor's argument that (II) is uncountable:

Suppose that s is a (countable) sequence of members of (II), say with initial element a. Let a' be a member of s, if any, such that $a < a'$; let a'' be a member of s, if any, such that $a' < a''$; and so forth. Then however long this process continues, the supremum of these numbers, or its successor, is not a member of s.

This argument was reminiscent of his [1874] argument that the reals are uncountable and suggested a correlation of the reals through their fundamental sequence representation with the members of (II) through associated cofinal sequences.[15] However, despite several announcements Cantor could never develop a workable correlation, an emerging problem in retrospect being that he could not *define* a well-ordering of the reals.

As for the approach through definable sets of reals, this evolved directly from Cantor's work on trigonometric series, the "symbols of infinity" used in the analysis of the P' operation transmuting to the transfinite numbers of the second number class (II).[16] In the *Grundlagen* Cantor studied P' for uncountable P and defined

---

[14]The "absolute infinite" is a varying but recurring explanatory concept in Cantor's work; see Jané [1995].

[15]After describing the similarity between ω and $\sqrt{2}$ as limits of sequences, Cantor [1887: 99] interestingly correlated the creation of the transfinite numbers to the creation of the irrational numbers, beyond merely breaking new ground in different number contexts: "The transfinite numbers are in a certain sense *new irrationalities*, and in my opinion the best method of defining the *finite* irrational numbers [via Cantor's fundamental sequences] is wholly similar to, and I might even say in principle the same as, my method of introducing transfinite numbers. One can say unconditionally: the transfinite numbers *stand or fall* with the finite irrational numbers: they are like each other in their innermost being [Wesen]; for the former like the latter are definite delimited forms or modifications of the actual infinite."

[16]Ferreirós [1995] suggests how the formulation of the second number class as a completed totality with a succeeding transfinite number emerged directly from Cantor's work on the operation P', drawing Cantor's transfinite numbers even closer to his earlier work on trigonometric

the key concept of a *perfect* set of reals (non-empty, closed, and containing no isolated points). Incorporating an observation of Ivar Bendixson [1883], Cantor showed in the succeeding [1884] that *any uncountable closed set of reals is the union of a perfect set and a countable set.* For a set A of reals, A has the *perfect set property iff* A is countable or else has a perfect subset. Cantor had shown in particular that *closed sets have the perfect set property.*

Since Cantor [1884; 1884a] had been able to show that any perfect set has the power of the continuum, he had established that "CH holds for closed sets": every closed set either is countable or has the power of the continuum. Or from his new vantage point, he had reduced the Continuum Problem to determining whether there is a closed set of reals of the power of the second number class. He was unable to do so, but he had initiated a program for attacking the Continuum Problem that was to be vigorously pursued (cf. 2.3 and 2.5).

1.3 Diagonalization and cardinal numbers

In the ensuing years, unable to resolve the Continuum Problem through direct correlations with transfinite numbers Cantor approached size and order from a broader perspective that *would* incorporate the continuum. He identified power with *cardinal number*, an autonomous concept beyond being *une façon de parler* about bijective correspondence, and he went beyond well-orderings to the study of linear *order types.* Cantor embraced a structured view of sets, when "well-defined", as being given together with a linear ordering of their members. Order types and cardinal numbers resulted from successive abstraction, from a set M to its order type $\overline{M}$ and then to its cardinality $\overline{\overline{M}}$.

Almost two decades after his [1874] result that the reals are uncountable, Cantor in a short note [1891] subsumed it via his celebrated diagonal argument. With it, he established: *For any set L the collection of functions from L into a fixed two-element set has a higher cardinality than that of L.* This result indeed generalized the [1874] result, since the collection of functions from the natural numbers into a fixed two-element set has the same cardinality as the reals. Here is how Cantor gave the argument in general form:[17]

Let M be the totality of all functions from L taking only the values 0 and 1. First, L is in bijective correspondence with a subset of M, through the assignment to each $x_0 \in L$ of the function on L that assigns 1 to x_0 and 0 to all other $x \in L$. However, there cannot be a bijective correspondence between M itself and L. Otherwise, there would be a function $\phi(x, z)$ of two variables such that for every member f of M there would be a $z \in L$ such that $\phi(x, z) = f(x)$ for every $x \in L$. But then, the "diagonalizing" function $g(x) = 1 - \phi(x, x)$ cannot be a member of M since for $z_0 \in L$, $g(z_0) \neq \phi(z_0, z_0)$!

series.

[17]Actually, Cantor took L to be the unit interval of reals presumably to invoke a standard context, but he was clearly aware of the generality.

In retrospect the diagonal argument can be drawn out from the [1874] proof.[18] Cantor had been shifting his notion of set to a level of abstraction beyond sets of reals and the like, and the casualness of his [1891] may reflect an underlying cohesion with his [1874]. Whether the new proof is really "different" from the earlier one, through this abstraction Cantor could now dispense with the recursively defined nested sets and limit construction, and he could apply his argument to any set. He had proved for the first time that there is a power higher than that of the continuum and moreover affirmed "the general theorem, that the powers of well-defined sets have no maximum."[19] The diagonal argument, even to its notation, would become method, flowing later into descriptive set theory, the Gödel Incompleteness Theorem, and recursion theory.

Today it goes without saying that a function from L into a two-element set corresponds to a subset of L, so that Cantor's Theorem is usually stated as: *For any set L its power set $\mathcal{P}(L) = \{X \mid X \subseteq L\}$ has a higher cardinality than L.* However, it would be an exaggeration to assert that Cantor was working on power sets; rather, he had expanded the 19th-Century concept of *function* by ushering in arbitrary functions.[20] In any case, Cantor would now have had to confront, in his function context, a general difficulty starkly abstracted from the Continuum Problem: *From a well-ordering of a set, a well-ordering of its power set is not necessarily definable.* The diagonal argument called into question Cantor's very notion of set: On the one hand, the argument, simple and elegant, should be part of set theory and lead to new sets of ever higher cardinality; on the other hand, these sets do not conform to Cantor's principle that every set comes with a (definable) well-ordering.[21]

---

[18]Moreover, diagonalization as such had already occurred in Paul du Bois-Reymond's theory of growth as early as in his [1869]. An argument is manifest in his [1875: 365ff] for showing that for any sequence of real functions $f_0, f_1, f_2, \ldots$ there is a real function g such that for each n, $f_n(x) < g(x)$ for all sufficiently large reals x.

Diagonalization can be drawn out from Cantor's [1874] as follows: Starting with a sequence s of reals and a half-open interval I_0, instead of successively choosing delimiting *pairs* of reals in the sequence, avoid the members of s *one* at a time: Let I_1 be the left or right half-open subinterval of I_0 demarcated by its midpoint, whichever does not contain the first element of s. Then let I_2 be the left or right half-open subinterval of I_1 demarcated by its midpoint, whichever does not contain the second element of s; and so forth. Again, the nested intersection contains a real not in the sequence s. Abstracting the process in terms of reals in binary expansion, one is just generating the binary digits of the diagonalizing real.

In that letter of Cantor's to Dedekind of 7 December 1873 (Ewald [1996: 845ff]) first establishing the uncountability of the reals, there already appears, quite remarkably, a doubly indexed array of real numbers and a procedure for traversing the array downward and to the right, as in a now common picturing of the diagonal argument.

[19]Remarkably, Cantor had already conjectured in the *Grundlagen* [1883: 590] that the collection of continuous real functions has the same power as the second number class (II), and that the collection of all real functions has the same power as the third number class (III). These are consequences of the later Generalized Continuum Hypothesis and are indicative of the sweep of Cantor's conception.

[20]The "power" in "power set" is from "Potenz" in the German for cardinal exponentiation, while Cantor's "power" is from "Mächtigkeit".

[21]This is emphasized in Lavine [1994: IV.2]. Cantor did consider power sets in a letter of 20

Cantor's *Beiträge*, published in two parts [1895] and [1897], presented his mature theory of the transfinite. In the first part he described his post-*Grundlagen* work on cardinal number and the continuum. He quickly posed *Cardinal Comparability*, whether

for cardinal numbers $\mathfrak{a}$ and $\mathfrak{b}$, $\mathfrak{a} = \mathfrak{b}$, $\mathfrak{a} < \mathfrak{b}$, or $\mathfrak{b} < \mathfrak{a}$,

as a property "by no means self-evident" and which will be established later "when we shall have gained a survey over the ascending sequence of transfinite cardinal numbers and an insight into their connection." He went on to define the addition, multiplication, and exponentiation of cardinal numbers primordially in terms of set-theoretic operations and functions. If $\mathfrak{a}$ is the cardinal number of M and $\mathfrak{b}$ is the cardinal number of N, then $\mathfrak{a}^{\mathfrak{b}}$ is the cardinal number of the collection of all functions $: N \to M$, i.e. having domain N and taking values in M. The audacity of considering arbitrary functions from a set N into a set M was encased in a terminology that reflected both its novelty as well as the old view of function as given by an explicit rule.[22] As befits the introduction of new numbers Cantor then introduced a new notation, one using the Hebrew letter aleph, $\aleph$. With $\aleph_0$ the cardinal number of the set of natural numbers Cantor observed that $\aleph_0 \cdot \aleph_0 = \aleph_0$ and that $2^{\aleph_0}$ is the cardinal number of continuum. With this he observed that the [1878] labor of associating the continuum with the plane and so forth could be reduced to a "few strokes of the pen" in his new arithmetic:

$$(2^{\aleph_0})^{\aleph_0} = 2^{\aleph_0 \cdot \aleph_0} = 2^{\aleph_0}\,.$$

September 1898 to Hilbert. In it Cantor entertained a notion of "completed set", one of the guidelines being that "the collection of *all subsets* of a completed set M is a completed set." Also, in a letter of 10 October 1898 to Hilbert, Cantor pointed out, in an argument focused on the continuum, that the power set $P(S)$ is in bijective correspondence with the collection of functions from S into $\{0,1\}$. But in a letter of 9 May 1899 to Hilbert, writing now "set" for "completed set", Cantor wrote: "... it is our common conviction that the 'arithmetic continuum' is a 'set' in this sense; the question is whether this truth is provable or whether it is an axiom. I now incline more to the latter alternative, although I would gladly be convinced by you of the former." For the first and third letters in context see Moore [2002: 45] and for the second, Ferreirós [2007: epilogue]; the letters are in Meschkowski-Nilson [1991].

[22] Cantor wrote [1895: 486]: "... by a '*covering* [Belegung] of N with M,' we understand a law by which with every element n of N a definite element of M is bound up, where one and the same element of M can come repeatedly into application. The element of M bound up with n is, in a way, a one-valued function of n, and may be denoted by $f(n)$; it is called a 'covering function [Belegungsfunktion] of n.' The corresponding covering of N will be called $f(N)$." A convoluted description! Arbitrary functions on arbitrary domains are now of course commonplace in mathematics, but several authors at the time referred specifically to Cantor's concept of covering, most notably Zermelo [1904]. Jourdain in his introduction to his English translation of the *Beiträge* wrote (Cantor [1915: 82]): "The introduction of the concept of 'covering' is the most striking advance in the principles of the theory of transfinite numbers from 1885 to 1895 ...."

With Cantor initially focusing on bijective correspondence [Beziehung] and these not quite construed as functions, Dedekind was the first to entertain an arbitrary function on an arbitrary domain. He [1888: §§21,36] formulated $\phi\colon S \to Z$, "a mapping [Abbildung] of a system S in Z", in less convoluted terms, but did not consider the totality of such. He quickly moved to the case $Z = S$ for his theory of chains; see footnote 36.

Cantor only mentioned

$$\aleph_0, \aleph_1, \aleph_2, \ldots, \aleph_\alpha, \ldots,$$

these to be the cardinal numbers of the successive number classes from the *Grundlagen* and thus to exhaust all the infinite cardinal numbers.

Cantor went on to present his theory of *order types*, abstractions of linear orderings. He defined an arithmetic of order types and characterized the order type η of the rationals as the countable dense linear order without endpoints, introducing the "forth" part of the now familiar back-and-forth argument of model theory.[23] He also characterized the order type θ of the reals as the perfect linear order with a countable dense set; whether a realist or not, Cantor the mathematician was able to provide a characterization of the continuum.

The second *Beiträge* developed the *Grundlagen* ideas by focusing on well-orderings and construing their order types as the *ordinal numbers*. Here at last was the general proof via order comparison of well-ordered sets that ordinal numbers are comparable. Cantor went on to describe ordinal arithmetic as a special case of the arithmetic of order types and after giving the basic properties of the second number class defined $\aleph_1$ as its cardinal number. The last sections were given over to a later preoccupation, the study of ordinal exponentiation in the second number class. The operation was defined via a transfinite recursion and used to establish a normal form, and the pivotal ϵ-numbers satisfying $\epsilon = \omega^\epsilon$ were analyzed.

The two parts of the *Beiträge* are not only distinct by subject matter, cardinal number and the continuum vs. ordinal number and well-ordering, but between them there developed a wide, irreconcilable breach. In the first part nowhere is the [1891] result $\mathfrak{a} < 2^{\mathfrak{a}}$ stated even in a special case; rather, it is made clear [1895: 495] that the procession of transfinite cardinal numbers is to be secured through their construal as the alephs. However, the second *Beiträge* does not mention any aleph beyond $\aleph_1$, nor does it mention CH, which could now have been stated as

$$2^{\aleph_0} = \aleph_1.$$

(Cantor did state this in an 1895 letter.[24]) Ordinal comparability was secured, but cardinal comparability was not reduced to it. Every well-ordered set has an aleph as its cardinal number, but where is $2^{\aleph_0}$ in the aleph sequence?

Cantor's initial [1874] proof led to the Continuum Problem. That problem was embedded in the very interstices of the early development of set theory, and in fact the structures that Cantor built, while now of intrinsic interest, emerged in significant part out of efforts to articulate and solve the problem. Cantor's [1891] diagonal argument, arguably a transmutation of his initial [1874] proof, exacerbated a growing tension between having well-orderings and admitting sets of arbitrary functions (or power sets). David Hilbert, when he presented his famous list of problems at the 1900 International Congress of Mathematicians at Paris,

[23]See Plotkin [1993] for an analysis of the emergence of the back-and-forth argument.
[24]See Moore [1989: 99].

made the Continuum Problem the very first problem and intimated Cantor's difficulty by suggesting the desirability of "actually giving" a well-ordering of the reals.

The next, 1904 International Congress of Mathematicians at Heidelberg was to be a generational turning point for the development of set theory. Julius Kőnig delivered a lecture in which he provided a detailed argument that purportedly established that $2^{\aleph_0}$ is not an aleph, i.e. that the continuum is not well-orderable. The argument combined the now familiar inequality $\aleph_\alpha < \aleph_\alpha^{\aleph_0}$ for α of cofinality ω with a result from Felix Bernstein's Göttingen dissertation [1901: 49] which alas does not universally hold.[25] Cantor was understandably upset with the prospect that the continuum would simply escape the number context that he had devised for its analysis.

Accounts differ on how the issue was resolved. Although one has Zermelo finding an error within a day of the lecture, the weight of evidence is for Hausdorff having found the error.[26] Whatever the resolution, the torch had passed from Cantor to the next generation. Zermelo would go on to formulate his Well-Ordering Theorem and axiomatize set theory, and Hausdorff, to develop the higher transfinite in his study of order types and cofinalities.[27]

2 MATHEMATIZATION

2.1 *Axiom of Choice and axiomatization*

Ernst Zermelo[28] (1871–1953), born when Cantor was establishing his trigonometric series results, had begun to investigate Cantorian set theory at Göttingen under the influence of Hilbert. In just over a month after the Heidelberg congress, Zermelo [1904] formulated what he soon called the *Axiom of Choice* (AC) and with it, established his Well-Ordering Theorem:

Every set can be well-ordered.

[25]The *cofinality* of an ordinal number α is the least ordinal number β such that there is a set of form $\{\gamma_\xi \mid \xi < \beta\}$ unbounded in α, i.e. for any $\eta < \alpha$ there is an $\xi < \beta$ such that $\eta \leq \gamma_\xi < \alpha$. α is *regular* if its cofinality is itself, and otherwise α is *singular*. There concepts were not clarified until the work of Hausdorff, brought together in his [1908], discussed in 2.6.

Kőnig applied Bernstein's equality $\aleph_\alpha^{\aleph_0} = \aleph_\alpha \cdot 2^{\aleph_0}$ as follows: If $2^{\aleph_0}$ were an aleph, say $\aleph_\beta$, then by Bernstein's equality $\aleph_{\beta+\omega}^{\aleph_0} = \aleph_{\beta+\omega} \cdot 2^{\aleph_0} = \aleph_{\beta+\omega}$, contradicting Kőnig's inequality. However, Bernstein's equality fails when α has cofinality ω and $2^{\aleph_0} < \aleph_\alpha$. Kőnig's published account [1905] acknowledged the gap.

[26]See Grattan-Guinness [2000, 334] and Purkert [2002].

[27]And as with many incorrect proofs, there would be positive residues: Zermelo soon generalized Kőnig's inequality to the fundamental Zermelo-Kőnig inequality for cardinal exponentiation, which implies that the cofinality of $2^{\aleph_\alpha}$ is larger than α, and Hausdorff [1904: 571] published his recursion formula $\aleph_{\beta+1}^{\aleph_\alpha} = \aleph_{\beta+1} \cdot \aleph_\beta^{\aleph_\alpha}$, in form like Bernstein's result.

[28]Ebbinghaus [2007] is a substantive biography of Zermelo. See Kanamori [1997; 2004] for Zermelo's work in set theory.

Zermelo thereby shifted the notion of set away from the implicit assumption of Cantor's principle that every well-defined set is well-ordered and replaced that principle by an explicit axiom about a wider notion of set, incipiently unstructured but soon to be given form by axioms.

In retrospect, Zermelo's *argument* for his Well-Ordering Theorem can be viewed as pivotal for the development of set theory. To summarize the argument, suppose that x is a set to be well-ordered, and through Zermelo's Axiom-of-Choice hypothesis assume that the power set $\mathcal{P}(x) = \{y \mid y \subseteq x\}$ has a choice function, i.e. a function γ such that for every non-empty member y of $\mathcal{P}(x)$, $\gamma(y) \in y$. Call a subset y of x a γ*-set* if there is a well-ordering R of y such that for each $a \in y$,

$$\gamma(\{z \mid z \notin y \text{ or } z \; R \; a \text{ fails}\}) = a\,.$$

That is, each member of y is what γ "chooses" from what does not already precede that member according to R. The main observation is that γ-sets cohere in the following sense: If y is a γ-set with well-ordering R and z is a γ-set with well-ordering S, then $y \subseteq z$ and S is a prolongation of R, or vice versa. With this, let w be the union of all the γ-sets, i.e. all the γ-sets put together. Then w too is a γ-set, and by its maximality it must be all of x and hence x is well-ordered.

The converse to this result is immediate in that if x is well-ordered, then the power set $\mathcal{P}(x)$ has a choice function.[29] Not only did Zermelo's argument analyze the connection between having well-orderings and having choice functions on power sets, it anticipated in its defining of approximations and taking of a union the proof procedure for von Neumann's Transfinite Recursion Theorem (cf. 3.1).[30]

Zermelo [1904: 516] noted without much ado that his result implies that every infinite cardinal number is an aleph and satisfies $\mathfrak{m}^2 = \mathfrak{m}$, and that it secured Cardinal Comparability — so that the main issues raised by Cantor's *Beiträge* are at once resolved. Zermelo maintained that the Axiom of Choice, to the effect that *every* set has a choice function, is a "logical principle" which "is applied without hesitation everywhere in mathematical deduction", and this is reflected in the Well-Ordering Theorem being regarded as a theorem. The axiom is consistent with Cantor's view of the finite and transfinite as unitary, in that it posits for infinite sets an unproblematic feature of finite sets. On the other hand, the Well-Ordering Theorem shifted the weight from Cantor's well-orderings with their residually temporal aspect of numbering through *successive* choices to the use of a function for making *simultaneous* choices.[31] Cantor's work had served to exacerbate a growing discord among mathematicians with respect to two related issues: whether infinite collections can be mathematically investigated at all, and how far the function concept is to be extended. The positive use of an arbitrary function

[29]Namely, with $\prec$ a well-ordering of x, for each non-empty member y of $\mathcal{P}(x)$, let $\gamma(y)$ be the the $\prec$-least member of y.

[30]See Kanamori [1997] for more on the significance of Zermelo's argument, in particular as a fixed point argument.

[31]Zermelo himself stressed the importance of simultaneous choices over successive choices in criticism of an argument of Cantor's for the Well-Ordering Theorem in 1899 correspondence with Dedekind, discussed in 2.2. See Cantor [1932: 451] or van Heijenoort [1967: 117].

operating on arbitrary subsets of a set having been made explicit, there was open controversy after the appearance of Zermelo's proof. This can be viewed as a turning point for mathematics, with the subsequent tilting toward the acceptance of the Axiom of Choice symptomatic of a conceptual shift in mathematics.

In response to his critics Zermelo published a second proof [1908] of his Well-Ordering Theorem, and with axiomatization assuming a general methodological role in mathematics he also published the first full-fledged axiomatization [1908a] of set theory. But as with Cantor's work this was no idle structure building but a response to pressure for a new mathematical context. In this case it was not for the formulation and solution of a *problem* like the Continuum Problem, but rather to clarify a specific *proof*. In addition to codifying generative set-theoretic principles, a substantial motive for Zermelo's axiomatizing set theory was to buttress his Well-Ordering Theorem by making explicit its underlying set existence assumptions.[32] Initiating the first major transmutation of the notion of set after Cantor, Zermelo thereby ushered in a new abstract, prescriptive view of sets as structured solely by membership and governed and generated by axioms, a view that would soon come to dominate. Thus, proof played a crucial role by stimulating an axiomatization of a field of study and a corresponding transmutation of its underlying notions.

The objections raised against Zermelo's first proof [1904] mainly played on the ambiguities of a γ-set's well-ordering being only implicit, as for Cantor's sets, and on the definition of the well-ordering being impredicative — defined as a γ-set and so drawn from a collection of which it is already a member. Largely to preclude these objections Zermelo in his second [1908] proof resorted to a rendition of orderings in terms of segments and inclusion first used by Gerhard Hessenberg [1906: 674ff] and a closure approach with roots in Dedekind [1888]. Instead of extending initial segments toward the desired well-ordering, Zermelo got at the collection of its final segments by taking an *intersection* in a larger setting.[33]

With his [1908a] axiomatization, Zermelo "started from set theory as it is historically given" to seek out principles sufficiently restrictive "to exclude all contradictions" and sufficiently wide "to retain all that is valuable". However, he would transform set theory by making explicit *new* existence principles and promoting a generative point of view. Zermelo had begun working out an axiomatization as early as 1905, addressing issues raised by his [1904] proof.[34] The mature presentation is a precipitation of seven axioms, and these do not just reflect "set theory as it is historically given", but explicitly buttress his proof(s) of the Well-Ordering Theorem.

[32]Moore [1982: 155ff] supports this contention using items from Zermelo's *Nachlass*.

[33]To well-order a set M using a choice function φ on $\mathcal{P}(M)$, Zermelo defined a Θ-*chain* to be a collection Θ of subsets of M such that: (a) $M \in \Theta$; (b) if $A \in \Theta$, then $A - \{\varphi(A)\} \in \Theta$; and (c) if $Z \subseteq \Theta$, then $\bigcap Z \in \Theta$. He then took the intersection I of all Θ-chains, and observed that I is again a Θ-chain. Finally, he showed that I provides a well-ordering of M given by: $a \prec b$ *iff* there is an $A \in I$ such that $a \notin A$ and $b \in A$. I thus consists of the final segments of the same well-ordering as provided by the [1904] proof. Note that this second proof is less parsimonious than the [1904] proof, as it uses the power set of the power set of M.

[34]This is documented by Moore [1982: 155ff] with items from Zermelo's *Nachlass*.

Zermelo's seven set axioms, now formalized, constitute the familiar theory Z, Zermelo set theory: Extensionality, Elementary Sets ($\emptyset$, $\{a\}$, $\{a,b\}$), Separation, Power Set, Union, Choice, and Infinity. His setting allowed for urelements, objects without members yet distinct from each other. But Zermelo focused on sets, and his Axiom of Extensionality announced the espousal of an extensional viewpoint. In line with this AC, a "logical principle" in [1904] expressed in terms of an informal choice function, was framed less instrumentally: It posited for a set consisting of non-empty, pairwise disjoint sets the existence of a *set* that meets each one in a unique element.[35] However, Separation retained an intensional aspect with its "separating out" of a new set from a given set using a *definite* property, where a property is "*definite* [*definit*] if the fundamental relations of the domain, by means of the axioms and the universally valid laws of logic, determine without arbitrariness whether it holds or not." But with no underlying logic formalized, the ambiguity of definite property would become a major issue. With Infinity and Power Set Zermelo provided for sufficiently rich settings for set-theoretic constructions. Tempering the logicians' extravagant and problematic "all" the Power Set axiom provided the provenance for "all" for subsets of a given set, just as Separation served to capture "all" for elements of a given set satisfying a property. Finally, Union and Choice completed the encasing of Zermelo's proof(s) of his Well-Ordering Theorem in the necessary set existence principles. Notably, Zermelo's recursive [1904] argumentation also brought him in proximity of the Transfinite Recursion Theorem and thus of Replacement, the next axiom to be adjoined in the subsequent development of set theory (cf. 3.1).

Fully two decades earlier Dedekind [1888] had provided an incisive analysis of the natural numbers and their arithmetic in terms of sets [Systeme], and several overlapping aspects can serve as points of departure for Zermelo's axiomatization.[36] The most immediate is how Dedekind's argumentation extends to Zermelo's [1908] proof of the Well-Ordering Theorem, which in the transfinite setting brings out the role of AC. Both Dedekind and Zermelo set down rules for sets in large part to articulate arguments involving simple set operations like "set of", union, and intersection. In particular, both had to argue for the equality of sets resulting after involved manipulations, and extensionality became operationally necessary. However vague the initial descriptions of sets, sets are to be determined solely by

---

[35] Russell [1906] had previously arrived at this form, his Multiplicative Axiom. The elimination of the "pairwise disjoint" by going to a choice function formulation can be established with the Union Axiom, and this is the only use of that axiom in the second, [1908] proof of the Well-Ordering Theorem.

[36] In current terminology, Dedekind [1888] considered arbitrary sets S and mappings $\phi\colon S \to S$ and defined a *chain* [Kette] to be a $K \subseteq S$ such that $\phi``K \subseteq K$. For $A \subseteq S$, the *chain of* $A$ is the intersection of all chains $K \supseteq A$. A set $N$ is *simply infinite iff* there is an injective $\phi\colon N \to N$ such that $N - \phi``N \neq \emptyset$. Letting 1 be a distinguished element of $N - \phi``N \neq \emptyset$ Dedekind considered the chain of $\{1\}$, the chain of $\{\phi(1)\}$, and so forth. Having stated an inherent induction principle, he proceeded to show that these sets have all the ordering and arithmetical properties of the natural numbers (that are established nowadays in texts for the (von Neumann) finite ordinals).

their elements, and the membership question is to be determinate.[37] The looseness of Dedekind's description of sets allowed him [1888: §66] the latitude to "prove" the existence of infinite sets, but Zermelo just stated the Axiom of Infinity as a set existence principle.

The main point of departure has to do with the larger issue of the role of proof for articulating sets. By Dedekind's time proof had become basic for mathematics, and indeed his work did a great deal to enshrine proof as the vehicle for algebraic abstraction and generalization.[38] Like algebraic constructs, sets were new to mathematics and would be incorporated by setting down the rules for their proofs. Just as calculations are part of the sense of numbers, so proofs would be part of the sense of sets, as their "calculations". Just as Euclid's axioms for geometry had set out the permissible geometric constructions, the axioms of set theory would set out the specific rules for set generation and manipulation. But unlike the emergence of mathematics from marketplace arithmetic and Greek geometry, sets and transfinite numbers were neither laden nor buttressed with substantial antecedents. Like strangers in a strange land stalwarts developed a familiarity with them guided hand in hand by their axiomatic framework. For Dedekind [1888] it had sufficed to work with sets by merely giving a few definitions and properties, those foreshadowing Extensionality, Union, and Infinity. Zermelo [1908a] provided more rules: Separation, Power Set, and Choice.

Zermelo [1908], with its rendition of orderings in terms of segments and inclusion, and Zermelo [1908a], which at the end cast Cantor's theory of cardinality in terms of functions cast as set constructs, brought out Zermelo's *set-theoretic reductionism.* Zermelo pioneered the reduction of mathematical concepts and arguments to set-theoretic concepts and arguments from axioms, based on sets doing the work of mathematical objects. Zermelo's analyses moreover served to draw out what would come to be generally regarded as set-theoretic out of the presumptively logical. This would be particularly salient for Infinity and Power Set and was strategically advanced by the relegation of property considerations to Separation.

Zermelo's axiomatization also shifted the focus away from the transfinite numbers to an abstract view of sets structured solely by $\in$ and simple operations. For Cantor the transfinite numbers had become central to his investigation of definable sets of reals and the Continuum Problem, and sets had emerged not only equipped with orderings but only as the developing context dictated, with the "set of" operation never iterated more than three or four times. For Zermelo his second, [1908] proof of the Well-Ordering Theorem served to eliminate any residual role that the transfinite numbers may have played in the first proof and highlighted the set-theoretic operations. This approach to (linear) ordering was to preoccupy his followers for some time, and through this period the elimination of the use of

[37] Dedekind [1888: §2] begins a footnote to his statement about extensional determination with: "In what manner this determination is brought about, and whether we know a way of deciding upon it, is a matter of indifference for all that follows; the general laws to be developed in no way depend upon it; they hold under all circumstances."

[38] Cf. the first sentence of the preface to Dedekind [1888]: "In science nothing capable of proof ought to be accepted without proof."

transfinite numbers where possible, like ideal numbers, was regarded as salutary.[39] Hence, Zermelo rather than Cantor should be regarded as the creator of abstract set theory.

Outgrowing Zermelo's pragmatic purposes axiomatic set theory could not long forestall the Cantorian initiative, as even $2^{\aleph_0} = \aleph_1$ could not be asserted directly, and in the 1920s John von Neumann was to fully incorporate the transfinite using Replacement (cf. 3.1).[40] On the other hand, Zermelo's axioms had the advantages of schematic simplicity and open-endedness. The generative set formation axioms, especially Power Set and Union, were to lead to Zermelo's [1930] cumulative hierarchy picture of sets, and the vagueness of the *definit* property in the Separation Axiom was to invite Thoralf Skolem's [1923] proposal to base it on first-order logic, enforcing extensionalization (cf. 3.2).

2.2 Logic and paradox

At this point, the incursions of a looming tradition can no longer be ignored. Gottlob Frege is regarded as the greatest philosopher of logic since Aristotle for developing quantificational logic in his *Begriffsschrift* [1879], establishing a logical foundation for arithmetic in his *Grundlagen* [1884], and generally stimulating the *analytic tradition* in philosophy. The architect of that tradition was Bertrand Russell who in his earlier years, influenced by Frege and Giuseppe Peano, wanted to found all of mathematics on the certainty of logic. But from a logical point of view Russell [1903] became exercised with paradox. He had arrived at Russell's Paradox in late 1901 by analyzing Cantor's diagonal argument applied to the class of all classes,[41] a version of which is now known as Cantor's Paradox of the largest cardinal number. Russell [1903: §301] also refocused the Burali-Forti Paradox of the largest ordinal number, after reading Cesare Burali-Forti's [1897].[42] Russell's Paradox famously led to the tottering of Frege's mature formal system, the *Grundgesetze* [1893, 1903].[43]

[39]Some notable examples: Lindelöf [1905] proved the Cantor-Bendixson result, that every uncountable closed set is the union of a perfect set and a countable set, without using transfinite numbers. Suslin's [1917], discussed in 2.5, had the unassuming title, "On a definition of the Borel sets without transfinite numbers", hardly indicative of its results, so fundamental for descriptive set theory. And Kuratowski [1922] showed, pursuing the approach of Zermelo [1908], that inclusion chains defined via transfinite recursion with intersections taken at limits can also be defined without transfinite numbers. Kuratowski [1922] essentially formulated Zorn's Lemma, and this was the main success of the push away from explicit well-orderings. Especially after the appearance of Zorn [1935] this recasting of AC came to dominate in algebra and topology.

[40]Textbooks usually establish the Well-Ordering Theorem by first introducing Replacement, formalizing transfinite recursion, and only then defining the well-ordering using (von Neumann) ordinals; this amounts to another historical misrepresentation, but one that resonates with how acceptance of Zermelo's proof broke the ground for formal transfinite recursion.

[41]Grattan-Guinness [1974], Coffa [1979], Moore [1988], and Garciadiego [1992] describe the evolution of Russell's Paradox.

[42]Moore-Garciadiego [1981] and Garciadiego [1992] describe the evolution of the Burali-Forti Paradox.

[43]See the exchange of letters between Russell and Frege in van Heijenoort [1967: 124ff]. Russell's Paradox showed that Frege's Basic Law V is inconsistent.

Russell's own reaction was to build a complex logical structure, one used later to develop mathematics in Whitehead and Russell's 1910-3 *Principia Mathematica.* Russell's *ramified theory of types* is a scheme of logical definitions based on *orders* and *types* indexed by the natural numbers. Russell proceeded "intensionally"; he conceived this scheme as a classification of propositions based on the notion of *propositional function*, a notion not reducible to membership (extensionality). Proceeding in modern fashion, we may say that the universe of the *Principia* consists of *objects* stratified into disjoint types T_n, where T_0 consists of the *individuals*, $T_{n+1} \subseteq \{Y \mid Y \subseteq T_n\}$, and the types T_n for $n > 0$ are further ramified into orders O_n^i with $T_n = \bigcup_i O_n^i$. An object in O_n^i is to be defined either in terms of individuals or of objects in some fixed O_m^j for some $j < i$ and $m \leq n$, the definitions allowing for quantification only over O_m^j. This precludes Russell's Paradox and other "vicious circles", as objects consist only of previous objects and are built up through definitions referring only to previous stages. However, in this system it is impossible to quantify over all objects in a type T_n, and this makes the formulation of numerous mathematical propositions at best cumbersome and at worst impossible. Russell was led to introduce his *Axiom of Reducibility*, which asserts that *for each object there is a predicative object consisting of exactly the same objects*, where an object is *predicative* if its order is the least greater than that of its constituents. This axiom reduced consideration to individuals, predicative objects consisting of individuals, predicative objects consisting of predicative objects consisting of individuals, and so on—the *simple theory of types.* In traumatic reaction to his paradox Russell had built a complex system of orders and types only to collapse it with his Axiom of Reducibility, a fearful symmetry imposed by an artful dodger.

The mathematicians did not imbue the paradoxes with such potency. Unlike Russell who wanted to get at everything but found that he could not, they started with what could be got at and peered beyond. And as with the invention of the irrational numbers, the outward push eventually led to the positive subsumption of the paradoxes.

Cantor in 1899 correspondence with Dedekind considered the collection Ω of all ordinal numbers as in the Burali-Forti Paradox, but he used it *positively* to give mathematical expression to his Absolute.[44] First, he distinguished between two kinds of multiplicities (Vielheiten): There are multiplicities such that when taken as a unity (Einheit) lead to a contradiction; such multiplicities he called "*absolutely infinite or inconsistent multiplicities*" and noted that the "totality of everything thinkable" is such a multiplicity. A multiplicity that can be thought of without contradiction as "being together" he called a "*consistent multiplicity* or a 'set [Menge]"'. Cantor then used the Burali-Forti Paradox argument to point out that the class Ω of all ordinal numbers is an inconsistent multiplicity. He proceeded to

[44]See footnote 3 for more about the 1899 correspondence. Purkert [1989: 57ff] argues that Cantor had already arrived at the Burali-Forti Paradox around the time of the *Grundlagen* [1883]. On the interpretations supported in the text *all* of the logical paradoxes grew out of Cantor's work — with Russell shifting the weight to paradox.

argue that every *set* can be well-ordered through a presumably recursive procedure whereby a well-ordering is defined through successive choices. The set must get well-ordered, else all of Ω would be injectible into it, so that the set would have been an inconsistent multiplicity instead.[45]

Zermelo found Russell's Paradox independently and probably in 1902,[46] but like Cantor, he did not regard the emergence of the paradoxes so much as a crisis as an overall delimitation for sets. In the Zermelian generative view [1908: 118], "... if in set theory we confine ourselves to a number of established principles such as those that constitute the basis of our proof — principles that enable us to form initial sets and to derive new sets from given ones – then all such contradictions can be avoided." For the first theorem of his axiomatic theory Zermelo [1908a] subsumed Russell's Paradox, putting it to use as is done now to establish that for any set x there is a $y \subseteq x$ such that $y \notin x$, and hence that there is no universal set.[47]

The differing concerns of Frege-Russell logic and the emerging set theory are further brought out by the analysis of the function concept as discussed below in 2.4, and those issues are here rehearsed with respect to the existence of the null class, or empty set.[48] Frege in his *Grundlagen* [1884] eschewed the terms "set [Menge]" and "class [Klasse]", but in any case the extension of the concept "not identical with itself" was key to his definition of zero as a logical object. Ernst Schröder, in the first volume [1890] of his major work on the algebra of logic, held a traditional view that a class is merely a collection of objects, without the { } so to speak. In his review [1895] of Schröder's [1890], Frege argued that Schröder cannot both maintain this view of classes and assert that there is a null class, since the null class contains no objects. For Frege, logic enters in giving unity to a class as the extension of a *concept* and thus makes the null class viable.

It is among the set theorists that the null class, *qua* empty set, emerged to the fore as an elementary concept and a basic building block. Cantor himself did not dwell on the empty set. At one point he did write [1880: 355] that "the identity of two pointsets P and Q will be expressed by the formula $P \equiv Q$"; defined disjoint sets as "*lacking intersection*"; and then wrote [1880: 356] "for the absence of points ... we choose the letter O; $P \equiv O$ indicates that the set P contains *no single point.*" (So, "$\equiv O$" is arguably more like a predication for being empty at this stage.)

Dedekind [1888: §2] deliberately excluded the empty set [Nullsystem] "for certain reasons", though he saw its possible usefulness in other contexts. Zermelo [1908a] wrote in his Axiom II: "There exists a (improper [uneigentliche]) set, the *null set* [*Nullmenge*] 0, that contains no element at all." Something of intension re-

[45] G.H. Hardy [1903] and Philip Jourdain [1904, 1905] also gave arguments involving the injection of Ω, but such an approach would only get codified at a later stage in the development of set theory in the work of von Neumann [1925] (cf. 3.1).

[46] See Kanamori [2004: §1].

[47] In 2.6 Hartogs's Theorem is construed as a positive subsumption of that other, the Burali-Forti Paradox.

[48] For more on the empty set, see Kanamori [2003a].

mained in the "(improper [uneigentliche])", though he did point out that because of his Axiom I, the Axiom of Extensionality, there is a single empty set. Finally, Hausdorff [1914] unequivocally opted for the empty set [Nullmenge]. However, a hint of predication remained when he wrote [1914: 3]: "... the equation $A = 0$ means that the set A has no element, vanishes [verschwindet], is empty." The use to which Hausdorff put "0" is much as "$\emptyset$" is used in modern mathematics, particularly to indicate the extension of the conjunction of mutually exclusive properties.

The set theorists, unencumbered by philosophical motivations or traditions, attributed little significance to the empty set beyond its usefulness. Although embracing both extensionality and the null class may engender philosophical difficulties for the logic of classes, the empty set became commonplace in mathematics simply through use, like its intimate, zero.

2.3 Measure, category, and Borel hierarchy

During this period Cantor's two main legacies, the investigation of definable sets of reals and the extension of number into the transfinite, were further incorporated into mathematics in direct initiatives. The axiomatic tradition would be complemented by another, one that would draw its life more directly from mathematics.

The French analysts Emile Borel, René Baire, and Henri Lebesgue took on the investigation of definable sets of reals in what was to be a paradigmatically constructive approach. Cantor [1884] had established the perfect set property for closed sets and formulated the concept of *content* for a set of reals, but he did not pursue these matters. With these as antecedents the French work would lay the basis for measure theory as well as *descriptive set theory,* the definability theory of the continuum.[49]

Soon after completing his thesis Borel [1898: 46ff] considered for his theory of measure those sets of reals obtainable by starting with the intervals and closing off under complementation and countable union. The formulation was axiomatic and in effect impredicative, and seen in this light, bold and imaginative; the sets are now known as the *Borel sets* and quite well-understood.

Baire in his thesis [1899] took on a dictum of Lejeune Dirichlet's that a real function is any arbitrary assignment of reals, and diverging from the 19th-Century preoccupation with pathological examples, sought a constructive approach via pointwise limits. His *Baire class 0* consists of the continuous real functions, and for countable ordinal numbers $\alpha > 0$, *Baire class* α consists of those functions f not in any previous class yet obtainable as pointwise limits of sequences $f_0, f_1, f_2, \ldots$ of functions in previous classes, i.e. $f(x) = \lim_{n\to\infty} f_n(x)$ for every real x. The functions in these classes are now known as the *Baire* functions, and this was the first stratification into a transfinite hierarchy after Cantor.[50]

[49]See Kanamori [1995] for more on the emergence of descriptive set theory. See Moschovakis [1980] or Kanamori [2003] for the mathematical development.

[50]Baire mainly studied the finite levels, particularly the classes 1 and 2. He [1898] pointed

Baire's thesis also introduced the now basic concept of *category*. A set of reals is *nowhere dense iff* its closure under limits includes no open set, and a set of reals is *meager* (or *of first category*) *iff* it is a countable union of nowhere dense sets — otherwise, it is *of second category*. Baire established the Baire Category Theorem: *Every non-empty open set of reals is of second category.* His work also suggested a basic property: A set of reals has the *Baire property iff* it has a meager symmetric difference with some open set. Straightforward arguments show that every Borel set has the Baire property.

Lebesgue's thesis [1902] is fundamental for modern integration theory as the source of his concept of measurability. Inspired in part by Borel's ideas but notably containing non-constructive aspects, Lebesgue's concept of measurable set through its closure under countable unions subsumed the Borel sets, and his analytic definition of measurable function through its closure under pointwise limits subsumed the Baire functions. Category and measure are quite different; there is a co-meager (complement of a meager) set of reals that has Lebesgue measure zero.[51] Lebesgue's first major work in a distinctive direction would be the seminal paper in descriptive set theory:

In the memoir [1905] Lebesgue investigated the Baire functions, stressing that they are exactly the functions definable via analytic expressions (in a sense made precise). He first established a correlation with the Borel sets by showing that they are exactly the pre-images of open intervals via Baire functions. With this he introduced the first hierarchy for the Borel sets, his *open sets of class* α not being in any previous class yet being pre-images of some open interval via some Baire class α function. After verifying various closure properties and providing characterizations for these classes Lebesgue established two main results. The first demonstrated the necessity of exhausting the countable ordinal numbers: *The Baire hierarchy is proper, i.e. for every countable* α *there is a Baire function of class* α*; correspondingly the hierarchy for the Borel sets is analogously proper.* The second established transcendence beyond countable closure for his concept of measurability: *There is a Lebesgue measurable function which is not in any Baire class; correspondingly there is a Lebesgue measurable set which is not a Borel set.*

The first result was the first of all hierarchy results, and a precursor of fundamental work in mathematical logic in that it applied Cantor's enumeration and diagonalization argument to achieve a transcendence to a next level. Lebesgue's second result was also remarkable in that he actually provided an explicitly defined set, one that was later seen to be the first example of a non-Borel analytic set (cf. 2.5). For this purpose, the reals were for the first time regarded as encoding something else, namely countable well-orderings, and this not only further embedded the transfinite into the investigation of sets of reals, but foreshadowed

---

out that Dirichlet's function that assigns 1 to rationals and 0 to irrationals is in class 2 and also observed with a non-constructive appeal to Cantor's cardinality argument that there are real functions that are not Baire.

[51]See footnote 11. See Hawkins [1975] for more on the development of Lebesgue measurability. See Oxtoby [1971] for an account of category and measure in juxtaposition.

the later coding results of mathematical logic.

Lebesgue's results, along with the later work in descriptive set theory, can be viewed as pushing the mathematical frontier of the actual infinite past $\aleph_0$, which arguably had achieved a mathematical domesticity through increasing use in the late 19th-Century, through Cantor's second number class to $\aleph_1$. It is somewhat ironic but also revealing, then, that this grew out of work by analysts with a definite constructive bent. Baire [1899: 36] viewed the infinite ordinal numbers and hence his function hierarchy as merely *une façon de parler*, and continued to view infinite concepts only in potentiality. Borel [1898] took a pragmatic approach and seemed to accept the countable ordinal numbers. Lebesgue was more equivocal but still accepting; recalling Cantor's early attitude Lebesgue regarded the ordinal numbers as an indexing system, "symbols" for classes, but nonetheless he worked out their basic properties, even providing a formulation [1905: 149] of proof by transfinite induction. All three analysts expressed misgivings about AC and its use in Zermelo's proof.[52]

As descriptive set theory was to develop, a major concern became the extent of the *regularity properties*, those properties indicative of well-behaved sets of reals of which Lebesgue measurability, the Baire property, and the perfect set property are the prominent examples. These properties seemed to get at basic features of the extensional construal of the continuum, yet resisted inductive approaches. Early explicit uses of AC through its role in providing a well-ordering of the reals showed how it allowed for new constructions: Giuseppe Vitali [1905] established that there is a non-Lebesgue measurable set of reals, and Felix Bernstein [1908], that there is a set of reals without the perfect set property. Soon it was seen that neither of these examples have the Baire property. Thus, that the reals are well-orderable, an early contention of Cantor's, permitted constructions that precluded the universality of the regularity properties, in particular his own approach to the Continuum Problem through the perfect set property.

2.4 Hausdorff and functions

Felix Hausdorff was the first developer of the transfinite after Cantor, the one whose work first suggested the rich possibilities for a mathematical investigation of the higher transfinite. A mathematician *par excellence*, Hausdorff took that sort of mathematical approach to set theory and extensional, set-theoretic approach to mathematics that would dominate in the years to come. While the web of 19th-Century intension in Cantor's work, especially his approach toward functions, now seems rather remote, Hausdorff's work seems familiar as part of the modern language of mathematics.

In [1908] Hausdorff brought together his extensive work on *uncountable* order types.[53] Deploring all the fuss being made over foundations by his contemporaries (p.436) and with Cantor having taken the Continuum Problem as far as

[52] See Moore [1982: 2.3].

[53] See Plotkin [2005] for translations and careful analyses of Hausdorff's work on ordered sets.

seemed possible, Hausdorff proceeded to venture beyond the second number class with vigor. He provided an elegant analysis of scattered linear order types (those having no dense subtype) in a transfinite hierarchy, and constructed the η_α sets, prototypes for saturated model theory. He first stated the *Generalized Continuum Hypothesis* (GCH), that $2^{\aleph_\alpha} = \aleph_{\alpha+1}$ for every α, clarified the significance of cofinality, and first considered (p.443) the possibility of an uncountable regular limit cardinal, the first *large cardinal.*

Large cardinal hypotheses posit cardinals with properties that entail their transcendence over smaller cardinals, and as it has turned out, provide a superstructure of hypotheses for the analysis of strong propositions in terms of consistency. Hausdorff observed that uncountable regular limit cardinals, also known now as *weakly inaccessible cardinals*, are a natural closure point for cardinal limit processes. In penetrating work of only a few years later Paul Mahlo [1911; 1912; 1913] investigated hierarchies of such cardinals based on higher fixed-point phenomena, the *Mahlo cardinals.* The theory of large cardinals was to become a mainstream of set theory.[54]

Hausdorff's classic text, *Grundzüge der Mengenlehre* [1914] dedicated to Cantor, broke the ground for a generation of mathematicians in both set theory and topology. A compendium of a wealth of results, it emphasized mathematical approaches and procedures that would eventually take firm root.[55] After giving a clear account of Zermelo's first, [1904] proof of the Well-Ordering Theorem, Hausdorff (p.140ff) emphasized its maximality aspect by giving synoptic versions of Zorn's Lemma two decades before Zorn [1935], one of them now known as Hausdorff's Maximality Principle.[56] Also, Hausdorff (p.304) provided the now standard account of the Borel hierarchy of sets, with the still persistent F_σ and G_δ notation. Of particular interest, Hausdorff (p.469ff, and also in [1914a]) used AC to provide what is now known as Hausdorff's Paradox, an implausible decomposition of the sphere and the source of the better known Banach-Tarski Paradox from Stefan Banach and Alfred Tarski's [1924].[57] Hausdorff's Paradox was the first, and a

[54] See Kanamori [2003] for more on large cardinals.

[55] Hausdorff's mathematical attitude is reflected in a remark following his explanation of cardinal number in a revised edition [1937:§5] of [1914]: "This formal explanation says what the cardinal numbers are supposed to do, not what they are. More precise definitions have been attempted, but they are unsatisfactory and unnecessary. Relations between cardinal numbers are merely a more convenient way of expressing relations between sets; we must leave the determination of the 'essence' of the cardinal number to philosophy."

[56] Hausdorff's Maximality Principle states that if A is a partially ordered set and B is a linearly ordered subset, then there is a $\subseteq$-maximal linearly ordered subset of A including B.

[57] Hausdorff's Paradox states that a sphere can be decomposed into four pieces Q, A, B, C with Q countable and A, B, C, and $B \cup C$ all pairwise congruent. Even more implausibly, the Banach-Tarski Paradox states that a ball can be decomposed into finitely many pieces that can be rearranged by rigid motions to form two balls of the same size as the original ball. Raphael Robinson [1947] later showed that there is such a decomposition into just five pieces with one of them containing a single point, and moreover that five is the minimal number. See Wagon [1985] for more on these and similar results; they stimulated interesting developments in measure theory that, rather than casting doubt on AC, embedded it further into mathematical practice (cf. 2.6).

dramatic, synthesis of classical mathematics and the Zermelian abstract view.

Hausdorff's reduction of functions through a defined ordered pair highlights the differing concerns of the earlier Frege-Russell logic and the emerging set theory.[58] Frege [1891] had two fundamental categories, *function* and *object*, with a function being "unsaturated" and supplemented by objects as arguments. A *concept* is a function with two possible values, the True and the False, and a *relation* is a concept that takes two arguments. The extension of a concept is its graph or course-of-values [Werthverlauf], which is an object, and Frege [1893: §36] devised an iterated or double course-of-values [Doppelwerthverlauf] for the extension of a relation. In these involved ways Frege assimilated relations to functions. As for the ordered pair, Frege in his *Grundgesetze* [1893: §144] provided the extravagant definition that the ordered pair of x and y is that class to which all and only the extensions of relations to which x stands to y belong.[59]

On the other hand, Peirce [1883], Schröder [1895], and Peano [1897] essentially regarded a relation from the outset as just a collection of ordered pairs. Whereas Frege was attempting an analysis of thought, Peano was mainly concerned with recasting ongoing mathematics in economical and flexible symbolism and made many reductions, e.g. construing a *sequence* in analysis as a *function* on the natural numbers. Peano from his earliest logical writings had used "(x, y)" to indicate the ordered pair in formula and function substitutions and extensions. In [1897] he explicitly formulated the ordered pair using "$(x; y)$" and moreover raised the two main points about the ordered pair: First, equation 18 of his Definitions stated the instrumental property which is all that is required of the ordered pair:

$$(*) \qquad \langle x, y\rangle = \langle a, b\rangle \ \textit{iff} \ x = a \text{ and } y = b\,.$$

Second, he broached the possibility of reducibility, writing: "The idea of a pair is fundamental, i.e. we do not know how to express it using the preceding symbols."

In Whitehead and Russell's *Principia Mathematica* [1910-3], relations distinguished in intension and in extension were derived from "propositional" functions taken as fundamental and other "descriptive" functions derived from relations. They [1910: ∗55] like Frege defined an ordered pair derivatively, in their case in terms of classes and relations, and also for a specific purpose.[60] Previously Russell [1903: §27] had criticized Peirce and Schröder for regarding a relation "essentially as a class of couples," although he did not mention this shortcoming in Peano.[61]

[58]For more on the ordered pair, see Kanamori [2003a].

[59]This definition, which recalls the Whitehead–Russell definition of the cardinal number 2, depended on Frege's famously inconsistent Basic Law V. See Heck [1995] for more on Frege's definition and use of his ordered pair.

[60]Whitehead and Russell had first defined a cartesian product by other means, and only then defined their ordered pair $x{\downarrow}y$ as $\{x\} \times \{y\}$, a remarkable inversion from the current point of view. They [1910: ∗56] used their ordered pair initially to define the ordinal number 2.

[61]In a letter accepting Russell's [1901] on the logic of relations for publication in his journal *Rivista*, Peano had pointedly written "The classes of couples correspond to relations" (see Kennedy [1975: 214]) so that relations are extensionally assimilated to classes. Russell [1903: §98] argued that the ordered pair cannot be basic and would itself have to be given sense, which

Commenting obliviously on *Principia* Peano [1911; 1913] simply reaffirmed an ordered pair as basic, defined a relation as a class of ordered pairs, and a function extensionally as a kind of relation, referring to the final version of his *Formulario Mathematico* [1905-8: 73ff.] as the source.

Capping this to and fro Norbert Wiener [1914] provided a definition of the ordered pair in terms of unordered pairs of classes only, thereby reducing relations to classes. Working in Russell's theory of types, Wiener defined the ordered pair $\langle x, y\rangle$ as

$$\{\{\{x\},\Lambda\},\{\{y\}\}\}$$

when x and y are of the same type and Λ is the null class (of the next type), and pointed out that this definition satisfies the instrumental property $(*)$ above. Wiener used this to eliminate from the system of *Principia* the Axiom of Reducibility for propositional functions of two variables; he had written a doctoral thesis comparing the logics of Schröder and Russell.[62] Although Russell praised Sheffer's stroke, the logical connective not-both, he was not impressed by Wiener's reduction. Indeed, Russell would not have been able to accept it as a genuine analysis. Unlike Russell, Willard V.O. Quine in a major philosophical work *Word and Object* [1960: §53] regarded the reduction of the ordered pair as a paradigm for philosophical analysis.

Making no intensional distinctions Hausdorff [1914: 32ff,70ff] defined an ordered pair in terms of unordered pairs, formulated functions in terms of ordered pairs, and the ordering relations as collections of ordered pairs.[63] Hausdorff thus made both the Peano [1911; 1913] and Wiener [1914] moves *in* mathematical practice, completing the reduction of functions to sets.[64] This may have been congenial to Peano, but not to Frege nor Russell, they having emphasized the primacy of functions. Following the pioneering work of Dedekind and Cantor Hausdorff was at the crest of a major shift in mathematics of which the transition from an intensional, rule-governed conception of function to an extensional, arbitrary one was a large part, and of which the eventual acceptance of the Power Set Axiom and the Axiom of Choice was symptomatic.

In his informal setting Hausdorff took the ordered pair of x and y to be

$$\{\{x,1\},\{y,2\}\}$$

would be a circular or an inadequate exercise, and "It seems therefore more correct to take an intensional view of relations ...".

[62]See Grattan-Guinness [1975] for more on Wiener's work and his interaction with Russell.

[63]He did not so define arbitrary relations, for which there was then no mathematical use, but he was the first to consider general *partial* orderings, as in his maximality principle. Before Hausdorff and going beyond Cantor, Dedekind was first to consider non-linear orderings, e.g. in his remarkably early, axiomatic study [1900] of lattices.

[64]As to historical priority, Wiener's note was communicated to the Cambridge Philosophical Society, presented on 23 February 1914, while the preface to Hausdorff's book is dated 15 March 1914. Given the pace of book publication then, it is arguable that Hausdorff came up with his reduction first.

where 1 and 2 were intended to be distinct objects alien to the situation.[65] In any case, the now-standard definition is the more intrinsic

$$\{\{x\}, \{x, y\}\}$$

due to Kazimierz Kuratowski [1921: 171]. Notably, Kuratowski's definition is a by-product of his analysis of Zermelo's [1908] proof of the Well-Ordering Theorem.[66]

2.5 Analytic and projective sets

A decade after Lebesgue's seminal paper [1905], descriptive set theory emerged as a distinct discipline through the efforts of the Russian mathematician Nikolai Luzin. He had become acquainted with the work of the French analysts while in Paris as a student and had addressed Baire's functions with a intriguing use of CH. What is now known as a *Luzin set* is an uncountable set of reals whose intersection with any meager set is countable, and Luzin established: CH *implies that there is a Luzin set.*[67] This would become a paradigmatic use of CH, in that a recursive construction was carried out in $\aleph_1$ steps where at each state only countable many conditions have to be attended to, in this case by applying the Baire Category Theorem. Luzin showed that the characteristic function of his set escaped Baire's function classification, and Luzin sets have since become pivotal examples of "special sets" of reals.

In Moscow Luzin began an important seminar, and from the beginning a major topic was the "descriptive theory of functions". The young Pole Wacław Sierpiński was an early participant while he was interned in Moscow in 1915, and undoubtedly this not only kindled a decade-long collaboration between Luzin and Sierpiński but also encouraged the latter's involvement in the development of a Polish school of mathematics and its interest in descriptive set theory.

Of the three regularity properties, Lebesgue measurability, the Baire property, and the perfect set property (cf. 2.3), the first two were immediate for the Borel sets. However, nothing had been known about the perfect set property beyond

[65] It should be pointed out that the definition works even when x or y is 1 or 2 to maintain the instrumental property $(*)$ of ordered pairs.

[66] The general adoption of the Kuratowski pair proceeded through the major developments of mathematical logic: Von Neumann initially took the ordered pair as primitive but later noted (von Neumann [1925:VI]; [1928: 338];[1929: 227]) the reduction via the Kuratowski definition. Gödel in his incompleteness paper [1931: 176] also pointed out the reduction. In his footnote 18, Gödel blandly remarked: "Every proposition about relations that is provable in [*Principia Mathematica*] is provable also when treated in this manner, as is readily seen." This stands in stark contrast to Russell's labors in *Principia* and his antipathy to Wiener's reduction of the ordered pair. Tarski [1931: n.3] pointed out the reduction and acknowledged his compatriot Kuratowski. In his recasting of von Neumann's system, Bernays [1937: 68] also acknowledged Kuratowski [1921] and began with its definition for the ordered pair. It is remarkable that Nicolas Bourbaki in his treatise [1954] on set theory still took the ordered pair as primitive, only later providing the Kuratowski reduction in the [1970] edition.

[67] Mahlo [1913a] also established this result.

Cantor's own result that the closed sets have it and Bernstein's that with a well-ordering of the reals there is a set not having the property. Luzin's student Pavel Aleksandrov [1916] established the groundbreaking result that *the Borel sets have the perfect set property*, so that "CH holds for the Borel sets".[68]

In the work that really began descriptive set theory another student of Luzin's, Mikhail Suslin, investigated the *analytic sets* following a mistake he had found in Lebesgue's paper.[69] Suslin [1917] formulated these sets in terms of an explicit operation $\mathcal{A}$[70] and announced two fundamental results: *a set B of reals is Borel iff both B and $\mathbb{R} - B$ are analytic;* and *there is an analytic set which is not Borel.*[71] This was to be his sole publication, for he succumbed to typhus in a Moscow epidemic in 1919 at the age of 25. In an accompanying note Luzin [1917] announced the regularity properties: *Every analytic set is Lebesgue measurable, has the Baire property, and has the perfect set property*, the last result attributed to Suslin.

Luzin and Sierpiński in their [1918] and [1923] provided proofs, and the latter paper was instrumental in shifting the emphasis toward the co-analytic sets, i.e. sets of reals X such that $\mathbb{R} - X$ is analytic. They used well-founded relations to provide a basic *tree representation* of co-analytic sets, one from which the main results of the period flowed, and it is here that well-founded relations entered mathematical practice.[72]

After the first wave in descriptive set theory brought about by Suslin [1917] and Luzin [1917] had crested, Luzin [1925a] and Sierpiński [1925] extended the domain of study to the *projective sets*. For $Y \subseteq \mathbb{R}^{k+1}$ and with ordered k-tuples defined

[68]After getting a partial result [1914: 465ff], Hausdorff [1916] also showed, in essence, that the Borel sets have the perfect set property.

[69]Sierpiński [1950: 28ff] describes Suslin's discovery of the mistake.

[70]A *defining system* is a family $\{X_s\}_s$ of sets indexed by finite sequences s of natural numbers. The result of the Operation $\mathcal{A}$ on such a system is that set $\mathcal{A}(\{X_s\}_s)$ defined by:

$$x \in \mathcal{A}(\{X_s\}_s) \ \textit{iff} \ (\exists f\colon \omega \to \omega)(\forall n \in \omega)(x \in X_{f|n})$$

where $f|n$ denotes that sequence determined by the first n values of f. For a set X of reals, X is *analytic iff* $X = \mathcal{A}(\{X_s\}_s)$ for some defining system $\{X_s\}_s$ consisting of closed sets of reals.

[71]Luzin [1925] traced the term "analytic" back to Lebesgue [1905] and pointed out how the original example of a non-Borel Lebesgue measurable set there was in fact the first example of a non-Borel analytic set.

[72]Building on the penultimate footnote, suppose that Y is a co-analytic set of reals, i.e. $Y = \mathbb{R} - X$ with $X = \mathcal{A}(\{X_s\}_s)$ for some closed sets X_s, so that for reals x,

$$x \in Y \ \textit{iff} \ x \notin X \ \textit{iff} \ (\forall f\colon \omega \to \omega)(\exists n \in \omega)(x \notin X_{f|n}).$$

For finite sequences s_1 and s_2 define: $s_1 \prec s_2$ *iff* s_2 is a proper initial segment of s_1. For a real x define: $T_x = \{s \mid x \in X_t$ for every initial segment t of $s\}$. Then:

$$x \in Y \ \textit{iff} \ \prec \text{ on } T_x \text{ is a well-founded relation},$$

i.e. there is no infinite descending sequence $\ldots \prec s_2 \prec s_1 \prec s_0$. T_x is a tree (cf. 3.5). Well-founded relations were explicitly defined much later in Zermelo [1935]. Constructions recognizable as via recursion along a well-founded relation had already occurred in the proofs that the Borel have the perfect set property in Aleksandrov [1916] and Hausdorff [1916].

from the ordered pair, the *projection of* Y is

$$pY = \{\langle x_1, ..., x_k\rangle \mid \exists y(\langle x_1, ..., x_k, y\rangle \in Y)\}.$$

Suslin [1917] had essentially noted that *a set of reals is analytic iff it is the projection of a Borel subset of* $\mathbb{R}^2$.[73] Luzin and Sierpiński took the geometric operation of projection to be basic and defined the projective sets as those sets obtainable from the Borel sets by the iterated applications of projection and complementation. The corresponding hierarchy of projective subsets of $\mathbb{R}^k$ is defined, in modern notation, as follows: For $A \subseteq \mathbb{R}^k$,

$$A \text{ is } \mathbf{\Sigma^1_1} \;\; \textit{iff} \;\; A = pY \text{ for some Borel set } Y \subseteq \mathbb{R}^{k+1} ,$$

i.e. A is analytic[74] and for integers $n > 0$,

$$\begin{aligned} A \text{ is } \mathbf{\Pi^1_n} \;\; &\textit{iff} \;\; \mathbb{R}^k - A \text{ is } \mathbf{\Sigma^1_n} , \\ A \text{ is } \mathbf{\Sigma^1_{n+1}} \;\; &\textit{iff} \;\; A = pY \text{ for some } \mathbf{\Pi^1_n} \text{ set } Y \subseteq \mathbb{R}^{k+1} , \text{ and} \\ A \text{ is } \mathbf{\Delta^1_n} \;\; &\textit{iff} \;\; A \text{ is both } \mathbf{\Sigma^1_n} \text{ and } \mathbf{\Pi^1_n} . \end{aligned}$$

Luzin [1925a] and Sierpiński [1925] recast Lebesgue's use of the Cantor diagonal argument to show that the projective hierarchy is proper, and soon its basic properties were established. However, this investigation encountered basic obstacles from the beginning. Luzin [1925a] emphasized that whether the $\mathbf{\Pi^1_1}$ sets, the co-analytic sets at the bottom of the hierarchy, have the perfect set property was a major question. In a confident and remarkably prophetic passage he declared that his efforts towards its resolution led him to a conclusion "totally unexpected", that "one does not know *and one will never know*" of the family of projective sets, although it has cardinality $2^{\aleph_0}$ and consists of "effective sets", whether every member has cardinality $2^{\aleph_0}$ if uncountable, has the Baire property, or is even Lebesgue measurable. Luzin [1925b] pointed out the specific problem of establishing whether the $\mathbf{\Sigma^1_2}$ sets are Lebesgue measurable. Both these difficulties were also pointed out by Sierpiński [1925]. This basic impasse in descriptive set theory was to remain for over a decade, to be surprisingly resolved by penetrating work of Gödel involving metamathematical methods (cf. 3.4).

2.6 Equivalences and consequences

In this period AC and CH began to be explored no longer as underlying axiom and primordial hypothesis but as part of mathematics. Consequences were drawn and even equivalences established, and this mathematization, like the development of non-Euclidean geometry, led eventually to a deflating of metaphysical attitudes and attendant concerns about truth and existence.

[73] Borel subsets of $\mathbb{R}^k$ are defined analogously to those of $\mathbb{R}$.

[74] Analytic subsets of $\mathbb{R}^k$ are defined as for the case $k = 1$ in terms of a defining system consisting of closed subsets of $\mathbb{R}^k$.

Friedrich Hartogs [1915] established an equivalence result for AC, and this was the first substantial use of Zermelo's axiomatization after its appearance. The axiomatization had initially drawn ambivalent response among commentators,[75] especially those exercised by the paradoxes, and its assimilation by structuring sets and clarifying arguments began with such uses.

As noted in 1.3, Cardinal Comparability had become a concern for Cantor by the time of his *Beiträge* [1895]; Hartogs showed in Zermelo's system *sans* AC that *Cardinal Comparability implies that every set can be well-ordered.* Thus, an evident consequence of every set being well-orderable also implied that well-ordering principle, and this first "reverse mathematics" result established the equivalence of the well-ordering principle, Cardinal Comparability, and AC over the base theory.

Hartogs actually established without AC what is now called *Hartogs's Theorem*: *For any set M, there is a well-orderable set E not injectible into M.* Cardinal Comparability would then imply that M is injectible into E and hence is well-orderable. For the proof Hartogs first worked out a theory of ordering relations in Zermelo's system in terms of inclusion chains as in Zermelo's [1908] proof.[76] He then used Power Set and Separation to get the set M_W of well-orderable subsets of M and the set E of equivalence classes partitioning M_W according to order-isomorphism. Finally, he showed that E itself has an inherited well-ordering and is not injectible into M.[77] Reminiscent of Zermelo's subsumption of Russell's Paradox in the denial of a universal set, Hartogs's Theorem can be viewed as a subsumption of the Burali-Forti Paradox into the Zermelian setting.

The first explicit uses of AC mostly amounted to appeals to a well-ordering of the reals, Cantor's preoccupation. Those of Vitali [1905] and Bernstein [1908] were mentioned in 2.3, and Hausdorff's Paradox [1914; 1914a], in 2.4. Georg Hamel [1905] constructed by transfinite recursion a basis for the reals as a vector space over the rationals; cited by Zermelo [1908, 114], this provided a useful basis for later work in analysis and algebra. These various results, jarring at first, broached how a well-ordering allows for a new kind of *arithmetical* approach to the continuum.

The full exercise of AC in ongoing mathematics first occurred in the pioneering work of Ernst Steinitz [1910] on abstract fields. This was the first instance of an emerging phenomenon in algebra and topology: the study of axiomatically given structures with the range of possibilities implicitly including the transfinite. Steinitz studied algebraic closures of fields and even had an explicit transfinite parameter in the transcendence degree, the number of indeterminates necessary for closure. Typical of the generality in the years to come was Hausdorff's [1932] result using well-orderings that *every vector space has a basis.* As algebra and

[75] See Moore [1982: 3.3].

[76] This is better done in Kuratowski [1921]. The Hausdorff [1914] approach with an ordered pair could have been taken, but that only became standard later when more general relations were considered.

[77] As with Zermelo's Well-Ordering Theorem, textbooks usually establish Hartogs's Theorem after first introducing Replacement and (von Neumann) ordinals, and this amounts to a historical misrepresentation.

topology developed however, such results as these came to be based on the maximal principles that Hausdorff had first broached (cf. 2.4) and began to dominate after the appearance of Zorn's Lemma [1935]. Explicit well-orderings seemed out of place at this level of organization, and Zorn's Lemma had the remarkable feature that its hypothesis was easily checked in most applications.

Poland since its reunification in 1918 featured an active school of mathematics establishing foundational results in mathematical logic, topology, and analysis, and at Warsaw Tarski and Kuratowski together with Sierpiński were making crucial contributions to set theory and the elucidation of its role in mathematics. The Polish school of mathematics carried out a penetrating investigation of the role of AC in set theory and analysis. Sierpiński's earliest publications, culminating in his survey [1918], not only dealt with specific constructions but showed how deeply embedded AC was in the informal development of cardinality, measure, and the Borel hierarchy (cf. 2.3), supporting Zermelo's contention [1904: 516] that the axiom is applied "everywhere in mathematical deduction". Tarski [1924], explicitly building his work on Zermelo's system, provided several propositions of cardinal arithmetic equivalent to AC, most notably that ($\mathfrak{m}^2 = \mathfrak{m}$ for every infinite cardinal $\mathfrak{m}$. Adolf Lindenbaum and Tarski in their [1926] gave further cardinal equivalents, some related to the Hartogs [1915] result, and announced that GCH, in the form that $\mathfrak{m} < \mathfrak{n} < 2^{\mathfrak{m}}$ holds for no infinite cardinals $\mathfrak{m}$ and $\mathfrak{n}$, implies AC. This study of consequences led to other choice principles, further implications and sometimes converses in a continuing cottage industry.[78]

The early mathematical study of AC extended to the issue of its independence. Abraham Fraenkel's first investigations [1922] directly addressed Zermelo's axioms, pointing out the need for the Replacement Axiom and attempting an axiomatization of the *definit* property for the Separation Axiom (cf. 3.1). The latter was motivated in part by the need to better articulate independence proofs for the various axioms. Fraenkel [1922a] came to the fecund idea of starting with urelements and some initial sets closing off under set-theoretic operations to get a model. For the independence of AC he started with urelements $a_n, \overline{a_n}$ for $n \in \omega$ and the set $A = \{\{a_n, \overline{a_n}\} \mid n \in \omega\}$ of unordered pairs and argued that for any set M in the resulting model there is a co-finite $A_M \subseteq A$ such that M is invariant if members of any $\{a_n, \overline{a_n}\} \in A_M$ are permuted. This immediately implies that there is no choice function for A in the model. Finally, Fraenkel argued that the model satisfies the other Zermelo axioms, except Extensionality because of the urelements.

Fraenkel's early model building emphasized the Zermelian generative framework, anticipated well-founded recursion, and foreshadowed the later play with models of set theory. That Extensionality was not to be had precluded settling the matter, but just as for the early models of non-Euclidean or finite geometries Fraenkel's achievement lay in stimulating interest in mathematical constructions despite relaxing some basic tenet. Fraenkel tried to develop his approach from time to time, but it needed the articulation that would come with the full espousal of the satisfaction relation. In the latter 1930s Lindenbaum and Andrzej Mostowski

[78]See Moore [1982], especially its 5.1, for other choice principles.

so cast and extended Fraenkel's work. Mostowski [1939] forged a method according to post-Gödelian sensibilities, bringing out the importance of groups of permutations leaving various urelements fixed, and the resulting models as well as later versions are now known as the *Fraenkel-Mostowski models.*

Even more than AC, Sierpiński investigated CH, and summed up his researches in a monograph [1934]. He provided several notable equivalences to CH, e.g. (p.11) the plane $\mathbb{R}^2$ is the union of countably many curves, where a *curve* is a set of form $\{\langle x, y\rangle \mid y = f(x)\}$ or $\{\langle x, y\rangle \mid x = f(y)\}$ with f a real function.

Moreover, Sierpiński presented numerous consequences of CH from the literature, one in particular implying a host of others: Mahlo [1913a] and Luzin [1914] had shown that CH *implies that there is a Luzin set*, an uncountable set of reals whose intersection with any meager set is countable (cf. 2.5). To state one consequence, say that a set X of reals has *strong measure zero iff* for any sequence $\epsilon_0, \epsilon_1, \epsilon_2, \ldots$ of positive reals there is a sequence of intervals $I_0, I_1, I_2, \ldots$ such that the length of I_n is less than ϵ_n for each n and $X \subseteq \bigcup_n I_n$. Borel [1919] conjectured that such sets are countable. However, Sierpiński [1928] showed that *a Luzin set has strong measure zero.* Analogous to a Luzin set, a *Sierpiński set* is an uncountable set of reals whose intersection with any Lebesgue measure zero set is countable. Sierpiński [1924] showed that CH *implies that there is a Sierpiński set*, and emphasized [1934] an emerging duality between measure and category.

The subsequent work of Fritz Rothberger would have formative implications for the Continuum Problem. He [1938] observed that if both Luzin and Sierpiński sets exist, then they have cardinality $\aleph_1$, so that the joint existence of such sets of the cardinality of the continuum implies CH. Then in penetrating analyses of the work of Sierpinski and Hausdorff on gaps (cf. 2.1) Rothberger [1939; 1948] considered other sets and implications between cardinal properties of the continuum *independent* of whether CH holds. It became newly clarified that absent CH one can still isolate uncountable cardinals $\leq 2^{\aleph_0}$ that gauge and delimit various recursive constructions, and this approach was to blossom half a century later in the study of *cardinal characteristics (or invariants)* of the continuum.[79]

These results cast CH in a new light, as a construction principle. Conclusions had been drawn from having a well-ordering of the reals, but one given by CH allowed for recursive constructions where at any stage only countably many conditions corresponding to as many reals had to be handled. The construction of a Luzin set was a simple recursive application of the Baire Category Theorem, and later constructions took advantage of the possibility of diagonalization at each stage. However, whereas the new constructions using AC, though jarring at first, were eventually subsumed as concomitant with the acceptance of the axiom and as expressions of the richness of possibility, constructions from CH clashed with that very sense of richness for the continuum. It was the *mathematical* investigation of CH that increasingly raised doubts about its truth and certainly its provability

[79]See Miller [1984] for more on special sets of reals and van Douwen [1984] as a trend setting paper for cardinal characteristics of the continuum. See Blass [2008] and Bartoszyński [2008] for recent work on cardinal characteristics.

(cf. end of 3.4).

3 CONSOLIDATION

3.1 Ordinals and Replacement

In the 1920s fresh initiatives structured the loose Zermelian framework with new features and corresponding developments in axiomatics: von Neumann's work with ordinals and Replacement; the focusing on well-founded sets and the cumulative hierarchy; and extensionalization in first-order logic. Von Neumann effected a counter-reformation of sorts: The transfinite numbers had been central for Cantor but peripheral to Zermelo; von Neumann reconstrued them as *bona fide* sets, now called simply the *ordinals*, and established their efficacy by formalizing transfinite recursion.

Von Neumann [1923; 1928], and before him Dimitry Mirimanoff [1917; 1917a] and Zermelo in unpublished 1915 work,[80] isolated the now familiar concept of ordinal, with the basic idea of taking precedence in a well-ordering simply to be membership. Appealing to forms of Replacement Mirimanoff and Von Neumann then established the key instrumental property of Cantor's ordinal numbers for ordinals: *Every well-ordered set is order-isomorphic to exactly one ordinal with membership.* Von Neumann in his own axiomatic presentation took the further step of ascribing to the ordinals the role of Cantor's ordinal numbers. Thus, like Kepler's laws by Newton's, Cantor's principles of generation for his ordinal numbers would be subsumed by the Zermelian framework. For this reconstrual of ordinal numbers and already to define the arithmetic of ordinals von Neumann saw the need to establish the Transfinite Recursion Theorem, the theorem that validates definitions by transfinite recursion. The proof was anticipated by the Zermelo 1904 proof, but Replacement was necessary even for the very formulation, let alone the proof, of the theorem. With the ordinals in place von Neumann completed the restoration of the Cantorian transfinite by defining the *cardinals* as the *initial ordinals*, those ordinals not in bijective correspondence with any of its predecessors. Now, the infinite initial ordinals are denoted

$$\omega = \omega_0, \omega_1, \omega_2, \ldots, \omega_\alpha, \ldots,$$

so that ω is to be the set of natural numbers in the ordinal construal, and the identification of different intensions is signaled by

$$\omega_\alpha = \aleph_\alpha$$

with the left being a von Neumann ordinal and the right being the Cantorian cardinal number.

Replacement has been latterly regarded as somehow less necessary or crucial than the other axioms, the purported effect of the axiom being only on

[80] See Hallett [1984: 8.1].

large-cardinality sets. Initially, Fraenkel [1921; 1922] and Skolem [1923] had independently proposed adjoining Replacement to ensure that $E(a) = \{a, \mathcal{P}(a), \mathcal{P}(\mathcal{P}(a)), \ldots\}$ be a set when a is the particular infinite set $Z_0 = \{\emptyset, \{\emptyset\}, \{\{\emptyset\}\}, \ldots\}$ posited by Zermelo's Axiom of Infinity, since, as they pointed out, Zermelo's axioms cannot establish this. However, even $E(\emptyset)$ cannot be proved to be a set from Zermelo's axioms,[81] and if his axiom of Infinity were reformulated to accommodate $E(\emptyset)$, there would still be many finite sets a such that $E(a)$ cannot be proved to be a set.[82] Replacement serves to rectify the situation by admitting new infinite sets defined by "replacing" members of the one infinite set given by the Axiom of Infinity. In any case, the full exercise of Replacement is part and parcel of transfinite recursion, which is now used everywhere in modern set theory, and it was von Neumann's formal incorporation of this method into set theory, as necessitated by his proofs, that brought in Replacement.

That Replacement became central for von Neumann was intertwined with his taking of function, in its full extensional sense, instead of set as primitive and his establishing of a context for handling *classes*, collections not necessarily sets. He [1925; 1928a] formalized the idea that a class is *proper*, i.e. not a set, exactly when it is in bijective correspondence with the entire universe, and this exactly when it is not an element of any class. This thus brought in another move from Cantor's 1899 correspondence with Dedekind (cf. 2.2). However, von Neumann's axiomatization [1925; 1928] of function was complicated, and reverting to sets as primitive Paul Bernays (cf. his [1976]) recast and simplified von Neumann's system. Still, the formal incorporation of proper classes introduced a superstructure of objects and results distant from mathematical practice. What was to be inherited was a predisposition to entertain proper classes in the mathematical development of set theory, a willingness that would have crucial ramifications (cf. 3.6).

3.2 Well-foundedness and the cumulative hierarchy

With ordinals and Replacement, set theory continued its shift away from pretensions of a general foundation toward a theory of a more definite subject matter, a process fueled by the incorporation of well-foundedness. Mirimanoff [1917]] was the first to study the well-founded sets, and the later hierarchical analysis is distinctly anticipated in his work. But interestingly enough well-founded relations next occurred in the direct definability tradition from Cantor, descriptive set theory (cf. 2.5).

In the axiomatic tradition Fraenkel [1922], Skolem [1923] and von Neumann [1925] considered the salutary effects of restricting the universe of sets to the well-founded sets. Von Neumann [1929: 231,236ff] formulated in his functional terms the Axiom of Foundation, that every set is well-founded,[83] and defined

[81]The union of $E(Z_0)$, with membership restricted to it, models Zermelo's axioms yet does not have $E(\emptyset)$ as a member.

[82]See Mathias [2001].

[83]$\forall x(x \neq \emptyset \longrightarrow \exists y \in x(x \cap y = \emptyset))$. This is von Neumann's Axiom VI4 in terms of sets. The

the resulting hierarchy of sets in his system via transfinite recursion: In modern notation, the axiom, as is well-known, entails that the universe V of sets is stratified into cumulative ranks V_α, where

$$V_0 = \emptyset;\ V_{\alpha+1} = \mathcal{P}(V_\alpha);\ V_\delta = \bigcup_{\alpha<\delta} V_\alpha \text{ for limit ordinals } \delta\,;$$

and

$$V = \bigcup_\alpha V_\alpha\,.$$

Von Neumann used this, *the cumulative hierarchy*, to establish the first relative consistency result in set theory via "inner models"; his argumentation in particular established the consistency of Foundation relative to Zermelo's axioms plus Replacement.

During this period mathematical logic gained new currency, and a tussle based on the different approaches of first- and second-order logic to set theory would lead to a substantial axiomatic development.[84] The prescient Skolem [1923] made the proposal of using for Zermelo's definite properties for the Separation Axiom those properties expressible in first-order logic with $\in$ as a binary relation symbol. After Leopold Löwenheim [1915] had broken the ground for model theory with his result about the satisfiability of a first-order sentence, Skolem [1920; 1923] had located the result solidly in first-order logic and generalized it to the Löwenheim-Skolem Theorem: *If a countable collection of first-order sentences is satisfiable, then it is satisfiable in a countable domain.* That Skolem intended for set theory to be a first-order system without a privileged interpretation for $\in$ becomes evident in the initial application of the Löwenheim-Skolem Theorem to get *Skolem's Paradox*: In first-order logic Zermelo's axioms are countable, Separation having become a *schema*, a schematic collection of axioms, one for each first-order formula; the theorem then implies the existence of countable models of the axioms although they entail the existence of uncountable sets. Skolem intended by this means to deflate the possibility of set theory becoming a foundation for mathematics. Exercised by this relativism and by the recent work of Fraenkel and von Neumann, Zermelo [1929] in his first publication in set theory in two decades proposed an axiomatization of his *definit* property in second-order terms. In direct response Skolem [1930] pointed out possible difficulties with this approach and reaffirmed his first-order formulation, completing the backdrop for a new axiomatic synthesis.

Zermelo in his remarkable [1930] offered his final axiomatization of set theory as well as a striking view of a procession of natural models that would have a modern resonance. While ostensibly a response to Skolem [1930], the dramatically new picture of sets in Zermelo [1930] reflects gained experience and the germination of ideas over a prolonged period. The main axiomatization incorporated Replacement but also the Axiom of Foundation. In contrast to Zermelo [1908a], while urelements

term "Foundation [Fundierung]" itself comes from Zermelo [1930].

[84]First-order logic is the logic of formal languages consisting of formulas built up from specified function and relation symbols using logical connectives and first-order quantifiers $\forall$ and $\exists$, quantifiers to be interpreted as ranging over the *elements* of a domain of discourse. Second-order logic has quantifiers to be interpreted as ranging over arbitrary subsets of a domain.

continued to be allowed, Infinity was eschewed and Choice was regarded as part of the underlying logic. Concerning Separation and Replacement it becomes evident from how Zermelo proceeded that he regarded their applicability in what we would now call a fully second-order context.

The now standard axiomatization of set theory

ZFC, *Zermelo-Fraenkel with Choice*,

is recognizable if we inject Infinity and AC, the main difference being that ZFC is a first-order theory (see below). "Fraenkel" acknowledges the early suggestion by Fraenkel to adjoin Replacement; and the Axiom of Choice is explicitly mentioned.

ZF, *Zermelo-Fraenkel*,

is ZFC without AC and is a base theory for the investigation of weak Choice-type propositions as well as propositions that contradict AC.

As described in above, Foundation in modern set theory ranks the universe of sets into the cumulative hierarchy $V = \bigcup_\alpha V_\alpha$. Zermelo substantially advanced this schematic generative picture with his inclusion of Foundation in an axiomatization. Replacement and Foundation focused the notion of set, with the first making possible the means of transfinite recursion and induction, and the second making possible the application of those means to get results about *all* sets. It is now almost banal that Foundation is the one axiom unnecessary for the recasting of mathematics in set-theoretic terms, but the axiom ascribes to membership the salient feature that distinguishes investigations specific to set theory as an autonomous field of mathematics. Indeed, it can be fairly said that modern set theory is at base a study couched in well-foundedness, the Cantorian well-ordering doctrines adapted to the Zermelian generative conception of sets.

In [1930] Zermelo described a range of models for set theory, each an initial segment of a cumulative hierarchy built on an initial set of urelements. Zermelo then established a categoricity of sorts for his axioms, one made possible by his second-order context. He showed that his models are characterized up to isomorphism by two cardinals, the number of their urelements and the height of their ordinals. Moreover, he established that if two models have the same number of urelements yet different heights, then one is isomorphic to an initial segment of the other's cumulative hierarchy. Grappling with Power Set and Replacement he characterized the heights of his models ("Grenzzahlen") as $\aleph_0$ or the *(strongly) inaccessible cardinals*, those uncountable regular cardinals κ that are strong limit, i.e. if $\lambda < \kappa$, then $2^\lambda < \kappa$.

Zermelo posited an endless procession of models, each a set in a next, advocating a dynamic view of sets that was a marked departure from Cantor's (and later, Gödel's) realist presumption of a fixed universe of sets. In synthesizing the sense of progression inherent in the new cumulative hierarchy picture and the sense of completion in the limit numbers, the inaccessible cardinals, he promoted the crucial idea of internal models of set theory. The open-endedness of Zermelo's original [1908a] axiomatization had been structured by Replacement and Foundation, but he advanced a new open-endedness with an eternal return of models approaching Cantor's Absolute.

In the process, inaccessible cardinals became structurally relevant. Sierpiński-Tarski [1930] had formulated these cardinals arithmetically as those uncountable cardinals that are not the product of fewer cardinals each of smaller power and observed that they are weakly inaccessible — the first large cardinal concept, from Hausdorff [1908: 443] (cf. 2.4). Be that as it may, in the early model-theoretic investigations of set theory the inaccessible cardinals provided the natural models as envisioned by Zermelo. Moreover, strong large cardinal hypotheses emerging in the 1960s were to be formulated in terms of these initial segments of the cumulative hierarchy.[85]

The journal volume containing Zermelo's paper also contained Stanisław Ulam's seminal paper [1930] on *measurable cardinals*, the most important of all large cardinals. For a set S, U is a (non-principal) *ultrafilter over S iff* U is a collection of subsets of S containing no singletons, closed under the taking of supersets and finite intersections, and such that for any $X \subseteq S$, either $X \in U$ or $S - X \in U$. For a cardinal λ, an ultrafilter U is *λ-complete iff* for any $D \subseteq U$ of cardinality less than λ, $\bigcap D \in U$. Finally, an uncountable cardinal κ is *measurable iff* there is a κ-complete ultrafilter over κ. Thus, a measurable cardinal is a cardinal whose power set is structured with a two-valued measure having strong closure property.

Measurability embodied the first large cardinal confluence of Cantor's two legacies, the investigation of definable sets of reals and the extension of number into the transfinite: Distilled from measure-theoretic considerations related to Lebesgue measure, the concept *also* entailed inaccessibility in the transfinite. Moreover, the initial airing generated a problem that was to keep the spark of large cardinals alive for the next three decades: *Can the least inaccessible cardinal be measurable?* In the 1960s consequences of, and a structural characterization of, measurability were established that became fundamental in the setting structured by the new Zermelian emphasis on well-foundedness (cf. 3.6).

3.3 *First-order logic and extensionalization*

The final structuring of set theory before it was to sail forth on its independent course as a distinctive field of mathematics was its full extensionalization in first-order logic.[86] However influential Zermelo's [1930] and despite his subsequent advocacy [1931; 1935] of infinitary logic, his efforts to forestall Skolem were not to succeed, as stronger currents were at work in the direction of first-order formalization.

Hilbert effected a basic shift in the development of mathematical logic when he took Whitehead and Russell's *Principia Mathematica*, viewed it as an uninterpreted formalism, and made it an object of mathematical inquiry. The book

[85]See Kanamori [2003: chap.5].

[86]See Goldfarb [1979] and Moore [1988a] for more on the emergence of first-order logic.

[1928][87] by Hilbert and Wilhelm Ackermann reads remarkably like a recent text. In marked contrast to the formidable works of Frege and Russell with their forbidding notation and all-inclusive approach, it proceeded pragmatically and upward to probe the extent of structure, making those moves emphasizing forms and axiomatics typical of modern mathematics. After a complete analysis of sentential logic it distinguished and focused on first-order logic ("functional calculus", and later "(restricted) predicate calculus") as already the source of significant problems. Thus, while Frege and Russell never separated out first-order logic, Hilbert through his mathematical investigations established it as a subject in its own right.

Hilbert in the 1920s developed proof theory, i.e. metamathematics, and proposed his program of establishing the consistency of classical mathematics. The issues here gained currency because of Hilbert's preeminence, just as mathematics in the large had been expanded in the earlier years of the century by his reliance on non-constructive proofs and transcendental methods and his advocacy of new contexts. Through this expansion the full exercise of AC had become a mathematical necessity (cf. 2.6) and arbitrary functions, and so Power Set, had become implicitly accepted in the extensive investigation of higher function spaces.

Hilbert-Ackermann [1928: 65ff,72ff] raised two crucial questions directed at the further possibilities for first-order logic: the completeness of its axioms and the Decision Problem [Entscheidungsproblem]. These as well as Hilbert's program for securing consistency were to be decisively informed by penetrating work that for set theory eventually led to its first sophisticated metamathematical result, the relative consistency of AC and GCH.

Kurt Gödel (1906–1978), born when Zermelo was devising his proofs of the Well-Ordering Theorem, virtually completed the mathematization of logic by submerging metamathematical methods into mathematics. The main vehicle was of course the direct coding, "the arithmetization of syntax", in his celebrated Incompleteness Theorem [1931]. Establishing a fundamental distinction between what is *true* about the natural numbers and what is *provable*, this theorem transformed Hilbert's consistency program and led to the undecidability of the Decision Problem from Hilbert–Ackermann [1928] and the development of recursion theory. Gödel's work showed in particular that for a (schematically definable) collection of axioms A, its *consistency*, that from A one cannot prove a contradiction, has a formal counterpart in an arithmetical formula Con(A) about natural numbers. Gödel's "second" theorem asserts that if A is consistent and subsumes some elementary arithmetic of the natural numbers, then Con(A) cannot be proved from A. But starting an undercurrent, the earlier Completeness Theorem [1930] from his thesis answered affirmatively a Hilbert–Ackermann [1928] question about semantic completeness, clarified the distinction between the formal syntax and semantics of first-order logic, and secured its key instrumental property with the Compactness Theorem.

[87] The historical development is clarified by the fact that while this book was published in light of the developments of the 1920s, it has a large overlap with unpublished lecture notes for a 1917-8 course given by Hilbert at Göttingen.

Tarski [1933; 1935] then completed the mathematization of logic by providing his definition of truth, exercising philosophers to a surprising extent ever since. Through Hilbert-Ackermann [1928] and Gödel [1930] the satisfaction relation had been informal, and in that sense completeness could be said to have remained inadequately articulated. Tarski simply extensionalized truth in formal languages and provided a formal, *recursive* definition of the satisfaction relation in set-theoretic terms. This new response to a growing need for a mathematical framework became the basis for model theory, but thus cast into mathematics truth would leave behind any semantics in the real meaning of the word. Tarski's [1933] was written around the same time as his [1931], a seminal paper that highlights the thrust of his initiative. In [1931] Tarski gave a precise mathematical (that is, set-theoretic) formulation of the informal concept of a (first-order) definable set of reals, thus infusing the intuitive notion of definability into ongoing mathematics. This mathematization of intuitive or logical notions was accentuated by Kuratowski-Tarski [1931], where second-order quantification over the reals was correlated with the geometric operation of projection, beginning the process of explicitly wedding descriptive set theory to mathematical logic. The eventual effect of Tarski's [1933] mathematical formulation of (so-called) semantics would be not only to make mathematics out of the informal notion of satisfiability, but also to enrich ongoing mathematics with a systematic method for forming mathematical analogues of several intuitive semantic notions.[88]

In this process of extensionalization first-order logic came to be accepted as the canonical language because of its mathematical possibilities as epitomized by the Compactness Theorem, and higher-order logics became downgraded as the workings of the power set operation in disguise. Skolem's early suggestion for set theory was thus taken up generally, and again the ways of paradox were positively subsumed, as the negative intent of Skolem's Paradox gave way to the extensive, internal use of Skolem functions from the Löwenheim-Skolem Theorem in set-theoretic constructions.

3.4 Relative consistency

Set theory was launched on an independent course as a distinctive field of mathematics by Gödel's construction of L [1938; 1939] leading to the relative consistency of the Axiom of Choice and the Generalized Continuum Hypothesis. Synthesizing all that came before, Gödel built on the von Neumann ordinals as sustained by Replacement to formulate a relative Zermelian universe of sets based on logical definability, a universe imbued with a Cantorian sense of enumerative order.

Gödel's advances in set theory can be seen as part of a steady intellectual development. In a lecture [1933] on the foundations of mathematics Gödel propounded the axiomatic set theory "as presented by Zermelo, Fraenkel and von Neumann"

[88]Incidentally, Tarski [1931] stated a result whose proof led to Tarski's well-known theorem [1951] that the elementary theory of real closed fields is decidable via the elimination of quantifiers.

as "a natural generalization of [Russell's simple] theory of types, or rather, what becomes of the theory of types if certain superfluous restrictions are removed." First, the types can be taken to be cumulative, and second, the process can be continued into the transfinite. As for how far this cumulative hierarchy of sets is to continue, "the first two or three [infinite] types already suffice to define very large [Cantorian ordinal numbers]" which can then serve to index the process, and so on. Implicitly referring to his incompleteness result Gödel noted that for a formal system S based on the theory of types a number-theoretic proposition can be constructed which is unprovable in S but becomes provable if to S is adjoined "the next higher type and the axioms concerning it."[89] Thus, although he never mentioned Zermelo [1930], Gödel was entertaining its cumulative hierarchies but as motivated by the theory of types.

It is to this initiative, separately fueled by Zermelo and Gödel, that one can date how the formation of sets out of sets iterated into the transfinite as embodied by the cumulative hierarchy can be regarded as a motivation for the subject matter of set theory. In a notable inversion, what has come to be regarded as the *iterative conception* became a heuristic for motivating the axioms of set theory generally.[90] The iterative conception of sets, like Tarski's definition of truth, has exercised philosophers to a surprising extent with respect to extrinsic justifications. This has opened the door to a metaphysical appropriation in the following sense: It is as if there is some notion of set that is "there", in terms of which the axioms must find some further justification. But set theory has no particular obligations to mirror some prior notion of set arrived at *a posteriori*. Replacement and Choice for example do not quite "fit" the iterative conception,[91] but if need be, Replacement can be "justified" in terms of achieving algebraic closure of the axioms, a strong motivation in the work of Fraenkel and the later Zermelo, and choice can be "justified" in terms of Cantorian well-ordering doctrines or as a logical principle as Zermelo did.

In his first announcement [1938] about L Gödel described it as a hierarchy "which can be obtained by Russell's ramified hierarchy of types, if extended to include transfinite orders." Indeed, with L Gödel had refined the cumulative hierarchy of sets to a cumulative hierarchy of *definable* sets which is analogous to the orders of Russell's *ramified* theory. Gödel's further innovation was to continue the indexing of the hierarchy through *all* the ordinals. Von Neumann's canonical

[89]Gödel was evidently referring to propositions like Con(S). In a prescient footnote, 48a, to his incompleteness paper [1931] Gödel had already written: "...the true reason for the incompleteness inherent in all formal systems of mathematics is that the formation of ever higher types can be continued into the transfinite ... while in any formal system at most denumerably many of them are available. For it can be shown that the undecidable propositions constructed here become decided whenever appropriate higher types are added (for example, the type ω to the system P [Peano Arithmetic]). An analogous situation prevails for the axiom system of set theory."

[90]Shoenfield [1967: 238ff][1977], Wang [1974a], Boolos [1971], and Scott [1974] motivate the axioms of set theory in terms of an iterative concept of set based on stages of construction. Parsons [1977] raises issues about this approach.

[91]See Boolos [1971] for Replacement and Scott [1974: 214] for Choice.

well-orderings would be the spine for a thin hierarchy of sets, and this would be the key to both the AC and GCH results.

In a brief account [1939] Gödel informally presented L essentially as is done today: For any set x let def(x) denote the collection of subsets of x first-order definable over $\langle x, \in \rangle$.[92] Then define:

$$L_0 = \emptyset;\ L_{\alpha+1} = \mathrm{def}(L_\alpha),\ L_\delta = \bigcup\{L_\alpha \mid \alpha < \delta\} \text{ for limit ordinals } \delta;$$

and the *constructible universe*

$$L = \bigcup\{L_\alpha \mid \alpha \text{ is an ordinal}\}.$$

Gödel pointed out that L "can be defined and its theory developed in the formal systems of set theory themselves." This is a remarkable understatement of arguably the central feature of the construction of L. L is a class definable in set theory via a transfinite recursion that could be based on the formalizability of def(x), the definability of definability, which was later reaffirmed by Tarski's systematic definition of the satisfaction relation in set-theoretic terms. With this, one can formalize the *Axiom of Constructibility* $V = L$, i.e. $\forall x(x \in L)$. In modern parlance, an *inner model* is a transitive class[93] containing all the ordinals such that, with membership and quantification restricted to it, the class satisfies each axiom of ZF. In summary terms, what Gödel did was to show in ZF that L is an inner model, and moreover that L satisfies AC and GCH. He thus established the relative consistency Con(ZF) implies Con(ZFC + GCH).

In the approach via def(x) it is necessary to show that def(x) remains unaltered when applied in L with quantifiers restricted to L. Gödel himself would never establish this *absoluteness of first-order definability* explicitly, preferring in his one rigorous published exposition of L to take an approach that avoids def(x) altogether.

In his monograph [1940], based on 1938 lectures, Gödel provided a specific, formal presentation of L in a class-set theory emanating from that of Bernays (cf. [1976]), a theory based in turn on a theory of von Neumann [1925]. Using eight binary operations producing new classes from old, Gödel generated L set by set via transfinite recursion. This veritable "Gödel numbering" with ordinals bypassed def(x) and made evident certain aspects of L. Since there is a direct, definable well-ordering of L, choice functions abound in L, and AC holds there. Of the other axioms the crux is where first-order logic impinges, in Separation and Replacement. For this, "algebraic" closure under Gödel's eight operations

[92]For a first-order formula $\varphi(v_1, \ldots, v_n)$ in $\in$, $\varphi^x(x_1, \ldots, x_n)$ is the restriction of the formula to x, i.e. each $\forall y$ is replaced by $\forall y \in x$ and each $\exists y$ is replaced by $\exists y \in x$ (with these abbreviations having the expected formal articulation). A set $y \subseteq x$ is *first-order definable over* $\langle x, \in \rangle$ if there is a first-order formula $\varphi(v_0, v_1, \ldots, v_n)$ and $a_1, \ldots, a_n$ all in x such that $y = \{z \in x \mid \varphi^x(z, a_1, \ldots, a_n)\}$.

[93]A class C is *transitive* if members of members of C are themselves members of C, so that C is "closed under membership".

ensured "logical" Separation for bounded formulas,[94] and then the full exercise of Replacement (in V) secured all of the ZF axioms in L.

Gödel's proof that L satisfies GCH consisted of two separate parts. He established the implication $V = L \to$ GCH, and, in order to apply this implication within L, that L as defined within L with quantifiers restricted to L is again L itself. This latter follows from the aforementioned absoluteness of def(x), and in [1940] Gödel gave an alternate proof based on the absoluteness of his eight binary operations.

Gödel's argument for $V = L \to$ GCH rests, as he himself wrote in [1939], on "a generalization of Skolem's method for constructing enumerable models." This was the first significant use of Skolem functions since Skolem's own to establish the Löwenheim-Skolem theorem, and with it, Skolem's Paradox. Ironically, though Skolem sought through his paradox to discredit set theory based on first-order logic as a foundation for mathematics, Gödel turned paradox into method, one promoting first-order logic. Gödel [1939] specifically established:

(*) For infinite α, every constructible subset of L_α belongs to some L_β for a β of the same cardinality as α.

It is straightforward to show that for infinite α, L_α has the same cardinality as that of α. It follows from (*) that in the sense of L, the power set of $L_{\aleph_\alpha}$ is included in $L_{\aleph_{\alpha+1}}$, and so GCH follows in L. To establish (*), Gödel actually iterated the Skolem closure procedure, and made the first use of the now familiar Mostowski collapse (cf. 3.6). In an incisive 1939 lecture Gödel announced the version of (*) for countable α as the crux of the consistency proof of CH and asserted that "this fundamental theorem constitutes the corrected core of the so-called Russellian axiom of reducibility."[95] Thus, Gödel established another connection between L and Russell's ramified theory of types. But while Russell had to *postulate* his ill-fated Axiom of Reducibility for his finite orders, Gödel was able to *derive*, with an important use of Replacement, an analogous form for his transfinite hierarchy that asserts that the types are delimited in the hierarchy of orders.

The synthesis at L extended to the resolution of difficulties in descriptive set theory (cf. end of 2.5). Gödel [1938] announced, in modern terms: *If $V = L$, then (a) there is a $\mathbf{\Delta}^1_2$ set of reals that is not Lebesgue measurable, and (b) there is a a $\mathbf{\Pi}^1_1$ set of reals without the perfect set property.* Thus, the descriptive set theorists were confronting an obstacle insurmountable in ZFC! Gödel [1938] listed each of these impossibility results on an equal footing with his AC and GCH results. Unexpected, they were the first instances of metamathematical methods resolving outstanding mathematical problems that exhibited no prior connection to such methods. When eventually confirmed and refined, the results were seen to turn on a $\mathbf{\Sigma}^1_2$ well-ordering of the reals in L defined via reals coding well-founded structures and thus connected to the well-founded tree representation of a $\mathbf{\Pi}^1_1$ set

[94] That is, those first-order formulas in which all the quantifiers can be rendered as $\forall x \in y$ and $\exists x \in y$.

[95] See Gödel [1939a: 141].

(cf. 2.5).[96]

Set theory had progressed to the point of establishing, in addition to a consistent resolution of CH, a consistent possibility for a definable well-ordering of the reals as Cantor had wanted, one that synthesizes the two historical sources of well-foundedness. Put into a broader historical context, formal definability was brought into descriptive set theory by Tarski [1931], and by Kuratowski-Tarski [1931] and Kuratowski [1931] which pursued the basic connection between existential number quantifiers and countable unions and between existential real quantifiers and projection and used these "logical symbols" to aid in the classification of sets in the Borel and projective hierarchies. Gödel's results (a) and (b) constitute the first real synthesis of abstract and descriptive set theory, in that the axiomatic framework is brought to bear on the investigation of definable sets of reals.

Gödel brought into set theory a method of construction and argument which affirmed several features of its axiomatic presentation. Most prominently, Gödel showed how first-order definability can be formalized and used in a transfinite recursive construction to establish striking new mathematical results. This significantly contributed to a lasting ascendancy for first-order logic which beyond its *sufficiency* as a logical framework for mathematics was seen to have considerable *operational efficacy*. Moreover, Gödel's construction buttressed the incorporation of Replacement and Foundation into set theory. Replacement was immanent in the arbitrary extent of the ordinals for the indexing of L and in its formal definition via transfinite recursion. As for Foundation, underlying the construction was the well-foundedness of sets. Gödel in a footnote to his 1939 note wrote: "In order to give A [the axiom $V = L$] an intuitive meaning, one has to understand by 'sets' all objects obtained by building up the simplified hierarchy of types on an empty set of individuals (including types of arbitrary transfinite orders)."

How Gödel transformed set theory can be broadly cast as follows: On the larger stage, from the time of Cantor, sets began making their way into topology, algebra, and analysis so that by the time of Gödel, they were fairly entrenched in the structure and language of mathematics. But how were sets viewed among set *theorists*, those investigating sets as such? Before Gödel, the main concerns were what sets *are* and how sets and their axioms can serve as a reductive basis for mathematics. Even today, those preoccupied with ontology, questions of mathematical existence, focus mostly upon the set theory of the early period. After

[96]When every real is in L, this $\mathbf{\Sigma}^1_2$ well-ordering is also $\mathbf{\Delta}^1_2$ and does not satisfy Fubini's Theorem for Lebesgue measurable subsets of the plane, and this is one way to confirm (a). What may have been Gödel's original argument for (b) is given in Kanamori [2003: 170].

Texts establish (b) indirectly via the Kondô $\mathbf{\Pi}^1_1$ Uniformization Theorem, and this leads to a historical point about Gödel the working mathematician. As 1938 correspondence with von Neumann makes evident, Gödel was working on one-to-one continuous images of Π^1_1 sets, and his [1938] actually states the results (a) and (b) in these terms. In a 1939 letter, von Neumann informed Gödel of Kondô [1939], the paper containing the uniformization result, from which it is immediate that the $\mathbf{\Sigma}^1_2$ sets are exactly the one-to-one continuous images of $\mathbf{\Pi}^1_1$ sets. In a replying letter to von Neumann of 20 March 1939 Gödel wrote: "The result of Kondô is of great interest to me and will definitely allow an important simplification in the consistency proof of [(a)] and [(b)] of the attached offprint." See Gödel [2003] for the Gödel–von Neumann correspondence.

Gödel, the main concerns became what sets *do* and how set theory is to advance as an autonomous field of mathematics. The cumulative hierarchy picture was in place as subject matter, and the metamathematical methods of first-order logic mediated the subject. There was a decided shift toward epistemological questions, e.g. what can be proved about sets and on what basis.

As a pivotal figure, what was Gödel's own stance? What he *said* would align him more with his predecessors, but what he *did* would lead to the development of methods and models. In a 1944 article on Russell's mathematical logic, in a 1947 article on Cantor's continuum problem (and in a 1964 revision), and in subsequent lectures and correspondence, Gödel articulated his philosophy of "conceptual realism" about mathematics. He espoused a staunchly objective "concept of set" according to which the axioms of set theory are true and are descriptive of an objective reality schematized by the cumulative hierarchy. Be that as it may, his actual mathematical work laid the groundwork for the development of a range of models and axioms for set theory. Already in the early 1940s Gödel worked out for himself a possible model for the negation of AC, and in a 1946 address he described a new inner model, the class of ordinal definable sets.

In later years Gödel speculated about the possibility of deciding propositions like CH with large cardinal hypotheses based on the heuristics of *reflection* and later, *generalization*. Already in that 1946 address he suggested[97] the consideration of "stronger and stronger axioms of infinity," and reflection as follows: "Any proof of a set-theoretic theorem in the next higher system above set theory (i.e. any proof involving the concept of truth ...) is replaceable by a proof from such an axiom of infinity." This ties in with the class of all ordinal numbers cast as Cantor's Absolute: A largeness property ascribable to the class might be used to derive some set-theoretic proposition; but any such property confronts the antithetical contention that the class is mathematically incomprehendable, fostering the synthetic move to a large cardinal posited with the property.

In the expository article [1947] on the Continuum Problem Gödel presumed that CH would be shown independent from ZF and speculated more concretely about possibilities with large cardinals. He argued that the axioms of set theory do not "form a system closed in itself" and so the "very concept of set on which they are based suggests their extension by new axioms that assert the existence of still further iterations of the operation of 'set of' ", citing Zermelo [1930] and echoing its theme. In an unpublished footnote 20 toward a 1966 revision of [1947] Gödel was to acknowledge[98] "extremely strong axioms of infinity of an entirely new kind", generalizations of properties of ω "supported by strong arguments from analogy". This heuristic of generalization ties in with Cantor's view of the finite and transfinite as unitary, with properties like inaccessibility and measurability technically satisfied by ω being too accidental were they not also ascribable to higher cardinals through the uniformity of the set-theoretic universe.[99]

---

[97]See Gödel [1990: 151].

[98]See Gödel [1990: 260ff].

[99]See Wang [1974: §§1,4] for more on Gödel's view on heuristics as well as the criteria of

Gödel [1947] at the end actually argued against CH by drawing on the work of Sierpinski and others (cf. 2.6) to exhibit six "paradoxical" consequences. One of them is the existence of a Luzin set of cardinality of the continuum, and three others actually follows from the existence of such a set. This brought to the fore Gödel's stance about what is true in set theory. Whether CH is proved consistent or independent of ZFC, he believed in a "truth of the matter" both from the point of view of intuitions about the continuum and from his philosophical standpoint. That CH is implausible because it led to various implausible conclusions became a prominent attitude, one that would stay with set theory through its subsequent development.

3.5 Combinatorics

Gödel's construction of L was both a culmination in all major respects of the early period in set theory and a source for much that was to follow. But for quite some time it was to remain an isolated monument in the axiomatic tradition. No doubt the intervening years of war were a prominent factor, but there was a continuing difficulty in handling definability within set theory and a stultifying lack of means for constructing models of set theory to settle issues of consistency and independence. It would take a new generation versed in emerging model-theoretic methods to set the stage for the next major methodological advances.

In the meantime, the direct investigation of the transfinite as extension of number was advanced, gingerly at first, by a new initiative. The seminal results of *infinite combinatorics* were established beginning in the 1930s. As for algebra and topology, it was natural to extend concepts over the transfinite, and significantly, the combinatorics that would have the most bearing there had their roots in the mathematization of logic.

Frank Ramsey [1930] established a special case of the Decision Problem of Hilbert-Ackermann [1928], the decidability of validity for the $\exists\forall$ formulas with identity. For this purpose he established a basic generalization of the pigeonhole principle. In a move that transcended purpose and context he also established an infinite version implicitly applying the now familiar Kőnig's Lemma for trees. Stated more generally for graphs by Dénes Kőnig [1927: 121] the lemma had also figured implicitly in Löwenheim [1915]. In what follows we affirm the general terminology for the formulation of Ramsey's results and then Kőnig's Lemma, anticipating extensions into the transfinite.

For ordinals α, β, and δ and $n \in \omega$ the *partition property*

$$\beta \longrightarrow (\alpha)^n_\delta$$

is the assertion that for any partition $f\colon [\beta]^n \to \delta$ of the n-element subsets of β into δ cells there is an $H \subseteq \beta$ of order type α *homogeneous* for the partition, i.e. all the n-element subsets of H lie in the same cell. Ramsey showed that for

intrinsic necessity and *pragmatic success* for accepting new axioms.

any k, n, and r all in ω, there is a $m \in \omega$ such that $m \longrightarrow (k)^n_r$. Skolem [1933] sharpened Ramsey's argument and thereby lowered the possibilities for the m's, but to this day the least such m's, the *Ramsey numbers*, have not been determined except in the simplest cases. Ramsey's infinite version is: $\omega \longrightarrow (\omega)^n_r$ for every $n, r \in \omega$. This partition property and its variants have been adapted to a variety of situations, and today *Ramsey theory* is a thriving field of combinatorics.[100]

A *tree* is a partially ordered set T such that the predecessors of any element are well-ordered. The *αth level* of T consists of those elements whose predecessors have order type α, and the *height* of T is the least α such that the αth level of T is empty. A *chain* of T is a linearly ordered subset, and an *antichain* is a subset consisting of pairwise incomparable elements. A *branch* of T is a maximal chain, and a *cofinal branch* of T is a branch with elements at every non-empty level of T. Finally, for a cardinal κ, a *κ-tree* is a tree of height κ each of whose levels has cardinality less than κ, and

κ has the *tree property iff* every κ-tree has a cofinal branch .

Finite trees of course are quite basic to current graph theory and computer science. With infinite trees the concerns are rather different, typically involving cofinal branches. Kőnig's Lemma asserts that *ω has the tree property.*

The first systematic study of transfinite trees was carried out in Đuro Kurepa's thesis [1935], and several properties emerging from his investigations, particularly for ω_1-trees, would later become focal in the combinatorial study of the transfinite.

An *Aronszajn tree* is an ω_1-tree without a cofinal branch ,

i.e. a counterexample to the tree property for ω_1. Kurepa [1935: §9,thm 6] gave Nachman Aronszajn's result that *there is an Aronszajn tree.*

A *Suslin tree* is an ω_1-tree with no uncountable chains or antichains .

Kurepa [1935: appendix] reduced a hypothesis growing out of a problem of Suslin [1920] about the characterizability of the order type of the reals to a combinatorial property of ω_1 as follows: *Suslin's Hypothesis holds iff there are no Suslin trees.*

A *Kurepa tree* is an ω_1-tree with at least ω_2 cofinal branches ,

and *Kurepa's Hypothesis* deriving from Kurepa [1942: 143], is the assertion that such trees exist. Much of this would be rediscovered, and both Suslin's Hypothesis and Kurepa's Hypothesis would be resolved three decades later with the advent of forcing, several of the resolutions in terms of large cardinal hypotheses.[101] Kurepa's work also anticipated another development from a different quarter:

Paul Erdős, although an itinerant mathematician for most of his life, was the prominent figure of a strong Hungarian tradition in combinatorics, and through

[100] See the text Graham-Rothschild-Spencer [1990] and the compendium Nešetřil-Rödl [1990] for the recent work on Ramsey Theory.

[101] See Todorčević [1984] for a wide-ranging account of transfinite trees.

some seminal results he introduced major initiatives into the detailed combinatorial study of the transfinite. Erdős and his collaborators simply viewed the transfinite numbers as a combinatorially rich source of intrinsically interesting problems, the concrete questions about graphs and mappings having a natural appeal through their immediacy. One of the earliest advances was Erdős-Tarski [1943] which concluded enticingly with an intriguing list of six combinatorial problems, the positive solution to any, as it was to turn out, amounting to the existence of a large cardinal. In a footnote various implications were noted, one of them being essentially that *for inaccessible* κ*, the tree property for* κ *implies* $\kappa \longrightarrow (\kappa)^2_2$, generalizing Ramsey's $\omega \longrightarrow (\omega)^2_2$ and making explicit the Kőnig Lemma property needed. While Kurepa was investigating distinctive properties of uncountable trees, Erdős-Tarski [1943] was evidently motivated by strong properties of ω to formulate direct combinatorial generalizations to inaccessible cardinals by analogy.[102] The situation would become considerably clarified, but only two decades later.[103]

The detailed investigation of partition properties began in earnest in the 1950s, with Erdős and Richard Rado's [1956] being representative. For a cardinal κ, let κ^+ denote its successor cardinal and set $\exp_0(\kappa) = \kappa$ and $\exp_{n+1}(\kappa) = 2^{\exp_n(\kappa)}$. What became known as *the* Erdős–Rado Theorem asserts: *For any infinite cardinal* κ *and* $n \in \omega$,

$$\exp_n(\kappa)^+ \longrightarrow (\kappa^+)^{n+1}_\kappa .$$

This was established using the basic tree argument underlying Ramsey's results, whereby a homogeneous set is not constructed recursively, but a tree is constructed such that its branches provide homogeneous sets, and a counting argument ensures that there must be a homogeneous set of sufficient cardinality. Kurepa [1937; 1939] in effect had actually established the case $n = 1$ and shown that $\exp_1(\kappa)^+$ was the least possible. The $\exp_n(\kappa)^+$ was also shown to be the least possible in the general case, and so unlike for the Ramsey numbers in the finite case an exact analysis was quickly achieved in the transfinite. This was to be a recurring phenomenon, that the gross features of transfinite cardinality make its combinatorics actually easier than in the analogous finite situation. And notably, iterated cardinal exponentiation figured prominently, so that shedding deeper concerns the power set operation became further domesticated in the arithmetic of combinatorics. In fact, assuming GCH simplified results and formulations, and this was often done, as in Erdős, András Hajnal, and Rado's [1965], representative of the 1960s. Increasingly, a myriad of versions have been investigated in the larger terrain without GCH.[104]

---

[102]On the other hand, Kurepa [1935: §10.3] did ask whether every inaccessible cardinal has the tree property, a question only resolved by work of Hanf (cf. 3.6).

[103]The details of implications asserted at the end of Erdős-Tarski [1943] were worked out in an influential seminar conducted by Tarski and Mostowski at Berkeley in 1958-9, and appeared in Erdős-Tarski [1961].

[104]The results of Erdős-Hajnal-Rado [1965] were extended in Byzantine detail to the general situation without GCH by the book Erdős-Hajnal-Máté-Rado [1984]. See Hajnal–Larson [2008] for recent work on partition relations.

Still among the Hungarians, Géza Fodor [1956] established the now familiar *regressive function lemma* for stationary sets: *If* λ *regular and uncountable,* S *is stationary in* λ,[105] *and* $f: S \to \lambda$ *is regressive (i.e.* $f(\xi) < \xi$ *for* $\xi \in S$*), then there is an* $\alpha < \lambda$ *such that* $\{\xi \in S \mid f(\xi) = \alpha\}$ *is stationary in* λ. It is a basic fact and a simple exercise now, but then it was the culmination of a progression of results beginning with a special case established in Aleksandrov-Urysohn [1929] and getting to the right largeness notion of stationarity. The contrast with how the lemma's earlier precursors were considered difficult and even paradoxical is striking, indicative of both the novelty of uncountable cofinality and the great leap forward that set theory has made.

3.6 Model-theoretic methods

Model theory began in earnest with the method of diagrams of Abraham Robinson's thesis [1951] and the related method of constants from Leon Henkin's thesis which gave a new proof [1949] of the Gödel Completeness Theorem. Tarski had set the stage with his definition of truth and more generally his casting of formal languages and structures in set-theoretic terms, and with him established at the University of California at Berkeley a large part of the development in the 1950s and 1960s would take place there. The construction of models freely used transfinite methods and soon led to new questions in set theory, but also set theory was to be decisively advanced by the infusion of model-theoretic methods.

The first relevant result was a generalization accreditable to Mostowski [1949] of the Mirimanoff–von Neumann result that every well-ordered set is order-isomorphic to exactly one ordinal with membership. A binary relation R is *extensional on* X *iff* for any $x \neq y$ both in X there is a $z \in X$ such that $z\,R\,x$ *iff* $\neg z\,R\,y$. Recall that x is *transitive iff* members of members of x are themselves members of x, so that x is "closed under membership". *If* R *is a well-founded relation on a set* X *and extensional on* X*, there is a unique isomorphism of* $\langle X, R\rangle$ *onto* $\langle T, \in\rangle$ *where* T *is transitive, i.e. a bijection* $\pi: X \to T$ *such that for any* $x, y \in X$, $x\,R\,y$ *iff* $\pi(x) \in \pi(y)$. T is the *transitive collapse* of X, and π the *collapsing isomorphism.* Thus, the linearity of well-orderings has been relaxed to well-foundedness and an analogue of the Axiom of Extensionality, and the transitive sets become canonical representatives as ordinals are for well-orderings. Gödel [1939: 222] had made the first substantial use of the transitive collapse; Mostowski [1949: 147] established the general result much later; and John Shepherdson [1951: 171] in a structured setting that brought out a further necessary hypothesis for classes X: R is *set-like,* i.e. for any $x \in X$, $\{y \mid y\,R\,x\}$ is a set. The initial applications in Mostowski [1949] and Shepherdson [1953] were to establish the independence of the assertion that there is a transitive set M which with $\in$ restricted to it is a model of set theory. While the Mirimanoff–von Neumann result was basic to the analysis of number in

[105] A set $C \subseteq \lambda$ is *closed unbounded in* λ *iff* C contains its limit points, i.e. those $0 < \alpha < \lambda$ such that $C \cap \alpha = \alpha$, and is cofinal, i.e. $\bigcup C = \lambda$. A set $S \subseteq \lambda$ is *stationary in* λ *iff* for any C closed unbounded in λ, $S \cap C$ is not empty.

the transfinite, the transitive collapse result grew in significance from specific applications and came to epitomize how well-foundedness made possible a coherent theory of models of set theory.

The relationship between ZFC and Bernays-Gödel (BG), the class-set theory brought into prominence by its use in Gödel [1940], was clarified during this period. As analyzed in Hao Wang [1949], BG can be construed as an extension of ZFC via the introduction of class variables intended to range over subcollections of V and correlative axioms, together with a comprehension principle asserting for each formula φ with just one set variable v free and no class variables quantified that there is a corresponding class $\{v \mid \varphi\}$. Ilse Novak [1950] and Barkley Rosser and Wang in their [1950] established that *if* ZFC *is consistent, then so is* BG by providing model-theoretic interpretations of BG relative to ZFC. Then Mostowski [1950] showed that BG is a conservative extension of ZF, i.e. any sentence σ without class variables provable in BG is already provable in ZFC. Subsequently, Joseph Shoenfield [1954] showed how to convert directly a proof of such a σ in BG into a proof in ZFC. These results reinforced the impression that, as far as the axiomatic tradition from Zermelo through Gödel is concerned, there is essentially one set theory, and one might as well work in the parsimonious ZFC.

Shepherdson [1951; 1952; 1953] studied "inner" models of set theory, with [1952] giving a rigorous first-order account of the results of Zermelo [1930]. The term is now reserved for the case mentioned in 3.4: A transitive class containing all the ordinals such that, with membership and quantification restricted to it, the class satisfies each axiom of ZF. The archetypal inner model is Gödel's L, and $L \subseteq M$ for any inner model M since the construction of L carried out in M is again L. Because of this Shepherdson [1953] observed that the relative consistency of hypotheses like the negation of CH cannot be established via inner models.

Hajnal [1956; 1961] and Azriel Levy [1957; 1960] developed generalizations of L that were to become basic in a richer setting. For a set A, Hajnal formulated the *constructible closure* $L(A)$ *of* A, i.e. the smallest inner model M such that $A \in M$, and Levy formulated the *class* $L[A]$ *of sets constructible relative to* A, i.e. the smallest inner model M such that for every $x \in M$, $A \cap x \in M$.[106] $L(A)$ realizes the algebraic idea of building up a model starting from a set of generators, and $L[A]$ the idea of building up a model using A construed as a predicate. $L(A)$ may not satisfy AC since e.g. it may not have a well-ordering of A, yet $L[A]$ always satisfies that axiom. This distinction was only to surface later, as both Hajnal and Levy took A to be a set of ordinals, when $L(A) = L[A]$. Hajnal and Levy (and also Shoenfield [1959], who formulated a special version of Levy's construction) used these models to establish conditional independence results of the sort: if the failure of CH is consistent, then so is that failure together with $2^\lambda = \lambda^+$ for sufficiently

[106]To formulate $L(A)$, define: $L_0(A)$ = the smallest transitive set $\supseteq \{A\}$ (to ensure that the resulting class is transitive); $L_{\alpha+1} = \text{def}(L_\alpha(A))$ (where def is as in 3.4); $L_\delta = \bigcup_{\alpha<\delta} L_\alpha(A)$ for limit $\delta > 0$; and finally $L(A) = \bigcup_\alpha L_\alpha(A)$. To formulate $L[A]$, first let $\text{def}^A(x)$ denote the collection of subsets of x first-order definable over $\langle x, \in, A \cap x\rangle$, i.e. $A \cap x$ is now allowed as a predicate in the definitions. Then define: $L_0[A] = \emptyset$; $L_{\alpha+1}[A] = \text{def}^A(L_\alpha[A])$; $L_\delta[A] = \bigcup_{\alpha<\delta} L_\alpha[A]$ for limit $\delta > 0$; and finally $L[A] = \bigcup_\alpha L_\alpha[A]$.

large cardinals λ.

After Richard Montague [1956; 1961] applied reflection phenomena to investigate finite axiomatizability for set theory, Levy [1960a; 1960b] also formulated reflection principles and established their broader significance. The *Reflection Principle* for ZF, drawn from Montague [1961: 99] and from Levy [1960a: 234], asserts: *For any (first-order) formula $\varphi(v_1, \ldots, v_n)$ and any ordinal β, there is a limit ordinal $\alpha > \beta$ such that for any $x_1, \ldots, x_n \in V_\alpha$,*

$$\varphi[x_1, \ldots, x_n] \text{ iff } \varphi^{V_\alpha}[x_1, \ldots, x_n],$$

i.e. the formula holds exactly when it holds with all the quantifiers restricted to V_α. Levy showed that this schema is equivalent to the conjunction of the Replacement schema together with Infinity in the presence of the other axioms of ZF. Moreover, he formulated reflection principles in local form that characterized cardinals in the Mahlo hierarchy (2.4), conceptually the least large cardinals after the inaccessible cardinals. Then William Hanf and Dana Scott in their [1961] posited analogous reflection principles for higher-order formulas, leading to what are now called the *indescribable cardinals*, and eventually Levy [1971] carried out a systematic study of the sizes of these cardinals.[107] The model-theoretic reflection idea thus provided a coherent scheme for viewing the bottom of an emerging hierarchy of large cardinals as a generalization of Replacement and Infinity, one that resonates with the procession of models in Zermelo [1930]. The heuristic of reflection had been broached in 1946 remarks by Gödel (cf. 3.4), and another point of contact is the formulation of the concept of *ordinal definable set* in those remarks. With the class of ordinal definable sets formalized by $\mathrm{OD} = \bigcup_\alpha \mathrm{def}(V_\alpha)$, the adequacy of this definition is based on some form of the Reflection Principle for ZF. With $\mathrm{tc}(y)$ denoting the smallest transitive set $\supseteq y$, let $\mathrm{HOD} = \{x \mid \mathrm{tc}(\{x\}) \subseteq \mathrm{OD}\}$. the class of *hereditarily ordinal definable sets*. As adumbrated by Gödel, HOD is an inner model in which AC, though not necessarily CH, holds. The basic results about this inner model were to be rediscovered several times.[108] In these several ways reflection phenomena both as heuristic and as principle became incorporated into set theory, bringing to the forefront what was to become a basic feature of the study of well-foundedness.

The set-theoretic generalization of first-order logic allowing transfinitely indexed logical operations was to lead to the solution of the problem of whether the least inaccessible cardinal can be measurable (cf. 3.2). Extending familiarity by abstracting to a new domain Tarski [1962] defined the *strongly compact* and *weakly compact* cardinals by ascribing natural generalizations of the key compactness property of first-order logic to the corresponding infinitary languages. These cardinals had figured in Erdős-Tarski [1943] (cf. 3.5) in combinatorial formulations that was later seen to imply that *a strongly compact cardinal is measurable,* and *a measurable cardinal is weakly compact.* Tarski [1962] pointed out that his student Hanf (cf. [1964]) established, using the satisfaction relation for infinitary

[107]See Kanamori [2003: §6].

[108]See Myhill-Scott [1971], especially p. 278.

languages, that *there are many inaccessible cardinals (and Mahlo cardinals) below a weakly compact cardinal. A fortiori, the least inaccessible cardinal is not measurable.* This breakthrough was the first result about the size of measurable cardinals since Ulam's original paper [1930] and was greeted as a spectacular success for metamathematical methods. Hanf's work radically altered size intuitions about problems coming to be understood in terms of large cardinals and ushered in model-theoretic methods into the study of large cardinals beyond the Mahlo cardinals.[109]

Weak compactness was soon seen to have a variety of characterizations; most notably in terms of 3.5, κ is weakly compact *iff* $\kappa \longrightarrow (\kappa)^2_2$ *iff* $\kappa \longrightarrow (\kappa)^n_\lambda$ for every $n \in \omega$ and $\lambda < \kappa$ *iff* κ is inaccessible and has the tree property. Erdős and Hajnal [1962] noted that the study of stronger partition properties had progressed to the point where a combinatorial proof that the least inaccessible cardinal is not measurable could have been given before Hanf came to his argument. However, model-theoretic methods quickly led to far stronger conclusions, particularly through the connection that had been made in Ehrenfeucht-Mostowski [1956] between partition properties and *sets of indiscernibles.*[110]

The concurrent emergence of the *ultraproduct construction* in model theory set the stage for the development of the modern theory of large cardinals in set theory. With a precursor in Skolem's [1933a; 1934] construction of a non-standard model of arithmetic the ultraproduct construction was brought to the forefront by Tarski and his students after Jerzy Łoś's [1955] adumbration of its fundamental theorem. This new method of constructing concrete models brought set theory and model theory even closer together in a surge of results and a lasting interest in ultrafilters. Measurable cardinals had been formulated (cf. 3.2) in terms of ultrafilters construed as two-valued measures; Jerome Keisler [1962] struck on the idea of taking the ultrapower of a measurable cardinal κ by a κ-complete ultrafilter over κ to give a new proof of Hanf's result, seeing the crucial point that the completeness property led to a well-founded, and so in his case well-ordered, structure.

Then Scott [1961] made the further, crucial move of taking the ultrapower of the universe V itself by such an ultrafilter. The full exercise of the transitive collapse as a generalization of the correlation of ordinals to well-ordered sets now led to an inner model $M \neq V$ and an elementary embedding $j\colon V \to M$.[111] With this Scott established: *If there is a measurable cardinal, then* $V \neq L$. Large cardinal hypotheses thus assumed a new significance as a means for maximizing possibilities away from Gödel's delimitative construction. Also, the Cantor-Gödel

[109]See Kanamori [2003: §4] for these results about strongly and weakly compact cardinals.

[110]See Kanamori [2003: §§7, 8, 9] for more on partition relations and sets of indiscernibles, particularly their role in the formulation the set of natural numbers $0^\#$ and its role of transcendence over L.

[111]That is, for any formula $\varphi(v_1, \ldots, v_n)$ and sets $x_1, \ldots, x_n$, $\varphi(x_1, \ldots, x_n) \longleftrightarrow \varphi^M(j(x_1), \ldots, j(x_n))$, i.e. the formula holds of the x_is exactly when it holds of the $j(x_i)$s with the quantifiers restricted to M. Thus elementary embeddings are just the extension of algebraic monomorphisms to the preservation of logical properties.

realist view of a fixed set-theoretic universe notwithstanding, Scott's construction fostered the manipulative use of inner models in set theory. The construction provided one direction and Keisler [1962a] the other of a new characterization that established a central structural role for measurable cardinals: *There is an elementary embedding* $j\colon V \to M$ *for some inner model* $M \neq V$ *iff there is a measurable cardinal.* This result is not formalizable in ZFC because of the use of the satisfaction relation and the existential assertion of a proper class, but technical versions are. Despite the lack of formalizability such existential assertions have been widely entertained since, and with this set theory in practice could be said to have overleaped the bounds of ZFC. On the other hand, that the existence of a class elementary embedding is equivalent to the existence of a certain set, the witnessing ultrafilter for a measurable cardinal, can be considered a means of formalization in ZFC, one that would be paradigmatic for such reductions.

Work of Petr Vopěnka, who started the active Prague seminar in set theory in the spring of 1963, would be closely connected to that of Scott. Aware of the limitations of inner models for establishing independence results Vopěnka (cf. [1965]) embarked on a systematic study of (mostly ill-founded) class models of Bernays-Gödel set theory using ultrapower and direct limit constructions. Vopěnka not only established [1962] Scott's result on the incompatibility of measurability and constructibility via different means, but he and his student Karel Hrbáček in their [1966] soon established a global generalization for inner models $L(A)$: *If there is a strongly compact cardinal, then* $V \neq L(A)$ *for any set* A.

Through model-theoretic methods set theory was brought to the point of entertaining elementary embeddings into well-founded models,[112] soon to be transfigured by a new method for getting well-founded *extensions* of well-founded models.

4 INDEPENDENCE

4.1 Forcing

Paul Cohen (1934–2007), born just before Gödel established his relative consistency results, established the independence of AC from ZF and the independence of CH from ZFC [1963; 1964]. That is, Cohen established that Con(ZF) implies Con(ZF + ¬AC) and Con(ZF) implies Con(ZFC + ¬CH). These results delimited ZF and ZFC in terms of the two fundamental issues at the beginnings of set theory. But beyond that, Cohen's proofs were soon to flow into method, becoming the inaugural examples of *forcing*, a remarkably general and flexible method for extending models of set theory. Forcing has strong intuitive underpinnings and reinforces the notion of set as given by the first-order ZF axioms with conspicuous uses of Replacement and Foundation. If Gödel's construction of L had launched set theory as a distinctive field of mathematics, then Cohen's method of forcing

[112] See Keisler-Tarski [1964] for a comprehensive account of the theory of large cardinals through the use of ultrapowers in the early 1960s.

began its transformation into a modern, sophisticated one.[113]

Cohen's approach was to start with a model M of ZF and adjoin a set G, one that would exhibit some desired new property. He realized that this had to be done in a minimal fashion in order that the resulting structure also model ZF, and so imposed restrictive conditions on both M and G. He took M to be a countable standard model, i.e. a countable transitive set that together with the membership relation restricted to it is a model of ZF.[114] The ordinals of M would then coincide with the predecessors of some ordinal ρ, and M would be the cumulative hierarchy $M = \bigcup_{\alpha<\rho} V_\alpha \cap M$. Cohen then established a system of terms to denote members of the new model, finding it convenient to use a ramified language: For each $x \in M$ let $\dot{x}$ be a corresponding constant; let $\dot{G}$ be a new constant; and for each $\alpha < \rho$ introduce quantifiers $\forall_\alpha$ and $\exists_\alpha$. Then develop a hierarchy of terms as follows: $\dot{M}_0 = \{\dot{G}\}$, and for limit ordinals $\delta < \rho$, $\dot{M}_\delta = \bigcup_{\alpha<\delta} \dot{M}_\alpha$. At the successor stage, let $\dot{M}_{\alpha+1}$ be the collection of terms $\dot{x}$ for $x \in V_\alpha \cap M$ and "abstraction" terms corresponding to formulas allowing parameters from $\dot{M}_\alpha$ and quantifiers $\forall_\alpha$ and $\exists_\alpha$. It is crucial that this ramified language with abstraction terms be entirely formalizable in M, through a systematic coding of symbols. Once a set G is provided from the outside, a model $M[G] = \bigcup_{\alpha<\rho} M_\alpha[G]$ would be determined by the terms, where each $\dot{x}$ is to be interpreted by x for $x \in M$ and $\dot{G}$ is to be interpreted by G, so that: $M_0[G] = \{G\}$; for limit ordinals $\delta < \rho$, $M_\delta[G] = \bigcup_{\alpha<\delta} M_\alpha[G]$; and $M_{\alpha+1}[G]$ consists of the sets in $V_\alpha \cap M$ together with sets interpreting the abstraction terms as the corresponding definable subsets of $M_\alpha[G]$ with $\forall_\alpha$ and $\exists_\alpha$ ranging over this domain.

But what properties can be imposed on G to ensure that $M[G]$ be a model of ZF? Cohen's key idea was to tie G closely to M through a system of sets in M called *conditions* that would approximate G. While G may not be a member of M, G is to be a subset of some $Y \in M$ (with $Y = \omega$ a basic case), and these conditions would "force" some assertions about the eventual $M[G]$ e.g. by deciding some of the membership questions, whether $x \in G$ or not, for $x \in Y$. The assertions are to be just those expressible in the ramified language, and Cohen developed a corresponding *forcing relation* $p \Vdash \varphi$, "p forces φ", between conditions p and formulas φ, a relation with properties reflecting his approximation idea. For example, if $p \Vdash \varphi$ and $p \Vdash \psi$, then $p \Vdash \varphi \,\&\, \psi$. The conditions are ordered according to the constraints they impose on the eventual G, so that if $p \Vdash \varphi$, and q is a stronger condition, then $q \Vdash \varphi$. Scott actually provided the now common forcing symbol $\Vdash$, and with Cohen having worked with prenex formulas, Scott showed how to proceed generally by separating out negation with: $p \Vdash \neg\varphi$ *iff* for no stronger condition q does $q \Vdash \varphi$. It was crucial to Cohen's approach that the forcing relation, like the ramified language, be definable in M.

[113]According to Scott (Bell [1985: ix]): "Set theory could never be the same after Cohen, and there is simply no comparison whatsoever in the sophistication of our knowledge about models of set theory today as contrasted to the pre-Cohen era."

[114]The existence of such a model is an avoidable assumption in formal relative consistency proofs via forcing.

The final ingredient is that the whole scaffolding is given life by incorporating a certain kind of set G. Stepping out of M and making the only use of its countability, Cohen enumerated the formulas of the ramified language in a countable sequence and required that G be completely determined by a countable sequence of stronger and stronger conditions $p_0, p_1, p_2, \ldots$ such that for every formula φ of the ramified language exactly one of φ or $\neg\varphi$ is forced by some p_n. Such a G is called a *generic* set. Cohen was able to show that the resulting $M[G]$ does indeed satisfy the axioms of ZF: Every assertion about $M[G]$ is already forced by some condition; the forcing relation is definable in M; and so the ZF axioms, holding in M, most crucially Power Set and Replacement, can be applied to derive corresponding forcing assertions about ZF axioms holding in $M[G]$.

The foregoing outline in its main features reflects how forcing was viewed by July 1963 and presented by Cohen himself in a course in Spring 1965.[115] He first described the case when $G \subseteq \omega$ and the conditions p are functions from some finite subset of ω into $\{0, 1\}$ and $p \Vdash \dot{n} \in \dot{G}$ if $p(n) = 1$ and $p \Vdash \dot{n} \notin \dot{G}$ if $p(n) = 0$. Today, a G so adjoined to M is called a *Cohen real over M*. Cohen established the independence of CH by adjoining a set which can be construed as a sequence of many Cohen reals. He established the independence of AC by a version of the above scheme where in addition to $\dot{G}$ there are also new constants $\dot{G}_i$ for $i \in \omega$, with $\dot{G}$ to be interpreted by a set X of Cohen reals, each an interpretation of some $\dot{G}_i$. The point is that X is not well-orderable in the extension.

The appeal to a countable model in Cohen's approach is a notable positive subsumption of Skolem's Paradox (cf. 3.2) into a new method. Remarkably, Skolem [1923: 229] had entertained the possibility of adjoining a new subset of the natural numbers to a countable model of Zermelo's system and getting a new model, adding in a footnote that "it is quite probable" that the Continuum Hypothesis is not decided by Zermelo's axioms. Just as starting with a countable standard model is not formally necessary for relative consistency results, other features of Cohen's argument would soon be reformulated, reorganized, and generalized, but the main thrust of his constructive approach through definability and genericity would remain. Cohen's particular achievement lies in devising a concrete procedure for extending well-founded models of set theory in a minimal fashion to well-founded models of set theory with new properties but without altering the ordinals.[116] Set theory had undergone a sea-change, and beyond how the subject was enriched, it is difficult to convey the strangeness of it.

The creation of forcing is a singular phenomenon in the development of set theory not only since it raised the level of the subject dramatically but also since it could well have occurred decades earlier. But however epochal Cohen's advance there was a line of development for which it did provide at least a semblance of con-

[115] See Cohen [1966].

[116] Scott continued (Bell [1985: ix]): "I knew almost all the set-theoreticians of the day, and I think I can say that no one could have guessed that the proof would have gone in just this way. Model-theoretic methods had shown us how many *non-standard* models there were; but Cohen, starting from very primitive first principles, found the way to keep the models *standard* (that is, with a well-ordered collection of ordinals)."

tinuity: Interest in independence results for weak versions of AC had been on the rise from the mid-1950's, with more and more sophisticated Fraenkel-Mostowski models being constructed.[117] Solomon Feferman, who was one of the first to whom Cohen explained his ideas for the independence proofs in the process of their development, was the first after him to establish results by forcing; Levy soon followed; and among their first results were new independences from ZF for weak versions of AC (Feferman-Levy [1963], Feferman [1965]). Cohen [1965: 40] moreover acknowledged the similarities between his AC independence result and the previous Fraenkel-Mostowski models. In fact, consistencies first established via Fraenkel-Mostowski models were soon "transferred" to consequence of ZF via forcing by correlating urelements with generic sets.[118]

After an initial result by Feferman [1963], Levy [1963; 1965; 1970] also probed the limits of ZFC definability, establishing consistency results about definable sets of reals and well-orderings and in descriptive set theory. Intriguingly, inaccessible cardinals were brought in to overcome a technical hurdle in this study; Levy [1963: IV] applied the defining properties of such a cardinal to devise its "collapse" to $\aleph_1$ by making every smaller ordinal countable, and this forcing is now known as the *Levy collapse.*

Forcing was quickly generalized and applied to achieve wide-ranging results, particularly by Robert Solovay. He above all epitomized this period of great expansion in set theory with his mathematical sophistication and fundamental results about and with forcing, and in the areas of large cardinals and descriptive set theory. Just weeks after Cohen's breakthrough Solovay [1963; 1965] elaborated the independence of CH by characterizing the possibilities for the size of 2^κ for regular κ and made the first exploration of a spectrum of cardinals. Then William Easton [1964; 1970] established the definitive result for powers of regular cardinals: *Suppose that* GCH *holds and F is a class function from the class of regular cardinals to cardinals such that for $\kappa \leq \lambda$, $F(\kappa) \leq F(\lambda)$ and the cofinality of $F(\kappa)$ is greater than κ. Then there is a forcing extension preserving cofinalities in which $2^\kappa = F(\kappa)$ for every regular κ.* Thus, as Solovay had seen locally, the *only* restriction beyond monotonicity on the power function for regular cardinals is that given by the Zermelo–Kőnig inequality.[119] Easton's result vitally infused forcing not only with the introduction of proper classes of forcing conditions but the now basic idea of a product analysis and the now familiar concept of *Easton support.* Through its reduction Easton's result focused interest on the possibilities for powers of *singular* cardinals, and this *Singular Cardinals Problem* together with the *Singular Cardinals Hypothesis* would stimulate the further development of set theory much as the Continuum Problem and the Continuum Hypothesis had stimulated its early development.[120]

[117]See Moore [1982: 5.1].

[118]See Felgner [1971] and Jech [1973] for more on the independence of weak versions of AC and transfers, and Pincus [1972] for a strong transfer theorem.

[119]See 1.3, especially footnote 27.

[120]The Singular Cardinal Hypothesis asserts that 2^λ for singular λ is the least possible with respect to the powers 2^μ for $\mu < \lambda$, as given by monotonicity and the Zermelo-Kőnig inequality.

In the Spring of 1964 Solovay [1965b; 1970] established a result remarkable for its mathematical depth and revelatory of what standard of argument was possible with forcing: *If there is an inaccessible cardinal, then in an inner model of a forcing extension every set of reals is Lebesgue measurable, has the Baire property, and has the perfect set property.* Like Cohen's results, this contextually decided issues dating back to the turn of the century and before; as described in 2.3 the regularity properties of sets of reals was a major concern of the early descriptive set theorists. Classical counterexamples show that Solovay's inner model cannot have a well-ordering of the reals, and so AC fails there. However, he established that the model satisfies the Principle of Dependent Choices, a principle sufficient for the formal rendition of the traditional theory of measure and category. Thus, Solovay's work vindicated the early descriptive set theorists in the sense that the regularity properties can consistently hold for all sets of reals in a *bona fide* model for mathematical analysis. For his result Solovay applied the Levy collapse of an inaccessible cardinal and built on its definability properties as first exploited by Levy [1963: IV]; for the Lebesgue measurability he introduced a new kind of forcing beyond Cohen's direct ways of adjoining new sets of ordinals or collapsing cardinals, that of adding a *random real*. Solovay's work not only opened the door to a wealth of different forcing arguments, but to this day his original definability arguments remain vital to descriptive set theory.

The perfect set property, central to Cantor's direct approach to the Continuum Problem through definability (1.2,2.3,2.5), led to the first acknowledged instance of a new phenomenon in set theory: the derivation of *equiconsistency* results with large cardinal hypotheses based on the complementary methods of forcing and inner models. A large cardinal hypothesis is typically transformed into a proposition about sets of reals by forcing that "collapses" that cardinal to $\aleph_1$ or "enlarges" the power of the continuum to that cardinal. Conversely, the proposition entails the same large cardinal hypothesis in the clarity of an inner model. Solovay's result provided the forcing direction from an inaccessible cardinal to the proposition that every set of reals has the perfect set property (and $\aleph_1$ is regular). But Ernst Specker [1957: 210] had in effect established that if this obtains, then $\aleph_1$ (of V) is inaccessible in L. Thus, Solovay's use of an inaccessible cardinal was actually necessary, and its collapse to $\aleph_1$ complemented Specker's observation. Other propositions figuring in the initial applications of inaccessibility to forcing turned out to require inaccessibility, further integrating it into the interstices of set theory.[121] The emergence of such equiconsistency results is a subtle transformation of earlier hopes of Gödel (cf. 3.4): Propositions can be positively subsumed if there

[121]The original application of the Levy collapse in Levy [1963: IV] also turned out to require an inaccessible cardinal (Levy [1970: 131ff]) – a remarkable turn of events for an apparently technical artifact at the beginning of forcing.

Many years later, Saharon Shelah [1980; 1984] was able to establish the necessity of Solovay's inaccessible for the proposition that every set of reals is Lebesgue measurable; on the other hand, Shelah also showed that the inaccessible is not necessary for the proposition that every set of reals has the Baire property.

are enough ordinals, how many being specified by positing a large cardinal.[122] On the other hand, forcing quickly led to the conclusion that there could be no direct implication for CH: Levy and Solovay (Levy [1964], Solovay [1965a], Levy-Solovay [1967]) established that measurable cardinals neither imply nor refute CH, with an argument generalizable to most inaccessible large cardinals. Rather, the subsumption for many other propositions would be in terms of consistency, the methods of forcing and inner models being the operative modes of argument. In a new synthesis of the two Cantorian legacies, hypotheses of length concerning the extent of the transfinite are correlated with hypotheses of width concerning sets of reals.

It was the incisive work of Scott and Solovay through this early period that turned Cohen's breakthrough into a general method of wide applicability. Scott simplified Cohen's original formulation as noted above; Solovay made the important move to general partial orders and generic filters; and they together developed, with vicissitudes, the formulation in terms of Boolean-valued models.[123] These models forcibly showed how to avoid Cohen's ramified language as well as his dependence on a countable model. With their elegant algebraic trappings and seemingly more complete information they held the promise of being the right approach to independence results. But Shoenfield [1971] showed that forcing with partial orders can get at the gist of the Boolean approach in a straightforward manner. Moreover, Boolean-valued models were soon found to be too abstract and unintuitive for establishing *new* consistency results, so that within a few years set theorists were generally working with partial orders. It is a testament to Cohen's concrete approach that in this return from abstraction even the use of ramified languages has played an essential role in careful forcing arguments at the interface of recursion theory and set theory.

4.2 Envoi

Building on his Lebesgue measurability result Solovay soon reactivated the classical descriptive set theory program (cf. 2.5) of investigating the extent of the regularity properties by providing characterizations for the $\mathbf{\Sigma}^1_2$ sets, the level at which Gödel established from $V = L$ the failure of the properties (cf. 3.4), and showed in particular that the regularity properties for these sets follow from the existence of

[122]There is a telling antecedent in the result of Gerhard Gentzen [1936; 1943] that the consistency strength of arithmetic can be exactly gauged by an ordinal ε_0, i.e. transfinite induction up to that ordinal in a formal system of notations. Although Hilbert's program of establishing consistency by finitary means could not be realized, Gentzen provided an exact analysis in terms of ordinal length. Proof theory blossomed in the 1960s with the analysis of other theories in terms of such lengths, the proof theoretic ordinals.

[123]Vopěnka had developed a similar concept in a reworking [1964] of the independence of CH. The concept was generalized and simplified in a series of papers on the so-called ∇-models from the active Prague seminar founded by Vopěnka (see Hájek [1971: 78]), culminating in the exposition Vopěnka [1967]. However, the earlier papers did not have much impact, partly because of an involved formalism in which formulas were valued in a complete lattice rather than Boolean algebra.

a measurable cardinal. Thus, although measurable cardinals do not decide CH, they do establish the perfect set property for $\mathbf{\Sigma}^1_2$ sets (Solovay [1969]) so that "CH holds for the $\mathbf{\Sigma}^1_2$ sets" – a vindication of Gödel's hopes for large cardinals through a direct implication. Donald Martin and Solovay in their [1969] then applied large cardinal hypotheses weaker than measurability to push forward the old tree representation ideas of the classical descriptive set theorists, with the hypotheses cast in the new role of securing well-foundedness in this context.

The method of forcing as part of the axiomatic tradition together with the transmutations of Cantor's two legacies, large cardinals furthering the extension of number into the transfinite and descriptive set theory investigating definable sets of reals, established set theory as a sophisticated field of mathematics, a study of well-foundedness expanded into one of consistency strength. With the further development of forcing through increasingly sophisticated iteration techniques questions raised in combinatorics and over a broad landscape would be resolved in terms of consistency, sometimes with equiconsistencies in terms of large cardinals. The theory of large cardinals would itself be much advanced with the heuristics of reflection and generalization and sustained through increasing use in the study of consistency strength. In the most distinctive and intriguing development of contemporary set theory, the investigation of the determinacy of games, large cardinals would be further integrated into descriptive set theory. They were not only used to literally incorporate well-foundedness of inner models into the study of tree representations, historically first context involving well-foundedness, but also to provide the exact hypotheses, with Woodin cardinals, for gauging consistency strength.[124]

Stepping back to gaze at modern set theory, the thrust of mathematical research should deflate various possible metaphysical appropriations with an onrush of new models, hypotheses, and results. Shedding much of its foundational burden, set theory has become an intriguing field of mathematics where formalized versions of truth and consistency have become matters for manipulation as in algebra. As a study couched in well-foundedness ZFC together with the spectrum of large cardinals serves as a court of adjudication, in terms of relative consistency, for mathematical statements that can be informatively contextualized in set theory by letting their variables range over the set-theoretic universe. Thus, set theory is more of an open-ended framework for mathematics rather than an elucidating foundation. It is as a field *of* mathematics that both proceeds with its own internal questions and is capable of contextualizing over a broad range which makes of set theory an intriguing and highly distinctive subject.

ACKNOWLEDGEMENTS

This is a revision with significant changes of the author's The Mathematical Development of Set Theory from Cantor to Cohen, *The Bulletin of Symbolic Logic*,

[124]See Kanamori [2003] for these recent developments.

volume 2, 1996, pages 1–71, and appears here with the permission of the *Bulletin*.

BIBLIOGRAPHY

[Addison *et al.*, 1965] J. W. Addison Jr., L. Henkin, and A. Tarski, eds. *The Theory of Models*, Proceedings of the 1963 International Symposium at Berkeley, North-Holland, Amsterdam, 1965.

[Aleksandrov, 1916] P. S. Aleksandrov. Sur la puissance des ensembles mesurables B, *Comptes Rendus Hebdomadaires des Séances de l'Académie des Sciences, Paris*, 162: 323–325, 1916.

[Aleksandrov and Urysohn, 1929] P. S. Aleksandrov and P. S. Urysohn. Mémoire sur les espaces topologiques compacts, *Verhandelingen der Koninklijke Nederlandse Akademie van Wetenschappen Tweed Reeks. Afdeling Natuurkunde*, 14: 1-96, 1929.

[Baire, 1898] R. Baire. Sur les fonctions discontinues qui se rattachent aux fonctions continues, *Comptes Rendus Hebdomadaires des Séances de l'Académie des Sciences, Paris*, 129:1621–1623, 1898.

[Baire, 1899] R. Baire. Sur les fonctions de variables réelles, *Annali di Matematica Pura ed Applicata*, 3:1-123, 1899.

[Banach, 1967] S. Banach. In S. Hartman and E. Marczewski, eds., *Oeuvres*, vol. 1, Państwowe Wydawnictwo Naukowe, Warsaw, 1967.

[Banach and Tarski, 1924] S. Banach and A. Tarski. Sur la décomposition des ensembles de points en parties respectivement congruentes, *Fundamenta Mathematicae*, 6: 244–277, 1924. Reprinted in [Banach, 1967, 118–148]. Reprinted in [Tarski, 1986, vol. 1, 121–154].

[Bar-Hillel *et al.*, 1961] Y. Bar-Hillel, E. I. J. Poznanski, M. O. Rabin, and A. Robinson, eds. *Essays on the Foundations of Mathematics*, Magnes Press, Jerusalem, 1961.

[Bartoszyński, 2008] T. Bartoszyński. Invariants of Meaure and Category, in [Foreman and Kanamori, 2008].

[Bell, 1985] J. L. Bell. *Boolean-Valued Models and Independence Proofs in Set Theory*, second edition, Oxford Logic Guides #12, Oxford University Press, Oxford, 1985.

[Benacerraf and Putnam, 1964] P. Benacerraf and H. Putnam, eds. *Philosophy of Mathematics. Selected Readings.* Prentice Hall, Englewood Cliffs, N.J., 1964.

[Benacerraf and Putnam, 1983] P. Benacerraf and H. Putnam, eds. Second edition of [1964]. Cambridge University Press, Cambridge, 1983.

[Bendixson, 1883] I. Bendixson. Quelques théorèmes de la théorie des ensembles de points, *Acta Mathematica*, 2:415–429, 1883.

[Bernays, 1937] P. Bernays. A system of axiomatic set theory, Part I, *The Journal of Symbolic Logic*, 2:65–77, 1937.

[Bernays, 1976] P. Bernays. A system of axiomatic set theory. In Gert H. Müller, ed., *Sets and Classes. On the Work of Paul Bernays*, pp. 1–119. North-Holland, Amsterdam, 1976. Individually appeared in various parts in *The Journal of Symbolic Logic*, 2:65–77, 1937; 6:1–17, 1941; 7:65–89 and 133–145, 1942; 8:89–106, 1943; 3:65–79, 1948; and 19:81–96, 1954.

[Bernstein, 1908] F. Bernstein. Zur Theorie der trigonometrischen Reihen, *Berichte über die Verhandlungen der Königlich Sächsischen Gesellschaft der Wissenschaften zu Leipzig, Mathematische-Physische Klasse*, 60:325–338, 1908.

[Blass, 1984] A. R. Blass. Existence of bases implies the Axiom of Choice. In J. E. Baumgartner, D. A. Martin, and S. Shelah, eds., *Axiomatic Set Theory*, pp. 31–33. Contemporary Mathematics vol. 31, American Mathematical Society, Providence, 1984.

[Blass, 2010] A. R. Blass. Combinatorial Cardinal Charateristics of the Continuum. In [Foreman and Kanamori, 2010].

[Boolos, 1971] G. S. Boolos. The iterative concept of set, *Journal of Philosophy*, 68:215–231, 1971. Reprinted in [Benacerraf and Putnam, 1983, 486–502], and in [Boolos, 1998, 13–29].

[Boolos, 1998] G. S. Boolos. *Logic, Logic, and Logic*, Harvard University Press, Cambridge, 1998.

[Borel, 1898] E. Borel. *Leçons sur la théorie des fonctions*, Gauthier-Villars, Paris, 1898.

[Borel, 1919] E. Borel. Sur la classification des ensembles de mesure nulle, *Bulletin de la Société Mathématique de France*, 47:97–125, 1919.

[Bourbaki, 1954] N. Bourbaki. *Eléments de Mathématique. I. Théorie des Ensembles*, Hermann, Paris, 1954.

[Bourbaki, 1970] N. Bourbaki. *Eléments de Mathematique. I. Théorie des Ensembles*, combined edition, Hermann, Paris, 1970.

[Brouwer, 1911] L. E. J. Brouwer. Beweis der Invarianz der Dimensionenzahl, *Mathematische Annalen*, 70:161–165, 1911. Reprinted in [Brouwer, 1976, 430–434].

[Brouwer, 1976] L. E. J. Brouwer. H. Freudenthal, ed., *Collected Works*, vol. 2, North-Holland, Amsterdam, 1976.

[Burali-Forti, 1897] C. Burali-Forti. Una questione sui numeri transfini, *Rendiconti del Circolo Matematico di Palermo*, 11: 154–164, 1897. Translated in [van Heijenoort, 1967, 104–111].

[Cantor, 1872] G. Cantor. Über die Ausdehnung eines Satzes aus der Theorie der trigonometrischen Reihen, *Mathematische Annalen*, 5:123–132, 1872. Reprinted in [Cantor, 1932, 92–102].

[Cantor, 1874] G. Cantor. Über eine Eigenschaft des Inbegriffes aller reellen algebraischen Zahlen, *Journal für die reine und angewandte Mathematik*, 77:258–262, 1874. Reprinted in [Cantor, 1932, 115–118]. Translated with commentary in [Ewald, 1996, 839–843].

[Cantor, 1878] G. Cantor. Ein Beitrag zur Mannigfaltigkeitslehre, *Journal für die reine und angewandte Mathematik* 84:242–258, 1878. Reprinted in [Cantor, 1932, 119–133].

[Cantor, 1879] G. Cantor. Über einen Satz aus der Theorie der stetigen Mannigfaltigkeiten, *Nachrichten von der Königlichen Gesellschaft der Wissenschaften und der Georg-Augusts-Universität zu Göttingen*, 127–135, 1879. Reprinted in [Cantor, 1932, 134–138].

[Cantor, 1880] G. Cantor. Über unendliche, lineare Punktmannigfaltigkeiten. II, *Mathematische Annalen* 17:355–358, 1880. Reprinted in [Cantor, 1932, 145–148], but excluding the referenced remark.

[Cantor, 1883] G. Cantor. Über unendliche, lineare Punktmannigfaltigkeiten. V. *Mathematische Annalen*, 21:545–591, 1883. Published separately as *Grundlagen einer allgemeinen Mannigfaltigkeitslehre. Ein mathematisch-philosophischer Versuch in der Lehre des Unendlichen*, B.G. Teubner, Leipzig, 1883. Reprinted in [Cantor, 1932, 165–209]. Translated with commentary in Ewald [1996], 878–920.

[Cantor, 1884] G. Cantor. Über unendliche, lineare Punktmannigfaltigkeiten. VI, *Mathematische Annalen*, 23:463–488, 1884. Reprinted in [Cantor, 1932, 210–246].

[Cantor, 1884a] G. Cantor. De la puissance des ensembles parfaits de points, *Acta Mathematica*, 4:381–392, 1884. Reprinted in [Cantor, 1932, 252–260].

[Cantor, 1887] G. Cantor. Mitteilungen zur Lehre vom Transfiniten, *Zeitschrift für Philosophie und philosophische Kritik*, 91:81–125 and 252–270, 1887. Reprinted in [Cantor, 1932, 378–439].

[Cantor, 1891] G. Cantor. Über eine elementare Frage der Mannigfaltigkeitslehre, *Jahresbericht der Deutschen Mathematiker-Vereinigung*, I:75–78, 1891. Reprinted in [Cantor, 1932, 278–280]. Translated with commentary in [Ewald, 1996, 920–922].

[Cantor, 1895] G. Cantor. Beiträge zur Begründung der transfiniten Mengenlehre. I, *Mathematische Annalen*, 46:481–512, 1895. Translated in [Cantor, 1915]. Reprinted in [Cantor, 1932, 282–311].

[Cantor, 1897] G. Cantor. Beiträge zur Begründung der transfiniten Mengenlehre. II, *Mathematische Annalen*, 49:207–246, 1897. Translated in [Cantor, 1915]. Reprinted in [Cantor, 1932, 312–351].

[Cantor, 1915] G. Cantor. *Contributions to the Founding of the Theory of Transfinite Numbers.* Translation of [Cantor, 1895] and [Cantor, 1897] with introduction and notes, by Philip E. B. Jourdain, Open Court, Chicago 1915. Reprinted Dover, New York, 1965.

[Cantor, 1932] G. Cantor. E. Zermelo, ed., *Gesammelte Abhandlungen mathematicschen und philosophischen Inhalts*, Julius Springer, Berlin 1932. Reprinted Springer, Berlin, 1980.

[Cavaillès, 1962] J. Cavaillès. *Philosophie Mathématique.* Paris, Hermann, 1962. Includes French translation of [Noether and Cavaillès, 1937].

[Coffa, 1979] J. A. Coffa. The humble origins of Russell's paradox, *Russell: The Journal of Betrand Russell Studies*, 99: 31–37, 1979.

[Cohen, 1963] P. J. Cohen. The independence of the Continuum Hypothesis. I, *Proceedings of the National Academy of Sciences USA*, 50:1143–1148, 1963

[Cohen, 1964] P. J. Cohen. The independence of the Continuum Hypothesis. II, *Proceedings of the National Academy of Sciences USA*, 51:105–110, 1964.

[Cohen, 1965] P. J. Cohen. Independence results in set theory, in [Addison *et al.*, 1965, 39-54].

[Dauben, 1979] J. W. Dauben. *Georg Cantor. His Mathematics and Philosophy of the Infinite,* Harvard University Press, Cambridge, 1979. Paperback edition, 1990.

[Dedekind, 1871] R. Dedekind. Über die Komposition der binären quadratischen Formen, supplement X to Dirichlet's *Vorlesungen über Zahlentheorie*, second edition, F. Vieweg, Braunschweig, 1871. Reprinted in [Dedekind, 1932, vol. 3, 399–407]. Excerpt translated with commentary in [Ewald, 1996, 762-765].

[Dedekind, 1872] R. Dedekind. *Stetigkeit und irrationale Zahlen,* F. Vieweg, Braunschweig, 1872. Fifth, 1927 edition reprinted in [Dedekind, 1932, vol. 3, 315–334]. Translated in [Dedekind, 1963, 1-27]. Translated with commentary in [Ewald, 1996, 765–779].

[Dedekind, 1888] R. Dedekind. *Was sind und was sollen die Zahlen?*, F. Vieweg, Braunschweig, 1888. Sixth, 1930 edition reprinted in [Dedekind, 1932, vol. 3, 335–390]. Second, 1893 edition translated in [Dedekind, 1963, 29-115]. Third, 1911 edition translated with commentary in [Ewald, 1996, 787–833].

[Dedekind, 1900] R. Dedekind. Über die von drei Moduln erzeugte Dualgruppe, *Mathematische Annalen*, 53:371–403, 1900. Reprinted in [Dedekind, 1932, vol. 2, 236–271].

[Dedekind, 1932] R. Dedekind. R. Fricke, E. Noether, and Ö. Ore, eds. *Gesammelte mathematische Werke*, F. Vieweg, Braunschweig 1932. Reprinted by Chelsea, New York, 1969.

[Dedekind, 1963] R. Dedekind. *Essays on the Theory of Numbers.* Translations by W. W. Beman, Dover, New York, 1963. (Reprint of original edition, Open Court, Chicago, 1901).

[du Bois-Reymond, 1869] P. du Bois-Reymond. Bemerkungen über die verschidenen Werthe, welche eine Function zweier reellen Variabeln erhält, wenn man diese Variabeln entweder nach einander oder gewissen Beziehungen gemäss gleichzeitig verschwinden lässt, *Journal für die reine und angewandte Mathematik*, 70:10–45, 1869.

[du Bois-Reymond, 1875] P. du Bois-Reymond. Über asymptotische Werthe, infinitäre Approximationen und infinitäre Auflösung von Gleichungen, *Mathematische Annalen*, 8:363–414, 1875.

[Dugac, 1976] P. Dugac. *Richard Dedekind et les fondements des mathématiques,* Collection des travaux de l'Académie Internationale d'Histoire des Sciences #24, Vrin, Paris, 1976.

[Easton, 1964] W. B. Easton. Powers of regular cardinals. Ph.D. thesis, Princeton University 1964. Abstracted as:Proper classes of generic sets, *Notices of the American Mathematical Society*, 11:205, 1964. Published in abridged form as [Easton, 1970].

[Easton, 1970] W. B. Easton. Powers of regular cardinals, *Annals of Mathematical Logic*, 1:139–178, 1970.

[Ebbinghaus, 2007] H.-D. Ebbinghaus. *Ernst Zermelo. An Approach to his Life and Work*, Springer, Berlin, 2007.

[Ehrenfeucht and Mostowski, 1956] A. Ehrenfeucht and A. M. Mostowski. Models of axiomatic theories admitting automorphisms, *Fundamenta Mathematicae,* 43:50–68, 1956. Reprinted in [Mostowski, 1979, 494–512].

[Erdős and Hajnal, 1962] P. Erdős and A. Hajnal. Some remarks concerning our paper "On the structure of set mappings", *Acta Mathematica Academiae Scientiarum Hungaricae*, 13:223–226, 1962.

[Erdős *et al.*, 1984] P. Erdős, A. Hajnal, A. Máté, and R. Rado. *Combinatorial Set theory:Partition Relations for Cardinals*, North-Holland, Amsterdam, 1984.

[Erdős *et al.*, 1965] P. Erdős, A. Hajnal, and R. Rado. Partition relations for cardinal numbers, *Acta Mathematica Academiae Scientiarum Hungaricae*, 16:93–196, 1965.

[Erdős and Rado, 1956] P. Erdős and R. Rado. A partition calculus in set theory, *Bulletin of the American Mathematical Society*, 62:427–489, 1956. Reprinted in [Gessel and Rota, 1987, 179–241].

[Erdős and Tarski, 1943] P. Erdős and A. Tarski. On families of mutually exclusive sets, *Annals of Mathematics*, 44:315–329, 1943. Reprinted in [Tarski, 1986, vol. 2, 591–605].

[Erdős and Tarski, 1961] P. Erdős and A. Tarski. On some problems involving inaccessible cardinals. In [Bar-Hillel, *et al.*, 1961, 50–82]. Reprinted in [Tarski, 1986 vol. 4, 79–111].

[Ewald, 1996] W. Ewald, ed. *From Kant to Hilbert:A Source Book in the Foundations of Mathematics*, two volumes, Clarendon Press, Oxford, 1996.

[Feferman, 1963] S. Feferman. Some applications of the notions of forcing and generic sets (summary), in [Addison *et al.*, 1965, 89–95].

[Feferman, 1965] S. Feferman. Some applications of the notions of forcing and generic sets, *Fundamenta Mathematicae*, 56:325–345, 1965.

[Feferman and Levy, 1963] S. Feferman and A. Levy. Independence results in set theory by Cohen's method. II (abstract), *Notices of the American Mathematical Society*, 10:593, 1963.

[Felgner, 1971] U. Felgner. *Models of ZF-Set Theory,* Lecture Notes in Mathematics #223, Springer, Berlin, 1971.

[Ferreirós, 1995] J. Ferreirós. "What fermented in me for years":Cantor's discovery of transfinite numbers, *Historia Mathematica*, 22:33–42, 1995.

[Ferreirós, 2007] J. Ferreirós. *Labyrinth of Thought:A History of Set Theory and its Role in Modern Mathematics*, second edition, Birkhauser, Basel, 2007.

[Fodor, 1956] G. Fodor. Eine Bemerking zur Theorie der regressiven Funktionen, *Acta Scientiarum Mathematicarum (Szeged)*, 17:139–142, 1956.

[Foreman and Kanamori, 2010] M. Foreman and A. Kanamori. *The Handbook of Set Theory,* Springer, Berlin, 2010.

[Fraenkel, 1921] A. A. Fraenkel. Über die Zermelosche Begründung der Mengenlehre (abstract), *Jahresbericht der Deutschen Mathematiker-Vereinigung*, 30II:97–98, 1921.

[Fraenkel, 1922] A. A. Fraenkel. Zu den Grundlagen der Cantor-Zermeloschen Mengenlehre, *Mathematische Annalen*, 86:230–237, 1922.

[Fraenkel, 1922a] A. A. Fraenkel. Über den Begriff 'definit' und die Unabhängigkeit des Auswahlaxioms, *Sitzungsberichte der Preussischen Akademie der Wissenschaften, Physikalisch-mathematische Klasse*, 253–257, 1922. Translated in [van Heijenoort, 1967, 284–289].

[Fraenkel, 1930] A. A. Fraenkel. Georg Cantor, *Jahresbericht der Deutschen Mathematiker-Vereinigung*, 39:189–266, 1930. Published separately as *Georg Cantor*, B.G. Teubner, Leipzig, 1930. Published in abridged form in [Cantor, 1932, 452–483].

[Fraenkel, 1953] A. A. Fraenkel. *Abstract Set Theory,* North-Holland, Amsterdam, 1953.

[Frege, 1879] G. Frege. *Begriffsschrift, eine der arithmetischen nachgëbildete Formelsprache des reinen Denkens,* Nebert, Halle, 1879. Reprinted Hildesheim, Olms, 1964. Translated in [van Heijenoort, 1967, 1–82].

[Frege, 1884] G. Frege. *Die Grundlagen der Arithmetik, eine logisch-mathematische Untersuchung über den Begriff der Zahl,* Wilhelm Köbner, Breslau, 1884. Translated with German text by J. L. Austin, as *The Foundations of Arithmetic, A logico-mathematical enquiry into the concept of number*; Blackwell, Oxford, 1950. Later editions without German text, Harper, New York.

[Frege, 1891] G. Frege. Function und Begriff, Hermann Pohle, Jena, 1891. Translated in [Frege, 1952, 21–41].

[Frege, 1893] G. Frege. *Grundgesetze der Arithmetik, Begriffsschriftlich abgeleitet.* Vol. 1, Hermann Pohle, Jena 1893. Reprinted Hildesheim, Olms, 1962. Partially translated in [Frege, 1964].

[Frege, 1895] G. Frege. Kritische Beleuchtung einiger Punkte in E. Schröders *Vorlesungen über die Algebra der Logik, Archiv für systematische Philosophie*, 1:433–456, 1895, 433–456. Translated in [Frege, 1952, 86–106].

[Frege, 1903] G. Frege. Vol. 2 of [1893]. Reprinted Hildesheim, Olms, 1962.

[Frege, 1952] G. Frege. P. Geach and M. Black, trans. and eds., *Translations from the Philosophical Writings of Gottlob Frege*, Blackwell, Oxford, 1952. Second, revised edition, 1960. Latest edition, Rowland & Littlewood, Ottowa, 1980.

[Frege, 1964] G. Frege. M. Furth, trans. and ed., *The Basic Laws of Arithmetic. Exposition of the System*, University of California Press, Berkeley, 1964.

[Garciadiego, 1992] A. Garciadiego. *Bertrand Russell and the Origins of the Set-theoretic "Paradoxes",* Birkhäuser, Boston, 1992.

[Gentzen, 1936] G. Gentzen. Die Widerspruchsfreiheit der reinen Zahlentheorie, *Mathematicshe Annalen*, 112:493–565, 1936. Translated in [Gentzen, 1969, 132–213].

[Gentzen, 1943] G. Gentzen. Beweisbarkeit und Unbeweisbarkeit von Anfangsfällen der transfiniten Induktion in der reinen Zahlentheorie, *Mathematicshe Annalen*, 119:140–161, 1943. Translated in [Gentzen, 1969, 287–308].

[Gentzen, 1969] G. Gentzen. M. E. Szabo, ed., *The Collected Papers of Gerhard Gentzen*, North-Holland, Amsterdam, 1969.

[Gessel and Rota, 1987] I. Gessel and G.-C. Rota, eds. *Classic Papers in Combinatorics,* Birkhäuser, Boston, 1987.

[Gödel, 1930] K. F. Gödel. Die Vollständigkeit der Axiome des logischen Funktionenkalküls, *Monatshefte für Mathematik und Physik*, 37:349–360, 1930. Reprinted and translated in [Gödel, 1986, 102–123].

[Gödel, 1931] K. F. Gödel. Über formal unentscheidbare Sätze der *Principia Mathematica* und verwandter Systeme I, *Monatshefte für Mathematik und Physik*, 38:173–198, 1931. Reprinted and translated with minor emendations by the author in [Gödel, 1986, 144–195].

[Gödel, 1933] K. F. Gödel. The present situation in the foundations of mathematics. In [Gödel, 1995, 45–53], and the page references are to these.

[Gödel, 1938] K. F. Gödel. The consistency of the Axiom of Choice and of the Generalized Continuum Hypothesis, *Proceedings of the National Academy of Sciences USA*, 24:556–557, 1938. Reprinted in [Gödel, 1990, 26–27].

[Gödel, 1939] K. F. Gödel. Consistency-proof for the Generalized Continuum-Hypothesis, *Proceedings of the National Academy of Sciences USA*, 25:220–224, 1939. Reprinted in [Gödel, 1990, 28–32].

[Gödel, 1939a] K. F. Gödel. Vortrag Göttingen, text and translation in [Gödel, 1995, 126–155], and the page references are to these.

[Gödel, 1940] K. F. Gödel. *The Consistency of the Axiom of Choice and of the Generalized Continuum Hypothesis with the Axioms of Set Theory,* Annals of Mathematics Studies #3, Princeton University Press, Princeton, 1940. Reprinted in [Gödel, 1990, 33–101].

[Gödel, 1947] K. F. Gödel. What is Cantor's Continuum Problem?, *American Mathematical Monthly*, 54:515–525, 1947. Errata, 55:51, 1948. Reprinted in [Gödel, 1990, 176–187]. Revised and expanded version in [Benacerraf and Putnam, 1964, 258–273]. This version reprinted with minor emendations by the author in [Gödel, 1990, 254–270].

[Gödel, 1986] K. F. Gödel. S. Feferman, *et al.*, eds., *Kurt Gödel, Collected Works*, vol. I, Oxford University Press, Oxford, 1986.

[Gödel, 1990] K. F. Gödel. S. Feferman, *et al.*, eds., *Kurt Gödel, Collected Works*, vol. II, Oxford University Press, New York, 1990.

[Gödel, 1995] K. F. Gödel. S. Feferman, *et al.*, eds., *Kurt Gödel, Collected Works*, vol. III, Oxford University Press, New York, 1995.

[Gödel, 2003] K. F. Gödel. S. Feferman and J. W. Dawson, eds., *Kurt Gödel, Collected Works*, vol. V, Clarendon Press, Oxford, 2003.

[Goldfarb, 1979] W. D. Goldfarb. Logic in the Twenties:the nature of the quantifier, *The Journal of Symbolic Logic*, 44:351–368, 1979.

[Graham *et al.*, 1990] R. L. Graham, B. L. Rothschild, and J. H. Spencer. *Ramsey Theory*, second edition, Wiley & Sons, New York, 1990.

[Grattan-Guinness, 1978] I. Grattan-Guinness. Wiener on the logics of Russell and Schröder. An account of his doctoral thesis, and of his subsequent discussion of it with Russell, *Annals of Science*, 32:103–132, 1978.

[Grattan-Guinness, 1978a] I. Grattan-Guinness. How Bertrand Russell discovered his paradox, *Historia Mathematica*, 5:127–137, 1978.

[Grattan-Guinness, 2000] I. Grattan-Guinness. *The Search for Mathematical Roots 1870–1940:Logics, Set Theories and the Foundations of Mathematics from Cantor through Russell and Gödel*, Princeton University Press, Princeton, 2000.

[Gray, 1994] R. Gray. Georg Cantor and Transcendental Numbers, *American Mathematical Monthly*, 101:819–832, 1994.

[Hájek, 1971] P. Hájek. Sets, semisets, models. In [Scott, 1971, 67-81].

[Hajnal, 1956] A. Hajnal. On a consistency theorem connected with the generalized continuum problem, *Zeitschrift für Mathematische Logik und Grundlagen der Mathematik*, 2:131–136, 1956.

[Hajnal, 1961] A. Hajnal. On a consistency theorem connected with the generalized continuum problem, *Acta Mathematica Academiae Scientiarum Hungaricae*, 12:321–376, 1961.

[Hajnal and Larson, 2010] A. Hajnal and J. A. Larson. Partition relations. In [Foreman and Kanamori, 2010].

[Hallett, 1984] M. Hallett. *Cantorian Set Theory and Limitation of Size*, Oxford Logic Guides #10, Clarendon Press, Oxford, 1984.

[Hamel, 1905] G. Hamel. Eine Basis aller Zahlen und die unstetigen Lösungen der Funktionalgleichung:$f(x+y) = f(x) + f(y)$, *Mathematische Annalen*, 60:459–462, 1905.

[Hanf, 1964] W. P. Hanf. Incompactness in languages with infinitely long expressions, *Fundamenta Mathematicae*, 53:309–324, 1964.

[Hartogs, 1915] F. Hartogs. Über das Problem der Wohlordnung, *Mathematische Annalen*, 76:436–443, 1915.

[Hardy, 1903] G. H. Hardy. A theorem concerning the infinite cardinal numbers, *The Quarterly Journal of Pure and Applied Mathematics*, 35:87–94, 1903. Reprinted in [Hardy *et al.*, 1979, vol. 7, 427–434].

[Hardy *et al.*, 1979] G. H. Hardy. I. W. Busbridge and R. A. Rankin, eds., *Collected Papers of G. H. Hardy*, Clarendon, Oxford, 1979.

[Hausdorff, 1908] F. Hausdorff. Grundzüge einer Theorie der geordneten Mengen, *Mathematische Annalen*, 76:436–443, 1908.

[Hausdorff, 1914] F. Hausdorff. *Grundzüge der Mengenlehre*, de Gruyter, Leipzig 1914. Reprinted by Chelsea, New York, 1965.

[Hausdorff, 1914a] F. Hausdorff. Bemerkung über den Inhalt von Punktmengen, *Mathematische Annalen*, 75:428–433, 1914.

[Hausdorff, 1916] F. Hausdorff. Die Mächtigkeit der Borelschen Mengen, *Mathematische Annalen*, 77:430–437, 1916.

[Hausdorff, 1932] F. Hausdorff. Zur Theorie der linearen metrischen Räume, *Journal für die reine und angewandte Mathematik*, 167:294–311, 1932.

[Hausdorff, 1937] F. Hausdorff. *Mengenlehre*, third, revised edition of [1914]. Translated by John R. Auman as *Set Theory*, Chelsea, New York, 1962.

[Hawkins, 1975] T. W. Hawkins. *Lebesgue's Theory of Integration. Its Origins and Development,* second edition, Chelsea, New York, 1975.

[Heck, 1995] R. G. Heck, Jr. Definition by induction in Frege's *Grundgesetze der Arithmetik.* In W. Demopoulos, ed., *Frege's Philosophy of Mathematics*, pp. 295–233. Harvard University Press, Cambridge, 1995.

[Henkin, 1949] L. Henkin. The completeness of the first-order functional calculus, *The Journal of Symbolic Logic*, 14:159–166, 1949.

[Hessenberg, 1906] G. Hessenberg. *Grundbegriffe der Mengenlehre,* Vandenhoeck & Ruprecht, Göttingen, 1906. Reprinted in *Abhandlungen der Fries'schen Schule, Neue Folge*, 1:479–706, 1906.

[Hilbert, 1890] D. Hilbert. Über die Theorie der algebraischen Formen, *Mathematische Annalen*, 36:473–534, 1890. Translated in [Hilbert, 1978, 143–224].

[Hilbert, 1900] D. Hilbert. Mathematische Probleme. Vortrag, gehalten auf dem internationalem Mathematiker-Kongress zu Paris, 1900. *Nachrichten von der Königlichen Gesellschaft der Wissenschaften zu Göttingen*, pp. 253–297, 1900. Translated in the *Bulletin of the American Mathematical Society*, 8:437–479, 1902.

[Hilbert, 1926] D. Hilbert. Über das Unendliche, *Mathematische Annalen*, 95:161–190, 1926. Translated in [van Heijenoort, 1967, 367–392]. Partially translated in [Benacerraf and Putnam, 1983, 183–201].

[Hilbert, 1978] D. Hilbert. M. Ackerman, trans., *Hilbert's Invariant Theory Papers*, Math Sci Press, Brookline, 1978.

[Hilbert and Ackermann, 1928] D. Hilbert and W. Ackermann. *Grundzüge der theoretischen Logik*, Julius Springer, Berlin, 1928. Second edition, 1938; third edition, 1949. Second edition was translated by L. M. Hammond, G. G. Leckie, and F. Steinhardt as *Principles of Mathematical Logic*, Chelsea, New York, 1950.

[Howard and Rubin, 1998] P. Howard and J. E. Rubin. *Consequences of the Axiom of Choice*, Mathematical Surveys and Monographs, vol. 59, American Mathematical Society, Providence, 1998.

[Jané, 1995] I. Jané. The role of the absolute infinite in Cantor's conception of set, *Erkenntnis*, 42:375–402, 1995.

[Jech, 1973] T. J. Jech. *The Axiom of Choice*, North-Holland, Amsterdam, 1973.

[Jech, 2003] T. J. Jech. *Set Theory*, the third millennium edition, revised and expanded, Springer, Berlin, 2003.

[Jourdain, 1904] P. E. B. Jourdain. On the transfinite cardinal numbers of well-ordered aggregates, *Philosophical Magazine*, 7:61–75, 1904.

[Jourdain, 1905] P. E. B. Jourdain. On a proof that every aggregate can be well-ordered, *Mathematische Annalen*, 60:465–470, 1905.

[Kac and Ulam, 1968] M. Kac and S. M. Ulam. *Mathematics and Logic*, Praeger, New York, 1968.

[Kanamori, 1995] A. Kanamori. The emergence of descriptive set theory. In J. Hintikka, ed., *Essays on the Development of the Foundations of Mathematics*, pp. 241–266. Kluwer, Dordrecht, 1995.

[Kanamori, 1997] A. Kanamori. The mathematical import of Zermelo's well-ordering theorem, *The Bulletin of Symbolic Logic*, 3:281–311, 1997.

[Kanamori, 2003] A. Kanamori. *The Higher Infinite. Large Cardinals in Set Theory from their Beginnings*, second edition, Springer, Berlin, 2003.

[Kanamori, 2003a] A. Kanamori. The empty set, the singleton, and the ordered pair, *The Bulletin of Symbolic Logic*, 9:273–298, 2003.

[Kanamori, 2004] A. Kanamori. Zermelo and set theory, *The Bulletin of Symbolic Logic*, 10:487–553, 2004.

[Kanamori, 2007] A. Kanamori. Gödel and set theory, *The Bulletin of Symbolic Logic*, 13: 153–188, 2007.

[Kechris and Louveau, 1987] A. S. Kechris and A. Louveau. *Descriptive Set Theory and the Structure of Sets of Uniqueness*, London Mathematical Society Lecture Note Series #128, Cambridge University Press, Cambridge, 1987.

[Keisler, 1962] H. J. Keisler. Some applications of the theory of models to set theory. In E. Nagel, P. Suppes, and A. Tarski, eds., *Logic, Methodology and Philosophy of Science*, Proceedings of the 1960 [and first] International Congress (Stanford, California), pp. 80–86. Stanford University Press, Stanford, 1962.

[Keisler, 1962a] H. J. Keisler. The equivalence of certain problems in set theory with problems in the theory of models (abstract), *Notices of the American Mathematical Society*, 9:339–340, 1962.

[Keisler and Tarski, 1964] H. J. Keisler and A. Tarski. From accessible to inaccessible cardinals, *Fundamenta Mathematicae*, 53:225–308, 1964. Corrections, 57:119, 1965. Reprinted in [Tarski, 1986, vol. 4, 129–213].

[Kennedy, 1975] H. C. Kennedy. Nine letters from Giuseppe Peano to Bertrand Russell, *Journal of the History of Philosophy*, 13:205–220, 1975.

[Kőnig, 1927] D. Kőnig. Über eine Schlussweise aus dem Endlichen ins Unendliche: Punktmengen. Kartenfärben. Verwandtschaftsbeziehungen. Schachspiel. *Acta Litterarum ac Scientiarum Regiae Universitatis Hungaricae Francisco-Josephinae, sectio scientiarum mathematicarum*, 3:121–130, 1927.

[Kőnig, 1905] J. (G.) Kőnig. Zum Kontinuum-Problem. In A. Krazer, ed., *Verhandlungen des Dritten Internationalen Mathematiker-Kongresses in Heidelberg vom 8. bis 13. August 1904*, pp. 144–147. B.G. Teubner, Leipzig, 1905. Reprinted in *Mathematische Annalen*, 60:177–180, 1905, and Berichtigung, 462.

[Kunen and Vaughan, 1984] K. Kunen and J. E. Vaughan, eds. *Handbook of Set-Theoretic Topology*, North-Holland, Amsterdam, 1984.

[Kuratowski, 1921] K. Kuratowski. Sur la notion de l'ordre dans la théorie des ensembles, *Fundamenta Mathematicae*, 2:161–171, 1921. Reprinted in [Kuratowski, 1988, 1–11].

[Kuratowski, 1922] K. Kuratowski. Une méthode d'élimination des nombres transfinis des raisonnements mathématiques, *Fundamenta Mathematicae*, 3:76–108, 1922.

[Kuratowski, 1931] K. Kuratowski. Evaluation de la class Borélienne ou projective d'un ensemble de points à l'aide des symboles logiques, *Fundamenta Mathematicae*, 17: 249–272, 1931.

[Kuratowski *et al.*, 1988] K. Kuratowski. K. Borsuk *et al.*, eds., *Selected Papers*, Państwowe Wydawnictwo Naukowe, Warsaw, 1988.

[Kuratowski and Tarski, 1931] K. Kuratowski and A. Tarski. Les opérations logiques et les ensembles projectifs, *Fundamenta Mathematicae*, 17:240–248, 1931. Reprinted in [Tarski, 1986, vol. 1, 551–559], and in [Kuratowski, 1988, 367–375]. Translated in [Tarski, 1983, 143–151].

[Kurepa, 1935] Đ. R. Kurepa. *Ensembles ordonnés et ramifiés*, Thèse, Paris. Published as:*Publications mathématiques de l'Université de Belgrade*, 4:1–138, 1935.

[Kurepa, 1942] Đ. R. Kurepa. propos d'une généralisation de la notion d'ensembles bien ordonnés, *Acta Mathematica*, 75:139–150, 1942.

[Kurepa, 1959] Đ. R. Kurepa. On the cardinal number of ordered sets and of symmetrical structures in dependence on the cardinal numbers of its chains and antichains, *Glasnik Matematičko-fizički i astronomski, Periodicum mathematico-physicum et astronomicum*, 14:183–203, 1959.

[Lavine, 1994] S. M. Lavine. *Understanding the Infinite*, Harvard University Press, Cambridge, 1994.

[Lebesgue, 1902] H. Lebesgue. Intégrale, longueur, aire, *Annali di Matematica Pura ed Applicata*, 7:231-459, 1902. Reprinted in [Lebesgue, 1972, vol. 1, 203–331].

[Lebesgue, 1905] H. Lebesgue. Sur les fonctions représentables analytiquement, *Journal de mathématiques pures et appliquées*, 1:139–21, 1905. Reprinted in [Lebesgue, 1972, vol. 3, 103–180].

[Lebesgue, 1972] H. Lebesgue. *Oeuvres Scientifiques*, Kundig, Geneva, 1972.

[Levy, 1957] A. Levy. Indépendance conditionelle de $V = L$ et d'axiomes qui se rattachent au système de M. Gödel, *Comptes Rendus Hebdomadaires des Séances de l'Académie des Sciences, Paris*, 245:1582–1583, 1957.

[Levy, 1960] A. Levy. Axiom schemata of strong infinity in axiomatic set theory, *Pacific Journal of Mathematics*, 10: 223–238, 1960.

[Levy, 1960a] A. Levy. Principles of reflection in axiomatic set theory, *Fundamenta Mathematicae*, 49:1–10, 1960.

[Levy, 1963] A. Levy. Independence results in set theory by Cohen's method. I, III, IV (abstracts), *Notices of the American Mathematical Society*, 10:592–593, 1963.

[Levy, 1964] A. Levy. Measurable cardinals and the continuum hypothesis (abstract), *Notices of the American Mathematical Society*, 11:769–770, 1964.

[Levy, 1965] A. Levy. Definability in axiomatic set theory I. In Y. Bar-Hillel, ed., *Logic, Methodology and Philosophy of Science*, Proceedings of the 1964 International Congress, Jerusalem, pp. 127–151. North-Holland, Amsterdam, 1965.

[Levy, 1970] A. Levy. Definability in axiomatic set theory II. In Y. Bar-Hillel, ed., *Mathematical Logic and Foundations of Set Theory*, pp. 129–145. North-Holland, Amsterdam, 1970.

[Levy and Solovay, 1967] A. Levy and R. M. Solovay. Measurable cardinals and the Continuum Hypothesis, *Israel Journal of Mathematics*, 5:234–248, 1967.

[Lindelöf, 1905] E. Lindelöf. Remarques sur un théorème fondamental de la théorie des ensembles, *Acta Mathematica*, 29:183–190, 1905.

[Lindenbaum and Tarski, 1926] A. Lindenbaum and A. Tarski. Communication sur les recherches de la théorie des ensembles, *Sprawozdana z Posiedzeń Towarzystwa Naukowego Warszawskiego, Wydział III, Nauk Matematyczno-fizycznych (Comptes Rendus des Séances de la Société des Sciences et des Lettres de Varsovie, Classe III, Sciences Mathématiques et Physiques)* , 19:299–330, 1926. Reprinted in [Tarski, 1986, vol. 1, 171–204].

[Liouville, 1844] J. Liouville. Des remarques relatives 1° à des classes très-étendues de quantités dont la valeur n'est ni rationelle ni même réductible à des irrationnelles algébriques; 2° à un passage du livre des *Principes* où Newton calcule l'action exercée par une sphère sur un point extérieur, *Comptes Rendus Hebdomadaires des Séances de l'Académie des Sciences, Paris*, 18:883–885, 1844.

[Liouville, 1851] J. Liouville. Sur des classes très-étendues de quantités dont la valeur n'est ni algébrique ni même réductible à des irrationnelles algébriques, *Journal de mathématiques pures et appliquées*, 16:133–142, 1851.

[Loś, 1955] J. Loś. Quelques remarques, théorèmes et problèmes sur les classes définissables d'algèbres. In T. Skolem, *et al.*, eds., *Mathematical Interpretation of Formal Systems*, pp. 98–113. North-Holland, Amsterdam, 1955.

[Löwenheim, 1915] L. Löwenheim. Über Möglichkeiten im Relativkalkul, *Mathematische Annalen*, 76:447–470, 1915. Translated in [van Heijenoort, 1967, 228–251].

[Luzin, 1914] N. N. Luzin. Sur un problème de M. Baire, *Comptes Rendus Hebdomadaires des Séances de l'Académie des Sciences, Paris*, 158:1258–1261, 1914.

[Luzin, 1917] N. N. Luzin. Sur la classification de M. Baire, *Comptes Rendus Hebdomadaires des Séances de l'Académie des Sciences, Paris*, 164:91–94, 1917.

[Luzin, 1925] N. N. Luzin. Sur un problème de M. Emile Borel et les ensembles projectifs de M. Henri Lebesgue; les ensembles analytiques, *Comptes Rendus Hebdomadaires des Séances de l'Académie des Sciences, Paris*, 180:1318–1320, 1925.

[Luzin, 1925a] N. N. Luzin. Sur les ensembles projectifs de M. Henri Lebesgue, *Comptes Rendus Hebdomadaires des Séances de l'Académie des Sciences, Paris*, 180:1572–1574, 1925.

[Luzin, 1925b] N. N. Luzin. Les propriétés des ensembles projectifs, *Comptes Rendus Hebdomadaires des Séances de l'Académie des Sciences, Paris*, 180:1817–1819, 1925.

[Luzin and Sierpiński, 1918] N. N. Luzin and W. Sierpiński. Sur quelques propriétés des ensembles (A), *Bulletin de l'Académie des Sciences de Cracovie, Classe des Sciences Mathématiques et Naturelles, Série A*, pp. 35–48, 1918. Reprinted in [Sierpiński, 1975, 192–204].

[Luzin and Sierpiński, 1923] N. N. Luzin and W. Sierpiński. Sur un ensemble non mesurable B, *Journal de Mathématiques Pures et Appliquées*, 2:53–72, 1923. Reprinted in [Sierpiński, 1975, 504-519].

[Mahlo, 1911] P. Mahlo. Über lineare transfinite Mengen, *Berichte über die Verhandlungen der Königlich Sächsischen Gesellschaft der Wissenschaften zu Leipzig, Mathematische-Physische Klasse*, 63:187–225, 1911.

[Mahlo, 1912] P. Mahlo. Zur Theorie und Anwendung der ρ_0-Zahlen, *Berichte über die Verhandlungen der Königlich Sächsischen Gesellschaft der Wissenschaften zu Leipzig, Mathematische-Physische Klasse*, 64:108–112, 1912.

[Mahlo, 1913] P. Mahlo. Zur Theorie und Anwendung der ρ_0-Zahlen II, *Berichte über die Verhandlungen der Königlich Sächsischen Gesellschaft der Wissenschaften zu Leipzig, Mathematische-Physische Klasse*, 65:268–282, 1913.

[Mahlo, 1913a] P. Mahlo. Über Teilmengen des Kontinuums von dessen Mächtigkeit, *Berichte über die Verhandlungen der Königlich Sächsischen Gesellschaft der Wissenschaften zu Leipzig, Mathematische-Physische Klasse*, 65:283–315, 1913.

[Martin and Solovay, 1969] D. A. Martin and R. M. Solovay. A basis theorem for Σ^1_3 sets of reals, *Annals of Mathematics*, 89:138–159, 1969.

[Mathias, 2001] A. R. D. Mathias. Slim models of Zermelo set theory, *The Journal of Symbolic Logic*, 66:487–496, 2001.

[Meschkowski, 1983] H. Meschkowski. *Georg Cantor. Leben. Werk und Wirkung*, Bibliographisches Institut, Mannheim, 1983.

[Meschkowski and Nilson, 1991] H. Meschkowski and W. Nilson. *Briefe / Georg Cantor*, Springer, Berlin, 1991.

[Miller, 1984] A. W. Miller. Special subsets of the real line. In [Kunen and Vaughan, 1984, 201–233].

[Mirimanoff, 1917] D. Mirimanoff. Les antinomies de Russell et de Burali-Forti et le problème fondamental de la théorie des ensembles, *L'Enseignment Mathématique*, 19:37–52, 1917.

[Mirimanoff, 1917a] D. Mirimanoff. Remarques sur la théorie des ensembles et les antinomies Cantoriennes. I, *L'Enseignment Mathématique*, 19:209–217, 1917.

[Montague, 1956] R. M. Montague. Zermelo-Fraenkel set theory is not a finite extension of Zermelo set theory (abstract), *Bulletin of the American Mathematical Society*, 62:260, 1956.

[Montague, 1961] R. M. Montague. Fraenkel's addition to the axioms of Zermelo. In [Bar-Hillel *et al.*, 1961, 91–114].

[Moore, 1982] G. H. Moore. *Zermelo's Axiom of Choice. Its Origins, Development and Influence*, Springer, New York, 1982.

[Moore, 1988] G. H. Moore. The roots of Russell's paradox, *Russell: The Journal of Bertrand Russell Studies*, 8:46–56, 1988.

[Moore, 1988a] G. H. Moore. The emergence of first-order logic. In W. Aspray and P. Kitcher, eds., *History and Philosophy of Modern Mathematics*, Minnesota Studies in the Philosophy of Science, vol. 11, pp. 95–135. University of Minnesota Press, Minneapolis, 1988.

[Moore, 1989] G. H. Moore. Towards a history of Cantor's Continuum Problem. In D. E. Rowe and J. McCleary, eds., *The History of Modern Mathematics*, vol. 1:*Ideas and their Reception*, pp. 79–121. Academic Press, Boston, 1989.

[Moore, 2002] G. H. Moore. Hilbert on the infinite:The role of set theory in the evolution of Hilbert's thought, *Historia Mathematics*, 29:40–64, 2002.

[Moore and Garciadiego, 1981] G. H. Moore and A. R. Garciadiego. Burali-Forti's paradox:A reappraisal of its origins, *Historia Mathematica*, 8:319–350, 1981.

[Moschovakis, 1980] Y. Moschovakis. *Descriptive Set Theory*, North-Holland, Amsterdam, 1980.

[Mostowski, 1939] A. M. Mostowski. Über die Unabhängigkeit des Wohlordnungssatzes vom Ordnungsprinzip, *Fundamenta Mathematicae*, 32:201–252, 1939. Translated in [Mostowski, 1979, vol. 1, 290–338].

[Mostowski, 1949] A. M. Mostowski. An undecidable arithmetical statement, *Fundamenta Mathematicae*, 36:143–164, 1949. Reprinted in [Mostowski, 1979, vol. 1, 531–552].

[Mostowski, 1950] A. M. Mostowski. Some impredicative definitions in the axiomatic set theory, *Fundamenta Mathematicae*, 37:111–124, 1950. Correction in 38:238, 1952.

[Mostowski, 1979] A. M. Mostowski. *Foundational Studies. Selected Works,* North-Holland, Amsterdam, 1979.

[Myhill and Scott, 1971] J. R. Myhill and D. S. Scott. Ordinal definability. In [Scott, 1971, 271–278].

[Nešetřil and Rödl, 1990] J. Nešetřil and V. Rödl. *Mathematics of Ramsey Theory*, Springer, Berlin, 1990.

[Noether and Cavaillès, 1937] E. Noether and J. Cavaillès, eds. *Briefwechsel Cantor-Dedekind*, Hermann, Paris, 1937. Translated into French in [Cavaillès, 1962].

[Novak, 1950] I. L. Novak. A construction for models of consistent systems, *Fundamenta Mathematicae*, 37:87–110, 1950.

[Oxtoby, 1971] J. C. Oxtoby. *Measure and Category. A Survey of the Analogies between Topological and Measure Spaces*, Springer, New York, 1971.

[Parsons, 1977] C. Parsons. What is the iterative concept of set? In R. E. Butts and J. Hintikka, eds., *Logic, Foundations of Mathematics and Computability Theory*. Proceedings of the Fifth International Congress of Logic, Methodology, and the Philosophy of Science (London, Ontario, 1975), The University of Western Ontario Series in Philosophy and Science, vol. 9, pp. 335–367. D. Reidel, Dordrecht 1977. Reprinted in [Benacerraf and Putnam, 1983, 503–529].

[Peano, 1897] G. Peano. Studii di logica matematica, *Atti della Accademia delle scienze di Torino, Classe di scienze fisiche, matematiche e naturali*, 32:565–583, 1897. Reprinted in [Peano, 1958, 201–217].

[Peano, 1905-8] G. Peano. *Formulario Mathematico*, Bocca, Torino 1905-8. Reprinted Edizioni Cremonese, Rome, 1960.

[Peano, 1911] G. Peano. Sulla definizione di funzione. *Atti della Accademia nazionale dei Lincei, Rendiconti, Classe di scienze fisiche, matematiche e naturali*, 20-I:3–5, 1911.

[Peano, 1913] G. Peano. Review of:A.N. Whitehead and B. Russell, *Principia Mathematica*, vols. I,II, *Bollettino di bibliografia e storia delle scienze matematiche Loria*, 15:47–53; 75–81, 1913. Reprinted in [Peano, 1958, 389–401].

[Peano, 1958] G. Peano. *Opere Scelte*, vol. 2, Edizioni Cremonese, Rome, 1958.

[Peckhaus, 1990] V. Peckhaus. "Ich habe mich wohl gehütet alle Patronen auf einmal zu verschiessen." Ernst Zermelo in Göttingen, *History and Philosophy of Logic*, 11:19–58, 1990.

[Peirce, 1883] C. S. Peirce. A theory of probable inference. Note B. The logic of relatives, *Studies in Logic by Members of the John Hopkins University*, pp. 187–203, Boston, 1883. Reprinted in C. Hartshorne and P. Weiss, eds., *Collected Papers of Charles Sanders Peirce*, vol. 3, pp. 195–209. Harvard University Press, Cambridge, 1933.

[Pincus, 1972] D. Pincus. Zermelo-Fraenkel consistency results by Fraenkel-Mostowski methods, *The Journal of Symbolic Logic*, 37:721–743, 1972.

[Plotkin, 1993] J. M. Plotkin. Who put the "back" in "back-and-forth?". In Crossley, Remmel, Shore and Sweedler, eds., *Logical Methods, In Honor of Anil Nerode's Sixtieth Birthday*, Birkhäuser, Boston, 1993.

[Plotkin, 2005] J. M. Plotkin. *Hausdorff on Ordered Sets*, American Mathematical Society, Providence, 2005.

[Purket, 1989] W. Purkert. Cantor's views on the foundations of mathematics. In D. E. Rowe and J. McCleary, eds., *The History of Modern Mathematics*, vol. 1:*Ideas and their Reception*, pp 49–65. Academic Press, Boston, 1989.

[Purket, 2002] W. Purket. Grundzüge der Mengenlehre – Historische Einführung. In F. Hausdorff, *Gesammelte Werke*, vol. 2, 1–89, 2002.

[Purket and Ilgauds, 1987] W. Purkert and H. J. Ilgauds. *Georg Cantor:1845–1918*, Birkhäuser, Basel, 1987.

[Quine, 1960] W. V. O. Quine. *Word and Object*, MIT Press, Cambridge, 1960.

[Ramsey, 1930] F. P. Ramsey. On a problem of formal logic, *Proceedings of the London Mathematical Society*, 30:264–286, 1930. Reprinted in [Gessel and Rota, 1987, 2–24].

[Robinson, 1947] R. M. Robinson. On the decomposition of spheres, *Fundamenta Mathematicae*, 34:246–270, 1947.

[Robinson, 1951] A. Robinson. *On the Metamathematics of Algebra*, North-Holland, Amsterdam, 1951.

[Rosser and Wang, 1950] J. B. Rosser and H. Wang. Non-standard models for formal logics, *The Journal of Symbolic Logic*, 15:113–129, 1950.

[Rothberger, 1938] F. Rothberger. Eine Äquivalenz zwischen der Kontinuumhypothese und der Existenz der Lusinschen und Sierpińskischen Mengen, *Fundamenta Mathematicae*, 30:215–217, 1938.

[Rothberger, 1939] F. Rothberger. Sur un ensemble toujours de première categorie qui est depourvu de la properté λ, *Fundamenta Mathematicae* ,32:294–300, 1939.

[Rothberger, 1948] F. Rothberger. On some problems of Hausdorff and Sierpiński, *Fundamenta Mathematicae*, 35:29–46, 1948.

[Rubin and Rubin, 1985] H. Rubin and J. E. Rubin. *Equivalents of the Axiom of Choice, II.* Amsterdam, North-Holland, 1985. Revised, expanded version of their *Equivalents of the Axiom of Choice*, North-Holland, Amsterdam, 1963.

[Russell, 1901] B. Russell. Sur la logique des relations avec des applications à la théorie des séries, *Revue de mathématiques (Rivista di matematica)* 7, 115–148; partly reprinted in G. H. Moore, ed., *The Collected Papers of Bertrand Russell*, vol. 3, pp. 613–627. Routledge, London, 1993; translated in same, pp. 310–349.

[Russell, 1903] B. Russell. *The Principles of Mathematics*, Cambridge University Press, Cambridge, 1903. Later editions, George Allen & Unwin, London.

[Russell, 1906] B. Russell. On some difficulties in the theory of transfinite numbers and order types, *Proceedings of the London Mathematical Society*, 4L 29–53, 1906. Reprinted in [Russell, 1973, 135–164].

[Russell, 1959] B. Russell. *My Philosophical Development*, George Allen & Unwin, London, 1959.

[Russell, 1973] B. Russell. D. Lackey, ed., *Essays in Analysis*, George Braziller, New York, 1973.

[Schröder, 1890] E. Schröder. *Vorlesungen über die Algebra der Logik (exakte Logik).* Vol. 1. Leipzig, B.G. Teubner, 1890. Reprinted in [Schröder, 1966].

[Schröder, 1895] E. Schröder. *Vorlesungen über die Algebra der Logik (exakte Logik).* Vol. 3:*Algebra und Logik der Relative.* Leipzig, B.G. Teubner, 1895. Reprinted in [Schröder, 1966].

[Schröder, 1966] E. Schröder. *Vorlesungen über die Algebra der Logik* (three volumes), Chelsea, New York, 1966.

[Scott, 1961] D. S. Scott. Measurable cardinals and constructible sets, *Bulletin de l'Académie Polonaise des Sciences, Série des Sciences Mathématiques, Astronomiques et Physiques*, 9:521–524, 1961.

[Scott, 1971] D. S. Scott. *Axiomatic Set Theory*, Proceedings of Symposia in Pure Mathematics vol. 13, part 1, American Mathematical Society, Providence, 1971.

[Scott, 1974] D. S. Scott. Axiomatizing set theory. In T. J. Jech, ed., *Axiomatic Set Theory.* Proceedings of Symposia in Pure Mathematics vol. 13, part 2, pp. 207–214. American Mathematical Society, Providence, 1974.

[Shelah, 1980] S. Shelah. Going to Canossa (abstract). *Abstracts of papers presented to the American Mathematical Society*, 1:630, 1980.

[Shelah, 1984] S. Shelah. Can you take Solovay's inaccessible away? *Israel Journal of Mathematics*, 48:1–47, 1984.

[Shepherdson, 1951] J. C. Shepherdson. Inner models for set theory — Part I, *The Journal of Symbolic Logic*, 16:161–190, 1951.

[Shepherdson, 1952] J. C. Shepherdson. Inner models for set theory — Part II, *The Journal of Symbolic Logic*, 17:225–237, 1952.

[Shepherdson, 1953] J. C. Shepherdson. Inner models for set theory — Part III, *The Journal of Symbolic Logic*, 18:145–167, 1953.

[Shoenfield, 1954] J. R. Shoenfield. A relative consistency proof, *The Journal of Symbolic Logic*, 19:21–28, 1954.

[Shoenfield, 1959] J. R. Shoenfield. On the independence of the axiom of constructibility, *American Journal of Mathematics*, 81:537–540, 1959.

[Shoenfield, 1967] J. R. Shoenfield. *Mathematical Logic*, Addison-Wesley, Reading, 1967.

[Shoenfield, 1971] J. R. Shoenfield. Unramified forcing. In [Scott, 1971, 357–381].

[Shoenfield, 1977] J. R. Shoenfield. Axioms of set theory. In J. K. Barwise, ed., *Handbook of Mathematical Logic*, pp. 321–344. North-Holland, Amsterdam, 1977.

[Sierpiński, 1918] W. Sierpiński. L'axiome de M. Zermelo et son rôle dans la Théorie des Ensembles et l'Analyse, *Bulletin de l'Académie des Sciences de Cracovie, Classe des Sciences Mathématiques et Naturelles, Série A*, 97–152, 1918. Reprinted in [Sierpińsi, 1975, 208–255].

[Sierpiński, 1924] W. Sierpiński. Sur l'hypothèse du continu ($2^{\aleph_0} = \aleph_1$), *Fundamenta Mathematicae*, 5:177–187, 1924. Reprinted in [Sierpiński, 1975, 527–536].

[Sierpiński, 1925] W. Sierpiński. Sur une class d'ensembles, *Fundamenta Mathematicae* , 7:237–243, 1925. Reprinted in [Sierpiński, 1975, 571–576].

[Sierpiński, 1928] W. Sierpiński. Sur un ensemble non dénombrable, dont toute image continue est de mesure nulle, *Fundamenta Mathematicae*, 11:302–304, 1928. Reprinted in [Sierpiński, 1975, 702–704].

[Sierpiński, 1934] W. Sierpiński. *Hypothèse du Continu*, Monografie Matematyczne vol. 4, Warsaw 1934. Second, revised edtion, Chelsea, New York, 1956.

[Sierpiński, 1950] W. Sierpiński. *Les ensembles projectifs et analytiques*, Mémorial des Sciences Mathématiques #112, Gauthier-Villars, Paris, 1950.

[Sierpiński, 1975] W. Sierpiński. S. Hartman *et al.*, eds., *Oeuvres Choisies*. vol. 2, Warsaw, Państwowe Wydawnictwo Naukowe, 1975.

[Sierpiński, 1976] W. Sierpiński. S. Hartman *et al.*, eds., *Ouevres Choisies*. vol. 3, Państwowe Wydawnictwo Naukowe, Warsaw, 1976.

[Sierpiński and Tarski, 1930] W. Sierpiński and A. Tarski. Sur une propriété caractéristique des nombres inaccessibles, *Fundamenta Mathematicae*, 15:292–300, 1930. Reprinted in [Sierpiński, 1976, 29-35], and in [Tarski, 1986 vol. 1, 289–297].

[Skolem, 1920] T. Skolem. Logisch-kombinatorische Untersuchungen über die Erfüllbarkeit oder Beweisbarkeit mathematischer Sätze nebst einem Theoreme über dichte Mengen, *Videnskapsselskapets Skrifter, I. Matematisk-Naturvidenskabelig Klass*, #4:1–36, 1920. Reprinted in [Skolem, 1970, 103–136]. Partially translated in [van Heijenoort, 1967, 252–263].

[Skolem, 1923] T. Skolem. Einige Bemerkungen zur axiomatischen Begründung der Mengenlehre, in: *Matematikerkongressen i Helsingfors den 4–7 Juli 1922, Den femte skandinaviska matematikerkongressen, Redogörelse,* pp. 217–232. Akademiska-Bokhandeln, Helsinki, 1923. Reprinted in [Skolem, 1970, 137–152]. Translated in [van Heijenoort, 1967, 290–301].

[Skolem, 1930] T. Skolem. Einige Bemerkungen zu der Abhandlung von E. Zermelo: "Über die Definitheit in der Axiomatik", *Fundamenta Mathematicae*, 15:337–341, 1930. Reprinted in [Skolem, 1970, 275–279].

[Skolem, 1933] T. Skolem. Ein kombinatorischer Satz mit Anwendung auf ein logisches Entscheidungsproblem, *Fundamenta Mathematicae*, 20: 254–261, 1933. Reprinted in [Skolem, 1970, 337–344].

[Skolem, 1933a] T. Skolem. Über die Unmöglichkeit einer vollständigen Charakterisierung der Zahlenreihe mittels eines endlichen Axiomensystems, *Norsk Matematisk Forenings Skrifter*, #10:73–82, 1933. Reprinted in [Skolem, 1970, 345–354].

[Skolem, 1934] T. Skolem. Über die Nicht-charakterisierbarkeit der Zahlenreihe mittels endlich oder abzählbar unendlich vieler Assagen mit ausschliesslich Zahlenvariablen, *Fundamenta Mathematicae*, 23':150–161, 1934. Reprinted in [Skolem, 1970, 355–366].

[Skolem, 1970] T. Skolem. J. E. Fenstad, ed., *Selected Works in Logic*, Univesitetsforlaget, Oslo, 1970.

[Solovay, 1963] R. M. Solovay. Independence results in the theory of cardinals. I, II (abstracts), *Notices of the American Mathematical Society*, 10:595, 1963.

[Solovay, 1965] R. M. Solovay. $2^{\aleph_0}$ can be anything it ought to be (abstract). In [Addison *et al.*, 1965, 435].

[Solovay, 1965a] R. M. Solovay. Measurable cardinals and the continuum hypothesis (abstract), *Notices of the American Mathematical Society*, 12:132, 1965.

[Solovay, 1965b] R. M. Solovay. The measure problem (abstract), *Notices of the American Mathematical Society*, 12:217, 1965.

[Solovay, 1969] R. M. Solovay. The cardinality of Σ^1_2 sets of reals. In J. J. Bulloff, T. C. Holyoke, and S. W. Hahn, eds., *Foundations of Mathematics*, pp. 58–73. Symposium papers commemorating the sixtieth birthday of Kurt Gödel. Springer, Berlin, 1969.

[Solovay, 1970] R. M. Solovay. A model of set theory in which every set of reals is Lebesgue measurable, *Annals of Mathematics*, 92:1–56, 1970.

[Specker, 1957] E. Specker. Zur Axiomatik der Mengenlehre (Fundierungs- und Auswahlaxiom), *Zeitschrift für mathematische Logik und Grundlagen der Mathematik*, 3:173–210, 1957.

[Steinitz, 1910] E. Steinitz. Algebraische Theorie der Körper, *Journal für die reine und angewandte Mathematik*, 137:167–309, 1910.

[Suslin, 1917] M. Y. Suslin. Sur une définition des ensembles mesurables B sans nombres transfinis, *Comptes Rendus Hebdomadaires des Séances de l'Académie des Sciences, Paris*, 164:88–91, 1917.

[Suslin, 1920] M. Y. Suslin. Problème 3, *Fundamenta Mathematicae*, 1:223, 1920.

[Tarski, 1924] A. Tarski. Sur quelques théorèmes qui équivalent à l'axiome du choix, *Fundamenta Mathematicae*, 5:147–154, 1924. Reprinted in [Tarski, 1986, vol. 1, 41–48].

[Tarski, 1931] A. Tarski. Sur les ensembles définissables de nombres réels, *Fundamenta Mathematicae*, 17:210–239, 1931. Translated in [Tarski, 1983, 110–142].

[Tarski, 1933] A. Tarski. Pojęcie prawdy w językach nauk dedukcyjnych (The concept of truth in the languages of deductive sciences), *Prace Towarzystwa Naukowego Warszawskiego, Wydział III, Nauk Matematyczno-fizycznych (Travaux de la Société des Sciences et des Lettres de Varsovie, Classe III, Sciences Mathématiques et Physiques*, #34, 1933. See also [Tarski, 1935].

[Tarski, 1935] A. Tarski. Der Wahrheitsbegriff in den formalisierten Sprachen (German translation of [Tarski, 1933] with a postscript), *Studia Philosophica*, 1:261–405, 1935. Reprinted in [Tarski, 1986, vol. 2, 51–198]. Translated in [Tarski, 1983, 152–278].

[Tarski, 1951] A. Tarski. *A Decision Method for Elementary Algebra and Geometry* (prepared by J.C.C. McKinsey), University of California Press, Berkeley, 1951. Second revised edition.

[Tarski, 1962] A. Tarski. Some problems and results relevant to the foundations of set theory, in:Nagel, Ernest, Patrick Suppes, and Alfred Tarski (editors). *Logic, Methodology and Philosophy of Science.* Proceedings of the 1960 [and first] International Congress (Stanford, California), pp. 125–135. Stanford University Press, Stanford, 1962. Reprinted in [Tarski, 1986, vol. 4, 115–125].

[Tarski, 1983] A. Tarski. *Logic, Semantics, Metamathematics. Papers from 1923 to 1938.* Translations by J.H. Woodger, second edition, Hackett, Indianapolis, 1983.

[Tarski, 1986] A. Tarski. S. R. Givant and R. N. McKenzie, eds., *Collected Papers*, Basel, Birkhäuser, 1986.

[Todorčević, 1984] S. Todorčević. Trees and linearly ordered sets. In [Kunen and Vaughan, 1984, 235–293].

[Ulam, 1930] S. M. Ulam. Zur Masstheorie in der allgemeinen Mengenlehre, *Fundamenta Mathematicae*, 16:140–150, 1930. Reprinted in [Ulam, 1974, 9–19].

[Ulam, 1974] S. M. Ulam. W. A. Beyer, J. Mycielski, and G.-C. Rota, eds., *Sets, Numbers, and Universes.* Selected Works, MIT Press, Cambridge, 1974.

[van Douwen, 1984] E. K. van Douwen. The integers and topology. In [Kunen and Vaughan, 1984, 111–168].

[van Heijenoort, 1967] J. van Heijenoort, ed. *From Frege to Gödel:A Source Book in Mathematical Logic, 1879-1931.* Harvard University Press, Cambridge, 1967.

[Vitali, 1905] G. Vitali. Sul problema della misura dei gruppi di punti di una retta, Tip. Gamberini e Parmeggiani, Bologna, 1905.

[von Neumann, 1923] J. von Neumann. Zur Einführung der transfiniten Zahlen, *Acta Litterarum ac Scientiarum Regiae Universitatis Hungaricae Francisco-Josephinae, sectio scientiarum mathematicarum*, 1:199–208, 1923. Reprinted in [von Neumann, 1961, 24–33]. Translated in [van Heijenoort, 1967, 346–354].

[von Neumann, 1925] J. von Neumann. Eine Axiomatisierung der Mengenlehre, *Journal für die reine und angewandte Mathematik*, 154:219–240, 1925. Berichtigung 155, 128. Reprinted in [von Neumann, 1961, 34–56]. Translated in [van Heijenoort, 1967, 393–413].

[von Neumann, 1928] J. von Neumann. Über die Definition durch transfinite Induktion und verwandte Fragen der allgemeinen Mengenlehre, *Mathematische Annalen*, 99:373–391, 1928. Reprinted in [von Neumann, 1961, 320–338].

[von Neumann, 1928a] J. von Neumann. Die Axiomatisierung der Mengenlehre, *Mathematische Zeitschrift*, 27:669–752, 1928. Reprinted in [von Neumann, 1961, 339–422].

[von Neumann, 1929] J. von Neumann. Über eine Widerspruchfreiheitsfrage in der axiomaticschen Mengenlehre, *Journal für die reine und angewandte Mathematik*, 160:227–241, 1929. Reprinted in [von Neumann, 1961, 494–508].

[von Neumann, 1961] J. von Neumann. A. H. Taub, ed., *John von Neumann, Collected Works*, vol. 1, Pergamon Press, New York, 1961.

[Vopěnka, 1962] P. Vopěnka. Construction of models of set theory by the method of ultraproducts (in Russian), *Zeitschrift für mathematische Logik und Grundlagen der Mathematik*, 8:293–304, 1962.

[Vopěnka, 1964] P. Vopěnka. The independence of the Continuum Hypothesis (in Russian), *Commentationes Mathematicae Universitatis Carolinae*, 5 Supplement, I:1–48, 1964. Translated in *American Mathematical Society Translations*, 57:84–112, 1966.

[Vopěnka, 1965] P. Vopěnka. Construction of a model for Gödel-Bernays set theory for which the class of natural numbers is a set of the model and a proper class in the theory. In [Addison *et al.*, 1965, 436–437].

[Vopěnka, 1967] P. Vopěnka. The general theory of ∇-models, *Commentationes Mathematicae Universitatis Carolinae*, 8:145–170, 1967.

[Wagon, 1985] S. Wagon. *The Banach-Tarski Paradox*, Encyclopedia of Mathematics and Its Applications, vol. 24, Cambridge University Press, Cambridge, 1985. Paperback edition, 1993.

[Wang, 1974] H. Wang. *From Mathematics to Philosophy,* Humanities Press, New York ,1974.

[Wang, 1974a] H. Wang. The concept of set. In [Wang, 1974, 181–223]. Reprinted in [Benacerraf and Putnam, 1983, 530–570].

[Whitehead and Russell, 1910] A. N. Whitehead and B. Russell. *Principia Mathematica*, vol. 1, Cambridge University Press, Cambridge, 1910.

[Whitehead and Russell, 1912] A. N. Whitehead and B. Russell. *Principia Mathematica*, vol. 2, Cambridge University Press, Cambridge, 1912.

[Whitehead and Russell, 1913] A. N. Whitehead and B. Russell. *Principia Mathematica*, vol. 3, Cambridge University Press, Cambridge, 1913.

[Wiener, 1914] N. Wiener. A simplification of the logic of relations, *Proceedings of the Cambridge Philosophical Society*, 17:387–390, 1914. Reprinted in [van Heijenoort, 1967, 224–227].

[Wittgenstein, 1956] L. Wittgenstein. G. H. von Wright, R. Rhees, and G. E. M. Anscombe, eds., *Remarks on the Foundations of Mathematics,* Basil Blackwell, Oxford, 1956. Second printing, 1967.

[Zermelo, 1904] E. Zermelo. Beweis, dass jede Menge wohlgeordnet werden kann (Aus einem an Herrn Hilbert gerichteten Briefe), *Mathematische Annalen*, 59:514–516, 1904. Translated in [van Heijenoort, 1967, 139–141].

[Zermelo, 1908] E. Zermelo. Neuer Beweis für die Möglichkeit einer Wohlordnung, *Mathematische Annalen*, 65:107–128, 1908. Translated in [van Heijenoort, 1967, 183–198].

[Zermelo, 1913] E. Zermelo. Über eine Anwendung der Mengenlehre auf die Theorie des Schachspiels. In E. W. Hobson and A. E. H. Love, eds., *Proceedings of the Fifth International Congress of Mathematicians*, pp. 501–504. Cambridge, 1912. Vol. 2, Cambridge University Press, Cambridge, 1913.

[Zermelo, 1929] E. Zermelo. Über den Begriff der Definitheit in der Axiomatik, *Fundamenta Mathematicae*, 14:339–344, 1929.

[Zermelo, 1930] E. Zermelo. Über Grenzzahlen und Mengenbereiche:Neue Untersuchungen über die Grundlagen der Mengenlehre, *Fundamenta Mathematicae* , 16:29–47, 1930. Translated with commentary by Michael Hallett in [Ewald, 1996, 1208–1233].

[Zermelo, 1931] E. Zermelo. Über Stufen der Quantifikation und die Logik des Unendlichen, *Jahresbericht der Deutschen Mathematiker-Vereinigung*, 41:85–88, 1931.

[Zermelo, 1935] E. Zermelo. Grundlagen einer allgemeinen Theorie der mathematicschen Satzsysteme, *Fundamenta Mathematicae*, 25:136–146, 1935.

[Zorn, 1935] M. Zorn. A remark on method in transfinite algebra, *Bulletin of the American Mathematical Society*, 41:667–670, 1935.

HISTORY OF THE CONTINUUM IN THE 20th CENTURY

Juris Steprāns

1 INTRODUCTION

The subject of this history is the cardinality of the continuum, as described by Georg Cantor, as well as the various other cardinal numbers defined by combinatorial structures associated with the continuum. Among these are the cardinals of various subsets of the ideals of meagre and Lebesgue null subsets of the reals. Typical examples of cardinals arising in this context are the least cardinal of a non-measurable set of reals or the least number of meagre sets whose union covers the reals. On the other hand, the combinatorial structures giving rise to cardinals are usually extremal objects embedded in simply definable equivalence relations or partial orders associated with the continuum. Typical examples include maximal families of subsets of $\mathbb{N}$ with all pairwise intersections being finite, or unbounded sets of sequences of natural numbers under the ordering of eventual dominance of minimal cardinality. Other examples of combinatorial structures associated with the continuum are sets of reals with unusual properties; *Lusin sets* — these are uncountable sets whose intersection with each meagre set is countable — provide typical examples of such sets. However, the theory of ultrafilters will play only a minor role in this discussion since these objects belong to a different realm than the continuum. As well, even though the descriptive set theory of the reals is very closely linked to the continuum, this topic would require its own chapter to be treated adequately.

Instead, this article will trace the growing understanding of the continuum, as well as its related cardinals and associated combinatorial structures, starting at the beginning of the twentieth century and ending at the start of the twenty-first century. However, it would be more accurate to describe this as a history of the documentation of the subject as recorded in the published literature. No attempt has been made to interview the surviving contributors to the development of the subject, nor will this article attempt to settle controversial questions of provenance. Instead, the focus of this history is on the questions that stimulated research on the continuum and the often long chains of incremental progress that eventually lead to greater understanding and, in many cases, solutions to longstanding, open problems.

Handbook of the History of Logic: Sets and Extensions in the Twentieth Century.
Volume editor: Akihiro Kanamori. General editors: Dov M. Gabbay, Paul Thagard and John Woods.

Moreover, this is not intended to be a comprehensive history providing a catalogue of all significant results obtained in the last century. The reader looking for that much detail is advised to consult [Blass, 2010] or [Bartoszyński and Judah, 1995] as a starting point for further investigations into the research literature. The reader with no knowledge of the subject would also benefit from at least a superficial reading of [Blass, 2010] before turning to this history. For the benefit of such readers much of the standard terminology and definitions have been provided throughout the text. However, definitions are provided at that point in the text where the explanation contributes most to the development of the story rather than at their first appearance. Entries to the volume index should help the reader searching for a definition or an explanation of some terminology. Also, the full names of individuals mentioned in this history are used only at their first occurrence in the story and the index can be used to find these occurrences.

2 HILBERT'S ADDRESS

It is difficult to imagine that the story of the continuum in the twentieth century could start anywhere other than with David Hilbert's 1900 address [1901] in Paris. The very first of his celebrated list of problems presented at the International Congress of Mathematicians is the Continuum Hypothesis, which, by this point in time, had already been well studied. As Hilbert says, "The investigations of Cantor on such assemblages of points suggest a very plausible theorem, which nevertheless, in spite of the most strenuous efforts, no one has succeeded in proving."[1] Hilbert goes on to emphasize a possible line of attack on the problem which, while not solving the problem, does serve as a suitable prelude for much of the developments in the study of the continuum that were to unfold throughout the coming century. In his address, Hilbert focusses his attention on "another very remarkable statement of Cantor's which stands in the closest connection with the theorem mentioned and which, perhaps, offers the key to its proof."[2] He is, of course, making reference here to the well-ordering of the reals, a subject that would occupy the minds of mathematicians concerned with the foundations of their subject well into the coming century. However, Hilbert's interest here was not with foundations, but with understanding how the new concept of well ordering could be used to deal with existing problems such as that of whether $2^{\aleph_0} = \aleph_1$.

Both the novelty of the concept as well as the importance Hilbert attaches to it are underlined by the detailed description of what a well-ordering of the reals would look like. This is to set up the question of "whether the totality of all numbers may not be arranged in another manner so that every partial assemblage

[1](The translation from the address here, as well as later, is from [Hilbert, 1902].) Die Untersuchungen von Cantor über solche Punktmengen machen einen Satz sehr wahrscheinlich, dessen Beweis jedoch trotz eifrigster Bemühungen bisher noch Niemanden gelungen ist ...

[2]Es sei noch eine andere sehr merkwürdige Behauptung Cantors erwähnt, die mit dem genannten Satze in engstem Zusammenhange steht und die vielleicht den Schlüssel zum Beweise dieses Satzes liefert.

may have a first element ... It appears to me most desirable to obtain a direct proof of this ... perhaps by actually giving an arrangement of numbers such that in every partial system a first number can be pointed out."[3] Later developments in descriptive set theory would have much to say on the extent to which Hilbert's plan could be carried out, but the importance from the point of view of the history of the continuum is Hilbert's clear vision of the potential connection between combinatorial structures, such as well-ordered sets, and the topological or set-theoretic continuum. His hope is that this will provide "a new bridge between the countable assemblage and the continuum."[4]

By the time of Hilbert's address it had already been more than 15 years since Cantor [1884] had concentrated his attention on perfect sets and had shown them all to have the same cardinality as the continuum. It would be another 8 years after Hilbert's address before Felix Bernstein [1908] would make use of the Axiom of Choice to construct what is now known as a Bernstein set, an uncountable set such that neither it nor its complement contains a perfect set. While Bernstein's construction put an end to any hope of establishing the Continuum Hypothesis using Cantor's theorem on the cardinality of perfect sets, it did provide another tool to be used in investigating the "new bridge" of Hilbert's address, between the countable and the continuum. This line of reasoning, looking for connections between infinite combinatorial objects and the continuum, can be traced throughout the twentieth century. On one hand, advances in understanding the continuum followed from results such as Pavel Alexandroff and Felix Hausdorff's independent proofs [Alexandroff, 1916; Hausdorff, 1916] that uncountable Borel sets contain perfect sets. On the other hand, important information about the combinatorial nature of ω_1 was obtained from constructions such as Hausdorff's (ω_1, ω_1^*)–gap [Hausdorff, 1909] or Stevo Todorčević's counterexample [Todorcevic, 1987] to $\aleph_1 \to [\aleph_1]^2_{\aleph_1}$. The motivation behind such results is often the hope of gaining better understanding of the connection between these two facets of the continuum, understanding such as that gained from Robert M. Solovay's coding by almost disjoint families. It is the purpose of this history of the continuum to examine this line of research and the critical developments along it.

While Hilbert's address might provide a dramatic starting point, the truth of the matter is that there is no convenient anchor for a history of twentieth century developments on the continuum. However, it is instructive to begin investigations into this history by considering those questions in analysis that had attracted the attention of mathematicians with a set-theoretic interest at the end of the nineteenth and start of the twentieth centuries, questions which would turn out to have a critical influence on developments in set theory and, eventually, would lead

[3] Denn, wenn wir als Teilmenge die Punkte einer endlichen Strecke mit Ausnahme des Anfangspunktes der Strecke ins Auge fassen, so besitzt diese Teilmenge jedenfalls kein frühestes Element. Es erhebt sich nun die Frage, ob sich die Gesamtheit aller Zahlen nicht in anderer Weise so ordnen läßt, daß jede Teilmenge ein frühestes Element, hat, d.h. ob das Continuum auch als wohlgeordnete Menge aufgefaßt werden kann, was Cantor bejahen zu müssen glaubt.

[4] der Beweis dieses Satzes würde mithin eine neue Brücke schlagen zwischen der abzählbaren Menge und dem Continuum.

to the detailed study of cardinal invariants[5] of the continuum. A focal point, if not a starting point, for these investigations is found a year before Hilbert's address in the thesis of René Baire [1899b]. Baire had set himself the task of analyzing pointwise limits of continuous functions. Among many other results, this thesis contains the Baire Category Theorem, that no interval is of the first category, a theorem which 70 years later could be interpreted as a weak form of Martin's Axiom. The developments leading from Baire's theorem through Martin's Axiom to the Proper Forcing Axiom and its variants will be discussed in §8.

His continuation of this work six years later would introduce Baire's classification of functions as a hierarchy of classes, each the pointwise limit of functions from previous classes. His sufficient condition for a function to belong to this classification inspired Nikolai Lusin's [1914] construction, assuming the Continuum Hypothesis, of what is now known as a Lusin set, even though Paul Mahlo's article [1913] with a similar construction appeared a year earlier. Almost a century later this remains a critical example delineating the behaviour of cardinal invariants of the continuum. It is a happy circumstance for the development of mathematics that Lusin's construction seems to have been ignored by mathematicians, such as Abram Besicovitch and Fritz Rothberger, working on related questions much later. While a Lusin set would have been sufficient for their purposes, the concentrated sets they constructed, assuming the same hypothesis as Lusin, were revealed half a century later to be subtly distinct from Lusin sets. Moreover, their method of construction lead to important developments in the study of the structure of $\mathbb{N}^{\mathbb{N}}$ under the $\leq^*$ relation and the related cardinal invariants $\mathfrak{b}$ and $\mathfrak{d}$ which will be examined in §5.

The study of $\mathbb{N}^{\mathbb{N}}$ under the $\leq^*$ has its roots in the work of Paul du Bois-Reymond [1870-71; 1872] on rates of convergence, referred to by du Bois-Reymond as pantachies. It would be more than thirty years before du Bois-Reymond's work would attract serious attention. G. H. (Godfrey Harold) Hardy [1924] provided an English exposition and considered the implication for power series and L-functions. Hausdorff, on the other hand, recognized the critical role set theory would play in the study of pantachies. This will be the starting point for §5.

While du Bois-Reymond, his contemporary Jean-Gaston Darboux and, of course, Cantor were engaged in the first, tentative examinations of questions in analysis that would lead to interesting set-theoretic developments it would be a generation later before the study of such questions found any real resonance among analysts. Writing in 1922, Henri Léon Lebesgue acknowledges (cf. [1972, page 92]) that Camille Jordan has established the usefulness of set theory and incorporated it into the mainstream of mathematics, even though he does refer to Darboux's counterexamples as "une sorte de musée de monstruosités". Indeed, Hélène Gis-

[5]The use of the term *invariant* to describe cardinals associated with various structures on the continuum is not universally accepted. For example, Andreas Blass [2010] uses the term *characteristic* instead, arguing that these cardinals are not at all invariant and can assume different values in different models of set theory. Nevertheless, since the term *invariant* has found widespread acceptance and usage in the literature, this survey will not deviate from this tradition.

pert [1995] argues that the French tradition of interest in the theory of functions was the source of the increased interest in set-theoretic questions, especially in the work of Émile Borel, Lebesgue and Baire. Given the influence of French mathematics, it is no surprise to see this interest reflected beyond France's borders. Indeed, the analysts in Moscow and Warsaw took up this direction so enthusiastically that it is fair to say that set-theoretic developments in these two centres soon eclipsed those of Paris.

The year 1920 witnessed the founding of the journal *Fundamenta Mathematicae* in Warsaw and even a casual scanning of the early editions reveals the intense interest in set-theoretic questions during those years, both in Warsaw and in Moscow. Alongside the Polish mathematicians Kazimierz Kuratowski, Stefan Banach, Hugo Steinhaus and, of course, Wacław Sierpiński one also sees the names of Lusin, Pavel Urysohn and Mikhail Suslin although Russian contributions are seen to have declined with the rise of Stalin by the end of the 1920's. An enlightening indication of the set-theoretic interests of these two groups can be found in the problems section at the end of each of the early issues. Perhaps the best known of these questions, due to Suslin, is recorded on page 223 of the section *Problèmes* of the first issue of *Fundamenta Mathematicae*:

QUESTION 1. Can the real line be characterized as the unique complete, dense in itself linear order with the countable chain condition?[6]

The ramifications of the answer to Suslin's question are profound and lead to mathematics far beyond what could have been imagined by Suslin and his contemporaries at the time. These will be discussed in §8. As well, one finds Sierpiński asking:

QUESTION 2. Is there a partition of the unit interval into $\aleph_1$ Borel sets?[7]

The key idea to the solution, Hausdorff's (ω_1, ω_1^*)–gap, had already appeared some ten years earlier, albeit in a rather obscure journal. In the same list of problems Sierpiński also asks:

QUESTION 3. Is there an uncountable set of reals every homeomorphic image of which is Lebesgue null?[8]

Almost as an afterthought he asks whether assuming that $2^{\aleph_0} = \aleph_1$ helps. The answers to both of these questions would be known considerably earlier than the answer to Suslin's question.

Other questions seem considerably ahead of their time, given the little that was known about the value of the continuum, but, nevertheless were asked by

[6]Un ensemble ordonné (linéairement, sans sauts at sans lacunes, possédant cette propriété que tout ensemble d'intervalles contenant plus qu'un élément) n'empiétant pas les un sur les autres est toujours au plus dénombrable, est il nécessairement un continu linéaire ordinaire?

[7]Existe-il une décomposition d'un intervalle en $\aleph_1$ ensembles mesurables (B), non vides, sans point commun deux à deux?

[8]Existe-il un ensemble linéaire non dénombrable E, tel que tout ensemble linéaire homéomorphe avec E soit de mesure lebesguienne nulle? Peut-on démontrer l'existence d'un tel ensemble même en admettant que $2^{\aleph_0} = \aleph_1$?

Sierpiński in the first issue of *Fundamenta Mathematicae*:

QUESTION 4. Is it possible, without assuming that $2^{\aleph_0} = \aleph_1$, to show that there are $\aleph_1$ Lebesgue null sets whose union is not null? What of meagre sets? What of analytic sets?[9]

Two issues later, in a similar spirit, Stanisław Ruziewicz asks:

QUESTION 5. Are sets of cardinality less than the continuum necessarily meagre?[10]

In the language of cardinal invariants, of course, Sierpiński is asking about the additivity of the null and meagre ideals, while Ruziewicz is asking about the uniformity invariant of the ideal of meagre sets. All these questions will be considered in §6.

Sierpiński's final question is about the additivity of analytic sets, while, in the same list Lusin asks what the cardinality of a co-analytic set is. This second question would soon be answered in [Lusin and Sierpiński, 1923], but this line of investigation leads away from the study of cardinal invariants. However, it is worthy of note that these sorts of questions seemed natural long before they could be answered.

The following question from the first issue of *Fundamenta*, due to Steinhaus, deserves special mention:

QUESTION 6. Is it possible, without assuming that $2^{\aleph_0} = \aleph_1$, to show that there is a subset of the plane which is null in all vertical sections, yet whose complement is null in all horizontal sections?[11]

Sierpiński would show that the answer is negative and this answer would play a central role in his book [1934a] on the Continuum Hypothesis. In the fourth issue Kuratowski asked:

QUESTION 7. Given a set which is nowhere meagre, is it possible to partition it into two disjoint sets, both of which are nowhere meagre?[12]

He mentions that this can be done by assuming either that the set in question is analytic or by assuming the Continuum Hypothesis but, that the answer for an arbitrary subset of the reals is not known. Partial answers would invoke large cardinal assumptions, even though Lusin's final answer would have no need of

[9]Peut-on démontrer sans l'hypthèse du continu ($2^{\aleph_0} = \aleph_1$) qu'une somme de $\aleph_1$ ensembles de mesure lebesguienne nulle n'est pas ncéssairement de mesure (L) nulle? Qu'une somme de $\aleph_1$ ensembles de première catégorie n'est pas nécessairement de première catégorie? Qu'un produit de $\aleph_1$ ensembles (A) n'est pas nécessairement un ensemble (A)?

[10]Un ensemble (linéaire) de puissance inférieure à celle du continu, est-il nécessairement de la première catégorie de M. Baire?

[11]Peut-on démontrer sans l'hypothèse du continu l'existence d'un ensemble plan qui est de la mesure lebesguienne nulle sur toute parallèle à l'axe d'abscisses et dont le complémentaire est de mesure nulle sur toute parallèle à l'axe d'ordonnées?

[12]A étant un ensemble de nombres réels qui n'est de 1 catégorie dans aucun intervalle, existe-il une décomposition: $A = B + C$, $B \times C = 0$ telle que ni B ni C ne soient de 1 catégorie dans aucun intervalle?

these. This question will be revisited in §6.

These questions, and ones like them, will play an important role in this effort to trace the central developments leading to our current understanding of the continuum since they can be viewed as giving meaning to questions such as the Continuum Hypothesis. Given the broad range of independence results concerning the continuum and its various invariants, there may be a tendency to view any results in this area as merely a formalist exercise. Indeed, Paul Cohen himself admits to such a perspective in an essay written shortly after obtaining the independence of the Continuum Hypothesis — the impact of the forcing technique he developed to do this will be discussed in §7 — and he acknowledges that Abraham Robinson's views on the subject have convinced him to accept the formalist position. On this Cohen [1971] says,

> It is a choice which carries with it certain heavy weights. Certainly one of the heaviest is the admission that CH, perhaps the first significant question about uncountable sets which can be asked, has no intrinsic meaning.

If, contrary to Cohen's view, there is any meaning to be found in statements about the continuum, then, at least part of this meaning will be found in the applications to which these statements are put. The applications of interest to set theorists shortly after the First World War have been outlined in this section. Subsequent sections will survey the developments stemming from these.

3 LUSIN'S SECOND CONTINUUM HYPOTHESIS

Lusin's [1914] construction using transfinite induction and assuming $2^{\aleph_0} = \aleph_1$ to enumerate the reals and perfect sets is now a standard technique and part of the arsenal of any set theorist. But it is of interest to reflect on his conclusion and the use to which it was to put. More than a decade earlier, Baire had examined the classification of functions as a hierarchy of classes, each the pointwise limit of functions from previous classes. His necessary condition for a function f to belong to this classification scheme was that given any perfect set P there is a set M which is meagre relative to P such that the restriction of f to $P \setminus M$ is continuous. It is easy to see that the characteristic function of a Lusin set has this property; indeed, given any perfect set P there is a countable set M such that the restriction of the characteristic function to $P \setminus M$ is constant. But how did Lusin show that the characteristic function of the set he constructed does not belong to Baire's classification scheme? Had he known that all Borel sets have the Perfect Set Property, this would have been immediate. But this fact was to be established only two years later, in 1916, independently by Alexandroff [1916] and Hausdorff [1916]. Along with the later result that Baire functions are Borel, it is easy to show that the characteristic functions of Lusin sets are not Baire. However, this was not available to Lusin in 1914. He had to rely on a less constructive pigeonhole argument based on $2^{\aleph_0} < 2^{\aleph_1}$. Various results throughout the coming century would provide further illustrations of the usefulness of this assumption.

It is interesting to note that in the section *Problèmes* of Volume 4 of *Fundamenta Mathematicae* Sierpiński asks in Question 24 about the cardinality of all functions satisfying Baire's condition. It seems very likely that the motivation for this question was rooted in considerations of the hypothesis used in Lusin's argument for constructing a function satisfying Baire's condition but not belonging to the Baire classification. It is worth noting though, that already in [1921] Lusin had shown that there are uncountable non-Borel sets that are meagre in every perfect set. In [1924] Sierpiński considers varous consequences of the Continuum Hypothesis, but also devotes some space to the weak Continuum Hypothesis $2^{\aleph_0} < 2^{\aleph_1}$ and, more than a decade later, it is clear that this sort of question continued to exercise both Lusin and Sierpinski. In [Sierpiński, 1935] Sierpiński returns to the hypothesis $2^{\aleph_0} < 2^{\aleph_1}$ since its failure is required for the following hypothesis of Lusin [1935] to hold: Every set of reals of cardinality $\aleph_1$ is co-analytic. Lusin [1935] refers to the equality $2^{\aleph_0} = 2^{\aleph_1}$ as the "seconde hypothèse du continu" and admits to not being aware of the first author to consider this hypothesis. Martin's Axiom is discussed in §8, and in the paper [Martin and Solovay, 1970] introducing this axiom it is shown that it implies that if $2^{\aleph_0} > \aleph_1$ then Lusin's hypothesis about sets of cardinality $\aleph_1$ being co-analytic sets holds if and only there is a real t such that $\aleph_1 = \aleph_1^{L[t]}$ and, hence, there are models of Martin's Axiom in which this hypothesis of Lusin is true.

However, the weak form of the Continuum Hypothesis, whose negation is referred to by Luzin as the Second Continuum Hypothesis, had already been examined by Hausdorff in 1907 and found to be useful. Hausdorff's work on maximal linear orderings of sequences of integers under eventual dominance is examined in more detail in §5. A key development in this early work occurred in [Hausdorff, 1907] in which Hausdorff established that these maximal linear orders — which, following du Bois-Reymond, he called *pantachies* — must have cardinality $2^{\aleph_0}$. Moreover, if such a pantachie happens to have no (ω_1, ω_1^*)–gaps — these are defined in §5 — then the cardinality of the pantachie must be $2^{\aleph_1}$. In other words, if $2^{\aleph_0} < 2^{\aleph_1}$ then there must be (ω_1, ω_1^*)–gaps in each pantachie. As it happens, this early application of the negation of the Second Continuum Hypothesis was superseded two years later by Hausdorff's critically important discovery [1909] that (ω_1, ω_1^*)–gaps actually exist without having to appeal to any hypothesis other than the Axiom of Choice.

Around the same time that Lusin and Sierpiński were considering the Second Continuum Hypothesis in the context of co-analytic sets of cardinality $\aleph_1$, across the ocean the same axiom was finding applications to the theory of metrizability of general topological spaces. In 1937 F. Burton Jones showed that if $2^{\aleph_0} < 2^{\aleph_1}$ then every separable, normal Moore space is metrizable. The point in the argument at which the weak Continuum Hypothesis plays a role is in showing that there are far more instances of normality — namely, disjoint closed sets that need to be separated by disjoint open sets — than can be handled by the $2^{\aleph_0}$ open sets that are available in a separable space.

About the same time as Jones was working on questions of metrizability, Fritz Rothberger was making some of his initial advances in the study of the structure $\mathcal{P}(\mathbb{N})$ under inclusion modulo a finite set as well as $\mathbb{N}^{\mathbb{N}}$ under eventual dominance which are described in more detail in §5. In the spirit of Hausdorff's results on the weak Continuum Hypothesis and (ω_1, ω_1^*)–gaps, Rothberger [1948] showed that this hypothesis implies that ω_1–limits exist. Recall that an ω_1–limit is a $\subseteq^*$ descending family $\{A_\xi\}_{\xi\in\omega_1}$ — in other words, if $\xi \in \eta$ then $A_\eta \subseteq^* A_\xi$, i.e. $A_\eta \setminus A_\xi$ is finite — for which there does not exist an infinite X such that $X \subseteq^* A_\xi$ for each $\xi \in \omega_1$. Once again the key step is to find a family of $2^{\aleph_1}$ descending families any two of which are eventually almost disjoint. Assuming $2^{\aleph_0} < 2^{\aleph_1}$, it impossible to have an infinite set below each of these.

After stating some open problems suggested by his results, Rothberger comments "Many propositions are known that *imply* Lusin's 'second continuum hypothesis' but very little is known about *consequences* of Lusin's hypothesis." He refers here explicitly to Sierpiński's paper [1935] dealing with Lusin's hypothesis that all sets of reals of cardinality $\aleph_1$ are co-analytic, but he is actually referring to the hypothesis $2^{\aleph_0} = 2^{\aleph_1}$.

In the same paper [1948], Rothberger considers the following two related notions:

- A *Q*–set is an uncountable set of reals every subset of which is a relative F_σ.
- A *denumerable base* for a family $\mathcal{G}$ of $\mathbb{R}$-valued functions on a set X is a countable family $\mathcal{F}$ of $\mathbb{R}$-valued functions on X such that every function in $\mathcal{G}$ is the pointwise limit of functions from $\mathcal{F}$.

He goes on to show that there is a Q–set if and only if there is an uncountable set $X \subseteq \mathbb{R}$ for which $\mathbb{R}^X$ has a denumerable base. Providing more evidence for his comment on page 43, he shows that both these assertions fail unless Lusin's Second Continuum Hypothesis $2^{\aleph_0} = 2^{\aleph_1}$ holds. R. H. Bing uses this idea in his [1951] to show that if there is a Q–set then there is a separable, normal, non-metrizable Moore space.

In spite of the extensive use made of the hypothesis that $2^{\aleph_0} < 2^{\aleph_1}$ in various contexts and by geographically diverse groups, there is a clear lack of new ideas until the revived interest in such problems in the forcing era. The new impetus for applications of the inequality $2^{\aleph_0} < 2^{\aleph_1}$ came from two sources, one was Ronald Jensen's detailed analysis of Kurt Gödel's constructible universe and the other was rooted in the theory of abelian groups.

One of the discoveries made during the renewed examination of inner models was reminiscent of work done by Jones, Rothberger, Bing and others who took advantage of the fact that $2^{\aleph_0} < 2^{\aleph_1}$ implies there are more uncountable sets of reals than there are reals, or objects coded by reals. Jensen and Solovay gained an even deeper insight into this phenomenon by devising (cf. their [1970]) a coding by almost disjoint sets. Given an almost disjoint family $\{A_\xi\}_{\xi\in\omega_1}$ with the property that for any $X \subseteq \omega_1$ there is a set $A \subseteq \mathbb{N}$ such that $A_\xi \cap A$ is finite if and only if

$\xi \in X$, it is natural to associate with each subset of ω_1 a subset of $\mathbb{N}$ which codes it in this way. Using this they are able to show that in a countable transitive model $\mathfrak{M}$ of set theory which shares the same $\aleph_1$ with L and $A \subseteq \omega_1$ there is a Cohen extension $\mathfrak{M}[x]$, with the same cardinals as $\mathfrak{M}$, such that x is a real from which A is constructible.

However, following a different line of enquiry, in studying Gödel's condensation arguments Jensen formulated the following principle known as $\diamondsuit$: There is an indexed family $\{X_\xi\}_{\xi\in\omega_1}$ such that

- $X_\xi \subseteq \xi$ for each countable limit ordinal ξ
- for any $X \subseteq \omega_1$ there is (a stationary set of) ξ such that $X \cap \xi = X_\xi$.

Jensen had shown that $\diamondsuit$ holds in the constructible universe and, even though it implies that $2^{\aleph_0} = \aleph_1$ the reverse implication fails. The fact that $\diamondsuit$ implies there is a Suslin tree will play a role in §8.

One of the first to realize the potential applications of $\diamondsuit$ was Saharon Shelah [1974], who showed that $\diamondsuit$ implies that every Whitehead group of cardinality $\aleph_1$ is free. On the other hand, Shelah [1974] had also shown that assuming Martin's Axiom and the failure of the Continuum Hypothesis — which is discussed in §8 — there is a non-free Whitehead group of cardinality $\aleph_1$. In his proof Shelah isolated the following combinatorial consequence $C(S)$ of Martin's Axiom and the failure of the Continuum Hypothesis: S is a stationary subset of ω_1 consisting of limit ordinals and for each $\xi \in S$ there is an increasing sequence $\{\eta^i_\xi\}_{i\in\omega}$ converging to ξ and if $c_\xi : \omega \to 2$ is a colouring of the integers by two colours for each $\xi \in S$ then there is a single colouring $c : \omega_1 \to 2$ such that for each $\xi \in S$ the equality $c(\eta^i_\xi) = c_\xi(i)$ fails for only finitely many i.

In the introduction to their [1978], Keith Devlin and Shelah recount the developments leading to a new consequence of the Second Continuum Hypothesis:

> Shelah conjectured that $C(\omega_1)$ was consistent with ZFC + GCH. Devlin refuted this conjecture by showing that CH implies $\neg C(\omega_1)$... Devlin's original proof used metamathematical techniques (precisely, inner models of set theory). Devlin, Jensen and Shelah all independently observed that the proof could be modified to eliminate the use of inner models, and that the assumption of CH could be weakened to $2^{\aleph_0} < 2^{\aleph_1}$. Shelah took this a step further by "extracting" from the proof the principle Φ ...

The principle Φ, which follows from the negation of Lusin's Second Continuum Hypothesis, is now known as weak $\diamondsuit$ and asserts the following: For any function F colouring the complete binary tree of height ω_1 with two colours, 0 and 1, there is a function $g : \omega_1 \to 2$ such that for any branch B of the binary tree the set of all ξ such that $F(B \restriction \xi) = g(\xi)$ is stationary. While the proof of Φ recorded in [Devlin and Shelah, 1978] is quite elementary and does not use concepts unavailable to Sierpiński, Rothberger or Lusin, it is clear from the description of the history leading to this proof that significant new ideas were needed to arrive at the elementary proof. As well, the formulation is very closely tied to the combinatorial problems considered by Shelah in dealing with Whitehead groups. It is no surprise then, that the weak $\diamondsuit$ implies there is a non-free Whitehead group. Applications to topology and model theory were soon also found and the principle continues to play a central role in set theory.

A more recent development along these lines is a family of parametrized versions of $\diamondsuit$ that are closely related to certain cardinal invariants of the continuum. One of these is introduced in [Hrušák, 2001] and is closely related to the cardinal $\mathfrak{d}$ examined in §5. Variations and generalizations of this to other cardinal invariants of the continuum are found in [Moore *et al.*, 2004] and some of these have their roots in a much earlier weakening of $\diamondsuit$ referred to as $\clubsuit$ introduced by Adam Ostaszewski [1976]. Just as Φ does not require the Continuum Hypothesis, neither does $\clubsuit$. But to follow this line of development would lead considerably away from the continuum. The reader interested in recent developments is encouraged to consult the introduction of [Brendle, 2006].

4 THE CONTINUUM HYPOTHESIS

This history chose as its starting point Hilbert's address to the International Congress in 1900 because the first question on his famous list was Cantor's continuum problem. Cantor 's unsuccessful search for a proof of his Continuum Hypothesis is well documented — for example see [Moore, 1989] or [Dauben, 1979] — and Hilbert himself was to devote considerable energies to this problem some twenty-five years after his address, with no more success than Cantor. However, there were some minor successes, and even more notable failures, around this time as well.

Four years after Hilbert's address in Paris the International Congress of Mathematicians met again in 1904 in Heidelberg. The Continuum Hypothesis was again the topic of at least one talk. Julius König presented a proof (cf. [1905]) in which he attempted to show that the Continuum Hypothesis is false on the grounds that $2^{\aleph_0} \neq \aleph_\xi$ for any ξ. The correct part of the proof is what is now known as König's Lemma: For any sequence of ordinals $\{\sigma_n\}_{n\in\omega}$ the inequality $\sum_{n\in\omega} \aleph_{\sigma_n} < \prod_{n\in\omega} \aleph_{\sigma_n}$ holds. It should be emphasized that it was the proof that was correct, not the statement. Indeed it was Hausdorff who was to make precise the notion of cofinality in his researches on order types which are discussed in §5. The error in König's argument was his use of an assertion of Bernstein that $\aleph_\alpha^{\aleph_0} = \aleph_\alpha 2^{\aleph_0}$ in which Bernstein had claimed, essentially, that if $\alpha > 0$ then the cofinality of $\aleph_\alpha$ is uncountable. Taking this into account one sees that what König's argument actually establishes is that the cofinality of $2^{\aleph_0}$ is uncountable.

Bernstein himself was also scheduled to give a talk on the continuum problem, resolving it in the opposite way from König. He did not give his talk though, but published a version [1905] a year later. He argued that the continuum has cardinality $\aleph_1$ because each real number can be calculated according to some rule and counting the set of all such rules was to yield $\aleph_1$. The strategy employed by Hilbert [1926] a quarter century after his 1900 address followed similar lines. However, such arguments did not lead to any resolution of the problem until Gödel's insightful analysis [1940] of such ideas lead to the consistency of $2^{\aleph_0} = \aleph_1$ as exemplified by the constructible universe.

It should be mentioned that the continuum problem was considered to have two variants in the early years of the twentieth century. The strong form was whether what is now known as the continuum hypothesis is true: Does $2^{\aleph_0} = \aleph_1$? The alternate form of the continuum hypothesis merely asserted that there are only two possible values for the cardinalities of subsets of the reals. Assuming the Axiom of Choice, of course, these problems are equivalent but, since this axiom was still the subject of much controversy at the time, it made some sense to consider both forms of the continuum problem.

Interest in the alternate version of the continuum problem was stimulated by Cantor's result that the alternate Continuum Hypothesis is true for closed sets. Extending this to other classes of sets seemed a natural course to follow. Indeed, Gregory Moore [1989, page 109] claims Hausdorff's interest in order types was motivated by a desire to solve the Continuum Problem. According to Moore, Hausdorff "put forward a new way of posing the problem, in the hope that this would lead to a solution. Namely, CH is true if the following proposition holds ...: There is a dense order-type of power $\aleph_1$ having no (ω, ω^*)–gaps." However, Hausdorff's own words[13] do not add as much weight to the result quoted by Moore:

> For simple reflection reveals that each everywhere dense type contains the countable everywhere dense type η; consequently, *each everywhere dense type without (ω, ω^*)–gaps contains the usual continuum as a subset.* As long as there is no success, for instance, in constructing an everywhere dense type of the second infinite cardinality without (ω, ω^*)–gaps and with that in verifying Cantor's hypothesis $2^{\aleph_0} = \aleph_1$, the questions of the existence of homogeneous types of the second infinite cardinality will be restricted to the 32 species with (ω, ω^*)–gaps.

It seems from this passage that Hausdorff is primarily interested in the classification of linear order types, but that he nevertheless recognizes the significance of his work to the Continuum Problem.

However, after the first decade of the century the role of the Continuum Hypothesis as a construction tool began to predominate. Lusin's and Mahlo's early construction of what is now known as a Lusin set has already been mentioned in

[13](Page 156 of [Hausdorff, 1906]; the translation is from [Hausdorff, 2005].) Denn eine einfache Überlegung lehrt, daßjeder überalldichte Typus den abzählbaren überalldichten Typus η, folglich *jeder überalldichte Typus ohne $\omega\omega^*$-Lücken das gewöhnliche Kontinuum als Teilmenge enthält.* Solange es also nicht etwa gelingt einen überalldichten Typus zweiter Mächtigkeit ohne $\omega\omega^*$-Lücken zu konstruieren und damit die CANTORsche Hypothese $2^{\aleph_0} = \aleph_1$ zu verifizieren, wird man die Frage nach der Existenz homogener Typen zweiter Mächtigkeit nur auf die 32 Spezies mit $\omega\omega^*$-Lücken beschränken.

§2. As Akihiro Kanamori [1996, page 25] points out, "[i]t was the mathematical investigation of CH that increasingly raised doubts about its truth and certainly its provability". Indeed, by 1934 such constructions had become sufficiently commonplace that Sierpiński was able to devote an entire book [1934a] to the equivalents and consequences of the Continuum Hypothesis, a hypothesis which he denoted "H".

Judging by its rank in the list of equivalents, the following proposition — answering Steinhaus' Question 6 — seemed to Sierpiński to be a critical equivalent of the Continuum Hypothesis:

P_1: The plane is the union of two sets, A and B, such that each vertical section of A and each horizontal section of B is countable.[14]

Sierpiński himself had shown [1919] this to be equivalent to the Continuum Hypothesis. From this follow various consequences which are also equivalent to the Continuum Hypothesis. For example, the plane is the union of countably many functions and their inverses if and only if the Continuum Hypothesis holds. As a remark, Sierpiński points out that a similar result holds for Euclidean space as well. Much later though, Sierpiński [1965] would show that the Continuum Hypothesis is equivalent to the following: There is a partition of Euclidean space $\mathbb{R}^3 = A_1 \cup A_2 \cup A_3$ such that if L is a line parallel to the x_i-co-ordinate axis then $A_i \cap L$ is not only countable, but actually finite. Moreover, there is a partition $\mathbb{R}^{k+2} = A_1 \cup A_2 \cup \ldots A_{k+2}$ such that if L is a line parallel to the x_i-co-ordinate axis then $A_i \cap L$ is finite if and only if $2^{\aleph_0} \leq \aleph_k$.

A related statement from [Morayne, 1987] is even more striking: the Continuum Hypothesis is equivalent to the existence of a Peano-like function $P : \mathbb{R} \to \mathbb{R}^2$ such that at each $x \in \mathbb{R}$ at least one of the co-ordinate functions of P is differentiable, but even continuity can fail at the other co-ordinate. Moreover, this was improved by Jacek Cichoń and Michał Morayne to show that $2^{\aleph_0} \leq \aleph_k$ is equivalent to the existence of a function $P : \mathbb{R}^k \to \mathbb{R}^{k+1}$ such that at each $x \in \mathbb{R}$ at least k of the co-ordinate functions of P are differentiable, but even continuity may fail at the other co-ordinate. It is interesting to note though, that even though this result seems to be totally unrelated to P_1 it, nevertheless, justifies the prominent role of P_1 emphasized by Sierpiński. For example, one can prove that the Continuum Hypothesis implies the existence of a function $P : \mathbb{R} \to \mathbb{R}^2$ such that at each $x \in \mathbb{R}$ at least one of the co-ordinate functions of P is differentiable by starting with a decomposition of the plane as described by P_1. One then defines the two co-ordinate functions of the Peano curve in such a way that the first co-ordinate function guarantees that the image of $(\infty, 1)$ covers one half of the decomposition of the plane, regardless of the definition of the second co-ordinate function, while the second co-ordinate function guarantees that the image of $(-1, \infty)$ covers the

[14] L'ensemble du tous les points du plan est une somme de deux ensembles dont l'un est au plus dénombrable sur toute parralèle à l'axe d'ordonnées et l'autre est au plus dénombrable sur toute parralèle à l'axe d'abscisses.

other half of the decomposition of the plane, regardless of the definition of the first co-ordinate function.

Other similar results stemming from Sierpiński's P_1 can be found, for example, in [Komjáth and Shelah, 1993/94]. However, it should be noted that considerably earlier, in 1967, Barbara Osofsky [1968] had discovered algebraic results equivalent to the Continuum Hypothesis. For example, she showed that the homological dimension of the field of rational polynomials in 3 variables with real coefficients over the ring of such polynomials has dimension 2 if and only if the Continuum Hypothesis holds.

Even in situations in which the connection between combinatorics in Euclidean spaces of different dimension and cardinal arithmetic is not immediately apparent, other phenomena related to dimension manifest themselves. For example, a subset $X \subseteq \mathbb{R}^n$ is said to have *distinct distances* if any two distinct pairs of points from X determine distinct distances. Kenneth Kunen [1987] generalized to $\mathbb{R}^n$ the following result of Paul Erdős and Shizuo Kakutani: Assuming the Continuum Hypothesis, there is a partition of $\mathbb{R}$ into countably many pieces such that each piece has distinct distances. The difficulty in generalizing results involving cardinal arithmetic from $\mathbb{R}$ to $\mathbb{R}^n$ appears in different contexts as well.

The following is the last in Sierpiński's section on the equivalences of the Continuum Hypothesis and was studied first by Alfred Tarski [1928].

P_{11}: There do not exist more than $\aleph_1$ distinct infinite subsets of ω_1 any two of which have finite intersection.[15]

The restriction to infinite sets here is crucial for obtaining the equivalence with the Continuum Hypothesis, even though asking for the sets to be uncountable seems more natural. However, there may have been a good reason for dealing only with this case since it was shown much later, in 1976, that such questions lead to independence results and were beyond the techniques available to Sierpiński and his contemporaries. James Baumgartner [1976] showed that the existence of a family of cardinality $\aleph_2$ of subsets of ω_1 all of whose pairwise intersections are finite is independent of the usual axioms of set theory.

The largest part of Sierpiński book deals with implications of the Continuum Hypothesis. The first of the properties listed by him that follow from the Continuum Hypothesis was the existence of a Lusin set and was denoted as C_1. The early work regarding Lusin sets is described in §2 and later work will be described in §5. However, a key result of Rothberger [1938a] certainly bears mentioning here: the Continuum Hypothesis is equivalent to the existence of both Lusin and Sierpiński sets, at least of one of which has cardinality $2^{\aleph_0}$. While this may not be the most elegant of statements, it is significant in that Rothberger's proof isolates a key translational property that has turned out to be extremely useful in many situations. He shows that if there is a second category set of cardinality κ then $\mathbb{R}$ is the union of κ meagre sets; while, on the other hand if there is a non-measurable

[15] Aucun ensemble de puissance $\aleph_1$ n'est une somme de plus que $\aleph_1$ ensembles infinis ayant deux à deux un nombre fini d'éléments communs.

set of cardinality κ then $\mathbb{R}$ is the union of κ null sets. He argues by decomposing $\mathbb{R}$ into a null and a meagre set and translating these. It follows easily that both the Sierpiński and Lusin sets must be of size $\aleph_1$. The requirement that at least one of the Sierpiński and Lusin sets have cardinality $2^{\aleph_0}$ seems awkward, but it is essential. Adding alternately Cohen and random reals to any model where $2^{\aleph_0} > \aleph_1$ produces a model with both Sierpiński and Lusin sets of size $\aleph_1$.

Also mentioned in §5 is the Property C whose central role Sierpiński acknowledges by introducing it right after the Lusin set existence hypothesis:

C_1: there is a set of cardinality the continuum whose intersection with each nowhere dense set is at most countable.[16]

He notes that the sets with Property C are invariant under continuous maps and then concludes that C_1 implies the following:

C_5: there is a set of cardinality the continuum no continuous image of which contains the unit interval.[17]

Sierpiński [1928a] had established this from the Continuum Hypothesis. However, this is considerably weaker than C_1 as was shown by Arnold W. Miller [1983]. In particular, he showed that in the iterated Sacks model every set of reals of cardinality $\aleph_2 = 2^{\aleph_0}$ can be mapped continuously onto the unit interval. This is discussed in §7. Moreover, he also remarks that it is consistent with any cardinal arithmetic that the real line can be partitioned into $\aleph_1$ pieces none of which can be mapped onto the unit interval.

An indication of the range of implications considered by Sierpiński is provided by the fact that the indexing of properties stops at C_{81}. All of these are now known to follow from Martin's Axiom and are discussed in the paper [Martin and Solovay, 1970] introducing this hypothesis. In [1997] Todorčević singles out Sierpiński's Property P_3: There is an uncountable subset of the Hilbert cube every uncountable subset of which is projected onto the interval $[0, 1]$ by all but finitely many of the co-ordinate projections. He says of this hypothesis:

> Proving this statement from CH did cause a considerable difficulties ... and we conjecture that this is because of a too strong preoccupation with getting a statement equivalent to CH rather than just its consequence (like C_1). In fact, P_3 is so made (perhaps artificially) that the implication $P_3 \to$ CH is a triviality.

Perhaps a similar "preoccupation" can be found in the result of Roy Davies [1974] who showed that the Continuum Hypothesis is equivalent to the assertion that each function from $\mathbb{R}^2$ to $\mathbb{R}$ can be represented as a sum of products of single variable functions. The last chapter of Sierpiński's book also contains several

[16] Il existe un ensemble linéaire de puissance du continu qui admet un ensemble au plus dénombrable de points communs avec tout ensemble (linéaire) parfait non-dense.

[17] Il existe un ensemble linéaire E de puissance du continu et tel que l'intervalle linéaire n'en est pas une image continue.

similar instances of this "preoccupation". The question of determining that an assertion is *not* a consequence of the Continuum Hypothesis however, is an even more delicate task.

The first result of this kind was obtained by Jensen. It had already been shown by Solovay and Stanley Tennenbaum [1971] that it was consistent that no Suslin trees exists. Moreover, Jensen had also shown that Suslin trees could be constructed by using his principle $\diamondsuit$ which holds in Gödel's constructible universe. This, however, did no preclude the possibility that the Continuum Hypothesis alone may be sufficient for the construction of a Suslin tree. Perhaps there was an argument involving some key idea that Sierpiński, Lusin, Duro Kurepa and the other early masters at using the Continuum Hypothesis had all missed. However, Jensen (cf. [Devlin and Johnsbråten, 1974]) provided a sophisticated iterated forcing construction that established the consistency of the Continuum Hypothesis along with no Suslin trees. Later Uri Abraham, Devlin and Shelah modified this approach [Avraham *et al.*, 1978] to obtain that some of the consequences of Martin's Axiom and $2^{\aleph_0} > \aleph_1$ are consistent with the Continuum Hypothesis.

However, Shelah later provided a general method for establishing such results as those of Chapter V of his [1982]. The completeness systems introduced there together with techniques for iterating proper partial orders allowed Jensen's result to be established with methods amenable to other problems. For example, Abraham and Todorčević [1997] study the following property of ideals on ω_1: every $\subseteq^*$-σ-directed ideal $\mathcal{I}$ on ω_1 is such that one of the following two alternatives holds:

- there exists an uncountable subset
- ω_1 can be decomposed into countably many subsets A_i such that $A_i \cap I$ is finite for all $i < \omega$ and $I \in \mathcal{I}$.

They show this to be consistent with the Continuum Hypothesis. While this statement is weaker than the Proper Forcing Axiom it suffices to imply that there are no Suslin trees — the history of these results will be discussed in §8 — and that all (ω_1, ω_1^*)–gaps are Hausdorff gaps, which will be discussed in §5.

The principle $\diamondsuit$ is discussed in §3 and it is mentioned that Shelah was one of the first to realize the potential applications of this principle and that he used it to great effect in studying the freeness of abelian groups. Another of the early applications of was to topology. Ostaszewski [1976] used to construct a perfectly normal, countably compact space that is not compact. In addition, the space constructed by Ostaszewski has the property that all closed sets are either countable or co-countable. The question of whether $\diamondsuit$ is necessary for establishing the existence of such a space, or whether the Continuum Hypothesis alone suffices had vexed a generation of topologists until Todd Eisworth and Judith Roitman [1999] showed that the Continuum Hypothesis alone does not suffice. The key difficulty here is to devise a forcing that destroys Ostaszewski type spaces without adding reals and, more, such that the iteration also does not add reals. Shelah's general framework from his [1982] provides the starting point for Eisworth and Roitman's work.

An important result of Woodin delineates the possibilities for obtaining results independent of the Continuum Hypothesis. If there is a proper class of measurable Woodin cardinals, and Ψ is a Σ_1^2 sentence — in other words, $\Psi = (\exists X)\Phi(X)$ where the quantifiers in Φ are all over reals, but X is allowed to be a set of reals — such that Ψ holds in some forcing extension of the set-theoretic universe, then it holds in every forcing extension that also satisfies the Continuum Hypothesis. For most purposes, this is almost the same as saying that the assertion actually follows from the Continuum Hypothesis. In other words, in order to establish that a statement asserting the existence of a set of reals with a Borel definable property follows from the Continuum Hypothesis, it will quite surely suffice to show only that it is consistent with set theory. Note that this result does not apply to the existence of Ostaszewski's space since describing this space requires a universal quantifier over sets of reals as well as an existential quantifier asserting the existence of the space. The original argument of Woodin was never published, but an accessible account of this can be found in [Farah, 2007].

5 CARDINAL INVARIANTS OF THE CONTINUUM ASSOCIATED WITH CONVERGENCE RATES

Investigations into the structure of sequences of natural numbers have their origins in the work of du Bois-Reymond [1870-71], work which was later continued by mathematicians such as Jacques Hadamard [1894] and Hardy [1910], who [1924] used du Bois-Reymond's ideas as a starting point for investigation into the theory of L-functions and what are now known as Hardy fields. But, much more relevant from the point of view of this history is the work of Hausdorff and its continuation by Rothberger leading to the modern theory of the invariants $\mathfrak{b}$, $\mathfrak{d}$ and related notions.

The question on which du Bois-Reymond had fixed his attention was that of classifying rates of convergence occurring in analysis by attaching to them some notion of order of infinity in analogy with Cantor's. He approached this problem by defining an ordering on positive, real-valued functions by declaring $f < g$ if $\lim_{x\to\infty} f(x)/g(x) = 0$, $f > g$ if $\lim_{x\to\infty} f(x)/g(x) = \infty$ and $f \sim g$ if $\lim_{x\to\infty} f(x)/g(x)$ exists and is finite but non-zero. If $f < g$ then du Bois-Reymond considered that f represented a smaller infinity than g whereas if $f \sim g$ then the two functions were considered as representing the same infinity. This structure was referred to by du Bois-Reymond as a pantachie and it was du Bois-Reymond's intention to find in it the cut point separating convergence from divergence. There are several problems here, the existence of the cut point being one of them.

The even more serious problem however, is that, unlike Cantor's cardinalities under the Axiom of Choice, du Bois-Reymond's "infinitary calculus" of pantachies suffers from the existence of incomparable orders of infinity. Unlike Cantor's cardinals, this problem persists even assuming the Axiom of Choice and under any conceivable modification of the definition of the ordering of the types considered by du Bois-Reymond. Indeed, Salvatore Pincherle and later Hausdorff [1907] had observed that replacing the quotient by a difference in du Bois-Reymond's definition leads to an equally reasonable definition. Hausdorff explains:[18]

> In addition, starting out from the mode of expression that x^α is infinite of order α, people have repeatedly endeavored to associate to the elements of such restricted function classes magnitude-like symbols with corresponding laws of combination; among these are Stoltz's moments, Thomae's complex numbers, and the symbols of Pincherle, Borel, Bortolotti et al. These magnitudes do not satisfy the Archimedean axiom and are thus "actually" infinitely large or infinitely small relative to each other, which says nothing against their logical admissibility; their usefulness can be debated since, on the one hand, they break down with respect to the intermediate levels of scale, and, on the other hand, they are more complicated than the functions whose infinity they are supposed to express.

One direction of research that has proved very productive is abstracting to the simple situation of comparing the growth rates of sequences term by term. This, of course, leads directly to the study of $\mathbb{N}^\mathbb{N}$ under $\leq^*$ which will be the subject of much of this history. Recall that if $f : \mathbb{N} \to \mathbb{N}$ and $g : \mathbb{N} \to \mathbb{N}$ then the relation $f \leq^* g$ is define to hold if and only if $f(n) \leq g(n)$ for all but finitely many n. However, it should be remarked that this abstracted, discrete version of the problem is not entirely equivalent to the original problem concerning functions defined on $[0, \infty)$ considered by du Bois-Reymond. For example, Vladimir Kanovei and Peter Koepke examine various possibilities of the discrete orderings as well as the continuous structure $\mathcal{C}([0, \infty))$ and mention that, while the existence of (κ, λ^*)–gaps in $\mathbb{R}^\mathbb{N}$ under the co-ordinate-wise ordering implies the existence of similar gaps in $\mathcal{C}([0, \infty))$, the converse is an open question. In their discussion of Gödel's argument that $2^{\aleph_0} = \aleph_2$ Jörg Brendle, Paul Larson and Todorčević (cf. their [2001]) examine this same structure in some detail, as does the survey article [Scheepers, 1993]. However, it is the discrete versions of problems on order types that have driven developments in the set-theoretic study of the continuum and its associated structures.

It was on the basis of the existence of incomparable rates of convergence that Cantor [1895] rejected du Bois-Reymond's work on this subject. On the other

[18](Page 108 of [Hausdorff, 1907]; the translation is from [Hausdorff, 2005]) Nebenbei hat man sich, von der Ausdrucksweise ausgehend, daßx^α von der Ordnung α unendlich werde, mehrfach bemüht, den Elementen solcher beschränker Funktionsklassen größentartige Symbole mit entsprechenden Verknüpfungsgesetzen zuzuordnen; hierher gehören die STOLZschen Momente, die THOMAEschen komplexen Zahlen, die Symbole von PINCHERLE, BOREL BORTOLOTTI u. A. Diese Größen erfüllen das Archimedische Axiom nicht und sind also, relativ zu einander, "aktuell" unendlich groß oder unendlich klein, womit nichts gegen ihre logische Zulässigkeit gesagt ist; über ihre Zweckmäßigkeit läst sich streiten, denn die einen versagen gegenüber den Zwischenstufen der Skala und die andern sin komplizierter als die Funktionen, deren Unendlich sie ausdrüchken sollen.

hand, Hausdorff was not at all discouraged by this realization and took up the challenge of analyzing the structure of rates of convergence in [Hausdorff, 1907]. Given that there are incomparable rates of convergence, Hausdorff's starting point was to take an arbitrary maximal, linearly ordered family of rates of convergence which, extending du Bois-Reymond's terminology, he also called a pantachie. In Hausdorff's words,[19] "if we designate it as our task to connect the infinitary rank ordering as a whole with Cantor's theory of order types, then nothing remains but to investigate the sets of pairwise comparable functions that are as comprehensive as possible." Hausdorff justifies the necessity of this approach by explaining that,[20] "all attempts to produce a simple (linearly) ordered set of elements in which each infinity occupies its specific place had to fail: the infinitary pantachie in the sense of du Bois-Reymond does not exist."

It is worth observing that Hausdorff was not alone is viewing pantachies, in this new and more precise formulation, as natural objects of study. In his treatise [1914] Borel argues[21] that the existence of pantachies should be an axiom, and enjoy the same status as the Archimedean Axiom. In particular, he feels that the transfinite process of constructing functions of ever increasing growth so as to form a well-ordered set under eventual dominance will, in fact, produce a set of functions which will eventually dominate any given function.

By the time he arrived at the study of pantachies, Hausdorff already had a wealth of experience in studying these abstract order types. Naturally, the same sort of questions which had appealed to his interest in studying abstract order types, figured prominently in forming his approach to pantachies. A significant part of Hausdorff's interest had to do with η–sets which, being generalizations of the rationals, had certain saturation properties. In contrast and conjunction to this study of saturated linear orders, it was reasonable for Hausdorff to consider gaps in linear orders and to attempt to use these as a classifying tool. Most of his [1906] is devoted to this pursuit. Here he isolates 50 possible uncountable order types according to which gaps and order types appear and tackles the problem of realizing these types by homogeneous orders. The culmination of this work is his [1909] construction of an (ω_1, ω_1^*)–gap. While the unresolved Continuum Hypothesis obviously thwarted Hausdorff's attempts at a complete solution to his classification problem, this seminal result must have provided a fair degree of satisfaction. Indeed, even in 1936, when Hausdorff republished it in a different

[19](Page 110 of [Hausdorff, 1907]; the translation is from [Hausdorff, 2005]) Bezeichen wir es also als unsere Aufgabe, die infinitäre Rangordnung als Ganzes mit der Theorie der CANTORschen Ordnungstypen in Verbindung zu setzen, so bleibt nichts anders übrig als *möglichst umfassende Mengen paarweise vergleichbarer Funktionen* zu untersuchen ...

[20](Page 107 of [Hausdorff, 1907]; the translation is from [Hausdorff, 2005]) ... alle Versuche, eine einfach (linear) geordnete Menge von Elementen herzustellen, in der jedes Unendlich seine bestimmte Stelle inne hat, mußten aus diesem Grunde scheitern: *die infinitäre Pantachie* im Sinne DU BOIS-REYMOND *existiert nicht.*

[21]Page 117 of [Borel, 1914]. In Borel's words, "... étant donné une fonction croissante quelconque, on finira par la dépasser en répétant transfiniment les procédés que nous avons indiqués. Il nous semble que c'est là un axiome qui doit être admis au même titre que l'axiome d'Archimède... ".

form in [1936], this argument still defined the state of the art.

However there is another path of enquiry, not entirely different, leading to the modern detailed studies of the structure $(\mathbb{N}^{\mathbb{N}}, \leq^*)$. This has its roots in the investigations of certain sets of reals and problems of category and measure. Borel [1919] undertook a study of sets of Lebesgue measure zero and a classification scheme that ranked null sets $X \subseteq \mathbb{R}$ by assigning to them sequences $\sigma : \mathbb{N} \to \mathbb{R}^+$ according to whether there exists a sequence of intervals $\{I_n\}_{n\in\mathbb{N}}$ such that $\bigcup_n I_n \supseteq X$ and the length of I_n is less than $\sigma(n)$. This, of course, is reminiscent of du Bois-Reymond's scheme for classifying convergence and Borel acknowledges the key fact established by du Bois-Reymond that countable families of sequences of natural numbers can be $\leq^*$-dominated in the limit. The genesis of strong measure zero sets can be traced to this paper and Borel's isolation of those sets which he describes by the phrase *"mesure asymptotique infèrieure a toute serie donné a l'avance"*. In particular, as set of reals X has *strong measure zero* if and only for any sequence of positive reals $\{\epsilon_n\}_{n=1}^{\infty}$ there is a corresponding sequence of intervals $\{I_n\}_{n=1}^{\infty}$ such that the length of I_n is less than ϵ_n and $X \subseteq \bigcup_n I_n$.

It was 15 years later that Besicovitch took up this study of the classification of null sets, apparently unaware of the earlier work of Borel. Much of Besicovitch's life's work was focused on questions of geometric measure theory and much of this theory examines measures obtained by assigning measure to certain open sets and then extending this to at least all Borel sets. The fractional Hausdorff measures are typical examples of this type of construction. It was therefore natural for Besicovitch to turn, at some point, to the simplest such example. This is obtained by taking a continuous, monotone function ϕ defined on the positive reals such that $\lim_{x\to 0^+} \phi(x) = 0$ and defining the measure of an interval (a, b) to be $\phi(b-a)$. This is then used to define an outer measure in the usual way. Besicovitch [1934a; 1934b] states that his interest is centred on determining those subsets of the real line which are null with respect to any such measure. As well, he is interested in the related problem of determining those sets on which the variation of any continuous, monotone function is zero. Relying on the *Perfect Set Property* for Borel sets — namely, that every uncountable Borel set contains a copy of the Cantor set — Besicovitch observes that the only Borel sets with either of these properties are the countable ones. He then poses himself the question of whether this is true in general.

His answer is that a set of reals is null with respect to all measures obtained from continuous, monotone functions if and only if it has the property currently referred to as strong measure zero and denoted by Sierpiński as Property C. Besicovitch's main result in [1934a] is that, assuming the Continuum Hypothesis, there is an uncountable set of reals which has strong measure zero; in other words, it is null with respect to all the measures Besicovitch considers and, furthermore, all continuous, monotone functions defined on the set have zero variation.

What Besicovitch actually constructs is what he calls a *concentrated set*, which he defines as an uncountable set $X \subseteq \mathbb{R}$ for which there is a countable set $H \subseteq \mathbb{R}$ such that any open set containing H contains all but countably many points of

X. He provides the simple argument that concentrated sets have strong measure zero and then shows how to construct a concentrated set assuming the Continuum Hypothesis. From the point of view of the study of $(\mathbb{N}^{\mathbb{N}}, \leq^*)$ the notion of a concentrated set will turn out to be a critical notion. Seemingly unaware of Besicovitch's work (although by 1942 he has met Besicovitch and the latter has introduced him to the Cambridge Philosophical Society) Rothberger will study the same family as well as the strong measure zero sets — although, using Sierpiński's [1928c] terminology of sets with Property C — establishing their relationship to $\mathfrak{b}$. An important realization, due to Sierpiński [1939], established the correspondence between open sets containing the rationals and the sequences of natural numbers obtained by thinking of the sequences of naturals as functions $\Phi : \mathbb{Q} \to \mathbb{N}$ and then sending Φ to the open set $\bigcup_{q\in\mathbb{Q}}(q - 1/\Phi(q), q + 1/\Phi(q))$. This mapping establishes that these two structures are cofinal in each other and, so, for many purposes equivalent.

Before following these later developments though, it is worth returning twenty years to 1914 since it was in this year that Lusin's article [1914] constructing a set with an even stronger property than that required by Besicovitch, but under the same assumption, appeared. These are now called Lusin sets and it is fortunate for the development of set theory that Besicovitch had ignored their existence, since the subtle nature of the differences between them and concentrated sets has led to interesting mathematical developments. The central role of concentrated sets in the evolution of the study of $\mathfrak{b}$ has already been noted. There is no indication that finding a weaker form of a Lusin set was Besicovitch's motivation for his construction, although we now do know, with the hindsight provided by close to a century of study, that concentrated sets can exist even when Lusin sets do not. (More on this can be found in the discussion of the random real model in §7.) Even though he does not refer to Lusin's much earlier work, Besicovitch must have been aware of the possibility of Lusin's construction since he also establishes, using the Continuum Hypothesis, the existence of what he calls *rarified sets*, but which are now usually referred to as Sierpiński sets. Since these are the measure theoretic analogue of Lusin sets — namely, their intersection with every null set is countable — it is hard to imagine that Besicovitch would have overlooked the possibility of using Lusin's topological construction for the purposes of the first half of [Besicovitch, 1934a]. On the other hand, Besicovitch's interest in the second half of that paper is showing that it is possible to have planar sets of positive measure that contain no subset of finite, but non-zero, measure. Further evidence for the view that Besicovitch was not concerned with the distinction between concentrated and Lusin sets is provided by the fact that he does not ask any questions along these lines. He could, for example, have asked whether the family of strong measure zero sets is the same as the family of concentrated sets, but it was Sierpiński who asked this question in print in his [1938]. However, the fact that it was Besicovitch [1942] who finally did answer this question indicates his interest in this line of enquiry. The related question of whether the existence of a set with Property C implies the existence of a concentrated set, of course, would have to wait for methods capable of producing models with large continuum. The same remark applies to the question of whether the existence of a concentrated set implies the existence of a Lusin set. This would have to wait for the arrival of forcing constructions and, in particular, the random model of Solovay that is discussed in §7.

The first hint of the connection between strong measure zero sets and the cardinal invariant $\mathfrak{b}$ appears in Lusin's article [1921]. His proof, in fact, shows that there is a set of reals of cardinality $\mathfrak{b}$ which is meagre in each perfect set, although Lusin was satisfied with simply producing an uncountable such set. This uncountable set has the same property that had motivated Lusin's earlier construction of a Lusin set using the Continuum Hypothesis; three years earlier he had used this to produce a counterexample to show that Baire's sufficient condition for belonging to the Baire classification was not necessary. In 1914, Lusin had not been able to use the 1916 result of Alexandroff and Hausdorff, independently, that any uncountable Borel set contains a perfect set and, so, he had had to rely on some trickery using cardinal arithmetic as described in §3. Now however, he had the same conclusion without having to rely on any hypothesis other than the Axiom of Choice.

Lusin's realization was that, by following Baire and associating with each sequence of integers $\sigma : \mathbb{N} \to \mathbb{N}$ the irrational $\nu(\sigma)$ obtained from the continued fraction

$$\nu(\sigma) = \frac{1}{\sigma(0) + \frac{1}{\sigma(1) + \frac{1}{\sigma(2)+\dots}}}$$

it is possible to take a scale $\{\sigma_\beta\}_{\beta\in\omega_1}$ and produce a set of irrationals $G = \{\nu(\sigma_\beta)\}_{\beta\in\omega_1}$. Lusin then uses the fact that, given any perfect set P one may as well assume that $G \cap P$ is dense in P and, hence, there is an ordinal $\gamma \in \mathfrak{p}$ — $\mathfrak{p}$ is the least cardinal of a family $\mathcal{F}$ of infinite subsets of $\mathbb{N}$ such that there is no infinite $X \subseteq \mathbb{N}$ such that $X \subseteq^* F$ for all $F \in \mathcal{F}$ — such that $G_\gamma \cap P$ is dense in P where $G_\gamma = \{\nu(\sigma_\beta)\}_{\beta\in\gamma}$. Note that du Bois-Reymond's observation that countable families of sequences can be dominated plays a key role here in guaranteeing that $\gamma \in \mathfrak{p}$. Lusin then realizes that $G \setminus G_\gamma$ must be relatively meagre in P. The fact that G_γ itself is relatively meagre in P is immediate if one is only interested in producing an uncountable set since initial segments are countable. However, the tools were at hand for Lusin to have proved that this can also be done for a set of size $\mathfrak{p}$. He could have chosen γ to be the least ordinal such that G_γ is not meagre in P and obtained a contradiction using diagonalization and the Baire Category Theorem. However, at this early stage in the developments of set theory this sort of result, asserting the existence of sets of a cardinality not known to be different from $\aleph_1$, may have seemed unjustifiable.

In any case, this slightly more delicate argument would have to wait for Rothberger and his construction of what he called a λ–set which is not universally meagre. Kuratowski had introduced the notion of a set with property λ on page 269 of his monograph [1948]. A λ–set is a subset of n-dimensional Euclidean space with the property that everyone of its countable subsets is a relative G_δ. Sierpiński

[1937] addresses the question of the additivity of the λ property and establishes that this family is countably additive if it is finitely additive and, moreover, to obtain additivity it suffices to show that the union of a λ-set with a countable set is always a λ-set. He defines a set $X \subseteq \mathbb{R}^n$ to have property λ' if $X \cup Y$ has property λ for every countable $Y \subseteq \mathbb{R}^n$ and asks whether every λ-set is also a λ'-set.

Rothberger [1939] describes a construction of a λ-set which is not a λ'-set using only the Axiom of Choice, thus answering Sierpiński's question in the negative. In the same issue of *Fundamenta* Sierpiński shows [1939] that Lusin's argument of 1917 — recorded in [Lusin, 1921] — provides an alternate proof and, indeed, the key ideas were already in place at that time. From the point of view of the development of the study of cardinal invariants though, it is interesting to note that both Rothberger and Sierpiński had introduced notation for the cardinal now known as $\mathfrak{b}$ and had isolated it as a mathematical object of study in its own right. Sierpiński names it ϕ whereas Rothberger's argument relies on defining the property he calls $\mathbf{B}(\aleph_\xi)$ and which is defined to hold if every family of sequences of natural numbers of cardinality $\aleph_\xi$ is bounded. He then defines $\aleph_\eta$ to be the least cardinal for which $\mathbf{B}(\aleph_\eta)$ fails; in other words, in contemporary notation $\aleph_\eta = \mathfrak{b}$. Both arguments, of course, rely on constructing sequences of reals well-ordered by $\leq^*$ of length $\mathfrak{b}$ or, to be precise, such that the sequences of integers obtained by the continued fraction expansions of the corresponding reals are well-ordered by $\leq^*$. However, Rothberger's interest in $\aleph_\eta$ is clear. He remarks, for example, that the cofinality of $\mathfrak{b}$ must be uncountable — but does not mention du Bois-Reymond — and, in a footnote, adds that this is close to all that is known about it. Do we know much more about $\mathfrak{b}$ today than Rothberger did in 1939? Answers to this question will be found when discussing Stephen Hechler's work on the subject in the seventies.

Both Rothberger's and Sierpiński's argument rely on showing that any G_δ containing the rationals must intersect any unbounded scale in the irrationals. Of course one can immediately conclude more: Any G_δ containing the rationals must contain all but a bounded subset of any scale in the irrationals. Recall that Besicovitch had defined a set X to be concentrated on the rationals if every G_δ containing the rationals contained all but countably many of the points of X. If an unbounded scale has length ω_1 then the result of Rothberger and Sierpiński shows that the scale must be concentrated on the rationals. Moreover, their result points to a useful generalization of Besicovitch's definition that can be phrased in terms of the cardinal invariant $\mathfrak{b}$. Recall that $\mathfrak{b}$ is the least cardinal of a family $\mathcal{B} \subseteq \mathbb{N}^\mathbb{N}$ such that for each $f \in \mathbb{N}^\mathbb{N}$ there is $b \in \mathcal{B}$ such that $b \not\leq^* f$.

This discussion has rushed ahead of many interesting and critical developments in the history of the $(\mathbb{N}^\mathbb{N}, \leq^*)$ to which we will return shortly. But, the point to be emphasized here is the motivation for studying one particular version of the questions surrounding gaps and limits in the various discrete structures that might be viewed as abstractions of du Bois-Reymond's pantachies. Hausdorff himself had deemed worthy of examination many of these variants in 1907 and these are examined in detail in [Kanovei and Koepke, 2000]. For example, two sequence of natural numbers a and b could be compared by asking whether $a(n) \lneq b(n)$ for all but finitely many n or whether it is simply the case that eventually $a(n) \leq b(n)$. Of course the gap structure of dyadic sequences was also of interest. However, the structure of sequences of natural numbers was the focus of the greatest attention of researchers such as Lusin, Sierpiński and Rothberger because of the natural map from the irrationals to $\mathbb{N}^{\mathbb{N}}$ using continued fractions already described by Baire [1899a; 1992]. It has already been mentioned that the usefulness of this map was exploited by Lusin [1914] and, later, Sierpiński and Rothberger. Indeed, many of the papers dealing with Lusin sets and λ–sets invoke the continued fraction map in an almost boiler plate fashion, the introductory remarks seeming very similar to each other. The point to realize though, is that it was the questions on the properties of subsets of the reals constructed using the Axiom of Choice which drove developments on subsets of the irrationals and, this, in turn, focused attention on $(\mathbb{N}^{\mathbb{N}}, \leq^*)$ because of Baire's continued fraction map.

Hausdorff however, was motivated by understanding the discrete structures themselves and devotes some energy in [Hausdorff, 1907] to comparing the possible variants. He spends several pages[22] discussing whether continuous or arbitrary real-valued functions should be considered, whether or not they should be monotone, and eventually arrives at the conclusion that studying discrete sequences of numbers — *Zahlenreihen* or *Zahlenfolge* — is the correct level of generality.

Recall that du Bois-Reymond's initial investigations were about the structure of real-valued functions defined on the positive reals ordered by limiting ratios. Hausdorff [1909] abstracted to the context of a discrete domain and his pantachies were maximal linearly ordered subsets of $\mathbb{R}^{\mathbb{N}}$ under the ordering $<^*$ defined by $f <^* g$ if and only if $f(n) < g(n)$ for all but finitely many $n \in \mathbb{N}$. For $\mathbb{N}^{\mathbb{N}}$ one can consider the two different orderings, $\leq^*$ and $<^*$, but, of course, for $2^{\mathbb{N}}$ only $\leq^*$ is interesting. In all of these structures it makes sense to examine the limit and gap structures considered by Hausdorff.

In Hausdorff's terminology, given a linearly ordered set $(L, \leq)$ and ordinals γ and λ the pair of subsets $\{a_\xi\}_{\xi\in\gamma}$ and $\{b_\xi\}_{\xi\in\lambda}$ will be said to form a (γ, λ^*)–gap if and only if

- $a_\xi \leq a_\eta \leq b_\alpha \leq b_\beta$ provided that $\xi \in \eta \in \gamma$ and $\beta \in \alpha \in \lambda$
- there is no $c \in L$ such that $a_\xi \leq c \leq b_\alpha$ and $a_\xi \neq c \neq b_\alpha$ for all $\xi \in \gamma$ and $\alpha \in \lambda$.

A γ–limit in $(L, \leq)$ is simply a $(\gamma, 1)$–gap. In this general setting of linear orders the notion of a *scale* corresponds to a cofinal, well-ordered subset of L. In a general partial order though, a scale is a cofinal, well-ordered family — for our purposes a *scale* is a subset of $\mathbb{N}^{\mathbb{N}}$ cofinal and well-ordered by $\leq^*$. An important point to keep in mind is that, since Hausdorff was motivated by putting du Bois-Reymond's work on a solid foundation, his notion of pantachie, a maximal linearly

[22] Pages 111 to 116.

ordered subset of $\mathbb{R}^{\mathbb{N}}$ with respect to $<^*$, was central to all his work, at least in 1909. For pantachies and, indeed, for all linearly ordered sets, the notions of unbounded and cofinal coincide; in particular, well-ordered cofinal or unbounded sets are both scales. It is not surprising, therefore, that these concepts do not figure prominently in Hausdorff's work. Rothberger on the other hand, who arrived at the same questions from a different direction, did formulate and analyze these different notions. In Rothberger's terminology a *tower* of length κ is simply a $(\kappa, 1)$–gap in $(2^{\mathbb{N}}, \leq^*)$. For Hausdorff's purposes this would be the same as a scale.

However, when dealing with problems about sets of reals, as in [Hausdorff, 1936] where he considers the problem of partitioning the reals into $\aleph_1$ Borel sets, Hausdorff chose to view the reals as the Cantor set instead of $\mathbb{N}^{\mathbb{N}}$. In 1909 he had established the existence of (ω_1, ω_1^*)–gaps in the structure $\mathbb{R}^{\mathbb{N}}$ whereas, in 1936, by using the same methods he established the analogous result for $2^{\mathbb{N}}$. While the formulation for $\mathbb{R}^{\mathbb{N}}$ may have caused some slight awkwardness in establishing that the annuli defined from his gap are Borel, this would have been very similar for $\mathbb{N}^{\mathbb{N}}$ and $2^{\mathbb{N}}$. In [1936], using the same argument Hausdorff establishes the very closely related result that there is an (ω_1, ω_1^*)–gap in the structure $(2^{\mathbb{N}}, \leq^*)$. Nevertheless the second paper quickly became widely cited and Lusin even published his own proof of the result in [1947] and Rothberger was inspired to continue work on the subject. Why was the earlier paper ignored while the later one so enthusiastically accepted? In a footnote to [1936], Hausdorff himself explains that he had the $\mathbb{R}^{\mathbb{N}}$ version of the result much earlier but that this paper is little known. Kanovei and Koepke [2000] offer the following explanation:

> Clearly the gap construction was far ahead of the level of development and, perhaps, even motivation of set theory in the early years of the century. In addition, the paper was published in a rather provincial journal.

The "level of development" seems unlikely — surely Lusin and Sierpiński would have understood the argument and embraced it. The "motivation" explanation is certainly more convincing. Keeping in mind the questions occupying Lusin and Sierpiński, questions leading to the discovery of Lusin sets and λ–sets, it seems more likely that Hausdorff's investigations into abstract order types did not seem germane at the time. As well, the motivation for Hausdorff's 1936 paper was Question 2 of Sierpiński. This was something more likely to attract the attention of Lusin and Sierpiński who had been working on these questions for some time, and Rothberger, who was just starting in the area.

Nevertheless, the key results were in place by 1909. The (ω_1, ω_1^*)–gap has already been mentioned, but Hausdorff also built on du Bois-Reymond's work to show that there are no (ω, ω^*)–gaps in $(\mathbb{R}^{\mathbb{N}}, <^*)$, indeed, this is a key component in showing that there are (ω_1, ω_1^*)–gaps. He also established some important, but not as surprising, results assuming the Continuum Hypothesis; for example, there are ω_1–limits, (ω_1, ω^*)–gaps as well as towers and scales.

The study of these order types in $(\mathbb{N}^{\mathbb{N}}, \leq^*)$ was systematically taken up as a subject in its own right only some thirty years later in the 1941 work of Rothberger, who came to it through an interest in properties of sets of reals. Of course, some five years earlier, Hausdorff had dealt with Question 2 and Question 4 of Sierpiński by showing that (ω_1, ω_1^*)–gaps produce partitions of the reals into $\aleph_1$ Borel sets. Of course, the relevance of scales to λ–sets and their variants had been known for quite some time. But now, in 1936, Hausdorff showed that his much more sophisticated construction of 1909 also had applications to the study of sets of reals and, from the modern perspective, to cardinal invariants of the continuum. It is interesting that Hausdorff cites Question 4 as his motivation, referring to Sierpiński's questions as (A) and (B):

A: Can one show, without assuming the Continuum Hypothesis, that there is a union of $\aleph_1$ first category sets which is of second category?[23]

B: Can one show, without assuming the Continuum Hypothesis, that there is a union of $\aleph_1$ null sets which has positive outer measure?[24]

He is able to show, that given a gap $\{A_\xi\}_{\xi\in\omega_1} \cup \{B_\xi\}_{\xi\in\omega_1}$ the set of all $X \subseteq \mathbb{N}$ such that $A_\xi \subseteq^* X \subseteq^* B_\xi$ is an F_σ in the Cantor set. Consequently, the family of differences of these sets provides the required Borel partition. However, Hausdorff's analysis of the situation, while providing a solution to Question 2, does not answer Question 4. The fact that this is a hopeless task would not become evident until after Cohen's work of the early sixties.

In any case, these were the developments that lead to Rothberger's [1939] in which he introduced his already mentioned concept $B(\aleph_\xi)$ to study λ–sets. By 1941, he had become aware of Besicovitch's work on concentrated sets and was able to more clearly outline the connection between these objects and $B(\aleph_\xi)$. In synthesizing earlier work of Sierpiński, as well as himself, Rothberger [1941] presents the following picture of the knowledge about λ–sets and unbounded families in $(\mathbb{N}^\mathbb{N}, \leq^*)$. The following are all equivalent:

- Any subset of $\mathbb{N}^\mathbb{N}$ of size $\aleph_\xi$ is $\leq^*$-bounded.
- Any subset of the irrationals of size $\aleph_\xi$ has Property Θ.[25]
- Given disjoint sets of reals A and B, with B an F_σ and A a set of cardinality no greater than $\aleph_\xi$, an F_σ can be found containing A and disjoint from B.
- The union of $\aleph_\xi$ sets with Property Θ has Property Θ.
- All sets of reals of cardinality $\aleph_\xi$ have property λ.
- All sets of reals of cardinality $\aleph_\xi$ have property λ'.

[23] Kann eine Summe von $\aleph_1$ Mengen 1. Kategorie von 2. Kategorie sein?
[24] Kann eine Summe von $\aleph_1$ Nullmengen von positivem äusserem Masse sein?
[25] A set or irrationals has Property Θ if it is contained in an F_σ contained in the irrationals.

- All sets of reals of cardinality $\aleph_\xi$ have property σ, in other words, all relative G_δ-subsets are also F_σ.
- Every real-valued function on a set of reals of cardinality $\aleph_\xi$ is of Baire class no greater than 1.

Rothberger also points out the special role of $\aleph_1$ in these problems. The key result here is: The failure of $B(\aleph_1)$ is equivalent to the existence of an uncountable concentrated set of reals.

Recall that a set X is concentrated on the rationals if every G_δ containing the rationals contains all but a countable subset of X. The role of countability allows Rothberger's equivalence in the case of $\aleph_1$, but not for other cardinals.

This special role of $\aleph_1$ would be more subtly elucidated in section 4 of Rothberger's paper. Here he returns to the question of gaps themselves. A preliminary key result here is that the failure of $B(\aleph_1)$ is equivalent to the existence of (ω_1, ω^*)–gaps in $(2^{\mathbb{N}}, \leq^*)$, even though his proof actually shows, in his notation, that there is an (ω_η, ω^*)–gap in $(2^{\mathbb{N}}, \leq^*)$.

However, in 1948 Rothberger shows that the existence of (ω_1, ω^*)–gaps in $(2^{\mathbb{N}}, \leq^*)$ implies the existence of ω_1–limits in $(\mathbb{N}^{\mathbb{N}}, \leq^*)$. He does this by setting up a bijection between $\mathbb{N} \times \mathbb{N}$ and $\mathbb{N}$ and then noting that if $f : \mathbb{N} \to \mathbb{N}$ and $g : \mathbb{N} \to \mathbb{N}$ and $f \leq^* g$ then $\{(n, m) \in \mathbb{N} \times \mathbb{N} \mid m \leq f(n)\} \subseteq \{(n, m) \in \mathbb{N} \times \mathbb{N} \mid m \leq g(n)\}$. Indeed, from the point of view of cardinal invariants, this correspondence takes gaps to gaps and can be used to establish an inequality between the associated cardinal invariants and need not be restricted to the case of $\aleph_1$.

In Lemma 7 of his [1948], however, Rothberger uncovers an essential use of $\aleph_1$ that stands besides Hausdorff's gap construction as a milestone in the understanding of the role of ω_1 in the combinatorics of the continuum. He begins by showing that the existence of an (ω_1, ω^*)–gap is equivalent to the following assertion: There are families $\{X_\alpha\}_{\alpha \in \omega_1}$ and $\{Y_m\}_{m \in \omega}$ of subsets of $\mathbb{N}$ such that $X_\alpha \subseteq^* Y_m$ for each $\alpha \in \omega_1$ and $m \in \omega$ yet there is no $Z \subseteq \mathbb{N}$ such that $X_\alpha \subseteq^* Z \subseteq^* Y_m$ for each $\alpha \in \omega_1$ and $m \in \omega$. In other words, Rothberger extends his interest to gaps that do not need to be linearly ordered, gaps that were not of interest to Hausdorff because his motivation was to make sense of the pantachies of du Bois-Reymond by considering maximal linear orders. He shows that, at least for the case of gaps which have $\aleph_1$ sets on one side and countably many on the other, there is no loss of generality in assuming linearity. He then uses this result to show that if there are no (ω_1, ω^*)–gaps then for every filter with base $\{Z_\alpha\}_{\alpha \in \omega_1}$ there is a $\subseteq^*$-descending family $\{B_\alpha\}_{\alpha \in \omega_1}$ such that $B_\gamma \subseteq^* Z_\gamma$. In the language of cardinal invariants the result can be stated as: If $\mathfrak{p} = \aleph_1$ then $\mathfrak{t} = \aleph_1$. But, from Rothberger's point of view, if $\mathfrak{p} = \aleph_1$ then there is an ω_1–limit in $(2^{\mathbb{N}}, \subseteq^*)$ and so, using his terminology, $\mathfrak{t}$ can be defined to be the least cardinal such that there is a $\mathfrak{t}$–limit.

Rothberger's idea here is that, given a filter base $\{Z_\alpha\}_{\alpha \in \omega_1}$ it is possible to first choose for each α a set A_α such that $A_\alpha \subseteq Z_\gamma$ for each $\gamma \in \alpha$. This is not difficult and was known already to du Bois-Reymond. However, there is no reason to believe that the family $\{A_\alpha\}_{\alpha \in \omega_1}$ is $\subseteq^*$-descending. To achieve this Rothberger chooses inductively $\subseteq^*$-descending $\{B_\alpha\}_{\alpha\in\omega_1}$ such that $Z_\gamma \supseteq^* B_\gamma \supseteq^* A_\beta$ for $\gamma \in \beta$. The reason he is able to do this is that failure at stage γ would guarantee that the family $\{B_\beta\}_{\beta\in\gamma} \cup \{A_\alpha\}_{\alpha\in\omega_1\setminus\gamma}$ is an (ω_1, ω^*)–gap, contradicting his hypothesis. This is a central result which sheds considerable light on one of the major outstanding problems concerning cardinal invariants of the continuum: Is $\mathfrak{t} = \mathfrak{p}$?

Before leaving the work of Rothberger it is worthwhile to look at some of his other results that have to do with the structure of $(\mathbb{N}^\mathbb{N}, \leq^*)$, some of which we will return to later. Results such as that $\mathfrak{p} = \aleph_1$ implies the existence of ω_1–limits in $(2^\mathbb{N}, \subseteq^*)$ or the existence of Hausdorff's gaps, to say nothing of Gódel's consistency results, must surely have provided encouragement to researchers in the thirties and forties to continue trying to prove the Continuum Hypothesis. Some of Rothberger's results must have seemed tantalizingly close. For example, Rothberger [1938b] shows that the Continuum Hypothesis is equivalent to the conjunction of two assertions of a cardinal invariant nature. The first is that there is a family $F \subseteq \mathbb{N}^\mathbb{N}$ of size $2^{\aleph_0}$ such that for each $g \in \mathbb{N}^\mathbb{N}$ the set $\{f \in F \mid f \leq^* g\}$ is countable. The second assertion is that there is a scale of cardinality $2^{\aleph_0}$ in $(\mathbb{N}^\mathbb{N}, \leq^*)$. The result is not at all difficult, but, in order to obtain it, it was crucial that Rothberger was looking at the whole structure $(\mathbb{N}^\mathbb{N}, \leq^*)$ rather than a pantachie within it such as Hausdorff would have done in 1909. It is also noteworthy that Rothberger begins to consider cofinal families in $(\mathbb{N}^\mathbb{N}, \leq^*)$ rather than just unbounded families. The cardinal $\mathfrak{d}$ is certainly implicit in this work — recall that $\mathfrak{d}$ is defined to be the least cardinal of a cofinal family in $(\mathbb{N}^\mathbb{N}, \leq^*)$, namely a family $\mathcal{D} \subseteq \mathbb{N}^\mathbb{N}$ such that for each $f \in \mathbb{N}^\mathbb{N}$ there is a $d \in \mathcal{D}$ such that $f \leq^* d$. However, it was first considered explicitly by Miroslav Katětov [1960] in the context of studying the character of points in topological spaces. Other cardinal invariants are also standing in the wings waiting to be introduced. Rothberger [1948] shows that the failure of the Lusin's weak Continuum Hypothesis, namely $2^{\aleph_0} = 2^{\aleph_1}$, implies the existence of ω_1–limits. In the same paper he shows that, in modern terminology, if $\aleph_1$ is equal to the minimum of $\mathbf{non}(\mathcal{N})$ and $\mathbf{non}(\mathcal{M})$ then there are ω_1–limits.

However, a lengthy hiatus of close to twenty years would be experienced in the study of the structure $(\mathbb{N}^\mathbb{N}, \leq^*)$ before it was known that the existence of ω_1–limits does not imply the Continuum Hypothesis, let alone what the nature of the relationship between these concepts is. So, before skipping these twenty years to begin looking at developments in the post-Cohen era, it is worthwhile reviewing the progress that had been made by 1952, the year that Rothberger published his last paper [1952a] on the subject. (The other paper Rothberger published that year, [1952b], was considered in §3.) Hausdorff's initial research on $(\mathbb{N}^\mathbb{N}, \leq^*)$ was motivated by examining the pantachies of du Bois-Reymond. He knew that the axiom of choice produced these maximal linearly ordered subsets, but had more detailed information about them only by assuming the Continuum Hypothesis. In particular, Hausdorff knew that (ω_1, ω_1^*)–gaps existed, but he did not know whether they existed in all pantachies. Lusin, on the other hand, was brought to studying unbounded families by his interest in λ–sets. He was able to construct a λ–set without assuming the Continuum Hypothesis; but, the existence of Lusin sets or concentrated sets still requires some extra hypothesis. Rothberger continued both lines of investigation. The relationship between λ–sets and the structure of $(\mathbb{N}^{\mathbb{N}}, \leq^*)$ is detailed by him in his list of equivalences to sets of a fixed cardinality being $\leq^*$-bounded. He is also able to show that $\mathfrak{p} = \aleph_1$ implies the existence of ω_1–limits in $(2^{\mathbb{N}}, \subseteq^*)$ or, in other words, $\mathfrak{t} = \aleph_1$. Furthermore, Besicovitch had shown that there is a set with Property C which is not concentrated, but the question of whether the existence of a set with Property C implies the existence of a concentrated set remained unanswered in 1952.

QUESTION 8. Does the existence of a concentrated set imply that there is a Lusin set?

What is the role of $\mathfrak{b}$ in these questions? It was known by Rothberger that $\mathfrak{b} = \aleph_1$ is equivalent to the existence of a concentrated set and that concentrated sets have Property C. However, the following question could have been asked:

QUESTION 9. Is there a set with Property C even if $\mathfrak{b} > \aleph_1$?

Rothberger explicitly asks in [1948] whether the implication in his Theorem 3 can be reversed, namely,

QUESTION 10. Does $\mathfrak{b} = \aleph_1$ imply that $\mathfrak{t} = \aleph_1$?

The answers to all of these questions would have to await the advent of forcing techniques.

The first of the results to address these unanswered questions about $(\mathbb{N}^{\mathbb{N}}, \leq^*)$ dealt, unknowingly, with questions arising from Rothberger's article [1939]. Recall that in a footnote Rothberger had remarked that the cofinality of $\mathfrak{b}$ is uncountable, but that no more is known about it. Hechler [1974] shows that Rothberger did indeed know all that it was possible to know at the time. However, before describing Hechler's result, it should be noted that history did not unfold as it should at this point. One might have expected that the newly available forcing techniques would have precipitated a re-examination of the intense work on set-theoretic questions recorded in *Fundamenta Mathematicae* from the twenties to the end of the forties. However, the early applications of forcing techniques produced their own questions in some cases, and answered questions from entirely different sources in others.

Hechler's main theorem in [1974] is the following: Given any partially ordered set $(P, \leq)$ such that:

- $|P| \leq 2^{\aleph_0}$
- for all countable subsets $X \subseteq P$ there is some $p \in P$ such that $x \leq p$ for all $x \in X$

there is a forcing extension in which there is a $\Phi : P \to \mathbb{N}^{\mathbb{N}}$ such that $p \leq q$ if and only if $\Phi(p) \leq^* \Phi(q)$ and, moreover, the image of Φ is cofinal; in other words, for each $f \in \mathbb{N}^{\mathbb{N}}$ there is some $p \in P$ such that $\Phi(p) \geq^* f$.

The result can, in fact, be strengthened as has been noted by Maxim Burke [1997] who provides a modern version of Hechler's argument. The strengthening

Burke notes is that one does not need to assume that $|P| \leq 2^{\aleph_0}$ as long as one realizes that in the generic extension the size of the continuum will be $|P|^{\aleph_0}$. As a particular corollary to Hechler's theorem it follows that any uncountable regular cardinal κ can play the role of $(P, \leq)$ in the theorem and, in this case, it follows that $\mathfrak{b} = \kappa$. So Rothberger's footnote is, indeed, accurate. Moreover, if $\kappa < \rho$ are two different uncountable regular cardinals then $\kappa \times \rho$ can play the role of $(P, \leq)$ in Hechler's theorem and, in this case, it follows that $\mathfrak{b} = \kappa$ even though there are also ρ–limits in $(\mathbb{N}^{\mathbb{N}}, \leq^*)$.

Hechler's construction also plays a key role in solving Question 10. R. C. Solomon [1977] solves this by first showing that the equalities $\mathfrak{b} = \aleph_2 = 2^{\aleph_0}$ and $2^{\aleph_1} = \aleph_3$ hold in a model obtained by Hechler's construction used to establish his consistency result for the case $(P, \leq) = (\omega_2, \in)$. Using Rothberger's result that the weak Continuum Hypothesis implies there are ω_1–limits — this was examined in §3 — it follows that $\mathfrak{t} = \aleph_1$ in this model. It is worthy of note that Solomon's discussion of sets with Property C mentions that they are the same as the sets which have μ^h-Hausdorff measure zero[26] for each function h. This clear link to the work of Besicovitch may explain why Solomon was more familiar with the work of Rothberger than many North American set theorists, such as Hechler, were at the time.

In the same paper Solomon points to the significance of the cardinal arithmetic of this model and a necessary condition for solving Question 9. He notes that Rothberger had shown that $\mathfrak{b} = \aleph_1$ implies there is a set which is concentrated and hence has Property C. On the other hand, Rothberger [1952a] also showed that if $X \subseteq \mathbb{R}$ has cardinality less than $\mathfrak{p}$ then X has Property C. So, in order to have a model of set theory without any uncountable set with Property C it is necessary that $\mathfrak{p} < \mathfrak{b}$. The fact that this can not be ruled out follows from the cardinal arithmetic of Hechler's model.

Not long afterwards, Laver [1976] was able to produce a model of set theory of the "Borel Conjecture": all sets with Property C, or, as he referred to them, all sets with *strong measure zero*, are countable. This was a milestone result in several respects. The consistency result itself, of course, was of great interest. Laver's motivation here is rooted in considerations about strengthenings of the property of being Lebesgue null. Strong Measure Zero is one such strengthening, but so is being universally null. Laver reminds his reader that Hausdorff had established, as a corollary to the existence of (ω_1, ω_1^*)–gaps, that there are universally null sets of size $\aleph_1$. Laver points out that, "Regarding universal measure zero sets, Hausdorff's theorem is best possible in the sense that there is a model of ZFC + $2^{\aleph_0} > \aleph_1$ in which there are no universal measure zero sets of power $\aleph_2$". He explains that adding ω_2 Sacks reals or Solovay reals to a model of the Continuum Hypothesis will provide the required counterexample and that this fact is an extension of unpublished results of Baumgartner establishing that there are no sets with Property C of cardinality $\aleph_2$ in these models.

The other milestone Laver's paper marks is a new level of sophistication in forc-

[26] See [Rogers, 1970] for the relevant definitions.

ing methods. Countable support iterations had previously been used for partial orders which are either countably closed or, at least, not too far removed from the countably closed domain. In [1976] (announced in [1975]) Laver showed that this sort of support had much broader applicability. Efforts to generalize Laver's arguments would lead to the notion of Axiom A[27] forcing but, eventually Shelah was to show that countable support forcing could be used in contexts not even hinted at by Laver's result, although the Axiom A partial orders continue to provide important examples of Shelah's "proper" partial orders. However, this is leading us to topics that are discussed in greater detail in §7 since the Borel Conjecture paper was not Laver's only contribution to settling questions left unanswered by the efforts of Hausdorff, Rothberger, Sierpiński and their contemporaries.

In [1979] Laver shows that it is possible to construct a model of set theory in which $2^{\aleph_0} > \aleph_1$ and the $2^{\aleph_0}$-saturated linear order — recall that Hausdorff had devoted considerable effort to analyzing precisely such orders some 70 years earlier — is embedded into the structure $(\mathbb{N}^{\mathbb{N}}, \leq^*)$. But, just as Hechler's motivation for his embedding result was not in answering the questions that were beyond the techniques available to Rothberger, Laver's embedding result was used to answer a question raised by Solovay and Woodin concerning the automatic continuity of homomorphisms of Banach algebras. So, while Laver's result is directly related to the pantachies of Hausdorff, these are not even mentioned in [Laver, 1979]. Using routine methods, Laver's embedding result can easily be modified so as to arrange that the linear order he embeds into $\mathbb{N}^{\mathbb{N}}$ is actually maximal, in other words, a pantachie. This then answers question α of Hausdorff's [1907] which asks whether there is a pantachie without (ω_1, ω_1^*)–gaps.

Since Hechler has already been mentioned, it may be useful to pause briefly to explain why Laver's result is not implied by Hechler's. After all, one might argue, the $2^{\aleph_0}$-saturated linear order satisfies the weakened hypothesis of Hechler's theorem noticed by Burke, so it follows that it can be embedded, in fact embedded cofinally. The subtle error in this argument is that being $2^{\aleph_0}$-saturated is not an absolute property. So, while it is true that Hechler's argument will embed the ground model $2^{\aleph_0}$-saturated linear order into $(\mathbb{N}^{\mathbb{N}}, \leq^*)$, the forcing which does this will destroy the saturation property. So this is the difficulty Laver had to overcome.

His approach was to embed the $2^{\aleph_0}$-saturated linear order by iteratively adding new generic elements to $\mathbb{N}^{\mathbb{N}}$ which would be the targets of the embedding. At the same time, it is possible to keep track of new gaps in the $2^{\aleph_0}$-saturated linear order which need to be mapped to $\mathbb{N}^{\mathbb{N}}$. This seems a natural strategy, but there is a critical obstacle to overcome and there is a degree of irony in the fact that, while Laver was not interested in Hausdorff's pantachies, the obstacle which he had to overcome was one discovered by Hausdorff in his study of pantachies, namely the (ω_1, ω_1^*)–gap. It may occur that the strategy used by Laver arrives at a situation where the embedding has already been defined on a subset of order type $\omega_1 + \omega_1^*$ in

[27]These should not be confused with Axiom A from [Martin and Solovay, 1970] which is now known as Martin's Axiom.

the saturated linear order and has been mapped to an (ω_1, ω_1^*)–gap in $\mathbb{N}^{\mathbb{N}}$. From the fact that the linear order is $2^{\aleph_0}$-saturated and $2^{\aleph_0} > \aleph_1$ in the final model, it follows that there must be an element A of the saturated linear order falling between both halves of the subset of order type $\omega_1 + \omega_1^*$. However, the entire strategy must fail at this stage since the (ω_1, ω_1^*)–gap can not, by definition, be filled by any image of A. Laver's key realization though, is that since the image of the embedding consists of generic elements of $\mathbb{N}^{\mathbb{N}}$, every potential (ω_1, ω_1^*)–gap in $\mathbb{N}^{\mathbb{N}}$ can be split by an element of $\mathbb{N}^{\mathbb{N}}$ obtained by forcing with an appropriate partial order.

Laver's realization that generic (ω_1, ω_1^*)–gaps can be split by suitable forcing unveils the fine structure of these objects which could not even have been formulated by Hausdorff. It was Kunen who provided (cf. [Kunen, 1975; Baumgartner, 1984]) an even clearer picture of the nature of Hausdorff's gaps as well as gaps of different cofinalities. To put Kunen's contribution in the proper context, it has to be noted that Lavers' embedded linear order, of course, has no (κ, λ^*)–gaps at all for κ and λ less than $2^{\aleph_0}$. On the other hand, there must be $(\mathfrak{c}, \mathfrak{c}^*)$–gaps in the embedded liner order since these can be constructed by a standard diagonalization argument using the $2^{\aleph_0}$-saturation property. However, Kunen was able to show that it is consistent with Martin's Axiom and $2^{\aleph_0} = \aleph_2$ that there is a linear order of cardinality $2^{\aleph_0}$ which does not embed into $(2^{\mathbb{N}}, \leq^*)$.

His key idea was to establish a relationship between the assertion that a subset of $(2^{\mathbb{N}}, \leq^*)$ of order type $\kappa + \lambda^*$ is a gap with the assertion that the natural order for forcing a set which splits the gap enjoys the countable chain condition. It will become apparent that this idea is very closely related to the construction of a Hausdorff gap. In fact, it shows that there is some justification in using the preposition "the" when referring to a Hausdorff gap. Given a family $\mathcal{Z} = \{Z_\gamma\}_{\gamma \in \kappa + \lambda^*}$ of subsets of $\mathbb{N}$ such that $Z_\gamma \subseteq Z_{\gamma'}$ if $\gamma < \gamma'$ let $\mathbb{P}(\mathcal{Z})$ be the partial order consisting of finite approximations to a set splitting $\{Z_\gamma\}_{\gamma \in \kappa}$ from $\{Z_\gamma\}_{\gamma \in \lambda^*}$. To be more precise, $p = (x, a, b) \in \mathbb{P}(\mathcal{Z})$ where x is a finite subset of $\mathbb{N}$, a is a finite subset of κ and x is a finite subset of λ^*. Moreover $\bigcap_{\gamma \in a} Z_\gamma \setminus \max(x) \supseteq \bigcup_{\gamma \in b} Z_\gamma \setminus \max(x)$. Kunen's key lemma[28] describes precisely when this partial order has the countable chain condition.

1. If $\mathcal{Z}$ is filled — in other words, if $\mathcal{Z}$ is not a gap — then $\mathbb{P}(\mathcal{Z})$ has the countable chain condition.
2. If the cofinality of either κ or λ is different from ω_1 then $\mathbb{P}(\mathcal{Z})$ has the countable chain condition.
3. If $\kappa = \lambda = \omega_1$ then there is a countable chain condition partial order which forces an uncountable set of pairwise incomparable elements, known as an antichain, in $\mathbb{P}(\mathcal{Z})$.

The key point to observe is that it is immediate that, assuming Martin's Axiom, the only possible (κ, λ^*)–gaps are

[28]Theorem 4.2 of [Baumgartner, 1984].

- $(\mathfrak{c}, \omega_1^*)$ and $(\omega_1, \mathfrak{c}^*)$–gaps
- $(\mathfrak{c}, 1)$ and $(1, \mathfrak{c}^*)$–gaps
- (ω_1, ω_1^*)–gaps
- $(\mathfrak{c}, \mathfrak{c}^*)$–gaps

and the $(\mathfrak{c}, 1)$ and $(1, \mathfrak{c}^*)$–gaps — or, limits in Hausdorff's and Rothberger's terminology — are easily constructed by induction under Martin's Axiom. However, the same argument does not work for $(\mathfrak{c}, \mathfrak{c}^*)$–gaps because an (ω_1, ω_1^*)–gap may arise at an intermediate stage bringing the induction to a halt. The same applies to $(\mathfrak{c}, \omega_1^*)$ and $(\omega_1, \mathfrak{c}^*)$–gaps.

Kunen was able to show that, in fact, it is consistent with Martin's Axiom that there are no $(\mathfrak{c}, \mathfrak{c}^*)$–gaps, nor are there $(\mathfrak{c}, \omega_1^*)$ or $(\omega_1, \mathfrak{c}^*)$–gaps. The strategy is to begin with a model of the Continuum Hypothesis and $\diamondsuit_{\omega_2}$ for the ordinals of cofinality ω_1. This version of $\diamondsuit_{\omega_2}$ is used to guess initial segments of the embedding, for example, of a $(\mathfrak{c}, \omega_1^*)$–gap. When an initial segment is trapped at stage δ, as

$$\mathcal{Z}_\delta = \{Z_\xi\}_{\xi \in \delta} \cup \{Z_\xi\}_{\xi \in \omega_1^*}$$

it follows that the cofinality of δ is ω_1. Hence the third clause of Kunen's lemma can be applied to it. In other words, there is a countable chain condition partial order which adds an uncountable antichain to $\mathbb{P}(\mathcal{Z}_\delta)$. Once there is an uncountable antichain in $\mathbb{P}(\mathcal{Z}_\delta)$ it will persist in all larger models and, hence, it is not possible in any later model to split $\mathcal{Z}_\delta$, by the first clause of Kunen's lemma. In other words, once a $(\mathfrak{c}, \omega_1^*)$–gap is trapped by the $\diamondsuit_{\omega_2}$-sequence, it is possible to force with a countable chain condition partial order so that the trapped gap can not be extended any further. It is easy to dovetail countable chain condition partial orders into this strategy to obtain a model of Martin's Axiom and no gaps other than those guaranteed by the construction of Hausdorff.

It is interesting to examine this proof in a bit more detail to see that it is even more closely connected to Hausdorff's construction than might be suggested by the statement of the result. Recall that the key idea of Hausdorff in his [1909], and later in his [1936], was to arrange that the gap $\mathcal{H} = \{Z_\xi\}_{\xi \in \omega_1} \cup \{W_\xi\}_{\xi \in \omega_1^*}$ constructed has the following property: Given any $\xi \in \omega_1$ and any $k \in \mathbb{N}$ there are only finitely many $\zeta \in \xi$ such that $Z_\zeta \subseteq k \cup W_\xi$. If there were a set $Y \subseteq \mathbb{N}$ such that $Z_\zeta \subseteq^* Y \subseteq^* W_\zeta$ for each $\zeta \in \omega_1$ then there would also be some $k \in \mathbb{N}$ and an uncountable set $\Gamma \subseteq \omega_1$ such that $Z_\zeta \subseteq Y \cup k \subseteq^* W_\zeta \cup k$ for each $\zeta \in \Gamma$. Any $\xi \in \Gamma$ with infinitely many predecessors in Γ then yields a contradiction. In particular, there are only finitely many $\zeta \in \xi \cap \Gamma$ such that $Z_\zeta \subseteq Y \cup k \subseteq^* W_\zeta \cup k$. In the terminology of the partial order $\mathbb{P}(\mathcal{H})$, the key property of the gap $\mathcal{H}$ constructed by Hausdorff is that for any uncountable set $\Gamma \subseteq \omega_1$ there are ζ and ξ in Γ such that $(\emptyset, \{\xi\}, \{\xi\})$ and $(\emptyset, \{\zeta\}, \{\zeta\})$ are incompatible. One may say that for the gap constructed by Hausdorff the partial order $\mathbb{P}(\mathcal{H})$ hereditarily fails to have the countable chain condition.

The success of Hausdorff's construction of an (ω_1, ω_1^*)–gap can be thought of as being the result of producing a family which is as chaotic as possible with respect to the finite defects allowed by the $\leq^*$ ordering. Todorčević has extended this approach by more carefully analyzing the defects in scales of length $\mathfrak{b}$. He defines an oscillation of two functions f and g in $\mathbb{N}^{\mathbb{N}}$, denoted by $osc(f, g)$, to be the cardinality of the set $\{n \in \mathbb{N} \mid g(n) \leq f(n) \text{ and } g(n+1) > f(n+1)\}$. Observe that if $f \leq^* g$ then $osc(f, g)$ is finite. In Corollary 1 of [1988] (see also his [1998]) Todorčević establishes the following: If $B \subseteq \mathbb{N}^{\mathbb{N}}$ is any $\leq^*$-unbounded family of monotonically increasing functions such that all countable subsets of B have a $\leq^*$ upper bound in B then $\{osc(f, g) \mid f, g \in B\} = \mathbb{N}$. This idea would later be used to great effect in the seminal paper [Todorcevic, 1987] establishing the negative partition relation

$$\aleph_1 \not\to [\aleph_1]^2_{\aleph_1}$$

But the key realization contained in the earlier result is that the behaviour of initial segments of unbounded scales is as chaotic as possible. So, while Rothberger's remark that all that can be said about $\mathfrak{b}$ is that it has uncountable cofinality is true, if one looks at the families of sequences exemplifying $\mathfrak{b}$ then one finds there a very complex structure.

Since Todorčević's results on unbounded sets in $(\mathbb{N}^{\mathbb{N}}, \leq^*)$ do not use methods that were unavailable to Rothberger, or even to Sierpiński and Hausdorff before him, it is natural to ask why it took so long before this deeper structure was uncovered. A part of the explanation is certainly the difference in motivation for looking at $(\mathbb{N}^{\mathbb{N}}, \leq^*)$. Except for Hausdorff, all the researchers working on $(\mathbb{N}^{\mathbb{N}}, \leq^*)$, from Lusin to Rothberger, were primarily concerned with properties of subsets of the reals; sets with Property C, λ–sets and concentrated sets were the focus of greatest attention and the properties defining these sets are one dimensional in nature. It is an accident of terminology that these sets are often referred to as "linear" by the early researchers, but it is an accident worth keeping in mind. Hausdorff's construction of the gaps named in his honour are the only hint of an interaction between pairs of elements of $\mathbb{N}^{\mathbb{N}}$. Todorčević, however, was primarily interested in higher dimensional phenomena as can be seen by examining the generality of statements in his [1988]. Indeed, while the statement of Kunen's lemma has been stated in a forcing context, it is often more profitable to view Kunen's results in the context of partition relations as is done, for example, in [Abraham *et al.*, 1985].

Before leaving the history of $\mathbb{N}^{\mathbb{N}}$ there is still the question of the implications concerning concentrated sets, Lusin sets and sets with Property C. Since Lusin sets are concentrated and concentrated sets have Property C, the two questions to answer are:

QUESTION 11. Does the existence of an uncountable concentrated set imply the existence of a Lusin set?

QUESTION 12. Does the existence of an uncountable set with Property C imply the existence of an uncountable concentrated set?

Since Rothberger showed that there is an uncountable concentrated set if and only if $\mathfrak{b} = \aleph_1$ the first question is equivalent to asking whether the equality $\mathfrak{b} = \aleph_1$ implies that there is a Lusin set. This is easily seen to be false in the model obtained by adding $\aleph_2$ random reals since all sets of size $\aleph_1$ are meagre in this model as was shown by Solovay. The second question is equivalent to asking whether $\mathfrak{b} > \aleph_1$ implies the Borel Conjecture. Martin's Axiom and $2^{\aleph_0} > \aleph_1$ easily shows this to be false too. However, these results may be better understood if considered at the same time as the historical developments surrounding the properties of measure and category.

6 CARDINAL INVARIANTS OF MEASURE AND CATEGORY

The discussion of the cardinal $\mathfrak{b}$ in §5 was quite narrowly focused and readers familiar with the subject will have noticed, and perhaps even objected to, the lack of reference to the notions of measure and category. It is time to rectify this situation since, indeed, the cardinals $\mathfrak{b}$ and $\mathfrak{d}$ that are central to the study of $(\mathbb{N}^{\mathbb{N}}, \leq^*)$ are just as central to the study of measure and category. Coincidentally, these two invariants are also central in the diagram attributed by David Fremlin [1984] to Cichoń, as well as Anastasis Kamburelis and Janusz Pawlikowski:

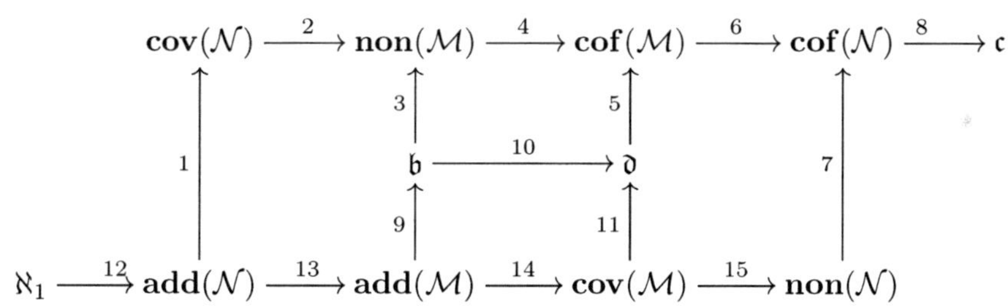

in which $\mathcal{M}$ represents the ideal of meagre subsets of the reals and $\mathcal{N}$ represents the ideal of Lebesgue null subsets of the reals. Furthermore, if $\mathcal{I}$ is either $\mathcal{M}$ or $\mathcal{N}$ then

- $\mathbf{cov}(\mathcal{I})$ represents the least cardinality of a set $\mathcal{A} \subseteq \mathcal{I}$ such that $\bigcup \mathcal{A} = \mathbb{R}$
- $\mathbf{add}(\mathcal{I})$ represents the least cardinality of a set $\mathcal{A} \subseteq \mathcal{I}$ such that $\bigcup \mathcal{A} \notin \mathcal{I}$
- $\mathbf{non}(\mathcal{I})$ represents the least cardinality of a set $A \subseteq \mathbb{R}$ such that $A \notin \mathcal{I}$
- $\mathbf{cof}(\mathcal{I})$ represents the least cardinality of a set $\mathcal{A} \subseteq \mathcal{I}$ such that for every $X \in \mathcal{I}$ there is $A \in \mathcal{A}$ such that $X \subseteq A$.

The definition of these four invariants can, of course, be applied to any ideal $\mathcal{I}$.

In fact, much of the development of the study of measure and category on the reals can be seen from the perspective of the cardinal invariants in Cichoń's diagram.

Just as importantly, the diagram marks a watershed in the study of the continuum by attaching names and bringing to the forefront the cardinal invariants associated with the reals in its various guises. It seems a simple enough idea that one may wonder why it had not appeared earlier. A probable reason is the early preoccupation with the Continuum Hypothesis. Recall that one of Sierpiński's questions in the first volume of *Fundamenta Mathematicae* was Question 4 asking whether it can be proved, without appealing to the Continuum Hypothesis of course, that there are $\aleph_1$ null sets whose union is not null or $\aleph_1$ meagre sets whose union is not meagre. While it is true that in the same volume Ruziewicz asks Question 5, whether sets of cardinality less than $2^{\aleph_0}$ are necessarily meagre, the focus at that time is certainly in $\aleph_1$. This remains true even later in 1936 when Hausdorff tries to answer Question 4 using his (ω_1, ω_1)–gap. Given this approach, drawing something similar to Cichoń's diagram would certainly have seemed senseless since the entire diagram collapses to a single point in this case. With the range of possibilities for the continuum unveiled by Cohen though, it makes perfect sense and similar diagrams seem to have been used by many researchers in the early years after Cohen's introduction of forcing techniques. For example, Miller [1981; 1982b] considers the same properties of measure and category as those that appear in Cichoń's diagram, but he is only interested in whether or not the cardinals are equal to the continuum. The organizational chart he uses is usually referred to as the Kunen–Miller chart.

So, the history of measure and category can begin by examining what could have been said about Cichoń's diagram in the years before the forcing era. It should be explained that an arrow from a cardinal invariant x to the invariant y means two things:

A: it is possible to prove that $x \leq y$ without appealing to extra axioms of set theory,

B: it is not possible to prove that $y \leq x$ without appealing to extra axioms of set theory.

In other words, each arrow corresponds to an inequality and none of the arrows can be replaced by an equality in the sense that for each of the 15 arrows there is a model of set theory where the cardinal invariant at the tail of the arrow is smaller than the one at its tip. This section will concentrate on the (A) component of the arrows in Cichoń's diagram, leaving the (B) component to §7. In fact, the (B) component is best described by the Kunen–Miller chart, a matrix whose entries correspond to possible combinations of cardinal invariants being equal or less than $\mathfrak{c}$. The final entries to this chart were added in [Judah and Shelah, 1990].

The (A) part of Arrow 8 is nothing more than that each null set is contained in a null G_δ and the number of such G_δ sets is no greater than $2^{\aleph_0}$ while the (A) part of Arrow 12 is just the additivity of Lebesgue measure. The (A) part of Arrows 1, 7, and 14 are immediate consequences of the definitions, as is the (A) part of Arrow 10. Recall that in Rothberger [1938a] showed that $\mathbf{non}(\mathcal{M}) \geq \mathbf{cov}(\mathcal{N})$ and

$\mathbf{non}(\mathcal{N}) \geq \mathbf{cov}(\mathcal{M})$ or, in other words, he justified the (A) parts of Arrow 2 and Arrow 15. But, of course, this was not Rothberger's primary goal since, in fact, he was concerned with the Continuum Hypothesis. The result he was able to prove with the use these two inequalities is that the Continuum Hypothesis is equivalent to the existence of Lusin and Sierpiński sets at least one of which has cardinality $2^{\aleph_0}$. However, this is simply a convoluted way of expressing his key result which, from the modern perspective seems more elegant than the equivalence: If there are both Sierpiński and Lusin sets then they both have cardinality $\aleph_1$.

As well, as Miller [1984b, page 205] has observed, Rothberger's result has direct implications for forcing constructions. For example, adding κ Cohen reals will yield a Lusin set of size κ consisting of the Cohen reals themselves. The same is true of random reals. Iteratively adding a pair of reals, one Cohen and the other random, to a model of $2^{\aleph_0} > \aleph_1$ will yield both a Lusin and a Sierpiński set. Rothberger's argument shows that there is no way of modifying the support or iteration strategy to get one of the sets to have cardinality greater than $\aleph_1$.

What can be said of the other arrows? Part (A) of Arrow 3 is essentially due to Sierpiński [1939] who showed that a subset of the irrationals is $\leq^*$-bounded if and only if it is σ-compact. It is important to recall that the representation of the irrationals as sequences of integers was already known to Lusin and Baire, but that the definition of the cardinal invariants associated with the meagre and null ideals are not specific to the metric space involved and, indeed, it can be shown that this is irrelevant for these invariants. For a result relating unboundedness to σ-compactness though, the role of the irrationals is, of course, crucial.

It has already been mentioned that in the same article Sierpiński established the cofinality preserving mappings between $\mathbb{N}^\mathbb{N}$ and the family of open sets containing the rationals. Using this, it is easy to establish the (A) part of Arrow 11. However, it is also worth observing that the nature of the map allows the definition of a Tukey function [Tukey, 1940]. A Tukey function between partial orders is any function, not necessarily order-preserving, which sends unbounded sets to unbounded sets. John Tukey defined these functions within a year or two of Sierpiński's results, but for quite different purposes. However it was to be only much later that the relevance of Tukey functions to cardinal invariants would become apparent, indeed, would become a guiding principle in their further study. Their relevance to the study of the continuum was realized by Peter Vojtas [1993] who called them Galois-Tukey functions, but some of the ideas are already apparent in the paper of Fremlin [1984] which gave Cichoń's diagram its name. For example, a dual argument using the Tukey function established by Sierpiński yields the (A) part of Arrow 9.

Excluding the arrows involving the cofinality invariants, this leaves only Arrows 13 to discuss, the theorem of Tomek Bartoszynski. Certainly the notions involved were familiar long before Inequality 13 was proved; one need look no further than Question 4 to see this. In order to discuss Bartoszynski's Theorem in its historical context, it is worthwhile using the same notation he did, notation borrowed from Miller [1981]. $A(m)$ represents the statement that all families of less than $2^{\aleph_0}$ null sets have null union and similarly for $A(c)$ and category and

$B(m)$ for covering. Earlier Miller [1981] had shown that $\mathbf{add}(\mathcal{N}) \leq \mathfrak{d}$, an inequality that Cichoń's diagram reveals was later refined. Also, Kunen had shown that $\mathbf{add}(\mathcal{N}) \lneq \mathbf{add}(\mathcal{M}))$ is possible.

Bartoszynski starts his [1984] by showing that $A(m)$ is equivalent to the statement that for any family of less than $2^{\aleph_0}$ convergent series there is a single convergent series dominating all of the members of the family; he calls this last property hD. It is interesting to recall, in this context, that Hausdorff had been interested in classifying the convergent series by finding a cut in a pantachie. The critical lemma then is one which establishes that hD is equivalent to a statement about what, in three years time, Bartoszynski [1987] would be calling *slaloms*. Part of this lemma establishes that hD is equivalent to:

STATEMENT 13. For every $F \subseteq \mathbb{N}^{\mathbb{N}}$ of size less than $2^{\aleph_0}$ there exist sets $I_n \subseteq \mathbb{N}$ such that $|I_n| < n^2$ and for every $f \in F$ there is some M such that $f(m) \in I_m$ for all $m \geq M$.

The key point here is that such statements are similar to combinatorial characterizations of the meagre ideal. The role of n^2 is simply that it produces a convenient convergent series $\sum_{n=1}^{\infty} n^{-2}$.

John Truss [1977] and Miller [1982b] had shown that $\mathbf{add}(\mathcal{M}) = \mathfrak{c}$ if and only $\mathbf{cov}(\mathcal{M}) = \mathfrak{b} = \mathfrak{c}$. Indeed, the arguments show that $\mathbf{add}(\mathcal{M})$ is the minimum of $\mathbf{cov}(\mathcal{M})$ and $\mathfrak{b}$, as might be conjectured by considering Arrows 9 and 14 of Cichoń's diagram. Miller [1982b] also shows that if $\mathfrak{b} = \mathfrak{c}$ then the equality $\mathbf{cov}(\mathcal{M}) = \mathfrak{c}$ can be characterized combinatorially by the statement:

STATEMENT 14. For every $F \subseteq \mathbb{N}^{\mathbb{N}}$ of cardinality less than $2^{\aleph_0}$ there is a $g \in \mathbb{N}^{\mathbb{N}}$ such that for every $f \in F$ there are infinitely many $n \in \mathbb{N}$ such that $f(n) = g(n)$.

Slightly earlier, Michel Talagrand had found a similar combinatorial approach to category in his study [1980] of measurable and non-measurable filters. A set $X \subseteq 2^{\mathbb{N}}$ is meagre if and only if there is $y : \mathbb{N} \to 2$ and an increasing sequence of natural numbers $\{n_i\}_{i=0}^{\infty}$ such that for every $x \in X$ there are only finitely many i such that $x \restriction [n_i, n_{i+1}) = y \restriction [n_i, n_{i+1})$.

Bartoszynski uses Statement 14 by adopting a different perspective on functions from $\mathbb{N}$ to $\mathbb{N}$ in order to show that Statement 13 implies Statement 14. Given a family of functions $F \subseteq \mathbb{N}^{\mathbb{N}}$ of size less than $2^{\aleph_0}$ he associates to $f \in F$ the function f' which maps n to the finite partial function $f \restriction [n^3, (n+1)^3)$. Statement 13 then allows the selection of a family I_n of size n^2 of functions defined on the interval $[n^3, (n+1)^3)$ such that for every $f \in F$ eventually $f \restriction [n^3, (n+1)^3)$ belongs to I_n. Since the domains of the functions in I_n are larger than $|I_n|$ it is easy to find a single $g_n : [n^3, (n+1)^3) \to \mathbb{N}$ which agrees with each member of I_n on at least one point in $[n^3, (n+1)^3)$. Stringing together the g_n produces the desired function g witnessing that Statement 14 holds. The argument for $\mathfrak{c}$ then converts without difficulty to a proof that Arrow 13 holds.

It is interesting that the same result along with strengthenings were obtained by Jean Raisonnier and Jacques Stern [1983; 1985] as well as Pawlikowski [1985] but in a much different spirit. Whereas Bartoszynski's arguments would have been comprehensible to Sierpiński or Rothberger, even the statement of Pawlikowski's result would have required considerable explanation: *Let M be a standard model of* ZFC. *If the union of all Borel sets of measure (Lebesgue) zero coded in M has measure zero then the union of all Borel sets of first category coded in M is of first category.* What has become evident by the 1980's is that even combinatorial statements that could have been considered and even proved in the 1930's are often more profitably viewed from the perspective of adding a generic real over a certain model of set theory.

It is instructive to look a bit more closely at Pawlikowski's results. In his [1985] his arguments are typical of this approach. He first recasts Bartoszynski's Theorem in the following form: If M is a standard model of ZFC then Statement 13 holds for $F = \mathbb{N}^{\mathbb{N}} \cap M$ if and only if the union of all Borel sets of measure zero in M has measure zero. He then shows that if M is a standard model of ZFC such that Statement 13 holds for $F = \mathbb{N}^{\mathbb{N}} \cap M$ then the union of all meagre Borel sets in M is meagre. Bartoszynski's Theorem follows from this by realizing that if $\kappa < \mathbf{add}(\mathcal{N})$ and $\mathcal{X}$ is a family of κ meagre sets then one can take a model of sufficiently much set theory containing $\mathcal{X}$ yet still having cardinality κ. It follows from the recast version of Bartoszynski's Theorem that Statement 13 holds for the functions of this model and hence the union of all meagre sets in the model is meagre. In particular, the union of $\mathcal{X}$ is meagre.

Moreover, the techniques used in Pawlikowski's argument lend themselves to arguments providing a deeper explanation of some of the inequalities of Cichoń's diagram. For example, in Section 2.3 of [Bartoszyński and Judah, 1995] it is shown that there is a Tukey function from $(\mathcal{M}, \subseteq)$ to $(\mathcal{N}, \subseteq)$. It follows directly from the existence of this Tukey function that $\mathbf{add}(\mathcal{N}) \leq \mathbf{add}(\mathcal{M})$, namely the (A) part of Arrow 13 holds. Moreover, a duality argument using the Tukey function establishes that $\mathbf{cof}(\mathcal{N}) \leq \mathbf{cof}(\mathcal{M})$ or, in other words, that the (A) part of Arrow 6 holds.

It was mentioned that the cardinal invariants related to the cofinality of the ideals $\mathcal{N}$ and $\mathcal{M}$ did not receive much attention before the forcing era, but this is not entirely true. An important counterexample is the Sierpiński-Tarski duality theorem assuming the Continuum Hypothesis. This uses cofinality as a key part of an argument — although under the Continuum Hypothesis, of course, many of these invariants become difficult to distinguish — to construct a function from $\mathcal{M}$ to $\mathcal{N}$ somewhat in the spirit of a Tukey function. The idea behind the result is that measure and category are, in many senses, very similar so it would be very useful if information about one structure could be transferred to the other. Which direction is more useful depends on your point of view though. John C. Oxtoby [1980] takes the view that, since "the theory of measure is more extensive and "important" than that of category, the service (of such a transfer) is mainly in the direction of measure theory." On the other hand, Shelah [2000b] comments that, "Mathematicians who are not set theorists generally consider "*null*" as senior to "*meagre*", that is, as a more important case; set theorists inversely, as set-theoretically Cohen reals are much more manageable than random reals ..." The two points of view differ not only in bias, but also in perspective. For Shelah measure and category are the study of random and Cohen reals, whereas Oxtoby's perspective is more traditional.

The duality was first formulated and proved by Sierpiński [1934b] and then later strengthened by Erdős [1943]. The basic idea of the proof is that, assuming the Continuum Hypothesis , there are cofinal families of order type ω_1 under inclusion in both $\mathcal{N}$ and $\mathcal{M}$. Constructing a bijection which respects annuli of these cofinal families will be the required mapping. Sierpiński concludes that, assuming the Continuum Hypothesis, any property of sets mentioning only the notions of null and meagre can be dualized by interchanging the roles of "meagre" and "null" and the dual statement will have the same truth value.

The question naturally arises of how similar the properties, meagre and null, actually are. One indication that full duality does not hold has already been seen in the inequalities $\mathbf{add}(\mathcal{N}) \leq \mathbf{add}(\mathcal{M})$, although consistency arguments for the (B) part of Arrow 13 are needed to make this more credible. A slightly earlier hint at this phenomenon was Shelah's proof [1984a] that the consistency of having all sets of reals exhibit the Baire Property did not require a large cardinal, whereas the analogous result for measure, due to Solovay [1970], does require an inaccessible.

But it would be very wrong to leave the impression that the role of inaccessible cardinals in the study of measure and category began to be appreciated only in the forcing era. This idea is quickly dispelled by considering Kuratowski's Question 7 about whether a set which is nowhere meagre can be partitioned into two disjoint sets, both of which are nowhere meagre. Stanisław Ulam [1933] had shown that if all weakly inaccessible cardinals are greater than $\mathfrak{c}$ then any second category set contains uncountably many disjoint second category sets. Assuming the same hypothesis Sierpiński [1934c] was then able to answer Question 7 in the positive. However, only a bit later, Lusin [1934] discovered an even more ingenious argument that dispensed with the extra hypothesis entirely.

While there are a few results — such as Zbigniew Piotrowski and Andrzej Szymański's inequality establishing $\mathfrak{t} \leq \mathbf{add}(\mathcal{M})$ — that link the cardinal invariants of measure and category with others, it is worth taking a bit of a detour to consider the cofinalities of some of these invariants. Two results deserve special attention in this context. The first is Miller's [1982a] proof that the cofinality of $\mathbf{cov}(\mathcal{M})$ is uncountable. In stark contrast to this, Shelah [2000a] showed that the cofinality of $\mathbf{cov}(\mathcal{N})$ can be countable in some models. It is interesting to speculate whether such questions were ever considered by Baire or Lebesgue. When Baire showed that $\mathbf{cov}(\mathcal{M})$ is uncountable the notion of cofinality had not yet been properly formulated. Within five years though, it had been clarified in response to the Bernstein and König controversy after the 1904 International Congress of Mathematicians mentioned in §4 and so it may have occurred to Baire whether he could improve his result in the way Miller eventually did. The same may have occurred to Lebesgue but the answer would certainly have surprised him, as it does even those who first encounter it today.

Even though this will be the topic of the next section, it is not possible to end this section without mentioning the great body of work on finding models exhibiting the entire spectrum of behaviour allowed by the various cardinal invariants related to measure and category that are not ruled out by Cichoń's diagram. Indeed, it is certainly true that much of the advances in developing forcing techniques were spurred on by the goal of trying to find a complete understanding of the possible relations among the invariants of Cichoń's diagram. The reader interested in the details of this development is encouraged to consult [Bartoszyński and Judah, 1995] or [Blass, 2010] which devote considerable space to the this topic. Part of the story will be examined in §7.

7 WHAT FORCING ARGUMENTS REVEAL ABOUT THE CONTINUUM

The discussion in §6 focused entirely on condition (A) in the explanation of the meaning of the arrows in Cichoń's diagram. The use of forcing to obtain independence results, thereby establishing the (B) component of the arrows in Cichoń's diagram, is the goal of this section. The tremendous effect that Cohen's introduction of forcing techniques in the 1960's had on the development of the study of the continuum has been mentioned several times already. However, Cohen's motivation for his work seems to have been equally based on establishing the independence of the Axiom of Choice from the other axioms of set theory as it was for showing that the negation of the Continuum Hypothesis is consistent. As it turns out, the same reals used to establish the failure of the Axiom of Choice could be used to control the size of the continuum; indeed, the arguments in this case are somewhat simpler.

Cohen's original arguments on the general framework of forcing were soon discovered to be more cumbersome than necessary, but the combinatorial arguments on adding κ Cohen reals to a model of the Continuum Hypothesis are essentially the same as those used today. A difference that is inessential from the combinatorial point of view is that, instead of finite functions from κ to 2, Cohen's original argument uses finite collections of forcing statements. In order to show that new countable subsets of the ground model are contained in old countable subsets of the ground model Cohen shows that his partial order has what is now known as the *countable chain condition*, namely every family of pairwise incompatible elements is countable. Cohen's Lemma 2 on page 131 of [Cohen, 1966] states exactly this and the proof can be recognized as a proof of what is now known as the Δ-system lemma. However, it would be Solovay [1970] who would realize that Cohen's partial order is equivalent, for the purposes of forcing, to the partial order of equivalence classes of Borel sets modulo the ideal of meagre sets. This is done as an afterthought in the context of proving the analogous result for random reals.

But before considering this it is worth noting that Cohen's interest in the Axiom of Choice resulted in him producing various models in [Cohen, 1966] exhibiting different types of failures of the Axiom of Choice. One of these is a model where ω_1 is the union of countably many countably ordinals. This is not of great interest from the point of view of the continuum, but it is of interest to note that the method of proof relies on collapsing cardinals with finite conditions. Cohen clearly saw this as a generalization of the original Cohen partial order. Around the same time as Cohen's proof became available, Azriel Levy produced generalizations, one of them being to apply this collapsing partial order to inaccessible cardinals. The resulting partial order collapses all uncountable cardinals below the inaccessible, but not the inaccessible itself.

Solovay realized that this partial order has important homogeneity and embedding properties that can be used to control the appearance of reals in the generic extension. His key innovation, already appearing in [Solovay, 1969], was the definition of what he called a "random real" over a model $\mathfrak{M}$ as a real which does not belong to any null set belonging to $\mathfrak{M}$. In order to have this definition make any sense at all, Solovay had had to develop the theory of codes for Borel sets. With the notion of a code for a Borel set it is possible to understand the notion of a real x not belonging to a null set in $\mathfrak{M}$ as meaning that there is no G_δ set with a code in $\mathfrak{M}$ which contains x. Establishing absoluteness for the concepts involved is a critical step in developing the theory of codes for Borel sets. In other words, Solovay had to show that if x does not belong to a Borel set with code c in some model containing both x and c then x does not belong to that Borel set in any other model containing both x and c.

Solovay [1970] describes his intuition for arriving at the definition for a random real as follows:

> Let x be a real random over $\mathfrak{M}$. An observer stationed in $\mathfrak{M}$ cannot have total knowledge about x (since x is not in $\mathfrak{M}$). However, he can have partial knowledge about x. For example, if B is a Borel set rational over $\mathfrak{M}$, then a natural question the observer can ask about x is "Is $x \in B$?" If $\mu(B) = 0$, then the answer is certainly no. On the other hand, if $\mu(B) > 0$, it is possible for x to be in B .... A similar discussion shows that if B_1 and B_2 are Borel sets rational over $\mathfrak{M}$, and $\mu((B_1 \Delta B_2) = 0$ (i.e., B_1 and B_2 are equal almost everywhere), then for x random over $\mathfrak{M}$, the questions "Is $x \in B_1$?" and "Is $x \in B_2$?" are equivalent.

With this discussion as motivation, Solovay then goes on to provide the now standard definition of random forcing as equivalence classes of Borel sets modulo the null ideal with the natural ordering. He notes that an analogue of a random real is obtained by replacing the null ideal by the ideal of meagre sets and he realizes that the associated partial order is precisely the forcing Cohen had used for his original consistency results on the failure of the Continuum Hypothesis. However, the random partial order plays a secondary role in Solovay's argument to the notion of a random real as one which avoids ground model null sets. It is this definition that plays a key role in establishing the consistency of all sets being Lebesgue measurable.

A precise discussion of Solovay's arguments in [1970] establishing the measurability of all sets, thus contradicting the Axiom of Choice, would lead a bit too far away from the main theme of this article, but some key points must be mentioned. Kanamori [1994, page 132] remarks that Cohen had hinted at the possibility of using forcing to obtain the measurability of all sets of reals but, given the complexity and sophistication of Solovay's argument, even by modern standards, it is hard to imagine that Cohen was relying on anything other than intuition. In exploiting Levy's inaccessible version of the collapsing functions considered by Cohen, Solovay observes that if the continuum is collapsed to be countable then the total number of codes for null sets in the ground model is also countable after the collapse and hence their union is also null. In other words, all but a null set of reals in the generic extension are random reals. (The same argument, of course, works for Cohen reals and the meagre ideal and this leads to the consistency of all sets having the Baire property.) Solovay would later in [1969][29] establish that this would yield the measurability of all Σ^1_2 if it was true with respect to all $L[a]$. The properties of the Levy collapse allowed precisely this.

Solovay [1969] showed that the measurability of all Σ^1_2 sets is equivalent to the statement that for any real a the set of reals not random over $L[a]$ is null and, similarly, that all Σ^1_2 sets having the Baire property is equivalent to the statement that for any real a the set of reals not Cohen over $L[a]$ is meagre. These ideas developed by Solovay eventually lead to the result of Raisonnier and Stern [1983; 1985] that if all Σ^1_2 sets are measurable then all Σ^1_2 sets have the Baire property. That the converse is false was established by Shelah and this will be discussed in a bit greater detail soon. This result is noteworthy in several ways, but it should be remarked here that it provided the first evidence that the duality hinted at by the results of Sierpiński [1934b] and Erdős fails. This has already been discussed in the section of Bartoszynski's work on additivity and this is closely related to the work of Raisonnier and Stern.

The work of Solovay and Tennenbaum and Martin on iterations of forcing partial orders will be discussed in §8 but the same ideas were critical for the Levy collapse. In particular, Solovay showed that given any real a in the generic extension by the Levy collapse it is possible to find a small partial order completely embedded in the collapse such that a already belongs to the generic extension by this partial order. In a sense, this was the reverse of the problem considered in the work of Solovay and Tennenbaum and Martin. The second key observation is that the quotient of the Levy collapse by the the small algebra is again a Levy collapse in this intermediate model. In other words, the observation about almost all reals being random over this model is valid in this intermediate extension.

It is significant that Solovay's focus is on the measurability question and the parallel result for the Baire Property is treated almost as a footnote. It is interesting in this context to recall the views of Oxtoby and Shelah on the relative importance of the meagre and null ideals. Shelah's comment that the meagre ideal is more natural and tractable was likely based, in part, on its appearance in Cohen's original forcing arguments. But Solovay's introduction to his paper lists

[29]Even though the results of [Solovay, 1969] were obtained later, this article was published before [Solovay, 1970]. See [Kanamori, 1994].

several consequences of the measurability of all sets to problems in analysis, such as Newtonian capacity, but none for having the Baire Property for all sets.

Given the very similar proofs for the two results, it came as a great surprise when Shelah [1984a] showed that the consistency of all sets having the Baire property could be established without assuming the existence of an inaccessible cardinal. The argument employs a similar strategy to Solovay's but the critical stumbling block is to obtain the embedding and homogeneity properties of the Levy collapse of an inaccessible without actually resorting to assuming that there is an inaccessible cardinal. The argument relies on a careful analysis of when countable partial orders can be completely embedded into a larger, homogeneous, partial order in such a way that all but a meagre set of later reals are Cohen over the countable partial order. As in Solovay's argument, the abundance of Cohen reals will guarantee the Baire Property for all sets of reals definable in a generic extension by the smaller order. The key point is the argument allowing the homogeneity arguments in the Levy collapse to be simulated. Shelah's ideas here allow various generalizations and lead to a large class of forcing partial orders known as "sweet". For example, the article of Andrzej Rosłanowski and Shelah [2004] combines these ideas with norms on possibilities to provide a classification of countable chain condition partial orders according to whether they are Cohen-like or random-like.

While the notion of a random real played a central role in Solovay's work on the measurability question, the random algebra which generates random reals was of less significance to this argument. However, in work following on the heels of his [1970], Solovay [1971] showed that, assuming the consistency of a measurable cardinal it is consistent that the continuum is real-valued measurable. This provided answers to questions with roots in the work of Ulam [1930]. Ulam had shown that if κ is not a weakly inaccessible cardinal then there is no κ-complete measure that measures all subsets of κ. This did not exclude the possibility that there might be such a measure on the continuum though. On the other hand, Solovay [1971] also showed that if there is a real-valued measurable cardinal then there is an inner model with a measurable cardinal. This result, in combination with work of Dana Scott [1961], showing that measurable cardinals do not exist if $V = L$ points the direction in which to look for models where measurable cardinals do exist.

Solovay's consistency result relied on forcing with the measure algebra of size κ where κ is a 2-valued measurable cardinal. The continuum then becomes κ and the κ-complete ultrafilter on κ becomes a κ-complete filter on $\mathbb{R}$. However, there are new subsets of κ not measured by the ground model ultrafilter. Solovay was able to exploit the measure on the forcing algebra by analyzing the names for sets of reals and assigning a measure to such names according the average probability that a real belongs to it. The probability is calculated in the forcing algebra, using its measure, while the average is calculated using the κ-complete measure on κ.

The generic model obtained by forcing with the measure algebra on 2^κ is usually known as the random real model. Since Solovay showed that the reals added are random, these are sometimes referred to as Solovay reals. Aside from producing a real-valued measure, the random real model is useful as an example exhibiting an interesting selection of values for various cardinal invariants. Moreover, the real-valued measurable cardinal imposes severe restrictions on the possible combinatorics at $\mathfrak{c}$, such as that $\diamondsuit_{\mathfrak{c}}$ holds. The article [Fremlin, 1993] can be consulted for more information. Even before Solovay's work, Petr Vopěnka and Karel Hrbáček [1967] had shown that if there is a Lusin set of size $2^{\aleph_0}$ then there can be no real-valued measure on the reals. The significance of this is that it points to the fact that Lusin sets can be considered as sets of Cohen reals. Indeed, it follows from the characterization of Cohen reals as those not belonging to any ground model meagre set — which was remarked on by Solovay, but even earlier also noticed by Vopěnka and Hrbáček [1967] — that an uncountable set of Cohen reals is a Lusin set. The Vopěnka–Hrbáček result shows that something very much like the measure algebra needs to be used to get a real-valued measure since no Cohen reals can be added, at least not cofinally.

Before continuing to look at other methods of adding generic reals it is worth pausing to consider the two models, Cohen and random (sometime known as Solovay). These are already sufficient to display a contrasting range of possibilities for $\mathfrak{b}$ and $\mathfrak{d}$ and the other invariants of Cichoń's diagram. In the model obtained by adding κ Cohen reals to a model of the Continuum Hypothesis, Vopěnka and Hrbáček showed that $\mathbf{cov}(\mathcal{M}) = \kappa$ while $\mathbf{non}(\mathcal{M}) = \aleph_1$. One therefore has that all the values of cardinals on the left half of Cichoń's diagram are $\aleph_1$ while those on the right hand side are all $\kappa = 2^{\aleph_0}$. In particular, this establishes the (B) component of Arrows 4, 10 and 14 in Cichoń's diagram. In the analogous model for random reals, Mathias showed that $\mathbf{cov}(\mathcal{N}) = \kappa$ while $\mathbf{non}(\mathcal{N}) = \aleph_1$. One therefore has that all the values of cardinals on the bottom row of Cichoń's diagram are $\aleph_1$ while those on the top row are all $\kappa = 2^{\aleph_0}$. Moreover $\mathfrak{b} = \aleph_1$ and $\mathfrak{d} = \aleph_1$. Hence, Arrows 1, 3, 5 and 7 in Cichoń's diagram are also fully justified.

A model where both $\mathfrak{b}$ and $\mathfrak{d}$ are the continuum but the Continuum Hypothesis fails can be obtained by Hechler's work, which has already been mentioned in the discussion of the structure of $(\mathbb{N}^{\mathbb{N}}, \leq^*)$. Indeed there is a simple version of Hechler's result, which corresponds to adding a κ-scale, that is very useful in distinguishing the behaviour of various cardinal invariants of the continuum. The partial order for adding a single real dominating all the ground model functions is usually referred to as the Hechler order or, simply, the partial order for adding a dominating real. In the generic extension obtained by iterating with finite support the Hechler partial with length κ, Hechler's results show that $\mathfrak{b} = \kappa$. Hechler did not have the general iteration techniques of Martin and Solovay [1970] available to him and, in any case, the general results he obtained yielding a wide spectrum of possible structures on $\mathbb{N}^{\mathbb{N}}$ required methods beyond the linear iteration considered by Martin and Solovay. Results of Truss [1977] establish that in the linearly iterated Hechler model $\mathbf{add}(\mathcal{M}) = \mathbf{cof}(\mathcal{M}) = \kappa$. It is interesting to note what Truss says of the motivation for his work, "We first became interested in these properties on reading the paper by K. Kunen and F. D. Tall (Between Martin's axiom and Suslins's hypothesis, to appear), and in particular, in the case of *calibre*, on learning of R. M. Solovay's result that, assuming $A_{\aleph_1}$ (Martin's axiom), the measure algebra has calibre $\aleph_1$."[30] In other words, he was lead to studying properties of the continuum, in part, by work with its roots in set-theoretic topology. Other examples of this phenomenon are discussed in §9 and §8.

Further investigations revealed that arguments of Miller, Truss and Solovay combine to show that $\mathbf{add}(\mathcal{N}) \neq \mathbf{add}(\mathcal{M})$ in Hechler's model, thereby establishing that Bartoszynski's inequality, Arrow 13, cannot be reversed. It has already been mentioned that Truss and Miller had shown that $\mathbf{add}(\mathcal{M})$ is the minimum of $\mathbf{cov}(\mathcal{M})$ and $\mathfrak{b}$. Iterating Hechler reals and Cohen reals — this is not realy needed, since the finite support iteration will add Cohen reals itself — with finite support ω_2 times over a model of the Continuum Hypothesis then produces a model in which both $\mathfrak{b}$ and $\mathbf{cov}(\mathcal{M})$ have value $\aleph_2$. Hence $\mathbf{add}(\mathcal{M}) = \aleph_2$ also. Once it is shown that no random reals have been added, Solovay's characterization of these reals yields that $\mathbf{cov}(\mathcal{N}) = \aleph_1$ in this model and the (A) part of Arrow 1 then implies that $\mathbf{add}(\mathcal{N}) \neq \mathbf{add}(\mathcal{M})$. The details can be found in [Miller, 1982b]. Duality of a Tukey function argument yields that Arrow 6 of Cichoń's diagram is also justified. Similar reasoning establishes that the (B) component for Arrows 2 and 15 is also justified by Hechler's model.

With the landscape of possibilities for values of the cardinal invariants having been revealed to be remarkably broad by the models of Cohen, Solovay and Hechler, the seventies witnessed a tremendous interest in exploring this landscape and determining its exact boundaries. A wealth of new forcing structures was developed in this period, most of them motivated as probes of the possibilities of cardinal invariants, but not all.

A notable exception is the Sacks real which was based on recursion theoretic arguments for constructing a minimal degree. Adding a Sacks real to Gödel's constructible universe produces a model that has exactly two degrees of constructibility. The behaviour of Sacks forcing [Sacks, 1971] is at the opposite end of the cardinal invariant spectrum from Hechler forcing. While increasing the size of the continuum with Hechler forcing will increase most invariants along with the continuum, Sacks forcing increases the continuum by keeping most other invariants small. In particular, the side-by-side Sacks model establishes that Arrow 8 of Cichoń's diagram is also fully justified; indeed, Miller attributes the inequalities $\mathbf{non}(\mathcal{M}) < \mathfrak{c}$ and $\mathbf{non}(\mathcal{N}) < \mathfrak{c}$ to Gerald Sacks. While, the behaviour of most invariants is the same in both constructions, Baumgartner and Laver [1979] showed how to iterate Sacks reals in based on Laver's earlier [1976] work on the Borel Conjecture. They point out that while many researchers had realized that Sacks reals could be added "side-by-side" or simultaneously, the iteration is a more delicate construction. It is shown in [Steprāns, 1999] that there are invariants distinguishing the two types of forcing.

An important realization of Miller is that the iterated Sacks model enjoys a certain type of homogeneity and he was able to exploit this in [Miller, 1983] to show that every set of reals of size $\aleph_2$ can be mapped continuously onto the reals.

[30] The article cited by Truss did appear later as [Kunen and Tall, 1979].

This shows that Sierpiński's Property C_5 mentioned in §4 is consistently false. Somewhat earlier, Miller [1980] had used a modification of Sacks forcing to get a partition of the reals into $\aleph_1$ closed sets together with the equality $\mathfrak{c} = \aleph_2$.

The notion of forcing now referred to as "Laver forcing" was introduced quite early as part of Laver's work on the Borel Conjecture, even before the work with Baumgartner on iterating Sacks forcing. Recall from §5 that Borel had been concerned with an analysis of null sets based on the rates of convergence of the size of intervals and that Laver had provided a model where there are no uncountable sets with Property C, in the terminology of Rothberger. Noteworthy about Laver's argument is that it is the first instance of a countable support iteration used to iterate partial orders that add reals at each stage of the iteration. It became apparent that the strategy used by Laver was far more widely applicable and this was crystallized by Baumgartner into an axiomatic form. A class of partial orders satisfying a property Baumgartner [1983] called Axiom A has turned out to be a very useful source of examples. As it turns out though, the iteration strategy devised by Laver was soon to be eclipsed by the far more general strategies for iterating proper partial orders, a class of partial orders including those satisfying Axiom A.

However, Laver reals continued to play a significant role in studying the continuum beyond the Borel Conjecture result. It was shown already in Laver's original article that $\mathfrak{b} = \aleph_2$ in the iterated Laver model. Haim Judah and Shelah [1990] showed that $\mathbf{non}(\mathcal{N}) = \aleph_1$ in this model as well, thus establishing even more than is needed for (B) of Arrows 9 and 11. The crucial contribution of this work was to the theory of iteration of proper partial orders, since knowing that adding a single Laver real preserves the outer measure of the ground model reals does not, by itself, yield the desired result.

Woodin [1990] showed that adding any number of random reals to Laver's model preserves the Borel conjecture. The significance here is that countable support iterations can only be used for iterations of length ω_2 without collapsing cardinals. This is the the key source of difficulties in trying to establish the consistency of $\mathfrak{p} < \mathfrak{t}$. Recall from §5 that Rothberger has shown that if $\mathfrak{p} = \aleph_1$ then, in fact, $\mathfrak{p} = \mathfrak{t}$. Hence a model where $\mathfrak{p} < \mathfrak{t}$ would require that $\aleph_3 \leq \mathfrak{t}$, putting this outside the realm of models constructed with countable support. Woodin's result provides at least some alternate method of creating models with $2^{\aleph_0} > \aleph_2$ in situations where countable support iterations seem to be essential, as is the case with the Borel Conjecture.

A variant of Laver forcing, due to Miller [1984a], has played a central role in furthering our understanding of forcing and the continuum. The partial order introduced by Miller is "intermediate between Sacks perfect set forcing and Laver forcing". While defined originally as the partial order of perfect subsets of the real line in which the rationals are dense, it is equivalent to forcing with infinitely branching trees. Unlike Laver's partial order though, the conditions in Miller's partial order do not have to branch at every node, this is only required to happen cofinally often. In the paper introducing this notion, Miller shows that the generic real added is not dominated by the ground model and is of minimal degree. Moreover, any new real in the extension is either dominated by a ground model real or is itself generic with respect to Miller's partial order. A key property is shared with Sacks forcing: Every subset of $\mathbb{N}$ in the extension either contains or is disjoint from some infinite subset of $\mathbb{N}$ in the ground model. This last property implies that ultrafilters are generated by ground model sets and so, it allows much more control of these objects in forcing extensions.

This sort of property is extremely useful when dealing with problems such as destroying P–points. What was missing from Miller's construction though, was an iteration lemma, an iteration lemma which would soon be supplied by Shelah [1987]. In that paper Blass and Shelah introduce a proper forcing notion which has the property that P–points[31] in the ground model generate P–points in the generic extension. The key point though is the iteration lemma asserting that in any countable support proper forcing iteration, if a ground model ultrafilter generates a P–point at every intermediate step of the iteration then it continues to do so at the limit. This then yields that iterating the partial order constructed by Blass and Shelah with countable support provides a model without P–points. However, not long after this construction was discovered, it was realized that Miller's rational perfect set forcing would also fit into the iteration scheme, considerably simplifying the argument.

This incident points to the critical role of iteration lemmas in sophisticated forcing arguments starting from the mid-1980's. A good illustration of this is provided in [Mathias, 1977] by the forcing now known as the Mathias partial order. It is designed to add a generic ultrafilter and then diagonalize it — or at least, this is modern perspective — and it was introduced by Adrian Mathias to deal with questions on Ramsey theory without the Axiom of Choice. Among the things he proves are that by collapsing a Mahlo cardinal one obtains a model where Dependent Choice holds yet there are no maximal almost disjoint families. (These are discussed in more detail in §9.) The questions Mathias considered did not require an elaborate iteration theory since he was able to extract a great deal of information by insightful arguments applied to a single stage forcing. However, Mathias forcing has become one of the standard partial orders in the theory of countable support iterations and the collection of cardinal invariant values in the iterated Mathias model are important for delineating the boundaries of the possible.

For example, recall from §5 that Rothberger had shown that $\mathfrak{t} \leq \mathbf{add}(\mathcal{M})$ and that the inequality $\mathfrak{t} \leq \mathfrak{h}$ is discussed in §9. This raises the obvious question of whether these results can be improved to show that $\mathfrak{h} \leq \mathbf{add}(\mathcal{M})$, but this fails in the Mathias model. In particular, $\mathfrak{t} = \mathbf{add}(\mathcal{M}) = \aleph_1$ in that model [Dordal, 1982], yet $\mathfrak{h} = \aleph_2$.

The Mathias forcing is also significant in that it is a prototype for the single stage forcing in the 1982 construction by Shelah [1984b], of a model for $\mathfrak{s} > \mathfrak{b}$. This is the first instance of norms on possibilities and creature forcing which has

[31] A P–point is an ultrafilter U on the integers such that for every countable subset of $C \subseteq U$ there is a pseudo-intersection $X \in U$; in other words, $X \subseteq^* Y$ for each $Y \in C$.

had a profound influence on the theory of forcing. The memoir [Rosłanowski and Shelah, 1999] describes the further developments along this line of research, while an excellent example of the strength of these methods can be found in the article [Rosłanowski and Shelah, 2006] in which it is shown to be consistent that every real-valued function agrees with a continuous function on a non-measurable set. However, it is not the purpose of this section to provide a full account of the developments in forcing techniques since their introduction by Cohen in the early sixties. However, the far more modest goal of justifying the arrows in Cichoń's diagram has almost been achieved. All that remains to be shown is that condition (B) for Arrow 12 holds. This is the topic of §8 in which this and far more will be discussed.

8 THE BAIRE CATEGORY THEOREM AND MARTIN'S AXIOM

Perhaps the longest and most remarkable chain of related ideas in the history of the continuum is that leading from Baire's work on continuous functions to Martin's Maximum. In his doctoral thesis [1899b] Baire provides a completely abstract definition of the notion of function, one based simply on existence rather than rules for calculating values. Moreover, he explains that he is interested in examining which properties of these abstract functions imply others. An example he provides foreshadows in a surprising way the uses to which the methods he developed in his thesis would be put. He reminds his readers of the existence of continuous functions without derivatives. In other words, continuous functions with derivatives are an exceptional subset of the family of all continuous functions and the one property does not imply the other.[32] Baire's work on classifying pointwise limits of continuous functions falls squarely into this quite modern, non-constructive view of functions.

His thesis contains what is now known as the Baire Category Theorem, that no interval in the real line is the union of countably many closed nowhere dense sets. While Baire makes very effective use of this result and was clearly aware of its significance, it would take some thirty years before Banach would distill the constructive essence of Baire's Theorem into a method for proving the existence of mathematical objects. In 1931 Banach [1931] and Stefan Mazurkiewicz [1931] independently employed the Baire Category Theorem to show how to prove the existence of a continuous but nowhere differentiable function from the reals to the reals. The key idea is that Baire's Theorem is true in a much broader context than just the real numbers since his proof can easily be modified to apply to any separable, complete metric space. Banach then exploits the fact that the

---

[32]Par example, on a reconnu, contrairement à ce qui avait été longtemps admis, qu'il existe des fonctions continues n'admettant pas de dérivée. Ce résultat doit être entenda de la manière suivante: le fait d'imposer à une fonction la continuité n'entraîne pas comme conséquence l'existence d'une dérivée; il en résulte que les fonctions continue qui admettent une dérivée ne forment qu'une classe particulière dans l'ensemble des fonctions continue; autrement dit, *c'est par exception qu'une fonction continue admet une dérivé.* (Page 2 of [Baire, 1899b])

continuous functions on the unit interval form a separable, complete metric space with respect to the uniform norm.

Banach's idea has since been used countless times to construct examples of what are now known as generic objects. The term "generic" refers to the fact that the Baire Category Theorem provides more information than just existence, it actually shows that most objects have the desired property. For example, not only is there a continuous nowhere differentiable function, but most continuous functions are nowhere differentiable in the sense that all but a meagre subset of the space of continuous functions with respect to the uniform norm will be nowhere differentiable. This method of proof is often described as non-constructive, but it should be realized that any such proof can be transformed into a constructive argument by suitably enumerating the dense open sets used in the argument and tracing the steps to create an inductive construction. The usefulness of Banach's approach is that it allows one to ignore these bookkeeping details in favour of the conceptual. However, some forty years later Banach's ideas began to be used in the context of Martin's Axiom in a way that is essentially non-constructive.

Before examining these ideas though, it is necessary to take what may seem to be a digression and return to the year 1920 in which the first issue of *Fundamenta Mathematicae* appeared, the one in which Suslin asked Question 1. The motivation for the question is clear. Suslin was aware that the real line could be characterized as the unique separable, complete, linear order with neither end points nor isolated points. So Suslin is asking whether the separability condition can be replaced by the seemingly weaker one that all disjoint families of open sets are countable. The question of why Suslin was interested in this particular weakening can only be guessed at, but the future development of the subject was to reveal that it is a central concept and so it may well be correct to say that it was a natural question for him to ask.

However the methods available to Suslin and his contemporaries were far from adequate for making much progress on the problem. The one bit of early progress worth noting is a reformulation of the problem due to Kurepa. He [1935] realized that by considering the set of intervals of a putative counterexample to Suslin's question one can create a tree structure capturing the relevant combinatorics. To be precise, Kurepa showed that the following are equivalent:

- The real line is not the only complete, linear order without endpoints or isolated points in which all disjoint families of intervals are countable.
- There is a tree of height ω_1 in which all nodes split, yet all incomparable sets of nodes are countable.

Trees satisfying the second alternative are called *Suslin trees.*

It is interesting to note that there is a tree structure similar to a Suslin tree that was studied for different reasons. It is easy to check that a Suslin tree can have no branches of length ω_1. However, trees satisfying just this weaker property are known as *Aronszajn trees* and the earliest recorded construction appears in

[Kurepa, 1938]. The point to remark on is that this construction has been shown by Todorčević to be equivalent to Hausdorff's construction of an (ω_1, ω_1^*)–gap. This construction itself is similar to Lusin's construction of an almost disjoint family no two uncountable subsets of which can be separated as described in §9. So, even though Hausdorff had described his construction in 1908 it seems that 30 years later the subject had matured to the point that similar constructions were now being discovered by various researchers.

The status of Suslin trees, however, would not be resolved until forcing techniques became available. In 1967 Tomáš Jech [1967] proved the consistency of the existence of a Suslin tree using an argument which could be rephrased as forcing with a countably closed partial order. The reviewer of the article for the AMS, Michael Morley makes the following comment in his review of this article:

> The author notes that he has heard that S. Tennenbaum had obtained similar results in 1964. The facts known to the reviewer are as follows: (1) Tennenbaum announced the main result of this paper in a lecture at Harvard in January,1964; (2) his proof has been presented at numerous seminars; (3) it has never been published even as an abstract (excluding mimeographed notes); and (4) the author's proof is significantly different from Tennenbaum's.

Indeed, Tennenbaum did subsequently publish [1968] another proof of the consistency of the existence of a Suslin tree and, as Morley suggests, his proof is quite different from Jech's, relying on finite conditions rather than Jech's countable ones. About the same time, Ronald Jensen showed that $V = L$ also implies that there is a Suslin tree. In his memorial address at a conference in honour of Tennenbaum, Kanamori [2007] includes a letter from Solovay recalling the consistency of a Suslin tree having actually been obtained in 1963. Indeed, Solovay's letter offers a detailed recollection of his collaboration with Tennenbaum on the key result in the opposite direction.

Solovay and Tennenbaum proved that it is consistent that there are no Suslin trees, thus showing that Suslin's question was not answerable by the methods available at the time he asked it. It is here that this seeming digression on Suslin's problem becomes relevant to the development stemming from the Baire Category Theorem. The important point to emphasize is that Solovay and Tennenbaum's proof relied on the new technique of iterated forcing. Tennenbaum had realized that thinking of the Suslin tree itself as a partial order allowed one to force with it. The result of this forcing is an ω_1 length path through the tree and it has already been noted that no Suslin tree can have such a path. In other words, forcing with a Suslin tree destroys that tree.

The critical property of a Suslin tree that allows it to be used as a forcing partial order without collapsing cardinals is its defining feature: There are no uncountable pairwise incompatible sets of elements of the tree considered as a partial order. This is the same condition Cohen had used to show that cardinals are not collapsed when increasing the size of the continuum. Solovay and Tennenbaum then developed the strategy of iteratively destroying each tree in turn. The new concept introduced here was that of creating a partial order which has the same effect as forcing to destroy one Suslin tree and then forcing over the resulting model to destroy the next Suslin tree and so on.

The next step was for Martin and Solovay to realize that the new iteration strategy allowed any partial order with the countable chain condition to be forced with in an iteration scheme. In its ultimate form, this meant that one could line up all partial orders with the countable chain condition and force with each in turn. Martin and Solovay [1970] explain:

> Martin observed that the construction of $\mathcal{N}$ depended only on very general properties of the Cohen extensions $\mathcal{M} \to \mathcal{M}^T$. He and, independently Rowbottom, suggested an "axiom" which asserts that all Cohen extensions having these very general properties can be carried out inside the universe of sets: that the universe of sets is — so to speak — *closed* under a large class of Cohen extensions.

Here $\mathcal{N}$ denotes the model constructed by Martin and Solovay.

The result is a strengthening of the Baire Category Theorem that applies to uncountably many dense sets and is valid in topological spaces satisfying the countable chain condition. Different variants of the countable chain condition define different classes of topological spaces to each of which correspond different versions of Martin's Axiom. The realization of the significance of the axiom to topological questions crystallized in the work of early researchers such as Franklin D. Tall [1969] and István Juhász [1970]. Of course, the work of Jensen on Suslin's Hypothesis can also be considered to be topological in nature.

An important step in unravelling the connections between various versions of the Baire Category Theorem was obtained by Murray Bell [1981], who showed that the intersection of less that $\mathfrak{c}$ dense open sets in σ-centred spaces being non-empty is equivalent to $\mathfrak{p}$ being equal to the continuum. Once again the motivation for Bell was topological since the main application of his result is the existence of first countable Dowker spaces assuming that $\mathfrak{p} = \mathfrak{c}$. However, the combinatorial equivalent of his theorem has proved to be very useful: Martin's Axiom restricted to σ-centred partial orders holds if and only if $\mathfrak{p} = \mathfrak{c}$. The usefulness of this equivalence stems from the the fact that there is a broad range of σ-centred partial orders and these can be designed according to the application in which one is interested. Bell's Theorem allows the general techniques of Martin's Axiom to be applied in the situation where one knows only that $\mathfrak{p} = \mathfrak{c}$.

It was the set-theoretic topologists who most vigorously applied Martin's Axiom in the early years after its appearance. Indeed, it can be said that, like the problems of analysis that drove set theoretic developments in the early years of the century, in the 1970's it was topological problems that provided the impetus for many new developments.

Central among these were the questions concerned with normality and metrizability. Recall from §3 that Jones [1937] had used the hypothesis that $2^{\aleph_0} < 2^{\aleph_1}$

to show that every separable, normal Moore space is metrizable. Bing [1951] had shown that a certain topological space constructed from an uncountable set of reals is a separable, normal, non-metrizable Moore space provided that the uncountable set of reals used is a Q–set. Recall also that Rothberger had shown that if there are no ω_1–limits, or equivalently that $\mathfrak{p} > \aleph_1$, then there is a Q–set. It would then seem a natural progression to conclude that Martin's Axiom implies there is a separable, normal, non-metrizable Moore space. Indeed, this was announced by Jack Silver (cf. [Tall, 1969]) and the proof was based on Bing's result, but with no reference to Rothberger's work. It was only later pointed out by David Booth (cf. [Tall, 1969]) that Silver's result follows immediately from Rothberger's and Bing's work. However, it is clear that the set-theoretic researchers of Europe and the general topologists of Texas had been unaware of each others work throughout the middle decades of the twentieth century. (A curious counterexample is provided by Émile Borel's address at the inauguration of Rice University in Houston, Texas in 1912 in which he developed some of the ideas leading to the Borel Conjecture.) The renewed interest in these questions resulting from the availability of the new forcing techniques provided by Cohen and Solovay as well as Tall's thesis [1969] on the normal Moore space problem finally rectified this situation and resulted in a lengthy period of intense activity.

The study of convergence properties as well as of $\beta\mathbb{N} \setminus \mathbb{N}$ proved to be particularly amenable to the newly available methods. Perhaps the most influential and far reaching development though, was the solution of the S–space problem by Todorčević [1989] using innovative proper forcing techniques. The problem had been the object of intense study by various authors and several partial results had been available earlier. For example, Kunen [1977] showed that Martin's Axiom and the failure of the Continuum Hypothesis implied that there are no regular spaces all of whose finite products are S–spaces. However, the hypothesis used by Kunen is especially amenable to Martin's Axiom since it lends itself to countable chain condition partial orders. The case of an arbitrary S–space, however, required the use of Todorčević's techniques and this focused attention on the Proper Forcing Axiom described by Baumgartner [1984].

Cohen, in his concluding remarks in [1966] says of the problem of determining the value of the continuum,

> It is unreasonable to expect that any description of a larger cardinal which attempts to build up that cardinal from ideas deriving from the Replacement Axiom can ever reach $\mathfrak{c}$. Thus $\mathfrak{c}$ is greater than $\aleph_n$, $\aleph_\omega$, $\aleph_\alpha$ where $\alpha = \aleph_\omega$ etc. This point of view regards $\mathfrak{c}$ as an incredibly rich set given to us by one bold new axiom, which can never be approximated by any piecemeal process of construction.

As it turns out, there are versions of Martin's Axiom sufficiently strong to establish the value of $\aleph_2$ for the continuum. However, as will be seen, while this does contradict Cohen's view that the continuum must be very large, it does not contradict his view that the continuum "can never be approximated by any piecemeal process of construction."

While many of the questions concerning cardinal invariants are settled by Martin's Axiom, indeed they are all equal to $2^{\aleph_0}$, other questions are not. For example, Martin's Axiom does not settle the question of whether there are $(\mathfrak{c}, \mathfrak{c}^*)$–gaps as was shown by Kunen [1975]. Even less is determined by assuming the following consequence of Martin's Axiom and $2^{\aleph_0} > \aleph_1$ which is often referred to as $MA_{\aleph_1}$: Given $\aleph_1$ dense open sets in a countable chain condition partial order, there is filter meeting each of them. Kunen and Tall [1979] consider variations on Martin's Axiom and remark on two different types of consequences of Martin's Axiom. They distinguish between what they call "combinatorial" consequences and "Souslin type" consequences. They point out that the combinatorial consequences "readily imply that $2^{\aleph_0} > \aleph_1$, while the Souslin type consequences do not." However, even the full $MA_{\aleph_1}$ does not determine the value of the continuum as is immediate from the construction of Solovay and Tennenbaum [1971].

In [1997] Todorčević explains the significance of this question.

> The increasing sophistication in the area of iterated forcing resulted in a profusion of models of Set Theory built for the purpose of proving the independence of statements like Souslin's Hypothesis, Borel's Conjecture, and so on. It turns out that many such proofs, especially the deeper ones involving other than the finite support iterations, could not produce models where the continuum would have value different from $\aleph_2$ ... It is therefore natural to ask whether the statements in question do pose some restriction on the continuum such as $\mathfrak{c} = \aleph_2$. This kind of question is especially interesting when the considered statements are some kind of maximality principles, such as, for example, various forcing axioms asserting that we can have a sufficiently generic filter for every member of a certain class of posets.

Therefore it was very surprising when Todorčević was able to show that the strengthening of $MA_{\aleph_1}$ known as the Proper Forcing Axiom actually implies that $2^{\aleph_0} = \aleph_2$. It will be seen that the argument here does indeed have two distinct parts that might be seen as corresponding to Kunen and Tall's division of Martin's Axiom into "combinatorial" and "Souslin type" consequences. However, the story is not quite so simple and the two parts of Todorčević's argument involve several new ingredients that were not yet available at the writing of [Kunen and Tall, 1979].

An early step in this direction was taken in [Todorčević, 1993]. Rado's Conjecture is a reflection type statement about linear orders: A family of intervals of a linearly ordered set is the union of countably many disjoint subfamilies if and only if every subfamily of size $\aleph_1$ has this property. Todorčević [1993] studies this conjecture and shows that it implies that the size of continuum is at most $\aleph_2$.

However, further progress in this direction hinges on an axiom that might be viewed as being at the core of the "Souslin type" consequences mentioned by Kunen and Tall in their [1979], an axiom now commonly known as OCA. However, there is room for considerable confusion here since a related family of axioms was studied earlier by Uri Abraham, Matatyahu Rubin and Shelah in [Abraham *et al.*, 1985].

The semi-open colouring axiom (SOCA) states that for any second countable space X with $|X| = \aleph_1$ and any open set $U \subseteq [X]^2$ — this can be identified with a symmetric open subset of $X \times X$ — there is $A \subseteq X$ with $|A| = \aleph_1$ such that either $[A]^2 \subseteq U$ or $[A]^2 \cap U = \emptyset$. However, another axiom Abraham, Rubin and Shelah call OCA is also defined. An open colouring of a second countable Hausdorff space X with $|X| = \aleph_1$ is a finite cover U of $[X]^2$. A set $A \subseteq X$ is said to be U-homogeneous if $[A]^2$ is contained in one member of U. The open colouring axiom (OCA) states that for every such X and U, X can be partitioned into countably many U-homogeneous sets.

The axiom currently known as OCA was introduced by Todorčević [1989] and is actually a strengthening of SOCA: If X is a separable metric space and $G \subseteq [X]^2$ is open then either there is a decomposition of X into countably many pieces $\{X_i\}_{i\in\omega}$ such that $[X_i]^2 \cap G$ is empty for each i or there is an uncountable $H \subseteq X$ such that $[H]^2 \subseteq G$. Observe that the two alternatives are not symmetric as they are in SOCA.

The first component of Todorčević's argument was to show that OCA, as formulated by him, implies that $\mathfrak{b} = \aleph_2$. A crucial step in the argument in [Todorčević, 1989] is to show that every gap in $(\mathbb{N}^{\mathbb{N}}, <^*)$ is either of type (ω_1, ω_1^*) or of type (ω, ω_1^*) or of type (ω_1, ω^*). (As discussed in §5 the first type of gap was first constructed by Hausdorff, while the other two were analyzed by Rothberger.) On the other hand, in [Todorcevic, 1988] it is shown that if $\mathfrak{b} > \aleph_2$ then there is a gap in $(\mathbb{N}^{\mathbb{N}}, <^*)$ which is neither of the Hausdorff nor Rothberger type. It is interesting to note that this argument depends on choosing a maximal linear order in $(\mathbb{N}^{\mathbb{N}}, <^*)$, in other words the pantachies as defined and studied by Hausdorff had not yet outlived their usefulness almost a century later.

The other essential step in the argument is to use the value of $\mathfrak{b}$ to code reals and put an upper bound on $\mathfrak{c}$. In this coding a well-ordered, unbounded subset of $\mathbb{N}^{\mathbb{N}}$ plays a central role and, of course, this set must have cardinality $\mathfrak{b}$. A scheme for having certain subsets of this well-ordered set code reals is described in [Todorčević, 1997] and this coding has the property that sets coding distinct reals have distinct suprema. Hence showing that each real is coded will establish that $\mathfrak{b}$ is a bound on the number of reals. Showing that each real actually is coded requires a relatively weak extension of Martin's Axiom, namely Martin's Axiom for partial orders of the form $P * Q$ where P is σ-closed and Q has the countable chain condition.

Before leaving the topic, it is interesting to note that Justin Moore [2002] has shown that the conjunction of two versions of OCA — the original OCA as defined by Abraham, Rubin and Shelah and the later version of OCA defined by Todorčević and strengthening SOCA — also implies that $2^{\aleph_0} = \aleph_2$. It must be remarked though, that finding bounds on the continuum using variations of Martin's Axiom are not an isolated phenomenon, but should really be placed within the context of discoveries, many of them quite unexpected, about combinatorial implications of variations of Martin's Axiom whose consistency is obtained by assuming some large cardinal hypothesis. As examples, one can mention Todorčević's proof [1984] that the Proper Forcing Axiom implies the failure of $\square_\kappa$ for all uncountable κ and the applications of Martin's Maximum in [Foreman *et al.*, 1988a; Foreman *et al.*, 1988b; Foreman *et al.*, 1989]. A very different approach to these axioms is provided by Woodin's P_{max} [1999] which is a canonical model for the failure of the Continuum Hypothesis. However, providing more detail here would require too long a diversion for the scope of this article.

9 CARDINAL INVARIANTS OF THE CONTINUUM ASSOCIATED WITH $\beta\mathbb{N}\setminus\mathbb{N}$

The discussion of the structure of $\mathbb{N}^\mathbb{N}$, the invariants associated to measure and category and the development of Baire Category like theorems have brought us to consider $\mathfrak{b}$, $\mathfrak{d}$, the invariants of the ideals $\mathcal{N}$ and $\mathcal{M}$ as well as $\mathfrak{p}$ and $\mathfrak{t}$. Little mention so far has been made of the combinatorial invariants such as $\mathfrak{a}$, $\mathfrak{i}$, $\mathfrak{s}$ and $\mathfrak{r}$. Indeed, all of these seem to have escaped the attention of set theorists and topologists for considerably longer than the invariants discussed in §5. Certainly part of the explanation for this must be found in the fact that these later invariants are much less closely connected to the notion of rate of convergence which were of interest to du Bois-Reymond and Hausdorff after him. The interest in the invariants studied in §6 is motivated by very different considerations of course.

The cardinal $\mathfrak{a}$ is defined to be the least cardinality of a maximal, infinite family $\mathcal{A}\subseteq\mathcal{P}(\mathbb{N})$ such that any two elements of $\mathcal{A}$ are almost disjoint, in the sense that their intersection is finite. The notion of an almost disjoint family was considered already in 1928 by both Sierpiński [1928b] and Tarski [1928] in considerable generality and Sierpiński established the result — a result he viewed as paradoxical — that in any infinite set of cardinality κ it is possible to find more than κ pairwise almost disjoint sets. In the case that $\kappa=\aleph_0$ there are, in fact, $\mathfrak{c}$ almost disjoint sets. As a result, until the era of forcing and the ability to construct models where $\mathfrak{c}$ is large it would not have made much sense to consider the question of the cardinality of *maximal* almost disjoint families of subsets of the integers. It was only in 1972, just shortly after the Martin–Solovay paper introducing Martin's Axiom, that articles of Stephen Hechler [1972] and Solomon [1977] appeared considering this question.

It is worth noting that Hechler's paper appeared in a topology journal and that the cardinal $\mathfrak{a}$ was looked at as the least cardinal of a maximal, infinite family of disjoint clopen sets in $\beta\mathbb{N}\setminus\mathbb{N}$. Moreover, in Eric van Douwen's very influential article [1984] the discussion of $\mathfrak{a}$ highlights the Ψ space constructed from an almost disjoint family. Given an almost disjoint family $\mathcal{A}$, the Ψ–space has as points $\mathbb{N}\cup\mathcal{A}$ and the neighbourhoods of $A\in\mathcal{A}$ are of the form $A\setminus k$ while each $n\in\mathbb{N}$ is isolated. The study of the topological properties of this space has motivated much of the developments in studying $\mathfrak{a}$ and almost disjoint families in general.

Hechler noted that Martin's Axiom implies that $\mathfrak{a} = \mathfrak{c}$ but, of course, the definition would not have been of much interest had he not been able to show that it is consistent that $\mathfrak{a} \neq \mathfrak{c}$. And, indeed, he was able to show that it is consistent that $\mathfrak{c} > \aleph_1$ and there are maximal almost disjoint families of each uncountable cardinality not greater than the continuum. By keeping the continuum smaller than the first singular, Hechler was able to avoid the question of singular cardinality for $\mathfrak{a}$. Much later, Jörg Brendle [2003a] was able to use a modification of a forcing technique due to Shelah to obtain the consistency of $\mathfrak{a} = \aleph_\omega$. The argument of Shelah which Brendle modified was originally used in 1999 to show that it is consistent that $\mathfrak{a} > \mathfrak{d}$, but the paper [Shelah, 2004] appeared only five years later. The fact that $\mathfrak{a} \geq \mathfrak{b}$ was noted by Solomon [1977] and a model where $\mathfrak{a} \neq \mathfrak{b}$ was obtained by Shelah in 1982, along with several other inequalities, in [Shelah, 1984b], a paper whose influence on forcing constructions was discussed at the end of §7. However, obtaining a model where $\mathfrak{a} \neq \mathfrak{d}$ seemed to be beyond the capabilities of the techniques whose development was initiated in [Shelah, 1984b]. And, in fact, the methods used to finally obtain such a model where entirely novel, but their influence on future developments is yet to be determined. It is worth noting though that the following question [Miller, 1993] of Judith Roitman is still open: If there is a dominating family in $\mathbb{N}^{\mathbb{N}}$ of size $\aleph_1$ does it follow that $\mathfrak{a} = \aleph_1$? The appearance of $\aleph_1$ here is not as a place holder for $\mathfrak{c}$ as it might have been in the questions posed in the early issues of *Fundamenta Mathematicae*, but because of a fundamental problem in modifying Shelah's argument to handle this cardinal.

In discussing the cardinal $\mathfrak{a}$ it is worth remarking that in the same year that Hausdorff published his second paper on constructing Hausdorff gaps — the paper which finally got set theorists to take notice of a result Hausdorff had obtained close to thirty years earlier — Lusin [1947] produced a very similar argument constructing an unusual almost disjoint family. What he did was to construct an almost disjoint family $\mathcal{A}$ such that no two disjoint, uncountable subfamilies can be separated: In other words, if $\mathcal{B} \subseteq \mathcal{A}$ and $\mathcal{C} \subseteq \mathcal{A}$ and $\mathcal{B} \cap \mathcal{A} = \emptyset$ then there is no $X \subseteq \mathbb{N}$ such that $B \subseteq^* X$ for all $B \in \mathcal{B}$ and $C \cap X$ is finite for all $C \in \mathcal{C}$. While it seems that Lusin was unaware of Hausdorff's work at the time — at least van Douwen expresses this opinion in his [1984] — it is at least clear that the work of du Bois-Reymond had motivated Lusin's interest in these questions, as it had Hausdorff's much earlier construction. Indeed, Lusin attributes the non-existence of non-separable countable families to du Bois-Reymond. In one sense, this is a stronger result than Hausdorff's gap construction because it yields $2^{\aleph_1}$ pairs that can not be separated. On the other hand, the pairs consist of almost disjoint sets rather than $\subseteq^*$-increasing towers as in Hausdorff's gaps. And, of course, the connection to the invariant $\mathfrak{a}$ is quite weak because there is no claim that Lusin's almost disjoint family is maximal. Moreover, using the Axiom of Choice to extend it to a maximal almost disjoint family is likely to destroy the key property of the Lusin family.

This very brittle behaviour of almost disjoint families is at the heart of some key questions still open in set theory. For example, Hechler [1971] raises the question of whether there is a *completely separable* maximal almost disjoint family. An almost disjoint family $\mathcal{A}$ is said to be completely separable if for every $X \subseteq \mathbb{N}$ either X is almost contained in the union of finitely many members of $\mathcal{A}$ or there is $A \in \mathcal{A}$ such that $X \supseteq A$. While this is still an open question, if the requirement of maximality is removed a solution was obtained by Petr Simon (cf. [Balcar and Simon, 1989]). The argument used by Simon has its roots in methods developed by Bohuslav Balcar, Jan Pelant and Simon in [Balcar *et al.*, 1980] in which a cardinal invariant quite different in nature to those studied by Rothberger or Sierpiński is examined. The notation for the cardinal now known as $\mathfrak{h}$ was introduced in [Balcar and Simon, 1989], but Balcar, Pelant and Simon designated it as κ — or, to be precise, $\kappa(\mathbb{N}^*)$ — in the earlier paper. It is defined to be the least cardinal of a *shattering matrix*, which is defined in [Balcar *et al.*, 1980] to be a collection of almost disjoint families $\{\mathcal{A}_\xi\}_{\xi\in\mathfrak{h}}$ such that for each infinite $X \subseteq \mathbb{N}$ there is some $\xi \in \mathfrak{h}$ such that there are at least two $A \in \mathcal{A}_\xi$ intersecting X on an infinite set.

The motivation behind the definition was firmly rooted in the study of $\beta\mathbb{N} \setminus \mathbb{N}$, the study of the number of nowhere dense sets required to cover $\beta\mathbb{N} \setminus \mathbb{N}$ to be precise. It did not take long, though, to realize that there is a close connection to distributivity properties of Boolean algebras. In the course of their analysis Balcar, Pelant and Simon show that $\mathfrak{h}$ is also the same as the minimal height of a tree π-base for $\beta\mathbb{N} \setminus \mathbb{N}$.

By the time of the survey article which revisited some of the results from [Balcar *et al.*, 1980] the article of van Douwen [1984] had already made its mark on the community of set-theoretic topologists and the position of $\mathfrak{h}$ among the cardinals considered in §5 seemed a natural question to answer. The following diagram describes the answer:

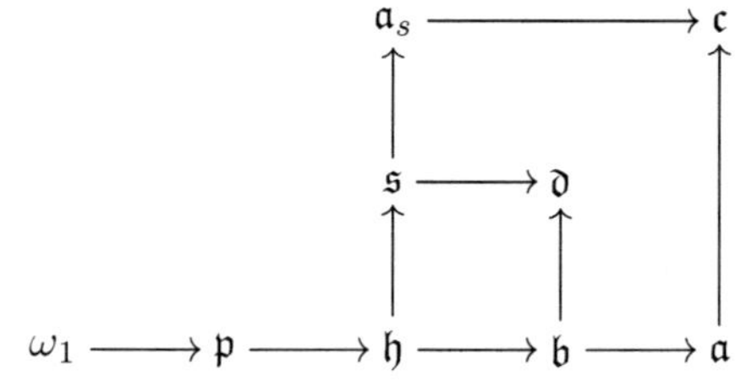

Here, as usual, an arrow signifies that the cardinal invariant at the tail of the arrow is less than or equal to the invariant at its tip and that it is consistent that inequality holds. The invariants $\mathfrak{b}$, $\mathfrak{d}$, $\mathfrak{h}$ and $\mathfrak{p}$ are treated in more detail in §5 but the invariants $\mathfrak{s}$ and $\mathfrak{a}_s$ have not yet been mentioned.

The invariant $\mathfrak{a}$ has a great many variants, some of which are known to be different from $\mathfrak{a}$ itself and others are not. The invariant $\mathfrak{a}_s$ is one of these and is defined to be the least cardinal of an almost disjoint family consisting of partial functions from $\mathbb{N}$ to $\mathbb{N}$. Therefore the inequality $\mathfrak{a} \leq \mathfrak{a}_s$ is immediate even though it is not indicated in the diagram. The cardinal $\mathfrak{s}$ is implicitly due to Booth in [1974] where he studies the sequential compactness of products of the two point discrete space. He shows that 2^λ is sequentially compact if and only if for every family $\mathcal{A}$ of subsets of $\mathbb{N}$ of size λ there is some infinite $X \subseteq \mathbb{N}$ such that for each $A \in \mathcal{A}$ either $X \subseteq^* A$ or $X \cap A$ is finite. The cardinal invariant is now usually defined as the least cardinal for which this combinatorial equivalence of the sequential compactness of 2^λ fails; namely $\mathfrak{s}$ is the least cardinal of a family $\mathcal{A}$ of subsets of $\mathbb{N}$ such that for every infinite $X \subseteq \mathbb{N}$ there is some $A \in \mathcal{A}$ such that $X \setminus A$ and $X \cap A$ are both infinite. Moreover, Booth also showed that $\mathbf{non}(\mathcal{N}) \geq \mathfrak{s}$ which can be compared to the inequality $\mathfrak{t} \leq \mathbf{add}(\mathcal{M})$ established much earlier by Rothberger and discussed in §5.

The inequalities $\omega_1 \leq \mathfrak{p} \leq \mathfrak{h}$ are immediate, but more interesting are $\mathfrak{h} \leq \mathfrak{b}$ and $\mathfrak{s} \leq \mathfrak{a}_s$. Since it involves $\mathfrak{b}$, it should be no surprise that the first of the last two inequalities is established by appealing to arguments used by Rothberger, but a structure similar to a shattering matrix also plays a role. The argument for the second inequality is quite subtle and can be found in [Balcar and Simon, 1989].

The problem of showing that none of these inequalities is actually an equality is the work of several authors. However, it is worth noting the simple fact that the invariant $\mathfrak{t}$ studied by Rothberger is bounded by $\mathfrak{h}$. The fact that $\mathfrak{t} < \mathfrak{h}$ is consistent is proved in [Dordal, 1982] so this yields the consistency of $\mathfrak{p} < \mathfrak{h}$. The consistency of $\mathfrak{s} < \mathfrak{b}$ is obtained by adding $\aleph_1$ of Solovay's random reals to a model of Martin's Axiom and $\mathfrak{c} = \aleph_2$ while arguments in a similar spirit (cf. [Balcar and Simon, 1989]) yield all the other arrows except for $\mathfrak{d} < \mathfrak{a}$ and $\mathfrak{a} < \mathfrak{s}$. However, the consistency of $\mathfrak{d} < \mathfrak{a}$ has already been discussed as a result of Shelah [2004] using techniques developed expressly for the purpose of obtaining this strict inequality. It had taken some twenty years to achieve this since the time of his seminal paper [Shelah, 1984b] in which he showed the consistency of $\mathfrak{a} \neq \mathfrak{b}$. In fact, in the same paper the consistency of $\mathfrak{a} < \mathfrak{s}$ was also proved. Noting that $\mathfrak{s} < \mathfrak{a}_s$, this also establishes the consistency of $\mathfrak{a} < \mathfrak{a}_s$, a quite unexpected result.

While it would not be correct to attribute the notion of an ultrafilter to Marshall Stone — indeed, Stanislaw Ulam [1929] showed how the Axiom of Choice can be used to construct these objects in 1932 and even before him, in 1909, Frederic Riesz [1909] had introduced the notion to little acclaim— it is certainly true that his work in [Stone, 1937] was a critical point in bringing to the attention of topologists the importance of this notion. While Stone's work on what is now known as the Čech–Stone compactification was motivated by his earlier research on the spectral theory of linear operators, Eduard Čech's seminal paper [1937] on the subject grew out of earlier work of Andrei Nikolaevich Tychonov in this direction, even to the extent of continuing to use Tychonov's β notation for the compactification. While Čech's construction, unlike Stone's, was not based on prime ideals in Boolean algebras, he was certainly aware of the connections to set theory as the following remark from [Čech, 1937] makes clear:

> Let I denote an infinite countable isolated space (e.g. the space of all natural numbers). It is an important open problem to determine the cardinal $\mathfrak{m}$ of $\beta(I)$. All I know about it is that $2^{\aleph_0} \leq \mathfrak{m} \leq 2^{2^{\aleph_0}}$

It would not be long before this question was answered though. Published in the same year, an article [1937] by Bedřich Pospíšil showed that in fact $\mathfrak{m} = 2^{2^{\aleph_0}}$.

However, had he used Stone's construction, Pospíšil's answer would have been immediately clear to Čech if he had also been aware of the following result of Gregory Fichtenholz and Leonid Kanterovich [1935]: There is an independent family of subsets of $\mathbb{N}$ of cardinality $\mathfrak{c}$. It follows that each function from $\mathfrak{c}$ to 2 corresponds to a distinct ultrafilter. These very early results on $\beta\mathbb{N}$ were a clear signpost for set theorists to follow in the era of independence results. It has already been mentioned that Hechler's results on $\mathfrak{a}$ as well as the later work of Balcar, Pelant and Simon were clearly focused on $\beta\mathbb{N} \setminus \mathbb{N}$. It was perfectly natural, therefore, to consider some of the earliest constructions through the lens of independence results.

For example, given that there is an independent family of cardinality $\mathfrak{c}$, in analogy with the definition of $\mathfrak{a}$ one can ask for the least cardinal of a maximal independent family. This appears for the first time in an appendix to the survey article [Vaughan, 1990] which provides a proof due to Shelah that $\mathfrak{d} \leq \mathfrak{i}$ and that it is consistent that $\mathfrak{d} \neq \mathfrak{i}$.

Another cardinal invariant found in [Vaughan, 1990] that has shown itself to be quite closely connected to various notions is $\mathfrak{u}$, the least cardinal of a base for a non-principal ultrafilter on $\mathbb{N}$. Of course, the nature of points in $\beta\mathbb{N} \setminus \mathbb{N}$ was one of the first and most obvious questions to be considered by topologists, but, one might argue that the structure of this space is really not that closely related to the continuum since even the cardinalities of the two objects are different. While this is true to some degree, the cardinal $\mathfrak{u}$ has played an important role in the development of our understanding of the cardinal invariants associated with the continuum. Already Pospíšil [1939] had considered the question of showing that there is an ultrafilter on $\mathbb{N}$ with no base of cardinality smaller than $\mathfrak{c}$. However, the results of [Baumgartner and Laver, 1979] already yielded that it is consistent that $\mathfrak{u}$ is less than the continuum. This is quite surprising since $\mathfrak{u}$ seems as if it should be quite a large cardinal. Indeed there is no upper bound known for it other than $\mathfrak{c}$.

While a great deal of the interest in ultrafilters comes from topologists studying the Čech–Stone compactification there are other fields that have influenced development in this area as well. The study of abelian groups is one of them. Motivated by very different questions than du Bois-Reymond, Ernst Specker [1950] defined growth types as ideals of non-decreasing functions from $\mathbb{N}$ to $\mathbb{N}$. By an ideal he meant a family of non-decreasing functions closed under pointwise sums and dominated functions. Rüdiger Göbel and Burkhard Wald [1979; 1980] defined an ordering on equivalence classes of these growth types and raised the question of their number. In providing answers to some of the questions raised by Göbel and Wald, Andreas Blass and Claude Laflamme [1989] found it useful to introduce a new cardinal invariant known as $\mathfrak{g}$.

A family $\mathcal{S}$ of infinite subsets of $\mathbb{N}$ which is downwards closed under the $\subseteq^*$ relation is known as *groupwise dense* if for every infinite family of pairwise disjoint finite subsets of $\mathbb{N}$ there is an infinite subfamily whose union belongs to $\mathcal{S}$. The cardinal $\mathfrak{g}$ is defined to be the least cardinal of a family of groupwise dense families whose intersection is empty. Comparing this definition to the definition of $\mathfrak{h}$ using shattering matrices will reveal several similarities. Indeed, Blass and Laflamme [1989] remark that the inequality $\mathfrak{h} \leq \mathfrak{g}$ is immediate. Not too much more difficult is the fact that $\mathfrak{g} \leq \mathfrak{d}$ which establishes a connection to growth types as considered more than a century earlier by du Bois-Reymond. In order to obtain the consistency of there being only four growth types, Blass and Laflamme employ the following hypothesis: $\mathfrak{u} \leq \mathfrak{g}$. Furthermore, they actually show that their hypothesis implies that for any two ultrafilters $\mathcal{U}$ and $\mathcal{V}$ there is a finite-to-one function f such that $f(\mathcal{U}) = f(\mathcal{V})$, a hypothesis known as NCF, the Near Coherence of Filters.

NCF was introduced in [Blass, 1986], which dealt with models of arithmetic. However, it soon turned out to be applicable to a wide variety of problems. For example it is shown under this assumption that the ideal of compact operators on the separable Hilbert space is the sum of two smaller ideals. In a similar fashion, the Čech–Stone compactification of the half line is shown to have only one composant and applications to the same sort of questions about abelian groups — dealt with later [Blass and Laflamme, 1989] together with Laflamme — are addressed here as well.

The consistency of NCF, however, was only established in by Blass and Shelah in their [1987], with a simpler proof provided soon after that in their [1989]. It is interesting to note that a key step on the road to this result was Shelah's construction [1982] of a model where there are no P–points. This is yet another example of a forcing model devised to answer a topological question, but the implications of whose techniques have had far reaching consequences on the study of independence results about the continuum.

Specker [1950] studied the group $\mathbb{Z}^{\mathbb{N}}$, the direct product of countably many copies of the integers. Letting $\{e_n\}_{n=0}^{\infty}$ be the natural generators for this group, Specker showed that for any homomorphism $H : \mathbb{Z}^{\mathbb{N}} \to \mathbb{Z}$ the value of $H(e_n)$ is 0 for all but finitely many of the generators e_n. Moreover, he also showed that this holds for many subgroups of $\mathbb{Z}^{\mathbb{N}}$, all of which have cardinality $\mathfrak{c}$. The question then arose of whether the cardinality of any such group must be maximal. However, Katsuya Eda [1983] showed that this question cannot be decided in the absence of some extra set-theoretic axioms since he could show that the minimal cardinality of a subgroup of $\mathbb{Z}^{\mathbb{N}}$ satisfying Specker's condition can be no less than $\mathfrak{p}$ and no greater than $\mathfrak{d}$. Later, Blass [1994] was able to improve this to show that the lower bound $\mathfrak{p}$ could be replaced by $\mathbf{add}(\mathcal{N})$ and the upper bound could be replaced by $\mathfrak{b}$. In the course of studying the cardinal invariants associated with the Specker phenomenon Blass introduced the notion of evasion and prediction.

A *predictor* consists of an infinite $D \subseteq \mathbb{N}$ and functions $\pi_d : \mathbb{N}^d \to \mathbb{N}$ for $d \in D$. Such a predictor is said to predict a function $f : \mathbb{N} \to \mathbb{N}$ if and only if for all but finitely many $d \in D$ the predictor guesses $f(d)$ in the sense that $\pi_d(f \restriction d) = f(d)$. Otherwise f has evaded the predictor. For any function $g : \mathbb{N} \to \mathbb{N}$ — to avoid trivialities we require that $g(i) > 2$ for all i — Blass defines $\mathfrak{e}_g$ to be the least cardinality of a family $\mathcal{E} \subseteq \prod_{i=0}^{\infty} g(i)$ such that for any predictor there is a function in $\mathcal{E}$ evading it. This cardinal has inspired a great deal of subsequent work by Blass as well as others such as Laflamme [1997], Brendle [1995; 2003b] and Brendle and Shelah [2003; 1996].

An interesting aspect of these evasion cardinals is that they are parametrized by the function g, which can be thought of as bounding the growth rates of the functions being predicted. This raises the possibility of infinite, indeed uncountable, families of cardinal invariants. The fact that this many invariants can actually be realized is a result due to Martin Goldstern and Shelah [1993] who considered invariants defined in a spirit similar to the evasion numbers of Blass. As with the evasion numbers, the invariants of Goldstern and Shelah are parametrized by growth rates and they show that it is consistent that uncountably many of these parametrized invariants are all different. Of course, this requires the continuum to be at least $\aleph_{\omega_1}$ in order to accommodate the uncountably many different cardinal values. Moreover, it is possible to exercise some control over the values of the invariants providing a vast landscape of models of set theory which differ for very concrete reasons, namely the value of certain cardinal invariants of the continuum.

10 EPILOGUE

Since this history began with a list of questions occupying set-theoretic researchers just after the first world war, it is fitting to provide a glimpse into the questions considered important at the end of the 20th century. We are fortunate to have a record of those questions considered important in the eyes of one of the foremost contributors to the study of the continuum in the last half of the last century, Saharon Shelah. In an issue of *Fundamenta Mathematicae* devoted to his work, Shelah provided an introductory article [2000b] listing questions he would have liked to have seen solved at that time. Some have been solved since then, but others persist.

Among those is the question of the consistency of $\mathfrak{p} < \mathfrak{t}$. It is interesting, and quite revealing, to see why Shelah would consider this seemingly technical question sufficiently important to include on his millennium list. Indeed, he provides a detailed discussion of this matter but, before seeing what he has to say on this it is worth recalling that Rothberger had shown that if $\mathfrak{p} = \aleph_1$ then $\mathfrak{t} = \aleph_1$. Shelah explains further:

> ...the advances in proper forcing make us "rich in forcing" for $2^{\aleph_1} = \aleph_2$, making the higher values more mysterious. ... So, because we know much more how to force to get $2^{\aleph_0} = \aleph_2$, the independence results on the problems of the interrelation of cardinal invariants of the continuum have mostly dealt with relationships of two cardinals, as their values are $\in \{\aleph_1, 2^{\aleph_0}\}$. Thus, having only two possible values $\{\aleph_1, \aleph_2\}$, among any three two are equal; the Pigeonhole Principle acts against us. As

> we are rich in our knowledge to force $2^{\aleph_0} = \aleph_2$, naturally we are quite poor concerning ZFC results. ... We are not poor concerning forcing for the Continuum Hypothesis (and are rich in ZFC). But for $2^{\aleph_1} \geq \aleph_3$ we are totally lost: very poor in both directions. We would like to have iteration theory for length $\geq \omega_3$. I tend to think good test problems will be important in developing such iterations.

So, by Rothberger's Theorem, the consistency of $\mathfrak{p} < \mathfrak{t}$ is a good test problem for iteration theory. Rephrased to avoid mention of consistency, the question of whether $\mathfrak{p} \geq \mathfrak{t}$ would have seemed natural and, very likely, have been an attractive problem to the early contributors to *Fundamenta.* The motivation of course, even if comprehensible, would have seemed entirely foreign. The questions these early set-theoretic researchers dealt with could mostly be motivated by their connection to problems in analysis. Eighty years later, though, motivation comes from a very different source.

Shelah mentions several other test problems. Among them is the very general question of finding relationships between triples of the well studied cardinal invariants. Certainly the iteration problem will need to be solved to have a general tool for dealing with independence results along these lines. However, Shelah believes that there are outright theorems waiting to be discovered, but that they "are camouflaged by the independent statements". Emphasizing once more the source of his motivation, he goes on to say that, "cardinal invariants from this perspective are excellent excuses to find iteration theorems".

However, it is perhaps true that the major set-theoretic question waiting to be resolved at the end of the twentieth century is the same one as at the start, the value of $2^{\aleph_0}$. The fact that this value can be calculated to be $\aleph_2$ under the Proper Forcing Axiom has already been discussed and there is a suggestion of Hugh Woodin [2001a; 2001b; 2002] that related types of arguments might be used to establish a true value for the continuum. A further discussion along these lines would be far from the topic of this history, but it may be worth reflecting on an earlier failed attempt in this direction by Gödel (cf. [1995]).

Gödel's idea, discussed in much greater detail in [Brendle *et al.*, 2001], was that the conjunction of four statements would imply that $2^{\aleph_0} = \aleph_2$. The first two of these axioms have to do with a generalization of the cardinal invariant $\mathfrak{d}$ to the context of the cardinals $\aleph_n$. In particular he asserted the existence of a family of functions from ω_n to ω_n of cardinality $\aleph_{n+1}$ which dominates all such functions with respect to the partial order of domination on a cofinal subset of ω_n, yet all of whose initial segments form a set of cardinality $\aleph_n$. While this structure is reminiscent of that studied by Hausdorff, Lusin, Sierpiński, Rothberger, Hechler and all those that followed this line of enquiry, its behaviour, in fact, is quite different. A discussion of this difference would lead too far away from the continuum, but the interested reader might consult [Koszmider, 2000] as a starting point for further investigation.

The last axioms Gödel invoked are familiar from the early work of Hausdorff on pantachies mentioned in §5. Gödel hypothesized a complete scale in $\mathbb{R}^\omega$ with no increasing or decreasing ω_2 sequences with what Gödel called the "Hausdorff Continuity Axiom". The exact meaning of this is not clear, but Brendle, Larson and Todorčević provide three axioms, too technical to be included here, and show that these follow from Gödel's hypotheses under the correct interpretation. Along with the complete scales mentioned, sets of strong measure zero play a central role. So does the assertion, in the spirit of Baire's Category Theorem, that any set of reals of cardinality $\aleph_3$ which is the intersection of $\aleph_1$ open sets contains a perfect set. Brendle, Larson and Todorčević then show that the three axioms — axioms which would have been recognizable to Hausdorff, Lusin and Sierpiński, but perhaps not to Baire — imply that $2^{\aleph_0} = \aleph_2$.

While Gödel's plan for proving that $2^{\aleph_0} = \aleph_2$ must be considered a failure, it may well be that more sophisticated arguments eventually succeed with the more ambitious goal of establishing that some form of the Proper Forcing Axiom is actually true. The effect of this on the study of cardinal invariants of the continuum would be profound since all these invariants would take on the value $2^{\aleph_0} = \aleph_2$. Even most questions regarding Hausdorff gaps, special sets of reals and related structures would be settled if this were to be the case. Of course, it could not be denied that the study of the cardinal structures associated with the continuum would have been crucial in coming to this new understanding. Nevertheless, it is possible that their study would likely become no more than a footnote in the history of set theory — and perhaps this article will serve as that footnote.

ACKNOWLEDGEMENTS

Research for this paper was partially supported by NSERC of Canada. The author is grateful to Francisco Kibedi for proofreading and providing valuable comments on a draft version of this article.

BIBLIOGRAPHY

[Abraham and Todorčević, 1997] U. Abraham and S. Todorčević. Partition properties of ω_1 compatible with CH. *Fundamenta Mathematicae*, 152(2):165–181, 1997.

[Abraham *et al.*, 1985] U. Abraham, M. Rubin, and S. Shelah. On the consistency of some partition theorems for continuous colorings, and the structure of $\aleph_1$-dense real order types. *Annals of Pure and Applied Logic*, 29(2):123–206, 1985.

[Alexandroff, 1916] P. Alexandroff. Sur la puissance des ensembles mesurables *B*. *Comptes Rendus des Séances de l'Académie des Sciences. Série I. Mathématique*, 162:323–325, 1916.

[Avraham *et al.*, 1978] U. Avraham, K. J. Devlin, and S. Shelah. The consistency with CH of some consequences of Martin's axiom plus $2^{\aleph_0} > \aleph_1$. *Israel Journal of Mathematics*, 31(1):19–33, 1978.

[Baire, 1899a] R. Baire. Sur la théorie des ensembles. *Comptes Rendus des Séances de l'Académie des Sciences. Série I. Mathématique*, 129:946–949, 1899.

[Baire, 1899b] R. Baire. Sur les fonctions de variables réelles. *Annali di matematica pur ed applicata*, 3:1–123, 1899.

[Baire, 1992] R. Baire. *Théorie des nombres irrationnels, des limites et de la continuité.* Éditions Jacques Gabay, Sceaux, 1992. Reprint of the 1905 original.

[Balcar and Simon, 1989] B. Balcar and P. Simon. Disjoint refinement. In *Handbook of Boolean algebras, Vol. 2*, pages 333–388. North-Holland, Amsterdam, 1989.

[Balcar *et al.*, 1980] B. Balcar, J. Pelant, and P. Simon. The space of ultrafilters on **N** covered by nowhere dense sets. *Fundamenta Mathematicae*, 110(1):11–24, 1980.

[Banach, 1931] S. Banach. Über die Bairesche Kategorie gewisser Funktionenmengen. *Studia Mathematica*, 3:174–179, 1931.

[Bartoszyński and Judah, 1995] T. Bartoszyński and H. Judah. *Set Theory. On the Structure of the Real Line.* A K Peters Ltd., Wellesley, MA, 1995.

[Bartoszyński, 1984] T. Bartoszyński. Additivity of measure implies additivity of category. *Transactions of the American Mathematical Society*, 281(1):209–213, 1984.

[Bartoszyński, 1987] T. Bartoszyński. Combinatorial aspects of measure and category. *Fundamenta Mathematicae*, 127(3):225–239, 1987.

[Baumgartner and Laver, 1979] J. E. Baumgartner and R. Laver. Iterated perfect-set forcing. *Annals of Mathematical Logic*, 17(3):271–288, 1979.

[Baumgartner, 1976] J. E. Baumgartner. Almost-disjoint sets, the dense set problem and the partition calculus. *Annals of Pure and Applied Logic*, 9(4):401–439, 1976.

[Baumgartner, 1983] J. E. Baumgartner. Iterated forcing. In *Surveys in set theory*, volume 87 of *London Mathematical Society Lecture Note Series*, pages 1–59. Cambridge University Press, Cambridge, 1983.

[Baumgartner, 1984] J. E. Baumgartner. Applications of the proper forcing axiom. In *Handbook of set-theoretic topology*, pages 913–959. North-Holland, Amsterdam, 1984.

[Bell, 1981] M. G. Bell. On the combinatorial principle $P(\mathfrak{c})$. *Fundamenta Mathematicae*, 114(2):149–157, 1981.

[Bernstein, 1905] F. Bernstein. Die Theorie der reellen Zahlen. *Deutsche Mathematische Vereinigung*, 14:447–449, 1905.

[Bernstein, 1908] F. Bernstein. Zur Theorie der trigonometrischen Reihe. *Berichte über die Verhandlungen der Königlich-Sächsischen Gesellschaft der Wissenschaften zu Leipzig, Mathematisch-Physische Klasse*, 60:325–338, 1908.

[Besicovitch, 1934a] A. S. Besicovitch. Concentrated and rarified sets of points. *Acta Mathematica*, 62:289–300, 1934.

[Besicovitch, 1934b] A. S. Besicovitch. Correction. *Acta Mathematica*, 62:317–318, 1934.

[Besicovitch, 1942] A. S. Besicovitch. Relations between concentrated sets and sets possessing Property *C*. *Mathematical Proceedings of the Cambridge Philosophical Society*, 38:20–23, 1942.

[Bing, 1951] R. H. Bing. Metrization of topological spaces. *Canadian Journal of Mathematics. Journal Canadien de Mathématiques*, 3:175–186, 1951.

[Blass, 2010] A. Blass. Combinatorial cardinal characteristics of the continuum. In *Handbook of Set Theory*, M. Foreman and A. Kanamori, eds. Springer, Berlin, 2010.

[Blass and Laflamme, 1989] A. Blass and C. Laflamme. Consistency results about filters and the number of inequivalent growth types. *The Journal of Symbolic Logic*, 54(1):50–56, 1989.

[Blass and Shelah, 1987] A. Blass and S. Shelah. There may be simple $P_{\aleph_1}$- and $P_{\aleph_2}$-points and the Rudin-Keisler ordering may be downward directed. *Annals of Pure and Applied Logic*, 33(3):213–243, 1987.

[Blass and Shelah, 1989] A. Blass and S. Shelah. Near coherence of filters. III. A simplified consistency proof. *Notre Dame Journal of Formal Logic*, 30(4):530–538, 1989.

[Blass, 1986] A. Blass. Near coherence of filters. I. Cofinal equivalence of models of arithmetic. *Notre Dame Journal of Formal Logic*, 27(4):579–591, 1986.

[Blass, 1994] A. Blass. Cardinal characteristics and the product of countably many infinite cyclic groups. *Journal of Algebra*, 169(2):512–540, 1994.

[Booth, 1974] D. Booth. A Boolean view of sequential compactness. *Fundamenta Mathematicae*, 85(2):99–102, 1974.

[Borel, 1914] É. Borel. *Leçons sur la théorie des fonctions. Deuxième édition.* Paris, Gauthier-Villars , 1914.

[Borel, 1919] É. Borel. Sur la classification des ensembles de mesure nulle. *Bulletin de la Société Mathématique de France*, 47:97–125, 1919.

[Brendle and Shelah, 1996] J. Brendle and S. Shelah. Evasion and prediction. II. *Journal of the London Mathematical Society. Second Series*, 53(1):19–27, 1996.

[Brendle and Shelah, 2003] J. Brendle and S. Shelah. Evasion and prediction. IV. Strong forms of constant prediction. *Archive for Mathematical Logic*, 42(4):349–360, 2003.

[Brendle *et al.*, 2001] J. Brendle, P. Larson, and S. Todorčević. Revisiting Gödel's argument for the true power of the continuum. Circulated notes, January 2001.

[Brendle, 1995] J. Brendle. Evasion and prediction—the Specker phenomenon and Gross spaces. *Forum Mathematicum*, 7(5):513–541, 1995.

[Brendle, 2003a] J. Brendle. The almost-disjointness number may have countable cofinality. *Transactions of the American Mathematical Society*, 355(7):2633–2649 (electronic), 2003.

[Brendle, 2003b] J. Brendle. Evasion and prediction. III. Constant prediction and dominating reals. *Journal of the Mathematical Society of Japan*, 55(1):101–115, 2003.

[Brendle, 2006] J. Brendle. Cardinal invariants of the continuum and combinatorics on uncountable cardinals. *Annals of Pure and Applied Logic*, 144(1-3):43–72, 2006.

[Burke, 1997] M. R. Burke. A proof of Hechler's theorem on embedding $\aleph_1$-directed sets cofinally into $(\omega^\omega, <^*)$. *Archive for Mathematical Logic*, 36(6):399–403, 1997.

[Cantor, 1884] G. Cantor. On the power of perfect sets of points. (De la puissance des ensembles parfaits de points.). *Acta Mathematica*, IV, 1884.

[Cantor, 1895] G. Cantor. Sui numeri transfiniti. Estratto di una lettera a G. Vivanti. *Rivista di Matematica Pura ed Applicata*, V, 1895.

[Čech, 1937] E. Čech. On bicompact spaces. *Annals of Mathematics. Second Series*, 38(4):823–844, 1937.

[Cohen, 1966] P. J. Cohen. *Set theory and the continuum hypothesis.* W. A. Benjamin, Inc., New York-Amsterdam, 1966.

[Cohen, 1971] P. J. Cohen. Comments on the foundations of set theory. In *Axiomatic Set Theory (Proc. Sympos. Pure Math., Vol. XIII, Part I, University of California, Los Angeles, Calif., 1967)*, pages 9–15. American Mathematical Society, Providence, R.I., 1971.

[Dauben, 1979] J. W. Dauben. *Georg Cantor: His mathematics and philosophy of the infinite.* Harvard University Press, Cambridge, Mass., 1979.

[Davies, 1974] R. O. Davies. Representation of functions of two variables as sums of rectangular functions. I. *Fundamenta Mathematicae*, 85:177–183, 1974.

[Devlin and Johnsbråten, 1974] K. J. Devlin and H. Johnsbråten. *The Souslin problem.* Springer-Verlag, Berlin, 1974. Lecture Notes in Mathematics, Vol. 405.

[Devlin and Shelah, 1978] K. J. Devlin and S. Shelah. A weak version of $\diamondsuit$ which follows from $2^{\aleph_0} < 2^{\aleph_1}$. *Israel Journal of Mathematics*, 29(2-3):239–247, 1978.

[Dordal, 1982] P. L. Dordal. *Independence results concerning some combinatorial properties of the continuum.* PhD thesis, Harvard University, 1982.

[du Bois-Reymond, 1870-71] P. du Bois-Reymond. Sur la grandeur relative des infinis des fonctions. *Annali di matematica pur ed applicata*, 4:338–353, 1870-71.

[du Bois-Reymond, 1872] P. du Bois-Reymond. A general theorem concerning the relative growth of the infinities of functions and of their derivatives. (Théorème général concernant la grandeur relative des infinis des fonctions et de leurs dérivées.). *Journal für die Reine und Angewandte Mathematik*, 1872.

[Eda, 1983] K. Eda. A note on subgroups of $\mathbf{Z}^{\mathbf{N}}$. In *Abelian group theory (Honolulu, Hawaii, 1983)*, volume 1006 of *Lecture Notes in Mathematics*, pages 371–374. Springer, Berlin, 1983.

[Eisworth and Roitman, 1999] T. Eisworth and J. Roitman. CH with no Ostaszewski spaces. *Transactions of the American Mathematical Society*, 351(7):2675–2693, 1999.

[Erdös, 1943] P. Erdös. Some remarks on set theory. *Annals of Mathematics. Second Series*, 44:643–646, 1943.

[Farah, 2007] I. Farah. A proof of the Σ^2_1-absoluteness theorem. In *Advances in logic*, volume 425 of *Contemporary Mathematics*, pages 9–22. American Mathematical Society, Providence, RI, 2007.

[Fichtenholz and Kantorovitch, 1935] G. Fichtenholz and L. Kantorovitch. Sur les opérations linéaires dans l'espace des fonctions bornées. *Studia Mathematica*, 5:69–98, 1935.

[Foreman *et al.*, 1988a] M. Foreman, M. Magidor, and S. Shelah. Martin's maximum, saturated ideals, and nonregular ultrafilters. I. *Annals of Mathematics. Second Series*, 127(1):1–47, 1988.

[Foreman *et al.*, 1988b] M. Foreman, M. Magidor, and S. Shelah. Martin's maximum, saturated ideals and nonregular ultrafilters. II. *Annals of Mathematics. Second Series*, 127(3):521–545, 1988.

[Foreman *et al.*, 1989] M. Foreman, M. Magidor, and S. Shelah. Correction to: "Martin's maximum, saturated ideals, and nonregular ultrafilters. I" *Annals of Mathematics. Second Series*, 129(3):651, 1989.

[Fremlin, 1984] D. H. Fremlin. Cichoń's diagram. *Publications Mathématiques de l'Université Pierre et Marie Curie, Sémin. Initiation Anal.*, 66(5), 1984.

[Fremlin, 1993] D. H. Fremlin. Real-valued-measurable cardinals. In *Set theory of the reals (Ramat Gan, 1991)*, volume 6 of *Israel Math. Conf. Proc.*, pages 151–304. Bar-Ilan University, Ramat Gan, 1993.

[Gispert, 1995] H. Gispert. Set theory in France before the crisis of 1905: Baire, Borel, Lebesgue ... and all the others. (La théorie des ensembles en France avant la crise de 1905: Baire, Borel, Lebesgue ... et tous les autres.). *Revue d'Histoire des Mathématiques*, 1(1):39–81, 1995.

[Göbel and Wald, 1979] R. Göbel and B. Wald. Wachstumstypen und schlanke Gruppen. In *Symposia Mathematica, Vol. XXIII (Conf. Abelian Groups and their Relationship to the Theory of Modules, INDAM, Rome, 1977)*, pages 201–239. Academic Press, London, 1979.

[Göbel and Wald, 1980] R. Göbel and B. Wald. Martin's axiom implies the existence of certain slender groups. *Mathematische Zeitschrift*, 172(2):107–121, 1980.

[Gödel, 1940] K. Gödel. *The Consistency of the Continuum Hypothesis*. Annals of Mathematics Studies, no. 3. Princeton University Press, Princeton, N. J., 1940.

[Gödel, 1995] K. Gödel. *Collected works. Vol. III*. The Clarendon Press Oxford University Press, New York, 1995. Unpublished essays and lectures, With a preface by Solomon Feferman, Edited by Feferman, John W. Dawson, Jr., Warren Goldfarb, Charles Parsons and Robert M. Solovay.

[Goldstern and Shelah, 1993] M. Goldstern and S. Shelah. Many simple cardinal invariants. *Archive for Mathematical Logic*, 32:203–221, 1993. math.LO/9205208.

[Hadamard, 1894] J. Hadamard. On the characters of convergence of series with positive terms and on indefinitely decreasing functions. (Sur les caractères de convergence des séries à termes positifs et sur les fonctions indéfiniment croissantes.). *Acta Mathematica*, XVIII, 1894.

[Hardy, 1910] G. H. Hardy. *Orders of infinity: The 'Infinitärcalül' of Paul du Bois-Reymond.* Cambridge: University Press. V u. 62 S. (Nr. 12 of the Cambridge Tracts in Mathematics and Mathematical Physics.) , 1910.

[Hardy, 1924] G. H. Hardy. *Orders of infinity. The "Infinitärcalcül" of Paul du Bois-Reymond. Cambridge Tracts in Mathematics and Mathematical Physics. Nr. 12. Second edition.* Cambridge, 77 S. 8° , 1924.

[Hausdorff, 1906] F. Hausdorff. Untersuchungen über Ordnungstypen. *Berichte über die Verhandlungen der Königlich-Sächsischen Gesellschaft der Wissenschaften zu Leipzig, Mathematisch-Physische Klasse*, 58:106–169, 1906.

[Hausdorff, 1907] F. Hausdorff. Untersuchungen über Ordnungstypen. *Berichte über die Verhandlungen der Königlich-Sächsischen Gesellschaft der Wissenschaften zu Leipzig, Mathematisch-Physische Klasse*, 59:84–159, 1907.

[Hausdorff, 1909] F. Hausdorff. Die Graduierung nach dem Endverlauf. *Abhandlungen der Mathematisch-Physischen Klasse der Königlich-Sächsischen Gesellschaft der Wissenschaften*, 31:297–334, 1909.

[Hausdorff, 1916] F. Hausdorff. Die Mächtigkeit der *Borel*schen Mengen. *Mathematische Annalen*, 77:430–437, 1916.

[Hausdorff, 1936] F. Hausdorff. Summen von $\aleph_1$ Mengen. *Fundamenta Mathematicae*, 26:241–255, 1936.

[Hausdorff, 2005] F. Hausdorff. *Hausdorff on ordered sets*, volume 25 of *History of Mathematics*. American Mathematical Society, Providence, RI, 2005. Translated from the German, edited and with commentary by J. M. Plotkin.

[Hechler, 1971] S. H. Hechler. Classifying almost-disjoint families with applications to $\beta N - N$. *Israel Journal of Mathematics*, 10:413–432, 1971.

[Hechler, 1972] S. H. Hechler. Short complete nested sequences in $\beta N \backslash N$ and small maximal almost-disjoint families. *General Topology and Applications*, 2:139–149, 1972.

[Hechler, 1974] S. H. Hechler. On the existence of certain cofinal subsets of ${}^{\omega}\omega$. In *Axiomatic set theory (Proc. Sympos. Pure Math., Vol. XIII, Part II, University of California, Los Angeles, Calif., 1967)*, pages 155–173. American Mathematical Society, Providence, R.I., 1974.

[Hilbert, 1901] D. Hilbert. Mathematische Probleme. Vortrag, gehalten auf dem internationalen Mathematiker-Kongreß zu Paris 1900. *Archiv der Mathematik und Physik*, 1:213–237, 1901.

[Hilbert, 1902] D. Hilbert. Mathematical problems. *Bulletin of the American Mathematical Society*, 8:437–479, 1902. Translation by Maby Winton Newson.

[Hilbert, 1926] D. Hilbert. Über das Unendliche. *Math. Ann.*, 95(1):161–190, 1926.

[Hrušák, 2001] M. Hrušák. Another ◇-like principle. *Fundamenta Mathematicae*, 167(3):277–289, 2001.

[Jech, 1967] T. Jech. Non-provability of Souslin's hypothesis. *Commentationes Mathematicae Universitatis Carolinae*, 8:291–305, 1967.

[Jensen and Solovay, 1970] R. B. Jensen and R. M. Solovay. Some applications of almost disjoint sets. In *Mathematical Logic and Foundations of Set Theory (Proc. Internat. Colloq., Jerusalem, 1968)*, pages 84–104. North-Holland, Amsterdam, 1970.

[Jones, 1937] F. Burton Jones. Concerning normal and completely normal spaces. *Bull. Am. Math. Soc.*, 43:671–677, 1937.

[Judah and Shelah, 1990] H. Judah and S. Shelah. The Kunen-Miller chart (lebesgue measure, the baire property, laver reals and preservation theorems for forcing). *The Journal of Symbolic Logic*, 55:909–927, 1990.

[Judah *et al.*, 1990] H. Judah, S. Shelah, and W. H. Woodin. The Borel conjecture. *Annals of Pure and Applied Logic*, 50(3):255–269, 1990.

[Juhász, 1970] I. Juhász. Martin's Axiom solves Ponomarev's problem. *Bulletin de l'Académie Polonaise des Sciences. Série des Sciences Mathématiques, Astronomiques et Physiques*, 18:71–74, 1970.

[Kanamori, 1994] A. Kanamori. *The higher infinite.* Perspectives in Mathematical Logic. Springer-Verlag, Berlin, 1994. Large cardinals in set theory from their beginnings.

[Kanamori, 1996] A. Kanamori. The mathematical development of set theory from Cantor to Cohen. *The Bulletin of Symbolic Logic*, 2(1):1–71, 1996.

[Kanamori, 2007] A. Kanamori. Tennenbaum and set theory. Expanded version of invited memorial address., August 2007.

[Kanovei and Koepke, 2000] V. Kanovei and P. Koepke. Gaps in partially ordered sets and related problems. Draft, November 2000.

[Katětov, 1960] M. Katětov. Remarks on characters and pseudocharacters. *Commentationes Mathematicae Universitatis Carolinae*, 1(1):20–25, 1960.

[Komjáth and Shelah, 1993/94] P. Komjáth and S. Shelah. On uniformly antisymmetric functions. *Real Analysis Exchange*, 19(1):218–225, 1993/94.

[König, 1905] J. König. Zum Kontinuumproblem. *Math. Ann.*, 60:177–180, 1905.

[Koszmider, 2000] P. Koszmider. On strong chains of uncountable functions. *Israel Journal of Mathematics*, 118:289–315, 2000.

[Kunen and Tall, 1979] K. Kunen and F. D. Tall. Between Martin's Axiom and Souslin's Hypothesis. *Fundamenta Mathematicae*, 102:173–181, 1979.

[Kunen, 1975] K. Kunen. (κ, λ^*)-gaps under MA. Unpublished notes, August 1975.

[Kunen, 1977] K. Kunen. Strong S and L spaces under MA. In *Set-theoretic topology (Papers, Inst. Medicine and Math., Ohio Univ., Athens, Ohio, 1975-1976)*, pages 265–268. Academic Press, New York, 1977.

[Kunen, 1987] K. Kunen. Partitioning Euclidean space. *Mathematical Proceedings of the Cambridge Philosophical Society*, 102:379–383, 1987.

[Kuratowski, 1948] C. Kuratowski. *Topologie. I. Espaces Métrisables, Espaces Complets.* Monografie Matematyczne, vol. 20. Warszawa-Wrocław, 1948. 2d ed.

[Kurepa, 1935] D. Kurepa. Ensembles ordonnes et ramifies. *Publications mathématiques de l'Université de Belgrade*, 4:1–138, 1935.

[Kurepa, 1938] D. Kurepa. Ensembles linéaires et une classe de tableaux ramifies (tableaux ramifies de M. de Aronszajn). *Publications mathématiques de l'Université de Belgrade*, 6/7:129–160, 1938.

[Laflamme, 1997] C. Laflamme. Combinatorial aspects of F_σ filters with an application to $\mathcal{N}$-sets. *Proceedings of the American Mathematical Society*, 125(10):3019–3025, 1997.

[Laver, 1975] R. Laver. On strong measure zero sets. In *Infinite and finite sets (Colloq., Keszthely, 1973; dedicated to P. Erdős on his 60th birthday), vol. II*, pages 1025–1027. Colloq. Math. Soc. Janós Bolyai, Vol. 10. North-Holland, Amsterdam, 1975.

[Laver, 1976] R. Laver. On the consistency of Borel's conjecture. *Acta Mathematica*, 137(3-4):151–169, 1976.

[Laver, 1979] R. Laver. Linear orders in $(\omega)^\omega$ under eventual dominance. In *Logic Colloquium '78 (Mons, 1978)*, volume 97 of *Studies in Logic and the Foundations of Mathematics*, pages 299–302. North-Holland, Amsterdam, 1979.

[Lebesgue, 1972] H. Lebesgue. *Oeuvres scientifiques (en cinq volumes). Vol. I.* Institut de Mathématiques de l'Université de Genève, Geneva, 1972.

[Lusin and Sierpiński, 1923] N. Lusin and W. Sierpiński. Sur un ensemble non mesurable *B*. *Journal de Mathématiques Pure et Appliquées*, 9:2:53–72, 1923.

[Lusin, 1914] N. Lusin. Sur un problème de *M. Baire*. *Comptes Rendus des Séances de l'Académie des Sciences. Série I. Mathématique*, 158:1268–1261, 1914.

[Lusin, 1921] N. Lusin. Sur l'existence d'un ensemble non dénombrable qui est de première catégorie dans tout ensemble parfait. *Fundamenta Mathematicae*, 2:155–157, 1921.

[Lusin, 1934] N. Lusin. Sur la décomposition des ensembles. *Comptes Rendus des Séances de l'Académie des Sciences. Série I. Mathématique*, 198:1671–1674, 1934.

[Lusin, 1935] N. Lusin. Sur les ensembles analytiques nuls. *Fundamenta Mathematicae*, 25:109–131, 1935.

[Lusin, 1947] N. Lusin. On subsets of the series of natural numbers. *Izvestiya Akademiya Nauk SSSR, Seriya Matematicheskaya*, 11:403–410, 1947.

[Mahlo, 1913] P. Mahlo. Über Teilmengen des Kontinuums von dessen Mächtigkeit. *Leipz. Ber.*, 65:283–315, 1913.

[Martin and Solovay, 1970] D. A. Martin and R. M. Solovay. Internal Cohen extensions. *Annals of Pure and Applied Logic*, 2(2):143–178, 1970.

[Mathias, 1977] A. R. D. Mathias. Happy families. *Annals of Pure and Applied Logic*, 12(1):59–111, 1977.

[Mazurkiewicz, 1931] S. Mazurkiewicz. Sur les fonctions non derivables. *Studia Mathematica*, 3:92–94, 1931.

[Miller, 1980] A. W. Miller. Covering 2^ω with ω_1 disjoint closed sets. In *The Kleene Symposium Proceedings, Madison, Wisconsin 1978*, volume 101 of *Studies in Logic and the Foundations of Mathematics*, pages 415–421. North-Holland, 1980.

[Miller, 1981] A. W. Miller. Some properties of measure and category. *Transactions of the American Mathematical Society*, 266:93–114, 1981.

[Miller, 1982a] A. W. Miller. A characterization of the least cardinal for which the Baire category theorem fails. *Proceedings of the American Mathematical Society*, 86(3):498–502, 1982.

[Miller, 1982b] A. W. Miller. Corrections and additions to "Some properties of measure and category". *Transactions of the American Mathematical Society*, 271:347–348, 1982.

[Miller, 1983] A. W. Miller. Mapping a set of reals onto the reals. *The Journal of Symbolic Logic*, 48:575–584, 1983.

[Miller, 1984a] A. W. Miller. Rational perfect set forcing. In *Axiomatic set theory (Boulder, Colo., 1983)*, volume 31 of *Contemporary Mathematics*, pages 143–159. American Mathematical Society, Providence, RI, 1984.

[Miller, 1984b] A. W. Miller. Special subsets of the real line. Handbook of set-theoretic topology, 201-233 (1984)., 1984.

[Miller, 1993] A. W. Miller. Arnie Miller's problem list. In *Set theory of the reals (Ramat Gan, 1991)*, volume 6 of *Israel Math. Conf. Proc.*, pages 645–654. Bar-Ilan Univ., Ramat Gan, 1993.

[Moore, 2002] J. Moore. Open colorings, the continuum and the second uncountable cardinal. *Proceedings of the American Mathematical Society*, 130(9):2753–2759 (electronic), 2002.

[Moore *et al.*, 2004] J. Moore, M. Hrušák, and M. Džamonja. Parametrized $\diamondsuit$ principles. *Transactions of the American Mathematical Society*, 356(6):2281–2306 (electronic), 2004.

[Moore, 1989] G. H. Moore. Towards a history of Cantor's continuum problem. In *The history of modern mathematics, Vol. I (Poughkeepsie, NY, 1989)*, pages 79–121. Academic Press, Boston, MA, 1989.

[Morayne, 1987] M. Morayne. On differentiability of Peano type functions. I, II. *Colloquium Mathematicum*, 53(1):129–132, 133–135, 1987.

[Osofsky, 1968] B. L. Osofsky. Homological dimension and the continuum hypothesis. *Transactions of the American Mathematical Society*, 132:217–230, 1968.

[Ostaszewski, 1976] A. J. Ostaszewski. On countably compact, perfectly normal spaces. *Journal of the London Mathematical Society. Second Series*, 14:505–516, 1976.

[Oxtoby, 1980] J. C. Oxtoby. *Measure and category. A survey of the analogies between topological and measure spaces. 2nd ed.* Graduate Texts in Mathematics, 2. New York - Heidelberg - Berlin: Springer-Verlag. X, 106 p. , 1980.

[Pawlikowski, 1985] J. Pawlikowski. Lebesgue measurability implies Baire property. *Bulletin des Sciences Mathématiques. 2e Série*, 109(3):321–324, 1985.

[Pospíšil, 1937] B. Pospíšil. Remark on bicompact spaces. *Annals of Mathematics. Second Series*, 38(4):845–846, 1937.

[Pospíšil, 1939] B. Pospíšil. On bicompact spaces. *Publ. Fac. Sci. Univ. Masaryk*, 270:1–16, 1939.

[Raisonnier and Stern, 1983] J. Raisonnier and J. Stern. Mesurabilité et propriété de Baire. *Comptes Rendus des Séances de l'Académie des Sciences. Série I. Mathématique*, 296(7):323–326, 1983.

[Raisonnier and Stern, 1985] J. Raisonnier and J. Stern. The strength of measurability hypotheses. *Israel Journal of Mathematics*, 50:337–349, 1985.

[Riesz, 1909] F. Riesz. Stetigkeitsbegriff und abstrakte Mengenlehre. In *Atti del IV Congresso Internazionale dei Matematici (Roma, 6-11 Aprile 1908)*, volume 2, pages 18–24. International Congress of Mathematicians, 1909.

[Rogers, 1970] C. A. Rogers. *Hausdorff measures.* Cambridge University Press, London, 1970.

[Rosłanowski and Shelah, 1999] A. Rosłanowski and S. Shelah. Norms on possibilities. I. Forcing with trees and creatures. *Memoirs of the American Mathematical Society*, 141(671):xii+167, 1999.

[Rosłanowski and Shelah, 2004] A. Rosłanowski and S. Shelah. Sweet & sour and other flavours of ccc forcing notions. *Archive for Mathematical Logic*, 43(5):583–663, 2004.

[Rosłanowski and Shelah, 2006] A. Rosłanowski and S. Shelah. Measured creatures. *Israel Journal of Mathematics*, 151:61–110, 2006.

[Rothberger, 1938a] F. Rothberger. Eine Äquivalenz zwischen der Kontinuumhypothese und der Existenz der Lusinschen und Sierpińskischen Mengen. *Fundamenta Mathematicae*, 30:215–217, 1938.

[Rothberger, 1938b] F. Rothberger. Une remarque concernant l'hypothèse du continu. *Fundamenta Mathematicae*, 31:224–226, 1938.

[Rothberger, 1939] F. Rothberger. Sur un ensemble toujours de premiere categorie qui est depourvu de la propriété λ. *Fundamenta Mathematicae*, 32:294–300, 1939.

[Rothberger, 1941] F. Rothberger. Sur les familles indenombrables de suites de nombres naturels et les problèmes concernant la propriété C. *Mathematical Proceedings of the Cambridge Philosophical Society*, 37:109–126, 1941.

[Rothberger, 1948] F. Rothberger. On some problems of Hausdorff and of Sierpiński. *Fundamenta Mathematicae*, 35:29–46, 1948.

[Rothberger, 1952a] F. Rothberger. On the property C and a problem of Hausdorff. *Canadian Journal of Mathematics. Journal Canadien de Mathématiques*, 4:111–116, 1952.

[Rothberger, 1952b] F. Rothberger. A remark on the existence of a denumerable base for a family of functions. *Canadian Journal of Mathematics. Journal Canadien de Mathématiques*, 4:117–119, 1952.

[Sacks, 1971] G. E. Sacks. Forcing with perfect closed sets. In *Axiomatic Set Theory (Proc. Sympos. Pure Math., Vol. XIII, Part I, University of California, Los Angeles, Calif., 1967)*, pages 331–355. American Mathematical Society, Providence, R.I., 1971.

[Scheepers, 1993] M. Scheepers. Gaps in ω^ω. In *Set theory of the reals (Ramat Gan, 1991)*, volume 6 of *Israel Math. Conf. Proc.*, pages 439–561. Bar-Ilan University, Ramat Gan, 1993.

[Scott, 1961] D. Scott. Measurable cardinals and constructible sets. *Bulletin de l'Académie Polonaise des Sciences. Série des Sciences Mathématiques, Astronomiques et Physiques*, 9:521–524, 1961.

[Shelah, 1974] S. Shelah. Infinite Abelian groups, Whitehead problem and some constructions. *Israel Journal of Mathematics*, 18:243–256, 1974.

[Shelah, 1982] S. Shelah. *Proper forcing*, volume 940 of *Lecture Notes in Mathematics*. Springer-Verlag, Berlin, 1982.

[Shelah, 1984a] S. Shelah. Can you take Solovay's inaccessible away? *Israel Journal of Mathematics*, 48:1–47, 1984.

[Shelah, 1984b] S. Shelah. On cardinal invariants of the continuum. In *Axiomatic set theory (Boulder, Colo., 1983)*, volume 31 of *Contemporary Mathematics*, pages 183–207. American Mathematical Society, Providence, RI, 1984.

[Shelah, 2000a] S. Shelah. Covering of the null ideal may have countable cofinality. *Fundamenta Mathematicae*, 166(1-2):109–136, 2000. Saharon Shelah's anniversary issue.

[Shelah, 2000b] S. Shelah. On what I do not understand (and have something to say). I. *Fundamenta Mathematicae*, 166(1-2):1–82, 2000. Saharon Shelah's anniversary issue.

[Shelah, 2004] S. Shelah. Two cardinal invariants of the continuum ($\mathfrak{d} < \mathfrak{a}$) and FS linearly ordered iterated forcing. *Acta Mathematica*, 192(2):187–223, 2004.

[Sierpiński, 1919] W. Sierpiński. Sur un théorème équivalent à l'hypothèse du continu. *Bulletin international de l'Académie des sciences de Cracovie, Classe des sciences mathématiques et naturelles.*, 1919:1–3, 1919.

[Sierpiński, 1924] W. Sierpiński. Sur l'hypothése du continu ($2^{\aleph_0} = \aleph_1$). *Fundamenta Mathematicae*, 5:177-187, 1924.

[Sierpiński, 1928a] W. Sierpiński. Sur un ensemble non dénombrable dont toute image continue est de premiere catégorie. *Bulletin de l'Academie Polonaise des Sciences (A)*, 1928:455–458, 1928.

[Sierpiński, 1928b] W. Sierpiński. Sur une décomposition d'ensembles. *Monatshefte für Mathematik*, 35:239–242, 1928.

[Sierpiński, 1928c] W. Sierpiński. Sur une ensemble non dénombrable, dont toute image continue est de mesure nulle. *Fundamenta Mathematicae*, 11:302–304, 1928.

[Sierpiński, 1934a] W. Sierpiński. *Hypothèse du continu.* Monogr. mat. 4. Warszawa-Lwow: Subwencji Funduszu Kultur. Narodowej. V, 192 S. , 1934.

[Sierpiński, 1934b] W. Sierpiński. Sur la dualité entre la premiere categorie et la mesure nulle. *Fundamenta Mathematicae*, 22:276–280, 1934.

[Sierpiński, 1934c] W. Sierpiński. Sur les ensembles partout de deuxième categorie. *Fundamenta Mathematicae*, 22:1–3, 1934.

[Sierpiński, 1935] W. Sierpiński. Sur une hypothése de M. Lusin. *Fundamenta Mathematicae*, 25: 132–135, 1935.

[Sierpiński, 1937] W. Sierpiński. Sur une propriété additive d'ensembles. *Comptes Rendus de la Société des Sciences de Varsovie*, 30:257–259, 1937.

[Sierpiński, 1938] W. Sierpiński. Remarque sur le problème de l'invariance topologique de la propriété (C). *Fundamenta Mathematicae*, 30:56–58, 1938.

[Sierpiński, 1939] W. Sierpiński. Sur les ensembles concentres. *Fundamenta Mathematicae*, 32:301–305, 1939.

[Sierpiński, 1965] W. Sierpiński. *Cardinal and ordinal numbers. (Monografie Matematyczne. Tom 34.) 2nd ed. revised.* Warszawa: Panstwowe Wydawnictwo Naukowe. 491 p. , 1965.

[Solomon, 1977] R. C. Solomon. Families of sets and functions. *Czechoslovak Mathematical Journal*, 27(102)(4):556–559, 1977.

[Solovay and Tennenbaum, 1971] R. M. Solovay and S. Tennenbaum. Iterated Cohen extensions and Souslin's problem. *Annals of Mathematics. Second Series*, 94:201–245, 1971.

[Solovay, 1969] R. M. Solovay. On the cardinality of $\sum_2^1$ sets of reals. In *Foundations of Mathematics (Symposium Commemorating Kurt Gödel, Columbus, Ohio, 1966)*, pages 58–73. Springer, New York, 1969.

[Solovay, 1970] R. M. Solovay. A model of set-theory in which every set of reals is Lebesgue measurable. *Annals of Mathematics. Second Series*, 92:1–56, 1970.

[Solovay, 1971] R. M. Solovay. Real-valued measurable cardinals. In *Axiomatic set theory (Proc. Sympos. Pure Math., Vol. XIII, Part I, University of California, Los Angeles, Calif., 1967)*, pages 397–428. American Mathematical Society, Providence, R.I., 1971.

[Specker, 1950] E. Specker. Additive Gruppen von Folgen ganzer Zahlen. *Portugaliae Mathematica*, 9:131–140, 1950.

[Steprāns, 1999] J. Steprāns. Decomposing Euclidean space with a small number of smooth sets. *Transactions of the American Mathematical Society*, 351(4):1461–1480, 1999.

[Stone, 1937] M. H. Stone. Applications of the theory of Boolean rings to general topology. *Transactions of the American Mathematical Society*, 41(3):375–481, 1937.

[Talagrand, 1980] M. Talagrand. Compacts de fonctions mesurables et filtres non mesurables. *Studia Mathematica*, 67:13–43, 1980.

[Tall, 1969] F. D. Tall. *Set-theoretic consistency results and topological theorems concerning the normal Moore space conjecture and related problems.* PhD thesis, University of Wisconsin, 1969.

[Tarski, 1928] A. Tarski. Sur la décomposition des ensembles en sous-ensembles presque disjoints. *Fundamenta Mathematicae*, 12:188–205, 1928.

[Tennenbaum, 1968] S. Tennenbaum. Souslin's problem. *Proceedings of the National Academy of Sciences of the United States of America*, 59:60–63, 1968.

[Todorčević, 1984] S. Todorčević. A note on the proper forcing axiom. In *Axiomatic set theory (Boulder, Colo., 1983)*, volume 31 of *Contemporary Mathematics*, pages 209–218. American Mathematical Society, Providence, RI, 1984.

[Todorcevic, 1987] S. Todorcevic. Partitioning pairs of countable ordinals. *Acta Mathematica*, 159:261–294, 1987.

[Todorcevic, 1988] S. Todorcevic. Oscillations of real numbers. In *Logic colloquium '86, (Hull, 1986)*, volume 124 of *Studies in Logic and the Foundations of Mathematics*, pages 325–331. North-Holland, 1988.

[Todorčević, 1989] S. Todorčević. *Partition problems in topology*, volume 84 of *Contemporary Mathematics*. American Mathematical Society, Providence, RI, 1989.

[Todorčević, 1993] S. Todorčević. Conjectures of Rado and Chang and cardinal arithmetic. In *Finite and infinite combinatorics in sets and logic (Banff, AB, 1991)*, volume 411 of *NATO Advanced Science Institutes Series C: Mathematical and Physical Sciences*, pages 385–398. Kluwer Acad. Publ., Dordrecht, 1993.

[Todorcevic, 1998] S. Todorčević. Oscillations of sets of integers. *Advances in Applied Mathematics*, 20(2):220–252, 1998.

[Todorčević, 1997] S. Todorčević. Comparing the continuum with the first two uncountable cardinals. Dalla Chiara, Maria Luisa (ed.) et al., Logic and scientific methods. Volume one of the proceedings of the tenth international congress of logic, methodology and philosophy of science, Florence, Italy, August 19–25, 1995. Dordrecht: Kluwer Academic Publishers. Synth. Libr. 259, 145-155 (1997)., 1997.

[Truss, 1977] J. Truss. Sets having calibre $\aleph_1$. In *Logic Colloquium '76 (Oxford, 1976)*, volume 87 of *Studies in Logic and Found. Math.*, pages 595–612. North-Holland, Amsterdam, 1977.

[Tukey, 1940] J. W. Tukey. *Convergence and Uniformity in Topology*. Annals of Mathematics Studies, no. 2. Princeton University Press, Princeton, N. J., 1940.

[Ulam, 1929] S. Ulam. Concerning functions of sets. *Fundamenta Mathematicae*, 14:231–233, 1929.

[Ulam, 1930] S. Ulam. Zur Maßtheorie in der allgemeinen Mengenlehre. *Fundamenta Mathematicae*, 16:140–150, 1930.

[Ulam, 1933] S. Ulam. Über gewisse Zerlegungen von Mengen. *Fundamenta Mathematicae*, 20:221–223, 1933.

[van Douwen, 1984] E. K. van Douwen. The integers and topology. In *Handbook of set-theoretic topology*, pages 111–167. North-Holland, Amsterdam, 1984.

[Vaughan, 1990] J. E. Vaughan. Small uncountable cardinals and topology. In *Open problems in topology*, pages 195–218. North-Holland, Amsterdam, 1990. With an appendix by S. Shelah.

[Vojtáš, 1993] P. Vojtáš. Generalized Galois-Tukey-connections between explicit relations on classical objects of real analysis. In *Set theory of the reals (Ramat Gan, 1991)*, volume 6 of *Israel Math. Conf. Proc.*, pages 619–643. Bar-Ilan Univ., Ramat Gan, 1993.

[Vopěnka and Hrbáček, 1967] P. Vopěnka and K. Hrbáček. The consistency of some theorems on real numbers. *Bulletin de l'Académie Polonaise des Sciences. Série des Sciences Mathématiques, Astronomiques et Physiques*, 15:107–111, 1967.

[Woodin, 1999] W. Hugh Woodin. *The axiom of determinacy, forcing axioms, and the nonstationary ideal*, volume 1 of *de Gruyter Series in Logic and its Applications*. Walter de Gruyter & Co., Berlin, 1999.

[Woodin, 2001a] W. Hugh Woodin. The continuum hypothesis. I. *Notices of the American Mathematical Society*, 48(6):567–576, 2001.

[Woodin, 2001b] W. Hugh Woodin. The continuum hypothesis. II. *Notices of the American Mathematical Society*, 48(7):681–690, 2001.

[Woodin, 2002] W. Hugh Woodin. Correction: "The continuum hypothesis. II". *Notices of the American Mathematical Society*, 49(1):46, 2002.

INFINITE COMBINATORICS

Jean A. Larson

CONTENTS

Handbook of the History of Logic: Sets and Extensions in the Twentieth Century.
Volume editor: Akihiro Kanamori. General editors: Dov M. Gabbay, Paul Thagard and John Woods.

1 INTRODUCTION

There is a tension in mathematics between generalization and description in concrete terms. Cantor's extension of number to the transfinite is an example of generalization, and the pull of generalization and the counterbalancing forces are described in the passage that concludes with a famous declaration [Cantor, 1883, 545], "das *Wesen* der *Mathematik* liegt gerade in ihrer *Freiheit.*"[1]

> These principles, I think, do not represent a danger for science as some have suggested: first, the conditions under which new numbers can be generated leave little space for arbitrariness; second, every mathematical construct comes with a natural corrective; if it is impractical or uninspiring it will be dropped due to lack of impact. Conversely, any unnecessary constraint on the mathematical research impetus carries the far greater risk—the more so since there is no scientific justification for it; for the *essence* of *mathematics* is its *lack of constraints.*[2]

[1]The quote has been variously translated, e.g. "the essence of mathematics resides in its freedom." This version is the one I first encountered as part of a mural by art students at the University of Florida, and is the one I will be using in the postscript.

[2](The above translation of the following paragraph was made by Jürg Peters of the University of Florida. When asked why he had translated *Freiheit* as *lack of constraints* rather than *freedom* he said he felt that Cantor's notion of *free* was like that of free groups and that *lack of constraints* better captured that sense.) Es ist, wie ich glaube, nicht nöthig in diesen Grundsätzen irgendeine

Concrete structure may be sought through classification schemes, in basis problems, in universal structures, and especially in the search for and expectation of local uniformity, regions of simplicity, and inescapable structure. This aspect is captured by the following quote, most often used in connection with Ramsey theory: "complete disorder is impossible". This short sentence is embedded in the following quote from Motzkin [1967]:[3]

> The influence on mathematics of its two neighbors, physics and logic, is sometimes opposite or, at least, complementary. Whereas the entropy theorems of probability theory and mathematical physics imply that, in a large universe, disorder is probable, certain combinatorial theorems show that complete disorder is impossible. Already the "pigeon-hole principle" —out of $t+1$ objects of t kinds, at least two must be of the same kind—can be interpreted as saying that, while no 2-set (set of two objects) in the given $(t+1)$-set needs to be composed of objects of different kind (however probable that may be), at least one 2-set of objects of the same kind must occur. A sophisticated generalization of this principle was initiated in [7] by the mathematical logician Ramsey.

The neighbors most relevant to our study are logic and other areas of mathematics. From logic we see that opposite or complementary results may be obtained by extending our basic set-theoretic assumptions in different ways to get consistency results for different aspects of the combinatorial enterprise. Motivated by the Todorcevic paper *Basis problems in combinatorial set theory* [1998a], we shall be especially interested in what he calls *critical members* of classes $\mathcal{S}$ of interest, of which Todorcevic wrote: "Critical objects are almost always some canonical members of $\mathcal{S}$ simple to describe and visualize."

The Pigeonhole Principle,[4] which was mentioned in the Motzkin quote, is our first example of an assertion of the lack of disorder. It is a fundamental idea that

Gefahr für die Wissenschaft zu befürchten, wie dies von Vielen geschieht; einerseits sind die bezeichneten Bedingungen, unter welchen die Freiheit der Zahlenbildung allein geübt werden kann, derartige, dass sie der Willkür einen äussertst geringen Spielraum lassen; dann aber trägt auch jeder mathematische Begriff das nöthige Correctiv in sich selbst einher; ist er unfruchtbar oder unzweckmässig, so zeigt er es sehr bald durch seine Unbrauchbarkeit und er wird alsdann, wegen mangelnden Erfolgs, fallen gelassen. Dagegen scheint mir aber jede überflüssige Einengung des mathematischen Forschungstriebes eine viel grössere Gefahr mit sich zu bringen und eine um so grössere, als dafür aus dem Wesen der Wissenschaft wirklich keinerlei Rechtfertigung gezogen werden kann; denn das *Wesen* der *Mathematik* liegt gerade in ihrer *Freiheit.*

[3]Hans Jürgen Prömel [2005, 3] used the short quote as a theme in the life of Walter Deuber. Theodore S. Motzkin (1908 – 1970) spent the later part of his career at the University of California, Los Angeles.

[4]An early use of the English "pigeonhole principle" occurs in Raphael Robinson' paper [1941] and the first mention found by searching MathSciNet on December 8, 2010 for the word "pigeon" is the Erdős-Rado paper [1956], where "pigeon-hole principle" is offered as a translation of "the box argument" or "the chest of drawers argument" with "Schubfachprinzip" in parenthesis, and attributed (incorrectly) to Dedekind. Those familiar with pigeon hole desks, sometimes used for sorting letters, recognize the image that almost surely inspired this translation, but those unfamiliar with them may think of the pigeon houses of Iran and Egypt, with their many entrances, as is suggested by one Italian translation, *principio da casa dos pombos*, which appears

has been around for a long time, and the principle itself is attributed to Dirichlet in 1834, where he used the name *Schubfachprinzip.*[5]

These themes play out in this decade-by-decade review of infinite combinatorics in the 20th Century. The foci are the study of orderings, especially trees, and Ramsey theory, especially the partition calculus. Along the way, results about almost disjoint sets, set mappings, transversals, and regressive functions are discussed, and a wide variety of combinatorial tools are mentioned as they come into play. Lists of a variety of results are briefly described to capture the breadth and flavors of different decades. Occasionally, conferences and applications to or from infinite combinatorics are described to indicate some of the ways in which the field has benefited by interactions with communities of mathematicians in finite combinatorics, model theory, topology, and other aspects of set theory, including large cardinals and forcing.

My goal in writing this chapter was to be inclusive, that is, to acknowledge multiple discoveries, rediscoveries,[6] and inventions of ideas. I apologize in advance for the omissions and incorrect statements that surely appear in this wide ranging survey. I have not concerned myself greatly about who had the first seed leading to each discovery, nor with assessing the quality of the variety of contributions. Let me state my criteria for attribution of results: (1) they must be directly stated; (2) they must be proved and published in a reputable journal, or in rare cases vetted by someone other than the author or explicitly ascribed by others, usually when the idea of proof has been generalized. This approach is informed by having listened to Paul Erdős describe how he decided at what point and to whom money should be awarded for the solutions to certain of his problems.

1.1 Overview of the history of orderings

A broad overview of the history of orderings starts with foundational work on cardinal and ordinal numbers, including the Schröder-Bernstein Equivalence theorem, fundamentals of cardinal and ordinal arithmetic, and Zermelo's Well-Ordering Theorem. An early treatment in English was Huntington's book on the continuum as a type of order [1905c].

on Miller's site (see the next footnote). It would not be surprising if a perusal of British popular articles on mathematics and mathematical puzzles turned up much earlier references. Literary references to the concept go back still further, with an example dating to the 1600s appearing on http://pballew.blogspot.com/2009/06/updating-history-of-pigeon-hole-theorem.html.

[5]It is difficult to ascertain the first usage of any idea, symbol or word. The web page for letter P http://jeff560.tripod.com/p.html of Jeff Miller's site on *Earliest Known Uses of Some of the Words of Mathematics*, downloaded on December 9, 2010 makes a case for Dirichlet's priority.

[6]At the Scottish Book conference held in Denton, Texas in 1979, Paul Erdős [1981, 36] shared his perspective on rediscovery: "Ulam and I settled many of these questions long ago, but we never got around to publishing our results. These results were rediscovered and published by the Indian mathematician B. V. Rao; when Rao sent me a preprint I urged him to publish. Naturally, I did not tell him that Ulam and I had already done the work. Eventually he found out, however, and asked me why I hadn't said anything about it. I replied that this was the one respect in which I did not want to imitate Gauss, who had a nasty habit of 'putting down' younger mathematicians by telling them he had long ago obtained their supposedly new results."

The first systematic study of linear and partial orderings was carried out by Hausdorff, who set the stage at the beginning of century with his first mathematical paper [1901] in which he generalized ordinals. He investigated the family ${}^{\alpha}M$ of functions from an ordinal number α into a linearly ordered set M under various orderings including the lexicographic ordering and eventual domination. Hausdorff identified critical elements, found bases, generating sets, universal linear orders[7] for scattered linear orderings of bounded cardinality, showed every partial order has a maximal well-ordered subset, and set up and studied a classification scheme for linear orders that led him to identify critical types, such as what have come to be known as Hausdorff and Rothberger gaps.[8]

In the decade 1910–1920, the Moscow school, founded by Egorov and Lusin, was put on the map by results like Lusin's CH construction of an uncountable set of reals whose intersection with every meager set is countable, Alexandroff's proof that every uncountable Borel set includes a perfect set (independently found by Hausdorff), and Suslin's construction of a non-Borel analytic set. Moore developed postulates for analysis-situ or point-set theory that led to the development of the Texas school of topology. They were the initial source of the axioms he used in his distinctive teaching style: he doled out axioms, a few at a time, and guided the students to deduce consequences from them. Sierpiński published *An Outline of Set Theory* [1912], an early systematic treatment of set theory, and worked on consequences and equivalences of the Continuum Hypothesis. Hausdorff wrote an influential set theory book, *Grundzüge der Mengenlehre* [1914], which included some of his work on linear and partial orders and introduced topology in terms of neighborhoods. Subsequent editions omitted most of the material on orderings in favor of Borel sets, analytic sets and Baire functions, and his pioneering work on ordered sets faded into obscurity for a time.

In the 1920's, mathematics flourished in Poland with schools in Lwów (Banach, Steinhaus, Ulam) and Warsaw, where Janiszewski, Mazurkiewicz, Sierpiński founded *Fundamenta Mathematicae*, an international journal which focused on foundations, point sets and functions and whose first issue appeared in 1920. That issue included a problem (rephrased below in modern language) that was influential in the theory of order and allied areas across the century.

> *Suslin's Problem:* Must every complete, dense linearly ordered set without endpoints for which every pairwise disjoint set of intervals is countable be a copy of the real line?

Also in the 1920's, Dénes Kőnig, working within a graph-theoretic perspective, proved the first result about infinite trees, now known as *König's Lemma* or *König's Infinity Lemma.*[9] Moore constructed his automobile road space essentially using the complete binary tree of sequences of zeros and ones of length $\leq \omega$.

[7] A linear order $(P, <)$ is *universal for a class* $\mathcal{K}$ of linear orders if $(P, <)$ belongs to $\mathcal{K}$ and every element of $\mathcal{K}$ is order-embeddable into $(P, <)$.

[8] See the chapters by Kojman and Steprāns for more on gaps.

[9] König's Theorem refers to the cardinal inequality of Dénes Kőnig's father Gyula, who published under the name Julius Kőnig.

Urysohn proved that every ordered set whose increasing chains and decreasing chains are at most countable has power at most the continuum. Sierpiński introduced almost disjoint sets which he and Tarski investigated, and von Neumann introduced his representation of Cantor's ordinal numbers and put transfinite recursion on a firm foundation.

In the 1930's, Marczewski proved that all partial orders can be extended to total (linear) orders; and Mostowski constructed a denumerable partial order universal[10] for countable partial orderings. Kurepa made a systematic study of uncountable partial orders, especially trees in the wider framework of ramified sets. He identified the key parameters for a structural analysis of uncountable trees: height, supremum of the sizes of the levels, and cardinality of the set of branches whose length is the height of the tree. In his thesis, he developed the framework within which Aronszajn constructed the first *Aronszajn tree.*[11] Rather than include Aronszajn's construction, Kurepa gave his own construction of an Aronszajn tree as a subtree of $\sigma\mathbb{Q}$, the non-empty bounded well-ordered sets of rationals under end-extension. He extracted the combinatorial content of Suslin's Problem when he showed that a negative answer was equivalent to the existence of a Suslin tree.[12] He constructed the first *special* Aronszajn tree, i.e. one which is a countable union of antichains. The distinction was later recognized as a critical one. Kurepa also showed that Aronszajn trees are special if and only if they have a strictly increasing map into the rational numbers. With his First Miraculous Problem, he started the study of isomorphism types of Aronszajn trees. He studied partial order more generally, in particular monotone functions on them, and proved his Fundamental Relation, an inequality for partial orders bounding the cardinality of the underlying set in terms of the powers of their chains and antichains.

A special Aronszajn tree is implicit in the 1933 construction by Jones of a Moore space which is not metrizable, but this construction was not announced or published until much later. Jones proved his metrization theorem and introduced the Normal Moore Space Conjecture in [1937], which stimulated research related to trees by topologists.

In the 1940's, Dushnik and Miller introduced a notion of dimension for partial orders, investigated endomorphisms of linearly ordered sets, constructed an uncountable rigid set of real numbers, and gave conditions that guarantee that a partial order has a large linearly ordered subset. Kurepa proved that every uncountable subset of $\sigma\mathbb{Q}$ includes an uncountable antichain. Sierpiński constructed a family of $2^{\mathfrak{c}}$ many pairwise non-isomorphic linear orderings each of the cardinality $\mathfrak{c}$ of the continuum. Erdős and Tarski generalized the notion of almost disjointness to partial orders and applied their results to questions related to Kurepa's work. The first steps were taken toward the development of a theory of well-quasi-orders

[10] A partial order $(P,<)$ is *universal for a class* $\mathcal{K}$ of partial orders if $(P,<)$ belongs to $\mathcal{K}$ and every partial order in $\mathcal{K}$ has an order and incomparability preserving embedding into $(P,<)$.

[11] An Aronszajn tree is a tree of height ω_1, all of whose levels are countable which has no uncountable branch.

[12] A *Suslin tree* is a tree of height ω_1 in which every branch and every antichain is countable.

(wqos). Fraïssé's conjecture that there is no infinite strictly decreasing sequence of denumerable order types under the relation of order embeddability became a motivating question in this area. Specker generalized the work of Aronszajn and Kurepa by using GCH to construct κ^+-Aronszajn trees for regular κ. Kurepa focused attention on trees of height ω_1 with more than $\aleph_1$ branches, now known as *Kurepa trees*, and continued the study of subtrees of $\sigma\mathbb{Q}$ and monotone mappings into $\mathbb{Q}$ and $\mathbb{R}$. Knaster introduced what has come to be known as the *Property of Knaster* and proved that for topological spaces this property is equivalent to being separable,[13] while Marczewski asked if the countable chain condition implies the Property of Knaster. Miller and Sierpiński came up with equivalences in terms of partial orders to the existence of a Suslin line, seemingly unaware of Kurepa's equivalence with the existence of Suslin trees, and Maharam used the existence or non-existence of Suslin continua in her investigation of the existence or non-existence of measures on countably complete Boolean algebras, in essence, constructing a Suslin tree from a Suslin continuum as part of one of her arguments.

In the 1950's, interest in Suslin lines and their applications continued: Rudin constructed a Dowker space from a Suslin line; Kurepa showed that the Cartesian square of a Suslin line (level-wise product) is not the countable union of antichains, i.e. does not have the countable chain condition. Sierpiński, Shepherdson, Ginsburg and Johnston, among others, investigated the structure of linear and partial orderings under embeddability. Bing made progress on the Normal Moore Space Conjecture by proving that if there is a Q-set, then there is a separable non-metrizable Moore space.

In the 1960's, the introduction of forcing, Martin's Axiom and the combinatorial principle $\diamondsuit$ and its variants led to an explosion of results, as did the detailed analysis of the fine structure of the constructible universe. By the end of the decade, there were consistency and independence results for Kurepa and Suslin trees, and it was known that, under CH, there are many isomorphism types of Aronszajn trees. The decade featured a catalog of examples of well-quasi-orders (wqos), the invention of better-quasi-orders (bqos) and the study of their closure properties. In particular, the following families of structures were shown to be wqos under structure-preserving embeddings: trees (Kruskal for finite, Nash-Williams for infinite) and σ-scattered linear orders (Laver). Hanf, building on work by Erdős, showed that the generalization to regular uncountable cardinals of König's Lemma, i.e. the tree property, is equivalent to a partition property. Halpern and Läuchli proved a partition theorem for products of finite trees, and an analogous theorem for infinite perfect trees was proved independently (and later) by Laver.

In the 1970's, questions about rigidity, automorphisms and other structural properties of Aronszajn trees and Suslin trees were investigated under various hypotheses. While the concept of special tree dates back to the 1930's, the use of the word "special" for this concept entered the language in the 1970's. Consistency and independence results were proved on the existence or non-existence of special Aronszajn trees and their generalizations to successor cardinals; such results were

[13] A space is separable if it has a countable dense subset.

also proved for trees which have a strictly increasing map into the reals, and for combinations of the various trees under consideration, including Kurepa trees. Milliken proved a Ramsey theorem for finitely branching trees. Basis questions for uncountable linear orders were studied and consistency was shown for the statement that all $\aleph_1$-dense sets of reals are isomorphic. A linear order of size $\aleph_1$ universal for orders of size $\aleph_1$ was shown to be consistent with $2^{\aleph_0} = \aleph_2$. Shelah constructed a Countryman line, an important example of an uncountable linear order.

In the 1980's, Todorcevic surveyed work on trees from combinatorial and set-theoretic perspectives. The elegance of his presentation drew a wide audience for this work. Work continued on exploring consistent possibilities for various types of trees, now looking for results for trees on multiple cardinals, or with required or forbidden types of subtrees. The structure of trees was examined through a variety of lenses: a notion of base, club isomorphism, club embeddability, nearness, existence of a universal Aronszajn tree for club embeddability. Shelah used a complex combinatorial argument to show that adding a Cohen real also adds a Suslin tree; Todorcevic later came up with an elegant construction as part of his introduction of a family of trees constructed from coherent mappings that have become standard examples in the subject. Existence and non-existence of κ^+-Suslin trees were proved for a variety of values of κ in the presence of different values for the continuum. Baumgartner surveyed work on uncountable linear orders, prompting further work in this area. For example, two $\aleph_1$-dense sets of reals were constructed for which provably there is no C^1 isomorphism.

In the 1990's, several consistency results for the existence and non-existence of λ-Aronszajn trees for $\lambda \geq \omega_2$ were found. Cardinality conditions and combinatorial assumptions were found that entail the existence of Suslin trees on successor cardinals larger than $\aleph_1$ which are immune to specialization in cardinal and cofinality preserving extensions. A Suslin tree was constructed from $\clubsuit$ and the assumption that the meager idea $\mathcal{M}$ has a base of size $\aleph_1$. The tree $\sigma\mathbb{Q}$ of all non-empty bounded well-ordered subsets of $\mathbb{Q}$ was shown to be universal (as a partial order) for all Hausdorff trees with strictly increasing embeddings into the reals. Todorcevic and Väänänen studied the quasi-order of all trees of cardinality ω_1 with no uncountable branches under the relation $S \leq T$ if and only if there is a strictly increasing map from S to T (it need not preserve the tree structure). Todorcevic announced a Ramsey-theoretic reformulation, under the Proper Forcing Axiom,[14] for the existence of a two-element basis for the class of Aronszajn lines. Abraham, Rubin and Shelah investigated consistent finite possibilities for the structure of order types of $\aleph_1$-dense homogeneous subsets of the reals under the embeddability relation. Kojman and Shelah investigated the existence problem for universal linear and partial orders in various cardinalities;[15] Todorcevic and Väänänen proved

[14]The Proper Forcing Axiom (PFA) is a generalization of Martin's Axiom which will be introduced in §8.

[15]A linear (partial) order $(P, <)$ is *universal in power* λ (alternatively *in cardinality* λ) if it has cardinality λ and is universal for the class of linear (partial) orders of cardinality $\leq \lambda$.

that the class of linear orderings without uncountable well-ordered or conversely well-ordered subsets has no universal element. The possibilities for order types of the form $\mathbb{R} \cap M$ for M a model of set theory were investigated.

1.2 Overview of the history of Ramsey theory

Now let us turn to an overview of the history of Ramsey theory in the 20th Century with a focus on the partition calculus. Conventionally the history is started with Ramsey's famous theorem proved in the late 1920's and published in 1930.[16] In the latter half of the 20th Century, the influence of the habit of thinking "complete disorder is impossible" has spread beyond the communities of combinatorial mathematicians, set theorists, and those studying order whether algebraically, combinatorially or as a special case of binary relations to include set-theoretic topology, semi-groups, dynamical systems, ergodic theory, and Banach space theory. It makes sense to include some of the historical roots of those directions, albeit with only cursory attempts to describe their mathematical interconnections, which can be quite complicated.[17]

Following Hindman in his article [2006] on the mathematics of Rothschild and Graham, we give a short prehistory of Ramsey theory, starting with the result of Hilbert [1892] that for any positive integer n, if the positive integers are partitioned into finitely many cells, there is a natural number a and a finite sequence $\vec{m} = \langle m_1, m_2, \ldots, m_n \rangle$ of n positive integers such that all integers of the form $a + b$ where b is a finite sum of some or all of the integers in $\vec{m}$ are in the same cell of the partition.[18] Next, we state Schur's Theorem [1917] which is Hilbert's Theorem without the a and with $n = 2$: whenever the positive integers are finitely colored, there are positive integers x and y with x, y and $x + y$ all the same color. We conclude the pre-history with van der Waerden's Theorem [1927] that whenever the positive integers are finitely colored, there are arbitrarily long (finite) monochromatic arithmetic progressions.

We start the history in the 1920's with Ramsey's own work by stating the theorem for which he is known by set theorists, and which we call *Ramsey's Theorem*: for finite r and any finite partition of the r-element subsets of an infinite set, there is an infinite subset all of whose r-element subsets belong to the same cell of the

[16] I remember hearing in the 1970's that Ramsey's paper was one of the most cited works in the mathematical literature.

[17] See, for example, the second chapter of *Combinatorics: the Rota way* [Kung *et al.*, 2009] on matching theory, which discusses some of the mathematical connections between Kőnig's Theorem of 1916, Hall's Marriage Theorem, and more of Rado's work on finite systems than is included in this chapter. Kung, Yan and Rota connect matching theory to graph theory, matroids, linear algebra, submodular functions, and transversal theory. It is the latter which is of particular interest in this chapter which includes beautiful work by Milner and Shelah on infinite and uncountable transversal theory and the deep theorem of Galvin and Hajnal on almost disjoint transversals and cardinal arithmetic which was one of the precursors of Shelah's pcf theory.

[18] This overview is the only place where Hilbert's result is mentioned since it predates the 20th Century.

partition. He actually used a finite version, which we include as well, naming it *Finite Ramsey's Theorem.* Urysohn and Alexandroff proved that every regressive function on the countable ordinals is constant on an uncountable set, which can be regarded as partition theorem for the singletons of ω_1 for a special type of partition.

In the 1930's, Dushnik generalized the Alexandroff-Urysohn result to successor cardinals; Skolem rediscovered the Finite Ramsey's Theorem in [1933]; and Erdős and Szekeres rediscovered it on their way to proving a theorem in geometry. Rado introduced the concept of *k-regular* linear systems, and, in what we call Rado's Theorem for Linear Systems with Positive Integer Coefficients, he generalized Schur's Theorem and van der Waerden's Theorem.

In the 1940's, Rado generalized his analysis of k-regularity of linear systems to ones with coefficients from a ring of complex numbers. We discuss the generalizations of Ramsey's Theorem to the uncountable in terms of graphs, the pairs being represented as edges. Dushnik and Miller, with help from Erdős for singular cardinals, proved that for every graph on κ vertices, either there is independent subset of size κ (i.e. no pairs joined) or it has an infinite complete subgraph (i.e. every pair is joined in the subgraph). This theorem grew out of their work on partial orders. Erdős proved that if the edges of the complete graph on more than κ^κ many vertices are colored with κ many colors, then there is a complete subgraph on κ^+ many vertices all of whose edges have the same color. He used GCH to prove that any graph on κ^{++} many vertices either has complete subgraph of size κ^{++} or has an independent set (no edges) of size κ^+. Kurepa proved a related result in 1950 which is implicit in his [1939]: any graph on more than κ^λ vertices either has a complete subgraph on κ^+ vertices or an independent set of λ^+ on vertices. Neither [Erdős, 1942] nor [Kurepa, 1939] received much attention in the 1940's, due to disruptions of World War II and because of the likely related fact that the Erdős paper was published in an obscure Latin American journal. The connection between partition relations and ramification methods (tree methods) was made explicit in a paper of Erdős and Tarski [1943] but proofs did not follow until much later.

In the 1950's, Erdős and Rado initiated Ramsey theory for equivalence relations by identifying families of *canonical partitions* for the n-element subsets of $\mathbb{N}$. Erdős and Rado published a series of papers on what became the partition calculus, summarizing their findings in their joint paper [1956], the first systematic study of the area. This paper delineated the main lines of study, detailed a wide variety of partial results, and provided the important Positive Stepping Up Lemma for transferring results for r-element sets to ones for $(r+1)$-element sets. The modern Erdős-Rado Theorem follows from this lemma. Investigation of variations such as ordinal and polarized partition relations were initiated. Rado's Rectangle Refining Theorem is a canonical polarized partition relation. Specker proved interesting partition relations for ordinals of the form ω^n for finite $n \geq 2$. With multiple individuals contributing, the Regressive Function Theorem reached its standard form. In 1958, Mostowski and Tarski had a seminar that revisited the problems and results for inaccessible cardinals from [Erdős and Tarski, 1943]. By the end of the 1950's the peripatetic Hungarian Erdős and German born Rado, whose academic career was spent in England, were joined by Hungarian Hajnal and by London born Milner who spent most of his career in Calgary, Canada. The group formed the core of a powerful collaborative team that welcomed the participation of a wide group of mathematicians around the world in the study of what was sometimes called "Hungarian" set theory.

In the 1960's, the Canonization Lemma and the Negative Stepping Up Lemma for the partition calculus joined the Positive Stepping Up Lemma in the "Giant Triple Paper" of Erdős, Hajnal and Rado. Years in the making, this classic paper included a nearly complete description, under GCH, of partition relations for cardinals for pairs, and major results for larger r-tuples. It introduced square bracket partition relations for strong counter-examples to ordinary partition relations. The visit by Hajnal to Berkeley in 1964, the appearance of the triple paper in 1965, and the first international meeting after the development of forcing held in 1967 at the University of California, Los Angeles, introduced problems in the partition calculus to an audience with new techniques to apply. The Milner-Rado Paradoxical Decomposition showed that for any uncountable ordinal $\beta < \omega_2$ there is a partition of β into countably many pieces so that no piece has a set of ordinals of order type ${\omega_1}^{\omega}$.

The Halpern-Läuchli Theorem, a beautiful partition theorem about finite products of finitely branching trees, appeared in 1967, making a new connection between two themes, trees and Ramsey theory. The Hales-Jewett Theorem on variable words generalized van der Waerden's Theorem. Nash-Williams, in his quest to prove that the collection of infinite (height at most ω) trees have no infinite antichains and no infinite descending sequences under embeddability, proved a lemma which became known as the Nash-Williams Partition Theorem. Square bracket and polarized partition relations were proved for η, the order type of the rationals. At the end of the 1960's, theorems for Borel and analytic partitions of the infinite subsets of ω were proved, and relative to the existence of an inaccessible cardinal, a Ramsey Theorem for all infinite subsets of ω was shown consistent.

In the 1970's, further extensions, alternate proofs, and applications of theorems on partitions of infinite subsets of ω proliferated, with Ellentuck giving a particularly nice topological approach. Hindman's Finite Sums Theorem published in 1974 was the beginning of the exploration by Hindman of semi-group colorings and another example of a theorem that grew from a result in finite combinatorics, namely the Graham-Rothschild Parameter Sets Theorem. Techniques pioneered in the 1960's and early 1970's were applied to questions about partition relations, with singular cardinals and ordinal partition relations receiving increased attention. The Baumgartner-Hajnal Theorem is a spectacular case: for any coloring of the edges of a graph on ω_1 with red and blue and any countable ordinal α, there is a monochromatic subgraph (all edges red or all edges blue) whose set of vertices has order type α. Chang proved that ω^ω is a partition ordinal, i.e. every triangle-free graph has an independent subset of the full order type, and Hajnal proved that under the Continuum Hypothesis (CH), ${\omega_1}^2$ and $\omega_1 \cdot \omega$ are not. Shelah proved a canonization theorem useful for proving partition theorems for singular cardinals, and Milliken and Blass proved partition theorems for trees.

In the 1980's, Erdős, Hajnal, Máté and Rado published a compendium of results in the partition calculus, the culmination of decades of work. Strengthenings and generalizations of the Baumgartner-Hajnal Theorem for ω_1 were proved consistent, and progress was made on partition relations for triples with a meta-mathematical proof. The Baumgartner-Hajnal Theorem for order types was generalized to trees and partial orders in general. Consistency results were proved for cardinal and ordinal partition relations, including the determination whether certain ordinal products $\kappa \cdot \lambda$ for $\omega \leq \lambda < \kappa$ were or were not partition ordinals under a variety of assumptions. Todorcevic proved a startling square bracket partition result for the uncountable and introduced new technology whose ramifications are still unfolding, and proved a stepping up lemma for negative square bracket partition relations. Extensions of the results of Galvin, Prikry, Silver, Mathias, Ellentuck, Hindman and Graham and Rothschild were established. Laver published his proof that the perfect subtree version of the Halpern-Läuchli Theorem holds for the product of infinitely many trees building on work that dated back to 1969.

In the final decade of the 20th Century, Schipperus proved that there is an uncountable family of countable partition ordinals. There were new results for triples in both ordinary and square bracket partition relations. Moreover continuous colorings witnessing negative square bracket partition relations were found. The Erdős-Rado Theorem was strengthened to the Baumgartner-Hajnal-Todorcevic Theorem in both balanced and unbalanced formulations, and consistency results about possible extensions for the Dushnik-Miller Theorem (Komjáth) and the Erdős-Rado Theorem (Foreman and Hajnal) were explored. The inequality $\kappa^* = \aleph_\omega < 2^{\aleph_1}$ for the least cardinal κ^* satisfying a surprisingly simple looking polarized partition relation was show to be equiconsistent with the existence of a measurable cardinal.

The introduction concludes with the remark that the next section covers three decades. This decision has the pleasant effect that the 1930's occur in Section 3, the 1940's in Section 4, and so forth. Moreover, subsections, for example of the 1970's, are numbered 7.1, 7.2. For an introduction to the history of order prior to 1900, see [Alvarez, 1999, 185-188].

2 1900-1930: BEGINNINGS

Fundamental understanding of functions and of "nice" sets (Borel sets, Borel and Baire functions, nowhere dense sets, perfect sets, measurable sets and measurable functions) was developed at the end of the 19th Century and the beginning of the 20th Century as detailed in Kanamori's first chapter.

Georg Cantor had introduced the *order type* of a linearly ordered set A by abstraction to $\overline{A}$, which denotes what remains when the set is considered only with respect to the order of its elements.[19] Cantor used $\overline{\overline{A}}$ to denote what remains when even the order of the elements is no longer considered.[20] Cantor's choices of η to symbolize the order type of the rationals and ω to symbolize the order type of the non-negative integers were in place, as was the notion of everywhere dense set.

In the first decade of the 20th Century, David Hilbert [1900] enumerated his famous list of problems, including Cantor's Continuum Problem, which is of particular relevance for infinite combinatorics, and Ernst Zermelo [1904] published his Well-Ordering Theorem, the foundation for myriad recursive constructions in set theory. Henri Lebesgue published his classic work [1905] in the theory of definability, described by Moschovakis [2009, 1] as follows:

> It introduced and studied systematically several natural notions of definable functions and sets and it established the first important hierarchy theorems and structure results for collections of definable objects.

Edward Huntington[21] [1905d] developed postulates for the order type of the continuum and wrote two expository papers [1905a], [1905b] which were reprinted by Harvard University Press as a short text [1905c] with a new title, presumably for use by students and other researchers.[22] Veblen opened his review of this work with the following statement [1906, 302]:

> The *Annals of Mathematics* has for some time followed a plan of printing articles expository of subjects which are little known or not easily accessible in the English language. Reprints of these articles are then placed on sale with the double and laudable purpose of making the circulation of the article wider than it would otherwise be and of helping solve the difficult problem of financing a mathematical journal.

---

[19]With this notation, two ordered sets A and B have the same order type, $\overline{A} = \overline{B}$, if and only if there is an order-isomorphism between them. Thus $\overline{A}$ may be regarded as the class of linear orders which are order isomorphic to A. In modern notation, the order type of a set A is usually denoted by one of the following: $\mathrm{ot}(A)$, $\mathrm{otp}(A)$, $\mathrm{tp}(A)$. In the classification of orderings, it is useful to work up to equivalence under the relation of being order-isomorphic, i.e. having the same order type.

[20]With this notation, for two sets A and B, the relation $\overline{\overline{A}} = \overline{\overline{B}}$ holds if and only if there is a bijection between A and B. In modern notation we write $|A| = |B|$.

[21]Edward Vermilye Huntington (April 26, 1874 – November 25, 1952) earned an A.B. in 1895 and an A.M. in 1897, both from Harvard University. He was an instructor in mathematics at Williams College 1897–1899, then studied in Europe, receiving his doctorate in 1901 from the University of Strasbourg (at the time it was part of Germany). He spent the remainder of his academic career at Harvard University, returning in 1901 as an instructor of mathematics. He was promoted to assistant professor in 1905, to associate professor in 1915, and became a full professor of mechanics in 1919. He retired in 1941. He was elected to the American Academy of Arts and Sciences in 1913. To learn more about his life and his work in the foundations of mathematics, see Michael Scanlan's biography [1999].

[22]There was sufficient demand that Huntington updated his brief book on order in [1917].

Veblen praised the clear style and balance between generality and detail, enumerated the main contents (ordinal theory of integers, rational numbers, and the continuum, with an appendix on the transfinite numbers of Cantor), and mentioned the price, fifty cents. He noted that the text was intended for non-mathematical readers as well as for mathematicians, and pointed out that while it assumes little background knowledge, "it requires for complete comprehension a considerable maturity in abstract reasoning."

Strides were made in working out cardinal arithmetic. Felix Bernstein,[23] as a gymnasium student, participated in Cantor's Halle seminar, and his interaction with Cantor led to his proof of what we shall call the *Schröder-Bernstein Equivalence Theorem*:[24] If there is an injection from a set A into a set B and an injection from B into A, then there is a bijection between them. In his dissertation [1901][25] at Göttingen under Hilbert, Bernstein gave two proofs that the set of countable order types has power the continuum and gave what Plotkin [2005, 8] speculates is the first proof that the order type of the rationals is universal in the sense that all countable order types embed in it. Bernstein [1908] also proved that any Euclidean space may be decomposed into two disjoint totally imperfect sets (i.e. uncountable sets which have no perfect subsets).

An early result in the theory of partial orders introduced the study of chains in partial orders. Gerhard Hessenberg[26] [1906] considered subfamilies of power sets ordered by inclusion and studied general ways to construct well-ordered subsets of them. He stated and proved a fixed point theorem in this connection.[27]

[23]Felix Bernstein (February 24, 1878 – December 3, 1956) became an extraordinary professor at the University of Göttingen in 1911, received a medical exemption for military service during World War I, and was made head of the statistical branch of the Office of Rationing in Berlin. After the war he continued in government service, and in 1921 became both Commissioner of Finance and ordinary (full) professor at Göttingen, where he founded the Institute of Mathematical Statistics. He visited Harvard University in 1928, and after he lost his position at Göttingen in 1934 during the Nazi regime, he emigrated to the United States, where he taught at a number of different institutions until his retirement in 1948, when he returned to Göttingen where he was appointed professor emeritus. In the United States his research input was reduced, his own employment irregular, and he sought to help other scientists who had fled Europe in the face of Nazi policies find jobs. While his early work was in set theory, his later work was in applied mathematics, especially statistics, and included work in the mathematical theory of genetics. For more information, see the biography at http://www-history.mcs.st-andrews.ac.uk/Biographies/Bernstein_Felix.html by J. J. O'Connor and E. F. Robertson and the brief article by M. Frewer [1981] which includes a bibliography of Bernstein's work.

[24]The Schröder-Bernstein Equivalence Theorem is sometimes called the Cantor-Bernstein Theorem (e.g. see [Sierpiński, 1965, 34], [Jech, 2003, 28]). See [Kanamori, 2004, Section 4] to learn more about this theorem, and see Dedekind [1968, 447–449] for an early unpublished proof.

[25]A memorable mistake in Bernstein's thesis led to trouble for Julius (Gyula) Kőnig at the 1904 International Congress of Mathematicians as detailed in Kanamori's first chapter and Kojman's chapter.

[26]Gerhard Hessenberg (August 16, 1874 – November 16, 1925) received his 1899 doctorate from the University of Berlin where he studied under Hermann Schwarz and Lazarus Fuchs. He is known for a proof in projective geometry that Desargues' Theorem follows from Pascal's Theorem, and, in linear algebra, Hessenberg matrices are named for him.

[27]This fixed point theorem was rediscovered by Bourbaki [1939, 37] for ordered sets in general with essentially the same proof. For details see [Felscher, 1962].

Hessenberg defined the *natural sum* and *natural product* of ordinals which are sometimes called the Hessenberg sum and Hessenberg product. These operations are commutative and associative and the product distributes over the sum. So for example, the natural sum of 2 and ω is $\omega + 2$ which is larger than $2 + \omega = \omega$, and the natural product of 2 and ω is $\omega \cdot 2 > 2 \cdot \omega = \omega$. Hessenberg proved a division algorithm for arbitrary ordinals and proved that every ordinal can be expressed uniquely in Cantor normal form. The *Cantor normal form* of an ordinal α is its unique representation as $\alpha = \omega^{\beta_0} c_0 + \omega^{\beta_1} c_1 + \cdots + \omega^{\beta_{k-1}} c_{k-1}$ where $\beta_0 > \beta_1 > \cdots > \beta_{k-1} \geq 0$ and $k, c_0, \ldots, c_{k-1}$ are positive integers.

2.1 *Hausdorff*

Felix Hausdorff[28] analyzed order types in the early part of his mathematical career[29] and touched on many themes that will recur: generalizations of well-orderings, bases for order types, order types universal for nicely defined families, and the study of well-ordered sequences of real numbers.

Hausdorff's first paper [1901] on order types examines a generalization of well-orderings called *graded order types*, where a linear order is graded if no two of its initial segments are *similar*[30] to each other. He analyzes types through decompositions of ordered sets into initial, middle and final pieces, a theme he returns to in later papers. In particular, he generalizes the result he calls the Cantor-Bernstein Theorem, namely that the collection of countable order types has the cardinality of the continuum,[31] to show that the collection of all graded types of cardinality $\mathfrak{m}$ where $\mathfrak{m}^2 = \mathfrak{m}$ has cardinality $2^{\mathfrak{m}}$. On his way to the proof, Hausdorff notes that given a set M of cardinality $\mathfrak{m}$, the power $2^{\mathfrak{m}}$ is the cardinality of the set of all two-valued functions (Cantor's "coverings") with domain M (Hausdorff uses $+1$ and -1) and is also the cardinality of the set of all subsets of M, via the correspondence matching a function f with the subset of points where f takes value $+1$.

[28]Felix Hausdorff (November 8, 1868 – January 26, 1942) completed his 1891 doctorate in applications of mathematics to astronomy at Leipzig University, and his Habilitation thesis was in the same area. He was also a writer with interests and a wide range of friends in art, literature, and philosophy, and had plays performed and published under the pseudonym Paul Mongré. Hausdorff began working in pure mathematics around the turn of the century, became an extraordinary professor at Leipzig in 1902, and moved to Bonn in 1910 where the mathematical environment was richer and he became friends with Eduard Study. Hausdorff became an ordinary professor in 1913 at Greifswald where Study had formerly been an ordinary professor. His final professional move was back to Bonn in 1921 where he worked until 1935 when he was forced to retire by the Nazi regime. With his suicide in 1942 in the face of a forced move to Endenich, the world lost a remarkable man.

[29]Hausdorff's beautiful work on order types has been made more accessible by a careful translation by Jacob Plotkin [2005].

[30]Two ordered sets are *similar* if there is an order-isomorphism from one to the other, i.e. they have the same order type.

[31]Cantor provided a sufficiently large collection of examples, and Bernstein, in his thesis, showed there are at most continuum many, which by the Schröder-Bernstein Equivalence Theorem gives the desired equality.

In a five-part series of articles sharing the title *Investigations into order types* [1906], [1907], Hausdorff begins a systematic study of ordered sets, moving beyond the study of ordinal numbers, cardinal numbers, their powers, and the order types of subsets of the reals. In the first part, *The powers of order types*, he develops a formal notion of power formation with a base μ and exponent α via sequences indexed by α of elements of μ ("coverings"), where α and μ are arbitrary order types. He studies subsets of these powers under various orderings, especially in the case that the exponent is a well-ordered set. One special case is ordering by first difference, i.e. lexicographic ordering. To ensure a linear order, Hausdorff fixes a principal element m in the base M (of type μ) and looks at the collection of sequences from the representative set A of type α that differ only finitely from m to get powers of the first class. He observes that if the exponent set A has no last element, the resulting type is everywhere dense, and that if the base M and exponent A are ordinal numbers and the principal element m is the least element of the base, then the result is the same as Cantor's power. He generalizes to higher order powers, by replacing the condition that the sequences have all but finitely many places taking value the principal element m to the requirement that the sequences have all but a well-ordered subset of power $\leq \aleph_v$ for some fixed ν that specifies how much higher the power is; ordering by first differences is a linear order on such collections. He matches the collection of subsets of a well-ordered set A with their characteristic functions in a suitably high power with base 2 and exponent A which is linearly ordered by first differences. He concludes the first part of the series with transfinite sums and products where not all the factors are the same.

In the second part, *The higher continua* [1906], he focuses on ordered sets of fundamental sequences, i.e. sequences with domain an initial ordinal number, ordered by first difference. He analyzes these powers by looking at properties of the base, e.g. being dense, and identifies a variety of types. He introduces the critical notions of *cofinal*[32] and *coinitial*, characteristics of segments that became fundamental to the classification scheme he eventually developed.[33] Hausdorff proved that a type cannot be cofinal with two different initial ordinals. Another concept introduced was of a type being *homogeneous* if all its segments are similar.

In the third part of the series, *Homogeneous types of the second infinite cardinality* [1906], Hausdorff sets out a classification scheme for homogeneous types of cardinality $\aleph_1$ based on the cofinality and coinitiality of the segments, types of gaps, and occurrence or non-occurrence of uncountable well-ordered or converse-well-ordered subsequences.

In the fourth part of the series, *Homogeneous types of the cardinality of the continuum* [1907], Hausdorff continues his construction of examples fitting his

[32]Hausdorff used *confinal* for what in English is *cofinal*.

[33]In a note at the end of his introduction to Hausdorff's paper [1906], Plotkin (cf. [Hausdorff, 2005, 42]) pointed out that Hessenberg [1906] had introduced a related notion when he called a set M a *Kern* of an ordinal μ if the set of ordinals less than μ was the same as the set of ordinals for which there is an ordinal in M at least as large. Every Kern of μ is cofinal in μ.

classification scheme with the constraint that they have cardinality at most the continuum, but in the final part, *Pantachie types*, this constraint is lifted. In Section 1 of this last part, he looks at monotonically increasing functions on ω which go to infinity, and shows that the cardinality of this collection is the cardinality of the continuum ([Hausdorff, 1907, 111] or [Hausdorff, 2005, 133]). Hausdorff investigates *final rank ordering* ([Hausdorff, 2005, 137]) of increasing sequences, i.e. ordering under eventual domination. He introduces generalizations of the order type of the rationals, denoted η, and uses the corresponding upper case Greek letter H for them. In Section 3 [Hausdorff, 2005, 145], he focuses on H-types, specifying that an η_1-type is unbounded, everywhere dense, neither cofinal with ω nor coinitial with ω^*; no fundamental sequence in it has a limit; and it contains no (ω, ω^*)-gaps ([Hausdorff, 1909, 323]).[34]

He proves the universality of an η_1-type for linearly ordered types of cardinality at most $\aleph_1$ and the uniqueness of an η_1-type of cardinality $\aleph_1$, if it exists. He generalizes the notion to η_ν-types for finite ν.[35] In Section 6, he introduces the scale problem and uses the Continuum Hypothesis (CH) to construct a scale [Hausdorff, 2005, 167]: a sequence of ω_1 increasing functions in ${}^\omega\omega$ which is increasing in the final rank order and has the property that each function in ${}^\omega\omega$ is less than some member of the sequence.

Hausdorff in his *Fundamentals of a theory of ordered sets* [1908] streamlined and extended his earlier work on order. He worked out the basic properties of cofinality in [1908, 440–444], introduced the distinction between regular and singular cardinals, and included a proof in Theorem IV that each set without a last element is cofinal with one and only one regular initial number. He described dense and scattered linear orderings,[36] gave $\{\omega, \omega^*\}$ as a basis[37] for the denumerable linear orders, $\{1, \omega, \omega^*, \eta\}$ as a generating set for all countable linear orders, and discussed type rings and the preservation of properties from the base of such a ring to all its elements (see [Hausdorff, 1908, 454–456], [Hausdorff, 2005, 213–215]). Here is a quote in which Hausdorff described linearly ordered sets of cardinality less than a fixed $\aleph_\alpha$ [Hausdorff, 1908, 458]:

> Satz XII. Jede Menge ist entweder selbst zerstreut order eine Summe zerstreuter Mengen mit dichtem Erzeuger. Die zerstreuten Mengen

[34] For an order type λ, λ^* denotes its converse (reversed) order type. A (κ, λ^*)-*gap* in an ordered set $(P, <)$ is a pair $A, B \subseteq P$ such that A is linearly ordered with order type κ, B is linearly order with order type λ^*, for all $a \in A$ and $b \in B$, $a < b$; and there is no $x \in P$ with $a < x < b$ for all $a \in A$, $b \in B$. [Scheepers, 1993] is a very readable survey on gaps in ${}^\omega\omega$ through the early 1990's which is described as "an invitation to study gaps" (see page 441) and has an extensive bibliography. See also the memoir by Farah [2000].

[35] Chang and Keisler [1973, 522], in notes for Chapter 5, date the notions of α-saturated and saturated models back to the η_α-sets of Hausdorff, citing the 1914 text. Plotkin (cf. [Hausdorff, 2005, 109]) quotes from an email of Michael Morley in which he indicated that 1954 was about the time when he set out to generalize the "eta set construction" to other algebraic structures.

[36] A *scattered linear order* is one that does not embed the order type of the rationals.

[37] A basis B for a class $\mathcal{C}$ of linear orders is a set of linear orders in the class which has the property that for any element $(L, \lessdot)$ of $\mathcal{C}$ there is some $(K, <)$ in B which is order-embeddable into $(L, \lessdot)$.

> der Mätigkeit $< \aleph_\alpha$) ($\aleph_\alpha$ regulär) bilden einen Ring, dessen Basis aus allen regulären Anfangszahlen $< \omega_\alpha$) (inklusive der Zahl 1) and deren Inversen besteht.[38]

The "dichtem Erzeuger" [dense generator] may be thought of as an *index set.* The ring in the above quote is a *type ring*, which is a system of types closed under sum of two types and by sums of types in the ring indexed by a type in the ring. On page 456, Hausdorff listed several collections that form a ring, including unbounded types, types with a last element, and types that are cofinal with a fixed type. Using transfinite recursion, one can transform the type ring of scattered sets of cardinality $< \aleph_\alpha$ into a hierarchy of sets that has become known as the *Hausdorff hierarchy.*

In Theorem XIV [1908, 472], Hausdorff proved that any sequence of length ω_ν under the lexicographic order on the ω_ν-sequences of elements of σ for an ordered set σ will have a cofinal subsequence of one of two types: one with initial segments that form a strictly increasing chain or one where all share a common initial segment and differ on the first place not in this common initial segment.

In a substantial part of the paper, Hausdorff detailed his classification of dense types from [1907] and provided existence proofs. He used collections of functions defined on well-ordered domains in interesting ways. Specifically, in Theorem XVIII ([Hausdorff, 1908, 488], [Hausdorff, 2005, 243]), he identified a type universal for ordered sets of cardinality at most $\aleph_\pi$ and in Theorem XIX proved that there is at most one such type of cardinality $\aleph_\pi$. He also showed that there are at least $\aleph_1$ distinct types of power at most the continuum ([1908, 505, footnote]).

Hausdorff [1909] continued his exploration of powers of the alephs, i.e. collections of functions defined on a well-ordered set partially ordered by final rank order, and maximal well-ordered subsets of this partial order. He constructs an (ω_1, ω_1^*)-gap inside ${}^\omega\omega$ ordered by eventual difference using the Axiom of Choice but not the Continuum Hypothesis. He also shows that under eventual domination, any subset of ${}^\omega\omega$ has a maximal linearly ordered set.[39]

In the summer of 1910, Hausdorff moved to the University of Bonn, where he gave a course on set theory. He gave a revised and expanded course in the summer of 1912 and at the same time began work on his set theory book *Grundzüge der Mengenlehre* [1914], [2002] which was completed in 1913, and appeared in April 1914, shortly before the outbreak of World War I.[40] In 1920, Henry Blumberg [1920, 116] gave the book a glowing review in the *Bulletin of the American Mathematical Society*:

[38]Theorem XII. Every set is either scattered or a sum of scattered sets with a dense generator. The scattered sets of cardinality $< \aleph_\alpha$ ($\aleph_\alpha$ regular) form a ring whose basis consists of all regular initial numbers $< \omega_\alpha$ (including the number 1) and their inverses (translation by Plotkin (cf. [Hausdorff, 2005, 216])).

[39]See Sections 2 and 5 of Steprāns' chapter on the continuum for more on Hausdorff's work on gaps in maximal linear orderings of sequences of integers under eventual domination.

[40]Purkert's *An introduction to Hausdorff* (downloaded from http://www.hausdorff-edition.de/wp/?page_id=20 on February 5, 2011) for the Hausdorff Edition included these details on the timing of Hausdorff's work and a broad overview of his life and mathematical work.

> Its most striking feature is that it is the work of art of a master. No one thoroughly acquainted with its contents could fail to withhold admiration for the happy choice and arrangement of subject matter, the careful diction, the smooth, vigorous and concise literary style, and the adaptable notation; above all things, however, for the highly pleasing unifications and generalizations and the harmonious weaving of numerous original results into the texture of the whole.

In *Grundzüge der Mengenlehre*, there are chapters on basic set operations (sum or union, section or intersection, difference used to get complements), functions and their usage in products and powers, cardinal numbers (Zermelo's Well-Ordering Theorem, non-denumerability of the continuum, existence of a one-to-one correspondence between the real line and the Euclidean plane), ordered sets and their types, well-ordered sets and the ordinal numbers (including a proof Blumberg calls elegant of the fact that $\aleph_\alpha \aleph_\alpha = \aleph_\alpha$), Hausdorff's own work on partially ordered sets, topological spaces (axiomatically described in terms of neighborhoods,[41] including the requirement that the space be what is now called Hausdorff), special topological spaces, metric spaces, and Lebesgue measure.

This text and its subsequent editions were influential in shaping the direction of set theory in the coming decades. Plotkin (cf. [Hausdorff, 2005, 111]) asserts that the "1914 text made *back-and-forth* part of the mathematical mainstream."[42] Hausdorff included his η_α sets, patterned on the rationals. In Chapter VI, §1, Hausdorff considered partially ordered sets and proved the existence of maximal linearly ordered subsets in them. Many of the results on order were omitted from this important book, including the work on gaps and the type ring with its sums over arbitrary elements of the class. As a consequence, a number of Hausdorff's pioneering results on order were rediscovered later in the century.

2.2 1910–1920: Emerging schools

In the decade 1910–1920, Poland and Russia developed schools active in set theory and function theory; R. L. Moore initiated the study of analysis-situ or point-set topology that led to the development of the Texas school of topology; and Schur had an early result in what is now known as Ramsey Theory. The work of Baire became more widely known through the book [1916] by de la Vallée Poussin.

Wacław Sierpiński[43] published *Zarys teorii mnogości* (*An Outline of Set The-*

[41]Gregory H. Moore [2008a] observed that Hausdorff's neighborhoods are what today would be called a neighborhood base for the topology. Moore's article on the emergence of open and closed sets situates Hausdorff's "neighborhood spaces" in historical context.

[42]See [Plotkin, 1993] for more on the history of the back-and-forth construction, and see [1907] for Hausdorff's earlier usage.

[43]Wacław Franciszek Sierpiński (March 14, 1882 – October 21, 1969) graduated from the University of Warsaw in 1903 and became a school teacher in Warsaw. He earned his doctorate in Kraków with Stanisław Zaremba at the Jagiellonian University in 1908 and was appointed to the University of Lwów (now Lviv, Ukraine) in 1910, and at that point his interests turned to set theory according to the entry on him in [Gowers *et al.*, 2008]. After a period in Russia during

ory) [1912], which Kuratowski [1972, 2] described as "one of the first synthetic formulations of this theory in the world," and was based on systematic lectures on set theory by Sierpiński in 1909 at University of Lwów.

Investigations of the Axiom of Choice and the Continuum Problem continued. An especially elegant application of the Continuum Hypothesis from this period is the construction from a well-ordering $<_\Omega$ of the real numbers of the subset $S \subseteq \mathbb{R}^2$ of the plane defined by $S = \{(r, s) : r <_\Omega s\}$; it has the property that every vertical section is co-countable and every horizontal section is countable. Sierpiński[44] [1919] published a proof that the existence of a set with these properties is equivalent to the Continuum Hypothesis.[45]

Results of Alexandroff, Lusin and Suslin put the emerging Moscow school[46] founded by Dimitri Egorov[47] and Nikolai Nikolaevich Lusin[48] on the map. Egorov [1911] proved that if a sequence of measurable functions converges pointwise in an interval except for a set of measure zero, then it converges uniformly on the interval except for a set of arbitrarily small measure. This result became known as Egorov's Theorem and then as the Severini-Egorov Theorem, when it was realized

World War I, he returned in 1918 to Lwów, and in 1919 became a professor at the University of Warsaw, where he spent the remainder of his professional life. For further background, see http://www.hpm-americas.org/wp-content/uploads/2010/04/PSMatHPMcorrected_2.pdf, an extended version of the paper by Roman Sznajder entitled *90th anniversary of emergence of the Polish School of Mathematics; Polish mathematics between the world wars* presented at the 2010 East Coast meeting of the History and Pedagogy of Mathematics: Americas Section.

[44]Kuratowski [1972, 2], in his obituary of Sierpiński, wrote "the outbreak of the war in 1914, interrupted Professor Sierpiński's didactic work. He was interned by the Tsarist authorities, first in Vyatka, and later in Moscow. There, the excellent Russian mathematicians Egorov and Luzin extended a most cordial reception to him and created very convenient conditions for his scientific work. This period brought important joint works by professors Sierpiński and Luzin which laid the foundations for their long co-operation in the domain of analytic and projective sets as well as the theory of real functions."

[45]Erdős [1981, 41] called this result a "very beautiful theorem" and commented that it was "very startling" when he first saw it. In Section 2 of the chapter on the continuum, Steprāns briefly discusses this result.

[46]See [Shields, 1987] and [Paul, 1997] for more information about the Moscow school.

[47]Dimitri Fedorovich Egorov (December 22, 1869 – September 10, 1931) became a Privat Docent in 1893, spent 1902-1903 abroad on paid leave visiting Berlin, Paris and Göttingen, and became a professor at Moscow University in 1903. Alan Shields [1987, 1] speculated that Egorov was one of the first foreign mathematicians to embrace the theory of Lebesgue. Egorov was director of the Institute of Mathematics and Mechanics at Moscow University from 1921 to 1930.

[48]Nikolai Nikolaevich Lusin (December 9, 1883 – January 1950) was a student of Egorov, entering Moscow University in 1901 when his family moved to Moscow from Tomsk. He started study under Bugaev, but it was Egorov, with whom he shared a religious outlook, who invited him to his home and gave him challenging problems that stimulated his creativity. Another influence on Lusin was P. A. Florensky, a fellow student in mathematics at Moscow University, who shifted his studies to theology and became a priest and spiritual advisor to Lusin. Lusin completed his undergraduate studies in the fall of 1905 and went to Paris for further education. He underwent a spiritual crisis in 1905–1906, studied medicine and theology as well as mathematics, and eventually focused on mathematics. He worked on a master's thesis with Egorov, but civil war disrupted his studies, and he spent 1910 to 1914 in Göttingen, continuing to correspond with Egorov and talking with Edmund Landau. He returned to Moscow and received his doctorate in 1915.

that Carlo Severini had independently and earlier published a proof [1910] of this result in a paper on sequences of orthogonal functions.

Lusin [1914] used the Continuum Hypothesis to construct what is now known as a *Lusin set*: an uncountable set of reals whose intersection with each meager set is countable, or equivalently, an uncountable set such that every uncountable subset is non-meager.

Lusin also worked on the Continuum Problem and on ordinals.[49] In a review of the book *Naming Infinity* [2009][50] Cooke [2010, 63] gave the following description of some Lusin's work:

> His notes, which are now in the archives of the Russian Academy of Sciences, are full of attempts to get a complete description of all the ordinals at once. In one valiant attempt, for example, he tried to imitate the method Dedekind had used to define real numbers as equivalence classes of cuts in the rational numbers. He took the class of subsets of the rational numbers between 0 and 1 that are well-ordered in their natural ordering, and introduced the natural order-equivalence relation on this class. Could one then *define* a countable ordinal number to be one of these equivalence classes? Certainly any such equivalence class will be order-isomorphic to a countable ordinal. But how could it be known that every countable ordinal is order-isomorphic to one of these equivalence classes? Luzin struggled with this problem of *naming* the countable ordinals for decades. One can almost hear him sigh as he wrote in one plaintive note that he left behind, "How many times must I write out the set of all ordinals numbers of types I and II?"

Pavel Sergeyevich Alexandroff[51] recalled reading Cantor in his first year (1913) at Moscow University [1979, 281]:

> In the comfortable round reading room of the university library I found Cantor's memoirs on set theory and I began to read them with delight. One of the last mathematics books Eiges had given to me was

[49]See Roger L. Cooke's survey [1993] of unpublished work and miscellaneous items in the Lusin archive.

[50]Graham and Kantor [2006], [2009] proposed that Lusin's religious practice of name-worshipping inspired his mathematics.

[51]Pavel Sergeyevich Alexandroff (November 1896 – May 1982) who grew up in a cultured family was a student of both Egorov and Lusin at Moscow University. He traveled to Europe in the summers from 1923 through 1932, visiting Göttingen, Bonn, Paris and Holland. He obtained his doctorate in 1927 and went on to a successful career in topology at Moscow University, had several students, including Tychonoff, and was elected to membership in a number of academies, including the Russian Academy of Sciences, the Göttingen Academy of Sciences, the National Academy of Sciences of the United States, the London Mathematical Society, and the Polish Academy of Sciences. A. N. Arhangels'kii and A. N. Dranishnikov [1997]: "Alexandroff placed a value on a result not according to technical difficulty of its proof, but, first of all, according to its position in a mathematical theory, according to the new harmony and beauty which it brings into the theory."

> Kowalewsky's course in analysis, and it had acquainted me, to some extent, with the rudiments of set-theoretical thinking. But when I began to read Cantor in the original and learned what transfinite numbers are, a new world opened up before me, just as had once happened when I first learned about non-Euclidean geometry, and I was in a state of excitement. I was in the same state of excitement when in Baire's book, which was given to me by V. V. Stepanov, I became acquainted with the Cantor perfect set, which I immediately saw and still see to this day as one of the greatest wonders (and I mean a wonder, nothing less) discovered by the human mind.[52]

Both Hausdorff [1916] and Alexandroff [1916] proved that every uncountable Borel set has a perfect subset. Thereafter, Lusin encouraged Alexandroff to work on proving that the collection of sets which included the Borel sets and whose uncountable members he showed all included perfect sets are exactly the Borel sets and set Mikhail Yakovlevich Suslin[53] to study the work of Henri Lebesgue [1905]. This assignment by Lusin is described by Alexandroff in his chatty autobiography [1979, 375].

In summer 1916, Suslin found an error in Lebesgue's work, and constructed a counter-example. He went on to construct an A-set which is not a B-set and the following year, he [1917] published his work on the these sets, which have variously been called A-sets, analytic sets, and Suslin sets. In this work he introduced operation $\mathcal{A}$, also known as the *Suslin operation*,[54] extracting a key combinatorial aspect of Alexandroff's proof. The collection of analytic sets is the closure of the closed sets under the Suslin operation.

Lusin introduced an alternate approach to analytic sets called the method of sieves. In a modern formulation by Moschovakis [2009, 203], a *sieve* is a map $r \mapsto F_r$ which assigns to each rational number r a subset F_r of a space X. The

[52]Kowalewsky possibly refers to the German mathematician Gerhard Kowalewski rather than Sonia Kovalevskaya.

[53]In his short biography of Suslin, Igoshin [1996] wrote that Suslin (November 1894 – October 1919), the only child of landowning peasants in the Saratov province of Russia, was recognized as special by his primary school teacher who persuaded some of the more wealthy people in the village to send him to a nearby city for grammar school. By tutoring children of the wealthy, Suslin eventually was able to pay for his studies himself, and finished in spring 1913. He then went to Moscow University where he attended Egorov's seminar, and in 1914, Lusin's seminar, and was awarded a Diploma of the first degree in 1917. He was given permission to stay an additional two years to prepare for a professorship. During the Russian Civil War, after a period of failing health, he returned to his home village, where he died in 1919 of typhus less than a month prior to his twenty-fifth birthday. His brief mathematical career included a stint of teaching at the Ivanovo-Voznesensk Polytechnic Institute as an Extraordinary Professor starting in 1918. His students appreciated him enough to petition for his return, when he asked for a leave of absence because of poor health.

[54]A modern description of a *Suslin scheme* is a family $\mathcal{P} = \{P_s : s \in {}^{\omega>}\omega\}$ of subsets of a set X indexed by finite sequences of non-negative integers. The *Suslin operation* applied to such a scheme produces the set $\mathcal{A}\mathcal{P} = \bigcup_{x \in {}^{\omega}\omega} \bigcap_{n \in \omega} P_{x \restriction n}$.

set *sifted* by the sieve is defined by

$$x \in \text{Sieve}_r\, F_r \quad \longleftrightarrow \quad \{r : x \in F_r \text{ is not well ordered}\},$$

where the standard ordering on the rationals is used. The Suslin operation $\mathcal{A}$ and the method of sieves appear in the study of partitions later in the century. We close this brief aside on analytic sets by noting that the paper by Suslin [1917] discussed above, one by Lusin [1917] which included work of Suslin, and a pair of papers by Lusin and Sierpiński [1918] and [1923] were important early papers in the theory of analytic sets.

Egorov and Lusin gathered an increasingly active group of mathematicians around them, and Lusin became a Professor of Pure Mathematics at Moscow University in 1917 prior to the Russian Civil War and the ensuing economic disruption. Lusin and Suslin, among others, moved to Ivanovo-Voznesensk Polytechnic Institute. Lusin returned to Moscow in 1920 and worked together with Egorov to revitalize the Moscow school, and the group of students and collaborators around Lusin came to be called the *Lusitania*.

In [1916] R. L. Moore[55] developed the postulates for what he [1916, 132] called "plane analysis situs or what may be roughly termed the non-metrical part of plane point-set theory". It marked his entry in the emerging field of point set topology. He used *point* and *region* as his basic (undefined) concepts; the concept of region evolved into open sets. He wrote the paper while at the University of Pennsylvania and only in 1920 moved back to the University of Texas at Austin, where he did his undergraduate work. David Zitarelli [2004, 471] has asserted that this paper "firmly solidified his reputation as a first-rate researcher and later formed the basis for the Moore Method." Sometimes called the father of the school of Texas topologists, he had 50 graduate students, the first completing his degree in 1916 and the last in 1969; on September 10, 2010, the Mathematics Genealogy Project site listed 2,661 descendants. In his Math Review MR2103548 (2006a:01005) of Parker's biography [2005] of Moore, Ivan Reilly highlights an interesting measure of the success for graduate students of Moore's teaching methods in which he gave his students axioms (starting with those from his [1916]) and asked them to prove theorems from them in an orchestrated way:

> One measure of this success is described in the Albert Report "A survey of research potential and training in the mathematical sciences",

[55]Robert Lee Moore (November 14, 1882 – October 4, 1974), a student of Oswald Veblen and E. H. Moore, earned his degree at the University of Chicago in 1905. He was a major figure in the mathematical life of the USA in the middle of the 20th Century, noted for his mathematics, his teaching style, and his service to the professional community (he was president of the American Mathematical Society in 1938). From the many resources available about his work and life, one learns that he was a tall Texan, opinionated, racist (see http://www.math.buffalo.edu/mad/special/RLMoore-racist-math.html, accessed September 10, 2010, especially quotes by Vivienne Malone), and a charismatic and inspiring teacher. One of many resources on him is a book by Parker [2005] published by the Mathematical Association of America, an organization that Moore once called the "Salvation Army of Mathematics."

> produced in 1957 for the National Science Foundation (page 238). Data was gathered from a very large number of people who received their Ph.D.s in mathematics between 1915 and 1954. This survey collated the work of Ph.D. students after graduation and categorized them by the amount of published mathematical research. Of the top 15 per cent, 25 per cent were graduates from UT (made up almost entirely of Moore students), 20 per cent from Princeton, 16 per cent from Harvard, and 8 per cent from Chicago.

Issai Schur[56] [1917] proved an early result in Ramsey theory.

> *Schur's Theorem:* When one divides the numbers $1, 2, \ldots, S(k)$ into k (color) classes for sufficiently large $S(k)$, there is always in one of the classes a triple of the form $a, b, a+b$, i.e. a monochromatic solution to the homogeneous linear system $x + y = z$.

Note that the triple $a, b, a+b$ can be thought of as a rather short monochromatic arithmetic progression as well as a monochromatic solution to a linear system. Schur used this result to prove that, for all m, if p is a sufficiently large prime, then the equation $x^n + y^n = z^n$ has a non-zero solution in the integers if you compute all the values modulo p.

2.3 1920–1930: Early structural results

The 1920's saw the flourishing of the Warsaw and Lwów schools and the founding of *Fundamenta Mathematicae*, whose first volume included Suslin's Problem. The decade saw the publication of Sierpiński's book on transfinite numbers and the second edition of Hausdorff's seminal text. New concepts for the decade included von Neumann ordinals and almost disjoint families. Two results related to infinite trees appeared, König's Lemma (it was expressed graph theoretically) and Moore's road space, a topological example with a tree implicit in its construction. Alexandroff and Urysohn proved their Regressive Function Theorem and van der Waerden proved his theorem on arithmetic progressions.

The Warsaw group in set theory, topology, and their applications prospered under the direction of Zygmunt Janiszewski, Stefan Mazurkiewicz and Wacław Sierpiński, founders of *Fundamenta Mathematicae*[57] (see [Kuratowski, 1980]).

A concept introduced by Sierpiński [1928b], that of an *almost disjoint* family[58] would become important in the study of partitions of infinite subsets of ω. Sierpiński proved that for any infinite set of cardinality κ, there is an almost

[56] Issai Schur (January 10, 1875 – January 10, 1941) was a student of Frobenius. He worked at the University of Berlin until 1935, and a few years later moved to Palestine. For more details on his life and mathematics see the biography by J. J. O'Connor and E. F. Robertson at http://www-history.mcs.st-andrews.ac.uk/Biographies/Schur.html.

[57] See [Kuzawa, 1970] on the founding and significance of *Fundamenta Mathematicae*.

[58] Sierpiński called a family $\mathcal{F}$ *almost disjoint* if for every two distinct sets $M, N \in \mathcal{F}$, the cardinality of the intersection $M \cap N$ is less than $|M|$ and less than $|N|$.

disjoint family of its subsets of cardinality 2^λ where λ is the least cardinal with $2^\lambda > \kappa$. Alfred Tarski [1928] refined Sierpiński's notion by defining the *degree of disjunction* $\delta(\mathcal{F})$ of an almost disjoint family $\mathcal{F}$ to be the minimum cardinal δ such that for all distinct $M, N \in \mathcal{F}$, $|M \cap N| < \delta$. Tarski [1928], [1929] used the GCH to prove various theorems on the existence of almost disjoint families of large size with specified cardinality of the subsets involved and specified maximal degree of disjointness.

The first volume of *Fundamenta Mathematicae* included the posthumous publication of Suslin's Problem [Suslin, 1920] on linear orders which, like Cantor's Continuum Problem, became a motivating force for a significant body of work in the 20th Century (Problème 3, page 223):

> Un ensemble ordonné (linéairement) sans sauts ni lacunes et tel que tout ensemble de ses intervalles (contenant plus qu'un élément) n'empiétant pas les uns sur les autres est au plus dénombrable, est-il nécessairement un continu linéaire (ordinare)?[59]

Here is a modern take on the problem. A linear order satisfies the *countable chain condition* (alternatively, it has countable *cellularity*) if every collection of pairwise disjoint non-empty open intervals is countable.[60] Suslin asked if every linearly ordered set without endpoints whose order is (1) dense, (2) *complete* in the sense that every non-empty bounded subset has a greatest lower bound and a least upper bound, and (3) has the countable chain condition (i.e. countable cellularity) is isomorphic to the real line. If the requirement that the linear order have the countable chain condition is replaced by the requirement of having a countable dense set (being *separable*), then such a linear order is isomorphic to the real line. Call a linearly ordered set without endpoints satisfying the three conditions but not isomorphic to the real line a *Suslin continuum.* Call a dense linear order without endpoints which satisfies the countable chain condition but is not separable a *Suslin line.* Note that the existence of a Suslin line implies the existence of a Suslin continuum by passage to the completion of the line. *Suslin's Hypothesis* (SH) asserts the non-existence of Suslin lines.

Sierpiński published his book *Leçons sur les Nombres Transfinite* [1928a]. Émile Borel wrote the preface, lauding the work of the author, welcoming the inclusion of Lusin's work, and calling attention to the modern material featured in *Fundamenta Mathematicae.* Borel remarked (cf. [Sierpiński, 1928a, VI]): "Ce qui le distingue surtout, c'est le fait que M. Sierpinski croit effectivement à la réalité de *tous* les nombres transfinis, et admet sans restriction les raisonnements tels que celui par

[59] Must a (linearly) ordered set which has neither jumps nor gaps and in which every set of disjoint [non-encroaching] intervals (all containing more than one element) is at most denumerable, must it, of necessity, be a (ordinary) linear continuum [copy of the real line]?

[60] Recall that the countable chain condition for a partial order P is the condition that every set of pairwise incompatible elements is countable. The countable chain condition in the restatement of Suslin's Problem is applied to the partial order of open intervals of order by inclusion.

lequel M. Zermelo a $\ll$démontré$\gg$ que le continu peut être bien ordonné."[61] In Chapter VII Sierpiński discussed order types and in the very last section, §64, he listed three properties (1° no end points, 2° continuous,[62] 3° has a countable dense set) that characterize the order type λ of the continuum. At the end of the section, he noted that ordered sets which have a countable dense subset also have the property 4° that all sets of non-empty pairwise disjoint intervals are at most denumerable, and concluded [1928a, 153]:

> Or, on ne sait pas si tout ensemble ordonné, jouissant des propriétés 1°, 2° and 4°, est nécessairement du type λ, et ce problème [dû à M. Souslin (^{2})] semble très difficile.[63]

The Lwów school of mathematics grew around Hugo Steinhaus[64] and his brilliant student, Stefan Banach.[65] At the end of the decade (in 1929 to be precise), Stefan Banach and Hugo Steinhaus founded the journal *Studia Mathematicae* in Lwów accepting papers in international languages, playing a role for the Lwów school comparable to that of *Fundamenta Mathematicae* for the Warsaw school. Among those publishing in the first year were Banach, Stanisław Mazur, Władysław Orlicz, Juliusz Schauder and Steinhaus. Other notable members of the Lwów school were Stanisław Ulam, Mark Kac, Antoni Łomnicki, Herman Auerbach and Stanisław Ruziewicz.

Banach and Kazimierz Kuratowski[66] looked at generalizations of Lebesgue's Measure Problem, and it was a question of Banach in their [1928] that led Ulam[67]

[61] What most distinguishes it [the book] is the fact that Mr. Sierpinski actually believes in the reality of all transfinite numbers, and allows, without restriction, reasoning such as that by which Mr. Zermelo has shown that the continuum can be well ordered.

[62] We would say complete where Sierpiński said continuous.

[63] However, we do not know if any ordered set, having the properties 1°, 2° and 4°, is necessarily of type λ, and this problem [due to Mr. Souslin (^{2})] seems very difficult. The reference (^{2}) is to the statement of Suslin's problem [1920].

[64] Steinhaus was a student of Hilbert who earned his degree in 1911 in Göttingen.

[65] The book [Kuratowski, 1980, 43] includes a quote from Stanisław Mazur to the effect that Banach's thesis, presented in June 1922 for the Ph.D. at University of Lwów and published in *Fundamenta Mathematicae* [Banach, 1922], was the beginning of functional analysis as a field. Kuratowski's book gives delightful descriptions of the mathematicians of the day and some of their ways of working. To learn more about the intellectual atmosphere of the time, see the wide variety of problems recorded in a notebook in the Scottish cafe in Lwów and recollections recorded from a conference on the this notebook [Mauldin, 1981].

[66] Kazimierz Kuratowski (February 2, 1896 – June 18, 1980) studied for one year at the University of Glasgow in Scotland before the outbreak of World War I in August 1914. He then moved to Warsaw University in 1915 and received his doctorate in 1921. On the Mathematics Genealogy Project site, Mazurkievic and Janiszewski are listed as his advisors; note that Janiszewski died before the thesis was awarded. Kuratowski taught at Warsaw University 1925–1927 and was then appointed to a professorship at Lwów University. In 1934, after the closure of the mathematics department in Lwów, he became a professor at Warsaw University where he remained until 1952, when he became a member of the Polish Academy of Sciences.

[67] Stanislaw (Stan) Marcin Ulam (April 13, 1909 – May 13, 1984), was born in Lwów, Poland. He received his bachelor's degree in 1931, his master's in 1932 and his doctoral degree in 1933, all from the Lwów Polytechnic Institute. He visited Princeton for three months in 1935 at the invitation of John von Neumann and returned the following year for a longer period. He was

to invent what are now called Ulam matrices in his ground-breaking paper [1930]. The definition of measurable cardinal traces back to this seminal work of Ulam; their properties as cardinals and their utility in combinatorial arguments took awhile to be fully recognized. See Kanamori's two chapters and Mitchell's chapter on inner model theory and large cardinals for more on the impact of this discovery. For more on the history of this accomplishment, see Chapter 1, Section 2 of [Kanamori, 1994].

Hausdorff published a second edition of his text under the new title of *Mengenlehre* in 1927, where he expanded the treatment of Borel sets, Suslin sets, Baire functions, continuous mappings and homeomorphisms. To keep the length reasonable, he had to cut material from the first edition and found it "easiest to sacrifice, of the topics treated in the first edition, most of the theory of ordered sets, a subject that stands somewhat by itself" (translation by John Aumann in [Hausdorff, 1957, 5]). The growing familiarity of the theory of ordered sets, as witnessed by Sierpiński's 1928 book, may have been one of the reasons that Hausdorff omitted this theory from the second and third editions of his textbook.

Next we turn to results related to trees. In 1925 Dénes Kőnig submitted *Sur les correspondences multivoques des ensembles* which includes a version of *König's Lemma*[68] described in terms of sequences and relations [1926, 120]:

> E) Soit $E_1, E_2, E_3, \ldots$ une suite dénombrable, d'ensembles finis, non vides, et soit R une relation telle qu'à chaque élément x_{n+1} de chaque ensemble E_{n+1} corresponde au moins un élément x_n de E_n, lié á x_{n+1} par la relation R, c'est ce que nous écrivons sous la forme $x_n \, R \, x_{n+1}$ $(n = 1, 2, 3, \ldots)$. Alors on peut choisir dans, chaque ensemble E_n un élément a_n de sorte que, pour la suite infinie $a_1, a_2, a_3, \ldots$, on ait $a_n \, R \, a_{n+1}$ $(n = 1, 2, 3, \ldots)$.[69]

a Harvard Junior Fellow and then a lecturer at Harvard during 1936–1940, visiting Poland in the summers until he left Poland in 1939 to return to the United States a mere month before World War II arrived in Poland. He was on the faculty of the University of Wisconsin, Madison 1941–1944, and then went to Los Alamos Laboratories 1944-1967, and remained a consultant for the remainder of his life. He continued his academic career 1965–1977 at the University of Colorado and at the University of Florida 1974–1984. His mathematical interests were broad, including set theory and abstract measures at the beginning of his career, topology, ergodic theory, computational approaches to mathematics, and bio-mathematics. He enjoyed conversations with mathematicians, physicists, and biologists. He held von Neumann in high regard, and valued their close association which lasted many years. He used to joke during visits by Erdős that the two had known each other since before I was born. I was privileged to enjoy the hospitality of Stan and his wife Françoise in Santa Fe where other guests and visitors included Gian-Carlo Rota, Dan Mauldin, Richard Laver, and the Mycielskis. For more information on Ulam, including further references, see his autobiography [1976], selected works [1974], the biography at http://www-history.mcs.st-andrews.ac.uk/Biographies/Ulam.html by J. J. O'Connor and E. F. Robertson and the brief article by Ciesielski and Rassias [2009].

[68]See [Franchella, 1997] for information of the history of this important result.

[69]E) Let $E_1, E_2, E_3, \ldots$ be a countable sequence of finite non-empty sets, and let R be a relation such that to each element x_{n+1} of each set E_{n+1} there corresponds at least one element x_n of E_n which is linked by the relation to x_{n+1}, which we will write as $x_n \, R \, x_{n+1}$ $(n = 1, 2, 3, \ldots)$. Then one can choose in each set E_n an element a_n such that, for $a_1, a_2, a_3, \ldots$, one has $a_n \, R \, a_{n+1}$

Kőnig remarked in a footnote to his paper that the proof is based on an idea he owes to M. St. Valkó.[70]

The modern formulation of König's Lemma is for trees considered as partial orders: $(T, <)$ is a *tree* if $<$ is a partial order such that for each element $t \in T$, the set of predecessors of t is well-ordered by $<$. By a common abuse of notation the tree is often referred to by the underlying set T alone. For an ordinal α, the αth *level* of the tree is the set of all elements $t \in T$ whose set of predecessors is order-isomorphic to α. The *height* of a tree T is the least α for which the αth level is empty. A *branch* is a maximal linearly ordered subset. With this terminology in hand, we can give the now usual formulation of König's Lemma: If T is a tree of height ω such that all levels of T are finite, then T has an infinite branch.

In the late 1920's, Moore started to revise his axioms and in the process constructed his famous road space. Moore presented a paper at the annual meeting of the American Mathematical Society in December, 1926 and his abstract [1927] gave a brief commentary on Axioms 1, 2, and 4 and Theorem 4 of his early work [1916], including a list of types of spaces in which either Axiom 1 or Axiom 4 failed to hold, together with a proposed Axiom 1'. He gave a series of four colloquium lectures on point set theory (we would call it point set topology) for the Summer Meeting of the American Mathematical Society in August, 1929 in Boulder, Colorado. The synopsis he passed out to the participants gave a variation of Axiom 1' which he further modified in the course of his lectures; he eventually ended up with a version that allowed him to prove what is known as the Moore Metrization Theorem [1935]. On his way home from the meeting, Moore discovered his automobile road space [1942].[71] Mary Ellen Rudin [1975b], in her personal survey on the metrizability of normal Moore spaces, described road spaces in terms of trees where a connecting road or copy of an interval of the real line is added between a node and each of its immediate successors, and any terminal nodes are also extended. The Moore automobile road space is then the space obtained from the complete binary tree of sequences of length $\leq \omega$.

Important tools for the understanding of the infinite were developed. John von Neumann[72] developed what are now known as the *von Neumann ordinals* and put

---

$(n = 1, 2, 3, \ldots)$.

[70]Stephan Valkó who was his co-author in [Kőnig and Valkó, 1926], a paper which was received by the journal in December 1924.

[71]See [Jones, 1997, 99–1000] for the timing, a description of the road space, and a graphical illustration. See [Fitzpatrick, 1997] for a carefully documented history of these lectures, their part in the development of the axioms for a Moore space, and their connections to the Moore metrization theorem and road spaces.

[72]John (János) von Neumann (December 28, 1903 – February 8, 1957), while at the Lutheran Gymnasium, was tutored by professors at the University of Budapest at the behest of his mathematics teacher László Rátz. In 1920, as a teenager, von Neumann submitted his first mathematical paper for publication. He studied chemistry at the University of Berlin 1921–1923 and then went to Zurich where he received his diploma in chemical engineering in 1926. He had an ongoing interest in mathematics, had been admitted to the University of Budapest in 1921, and did well on mathematics examinations there without taking any courses there. In Zurich he talked with Weyl and Pólya about mathematics. In the same year he received his chemical engineering degree he was awarded his doctorate with a thesis in set theory from the University

transfinite recursion on a firm foundation in the 1920's as detailed in Kanamori's first chapter. With concrete representatives of Cantor's ordinal numbers, the order types of well-orderings now had canonical representatives, and Hausdorff's initial ordinals received the names $\omega = \omega, \omega_1, \ldots$. Specifically, the order type of the non-negative integers, which Cantor denoted by ω, was now the order type of the von Neumann ordinal ω, which minimized the distinction between the order type of an ordinal number and its representative as a von Neumann ordinal.

In response to a question of Sierpiński about the size of an ordered set whose increasing chains and decreasing chains are known to be at most denumerable, Pavel Urysohn[73] [1924] built on work of Hausdorff [1914] to prove that the cardinality of such an ordered set is at most that of the continuum. The proof generalizes to the inequality $|S| \leq 2^{U(S)}$ where $U(S)$ is the supremum over the cardinalities of increasing and decreasing well-ordered chains, which we will call the *Hausdorff-Urysohn Inequality.*

Alexandroff and Urysohn, in the supplemental material of [1929],[74] made a simple but fundamental observation about *regressive functions*[75] which we will call the *Regressive Function Theorem* (its generalizations will be indexed under this name):[76]

---

of Budapest. The Mathematics Genealogy Project site lists the harmonic analyst Lipót Fejér (he is particularly well known for *Fejér's Theorem* on convergence of Fourier series) as his doctoral advisor. von Neumann held a Rockefeller Fellowship 1926–1927 which enabled him to study with Hilbert in Göttingen. He lectured at the University of Berlin 1926–1929, and in Hamburg from 1929–1930. He became a visiting lecturer at Princeton University in 1930, was appointed professor there in 1931, and in 1933 became one of the first six mathematics professors in the newly founded Institute for Advanced Study in Princeton. This remarkable man contributed to many fields of pure and applied mathematics, physics, computer science and received wide recognition for his work. Since the work of von Neumann related to the topics of this chapter occurred when he was young, we have restricted this brief sketch of his academic career to the early years.

[73]Pavel Samuilovich Urysohn (February 3, 1898 – August 17, 1924) started studying physics at the age of fifteen under the direction of P. Lasareft, a member of the Academy of St. Petersburg, and Urysohn already had a publication in 1915. Influenced by courses taken with Egorov and Lusin, he switched to mathematics and finished his undergraduate degree at Moscow University where his focus was analysis. Lusin persuaded him to continue with graduate studies, and Urysohn was awarded a doctorate in 1921 under Lusin's direction, after which he became an assistant professor there. He is perhaps best known for his work in topology, specifically for Urysohn's Lemma and the metrization theorem bearing his name. He died after swimming in stormy seas off the coast of Brittany near Batz-sur-Mer, and is buried in the cemetery there. For more information, see the brief biography at http://www-history.mcs.st-andrews.ac.uk/Biographies/Urysohn.html by J. J. O'Connor and E. F. Robertson, and the obituary by Alexandroff [1925]. Another resource of possible interest to those reading Russian is a memoir entitled *Joy of Discovery* written by his sister, Lina Neiman, which discusses his life, and includes articles by mathematicians on his mathematics.

[74]See Kojman's chapter for more on the Alexandroff and Urysohn paper.

[75]A function f on ordinals into ordinals is *regressive* if $f(\alpha) < \alpha$ for all non-zero α.

[76]In the preface to [Alexandroff and Urysohn, 1929], Alexandroff noted that the research was done in the summer of 1922, written up in winter 1922-1923, and accepted in 1923 by the editors of *Fundamenta Mathematicae*, but did not appear there for technical reasons. Math Review MR0370492 (51 #6719) of a reissue in Russian of the monograph noted that its first appearance was delayed for six years "mainly because its length posed problems for the editors of *Fundamenta*

> *Théorème:* Supposons qu'on ait défini, pour tout nombre ordinal a de la seconde classe, un nombre ordinal $\mu(\alpha)$ sous la seule condition $\mu(\alpha) < \alpha$.
>
> Il existe alors un ensemble indénombrable de nombres ordinaux
>
> $$\alpha_1, \alpha_2, \ldots, \alpha_n, \ldots, \alpha_\omega, \ldots, \alpha_\lambda, \ldots \quad |\Omega$$
>
> tels que les $\mu(\alpha)$ correspondant sont tous identiques:
>
> $$\mu(\alpha_1) = \mu(\alpha_2) = \cdots = \mu(\alpha_n) = \cdots = \mu(\alpha_\omega) = \cdots = \mu(\alpha_\lambda) = \ldots \quad |\Omega$$ [77]

In modern language, every regressive function on $\omega_1 \setminus \omega$ is constant on an uncountable set. They show that there cannot be a function ν with the property that for all α, $\nu(\alpha)$ is the least ordinal number greater than α such that $\mu(\beta) > \alpha$ when $\beta > \nu(\alpha)$. Thus there is some α such that for unboundedly many β, $\mu(\beta) \leq \alpha$ and, by cardinality considerations, uncountably many of these β have the same value for $\mu(\beta)$. They used this theorem to prove that ω_1 with the order topology is a non-metrizable topological space.

The Regressive Function Theorem can be regarded as a generalized pigeonhole principle in which one is required to distributes items in a list into indexed pigeonholes so that each item is placed in a pigeonhole with a smaller index.

Mathematicae, and it was the intervention of L. E. J. Brouwer that led to its publication, some years after the appearance of the six articles in which the results first appeared in print."

77

> *Theorem:* Suppose one has defined for every ordinal number α of the second number class, an ordinal number $\mu(\alpha)$ satisfying the sole condition $\mu(\alpha) < \alpha$.
>
> Then there exists an uncountable set of ordinal numbers of the second class
>
> $$\alpha_1, \alpha_2, \ldots, \alpha_n, \ldots, \alpha_\omega, \ldots, \alpha_\lambda, \ldots \quad |\Omega$$
>
> such that the corresponding $\mu(\alpha)$[s] are all identical:
>
> $$\mu(\alpha_1) = \mu(\alpha_2) = \cdots = \alpha_\lambda = \ldots \quad |\Omega$$

Todorcevic, after noting that for limit ordinals $\alpha < \varepsilon_0$, there is a natural recursive way to choose an increasing sequence of ordinals with limit α, made the following comment about this result of Alexandroff and Urysohn in a footnote [2007b, 19]: "one is tempted to believe that the recursion can be stretched all the way up to ω_1 and this is probably the way P. S. Alexandroff discovered the phenomenon that regressive mappings on ω_1 must be constant on uncountable sets."

2.4 Ramsey and van der Waerden

Bartel Leendert van der Waerden[78] was an algebraist, whose textbook *Modern Algebra* [1930][79] has been translated into English and Russian, was reissued in 1991, and is cited to this day. Van der Waerden [1927] (cf. [Graham *et al.*, 1990]) used a double induction argument to prove that for every partition of the positive integers into finitely many pieces there are arbitrarily long arithmetic progressions $(a+b, a+2b, a+3b, ..., a+\ell b)$ in one of the cells, and thereby solved a conjecture he attributed to a Dutch mathematician Baudet.[80] Van der Waerden wrote an account[81] of how he came to prove the theorem. It started with a lunch time conversation with Emil Artin and Otto Schreier, continued in Artin's office, where the trio exchanged questions and ideas, with lots of diagrams on the board, and ended with van der Waerden having an insight into a method for finding the arithmetic progressions based on an example about which he said [1927, 259]: "This final idea was accompanied by a feeling of complete certainty. I felt quite sure that this method of proof would work for arbitrary k and ℓ. I cannot explain this feeling; I can only say that mathematicians often have such conviction. When a decisive idea comes to our mind, we feel that we have the whole proof we are looking for: we have only to work it out in detail."

A critical theorem in infinite combinatorics is Ramsey's Theorem. In 1928 Frank P. Ramsey[82] submitted a paper [1930] addressing a problem in formal logic

[78]Bartel Leendert van der Waerden (February 2, 1903 – January 12, 1996) started studying mathematics at the University of Amsterdam in 1919 when he was sixteen, where Brouwer was the teacher he most respected. Most of his graduate study was at the University of Amsterdam where, in 1925, he defended his doctoral thesis written with Hendrik de Vries, and he also studied in Göttingen where he shifted to algebra after contact with Hellmuth Kneser and Emmy Noether. After a postdoctoral visit to Göttingen, he went to Hamburg for a semester with Hecke, Artin and Schreier and started work on the first volume of *Moderne Algebra* [1930] (in German) based on lectures by Artin and Noether. Van der Waerden held positions in Groningen and Göttingen before becoming a professor at the University of Leipzig in 1931, where he continued working during the war until his house was bombed in 1943. He and his family stayed in various places with friends and relatives and in 1945 returned to the Netherlands. Van der Waerden visited Johns Hopkins University in 1947, returned in 1948 to a chair in mathematics at the University of Amsterdam, and in 1951 moved to the University of Zurich where he spent the remainder of his academic career.

[79]*Moderne Algebra. Teil 1* was published as volume 33 in the first Springer series of higher mathematics texts, started by Richard Courant in 1920. In the preface to the translation of the second edition, van der Waerden shared his attitude toward set theory in [1937], translated to English in [1953, v–vi]: "I have tried to avoid as much as possible any questionable set-theoretical reasoning in algebra. ... I completely omitted those parts of the theory of fields which rest on the axiom of choice and the well-ordering theorem. Other reasons for this omission were the fact that, by the well-ordering principle, an extraneous element is introduced into algebra, and furthermore, the consideration that for virtually all applications the special case of countable fields, in which the counting replaces the well-ordering, is wholly sufficient."

[80]Schur independently made the conjecture and shared it with his students (cf. Soifer [2009]).

[81]For van der Waerden's account, see [1954], reprinted in [1998] and, in English, see [1971].

[82]Frank Plumpton Ramsey (February 22, 1903 – January 19, 1930) was the son of Arthur Ramsey, President of Magdalene College, Cambridge and a mathematics tutor. He completed his undergraduate work at Trinity College in 1923 and was elected a fellow of King's College in 1924. The Mathematical Genealogy Project site lists him as having a University of Cambridge

which included a proof of his oft-cited theorem as a lemma.[83] For a photograph of Ramsey and a discussion of the beginnings of finite Ramsey theory, see Harary [1983] and Spencer [1983]. Below is his statement of what he labeled *Theorem A* [1930, 264].

> *Ramsey's Theorem:* Let Γ be an infinite class, and μ and r positive integers; and let all those sub-classes of Γ which have exactly r members, or, as we may say, let all r-combinations of the members of Γ be divided in any manner into μ mutually exclusive classes C_i $(i = 1, 2, \ldots, \mu)$, so that every r-combination is a member of one and only one C_i; then, assuming the Axiom of Selections, Γ must contain an infinite sub-class Δ such that all the r-combinations of the members of Δ belong to the same C_i.[84]

Ramsey's Theorem is an example of a *balanced* partition theorem, since the size (infinite) of the set all of whose r-combinations belong to the same cell of the partition does not depend on which cell it is.

Ramsey's proof uses induction on the number of cells μ in the partition, with the critical case being $\mu = 2$, and induction on the size r of the combinations. He observed that the theorem is clearly true for $r = 1$ and for $\mu = 1$. Fixing $\mu = 2$, he assumes the statement is true for $r = \rho - 1$, and proceeds to the induction step with $r = \rho$. Having fixed a partition C_1 and C_2 of the ρ-combinations of Γ, he defines as long as possible sequences $x_1 < x_2 < \ldots$ and $(\Gamma \supseteq)\Gamma_1 \supseteq \Gamma_2 \supseteq \ldots$ so that $x_1 \in \Gamma$, $x_i \in \Gamma_{i-1}$ for $i > 1$ and every ρ-combination obtained by adding x_i to $\rho - 1$ members of Γ_i is in C_1. If the process goes on infinitely, he concludes that $\Delta = \{x_1, x_2, \ldots\}$ is the desired set. If the process stops after finitely many steps, say with the definition of Γ_n, then he initiates a new process of choosing $y_1 < y_2 < \ldots$ and $\Gamma_{n-1} \supseteq \Delta_1 \supseteq \Delta_2 \supseteq \ldots$ with $y_1 \in \Gamma_{n-1}$, $y_i \in \Delta_{i-1}$ for $i > 1$, and every ρ-combination obtained by adding y_i to $\rho - 1$ members of Δ_i is in the same C_i, which can be arranged by applying the induction hypothesis for $\rho - 1$. Since the first process stopped after finitely many steps, it must be the case that $C_i = C_2$ at every stage in the second process, so that $\Delta = \{y_1, y_2, \ldots\}$ is the desired set. He then completes the induction on r. He reduces the case for $\mu > 2$ many cells to at most $\mu - 1$ cells, by combining the last two cells together to get a new partition and applying the induction hypothesis to it to get an infinite set Δ and a specified cell containing all r-combinations from it. If the specified cell is not

doctorate with no mention of date or advisor. With his first major work (*The Foundations of Mathematics*) published in 1925, it is likely that his doctorate was awarded after he became a fellow of King's College and prior to being appointed a university lecturer in mathematics at King's College in 1926. His interests ranged widely and included foundations, philosophy and economics.

[83]Ramsey's paper was published posthumously.

[84]The Axiom of Selections is presumably another name for the Axiom of Choice. Bertrand Russell discussed the "theory of selections" in Chapter XII of his book *Introduction to Mathematical Philosophy* [1920, 117ff], including Zermelo's Theorem.

the new cell obtained by joining two old ones, Δ is the desired set, and otherwise he applies the theorem again with $\mu = 2$ to get the required infinite $\Delta' \subseteq \Delta$.

We will call a sequence $\{z_1, z_2, \dots\}$ where the combination of any element z_i with any $\rho - 1$ later elements lands in the same cell C_i *begin-homogeneous*.[85] With this language, Ramsey's basic induction step argument starts with an to attempt to construct a sequence begin-homogeneous for which all the ρ-combinations end in the first cell, and if it does not work, repeat starting from the point of failure to get a sequence begin-homogeneous and verify that all the ρ-combinations end in the second cell.

Corresponding to the theorem discussed above, sometimes also called the *Infinite Ramsey's Theorem*, Ramsey proved a finite version quoted below [1930].

> *Finite Ramsey's Theorem:* Given any r, n and μ we can find an m_0 such that, if $m \geq m_0$ and the r-combinations of any Γ_m are divided in any manner into μ mutually exclusive classes C_i $(i = 1, 2, \dots, \mu)$, then Γ_m must contain a sub-class Δ_n such that all the r-combinations of the members of Δ_n belong to the same C_i.[86]

Ramsey followed the statement of the theorem with the following remarks: "This is the theorem which we require in our logical investigations, and we should at the same time like to have information as to how large m_0 must be taken for any given r, n, and μ. This problem I do not know how to solve, and I have little doubt that the values for m_0 obtained below are far larger than is necessary." The study of this question and its generalizations is the core of finite Ramsey theory.

3 1930-1940: EARLY RAMIFICATIONS

In the 1930's Erdős and Rado, two pioneers in the partition calculus, forged their initial results; Sierpiński constructed a key example which shows that there is no direct uncountable analog of Ramsey's Theorem; Jones proved his metrization theorem and constructed a special Aronszajn tree[87] but did not publish the work until later; and Kurepa made the first systematic study of uncountable trees, giving an equivalent formulation of Suslin's Problem in terms of trees. Partial orders were shown to be extendible to linear orders, and a denumerable partial order universal for countable partial orders was constructed. The Regressive Function Theorem was generalized, and large families of almost disjoint sets were constructed. Equivalence relations became part of the tool kit for analysts;[88] graph theory became

---

[85]The use of *begin-homogeneous* here is designed to parallel *end-homogeneous* used in later constructions.

[86]Here Γ_m denotes a class with m members.

[87]Recall that a *special Aronszajn tree* is a tree of height ω_1 all of whose levels are countable and which is the union of countably many antichains.

[88]Equivalence relations were one of the topics in [Bourbaki, 1939], a 51-page overview of set theory designed to prepare individuals for the study of analysis.

a subject in its own right[89] with the publication of *Theorie der endlichen und unendlichen Graphen* [Kőnig, 1936], [1990], the first book to present graph theory as a subject.

There were ups and downs in mathematical publishing. *Zentralblatt für Mathematik* was founded in 1931 by Springer Verlag in cooperation with the Prussian Academy of Science.[90] *The Journal of Symbolic Logic* was founded in 1936 (cf. [Ducasse and Curry, 1962]), and Alonzo Church included a bibliography of symbolic logic in the first issue. However, on September 1, 1939 Germany invaded Poland and volume 33 of *Fundamenta Mathematicae* was not published.[91]

Hausdorff published the third edition of his set theory text; Moore [1932] published his monograph *Foundations of Point Set Theory* with the revised version of his axioms that he had been working toward in the 1920's.[92]

Sierpiński published a monograph [1934] on the Continuum Hypothesis. It naturally included his own work, and also included that of Lusin, Tarski, Kuratowski and others of the Polish school. It started with equivalences of the statement and continued with discussions, all the while assuming CH, of Lusin sets, the duality between sets of measure zero and sets of first category, the proof that there is no measure on all subsets of the real numbers which is (1) not identically zero; (2) zero on singletons; (3) countably additive.

The founders of the Moscow school, Egorov and Lusin, fell on hard times in the 1930's. In September 1930 Egorov was arrested for his religious beliefs, Name Worshipping, and exiled to a camp near Kazan on the Volga River, and he died after a hunger strike initiated because the prison guards would not let him practice his religious beliefs.[93] The Moscow Mathematical Society had elected Kolman, a loyal Bolshevik, as president. Lusin was alarmed by what happened to Egorov, so he left Moscow University for the Steklov Institute of the Central Aero-Hydrodynamics Institute in Leningrad, but kept his position as the head of the mathematical group of the Russian Academy of Sciences. Kolman's denunciation of Lusin in print prevented Lusin from attending the International Congress of 1932. The Russian Academy of Sciences and the Steklov Institute were moved to Moscow in 1934. In July 1936, articles in Pravda appeared accusing Lusin, in particular, of claiming results of his students as his own and of publishing his best papers in the West. These articles led to a series of meetings of the Moscow Mathematical Society which drafted a resolution against Lusin that led to the loss of his university position and loss of good will from some of his colleagues, but not his position as

[89]See the AMS Math Review by Robin Wilson [1991].

[90]See the *History of Zentralblatt* page whose url follows: http://www.zentralblatt-math.org/zmath/en/about/history/ downloaded September 7, 2010.

[91]See [Kuzawa, 1970, 489–490] for an overview of the loss of life and property of the mathematical community of Poland during the war period 1939–1945.

[92]Jones [1997, 97], in his article *The beginning of topology in the United States and the Moore school*, assessed this work of Moore and *Topology* by Solomon Lefschetz [1930] as follows: "While definitely limited, these books were of considerable help in formulating the knowledge and tools necessary for research."

[93]See [Graham and Kantor, 2007] for details.

an academician. Lusin made a statement in response to the resolution promising, in part, to publish primarily in the Soviet Union. What became known as the *Lusin Affair* had a chilling effect: mathematicians in the Soviet Union basically stopped publishing their results abroad.[94]

The study of transversal theory was launched when Philip Hall [1935] proved his celebrated theorem on matching:

> *Hall's Marriage Theorem:* Given n boys, if, for each positive $k \leq n$, any k of the boys between them know at least k girls, then it is possible to marry each boy to a girl that he knows.

The result can also be formulated as an existence theorem for a system of distinct representatives: For each boy b, there is a set G_b of girls he knows, and a selection of distinct girls from each of these sets yields the desired outcome. D. Kőnig [1931] had an equivalent but initially lesser known result published in Hungarian. Transversals are regarded in a variety of ways: as injective choice functions for a sequence of sets; as sets whose intersection with every set in a family is a singleton; as systems of unique representatives.

3.1 Extensions

This section includes extensions of results from earlier decades for linear and partial orderings, almost disjoint families and regressive functions. Also included is the following result which discusses an extension of a different kind.

Edward Marczewski[95] [1930] proved a theorem that is known in the literature by his birth name:

> *Szpilrajn's Theorem:* Every partial order has an extension to a total (linear) order.

For Marczewski, a linear order was a transitive, irreflexive, trichotomous binary relation. Marczewski first proved a lemma that any partial order with an incomparable pair can be extended to a partial order in which the pair has a specified order. He next observed that the union of an increasing well-ordered chain of partial orders under extension is a partial order. He concluded that there must be a maximal extension, hence one which is a linear order, citing [Kuratowski, 1922] and [Hausdorff, 1914]. In his paper, he remarked that the theorem was already known, citing the third edition of Sierpiński's book, *Zarys teorji mnogości*, but that the proofs of Banach, Kuratowski and Tarski had not yet been published.

[94]See [Demidov and Levshin, 1999] and the Math Review MR1790419 (2001k:01066) by F. Smithies of it for more information.

[95]Marczewski had the surname Szpilrajn until 1940. Kuratowski directed his thesis at the University of Lwów. Other influential colleagues there included Mazurkiewicz and Sierpiński.

Andrzej Mostowski[96] [1938] constructed a denumerable partial order universal for denumerable partial orders,[97] extending the Bernstein result for countable linear orders which Hausdorff had extended under CH to larger size linear orders.

Hausdorff [1936] revisited ordered sets and constructed an (ω_1, ω_1^*)-gap inside the "dyadic number sequences." He commented on its relationship to his earlier work in a footnote [Hausdorff, 1936, 244, footnote 1]:

> Für Folgen reeler order rationaler Zahlen habe ich dies bereits 1909 gemacht (*Graduierung nach dem Endverlauf*, Abh. Sächs. Ges. d. Wiss. 31). Da dieser Arbeit wohl wenig bekannt is und die Konstruktion für dyadische Folgen, die zum vollen Beweis des Satzes I erforderlich ist, doch einige Modifikationen velangt, möchte ich die Sache vollständig darstellen.[98]

Sierpiński [1931] investigated the collection of all subsets of a denumerable set D partially ordered by the subset relation. He proved that an order type φ is realizable as a family of subsets of D linearly ordered by the subset relation if and only if it is a denumerable real type, where a *real type* is the order type of some set of real numbers with their usual order. At the end of his paper he asked the following question.

> *Question:* What are necessary and sufficient conditions for an order type φ to be realizable as a family of denumerable sets linearly ordered by the subset relation?

He concluded with the comment that all such types are necessarily realizable as subsets of $(1 + \lambda)\,\omega_1$, where λ is the order type of the set of real numbers.

Sierpiński [1937], who had introduced the notion of almost disjoint families in [1928b], published his proof that for any cardinal κ there is a family of cardinality

[96] Andrzej Mostowski (November 1, 1913 – August 22, 1975) wrote up his doctoral thesis, *On the Independence of the Definition of Finiteness in the System of Logic* in 1938 and defended it at Warsaw University in February 1939 (recall that Poland was invaded in September 1939). Kuratowski was the official supervisor, but Mostowski counted himself a student of Tarski. Lindenbaum had suggested that he work on the method of independence proofs sketched by Fraenkel (cf. [Crossley, 1975, 44]). This conversation was the starting point of the main content of Mostowski's dissertation, namely, what came to be known as Fraenkel-Mostowski permutation models. In conversation with Crossley [1975], Mostowski shared some memories of his three escapes during World War II, and his decision to take some bread instead of his "nice, very big, wonderful notebook with all these discoveries" which subsequently was burned. For more on his life and his interactions with mathematicians inside and outside of Poland, see [Ehrenfeucht *et al.*, 2008].

[97] Recall that a partial order $(P, <)$ is *universal for a class* $\mathcal{K}$ of partial orders if $(P, <)$ belongs to $\mathcal{K}$ and every partial order in $\mathcal{K}$ has an order and incomparability preserving embedding into $(P, <)$. I only know of this result from the reference to it in [Johnston, 1956].

[98] I already did this in 1909 for sequences of real or rational numbers (*Graduierung nach dem Endverlauf*, Abh. Sächs. Ges. d. Wiss. 31). Since this work is undoubtedly little known and since the construction for dyadic sequences, which is necessary for the full proof of Theorem 1, needs some modifications, I would like to present the matter completely. (Translation by Plotkin cf. [Hausdorff, 2005, 307])

at least κ^+ of increasing κ-sequences which, pairwise, have disjoint tails, i.e. there is an almost disjoint family of size 2^λ of subsets of κ where λ is the least cardinal μ with $2^\mu > \kappa$.

Ben Dushnik[99] [1931] extended the Regressive Function Theorem of Alexandroff and Urysohn to non-limit [successor] cardinals κ using the well-foundedness of the ordinals and basic cardinal arithmetic to find a subset of size κ on which the function is constant. He gave the example of the function $F : \aleph_\omega \to \omega$ defined by $F(\alpha) = 0$ if α is finite and otherwise $F(\alpha) = n$ if and only if $|\alpha| = \aleph_n$ to show some limitation was necessary.

3.2 Erdős and Rado

Now we turn to Ramsey-theoretic results of Erdős, Rado and Sierpiński.

Richard Rado[100] was a student of Schur in Berlin. Rado in his thesis [1931] (see also his [1933]) studied partitions of positive integers into finitely many classes and homogeneous systems of linear equations with positive integer coefficients. He called such a system of linear equations *k-fold regular* (now shortened to *k*-regular) if for every partition of the positive integers into k classes, there is one class which has a solution to the system. Homogeneous systems which are k-regular for all k are called *partition regular*, and Rado classified partition regular systems, generalizing van der Waerden's Theorem [1927] and Schur's Theorem [1917]. We will call this *Rado's Theorem for Linear Systems with Positive Integer Coefficients.*

Here is a modern take on the theorem (see Hindman [2006, 4] for example). Say that an $n \times n$ matrix A with rational entries and columns $A = [\vec{c}_0, \vec{c}_1, \ldots, \vec{c}_n]$ satisfies the *columns condition* if and only if there is a partition $\mathcal{I} = \{I_0, I_1, \ldots, I_{m-1}\}$ of the indices such that

1. $\sum_{k \in I_0} \vec{c}_k = \vec{0}$; and

2. for all $j < m$, if j is positive, then the sum $\sum_{k \in I_j} \vec{c}_k$ is a linear combination of the vectors in $\bigcup_{\ell < j} \{\vec{c}_m : m \in I_\ell\}$.

[99] In the middle 1920's, Ben Dushnik joined the faculty of the University of Michigan (see [Kaplan, 1971, 184-185]). He received his Ph.D. from the University of Michigan under Theophil H. Hildebrandt in 1931.

[100] Richard Rado (April 28, 1906 – December 23, 1989), son of Leopold Rado, a Hungarian from Budapest, was born in Germany, and met and married Luise Zadek there. Since opportunities for those of Jewish descent were slim under Hitler, they went to Cambridge, England to study further. After earning a second doctorate under Hardy in 1935 (see [Richards, 1971] for details), he was a lecturer at Cambridge and then at the University of Sheffield. He was appointed Reader in Mathematics at King's College, London, before becoming a Professor of Mathematics at Reading in 1954. He was elected a Fellow of the Royal Society in 1978. He worked in a variety of fields and was particularly known for his work in combinatorics, including abstract independence structures, transversal theory, and Ramsey theory, especially the partition calculus. For more on his life and work, see [Rogers, 1998].

Call a matrix A with rational entries *partition regular* over $\mathbb{N}$ if and only if for any finite coloring of $\mathbb{N}$, there is a solution $\vec{e}$ to $A\vec{x} = \vec{0}$ such that all entries of $\vec{e}$ receive the same color. Here is Rado's Theorem reformulated this terminology.

> *Rado's Theorem for Linear Systems with Positive Integer Coefficients:* A is partition regular if and only if A satisfies the columns condition.

At a celebration of his 65th birthday, Richard Rado spoke of his beginning steps in mathematics ([Richards, 1971]):

> Having been registered as a research student at Berlin University for some time, I was in the habit of staring gloomily at a blank sheet of paper wondering how I could ever cover it with any worthwhile mathematics. One day I attended a seminar conducted by a group of distinguished mathematicians. A fellow research student gave a lecture on van der Waerden's theorem on arithmetical progressions. To me the theorem sounded quite incredible and the proof a string of fallacies. I went away determined to shatter whatever belief there could exist in the truth of such a theorem. But on studying the matter more closely I had to admit that the theorem was true and the proof sound. This gave me my start in mathematics and I have never looked back.

In 1931, Rado's future collaborator, Paul Erdős[101] was a first-year undergraduate taking a graph theory course from Dénes Kőnig, when he proved a generalization of Menger's Theorem[102] for infinite graphs, which only appeared in 1936 at the end of [Kőnig, 1936]. While at the university, Erdős interacted with a dozen or so young mathematicians, including György Szekeres and Eszter Klein. Sós [2002, 88] speculated:

> It might be that this intensive collaboration, which was quite unusual at the time, has been influential in shaping the future character of the mathematical life in Hungary.

Erdős received his Ph.D. from Péter Pázmaány University[103] in 1934. About his university professors, Erdős said in an interview:[104] "I learned a lot from Lipót

[101]Paul Erdős (March 26, 1913 - November 9, 1996) moved out of Hungary because of World War II and eventually became an itinerant mathematician whose travels knit together a world-wide community of mathematicians. See [Bollobás, 1997] for a short biography; see [Graham and Nešetřil, 1997] for personal memories of many close to him, and [Hajnal, 1997] for a discussion of his work in the partition calculus by a close collaborator.

[102]Menger's Theorem [Menger, 1927] says that for any vertices x and y in a graph, the minimum number of vertices that need to be removed from a graph to end up with x and y not connected by a path is the same as the maximum size of a family of paths from x to y with no two paths sharing an edge. For a strong version of Menger's Theorem that answers a question of Erdős, see [Aharoni and Berger, 2009].

[103]The university was renamed Loránd Eötvös University in 1948.

[104]The interview was published in 1979 in Hungarian [Staar, 1979] in the monthly *Terméózet Világa* and quoted by Sós [2002] in an English translation.

Fejér, but very probably, I learned the most from László Kalmár." However, Erdős had learned the basics of set theory from his father, who was a high school teacher, and Erdős "soon became fascinated with 'Cantor's paradise' " according to András Hajnal [1997, 352]. Erdős and Szekeres [1935][105] (see also [Spencer, 1973]) rediscovered the Finite Ramsey's Theorem.[106] They comment that Thoralf Skolem published a proof of the Finite Ramsey's Theorem [1933]. Erdős and Szekeres developed what Graham and Nešetřil [2002] called an *ordered pigeonhole principle.*

> *Ordered Pigeonhole Principle:* Let m, n be positive integers. Then every sequence of $(m-1)(n-1)+1$ distinct integers contains either a monotone increasing m-set or monotone decreasing n-set.

They then used their version of the Finite Ramsey's Theorem for quadruples to prove a Euclidean partition property.

After his doctoral degree, Erdős went to Manchester, England at the invitation of Mordell, where he met Rado and Ulam, among others. Erdős' interest in generalizing Ramsey's Theorem began early, as is illustrated in the obituary of Rado by Rogers which included a sentence from a letter to Rado from early in 1934 [1998, 190]: "Erdős asked: 'Is it true that when S is a set of infinite cardinal[ity], and the countable subsets of S are split into two classes, then there is always an infinite subset S^* of S all of whose countable subsets are in the same class?' Rado replied with a counterexample almost immediately." In 1938 Erdős was offered a fellowship at the Institute for Advanced Study in Princeton, where he spent a year and a half, and after that he became an itinerant mathematician, fostering collaborations everywhere he went. For more on his life, including passport and visa troubles with Hungary and the United States and the story of how he became a dual citizen of Israel, I recommend the article by Bollobás [1997].

Ramsey's Theorem is called a *positive* result, since it guarantees the existence of an infinite set whose n-element subsets are indistinguishable with respect to a given partition, coloring, or n-ary symmetric relation. An important early *negative* result (published in 1933) is the *Sierpiński partition* which shows that there is no direct uncountable analog of Ramsey's Theorem. Bronisław Knaster[107] had asked the following question (cf. [Sierpiński, 1933, 285]):

> Existe-t-il une relation symétrique R, dont le champ E est non dénombrable, telle que dans tout sous-ensemble non dénombrable de E existent deux éléments différents α et β, tels que $\alpha R \beta$, et deux éléments différents γ et δ, tels que γ non $R\delta$.[108]

[105]The paper was submitted December 7, 1934.

[106]In [Erdős, 1942, 363] and [Erdős, 1996, 113] the rediscovery of the Finite Ramsey Theorem is attributed to Szekeres.

[107]Knaster (May 22, 1893 – November 3, 1990) received his Ph.D. degree from the University of Paris in 1925 under the supervision of Stefan Mazurkiewicz. He was a professor in Lwów starting in 1939, and in Wrocław from 1945.

[108]Is there a symmetric relation R, of which the domain E is uncountable, such that in every

For the domain E, Sierpiński selected the set of finite and denumerable ordinals.[109] He applied the Axiom of Choice to obtain a one-to-one function $r : E \to \mathbb{R}$, and defined $\alpha R \beta$ if and only if $r(\min(\alpha, \beta)) < r(\max(\alpha, \beta))$. The symmetric binary relation R may be cast as a graph (without loops) by letting E be the set of vertices and letting the edges be the unordered pairs $\{\alpha, \beta\}$ for which $\alpha R \beta$ (and $\beta R \alpha$). Thus this example shows that Ramsey's Theorem cannot be directly extended to the uncountable.

3.3 Jones

F. Burton Jones,[110] in a one-page abstract for a paper presented at the October 28, 1933 meeting of the American Mathematical Society at Columbia University, included the statement of a metrization theorem [Jones, 1933]: "a separable normal space satisfying the first three parts of Axiom 1 of R. L. Moore's *Foundations of Point Set Theory* is completely separable and therefore metric, provided that in this space every uncountable point set contains a subset of power $\mathfrak{c}$." The reference is to Moore's book [1932]. Spaces which satisfy Axiom 0 and the first three parts of Axiom 1 were named *Moore spaces* by Jones [1937, 675]. A space is *normal* if disjoint closed sets P and Q can always be separated by disjoint open supersets $U \supseteq P$ and $V \supseteq Q$.

After a four-year gap, Jones published the proof of his Theorem 5 [1937, 676]:

> *Jones Metrization Theorem:* If $2^{\aleph_1} > 2^{\aleph_2}$, then every separable metric space M is completely separable and metric.[111]

Below this theorem and on the same page, Jones commented: "The author has tried for some time without success to prove that $2^{\aleph_1} > 2^{\aleph_0}$." He then stated what has come to be known as the Normal Moore Space Conjecture: "Is every normal Moore space M metric?"

Jerry Vaughan [1980] asked Jones why he thought that $2^{\aleph_0} < 2^{\aleph_1}$ was true, and here is the response Vaughan recorded from their October 12-13, 1979 conversation [1980, 40]: "Think of a countable sequence, say ω, as being an initial segment of an uncountable set, say ω_1. Well, it is perfectly obvious to even the most stupid

---

uncountable subset of E there are two different elements α and β, such that $\alpha R \beta$ and two different elements γ and δ such that not $\gamma R \delta$ (translation by the author).

[109] Formally, he let E be the set of all ordinal numbers of the first and second class of Cantor.

[110] Floyd Burton Jones (November 22, 1910 – April 15, 1999) was a student of R. L. Moore who received his degree at the University of Texas at Austin in 1935, where he taught until 1950 with the exception of a couple years during the war at the Harvard Underwater Sound Laboratory. For information on Burton Jones, see the dedication by Jones' first doctoral student [Transue, 1969] and the overview of Jones' work in the conference proceedings [McAuley and Rao, 1981]. Both were recommended by M. E. Rudin [1997]. Other resources include [Rogers, 1999] and the brief biography at http://www-history.mcs.st-andrews.ac.uk/Biographies/Jones_Burton.html by J. J. O'Connor and E. F. Robertson.

[111] Peter Nyikos [1980a] pointed out that Jones and Moore used *completely separable* for *second countable*, i.e. there is a countable family of open sets such that every open set can be represented as a union of some of them.

person, and I guess maybe it helps to be a little stupid, if you take all the sequences which run through the first infinity of ordinals, which gives a set of cardinality $2^{\aleph_0}$, look how many more you can get by running the rest of the way through ω_1 with uncountable sequences. It is clear that there ought to be a whole lot more of them. I thought surely that I could prove that overnight, maybe even in an hour, but I could not get anywhere on it. I remember one week I worked on that problem pretty hard and could not see it. I happened to run into Moore walking across campus and told him that I was losing my mind: I could not prove $2^{\aleph_0} < 2^{\aleph_1}$ even though it ought to be true. Well, he just laughed and said 'Neither could Sierpinski.' It had not occurred to me to look in Sierpinski's book; so I looked in there and found some things equivalent to this. He did not do much with it, but he did enough to convince me that I was wasting my time." Peter Nyikos [1980a, 28] completes the story: "The real cause of the delay, as Jones has informed us, is that he was hoping to settle the entire normal Moore space problem $\cdots$."

Jones [1980][112] shared reactions of his thesis advisor Moore to this work:

> When I was a student at Texas, Moore apparently didn't consider work in abstract spaces very highly. $\cdots$ I think he just felt there was more substance and beauty in other less abstract kinds of problems - - continua, decompositions, etc. So when I proved that if $2^{\aleph_0} < 2^{\aleph_1}$ then every separable, normal Moore space was metrizable [1], I wasn't asked to present the proof in class but rather to the Mathematics Club (as it was called). And when I wanted to use it together with other things for a thesis, Moore didn't like the idea at all but chose instead a couple of embedding theorems I had gotten. With hindsight he was certainly right. Since I thought that one ought to be able to prove $2^{\aleph_0} < 2^{\aleph_1}$, I might still be there!

Jones made a presentation entitled *A false proposition of logic* to the Mathematics Club of the University of Texas in the spring of 1933 (see footnote 6 of [1953]) in which he presented the construction/example of an Aronszajn tree as he much later recounted for the American Mathematical Society on December 29, 1952 [1953].[113] A modern description of his example follows [1953, 32–33]:

> *Jones' Example 1:* There exists a well-ordered family F of collections $H_1, H_2, \ldots, H_z, \ldots (z < \omega_1)$ such that (1) for each ordinal x, $x < \omega_1$, H_x is a countable collection of mutually exclusive sets, (2) if $x < y < \omega_1$, each set of H_y is a proper subset of H_x, (3) if $x < y < \omega_1$, and h_x is an element of H_x, there exists a sequence $h_x, h_{x+1}, h_{x+2}, \ldots, h_y$

[112]The paper [Jones, 1980] appeared in a volume dedicated to his mathematical work and influence and containing some of the papers presented at a conference held in Greensboro, North Carolina, in October, 1979.

[113]While Jones' example was published two decades after its construction (the paper included suitable references to related work that had been published in the meantime), it is included here because of its influence on students in the Texas school of topology, who heard him talk about it and the related road spaces constructed from it.

> such that $h_x \supset h_{x+1} \supset h_{x+2} \supset \cdots \supset h_y$, and for each z, $x \le z \le y$, H_z contains exactly one of the sets $h_x, h_{x+1}, h_{x+2}, \ldots, h_y$, but (4) there exists no sequence $h_1, h_2, h_3, \ldots$ such that $h_1 \supset h_2 \supset h_2 \supset \ldots$ and for each x, $x < \omega_1$, H_x contains exactly one of the sets $h_1, h_2, h_3, \ldots$.

Jones used his example to construct a non-metrizable Moore space [1954, 31–33]. He compared his example with the one in Kurepa's thesis, and described it as follows [1954, 34]: "Aronszajn's example is similar to the one I have constructed. It yields Example 1 but is not as suitable for the application to Moore spaces."

3.4 Kurepa

Đuro Kurepa[114] made a systematic study of uncountable partial orders, especially trees in the wider framework of ramified sets, starting in his thesis, *Ensembles et Tableaux Ramifiées*. This section starts with a detailed review of his thesis which includes the definition of a Suslin tree;[115] the equivalence of Suslin's Problem with the non-existence of a Suslin tree; the construction of an Aronszajn tree as a subtree of the tree of all non-empty bounded well-ordered sequences of rationals under end-extension; and the definition of the key parameters for the structural analysis of uncountable trees: height, supremum of the sizes of the levels, and cardinality of the set of branches whose length is the height of the tree. The review of his thesis is followed by an overview of results from the 1930's beyond the thesis: his construction of a special Aronszajn tree; his initial results on $\mathbb{Q}$-embeddable and $\mathbb{R}$-embeddable[116] Aronszajn trees; his Fundamental Relation

[114] Đuro Kurepa (August 16, 1907 – November 2, 1993) was a proud, tall, energetic individual from a region known for tall people. He spent a substantial portion of his career at the University of Zagreb, an ethnic Serb in Croatia. Specifically, he got his diploma there in 1931 and became an assistant in mathematics there. He spent 1932-1935 in Paris at the Faculté des Sciences and the Collège de France. He obtained his doctoral degree at the Sorbonne in 1935 where the president of his committee was Paul Montel, his supervisor (we would say advisor) was Maurice Fréchet, and whose third member was Denjoy. Knaster was one of the reviewers of his thesis (cf. Kurepa's Math Review MR0423284 (54 #11264)). He spent some postdoctoral time in Paris and Warsaw. Then he returned to the University of Zagreb, where he became an assistant professor in 1937, was promoted to associate professor 1938, and full professor in 1948. According to Hrovje Šikić, a Professor at the University of Zagreb and currently editor-in-chief of *Glasnik Matematički*, Kurepa was interested not only in mathematics, but also in languages, and would sometimes include in his lectures comments about idiosyncrasies in Serbian and Croatian word formation. An example Šikić recalled was the words for curve, which is *krivulja* in Croatian and *kriva* in Serbian. They came up a semi-popular seminar, about which Kurepa said neither should be used since they failed to follow the rule of names ending in "un", and he recommended that *krivun* be used instead to great laughter in his audience. Kurepa moved to the University of Belgrade in 1965 and remained there until his retirement in 1977. Kurepa was instrumental in the introduction in 1966 of *Glasnik Matematički* Series 3 as the mathematical successor to *Glasnik matematičko-fizički i astronomski* Series 2. He died tragically in the time of Milošević and rampant inflation after being beaten by thugs and hidden from view under stairs.

[115] Recall that a *Suslin tree* is a tree of height ω_1 in which every branch and every antichain is countable.

[116] A partially ordered set is $\mathbb{Q}$-*embeddable*, respectively $\mathbb{R}$-*embeddable*, if it has a strictly increasing map into $\mathbb{Q}$, $\mathbb{R}$ respectively.

bounding the size of a partial order in terms of the sizes of its well-ordered and conversely well-ordered subsets and the sizes of its antichains; and his use of a super-position of two linear orders to define a witness to the sharpness of the bound in his Fundamental Relation.

In the opening paragraph of the published version of his thesis, Kurepa [1935, 1] listed two logical problems that the theory of ordered sets had run up against: (1) the existence of a well-ordering of an arbitrary set,[117] and (2) the existence of a cardinal number between any cardinal κ and its power 2^κ.[118] They shaped much of the research in set theory in the 20th Century, and provided a context for Suslin's problem [1920], the motivating force in Kurepa's thesis.

While working on his thesis, Kurepa shared his progress in a series of four announcements in *Comptes Rendus*: [1934a], [1934b], [1934d], [1934c].[119] Kurepa [1934a] announced a positive solution to Suslin's Problem in February with a theorem asserting the equivalence of seven statements about a linearly ordered set E, the first of which posits that E is a continuum isomorphic to each of its intervals and not the product of two sub-continua and the last of which asserts that the set E is dense, it has neither a first nor a last element, is metrizable and is complete. This note introduced the notion of the development or atomization of a linearly ordered set. In modern language this is the process of repeatedly partitioning into subintervals to end up with a family of intervals which are pairwise either disjoint or have one a subset of the other, and so that the family forms a tree under the subset relation. Kurepa discussed developments in §12 of his thesis, and the development of a Suslin line led to the equivalence of the existence of a Suslin line and the existence of a Suslin tree.

The second announcement, [1934b] in March, also claimed a positive solution to Suslin's problem, and as a step toward the proof introduced the two cardinals:

- $p_1E = \inf\{|F| : F \subseteq E \text{ is a dense subset of } E\}$, and
- $p_2E = \inf\{|\mathcal{F}| : F \text{ is a family of disjoint and non-empty open intervals of } E\}$.

Now p_1E is written $d(E)$ and called the *density*, and p_2E is written $c(E)$ and called the *cellularity*. These cardinals represent the beginning of Kurepa's generalization of the Suslin problem to a wider context, and they figure in Kurepa's Fundamental Relation discussed at the end of this section. In §7 [1996, 58–64] of his thesis, Kurepa defined and investigated p_1E and p_2E for a linearly ordered set E. He focused on the relationship between these cardinals and proved in Lemma 10 that to prove $p_1E = p_2E$ it sufficed to prove it for linearly ordered continua. He also explored when the supremum of p_2E is attained, and proved in Lemma 10′ that if the supremum p_2E is attained for every linearly ordered continua, then it is attained for every linearly ordered set. In §12.C.3, Theorem 2, Kurepa proved

[117]This problem was addressed by Zermelo and Fraenkel as discussed in Kanamori's first chapter.

[118]On this problem, Kurepa wrote [1935, 1, footnote 2], [1996, 11, footnote 2] "La résponse présumée negative à cette problème est appelée l'Hypothèse de Cantor."

[119]Carlos Alvarez [1999] has written on the history Suslin's Problem, and his discussion of these announcements is particularly interesting and detailed.

that $p_2E \leq (p_1E)^+$.[120] In §12.C.4, Lemma 5, he proved that if the density p_1E is singular, then the cellularity $p_2E = p_1E$ is attained. Todorcevic (cf. [Kurepa, 1996, 299]) combined Lemma 4 of §12.C.3, Lemma 5 of §12.C.4 and Theorem 3 of §11.4 [Kurepa, 1935, 110], [Kurepa, 1996, 100] to point out that Kurepa showed that if the cellularity $c(E) = p_2E$ is singular then it is attained. In footnote 11 [1935, 131][121] (the degree of) cellularity is extended to abstract spaces as the supremum of the sizes of disjoint families of open sets.

Now we turn to Kurepa's systematic study of set-theoretic trees[122] (ramified tables) within the context of ramified sets and ramified tables. Since the definition of a ramified table essentially filled six pages of the published version of the thesis as printed in [1996], we look at an important family of examples.

Suppose that $(P, <)$ is a strict partial ordering and that the relation $>$ is defined from it in the usual way. Let $\|$ be defined on P by $a \parallel b$ if and only $a \not< b$, $b \not< a$ and $a \neq b$. Let $\equiv$ be the relation of identity. Then $* = (\equiv, <, >, \|)$ is a *ramification relation* if for all points $a, b, c \in P$, if $a < c$ and $b < c$, then a and b are comparable, i.e. the set of predecessors of any point forms a chain.[123] Since all pairs are related by one of $\equiv$, $<$, $>$ and $\|$, the set P is a ramified set with respect to this ramification relation.

Kurepa gave, as his first example of a ramified set, a genealogical tree [1996, 67], and his first example of a ramified table was the genealogical tree again [1996, 70], so it is clear that he intended to generalize the notion of tree.

For Kurepa, a ramified table was the result of the process of *ramification* applied to a partial order P, i.e. the stratification of the elements of the partial order into levels, by identifying a first level, removing the first level, letting the second level be the first level of the reduced set, and continuing in this fashion. Some partial orders fail to be suitable for ramification. A set $E \subseteq P$ *has a first level* [première rangée] if for every point $a \in E$, the set of predecessors of a in E has a least element; in such a case, the *first level of* E is the set of the least elements of the sets of predecessors of elements of E. Kurepa attributed this version of the definition to Fréchet; his original definition of first level was that it was the unique set F, if it existed, such that every element of $E \setminus F$ was preceded by one and only one element of F.

A ramified set is a *ramified table* if each of its non-empty subsets has a first level. In the lemma immediately following the definition, he showed that it is equivalent

[120] Todorcevic [1996, 299] called this the first important result on the relationship between density $d(E) = p_1E$ and cellularity $c(E) = p_2E$.

[121] Unfortunately, the supplementary material in pages 127–135 of [Kurepa, 1935] entitled *Complément des résultats précédents* were omitted from the version of Kurepa's thesis in [Kurepa, 1996].

[122] Arthur Cayley studied finite trees, especially rooted finite trees, coming up with the generating function to list the number of rooted trees with n nodes in [1857], made connections with isomers in chemistry in [1875], and counted the number of distinct rooted trees with n nodes in [1881].

[123] The notation and splitting of all pairs recalls the formal splitting by Hausdorff [1909, 300], [2005, 273] of real-valued functions with a common ordered domain by final rank ordering into $f < g$, $f = g$, $f > g$, $f \parallel g$.

to require that each linearly ordered subset be well-ordered and commented in footnote 10 (see [1996, 70]) that Fréchet used this statement as the definition and that part of the proof of the lemma of equivalence of these definitions is due to Fréchet. Thus in the example we are considering, the definition is equivalent to and, in the alternate form, remarkably close to, the current standard definition of a set-theoretic tree as a partially ordered set in which the predecessors of each point are well-ordered.

Immediately after the discussion of the definition of ramified table T, Kurepa recursively defined the αth level [rangée], which he denoted $R_\alpha T$, and used the levels to define the *height* [rang] as the order type of the set of α for which $R_\alpha T$ is non-empty, the *length* [longeur] as the cardinality of the height, the *cofinality* [longeur réduite] (the cofinality of the height), and the *width* [largeur] as the upper bound of the cardinalities of the $R_\alpha T$s.

The name ramified table suggests that Kurepa envisioned a distribution of the elements of T into levels, with boxes to contain the elements at each level and immediate successor boxes all stacked above the box that is their common immediate successor, as might be drawn to illustrate the atomization of a linear order by intervals into a tree. Indeed, in the opening paragraph of the second chapter of his thesis Kurepa cited the idea of subdivision as motivation for his generalization of the relation of order to the relation of ramification. He used this approach to the construction of ramified tables and thus was able to show that if all ramified tables of cardinality ω_1 have either a chain or an antichain of cardinality ω_1, then there is no Suslin line.

Many graduate students have been challenged by their dissertation advisors to give examples that show that their theoretical approaches are substantive, and Kurepa was not the only one who received help from a fellow graduate student in producing a desired example. Nachman Aronszajn,[124] also a student of Maurice Fréchet, constructed the first tree of uncountable height with all levels countable. The Aronszajn construction was announced in an abstract of Kurepa [1934c] and is briefly described in a footnote to Theorem 6 of §9 of Kurepa's thesis [1935, 96], [1996, 88]. Essentially, he built a suitable subtree of the set of one-to-one sequences of countable ordinal length of rationals in $\mathbb{Q} \cap (0, 1)$ whose sums are finite and partially ordered them by end-extension. Aronszajn may have constructed such

[124]Nachman Aronszajn was born July 26, 1907 in Warsaw and died February 1980 in Corvallis, Oregon (cf. [Aronszajn, 2011], [Mardešić, 1999]). He received a University of Warsaw doctorate in 1930 under the direction of Stefan Mazurkiewicz, and a second doctorate [1935] from Université de Paris in 1935 where Fréchet was his advisor. In his Math Review MR0746781 (85m:01086a), Gerlach describes Aronszajn's working life as follows: 1930–1940 in France, 1940–1945 in England, 1945–1948 in France again, and the United States after 1948. He spent time at Oklahoma Agricultural and Mechanical University, leaving after the university fired Ainsley Diamond for failing to sign a loyalty oath (cf. [Magill, 1996]), and moved to the University of Kansas where he was a professor from 1951 until his retirement in 1977. He then moved to Corvallis where he was appointed an adjunct professor. Aronszajn was an analyst known for the Aronszajn-Smith Theorem that every compact operator on any Banach space of dimension at least two has an invariant subspace [1954] and for his work in reproducing kernel Hilbert spaces [1950]. For more on his life see [Szeptycki, 1983]. For a list of his publications, see [Szeptycki, 1980].

a tree prior to Kurepa, but Kurepa came up with his own construction of what is now known as an Aronszajn tree[125] inside the tree $\sigma\mathbb{Q}$ of bounded well-ordered subsets of the rationals ordered by end-extension. He started with an initial level consisting of all rational singletons. At successor levels of the construction, he extended every node on the immediately preceding level in all possible ways. At limit levels α, he took any denumerable subset F of the collection of well-ordered subsets of $\mathbb{Q}$ all of whose initial segments were in levels already constructed which had the property that it and the tree up to that point were dense in each other. More generally, one can build trees inside the ramified table σE, which consists of the collection of non-empty bounded well-ordered subsets of a ordered set E ordered by end-extension.

In §10.1, Kurepa [1935, 98], [1996, 89] introduced the notion of *similarity* for ramified tables and called the similarity classes types of ramification. He showed that if E and F are similar linearly ordered sets, then so are the ramified tables σE and σF of the well-ordered bounded non-empty subsets of E, respectively F, ordered by end-extension. In §10.2, Kurepa sketched the proof that $\sigma\mathbb{Q}$ is *homogeneous*, i.e. that for every element $s \in \sigma\mathbb{Q}$, the set of $t \in \sigma\mathbb{Q}$ properly extending s is similar to $\sigma\mathbb{Q}$.

In §10.3, Kurepa introduced *distinguished ramified suites* [suites ramifiées distinguées]. A ramified table T is a *ramified suite* if for every $a \in T$, the set of $t \in T$ for which either $t <_T a$ or $a <_T t$ has cardinality $|T|$ (see §8.12 [1935, 76], [1996, 82]). A *distinguished ramified suite* is a ramified table S whose height κ is a regular initial ordinal and satisfies additional conditions, which, in the case $\kappa = \lambda^+$ is a successor cardinal may be phrased as follows: S has width λ [ambigu, i.e. ambiguous, neither narrow, nor wide]; S does not have a cofinal branch [descente monotone] and for every element $s \in S$, the set of all $x \in S$ with the same set of predecessors as s has cardinality λ. Let us call S *uniformly branching* if the last condition together with the condition that every element s has successors on every larger level both hold. Then in modern language, the distinguished ramified suites of height λ^+ are the uniformly branching λ^+-Aronszajn trees.[126] At the end of §10.3, he asserted the existence of uniformly branching $\omega_{\beta+1}$-Aronszajn trees without stating the hypotheses (GCH and ω_β regular) that would make his proof generalize. He also raised the question of the existence of a κ^+-Aronszajn tree for an inaccessible cardinal.[127]

Kurepa opened §10.4 with the statement that we do not know if every uniformly branching Aronszajn tree is necessarily similar to a subset of $\sigma_0 = \sigma\mathbb{Q}$, and then stated the following questions:

---

[125]According to Aronszajn's colleague Pawel Szeptycki [1983] at the University of Kansas, Aronszajn constructed the tree associated with his name "in an attempt to fill a gap in a work of G. Kurepa."

[126]For regular κ, a *κ-tree* is a tree of height κ all of whose levels have cardinality $< \kappa$, and a *κ-Aronszajn tree* is a κ-tree with no branch of length κ.

[127]Kurepa's definition of inaccessible is regular limit cardinal [1996, 22]. He does not mention GCH, under which these cardinals and the (strongly) inaccessible cardinals coincide.

Kurepa's First Miraculous Problem: Must every two uniformly branching Aronszajn trees necessarily be isomorphic?[128]

He pointed out that the problem is equivalent to showing that every uniformly branching Aronszajn tree is homogeneous, and in [1935, 100, footnote 4], [1996, 91, footnote 26] posed the question of the existence of a homogeneous uniformly branching Aronszajn tree.

Next we examine results of Kurepa beyond his thesis. Recall that a tree is *special* if it is the union of countably many antichains. Kurepa constructed a special Aronszajn tree inside $\sigma\mathbb{Q}$ not long after his thesis in [1937a, 147–156], [1996, 130–138] introducing a class of trees that has resonated through the century.[129] Note that any Suslin tree is an Aronszajn tree, but since any countable collection of antichains whose union is uncountable must include an uncountable antichain, no Suslin tree can be special. To construct his special Aronszajn tree, Kurepa started with a function $g : \omega_1 \to \mathbb{R}$ with the property that for all $\beta < \omega_1$, the restriction of g to the interval of ordinals $[\omega_2\beta, \omega^2(\beta+1))$ is one-to-one and has range dense in $\mathbb{R}$. He used recursion to define $\varphi : \omega_1 \to \sigma\mathbb{Q}$ so that for all $\alpha < \omega_1$, $g(\alpha) = \sup \varphi(\alpha)$, and so that for each $\beta < \omega_1$, the image under φ of $[\omega^2\beta, \omega^2(\beta+1))$ is the βth level of the tree he is building, requiring essentially the same properties he required in his thesis, with the additional constraint imposed by g, and the requirement that for all $\xi < \zeta$, if $\varphi(\xi)$ is an initial segment of $\varphi(\zeta)$, then $g(\xi) < g(\zeta)$.[130] Most of the work in the proof is to ensure that the values assigned by g to limit levels are attained and are assigned so that the last mentioned inequality is always satisfied. At the end of the proof he considers two alternative restrictions on the functions: (a) for all β, the range of g restricted to $[\omega_2\beta, \omega^2(\beta+1))$ is the same as the range restricted to $[0, \omega^2)$; and (b) g is one-to-one. In case (a) with the range being the set of rational numbers, his proof gives a strictly increasing map into the rationals, i.e. it is $\mathbb{Q}$-embeddable, but he does not mention this fact explicitly. Instead he observes that for any x the set of elements of his tree whose sup is x is an antichain, so in this case the tree is a countable union of antichains. He also proved (see the discussion of [1996, 118] and Théorème 3bis of Section 29 [1996, 140]) that there is an $\mathbb{R}$-*embeddable* Aronszajn tree, that is, one with a strictly increasing mapping into the set of real numbers. The paper also includes a proof by Aronszajn of the existence of an Aronszajn tree whose elements are closed subsets of a compact subset of a metric space ordered by the superset relation (see Théorème 2 [Kurepa, 1937a, 135],[131] [1996, 121]).

---

[128]For the original statement of the "problème miraculeux 1," see [1935, 100] or [1996, 91].

[129]Kurepa only noted in passing (see [Kurepa, 1996, 138]) that he was able to construct an Aronszajn tree which is a denumerable union of antichains; that is, this fact is not one of the numbered theorems, corollaries and lemmas in the paper.

[130]It is known today that care must be taken at the limit levels to ensure that such a tree is special. See [Devlin, 1972] discussed in §7.7.

[131]Unfortunately, pages 135-138 are omitted from the copy of this paper posted on the site http://elib.mi.sanu.ac.rs/ of the eLibrary of the Mathematical Institute of the Serbian Academy of Sciences and Arts.

Kurepa [1940a, 837] proved that a partially ordered set admits a strictly increasing mapping into the rationals if and only if it is the union of countably many antichains. Let us expand the scope of the concept of *special* to apply to partially order sets in general. Then Kurepa's result may be restated as follows: a partial order is $\mathbb{Q}$-embeddable if and only if it is special.

Kurepa announced in December [1937b] an upper bound on the cardinality of a partially ordered set E. He used pE to denote the cardinality of E (p for puissance, power). For a non-empty E, he denoted the supremums of the cardinalities of several families of subsets of E as follows: p_cE is the sup of sizes of well-ordered subsets (c for croissant, increasing); p_dE is the sup of sizes of conversely well-ordered subsets (d for decroissant, decreasing); and p_sE is the sup of sizes of subsets of pairwise incomparable elements. He then set $p_0E = \max(p_cE, p_dE)$ and $bE = \max(p_0E, p_sE)$, prior to introducing his *fundamental relation* [1937b, 1196], [1996, 387]:

$$\textit{Kurepa's Fundamental Relation:} \quad bE \leq pE \leq (2p_sE)^{p_0E}.$$

Kurepa also showed that any well partial order in which all well-ordered subsets and all subsets of pairwise incomparable elements are countable must itself be countable [Kurepa, 1937b, Section V].

The proof of Kurepa's Fundmental Relation [1939] was published just before World War II, overlooked by reviewing journals, and reprinted in [1959b] in order to distribute it more widely. Note that the Fundamental Relation for partial orders generalizes the Hausdorff-Urysohn Inequality [Urysohn, 1924] for linearly ordered sets (i.e. those with $p_sE = 1$): $pE \leq 2^{p_0}$, using the notation given above. The proof is divided into two cases. In case A, Kurepa proved the theorem for partial orders in which all linearly ordered subsets are well-ordered, and in case B, Kurepa considered an arbitrary partial order, superimposed a well-order of the underlying set on it to get a new partial order in which all linearly ordered subsets are well-ordered, and applied case A to conclude the inequality held in general.

To show the sharpness of the bound in his Fundamental Relation, Kurepa gave the example of the lexicographic ordering on the set E of sequences of 0's and 1's of length $|n|$ since for any cardinal n (which we may assume is an initial ordinal), one has $p_0E = n$, $p_sE = 1$ and $pE = 2^n$. Thus the partial order considered in case B to prove the Fundamental Relation is the superposition of a lexicographic order and a well-order, which is now known as the Sierpiński poset.[132]

In his Math Review MR0058687 of [Erdős and Rado, 1953], Kurepa asserted that "the idea of using superposition of orderings of a same set was, as far as we know, first introduced in 1937 by the reviewer." He referred the interested reader to the paper discussed above and to the announcement in [1937c] of a two-part paper [1940a], [1941], specifically pointing to page 487 in the latter (see also

[132]For example, Fraïssé [2000, 47] defined the *Sierpiński poset* to be the superposition of the usual ordering on the countable ordinals with an order induced from the real numbers via an injection of the set of countable ordinals into the set of real numbers. Recall that Sierpiński [1933] used this approach to define a symmetric binary relation, which via translation is now known as the Sierpiński partition, but it is not a partial order.

footnote 28 on page 179 of [1996]), where he had given the superposition of a well-order on another partial order.

Kurepa was a pioneer in the study of uncountable partial orders and what have come to be called trees, and, as with many pioneers, his ground-breaking work was not immediately recognized.

4 1940-1950: PIONEERING PARTITION RESULTS

The 1940's featured pioneering results in the Ramsey theory of the uncountable and the development and refinement of some basic tools of the trade. Progress was made on understanding the structure of partial orders and trees, and Maharam used a Suslin line in her exploration of the theory of measure algebras. Specific combinatorial results for this decade include the Δ-system lemma, the first unbalanced generalization of Ramsey's Theorem to the uncountable, and foundational results in partitions of pairs from a set. König's Lemma was generalized to measurable cardinals, and generalizations of Aronszajn trees to cardinals above ω_1 were made. Questions were raised that stimulated interest in quasi-orders and in ω_1-trees with many branches.

In this decade, Gödel [1940] published his monograph on the constructible universe, establishing the consistency of the Axiom of Choice and the Generalized Continuum Hypothesis with the axioms of set theory. Garrett Birkhoff's *Lattice Theory* [1940] became an oft-cited reference in the theory of orders. Mostowski and Tarski [1949] announced that the first-order theory of well-ordered sets is decidable.[133]

World War II had a devastating effect on mathematical life with many mathematicians killed outright or sent to concentration camps, and others decamping: Erdős left Hungary; Rado emigrated to England; and Tarski emigrated to the United States. After Germany invaded Poland and disrupted intellectual life there, an underground university sprang up in Warsaw, and Kuratowski, Mostowski and Sierpiński were among those who taught in it. After the end of the war, they rebuilt the group in logic and set theory. Publication of journals and books were slowed or disrupted as in the case of *Fundamenta Mathematicae*, and authors shifted to journals they had not used before and did not return to after the war. With the German occupation of France during World War II, the publication of the treatise *L'Énumération transfinite, Livre I* [1946] by Arnaud Denjoy[134] was postponed

[133]Slomson in his Math Review MR0519800 (80d:03027a) observed that the decidability of the first-order theory of well-ordered sets was first obtained by Mostowski and Tarski in 1938–1941. It was one of the items in Mostowski's notebook that was burned during World War II. The proof via an axiomatization and quantifier elimination argument appears in a paper joint with John E. Doner, a student of Tarski, in [1978]. Slomson noted that he had an essentially similar axiomatization in [1972].

[134]Arnaud Denjoy received his doctorate in 1909 from École Normale Supérieure Paris, where Baire was his advisor. His main focus was the theory of functions of a real variable.

to 1946.[135] This book included a bibliography compiled by Gustave Choquet[136] (cf. [Denjoy, 1946, xxv–xxxvii]) consisting mainly of references appearing after 1920 and organized by topics, including the following most closely related to the subject of this chapter: (a) linearly ordered sets; the general case; (b) characterization of the linear continuum; Suslin's Problem; (c) well-ordered sets; ordinal numbers; (d) ordinals of the second number class; enumeration; (e) inaccessible ordinal numbers. It included Erdős' paper of [1942] with the foundational results in the partition calculus and papers by Dushnik and Miller prior to their [1941], and listed Kurepa's work.

4.1 Combinatorial tools

Before we get into the specifics for order and Ramsey theory, we review advances in the basic tools of combinatorial set theory in a short subsection for ease of reference. A collection of sets is a Δ-*system* if there is a set K such that any two distinct sets in the collection have K as their intersection. Nikolaĭ Šanin[137] [1946] proved the following lemma, which I have restated in modern terminology.

> *The Δ-System Lemma for Finite Sets:* Any uncountable collection of finite sets has an uncountable subcollection which is a Δ-system.

Another tool in combinatorial set theory is the use of equivalence relations, which were discussed by Stoll [1949] in an *American Mathematical Monthly* article, particularly in connection with the study of algebra. We quote from page 377: "we mention that the set E of all possible equivalence relations definable over a given set S has been made an object of study," and note his examples: [Birkhoff, 1940] and [Ore, 1942].

Rado [1949a] introduced a useful principle, which he called "a combinatorial lemma" in his article extending Whitney's notion of rank for finite sets to rank for sets of arbitrary cardinality. Below is a modern formulation using functions.

> *Rado Selection Principle:* Let $\mathcal{F} = \{F_i : i \in I\}$ be a finite family of sets and for each finite $J \subseteq I$, let $g_J : J \to \bigcup \mathcal{F}$ be such that $g_J(j) \in F_j$ for all $j \in J$. Then there is a function $h : I \to \bigcup \mathcal{F}$ such that for all finite $M \subseteq I$ there is a set N with $M \subseteq N \subseteq I$ so that for all $n \in M$, $h(n) = g_N(n)$.

[135] In a footnote to the preface dated March 1946, Denjoy [1946, vi] wrote "Les conditions créées par l'occupation allemande en France de 1940 à 1944 ont fait que de ce Traité le Livre I, composé typographiquement en première épreuve dès les premier mois de 1942, n'a pu être édité qu'en 1946.

[136] Choquet earned his doctorate at École Normale Supérieure Paris in 1946 with Arnaud Denjoy as his advisor.

[137] Nikolaĭ Aleksandrovič Šanin (May 25, 1919 –) was a student of A. A. Markov at Leningrad State University. His Ph.D. dissertation *Extensions of Topological Spaces* was defended in 1942 and his D.Sci. dissertation was defended in 1946. (See G. Mints, http://logic.pdmi.ras.ru/šhanin/biogr2.html downloaded on August 11, 2009.)

The Rado Selection Principle was the compactness principle of choice for combinatorialists of his day and played an important role in transversal theory.

4.2 Ordered sets and their structure

We start this section with a result of Rothberger which is pertinent to the Normal Moore Space Conjecture, followed by two questions of Erdős and Tarski.

The *pseudointersection number* $\mathfrak{p}$ is the minimal cardinality of a family of subsets of ω with the strong finite intersection property (all intersections of its finite subfamilies are infinite) which has no infinite pseudointersection, i.e. no single set that is almost contained in every member of the family. Fritz Rothberger [1948] proved that if the pseudointersection number is greater than $\aleph_1$, then there is a *Q-set*, i.e. an uncountable subset of the real line every subset of which is a G_δ[138] in the relative topology.

The last part of the paper [1943] by Erdős and Tarski is subtitled *General remarks on inaccessible numbers* and included a list of six problems concerning a cardinal $\mathfrak{n}$, of which the last two (see [1943, 327–328]) are most pertinent to this section.

> Problem 5. (*The Ordering Problem.*) Is it true that every ordered set N of power $\mathfrak{n}$ contains a subset X of power $\mathfrak{n}$, which is either well ordered, or becomes well ordered if we invert the ordering relation?
>
> Problem 6. (*The Ramification Problem.*) Let ν be the smallest ordinal number such that the power of all ordinals $\xi < \nu$ is $\mathfrak{n}$. Is it true that every ramification system of the νth order has power $< \mathfrak{n}$ for every $\xi < \nu$, contains a well-ordered subset of the type ν. (By a ramification system S we understand a partially ordered set which has the property that, for every $x \in S$, the set $S(x)$ of all elements $y \leq x$ is well ordered; If the set $S(x)$ is of the type ξ the element x is said to be of the ξth order. The order of the whole set S is the smallest ordinal number greater than the order of all elements of S.)

For the Ordering Problem, Erdős and Tarski noted that the answer is obvious for $\mathfrak{n} = \aleph_0$, and for uncountable $\mathfrak{n}$ which are not inaccessible they credited Hausdorff [1914, 145–146] with the solution, noting [Erdős and Tarski, 1943, 328]: "he does not state the solution explicitly, but it can be deduced easily from his results." For the Ramification Problem, the authors noted that a solution for $\mathfrak{n} = \aleph_0$ was given by Dénes Kőnig [1927]. Erdős and Tarski continued: "for numbers $\mathfrak{n} > \aleph_0$ which are not inaccessible it was given by Aronsajn" (sic). It is surprising that the authors did not mention Kurepa, who brought the word *ramification* into mathematics for this type of system. Tarski presumably was acquainted with Kurepa's work after Kurepa's visit to Warsaw in 1937 (see [Kurepa, 1937a]), but

[138] A set is a G_δ if it is the countable intersection of open sets.

Erdős was not. In his prior [1942], Erdős had already used the approach he came to call the *ramification method* without using this name.

Disjointness

In a paper received by the journal in 1946, Rado [1949b] investigated the circumstances under which it is possible to divide a system of intervals $\langle I_\nu \mid \nu \in N \rangle$ into k (k is finite) pairwise disjoint families, where N is an abstract index set. At the beginning of Section 2 on page 510, he sets the scene: "Sets occurring in this note may have any arbitrary cardinality. Zermelo's axiom will be used freely." He proved that a necessary and sufficient condition is that no $k+1$ intervals I_ν belonging to $k+1$ distinct indices have a point in common. He gave a different characterization in the following proposition.

> *Rado's Disjoint Interval Theorem:* If any subsystem of $k+1$ distinct interval $I_{\nu_0}, I_{\nu_1}, \ldots, I_{\nu_k}$ can be subdivided into k groups of pairwise disjoint intervals, then the same kind of subdivision is possible for the whole system.

Erdős and Tarski in [1943] generalized almost disjointness to partial orders and investigated sizes of mutually exclusive subsets of a partial order $(P, \leq)$. They declared y and z from $(P, \leq)$ to be *disjoint* if there is no non-null element x with $x \leq y$ and $x \leq z$, and called a subset of P *mutually exclusive* if its elements are pairwise disjoint in this generalized way. For $F \subseteq P$, they set $\mathfrak{d}(F)$ to be the least cardinal greater than the power of every mutually exclusive subset of F. Define a closely related notion by setting $m(F)$ to be the least upper bound of the set of cardinalities of mutually exclusive subsets of F. Erdős and Tarski showed that for any partial order P, if $m(P)$ is not an uncountable inaccessible cardinal, then P contains a mutually exclusive set of power $m(P)$. Their proof involved an analysis of non-decreasing functions from P to the cardinal numbers. They applied their results to Boolean algebras, rings of sets and fields of sets. In particular, they showed that if F is a field of sets and $m(F)$ is not a regular uncountable inaccessible cardinal, then there is a mutually exclusive subset of power $m(P)$; and if κ is a regular uncountable inaccessible cardinal, then there is a field G with $m(G) = \kappa$ which does not contain a mutually exclusive subset of power κ. Erdős and Tarski also showed in Corollary 3 and its proof that in a topological space X with uncountable cellularity which is not a regular limit cardinal, the supremum $c(X)$ is realized as the size of a pairwise disjoint family of $c(X)$ many open sets of X. This result extends Kurepa's result [1935] for spaces with singular cellularity.

Dushnik and Miller

Dushnik and Edwin Miller[139] [1940] examined linearly ordered sets and order-preserving embeddings. Recall that every infinite well-ordering is order-isomorphic to a proper subset. The first question that Dushnik and Miller asked was if the same is true for every infinite linear order, and they proved that it is true for denumerable linear orders. They observed that for a well-ordering $(A, <)$, every order-preserving embedding $f : A \to A$ has the property that $f(a) \geq a$ for all $a \in A$, and asked if this property characterizes well-orderings among linear orderings. They show that this property does characterize the well-orderings among the denumerable linear orderings.

In the same paper, they go on to construct, using the Axiom of Choice, an uncountable subset E of the reals which is rigid, i.e. the only order-endomorphism is the identity. They appear not to be aware of the type ring for scattered linear order types discussed in [Hausdorff, 1908] or the fact that he showed that every type is either scattered or a sum over a set of dense order type of scattered types.

In a subsequent paper, Dushnik and Miller [1941] introduced a notion of dimension for partial orders, namely the minimal size of a family of linear orderings whose intersection results in the partial order. They constructed examples showing that for every cardinal κ, there is a partial order whose underlying set has cardinality κ and whose dimension is κ. They revisited their rigidity result from [1940] by constructing an example N that has the property that for any uncountable subset $M \subseteq N$, M is not similar to any proper subset of itself which differs from M in more than a denumerable infinity of points.

In Theorems 5.24 and 5.25, Dushnik and Miller showed that in a infinite partial order P, on the one hand, if every subset of power $|P|$ has two incomparable elements, then P has a denumerable linearly ordered subset, and on the other hand, if every denumerable subset has two incomparable elements, then P has a linearly ordered subset of power $|P|$. Note that the above results of Dushnik and Miller for uncountable partial orders follows from the approach Kurepa took to showing that any partial order in which all well-ordered subsets and all subsets of pairwise incomparable elements are countable must itself be countable in [1937b, Section V], and from his proof that any narrow tree has a linearly ordered subset of cardinality the power of the entire tree in Théorème 5 [1935, 80], [1996, 76]. A proof appears in [Kurepa, 1948a, 30, Lemma 5], [1996, 209] that every uncountable well-founded partial order with all finite levels in its stratification has an uncountable chain.

---

[139] Edwin Wilkinson Miller and Dushnik were fellow students at the University of Michigan: Miller received his Ph.D. from Raymond Wilder in 1930 (Raymond Wilder's doctorate was from the University of Texas at Austin, where he was student of R. L. Moore). In a letter dated October 15, 1929 (parts of this letter are quoted in [Fitzpatrick, 1997, 47]), Hildebrandt [1929] wrote to Moore about Miller's work on whether two of Moore's axioms from his 1929 Colloquium Lecture for the Summer Meeting of the AMS in Boulder define a metric space. Thus Miller was familiar with families of sets which were nested or disjoint. In 1934, when Hildebrandt, began a term as chair of the department, Miller became a member of the department and continued there until his death of a heart attack in 1942 (see [Kaplan, 1971, 184–185]).

Further work of Miller on order is discussed in §4.3, and additional Ramsey related work of Dushnik and Miller is discussed §4.4.

Order types

Sierpiński [1946] showed there are at least $2^{\mathfrak{c}}$ many real types by defining a family of that many pairwise incomparable order types of sets of real numbers of power $\mathfrak{c} = 2^{\aleph_0}$ (he did not use the Continuum Hypothesis).

Philip Carruth [1942] showed that the maximum order type of the union of sets of ordinals A and B of order type α and β respectively is the Hessenberg sum of α and β.

Noberto Cuestra Dutari [1943, 243] showed in Theorem 15 that the collection of dyadic sequences of length ω_α which are eventually zero, ordered lexicographically, is universal for linearly ordered sets of cardinality $\aleph_\alpha$, reproving a result known to Hausdorff. Recall that in the case of countable linear orders, the set of rationals is a countable linear order universal for all countable linear orders.

Kurepa [1948b] characterized in [1996, 200, Théorem I] the denumerable scattered linear order types as the denumerable order types that do not embed the order type of some countable ordinal. He noted that the supremum of the order types of the well-ordered subsets of η, the order type of the rationals, is $\aleph_1$, citing [Hausdorff, 1914, 99].

We continue the discussion of order types in the next section with Fraïssé's paper which includes results for order types so as not to split the discussion of it between two subsections.

Well-quasi-order theory

First steps toward the development of a theory of well-quasi-orders[140] were taken. Roland Fraïssé,[141] in his paper [1948] on the comparison of order types, wrote $S < T$ to indicate there is a subset of T of type S and $S \ll T$ for $S < T$ and not $T < S$, and called a type T poor (pauvre) if $T \ll \eta$, i.e. T is countable and scattered. He used Sierpiński [1928a] as a reference and announced several results, a few of which are listed below, including an independent discovery of the characterization of countable non-scattered ordinals given by Kurepa above:

[140] A *quasi-order* is a reflexive transitive binary relation $\leq_Q$ on a set A; it is a well-quasi-order (wqo) if for every infinite sequence $\langle a_i : i < \omega \rangle$ of elements of A, there are $i < j$ such that $a_i <_Q a_j$.

[141] Roland Fraïssé (1920 – March 30, 2008) received his doctorate from Université de Paris in 1953 under René de Possel (cf. http://www-history.mcs.st-andrews.ac.uk/Biographies/Possel.html). Denjoy was de Possel's advisor (cf. [Pouzet, 1992]). In his thesis Fraïssé [1953], [1954] used the back-and-forth method to determine whether two model-theoretic structures are isomorphic. He introduced amalgamation and developed a method for constructing countable homogeneous models from finite structures via amalgamation (see his influential book [2000]). Fraïssé held a chair of mathematics in Algiers 1941–1959, and then spent most of his career at Université de Provence in Marseille, France. He was one of the founders of Bourbaki (cf. [Delsarte, 2000]).

1. A necessary and sufficient condition for a type to be non-scattered is that it embeds every denumerable well-ordered set.

2. For every finite or denumerable set of scattered types T_i, there corresponds a scattered type T with $T_i \ll T$ for all i.

3. Every scattered type T is the sum of a finite sequence, an ω-sequence or an ω^*-sequence of scattered types T_i with $T_i \ll T$.

Here are the four *Fraïssé Conjectures* he made in [1948, 1331]:

> *Hypothèses.* – Nous considérons comme vraisemblables les propositions suivantes:

I. Il n'existe pas de suite infinie de types dénombrables $T_1, T_2, \ldots, T_i, \ldots$ avec $T_{i+1} \ll T_i$ pour tout i.

II. Si T est pauvre, le nombre des classes d'équivalence définies par la relations $\approx$ sur l'ensemble des types $< T$ est au plus dénombrable.

III. Soient E une famille, finie ou dénombrable, de types dénombrables t; $T(E)$ un type quelconque $> t$ quel que soit t de E; $\mathcal{T}$ un type inférieur à tout $T(E)$; alors $\mathcal{T}$ est inférieur à l'un des types t.

IV. Tout type indécomposable pauvre T est similaire au type d'une ω-suite ou d'une ω^*-suite de types $T_i \ll T$, ces types T_i étant soit indécomposables, soit de puissance 1.[142]

Erdős [1949] asked the following problem in the *American Mathematical Monthly* that is related to the first Fraïssé Conjecture:

> If a set A of integers has the *divisor property* (i.e. for every sequence $\langle a_i : i < \omega \rangle$ of elements of A, there are $i < j$ such that $a_i | a_j$), then does it follow that the set of all products of n-elements from A has the divisor property.

[142] *Conjectures.* – We consider the following propositions plausible:

I. There is no infinite sequence of denumerable types $T_1, T_2, \ldots, T_i, \ldots$ with $T_{i+1} \ll T_i$ for all i.

II. If T is poor $[T \ll \eta]$, the number of equivalence classes defined by $\approx$ of the set of types $< T$ is at most denumerable.

III. Let E be a family, finite or denumerable, of denumerable types t; let $T(E)$ be an arbitrary type $> t$ for every t in E; let $\mathcal{T}$ be a type strictly below $T(E)$; then $\mathcal{T}$ is below one of the types t.

IV. Every indecomposable poor T is similar to the type of an ω or ω^* sequence of type T_i strictly below T, with the types of the T_i being indecomposable or of power 1.

This question stimulated interest in the notions of well-quasi-order (wqo) and well partial order.[143] Eric C. Milner [1985] attributes the first development of wqo theory to Graham Higman [1952], in his analysis of divisibility in algebras. Higman used the equivalent notion of having the *finite-basis property.*[144]He proved the equivalence of the finite-basis property with the non-existence of both infinite antichains and infinite descending sequences, and referred to parallel work by Erdős and Rado (see [1952b], [Rado, 1954b][145]).

4.3 Around Suslin's Problem

The definitions of Aronszajn and Kurepa trees grew out of Kurepa's investigation of Suslin's Problem.

Aronszajn and Kurepa trees

Kurepa [1940b] revisited the family $\sigma_0 = \sigma\mathbb{Q}$ of all (non-empty bounded) well-ordered subsets of the rationals introduced in his thesis and proved (see Lemma 1 on page 148) that every uncountable subset of this tree under end-extension includes an uncountable antichain. On the same page, in footnote 7, he commented that we do not know if there is a Suslin line which is universal for Suslin lines. He also proved that every $\mathbb{R}$-embeddable uncountable ramified table T has an antichain of cardinality $|T|$.

Note that the result described in §3.4 for partial orders of the equivalence of being the countable union of antichains and being $\mathbb{Q}$-embeddable applies to Aronszajn trees: they are special if and only if they are $\mathbb{Q}$-embeddable.

Kurepa continued his work on the classification of trees with a paper [1942][146] that focused on suites T[147] of height ω_1 in which every level is countable and has the further property that if the set of predecessors of an element is a successor ordinal or empty, then there are denumerably many elements with this set of predecessors; while if its set of predecessors has limit order type, then it is the unique element with this set of predecessors. That is, he focused on those T which are a denumerable forest of uniformly branching ω_1-trees. As Kurepa's earlier work was missed by many, we note that in this paper Kurepa referred to his thesis and the proof of the existence of an Aronszajn tree and the equivalence of a negative answer to Suslin's Problem to the existence of a Suslin tree. He formulated an

[143]A *well partial order* is a well-founded antisymmetric partial order $(P, <)$ with no infinite antichains.

[144]The *closure* of a subset A of a quasi-order $(P, <)$ is the set $\mathrm{cl}(A)$ of all $q \in P$ for which there is some $a \in A$ with $a \leq q$. A *closed subset of P* is a set which is its own closure. A quasi-order has the *finite-basis property* if every closed subset is the closure of a finite subset.

[145]Rado [1954b] used the concept of *partial well-order*, a well-founded relation with the finite-basis property but not necessarily transitive.

[146]Todorcevic (cf. [Kurepa, 1996, 8]) referred to [Kurepa, 1942] as "one of Kurepa's best written and most frequently cited papers."

[147]Recall that for Kurepa, a suite is a particular kind of ramified table.

equivalence of the Continuum Hypothesis in terms of a particular kind of ramified suite.

Kurepa proved that every denumerable forest T of normal ω_1-trees with the property that every element of T is in an uncountable branch must have uncountably many uncountable branches, and pointed out that we do not know how to evaluate the cardinality of the set of uncountable branches. The existence of an ω_1-tree with more than $\aleph_1$ branches is now known as *Kurepa's Hypothesis*, abbreviated KH and such trees are called *Kurepa trees*. Kurepa constructed an example μ of a denumerable forest of ω_1-trees in which every element is in an uncountable branch as follows: the elements of μ are finite sequences of countable ordinals ordered by $\langle\alpha_0, \alpha_1, \ldots, \alpha_n\rangle < \langle\beta_0, \beta_1, \ldots, \beta_k\rangle$ if and only if $n \leq k$, $\alpha_n < \beta_n$ and for all $i < n$, $\alpha_i = \beta_i$. Note that s and $s^\frown\langle 0\rangle$ are incomparable and have the same set of predecessors, so each element has infinitely many companions with the same set of predecessors. Also, the sequence of all $s^\frown\langle\xi\rangle$ for $\xi < \omega_1$ is a branch, so that every node is on an uncountable branch. Kurepa proved that every denumerable forest of ω_1-trees contains a copy of μ.

While Ernst Specker[148] was a graduate student, Roman Sikorski visited and asked the following question. For a regular cardinal κ, suppose collections D_α for $\alpha < \kappa$ of sequences of length α of 0's and 1's are given which enjoy the following four properties: (1) D_1 is non-empty; (2) if $\alpha < \beta < \kappa$, then every element of D_β is an extension of an element of D_α; (3) if $\alpha < \beta < \kappa$, then every element of D_α has an extension in D_β; and (4) for all $\alpha < \kappa$, $|D_\alpha| < \kappa$; must there be a sequence of length κ which, for all $\alpha < \kappa$, extends an element of D_α? The condition that the sequences be binary was dropped to make the question more interesting. Henry Helson [1949] had already shown the existence of a family of sets of sequences with the first three properties and no sequence of length κ whose initial segments all appear in the appropriate D_α. Specker set himself the task of constructing a counter-example, that is, the levels of a κ-Aronszajn tree for κ the successor of a regular cardinal. He succeeded by assuming the GCH and using increasing functions (see [Specker, 1949]).

Suslin Problem variations

Suslin's Problem continued to stimulate and influence research. During this decade various topological questions related to Suslin's problem were investigated; Miller and Sierpiński came up with equivalences in terms of partial order; and Maharam used Suslin's Problem in her investigation of measure for Boolean algebras. The lack of references to Kurepa by Miller, Maharam, and Sierpiński show that Kurepa's [1935] formulation in terms of a ramified table (Suslin tree) was not

[148]Specker was born February 11, 1920 in Zurich, received his doctorate from Eidgenössische Technische Hochschule (ETH) in 1949, and spent his professional life there. He worked on Quine's set theory New Foundations, and is perhaps best known for the Kochen-Specker Theorem in quantum mechanics. On August 14, 2009, the Mathematics Genealogy Project site listed 42 students for him, including Walter Deuber and Hans Läuchli, and selected works have been collated in [Specker, 1990].

well-known at the time.

We note in passing that Y. S. Ochan[149] [1941] investigated two strengthenings of a topological variation of Suslin's problem and showed that they were equivalent.

Say a topological space T has the *Property of Knaster* if every uncountable family of open sets of T contains an uncountable subfamily of sets which pairwise have non-empty intersection. In April 1941,[150] Knaster [1945] proved that for complete linearly ordered sets, the Property of Knaster is equivalent to separability, i.e. having a countable dense set. Thus Suslin's Problem is equivalent to the question of whether, every Suslin continuum has the Property of Knaster. In May 1941, Knaster and Marczewski (under the name of Szpilrajn) asked in problem 192 of the *Scottish Book* [Mauldin, 1981, 265–266] if there is a topological space with the countable chain condition which does not have the Property of Knaster. Marczewski went on to ask whether the product of two partial orders with the countable chain condition necessarily has the countable chain condition. Both questions turned out to be independent of set theory; for details see [Mauldin, 1981] and Marczewski [1945], [1947].

Edwin Miller [1943][151] described a partial order whose existence is equivalent to the existence of a Suslin line: P has power $\aleph_1$; every subset of power $\aleph_1$ contains two comparable elements and two non-comparable elements; and if x and y are non-comparable elements, then there exists no z such that $x < z$ and $y < z$.

To prove the existence of such a partial order from a Suslin line, Miller recursively defined a sequence of non-overlapping (closed) intervals of length ω_1 and ordered them by reverse inclusion.[152]

To prove the existence of a Suslin line from such a partial order, he used two results that follow from the Dushnik-Miller Theorem (see below) to build a tree where each node has denumerably many successors; he linearly ordered the successors in type $\omega^* + \omega$; he passed to the lexicographic order, noted that there are no jumps, and filled any gaps by adding new elements to get a complete linear order.

Sierpiński [1948, 165] also gave an equivalence to the existence of a Suslin line in terms of families of infinite subsets partially ordered by the subset relation, which is, in essence, another description of a Suslin tree:

> *Problème P.* Soit F une famille infinie d'ensembles jouissant de trois propriétés suivantes:
>
> 1. X et Y étant deux ensembles de la famille F, on a ou bien $X \subset Y$, ou bien $Y \subset X$, ou bien $XY = 0$.

[149] Yuri Semyonovich Ochan is perhaps best known for his book of problems in mathematical physics [1967]. His name has been variously transliterated as Ju. S. Očan, Y. S. Ochan, G. Otchan.

[150] The date for Knaster's proof is in the *Scottish Book* [Mauldin, 1981].

[151] A footnote in Miller's paper noting that the paper was received July 10, 1942 also notes that the author died on July 23, 1942.

[152] The idea of using a sequence of non-overlapping intervals to study a space is natural to those familiar with Moore spaces.

2. Chaque sous-famille de F formée d'ensembles disjoints est au plus dénombrable.
3. Chaque sous-famille F_1 de F telle que, dans chaque couple de ses éléments, il existe un élément qui est contenu dans l'autre, est au plus dénombrable et a un ensemble maximal (c. à d. contenant tous les autres ensembles de F_1).

La famille F est-elle nécessairement dénombrable?[153]

On the one hand, Sierpiński appeared unaware of Kurepa's work on Suslin's Problem, while on the other, he referenced [Denjoy, 1946] for literature on the problem, and this reference includes a bibliography complied by Choquet, as mentioned at the beginning of Section 4, which listed Kurepa's work. Sierpiński, in a note added in proof, cited work of Miller [1943] and Maharam [1947] on Suslin's Problem.

Dorothy Maharam[154] [1947] provided a set of necessary and sufficient conditions for a countably complete Boolean algebra to have a measure.[155] In a discussion of whether one of her conditions for the existence of a continuous outer measure can be replaced by the countable chain condition, she indicated the thinking of the times about Suslin's hypothesis [1947, 164]: "We now show that an affirmative answer is unlikely—or at all events can hardly have an easy proof—by showing

[153] *Problem P.* Let F be an infinite family of sets enjoying the following three properties:

1. If X and Y are two sets of the family F, then one has pairwise $X \subset Y$, or $Y \subset X$, or $X \cap Y = \emptyset$.
2. Each subfamily of F of pairwise disjoint sets is at most denumerable.
3. Each subfamily F_1 of F such that, in each pair of its elements, there is one element which is contained in the other, is at most denumerable and has a maximal set (that is to say, containing all the other sets in F_1).

Is the family F necessarily denumerable?

[154] Dorothy Maharam was born in West Virginia in 1917. She did her undergraduate work at Cornell University and obtained her doctorate at Bryn Mawr College under the supervision of Anna Pell Wheeler in 1940 with a dissertation titled *On Measure in Abstract Sets.* Next she held a postdoc at the Institute for Advanced Study at Princeton, where she met Arthur H. Stone, whom she married in April 1942. Stone was at Princeton University during 1941–1942 and held a series of short appointments (Purdue 1942–1944, Geophysical Laboratory of the Carnegie Institute in Washington, D.C., Trinity College) before becoming a lecturer at Manchester University 1948–1957, where Maharam joined the staff in 1952. They both obtained professorships at the University of Rochester in 1961 where they stayed until retirement (cf. [Cohn, 2000]). For a brief biography of Maharam, see [Oxtoby, 1989] and for an overview of her work, especially on measure theory, see [Choksi *et al.*, 1989]. David Fremlin [1989] declared her "the chief architect of the present theory of measure algebras" in his masterful exposition of measure algebras in the *Handbook of Boolean Algebras.* A recent paper of Boban Veličković includes a survey of recent work on Maharam algebras, i.e. complete Boolean algebras that admit probability measures. Arthur Stone described working with her in an interview with W. Wistar Comfort [1997].

[155] She thanked Arthur H. Stone in a footnote at the bottom of the first page, where she noted that the paper is based in part on work done during 1941-1942 when she held the Alpha Delta Xi Fellowship of the American Association of University Women and that she had presented a preliminary version of the paper to the American Mathematical Society in November 1942.

that it would imply the truth of a well-known hypothesis of Souslin." She called a family of sets A a *Souslin system* if

1. every pair is either comparable under $\subseteq$ or disjoint,
2. every subfamily whose sets are pairwise disjoint is countable,[156] and
3. every subfamily whose sets are pairwise comparable is countable.

Maharam stated that "it was conjectured by Souslin that every Souslin system is countable." By suitable thinning and selection of maximal disjoint families a level at a time, Maharam transformed an uncountable Souslin system into a Suslin tree (or forest of trees) under the subset relation. She used this tree to construct a Boolean σ-algebra with no non-trivial outer measure (see Theorem 5).

The following year, Maharam [1948] proved that all Souslin subsystems of Boolean algebras are countable if and only if on each non-atomic Boolean algebra satisfying the countable chain condition there exists a real-valued function f such that (1) f is monotone non-decreasing, (2) f vanishes only at 0 and (3) every non-zero element contains non-zero subelements on which f is arbitrarily small. She further showed that for a fixed non-atomic Boolean algebra E with the countable chain condition, to show that all its Souslin subsystems are countable it is sufficient to have such functions satisfying the three conditions for every such subsystem, but it is not sufficient to have a single function for E since under the assumption that Suslin's Hypothesis is false, she gave a counter-example.

In setting the context for the paper, Maharam recalled her result that Suslin's Hypothesis is implied by the existence of a non-trivial outer measure satisfying certain conditions and in a footnote on the first page attributed the implication from the existence of such a measure to Gödel. She asserted that the equivalence of Suslin's Hypothesis to the statement that all Souslin systems are countable was well-known, citing [Miller, 1943] together with some results from [Dushnik and Miller, 1941].

4.4 Ramsey theory

Marshall Hall [1948] extended Philip Hall's Marriage Theorem to an infinite system of finite subsets of a given set. Independently, C. J. Everett and G. Whaples [1949] extended Hall's Marriage Theorem to infinite sets.[157] Other versions include one due to Paul Halmos and Herbert Vaughan [1950] using Tychonoff's Theorem, and another due to Walter Helbig Gottschalk [1951], who derived it as a corollary to the Rado Selection Principle (see Subsection 4.1).

Rado [1943] extended his classification of which systems of linear equations with positive integer coefficients are k-regular to systems of homogeneous linear equations whose coefficients are in a ring of complex numbers.

[156] Maharam [1948] called this constraint the *countable chain condition* where Kurepa [1935] used *countable degree of cellularity*.

[157] In a footnote, Everett and Whaples credit the basic idea of the proof to Erdős.

We now turn to the first steps in extending Ramsey's Theorem for partitions of pairs to the uncountable, and we will look at three papers: the Dushnik and Miller paper [1941] discussed above for its results on order types, a less well-known paper [1942] by Erdős published in a Latin American journal, and a related paper [1953] by Kurepa who obtained the results in 1950, extending previous work from 1939. We discuss the mathematics of Kurepa's paper here and postpone comment on its publication date in 1953 to the section on the 1950's.

Dushnik and Miller [1941] formulated a positive generalization of Ramsey's Theorem for pairs from an uncountable set of power κ partitioned into two color sets, K_0 and K_1, by weakening the conclusion to ask for either a set of size κ all of whose pairs are in K_0 or a countable set all of whose pairs are in K_1. They discussed graphs and partitions of sets of pairs in terms of the structure theory for order, having observed that any partial order gives rise to a graph (binary, symmetric relation) under the relation of comparability. Here is the statement of their Theorem 5.22 [1941, 606] under the name by which it is now known.[158]

> *Dushnik-Miller Theorem:* If G is a graph of power $\mathfrak{m}$, where $\mathfrak{m}$ is a transfinite cardinal, and if every subset of G of power $\mathfrak{m}$ contains two connected elements, then G contains a complete subgraph of power $\aleph_0$.

Dushnik and Miller acknowledge help from Erdős [1941, 606, footnote 6]:

> We are indebted to P. Erdős for suggestions in connection with Theorems 5.22 and 5.23. In particular, Erdős suggested the proof of 5.22 for the case in which $\mathfrak{m}$ is a singular cardinal.[159]

The above result is the first *unbalanced* generalization of Ramsey's Theorem. Theorem 5.23 is dual of 5.22; it showed that if every subset of G of power $\aleph_0$ contains two connected elements, then G contains a complete graph of power $\mathfrak{m}$. These two theorems are translations to the language of graphs of Theorems 5.24 and 5.25 discussed in §4.3.

Dushnik and Miller [1941, 608] exhibit a counter-example to a possible extension of these theorems in the case when $\mathfrak{m} = \aleph_1$, namely the comparability graph of the superposition, on a set of cardinality $\aleph_1$, of a linear order obtained by mapping it into the reals and a linear order obtained by mapping it into "the well-ordered series consisting of all the ordinals of the first and second class" [finite and countable ordinals]. That is, they used the comparability graph of what is now known as the Sierpiński poset.

Soon after his work for [Dushnik and Miller, 1941],[160] Erdős [1942] stated and proved some of the foundational results of what has become known as the partition calculus. Here he used, as was common at the time, "sum" for "union" [1942, 364]:

[158] This theorem is sometimes called the Erdős-Dushnik-Miller Theorem.

[159] In [Hajnal, 1997] the proof by Erdős is described as requiring a good technical knowledge of the set theory of the time.

[160] Baumgartner [1997, 71] also used "soon after" to indicate when Erdős proved the results of [1942].

> *Theorem I:* Let a and b be infinite cardinals such that $b > a^a$. If we split the complete graph of power b into a sum of a subgraphs at least one of them contains a complete graph of power $> a$.
>
> In particular: If $b > c$ (the power of the continuum) and we split the complete graph of power b into a countable sum of subgraphs; at least one subgraph contains a non-denumerable complete graph.
>
> Theorem I is best possible. As a matter of fact, if $b = a^a = 2^a$ we can split the complete graph of power b into the sum of a subgraphs, such that no one of them contains a triangle. For the sake of simplicity we show this only in the case $b = c = 2^{\aleph_0}$.

Erdős described the example[161] as the complete graph G on points of the interval $(0,1)$ and for each positive integer k, the edges of G_k are those pairs $\{x,y\}_<$ such that

$$\frac{1}{2^{k-1}} \geq y - x = \frac{1}{2^k}.$$

Erdős, in his Theorem II, assumed the Generalized Continuum Hypothesis and supposed that the complete graph on $m = \aleph_{\alpha+2}$ many points is the sum $G = G_1 + G_2$, and proved that if G_1 does not contain a complete graph of power m, then G_2 contains a complete graph of power $\aleph_{\alpha+1}$.

Prior to his proofs of the two theorems, he wrote "Tukey and I have shown by a result of Sierpiński that the complete graph of power $\aleph_1$ can be decomposed into the countable sum of trees. Without assuming the continuum hypothesis we can not decide whether this also holds for the complete graph of power $\aleph_2$." (See [1942, 365]; the result that Tukey and Erdős used is in [Sierpiński, 1924].)

Both proofs used a construction process that used a graph G to build a tree or *ramification system*, as it was called in his paper [1943] with Tarski; we describe only the proof for Theorem I. If G is a graph on $b > a^a$ points and the edges of a G are partitioned into a subgraphs enumerated as $\langle G_\alpha : \alpha < |a| \rangle$, then the process starts by choosing an arbitrary p, splitting the remaining points into sets Q_α where $q \in Q_\alpha$ if α is the least index such that the pair $\{p,q\}$ is in G_α. At successor stages $\xi = \eta + 1$, this approach is repeated; if $s : \eta \to |a|$ is the index of a non-empty set Q_s, then p_s is chosen as an arbitrary element of Q_s and the remaining points are split into sets $Q_{s \frown \langle \alpha \rangle}$ where $q \in Q_{s \frown \langle \alpha \rangle}$ if α is the least index such that the pair $\{p_s, q\}$ is in G_α. At limit stages, intersections are taken. The construction stops when all the points have taken on the role of being the chosen point. By cardinality considerations, the number of chosen points whose index is a sequence of length less than $|a|^+$ is at most $|a| \cdot |a|^+ = |a|^+$, so there must be a point r not chosen at any level of the construction of ordinal height $< |a|^+$. Since at the ηth level, every point of G has either been chosen or is in one of the sets with an indexing sequence of that length, it follows that r is in the intersection $\overline{G}$ of the level sets Q_s to which it belongs. For some α^* there must be $|a|^+$ many

[161] Hajnal [1997, 354] reported that Erdős mentioned that the example was pointed out to him by Gödel.

sequences s with $\overline{G} \subseteq Q_{s^\frown\langle\alpha^*\rangle}$ and the set of points p_s for which this is true is a complete subgraph of G_{α^*} of order type $|a|^+$. Erdős did not point out that r may be added to get a sequence of length $|a|^+ + 1$.

Erdős [1942] attributed the result that there is a graph on $2^{\aleph_\alpha}$ which does not have a complete subgraph of power $\aleph_{\alpha+1}$ nor an independent[162] subset of that power to Sierpiński, who proved it for graphs of size $\aleph_1$.

The last part of Erdős-Tarski [1943], subtitled *General remarks on inaccessible numbers*, included a list of six problems, of which problems 5 and 6 were discussed in §4.3. The following is a problem for an inaccessible cardinal $\mathfrak{n}$.

> Problem 4. (*The Graph Problem.*) Is it true that if a complete graph G of power $\mathfrak{n}$ is split into two graphs G_1 and G_2, at least one of the contains a [complete] subgraph of power $\mathfrak{n}$? (A graph is to be defined as an arbitrary set of non-ordered couples (x, y) with $x \neq y$. By a complete graph of power $\mathfrak{n}$ we mean the set of all such couples formed from the elements of a set N of power $\mathfrak{n}$.)

In the footnote (page 328) with information on the Graph Problem, the authors cited Ramsey's Theorem for $\mathfrak{n} = \aleph_0$ (where the answer is positive), and a paper by Erdős to appear in *Revista de Tucumán*[163] for uncountable $\mathfrak{n}$ which are not inaccessible (where the answer is negative).

The final paragraph of the footnote listed implications of the form positive solution to problem a implies positive solution to problem b. Of particular note is the fact that a positive solution of the Ramification Problem discussed in §4.3 leads to a positive solution of the Graph Problem. Proofs of these implications were postponed.

Some connections between graphs and partial orders were investigated. Marczewski [1945, 307] had shown that every graph can be represented by the disjunction relation[164] on some collection of sets. Kazimierz Zarankiewicz [1947] connected Marczewski's representation of graphs via the disjunction relation to symmetric relations.[165]

At some point in the late 1940's and early 1950's, Kurepa expanded his focus beyond ramified tables, partial orders and the study of monotone functions of them, to include graphs. Since a disjunction relation was one of the constituent parts of Kurepa's relation $* = (<, >, \equiv, \|)$ of ramification,[166] as explicated in his thesis

[162] An *independent* subset of a graph is one in which no pair of points is joined by an edge of the graph.

[163] The paper by Erdős mentioned here is his [1942] discussed above.

[164] Given a graph represented as $G = (V, E)$, Marczewski observed that the collection $\mathcal{D}$ of non-empty sets $D_x = \{\{x, y\} \notin V \mid y \in V \setminus \{x\}\}$ has the property that $\{a, b\} \in E$ if and only if $D_a \cap D_b = \emptyset$.

[165] Kurepa cited Zarankiewicz' paper in [Kurepa, 1959a]. Erdős, in Math Review MR0023047 (9,297c), translated the Zarankeiwicz results for symmetric relations into the language of finite graph theory and noted that the author's results can be deduced from [Turán, 1941].

[166] Kurepa used $\|$ as the incomparability relation; $\equiv$ was an equivalence relation, i.e. equality in the case of trees; and $>$ was the converse of $<$.

[1935], the paper [Marczewski, 1945], cited in the paper of Kurepa discussed below, may have been the one that led Kurepa to think about applying his Fundamental Relation (see the end of §3) to graphs.

In May 1950 Kurepa proved the following theorem for a binary symmetric relation[167] $G = (G; \rho)$, where kG is the cardinality of the vertex set, k_cG is the supremum of the sizes of connected subsets of G,[168] and k_sG is the supremum of the sizes of disconnected subsets of G.[169] We will call this theorem *Kurepa's Graph Relation* ([1953], [1996, 399]):

> *Theorem 0.1* For any graph $G = (G; \rho)$ one has
>
> (0.4) $\quad kG \leq (2k_sG)^{k_cG}$.
>
> For any $\aleph_\alpha$ and any cardinal number $n \leq 2^{\aleph_\alpha}$ there is a graph $g = g(\aleph_\alpha, n)$ such that
>
> (0.5) $\quad k_cg \leq \aleph_\alpha,\ k_sg \leq \aleph_\alpha,\ kg = n;$
>
> in particular (for $n = 2^{\aleph_\alpha}$) there is a graph M_α such that
>
> (0.6) $\quad k_cM_\alpha = \aleph_\alpha k_sM_\alpha,\ kM_\alpha = 2^{\aleph_\alpha}$.

With his orientation toward the study of partial order and the fact that he was updating his proof of the Fundamental Relation from [1939], Kurepa referred to the totally disconnected sets as antichains and the pairwise connected sets as chains.

Toward an overview of Kurepa's proof, for $\alpha < (k_cG)^+$, let $T_\alpha = {}^\alpha G$ be the set of functions (sequences) from α into G. Note that $T = \bigcup_\alpha T_\alpha$ is a tree under end-extension. Let $\delta(X)$ denote a maximal disconnected subset of X. Kurepa used δ to define sets $D_\alpha \subseteq T_\alpha$ (see below) by recursion on α such that $D = \bigcup\{D_\alpha : \alpha < (k_cG)^+ \wedge D_\alpha \neq \emptyset\}$ is a subtree of T closed under initial segments that has the properties that (1) each element is an enumeration of a chain (complete subgraph) in G, and (2) each element with domain a limit ordinal is the union of its initial segments. It follows that each branch through D may be thought of as an enumeration of a chain of G, and hence has cardinality at most k_cG. Moreover, for each non-terminal element $s \in D$, the set of $a \in G$ such that $s^\frown\langle a\rangle \in D$ is an antichain (disconnected subset, independent subset) of G. Thus for all ξ with $D_\xi \neq \emptyset$, the subtree $\bigcup_{\eta\leq\xi} D_\eta$ is injectively embeddable into the tree ${}^{\xi\geq}(k_sG)$. Consequently, for positive $\xi < (k_sG)^+$, $|\bigcup_{\eta\leq\xi} D_\eta| \leq (k_sG)^{|\xi|} \leq (2\cdot(k_sG))^{(k_cG)}$. Since $D \subseteq T$ has no branches of length $(k_cG)^+$ and $(k_cG)^+ \leq 2^{(k_cG)}$, it follows that $|D| \leq (2\cdot(k_sG))^{(k_cG)}$. Kurepa ensured that for each $a \in G$ there are $\alpha < (k_cG)^+$

[167]Kurepa followed Sierpiński [1933] in his usage of binary symmetric relations.

[168]The *connected subsets* may be regarded as pairwise related subsets, complete subgraphs, or cliques.

[169]The disconnected subsets may be regarded as pairwise unrelated subsets, subgraphs in which no pairs are joined by an edge, or independent subsets.

and a element $s \in D_{\alpha+1}$ with $s(\alpha) = a$. Therefore $|G| \leq |D|$ and the theorem follows.

For the interested reader, we describe the construction of the sets D_α from $G = (G; \rho)$ where ρ is a reflexive symmetric relation. It uses a function $E : \wp(G) \to \wp(G)$ defined by $E(X) = \{y \in G \mid \forall x \in X \rho(x, y)\}$. Note that since ρ is symmetric, $X \subseteq E(X)$. By an abuse of notation, for $s \in T$, write $E(s)$ for $E(\text{ran}(s))$ and $E(s) \setminus s$ for $E(\text{ran}(s)) \setminus \text{ran}(s)$. To start the recursion, $D_0 = \{\emptyset\}$. For $\xi = \eta + 1$ successor, D_ξ is the set of all $s^\frown\langle a\rangle$ for $s \in D_\eta$ and $a \in \delta(E(s) \setminus s)$. For ξ limit, D_ξ is the set of all $s \in T_\xi$ such that $E(s) \setminus s \neq \emptyset$ and all proper initial segments of s are in $\bigcup\{D_\eta \mid \eta < \xi\}$.

Kurepa showed that his graph inequality is sharp for graphs of size $2^{\aleph_\alpha}$ as follows. He let M_α be the collection of all ω_α-sequences of rational numbers, and defined a partial order $\leq$ on M_α as the superposition of a well-ordering of M_α and the lexicographic ordering. His example is the comparability graph of this partial order, $(M_\alpha, \leq)$.[170]

Kurepa used Hausdorff's language for the ordering "by first difference" to describe the lexicographic order, and used Hausdorff's Theorem XIV [1908, 472] in the proof of the lemma that states that the subsets of the collection of ω_α-sequences of rationals which are well-ordered or converse well-ordered under the first difference order have power at most ω_α, and that there is a subset S of such sequences which either has order type ω_α or has order type ω_α^*.

Kurepa derived two corollaries and applied his results to a question of Sierpiński [1933]. Kurepa's Corollary 0.1 [1953, 400] is the assertion that if a graph has cardinality at least $(2^{\aleph_0})^+$, then either it contains an uncountable connected (complete) induced subgraph or it contains an uncountable totally disconnected induced subgraph (independent set), which is an independent rediscovery of part of Theorem I by Erdős [1942].[171]

The corollary may be obtained by expressing the graph relation as an implication and taking the contrapositive. That is, the graph relation is equivalent to the following conditional statement: for any graph $G = (V, E)$, if the supremum of the cardinalities of its complete subgraphs is κ and the supremum of the cardinalities of its independent subsets is λ, then $|V| \leq (2\kappa)^\lambda$. While Kurepa did not do so, the modern reader, taking the contrapositive of the above conditional statement in the case where κ and λ are infinite cardinals and using basic cardinal arithmetic, derives the statement that if a graph (G, ρ) has at least $(\kappa^\lambda)^+$ points, then either there is a complete subgraph of size κ^+ or there is an independent set of size λ^+. For the situation of Erdős' Theorem II [1942] (and others like it with different size goals), this modern interpretation does not require the GCH which

[170] Ginsburg, in his Math Review MR0071485 (17,135b) noted that $(M_\alpha; \leq)$ given on page 91 of [Kurepa, 1953] is a generalization of an example in [Dushnik and Miller, 1941]. Recall that at the end of §3, we discussed the fact that in the late 1930's Kurepa used a partial order obtained as the superposition of a well-order and the lexicographic order on ${}^\omega 2$ to show that his Fundamental Relation for partial orders is sharp.

[171] Apply the Erdős result to the partition of the complete graph, i.e. the set of all pairs of vertices, into K_0, the set of connected ones, and K_1, the set of disconnected ones.

Erdős used, instead the guarantee that any graph on at least $(\aleph_{\alpha+1}{}^{\aleph_\alpha})^+$ vertices will have either a complete subgraph on $\aleph_{\alpha+2}$ vertices or an independent set of size $\aleph_{\alpha+1}$. On the other hand, for the situation with equal size outcomes of size κ^+, neither Kurepa nor Erdős used GCH; both start with the same size set but Erdős' Theorem I [1942] works for κ many colors.

Kurepa applied his theorem to a question of Sierpiński on a problem of Knaster. Kurepa's description of Knaster's problem appears below, translated by the author:

> Does there exist a non-denumerable infinite graph G so that each infinite nondenumerable subgraph of G contains both: two distinct connected points and two distinct disconnected points.

Sierpiński came up with an example of a graph on $\aleph_1$ points meeting Knaster's requirement, and asked if one could have an example with $\aleph_2$ points. Kurepa pointed out that the graph relation means that the cardinality of any graph satisfying Knaster's condition has power at most the continuum, giving a negative answer to Sierpiński's question if the Continuum Hypothesis holds. Since his counterexample M_0 (essentially the generalization of Sierpiński's example to the continuum) is a graph of cardinality the continuum in which every uncountable subset contains both connected and disconnected pairs, a positive answer to Sierpiński's question is equivalent to the negation of the Continuum Hypothesis.[172]

Kurepa's Corollary 0.2 is the inequality for partial orders corresponding to the graph relation, i.e. equation (0.4) in Theorem 0.1. It is an updated version of the Fundamental Relation for partial orders announced by Kurepa in [1937b] and published in [1939], obtained by using the supremum of the cardinalities of *all* chains rather than restricting attention to increasing and decreasing well-ordered chains.

To place Kurepa's paper [1953] in context, note that the results on graphs are closely related to those of Dushnik and Miller [1941] and Erdős [1942], discussed above. Moreover, the proof of the Fundamental Relation that Kurepa gave in [1939] can be immediately adapted to prove his graph relation (0.4). Kurepa had the tools in hand early, but missed the opportunity to be the first to state and prove these fundamental results in Ramsey theory.

5 1950-1960: FOUNDATION OF THE PARTITION CALCULUS

In the 1950's, a systematic study of partitions of finite subsets of an ordered set was initiated, the main lines of study delineated, and the useful arrow notation for partition relations was introduced. In a classic paper, Erdős and Rado proved a Positive Stepping Up Lemma from which the modern Erdős-Rado Theorem follows. Erdős and Rado generalized partition relations from cardinals to ordinals, from

[172]Ginsburg, in his Math Review MR0071485 (17,135b) noted that Erdős [1942] already answered the question of Sierpiński.

colorings of r-element sets to products of them, and introduced canonical partition relations as a way to treat partitions with potentially many classes. Kurepa gave an extended treatment of his approach to graphs and hypergraphs. Late in the decade, a seminar of Mostowski and Tarski revisited the Erdős and Tarski results for inaccessible cardinals which had been announced in 1943. By the end of the decade, all the members of the synergistic group consisting of Erdős, Hajnal, Milner and Rado had met one another and were working on partition relations.

The Regressive Function Theorem reached its modern form, and general Δ-system lemmas were proved. Interest in Suslin lines and objects that can be derived from them continued. Progress was made on the Normal Moore Space Conjecture: Bing proved that the existence of a Q-set implies the existence of a separable non-metrizable normal Moore space. Investigations into the basic structure of partial orders and their mappings were made and a partial order of size 2^κ universal for partial orders of size κ was constructed.

Sierpiński published *Cardinal and Ordinal Numbers* [1958], an expanded version his *Leçons sur les Nombres Transfinis* [1950]. In his Math Review MR650095787, Gillman commented "Professor Sierpiński writes in his usual kindly, patient style which makes reading a pleasure." Rodolfo A. Ricabarra published *Ordered and Ramified Sets (Contribution to the Study of Suslin's Problem)* [1958]. This book is an extended look at trees and linear order types related to them using Kurepa's ramified sets and tables, and has a chapter on equivalences and implications of Suslin's Problem.

We conclude this introduction with some results about combinatorial tools.

In answer to a question of Tarski, Robert Vaught [1952] showed that the statement "every family of sets has a maximal disjointed family" is equivalent to the Axiom of Choice. The statement can be derived by an application of Zorn's Lemma, and Vaught gave a simple argument that from it, one can find a transversal for any family of pairwise disjoint non-empty sets. Kurepa [1953], [1996, 401] used Vaught's result to show that the statement "every graph contains a maximal disconnected subgraph" is equivalent to Zermelo's axiom.

In 1958, Erdős and Rado submitted the paper [1960] in which they prove the general Δ-system lemma, with an interesting bound in the case of finite collections of finite sets. That year, Ernest Michael [1962] submitted his independently proved result equivalent to the general Δ-system lemma:[173] he generalized the result of Šanin [1946] to higher cardinals. In a footnote added in proof, Michael indicated that Erdős had alerted him to the equivalent Erdős-Rado result and its longer proof. It may be that Michael's paper alerted Erdős to the existence of Šanin's paper as well. It is unsurprising that such a useful result would be discovered multiple times.

András Hajnal[174] met Erdős in 1956 and in [1997, 361–362] recounted the de-

[173]Michael used *quasi-disjoint* for Δ-system. Michael is best known for the Michael Selection Principle [1956]; see also volume 155 of *Topology and its Applications*, a special issue dedicated to the theory of continuous selections of multivalued mappings and Michael's 80th birthday.

[174]András Hajnal (May 13, 1931 –) was awarded his doctorate in 1957 at Jószef Attila Uni-

lightful story of how they came to starting working together on set mappings at that very first meeting (it included climbing 300 odd stairs to the top of the local cathedral). The resulting paper is [Erdős and Hajnal, 1958]. Set mappings are an important part of infinite combinatorics and useful for proving partition relations, so I take this opportunity to quote a result by Hajnal under the name by which it is now known [1961, 123]:

> *Hajnal's Set Mapping Theorem:* Assume κ is an infinite cardinal, $\kappa > \lambda$, and let $f : \kappa \to \mathcal{P}(\kappa)$ be a set mapping with the property that $|f(\alpha)| < \lambda$ for every $\alpha < \kappa$. Then there is a set $X \subseteq S$ of cardinality κ free with respect to f.[175]

5.1 The partition calculus

This subsection details chronologically the work of Erdős, Hajnal, Kurepa, Rado, and others as they explored a variety of approaches to finding subsets with uniform behavior in graphs, binary symmetric relations, and in distributions, partitions and colorings of r-tuples. A chronological presentation was chosen to place in historical context the May 1950 results of Kurepa [1953] whose mathematical context was discussed in §4.4.

The Erdős-Rado paper [1950] has been frequently cited as the beginning of the study of *canonical partitions.*[176] Any decomposition or coloring function f defined on r-tuples of a set A induces the "same color" equivalence relation on the r-tuples of A. This equivalence relation is defined even when the number of colors is infinite. Erdős and Rado proved that for each finite r, there is a finite set $\mathfrak{C}_r$ of equivalence relations on the r-element subsets of $\mathbb{N}$ with the property that for every equivalence relation E on the r-tuples of A there is some infinite $B \subseteq \mathbb{N}$ and some $F \in \mathfrak{C}_r$ such that the restrictions of E and F to the r-tuples of B are identical. We will call such a collection $\mathfrak{C}_r$ a *canonical basis* for the equivalence relations on the r-tuples of A. For concreteness, we note that for pairs, the canonical equivalence relations are $s \equiv_{\min} t$ iff $\min(s) = \min(t)$; $s \equiv_{\max} t$ iff $\max(s) = \max(t)$ and $s \equiv_{=} t$ iff $s = t$; and the equivalence relation in which all s and t are equivalent. For each finite positive r and each $I \subseteq r$, let E_I be the equivalence relation on the

versity in Szeged, Hungary where his advisor was László Kalmar, a logician. His thesis was on relative constructibility (see Kanamori's first chapter for more on Hajnal's thesis work). The Mathematical Genealogy Project site lists 1956 for the year of his doctorate, but a curriculum vitae accessed September 9, 2010 from Hajnal's Hungarian website at http://www.renyi.hu/ ahajnal/hajn2006.pdf lists 1957.

[175]A *set mapping* is a function $f : S \to \mathcal{P}(S)$ so that $f(s) \notin s$ for every $s \in S$. A subset $X \subseteq S$ is *free* with respect to a set mapping f on S if for all $x, y \in X$, $x \notin f(y)$. See [Komjáth and Shelah, 2000] for the history of problems on set mappings and for some set mapping results about $\aleph_n$ at the end of the century. The history dates back to the thirties and a question of Paul Túran, and includes results by Sophie Piccard and Stanisław Ruziewicz, one of the founders of the Lwów School.

[176]The seminal paper [Ramsey, 1930] has an earlier classification of r-ary relations.

r-element subsets of $\mathbb{N}$ defined by

$$\{x_0, x_1, \ldots, x_{r-1}\}_< \; E_I \; \{y_0, y_1, \ldots, y_{r-1}\}_< \iff x_i = y_i \text{ for all } i \in I.$$

Here is a precise statement of Erdős and Rado's oft-quoted canonical partition relation for r-tuples of natural numbers:

> *Canonical Partitions for* $\mathbb{N}$: For every positive $r < \omega$ and every equivalence relation E on the r-element subsets of $\mathbb{N}$, there is an infinite set $M \subseteq \omega$ and an index set $I \subseteq \{0, 1, \ldots, k-1\}$ such that the restrictions of E and E_I to r-tuples from M are identical.

In [Erdős and Rado, 1952a], the authors looked at partitions of order types and proved (cf. Theorem 4, pp. 428-9) that if the set of pairs of rational numbers is partitioned into two pieces, then either there is an infinite monotonic sequence whose pairs are in the first cell or there is a subset dense in an open interval all of whose pairs are in the second cell.[177] Rado,[178] in his joint work with Erdős [1952a], used the Axiom of Choice to define a partition of the collection T of all infinite subsets of an infinite set S into two classes, $T = K_0 \cup K_1$, so that there is no infinite set all of whose infinite subsets are in the same class as follows. Start with a well-ordering $<$ of T. Let K_0 be the set of all $A \subseteq T$ for which there is an $A' < A$ with $A' \subseteq A$. Set $K_1 = T \setminus K_2$. Observe that if S' is in T and A is its $<$-least infinite subset, then A is in K_2. Since A is infinite, using the Axiom of Choice, we may partition $A = \{b_i \mid i < \omega\} \cup C$, where the b_j are all distinct, none of them is in C, and C is infinite. Let $B_j = \{b_i \mid i < j\} \cup C$. Then for some m_0, B_{m_0} is the $<$-least among the B_j's. In particular, B_{m_0+1} is a proper subset of B_{m_0} with $B_{m_0} < B_{m_0+1}$, so B_{m_0+1} is in K_1.

Erdő and Rado called attention to the following problem [1952a, 418]:

> *All Finite Subsets Problem:* Given an infinite set S, is it possible to divide all finite subsets of S into two classes in such a way that every infinite subset of S contains two finite subsets of the same number of elements but belonging to different classes?

They indicate that such partitions exist if $|S| \leq 2^{\aleph_0}$,[179] but indicate that nothing is known for sets of larger cardinality. We conclude with the comment that Erdős and Rado proved early results for polarized partitions in this paper, since we will refer to that fact later.

Kurepa gave several talks on his results about graphs from May 1950:[180] his September 6, 1952 talk in Munich at the Kongress der Deutschen Mathematiker-Vereinigung was entitled *Über binäre symmetrische Relation*; his December 3, 1952

[177] Erdős mentioned this result in his paper [1996, 120] on his favorite theorems.

[178] The attribution of this result to Rado was made in [Hajnal, 1997, 354], and it dates back to 1934, as discussed in §3.2.

[179] In Example 2 [1952a, 435], Erdős and Rado showed there is such a partition of the finite subsets of a denumerable set S, which shows in modern terminology that $\aleph_0$ is not a *Ramsey cardinal*.

[180] The details given in this paragraph were found in [Kurepa, 1954] and [Kurepa, 1959a].

talk for a Colloquium of the Society of Mathematicians and Physicists of Croatia in Zagreb was entitled *O simetričnim relacijama i grafovima*; and his February 24, 1953 talk for the Faculty of Sciences in Paris was entitled *Sur les relations binaires.* We learn later that Kurepa had intended to submit his work on graphs to *Journal für die reine und angewandte Mathematik* after his talk in Munich, but postponed publication in the expectation that he would be able to generalize the results. He submitted a preliminary report [1953] prior to the talk in Paris, probably in connection with his December 3, 1952 talk.[181] He learned from G. Riguet[182] during his visit to Paris about the work of others in Ramsey theory, citing in his preliminary report (1) Ramsey's seminal paper [1930], (2) Erdős' follow-up [1942] to the Dushnik-Miller Theorem, (3) the joint paper of Erdős and Rado [1953] discussed below, and (4) a paper of F. Burton Jones [1952].

In a paper with Erdős, Rado introduced the basic version of the arrow notation, "relation (5) $a \to (b_1, b_2)^2$." He called it the *decomposition relation* between cardinals a, b_1, b_2, and defined it as follows [1953, 427]:

> For any set S we denote by $\Omega_2(S)$ the set of all sets $S' \subset S$ such that $|S'| = 2$. Then we say that (5) holds if, and only if, the following statement is true. Whenever
>
> $$|S'| = \alpha; \quad \Omega_2(S) = K_1 + K_2,$$
>
> then there is $S' \subset S$ and $\lambda \in \{1, 2\}$ such that
>
> $$|S'| = b_\lambda; \quad \Omega_2(S') \subseteq K_\lambda.$$
>
> The relation (5) is fundamental in many investigations in set theory. The authors hope to deal in another paper with its numerous interesting properties and generalizations. In the present note it only serves as a convenient abbreviation.

The partition approach turned out to be remarkably well-suited to generalization, and Erdős spoke on his work with Rado [1956, 427] on some of these generalizations on October 24, 1953 in New York during his invited hour address, *Combinatorial problems in set theory*, for the Eastern Sectional Meeting of the American Mathematical Society (cf. [Cohen, 1954, 21]).

Rado [1954a] tackled the question of canonical polarized partition relations and proved two theorems, using his joint work with Erdős [1950] on canonical partitions and the approach to polarization in their [1952a]. Rado used the notation $[N]^n$ for the n-element subsets of a set N, which has been widely adopted. He wrote $[A_1, A_2, \ldots, A_\ell]^{n_1, n_2, \ldots, n_\ell}$ for the collection of all sequences $(X_1, \ldots, X_\ell)$ where each X_i is in $[A_i]^{n_i}$. The simplest case of the second theorem is Proposition B

[181] Kurepa described the December 1952 talk in a footnote [1953, 65] as a presentation "during the session of 23.10.1952 of the Slovenian Academy of Sciences and Art."

[182] Almost surely Kurepa is referring to Jacques Riguet, who used order in his study of binary relations, closures, and Galois correspondences [1948].

from page 74, which states that every infinite matrix all of whose elements are either 0 or 1 contains an infinite submatrix equal to one of the following four matrices: the matrix of all zeros; the matrix with zeros on and below the main diagonal and ones above it; the matrix of all ones on and below the main diagonal with zeros above it; and the matrix of all ones. Below, I have stated Rado's first theorem with a descriptive name for easy reference:

> *Rectangle Refining Theorem:* Given positive integers ℓ, $n_1, \ldots, n_\ell$, p, there is a positive integer $f = f_\ell(n_1, \ldots, n_\ell, p)$ which has the following property. If Δ is any partition of $[N, N_f, N_f, \ldots, N_f]^{n_1,\ldots,n_\ell}$ then there are sets N'_λ, such that N'_1 is infinite, and $N'_\lambda \in [N_f]^p$ for $1 \le \lambda \le \ell$, and
>
> $$\Delta(X_1, X_2, \ldots, X_\ell) = \prod_\lambda^\lambda \Delta_\lambda(X_\lambda) \qquad (X_\lambda \in [N'_\lambda]^{n_\lambda}),$$
>
> where each Δ_λ is a canonical partition.[183]

The displayed equation means that two sequences $\vec{X} = (X_1, X_2, \ldots, X_\ell)$ and $\vec{X}' = (X'_1, \ldots, X'_\ell)$ are in the same cell of the partition Δ of $[N, N_f, N_f, \ldots, N_f]^{n_1,\ldots,n_\ell}$ if and only if for each index λ, the components X_λ and X'_λ are in the same cell of Δ_λ.

In May 1954, Kurepa published his Math Review MR0058687 (15,410b) of [Erdős and Rado, 1953], in which the first half delineated the contents of the paper, and the second half listed references he felt should have been included, citing five of his own papers and Theorem 14 of [Hausdorff, 1908]. Of relevance to the partition calculus, he wrote "especially the proof and the result in the reviewer's paper [his Fundamental Relation proved in [1939]] $\cdots$ are connected with lemmas 3 and 4 of the paper under review." For infinite m, n, the contrapositive of the implication derived from the inequality for graphs in Kurepa's Theorem 0.1 of May 1950 ([1953], [1996, 399]) expressed in Rado's notation is $(m^n)^+ \to (m^+, n^+)^2$. This partition relation is closely related to the Erdős-Rado lemmas, since Lemma 3 is $a \to (b, a)^2$ for a regular limit cardinal a and $b < a$; and Lemma 4 is $a^+ \to (b, a^+)^2$ if a is infinite and b is the least cardinal such that $a^b > a$.

On November 11, 1954, Kurepa added a short note to [1953], [1996, 409]: "The results were obtained in [M]ay 1950 (and were to be presented in a talk in USA) when the author was anxious to give a new proof of a theorem he published earlier [v. Kurepa C.R. Acad. Sci. Paris **205** (1937), 1196-1198 "la relation fondamentale (1)" and Kurepa [1], [3, §20]."[184] That is, Kurepa added the announcement of his Fundamental Relation to the references and emphasized the starting point of the article was Corollary 0.2, his chain-antichain inequality for partial orders, as he

[183] See [Rado, 1954a, 76]. The notation N_f denotes a fixed set with f many elements; N denotes an infinite set, not necessarily the natural numbers.

[184] The references are to [Kurepa, 1939] and his book on set theory, *Teorija skupova*, published in 1951. Kurepa did make a lecture tour in the United States of America in 1950, as one can learn from [Kurepa, 1952, 97, footnote 1], but spoke on a different topic.

had noted in the body of the paper. Recall that the first theorem of [Kurepa, 1953] gives an inequality on the cardinality of a graph in terms of the supremum of the cardinalities of the sizes of its complete subgraphs and the supremum of the cardinalities of the sizes of its independent subsets, and addresses the sharpness of this inequality. Kurepa prefaced its statement with the remark that the purpose of the paper was to prove the first theorem.

Notice the shift in the way Kurepa presented his work and its connection with graphs: Kurepa described the first theorem of May 1950 as new at the time he found it [1953], but in the Math Review of [Erdős and Rado, 1953], Kurepa has, in essence, said it is implicit in his Fundamental Relation.

Erdős and Rado [1956][185] generalized the decomposition relation of their [1953] paper from cardinals to order types, from partitions into two cells into many cells, and from 2-tuples to r-tuples for finite values of r. They even mentioned, but did not treat, "more general partition relations referring to partial orders" [Erdős and Rado, 1956, 430]. In this seminal paper, Rado introduced, in full generality, the paradigm shifting definition of the *ordinary partition symbol* [1956, 429]:[186]

> Fundamental throughout this paper is the *partition relation* $a \to (b,d)^2$ introduced in [6]. More generally, for any a, b_ν, r the relation
>
> $$(1) \qquad a \to (b_0, b_1, \dots)^r_k$$
>
> is said to hold if, and only if, the following statement is true. The cardinals b_ν are defined for $\nu < k$. Whenever $|S| = a$; $[S]^r = \sum_{\nu<k} K_\nu$, then there are $B \subset S$; $\nu < k$ such that $|B| = b_\nu$; $[B]^r \subset K_\nu$.

When discussing $a \to (b_0, b_1, \dots)^r_k$, let us call a the *resource* and the b_i's the *goals*, and call a set *homogeneous for color* i or *monochromatic for color* i if all its r-tuples are in the ith cell of the partition. Erdős and Rado abbreviated $a \to (b_0, b_1, \dots)^r_k$ to $a \to (b)^r_k$ in the special case with $b_\nu = b$ for all $\nu < k$. Partition relations in which all the goals are the same have come to be called *balanced*; if not all the goals are the same, they are called *unbalanced*. This distinction, already mentioned in special cases earlier, proved useful since proof techniques varied for

[185] The paper was received by the editors on May 17, 1955. The authors did not mention the results of Kurepa [1953] which appeared very late in 1954 or possibly in 1955. A later paper, [Erdős *et al.*, 1965], reported that Erdős and Rado were unaware of Kurepa's work when they submitted their paper.

[186] As Hajnal said in [Hajnal and Larson, 2010], "There are cases in mathematical history when a well-chosen notation can enormously enhance the development of a branch of mathematics and a case in point is the ordinary partition symbol." A contrasting view is provided by Rudin [1975a, 7], who, after starting a section entitled *Ramification arguments and partition calculus* with eight statements, the first four of which are arrow relations and the second four of which are inequalities: "The gibberish above is a set of really quite important theorems, the first four from the partition calculus and the last four from topology. The topological theorems are proved using partition calculus theorems or techniques which often means using tree type proofs." In the next paragraph, she continues: "In this chapter we first introduce the partition calculus notation. This notation is a stone wall barrier which must be crossed in order to make the rich literature of partition calculus available."

the two types. They reviewed previous results, proved monotonicity results for the ordinary partition symbol and delimited the trivial cases (see Theorem 22, page 438). For example, with the arrow notation, Ramsey's Theorem is reduced to the compact statement $\omega \rightarrow (\omega)^r_\gamma$ for finite r and $1 \leq \gamma < \omega$. Sierpiński's partition generalizes easily to $2^{\aleph_0} \not\rightarrow (\aleph_1, \aleph_1)^2$. Theorem I from [Erdős, 1942] is the statement that if $b > a^a$, then $b \rightarrow (a^+)^2_a$. The core of Rado's result from his paper with Erdős [1952a] on decompositions of the infinite subsets into two classes may be written $\kappa \not\rightarrow (\aleph_0)^{\aleph_0}_2$, and their All Finite Subsets Problem may be written $\kappa \rightarrow (\kappa)^{<\omega}_2$, where here we are introducing a generalization of the ordinary partition relation that appears in [Erdős and Hajnal, 1958], which will be discussed below.

Erdős and Rado [1956] explored the partition calculus using the ramification method. They proved the following ordinal partition relation for $m, n, \ell \geq 2$: if $\ell \rightarrow (m,n)^2$, then $\omega \cdot \ell \rightarrow (m, \omega \cdot n)^2$ (see Theorem 25, page 440). They investigated the partition properties of λ, the order type of the set of real numbers, and proved that $\lambda \rightarrow (\alpha, n)^3$ for every $n < \omega$ and $\alpha < \omega_1$ (see Theorem 31, page 447). They proved that the same is true for any uncountable order type Φ such $\omega_1 \not\leq \Phi$ and $\omega_1^* \not\leq \Phi$, i.e. neither ω_1 nor its converse order ω_1^* is order-embeddable into Φ, and conjectured that every positive partition relation for λ generalized to such Φ, but in footnote 7 on page 443 they note that Specker disproved this conjecture. They also proved $\omega_1 \rightarrow (\omega + m, 4)^3$. Partition relations for triples would turn out to be particularly challenging. They focused attention on the question of whether $\omega_1 \rightarrow (\omega \cdot 2, \omega \cdot 2)^2$ after proving for $\alpha < \omega \cdot 2$ that $\omega_1 \rightarrow (\alpha, \alpha)^2$. In Theorem 44, they reproved the Dushnik-Miller Theorem by reproving that $\omega_1 \rightarrow (\omega_1, \omega + 1)^2$ (see [Erdős and Rado, 1956, 459]), and more generally if κ is regular of uncountable cofinality, by proving that $\kappa \rightarrow (\kappa, \omega + 1)^2$.

A key result of this paper is their Theorem 39 (i) quoted below with the name by which it has come to be known ([Erdős and Rado, 1956, 467–468]):

Positive Stepping Up Lemma: If

$$\ell \rightarrow (\alpha_0, \alpha_1, \ldots)^r_k, \tag{81}$$

$$|m| > \textstyle\sum_{\lambda < \ell} |k|^{|\lambda|^r}, \tag{82}$$

then

$$m \rightarrow (\alpha_0, \alpha_1, \ldots)^{r+1}. \tag{83}$$

One corollary of Theorem 39, which includes various consequences of the Positive Stepping Up Lemma is the fact that for every positive r and finite sequence of ordinal goals $\beta_0, \beta_1, \ldots, \beta_{m-1}$, there is a resource α sufficiently large so that $\alpha \rightarrow (\beta_0, \beta_1, \ldots, \beta_{m-1})^r$ (see page 470).

The proof of the Positive Stepping Up Lemma involved a new concept. An *end-homogeneous set* for a partition $g : [m]^{r+1} \rightarrow k$ is a set on which the color assigned by g to an $(r+1)$-element set does not depend on the largest element of the set.[187]

Let us assume m is a cardinal, or more precisely, the inital ordinal of that cardinality. For each ordinal $x < m$, define a sequence $\langle f_\beta(x)\rangle$ by recursion: $f_0(x) = 0$, and if $f_\beta(x)$ has been defined for $\beta < \gamma$, then $f_\gamma(x)$ is the least element y of $x\backslash\{f_\beta(x) : \beta < \gamma\}$ with the property that for all r-element $u \subseteq \{f_\beta(x) : \beta < \gamma\}$ one has $g(u \cup \{y\}) = g(u \cup \{x\})$, if there is such a $y < x$, and otherwise $f_\beta(x) = x$ and the recursion stops. Let $\rho(x)$ be the index of the final element of the sequence. Note that by construction, the sequence is end-homogeneous for g.

Erdős and Rado use an induction argument to show that if $\rho(x) = \rho(z)$ and for all indices $\beta_0 < \cdots < \beta_{r-1}$, $g(f_{\beta_0}(x), \ldots, f_{\beta_{r-1}}(x), x) = g(f_{\beta_0}(z), \ldots, f_{\beta_{r-1}}(z), z)$, then $x = z$. Note that if $\eta = f_\beta(x)$, then the same argument shows the sequence $\langle f_\zeta(\eta) : \zeta \leq \beta\rangle$ is an initial segment of $\langle f_\zeta(x) : \zeta \leq \rho(x)\rangle$. That is, these sequences form a tree under end-extension and the argument here is essentially a ramification argument.

A consequence of the induction arguments is that the number of such sequences of length σ is bounded by the cardinal of the set of functions $h : [\sigma]^r \to k$, which is at most $|k|^{|\sigma|^r}$. By the cardinality hypothesis that $|m| > \sum_{\lambda<\ell} |k|^{|\lambda|^r}$, for some x_0 one has $\rho(x_0) \geq \ell$ so the induction hypothesis may be applied to $S_0 = \{f_\beta(x_0) : \beta \leq \ell\}$ and the coloring $g' : [S_0]^r \to k$ defined by $g'(u) = g(u \cup \{x_0\})$, to get a subset S_1 of size α_i monochromatic for color i for g', and $S_1 \cup \{x_0\}$ is the required witness to the partition relation for the original coloring g.

In [Erdős and Rado, 1956] two special cases of what has come to be called the Erdős-Rado Theorem were derived, using weakenings of GCH to get around their lack of good notation for iterated exponentiation. From the assumptions that $2^{\aleph_\nu} \leq \aleph_n$ for $\nu < n$ and $\omega_n \to (\alpha_0, \alpha_1, \ldots)^r_k$, they derived $\omega_{n+1} \to (\alpha_0 + 1, \alpha_1 + 1, \ldots)^{r+1}_k$ (note that n, k need not be finite). From the assumptions that $|k| < \mathrm{cf}(\aleph_m)$, $r \geq 0$ and $2^{\aleph_\nu} \leq \aleph_n$ for $m \leq n < m + r$ and $\nu < n$, they derived $\omega_{m+r} \to (\alpha_0 + r, \alpha_1 + r, \ldots)^{r+1}_k$.

The Positive Stepping Up Lemma is the key ingredient in the proof of the modern version of the Erdős-Rado Theorem. For non-triviality, let us assume that $k \geq 2$ and each $\alpha_i \geq 3$. If $\ell = \omega$, then the assumption that $\ell \to (\alpha_0, \alpha_1, \ldots)^r_k$ means each $\alpha_i \leq \omega$, $k < \omega$, and in this case, Ramsey's Theorem already yields $\ell = \omega \to (\alpha_0, \alpha_1, \ldots)^{r+1}_k$. So assume $\ell > \omega$. Then $|k| < \ell$, else, since $\ell^r = \ell$, there would be a one-to-one mapping from $[\ell]^r$ into k. Also $\sum_{\lambda<\ell} |k|^{|\lambda|^r} = \sum_{\lambda<\ell} |k|^{|\lambda|} = \sum_{\lambda<\ell} 2^{|\lambda|} = 2^{<\ell}$. Note that $\sum_{n<\omega} 2^n = \omega$ yielding the following theorem.

> *Positive Stepping Up Lemma* (modern form): For all infinite cardinals κ, all γ with $2 \leq \gamma < \kappa$, all finite r, and all cardinals $\langle \lambda_\nu : \nu < \gamma\rangle$, if $\kappa \to (\lambda_\nu)^2_{\nu<\gamma}$, then $(2^{<\kappa})^+ \to (\lambda_\nu + 1)^{r+1}_{\nu<\gamma}$.

[187] The concept of an end-homogeneous set appears in the proof of the Positive Stepping Up Lemma, but the name appears only later e.g. in Chapter IV, Section 15 of the Erdős-Hajnal-Máté-Rado compendium [1984]. The idea also goes under the name *pre-homogeneous* which was used by Kunen [1977].

Let $\exp_n(\kappa)$ denote n-times iterated exponentiation, that is, $\exp_0(\kappa) = \kappa$ and $\exp_{n+1}(\kappa) = 2^{\exp_n(\kappa)}$. With this and the cardinal arithmetic above, we state below the modern version of their Theorem 39, obtained using the Positive Stepping Up Lemma repeatedly and starting from the clear fact that $\exp_0(\kappa) = \kappa \to (\kappa)^1_\gamma$ for $\gamma < \mathrm{cf}(\kappa)$.

> *Erdős-Rado Theorem* (modern form): For every infinite cardinal κ, every finite $r \geq 2$, and all $\gamma < \mathrm{cf}(\kappa)$, $(\exp_r(2^{<\kappa}))^+ \to (\kappa + (r-1))^r_\gamma$.

Since $\kappa^+ \to (\kappa^+)^1_\kappa$, by the Positive Stepping Up Lemma, $(2^{<\kappa^+})^+ \to (\kappa^+)^2_\kappa$. Use $2^{<\kappa^+} = 2^\kappa = \exp_0(\kappa)$ to rewrite this partition relation as $(\exp_0(\kappa))^+ \to (\kappa^+)^2_\kappa$. Repeated application of the Positive Stepping Up Lemma yields the most frequently quoted version of the Erdős-Rado Theorem: $(\exp_n(\kappa))^+ \to (\kappa^+)^{n+1}_\kappa$.

Erdős and Rado used the notation $[S_0, S_1, \ldots, S_{t-1}]^{r_1, r_2, \ldots, r_{t-1}}$ for the Cartesian product of the t sets $[S_i]^{r_i}$ from Rado's paper [1954a] and introduced the *polarized partition relation* (see [Erdős and Rado, 1956, 480]), denoted

$$\begin{pmatrix} a_0 \\ a_1 \\ \vdots \\ a_{t-1} \end{pmatrix} \longrightarrow \begin{pmatrix} b_{0,0} & b_{0,1} & \cdots \\ b_{1,0} & b_{1,1} & \cdots \\ \vdots & & \\ b_{t-1,0} & b_{t-1,1} & \cdots \end{pmatrix}^{r_0, r_1, \ldots, r_{t-1}}_k$$

for the assertion that whenever $|S_i| = a_i$ for $i < t$, and $[S_0, \ldots, S_{t-1}]^{r_1, \ldots, r_{t-1}}$ is partitioned into k classes, then there is some $j < k$ and sets $B_i \subseteq S_i$ of cardinality $b_{i,j}$ so that $[B_0, \ldots, B_{t-1}]^{r_1, \ldots, r_{t-1}}$ is a subset of the jth cell of the partition. They noted on page 480 that "the passage from our former type of partition relation i.e. the case $t = 1$, to the more general kind ... bears a certain resemblance to the process of polarization used in the theory of algebraic forms, which accounts for the name *polarized partition relation.*" They go on to prove some results expressed with this notation but do not give an extended study of it.

On a trip to Israel, Erdős stopped in Switzerland where he visited Specker and shared some of the partition problems with him. During the visit Specker was able to prove that for all $m < \omega$, the ordinal partition relation $\omega^2 \to (\omega^2, m)^2$ holds. Erdős continued to Israel and on his return learned that Specker had a counter-example witnessing $\omega^3 \not\to (\omega^2, 3)^2$ (see [Specker, 1957] for both results).

In a paper [1958] received by the journal in 1957, Erdős and Hajnal proved the first positive result in the study of partitions of all finite subsets of a set.[188] They introduced the notation $\kappa \to (\alpha)^{<\omega}_2$ which means that for any coloring of the finite subsets of κ with two colors, there is a subset A of order type α such that for all $n < \omega$, f is monochromatic on the collection of all n-element subsets of A. They proved that if κ is a measurable cardinal, then $\kappa \to (\kappa)^{<\omega}_2$. Thus if there is a measurable cardinal, then there is a set S for which the answer to the All Finite

[188] Erdős and Hajnal used $[S]^{<\aleph_0}$ to denote the collection of all finite subsets of S, where $[S]^{<\omega}$ is now used.

Subsets Problem of Erdős and Rado [1952a] is negative. Erdős and Hajnal also include in their paper [1958] a proof of a result due to Fodor that for all cardinals κ smaller than the least inaccessible cardinal, $\kappa \nrightarrow (\kappa)_2^{<\omega}$ holds.

Erdős and Hajnal met Eric Milner[189] at the International Congress of Mathematicians held in Edinburgh in 1958. Hajnal [1997, 374] recalled that Rado interested Milner in partition problems at the meeting and Milner settled a problem about countable ordinals which he [1969] only published much later. Erdős then visited him and starting working with him in Singapore, and they continued to work together throughout his career. At the suggestion of Erdős, Milner spent a sabbatical year 1958–1959 at the University of Reading working with Rado, and moved to Reading in 1961.

The finite powers of ω are all *additively indecomposable*, i.e. they cannot be written as the sum of two strictly smaller ordinals. It is well-known that the additively indecomposable ordinals are exactly those of the form ω^γ.[190] For specified indecomposable β and finite m, it is possible to determine an upper bound for the resource α needed to ensure that the positive partition relation holds. In particular, Erdős and Milner showed that $\omega^{1+\mu m} \rightarrow (\omega^{1+\mu}, 2^m)^2$. This result dates back to 1959 and a proof appeared in Milner's thesis in 1962. See also [Williams, 1977, 165–168] where the proof is given via the following stepping-up result:

> *Ordinal Stepping Up Lemma*: Suppose that γ and δ are countable and k is finite. If $\omega^\gamma \rightarrow (\omega^{1+\delta}, k)^2$, then $\omega^{\gamma+\delta} \rightarrow (\omega^{1+\delta}, 2k)^2$.
>
> *Corollary* [Erdős and Milner [1972]]: If $m < \omega$ and $\mu < \omega_1$, then $\omega^{1+\mu\cdot\ell} \rightarrow (\omega^{1+\mu}, 2^\ell)^2$.

During the 1958-1959 academic year, Andrzej Mostowski and Alfred Tarski[191] gave a seminar on the foundations of set theory at the University of California,

[189]Eric Charles Milner (May 17, 1928 – July 20, 1997) had an irregular start to his mathematical career, having gone from an undergraduate degree at the University of London to Malaysia in 1951 where he worked as a tin assayer for the Straits Trading Company, before joining the mathematics department of the University of Malaya in Singapore. Milner earned his doctorate as an external student at the University of London in 1963, with Richard Rado and Roy Davies serving as his examiners. Richard Guy, a former colleague from Singapore, became head of the department at the University of Calgary, where Milner became a professor in 1967 and continued for the remainder of his career, where, in addition to working on his own mathematics, he served as chair of the department, organized meetings and colloquia, and was a genial host for visits by the Hajnals and Erdős among many others. For more on his life and work, see [Nash-Williams, 2000] and [Hajnal, 1997].

[190]See Exercise 5 on page 43 of Kunen [1983].

[191]Tarski facilitated the expansion of the logic group at the University of California, Berkeley, in the 1950's. Leon Henkin was hired in 1953, having refused an earlier offer due to a loyalty oath previously required by the University of California; Robert Vaught arrived in 1958 (he had been a student of Alfred Tarski, receiving his doctorate in 1954). The interdepartmental graduate degree program in Logic and Methodology was established in 1959. Anil Nerode, who visited Berkeley 1958-1959 spoke of the state of logic in the United States in the 1950's in his preface [2004] to a volume for a Tarski centennial conference, and [Feferman, 2004] described the expansion of the logic group at Berkeley in the same volume.

Berkeley. Among the topics were the results announced by Erdős and Tarski in [1943], i.e. five problems involving inaccessible cardinals and other structures including the partition property $\kappa \to (\kappa)^2_2$, the tree property (every tree with height κ all of whose levels have size less than κ has a branch), the existence of non-principal κ-complete prime ideal in the set algebra $\mathcal{P}(\kappa)$ (the measurability of κ), and a representation problem for certain Boolean algebras. Erdős proved that if $\kappa \to (\kappa)^2_2$, then κ is inaccessible (see [Erdős and Tarski, 1961, 59, footnote 5], where the result is attributed to [Erdős, 1942], in which it is implicit). Later they [1961] published proofs, in part based on the seminar, noting (pp. 51-52) "some of the proofs are in the form in which they were reconstructed by some members of the seminar, Andrzej Ehrenfeucht, James D. Halpern, and Donald Monk, and may differ from those originally found by the authors".[192]

Near the end of the decade, Kurepa [1959a] gave a full exposition of his distinctive approach to the partition calculus. Conceived in 1952 when he submitted a preliminary report [1953] discussed earlier in this section, he had postponed the completion until well after [Erdős and Rado, 1956] since he [1996, 411] "thought that the same result should hold for n-ary symmetrical relations". His paper started from his thesis and his Fundamental Relation, whose proof he arranged to have reprinted in [Kurepa, 1959b]. It gave a gentle introduction to his approach to trees and ramification. In §7, Kurepa proved that if $|S| > 2^{\aleph_\alpha}$ and f is a coloring of the pairs of S with finitely many colors, then there is an $X \subseteq S$ with $|X| > \aleph_\alpha$ on which f is constant. In the arrow notation, $\left(2^{\aleph_\alpha}\right)^+ \to ({\aleph_\alpha}^+)^2_n$ for $n < \omega$. The proof is by induction on n. In a brief remark at the end of §7, Kurepa said he had expected this positive argument to generalize to larger tuples, which he then treated in §8 and §9 using a different approach. He introduced a compact notation for exponentiation that allowed him to state theorems with iterated use of exponentiation in a compact way. The results up to and including §7 appear to predate the appearance of [Erdős and Rado, 1956], while those after, that is, results for larger tuples, appear to be influenced by it. He reported learning from Erdős in 1959 that Hajnal had shown that the positive results were optimal, and Kurepa had decided to publish his recent work together with a reprinting of his [1939].

5.2 Applications of Suslin lines

Interest in Suslin's Problem and ordered sets continued in the 1950's, and new equivalences and applications of Suslin lines were found. Ricabarra published a book on the subject described briefly at the beginning of §5; Bourbaki added material on Suslin sets to the second edition of *Elements of Mathematics*. Yesenin-Volpin [1954] showed that Suslin's Hypothesis cannot be deduced in an axiom system which was used by Mostowski [1939].[193]

[192]Hajnal [1997, 356] reported that Monk wrote [Erdős and Tarski, 1961].

[193]Mostowski used his axiom system to show the independence of the Well-Ordering Principle from the Ordering Principle. This axiom system is closely related to the axiom systems of

Kurepa [1950] announced his result [1952] that the Cartesian square of a Suslin line (a level-wise product) does not have the countable chain condition. In fall 1950, Kurepa spoke on his work on Suslin's Problem at Harvard University, the University of Michigan at Ann Arbor, and the University of California, Berkeley. The following January, he spoke on it in Lausanne [1952, 97, footnote 1]. Josef Novák[194] [1950] investigated Suslin's Problem with closed dyadic [binary] partitions[195] seemingly unaware of Kurepa's work on such partitions. He [1952] also showed that a necessary and sufficient condition for a linear continuum of cardinality κ to have a "an element of character c_{00}" (limit both from below and above of sequences of length ω) is that $\kappa < 2^{\aleph_1}$.[196]

Mary Ellen Rudin, publishing under her maiden name of Estill,[197] investigated E spaces, that is, Moore spaces which do not have a countable dense set and do not contain uncountably many disjoint domains, i.e. open sets. In her thesis, Rudin [1950] constructed the first example of an E space satisfying the countable chain condition. Then she [1951] constructed an example which, in addition, is locally connected. Next she [1952] proved that the existence of a nonseparable locally connected Moore space satisfying the countable chain condition in which any two points can be separated by a countable set[198] if and only if there is a Suslin line.[199]

Bernays and Gödel. Jech [1967] described it as Gödel-Bernays set theory without the axioms of regularity and of choice.

[194]Josef Novák (April 19, 1905 – August 12, 1999) was a noted Czech topologist. He studied with Karl Menger in Vienna 1935–1936. He was a member of the Czechoslovak Academy of Sciences from its inception in 1952. For more on his life and work, see [Frolík and Zítek, 1975], [Frolík and Koutník, 1985], [Frič and Kent, 2000].

[195]A *closed dyadic partition* of a linearly ordered set C is a family P of closed intervals with the following properties: (1) the intersection of any pair of intervals is either empty, a singleton, or one of the intervals; (2) $C \in P$; (3) for every interval $X \in P$ there are X_1 and X_2 in P such that $X = X_1 \cup X_2$ and $X_1 \cap X_2$ is in P; and (4) the intersection of every decreasing set of intervals from P is either in P or a single element of C. That is, a dyadic partition gives a binary tree of intervals of the original linear order.

[196]However, as Bagemihl pointed out in his Math Review MR0056049 (15,17e), Hausdorff [1907] had already proved sufficiency.

[197]Mary Ellen Rudin (December 7, 1924 –) received her Ph.D. from the University of Texas at Austin in 1949. Her dissertation, supervised by R. L. Moore, was entitled *Concerning Abstract Spaces*. See [Albers and Reid, 1988] to learn more about her graduate student experience. Another early influence was F. Burton Jones, who had studied with Moore and was a professor at the University of Austin when Mary Ellen Rudin was a student. He moved to the University of North Carolina (UNC) the year after she moved to Duke University, and she attended weekly colloquia at UNC and talked frequently with Jones (see [Jones, 1993] for a description of their interaction and a glimpse of her working methods as she constructed a Dowker space from a Suslin line). It was at Duke that she met her husband Walter, a prominent analyst who was an instructor at the time in a neighboring cubical. Walter Rudin is known, for example, for his proof under CH there are $2^{\mathfrak{c}}$ many P-points (a type of ultrafilter), and for his textbook *Principles of Mathematical Analysis* [1953] with second and third editions in 1964 and 1976.

[198]Two points a and b are *separated by a countable subset* S of a space X if $X \setminus S$ can be partitioned into two disjoint clopen sets $X \setminus S = U \cup V$ with $a \in U$ and $b \in V$.

[199]This brief description of the early work of M. E. Rudin is informed by articles [Tall, 1993] and [Watson, 1993] from a conference in honor of Mary Ellen Rudin held June 26–29, 1991 in Madison, Wisconsin. Watson gives modern versions of some of her proofs and makes connections between this early work and Skolem functions, MAD families, the Pixley-Roy topology, and asserts that

Rudin had learned about Suslin's Problem from Moore.[200]

R. H. Bing[201] [1951] strengthened the notion of *normal* to *collectionwise normal*[202] and proved that a space is metrizable if and only if it is a collectionwise normal Moore space. He also showed that if there is a Q-set (defined in §4.2), then there is a separable non-metrizable normal Moore space (see Example E [1951, 182]).

C. Hugh Dowker[203] [1951] introduced the concept of countable paracompactness (every covering by countably many open sets has a locally finite refinement) and proved that for a topological space X, the property of X being countably paracompact and normal is equivalent to $X \times I$ being normal, where $I = [0, 1]$ is the unit interval. In footnote 4 [1951, 221], he asserted: "it would be interesting to have an example of a normal Hausdorff space which is not countably paracompact."

Rudin [1955] showed that if there is a Suslin line, the answer is negative by constructing from a Suslin space[204] a *Dowker space*, i.e. a normal space X whose product with the unit interval, $X \times [0, 1]$, is not normal.

John L. Kelley[205] [1959] investigated the regular open algebra of a Suslin line

---

[Watson, 1993, 175] her use of an uncountable almost increasing family $\mathcal{F}$ of functions from ω to ω as a combinatorial structure each uncountable family of which is likely to possess "bad" behavior is an early ancestor of Todorcevic's Theorem 1.1 in [8] $\cdots$." The reference is to [Todorcevic, 1989].

[200]Tall [1993, 3] reported that the one time Moore lectured to Rudin's class was when he (mistakenly) thought that he had solved Suslin's Problem.

[201]R. H. Bing (October 20, 1914 – April 28, 1986) earned his Ph.D. in 1945 as a student of R. L. Moore. In 1979, Jones [1980, 2] recalled returning to Texas after the war as Bing was finishing his degree, and described him as "solving all the topological world's unsolved problems." Jones showed Bing his tin can example [a special Aronszajn tree with the tree topology], and told him about the metrization problem he could not solve.

[202]A collection $\mathcal{D}$ of subsets of a topological space X is *discrete* if every set $D \in \mathcal{D}$ is a subset of an open set disjoint from $\bigcup(\mathcal{D} \setminus \{D\})$. A space X is *collectionwise normal* if every discrete collection $\mathcal{D}$ of closed sets expands to a disjoint collection $\mathcal{U}$ of open sets, i.e. each $D \in \mathcal{D}$ is in one and only one $U \in \mathcal{U}$.

[203]Clifford Hugh Dowker (March 2, 1912 – October 14, 1982) grew up in Canada, earned his doctorate in 1938 with Lefschetz at Princeton University, and held a variety of academic jobs before working for the MIT Radiation Laboratory 1943–1946. Afterwords he held short-term positions at Tufts, Princeton and Harvard, then decided to leave North America during the period of McCarthyism, when several of his mathematical friends were harassed. He joined the faculty of Birkbeck College in London in 1950 where he remained until his retirement in 1979. For more on his life and work, see [Strauss, 1984].

[204]Rudin's definition of a *Suslin space* is a linearly ordered space S in which neighborhoods are segments which has the properties that S is not separable but every collection of disjoint segments is countable. Her only references were the paper of Dowker [1951] for the question and Suslin's statement of the problem [1920].

[205]John Kelley (December 6, 1916 – November 26, 1999) did his undergraduate work and earned his master's degree at the University of California, Los Angeles, and obtained his doctorate in 1940 at the University of Virginia under the direction of Gordon T. Whyburn. He spent most of his career at the University of California, Berkeley. His textbook, *General Topology* [1955], based on lectures given at the University of Chicago 1946–1947, the University of California, Berkeley, 1948–1949 and Tulane University 1950–1951, has served generations of students. His autobiographical sketch [1989] is humorous and includes sketches both personal and mathematical

and developed some Boolean algebraic equivalents of the existence of a Suslin line. His work built on that of [Maharam, 1947].

5.3 *Ordered sets, structure and mappings*

A question of Fraïssé [1948] inspired Sierpiński [1950] to work on the structure of subsets of the reals under order embeddability. He developed a technique for identifying a subset H of a continuum-size linear order E such that no order-preserving map from E to H is onto. He proved that the class of uncountable linear orders has a decreasing chain of length the continuum, an increasing chain of length the continuum, and an antichain of size the continuum. He indicated in a footnote that the construction of a rigid real type by Dushnik and Miller [1941, Theorem 3, p. 325] was an inspiration for his first key lemma.

John C. Shepherdson[206] [1951a] examined the structure of ordered sets by computing the *index*, the least ordinal number strictly greater than the ordinals of all well-ordered subsets. He was able to give a recursive characterization of the general form of an order type except when the index is a regular cardinal. He showed that if the index of an ordered set is not of the form ω^β or $\omega^\beta + 1$, then it can be expressed as the sum of ordered sets with smaller indices. The representation for sets whose index has normal form $\sum_{i<k} \omega^{\beta_i} n_i$ is quite complex. In Theorem 3.2, he proved that among ordered sets of which every proper initial segment has index greater than ω_β, there is a minimal order, $\mathbb{Z}_\beta$, which is the lexicographic order on the set of functions from ω to ω_β which are non-zero only finitely often. His analysis built on work of Hausdorff [1908] and Gleyzal [1940].

Shepherdson [1954] published his construction of a denumerable ramified set F with the properties that no maximal antichain of F meets every maximal chain of F and that no maximal chain of F meets every maximal antichain of F. This example provided an answer to two questions of Kurepa at the end of his paper giving the following statement equivalent to a non-empty partially ordered set $(S, \leq)$ being finite: in every ramification $T(S)$ of S, every maximal antichain has non-empty intersection with every maximal chain.

Seymour Ginsburg,[207] a student Dushnik at University of Michigan, studied real types and interpolation theorems for them in the 1950's (see [1953a], [1953b], [1955]). He investigated partitions of real types into various nice sets, and proved [1953b, 531, Theorem 3.3] that if f is an order-preserving map of E into itself with no fixed points, then E is the union of countably many pairwise disjoint sets all of which are mutually embeddable in one another. Ginsburg [1953b, 516] noted that he knew of no order type other than λ, the order type of the real numbers, that

of his education and work.

[206] J. C. Shepherdson was on the faculty of the University of Bristol from 1946–1991, when he reached emeritus status. He was elected to fellowship in the philosophy section of the British Academy in 1990 and became a fellow of the Royal Society in 1999. Set theorists may know him best for his work on inner models [1951b], [1952], [1953], and those in computability for his work with Howard Sturgis on the unlimited register machines.

[207] Ginsburg (1928-2004) became a well-known computer scientist.

had a *similarity decomposition*, i.e. a partition of a set of that type into continuum many (pairwise disjoint) subsets all mutually embeddable with one another and with the original set.[208]

For every infinite cardinal κ, John B. Johnston [1956] constructed a partial order of cardinality 2^κ universal for partial orders of cardinality κ and dimension κ in his Theorem 3. Theorem 3 has an additional part generalizing the result of Mostowski [1938] that there is a countable partial order universal for countable partial orders.

Bjarni Jónsson [1956] gave a list of six axioms, which sufficed with the Continuum Hypothesis, to guarantee that structures satisfying the axioms have universal structures in all uncountable powers.

In a preliminary report [1957], Andrzej Ehrenfeucht[209] formulated two-player games as an organizing strategy,[210] inspired by Fraïssé's back-and-forth strategy for proving elementary equivalence of models.[211] Ehrenfeucht [1961] applied the technique to the analysis of properties of the arithmetic of ordinals.

The Dilworth [1950] chain decomposition theorem is the statement that if every antichain in a finite partially ordered set P has size $\leq k$, then P is the union of $\leq k$ many chains. In the proceedings of a symposium on combinatorial analysis held at Columbia University in 1958, Dilworth [1960] made connections between his decomposition theorem and Hall's Marriage Theorem [1935].

5.4 The Regressive Function Theorem

The seminal 1929 Alexandroff–Urysohn Regressive Function Theorem was steadily extended, getting to Fodor's 1956 widely applied version for stationary sets. We had already noted that Alexandroff and Urysohn [1929] showed that functions regressive on the set of infinite countable ordinals are constant on an uncountable set, and that Dushnik [1931] extended the result to regressive functions on the set of infinite ordinals smaller than a successor cardinal κ.

In a paper [1950, 140–141] received by the editors in November 1948, Erdős revisited Dushnik's version of the Regressive Function Theorem [1931] and proved that if Ω_k is an (uncountable) initial ordinal which is not cofinal to ω and f is a regressive function on the set of ordinals less than Ω_k, then there exists an ordinal

[208]Harzheim [2008] constructed a subset of the real numbers which does not order embed the real numbers but does have a similarity decomposition.

[209]Andrzej Ehrenfeucht (August 8, 1932 –) was a student of Andrzej Mostowski at the University of Warsaw, where he earned his master's degree in 1955. In 1961, he received his doctorate from the Institute of Mathematics of the Polish Academy of Sciences in Warsaw. After being an assistant professor at the University of California, Berkeley, and Stanford University, and an associate professor at the University of Southern California, he joined the Department of Computer Science at the University of Colorado in Boulder.

[210]For more on the early work of Ehrenfeucht on model theory, see [Vaught, 1997].

[211]Ehrenfeucht may have heard about Fraïssé's work from the mathematical logic colloquium in Paris in 1952 by Fraïssé titled *Sur les rapports entre la théorie des relations et la semantique au sens de A. Tarski*, since it is in the bibliography of [Ehrenfeucht, 1961] but otherwise unreferenced. The games Ehrenfeucht formulated are now called Ehrenfeucht-Fraïssé games.

β and a sequence α_i cofinal with Ω_k such that $f(\alpha_i) \leq \beta$. He noted that for regular Ω_k it is immediate from Dushnik's Theorem, and if Ω_k is cofinal with ω it does not hold for all such f. He pointed out that Dushnik's proof is easily adapted to prove the result for singular cardinals of uncountable cofinality, citing an oral communication, but he provided his own short proof.

As a tool for his investigation of Suslin continua with dyadic partitions, Novák [1950] proved a generalization of the Alexandroff–Urysohn Regressive Function Theorem:

> B being an uncountable closed subset of the set C of all countable ordinal numbers, let $f(x)$ be a single-valued transformation of B onto $A \subset C$ having the property that $f(x) < x$ for all $x \in B$. Then there exists a countable ordinal number $\alpha \in A$ and an uncountable subset $B^* \subset B$ such that $f(x) = \alpha$ for all $x \in B^*$.[212]

Walter Neumer[213] [1951] published a generalization of the Regressive Function Theorem that stated in modern language asserts that for all regular cardinals κ, any function regressive on a stationary[214] subset of κ is constant on an set of size κ. For Λ an initial ordinal, he wrote $W(\Lambda)$ for the set of ordinals smaller than Λ. He called a subset $M \subseteq W(\Lambda)$ a *full part* [Vollteil] if it is similar to $W(\Lambda)$; he called it *closed* if it contained all limits of well-ordered sequences; and a closed, full part of $W(\Lambda)$ was called a *band* [Band]. In particular, he proved that if the complement of M contains no band and f is regressive on M, then there is a value that f takes Λ many times. Moreover, he showed the necessity of this condition, proving that if the complement of M does contain a band, then there is a regressive function for which every value is realized fewer that Λ many times. Together these results of Neumer characterize stationary sets of a regular uncountable cardinal Λ as those for which every regressive function is constant on a subset of cardinality Λ.[215] He went on to explore basic features of stationary sets, such as the fact that stationary co-stationary sets exist for regular cardinals greater than ω_1.

Gérard Bloch[216] also investigated stationary sets. He independently proved Neumer's Theorem for ω_1 in his Theorem I: A necessary and sufficient condition for

---

[212]Denjoy [1953] independently proved Novák's result.

[213]Walter Neumer (March 9, 1906 – January 25, 1981) received his doctorate in 1929 From Justus-Liebig-Universität Giessen where his advisors were Friedrich Engel (primary) and Ludwig Schlesinger (secondary). Most of his academic career was at Johannes Gutenberg-Universität Mainz.

[214]A subset of κ is *stationary* if every closed unbounded subset of κ meets it; Bloch [1953] is credited with the first use of the term in connection with the Regressive Function Theorem; Kanamori [2009, Chapter 1] credits Mahlo with the first use of this concept.

The word *stationnaire* was used by Lusin [1934] to mean a sequence which is eventually constant, in a monograph where he extended the result of Réne Baire that all decreasing sequences of closed sets are countable to show that all decreasing sequences which are continuous at limits of sets which are countable unions of closed sets are eventually constant.

[215]Hajnal and Hamburger [1999] refer to this characterization as *Neumer's Theorem*, give two different proofs of it, and discuss its relationship with the Regressive Function Theorem of Fodor (see below).

[216]His note was presented by Denjoy, and is the only item by him listed in Math Reviews.

a subset to be stationary is that every closed subset of its complement be at most denumerable (translated from the original French). Given a set A of countable limit ordinals, he used *suite distinguée* for a sequence of regressive functions $f_n(x)$ ($n \in \mathbb{N}$) defined on A such that, for all $x \in A$, $f_1(x) \leq f_2(x) \leq \cdots \leq f_n(x) \leq \ldots$ and $\lim_n f_n(x) = x$. He used this concept to prove that stationary sets can be split into disjoint stationary sets. In a brief postscript, he reported that Walter Neumer had some of the results for stationary sets (a reference to Neumer's paper [1951] was included), and that he, Bloch, had not known of Neumer's article when he submitted his note.

Géza Fodor,[217] published in [1955] his rediscovery of Dushnik's generalization of the Alexandroff–Urysohn Regressive Function Theorem.[218] The very next year, Fodor [1956] proved that for any regular uncountable cardinal κ, every function regressive on a stationary subset of κ is constant on a stationary subset. This result is known by various names, including the Regressive Function Theorem, Fodor's Lemma or Fodor's Theorem (see [Jech, 2003, 93], [Hajnal and Hamburger, 1999, 151]), and the Pressing-Down Lemma (see [Kunen, 1983, 80]). The special case for $\kappa = \aleph_1$ of this useful result also appeared in Spanish in 1958 in [Ricabarra, 1956].[219] Neumer [1958], apparently unaware of Fodor's [1956], also published a proof of this important theorem.

6 1960-1970: FORCING, TREES AND PARTITIONS

The 1960's saw a sea-change in set theory with interactions with the growing field of model theory, the introduction of forcing, Martin's Axiom and diamond principles with multiple applications to combinatorial problems, and the development of

[217] Géza Fodor (May 6, 1927 – September 28, 1977) was born in Szeged, Hungary, educated at the University of Szeged, and received a degree in 1954 equivalent to the American doctorate under the supervision of B. Szőkefalvi-Nagy in the field of combinatorial set theory. Starting in 1954 he was a research associate and then an associate professor for Professor Laszlo Kalmár at the University of Szeged. He became a full professor in 1967 after receiving his Academy Doctor degree (like the Habilitation degree in German universities) and led the Department of Set Theory and Mathematical Logic from its creation in 1971. He was elected a corresponding member of the Hungarian Academy of Sciences in 1973 and served as Rector of the University of Szeged 1973 – 1967. This brief biography is based on [Editors, 1977], which also noted that Fodor was the originator of the question of whether a stationary subset of size κ could be split into κ disjoint subsets. For an overview of his mathematics in Hungarian, see Komjáth [1982/86].

[218] F. Bagemihl, in Math Review MR0076000, said Dushnik's result generalizes one in [Bachmann, 1955, 43].

[219] The international journal *Revista de Matemáticas Cuyana* was started by Mischa Cotlar who became director of the Instituto de Investigaciones Mathematicas de Cuyo, new at the time he arrived after earning his Ph.D. under Zygmund in 1953. In 1955, the institute was closed down by a military administration which claimed that it was a Communist organization financed by another Communist organization, UNESCO, according to testimony by Cora Sadowsky in conjunction with a conference honoring Mischa Cotlar held October 12, 2007 at the University of New Mexico in Albuquerque, New Mexico (see http://www.math.unm/conferences/10thAnalysis/resources/cotlar/sadosky.pdf accessed August 17, 2007). Ricabarra's paper was in an issue of the journal nominally for the year 1956, but only appeared in 1958.

wqo and bqo theory. Cohen's proofs that set theory was consistent with the negation of the Continuum Hypothesis and also with the negation of the Axiom of Choice set off a wave of new results. We mention Easton's work on the values the exponentiation function could have on regular cardinals, and Solovay's proof of the consistency of "all sets of reals are Lebesgue measurable" with Dependent Choice (DC).

Classical results for the partition calculus of cardinals were published in the middle of the decade in a massive paper that had been years in the making, and open questions about them were shared with the wider logic and set theory community. Suslin's Hypothesis was shown to be both consistent and independent of the usual axioms of set theory, and one might say a new section of Cantor's paradise was opened where realms were to be found representing different possibilities for Aronszajn, Kurepa, and Suslin trees.

The San Francisco Bay Area was a nexus for research during this decade with a wealth of graduate students, postdoctoral researchers and visitors in addition to the regular faculties of Stanford University and the University of California, Berkeley. Dana Scott was first on the Berkeley faculty and then the Stanford one.[220] Paul Cohen, Harvey Friedman,[221] and Ehrenfeucht were on the faculty at Stanford, where Jensen visited in the late 1960's with his student Adrian Mathias, who took a course with Scott. The following individuals (listed with their advisors) earned doctorates at Stanford:

1966	Platek	(Kreisel)	1968	Kunen	(Scott)
1967	Barwise	(Feferman)	1970	Mansfield	(Scott)

The mathematical life in Berkeley was enriched by the following visitors. Lévy was a visiting assistant professor 1959–1961; Jerzy Łoś visited 1959–1960 and 1962–1963;[222] and Hajnal visited for the calendar year 1964, lecturing on the partition calculus.[223] Fred Galvin held a post-doctoral position 1965–1968.[224] after finishing the research for his 1967 doctorate from the University of Minnesota under Jónsson.[225] The individuals listed below with their advisors earned doctorates at Berkeley:

[220]Dana Scott was a Miller Institute Fellow and assistant professor in 1960, was promoted to associate professor in 1962, and moved to Stanford University in 1963.

[221]Harvey Friedman was at Stanford 1967–1973, arriving as an assistant professor of Philosophy and promoted to associate professor in 1969. He spent 1970–1971 at the University of Wisconsin, Madison, and 1972–1973 as a Visiting Professor at the State University of Buffalo, where he was a Professor from 1973–1977. He has since been a professor at Ohio State University.

[222]Los [2000] gave a seminar jointly with Tarski in 1959–1960 and a seminar jointly with Jońsson 1962–1963.

[223]Hajnal noted [1997] that the persistent attendees of his 1964 lectures were Donald Monk, William Reinhardt and Jack Silver.

[224]From Berkeley, Galvin moved to the University of California, Los Angeles, and then in 1974 to the University of Kansas.

[225]Bjarni Jónsson moved to Vanderbilt University in 1966.

1962	Gaifman	(Tarski)	1967	Reinhardt	(Vaught)
1962	Halpern	(Lévy, Scott)	1967	Prikry	(Silver)
1966	Silver	(Vaught)	1969	Laver	(McKenzie)
1966	Malitz	(Vaught)	1970	Baumgartner	(Vaught)
1966	McAloon	(Solovay)	1970	Mitchell	(Silver)
1967	Hechler	(Solovay)			

The Prague Set Theory Seminar was founded by Petr Vopěnka after the death in 1963 of his advisor Ladislav S. Rieger,[226] the first Czech mathematician to work in the foundations of set theory. Petr Hájek [1971] listed members of the seminar: Bohuslav Balcar, Lev Bukovský, Karel Hrbáček, Thomas Jech, Antonín Sochor, Petr Štěpánek, Petr Vopěnka, and Hájek himself. Hájek closed his paper with a 55-item bibliography of the Prague seminar.[227]

The decade included the first international set theory conference after the introduction of forcing. Officially it was the Fourteenth Annual Summer Research Institute sponsored jointly by the American Mathematical Society and the Association for Symbolic Logic, with financial support from the National Science Foundation. In the preface to the proceedings, Scott [1971, v] reported that the four-week conference held at the University of California, Los Angeles (UCLA), July 10 - August 5, 1967, featured tutorials by himself and Joseph Shoenfield (10 lectures each), together with individual contributions (usually in hour talks) at the rate of four per day but by the last week reduced to three per day, "as the strength of the participants had noticeably weakened." Two other conferences of the decade were the International Symposium on Topology and its Applications in Herceg-Novi in 1968, whose organizing committee was chaired by Kurepa, and the first British Combinatorial Conference held in Oxford in 1969, whose invited speakers included Erdős and Marc Kac, which signaled the growing sense of community among British combinatorial mathematicians.

We mention in passing that in this decade Rado [1964] constructed the Rado graph (the denumerable random graph), which is universal for countable graphs.[228] Michael Morley and Robert Vaught [1962] extended Jónsson's work on the existence of universal structures, and Brian Rotman [1971] showed that most of Rado's results about the Rado graph can be deduced from the general model-theoretic framework of Jónsson, Morley and Vaught.

Cardinal transfer questions raised in this decade in model theory would turn out to be pivotal for questions about trees and about the partition calculus. Say a

[226] Ladislav Svante Rieger (1916–1963) received his doctorate from Charles University, Prague and in 1959 defended his Habilitation thesis, where the referee was Mostowski. Rieger founded the first seminar in Prague on mathematical logic in 1953 with a focus on Gödel's work on the consistency of the Axiom of Choice and the Continuum Hypothesis. For more on Rieger, see [Hájek *et al.*, 2008] and the thesis of E. Kozkíková, which is described as "a detailed monograph on L. S. Rieger (in Czech)".

[227] Karel Prikry is listed as participant in the Prague seminar by Hájek, Marek and Vopěnka in [2008].

[228] Rado's graph started a line of research which extends beyond the scope of this paper, for example with work by Shelah [1984b], [1990b], joint work with Komjáth [1995] and by Kojman [1998].

theory in a countable language with identity and distinguished unary predicate R *admits a pair* (α, β) if there is a model $M = \langle A, U, \cdots \rangle$ of T such that $|A| = \alpha$ and $|U| = \beta$, where U is the interpretation of R. Morley and Vaught [1962] proved that if T admits (κ, λ), then T admits $(\aleph_1, \aleph_0)$. Vaught restated the result as Theorem 14 and raised the following question in [1963, 308–309]:

> *Cardinal Transfer Question:* If T admits (κ, λ), what other pairs (κ', λ') must T admit?

Chang asked if Theorem 14 could be improved to the following question (see Vaught [1963, 309]): Does every model $\mathfrak{A}$ of T with $\overline{R_0^{\mathfrak{A}}} \neq \overline{\mathfrak{A}}$ have an elementary submodel $\mathfrak{B}$ with $\overline{\mathfrak{B}} = \aleph_1$ and $\overline{R^{\mathfrak{B}_0}} = \aleph_0$? The assertion that the answer to Chang's question is "yes" is now known as *Chang's Conjecture*. Vaught went on to connect the question with one in set theory: "A negative answer to the problem is an easy consequence of a positive answer to the following question: *Does there exist a family W of subsets of a set A of power $\aleph_1$ such that $\overline{W} = \aleph_2$ while, for any countable $B \subseteq A$, $\{X \cap B \mid B \in W\}$ is countable*?" Vaught pointed out that this question was a version of one raised by Kurepa [1942] (i.e. the questions of existence of a Kurepa tree) and quoted Ricabarra [1958, 344] as indicating it was still unresolved.

Chang proved the following theorem giving a partial answer under the Generalized Continuum Hypothesis [1965]:

> *Chang's Cardinal Transfer Theorem:* Assume GCH. If $\alpha > \beta \geq \omega$ are cardinals and a theory T in a countable language with a distinguished unary predicate R admits (α, β), then for all regular infinite cardinals δ, T admits (δ^+, δ).

We denote the cardinal transfer property "if T, R admits (α, β), then it admits (δ^+, δ)" by $(\alpha, \beta) \to (\delta^+, \delta)$.

We conclude this introduction with a brief discussion of the Normal Moore Space Conjecture. Robert Heath [1964] showed that there is a separable non-metrizable normal Moore space if and only if there is a Q-set, a result which some describe as the solution to the Separable Normal Moore Space Conjecture. During his 1967–1968 visit to the University of Wisconsin, encouraged by graduate student Franklin Tall, Silver showed the consistent existence of a Q-set (cf. [Tall, 1969], [Tall, 1977]). Recall that Rothberger [1948] proved that if the pseudointersection number is greater than $\aleph_1$, then Q-sets exist. Thus the combination of the work of Bing [1951] and Rothberger [1948] gives a consistency result. However, Rothberger's result had been forgotten in the sixties, and was rediscovered in the seventies by David Booth. Tall, in his 1969 thesis, part of which is published in [1977],[229] focused attention on the "real question .. when is a normal space collection-wise

[229] Eric van Douwen reported in his Math Review MR0454913 (56 #13156) that the material in Tall's monograph is unchanged from his thesis except for some brief mentions of results obtained after the thesis. van Douwen went on to assert that this paper changed the general feeling of topologists that results proving consistency and independence from ZFC are not really part of

normal?" According to Nyikos [2001, 1196], Tall, while a graduate student, proved the consistency of the statement that every first countable normal space is $\aleph_1$-collection-wise Hausdorff. His advisor, Rudin, observed that a corollary of Tall's result is the consistency of no Aronszajn tree being normal in the tree topology, so that in particular, it is consistent that the road spaces of Jones are not normal.

6.1 *The Halpern-Läuchli Theorem*

We start with the Halpern-Läuchli Theorem, a fundamental Ramsey-theoretic result for partitions of finite products of trees that generalized the rectangle refining theorem of Rado [1954a]. James D. Halpern[230] and Hans Läuchli[231] [1966] expressed their ground-breaking combinatorial result in terms of trees following a suggestion of a referee of their paper. The result was a key piece in the plan of Halpern and Lévy [1964], which they [1971] successfully carried out, to prove that in ZF, the Boolean Prime Ideal Theorem does not imply the Axiom of Choice. Halpern and Lévy [1971, 97] use Theorem 2 [Halpern and Läuchli, 1966, 362] in their proof, which is stated below in modern language, where a tree is *rooted* if there is a single element, called the *root*, in level 0:

> *Halpern-Läuchli Theorem:* For any finite sequence $\langle (T_i, <_i) \mid i < d\rangle$ of finitely branching rooted trees without treetops of height ω and any positive $q < \omega$ there is a positive integer n such that for any q-ary partition of the product $\prod_{i<d}(T_i \restriction n) = Q_0 \cup \cdots \cup Q_{q-1}$ there are $j < q$, $h < n$, $\langle x_i \mid i < d\rangle$ and $\langle A_i \subseteq (T_i \restriction n) \mid i < d\rangle$ such that $\prod_{i<d} A_i \subseteq Q_j$ and for all $i < d$, x_i is on level h of T_i and every immediate successor of x_i has an extension in A_i.

Included below, in modern language, is their immediate corollary [1966, 363] about level products which may look more familiar to some readers.

topology, toward recognition that they are with the discovery of consistency and independence results about "down-to-earth objects like first countable normal spaces," the subject of the monograph.

[230] James Daniel Halpern was a student at the University of California, Berkeley, in the late 1950's and early 1960's. In his thesis, Halpern [1962] proved the independence of the Axiom of Choice from the Boolean Prime Ideal Theorem (BPI) over the theory ZF minus the Axiom of Foundation by showing BPI holds in a Fraenkel-Mostowski model in which the Axiom of Choice did not. The result was announced in an abstract in the *Notices of the American Mathematical Society* in 1961; a revised version was published in [Halpern, 1964]. His dissertation work was done under the direction of Lévy, who was visited Berkeley 1959-1961, and written up under the supervision of Dana Scott. See page 58 of [Halpern, 1964] for Halpern's remarks on the supervision of his work. See [Kanamori, 2004] for more information about Lévy, and, in particular, his interactions with Halpern. The proof uses a combinatorial lemma that Halpern calls the Rado Corollary (Scott is credited with the proof included in the paper) and is a generalization of Ramsey's Theorem which may be simply expressed as a finitary polarized partition relation. Halpern spent 1965–1966 at the Institute for Advanced Study, and 1974–1981 at the University of Alabama, Birmingham.

[231] Hans Läuchli (1933 - August 13, 1997) was a student of Ernst Specker at Eidgenössische Technische Hochschule Zürich (ETH) and received his degree in 1962 with a dissertation on *Auswahlaxiom in der Algebra.*

Halpern-Läuchli Corollary: For any finite sequence $\langle (T_i, <_i) \mid i < d \rangle$ of finitely branching rooted trees without treetops of height ω and any positive $q < \omega$ there is a positive integer n such that for any q-ary partition of the product of the nth levels of the trees, $\prod_{i<d} T_i(n) = Q_0 \cup \cdots \cup Q_{q-1}$, there are $j < q$, $h < n$, $\langle x_i \mid i < d \rangle$ and $\langle A_i \subseteq T_i(n) \mid i < d \rangle$ such that $\prod_{i<d} A_i \subseteq Q_j$ and for all $i < d$, x_i is on level h of T_i and every immediate successor of x_i has an extension in A_i.

Most of the work to prove the above results is in the proof of Theorem 1, which I have relabeled but otherwise quote from [Halpern and Läuchli, 1966, 361–362].

Halpern-Läuchli Core: Let $\mathcal{T}_i = \langle T_i, \leq_i \rangle$, $1 \leq i \leq d$ be finitistic trees without tree tops and let $Q \subseteq \prod_1^d T_i$. Then either

a. for each k, Q includes a k-matrix or

b. there exists h such that for each k, $\left(\prod_1^d T_i\right) - Q$ includes an (h, k)-matrix.

A *finitistic tree without tree tops* is a finitely branching tree of height ω and no terminal nodes. An (h, k)-*matrix* is a product $\prod_1^d A_i$ where for each $1 \leq i \leq d$, $A_i \subseteq T_i$ and for some $\langle x_1, x_2, \ldots, x_d \rangle \in \prod_1^d T_i(h)$, for each $y \in T_i(k)$, if $x_i \leq_i y$, then there is some $z \in A_i$ with $y \leq_i z$. A k-*matrix* is a $(0, k)$-matrix, i.e. the special vector lists the roots of the trees.

In their proof of the Halpern-Läuchli Core Theorem, Halpern and Läuchli used metamathematical, heavily syntactic techniques. They comment "We have tried to eliminate the use of metamathematics without success and would welcome a simplification in this direction" [1966, 360].

In the late 1960's, Galvin (cf. [Erdős and Hajnal, 1974, 275]) proved that if you color the rationals with finitely many colors, then there is a subset of the same order type η which is at most *bi-colored*, i.e. realizes at most 2 colors. He conjectured that if you color r-element sequences of rationals with finitely many colors, then there are subsets $X_0, X_1, \ldots, X_{r-1}$, each of order type η so that $\prod_{i<r} X_i$ receives at most $r!$ colors. He proved this conjecture for $r = 2$,[232] and showed that for every positive r, the bound $r!$ cannot be improved. This conjecture, proved by Richard Laver in 1969 (cf. [Erdős and Hajnal, 1974, 275]), is expressed using a generalization of the polarized partition relation as follows:

$$\begin{pmatrix} \eta \\ \eta \\ \vdots \\ \eta \end{pmatrix} \longrightarrow \begin{pmatrix} \eta \\ \eta \\ \vdots \\ \eta \end{pmatrix}^{\overbrace{1,1,\ldots,1}^{r}}_{<\omega, r!}$$

[232]The Sierpiński partition restricted to the rationals is a partition in which every subset of the rationals of full order type meets both cells of the partition.

As Laver later recalled, he [1984] proved the polarized partition relation essentially by identifying the rationals with the dyadic rationals which can be regarded as represented by the complete binary tree $({}^{<\omega}2, <_q)$ of all finite sequences of 0's and 1's under a suitable ordering $<_q$, and proving the theorem in this alternative setting. More generally, Laver showed that there is a function $f : {}^n\omega \to \omega$ so that for any $\vec{r} = \langle r_0, r_2, \ldots, r_{n-1}\rangle$, and any finite coloring of $\prod_{i<n}[\mathbb{Q}]^{r_i}$, there is an $\langle X_0, X_1, \ldots, X_{n-1}\rangle$ with each X_i a subset of $\mathbb{Q}$ of order type η, such that $\prod_{i<n}[X_i]^{r_i}$ realizes at most $f(\vec{r})$ colors.

Laver was unaware of the Halpern-Läuchli Theorem at the time, so in order to prove the polarized partition theorem, he proved a perfect subtree version of the Halpern-Läuchli Theorem which I name and state below using $\prod^A \vec{T}$ as an abbreviation for $\bigcup_{n\in A}\prod_{i<d} T_i(n)$:

> *Laver HL_d Theorem:* If $\vec{T} = \langle T_i : i < d\rangle$ is a finite sequence of rooted finitely branching perfect trees of height ω and $\prod^\omega \vec{T} = G_0 \cup G_1$, then there are $\delta < 2$, an infinite subset $A \subseteq \omega$ and downwards closed perfect subtrees T'_i of T_i for $i < d$ with $\prod^A \vec{T}' \subseteq G_\delta$.

Call $\vec{X} = \langle X_i : i < d\rangle$ *n-dense* in $\vec{T} = \langle T_i : i < d\rangle$ if there is some $m \geq n$ such that $X_i \subseteq T_i(m)$ for all $i < d$ and for each $i < d$ and $t \in T_i(n)$, there is some $u \in X_i$ with $t \leq_i u$. For $\vec{x} \in \prod^\omega \vec{T}$, call $\vec{X}$ *$\vec{x}$-n-dense* in $\vec{T}$ if $\vec{X}$ is n-dense in the sequence $\langle (T_i)_{x_i} : i < d\rangle$ where $(T_i)_{x_i}$ is the subtree of T_i of all elements comparable with x_i. Most of the work of proving HD_d for $d < \omega$ is in the following result, which I have named for easy reference and appears as Theorem 1 in [Laver, 1984, 386]:

> *Laver HL_d Core:* Suppose that $\vec{T} = \langle T_i : i < d\rangle$ is a finite sequence of finitely branching tress of height ω with no terminal nodes, and $\prod^\omega \vec{T} = G_0 \cup G_1$. Then either
>
> a. for all $n < \omega$ there is an n-dense $\vec{X}$ with $\prod^\omega \vec{X} \subseteq G_0$, or
>
> b. for some $\vec{x} \in \prod^\omega \vec{T}$ and for all $n < \omega$ there is an $\vec{x}$-n-dense $\vec{X}$ with $\prod^\omega \vec{X} \subseteq G_1$.

6.2 *Countable height trees*

In 1960 Joseph B. Kruskal[233] published his celebrated tree theorem in an expanded version of his doctoral dissertation [1954] under the direction of Roger Lyndon at Princeton University:[234]

---

[233]Joseph Bernard Kruskal (January 29, 1928 – September 19, 2010) was an instructor at Princeton and the University of Wisconsin at Madison before becoming an assistant professor at the University of Michigan in Ann Arbor in 1958. In 1959 he joined the technical staff at Bell Telephone Labs. He has held visiting positions at Yale, Columbia and Rutgers.

[234]Erdős is listed as a second advisor for Kruskal at the Mathematics Genealogy Project site.

> *Kruskal's Tree Theorem* [1960]: If X is a wqo, then $T(X)$, the collection of finite trees with labels from X, is a wqo under tree embeddings $f : S \to T$ which respect the labels in the sense that for all $s \in S$, the label of s is less than or equal to the label of $f(s)$.

Independently, S. Tarkowski [1960] proved a theorem from which the above result follows. Tadashi Ohkuma announced [1960] and proved [1961] that under order-preserving maps, scattered trees[235] are linearly ordered under order preserving maps (they need not be injective). He also showed that every scattered tree is comparable with every tree under this ordering. He showed under CH that there are non-comparable trees.

Kruskal [1954] proved the closure of the family of wqos under finite unions and Cartesian products (quasi-ordered componentwise). He conjectured at the end of the paper that the set of all (connected) trees, finite or infinite, is a wqo.

Kruskal was motivated by a conjecture of Vázsonyi[236] that for any infinite sequence T_i of (finite) trees, there are two T_k and T_ℓ such that T_ℓ contains a homeomorphic image of T_k, or equivalently, T_ℓ is a subdivision of T_k.[237] Kruskal heard about the problem from Erdős.[238]

Connections between graphs and partial orders were investigated in the 1960's. Elliot S. Wolk [1962] came up with a characterization of when a graph is the comparability graph of a tree: on each path in between two vertices there is a vertex adjacent to both the initial and terminal vertex of the path. In 1962 Paul C. Gilmore and Alan J. Hoffman announced their characterization of when a graph is the comparability graph of a partial order [1964] at the International Congress of Mathematicians held in Stockholm in 1962: G is a comparability graph if and only if each of its odd cycles has at least one triangular cord (an edge which joins two consecutive edges of the cycle to form a triangle). Alain Ghouila-Houri [1962] gave an alternate description (and a proof for the finite case). Wolk [1965] proved that a graph is the comparability graph of a partial order if and only if all of its finite induced subgraphs are. Tibor Gallai [1967] gave a characterization in terms of forbidden subgraphs, which has proved very useful.

---

[235] A tree is *scattered* if it does not embed the tree of all finite sequences of zeros and ones.

[236] Erdős dates Vázsonyi's Conjecture to 1937 in Math Review MR0123491 (23 #A816).

[237] We say S is a subdivision of T if S can be obtained from T by changing some of the edges of T into finite paths by insertion of vertices of degree two.

[238] Erdős is listed as a second advisor to Kruskal on the Mathematics Genealogy Project site.

Crispin St. J. A. Nash-Williams[239] became interested in quasi-orders in 1957, on learning of Vázsonyi's Conjecture. He set out to extend Kruskal's Tree Theorem to infinite trees, and, as a side benefit, he [1963] came up with a simpler proof of it. He generalized Ramsey's Theorem to the *Nash-Williams Partition Theorem*. In his notation, P stands for the set of positive integers and $A(I)$ stands for the set of ascending finite sequences of elements of $I \subseteq P$, including the empty sequence. A subset of $A(P)$ is *thin* if it does not include two distinct sequences s, t such that one is a proper initial segment of the other. With this notation in hand, we are ready to quote his theorem [1965b, 33]:

> *Nash-Williams Partition Theorem:* Let $m \in P$ and I be an infinite subset of P and $\{T_1, \ldots, T_m\}$ be a partition of a thin subset T of $A(I)$ into m disjoint subsets. Then there exists an infinite subset K of I such that $T \cap A(K)$ is contained in a single T_j.

Lemma 4, which is the statement for partitions into two parts, is the heart of the theorem, which then follows from a simple induction. Nash-Williams included a note that Erdős sketched an alternate proof using a finitely additive probability measure μ on the subsets of P that can be constructed using the Axiom of Choice so that $\mu(X) = 0$ if X is finite and $\mu(Y) = 1$ if Y is cofinite.

While the paper is particularly well-known for the Nash-Williams Partition Theorem, the main objective of the paper, according to the author, was the proof of the conjecture of Rado [1954b] that given a well-quasi-order Q, the collection $\mathbb{R}(Q)$ of sequences with domain an ordinal and range a finite subset of Q form a well-quasi-order under the ordering $f \leq g$ if and only if there is a one-to-one order-preserving function $\mu : \text{dom}(f) \to \text{dom}(g)$ such that for all $\alpha \in \text{dom}(f)$, $f(\alpha) \leq f(\mu(\alpha))$. Nash-Williams noted that Higman [1952] had proved that the collection of finite sequences from a wqo form a wqo; Rado [1954b] proved it for sequences of length less than ω^3; and both Kruskal (unpublished) and Erdős and Rado in their joint work [1959] proved it for sequences of length less than ω^ω. Nash-Williams [1965b, 34] also indicated that the proof used an idea underlying Theorem 3 of [Rado, 1954b], and that he was "indebted to Prof. R. Rado for some improvements in the presentation of this paper."

[239]Crispin St. J. A. Nash-Williams earned his Ph.D. from Cambridge University in 1959. The obituary by Welsh [2003] lists as his advisors Shaun Wylie and Davis Rees of Cambridge, and N. E. Steenrod of Princeton University, whom he met while on a fellowship at Princeton in 1956-57. Richard Rado was one of the readers of his 500 page dissertation. Welsh recounted Luise Rado's dismay at the amount of time her husband spent during a walking vacation reading Nash-Williams's dissertation. When Rado formally retired in 1975 (he remained active mathematically), Nash-Williams moved into his chair at the University of Reading. For a further introduction to the life and work of Nash-Williams, this obituary is warmly recommended, especially for the overviews of Nash-Williams's work on decompositions of infinite graphs, which is outside the scope of this paper. Nash-Williams liked both finite and infinite graph theory, and introduced one of his survey articles as follows: "One of my own reasons for becoming interested in graph theory was that I was intrigued by the possibility of developing non-trivial and fairly deep mathematics from a very simple concept. The ultimate illustration of this might be set theory, since sets are presumably the simplest and most basic of all mathematical structures and yet some aspects of set theory are both deep and substantial" [Welsh, 2003, 177].

Nash-Williams [1965a] went on to prove that infinite trees form a wqo, but to do so, he invented the notion of *better-quasi-order* or bqo, which have nice closure properties. The definition of better-quasi-order is involved. Maurice Pouzet[240] and Norbert Sauer[241] set the concept of better-quasi ordering in historical context as follows [2006, §1.3]:

> A basic result due to G.Higman, see [10], asserts that a poset P is wqo if and only if $I(P)$, the set of initial segments of P, is well-founded. On the other hand, Rado [24] has produced an example of a well-founded partial order P for which $I(P)$ is well-founded and contains infinite antichains. The idea behind the bqo notion is to forbid this situation: $I(P)$ and all its iterates, $I(I((I(P)))$ up to the ordinal ω_1, have to be well-founded and hence wqo.[242]

In 1968, as part of his thesis work with Ralph McKenzie,[243] Laver [1969], [1971] proved Fraïssé Conjecture I [1948] (stated near the end of §4.2) when he proved that if $(A_0, \leq)$, $(A_1, \leq)$, ... is a sequence of countable ordered sets, then for some $i < j$ $(A_i, \leq)$ isomorphically embeds in $(A_j, \leq)$. To prove this celebrated result, Laver used the Hausdorff hierarchy [1908], [Erdős and Hajnal, 1962a],[244] the foundational work on bqo theory by Nash-Williams [1965b], and a characterization of σ-scattered linear order types[245] by Galvin to prove that the σ-scattered types form a bqo. In the same paper, he proved Fraïssé Conjecture II that for any countable scattered order type τ there are most countably many order types up to equivalence under mutual embeddability that are embeddable in but not equivalent to τ.

In his thesis, Pierre Jullien [1969] gave a counter-example to Fraïssé Conjecture III: the set $E = \{\omega \cdot \omega^*, \omega^* \cdot \omega\}$ has the property that $\omega + \omega^*$ is embeddable in every order type which embeds both members of E, but $\omega + \omega^*$ is not embeddable in either of the elements of E.

A positive answer to Fraïssé Conjecture IV follows from work of Laver [1973]. See Chapter 10 of [Rosenstein, 1982, 178, 179, 196, 197] for a discussion of the solution to this conjecture.

---

[240]Maurice Andre Pouzet received his doctorate from Université Claude Bernard, Lyon 1 in 1978 with Ernest Corominas as his advisor. He is known for his work in order, lattices and ordered algebraic structures.

[241]Nobert Werner Sauer received his doctorate in 1966 from the University of Vienna. A professor and long-time colleague of Eric Milner at the University of Calgary, Sauers interests range over set theory and universal algebra, especially Ramsey theory of partial orders and homogeneous structures.

[242]The citations [10] and [24] refer to Higman [1952] and Rado [1954b] respectively.

[243]Laver reputedly spent more time rock-climbing with McKenzie than talking mathematics with him. For Laver, rock climbing and mathematics were compatible. For example, he admitted to having proved a small theorem during an afternoon and overnight wait for a few minutes of assistance on a small ledge in Yosemite valley in the 1970's.

[244]Erdős and Hajnal [1962a] turned Hausdorff's type rings for scattered sets into a hierarchy.

[245]Recall that a scattered linear order is one that does not embed the order type of the rationals; σ-scattered linear orders are countable unions of scattered linear orders.

Galvin [1968b] explored partitions of the n-element subsets of the tree $C = {}^{\omega}2$ ordered lexicographically and with the topology that comes from identifying it with a subset of C^n, the n-fold product with the product topology. He proved what we will call Galvin's Partition Theorem for Perfect Sets: for $2 \leq n \leq 3$ and for any perfect set $P \subseteq C$, if $[P]^n$ is partitioned into finitely many open sets, then there is a perfect set (perfect subtree) $Q \subseteq P$ so that $[Q]^n$ meets at most $(n-1)!$ of the open sets. He conjectured that the theorem was true for all larger n.

Gerald Sacks spoke at the 1967 UCLA set theory conference on forcing with perfect closed sets. In his paper for the conference volume [1971, 335], he used the forcing to show the consistency of there being precisely two degrees of non-constructibility. He placed the origin[246] of the method in Cohen's forcing with finite conditions and Spector's [1956] construction of a minimal Turing degree. Sacks called the critical lemma which he used for proving that cardinals are preserved the *sequential lemma*, but noted in the footnotes that Mathias formulated the lemma more abstractly and called the result the *fusion lemma*.[247] Sacks noted that Shoenfield in an appendix to the 1967 Recursion theory seminar notes invoked the term "splitting" to describe the proof. Sacks cites the result in [Kleene and Post, 1954] of a minimal pair of upper bounds for the degrees of the arithmetic sets as the "germ of forcing with perfect conditions" and he references his joint paper with Robin Gandy [1967] where he used forcing with hyperarithmetic perfect closed sets to prove the existence of a minimal hyperdegree.

6.3 Suslin's Problem revisited

In the early days of forcing, Tennenbaum adapted Cohen's forcing to prove the relative consistency of the existence of a Suslin tree; Jech, working independently, gave an alternate proof; and Solovay and Tennenbaum proved the relative consistency of Suslin's Problem, i.e. the non-existence of Suslin trees.

According to Morley in his Math Review MR0215729 (35 #6564)), as early as 1964 Stanley Tennenbaum[248] announced that he was able to construct a model with a Suslin tree; Tennenbaum presented his proof in numerous seminars, but did not publish even an abstract prior to 1968, when his result appeared in [Tennenbaum, 1968]. In that paper, Tennenbaum, in his acknowledgment of support,

[246]See also the recursion-theoretic memories article [1999, 368–369] where he discusses the connections between forcing and recursion theory.

[247]Sacks expected Mathias's survey to appear in the second volume of the proceedings of 1967 UCLA set theory conference. It was distributed in 1968 as a preprint *A survey of recent results in set theory* from Stanford, and was later updated and published as [Mathias, 1979].

[248]Stanley Tennenbaum (April 11, 1927 – May 4, 2005) received a University of Chicago Ph.B. (an undergraduate degree) in 1945, and took graduate courses at the University of Chicago without earning a further degree. His academic career included many moves. Melvyn Nathanson [2008] recalled meeting Tennenbaum in fall 1964 when he was a visiting professor in the Department of Philosophy. Nathanson went to the University of Rochester for graduate work because Tennenbaum was teaching there at that time. In addition to his work in set theory, he is known for *Tennenbaum's Theorem* [1959]: there is no nonstandard recursive model of Peano arithmetic. A conference in his memory was held April 7, 2006 at the Graduate Center of the City University of New York.

mentioned summer 1963 support at Cornell University and listed two lecture series at which he presented the proofs of Theorems 1 and 2: Harvard University, January 27–29, 1964 and Symposium on Foundations of Set Theory, International Congress for Logic, Methodology and Philosophy of Science, Hebrew University, September 1, 1964. In Theorem 1 he proved the consistency of the existence of a Suslin tree together with GCH, and in Theorem 2 he proved the consistency of the existence of a Suslin tree with $\neg$CH. His reference for an equivalence of Suslin's Problem is [Miller, 1943]. Tennenbaum defined a tree to be a partial order in which the set of elements preceding any given element forms a chain. He listed Miller [1943] as his reference for the definition of Suslin tree and cited the original statement of the Suslin's Problem, Dowker's paper [1951] in which he posed the question of the existence of what have become known as Dowker spaces, Rudin's paper [1955], in which she showed the existence of a Suslin line implies the existence of a Dowker space, Cohen's article introducing forcing, and a forthcoming paper with Solovay, *Souslin's Problem II*, which is listed "to be submitted to *Fundamenta Mathematicae*."

Jech [1967] independently constructed a proof of the consistency of the existence of a Suslin continuum using ∇-models[249] of Vopěnka [1965] and ideas from Yesenin-Volpin's paper [1954].

Next, background material on Solovay is given to set the stage for a discussion of the consistency of the non-existence of Suslin trees by Solovay and Tennenbaum.

Robert Solovay[250] [2010b] recalled his explorations in logic and set-theory prior to his hearing of Cohen's work. He went to the University of Chicago with the clear idea that he wanted to be a theoretical physicist.

> Kaplansky did my entrance interview and I rattled off some 20 to 30 graduate texts I had worked through on my own. He clearly was unsure if I really had mastered the material in question but put me in two graduate courses with the comment that "if you really understand all that, these should be easy." I did, and they were.

Kaplansky went on to recommend that Solovay first get a Ph.D. in mathematics and then pursue his interest in physics. Solovay recalled how he came to know more logic than a mathematician should know [2010b]:

[249]Mostowski in his Math Review MR0182571 of Jech's paper indicated how to translate the proof from ∇-models to forcing.

[250]Robert Martin Solovay received his doctorate in 1964 from the University of Chicago where Saunders Mac Lane was his advisor. He was a lecturer at Princeton University 1962–1964, and a visiting member of the Institute for Advanced Study 1964–1965. Then he moved to the University of California, Berkeley, where he spent most of his academic career until retirement from teaching in 1994. He was at the Rockefeller University 1967–1968, at the IBM Thomas J. Watson Research Center in New York 1974–1976, and a Sherman Fairchild Scholar at the California Institute of Technology 1976–1977. His mathematical interests include set theory, new foundations, algorithms, and quantum computing. He became a member of the National Academy in 1986. In 2003 Solovay received jointly with Gary Miller, Michael Rabin and Volker Strassen the 2003 Association for Computing Machinery Paris Kanellakis Theory and Practice Award for the Solovay-Strassen and Miller-Rabin randomized algorithms for testing primality.

> Somehow, or other, the senior faculty got the impression that (though obviously not completely lacking in talent) I didn't quite know what a proof was. ...
>
> Mac Lane was my thesis advisor and he gave me the "sage advice" that it would help if I studied some logic to help with this supposed defect. And I retorted to him that "I know more logic than it is decent for a mathematician to know."
>
> At the time, I had worked through much of Kleene's *Introduction to Metamathematics* and certainly knew the Gödel Completeness and Incompleteness Theorems as well as the Orange Monograph presentation of the results on $V = L$. I knew more than this since I figured out on my own that the Skolem-Lowenheim theorem was at the root of Gödel's proof of GCH in L, but there certainly were significant gaps in my knowledge that I filled in later.

Solovay had finished the research for his dissertation around June 1962 and that fall became lecturer at Princeton University. He attended Cohen's general audience lecture on forcing in Princeton on May 3, 1963, a critical turning point in his mathematical career. Here is his description [2010a] of how he became a set theorist:

> I had been interested in set theory (and in particular the question of whether $V = L$ was a theorem of ZFC) from my (brief) undergraduate career at Harvard. (I dropped out during my sophomore year.) Cohen's talk (which I attended) got me intensely interested in it again. Either at that talk or his subsequent talk at Berkeley (on July 4, 1963) he posed the question of proving the consistency with ZF of "All sets Lebesgue Measurable". (I don't think he mentioned DC. I actually independently rediscovered that (and the name!) by reading through Halmos' book on measure theory and seeing what forms of choice were needed for the positive results. For the most part, countable choice suffices. But it seemed to me at the time that DC was needed for the proof of the Radon-Nikodym theorem. (I certainly had no proof that it was, in fact needed.)) After many twists and turns, I did, of course, indeed succeed in solving that problem. And by that point I was deep into set-theory.[251]

By July 1964, Solovay [1964], [1970] had solved the problem described above and spoke on it at a July meeting in Bristol,[252] as he recalled in [Jan 2011]:

[251]Based on an interview with Solovay in July 1981, G. Moore [1988] reported that Cohen posed the problem of the consistency of ZF with every set of reals being Lebesgue measurable on July 4, 1963.

[252]Solovay announced the results in an abstract [1965] received by the editors in October 1964 and, in [1970], noted that the main results of the paper were proved March–July, 1964, and were presented at the July meeting of the Association for Symbolic Logic at Bristol, England.

> At the time when I had to submit the abstract of my talk for the Bristol meeting I only had the weaker result that there was a model of ZF + countable axiom of choice + "There is a translation invariant countably additive measure defined on the power set of the reals and which extends Lebesgue measure". But in the interval between submitting the abstract and going to Bristol, I derived the full result. As a minor point, I knew that the measure described in the earlier abstract definitely was not Lebesgue measure.[253]

The model he created in which all sets of reals are Lebesgue measurable had many other nice properties and became known as *Solovay's model.*

During the academic year 1964–1965, Tennenbaum visited to discuss set theory, and in particular, his attempts to prove the consistency of SH, the non-existence of Suslin trees. Solovay recalled their discussions and his role in the proof in an email to Kanamori [2011a] (below, c.c.c. is an abbreviation for the countable chain condition introduced in §2.3).

> His [Tennenbaum's] attempt/plan for the consistency of SH had the following ingredients:
>
> 1. it was to be an iteration in which at each step another Souslin tree would be killed.
> 2. The steps in the iteration were to be forcing with Souslin trees, ...
> 3. Stan knew that forcing with a Souslin tree killed it and that the forcing was c.c.c.
>
> For much of the year, Stan was trying to prove that the iteration did not collapse cardinals. And he was considering iterations of length 2 and 3. My role was to passively listen to his proofs and spot the errors in them. With unjustified prescience, I kept saying that I was worried about "killing the same tree twice". Of course, there are now examples due to Jensen that show that forcing with the product of a Souslin tree with itself can collapse cardinals. But these weren't available then.
>
> At one of those meetings one of us (probably Stan) made progress and finally found reasonable conditions under which a two stage iteration did not collapse cardinals.
>
> Somehow this got me seriously thinking about the problem and by the time of our next meeting I had a proof of the theorem. This proof had the following ingredients which were new:

[253] G. Moore [1988], drawing on four interviews with Solovay in the period 1981–1984, reported that Solovay regarded the two weeks prior to the Bristol meeting as the most productive in his life. For more history of this result, see Moore's article.

(a) defining a transfinite sequence of forcing notions P_α where $P_{\alpha+1} = P_\alpha * Q_\alpha$.

This involves:

(a1) defining the operation $*$ where $P * Q$ is defined if Q is a poset in V^P;

(a2) defining what to do at limits.

(b) proving that if the component forcings are c.c.c. then the limit forcing is c.c.c.;

(c) exploiting the c.c.c.ness to see that if the length of the iteration has cofinality greater than ω_1 then all subsets of $\aleph_1$ in the final model appear at some proper intermediate stage;

(d) setting up the bookkeeping so that all Souslin trees are killed.

Solovay and Tennenbaum [1971] obtained the main results of their paper in June 1965, as they report in a footnote on the first page of their published paper. At the end of the paper, a brief note indicates that the paper was received by November 21, 1969. In footnote 2 on the first page of their article, Tennenbaum expressed regret that he had not mentioned, in [1968], Jech's work [1967]. In footnote 1, Solovay and Tennenbaum noted that Suslin's Problem had been studied by others leading to various formulations in terms of Boolean algebras and trees. In this discussion, they single out three papers: [Kurepa, 1952], [Miller, 1943], and [Sierpiński, 1948]. Solovay and Tennenbaum used a reformulation in terms of trees of the version of Suslin's Hypothesis due to Miller. They used the definition of tree used by Tennenbaum in [1968], but for the definition of *Suslin tree*, they extended the definition from Tennenbaum's [1968] by adding the condition that each element x has $\aleph_1$ elements in the tree above it, and in a brief appendix, in which they acknowledge that their reference to Miller is "not quite accurate," they show that if there is a tree without the additional condition, then there is one with it.

6.4 *Martin's Axiom and diamond principles*

The groundbreaking proofs of Tennenbaum [1968] and Solovay and Tennenbaum [1971] led to the extraction of combinatorial principles which opened the door to the application of set-theoretic methods to a wide variety of problems in set theory and set-theoretic topology.

Martin's Axiom

Around 1967, D. Anthony Martin[254] extracted the axiom which has become known as Martin's Axiom from work of Solovay for the Solovay-Tennenbaum paper. Rowbottom had independently suggested a similar axiom, but Solovay specified Martin in his paper with Tennenbaum and used the name *Martin's Axiom* for the following statement on page 233:

> *Martin's Axiom:* Let $\mathcal{P}$ be a reflexive partially ordered set satisfying c.a.c. Let $\mathcal{F}$ be a family of subsets of $\mathcal{P}$. Suppose that
>
> $$\max(\text{card}(\mathcal{P}),\ \text{card}(\mathcal{F})) < 2^{\aleph_0}.$$
>
> Then there is an $\mathcal{F}$-generic filter, G on $\mathcal{P}$.

Here *c.a.c.* is the *countable anti-chain condition*, i.e. the assertion that every antichain is at most countable, which is now universally referred to as the c.c.c. (sometimes written ccc), even though the terminology is less accurate. In what follows, we will write MA as an abbreviation for Martin's Axiom, and write MA_κ for Martin's Axiom restricted to families of dense sets of size $\leq \kappa$ together with the assumption that $\mathfrak{c} > \kappa$.

In the footnote to page 232, Solovay and Tennenbaum report on Martin's work:

> In addition to formulating his axiom, Martin observed that it could be proved consistent by the same techniques used by the authors to get a model of SH. He also noted that his axiom, in the presence of the inequality $2^{\aleph_0} > \aleph_1$, implied SH. This material is included here with his kind permission.

In their §7.2 they indeed observe that MA + $\neg$CH implies SH, and in the remainder of the paper they establish the consistency of MA + $\neg$CH.

Martin and Solovay [1970] made a case that an alternative to CH is needed in the presence of Cohen's proof that CH cannot be proved in ZF and the result of Lévy and Solovay [1967] that it cannot be proved even with the assumption of the existence of a measurable cardinal. Martin and Solovay pointed to Sierpiński's book *Hypothèse du Continu* [1934] with many consequences of CH which are not known to be decided by $\neg$CH. Then they proposed "axiom" **A**, now Martin's Axiom. Rowbottom is acknowledged as independently suggesting such an axiom, and some of Kunen's[255] results from his thesis are included in this substantial report on the first consequences of the new axiom. The main results of the paper [1970, 144] are collected into the following multipart statement:

[254] Donald Anthony Martin (December 24, 1940 –) received a bachelor's degree from the Massachusetts Institute of Technology, was a Harvard Junior Fellow, but never earned a doctorate. He taught at Rockefeller and now has a joint appointment in mathematics and philosophy at the University of California, Los Angeles.

[255] Kenneth Kunen received his doctorate in 1968 from Stanford University where his advisor was Scott. He has spent his career at the University of Wisconsin, Madison, with Keisler, Miller and the Rudins. Proofs of a number of his results were published by others. See Kanamori [2011b] to learn more about his set-theoretic work.

Theorem. If **A** then

1. $2^{\aleph_0} > \aleph_1 \rightarrow$ Souslin's hypothesis [22];
2. If $\aleph$ is an infinite cardinal number $< 2^{\aleph_0}$, then $2^{\aleph} = 2^{\aleph_0}$;
3. If $2^{\aleph_0} > \aleph_1$, every set of real numbers of cardinality $\aleph_1$ is $\mathbf{\Pi}^1_1$ if and only if every union of $\aleph_1$ Borel sets is $\mathbf{\Sigma}^1_2$ if and only if there is a real t with ${\aleph_1}^{L[t]} = \aleph_1$;
4. The union of $< 2^{\aleph_0}$ sets of reals of Lebesgue measure zero (respectively, of the first category) is of Lebesgue measure zero (of the first category);
5. If $2^{\aleph_0} > \aleph_1$, every $\mathbf{\Sigma}^1_2$ set of reals is Lebesgue measurable and has the Baire property;
6. $2^{\aleph_0}$ is not a real-valued measurable cardinal (see also [8]).[256]

Diamond principles

Tennenbaum's construction of a Suslin tree sparked others to work on related questions. In particular, Ronald Jensen, Jack Silver and Solovay analyzed his proof, according to Jensen in a private conversation with Larson and Mitchell in September 2008. Jensen [1968] announced the incompatibility of Suslin's Hypothesis with $V = L$. Here is a report from Larson's notes [2008]:

> While still at Bonn, Jensen thought that there must be a Suslin tree in L. His proof of the existence of a Suslin tree in L came out of Tennenbaum's forcing with finite trees, where the thing that made it work was that whenever you cut off at a limit α, every branch was generic over that and would hit every anti-chain in that model. Like many other proofs, this one occurred to him at night when he was trying unsuccessfully to fall asleep. He reported that he knew enough to think out the proof before falling asleep, and was able to get up the next morning and just write it up.
>
> Jensen analysed the Tennenbaum proof to see just what was needed combinatorially to construct such a tree and formulated his Diamond Principle $\diamondsuit$, which he then showed consistent by proving it held in L. Kunen independently extracted $\diamondsuit$ from Tennenbaum's proof, and Silver extracted $\diamondsuit^*$, a stronger principle. The principle $\diamondsuit^+$ was extracted from Solovay's construction of a Kurepa family.

Jensen was an Assistent at the University of Bonn from the time of his Ph.D. in 1964, through the time of his Habilitation in 1967, and until some point in 1969 when he became a professor at University of Oslo and simultaneously started

[256] The citation [22] refers to the Solovay-Tennenbaum paper, and [8] refers to Kunen's thesis [1968].

periodically lecturing at Rockefeller University. (See [Jensen, 1969b] for details of work on versions of the Diamond Principle and on Kurepa's Hypothesis asserting the existence of a Kurepa tree.) In an abstract outlining results on automorphism properties of Suslin continua, Jensen included an early published statement of his Diamond Principle [1969a, 576]:[257]

> $\diamondsuit$: There exists $\Gamma \subset \bigcup_{\alpha<\omega_1} {\omega_1}^\alpha$ such that $\Gamma \cap {\omega_1}^\alpha$ is countable for each α and if $f \in {\omega_1}^{\omega_1}$, then $\bigvee_\alpha f{\restriction}\alpha \in \Gamma$.

In other words, there is a family Γ of functions into ω_1 whose domains are countable ordinals such that for each $\alpha < \omega_1$, there are only countably many functions with domain α, and for every $f : \omega_1 \to \omega_1$, there is some α so that the restriction of f to α is in the family Γ.

This version of $\diamondsuit$ looks quite different from the general version introduced by Jensen [1972] in his master work on the constructible universe for each stationary subset $E \subseteq \kappa$:

> $\diamondsuit_\kappa(E)$: there is a sequence $\langle S_\alpha : \alpha \in E\}$ such that $S_\alpha \subseteq \alpha$ and for each $X \subseteq \kappa$, the set $\{\alpha : X \cap \alpha = S_\alpha\}$ is stationary in κ.

Jensen published a proof that if $V = L$ and κ is regular, then $\diamondsuit_\kappa(E)$ holds for every stationary set $E \subseteq \kappa$, and today we write $\diamondsuit$ for $\diamondsuit_{\aleph_1}(\omega_1)$.

To complete the overview of the basic diamond principles, we include the definitions of $\diamondsuit^+$ and $\diamondsuit^*_\kappa(E)$ for a stationary subset $E \subseteq \kappa$.

> $\diamondsuit^+$: There is a sequence $\langle \mathcal{S}_\alpha \subseteq \mathcal{P}(\alpha) : \alpha \in \omega_1\rangle$ of countable sets such that for all $A \subseteq \omega_1$, there is a closed unbounded set $C \subseteq \kappa$ such that the following properties hold:
>
> a. for all $\alpha \in C$, $A \cap \alpha \in \mathcal{S}_\alpha$;
>
> b. for all $\alpha \in C$, $C \cap \alpha \in \mathcal{S}_\alpha$.

> $\diamondsuit^*_\kappa(E)$: There is a sequence $\langle \mathcal{S}_\alpha \subseteq \mathcal{P}(\alpha) : \alpha \in E\rangle$ with $|\mathcal{S}_\alpha| \leq \alpha$ for all but a bounded set of $\alpha \in E$ such that for all $A \subseteq \kappa$, there is a closed unbounded set $C \subseteq \kappa$ with $A \cap \alpha \in \mathcal{S}_\alpha$ for all $\alpha \in C \cap E$.

Kunen, in work with Jensen [1969b], proved that $\diamondsuit^+$ implies $\diamondsuit$ by proving that $\diamondsuit$ is equivalent the following principle $\diamondsuit'$.

> $\diamondsuit'$: There is a sequence $\langle \mathcal{S}_\alpha : \alpha < \omega_1\rangle$ with each $\mathcal{S}_\alpha$ a countable set of $\mathcal{P}(\alpha)$ such that for each $X \subseteq \omega_1$ the set $\{\alpha < \omega_1 : X \cap \alpha \in S_\alpha\}$ is stationary.

---

[257] The abstract was received by the *Notices of the American Mathematical Society* on December 2, 1968.

The version of $\diamondsuit$ of the Jensen abstract can be made to look more like $\diamondsuit'$ if one replaces the functions by their ranges, and shrinks to a closed unbounded set E of α where the ranges of functions in Γ with domain α are also subsets of α. Alternatively, think of the functions as subsets of $\omega_1 \times \omega_1$ and use a bijection between $\omega_1 \times \omega_1$ and ω_1 to obtain the function version from the usual one.

It is clear that $\diamondsuit^+$ implies $\diamondsuit^* = \diamondsuit^*_{\aleph_1}(\omega_1)$, which in turn implies $\diamondsuit'$, so $\diamondsuit$ follows from all of the other variations listed. Presumably it was known to Kunen and Jensen in 1969 that $\diamondsuit$ does not imply Kurepa's Hypothesis (KH), that there is a Kurepa tree, since the model of $\neg$KH constructed by Silver in the 1960's (see [1971b]) from an inaccessible cardinal satisfies $\diamondsuit$. These diamond principles were analyzed in notes [1969b] prepared by Jensen when he was at Rockefeller University in 1969. Most of the results from the notes appeared in Devlin [1973], although the definition of $\diamondsuit^+$ was changed from asserting the existence of a closed unbounded set C for whose elements $C \cap \alpha$ and $A \cap \alpha$ in $\mathcal{S}_\alpha$, to asserting the existence of an unbounded set B for whose limit points α one has $C \cap \alpha$ and $A \cap \alpha$ in $\mathcal{S}_\alpha$. The principle $\diamondsuit^*$ was not mentioned in Devlin's book, but it is of little import, since the implications $\diamondsuit^+ \implies \diamondsuit^*$ and $\diamondsuit^* \implies \diamondsuit'$ follow from the definitions of these principles. There is a remark in the Jensen notes [1969b] that $\diamondsuit^+$ is stronger than $\diamondsuit^*$ since the former implies that there are Kurepa trees and the latter does not. Jensen's proof that $\diamondsuit^+$ implies KH appears later in the notes, but a proof $\diamondsuit^*$ does not imply KH only appears in the 1970's (see §7.1 where the discussion of diamond principles continues).

6.5 *More on uncountable trees*

We start by discussing terminology, mainly from [Kunen, 1983], to facilitate our discussion of results for Aronszajn, Suslin and Kurepa trees. Recall that for regular κ, a κ-tree $(T, <)$ is a tree of height κ all of whose levels have cardinality $< \kappa$; in particular, an ω_1-tree is a tree of height ω_1 all of whose levels are countable. With this terminology for regular κ, a κ-Aronszajn tree is a κ-tree with no branches of length κ, a special κ-Aronszajn tree is a κ-Aronszajn tree which is the union of fewer than κ antichains, and a κ-Suslin tree is a κ-Aronszajn tree with no antichains of size κ (it is definitely not a special κ-Aronszajn tree since special κ-Aronszajn trees have antichains of size κ).

If $s, t \in T$ and $s < t$, then we say that s is a *predecessor* of t and t is a *successor* of s. Call T *well-pruned* if it is rooted and every element of the tree has successors on all higher levels. Any subset of a tree is a *subtree* under the induced partial order, though we note that the height, width, and level structure need not be retained. One can show that every κ-tree has a well-pruned subtree which is a κ-tree. A tree is *Hausdorff* if different nodes on the same limit level have different sets of predecessors. A tree is *full* if every non-maximal element has infinitely many immediate successors, and λ-*full* if every non-maximal element has λ-many immediate successors. Note that the concept due to Kurepa that we called *uniformly branching* has been refined to *full* and we have imposed the requirements that the trees of interest in this case be rooted and Hausdorff. Note that the notion of *normal* ω_1*-tree* from [Jech, 2003] is the same as full well-pruned Hausdorff ω_1-tree.

Haim Gaifman[258] and Specker [1964] proved that there are at least 2^{κ^+} many isomorphism types of κ-full well-pruned Hausdorff κ^+-Aronszajn trees if $\kappa^\lambda = \kappa$ for all $\lambda < \kappa$.[259] In their introduction they mention Kurepa's problème miraculeux 1 (First Miraculous Problem stated in §3.4), i.e. whether or not any two uniformly branching $\aleph_1$-Aronszajn trees are isomorphic, which they solve in the negative. Gaifman and Specker showed that κ-full well-pruned Hausdorff κ^+-Aronszajn trees could be represented by *sequential trees*, which are trees of transfinite sequences closed under initial segment and ordered by end-extension. They referred to both Aronszajn's construction (cf. [Kurepa, 1935]) and Specker's construction [1949] of these trees. The proof of existence of many non-isomorphic trees used the formation of new trees from known ones via a process much like a shuffle of their underlying sequences.

Toward the end of the decade, conferences and surveys helped share the growing body of results about trees.

Kurepa [1969] surveyed results related to Suslin's Problem, from its statement through the independence results of Tennenbaum and Jech, at the International Symposium on Topology and its Applications held in Herceg-Novi in the summer of 1968.

Rudin [1969] popularized the use of Aronszajn, special Aronszajn and Suslin trees with her *American Mathematical Monthly* article. In this expository article, she described the Suslin problem as sounding easy (see page 1113): "Anyone who understands countable and uncountable can 'work' on it." She went on to point out that there are standard patterns, standard errors and "there are standard not-quite-counter-examples which almost everyone who looks at the problem happens upon." After noting the growing collection of results about Suslin trees, she described her goals for the paper: describe the standard pattern (development as a way to construct a tree from a linear order), prove an elementary result (the construction of a Suslin line from a Suslin topological space) and give one of the not-quite-counter-examples (the construction of an Aronszajn tree). She concluded her article with a brief review of recent research, which in part documents some results of Jensen on the existence of a variety of Suslin trees of cardinality κ in the constructible universe.

Jech [1971] expanded a May 1969 talk on trees given at a conference in Ober-

[258]Haim Gaifman earned a master's degree at Hebrew University in 1958 and a doctorate at the University of California, Berkeley, in 1962, where Tarski was his advisor. He directed the program in History and Philosophy of Science at the Hebrew University while a professor of mathematics prior to moving to Columbia University in 1990 as a professor of philosophy.

[259]Gaifman and Specker call the trees of interest *normal* κ*-trees*, but for $\kappa = \aleph_0$, their definition matches the definition of *normal* ω_1-tree in [Jech, 2003, 114] except that they have added the condition that there is no branch of length κ^+. This paper may well be the source of the word *normal* as used by Jech. I have stayed away from *normal* for trees because of its use in topology, and because different authors have different definitions.

wolfach into a survey article. He included brief proofs of classical as well as recent results, so his paper gives a snapshot of what was known at the end of the 1960's about uncountable trees. He straightforwardly pointed out in footnote 1 that Kurepa's [1968] assertion[260] that κ^+-Aronszajn trees exist for every κ is false. Jech showed how to generalize the construction of ω_1-Aronszajn trees to larger cardinals, and in his footnote 2, pointed out that Rowbottom and Silver observed that the existence of κ^+-Aronszajn trees follows immediately from the existence of ω_1-trees using Chang's cardinal transfer property $(\omega_1, \omega) \to (\kappa^+, \kappa)$, which is known to be true for regular κ assuming GCH. Jech showed the necessity of the assumption of GCH. He briefly discussed the tree property, the generalization of König's Lemma to trees with small levels and large heights, and reported that Tarski and his school had shown that for inaccessible cardinals, the tree property is equivalent to several other properties including $\kappa \to (\kappa)^2_2$ and Π^1_1-indescribability and being weakly compact [Keisler and Tarski, 1964]. In terms of trees, κ has the tree property exactly when there are no κ-Aronszajn trees.

Jech noted that Silver [1966] (see also [1971b]) observed that every real-valued measurable cardinal has the tree property, gave a sketch of the proof, and concluded that since the statement that $2^{\aleph_0}$ is real-valued measurable is consistent relative to some large cardinal assumption, so is the statement that the continuum has the tree property. This result also showed the consistency of the statement that not all cardinals with the tree property are weakly compact. Jech pointed out that a κ-complete λ-saturated non-principal ideal on κ for some $\lambda < \kappa$ is sufficient to get the tree property at κ.

Jech reported that Prikry [1968] proved the consistency of the statement that $\aleph_{\omega+1}$ does not have the tree property relative to a large cardinal assumption. He started with a measurable cardinal κ in the ground model, thus starting with a κ^+-Aronszajn tree T built from bounded increasing sequences from $Q = {}^{<\kappa}\kappa$, and constructed a generic extension in which $\kappa = \aleph_\omega$ and all cardinals above κ are preserved. In this extension T is an $\aleph_{\omega+1}$-tree whose levels have power at most $\aleph_\omega$, and there is no $\aleph_{\omega+1}$ branch in T, since otherwise there would be an increasing sequence of that length in Q which has cardinality $|Q| = \kappa = \aleph_\omega$ in the extension.

Next, Jech turned to the question of the existence of a Suslin tree. He cited Kurepa [1937b], Miller [1943], and Specker [1949] as having made translations of the problem to one about trees. Jech discussed the model of [Tennenbaum, 1968] for the existence of a Suslin tree and his own construction [1967].

Jech sketched the Solovay-Tennenbaum proof of the consistency of non-existence of Suslin trees we discussed in §6.3, noting that the continuum in their model is $2^{\aleph_0} = \aleph_2$. Jech announced the remarkable proof by Jensen of the consistency of non-existence of Suslin trees with the Continuum Hypothesis discussed in §7.7.

[260] In his Math Review MR0255410 (41 # 72) of Kurepa's paper entitled *On A-trees*, N. C. A. da Costa, noted that the *A*-trees of the title of have been considered by the author on several occasions, specifically in [Kurepa, 1937a]. Hence the title reference is clearly to Aronszajn trees. In the review, da Costa states that Kurepa proved the existence of an Aronszajn tree on every successor cardinal without the use of the Continuum Hypothesis.

His citation of mimeographed notes at the very end of his bibliography for this landmark result suggests that the announcement was a last minute addition.

In his discussion of Kurepa trees, Jech singled out as the first result that if ω_1 is inaccessible in L and if no ordinal between ω_1 and ω_2 is a cardinal in L, then there is a Kurepa tree; Jech indicated that several people found this result and mentioned Lévy, Rowbottom [1964],[261] and Bukovský [1966]. Later, Rowbottom and Stewart[262] [1966] proved the consistency of the existence of a Kurepa tree relative to ZFC. Next, Jech sketched the proof by Silver [1971a], only published in 1971, that the non-existence of Kurepa trees is consistent relative to the existence of an inaccessible cardinal. Jech then noted "recently," Solovay proved that there is a Kurepa tree whenever $V = L[X]$ for some $X \subseteq \omega_1$; Silver observed that from this result it follows that if ω_2 is not inaccessible in L, then there is a Kurepa tree, so that the assumption of inaccessibility in Silver's proof cannot be dropped.

In the fourth and final section, Jech presented what he called the "cream of the article", namely the proof that if $V = L$ both Suslin and Kurepa trees exist. We have already discussed the existence of Suslin trees in L due to Jensen; Jech credited Solovay with the construction of a Kurepa tree.

Like a spring that bubbles up from many tiny sources building up to a river, many independently generated ideas and approaches firmed up into our notions of ω_1-tree, Aronszajn tree, special Aronszajn tree, Suslin tree and Kurepa tree.

6.6 Transversals and decidability

Micha Perles [1963] showed that Dilworth's chain decomposition theorem (see the end of §5.3) does not generalize to the infinite, using $\omega_1 \times \omega$ ordered coordinatewise as his example of a partial order in which all antichains are countable but it is not the union of countably many chains.

H. A. Jung and Rado [1967] generalized Hall's Marriage Theorem to families of subsets of a set E which include a single infinite subset. They proved that a necessary and sufficient condition for a family $\langle A_i : i \in I\rangle^\frown\langle B\rangle$ of subsets of E with each A_i finite and B infinite to have a transversal is that $\langle A_i : i \in I\rangle$ have a transversal and B is not a subset of the union of all sets $A[T] := \bigcup_{i \in T} A_i$ over finite $T \subseteq I$ for which $|A[T]| = |T|$.

There were extensions of the generalization of Jung and Rado to allow finitely many infinite sets by several authors in the late 1960's and early 1970's. For example, here is the condition by Jon H. Folkman [1970] for a family $A = \langle A_i \mid i \in I\rangle$ of subsets of a set E having a transversal: A has a transversal if and only if the following condition is satisfied where for $J \subseteq I$, $A(J) = \langle A_j : j \in J\rangle$: for all $n \in \omega$ and $K \subseteq J \subseteq I$ such that $A(J) - A(K)$ is finite and $|J - K| \geq n + |A(J) - A(K)|$, there is a finite set $S \subseteq K$ such that $|A(T)| \geq n + |T|$ for all sets T such that $S \subseteq T \subseteq K$. Folkman gave the example of the set of denumerable ordinals each regarded as the set of its predecessors as an example of an uncountable family

[261] The reference to Rowbottom also includes a cryptic "(notes for 1967)" [Jech, 1971, 14].
[262] Stewart's proof is sketched in Jech's paper [1971].

which has no transversal (apply the Regressive Function Theorem). All three, [Brualdi and Scrimger, 1968], [Folkman, 1970] and [Woodall, 1972], gave equivalent characterizations, as shown in [Bry, 1981].

Milner became interested in transversals in the 1960's and wrote a series of papers on them, some coauthored with Erdős or Hajnal[263] or both. Of particular note is the triple paper [1968] in which they proved that given a sequence of κ^+ many disjoint sets, each of cardinality κ, there is a collection of κ^+ many almost disjoint transversals. L. Mirsky [1971] codified what was known about transversal theory up through the 1960's in his book of that name, with attention to infinite combinatorial problems using the Rado Selection Principle.

Next we consider an application of the classification of order types. Läuchli and John J. Leonard [1966] proved the decidability of the elementary theory of order through the analysis of the class M of order types such that $1 \in M$, M is closed under pairwise sum, multiplication by ω and by ω^*, and the shuffling operation σF for finite $F \subseteq M$, where σF is a linearly ordered set which can be partitioned into order type η many segments such that every segment has a type in F and between any two different types there are parts in each type from F. Läuchli and Leonard credit Skolem [1920], [1970, 132, Satz 2] with the uniqueness of the shuffling operation, i.e. between any two sets A and B with witnessing partitions $\langle C_\xi \in F : \xi \in \eta \rangle$ and $\langle D_\xi \in F : \xi \in \eta \rangle$ showing that A and B fit the definition of σF, there is an order-preserving map $\Phi : A \to B$ so that each C_ξ is taken to some D_ζ where C_ξ and D_ζ represent the same type from F. They contrasted their proof with that of Ehrenfeucht [1959], [1961] and note that a proof similar to theirs had been found independently by Fred Galvin. The Läuchli-Leonard approach recalls Hausdorff's analysis of countable linear order types in [1908], and Skolem cited this paper for terminology.

6.7 *Partition calculus classics*

The late 1950's and early 1960's were a fruitful time for those working in the partition calculus, particularly with the appearance of the landmark paper, *Partition relations for cardinal numbers* by Erdős, Hajnal and Rado [1965], and the proof of the Hales-Jewett Theorem, a result from finite combinatorics that would impact both finite and infinite combinatorics.

Let A be a finite alphabet (i.e. a set) and let v be a "variable" not in A. Let $\Sigma^* A$ be the set of all "words" over the alphabet A. A "variable word" is a word over $A \cup \{v\}$ in which v actually occurs. Alfred Hales and Robert Jewett [1963] used variable words to investigate partitions of H-fold products of an n-element set, a dimensional approach to partition theory. Their theorem implies van der Waerden's Theorem.

> *Hales-Jewett Theorem:* If $\Sigma^* A$ is finitely colored, there must exist a variable word $w(v)$ such that $\{w(a)\colon\ a \in A\}$ is monochromatic.

[263] Hajnal [1997, 374–375] recalled that his first joint paper with Milner was on transversals.

Before turning to a discussion of partition relations for cardinals, ordinals, and the reals, we remind the reader that partition results of Galvin for real types and rationals and of Laver for the rationals were discussed earlier in §6.1, since they will not be repeated in this section.

Early finite exponent results

Hajnal used the Continuum Hypothesis to show that the sharpening of Erdős and Rado of the Erdős-Dushnik-Miller Theorem could not be improved for $\kappa = \omega_1$ from $\omega_1 \to (\omega_1, \omega+1)^2$ [Erdős and Rado, 1956, Theorem 39] to $\omega_1 \to (\omega_1, \omega+2)^2$ [Hajnal, 1960, 283, Theorem 5]. The proof actually yields a bit more, namely $\omega_1 \not\to (\omega_1, \omega : 2)^2$, i.e. there is a coloring of the pairs of ω_1 which has no independent set of size ω_1 and also has no complete bipartite graph consisting of ω points in a set B all of which are below two points in a set C with all pairs between them joined. The main tool was a set mapping theorem (Theorem 1 [1960, 278]) he had proved using the Continuum Hypothesis in order to answer negatively a question of Erdős and Fodor. Hajnal also proved positive partition relations for ω_1 and for λ, the order type of the real numbers. For example, he proved that for any uncountable order type ϕ which embeds neither ω_1 nor ω_1^* (e.g. λ) that $\phi \to (\omega \cdot n, \alpha)^2$ for any $n < \omega$ and any countable ordinal α. Moreover, he proved (see Theorem 8 on page 290) that $\omega_1 \to (\omega \cdot n, \omega \cdot 2)^2$ for every finite n and stated as the simplest unsolved problems: Does $\omega_1 \to (\omega^2, \omega \cdot 2)^2$? Does $\omega_1 \to (\omega \cdot 3, \omega \cdot 3)^2$? His method of proof gives more (see [1960, 294]): $\omega_1 \to (\omega \cdot n, (\omega : \omega_1)_<)^2$,[264] where the $(\omega : \omega_1)_<$ stands for a complete bipartite graph on a pair of sets $B, C \subseteq \omega_1$with $B < C$, where B has type ω and C is uncountable, i.e. the graph of all pairs meeting both B and C.

Erdős and Hajnal took the approach of their [1958] a step further in [1962b] to prove that the least inaccessible cardinal κ satisfies the negative partition relation $\kappa \not\to (\aleph_1)_2^{<\omega}$, which we would now describe by saying the least inaccessible cardinal is not an $\aleph_1$-*Erdős cardinal.*

Building on the work of the late 1950's seminar organized by Mostowski and Tarski that led to [Erdős and Tarski, 1961], William Hanf[265] [1964] showed that $\kappa \to (\kappa)_2^2$ implies the tree property on κ, and hence by [Erdős, 1942] (the attribution is given in [Erdős and Tarski, 1961, 59, footnote 5]), the two properties are equivalent for regular uncountable κ.

Hajnal [1964], responding to this work, proved that if κ is a regular cardinal

[264]The notation $\omega_1 \to (\omega \cdot n, (\omega : \omega_1)_<)^2$ comes from [Erdős *et al.*, 1984, 62].

[265]William P. Hanf worked for IBM 1958–1968. He was supported by IBM during the year he completed his degree in 1963 as a student of Tarski with a dissertation titled *Some Fundamental Problems Concerning Languages with Infinitely Long Expressions.* He was a visiting lecturer at the University of California, Berkeley, for two years before joining the faculty at the University of Hawaii in 1970, where the William P. Hanf Memorial Fund now supports cash awards for winners of the Problem of the Month for Manoa Undergraduates, according to the website http://www.math.hawaii.edu/home/ProblemMonth_Hanf.html downloaded August 17, 2010. For more information on Hanf, see the eulogy by Dale Meyers [1991].

which supports a linear ordering with no well-ordered or conversely well-ordered subset of length κ, then for all $r \geq 3$, $\kappa \nrightarrow (\kappa, r+1)^r$. As he pointed out, it is enough to prove this for $r = 3$, since if f witnesses the relation for 4 and $r > 4$, then the function g given by $g(\alpha_0, \ldots, \alpha_{r-1}) = g(\alpha_0, \alpha_1, \alpha_2)$ witnesses the relation for r. He gave a simpler proof of Hanf's result[266] and noted that the four statements from [Erdős and Tarski, 1961] that inspired Hanf's work are equivalent, which he attributed to Hanf [1964] with prior work by [Erdős and Tarski, 1961] and [Monk and Scott, 1964].

Erdős and Hajnal [1962a] proved a partition relation for countable order types: $\theta \rightarrow (\theta, \omega)^2$ if and only if $\theta = \omega$, $\theta = \omega^*$ (the converse well-ordering) or θ is not scattered. A major tool was a hierarchy, the stratification of countable scattered order types into ranked layers as they are built up from 0 and 1 by sums indexed by ω or ω^* over sets of lower rank. In a footnote, Erdős and Hajnal shared their belief that this approach must be in the literature, but that they have not found it. Indeed, this hierarchy is a repackaging of a type ring of Hausdorff [1908], now known as the Hausdorff hierarchy. While the Hausdorff type rings were for scattered sets of cardinality less than $\aleph_\alpha$, the modern Hausdorff hierarchy includes scattered sets of every cardinality.

Milner and Rado [1965] investigated an infinitary generalization for ordinals of the Pigeonhole Principle. In terms of the arrow notation introduced by Rado, given ordinals $\alpha_0, \alpha_1, \ldots$, they sought the least ordinal α for which $\alpha \rightarrow (\alpha_0, \alpha_1, \ldots)^1$ holds. Of particular note was the discovery of what came to be called the *Milner-Rado Paradoxical Decomposition* of ordinals $\beta < \omega_2$ witnessing the surprising failure of $\beta \rightarrow (\omega_1, {\omega_1}^2, \ldots, {\omega_1}^n, \ldots)^1_\omega$, More generally, for any uncountable cardinal κ and every $\beta < \kappa^+$, the negative relation $\beta \nrightarrow (\kappa^\omega)^1_\omega$ holds.

The Giant Triple Paper

The triple paper [1965] of Erdős, Hajnal and Rado, finally appeared.[267] As Hajnal recounted in his paper *Paul Erdős' set theory* [1997, 363]: "Starting around 1957 or 58, we decided to write a triple paper with Rado on the partition calculus and the three of us set aside everything which we thought belonged there. Already in 1960 I visited Rado in Reading to work on the triple paper, carrying with me an almost completed manuscript." They succeeded in giving an almost complete discussion of the partition relation $\kappa \rightarrow (\lambda, \mu)^r$ for cardinal numbers in this massive paper under the assumption of GCH (theorems marked with (*) used this assumption). The Positive Stepping Up Lemma of [Erdős and Rado, 1956] was joined by the Negative Stepping Up Lemma: if $r \geq 2$, $\kappa \geq \omega$ and $\kappa \nrightarrow (\lambda)^r_{\nu<\gamma}$, then $2^\kappa \nrightarrow (\lambda_\nu + 1)^{r+1}_{\nu<\gamma}$, provided one of several situations holds as in the various parts of Lemma 5. Sections 13 and 14 (pages 118–129) include an analysis of counter-examples. The *Canonization Lemma* (Lemma 3 on page 110) was used to get positive results for singular strong limit cardinals using GCH. A sample is Theorem

[266]In a footnote Hajnal remarked that the original form of the proof was simplified by Hanf.

[267]Erdős reported that Hajnal used to refer to it as the "Giant Triple Paper" [Erdős, 1996, 121].

6 from page 114 (GCH): Let a be a singular cardinal of cofinality $a' = b^+$ and $c < a'$. Then for b' the cofinality of b, $a \to (a, (b')_c)^2$.

Among the variations of the partition symbol introduced in the Erdős-Hajnal-Rado classic paper is the square bracket partition relation (see 18.1, 18.2, 18.3 on page 144 of [1965]) $\kappa \to [\lambda]^n_\nu$, often used to express the existence of strong counter-examples. This partition relation holds if for every set S of cardinality κ and every $f : [S]^n \to \nu$, there is a subset $X \subseteq S$ of size λ so that the restriction of f to $[X]^n$ does not map onto ν; colloquially, f omits a color on $[X]^n$. For example, the Sierpiński partition [1933] may be expressed with this notation as $\omega_1 \not\to [\omega_1]^2_2$. Erdős, Hajnal and Rado assume GCH to prove in [1965, Theorem 17, 141] that for $\kappa = \lambda^+$, $\kappa \not\to [\kappa]^2_\kappa$.

Erdős and Hajnal quickly found an application of Theorem 17 to the problem of Jónsson in their [1966],[268] where they show that under the assumption of GCH no successor cardinal is Jónsson. A *Jónsson algebra* is an algebra without a proper subalgebra of the same cardinality and a cardinal κ is *Jónsson* if there are no Jónsson algebras with universe of size κ. Keisler and Rowbottom [1965] proved that κ is Jónsson if and only if every structure for a countable first-order language with universe of size κ has a proper elementary substructure whose universe also has size κ. They showed that under the assumption of $V = L$, no cardinal is Jónsson. In Theorem 4 of [1966], Erdős and Hajnal prove that every algebra whose universe has cardinality a measurable cardinal has a proper subalgebra with universe of the same cardinality, i.e. every measurable cardinal is a Jónsson cardinal. They went on to use a set mapping theorem from the joint paper [1958] that came out of their first meeting to prove in Theorem 5 that for every infinite cardinal λ there is a Jónsson algebra of cardinality λ with one ω-ary operation.

This brief overview gives only a taste of this paper about which one of the authors (Erdős) [1996, 120] said: "Hajnal, Rado and I nearly completely settled $m \to (n, q)^2$ but the results are very technical and can be found in our joint triple paper [69] and in our book [68]."

All Finite Subsets Problem

Results related to the All Finite Subsets Problem of Erdős and Hajnal discussed in §5.1 were proved in the 1960's.

Frederick Rowbottom [1964] received his doctorate from University of Wisconsin, Madison, and in his thesis he used partition properties of the form $\alpha \to (\beta)^{<\omega}_2$[269] to investigate the constructible universe. Specifically he showed that the existence of a cardinal κ such that $\kappa \to (\omega_1)^{<\omega}_2$ implies $V \neq L$. He also showed that measurable κ are *Ramsey cardinals* in the sense that $\kappa \to (\kappa)^{<\omega}_2$ but are

[268] The paper by Erdős and Hajnal comes with the notation that it was "presented by A. Mostowski on August 27, 1965".

[269] Recall that the partition relation $\alpha \to (\beta)^{<\omega}_2$ holds if and only if for every coloring $c : [\alpha]^{<\omega} \to \{0, 1\}$ of the finite subsets of α with two colors there is a subset $X \subseteq \alpha$ of order type β and a function $f : \omega \to \{0, 1\}$ such that for all $n < \omega$, for all n-element subsets $A \subseteq X$, $c(A) = f(n)$.

not the least cardinals with this property. The proof involved showing that for a normal ultrafilter U over a measurable cardinal κ, $\kappa \longrightarrow (U)^n_2$ for all $n < \omega$ by induction and using the closure of U under countable intersections to get the desired result.[270] Rowbottom showed that, in the presence of a cardinal κ satisfying $\kappa \longrightarrow (\kappa)^{<\omega}_2$, the reals of the constructible universe are countable.

Jack Silver[271] in Chapter 4 of his thesis [1966], [1970b] showed that $\kappa \longrightarrow (\omega)^{<\omega}_2$ is consistent with $V = L$. He further proved that if $\kappa \longrightarrow (\alpha)^{<\omega}_2$ and $\alpha < {\omega_1}^L$, then the statement $\kappa \longrightarrow (\alpha)^{<\omega}_2$ is true in L. He drew as a corollary the conclusion that if $\kappa \longrightarrow (\alpha)^{<\omega}_2$ for all countable α, then the same is true in the constructible universe. Silver, in his thesis [1966] and [1971b] showed the existence of $0^{\#}$ (named in [Solovay, 1967]) follows from the existence of a cardinal κ for which $\kappa \longrightarrow (\omega_1)^{<\omega}_2$.[272]

6.8 *Infinitary partition relations*

During the academic year 1967–1968 (including the summer of 1968), Mathias, Silver, and the team of Galvin and Prikry proved seminal results on partitions of infinite subsets of ω. Adrian Mathias[273] proved the consistency of $\omega \longrightarrow (\omega)^{\omega}$ in the absence of the Axiom of Choice and introduced what is now known as Mathias forcing. Galvin and Prikry[274] proved a remarkable generalization of the Nash-Williams Partition Theorem, subsequently generalized by Silver in the beginning of a chain of results that have led to a wide variety of applications. What follows is a mainly chronological account of how these theorems were proved.

Recall that Erdős and Rado, as early as 1935 and in their Example 1 of [1952a, 431], had looked at generalizations of Ramsey's Theorem to partitions of the infinite subsets of a set: they used a well-ordering of the collection of infinite subsets of an infinite set, say ω for concreteness, to construct a partition $\mathcal{P}(\omega) = K_0 \cup K_1$ such that neither cell contains all the subsets of an infinite subset of ω. Let us say family $\mathcal{F}$ of infinite subsets of ω has the *Ramsey property* or is *Ramsey* if there is an infinite subset $M \subseteq \omega$ such that either $[M]^{\omega} \subseteq \mathcal{F}$ or $[M]^{\omega} \cap \mathcal{F} = \emptyset$.

In fall 1967, Jensen was visiting Stanford and Scott had arranged for Mathias,

---

[270] Drake singled out this property and its generalization to n-tuples for special attention in his book *Set Theory* [1974, 189]: "The most important property of normal measures for us is usually called the *Ramsey* property, by analogy with Ramsey's Theorem of ch. 2 §8."

[271] Jack Howard Silver completed his doctoral thesis at the University of California, Berkeley, in 1966 with Vaught, spent a year at the University of Wisconsin, Madison, and like his advisor, came to be a member of the faculty at Berkeley where he had earned his degree.

[272] See Kanamori's second chapter and Mitchell's chapter for information on $0^{\#}$.

[273] Adrian Mathias graduated from the University of Cambridge in 1965 and wrote his fellowship and Ph.D. dissertation in 1968, for which the bulk of the work was done in Bonn and Stanford University under Jensen's supervision. He taught at Cambridge for twenty years without obtaining tenure; he then spent ten years teaching and researching in various places [Berkeley, Warsaw, Oberwolfach, Caen, the CRM at Barcelona, and Bogota], and finally was elected to a tenured position in Réunion in 2000.

[274] Karel Prikry studied at Charles University in Prague, and went to the University of California, Berkeley, in fall 1965 with what he [2010a] described as the equivalent of a master's degree including a short publication. He earned his doctorate with Jack Silver at Berkeley in 1968. He has spent most of his career at the University of Minnesota.

as Jensen's student, to be a research associate. Here is how Mathias described the initial impetus to his work on partitions of the infinite subsets of ω [1977, 110]:

> The author's interest in the problem of refuting the relation $\omega \to (\omega)^\omega$ without the axiom of choice was aroused by [Harvey] Friedman during Scott's seminar on partition theorems conducted at Stanford in 1967.

In a letter to Odell dated November 11, 1997, Mathias listed a sequence of events as he recalled them on the question of refuting $\omega \to (\omega)^\omega$ without choice. The first two items of the list are omitted since their core content is covered by the above quote, to which we add the remark that the seminar was held in the fall quarter.

3. In response to Friedman's enthusiasm, Paul Cohen and, independently, Ehrenfeucht find proofs that all open sets are Ramsey. Cohen uses this result to answer another question of Friedman by building an infinite subset X of ω not recursive in any $Y \subseteq X$ with $X \setminus Y$ infinite.
4. Whilst searching the literature in another connection, I find Nash-Williams' proof of a similar result.
5. Friedman gives a talk at Berkeley, with Silver, then an associate professor, and Prikry, then a graduate student, in the audience, describing his attempts to build a non-Ramsey set without using choice. He sketches a proof that if one Sacks real is added to a model of ZFC, every new infinite subset of ω contains or is disjoint from an old infinite subset of ω. In a discussion after his talk, Prikry points out that Friedman's argument is erroneous, and I give a correct proof.[275]

Next we turn to Galvin and Prikry [1973] who apparently proved their theorem immediately on hearing Scott's idea that infinitary partition theorems should be possible for nicely definable partitions, which they may have learned from the above described lecture of Friedman. In an email to Galvin and Larson, Prikry [2010a] recalled that he and Galvin "came up with the proof the same day or night," and described the circumstances.

> Fred did it while he was brushing his teeth before going to bed (plus the few minutes before he fell asleep). I spent the whole night writing up the proof in my office in T4 since I never liked claiming a theorem before I wrote it up completely. I think it would have been in the fall 1967. My motivation was Scot[t]'s talk. The notion of Ramseyness for infinite sets has been around – it has to do with infinitary Jonsson algebras and so forth.

[275]The Nash-Williams Partition Theorem [1965b] is the likely candidate for what Mathias found in the literature. I have omitted an unmatched [from this item.

Galvin and Prikry both had arrived at the University of California, Berkeley, in fall 1965, Prikry as a graduate student and Galvin as an instructor.[276] Prikry [2010b] recalled that he had gotten to know Silver during the year that they had overlapped in graduate school. Silver had finished his degree in 1966 and had spent a year at the University of Wisconsin, Madison. Already in the summer of 1967, Prikry had came up with his cofinality changing forcing, now known as *Prikry forcing.*

Galvin and Prikry identified the collection of subsets of ω with the product space 2^ω with the product topology. Their one-page proof that open sets are Ramsey is elegant and uses the following definition. Given a subset $S \subseteq 2^\omega$, say $M \subseteq \omega$ *accepts* X if every $N \subseteq M$ for which $X < N$ has $X \cup N \in S$, and M *rejects* X if there is no $N \subseteq M$ such that N accepts X.[277]

Galvin and Prikry introduced the notion *completely Ramsey* for subsets $S \subseteq 2^\omega$ with the property that for every continuous function $f : 2^\omega \to 2^\omega$, the set $f^{-1}(S)$ is Ramsey. They showed that the collection of completely Ramsey subsets of 2^ω includes all open sets and is closed under complements and countably unions, concluding that Borel sets are completely Ramsey. After deriving several corollaries including Ramsey's Theorem and the Nash-Williams Partition Theorem, they concluded with counter-examples to possible extensions, including a nowhere dense set of measure zero which is not Ramsey.

In January, Galvin [1968a][278] had already stated a generalization of the Nash-Williams Partition Theorem which can be translated to "open sets are Ramsey". Recently Galvin [2010] commented on this abstract:

> Of course, I was reinventing the wheel in that abstract; "open sets are Ramsey" had already been proved ... Also, at the time I submitted the abstract, I wouldn't have recognized it in the form "open sets are Ramsey".

Galvin and Prikry [1973] noted that their initial proofs that Borel sets are Ramsey used induction on the ordinals. Their published proof in [1973] instead uses ideas of Nash-Williams [1965b], "as they were explained to us by Richard Laver".[279] Documentably by the end of March, Galvin and Prikry had proved that Borel sets are Ramsey, since Robert Soare quoted this result in an abstract [1968a] received April 1, 1968, and in the paper [1969], Soare expressed gratitude to Galvin for the prompt reporting of his results.[280]

---

[276] Prikry recalled getting to know Galvin: "We had offices next to each other and had somewhat similar schedules, so we ended each day together around 1 am at a pizza place on Euclid [Avenue]."

[277] Galvin and Prikry noted that the fact that open sets are Ramsey was independently discovered by Ehrenfeucht and Cohen among others (cf. Mathias' point 3 above), but was not published prior to their paper [1973] in *The Journal of Symbolic Logic.*

[278] Galvin's abstract [1968a] was received January 26, 1968.

[279] Galvin and Laver played speed chess as well as talking mathematics together.

[280] Soare submitted a preliminary report [1968b] which was received February 8, 1968 (it mentioned Galvin for the proof of a combinatorial result formulated by Jockusch) and published

Silver [1970a], building on work of Galvin and Prikry, proved that analytic sets are completely Ramsey, and showed that under the assumption of the existence of a measurable cardinal that every Σ^1_2 set is Ramsey. For $X \subseteq \omega$ and $S \subseteq [X]^\omega$, the family S is called *X-Ramsey* if there is an infinite set $Y \subseteq X$ such that either $[Y]^\omega \subseteq S$ or $[Y]^\omega \cap S = \emptyset$. A family $S \subseteq [\omega]^\omega$ is *completely Ramsey* if, for every finite $F \subseteq \omega$ and every $X \subseteq \omega \setminus F$, the set $\{Z \subseteq X : F \cup Z \in S\}$ is X-Ramsey.[281] Silver's proof started with the formula $\exists r\phi(r, s, t)$ defining an analytic family, used forcing to extend a given model of a fragment of set theory to one in which MA_{ω_2} holds, proved that the family defined by the same formula in the extension is completely Ramsey, and applied an absoluteness argument to conclude that the original family was completely Ramsey.[282]

Mathias [1997, items 14–15] recalled an important step in his proof:

> Somewhere around the end of June 1968 I realise that there would be a good chance of proving that in Solovay's model all sets are Ramsey, provided that the Jensen-Friedman adaptation of Prikry forcing had the property that every infinite subset of a generic set is also generic. This is my main claim to originality: that such a phenomenon might be possible had never previously been suggested. ...
>
> Two weeks later, I finally reach the concept that I called *capturing*, and with its help establish my "all subsets generic" property, prove that in Solovay's model all sets are Ramsey, prove that my notion of forcing has the strong Cohen-type property that a Mathias-generic set X of integers is not constructible from any $Y \subseteq X$ with $X \setminus Y$ infinite, and, as easy corollaries, give new proofs of Silver's theorems.
>
> I write all this up in four weeks, and the resulting document, nearly 170 pages of A4, mainly in manuscript, is submitted in August 1968 to Trinity College, Cambridge in the (unfulfilled) hope of obtaining a Fellowship there under Title A.

Prior to the submission of the above document, Mathias sent in an abstract [1968] to the *Notices of the American Mathematical Society* which was received July 18, 1968. He gave a description of the forcing, now known as Mathias forcing: a *condition* is a pair $\langle s, S\rangle$ where $s \cup S \subseteq \omega$, s is finite and S infinite, and $n \in s \wedge m \in S \rightarrow n < m$; a condition (t, T) is stronger than (s, S) if $s \subseteq t \subseteq s \cup S$ and $T \subseteq S$. He included the statement of a lemma that subsets of a generic set are themselves generic, but did not mention the Solovay model, about which nothing

in the April issue of the *Notices of the American Mathematical Society*. Soare spoke on his results for the April 16–20, 1968 meeting of the American Mathematical Society in Chicago, so he welcomed the timely update by Galvin which allowed him to strengthen his theorem from continuous partial functions to Borel measurable functions.

[281] In their paper [1973], Galvin and Prikry indicated that they were the authors of both notions of *completely Ramsey*, and that every family satisfying the version they used in their paper also satisfies the version used by Silver.

[282] Mathias [1997, 2] noted that Silver had these results prior to Mathias' departure from California to Germany at the end of the winter quarter, in March or April of 1968.

had been published at the time. Here is how Mathias stated his theorem in the abstract:

> *Theorem:* Con(ZFC + there is an inaccessible cardinal) implies Con(ZF + DC + $\omega \to (\omega)_2^\omega$).

He included a note that his work yielded easy proofs of Silver's results.

Mathias spent 1968–1969 as a visiting lecturer at the University of Wisconsin, Madison, where he interacted with Kunen, his student Booth, and M. E. Rudin. Ultrafilters, especially on ω, were a topic of discussion and Mathias also thought about MAD families. In June he traveled to Australia where, in a very productive spurt, he proved several theorems and came up with the notion of *happy family*[283] for families $A \subseteq \mathcal{P}(\omega)$ whose complement $\mathcal{P}(\omega) - A$ is a free ideal (i.e. it contains $\{n\}$ for all n)[284] and for any decreasing sequence $\langle X_n : n \in \omega \rangle$ of elements of A there is $X \in A$ such that, for all n, if $n \in X$ then $X - X_n \subseteq \{0, 1, \cdots, n\}$. Examples of happy families include the family of all infinite subsets of ω, the complement of the ideal generated by a maximal family of almost disjoint sets, and a selective ultrafilter.[285] Mathias generalized Silver's result, and we present below a formulation by Andreas Blass [2009, 2583, Theorem 7]:[286]

> If U is a selective ultrafilter and if $[\omega]^\omega$ is partitioned into an analytic piece and a co-analytic piece, then there is a set $X \in U$ such that $[X]^\omega$ is included in one piece.

While Silver [1970a] published his result that analytic sets are Ramsey reasonably promptly, the publication of the work of Galvin and Prikry [1973] took longer to appear.[287] As noted above, Mathias submitted an extended version of his initial proofs to Cambridge University for his fellowship and Ph.D. dissertation. The above results were included in the published version [1977], but includes additional results, including a 1971 proof modeled on the classical proof that all analytic sets

[283] The name *happy family* was suggested by Crossley.

[284] This formulation of *free* comes from Math Review MR0491197 (58 #10462) by Baumgartner. Todorcevic [2010] calls such families non-principal co-ideals.

[285] A non-principal ultrafilter U on ω is *selective* if every partition of ω into disjoint sets not in U has a selector (transversal) in U. Equivalently, U is selective if every function $f : \omega \to \omega$ becomes one-to-one or constant when restricted to some set in U. Kunen (cf. [Booth, 1970]) showed that selective ultrafilters are *Ramsey*, i.e. whenever $[\omega]^2$ is partitioned into two pieces, there is a set $X \in U$ included in one piece. Booth and Kunen (cf. [Booth, 1970]) showed several conditions on ultrafilters were equivalent, including selective and Ramsey. Mathias used Ramsey in his paper.

[286] The Blass statement is very close to one direction of the theorem stated but not proved by Mathias in [1972, 210] which appears in a volume whose preface by Hodges was written November 1971, and a proof of the theorem may be extracted from Section 4 of his [1977]. Mathias' Theorem 2.2 [1973, 412] is similar to the Blass statement but in the context of Prikry sequences and ultrafilters on measurable cardinals; it was proved during a June 1969 stay at Monash University, and the resulting paper was received by the journal in September 1969, and revised the following August. Mathias mentioned in [1977] that these two theorems were proved at the same time.

[287] The Galvin-Prikry paper [1973] was received by the journal on October 12, 1971.

of real numbers are Lebesgue measurable, which he was inspired to seek after a conversation with Moschovakis at the Cambridge Summer School in Logic.[288]

7 1970-1980: STRUCTURES AND FORCING

The talented individuals who passed through the San Francisco Bay Area in the 1960's enriched mathematical life there and took away new perspectives and techniques to share around the world. Kunen moved to the University of Wisconsin, Madison, joining M. E. Rudin and graduate students Tall and Booth. Galvin moved to the University of California, Los Angeles, in 1968 and to the University of Kansas in 1974. Laver had a post-doctoral position at the University of Bristol and spent time at UCLA before moving to the University of Colorado, Boulder, where he joined a large and friendly working group with Malitz, Monk, Mycielski, Reinhardt, and Walter Taylor. Mathias moved to the University of Cambridge with a fellowship at Peterhouse, and Jensen spent time in Oslo, Rockefeller, and Oxford. Baumgartner moved to Dartmouth College as a John Wesley Young Instructor, and joined the faculty at Dartmouth with a one-year leave of absence spent at Caltech. Mitchell had a post-doctoral position at the University of Chicago and spent a few years at Rockefeller with Hao Wang and Martin, before moving to Pennsylvania State University with its longstanding logic group, which included Jech and Mansfield.

The two problem papers by Erdős and Hajnal [1971b], [1974] were valuable both for their problems and for their comments indicating what had been proved. The papers grew out of the presentation by Erdős at the 1967 UCLA set theory conference, and were circulated in preprint form. The results from the late 1960's on partitions of infinite subsets of ω by Galvin, Prikry, Mathias, and Silver finally appeared and were extended by Ellentuck and others in the 1970's. A version of the Halpern-Läuchli Theorem was used to prove a Ramsey theorem for trees. Partition relations with ordinal goals received increased attention, with a meta-mathematical approach yielding the Baumgartner-Hajnal Theorem, that $\omega_1 \rightarrow (\alpha)^2_2$ holds for all $\alpha < \omega_1$. Interesting ordinal partition relations were proved, i.e. ones where the resource or set being partitioned has order type a non-initial ordinal. A series of results about square bracket partitions provided strong counter-examples to ordinary partition relations and revealed additional structural information about partitions.

The word "special" entered the language of mathematics for trees and partial orders which are countable unions of antichains. Consistency was shown for combinations of existence and non-existence of various kinds of trees. Embeddings of trees into the rationals and into the reals were studied, adding to the understanding of their fundamental structure. The structure of Suslin trees in terms of their order was examined in terms of their order embeddings. Milliken proved

[288]The Cambridge Summer School in Logic (Mathias was the secretary of the organizing committee, i.e. the main organizer) was my first international meeting. There I met Erdős, Hajnal, Kunen, Milner, Prikry, Shelah, Silver and many others for the first time.

a partition theorem for strongly embedded trees. Shelah constructed a Countryman line and increased understanding of linear orders arising as the lexicographic orderings of Aronszajn trees. Basis questions were examined for linear orderings which are not a countable union of well-orderings. The statement that all $\aleph_1$-dense sets of reals are isomorphic was shown to be consistent as was the statement that $2^{\aleph_0} = \aleph_2$ and there is a linear order of cardinality $\aleph_1$ universal for linear orders of size $\aleph_1$.

M. E. Rudin [1975b], in a history of attempts to prove the Normal Moore Space Conjecture through the mid-1970's, highlighted connections between Aronszajn trees and Moore spaces, a thread explored by William Fleissner [1975],[289] Devlin and Shelah [1979a], [1979b], Taylor [1981][290] and others. In 1980, Peter Nyikos [1980b] used the *Product Measure Extension Axiom* (PMEA) to prove the consistency of the Normal Moore Space Conjecture. Kunen had shown that the consistency of PMEA follows from the consistent existence of a strongly compact cardinal (see [Fleissner, 1984] for a proof). There is more to the history of the Normal Moore Space Conjecture but at this point the story moves away from our main themes. We encourage the interested reader to continue the history with the excellent paper by Peter Nyikos [2001] and learn about further consistency and independence results there.

7.1 Combinatorial principles

John Truss observed that König's Lemma is equivalent to the principle that every countable set of finite sets has a choice function, in a paper exploring consequences of further restriction on the type of branching allowed.

In addition to diamond principles, Jensen [1972] introduced square principles for the "purpose of carrying out inductions which would otherwise breakdown" (see page 271) for infinite cardinals κ:[291]

> $\square_\kappa$: There is a sequence $\langle C_\lambda : \lambda < \kappa^+\rangle$ defined on limit ordinals $\lambda < \kappa^+$ such that
>
> (i). C_λ is a closed unbounded subset of λ;
>
> (ii). if $\mathrm{cf}(\lambda) < \kappa$, then $|C_\lambda| < \kappa$;
>
> (iii). if γ is a limit point of C_λ, then $C_\gamma = \gamma \cap C_\lambda$.

[289]The topologist Reed, in his Math Review MR0394583 (52 #15384) of Fleissner's paper [1975], discussed Jones' road space S [1953] which he said was introduced as a potential example of a normal, non-metrizable Moore space. Reed noted "The construction of S was discovered independently by N. Aronszajn, and it is known to set-theorists as a special Aronszajn tree," providing evidence that topologists were more familiar with the version of this construction from Jones.

[290]Taylor's article [1981] was received by the journal in 1979.

[291]I have changed the notation for cardinality of a set from $\overline{\overline{C}}$ to $|C|$ in the statement of $\square_\kappa$ for typographical simplicity.

Jensen proved in Theorem 5.2 that if $V = L$, then $\square_\kappa$ holds for every κ. Moreover, he proved that if $V = L$ and κ is regular, then the weak compactness of κ is equivalent to *stationary reflection*: if $A \subseteq \kappa$ is stationary in κ, then $A \cap \beta$ is stationary in β for some $\beta < \kappa$ (see [1972, 287]). Combinatorial consequences of the existence of a *non-reflecting* stationary subset of κ, i.e. a stationary subset $A \subseteq \kappa$ such that $A \cap \beta$ is non-stationary for all $\beta < \kappa$, were quickly found. For example, under the assumption of $V = L$, Shelah[292] used a non-reflecting stationary set to prove in Lemma 3.1 of [1975] that if λ is a regular uncountable non-weakly compact cardinal, then there is a graph G on λ so that the *coloring number*[293] of the restriction of G to any bounded set is $\leq \aleph_0$ but the coloring number of G is uncountable.

The remainder of §7.1 is devoted Jensen's Diamond Principle and its variants, and many of the results were found by Keith Devlin.[294]

Devlin [1979b] proved that $\diamondsuit^*$ is strictly stronger than $\diamondsuit$. Recall there was a remark in the handwritten notes of Jensen [1969b] that $\diamondsuit^+$ was stronger than $\diamondsuit^*$ since the former implied the existence of a Kurepa tree and the latter did not, but it was somewhat cryptic, since no proof was given that $\diamondsuit^*$ failed to guarantee the existence of Kurepa trees. The missing piece was filled in by Devlin [1978]. Using an idea of Jensen and starting from an inaccessible cardinal, Devlin gave a forcing construction of a model of $\diamondsuit^*$ which has no Kurepa trees. He said on page 292 that Jensen had used the idea to "convince himself that $\diamondsuit^*$ does not imply KT, [the existence of a Kurepa tree], but never wrote up the proof."

In the 1970's two variants weaker than $\diamondsuit$ were introduced. Adam J. Ostaszewski[295] [1976] published his proof of the existence of a non-compact, hereditarily separable, locally compact, perfectly normal, countably compact space. While he said in the introduction to his paper that the construction was based

[292]Saharon Shelah received his 1969 doctorate from Hebrew University where Michael O. Rabin was his advisor. He is a professor at Hebrew University and also at Rutgers University. While his main mathematical interests are in logic, especially model theory and set theory, he has broad interests. Combinatorial set theory has benefited from his deep knowledge of the behavior of singular cardinals, his development of myriad forcing techniques, and his powerful problem solving capabilities. His output is dauntingly prodigious, with gems that sparkle for many different groups of mathematicians, including his more that 200 co-authors. He was the first recipient of the Erdős Prize in 1977; he received the Israel Prize for Mathematics in 1998, the Bolyai Prize in 2000 and the Wolf Prize in Mathematics in 2001.

[293]The *coloring number* of a graph G on λ is the minimal cardinal μ such that we can order the vertices of G as $\langle a_i : i < \ell_0 \rangle$ such that each a_i is joined to fewer than μ many a_j's for $j < i$.

[294]Keith J. Devlin received his Ph.D. from the University of Bristol in 1971, where his advisor was Frederick Rowbottom. He was an assistant professor at the University of Bonn 1974–1976, a reader in mathematics at Lancaster University 1977–1987, visiting associate professor at Stanford University 1987–1989, chair of the Department at Colby College in Maine 1989–1993, dean at Saint Mary's College 1993-2001, and has been at Stanford University since 2001. In addition to his academic mathematics, he is widely known for sharing mathematics with the public, was winner of the 2007 Carl Sagan Award for Science Popularization, and is known as "the Math Guy" on National Public Radio.

[295]Adam J. Ostaszewski received his doctorate in 1973 from the University of London where his advisor was Claude Ambrose Rogers. He is a reader in the Department of Mathematics at the London School of Economics where his current research focuses on mathematical finance.

on Jensen's $\diamondsuit$, he actually introduced an ostensibly weaker principle, $\clubsuit$, which sufficed for his proof (see [1976, 507]):

> $\clubsuit$: There is an increasing sequence of countable limit ordinals λ_α indexing a sequence $\langle s(\lambda_\alpha) : \omega \le \alpha < \omega_1 \rangle$ where $s(\lambda_\alpha)$ is an ω sequence cofinal in λ_α, such that every uncountable subset of ω_1 contains one of them.

His paper includes a proof that $\diamondsuit$ is equivalent to CH + $\clubsuit$, which he attributes to Devlin in private correspondence (see also Devlin [1974][296]). Ostaszewski stated that it was not known whether $\clubsuit$ was consistent with the negation of the Continuum Hypothesis, and he pointed out that it is not difficult to show that it is false in the presence of MA + $\neg$CH.

At the end of the 1970's, Baumgartner and Shelah gave different proofs that $\clubsuit$ is weaker than $\diamondsuit$ by forcing over models of $V = L$ to get a model in which $\clubsuit$ holds but CH does not. Shelah [1980c, §5],[297] [1998, Chapter 1 §7] first added $\aleph_3$ Cohen subsets of ω_1 and then used finite conditions to collapse ω_1 to ω. Baumgartner, in an unpublished note,[298] added many Sacks reals side-by-side with countable support. See [Hrušák, 2001] for a proof of that $\clubsuit$ holds in this model.

Devlin and Shelah [1978] introduced the *Weak Diamond Principle*,[299] which states that for each $f : {}^{\omega_1>}2 \to 2$ there is a $g \in {}^{\omega_1}2$ such that for any $h \in {}^{\omega_1}2$, the set $\{\alpha \in \omega_1 : f(h\restriction\alpha) = g(\alpha)\}$ is stationary. This principle is a consequence of Jensen's $\diamondsuit$, and Devlin and Shelah showed it follows from the inequality $2^{\aleph_0} < 2^{\aleph_1}$. Their motivating application was Whitehead's problem on abelian groups, and they also applied it to ladder systems.

Several people, including Baumgartner, independently showed that the Weak Diamond Principle is equivalent to the cardinal inequality $2^{\aleph_0} < 2^{\aleph_1}$ from which it had been derived. A proof appears in a paper by Taylor [1981, §3] which was received by the journal in 1979 and discusses weak diamond type statements and the Normal Moore Space Conjecture.

Next, the discussion turns to the diamond variations that live on stationary sets. Recall that for a cardinal λ and a stationary subset $E \subseteq \lambda$, $\diamondsuit^*(E)$ is the statement that there is a sequence $\langle W_\alpha \subseteq \mathcal{P}(\alpha) : \alpha \in E \rangle$ with $|W_\alpha| \le \alpha$ for all but a bounded set of $\alpha \in E$ such that for all $X \subseteq \lambda$, there is a closed unbounded set $C \subseteq \lambda$ with $X \cap \alpha \in W_\alpha$ for all $\alpha \in C \cap E$. John Gregory [1976, 666] in Lemma 2.1 proved the following:

---

[296] Shelah [1980c] credits both Burgess and Devlin with the equivalence of $\diamondsuit$ with $\clubsuit$ + CH.

[297] The paper [Shelah, 1980c] was received by the journal December 24, 1978 and, in revised form, December 16, 1979. In an email of January 13, 2011, he recalled that he proved the result in 1977 shortly after learning of the problem from a list by Fleissner.

[298] Shelah and Džamonja [1999] briefly discuss the history of the problem, noting that Baumgartner's proof was given shortly after Shelah's. They list further extensions and answer questions of S. Fuchino and M. Rajagopalan on variants of $\clubsuit$.

[299] The principle, denoted Φ in [1978], was named the *Weak Diamond Principle* in Devlin's survey [1979b].

> *Gregory's Theorem:* For regular cardinals κ and μ, if $\kappa^\mu = \kappa$ and $2^\kappa = \kappa^+$, then for the stationary set $E = E(\kappa^+, \mu)$ of all ordinals less than κ^+ of confinality μ, the principle $\diamondsuit^*(E)$ holds.[300]

Specifically, if GCH holds and $\mathrm{cf}(\kappa) > \mu$ for μ regular, then $\diamondsuit^*(E(\kappa^+, \mu))$. Thus, if GCH holds, for all regular uncountable cardinals κ, $\diamondsuit(\kappa^+)$ holds.

In his survey article [1979b], Devlin reported on Shelah's result [1977] that it is consistent with GCH to have disjoint stationary sets $E, F \subseteq \omega_1$ such that $\diamondsuit(E)$ is true and $\diamondsuit(F)$ is false.

7.2 Transversals and cardinal arithmetic

At the 1974 International Congress of Mathematicians in Vancouver, individuals mentioned in this history were well-represented among the invited speakers: Barwise, Hajnal, Jónsson, Milner (plenary), Rado, Shelah, Silver. In addition, Yiannis Moschovakis was an invited speaker, and Leo Harrington, Martin and Prikry attended (see [Shelah, 1976]).

Milner [1975] spoke on transversals and included a discussion on various infinite extensions of Hall's Theorem. Milner had started working on transversals in the 1960's. He and Shelah [1975] extended transversal theory to regular uncountable cardinals in a paper for the proceedings of the colloquium held at Lake Balatón in 1973 dedicated to Erdős on his 60th birthday. They showed that if κ is weakly compact, $\mathcal{F}$ is an indexed family of κ sets each of cardinality less than κ, and every subsequence of $\mathcal{F}$ of smaller cardinality has a transversal, then so does $\mathcal{F}$. They also proved that the converse of this statement holds in L. They went on to prove a negative stepping up theorem for transversals.

Shelah [1975], in the same volume, tackled the case for strong limit cardinals. For example, he showed (see Theorem 4.6) that if κ is a strong limit cardinal of cofinality $\aleph_0$ and $\lambda < \kappa$, and if $\mathcal{F} = \langle F_i \colon i \in I \rangle$ is a system of κ sets each of cardinality $< \lambda$, and if every subsystem of $< \kappa$ sets has a transversal, then so does $\mathcal{F}$. Shelah also gave a criterion of an inductive kind which provided necessary and sufficient conditions for an arbitrary family of countable sets to have a transversal.

At the Vancouver meeting, Silver [1975] spoke on his groundbreaking new result on the singular cardinals problem, that if $2^\nu = \nu^+$ holds on a stationary set of cardinals below κ, a singular cardinal of uncountable cofinality, then it also holds at κ. Since this result is discussed in Kojman's chapter, we only include here a few remarks about connections with individuals working in combinatorial set theory.

Several mathematicians were involved in finding alternative proofs and generalizations of Silver's result. An elegant and purely combinatorial proof was found independently by Prikry, Baumgartner, and Jensen. The proof by Baumgartner and Prikry [1976], [1977] used a result of Erdős, Hajnal, Milner [1968] as the base case.

[300]Note that Gregory is using *confinality* as Hausdorff did, rather than the now standard *cofinality*.

Galvin and Hajnal[301] [1975], working independently of one another, proved what has become known as the Galvin-Hajnal Theorem.

> *Galvin-Hajnal Theorem* [1975, 492, Corollary 5]: Let $\aleph_\alpha$ be a singular strong limit cardinal of uncountable cofinality. Then $2^{\aleph_\alpha} < \aleph_\gamma$ where $\gamma = (2^{|\alpha|})^+$.

This result is a corollary of their Theorem 1, which relied on the notion of a family $\mathcal{F}$ of *almost disjoint transversals* of a sequence $\langle A_\alpha : \alpha < \kappa \rangle$, where transversals f and g are *almost disjoint* if $\{\alpha : f(\alpha) = g(\alpha)\}$ is bounded. Their Theorem 2 is a stronger result, obtained by replacing the ideal of bounded subsets with the non-stationary ideal to define a well-founded partial order, $f <_{\mathrm{NS}_\kappa} g$ iff $\{\alpha : f(\alpha) \geq g(\alpha)\}$ is non-stationary. The rank function associated with this partial order is now known as the *Galvin-Hajnal rank*.

Theorem 2 has as corollaries Silver's result and its immediate generalizations. It stimulated further work[302] by Shelah [1980b, 56], who wrote that he was continuing [Galvin and Hajnal, 1975]. Shelah [1978b] applied the Galvin-Prikry ideas to Jónsson algebras, and he [1994, xi] said the following:

> My interest in cardinal arithmetic started in 1975, having heard the theorems of Silver, Galvin, Hajnal and others, on bounds, proved in ZFC, for cardinal exponentiation, and Jensen's discovery of the covering theorem. I felt then that I had come too late to the game (later I have felt to be alone actively interested in it.)

This story is continued in Kojman's chapter on cardinal arithmetic.

7.3 *Partition relations on cardinals and ordinals*

This section starts with a discussion of the Baumgartner-Hajnal Theorem and related results. Then the following topics are discussed in turn: ordinary partition relations in which the resource is a cardinal, ordinal partition relations, and square bracket partition relations.

The Baumgartner-Hajnal Theorem

We begin with a special case:

---

[301]The Galvin-Hajnal paper was received September 25, 1974, a mere month after the August 21-29, 1974 Vancouver meeting at which Silver spoke. Galvin and Hajnal thanked Silver and Prikry for communicating their results to them, and point out that they use Prikry's ideas (see [Baumgartner and Prikry, 1976]). For a nice exposition of their work, see the teaching blog of Andrés Caicedo or his forthcoming book.

[302]See [Jech, 1995], Kojman's chapter on cardinal arithmetic, and the final chapter of [Holz *et al.*, 1999] for more on the relationship between the Galvin-Hajnal theorems and Shelah's work on cardinal arithmetic.

Baumgartner-Hajnal Theorem for ω_1: For all $\alpha < \omega_1$ and $k < \omega$,

$$\omega_1 \to (\alpha)^2_k.$$

The full Baumgartner-Hajnal Theorem solved a series of questions raised by Erdős, Hajnal and Galvin[303] about partitions of pairs from ω_1 and partitions of pairs from λ, the order type of the real numbers. Before stating the full theorem, we review these problems and partial results leading up to it.

Problem 10 of [Erdős and Hajnal, 1971b] asks, under the assumption of GCH, whether $\omega_{\rho+1} \to (\xi, \xi)^2$ for all $\xi < \omega_{\rho+1}$ and all ρ, and Problem 11 asks a parallel partition relation for λ, the order type of the reals: does $\lambda \to (\alpha)^2_k$ hold for all finite k and countable α? In addition, Problem 10/A asks three specific instances of $\omega_1 \to (\alpha_i)^2_k$ for ordinals $\alpha_i \leq \omega^2$ and finite k, and Problem 11/A asks a parallel series of specific questions about instances of $\lambda \to (\alpha_i)^2_k$.

Hajnal [1960, Theorem 8] had shown that, in the presence of CH, $\omega_1 \to (\omega \cdot 2, \omega \cdot k)^2$ for every finite k. In a letter to Erdős and Hajnal cited in a note added in proof to [1971b], Galvin informed them that he had proved $\omega_1 \to (\omega \cdot 3)^2_2$. Prikry [1972] proved $\omega_1 \to (\omega^2, \alpha)^2$ for $\alpha < \omega_1$, which completely solved Problem 10/A, since for any finite sequence $\langle \alpha_i : i < k \rangle$ of countable ordinals, there is a countable ordinal α such that $\alpha \to (\alpha_i)^2_k$.

Laver [1975], during his time at Bristol 1969-1971, used $\mathrm{MA}_{\aleph_1}$ to prove a partition relation equivalent to

$$\omega_1 \to (\omega_1, (\omega : \omega_1))^2.$$

This means that every graph on ω_1 either has an uncountable independent set or has a pair of sets B and C where B has type ω, C has type ω_1, and all pairs $\{x, y\}_{\neq}$ with $x \in B$ and $y \in C$ are joined. Laver's proof was the first use of Martin's Axiom to prove a partition relation, and a forerunner of the proof of the Baumgartner-Hajnal Theorem.

Prikry [1972] in the Corollary on page 78 proved (in ZFC) a partition relation closely related to Laver's result from $\mathrm{MA}_{\aleph_1}$. Below I have named it and restated it in an equivalent form.

Prikry's Theorem: For all countable ordinals α, $\omega_1 \to (\alpha, (\omega : \omega_1))^2$.

By the time Erdős and Hajnal were correcting the proofs of their problem paper [1971b], Galvin had already shown that for every real type φ, the partition relation $\varphi \to (\alpha)^2_2$ holds for all countable α. Erdős and Hajnal, after stating Galvin's result, commented that all the known results for λ, the order type of the reals, hold for order types in which neither ω_1 nor $\omega_1{}^*$ is embeddable (let us temporarily call

[303] In a November 2010 telephone call, Baumgartner recalled visiting Galvin in his Berkeley office in the 1960's where he had three boxes of papers on the floor. Galvin would occasionally open one of the boxes and reveal a paper on Erdős type problems. Baumgartner said that for many years, Galvin was the one he went to in order to learn the status of such problems.

such types *pseudo-real types*). Erdős and Hajnal asked in Problem 12 if one can prove a partition relation of the form $\lambda \to (\theta_1, \theta_2)^2$ which does not hold for some uncountable pseudo-real type ψ and some order types θ_1 and θ_2.

Erdős and Hajnal review some of the history of the generalization of the question to order types in their follow-up problem paper [1974, 272], noting that the topic was extensively studied by Galvin, "and it is fair to say that even the general conjecture is due to him." Among Galvin's "old theorems" they list the implication that if $\varphi \not\to (\omega)^1_\omega$, then $\varphi \not\to (\omega, \omega+1)^2$. Consequently, the order types φ satisfying $\varphi \to (\alpha)^2_2$ for all countable α, must satisfy $\varphi \to (\omega)^1_\omega$.

This property of order types, singled out by Galvin, turned out to be the key:

> *Baumgartner-Hajnal Theorem for order types:* For all $\alpha < \omega_1$ and $k < \omega$, if $\varphi \to (\omega)^1_\omega$, then $\varphi \to (\alpha)^2_k$.

Since λ satisfies the partition property $\lambda \to (\omega)^1_\omega$, these results provided solutions to Problems 10/A and 11 of the Erdős-Hajnal problem paper [1971b].

Baumgartner and Hajnal [1973] used a meta-mathematical argument. They started by using Martin's Axiom to prove the desired partition relation and then used an absoluteness argument to show that the use of MA is inessential. Baumgartner and Hajnal cited Silver [1966], [1970b] as the source of the absoluteness argument. Baumgartner and Hajnal did not use MA directly, but instead cited two lemmas. Lemma 1 [1973], due to Solovay (cf. [1971]), is the fact that under MA_β, for any finite k and any indexed family $\{A(i,\xi) : i < k \wedge \xi < \beta\}$ of partitions $\bigcup_{i<k} A(i,\xi) = \omega$ of ω, there are an infinite set $X \subseteq \omega$ and a function $f : \beta \to k$ so that X is almost included in every $A(f(\xi), \xi)$. Lemma 2 [1973], due to Kunen [1968], is the fact that under MA_β, any family of at most β many functions from ${}^\omega\omega$ can be eventually dominated.

Galvin [1975] presented a combinatorial proof of the Baumgartner-Hajnal Theorem for ω_1 in the clear and elegant style for which he is known. He first isolated in his Theorem 1 a combinatorial statement about a fixed partition of 2-element chains of a partially ordered set stating that there is a family of its subsets with sufficient closure properties to derive the Baumgartner-Hajnal Theorem in both its cardinal and order type forms. In the same paper he used Theorem 1 to prove that if φ is a partial order type and $\varphi \to (\eta)^1_\omega$, where η is the type of the rationals, then $\varphi \to (\alpha)^2_k$ for all $k < \omega$ and $\alpha < \omega_1$. Note that the partitions under consideration are partitions of 2-element chains, rather than arbitrary 2-element subsets. This result led to increased interest in order types with this property. Galvin stated some open problems, including the question of whether $\varphi \to (\omega)^1_\omega$ implies $\varphi \to (\alpha)^2_k$ for all $k < \omega$, all $\alpha < \omega_1$ and all partial order types φ. He also generalized the result of Erdős and Rado [1956, 472] that $\omega_1 \to (\omega+1)^r_k$ for finite r, k, by replacing ω_1 by a partial order P such that $P \to (\omega)^1_\omega$.

Cardinal resources

While the classical work on ordinary partition relations for cardinals was essentially done in the sixties [Erdős *et al.*, 1965] and the Baumgartner-Hajnal Theorem came in the first half of the decade, additional results appeared in the 1970's.

Shelah [1973] used GCH to partially extend the Baumgartner-Hajnal Theorem to successor cardinals by showing that if $|\gamma|^+ < \kappa$ and κ is regular, then $\kappa^+ \to (\kappa+\gamma)^2_2$. Rebholz [1974] applied morasses to show that Shelah's result was optimal for successors of successor cardinals under GCH by demonstrating that

$$\aleph_{\alpha+2} \not\to (\aleph_{\alpha+1} + \aleph_\alpha)^2 \text{ holds if } V = L.$$

Consistently, the continuum may have rather different partition properties from ω_1. Kunen [1971] announced that if κ is real-valued measurable, then $\kappa \to (\kappa, \alpha)^2$ for all $\alpha < \kappa$. Thus, since Solovay [1971] established the consistency of the continuum being a real-valued measurable cardinal, $2^{\aleph_0} \not\to (2^{\aleph_0}, \alpha)$ cannot be proved for $\omega + 2 \leq \alpha < \omega_1$ without some hypothesis on the size of the continuum.

Ideals were used to prove other partition relations of the form $2^\kappa \to (2^\kappa, \alpha)^2$ for some $\alpha < \kappa$. A *κ-ideal* is a non-trivial ideal of subsets of κ which is κ-complete. Laver called a κ-ideal $\mathcal{I}$ *(λ, μ, ν)-saturated* if every family of λ many $\mathcal{I}$-positive sets (i.e. ones not in the ideal) has a subfamily of size μ such that the intersection of any subset of the subfamily of size at most ν is also $\mathcal{I}$-positive. Laver [1978b] proved that if there is a (κ, κ, γ)-saturated κ-ideal and $\beta^\gamma < \kappa$ for all $\beta < \kappa$, then $\kappa \to (\kappa, \alpha)^2$ for every $\alpha < \gamma^+$. He further showed how to force the existence of such ideals starting with a measurable cardinal, with a sample consequence being the consistency, relative to a measurable cardinal, of the partition relation $2^{\omega_1} \to (2^{\omega_1}, \alpha)$ for all $\alpha < \omega_2$.

Laver [1975, 1936] proved that if there is a Mahlo cardinal κ, then there is a forcing extension in which $2^{\aleph_0} = \kappa$ is weakly Mahlo and $2^{\aleph_0} \not\to (2^{\aleph_0}, (\omega : 2))^2$.

Baumgartner [1976a, 434, Theorem 7.1] started with a model in which κ and λ are regular cardinals with $\kappa \leq \lambda$, and $\diamondsuit_\kappa$ if κ is not inaccessible, and forced to get $2^\kappa \geq \lambda$ while preserving cofinalities and arranging that in the extension the following partition relation holds for all regular $\mu \geq \kappa^{++}$ and every $\beta_0 < \kappa^+$:

$$\mu \to (\text{stationary } \mu, (\beta_0 : \text{stationary } \mu))^2.$$

Note the close relationship with Prikry's Theorem.

Recall that Hajnal [1960] had shown with CH that $\omega_1 \not\to (\omega_1, (\omega : 2))^2$, a refutation of a possible extension of the Dushnik-Miller Theorem. Baumgartner [1976a, 436, Theorem 7.2] used forcing to prove the consistency of $\kappa^+ \not\to (\kappa^+, (\nu : 2))^2$, where $2^\nu = \kappa^+$. Laver [1975, 1033] proved that if Martin's Axiom holds and $2^{\aleph_0} = \aleph_2$, then $2^{\aleph_0} \not\to (2^{\aleph_0}, (\omega : 2))^2$ (this result was later and independently found by Hajnal).

Stephen G. Simpson[304] [1970] came up with a model-theoretic proof of the Erdős-Rado theorem which was incorporated into the classic text *Model Theory*

[304]Stephen George Simpson (1945 –) received his doctorate in 1971 from the Massachusetts Institute of Technology where his advisor was Gerald Sacks. He is currently a professor at Pennsylvania State University.

by C. C. Chang[305] and H. Jerome Keisler[306] [1973].

Baumgartner [1975a] updated the Erdős-Rado Theorem by proving a canonical version of the Erdős-Rado Theorem: Write $\kappa \to *(\lambda)^n$ if for every f defined on $[\kappa]^n$, there is a finite set $\Delta \subseteq n$ and a subset $B \subseteq \kappa$ of order type λ such that $f(x) = f(y)$ for all subsets $x = \{x_0, x_1, \ldots, x_{n-1}\}_<$ and $y = \{y_0, y_1, \ldots, y_{n-1}\}_<$ of B for which $x_i = y_i$ for all $i \in \Delta$. Baumgartner proved that

$$\left(\exp_r(2^{<\kappa})\right)^+ \to *(\kappa + (r-1))^r,$$

using regressive functions together with ideas from the proof of the Erdős-Rado Theorem [1956].

Shelah [1975] solved Problem 3 of [Erdős and Hajnal, 1971b] by showing that if $n(i)$ $(i < \omega)$ is a sequence of integers such that $\aleph_\omega < \kappa_0 < \kappa_1 < \cdots$, where $\kappa_i = 2^{\aleph_{n(i)}}$, then $\sum\{\kappa_i \colon i < \omega\} \to (\aleph_\omega, \aleph_\omega)^2$. To do so, Shelah proved a Canonization Lemma, a special case of which is given below in a modern formulation.

> *Shelah's Canonization Lemma:* Suppose that $\langle \kappa_\xi : \xi < \mu\rangle$ is an exponentially increasing sequence[307] of infinite cardinals with $\kappa_0 \geq \tau, \mu, \omega$ for a cardinal $\tau \geq 2$. Then for any disjoint union $A = \dot{\bigcup}\{A_\nu : \nu < \mu\}$, any sequence $\langle F_\nu \subseteq \mathcal{P}(A_\nu) : \nu < \mu\rangle$, and any coloring $f : [A]^2 \to \tau$, if $|A_\nu| > 2^{\kappa_\nu}$ and F_ν sustains[308] A_ν over κ_ν for all $\nu < \mu$, then there is a sequence $\langle B_\nu \in F_\nu : \nu < \mu\rangle$ *weakly canonical* with respect to f with $|B_\nu| = {\kappa_\nu}^+$ for all $\nu < \mu$, i.e. $f(u) = f(v)$ whenever $u, v \in \left[\bigcup\{B_\nu : \nu < \mu\}\right]^2$ and $|u \cap B_\nu| = |v \cap B_\nu| \leq 1$ for every $\nu < \mu$.

Given a disjoint union $A = \dot{\bigcup}\{A_\nu : \nu < \mu\}$, a coloring $f : [A]^2 \to 2$ and an exponentially increasing sequence $\langle \kappa_\xi : \xi < \mu\rangle$ with $\kappa_0 \geq \mu, \omega$ and $|A_\nu| > 2^{\kappa_\nu}$ for all $\nu < \mu$, the sequence $\langle F_\nu : \nu < \mu\rangle$ where F_ν is the set of $X \subseteq A_\nu$ of size ${\kappa_\nu}^+$ which are homogeneous for f has the property that each F_ν sustains A_ν, so by Shelah's Canonization Lemma, there are $B_\nu \subseteq A_\nu$ of size ${\kappa_\nu}^+$ for $\nu < \mu$ such that $\langle B_\nu : \nu < \mu\rangle$ is weakly canonical with respect to f. Moreover it is *canonical with*

[305] Chen Chung Chang (October 13, 1927 –) received his doctorate from the University of California, Berkeley, in 1955 where his advisor was Tarski. The title of his dissertation was *Cardinal and Ordinal Factorization of Relation Types.* He spent a postdoctoral year 1955–1956 at Cornell University where a spring seminar by J. Barkley Rosser led him to work on MV-algebras, moved to University of Southern California in Los Angeles the following academic year where he was an assistant professor, and in July 1958, joined the faculty of the University of California, Los Angeles, retiring in 1990. For more on his early career and working habits, see [Chang, 1998].

[306] Howard Jerome Keisler (December 3, 1936 –) received his doctorate from the University of California, Berkeley, in 1961 where his advisor was Tarski. He is now an emeritus professor at the University of Wisconsin, Madison. He is known for his work in model theory and non-standard analysis, especially for his calculus text book [1976] using non-standard analysis, which is currently available from his homepage.

[307] A sequence $\langle \kappa_\xi : \xi < \mu\rangle$ is *exponentially increasing* if $\xi < \nu < \mu$ implies $2^\xi < 2^\nu$.

[308] We say $G \subseteq \mathcal{P}(C)$ *sustains C over κ* if for every subset X of C of cardinality $(2^\kappa)^+$, there is a $Y \in G$ so that $Y \subseteq X$ and $|Y| = \kappa^+$.

respect to f since $f(u) = f(v)$ whenever $u, v \in \bigcup_{\nu<\mu} B_\nu$ and $|u \cap B_\nu| = |v \cap B_\nu|$ for all $\nu < \mu$.

Since μ is a cardinal, there is some subset $M \subseteq \mu$ of full cardinality such that $B' = \bigcup_{\nu \in M} B_\nu$ has the property that every 2-element subset of B' with two elements from a single block among $\{B_\nu : \nu \in M\}$ receive the same color. With this final reduction, the same color equivalence relation has a nice description with respect to the block structure: a 2-element set has the *same block color* if both elements come from the same block, and have the (ξ, ζ)*-block color* if one element is from the ξth block and the other from the ζth block. Clearly this approach can be extended to more colors, and one could ask that the coloring of pairs across blocks be induced by a canonical equivalence relation on μ, if that makes sense. It does not have the simplicity of the Erdős-Rado canonization of equivalence relations on 2-element subsets of ω, where any same color equivalence relation was equal to one on their list when restricted to some infinite subset of ω.

However, the point of the canonization lemma was not simply to show that the partition could be made canonical. Below we mildly paraphrase the theorem Shelah derived from his canonization lemma [1975] to make the names of the cardinals better match our statement of his canonization lemma:

> Shelah's Singular Cardinal Partition Relation: If $\mu \to (\mu)^2_2$, $\mu = \operatorname{cf} \lambda$, $\langle 2^\kappa : \kappa < \lambda \rangle$ is not eventually constant, but is eventually $\geq \lambda$, and $\chi = \sum_{\kappa<\lambda} 2^\kappa$, then $\chi \to (\lambda)^2_2$.

Ordinal partition relations

In the late sixties and early seventies, a variety of partition results of the form $\omega^n \to (\omega^m, k)^2$ for finite values of n, m, k were found by Galvin, Hajnal,[309] Milner, and Labib Haddad and Gabriel Sabbagh [1969b], [1969a], [1969c]. The article by Milner [1971] gave a good exposition of this work including its history, and even included a proof of $\omega^5 \to (\omega^3, 6)^2$, which was then unknown. While open questions remain, Milner's student, Eva Nosal settled more cases [1974], [1979]]:

Theorem:

1. If $1 \leq \ell < \omega$, then $\omega^{2+\ell} \to (\omega^3, 2^\ell)^2$ and $\omega^{2+\ell} \not\to (\omega^3, 2^\ell + 1)^2$.
2. If $1 \leq \ell < \omega$ and $4 \leq r < \omega$, then $\omega^{1+r\cdot\ell} \to (\omega^{1+r}, 2^\ell)^2$ and $\omega^{r+r\cdot\ell} \not\to (\omega^{1+r}, 2^\ell + 1)^2$.

Galvin exploited Specker's counter-example $\omega^3 \not\to (\omega^3, 3)^2$ to the full. For ordinals α and β, α *can be pinned to* β, in symbols, $\alpha \to \beta$, if there is a mapping $\pi : \alpha \to \beta$ such that for every $X \subseteq \alpha$ of order type α, its image, $f[X]$, has order

[309] Erdős and Hajnal, in their problem paper [1971b, 21–22], attribute to Galvin and Hajnal independently a uniformization result which implies that for all $3 \leq k, n < \omega$, there is a least $f(k, n)$ such that $\omega^n \not\to (\omega^k, f(k, n))^2$.

type β.[310] Using pinning onto ω^3 in a paper with Jean Larson, he [1975] proved that $\alpha = \omega^\beta$ satisfies the relation $\alpha \not\to (\alpha, 3)^2$ for every decomposable countable ordinal $\beta \geq 3$.

The first additively indecomposable countable ordinal not covered by the Specker and Galvin results is ω^ω. Chang proved [1972] that ω^ω satisfies the partition relation $\omega^\omega \to (\omega^\omega, 3)^2$. This was the first major breakthrough in partition relations for countable ordinals since Specker's initial positive result for ω^2 and negative result for ω^3 in the mid-1950's. Chang's result was reported in the Erdős-Hajnal problem paper [1971b, 22] in a note added in proof and dated May 1970. Chang's 58-page paper [1972] in which this result appeared included a note that Milner had generalized the proof to show $\omega^\omega \to (\omega^\omega, m)^2$ for all $m < \omega$ and, added in proof, that Larson had a much shorter proof.[311] The $250 dollars that Erdős awarded for this result was the highest he had awarded to that time; Chang used the money to purchase a motorcycle.

Erdős and Hajnal [1971a] stimulated interest in partition relations for uncountable ordinals with a paper in which they investigated partition relations for small ordinal powers of uncountable cardinals. They asked in Problem 13 of their problem paper [1971b] whether the following partition relations hold in the presence of GCH:

$$\omega_1{}^2 \to (\omega_1{}^2, 3)^2 \text{ and } \omega_1 \cdot \omega \to (\omega_1 \cdot \omega, 3)^2.$$

Hajnal [1971] published elegant proofs from the Continuum Hypothesis that the answer is no by proving that

$$\omega_1{}^2 \not\to (\omega_1{}^2, 3)^2 \text{ and } \omega_1 \cdot \omega \not\to (\omega_1 \cdot \omega, 3)^2.$$

Larson[312] [1998] showed the partition relation is consistent with the failure of CH by proving it from the assumption that the *dominating number*[313] $\mathfrak{d}$ is ω_1. More generally, Hajnal showed that if $\kappa = \lambda^+$ is the successor of a regular cardinal and $2^\lambda = \lambda^+ = \kappa$, then $\kappa^2 \not\to (\kappa^2, 3)^2$ and $\kappa \cdot \lambda \not\to (\kappa \cdot \lambda, 3)^2$.

Baumgartner [1975b] then extended the first result to the successor κ of a singular cardinal λ for which $2^\lambda = \lambda^+ = \kappa$ by proving $\kappa^2 \not\to (\kappa^2, 3)^2$. He also showed that for a strong limit cardinal κ, the partition relation $\kappa^2 \to (\kappa^2, m)^2$ for finite m holds if and only if $(\operatorname{cf} \kappa)^2 \to ((\operatorname{cf} \kappa)^2, m)^2$ holds.

---

[310] Rotman [1970] introduced this notation for the notion used by Specker [1957] to transfer his negative partition result for the resource ω^3 to ω^n for all finite $n \geq 3$. Rotman proved that no countable ordinal pinned to a larger countable ordinal and conjectured that no ordinal could be pinned to a larger ordinal. In a surprising result on the structure of ordinals, Carlson [1984] showed that in the presence of the Continuum Hypothesis, $\omega_1^{\omega+2}$ pins to cofinally many ordinals in ω_2.

[311] See Larson's thesis [1972], [1973], and, for a particularly nice exposition, Neil Williams's book [1977, Chapter 7, Section 3].

[312] Larson remembers hearing Hajnal's beautiful proof from Galvin at UCLA in the early 1970's.

[313] A family $Y \subseteq {}^\omega\omega$ is *dominating* if for every function $f \in {}^\omega\omega$ there is a function $g \in Y$ such that $f \leq^* g$. The *dominating number* $\mathfrak{d}$ is the minimum size of a dominating subfamily of ${}^\omega\omega$ and can consistently be less than $2^{\aleph_0}$. For more on this cardinal invariant, see §5 of the chapter by Steprāns.

Kunen (cf. [Erdős and Hajnal, 1974, 274]) proved that if κ is real-valued measurable, then $\kappa^2 \to (\kappa^2, 3)^2$.

Baumgartner and Larson (cf. [Larson, 1975]) proved that if κ is weakly compact and $\alpha < \kappa$ satisfies $\alpha \to (\alpha, m)^2$, then $\kappa \cdot \alpha \to (\kappa \cdot \alpha, m)^2$.

Square bracket partition relations

While square bracket partition relations were introduced in the 1960's,[314] we have postponed discussion of this relation to the 1970's to consolidate the results of the two decades.

A nice positive result by Erdős, Hajnal and Rado is Theorem 22 (iii) [1965, 149] of which we state a special case that introduces a useful variant of the basic notation: For κ a singular strong limit cardinal of cofinality ω and $2 < \gamma < \omega$, $\kappa \to [\kappa]^r_{\gamma,2}$, i.e. for any coloring of the pairs of κ with γ colors, where $2 < \gamma < \omega$, there is a subset of size κ so that all but at most two colors are omitted.

Recall that we introduced, in §6.7, the following GCH negative result by Erdős, Hajnal and Rado Theorem 17 [1965, 145]:

$$\aleph_{\alpha+1} \not\to [\aleph_{\alpha+1}]^2_{\aleph_{\alpha+1}}.$$

Hajnal [1997, 365] called Theorem 17 a test result and noted that they later discovered that this "was known to Sierpiński in a different context." Erdős, Hajnal and Rado also proved stronger Theorems 17a, 18 [1965, 145–146].

In the proceedings of the 1967 UCLA set theory conference, Solovay [1971] proved a result with a square bracket consequence: every stationary subset of a regular cardinal κ can be split into κ many pairwise disjoint stationary sets. Thus if $\kappa > \aleph_0$ is regular then a witness that $\kappa \not\to [\kappa]^\omega_\kappa$ can be obtained by taking a partition $S = \bigcup\{S_\alpha \mid \alpha < \kappa\}$ of a stationary set S into disjoint stationary sets, and defining the coloring $f : [\kappa]^\omega \to \kappa$ by letting $f(s)$ be the unique α such that $\sup(s) \in S_\alpha$.[315]

Erdős and Hajnal asked, in Problem 15 of their problem paper [1971b], if one can prove without the Continuum Hypothesis that $2^{\aleph_0} \to [\aleph_1]^2_3$. Blass [1972][316] and Laver[317] independently proved, without using CH, that $\aleph_1 \not\to [\aleph_1]^2_3$. The Blass counter-example, in essence, superposed three linear orders of size $\aleph_1$: a well-ordering of length ω_1, the lexicographic ordering of the elements of an Aronszajn tree, $(T, <_{\text{lex}})$, and the usual ordering $<$ of the reals on a subset A of size $\aleph_1$ chosen such that $(A, <)$ is order-isomorphic to $(A, >)$ and such that every subset $Y \subseteq A$ has a countable subset $X \subseteq Y$ for which $X \cap [a, b) \neq \emptyset$ for all $a < b$ from Y. Blass noted that subsequently Galvin and Shelah [1973] improved the result to

[314]See §6.7 for the definition of the basic square bracket partition relation.

[315]This idea is used in Theorem 2 of the paper [1976, 370] by Galvin and Prikry who attribute it to Solovay and list it as unpublished.

[316]The paper was received April 25, 1971, and in revised form, February 25, 1972.

[317]Laver proved this partition relation in 1969 but did not publish it; it is attributed to him by Dow in his review [1989, 637] of several articles of Todorcevic.

$\aleph_1 \nrightarrow [\aleph_1]^2_4$, using the same partition. He also generalized the result to show that if κ is regular and accessible and there is a κ-Aronszajn tree, then $\kappa \nrightarrow [\kappa]^2_3$.

Recall that for a regular cardinal κ, a κ-*Suslin tree* is a tree of height κ all of whose chains and antichains have cardinality $< \kappa$. Jensen [1972, Lemma 6.6, 296] reported that Soare (corrected to Shore in the erratum) proved that if κ is a regular uncountable cardinal and there is a κ-Suslin tree, then $\kappa \nrightarrow (\kappa)^2_{r,\gamma}$ for $r < \gamma < \kappa$, and attributed the case $r = 2$ and $\gamma = 3$ to Martin. Erdős and Hajnal [1974, 274] reported that Richard Shore [1974] proved that if κ is regular and there is a κ-Suslin tree, then $\kappa \nrightarrow [\kappa]^2_\kappa$.[318] Since Jensen showed that if $V = L$, then for regular uncountable cardinals κ, there is a κ-Suslin tree if and only if κ is not weakly compact, Shore's result gives a characterization, under the assumption of $V = L$, of the regular uncountable cardinals for which $\kappa \nrightarrow (\kappa)^2_{r,\gamma}$ holds for $r < \gamma < \kappa$. Erdős and Hajnal [1974, 274] also reported that Shore proved that $\omega_2 \nrightarrow [\omega_2]^3_{\omega_1}$ holds in L.

Baumgartner [1975b] proved that for regular κ, if $\kappa^2 \rightarrow (\kappa^2, 3)^2$, then there is no κ-Suslin tree.

Galvin and Prikry [1976] gave a short proof of the equivalence of a cardinal κ being Jónsson and the square bracket partition relation $\kappa \rightarrow [\kappa]^{<\omega}_\kappa$, updating the result of Erdős and Hajnal [1966].

Galvin [1971] (see also [Erdős and Hajnal, 1974, 275]) proved that if $\varphi \geq \omega^\omega$ is an order type which embeds neither ω_1 nor the order type η of the rationals, then

$$\varphi \nrightarrow [\omega, \omega^2, \omega^2, \omega^3, \omega^3, \dots]^2_\omega.$$

All the order types in question are scattered, so Galvin used Hausdorff's result that the scattered order types are the closure of $\{0, 1\}$ under well-ordered and converse well-ordered sums. Galvin came up with an especially nice representation in terms of sets of finite sequences of non-negative integers and negative ordinals under the lexicographic order, with an additional property of being *good*, which will not be defined here. Then he used the coloring that is now known as the *oscillation map*: it takes a pair of sequences $\vec{a} = \langle a_0, a_1, \dots, a_{k-1}\rangle$, $\vec{b} = \langle b_0, b_1, \dots, b_{\ell-1}\rangle\}$, to the number of alternations between blocks of elements from $\{a_i : i < k\} \setminus \{b_j : j < \ell\}$ and blocks of elements from $\{b_j : j < \ell\} \setminus \{a_i : i < k\}$ as they occur in the list in increasing order of the union of these two sets. He proved that for finite positive n and k with $1 \leq k \leq 2n - 1$, any good subset of type ω^n contains a pair with k alternations. As reported in [Erdős and Hajnal, 1974, 275], the above partition relation negatively solved Problem 18 [Erdős and Hajnal, 1971b, 26], which asked if $\omega^\omega \rightarrow [\omega^\omega]^2_{\aleph_0}$ holds. Given that both η and ω_1 embed all countable ordinals, Galvin's result displayed above is optimal.

The discussion of square bracket relations is continued in the next subsection where we turn our attention to structural partitions, where the underlying set is

[318]In the book [Erdős *et al.*, 1984, 318], Jensen [1972] and Shore [1974] are both given credit for $\kappa \nrightarrow [\kappa]^2_\kappa$. Shore [1974] also credits Martin for the proof that for regular κ, $\kappa \rightarrow [\kappa]^2_3$ implies Suslin's Hypothesis, and in a footnote remarks that after Martin's discovery, both Jensen and Silver independently discovered the implication as well as the implication due to Shore that $\kappa \rightarrow [\kappa]^2_\kappa$ implies Suslin's Hypothesis.

either a tree or the set of rational numbers or real numbers, which have natural tree representations.

7.4 Ramsey theory for trees

While it was known in the 1960's that every real in a model obtained by adding a single Sacks real either contains or is disjoint from a ground model real, Baumgartner (cf. [Laver, 1984]) used HL_d for finite d (see §6.1) to show that the same is true when d side-by-side Sacks reals are added. He asked whether a theorem like HL_d is true for a product of infinitely many trees, because he was able to get the forcing consequence from such a statement.

In the late 1970's, Harrington (unpublished, see [Laver, 1984, 386] for the timing) gave an elegant proof of the Laver HL_d Core using forcing. Todorcevic [1995, 54–55] used Harrington's proof in showing that the appropriate forcing statement is equivalent to the Laver HL_d Core.

In his thesis [1979a], Denis Devlin generalized Galvin's result $\eta \to [\eta]^2_{<\omega,2}$ to the result $\eta \to [\eta]^r_{<\omega,t_r}$ and $\eta \not\to [\eta]^r_{t_r}$, where $\langle t_n \mid n \in \omega\rangle$ is the sequence of tangent numbers, $t_n = \tan^{(2n-1)}(0)$, which starts $t_1 = 1$, $t_2 = 2$, $t_3 = 16$, $t_4 = 272$.[319]

Keith Milliken[320] [1975b], [1979] developed a very useful tool in partition theory for countable structures in his Ramsey theorem for strongly embedded subtrees. Suppose that $(T, <_T)$ is a rooted, finitely branching tree whose levels have order type α and $S \subseteq T$ is such that $(S, <_S)$ is a rooted tree where $<_S$ is the restriction of $<_T$ to S. Let β be the order type of the levels of S. Then S is *strongly embedded in T* if there is an order-preserving embedding $\ell : \beta \to \alpha$ such that (1) if $s \in S$ is on level γ in S, then s is on level $\ell(\gamma)$ in T; and (2) if u is an immediate successor of $s \in S$ when s is regarded as a node of T, then there is a v, an immediate successor in S of s, such that $u = v$ or $u <_T v$. For (uniformly) d-branching trees of limit ordinal height α, this is equivalent to the existence of a level preserving map from the d-branching tree all of whose branches have order type γ.

> *Milliken's Ramsey Theorem for Trees:* If $(T, <_T)$ is a finitely branching tree of height ω and, for some $1 \le k < \omega$, the height k strongly embedded subtrees of T are partitioned into finitely many pieces, then there is a strongly embedded tree $S \subseteq T$ all of whose strongly embedded height k subtrees are in the same cell of the partition.

Milliken's proof used a pigeonhole principle which he described as a reformulation of the Halpern-Läuchli Theorem due to Laver in about 1969. He included a

[319] A sketch by Todorcevic of the proof of Devlin's result with some details for $n = 2$, $n = 3$ appeared in [1995, 45-47]. Vojkan Vuksanovič [2002] published a beautiful proof of it using Milliken's Ramsey Theorem for Trees (discussed later in the section). Devlin's result was generalized by Džamonja, Larson and Mitchell [Džamonja *et al.*, 2009] to κ-dense linear orders for suitably chosen uncountable cardinals κ. For $n > 2$, the critical numbers r_n for these κ-dense linear orders are larger than those for η.

[320] Keith Milliken received his doctorate from the University of California, Los Angeles, in 1975 where he was advised by Chang.

proof of it from the Halpern-Läuchli Theorem.

> *Milliken Pigeonhole Principle:* Suppose that d and r are positive integers and $\langle T_i : i \in d\rangle$ is a sequence of finitely branching trees each branch of which has order type ω. If
>
> $$\bigcup_{n\in\omega}\left(\prod_{i\in d} T_i(n)\right) = \bigcup_{j<r} C_j,$$
>
> then there must be a $k \in r$ and a sequence $\langle S_i : i \in d\rangle$ of subtrees with S_i strongly embedded in T_i for $i < d$ for which
>
> $$\bigcup_{n\in\omega}\left(\prod_{i\in d} S_i(n)\right) \subseteq C_k.$$

The Laver HL_d Theorem (see §6) is about perfect subtrees rather than strongly embedded trees. Both versions follow from the Laver HL_d Core, and, if starting from perfect trees, one can obtain perfect strongly embedded trees as well.

Blass [1981] used the Halpern-Läuchli Theorem to verify a conjecture of Galvin [1968b] by proving the following theorem (due to Galvin for $n \le 3$):[321]

> *Blass Perfect Set Partition Theorem:* If P is a perfect subset of Cantor space and $[P]^n$ is partitioned into finitely many open pieces, then there is a perfect set $Q \subseteq P$ such that $[Q]^n$ intersects at most $(n-1)!$ pieces.

Blass noted that the result generalizes to partitions of perfect subsets of the real line. He fit his result into a wider context by pointing out that Mycielski [1964], [1967] had shown that any meager set or any set of measure zero in $[\mathbb{R}]^n$ is disjoint from $[P]^n$ for some perfect set $P \subseteq \mathbb{R}$ and derived the corollary that if $[\mathbb{R}]^n$ is partitioned into finitely many pieces that all have the Baire property or are all measurable, then there is a perfect set $Q \subseteq \mathbb{R}$ such that $[Q]^n$ meets at most $(n-1)!$ pieces. Blass noted that some limitation on the partitions is necessary by the Galvin and Shelah result [1973] that there is a partition of the pairs of real numbers into ω many pieces such that the set of pairs of every subset isomorphic to $\mathbb{R}$ meets all the pieces (in the partition calculus notation, $\lambda \not\to [\lambda]^2_\omega$, where λ is the order type of the set of real numbers).

7.5 Hindman's Finite Sums Theorem

A Ramsey theorem for finite sums of a finite set was independently proved in the late 1960's by Jon Folkman (unpublished, see [Graham and Rothschild, 1971]), Rado [1970], and Jon Sanders [1968] which I will call the FRS Finite Sums Theorem, but in the textbook *Ramsey Theory* [1990] whose statement of the theorem I paraphrase, it is called Folkman's Theorem.

---

[321] Blass presented his perfect set theorem at a meeting on combinatorial set theory in Aachen in June 1976.

> *FRS Finite Sums Theorem:* If $\mathbb{N}$ is finitely colored, then there are arbitrarily large finite sets S such that the set of all non-repeating sums from S is monochromatic.[322]

This theorem is a corollary to Rado's Theorem for Linear Systems with Positive Integer Coefficients discussed in §3.2. It is also a corollary of the Graham-Rothschild[323] Parameter Sets Theorem [1971] for which I give a simplified version rephrased from [2006].

> *Graham-Rothschild Parameter Sets Theorem:* Suppose A is a non-empty finite alphabet. For all $m < k < \omega$ and positive r, there is a positive n such that whenever $[A]\binom{n}{m}$ is colored with r colors, there is a word $w \in [A]\binom{n}{k}$ such that $\{w\langle u\rangle : u \in [A]\binom{k}{m}\}$ is monochromatic.

Here, for $k = 0$, $[A]\binom{n}{k}$ is simply ${}^{n}A$ and, for $k > 0$, $[A]\binom{n}{k}$ is the set of all words w of length n over the enlarged alphabet $A \cup \{v_0, v_1, \ldots, v_{k-1}\}$ such that each variable v_i for $i < k$ occurs in w and the first occurrence of v_i precedes the first occurrence of v_{i+1} if $i + 1 < k$. Given a word w in $[A]\binom{n}{k}$ and a word u of length k on the alphabet $A \cup \{v_0, \ldots, v_k\}$, let $w\langle u\rangle$ be the word of length n such that for all $i < n$, $w\langle u\rangle(i) = w(i)$ if $w(i) \in A$, and $w\langle u\rangle(i) = u(j)$ if $w(i) = v_j$. Prömel and Voight [1990] in their discussion of parameter sets point out that the formal calculus of words, which we used in the above presentation, was introduced by Klaus Leeb [1973].

Graham and Rothschild [1971] gave many corollaries of their theorem including Schur's Theorem, Rado's Theorem for Linear Systems with Positive Integer Coefficients, Ramsey's Theorem, van der Waerden's Theorem, the Hales-Jewett Theorem, the FRS Finite Sums Theorem, and various theorems for affine and vector spaces. Graham and Rothschild were motivated by a problem of Rota about vector spaces and in a paper with Leeb [1972] they solve Rota's problem. In a survey of developments in Ramsey theory, Graham and Rothschild [1974] attribute to Leeb the notion of a category having the Ramsey property, extending the range of structures to which Ramsey theory is applied.

This brief account was informed by Hindman's account [2006] and the paper of Prömel and Voigt [1990]. For extensions and other perspectives, Todorcevic's book [2010] and the paper [2006] by Carlson, Hindman and Strauss are recommended.

At the end of their paper [1971, 290–291], Graham and Rothschild asked (see §9 Concluding Remark (ii)) if the various infinite versions of certain of their corollaries

[322]For $S = \{1, 2, 5\}$, the non-repeating sums are $1, 2, 5, 1 + 2, 1 + 5, 2 + 5, 1 + 2 + 5$.

[323]Ronald Lewis Graham (October 31, 1935 –) received his 1962 doctorate from the University of California, Berkeley, where his advisor was Derrick Lehmer. He had nearly 30 joint papers with Erdős. In addition to being a mathematician of broad interests, he is juggler and a person who learned to flip a quarter so that it comes up heads on demand. There is a comprehensive archive of his papers at http://www.math.ucsd.edu/~ronspubs/ which also includes copies of some articles about him, including [Kolata, 1981] which was my source for his birth date.

Bruce Lee Rothschild (1941 –) received his doctorate from Yale University in 1967 where his advisor was Öystein Ore. He is a professor at the University of California, Los Angeles. To learn more about his mathematics, see [Hindman, 2006].

are valid, and specifically asked if the positive integers are 2-colored, does there always exist an infinite subset A such that the sums of all (non-empty) finite subsets of A are the same color? Hindman proved the infinite version and the remainder of this section is focused on Hindman's solution.

> *Hindman's Finite Sums Theorem:* For any positive integer k and any partition of the natural numbers into k classes, $\mathbb{N} = \bigcup_{i<k} A_i$, there are $i < k$ and an infinite subset $X \subseteq N$ such that the sum of every finite set of distinct members of X is in A_i.

Hindman gave a delightful account in [2005] of how he came to prove his theorem.[324] He reported that around 1971, Galvin asked Erdős whether there were any (downward) almost translation invariant ultrafilters on $\mathbb{N}$. Erdős asked the question of W. Wistar Comfort, who had been Hindman's dissertation advisor,[325] and Comfort passed the question on to Hindman. Along the way, (downward) almost translation invariant ultrafilter became (upward) almost translation invariant, a simpler problem, which Hindman tackled when he heard that Erdős would be visiting the Claremont Colleges at a time when Hindman was teaching at Cal State Long Beach. Hindman learned from Erdős that the question originated with Galvin, so Hindman telephoned Galvin to share the proof. Galvin congratulated him and let him know that he had answered a question different from the one originally asked. The point of the original question was that a positive answer would provide a simple proof of the truth of a conjecture of Graham and Rothschild which is now Hindman's Finite Sums Theorem. Hindman then proved that under GCH, the Graham-Rothschild conjecture implied the existence of a (downward) almost translation invariant ultrafilter, and "with a great deal more effort" [2005, 322] he established the validity of the conjecture by "an elementary but very complicated proof" [2005, 322] (see [Hindman, 1974] for Hindman's original proof of the finite sums theorem). Hindman continues [2005, 323]:

> At ... the end of 1972 ... Galvin's (downward) almost translation invariant ultrafilters were figments of the continuum hypothesis. Galvin wanted to know if they really existed, that is, whether their existence could be proved in ZFC. In 1975 he [Galvin] encountered Steven Glazer and asked him whether such ultrafilters could be shown to exist. Glazer immediately answered "yes".

Glazer went on to explain that (1) any compact right topological semi-group has idempotents; (2) $\beta\mathbb{N}$[326] has a natural operation $+$ extending addition on $\mathbb{N}$ which makes $(\beta\mathbb{N}, +)$ a right topological group; and (3) for $p, q \in \beta\mathbb{N}$ and $A \subseteq \mathbb{N}$, $A \in p + q$ if and only if $\{x \in \mathbb{N} \mid -x + A \in q\} \in p$. Thus an idempotent in $\beta\mathbb{N}$ is

[324] The account is part of a printed lecture presented at the international meeting of the Japanese Association of Mathematical Science on the receipt of its 2003 International Prize.

[325] Comfort reminisced about Hindman's graduate student days and their times together over the years in [Comfort, 2005].

[326] $\beta\mathbb{N}$ is the Stone-Čech compactification of $\mathbb{N}$, identifiable with the space of ultrafilters on $\mathbb{N}$.

a (downward) almost translation invariant ultrafilter. Galvin told Hindman this proof in fall 1975, which Hindman identified in [2005] as the time he fell in love with the algebraic structure of the Stone-Čech compactification of a discrete semigroup, whose structure and applications to Ramsey theory have been the focus of the remainder of his mathematical career.

Hindman's Finite Sums Theorem received multiple proofs. In addition to the original proof and the Galvin-Glazer proof (see [Comfort, 1977]) described above, Baumgartner [1974] published a short proof of Hindman's Finite Sums Theorem in the same journal and the same year as Hindman's original proof, and proved an equivalent version in terms of finite sets and their unions: if $f : [\mathbb{N}]^{<\omega} \to k$ is partition into finitely many parts, there is an $i < k$ and an infinite family S of disjoint non-empty finite subsets so that all finite unions of members of S receive color i. This theorem is sometimes call the *Finite Unions Theorem*.

7.6 Infinitary partition relations

We start this subsection with results on infinitary partition relations by Kleinberg and Kunen and those partition relations growing out of the work of Nash-Williams, Galvin-Prikry, Silver and Mathias.

Eugene Kleinberg[327] studied partition relations for infinite exponents, particularly those of the form $\kappa \to (\kappa)^\alpha$ for κ a cardinal and α an ordinal in his thesis (see [1970] based on his thesis). By a result of Kunen quoted in Kleinberg's paper in a note apparently added in proof, the Axiom of Determinacy (AD) implies that $\aleph_1 \to (\aleph_1)^\alpha$ for all $\alpha < \aleph_1$. Moreover Kleinberg proved that if λ is a regular cardinal smaller than κ and $\kappa \to (\kappa)^{\lambda+\lambda}$, then κ has a normal ultrafilter concentrated on ordinals of cofinality λ. The ultrafilter in this case is generated by the subsets of κ that are unbounded and closed under sups of increasing λ-sequences. Kleinberg went on to investigate the conditions under which infinite-exponent partitions admit homogeneous sets of measure one for these measures.

Kleinberg wrote a monograph [1977] on infinitary combinatorics in the theory ZF + AD + DC with an emphasis on infinite exponent partition relations, the measures these generate, and ultraproducts by such measures. For more on the story of infinitary partitions see the chapter by Paul Larson.

Erik Ellentuck [1974] gave a topological proof of Silver's result that analytic sets are completely Ramsey, a result that he incorrectly attributed to both Mathias and Silver,[328] by characterizing completely Ramsey sets as those having the Baire property with respect to a certain topology on $[\omega]^\omega$, now called the *Ellentuck topology*[329] or the *exponential topology*, and is also known as the *Vietoris topol-*

[327]Kleinberg was a student of Martin and graduated from Rockefeller University in 1969. He was on the faculty of Massachusetts Institute of Technology, and then of the State University of New York, Buffalo, where he is now emeritus.

[328]Recall Mathias [1977] gave new proofs and generalized Silver's work.

[329]A neighborhood (s, A) in the Ellentuck topology for finite part $s \subseteq \omega$ and infinite part $A \subseteq \omega$ with $\max s < \min A$ is the collection of infinite subsets $X \subseteq \omega$ with the property that $s \subseteq X \subseteq s \cup A$.

ogy (cf. Todorcevic [2010]). Ellentuck characterized the Silver and Mathias proofs as using "relatively deep metamathematical notions involving forcing," while he characterized his own proof as "using nothing more than the methods of Galvin-Prikry." The paper includes a postscript in which the observation of Galvin that the proof actually shows that all sets in the classical Lusin hierarchy are completely Ramsey. Independently, Louveau [1974] gave a similar topological proof of the theorems of Silver and Mathias; Louveau used a Ramsey ultrafilter to show analytic sets are Ramsey.

Milliken [1975a] proved a theorem that generalized both Silver's analytic sets are completely Ramsey theorem and Hindman's Finite Sums Theorem. He also gave a short proof of the theorem of Mathias that if $\mathcal{U}$ is a selective ultrafilter on ω and $R \subseteq [\omega]^\omega$, then R is completely $\mathcal{U}$-Ramsey. Taylor [1975], [1976] independently generalized Hindman's Finite Sums Theorem in the Finite Unions Theorem form and found canonical equivalence relations for the "same color" equivalence relation. Here is a modern statement of what has come to be known as the Milliken-Taylor Theorem. For finite sets $A, B \subseteq \omega$, let us write $A \ll B$ if and only if $\max A < \min B$.

> *Milliken-Taylor Theorem:* For any finite partition $[\mathbb{N}]^r = C_0 \cup \cdots \cup C_{\ell-1}$, there is an infinite sequence $\langle a_n : n < \omega \rangle$ of natural numbers such that for some $i < \ell$,
>
> $$\left\{ \left\{ \sum_{i \in I_0} a_t, \ldots, \sum_{i \in I_{r-1}} a_i \right\} : I_0 \ll \cdots \ll I_{r-1} \in [\mathbb{N}]^{<\omega} \right\} \subseteq C_i.$$

Jacques Stern [1978] proved an extension of Silver's result to colorings of the infinite chains of a complete binary tree and applied his result to a problem of Brunel and Sucheston for real Banach spaces.

Hillel Furstenberg and Benjamin Weiss [1978] applied dynamical theorems to product spaces of countably many copies of a finite set to give new proofs of a series of combinatorial results including Rado's generalization of van der Waerden's Theorem and Hindman's Finite Sums Theorem.

Taylor [1978] investigated partitions of the pairs from the Cantor space ${}^\omega 2$. He ascribed to Galvin, Mycielski and Silver the statement that every partition of $[{}^\omega 2]^2$ into two pieces having the property of Baire has a homogeneous perfect set. He had learned about it from Baumgartner who proved it independently from Galvin, Mycielski and Silver, and Taylor noted that Burgess rediscovered it still later. Taylor proved that if $f : [{}^\omega 2]^2 \to \omega$ is a coloring for which $f^{-1}\{i\}$ has the property of Baire for all $i < \omega$, then there is a perfect set $P \subseteq {}^\omega 2$ such that either f is constant on $[P]^2$ or else f induces the same equivalence relation on $[P]^2$ as the discrepancy partition $\delta : [{}^\omega 2]^2 \to \omega$ defined by $\delta(\vec{x}, \vec{y}) = \min\{n : \vec{x}(n) \neq \vec{y}(n)\}$.

7.7 Structure of trees

The structure of trees of height ω_1 with countable levels and their embeddings continued to be an area of intense interest in the 1970's.

Embedding trees in the reals

Baumgartner, Jerome Malitz, and William Reinhardt[330] [1970] carried out an iterated c.c.c. forcing with finite conditions to prove the consistency of the statement that every Aronszajn tree and every tree-like order of size $\aleph_1$ is embeddable in the rationals, i.e. is $\mathbb{Q}$-embeddable. Baumgartner, Malitz and Reinhardt had shown in particular that MA + $\neg$CH implies that all Aronszajn trees are special.[331]

While they cited the paper [1937a] in which Kurepa constructed a special Aronszajn tree as an example of a paper on embeddings of trees into linear orders, they seemed unaware that he had shown the equivalence between being special and having an embedding into the rationals (see §3.4 for details). Most of the work for their consistency result is in Theorem 3 in which they show that their forcing satisfies the countable chain condition. In Theorem 4 they show under the assumption of MA + $\neg$CH that every tree-like partial order[332] T of cardinality less than the continuum in which every chain is countable is $\mathbb{Q}$-embeddable. The paper included announcements of a result by Laver of the existence of a tree of power the continuum which is $\mathbb{R}$-embeddable but not $\mathbb{Q}$-embeddable such that each uncountable subset contains an uncountable antichain, and a result by Galvin of the existence of such a tree which is not $\mathbb{R}$-embeddable (Baumgartner's thesis is given as a reference). Baumgartner, Malitz and Reinhardt pointed out that these examples show the independence of their result over ZFC, since in ZFC + $\neg$CH, it is not the case that every tree-like partial order without an uncountable chain is $\mathbb{Q}$-embeddable. Laver's example showed that one cannot hope to prove the consistency of the parallel embedding result for tree-like orders of size the continuum, and the Sierpiński poset (see §3.4) showed that one cannot hope to extend the embedding consistency result to arbitrary partial orders. The paper ended with a trio of open problems [1970, 1753]:

1. Is it consistent with ZFC to assume that for any partial order P, if card $P = \aleph_1$ and every uncountable subset of P contains an uncountable antichain, then P can be embedded in the rationals?

2. (Galvin, unpublished) Is it consistent with ZFC to assume that for any partial order P, if card $P = \aleph_2$ and if every $P' \subseteq P$ with card $P' \leq \aleph_1$ can be embedded in the rationals, then P can be embedded in the rationals?

3. (Baumgartner) Is it consistent with ZFC + $2^{\aleph_0} = \aleph_1$ to assume that every Aronszajn tree is embeddable in the rationals?

[330] Baumgartner, Malitz and Reinhardt were all students of Vaught, and Malitz and Reinhardt both had careers at the University of Colorado, Boulder.

[331] M. E. Rudin [1975b, 513] credited Kunen with having shown that MA + $\neg$CH implies that all Aronszajn trees are embeddable in the rationals, and the same is true for tree-like partial orders. This result is credited to Baumgartner in [Devlin, 1972]. In [Devlin and Shelah, 1978, p. 289], both Baumgartner and Kunen are given credit for proving that under MA + $\neg$CH, every Aronszajn tree is embeddable in the rationals.

[332] A partial order T is *tree-like* if the set of predecessors of each element is linearly ordered.

For future reference, we note that one can easily use their forcing to arrange for every Aronszajn tree to have a continuous strictly increasing embedding into the rationals.

Baumgartner [1970] announced two structural theorems (paraphrased below). The first theorem is that under MA + $\neg$CH, for every pair of full well-pruned Hausdorff trees $(S, \leq_S)$ and $(T, \leq_T)$ of power $\aleph_1$, there are countable sequences $\langle S_n \mid n \in \omega\rangle$ and $\langle T_n \mid n \in \omega\rangle$ of cofinal subtrees whose unions give back the original trees and which are term-by-term isomorphic. The second theorem is that if $V = L[A]$ for some $A \subseteq \omega_1$, then there are non-Suslin Aronszajn trees (a) which are $\mathbb{R}$-embeddable but not $\mathbb{Q}$-embeddable and (b) which are not $\mathbb{R}$-embeddable.[333]

Devlin, then a student at University of Bristol, heard about Baumgartner's results from Laver, who visited Bristol 1969-1971. After stating Theorem 1 with a review of some known results, in Theorem 2 [1972, 256], Devlin characterized what he called ω_1-trees (but we call full well-pruned Hausdorff ω_1-trees) that have strictly increasing embeddings into the rationals as the Aronszajn trees that are the union of countably many antichains, i.e. special Aronszajn trees. This result is a reiteration of one from [Kurepa, 1940a]. In the same theorem, he proved that if a full well-pruned Hausdorff ω_1-tree is $\mathbb{R}$-embeddable, then every uncountable subset contains an uncountable antichain. The proof used the fact that the union of the successor levels of any $\mathbb{R}$-embeddable tree is $\mathbb{Q}$-embeddable, which was also observed by Kurepa [1940a]. Devlin generalized both Baumgartner's result on the consistent existence of $\mathbb{R}$-embeddable non-$\mathbb{Q}$-embeddable Aronszajn trees and the Gaifman-Specker [1964] result that under CH there are $2^{\aleph_1}$ many non-isomorphic Aronszajn trees by proving that under $\diamondsuit$ there are $2^{\aleph_1}$ non-isomorphic $\mathbb{R}$-embeddable non-$\mathbb{Q}$-embeddable Aronszajn trees. Devlin (see pages 256–257) described his own approach as an adaptation of Baumgartner's argument "at the cost of some messy combinatorics" under the inspiration of arguments from Jech's paper [1972].

Jech [1972] investigated the automorphism group of ω_1-trees and used $\diamondsuit$ to proved the existence of $2^{\aleph_1}$ pairwise non-isomorphic Suslin trees. He also proved that it was consistent to assume the existence of $2^{\aleph_1}$ many pairwise non-isomorphic Kurepa trees.

Jensen on Suslin trees

Devlin and Håvard Johnsbråten[334] wrote on Jensen's work on Suslin's Problem in their [1974]. In the introduction, the authors noted that the first five chapters were based on a set of lecture notes written by Jensen in Kiel in 1969. The remaining five chapters were based on a set of notes written by Devlin in 1972, which were based on a set of notes written by Jensen some years earlier.

[333] Kunen [1980, 90] cited this abstract in his remark that $\diamondsuit$ implies that there is an Aronszajn subtree of the tree of one-to-one functions from countable ordinals into ω which is not special.

[334] Johnsbråten was a student of Fensted and studied with Jensen but did not complete his doctoral degree. He is now on the faculty of Telemark University College, in Norway.

The first half details Suslin's Problem, the connection with trees, and Jensen's $\diamondsuit$ and gives the construction of a Suslin tree first in L using the canonical well-ordering and then from $\diamondsuit$. Then $\diamondsuit$ is shown consistent both via forcing and via holding in the constructible universe. Various consistency results are proved assuming the consistency of ZF, including ZFC + GCH + $\neg$SH and ZFC + $\neg$CH + $\neg$SH.

In Chapters 4 and 5, the possible structure of Suslin trees was explored. Under $\diamondsuit$ there is a homogeneous Suslin tree, there is one with exactly ω_1 automorphisms, there is a non-reversible (i.e. not isomorphic to its converse) homogeneous Suslin tree with exactly ω_1 automorphisms, a reversible (i.e. isomorphic to its converse) homogeneous Suslin tree with exactly ω_1 automorphisms, a rigid[335] Suslin tree, and a rigid Suslin tree which is reversible. Jensen's proof that from $\diamondsuit^+$ there is a Kurepa tree is sketched.

In Chapter 6,[336] work of Solovay and Tennenbaum on the consistency of Suslin's hypothesis is given, along with an introduction to Martin's Axiom.

The remainder of the book is devoted to the remarkable proof of the consistency of Suslin's Hypothesis with CH. Jensen used a method of iterated forcing with conditions which were not countably closed but did not collapse cardinals and did not add reals. Todorcevic (cf. [Kurepa, 1996, 8]) called this result "one of the deepest applications of the method of forcing since its original invention".

Mix and match

Early in the decade, Jech [1972] asked about the existence of a rigid Aronszajn tree. We discussed the results of Jensen on rigid Suslin trees above. At the end of the decade, Uri Abraham[337] published his ZFC construction of a rigid Aronszajn tree[338] using non-isomorphic Aronszajn trees constructed by Gaifman and Specker [1964]. The same year, motivated by a question in Kurepa's thesis [1935], Todorcevic published his construction of a rigid Aronszajn tree.

By the end of the decade, the consistency of combinations of existence or non-existence of Suslin trees, existence or non-existence of Kurepa trees, together with either CH or MA + $\neg$CH were established.

Devlin had asked whether $V = L$ implies the existence of such a tree in 1969 (see [1983]). In 1971 Jensen (cf. Theorem 4 in [Devlin, 1974]) proved that if $V = L$, then there exists a Kurepa tree without an Aronszajn subtree. In the summer of 1980, while at the University of Toronto, Devlin [1983] came up with a simpler

[335] A structure is *rigid* if it has no non-trivial automorphism.

[336] Chapter 6 is essentially the only place where results appear that are not due to Jensen

[337] Uri Abraham, whose name is sometimes transliterated Avraham, received his 1979 doctorate [1979b] from Hebrew University where his advisors were Lévy and Shelah. Abraham has a joint appointment split between Computer Science and Mathematics at Ben Gurion University, Be'er-Sheva. His interests in set theory and logic include Boolean algebras, forcing and axiomatics as well as a variety of combinatorial topics. For a sample publication in computer science, see [1999].

[338] Baumgartner [1975c] knew of the existence of such a tree already in 1975, but the work was neither published nor widely publicized.

proof after talking with Fleissner about his work and Kunen's work on the Normal Moore Space Conjecture.

The *weak Kurepa Hypothesis* (wKH) asserts that there is a tree of height ω_1 of cardinality $\aleph_1$ which has more than $\aleph_1$ uncountable branches. Mitchell [1972] proved that the failure of the weak Kurepa Hypothesis is equiconsistent with the existence of an inaccessible cardinal. In his 1978 Master's Thesis, Stevo Todorcevic used techniques of [Mitchell, 1972] and [Devlin, 1978] to construct a model of MA + $\neg$wKH and did so in a way that allowed him to make the continuum any regular cardinal (see [Todorcevic, 1981c]). Todorcevic went on to derive a variety of topological consequences of MA + $\neg$wKH.

In a 63-page paper, Devlin [1978] reviewed known work and noted that models of CH had been constructed (a) with a Suslin tree but no Kurepa tree, (b) with a Kurepa tree but no Suslin tree, (c) with both a Suslin tree and a Kurepa tree, and that models of MA + $\neg$CH for the above possibilities had also been constructed. The two remaining possibilities were simultaneous non-existence of Suslin trees and Kurepa trees, either with CH or with MA + $\neg$CH. As for the latter, Devlin started with Silver's model of no Kurepa trees and iterated over it in such a way so as to produce a model of MA + $\neg$CH, adapting Silver's proof that σ-closed forcing does not add branches to control the addition of branches in this process, so that no Kurepa trees are added. As in any model of MA + $\neg$CH, all Aronszajn trees are special, so there are no Suslin trees. The last half of the paper is devoted to getting a model of CH with neither Suslin nor Kurepa trees, and Devlin used an idea of Jensen's for forcing the consistency of $\diamondsuit^*$ and adapted Jensen's proof of the consistency of no Suslin trees with CH.

Larger trees

Starting with a Mahlo cardinal, William Mitchell[339] [1972] constructed a model of set theory with no special $\aleph_2$-Aronszajn trees. He introduced the term *special* for κ^+-Aronszajn trees which are subtrees of the collection of one-to-one functions into κ whose domain is an ordinal $< \kappa^+$ ordered by end-extension. Mitchell's definition differs from our use of *special* for κ^+-trees that are the union of κ many antichains, which was introduced for $\kappa = \omega$ in §3.4 in the discussion of Kurepa's construction of a special Aronszajn tree.[340]

Mitchell [1972] proved that the non-existence of special $\aleph_2$-Aronszajn trees is equiconsistent with the existence of a Mahlo cardinal. He used a result, independently due to Silver and Rowbottom, which asserted the existence of a sentence

[339] William J. Mitchell (December 30, 1943 –) received his doctorate from the University of California, Berkeley, where Silver was his advisor. In [Mitchell, 1972, 22], where he published revised versions of his thesis work, he also thanked Baumgartner and Solovay for "helpful discussions and suggestions."

[340] Mitchell's definition of special recalls the original Aronszajn construction of a subtree of the one-to-one functions into ω. Such trees need not be special by the modern definition. Our definition (also used by Kunen [1980]) is based on Kurepa's initial construction of a special Aronszajn tree; the equivalence of being an ω_1-tree which is $\mathbb{Q}$-embeddable is used by Jech [2003].

which has a model of type (κ^+, κ) if and only if there is a special κ^+-Aronszajn tree. Since κ^+-Aronszajn trees are known to exist for regular κ with $2^{<\kappa} = \kappa$, it follows that any model of ZFC which has no special κ^+-Aronszajn tree fails to satisfy the cardinal transfer property $(\aleph_1, \aleph_0) \to (\kappa^+, \kappa)$. Silver extended Mitchell's result by showing that for regular κ, there are no κ^+-Aronszajn trees if and only if κ is weakly compact in L. Mitchell's construction involved a mixed iteration of length κ adding Cohen reals, so that eventually $2^{\aleph_0} = \kappa$, and collapsing cardinals below κ to $\aleph_1$, so that eventually $\kappa = \aleph_2$, but κ retains enough of its weak compactness to guarantee that every tree on ω_2 of height ω_2 with all levels of size $\leq \aleph_1$ has a branch. For a model created in the presence of a weakly compact cardinal with iterated Sacks forcing and with no ω_2-Aronszajn trees, see [Baumgartner and Laver, 1979].

In his [1976], Gregory combined work of Jensen and Solovay with Gregory's Theorem discussed in §7.1 to construct higher Suslin trees. After stating Theorem 1.3 attributed to Solovay, Gregory included commentary which concluded with the assertion of the existence of a non-reflecting stationary subset of κ from the existence of a $\Box_\kappa$ sequence. The commentary was followed with a proof of the result of Solovay "given in place of a reference." Gregory stated a multipart Theorem 1.10, attributed to Jensen,[341] which among other things says that if $\kappa^{<\kappa} = \kappa$ and there is a non-reflecting stationary set $E \subseteq \kappa^+$ for which $\diamondsuit(E)$ holds, then there is a κ^+-Suslin tree and it may be constructed with additional properties as well. Specifically, Gregory constructed an $\aleph_2$-Suslin tree using a non-reflecting stationary set consisting of ordinals of cofinality ω.

In his doctoral thesis [1979a], [1981d, Theorem 4.1, 240], Todorcevic[342] used CH + $\Box$ to construct an ω_2-Aronszajn tree with no ω_1-Aronszajn subtrees and no Cantor subtrees.[343] Devlin asked if CH or GCH was sufficient. Todorcevic showed the answer was no by proving the consistency, relative to the existence of a weakly compact cardinal, of GCH holds and every ω_2-Aronszajn tree contains a complete binary ω_1-tree. He also showed that the proof of Theorem 4.1 generalized to show that under the assumption of $\Box_\kappa$, there is a κ^+-Aronszajn tree with no λ-Aronszajn subtrees for every $\lambda \neq \kappa$ and no ν-Cantor trees for every ν. Moreover, in Theorem 5.5 [1981d, 249] he showed the consistency, via forcing over a model of $V = L$, that for every regular uncountable non-weakly compact cardinal κ, there is a κ-Suslin tree and there is a κ-Kurepa tree with no λ-Aronszajn subtrees for any λ and no ν-Cantor subtrees for any ν. Istvan Juhász and William Weiss [1978] had shown that a question of Sikorski[344] [1950] of the existence of a κ-metrizable κ-

[341]See the top of page 667 of [Gregory, 1976] for the attribution of Theorem 1.10 to Jensen.

[342]Stevo B. Todorcevic earned his doctoral degree in 1979 at the University of Belgrade with Kurepa as advisor and K. Devlin as outside reader. Devlin attended the defense; his encouragement led Todorcevic to visit Jerusalem where he sat in on Shelah's lectures on forcing.

[343]A tree is a κ-Cantor tree if it has height $\lambda + 1$ for some cardinal $\lambda \leq \kappa$, the first λ levels have size $\leq \kappa$, and the last level has size $> \kappa$.

[344]Roman Sikorski (July 11, 1920 – September 12, 1983) was a student of Mostowski, receiving his doctorate in 1949 and habilitation in 1950 at Warsaw University. He was a professor at Warsaw University 1952–1982, became a corresponding member of the Polish Academy of Sci-

compact space of power μ was equivalent to the existence of a κ-Kurepa tree with μ branches and no κ-Aronszajn subtree. Their result combined with Todorcevic's Theorem 5.5 gives a consistently positive answer to Sikorski's question.

7.8 Linear and quasi-orders

We start with a review of results about wqos and bqos. Dick H. J. de Jongh and Rohit Parikh[345] [1977] proved that the ordinal length of a well-quasi-ordered set[346] can always be realized in a linear extension.

In a series of papers, Laver [1971], [1973], [1976], [1978a] explored the theory of bqos.[347] In an early example, Laver [1971] proved there are $\aleph_1$ many countable scattered linear orders with respect to bi-embeddability.[348] In a later example, Laver [1978a] extended the Nash-Williams [1965a] result that infinite trees form a bqo. Laver [1978a] called a tree $(T, \leq)$ σ-scattered if it is representable as $T = \bigcup_{n<\omega} T_n$ where each T_n is a scattered[349] downwards closed subtree of T, and denoted the class of these trees by $\mathfrak{M}$.[350] He proved that $\mathfrak{M}$ is a bqo under order-preserving and meet preserving maps (the meet of two elements of a tree is the largest element which is less than or equal to both of them). He observed that Aronszajn trees do not form a wqo under $\diamondsuit$ and raised the question of whether it is consistent that Aronszajn trees form a wqo.[351] Galvin had shown that the class of trees of height $\omega + 1$ is not well-quasi-ordered, which led him to ask the of whether σ-scattered trees are a wqo. Laver proved closure properties for bqos, proving that if Q is a bqo, then the collection of Q-labeled $\mathcal{M}$-trees is a bqo.

ences in 1962 and a full member in 1967. He was president of Polish Mathematical Society from 1965–1977. He is known for the Rasiowa-Sikorski Lemma. For more on his life, see [Anonymous, 1984].

345 Dick de Jongh received his Ph.D. at the University of Wisconsin, Madison, with Kleene as advisor, and and Rohit Parikh received his Ph.D. at Harvard University with Hartley Rogers, Jr. and Burton S. Dreben as advisors.

346 The *ordinal length* of a well-quasi-ordered set is the maximum of the order types of its linear extensions.

347 Larson gave a graduate course on Laver's bqo work at UCLA when she was a Hedrick Assistant Professor (1972-1974), and remembers enjoyable conversations with him about bqo theory in the botanical garden.

348 Arnold Beckmann, Martin Goldstern and Norbert Preining [2008] generalize Laver's method to extend this result of Laver by showing that there are only $\aleph_1$ many countable closed linear orders with respect to bi-embeddability with continuous monotone maps and apply their result to show there are only denumerably many Gödel logics. More specifically, they show that the set of countable closed linear orderings is better-quasi-ordered by strictly monotone continuous embeddability, and the result extends to the case in which the orders are labeled. They defined "countable closed linear order" as a countable closed subset of the real line.

349 Recall from §6.2 that a tree T is *scattered* if it does not embed the tree of all finite sequences of zeros and ones. Note that Laver's definition departs from the usual usage of σ as a prefix to denote countable unions taken freely.

350 Todorcevic [2007a] writes $\mathcal{S}$ for the class of scattered trees and $\mathcal{S}_\sigma$ for the class of countable unions of scattered trees, which he calls σ-scattered trees.

351 Todorcevic answered Laver's question negatively in [2007a] in which he showed that there is an infinite, strictly decreasing sequence of Aronszajn trees (he used Lipschitz trees) under the weaker notion where $S \leq T$ if and only if there is a strictly increasing map from S into T.

The theory of wqos and bqos has a wide variety of applications. For example, Dexter Kozen [1988] used one of Laver's wqo results [1976] to establish the finite model property for the μ-propositional calculus. Montalbán [2007] used theorems from [Laver, 1971] in his analysis, mainly from the viewpoints of computability theory and reverse mathematics, of equimorphism types[352] of linear orders.

Laver [1979] examined the question of which linear orders can be embedded in $({}^{\omega}\omega, <^*)$,[353] i.e. the functions with domain and codomain ω under eventual domination, in the absence of the Continuum Hypothesis. CH implies that every linear order of power continuum is embeddable in $({}^{\omega}\omega, <^*)$. Laver pointed out that it was well-known that one can force the continuum to be large and ω_2 not embeddable in $({}^{\omega}\omega, <^*)$. Kunen however noted that MA is consistent with the existence of a linear order of power $2^{\aleph_0}$ which is not embeddable in $({}^{\omega}\omega, <^*)$.[354] Laver [1979, 299] pointed out that "the problem with trying to inductively embed, using Martin's Axiom, any linear ordering of power $2^{\aleph_0}$ into $(\omega)^{\omega}$, is the possibility of creating in the course of construction a 'Hausdorff gap'." In connection with his work with W. Hugh Woodin on homomorphisms of Banach algebras, Solovay had asked if it is consistent that the continuum is large and every linear order of power continuum is embeddable in $({}^{\omega}\omega, <^*)$. Laver answered the question positively by showing that if $\kappa^{<\kappa} = \kappa > \omega$, then there is a c.c.c. generic extension in which $2^{\omega} = \kappa$ and the saturated linear ordering of power 2^{ω} is embeddable into $\langle {}^{\omega}\omega, <\rangle$. As in the consistency proof for MA, generic functions $f_{\alpha}\colon \omega \to \omega$ for $\alpha < \kappa$ are added iteratively so that all cuts in their ordering via $<^*$ are eventually filled.

Mitchell [1972] used trees to answer negatively a question of Malitz [1968] whether for regular cardinals κ, the existence of a linear ordering of power 2^{κ} with a dense subset of cardinality κ is true in ZFC.

In answer to a question of Erdős and Hajnal [1974, 281–282], Devlin [1974] proved that if GCH holds and there is a Kurepa tree with no Aronszajn subtree, then the following statement Φ holds: there is an order type of cardinality ω_2 which embeds neither ω_2 nor $\omega_2{}^*$, for which every subtype of cardinality ω_1 embeds one of ω_1 and $\omega_1{}^*$. Hence Φ is true in L, by this theorem combined with Jensen's construction in L of a Kurepa tree with no Aronszajn subtree, mentioned above in §7.7. In his thesis [1979a], [1981d], Todorcevic reduced Devlin's use of $V = L$ to $\Box$. Recall that he had used $\Box$ to construct an ω_2-Aronszajn tree with no ω_1-Aronszajn subtrees and no Cantor subtrees, as mentioned above in the subsection of §7.7 on larger trees. He showed in Theorem 7.1 [1981d, 254] that Φ is equivalent to the existence of either a Kurepa tree with no Aronszajn subtree or an ω_2-Aronszajn tree with no ω_1-Aronszajn subtrees and no Cantor subtrees. Thus he reduced the use of $V = L$ to the hypothesis $\Box$ in Devlin's result. He also showed one can use forcing to construct a Kurepa tree with no Aronszajn subtree.

In 1971 Baumgartner announced his result [1973] that it is consistent that all

[352]Two linear orders are *equimorphic* if each can be embedded in the other.

[353]Laver used $(\omega)^{\omega}$ to represent this collection of functions. Currently, this work would be cast as a question about $\mathcal{P}(\omega)/\mathrm{fin}$.

[354]To learn more about Kunen's result, see the chapter by Steprāns.

subsets of the real numbers of cardinality $\aleph_1$ which are $\aleph_1$-dense[355] have the same order type. Under the assumption of CH, given a pair of $\aleph_1$-dense sets A and B, he used diagonal arguments like those of Sierpiński [1950] and back-and-forth arguments to construct partitions of A and B into continuum many countable sets each dense in the appropriate one of A, B so that the following partial order satisfies the countable chain condition: P is the set of finite, order-preserving functions from A to B which send elements of the αth cell of the partition of A to elements of the αth cell of the partition of B, ordered by inclusion.

Baumgartner iterated this forcing to build a model in which all pairs of $\aleph_1$-dense subsets of the reals are isomorphic and $2^{\aleph_0} = \aleph_2$. He then observed that in this model, a singleton whose only element is an $\aleph_1$-dense set of reals is a basis for the set of all uncountable real types.

Baumgartner [1973] also posed the following two questions:

- Is it consistent to have all $\aleph_2$-dense subsets of the reals isomorphic?
- Does the statement that all $\aleph_1$-dense subsets of the reals are isomorphic follow from the assumption of Martin's Axiom and $2^{\aleph_0} > \aleph_1$?

Shelah [1980a] used oracle forcing to establish the consistency of $2^{\aleph_0} = \aleph_2$ together with the existence of a linear order universal in power $\aleph_1$.[356] In his paper Shelah recalled the Baumgartner [1973] proof of the consistency of all $\aleph_1$-dense sets of reals being isomorphic with a quick sketch and compared and contrasted his own proof with it. Shelah noted, without proof, that in a model in which $2^{\aleph_0}$ Cohen reals have been added, no such linear order universal in power $\aleph_1$ exists. These results complement the CH construction of a universal linear order of power $\aleph_1$ by Hausdorff in [2005]; see also [1908, 488], [2005, 243]. Shelah's consistency result should also be situated with earlier work of model theorists who had shown that saturated models are universal and that such models of cardinality λ exist for theories of cardinality $< \lambda$ when $\lambda = \lambda^{<\lambda}$. (See Chapter 5 of [Chang and Keisler, 1973]), and Shelah's Theorem 0.6 in Chapter VIII [1978a]: T has a saturated model of cardinality λ if and only if $\lambda^{<\lambda} + |D(T)|$ or T is λ-stable.)

In a paper received by the journal in January of 1972, Baumgartner [1976b] discussed basis problems[357] for the class Φ of linear orderings which cannot be represented as the countable union of well-orderings. He set $\Phi_1 = \{\omega_1{}^*\}$, and let Φ_2 be the collection of uncountable real types. Baumgartner noted that Erdős and Rado [1956, 443] had asked if there were uncountable order types which embed neither ω_1 nor $\omega_1{}^*$ and do not embed an uncountable real type, and pointed

[355] A linearly ordered set $(P, <)$ is $\aleph_1$-*dense* if it has no first or last element and between any pair of points there are exactly $\aleph_1$ elements.

[356] Recall that a linear (partial) order U is *universal in power* λ if it is a linear (partial) order of cardinality λ that homomorphically embeds all linear (partial) orders of cardinality λ.

[357] Recall that a basis B for a class $\mathcal{C}$ of linear orders is a set of linear orders in the class which has the property that for any element $(L, \lessdot)$ of $\mathcal{C}$ there is some $(K, <)$ in B which is order-embeddable into $(L, \lessdot)$.

out that Specker had answered the question in the positive.[358] Baumgartner called types with these properties *Specker types* and let Φ_3 be the class of them. Baumgartner showed that every ordering witnessing a Specker type embeds a linear ordering obtained from an Aronszajn tree under a suitable ordering (such linear orderings are now called *Aronszajn lines*, or *A-lines*). Baumgartner reported that Galvin had asked whether or not $\Phi_1 \cup \Phi_2 \cup \Phi_3$ constitutes a basis for Φ, and Baumgartner identified a collection he called Φ_4 of order types φ such that every uncountable subtype embeds ω_1 but φ is not the union of countably many well-orderings. He answered Galvin's question negatively when he proved that Φ_4 is non-empty. Baumgartner also proved that Φ_4 is not a well-quasi-order, does not have a finite basis, and contains dense rigid types. Baumgartner pointed out a partition calculus application which was likely the motivation for Galvin's question: there is an ordering φ different from both ω_1 and λ (any ψ^* with $\psi \in \Phi_4$ works) such that $\varphi \to (\omega)^1_\omega$. If φ is such a type, then, by the Baumgartner-Hajnal Theorem for order types, $\varphi \to (\alpha)^2_k$ for all $\alpha < \omega_1$ and $k < \omega$, so the extra work for that theorem is justified. Baumgartner noted out that $\Phi_1 \cup \Phi_2 \cup \Phi_3 \cup \Phi_4$ is a basis for Φ. Let N be the class of all continuous ordinal-valued strictly increasing functions whose domains are countable successor ordinals, and for $f, g \in N$, write $f \sqsubset_{\text{lex}} g$ if either f has g as a proper end-extension or f precedes g in the lexicographic order. His main technique was to represent order types as subsets of N under $\sqsubset_{\text{lex}}$. In Theorem 2.1, he showed that every type φ which fails to embed $\omega_1{}^*$ can be represented by $(S, \sqsubset_{\text{lex}})$ for some $S \subseteq N$ such that the tree $T(S) := \{f \in N : (\exists g \in S)(f \sqsubset_{\text{lex}} g)\}$ contains no path of length ω_1, and that if $|\varphi|$ is a cardinal of uncountable cofinality, then the desired S may be chosen as a subset of ${}^{<\omega_1}|\varphi|$. In Proposition 1.4, he showed that if $T(S)$ contains an Aronszajn tree, then the order type of $(S, \sqsubset_{\text{lex}})$ contains a Specker type. The paper closed with a list of problems, of which we give one below [1976b, 221]:

Problem 5:

1. Is ZFC + "Φ_3 has a finite basis" consistent?
2. Is ZFC + "$\Phi_2 \cup \Phi_3$ has a finite basis" consistent?

Note that the first of these problems asks for the consistent existence of a finite basis for Specker types, and implicitly asks for a finite basis for Aronszajn lines, since, as noted above, every ordering witnessing a Specker type embeds an Aronszajn line. Since any uncountable linear order which is the union of countably many well-orderings must embed ω_1, and by definition every linear order in Φ_4 embeds ω_1, the addition of ω_1 and its converse $\omega_1{}^*$ to a finite basis for $\Phi_2 \cup \Phi_3$ would give rise to a finite basis for all uncountable linear orderings. So the second of the above questions implicitly asks if it is consistent that there is a finite basis for uncountable linear orders.

[358] As discussed earlier, in footnote 7 on page 443, Erdős and Rado [1956] noted that Specker had disproved their conjecture after the submission of their paper. Baumgartner listed Specker's work as unpublished, and likely the Erdős-Rado paper was the source.

In 1974 during the International Congress in Vancouver, Shelah [1976] constructed what is now known as a Countryman line,[359] namely an uncountable linearly ordered set, whose square, ordered lexicographically, is the union of countably many chains. Shelah first carefully and inductively constructed an Aronszajn subtree T of the tree $S = {}^{\omega_1>}\mathbb{Q}$ of all sequences of rationals, partially ordered by end-extension, which satisfied suitable constraints and then passed to a linear order imposed on T essentially by lexicographically ordering its nodes to obtain a Countryman line.

Shelah closed his paper with several observations, a sample of which follows. He noted that if $\lambda = \lambda^{<\lambda}$, then one could lift his result with λ and λ^+ in place of $\aleph_0$ and $\aleph_1$ by a similar construction. He observed that one can construct the tree so that it will be special. He commented that any uncountable linear order whose square is a countable union of chains is a Specker order (i.e. an Aronszajn line). Consequently it has cardinality $\aleph_1$, something Shelah reported that Countryman already knew.

Prior to Shelah's construction of a Countryman line, Galvin had observed that if I is Countryman, then for all $n > 2$, the set I^n can also by covered by countably many chains defined from those used to cover I^2 (Shelah and the referee observed this later), and that I cannot contain two anti-isomorphic uncountable subsets (Shelah and Abraham observed this later).

Shelah reported that Harvey Friedman[360] asked "for the existence of an infinite complete order I, such that any open interval of I is isomorphic to I, but I is not anti-isomorphic to itself." He then indicated how, with some care, one could construct a completion of his example with these properties.

Shelah called two linear orders *near* if they have isomorphic uncountable subsets and noted that one can use $\diamondsuit$ to prove that there are $2^{\aleph_1}$ pairwise not near orders each of which is uncountable and whose square can be covered by countably many chains.

In his ninth and final observation, Shelah conjectured the consistency of the following pair of statements [1976, 114]:

(A) For every Specker order I there is a J near to it which is an uncountable order whose square is the union of countably many chains.

(B) If I and J are both uncountable orders whose squares are the union of countably many chains, then I is near to J or its converse order J^*.

Shelah [2000b] later reflected back on his mid-seventies interest in the consistent existence of a finite basis for the uncountable linear orders, the necessity of such a basis having ω_1, $\omega_1{}^*$, a representative real type, and at least two Aronszajn lines because of his construction of a Countryman line which is not isomorphic to its

[359]The name Countryman line refers to R. S. Countryman, Jr., whose unpublished typescript, dated to 1970 in [Todorcevic, 1984c], included the question of the existence of such a type.

[360]Shelah included in his references the problem paper [Friedman, 1975], but did not cite it in the body of the paper.

reverse. This problem has become known as *Shelah's Finite Basis Problem* or *Shelah's 5-Element Basis Problem.*

8 1980-1990: CODIFICATIONS AND EXTENSIONS

New perspectives were reached in the 1980's. Shelah [1980a], [1982b] introduced a generalization of countable chain condition forcings called *proper forcing.* A partial order P is *proper* if it preserves stationary subsets of $[\lambda]^{\leq\omega}$[361] for all uncountable cardinals λ.[362] The use of iterated forcing spread, helped in part by a survey article by Baumgartner [1983a]. In particular, set-theoretic topologists used forcing in an increasing number of ways. Foreman, Magidor and Shelah [1988] introduced a maximal form of Martin's Axiom, called Martin's Maximum, and proved new consistency results that led to a paradigm shift in the understanding of large cardinal strength.

Todorcevic and Boban Veličković[363] [1987] reinterpreted many questions in combinatorial language leading to new perspectives, including giving an equivalence of $MA_{\aleph_1}$ to a statement about c.c.c. partitions. Abraham, Rubin and Shelah and, separately, Todorcevic, introduced distinct open coloring axioms. New ways of obtaining Suslin trees were discovered, and partition theorems for products of trees moved into the transfinite.

Conferences promoted interactions between set theorists and the communities built around set-theoretic topology, order theory and (finite) combinatorics. For example, the survey by Baumgartner [1982] on uncountable linear order types, based on a talk at the *Ordered Sets* conference held in Banff in 1981,[364] consolidated understanding of an area that had been spread out in the literature; another new resource for people interested in order was the book *Linear Orderings* by Joseph Rosenstein [1982].

Boolean algebras as a field of study matured with the appearance of the three-volume *Handbook of Boolean Algebras*: Sabine Koppelberg edited the first volume

[361] The the set of all countable subsets of λ is variously denoted in the literature by $[\lambda]^{\leq\omega}$, $\mathcal{P}_{\omega_1}(\lambda)$ and $\mathcal{P}_{\aleph_1}(\lambda)$. See [Jech, 2003] for the concepts of closed unbounded subsets and stationary subsets of $[\lambda]^{\leq\omega}$.

[362] See [Jech, 2003] and [Abraham, 2010] for more on proper forcing.

[363] Boban Veličković received his doctorate in 1986 from the University of Wisconsin, Madison, where his advisor was Kunen. He was a Bateman Research Instructor at Caltech 1986–1987, a C. Morrey Assistant Professor at the University of California 1988–1989 and 1990–1991, with the intervening year as a postdoctoral fellow at the Mathematical Sciences Research Institute in Berkeley, was an NSERC International Fellow at York University 1992–1994, an associate professor 1994–1995 before becoming a professor at University of Paris VI, where he is the head of the logic group.

[364] This two-week conference organized by Ivan Rival was, as Duffus [2003] pointed out in his obituary of Rival, a "significant point in the creation of an 'order theory community'... with links to lattice theory, combinatorics, set theory, and computer science". Subsequent conferences included *Graphs and Order*, a NATO Advanced Study Institute held in Banff in 1984, *Combinatorics and Ordered Sets*, an AMS-IMS-SIAM Joint Summer Research Conference in Arcata, California in 1985.

[Koppelberg, 1989], and Monk and Robert Bonnet edited the second and third volumes [1989a], [1989b]. Connections were made between the study of Boolean algebras and pcf theory in the following decade (see Kojman's chapter).

Fraïssé published *Theory of Relations* [1986], bringing together work on order and work on relational structures. It included in the first eight chapters a brief review of set theory, fundamental results for the theory of linear, partial and quasi orders, especially well-quasi-orders, widely applied results in Ramsey theory including Ramsey's Theorem, the Nash-Williams Partition Relation, the Dushnik-Miller Theorem, the Erdős-Rado Theorem and Galvin's work on canonical partitions for pairs of rationals. The second half of the book was devoted to relational structures and included Fraïssé's method of building countable universal structures through a process of amalgamation of finite structures that leads to what is now known as the *Fraïssé limit.*

The *Handbook of Set-Theoretic Topology*, edited by Kunen and Vaughan [1984], popularized many techniques and results of set theory in the set-theoretic community and made them more readily available to young researchers in both areas. Juhász, in his article on cardinal functions, encouraged other topologists to follow him in using forcing directly rather than combinatorial consequences in their work. Baumgartner, in his expository article on applications of proper forcing aimed at set-theoretic topologists, facilitated the spread of this powerful technique and introduced the *Proper Forcing Axiom* (PFA), a generalization of Martin's Axiom in which the role of c.c.c. partial orders is taken on by proper partial orders. Models of PFA are non-CH models which have nice combinatorial structure.[365] Kunen focused on ideals in his handbook article on random and Cohen reals and showed that their effects differ markedly on the meager ideal and the null ideal. Weiss discussed versions of Martin's Axiom, its influence on partial orders, and variants of the axiom where the partial orders considered are restricted, an area of research that continued. The volume also presented problems and results of set-theoretic topology in a way that invited others to work in the area. For example, there were reports on the Normal Moore Space Conjecture and work coming out of that line of research (see the chapters by Fleissner, Nyikos, and Tall). Judith Roitman's article, *Basic S and L*, was designed to introduce the reader to an area she described as one of the most active in set-theoretic topology for the preceding ten years, whose key questions are the following.

(S) Is every hereditarily separable space hereditarily Lindelöf?

(L) Is every hereditarily Lindelöf space hereditarily separable?

A space is *separable* if it has a countable dense subset and *hereditarily separable* if every subspace is separable. A space is *Lindelöf* if every open cover has a countable subcover and *hereditarily Lindelöf* if every subspace is Lindelöf. Roitman was the first to formulate these questions in terms of partitions [Roitman, 1978], and counter-examples will be soon discussed below.

[365] Todorcevic showed in [1984b] that PFA implies $\mathfrak{c} = \aleph_2$.

Of particular significance for this chapter is the survey by Todorcevic [1984c] of work on trees and linearly ordered sets from set-theoretic, topological and algebraic perspectives. Elegantly written and comprehensive for its time, it included many results from Kurepa's thesis and later works that had been largely inaccessible due to their appearing in journals not widely distributed.

8.1 Set-theoretic topology

In the 1970's, set-theoretic topology as a discipline became increasingly visible. Rudin was an invited speaker at the 1974 International Conference of Mathematicians[366] in Vancouver where she spoke on the normality of products,[367] and she gave a series of expository talks at the CBMS Regional Conference held in Laramie, Wyoming in August 1974 (see Rudin [1975a]).

Rudin [1972] used a Suslin tree to produce the first example of an *S-space*, a hereditarily separable space which is not hereditarily Lindelöf. It was well-known that a Suslin line is an example of an *L-space*, a hereditarily Lindelöf space which is not hereditarily separable.

In 1980, W. Weiss arranged for Todorcevic to come to North America for the six-week summer school called Settop held in Toronto during July and the first half of August. The summer school featured three two-week sessions each with a ten-lecture series by a single speaker: Baumgartner spoke on forcing (see his *Handbook of Set-Theoretic Topology* exposition of proper forcing [1984]); Devlin spoke on the constructible universe and combinatorial principles derived from L; Juhász spoke on topology. Todorcevic shared a room with Abraham. Both had proved the existence of rigid Aronszajn trees (see [Abraham, 1979a] and [Todorcevic, 1979b]) in response to a question by Jech [1971]. While they were at the conference, Abraham and Todorcevic [1984] proved the consistency of MA + $\neg$CH + there exists a first countable S-space.[368] This theorem answered a question of Hajnal and Juhász that the existence of a first countable S-space does not imply the existence of a first countable L-space, since by a result of Szentmiklóssy, none exist in models of MA.

After the Settop conference, Todorcevic visited Dartmouth College for a few months during the academic year 1980-1981.[369] During that time he [1981a],

---

[366] Tall attributed the invitation to Rudin's construction [1971] of a Dowker space without using a Suslin tree. See [Rudin, 2008] for recollections of this discovery.

[367] Rudin [2008] expressed her perspective on normality as follows: "I admit I most enjoy tearing down structures: proving that reasonable sounding conjectures are false: constructing counter-examples. Normality is my thing since it seldom has the strength to guarantee your space is not miserable is some other way."

[368] Footnote 1 of [Abraham and Todorcevic, 1984] locates this result in time and space. Roitman [1984] referenced handwritten notes by Abraham and Todorcevic titled *MA and strong properties of S-spaces* and dated 1980. Justin Moore [2006b] constructed a ZFC example of an L-space, showing that the S-space and L-space problems are surprisingly different.

[369] See the note of gratitude to Baumgartner and thanks to Galvin for a "stimulating correspondence concerning a class of problems about strong partition relations on ω_1, a small part of which is considered in this paper" in [Todorcevic, 1983a, 705].

[1983a] showed the consistency with $MA_{\aleph_1}$ of the statement "every regular hereditarily separable topological space is Lindelöf", i.e. the consistency of "there are no S-spaces", which fulfilled the expectations of Hajnal and Juhász [1974, 837] who asserted that "it is quite likely that the existence of such spaces is independent of the usual axioms of set theory. In fact, Martin's axiom + the negation of CH is a candidate to imply that no such spaces exist."[370]

At the end of the decade, Todorcevic published a monograph, *Partition Problems in Topology* [1989]. He observed that proof techniques developed for solving the S-space problem and the L-space problem turn out to be useful in many other problems in general topology, writing "this is so because Ramsey-type theorems are basic and so much needed in many parts of mathematics and (S) and (L) happen to be Ramsey-type properties of the uncountable most often needed by the topologist" [Todorcevic, 1989, 1].

A key part of Suslin's Problem is the chain condition, an important concept for the 20th Century with many close ties with infinite combinatorics, so the book *Chain Conditions in Topology* [1982] by Comfort and Negrepontis was a welcome new resource. They gave background in infinite combinatorics, in particular, examined generalized chain conditions useful for products of spaces among other topological concepts. Specifically, the book includes a proof of Kurepa's result that the product of two Suslin continua does not satisfy the countable chain condition, and the Galvin construction, under the assumption of $2^\kappa = \kappa^+$, of pairs of partial orders on κ^+ with all antichains of size at most κ but whose product has an antichain of size κ^+. Comfort and Negrepontis worked largely in ZFC, with occasional use of the Generalized Continuum Hypothesis.[371]

In 1941, Marczewski and Knaster wrote in the *Scottish Book* what it means for a topological space T to have the property (S) (of Suslin): every family of disjoint sets, open in T, is at most countable. Then, under his original name of Szpilrajn, he inscribed a problem in the *Scottish Book* (see [Mauldin, 1981, 265–266]) which I have named and quoted below:

> *The Szpilrajn Problem:* Is the property (S) an invariant of the Cartesian product of two factors?

Recall that Kurepa [1952] showed that the square of a Suslin line fails to have the countable chain condition. Thus, if the product of two c.c.c. partial orders is

[370]Gary Gruenhage, in his 1981 Math Review MR0588816 (81j:54001) of *A survey of S- and L-spaces* by Juhász [1980], stated "The question of the existence of S- and L-spaces is one of the most important open problems in set-theoretic topology." Rudin also surveyed the subject in [1980]. In her article for the *Handbook of Set-Theoretic Topology*, Roitman [1984, 297] referred to the S-space and L-space existence problems as "a natural outgrowth of Suslin's hypothesis and of investigations of properties of metric, semi-metric, and Fréchet spaces," and attributes the modern form independently to Countryman and to Hajnal and Juhász, without giving references. Todorcevic [1989, 12–13] briefly discussed the history of the problem, and pointed to a series of papers by Hajnal and Juhász [1967], [1968], [1969] from which an early modern statement of the problem may have arisen.

[371]In his Math Review MR0665100 (84k:04002), Blass expressed regret that the book did not discuss independence results.

always c.c.c., then there are no Suslin trees.

Galvin [1980] proved that if λ is a regular cardinal and the square bracket partition relation $\lambda \to [\lambda]^2_3$ holds, then for any two partially ordered sets satisfying the λ-chain condition, their product also satisfies the λ-chain condition.[372] $MA_{\aleph_1}$ implies that the product of c.c.c. partially ordered sets is also c.c.c. (see Theorem 2.24 in [Kunen, 1980, 61].) Galvin also proved that if κ is an infinite cardinal such that $2^\kappa = \kappa^+$, then there are two partially ordered sets satisfying the κ^+-chain condition whose product does not. The two papers of Todorcevic [1985b], [1986b] contain the first ZFC examples of pairs of partial orders which satisfy the κ-chain condition but whose product does not, in particular for $\kappa = \mathrm{cf}(2^{\aleph_0})$, $\kappa = \mathfrak{b}$ (the bounding number[373]) and $\kappa = (2^{\aleph_0})^{+\omega+1}$. Shelah [1988a] gave such examples for every successor cardinal $> \aleph_1$.

8.2 Partition relations

The authors, Erdős, Hajnal and Rado, of the classic paper [1965] on the partition calculus were joined by Attila Máté in writing a compendium [1984] on the topic. It is the place to go for the definitions of the various partition symbols; care was taken to use modern notation and to express what is known without the assumption of the GCH. In the preface, the authors delimit the scope [Erdős *et al.*, 1984, 6]:

> There are many interesting assertions that are consistent with set theory without the axiom of choice but contradict this latter, and there are many important theorems of set theory plus some interesting additional assumptions, e.g. the axiom of determinacy, that is known to contradict the axiom of choice. We did not include any of these; unfortunate though this may be, we had to compromise; we attempted to discuss infinity, but had to accomplish our task in finite time.

Cardinal resources

Todorcevic [1983a] proved the consistency of a strong partition relation for ω_1, of which $\omega_1 \to (\omega_1, \alpha)^2$ for all $\alpha < \omega_1$ is a special case. More specifically, Todorcevic introduced the partition relation $\omega_1 \to (\omega_1, (\omega_1; \mathrm{fin}\ \omega_1))^2$, used forcing to prove the consistency of this partition relation with $MA_{\aleph_1} + 2^{\aleph_0} = \aleph_2$. He derived multiple consequences, the most well-known of which is the consistency of the non-existence of S-spaces. More pertinent to this chapter is that fact that from the assumption of $MA_{\aleph_1}$ and $\omega_1 \to (\omega_1, (\omega_1; \mathrm{fin}\ \omega_1))^2$, he proved in Theorem 4 [1983a, 716] that for every partition $[\omega_1]^2 = K_o \cup K_1$, either there is an uncountable subset $A \subseteq \omega_1$ all of whose pairs are in K_0 or for every $\alpha < \omega_1$, there are $B, C \subseteq \omega_1$ such that

[372] Recall that Galvin and Shelah [1973] had shown that $\lambda \to [\lambda]^2_3$ does not hold for $\lambda \in \{\aleph_1, 2^{\aleph_0}, \mathrm{cf}(2^{\aleph_0})\}$.

[373] A family $Y \subseteq {}^\omega\omega$ is *unbounded (under eventual domination)* if for every function $f \in {}^\omega\omega$ there is a function $g \in Y$ such that $g \not\leq^* f$. The *bounding number* $\mathfrak{b}$ is the minimum size of an unbounded subfamily of ${}^\omega\omega$. For more on this cardinal invariant, see §5 of the chapter by Steprāns.

B has order type α, C is uncountable, and all pairs $\{\beta, \gamma\}_<$ with $\beta \in C$ and $\gamma \in B \cup C$ are in K_1. This consistent partition relation is considerably stronger than its easily stated consequence $\omega_1 \to (\omega_1, \alpha)^2$ for all $\alpha < \omega_1$. These results shed light on possible consistent extensions of the Baumgartner-Hajnal Theorem for ω_1, that $\omega_1 \to (\alpha)^2_k$ for all $\alpha < \omega_1$ and $k < \omega$.

Laver [1982], starting from a huge cardinal, forced to get a model in which ω_1 carries an $(\aleph_2, \aleph_2, \aleph_0)$-saturated ideal (cf. §7.3) and derived interesting consequences for the partition calculus, including the consistency of

$$\omega_2 \to (\omega_1 + \omega_1 + 1, \alpha)^2.$$

Kanamori [1986] surveyed partition relations of the form $\kappa^+ \to (\alpha)^2_\delta$ for regular cardinals κ, with an emphasis on $\delta = 2$. First he reported that Donder [1982] had used Jensen's covering lemma to show that if $\kappa > \omega_2$ is a successor cardinal with κ^- as its cardinal predecessor, $2^{\kappa^-} = \kappa$ and $0^\#$ does not exist, then $\kappa^+ \not\to [\kappa : \kappa^-]^2_\kappa$ (recall the colon notation in a partition relation was defined in §6.7). Next, given a weakly compact cardinal κ, Kanamori [1986] noted the ordinary partition relation $\kappa^+ \to (\kappa : \eta)^2_2$ holds by applying the square bracket polarized partition relation

$$\begin{pmatrix} \kappa^+ \\ \kappa \end{pmatrix} \to \begin{bmatrix} \eta \\ \kappa \end{bmatrix},$$

which was stated without proof by Chudnovsky [1975], with proofs supplied by [Wolfsdorf, 1980], Shelah, and [Kanamori, 1982]. Kanamori [1986, 156] then asked:

Question 2.2: If κ is weakly compact, does $\kappa^+ \to (\kappa + \kappa)^2_2$?

Kanamori also considered successors of measurable cardinals, but found that the existence of a sufficiently saturated ideal sufficed for his purposes. Building on a definition in the Cardinal resources subsection of §7.3, call a κ-ideal $\mathcal{I}$ a *Laver ideal* if every family of κ^+ many $\mathcal{I}$-positive sets has a subfamily of size κ^+ such that the intersection of any subset of the subfamily of *fewer* than κ sets is also $\mathcal{I}$-positive. Then, if $\kappa^{<\kappa} = \kappa$ and there is a Laver κ-ideal, then for every $\alpha < \kappa$:

$$\kappa^+ \to (\kappa + \kappa + 1, \alpha)^2.$$

Since measurable cardinals carry such ideals, they satisfy the positive partition relation. Laver [1982] had only outlined a proof in his paper of the consistency of the existence of such ideals; Kanamori [1986] rediscovered the proof, and included a proof with permission from Laver. At the end of the paper, Kanamori noted that the positive partition relation could consistently hold at a non-weakly compact inaccessible cardinal, since it can be verified that a saturated ideal constructed by Kunen [1978] on such a cardinal is a Laver ideal. Kanamori [1986, 161] focused attention on the boundary of knowledge about these partition properties:

Question 2.5: If κ is measurable, does $\kappa^+ \to (\kappa + \kappa + 2)^2_2$?[374]

[374]Foreman and Hajnal have proved that the answer is yes in Theorem 39 of their [2003], which gives $\kappa^+ \to (\rho)^2_m$ for finite m and ρ below a large bound.

Todorcevic [1986a] showed the consistency relative to the existence of a weakly compact cardinal of the partition relation $2^{\aleph_0} \to (2^{\aleph_0}, \alpha)^2$ for all $\alpha < \omega_1$, and indicated how to modify the argument to obtain the consistency of $2^{\aleph_1} \to (2^{\aleph_1}, \alpha)^2$ for all $\alpha < \omega_2$. He also gave a forcing argument for cardinals κ of uncountable cofinality that $\kappa \to (\kappa, \omega + 2)^2$ may hold. The context for this relation includes Hajnal's proof [1960] that under CH, $2^{\aleph_0} \nrightarrow (2^{\aleph_0}, (\omega\colon 2))^2$ and consistency results in [Baumgartner, 1976a], [Laver, 1975] that this result holds under stronger hypotheses, for various values of the continuum.

Milner and Prikry [1986] used MA + ¬CH to prove

$$\omega_1 \to (\omega + m, 4)^3.$$

By absoluteness arguments like those used in the metamathematical proof of the Baumgartner-Hajnal Theorem, the use of MA + ¬CH may be eliminated. This positive partition relation was the first significant progress on partitions of triples of countable ordinals since the paper of Erdős and Rado [1956].

Ordinal partition relations

Erdős continued to focus attention on partition relations of the form $\alpha \to (\alpha, m)^2$ through offering money. In 1985, he [1987] offered \$1000 for a complete characterization of those countable ordinals α for which $\alpha \to (\alpha, 3)^2$. Ordinals α, countable or not, which satisfy this partition relation have come to be called *partition ordinals* (see [Schipperus, 1999]).

Larson [1980] used the combinatorial structure of a lexicographic representation of the ordinal ${\omega_1}^{\omega+1}$ to prove a partition relation for an ordinal with infinite exponent:

$${\omega_1}^{\omega+1} \nrightarrow ({\omega_1}^{\omega+1}, 3)^2$$

In connection with a 1985 meeting held at Humboldt State University, Baumgartner and Hajnal [1987] proved a variety of ordinal partition relations involving simple ordinal products of cardinals, some of which are listed below:

- If κ is a regular cardinal and $2^\kappa = \kappa^+$, then $(\kappa^+)^2 \nrightarrow (\kappa^+ \cdot \kappa, 4)^2$.
- If κ is a singular cardinal of cofinality τ and $2^\kappa = \kappa^+$, then $\kappa^+ \cdot \tau \nrightarrow (\kappa^+ \cdot \tau, 3)^2$ (i.e. $\kappa^+ \cdot \tau$ is not a partition ordinal).
- If κ is a regular cardinal and $\kappa^{<\kappa} = \kappa$, then $(\kappa^+)^2 \to (\kappa^+ \cdot \kappa, 3, 3)^2$.

In a 1987 meeting held in Toronto, Baumgartner [1989] focused attention on partition ordinals by showing that the existence of uncountable partition ordinals is consistent. Under the assumption of MA + ¬CH, he proved two simple products were partition ordinals and asked the simplest open question suggested by these results:

- $\omega_1 \cdot \omega \to (\omega_1 \cdot \omega, 3)^2$.
- $\omega_1 \cdot \omega^2 \to (\omega_1 \cdot \omega^2, 3)^2$.
- *Question:* Is it consistent that ${\omega_1}^2 \to ({\omega_1}^2, 3)^2$?

Shelah and Lee J. Stanley [1987] gave proofs of several partition relations for ordinal products of cardinal numbers and we list some representative examples.

- Assume CH. Then $\omega_2 \cdot \omega \to (\omega_2 \cdot \omega, 3)^2$.
  This is a solution to Problem 13 [Erdős and Hajnal, 1971b, 24].
- It is consistent to have $\neg$CH + $\omega_2 \cdot \omega \nrightarrow (\omega_2 \cdot \omega, 3)^2$.
  The proof used forcing with finite conditions.
- It is consistent to have GCH + $\omega_3 \cdot \omega_1 \nrightarrow (\omega_3 \cdot \omega_1, 3)^2$.
  The proof used historic forcing.[375]
- Assume that the existence of a weakly compact cardinal is consistent. Then so is GCH + $\omega_3 \cdot \omega_1 \to (\omega_3 \cdot \omega_1, k)^2$ for all $k < \omega$.
- *Question:* Does $\mathfrak{c} \cdot \omega \to (\mathfrak{c} \cdot \omega, 3)^2$?

In his thesis, Tadatoshi Miyamoto [1988, 67-94] gave a morass construction of a partition witnessing $\omega_3 \cdot \omega_1 \nrightarrow (\omega_3 \cdot \omega_1, 3)^2$, which had earlier been proved by forcing by Shelah and Stanley [1987]. The existence of such a construction had been conjectured by Stanley and Daniel Velleman among others. Miyamoto's morass was a simplified morass, but had some extra structure, similar to linear limits. During a July 1988 meeting in Oberwolfach, Stanley and Velleman tried to understand Miyamoto's proof and, in the process, they came up with their own construction of a witness to the negative partition from a simplified $(\omega_2, 1)$-morass with linear limits, but used a complete amalgamation system in place of the rest of Miyamoto's extra structure. Later they learned that Charles Morgan had independently found essentially the same proof which they published together in [1991]. Assuming $2^{\aleph_1} = \aleph_2$, it is known that if both $\aleph_2$ and $\aleph_3$ are successor cardinals in L, then there is such a simplified $(\omega_2, 1)$-morass with linear limits. It follows that, assuming $2^{\aleph_1} = \aleph_2$, the positive partition relation $\omega_3{\cdot}\omega_1 \to (\omega_3{\cdot}\omega_1, 3)^2$ holds only if either $\aleph_2$ or $\aleph_3$ is inaccessible in L.

Square bracket partition relations

Todorcevic [1981d] developed methods for constructing uncountable trees while controlling the type of their subtrees and applied his tree results to prove consistency results for the partition calculus. In particular, he proved from $\square_\kappa$ the

[375]The process of historicization of naïve conditions was introduced in [Baumgartner and Shelah, 1987] to prove consistency results for superatomic Boolean algebras, without use of the name. The Shelah-Stanley paper was the first to introduce it officially according to Rosłanowski and Shelah [2001].

stepping up negative square bracket partition relations as given below when $\kappa > \omega$ is regular, $2 \leq r < \omega$, $r < \lambda_0 \leq \kappa$, $\omega < \lambda_\xi \leq \kappa$ and λ_ξ regular for $0 < \xi < \nu$:

$$\kappa \not\to [\lambda_\xi]^r_{\xi<\nu} \quad \text{implies} \quad 2^\kappa \not\to [\lambda_\xi + 1]^{r+1}_{\xi<\nu} \,.$$

The statement of this square bracket result was added in proof to the Erdős, Hajnal, Máté, Rado [1984, 323] compendium.

In the early 1980's the expectation was that ZFC results like the Galvin-Shelah result [1973] that $\lambda \not\to [\lambda]^2_\omega$ for λ the order type of the reals and $\aleph_1 \not\to [\aleph_1]^2_4$ discussed above would not extend all the way to a GCH result that $\aleph_1 \not\to [\aleph_1]^2_{\aleph_1}$ (see Theorem 17 in [Erdős *et al.*, 1965, 145]). Thus the proof of the partition relation $\aleph_1 \not\to [\aleph_1]^2_{\aleph_1}$ by Todorcevic [1987] was a wonderful shock that introduced a wide audience to the walks on ordinals and the oscillation function.

Todorcevic was a Miller Fellow at the University of California, Berkeley, when in May 1984 his came up with a new proof of the existence of a Countryman line.[376] A key ingredient was the method of walks on ordinals, which Todorcevic devised walking on stairs in one of the libraries on the Berkeley campus, trying to think of a way to attach something finite to a pair of countable ordinals. We think of ladders as a way to climb up. Imagine having a nice ladder system on ω_1, say a $\diamondsuit$-sequence $\langle S_\alpha : \alpha < \omega_1 \wedge \alpha \text{ limit}\rangle$. Descending sequences of ordinals are finite. One can "walk down" from a large δ to a smaller γ in stages $\delta_0 = \delta, \delta_1, \ldots, \delta_k = \gamma$, where at each stage $i+1$, you move from δ_i to the least ordinal δ_{i+1} in $S_{\delta_i} \setminus \gamma$. But if $\delta = \gamma + \omega$, the set $S_\delta \setminus \gamma$ is a subset of $\{\gamma + n : n < \omega\}$, and if $\gamma \notin S_\delta$, then the recursive definition breaks down in an easily fixable way: from $\eta + 1$, walk to η. Todorcevic called a sequence $\vec{C} = \langle C_\xi : 0 < \xi < \kappa\rangle$ a *c-sequence* if each C_ξ is a closed unbounded subset of ξ. In practice, $C_{\eta+1} = \{\eta\}$ for every successor ordinal. If the cofinal subsets for limit ordinals consist of successor ordinals, then the walk from a limit ordinal to a smaller one has at least one intermediate step.

Todorcevic obtained the partition relation $\aleph_1 \not\to [\aleph_1]^2_{\aleph_1}$ in September 1984, lectured on it in the Berkeley seminar that fall, wrote up the notes of his lectures and circulated them in January 1985.

Todorcevic produced colorings of the pairs of countable ordinals with countable ordinals that had large range on every uncountable set, for three notions of large: contains a club, contains a cobounded set, is all of ω_1. He showed that from one such mapping, one could define mappings of the other kinds, and the same interdefinability was true for mappings from $[\omega_1]^2$ to ω whose range on every uncountable set is all of ω.

To define a function from pairs of countable ordinals into ω that realizes all colors on every uncountable subset, Todorcevic used the *oscillation function* osc : ${}^\omega\omega \to \omega \cup \{\omega\}$ defined by letting $\mathrm{osc}(x, y)$ be the cardinality of the set of all n for which $x(n) \leq y(n)$ and $x(n+1) > y(n+1)$. Recall that Galvin [1971] had

[376]Recall that the original construction of a Countryman line by Shelah [1976] also involved construction of an Aronszajn tree. This is not an accident since given any Countryman line L, every countably branching partition tree (development) of intervals under $\subseteq$ whose leaves are singletons must be an Aronszajn tree. See Baumgartner's survey [1982] to be discussed in §8.6.

used an instance of this mapping as discussed in the subsection of §7.3 on square bracket partition relations. Todorcevic [1984a] had used this function to define interesting partitions of nice subsets of the Baire space ${}^{\omega}\omega$. His article [1988b] for the proceedings for the Logic Colloquium 1986 held in Hull is an exposition of the use of the oscillation function, and he proved that on any unbounded subset of the monotonically increasing functions in ${}^{\omega}\omega$ linearly ordered under eventual domination, the oscillation function attains all possible values.

In his concluding remarks, Todorcevic [1987, 287] wrote: "The key idea of our coloring can roughly be stated as follows: If the set of interesting places is stationary then in any unbounded set we can find $\alpha < \beta$ such that walking from β to α along the C_ξ's we pass through at least one interesting place." He pointed out that several papers had already been written using the methods of his circulated notes including [Hajnal *et al.*, 1987], [Shelah, 1988b], [Shelah, 1990a], and [Shelah and Steprāns, 1987] (see [Todorcevic, 1987, 288]). The applications of the ideas in this seminal paper are manifold; see the densely packed book by Todorcevic [2007b] for some of them.

Shelah, in a paper entitled *Was Sierpiński right? I* [1988b] built on the work of Todorcevic to show the consistency of $2^{\aleph_0} \rightarrow [\aleph_1]^2_3$ relative to the existence of an Erdős cardinal. Shelah's result answered a question of Erdős and Hajnal in the negative: the Sierpiński partition witnessing $2^{\aleph_0} \nrightarrow [\aleph_1]^2_2$ cannot be generalized from two colors to three colors in ZFC.

Shelah [1989] proved that it is consistent that for every order type θ and cardinal κ, there is an order type ψ such that $\psi \rightarrow [\theta]^2_{\kappa,2}$.

8.3 Structural partition relations

In a paper dedicated to the celebration of the 80th birthday of Deuber, Rado [1986] published a new proof of the the Canonical Ramsey Theorem for $[\mathbb{N}]^n$, showing that it is a corollary of the original Ramsey Theorem. Deuber [1989] surveyed results on partition regular linear systems that flowed from Rado's dissertation [1933], keeping this research area alive. Interesting generalizations to other types of partitions and the equivalence relations arising from them were proved, but because they tend to be difficult to state, no others will be included in this history.

Confirming a conjecture of Galvin [1975],[377] Todorcevic [1981b], [1985a] simultaneously generalized the Baumgartner-Hajnal theorem that for linear orders, $\varphi \rightarrow (\omega)^1_\omega$ implies $\varphi \rightarrow (\alpha)^2_k$ for all $\alpha < \omega_1$ and $k \in \omega$, and the Galvin [1975] partial order result that $P \rightarrow (\eta)^1_\omega$ implies $P \rightarrow (\alpha)^2_k$ for all $\alpha < \omega_1$ and $k \in \omega$.

> *Todorcevic Poset Partition Theorem:* $P \rightarrow (\omega)^1_\omega$ implies $P \rightarrow (\alpha)^2_k$ for all $\alpha < \omega_1$ and $k \in \omega$.

The first step in the proof was a reduction from partition problems for partial orders $(P, <_P)$ to partition relations for trees via passage to one of σP and $\sigma^* P$,

[377] In a brief note, Todorcevic [1985a, 4] thanked Galvin for many valuable communications concerning the problems listed in Galvin's paper [1975].

where σP is the collection of all increasing maps from ordinals into P and $\sigma^* P$ is the collection of maps from σP whose domain is a successor ordinal. Both σP and $\sigma^* P$ are trees under end-extension. The operation σP was first considered by Kurepa [1935], who observed that σP is a tree and constructed his first Aronszajn tree as a subtree of $\sigma\mathbb{Q}$. The hypothesis $P \to (\omega)^1_\omega$, when $P = T$ is a tree, becomes the condition that T is non-special. The second step consisted of showing that for non-special trees T of cardinality less than the pseudointersection number $\mathfrak{p}$, the desired partition relation holds. The final step is elimination of the hypothesis on the cardinality of the trees via a forcing argument.

The paper included proofs of generalizations of classical partition relations to higher cardinals, trees and partial orders. For example, the Dushnik-Miller Theorem for $\aleph_1$ is generalized to non-special Aronszajn trees: non-special tree $\to$ (non-special tree, $\omega+1)^2$. Also, the classical partition theorem $\mathfrak{c}^+ \to (\omega_1+2)^2_{\aleph_0}$ is generalized, for partial orderings P satisfying $P \to (\omega_1)^1_{\mathfrak{c}}$, to $P \to (\omega_1+1)^2_{\aleph_0}$.

In response to a question by Baumgartner, Laver [1984] generalized HL_d for finite d to HL_ω, a perfect subtree version of the Halpern-Läuchli Theorem for infinitely many trees. The notation here was introduced in §6.1.

> *Laver HL_ω Theorem:* If $\vec{T} = \langle T_i : i < \omega \rangle$ is a sequence of rooted finitely branching perfect trees of height ω and $\prod^\omega \vec{T} = G_0 \cup G_1$, then there are $j < 2$, an infinite subset $A \subseteq \omega$, and downwards closed perfect subtrees T_i' of T_i for $i < \omega$ with $\prod^A T_i' \subseteq G_i$.

In addition to proving this theorem, Laver showed that it is equivalent to the following two statements:

1. If $\mathbb{P}_\kappa$ is the partial ordering for adding κ side-by-side Sacks reals with countable supports, then every subset of ω in the extension contains or is disjoint from some infinite subset of ω in the ground model.

2. If $\langle f_i : i < \omega \rangle$ are continuous functions from the Hilbert cube $[0,1]^\omega$ into $[0,1]$, then there exist non-empty perfect sets $P_i \subseteq [0,1]$ for $i < \omega$ and an infinite subset $A \subseteq \omega$ such that, on $\prod^A P_i$, $\langle f_i : i \in A \rangle$ is monotonic (and uniformly convergent).

Continuous colorings were examined. After noting that an open coloring of pairs from X can be regarded as a continuous function on $X \times X$ (minus the diagonal), Abraham, Rubin and Shelah observed [1985, 153] that "it seems thus natural to examine these partition theorems for general continuous functions. We did not investigate these questions thoroughly ..." and concluded with a theorem which we omit. Suppose X and Y are topological spaces and X is Hausdorff and τ is a topology on X^n.[378] Let $D(X) = \{(x_0, x_1, \ldots, x_{n-1} : (\exists i < j < n)(x_i = x_j)\}$.

[378] A natural topology τ on X^n is the topology generated by the sets $[O_0, O_1, \ldots, O_{n-1}] = O_0 \times O_1 \times \cdots \times O_{n-1}$ where $\{O_0, O_1, \ldots, O_{n-1}\}$ is a family of pairwise disjoint open subsets of X.

A coloring $c : [X]^n \to Y$ is *continuous* if, when viewed as a symmetric function on $X^n \setminus D$, it is continuous with respect to the topology τ on X^n. We are most interested when X is a subset of the reals or a related topological space, e.g. $\mathbb{Q}$, Baire space ω^ω, and Cantor space 2^ω. For colorings c of pairs of reals, being continuous means that to determine a finite amount of information about $c(x, y)$ one only needs a finite amount of information about x and about y. Todorcevic [1985b] proved the existence a continuous coloring witnessing $\aleph_1 \to [\aleph_1]_{\aleph_0}$,[379] strengthening a result of Galvin and Shelah [1973].

Furstenberg [1981] investigated applications of recurrence in ergodic theory and topological dynamics to combinatorics and number theory, giving proofs of multi-dimensional versions of van der Waerden's Theorem on monochromatic arithmetic progressions, Szemerédi's theorem on arbitrarily long arithmetic progressions in sequences of integers with positive density, Hindman's Finite Sums Theorem and Rado's generalization of van der Waerden's Theorem. Taylor [1982] gave a combinatorial version of the Furstenberg-Weiss topological proof of van der Waerden's Theorem.

For a rooted finitely branching tree T without treetops, Milliken [1981] introduced a tree topology on the family $\mathrm{Str}^\omega(T)$ of strongly embedded subtrees of T. He generalized Silver's theorem to prove what we will call *Milliken's Complete Ramseyness for Trees Theorem* by showing that if $R \subseteq \mathrm{Str}^\omega(T)$ is analytic in the tree topology on $\mathrm{Str}^\omega(T)$, then it is completely T-Ramsey, i.e. there is a strongly embedded subtree S of T such that the collection of its strongly embedded subtrees is contained in or disjoint from R. He generalized Ellentuck's approach to Silver's theorem and used his own work on strongly embedded subtrees of finite height [1979].

In [1984], Timothy Carlson[380] and Simpson proved *dual Ramsey theorems*, i.e. Ramsey theorems about the set of all partitions. For $k \leq \omega$, let $(\omega)^k$ be the set of all partitions of ω into k parts. Their *Dual Ramsey Theorem* is the statement that if $k < \omega$, $(\omega)^k = C_1 \cup \cdots \cup C_n$ and each of the C_i's is Borel, then there is some i and some k-cell partition X such that all its coarsenings lie in C_i, where the relevant topology is the product topology on $2^{\omega\times\omega}$ and the partitions are regarded as equivalence relations on ω, i.e. subsets of $\omega \times \omega$. Their *Dual Galvin-Prikry Theorem* is obtained by setting $k = \omega$. Their *Dual Ellentuck Theorem* states that a subset D of $(\omega)^\omega$ is *Ramsey* if and only if it has the property of Baire and D is *Ramsey null* if and only if it is meager, where the relevant topology is the Dual Ellentuck topology.

At the end of the introduction to their paper they included some historical remarks which indicate that Simpson originated the idea that dual Ramsey theorems should be provable with a series of conjectures in August 1981 after conversations with Leeb,[381] which became their Theorem 2.2 (a generalization of the Dual Ram-

[379] See J. Moore [2000] for the attribution of this continuous coloring result to Todorcevic [1985b].

[380] In 1978, Carlson earned his doctorate at the University of Minnesota, Minneapolis where Prikry was his advisor. Carlson is now a professor at the Ohio State University.

[381] Recall that in §7.5 the discussion following the statement of the Parameter Sets Theorem

sey Theorem), the Dual Ellentuck Theorem, and two more theorems more technical than I have chosen to state. Simpson was able to prove the Dual Ramsey Theorem in the special case in which the partitions have exactly three cells and his proof used Hindman's Finite Sums Theorem. Simpson also isolated the combinatorial core (Lemma 2.4), which we here call the Key Lemma, that would suffice to prove these theorems.[382] The introductory sections lists the following theorems as Simpson's "chief" sources of inspiration: the Galvin-Prikry Theorem, The Graham-Rothschild Parameter Sets Theorem, and the Paris-Harrington Theorem [1977]. Simpson's reductions of these theorems to the Key Lemma are included in Sections 2–5. Section 2 presents the proof of the Dual Ramsey Theorem from the Key Lemma. (See also the contemporaneous independent work by Prömel and Voigt [1985a] which strengthens the Carlson-Simpson Lemma 2.3 from Borel partitions to partition's with the property of Baire.) Section 3 gives corollaries of the Dual Ramsey Theorem, including Ramsey's Theorem, the finite version of the Graham-Rothschild Parameter Sets Theorem, the Laver HL_d Theorem and the Laver HL_ω Theorem; the first two presumably were found by Simpson, and the last two were explicitly found by Carlson. Section 4 introduces the dual Ellentuck topology, presents the proof of the Dual Ellentuck Theorem from the Key Lemma, and derives the Ellentuck Theorem as a corollary. Section 5 introduces dual Mathias forcing and results proved with it. Section 6 is devoted to the proof of the Key Lemma due to Carlson, who learned about Simpson's work in July 1982 when they met at the American Mathematical Society Recursion Theory Institute at Cornell University and at an American Mathematical Society Meeting held in Toronto in August 1982, and found his proof shortly thereafter along with a stronger version which appears as Theorem 6.3.

Before we state the Key Lemma we need to introduce some additional definitions. For a finite alphabet A and some $\alpha \le \omega$, an *A-partition of α* is a collection of pairwise disjoint, non-empty *blocks* whose union is all of $A \cup \alpha$ and such that each block has at most one element of A. The blocks with no element of A are called *free*. If $\alpha < \beta \le \omega$ and X is an A-partition of β, let $X[\alpha] = \{x \cap (A \cup \alpha) : x \in X\} \setminus \{\emptyset\}$. Observe that $X[\alpha]$ is an A-partition of α. If X is an A-partition of ω, let X^*_A denote the collection of all $s \cup \{\{n\}\}$ such that $0 < n < \omega$, s is an A-partition of n with no free blocks and $s = Y[n]$ for some A-partition Y of ω which is a coarsening of X. With this notation in hand, we state Lemma 2.4 of [1984, 270]:

> *Carlson-Simpson Key Lemma:* Let A be a finite alphabet. If Y is an A-partition of ω and $Y^*_A = C^*_0 \cup \cdots \cup C^*_{\ell-1}$, then there exists an A-partition Z which is a coarsening of Y such that $Z^*_A \subseteq C^*_i$ for some $i < \ell$.

The statement of the stronger Theorem 6.3 proved by Carlson in their [1984] requires the following notation. Let $(\omega)^\omega_A$ denote the family of A-partitions for

included the Graham and Rothschild [1974] report that Leeb was the one who looked at the Ramsey property for categories.

[382]The conjecture that became Lemma 2.4 is described as "an infinitary Hales-Jewett [13] type conjecture" and the reference [13] is to [Hales and Jewett, 1963].

which any two free blocks x and y satisfy $\max x < \min y$ or $x = y$ or $\max y < \min x$. Theorem 6.3 states that for a finite alphabet A, if $(\omega)^*_A = C_0 \cup \cdots \cup C_{\ell-1}$, then there is some $X \in \langle\omega\rangle^\omega_A$ such that $(X)^*_A \subseteq C_i$ for some i. In Section 6 the authors announce what they call *Carlson's Theorem:* A set $C \subseteq \langle\omega\rangle^\omega_A$ is Ramsey if and only if it has the property of Baire with respect to the Ellentuck topology on $\langle\omega\rangle^\omega_A$. They report that Carlson found the proof of this theorem in October 1982 along with a still stronger result which gives a common generalization of the key lemma, Hindman's Finite Sums Theorem and the work of Milliken [1975a] and Taylor [1976], and that Carlson has derived results for this space analogous to the Dual Ramsey Theorem.[383] A note added in proof records the March 1983 verification by Carlson of a conjecture by Simpson on finite Borel colorings of k-dimensional affine subspaces of the infinite dimensional space F^ω over a finite field F which appears in [Carlson, 1987].

Carlson [1988] looked at the series of Ramsey results about partitions from the more abstract perspective of *Ramsey spaces*, which he described as structures which satisfy an analog of Ellentuck's Theorem. A *Ramsey space* is a space which is a collection of infinite sequences with a topology such that every set which has the property of Baire is Ramsey and no open set is meager. In Section 2, he sets up a framework for this study; discusses the notion of *finitization* which is a sequence of approximations; and introduces a neighborhood base defined from the approximations, which allows him to prove critical topological closure results in his reduction of the question of whether a topological space is a Ramsey space to a more combinatorial question. He introduces axiom-like statements (see page 125 and statements A1-A3) and proves an abstract Ellentuck Theorem. His approach generalizes a large part of quantitative Ramsey theory. Hindman and McCutcheon [2002, 2559] single out Carlson's Lemma 5.9: "Experience suggests that Carlson's 'main lemma' ... implies those Ramsey-theoretic corollaries of his theorem in which a finite collection of finite objects is partitioned into finitely many classes."

During the 1982 Prague Symposium on Graph Theory, several people suggested collecting papers to showcase the new Ramsey theory of the time. Nešetřil and Rödl took up the challenge, wrote an extended introduction, solicited and selected appropriate papers and edited the volume *Mathematics of Ramsey Theory* [1990]. This warmly recommended survey includes Ramsey theory from a wide variety of perspectives including many not touched upon in this chapter. Of particular interest are the paper by Weiss titled *Partitioning Topological Spaces* and the paper [1990] in which Carlson and Simpson survey topological Ramsey theory, i.e. work related to their seminal paper [1984], the two Carlson papers discussed above [1987], [1988] and related work of Prömel and Voigt, including some of their canonical partition theorems [1983], [1985b]. Carlson and Simpson define a Ramsey space to be a partial ordering with approximations which satisfies an Ellentuck type theorem. There is a list of eight assumptions, A1–A8. Theorem 1 (due to Carlson) states that a partial ordering with approximation which satisfies

[383] Wolfsdorf [1988] in an article received by the editors January 28, 1983, independently proved some results of the kind found in Section 6 of the Carlson and Simpson paper [1984].

A1–A7 is a Ramsey space if and only if it satisfies A8. The definition of partial order with approximations and eight conditions are somewhat technical to state, so we have omitted them, but the authors assert on page 175 that "Generally, when proving a particular partial ordering with approximations is a Ramsey space assumptions A1–A7 are immediate." They describe Theorem 1 as an abstract Ellentuck Theorem. They discuss finitary consequences, finite dimensional analogs of Ellentuck type theorems, and canonical partitions as well.

Arnold Miller [1989] investigated Borel versions of infinite combinatorial theorems, and proved a parameterized version of the Galvin-Prikry Theorem.

Many more results were found in the 20th Century, but are beyond the scope of this chapter. For a detailed look at infinitary partition relations and Ramsey spaces from a modern perspective, see Todorcevic [2010]. To learn about their applications to Banach space theory, see [Argyros and Todorcevic, 2005] and [Gowers, 2003].

8.4 Partial order, not trees

Kunen and Miller (cf. [Fraïssé, 2000, 164]) showed that the class of denumerable partial orders includes a strictly decreasing ω_1-sequence.

Abraham [1987] revisited the Perles counter-example to Dilworth's chain decomposition theorem holding in the infinite and showed that if a partial order P has no infinite antichain and the length of the well-founded poset of antichains ordered by reverse inclusion has height $< {\omega_1}^2$, then P is the union of countably many chains.

In answer to a question of Galvin, Milner [1982] proved another infinitary analog of Dilworth's theorem by showing that every partially order set P in which every antichain has cardinality less than μ can be decomposed into fewer than $\mu^{<\mu}$ many directed sets.

Todorcevic [1988a] proved that every σ-chain complete poset with the finite cutset property is the union of countably many chains.[384] For context, he listed several earlier papers that studied posets with finite cutsets.

Rado [1981] conjectured that a family of intervals of a linearly ordered set is the union of countably many disjoint subfamilies if and only if every subfamily of size $\aleph_1$ has this property. Todorcevic [1983b] heard about Rado's conjecture, and investigated its non-trivial consistency strength.

8.5 Tree results

Shelah [1984a] proved what we will call his *Cohen Real to Suslin Tree Theorem*: Forcing to add a Cohen real also adds a Suslin tree. Shelah and Stanley [1982]

[384] A partial ordering $(P, <)$ is *σ-chain complete* if every bounded countable chain in P has both a supremum and an infimum. It has the *finite cutset property* if for every $p \in P$, there is a finite set S with $p \in S \subseteq P$ such that S intersects every maximal chain of P and every element of $S \setminus \{p\}$ is incomparable with p.

generalized the result to larger successor cardinals by proving that if $2^\kappa = \kappa^+$ and κ^{++} is inaccessible in L, then there is a κ^{++}-Suslin tree. Velleman proved that if κ is a regular limit cardinal, $2^{<\kappa} = \kappa$, and κ^+ is a regular limit cardinal in L, then forcing to add a Cohen subset of κ also adds a κ^+-Suslin tree. In his Theorem 1.11, Velleman proved that if there is a simplified $(\kappa, 1)$-morass with a complete amalgamation system, then there is a κ^+-Suslin tree. Velleman proved in [1984a] that a simplified $(\omega, 1)$-morass can be constructed in ZFC and in [1984b] that the existence of a simplified $(\kappa, 1)$-morass for regular uncountable κ is equivalent to the existence of an ordinary $(\kappa, 1)$-morass. In Theorem 2.1 of [1984c], he proved that if κ is a regular cardinal and $2^{<\kappa} = \kappa$, then forcing over a model with a simplified $(\kappa, 1)$-morass to add a Cohen subset of κ also adds a complete amalgamation system to the simplified $(\kappa, 1)$-morass. With these results in hand, he derived as corollaries Shelah's Cohen Real to Suslin Tree Theorem and the Shelah-Stanley results quoted above. In the introduction he commented that his proof of the former is a "modified form of a proof by Mark Bickford,[385] and his proof of the latter was based on Shelah and Stanley's original proof.

Todorcevic [1987] gave a different proof of Shelah's Cohen Real to Suslin Tree Theorem. Call a function $e : [\omega_1]^2 \to \omega$ a *coherent mapping* if for all $\alpha < \beta < \omega$, there are only finitely many ξ with $e(\xi, \alpha) \neq e(\xi, \beta)$. Note that in this case, the restriction of the function $e(\cdot, \beta)$ to α is almost equal to $e(\cdot, \alpha)$, in symbols, $e(\cdot, \alpha) =^* e(\cdot, \beta)\restriction\alpha$. It is not difficult to construct a coherent mapping by transfinite induction. Todorcevic observed that if e is a coherent mapping, then the following set is an Aronszajn tree under $\sqsubseteq$:

$$T(e) := \{e_\beta\restriction\alpha : \alpha \leq \beta < \omega_1\}.$$

Note that if e is a coherent mapping, then the *stretching* $\overline{e} : [\omega_1]^2 \to \omega$ of e which Todorcevic defined on page 271 of [1987] by setting

$$\overline{e}(\alpha, \beta) = 2^{e(\alpha,\beta)}(1 + 2|\{\xi < \alpha : e(\xi, \alpha) = e(\xi, \beta)\}|)$$

is coherent and each $\overline{e}_\beta$ is one-to-one. He used such a map to prove the following [1987, 292–293]:

> *Todorcevic's Cohen Real to Suslin Tree Construction:* If $r : \omega \to \omega$ is a Cohen real and $e : [\omega_1]^2 \to \omega$ is a coherent mapping such that each e_β is one-to-one, then $T(r \circ e)$ is a Suslin tree.

The most common constructions of κ-Aronszajn trees to that point had been subtrees of the collection of functions ${}^{<\kappa}L$ for some linear order L under end-extension. Todorcevic [1987] associated various *rho functions* or *characteristics* with the walks on ordinals discussed in the subsection on partition properties. He proved several "metric" properties of his rho functions, and used them to define trees whose properties could be described in terms of features of the rho functions

[385] Mark Bickford received his 1983 doctorate from the University of Wisconsin, Madison, where his advisor was Terrence Millar.

from which they came. Define the *right lexicographic ordering of* ${}^{<\kappa}L$ by $s <_0 t$ if and only if s properly end-extends t or for the least α on which s and t disagree, $s(\alpha) < t(\alpha)$. Define the *left lexicographic ordering of* ${}^{<\kappa}L$ similarly by $s <_1 t$ if and only if t properly end-extends s or for the least α on which s and t disagree, $s(\alpha) < t(\alpha)$. Call an arbitrary tree T of height ω_1 a *coherent tree* if it can be represented as a downwards closed subtree of some tree of the form $T(a)$, where $a : [\omega_1]^2 \to \omega$ is a coherent mapping.[386]

> *Todorcevic's Coherent Tree Theorem:* Suppose that $e : [\omega_1]^2 \to \omega$ is a coherent mapping with each e_β finite-to-one. Further suppose that $\vec{C} = \langle C_\alpha : \alpha < \omega\rangle$ is a c-sequence and ρ_0 and ρ_1 are defined from C. Then $T(e)$, $T(\rho_0)$ and $T(\rho_1)$ are all coherent Aronszajn trees, $T(\rho_0)$ is special, and $T(\rho_0, <_0)$ and $T(\rho_1, <_1)$ are Countryman lines.

Here $\rho_0 : [\omega_1]^2 \to {}^{<\omega}\omega$ is recursively defined by

$$\rho_0(\alpha, \beta) = \langle \mathrm{ot}(C_\beta \cap \alpha)\rangle^\frown \rho_0(\alpha, \min(C_\beta \setminus \alpha),$$

and $\rho_1 : [\omega_1]^2 \to \omega$ is recursively defined by

$$\rho_1(\alpha, \beta) = \max\{\mathrm{ot}(C_\beta \cap \alpha), \rho_1(\alpha, \min(C_\beta \setminus \alpha))\}.$$

The definitions of ρ_0 and ρ_1 extend to larger successor cardinals. When working with walks on cardinals larger than ω_1, it is useful to have c-sequences with coherence properties like those of $\square_\kappa$-sequences. Todorcevic used his new approach to revisit some results on large cardinals and about combinatorial properties of L. For example, he generalized the notion of special tree to limit cardinals as follows. Suppose $(T, <_T)$ is a tree of height κ. A function $f : T \to T$ is *regressive* if for all non-minimal $t \in T$, $f(t) <_T t$. T is *special* if there is a regressive function $f : T \to T$ such that for every $t \in T$, $f^{-1}(\{t\})$ is the union of fewer than κ many antichains. Todorcevic [1987, 266] proved that among inaccessible cardinals κ, there are no special κ-Aronszajn trees if and only if κ is Mahlo.

Laver [1987] proved that Shelah's Cohen Real to Suslin Tree Theorem does not extend to forcing with random reals. By adding many random reals to a universe satisfying $\mathrm{MA}_{\aleph_1}$, Laver obtained a model of "$2^{\aleph_0}$ is singular with cofinality ω_1" with no Suslin tree.

In chapter IX of [1982b], Shelah came up with some new proper forcing techniques for specializing an Aronszajn tree, including one that did not add reals, and some that specialized an Aronszajn tree on a stationary set but not on the rest of the tree. Thus he was able to build a model with no Suslin trees without making all Aronszajn trees special.

Several consistency results for the existence or non-existence of Aronszajn trees on $\aleph_n$'s were proved in the 1980's. Note that the non-existence of κ-Aronszajn

[386] The definition of coherent tree given here is from Todorcevic [2007b].

trees for regular κ means that κ has the *tree property*, i.e. every κ-tree has a cofinal branch.

Abraham [1983] used a supercompact cardinal with a weakly compact cardinal above to show the consistency of having no $\aleph_2$-Aronszajn trees and no $\aleph_3$-Aronszajn trees. Included in his paper is a proof by Magidor that if there are no $\aleph_2$-Aronszajn trees and $\aleph_3$ is inaccessible in L (as would be the case if there were no $\aleph_3$-Aronszajn trees), then $0^\#$ exists, and that two weakly compact cardinal would be insufficient.[387]

Shelah [1982a] generalized bqo theory to uncountable cardinals and identified the least cardinal κ such that the class of unions of $\leq \lambda$ scattered orderings is κ-well-ordered (a generalization of wqo) and called these cardinals *beautiful cardinals*. He also gave a construction for rigid families of colored trees, and generalized the method to other structures.

Laver and Shelah [1981], starting from a weakly compact cardinal, used iterated forcing to construct a model of CH in which there are no $\aleph_2$-Suslin trees.

Baumgartner [1983b], as an application of Axiom A forcing,[388] proved that if the existence of an inaccessible cardinal is consistent, then so is MA + $2^{\aleph_0} = \aleph_2$ + every $\aleph_1$-tree has at most $\aleph_1$ uncountable branches.

Shelah and Stanley [1988] proved that a successor cardinal $\kappa = \lambda^+$ has a non-special Aronszajn tree when either $\square_\lambda$ holds or κ is not weakly compact in L.

Shai Ben-David and Shelah [1986] proved the consistency, relative to the existence of a supercompact cardinal, of the existence of λ^+-Suslin trees for every singular λ. Under the same assumption, they proved the consistency of the existence of an $\aleph_{\omega+1}$-Suslin tree together with the non-existence of any special $\aleph_{\omega+1}$-Aronszajn tree.

Chaz Schlindwein [1989], starting from an inaccessible cardinal, constructed a model of CH + SH (no Suslin trees) + KH (there is a Kurepa tree) in which not all Aronszajn subtrees are special.[389]

Masazumi Hanazawa [1982] defined a non-Suslin base for an Aronszajn tree $(T, <_T)$ to be a collection C of uncountable antichains such that for any uncountable $S \subseteq T$ there is an $X \in C$ such that for all $s \in X$ there is a $t \in S$ with $s \leq_T t$. Note that if T has a non-Suslin base, no subtree of T is Suslin. Hanazawa used Martin's Axiom to prove that no Aronszajn tree has a non-Suslin base of cardinality less than the continuum. He used $\diamondsuit^+$ to prove that there is a special Aronszajn tree which has no base of size $\aleph_1$, and asked whether $\diamondsuit^+$ was necessary.

Baumgartner [1985] defined *base for a tree* T of height ω_1 to be a collection $\mathcal{B}$ of subtrees of T with the property that every subtree of T contains one of the subtrees in $\mathcal{B}$. Note that the antichains used by Hanazawa can be turned into subtrees by downward closure, so the notion of Baumgartner generalizes the notion

[387] Recall that Mitchell and Silver [Mitchell, 1972] showed that non-existence of $\aleph_2$-Aronszajn trees and $\aleph_4$-Aronszajn trees is equiconsistent with the existence of two weakly compact cardinals.

[388] Axiom A forcings are proper and were a precursor to proper forcings.

[389] In his Math Review MR1031772 (91i:03101) of this article, Di Prisco commented that models of ZFC exist satisfying every combination of the form $\pm$ CH $\pm$ SH $\pm$ KH $\pm$ EATS that are not ruled out by the fact that EATS (every Aronszajn tree is special) implies SH.

of Hanazawa. Baumgartner answered Hanazawa's question by showing that, in the model obtained from the Levy collapse of an inaccessible cardinal to ω_2, every Aronszajn tree has a base of size $\aleph_1$ and $\diamondsuit$ holds. He defined an anti-Suslin tree to be a tree with no Suslin subtree and called a Specker order (also called Aronszajn line) *anti-Suslin* if it has no Suslin suborder. Such orderings arise from anti-Suslin trees. From the above theorem and the equivalence of bases with non-Suslin bases for anti-Suslin trees, the following conclusion may be drawn: if it is consistent that there is an inaccessible cardinal, then it is consistent that for every anti-Suslin Specker order S there is a collection C of subsets of S such that C has cardinality $\leq \aleph_1$ and every uncountable subset of S contains an order-isomorphic copy of an element of C.

Moreover, Baumgartner showed that if there is a Kurepa tree with at least κ branches, then there is a special Aronszajn tree for which every non-Suslin base has cardinality at least κ. The proof he gave, due to Todorcevic and included with his permission, is the observation that the level-wise product of a Kurepa tree $(K, \leq_K)$ with κ branches and a special Aronszajn tree $(T, \leq_T)$, namely $KT = \{(s,t) : s \in K, t \in T, \text{ and } \ell(s) = \ell(t)\}$[390] under the coordinate-wise ordering, is the required special Aronszajn tree in which every base has cardinality $\geq \kappa$. Baumgartner asked whether it is consistent that both CH holds and no Aronszajn tree has a base of cardinality $\aleph_1$.

Abraham and Shelah [1985][391] investigated Aronszajn trees and embeddings of their subtrees, particularly those subtrees obtained by choosing all nodes from a closed unbounded (club) set of levels. They strengthened some of the notions of structure considered earlier. Recall that in [Shelah, 1976], two Aronszajn trees were said to be *near* if for some club $C \subseteq \omega_1$ they have Aronszajn subtrees which are isomorphic when restricted to C. Abraham and Shelah called S and T *club isomorphic* if for some club C, the subtrees $S \restriction C$ and $T \restriction C$ are isomorphic, and said S is *club embeddable* in T, if for some club C, $S \restriction C$ embeds in $T \restriction C$. Say that two Aronszajn trees are *really different* if for every club $C \subseteq \omega_1$ their restrictions to C are non-isomorphic. They deemed an Aronszajn tree to be *really rigid* if for any club C its restriction to C has no order automorphisms other than the identity. They proved a series of consistency results using forcing and cardinality considerations:

1. It is consistent to have GCH + there is no Suslin tree, every two Aronszajn trees are near and there is an Aronszajn tree U universal in the sense that every Aronszajn tree is club embeddable in U.

2. If $2^\omega < 2^{\omega_1}$, then there are 2^{ω_1} really rigid pairwise really different Aronszajn trees.

3. If $2^\omega < 2^{\omega_1}$, then there is no minimum Aronszajn tree P in the sense that P is club embeddable into every Aronszajn tree.

[390] Here $\ell(s)$ is the level of s.

[391] The paper was received July 1, 1983; an abstract for it dates back to 1979.

4. It is consistent to have GCH + "there are exactly ω_1 non-isomorphic Suslin trees and there is an Aronszajn tree U which is almost universal in the sense that every Aronszajn tree without Suslin subtrees is club embeddable in U."

5. MA + $\neg$CH is independent of "every two Aronszajn trees are isomorphic on a club".[392]

6. It is consistent to have MA + "2^ω is as large as you like and no two Aronszajn trees are really different."

They asked if the existence of a Suslin tree implied the existence of two really different Suslin trees [1985, 78], and they asked a basis problem [1985, 79]:

Problem. It is not clear whether we can get the consistency of:

(*) Any Specker order contains a suborder in K'_s.[393]
This is equivalent to the following consistency question.

(**) If T is an Aronszajn tree, $T = A_1 \cup A_0$, then there is an unbounded $B \subseteq T$ and $\ell \in \{0, 1\}$ such that $x, y \in B \implies$ g.l.b.$\{x, y\} \in A_\ell$.[394]

Ingrid Lindström,[395] in a paper based on her 1979 Stanford dissertation, investigated a pre-order $\leq$ defined on rooted, Hausdorff, well-pruned $\geq$ 2-full ω_1-trees by $S \leq T$ if and only if for some closed unbounded set $C \subseteq \omega_1$ there is an order-preserving and level-preserving mapping carrying the restriction of T to levels indexed by C, briefly $T \restriction C$ to $S \restriction C$. In the language of Abraham and Shelah, $S \leq T$ if T is club-embeddable in S. Intuitively, the relation $S \leq T$ holds for Suslin trees if forcing with T adjoins a branch through S. She points out that the equivalence classes (degrees) of Aronszajn trees under $\leq$ with the level-wise product and disjointed union with added root operations forms a distributive lattice, and uses $\diamondsuit$ to prove that all countable linear order types embed in it and that there is no maximal Aronszajn degree. Recall that Jech [1972] had used $\diamondsuit$ to proved the existence of $2^{\aleph_1}$ pairwise non-isomorphic Suslin trees. Lindström uses $\diamondsuit$ to proved the existence of $2^{\aleph_1}$ pairwise $\leq$-incomparable families of several types of ω_1-trees: Suslin trees, special Aronszajn trees, non-special $\mathbb{R}$-embeddable Aronszajn trees.

---

[392]Shelah proved the consistency relative to Martin's Axiom for stable posets (see his paper with Abraham [1982] for the definition of stable). Then Abraham used generic reals to extend the result to the full Martin's Axiom. Shelah constructed a proper forcing with which to force two Aronszajn trees to be isomorphic on a club, using finite conditions. In §5 where the construction is given, the authors [1985, 100] remark "At the present state of knowledge, we do not know how to go with proper forcing beyond $2^{\aleph_0} = \aleph_2$".

[393]K'_s is the collection of the Countryman lines, so (*) may be rephrased as "every Aronszajn line embeds a Countryman line."

[394]The ordering of the indices $T = A_1 \cup A_0$ is the way it appears in [Abraham and Shelah, 1985]. The statement (**) is called the *Coloring Axiom for Trees* or CAT in [Todorcevic, 2007a]. In current usage one writes $x \wedge y$ for greatest lower bound of x and y.

[395]Ingrid Birgitta Lindström received her doctorate from Stanford University in 1979 under the direction of Ketonen. She is currently a senior lecturer at Uppsala University.

8.6 Linear orders

Baumgartner [1980] used iterated forcing to build a model of Martin's Axiom + $2^{\aleph_0} = \aleph_2$ in which (A) every uncountable subset of $\mathcal{P}(\omega)$ contains an uncountable chain or antichain and (C) any two $\aleph_1$-dense sets of reals are order-isomorphic. Then he proved that (A) + (C) implies that every uncountable Boolean algebra has an uncountable antichain.[396]

Abraham and Shelah [1981] answered a question of Baumgartner [1973] (see §7.8 for the statement of Baumgartner's question) when they proved that Martin's Axiom does not imply that all $\aleph_1$-dense sets of reals are isomorphic.

At the 1981 conference on ordered sets, Pouzet (cf. [Bonnet and Pouzet, 1982]) gave a series of lectures on what he called the most popular connection between partial orders and linear orders, namely that every partial order can be extended to a linear order (Szpilrajn's Theorem [Marczewski, 1930]). One topic was the question of whether a partial order with a given property P can be extended to a linear order with the same property P or whether all extensions to a linear order will have this property.

At the same conference, Baumgartner [1982] surveyed and extended known results for uncountable linear orders. He reviewed Specker types, and proved that every Specker order is isomorphic to a lexicographically ordered Aronszajn tree and that there are $2^{\aleph_1}$ non-isomorphic $\aleph_1$-dense Specker types. He highlighted the fact that the Countryman line constructed by Shelah with its converse are incompatible, no uncountable type is below both, and asserted that under the assumption of $\diamondsuit$, for any Aronszajn tree T there is an Aronszajn tree U such that no Aronszajn tree is embeddable in both T and U. He also proved under the stronger assumption of $\diamondsuit^+$ that there is a Specker type φ which is minimal, i.e. if ψ is embeddable in φ then φ is also embeddable in ψ.

Abraham, Matatyahu Rubin and Shelah [1985] presented a variety of consistency results, including the consistency of ZF + $2^{\aleph_0} < 2^{\aleph_1}$ + "for any two $\aleph_1$-dense subsets $A, B \subseteq \mathbb{R}$, there is an $\aleph_1$-dense $C \subseteq \mathbb{R}$ which is order-embeddable in both A and B." They also produced a ZFC example of two $\aleph_1$-dense sets for which the order-isomorphism cannot be taken to be C^1. Let $(K^H, \leq)$ be the set of order types of $\aleph_1$-dense homogeneous subsets of the reals the embeddability relation. They prove that for every finite model $(L, \leq)$, the statement $(K^H, \leq)$ is isomorphic to $(L, \leq)$ is consistent if and only if $(L, \leq)$ is a distributive lattice.

8.7 Other combinatorial results

In the paper just discussed above, Abraham, Rubin and Shelah introduced the *Semi-Open Coloring Axiom* (SOCA) and the *Open Coloring Axiom*, which we here will abbreviate $\mathrm{OCA}_{\mathrm{ARS}}$, along with a variety of other axioms. For a subset $Y \subseteq X$ of a topological space X, let $D(Y) = Y \times Y \setminus \{(y, y) : y \in Y\}$.

[396] Maria Losada and Todorcevic [2000] showed that $\mathrm{MA}_{\aleph_1}$ suffices.

> *Open Coloring Axiom (OCA_{ARS}):* If X is a second countable space of power $\aleph_1$ and $\{U_0, \ldots, U_{n-1}\}$ is a cover of $D(X)$ consisting of symmetric open sets, then there is a partition of $X = \bigcup_{i<\omega} X_i$ such that for every $i < \omega$, there is an $\ell < n$ such that $D(X_i) \subseteq U_\ell$.

The trio of authors proved, relative to ZFC, the consistency of MA + SOCA and moreover the consistency of MA + SOCA + $\neg$$\mathrm{OCA_{ARS}}$. They also prove that MA + $\mathrm{OCA_{ARS}}$ implies $2^{\aleph_0} = \aleph_2$.

The Abraham-Rubin-Shelah paper [1985] was received by the journal March 3, 1982. That spring, Todorcevic [1989] formulated a different open coloring axiom as a proposition using PFA, and I have reformulated it below using coloring instead of partitions.

> *Open Coloring Axiom (OCA_T):* For every coloring $f : [\mathbb{R}]^2 \rightarrow \{0, 1\}$ with $f^{-1}(\{0\})$ open, either there is an uncountable 0-homogeneous set Y, or $\mathbb{R}$ is the union of countably many 1-homogeneous sets.[397]

Todorcevic listed some straightforward consequences: (1) every uncountable subset of $\mathcal{P}(\omega)$ contains an uncountable chain or antichain, and (2) every function f from an uncountable set of reals into the reals is monotonic on some uncountable set.[398] In remarks at the end of chapter 8, Todorcevic credited Baumgartner [1973], [1980] with "the main advance towards the Open Coloring Axiom for sets of reals" and continued "the discovery of Baumgartner ... is that a CH diagonalization argument can be used in building [a] c.c.c. poset which forces homogeneous sets for certain open partitions on sets of reals" [Todorcevic, 1989, 86–87].

Moore [2002] proved that $\mathrm{OCA_{ARS}}$ + $\mathrm{OCA_T}$ implies $2^{\aleph_0} = \aleph_2$.

Aharoni, Nash-Williams and Shelah [1983] devised a general criterion for the existence of transversals. They used *critical families* $\mathcal{F}$, indexed families (functions) with a transversal that have the property that for every transversal φ, the range of φ is the union of the range of $\mathcal{F}$. They use critical families to identify a collection of *obstructions* and prove that a family has a transversal if and only if it does not have a subfamily in the collection of obstructions. In a follow-up, the authors [Aharoni *et al.*, 1984] gave a different criterion based on an idea of [Nash-Williams, 1978] that avoided use of critical families.

Shelah [1985] proved that if S is a stationary co-stationary subset of ω_1, then the Weak Diamond Principle for S does not follow from $2^{\aleph_0} < 2^{\aleph_1}$.

---

[397] Moore [2002] used the following reformulation of $\mathrm{OCA_T}$: if X is a separable metric space and $G \subseteq [X]^2$ is open, then either G is countably chromatic (there is a decomposition of X into countably many pieces X_i such that each $X_i \cap G = \emptyset$) or there is $H \subseteq X$ such that $[H]^2 \subseteq G$. That is, either X is the union of countably many independent sets or G includes an uncountable complete subgraph.

[398] The Axiom $\mathrm{OCA_T}$ has applications in analysis, e.g. Ilijas Farah [2011] used it to establish the consistency of all automorphisms of the Calkin algebra being inner. His paper complements a 2006 construction of Phillips and Weaver [2007] of an outer automorphism using CH. Together these results solve a problem of Brown, Douglas and Fillmore [1977].

9 1990–2000: A SAMPLING

For the final decade of the 20th Century, I am presenting a miscellaneous collection of results I found interesting and make no claims that they are representative for this time. In particular, singular cardinal combinatorics became an increasingly important part of the subject, but it is a subject of which I know little. I warmly recommend the survey article by James Cummings [2005], a recent paper [2009] by Todd Eisworth and his chapter [2010] for the *Handbook of Set Theory*, especially Section 5 on square bracket partition relations,

Alan Dow [1992] provided a valuable update to the *Handbook of Set-Theoretic Topology* [Kunen and Vaughan, 1984] in a survey paper written based on a talk at the Symposium on Topology (Topsym) held in Prague.[399] Another valuable resource that came out of that meeting is the compilation *Open Problems in Topology* [van Mill and Reed, 1990].[400]

In 1998, Hindman and his frequent collaborator, Dona Strauss, made widely available their work on Ramsey theory, especially semi-group colorings, in their book [1998],[401] which includes interesting historical remarks.

Todorcevic [1998a] spoke on *Basis problems in combinatorial set theory* at the 1998 International Congress of Mathematicians in Berlin. Of particular note for this chapter was his discussion of c-sequences and their related distance functions as critical objects for the collection of uncountable subsets of ω_1. The variety of applications is astounding, and his recent book, *Walks on Ordinals and Their Characteristics* [2007b], is an important new resource. Todorcevic focused on five general areas: (1) uncountable subsets of ω_1, (2) binary relations on ω_1 under Tukey reducibility, (3) transitive relations both on ω_1 and on the reals, (4) uncountable linear orderings, and (5) topological spaces. In each case, he identified a list of critical objects and either indicated why it is known to be sufficiently rich that one can relate any object in the class to one of the critical objects or made a conjecture to that effect under the assumption of PFA. He referred readers interested in metamathematical aspects to the 1,000 page book by Woodin, saying that they "will find a satisfactory explanation in the recent monograph of Woodin" (see [Woodin, 1999]).

In his examination of uncountable subsets of ω_1, Todorcevic focused on the delightful fact that many critical objects in families of uncountable structures whose domain is a subset of ω_1 can be defined in terms of a c-sequence.[402] Todorcevic reviewed walks on ordinals, the trace function of places visited on a walk

[399] The Prague Topsym conference meets every five years.

[400] In his Math Review MR1078636 (92c:54001) of the problem book, John Walsh noted "The journal *Topology and its Applications* has agreed to provide a readily available source for determining the status of the various problems. Each issue of the journal will have space devoted to updating the status as progress is made on individual problems. A second book of problems [Pearl, 2007] has followed.

[401] This book was cited in the 2003 JAMS award citation honoring Hindman's work.

[402] For the general audience, Todorcevic omitted the term c-sequence in favor of describing one on ω_1 with minimal jargon as a sequence C which assigns to each ordinal α a subset c_α of smaller ordinals of minimal order type with $\alpha = \sup c_\alpha$.

from β to α, distance functions, ρ_0, and Todorcevic's square bracket function, $[\alpha\beta] = \min(\mathrm{Tr}(\xi, \beta) \setminus \alpha)$ where $\xi = \min\{\zeta : \rho_0(\zeta, \alpha) \neq \rho_0(\zeta, \beta)\}$. He connected an uncountable set $X \subseteq \omega_1$ with one which contains a closed unbounded subset using the square bracket operation (see [Todorcevic, 1987]), namely

$$\{[\alpha\beta] : \alpha \in X \wedge \beta \in X \wedge \alpha < \beta\}.$$

He pointed to the utility of this approach through citation of others' work: the construction by Saharon Shelah and Juris Steprāns [1987] of Ehrenfeucht-Faber groups, that is, extraspecial p-groups G in which each Abelian subgroup has cardinality less than $|G|$; the construction by Shelah and Steprāns [1988] of a Banach space on which every bounded linear operator is the sum of a multiple of the identity operator and an operator with a separable range; and the construction by Erdős, R. Daniel Mauldin, and Steve Jackson [1997] of a family H of ω_1 hyperplanes in $\mathbb{R}^n$, a family X of ω_1 points in $\mathbb{R}^n$ and a coloring $P : H \to \omega$ such that any n hyperplanes of distinct colors meet in at most one point, and there is no coloring $Q : X \to \omega$ such that each hyperplane $h \in H$ meets X in at most $n-1$ points of its color, $P(h)$.

9.1 *Partition calculus results*

We split the discussion of partition calculus results into the same subsections as before, with the addition of one for polarized partitions so that this active area of research dating back to the 1950's has some representation in this chapter.

Cardinal resources

The earliest uncountable partition calculus result stated in the literature is the Dushnik-Miller Theorem

$$\kappa \to (\kappa, \omega)^2.$$

Implicit in the original proof and explicit in [Erdős and Rado, 1956] is the relation $\kappa \to (\kappa, \omega+1)^2$. Erdős and Hajnal [1971b, 22] asked in their problem paper arising from the 1967 UCLA set theory conference the following natural question about the limitations on the possible extension of the Dushnik-Miller Theorem:

> *Problem 8:* Can one prove without using the continuum hypothesis that $\omega_1 \not\to (\omega_1, \omega+2)^2$ holds?

Péter Komjáth [1998], [1999] examined what was known about this foundational theorem, derived some additional ZFC results, and clarified the boundary between what can be proved in ZFC with additional consistency results. Recall that Hajnal [1960] used CH to prove that $\omega_1 \not\to (\omega_1, \omega+2)$ and the proof gave $\omega_1 \not\to (\omega_1, (\omega{:}2))$ as he later pointed out in [Erdős and Hajnal, 1974]. Komjáth [1998] showed these two partition relations were different by forcing to produce a model in which the bipartite partition relation holds but the complete subgraph relation does not:

$\omega_1 \to (\omega_1, (\omega{:}2))^2$ and $\omega_1 \not\to (\omega_1, \omega+2)^2$. Recall that Baumgartner [1976a] proved the consistency of $\kappa \not\to (\kappa, (\kappa, (\omega{:}2))^2$ for every cardinal κ which is the successor of a regular cardinal. Laver [1975] proved that if κ is Mahlo, then one can force to make $\kappa \not\to (\kappa, (\omega{:}2))^2$ true and preserve the fact that κ is Mahlo. Komjáth [1999] proved that if λ is singular, then $\lambda \to (\lambda, (\mu{:}n))^2$ holds for $\mu < \lambda$ and $n < \omega$.

Recall that Todorcevic [1983a] proved that if λ is a cardinal of uncountable cofinality, there is a c.c.c. forcing for adding a witness to $\lambda \not\to (\lambda, \omega+2)^2$. Komjáth [1999] proved that under GCH, if κ is a regular cardinal and λ is a cardinal whose cofinality is greater than κ, then there is a κ-closed, κ^+-chain condition forcing which adds a witness to $\lambda \not\to (\lambda, \kappa+2)^2$. Recall that Erdős and Hajnal [1974] reported that Galvin had proved $\omega_2 \to (\omega_1, \omega+2)^2$ and that he had mentioned as open problems the questions about whether the following hold without the assumption of GCH: Does $\omega_2 \to (\omega_1, \omega+3)^2$? Does $\omega_2 \to (\omega_1, (\omega+2)_2)^2$? Komjáth pointed out that Galvin's proof generalizes to give $\omega_n \to (\omega_1, \omega+n)^2$ for positive $n < \omega$. Komjáth proved that for all $n < \omega$,

$$\omega_3 \to (\omega_1, \omega+n)^2.$$

He also showed that if the resource ω_3 shrinks to ω_2, then it is consistent that

$$\omega_2 \not\to (\omega_1, \omega+3)^2.$$

In a 1997 lecture series at Rutgers University, Shelah commented on a variety of problems he found interesting. A sample is Question 8.6 [2000b, 61] below:

> For which cardinals λ and which ordinals α with $2 \leq \alpha < \lambda$ does the partition relation $\lambda^{++} \to (\lambda^+ + \alpha)^2_{\aleph_0}$ hold? GCH may be assumed for simplicity.

Recall that Erdős and Hajnal [1971b, 23] asked the following natural question about the limitations on possible extensions of the Erdős-Rado Theorem:

> *Problem 10:* Assume GCH. Does then $\omega_{\rho+1} \to (\xi, \xi)^2$ hold for every $\xi < \omega_{\rho+1}$ and every ρ?

Baumgartner, Hajnal and Todorcevic [1993, 5], all of whom had given invited talks at the NATO Advanced Science Institute on Finite and Infinite Combinatorics in Sets and Logic which had been held in Banff in 1991, published their generalization of the Erdős-Rado theorem, now known as the Baumgartner-Hajnal-Todorcevic Theorem, in the conference volume using the notation $\log(\kappa)$ for the least cardinal μ with $2^\mu \geq \kappa$:

> *Balanced Baumgartner-Hajnal-Todorcevic Theorem:* If κ is a regular cardinal and $\lambda = 2^{<\kappa}$, then for $k < \omega$ and $\xi < \log(\kappa)$
>
> $$\lambda^+ \to (\kappa+\xi)^2_k.$$

Unbalanced Baumgartner-Hajnal-Todorcevic Theorem:
If κ is a regular cardinal and $\lambda = 2^{<\kappa}$, then for all $n < \omega$

$$\lambda^+ \to (\kappa^{\omega+2} + 1, \kappa + n)^2.$$

For $\kappa = \omega$, the results follow from the Baumgartner-Hajnal Theorem $\omega_1 \to (\alpha)^2_k$ for $k < \omega$ and $\alpha < \omega_1$, so the main interest is for κ uncountable. The authors note that in unpublished work, Hajnal proved for $\lambda = 2^{<\omega_1}$ that $\lambda^+ \to (\kappa + m)^2_2$ for $m < \omega$ "about thirty years ago" [Baumgartner *et al.*, 1993, 5], and Shelah proved the balanced version for $k = 2$ in his [1973]. The trio of authors used what has come to be called the *elementary submodel method* for their proof, which was different from the approach of Simpson [1970] in his model-theoretic proof of the Erdős-Rado Theorem. Later, Hajnal [2010] commented that this paper was the one which introduced non-reflecting ideals as a tool in the study of partition relations. Baumgartner, Hajnal and Todorcevic made use of elementary substructures of the form $(H(\mu), \in)$ where $H(\mu)$ is the set of all sets hereditarily of cardinality $< \mu$, and ideals on ordinals generated from such elementary substructures. They wrote: "it is possible to recast our arguments in such a way as to fit them under the heading of ramification arguments but we choose not to do so, both because the proofs remain clearer this way, and because we believe that this may generally be a better approach to ramification arguments" [Baumgartner *et al.*, 1993, 2]. They also used the metamathematical approach taken in the proof of the Baumgartner-Hajnal Theorem of using forcing to get the desired result in a generic extension, and then using an absoluteness argument to prove the statement is outright true.

Milner [1994] emphasized the utility of using elementary substructures in an expository article designed to encourage more people to use this approach, in place of the ramification arguments of many earlier proofs of combinatorial results.

For the proceedings of the Hajnal conference of October 1999, Baumgartner wrote about the use of ideals in arguments involving elementary substructures, which were a feature of Hajnal's work. He started with a treatment of the Erdős-Rado Theorem using the non-stationary ideal. In this paper, he attributed to Hajnal the following unbalanced extension of the Erdős-Rado Theorem: for regular uncountable κ, all ξ with $2^\xi < \kappa$, and all $k < \omega$, the following partition relation holds:

$$(2^\kappa)^+ \to (\kappa + \xi)^2_k,$$

whose proof appeared in [Baumgartner *et al.*, 1993]. Baumgartner revisited the proof using many elements of the proof from that paper, and the perspective gained from nearly a decade.

Todorcevic [1998b] proved a common generalization of the Baumgartner-Hajnal Theorem for ω_1 [1973] and Prikry's Theorem [1972] when he showed that for all finite k and countable ordinals α the following partition relation holds:

$$\omega_1 \to ((\alpha; \omega_1), (\alpha)_k)^2.$$

That is, if $[\omega_1]^2$ is partitioned into $k+1$ pieces, then either the first piece includes $\{\{\xi,\eta\}_< : \xi \in A \wedge \eta \in A \cup B\}$ for some A of order-type α and some uncountable B or one of the other pieces includes $[A]^2$ for some A of order-type α.

At the end of the decade, Matthew Foreman[403] and Hajnal [2003] were writing up their results on partition properties of large successor cardinals, which gave further insight into possibilities for Problem 10. They showed that if the Continuum Hypothesis holds and there is an $\aleph_1$-dense ideal on ω_1, then, for $\alpha < \omega_2$,

$$\omega_2 \rightarrow ({\omega_1}^2 + 1, \alpha)^2.$$

At the International Congress of Mathematicians held in Berlin, Foreman [1998b] spoke on *Generic large cardinals: new axioms for mathematics?* In part of his talk, he discussed his work with Hajnal and pointed out that the assumption of the existence of an $\aleph_1$-dense ideal on $\aleph_2$ implies both hypotheses that they used to prove their positive partition relation hold.[404] Note that Foreman [1998a] had shown the consistency of the existence of such an $\aleph_1$-dense ideal on ω_2 as well as the consistency of the existence of an $\aleph_1$-dense ideal on ω_1.

Now we shift attention to partitions of triples. Recall that Todorcevic [1983a] proved that under $\mathrm{MA}_{\aleph_1} + 2^{\aleph_0} = \aleph_2$, the following positive partition relation holds for all $\alpha < \omega_1$:

$$\omega_1 \rightarrow (\omega_1, \alpha)^2.$$

Using this result, Milner and Prikry [1991] proved that $\omega_1 \rightarrow (\omega \cdot 2 + 1, 4)^3$ also holds, using $\mathrm{MA}_{\aleph_1}$ and absoluteness as in the Baumgartner-Hajnal Theorem.

Albin Jones,[405] in his thesis [1999], [2000b], took full advantage of the model-theoretic approach to partition relations to prove the following partition relation for all $m, n < \omega$:

$$\omega_1 \rightarrow (\omega + m, n)^3.$$

> *Open Question:* Is it true that for all $\alpha < \omega_1$ and $n < \omega$, the following partition relation holds:
>
> $$\omega_1 \rightarrow (\alpha, n)^3?$$

Ordinal partition relations

Progress was made on the problem of determining which countable ordinals are partition ordinals.[406]

---

[403] Matthew Foreman received his doctorate in 1980 from the University of California, Berkeley, where his advisor was Solovay.

[404] In the mid-1990's, Woodin showed that if you have an $\aleph_1$-dense ideal on ω_2, then CH hold. See [Foreman, 2010] for a proof.

[405] Albin Jones received his doctorate from Dartmouth College, where Baumgartner was his advisor. His academic career, which included some time at the University of Kansas, was fairly short, but mathematics remains an avocation.

[406] Recall that α is a *partition ordinal* if it satisfies $\alpha \rightarrow (\alpha, 3)2$.

Galvin, in a paper with Larson [1975], had shown that no countable ordinal greater than ω^2 which is the ordinal product of two strictly smaller ordinals is a partition ordinal. Ordinals which do not have this property are called *multiplicatively indecomposable*, and one can show that they have the form ω^α for some additively indecomposable ordinal α. Since additively indecomposable ordinals can be written in the form $\alpha = \omega^\beta$ and Chang [1972] had shown that ω^ω is a partition ordinal, ω^{ω^2} was the next test case.

Carl Darby[407] and Rene Schipperus independently came up with proofs that ω^{ω^2} is a partition ordinal in the 1990's. I remember hearing Darby's proof during a visit to Boise State University, possibly as early as 1991, the year of the first BEST meeting. Darby's work used Ramsey's Theorem and the *good sequences* developed by Galvin [1971] to represent the ordinal ω^{ω^2} and a careful analysis of the avoidable and unavoidable ways two such disjoint sequences could be interlaced to prove ω^{ω^2} is a partition ordinal, though his proof remains unpublished.

Rene Schipperus recalled enjoyable conversations with Baumgartner about partition problems and partition ordinals in the 1991 meeting in Banff. He worked for a long time on his own, spent some time speaking in a seminar at the University of Calgary with Claude Laflamme[408] and Milner[409] about this work. Eventually, Schipperus wrote a thesis [1999; 2010] with Laver for which he won the Sacks prize for his proof that every denumerable ordinal of the form ω^{ω^β} where β is the sum of one or two indecomposable ordinals is a partition ordinal. In place of the lexicographically ordered good sequences used by Darby for his underlying set, Schipperus used finite labeled trees, carefully constructed to enable him to build his large subsets in a highly restricted way. The heart of his argument is a Ramsey dichotomy, proved using a game and the Nash-Williams Partition Theorem. Darby and Schipperus independently came up with new families of examples witnessing $\alpha \not\to (\alpha, k)$ for different values of k. All the known examples use oscillation type behavior where, for larger values, the blocks can be thought of as having a finite number of possible weights.

Darby [1999] developed a general method to show how to build finite sets witnessing a particular pattern of weighted blocks inside a large subset of an ordinal, depending on the order types of the original set and the large subset. This allowed him to reduce the question of counter-examples to one of finite combinatorics. Schipperus extended the notion of having a large "free" subset (see Lemma 7.2.2 of [Williams, 1977]) to enable him to construct the necessary finite sets.

These negative results are best expressed using the *additive normal form* for the exponent β for an ordinal of the form ω^{ω^β}, that is, its unique representation as a sum of non-decreasing additively indecomposable ordinals. Larson built on

[407]Carl Darby was a student of Laver at the University of Colorado, who earned his doctorate in 1990.

[408]In 1987, Claude Laflamme received his doctorate from the University of Michigan, Ann Arbor, where his advisor was Andreas Blass.

[409]Recall that Milner was the careful reader of Chang's proof who generalized it from 3 to all m; Milner died in 1997 and did not see the final version of the proof by Schipperus.

their work to improve one of the results obtained by both of them.

1. (Darby unpublished) If $\beta = \omega^{\alpha+1}$ and $m \to (4)^3_{2^{32}}$, then
$$\omega^{\omega^\beta} \not\to (\omega^{\omega^\beta}, m)^2.$$

2. (Darby [1999] for 6; Schipperus [1999] for 6; Larson for 5 [2000]) If $\beta \geq \gamma \geq 1$, then
$$\omega^{\omega^{\beta+\gamma}} \not\to (\omega^{\omega^{\beta+\gamma}}, 5)^2.$$

3. (Darby [1999]; Schipperus [1999]) If $\beta \geq \gamma \geq \delta \geq 1$, then
$$\omega^{\omega^{\beta+\gamma+\delta}} \not\to (\omega^{\omega^{\beta+\gamma+\delta}}, 4)^2.$$

4. (Schipperus [1999]) If $\beta \geq \gamma \geq \delta \geq \varepsilon \geq 1$, then
$$\omega^{\omega^{\beta+\gamma+\delta+\varepsilon}} \not\to (\omega^{\omega^{\beta+\gamma+\delta+\varepsilon}}, 3)^2.$$

The Ramsey game $R(\varphi, n, \psi)$ is played by two players, the Builder and the Spoiler, alternating choosing n-element subsets of a linear order L of order type φ, the resource. The Builder goes first and wins if and only if there is a set $L' \subseteq L$ of order type ψ, the goal, such that $[L']^n$ is a subset of the Builder's moves. Using Galvin's result from §7.3 that if φ is scattered and does not embed ω_1, then
$$\varphi \not\twoheadrightarrow [\omega, \omega^2, \omega^3, \dots]^1_\omega.$$
Darby and Laver [1998] showed that the Spoiler has a winning strategy in the Ramsey game $R(\omega^\omega, 3, \omega^\omega)$

We conclude this section with two open questions on ordinal partition relations, the first asked many times by Erdős and the second inspired by the proof by Hajnal [1960] of ${\omega_1}^2 \not\twoheadrightarrow ({\omega_1}^2, 3)^2$ and explicitly asked by Baumgartner.

Open Questions:

1. Characterize the set of $\alpha < \omega_1$ for which $\alpha \to (\alpha, 3)^2$ holds.
2. Is it consistent that ${\omega_1}^2 \to ({\omega_1}^2, 3)^2$?

Erdős and Hajnal [1974] specifically asked about the consistency of the partition relation ${\omega_1}^2 \to ({\omega_1}^2, 3)^2$ with $\mathrm{MA}_{\aleph_1} + 2^{\aleph_0} = \aleph_2$.

Polarized partitions

Polarized partitions may be interpreted in different ways. One approach is to think of them as treating partitions of a product of two sets that can be thought of as a rectangle for which you seek a homogeneous set which is a product of subsets of the given factors. Another approach is to think of one of the factors parameterizing a family of partitions. For each of the examples below, we specify the partition relation.

Example 1: Erdős, Hajnal and Rado had used GCH to prove that for every infinite cardinal κ, the following partition relation holds (see Theorem 39 [1965, 172–176]):

$$\begin{pmatrix} \kappa^+ \\ \\ \kappa^+ \end{pmatrix} \to \begin{pmatrix} \kappa^+ & & \kappa & \kappa \\ & \vee & & \\ \kappa & & \kappa^+ & \kappa \end{pmatrix}^{1,1}.$$

Here the partition relation asserts that if $f : \kappa^+ \times \kappa^+ \to \{0,1\}$ is a partition, then there are sets $A, B \subseteq \kappa^+$ so that one of the following happens:

1. $f \restriction A \times B$ is constantly 0 and $|A| = \kappa^+$ and $|B| = \kappa$;
2. $f \restriction A \times B$ is constantly 0 and $|A| = \kappa$ and $|B| = \kappa^+$;
3. $f \restriction A \times B$ is constantly 1 and $|A| = \kappa = |B|$.

Albin Jones [2000a] used elementary submodels to prove a ZFC version:

$$\begin{pmatrix} (2^{<\kappa})^+ \\ \\ (2^{<\kappa})^+ \end{pmatrix} \longrightarrow \begin{pmatrix} (2^{<\kappa})^+ & & 2^{<\kappa} & \kappa \\ & \vee & & \\ 2^{<\kappa} & & (2^{<\kappa})^+ & \kappa \end{pmatrix}^{1,1}$$

Baumgartner and Hajnal [2001] added a plus to get a balanced partition relation in a paper received by the journal in 1999.

$$\begin{pmatrix} (2^{<\kappa})^{++} \\ \\ (2^{<\kappa})^+ \end{pmatrix} \longrightarrow \begin{pmatrix} \kappa \\ \\ \kappa \end{pmatrix}^{1,1}$$

Example 2: Di Prisco and Todorcevic [1999] considered κ^*, the least cardinal κ satisfying the following partition relation:

$$\kappa \longrightarrow \begin{pmatrix} 2 \\ 2 \\ 2 \\ \vdots \end{pmatrix}^{<\omega}$$

This partition relation holds if for every $f \colon [\kappa]^{<\omega} \to 2$, there is a sequence $\langle H_n \in [\kappa]^2 : n < \omega\rangle$ such that for all $n < \omega$, the function f is constant on $\prod_{k<n} H_k)$. Di Prisco and Todorcevic [1999] prove that $\kappa^* = \aleph_\omega < 2^{\aleph_1}$ is equiconsistent with the existence of a measurable cardinal. The proof builds on the work of Peter Koepke [1984], [1983], who used core model arguments to prove that the free-subset property for ω_ω is equiconsistent with the existence of a measurable cardinal.[410]

[410] A subset X of a structure S is called *free* in S if no element x of X is in the structure generated inside S by $X \setminus \{x\}$. The *free-subset property for* ω_ω is the statement that for every structure S with $\omega_\omega \subseteq S$ which has at most countably many functions and relations there is an infinite subset $X \subseteq S$ free in S.

Square bracket partition relations

Velleman, in Theorem 2 of [1990], gave a short proof of the result of Todorcevic that $\aleph_1 \not\rightarrow [\aleph_1]^2_{\aleph_1}$ which avoided the use of walks, but which Velleman described as "essentially the same as the one given by Todorcevic."

Todorcevic [1994] examined square bracket relations for triples. He proved, for example, that $2^{\aleph_1} \not\rightarrow [\aleph_1]^3_{10}$. This is a special case of a general phenomenon. He showed that Chang's Conjecture is equivalent to $\aleph_2 \rightarrow [\aleph_1]^3_{\aleph_1}$. He proved that the partition relation $\aleph_2 \not\rightarrow [\aleph_1]^3_{\aleph_0}$ holds, and discussed the optimality of the result, showing the size of the tuples cannot be reduced to 2 under GCH, and the number of colors cannot be increased to $\aleph_1$ by the Chang's Conjecture result. His proof used the same basic ingredients as those for the result $\aleph_1 \not\rightarrow [\aleph_1]^2_{\aleph_1}$ (see [1987]), but the new setting required significant reworking. He also proved that for every continuous coloring $f : [\mathbb{Q}]^2 \rightarrow 2$, there is a monochromatic homeomorphic copy of $\mathbb{Q}$. On the other hand, he proved that there is a continuous coloring $d : [\mathbb{Q}]^3 \rightarrow \mathbb{Q}$ which is *irreducible* in the sense that d realizes every color on the pairs of every subset $X \subseteq \mathbb{Q}$ homeomorphic to $\mathbb{Q}$. He used almost disjoint coding of the coloring $f : [\omega_1]^2 \rightarrow \omega$ from his seminal [1987] which witnesses $\aleph_1 \not\rightarrow [\aleph_1]^2_{\aleph_0}$ to construct a set $X \subseteq \mathbb{R}$ of cardinality $\aleph_1$ and a continuous coloring $g : [X]^3 \rightarrow X$ which is irreducible.

Todorcevic's Remark 4.3.(d) of [1994] on partitions of pairs is useful for applications of the oscillation map on pairs. A (continuous) order-preserving map from a partial order into the rationals composed with osc will take all possible values $< 2k - 1$ on chains of order type ω^k. Recall that in §7.7 we noted that the Baumgartner-Malitz-Reinhardt proof of the consistency of all Aronszajn trees being embeddable in the rationals could be modified to show that they all have strictly increasing continuous maps into the rationals and the composition of such a function applied to each element of a pair and followed with the oscillation map yields a partition into ω many pieces for which every chain of order type ω^ω has pairs of every color.

Justin Moore[411] [2000] modified the oscillation map of Todorcevic to produce a continuous coloring from the family $\mathbb{P}$ of strictly increasing functions into ω which is irreducible in the sense that every unbounded and well-ordered subset of $\mathbb{P}$ receives all the colors. He introduced an ideal $\mathcal{N}_*$ of measure zero sets and proved that there is a subset $X \subseteq {}^\omega\omega$ of size $\text{add}(\mathcal{N}_*)$[412] and a coloring $c : [X]^2 \rightarrow \omega$ which is both continuous and irreducible on subsets of the same size. The proof uses lower bounds for finite Ramsey numbers.

[411]Justin Tatch Moore received his doctorate in 2000 at the University of Toronto under the direction of Todorcevic, held a post-doctoral position at the University of East Anglia where he worked with Džamonja, spent some years at Boise State University and is now at Cornell University.

[412]For any ideal $\mathcal{I}$ of subsets of a set X, the *additivity* of $\mathcal{I}$, denoted $\text{add}(\mathcal{I})$, is the smallest number of sets in $\mathcal{I}$ whose union is not in $\mathcal{I}$. Additivity is one of the well-studied cardinal invariants of ideals. Steprāns, in his chapter, pointed out that Sierpiński asked about the additivity of the null and meager ideals in the first issue of *Fundamenta Mathematicae*.

Galvin, Hajnal and Komjáth [1995] introduced a strengthening of $\lambda \nrightarrow [\kappa]^2_\mu$: for cardinals $\lambda \geq \kappa$ and μ: $Q(\lambda, \kappa, \mu)$ holds if whenever $X \subseteq [\lambda]^2$ is a graph on λ which has no independent set of cardinality κ, then X can be written as the disjoint union of μ graphs, $X = \dot{\bigcup}\{Y_\alpha : \alpha < \mu\}$, none of which has an independent set of cardinality κ. They prove the following:

1. If $2^\kappa = \kappa^+$, then $Q(\kappa^+, \kappa^+, \kappa^+)$ holds.
2. If more than $\aleph_1$ Cohen reals are added to a model of ZFC, then $Q(\omega_1, \omega_1, \omega_1)$ holds in the resulting model.
3. It is consistent that $Q(\omega_1, \omega_1, 2)$ fails.
4. If $2^\omega < \omega_\omega$, then $Q(2^\omega, 2^\omega, \omega)$ holds.
5. Assuming $V = L$, if κ is a regular uncountable non-weakly compact cardinal, then $Q(\kappa, \kappa, \kappa)$.

Hajnal and Komjáth [1997] show that there is a forcing of size $\aleph_1$ which adds an order type θ of size $\aleph_1$ with the property that $\psi \not\rightarrow [\theta]^2_{\omega_1}$ for every order type ψ, regardless of its size. This is in sharp contrast to the consistency result of Shelah [1989] discussed in §8.2.

Structural results

Louveau, Shelah and Veličković [1993] generalized both the Galvin-Prikry Theorem and the Galvin and Blass Perfect Set Partition Theorems. They defined a notion of type of a perfect tree, and showed that for any type τ if the set of all subtrees of a given perfect tree T which have type τ is partitioned into two Borel classes, there there is a perfect subtree $S \subseteq T$ such that all subtrees of S of type τ belong to the same class.

Pierre Matet [1993] used games to characterize sets which are completely Ramsey with respect to a happy family $\mathcal{F}$.

Veličković and Woodin [1998] use a variation of the Todorcevic oscillation function for reals to define a (partial) continuous coloring of triples of real numbers (regarded as infinite non-decreasing sequences of integers) with infinite binary sequences, such that every infinite binary sequence is obtained on the set of triples from every superperfect set.[413] Specifically, given (x, y, z), a triple of distinct non-decreasing functions in ${}^\omega\omega$, if

$$d(x, y, z) = \{n \in \omega : z(n-1) \leq \min(x(n-1), y(n-1)) \wedge \max(x(n), y(n)) < z(n)\}$$

is infinite and enumerated as $\langle n_k : k < \omega \rangle$, then $o(x, y, z)$ is the function $k \mapsto 0$ if and only if $x(n_k) \leq y(n_k)$.

[413] Recall that a subtree of ${}^{<\omega}\omega$ is *superperfect* if every node has an infinitely branching successor (it need not be an immediate successor); and a subset of ${}^\omega\omega$ is superperfect if the subtree of ${}^{<\omega}\omega$ it generates is superperfect.

9.2 *Linear and partial orders*

Galvin [1994] gave a short, elegant proof of the Dilworth chain decomposition theorem. For more on the impact of this theorem see [Bogart *et al.*, 1990].

Gary Gruenhage and Joe Mashburn continued the work of Kurepa on decompositions of partially order sets into chains and antichains by analyzing countable width partial orders $(X, <)$ which are *order-separable*, i.e. there is a countable set D such that if $p < q$, then there is an $r \in D$ with $p < r < q$. They [1999] showed that such partial orders can be written as a union of countably many chains under the Open Coloring Axiom $\mathrm{OCA_T}$, but not if there is a 2-entangled subset.[414]

Kunen and Tall [2000] examined the possibilities for the order types of the intersection of the reals with models of set theory, $\mathbb{R} \cap M$. Under CH they found there are exactly two possibilities: $\mathbb{Q}$ and $\mathbb{R}$. They also show the consistency of several other possibilities, including consistency of exactly three types. If the existence of a Ramsey cardinal is consistent, then so is the existence of $2^{2^{\aleph_0}}$ many non-order-isomorphic order types of cardinality continuum of the form $\mathbb{R} \cap M$. At the end of the paper, they recall that it is a long-standing open question whether it is consistent that $2^{\aleph_0} > \aleph_2$ and all $\aleph_2$-dense sets of reals are order-isomorphic. Then they pose [2000, 691]: "a potentially less demanding question: is it consistent that with $2^{\aleph_0} > \aleph_2$ that all the $\mathbb{R} \cap M$s of size $\aleph_2$ are order-isomorphic?"

Abraham and Bonnet [1999] generalized the Hausdorff characterization of the class of scattered linear orders to FAC partial orders, i.e. those in which every antichain is finite. Building on work of Bonnet and Pouzet [1969], they proved that the class of scattered FAC partial orderings may be characterized as the closure under lexicographic sums, inverses and augmentations (keep the domain the same but add new pairs to the partial order). They use the natural sum (Hessenberg sum) and introduce a Hessenberg based product and a Hessenberg based exponentiation which make their proofs go smoothly.[415]

Recall that Hausdorff used CH to prove the existence of a linear order universal in power $\aleph_1$[416] and Shelah [1980a] proved the consistent existence of a linear order universal in power $\aleph_1$ in a model of $2^{\aleph_0} = \aleph_2$. Kojman and Shelah [1992] proved the non-existence of universal orders and universal partial orders in power λ for a variety of cardinals λ. They noted that the problem of existence of a universal model for a first-order theory remains "unsettled in classical model theory only for cardinals $\lambda < 2^{<\lambda}$" and cite [Chang and Keisler, 1973] and VIII.4 of [Shelah, 1978a] on saturated, special and universal models. In Theorem 3.6 they used club guessing to prove that for every regular cardinal λ for which $\aleph_1 < \lambda < 2^{\aleph_0}$, there is no universal linear order in power λ. The first item of Conclusions 5.7 is that under the hypotheses on λ of Theorem 3.6 there are no universal partial orders

[414] Entangled subsets of partial orders were introduced by Shelah, shown to follow from CH and to be consistent with $\mathrm{MA}_{\aleph_1}$ in [Abraham *et al.*, 1985], and used by Roitman [1978] in her combinatorial reformulation of the questions (S) and (L).

[415] See [Džamonja and Thompson, 2006] for an extension to κ-scattered partial orders.

[416] Recall that a linear (partial) order $(P, <)$ is *universal in power* λ (alternatively *in cardinality* λ) if it has cardinality λ and is universal for the class of linear (partial) orders of cardinality λ.

in power λ. In Theorem 3.10, they generalized Theorem 3.6 by showing that if λ is a regular cardinal and there is some $\kappa^+ < \lambda < 2^\kappa$, then there is no universal linear order in power λ. They start Section 4 by noting that for a strong limit cardinal μ, any first-order theory T with $|T| < \mu$ has a universal model in power μ. They prove in Theorem 4.3 that if θ and $\kappa > \theta^+$ are regular cardinals, $\kappa < \mu$ and μ^* is the minimum size of a family $A \subseteq [\mu]^{\leq\kappa}$ which has the property that for all $X \in [\mu]^{\leq\kappa}$ there are fewer than κ members of A whose union covers A, and if there is a binary tree $T \subseteq {}^{<\theta}2$ of size $< \kappa$ with more than μ^* branches of length θ, then there is no linear order of size μ which is universal for linear orders of size κ. They show that if μ is a fixed point of the aleph function of the first order but not the second order[417] and $\sigma + \mathrm{cf}(\lambda) < \lambda$, then $\mu^* = \mu$ in the theorem stated above. They draw two corollaries, proving that there is no universal linear order of power μ for a singular cardinal if (1) μ is not a second order fixed point, and there is some $\sigma < \mu$ such that $2^{<\sigma} < \mu < 2^\sigma$, and if (2) $\aleph_\mu > \mu$ or μ is not a second order fixed point, $\mathrm{cf}(\mu) = \aleph_0$ or $2^{<\mathrm{cf}(\mu)} < \mu$, and $\mu \neq 2^{<\mu}$.

9.3 Trees

We discuss a variety of results, many of them consistency proofs, on existence or non-existence of trees. Later we look at the structure of families of trees under embeddings.

Existence results

Kojman and Shelah [1993] proved that if $\mu^{<\mu} = \mu$, $2^\mu = \mu^+$, and the set $\{\alpha < \mu^+ : \mathrm{cf}(\alpha) < \mu\}$ contains a non-reflecting stationary set, then there is a Suslin tree on μ^+ in which every ascending chain of size $< \mu$ has an upper bound, i.e. is closed in the tree topology. They use a principle they describe as "closely related to club guessing", and recalled Gregory's result [1976] that GCH together with the existence non-reflecting stationary set of ω-cofinal elements of ω_2 implies the existence of an $\aleph_2$-Suslin tree (see §7.7).

Magidor and Shelah [1996] proved that if a singular cardinal κ is the limit of strongly compact cardinals, then there are no κ^+-Aronszajn trees. Recall that for regular cardinals, Jensen proved that in L, there is no λ-Aronszajn tree if and only if λ is weakly compact. Magidor and Shelah went on to show the consistency relative to large cardinals of there being no $\aleph_{\omega+1}$-Aronszajn tree, shedding light on the difficulty of determining whether or not successors of singular cardinals must support Aronszajn trees.

Cummings [1997] showed that, under certain cardinal and combinatorial assumptions, there are Suslin trees on successor cardinals larger than $\aleph_1$ which are immune to specialization in cardinal and cofinality preserving forcing extensions.

[417] A cardinal λ is a *fixed point of the aleph function of the first order* if $\lambda = \aleph_\lambda$. It is not a fixed point of *the second order* if $|\{\nu < \lambda : \nu = \aleph_\nu\}| = \sigma < \lambda$.

Cummings and Foreman [1998] derived the consistency of no $\aleph_n$-Aronszajn trees for $2 \leq n < \omega$ from the consistency of infinitely many supercompact cardinals, and also derived the consistency of the existence of a strong limit cardinal κ of countable cofinality for which there are no κ^{++}-Aronszajn trees from the consistency of a supercompact cardinal with a weakly compact cardinal above.

In a paper submitted in 1998, Foreman, Magidor and Ralf Schindler [2001] investigated the consistency strength of the above result of Cummings and Foreman [1998] and from it prove the existence of sharps for inner models with n Woodin cardinals. In their proof they use a method of induction due to Woodin.

In an unpublished manuscript from 1994, Miyamoto showed that $\text{cov}(\mathcal{M}) \geq \aleph_2$ and the Stick Principle[418] were sufficient to get a complete amalgamation system for every ω-morass which by work of Velleman in his [1990] is enough to build a Suslin tree. Brendle [2006] adapts Todorcevic's Cohen Real to Suslin Tree Construction and Miyamoto's proof to show that ♣ together with the assumption that the meager idea $\mathcal{M}$ has a base of size $\aleph_1$ implies there is a Suslin tree.

Structure under embeddings

Taneli Huuskonen [1995] called a tree of height ω_1 *persistent* if it has a non-empty subtree in which each node has extensions of all countable heights. He introduced and identified the $\leq$-smallest persistent tree T^0 whose nodes are sequences $(t_0, \ldots, t_n)$ of countable ordinals ordered by $(t_0, \ldots, t_n) \leq (s_0, \ldots, s_m)$ if $n \leq m$, $t_n \leq s_n$, and $i < n$ implies $t_i = s_i$.

Todorcevic and Jouko Väänänen [1999] studied the quasi-order $(\mathbb{T}, \leq)$ where $\mathbb{T}$ denotes the class of all trees of cardinality ω_1 with no uncountable branches, and $S < T$ if and only if there is a strictly increasing map from S into T. Note that there is no requirement that the maps be one-to-one, so they are not necessarily homomorphisms. They proved a variety of results about the structure of $(\mathbb{T}, \leq)$, with one set of results for those trees below T^0 and another set for those at or above T^0 and $(\mathbb{T}/\equiv, \leq)$. For example, they showed in Theorem 33 that if $2^{\aleph_0} < 2^{\aleph_1}$, then there are incomparable special Aronszajn trees under order-preserving maps. Recall that Abraham and Shelah [1985] had a similar result for homomorphic embeddings.

In the same paper, Todorcevic and Väänänen identified the tree $\sigma\mathbb{Q}$ as the universal element in the class of $\mathbb{R}$-embeddable Hausdorff trees. They showed that the class LO_1 of linear orders without uncountable well-ordered or conversely well-ordered sub-orderings has no universal element, i.e. there is no $K \in \text{LO}_1$ such that every element of LO_1 embeds in it. They also showed that there is no universal element for the class PO_1 of all those partial orders of cardinality at most continuum which have no increasing ω_1-sequence.[419]

[418] The *Stick Principle* asserts the existence of a family $\mathcal{A}$ of $\aleph_1$ many countable subsets of ω_1 such that every uncountable subset of ω_1 contains a member of $\mathcal{A}$.

[419] Džamonja and Katherine Thompson [2005] showed that the least size of a universal family for well-founded partial orders of power λ under order preserving embeddings is the least size of a universal family for well-orders of size λ, which is λ^+. They used club guessing arguments

Todorcevic [1998a] announced a Ramsey theoretic reformulation, under PFA, of the existence of a two-element basis for the class of Aronszajn lines.[420]

9.4 Combinatorial principles

In the early 1990's, Shelah [2000a] used his pcf theory to make a remarkable connection between instances of Jensen's Diamond Principle for successor cardinals and instances of the GCH:[421] for cardinals $\lambda \geq \beth_\omega$,[422] $2^\lambda = \lambda^+$ if and only if $\diamondsuit_{\lambda^+}$.[423]

In a paper received October 19, 1998, Cummings, Foreman, and Magidor [2001] considered, in the context of singular cardinals, implications between square and the scales of pcf theory, and proved consistency results between relatively strong forms of square and stationary set reflection.

Abraham and Todorcevic [1997] focused attention on P-ideals on ω_1.[424] They introduced a new property of ideals which they called $(*)$: for every P-ideal, either

1. there exists an uncountable subset $A \subseteq \omega_1$ such that $[A]^\omega \subseteq \mathcal{I}$; or
2. ω_1 may be decomposed into countably many subsets A_i such that $A_i \cap I$ is finite for all $i < \omega$ and $I \in \mathcal{I}$ (they say A_i is *orthogonal* to $\mathcal{I}$).

They observed that PFA implies $(*)$, but in this paper, showed that $(*)$ is consistent with CH. They proved that if $(*)$ holds, then there are no Suslin trees, and they drew many other consequences.

In a paper [Todorcevic, 2000] submitted in November 1999, Todorcevic simplifies the statement of $(*)$ to get an optimal P-ideal dichotomy, also called $(*)$: for every P-ideal $\mathcal{I}$ of countable subsets of some set S, either

1. there exists an uncountable subset $A \subseteq S$ such that $[A]^\omega \subseteq \mathcal{I}$; or
2. S can be decomposed into countably many sets orthogonal to $\mathcal{I}$.

He used this version to prove, among other results, that Jensen's $\square_\kappa$ fails for any regular $\kappa > \omega_1$. Todorcevic proved that this P-ideal dichotomy follows from PFA and that the consistency of the existence of a supercompact cardinal gives the consistency of $(*)$ + GCH.

---

to prove that for regular λ strictly between $\aleph_1$ and $2^{\aleph_0}$, the least size of a universal family for well-founded partial orders of power λ under rank and order preserving embeddings is at least $2^{\aleph_0}$.

[420] Moore [2006a] showed that under the assumption of PFA, Aronszajn lines have a two-element basis and the uncountable linear orders have a 5-element basis. This resolved Shelah's Finite Basis Problem (cf. end of Section 8); see our Postscript for a discussion.

[421] The paper was received February 15, 1994, revised November 16, 1998, and appear in the December 2000 issue of the Israel Journal.

[422] Recall that $\beth_\omega$ is the least strong limit cardinal.

[423] Shelah [2010] improved his result to all uncountable λ.

[424] An ideal $\mathcal{I}$ on ω_1 is a *P-ideal* if all its members are countable and it is $\subseteq^*$-σ-directed.

10 POSTSCRIPT

In this final section we look back at some threads woven through this decade-by-decade account, and we hark back to the opening of the chapter with the quotes by Cantor and Motzkin. Inspired by Cantor, we look at *freedom* as shown in the generalization of notions of largeness, first examining those used in successive versions of the Regressive Function Theorem, then in generalizations of the Pigeonhole Principle, and in applications to partition relations and trees. We then examine the *winnowing* process mentioned by Cantor that shaped the notion of uncountable tree from the ramified sets, tables, and suites of Kurepa to the family of ω_1-trees classified as to whether they are special Aronszajn trees, non-special Aronszajn trees, Suslin trees, or Kurepa trees. The Motzkin quote inspired a look at structural results more generally, and we review the possibilities for describing the structure of uncountable linear orders that appeared in this century with a brief peek at one in the next century. We wrap up the postscript with a few remarks on interactions with the near neighbors of logic, topology and analysis.

First we look at the *freedom* to generalize notions of largeness for subsets of a given set. Our first lens is on the series of generalizations of the Regressive Function Theorem of Alexandroff and Urysohn [1929] that any regressive function on the set of infinite countable ordinals is constant on an uncountable set. Two aspects were generalized, the possibilities for the domain of the regressive function and the notion of largeness of the set on which the function is constant. Dushnik [1931] extended the family of possible domains to include the set of infinite ordinal numbers less than an uncountable successor cardinal κ, with cardinality κ the notion of largeness. Erdős [1950] then extended the domains to include the set of infinite ordinal numbers less than a singular cardinal κ of uncountable cofinality and used being cofinal in κ as the notion of largeness, but relaxed the constraint that the function be constant there to only require the values of the function to be bounded. Novák [1950] proved that a closed unbounded subset of the countable ordinals was a suitable domain and used being uncountable as the notion of largeness. Neumer [1951], apparently unaware of Novák's work, weakened the domain to stationary subsets of a regular cardinal κ, and used cardinality κ as the notion of largeness. Finally Fodor, after a rediscovery in [1955] of the work of Dushnik, strengthened Neumer's Theorem to show any function defined on a stationary set is constant on a stationary subset. Fodor was not alone in rediscovering work by others related to this theorem: Bloch [1953], Denjoy [1953], and Neumer [1958] did so as well. The history of this theorem exemplifies the sort of forgetting and remembering of theorems that happens over time, the independent discovery of generalizations and proofs with variations that evolve into an especially beautiful theorem.

The Regressive Function Theorem has been generalized to other contexts, but we will stop here to recall that the original theorem can be regarded as a kind of pigeonhole principle with $\aleph_1$ pigeons and $\aleph_1$ pigeonholes both labeled with infinite countable ordinals but a requirement that pigeons come to roost in pigeonholes with a smaller label than their own label and a conclusion that an uncountable number end up in the same pigeonhole. The generalizations may be similarly recast.

As we turn to look at various pigeonhole principles and how they are used to prove partition theorems, particularly for pairs, we keep in mind the slogan that is embedded in the Motzkin quote: *complete disorder is impossible.*

Recall the *Schubfachprinzip*, known in English as the *Pigeonhole Principle*, was identified as a mathematical principle by Dirichlet in the 1800s and its generalization to the transfinite has proved widely applicable. Basic cardinality arguments, with infinitely many objects deposited in finitely many places with at least one receiving infinitely many, or the extension with uncountable versus countable are now fundamental in mathematical practice.

The important concept of *cofinality* allowed more nuanced generalizations. This concept was brought into focus when a mistake in Bernstein's thesis lead to the (false) "proof" by Julius (Gyula) Kőnig that the continuum cannot be well-ordered at the 1904 International Congress in Heidelberg. Hausdorff introduced the concept of cofinality [confinality] in his paper [1906], worked out the basic properties in his paper [1908], and included this material in his book [1914]. With the notion of cofinality, one can say more: if κ objects are distributed among fewer than cofinality of κ many cells, then one cell must have κ objects.

Now we turn to the use of pigeonhole principles in proofs of partition relations. Ramsey specified the *Axiom of Selections* as his pigeonhole principle, i.e. the Axiom of Choice. Recall that he constructed a sequence begin-homogeneous in the inner induction step of his proof.

Many proofs of Ramsey's Theorem exist. One example is an ultrafilter proof. Building an end-homogeneous set is another natural approach. In their rediscovery of Ramsey's Theorem, Erdős and Szekeres used an Ordered Pigeonhole Principle in which a suitably long sequence was required to have either a large monotone increasing subsequence or a large monotone decreasing subsequence.

Recall that Erdős, in his proof [1942] that if $b > a^a$, then $b \to (a^+)^2_a$, built a tree of subsets of the vertices of the complete graph on his underlying set: he built a tree of subsets and a parallel tree of representatives with the entire set of vertices at the root of the first and a representative at the root of the second, and at each successive stage through the transfinite, selected from each set on the highest level a representative for the tree of vertices, and split the remaining vertices in each subset according to the color class or graph to which the vertex and the representative belonged. The process is quite like the one used by Ramsey, but proceeds on all fronts simultaneously to create two trees: one of nested subsets and another of the representative vertices with the set of vertices along any branch forming a set which is end-homogeneous. One could say that he is using color classes determined by specific representatives as his pigeonholes.

In the proof of the Positive Stepping Up Lemma, Erdős and Rado [1956] essentially built a tree of sequences of ordinals with a largest element, where the recursive definition of each sequence started with its top element and was so constructed to be end-homogeneous for r-element subsets for $r \geq 2$. They used the functions $h : [\sigma]^r \to k$ for $\sigma < \ell$ as pigeonholes and determined that they had more sequences than such pigeonholes so one must be of length at least ℓ and they could use it to finish their proof.

Milner and Rado [1965] investigated the generalization of the Pigeonhole Principle to ordinals that is encompassed in the partition relation $\alpha \to (\alpha_0, \alpha_1, \dots)^1$, and discovered their paradoxical decomposition: for any uncountable cardinal κ and every $\beta < \kappa^+$, the negative relation $\beta \not\to (\kappa^\omega)^1_\omega$ holds.

A variation of Laver's generalization of the Halpern-Läuchli Theorem was used by Milliken as his pigeonhole principle in the proof of Milliken's Ramsey Theorem for strongly embedded trees.

In Galvin's investigation of partition relations of the form $\varphi \to (\alpha)^2_2$ for all countable ordinals α, he (see [Erdős and Hajnal, 1974, 272] for a discussion of Galvin's "old theorems") focused attention on those uncountable linear orders φ that satisfy the pigeonhole principle represented by the relation $\varphi \to (\omega)^1_\omega$, where Erdős and Hajnal in their paper with Rado [1965] and in their problem paper [1971b] had focused attention on order types in which neither ω_1 nor ${\omega_1}^*$ is embeddable. The Baumgartner-Hajnal Theorem for order types validated Galvin's conjecture that this was the right approach. Galvin then tackled partial orders and conjectured the same condition would be critical (see [Erdős and Hajnal, 1974, 272]). He was able to show that the hypothesis $\varphi \to (\eta)^1_\omega$ sufficed in [1975]. Todorcevic [1985a] validated Galvin's conjecture with a proof that first showed the result was true for non-special Aronszajn trees and then used that result to derive his Poset Partition Theorem.

At the end of the century, co-ideals (the positive sets for their corresponding ideals) were used together with elementary submodels to prove partition relations for pairs starting with a cardinal resource, including the Balanced and Unbalanced Baumgartner-Hajnal-Todorcevic Theorems and the consistency result of Foreman and Hajnal.

Next we look at winnowing to enduring concepts, especially for trees. The first infinite trees were graph-theoretic trees used by Kőnig in his Infinity Lemma. The first systematic study of uncountable trees was by Kurepa who introduced ramified sets, ramified tables, and suites. His basic notion was more general than what we use today, allowing equivalent elements. His notion of homogeneous required a constant size across the entire ramified table for every set consisting of all elements with the same set of predecessors.

Eventually, these special partial orders became known as trees, and the key feature that the set of predecessors be well-ordered became standard. Further winnowing followed. Now the most studied trees are those whose levels have smaller cardinality than their height and for which nodes whose set of predecessors have limit order type are the unique elements with that set of predecessors, a feature which means that the tree topology is Hausdorff. These constraints allow trees of height ω_1 in which all levels are countable to be embedded in the collection of functions from countable ordinals into ω ordered by end-extension. This collection of trees includes Aronszajn trees, special Aronszajn trees, Suslin trees, and Kurepa trees, with special examples such as those coming from Countryman lines and from the rho functions of Todorcevic. For those interested in learning more about Aronszajn trees and their structure, the survey by Moore [2008b] is warmly recommended.

As one way to look at the interaction of logic and infinite combinatorics, we revisit the basis question for uncountable linear orders. There we find consistency results via forcing, new axioms, and representation theorems.

Sierpiński [1946] showed that under the Continuum Hypothesis, there are $2^{\mathfrak{c}}$ many pairwise incomparable order types of sets of real numbers of power $\mathfrak{c} = 2^{\aleph_0}$, so consistently any basis for the uncountable linear orders is very large.

Moore [2006a] proved the consistent existence of a finite basis for uncountable linear orders, thereby resolving Shelah's Finite Basis Problem (cf. end of Section 8). His accomplishment offers us the opportunity to review the steps leading up to his proof.

Recall that Baumgartner [1973] had given a forcing proof of the consistency that all $\aleph_1$-dense sets of reals are isomorphic, and Abraham and Shelah [1981] had shown that Martin's Axiom does not suffice. However, the Proper Forcing Axiom does suffice.

Baumgartner [1976b] had investigated basis questions, mostly due to Galvin, for the class of uncountable order types which are not the union of countably many well-orderings. Galvin had asked if the collection consisting of ${\omega_1}^*$ together with all real types and all Specker types, was a basis for the class of uncountable order types which are not a union of countably many well-orderings. Baumgartner had shown there were more, but they all embed ω_1. Baumgartner had shown in [1976b] that every Specker type embeds a lexicographically ordered Aronszajn tree, and in [1982] had showed that every Specker type is isomorphic to a lexicographically ordered Aronszajn tree.

Baumgartner singled out the Aronszajn lines (Specker types) for special attention with his Question 5(ii) [1976b, 221] which asked for the consistency of a finite basis for them, and asked with his fourth problem on page 275 [1982] on the consistency of the existence of a finite basis for uncountable linear orders. Under PFA, the questions are the same, since the union of a finite basis for the Aronszajn lines with $\{\omega_1, {\omega_1}^*, X\}$ where X is an $\aleph_1$-dense set of reals, is the desired basis.

As this chapter comes to a close, we note that just as finite combinatorics benefited by interaction with probability as indicated in the Motzkin quote, infinite combinatorics benefited by interaction with logic and other areas of mathematics.

From logic and other parts of set theory, the field benefited by research on forcing, large cardinals, ideals, model-theoretic methods, and an axiomatic approach that has led to a wide variety of additional axioms for consistency results.

From topology came the notion of compactness (Tychonoff's Theorem), closure operators, notions of largeness and smallness related to measure and category, interest in products, Suslin's Problem, the Normal Moore Space Conjecture and S-space and L-space problems that stimulated work in set theory. Since trees are the foundation of many interesting topological spaces, set-theoretic topologists have motivations and intuitions that lead to results interesting to both areas.

Finite combinatorics has provided tools applied applied in many proofs and has been a source of inspiration, with many finite results, especially partition relations, generalized to the infinite and uncountable in a variety of ways.

Dynamical systems and ergodic theory have given new approaches to Ramsey theory and interactions between Ramsey theory and Banach spaces has benefited both areas.

ACKNOWLEDGMENTS

I have benefited greatly from the many people who reviewed papers discussed here in the Mathematical Reviews of the American Mathematical Society. I have used the Mathematics Genealogy Project (http://genealogy.math.ndsu.nodak.edu/) as a resource multiple times in this chapter. I also benefited from the hospitality of the University of Münster in June 2010, received help from the library of the University of Wisconsin in summer 2010, enjoyed the excellent library of Oberwolfach in January 2011, and benefited from the hospitality of the University of California, Berkeley, during March and April, 2011. My thanks to my husband, Bill Mitchell, for his encouragement and advice, to András Hajnal, for friendship over the years and all he taught me as we wrote our chapter for the *Handbook of Set Theory*, and to Rich Laver for sharing his joy in doing mathematics and reading the paper and offering suggestions and comments. I thank Peter Koepke for help with Hausdorff's German and Yuli Rudyak for help with Russian history and language. My thanks to Stevo Todorcevic for stimulating conversations; I respect his opinions both expressed in person and in his writings and find them thought-provoking, even if in some cases we disagree. My thanks to Andrés Caicedo, James Cummings and Justin Moore for comments on earlier versions of this paper, and to the many people who shared personal knowledge of the history. Finally, my thanks to Aki Kanamori whose patient encouragement and numerous suggestions have been invaluable.

BIBLIOGRAPHY

[Abraham and Bonnet, 1999] U. Abraham and R. Bonnet. Hausdorff's theorem for posets that satisfy the finite antichain property. *Fund. Math.*, 159(1):51–69, 1999.

[Abraham and Shelah, 1981] U. Abraham and S. Shelah. Martin's Axiom does not imply that every two $\aleph_1$-dense sets of reals are isomorphic. *Israel J. Math.*, 38(1-2):161–176, 1981. Published under the transliteration Avraham.

[Abraham and Shelah, 1982] U. Abraham and S. Shelah. Forcing with stable posets. *J. Symbolic Logic*, 47(1):37–42, 1982. Published under the transliteration Avraham.

[Abraham and Shelah, 1985] U. Abraham and S. Shelah. Isomorphism types of Aronszajn trees. *Israel J. Math.*, 50(1-2):75–113, 1985.

[Abraham and Todorcevic, 1984] U. Abraham and S. Todorcevic. Martin's Axiom and first countable S- and L-spaces. In *Handbook of Set-Theoretic Topology*, pages 327–346. North-Holland, Amsterdam, 1984.

[Abraham and Todorcevic, 1997] U. Abraham and S. Todorcevic. Partition properties of ω_1 compatible with CH. *Fund. Math.*, 152(2):165–181, 1997.

[Abraham *et al.*, 1985] U. Abraham, M. Rubin, and S. Shelah. On the consistency of some partition theorems for continuous colorings, and the structure of $\aleph_1$-dense real order types. *Ann. Pure Appl. Logic*, 29(2):123–206, 1985.

[Abraham, 1979a] U. Abraham. Construction of a rigid Aronszajn tree. *Proc. Amer. Math. Soc.*, 77(1):136–137, 1979. Published under the transliteration Avraham.

[Abraham, 1979b] U. Abraham. *Isomorphisms of Aronszajn Trees and Forcing without the Generalized Continuum Hypothesis.* PhD thesis, Hebrew University, 1979.

[Abraham, 1983] U. Abraham. Aronszajn trees on $\aleph_2$ and $\aleph_3$. *Ann. Pure Appl. Logic*, 24(3):213–230, 1983.

[Abraham, 1987] U. Abraham. A note on Dilworth's theorem in the infinite case. *Order*, 4(2):107–125, 1987.

[Abraham, 1999] U. Abraham. *Models for Concurrency*, volume 11 of *Algebra, Logic and Applications.* Gordon and Breach Science Publishers, 1999.

[Abraham, 2010] U. Abraham. Successors of singular cardinals. In *Handbook of Set Theory, I*, chapter 5, pages 333–394. Springer, Berlin, 2010.

[Aharoni and Berger, 2009] R. Aharoni and E. Berger. Menger's theorem for infinite graphs. *Invent. Math.*, 176(1):1–62, 2009.

[Aharoni *et al.*, 1983] R. Aharoni, C. Nash-Williams, and S. Shelah. A general criterion for the existence of transversals. *J. London Math. Soc.(3)*, 47:43–68, 1983.

[Aharoni *et al.*, 1984] R. Aharoni, C. Nash-Williams, and S. Shelah. Another form of a criterion for the existence of transversals. *J. London Math. Soc. (2)*, 29(2):193–203, 1984.

[Albers and Reid, 1988] D. Albers and C. Reid. An interview with Mary Ellen Rudin. *College Math. J.*, 19(2):114–137, 1988.

[Alexandroff and Urysohn, 1929] P. Alexandroff and P. Urysohn. Mémoire sur les espaces topologiques compacts. *Verh. Nederl. Akad. Wetensch. Afd. Natuurk. Sect. 1*, 14(1):1–96, 1929. Note supplémentaire, pp. 95–96.

[Alexandroff, 1916] P. Alexandroff. Sur la puissance des ensembles mesurables *B. C. R. Acad. Sci. Paris*, 162:323–325, 1916.

[Alexandroff, 1925] P. Alexandroff. Paul Urysohn (Notice nécrologique). *Fund. Math.*, 7:138–140, 1925.

[Alexandroff, 1979] P. Alexandroff. Pages from an autobiography. *Uspekhi Mat. Nauk*, 34(6(210)):219–249, 1979. References are to the translation in Russian Math. Surveys 34 (1979), no. 6, 267–304.

[Alvarez, 1999] C. Alvarez. On the history of Souslin's problem. *Arch. Hist. Exact Sci.*, 54(3):181–242, 1999.

[Anonymous, 1984] Anonymous. Roman Sikorski (11 July 1920–12 September 1983). *Studia Math.*, 78(2):105, 1984.

[Argyros and Todorcevic, 2005] S. Argyros and S. Todorcevic. *Ramsey Methods in Analysis.* Advanced Courses in Mathematics. CRM Barcelona. Birkhäuser Verlag, Basel, 2005.

[Arhangel′skii and Dranishnikov, 1997] A. V. Arhangel′skii and A. N. Dranishnikov. P. S. Alexandroff and topology: an introductory note. *Topology Appl.*, 80(1-2):1–6, 1997.

[Aronszajn and Smith, 1954] N. Aronszajn and K. Smith. Invariant subspaces of completely continuous operators. *Ann. of Math. (2)*, 60:345–350, 1954.

[Aronszajn, 1935] N. Aronszajn. Sur les décompositions des fonctions analytiques uniformes et sur leurs applications. *Acta Math.*, 65(1):1–156, 1935.

[Aronszajn, 1950] N. Aronszajn. Theory of reproducing kernels. *Trans. Amer. Math. Soc.*, 68:337–404, 1950.

[Aronszajn, 2011] M. Aronszajn. Death Date Nachman Aronszajn. Email message to Larson dated March 2, 2011.

[Bachmann, 1955] H. Bachmann. *Transfinite Zahlen.* Ergebnisse der Mathematik und ihrer Grenzgebiete (N.F.), Heft 1. Springer-Verlag, Berlin-Göttingen-Heidelberg, 1955.

[Balcerzyk *et al.*, 2000] S. Balcerzyk, W. Bartol, E. Orłowska, A. Wieczorek, and A. Wojciechowska-Waszkiewicz. Jerzy Łoś (1920–1998). *Studia Logica*, 65(3):301–314, 2000.

[Banach and Kuratowski, 1928] S. Banach and K. Kuratowski. Sur une généralisation du problème de la mesure. *Fund. Math.*, 14(1):127–131, 1928.

[Banach, 1922] S. Banach. Sur les opérations dans les ensembles abstraits et leur application aux équations intégrales. *Fund. Math.*, 3:131–181, 1922.

[Baumgartner and Hajnal, 1973] J. Baumgartner and A. Hajnal. A proof (involving Martin's Axiom) of a partition relation. *Fund. Math.*, 78(3):193–203, 1973.

[Baumgartner and Hajnal, 1987] J. Baumgartner and A. Hajnal. A remark on partition relations for infinite ordinals with an application to finite combinatorics. In *Logic and Combinatorics (Arcata, Calif., 1985)*, volume 65 of *Contemp. Math.*, pages 157–167. Amer. Math. Soc., Providence, RI, 1987.

[Baumgartner and Hajnal, 2001] J. Baumgartner and A. Hajnal. Polarized partition relations. *J. Symbolic Logic*, 66(2):811–821, 2001.

[Baumgartner and Laver, 1979] J. Baumgartner and R. Laver. Iterated perfect-set forcing. *Ann. Math. Logic*, 17(3):271–288, 1979.

[Baumgartner and Prikry, 1976] J. Baumgartner and K. Prikry. On a theorem of Silver. *Discrete Math.*, 14(1):17–21, 1976.

[Baumgartner and Prikry, 1977] J. Baumgartner and K. Prikry. Singular cardinals and the generalized continuum hypothesis. *Amer. Math. Monthly*, 84(2):108–113, 1977.

[Baumgartner and Shelah, 1987] J. Baumgartner and S. Shelah. Remarks on superatomic Boolean algebras. *Ann. Pure Appl. Logic*, 33(2):109–129, 1987.

[Baumgartner *et al.*, 1970] J. Baumgartner, J. Malitz, and W. Reinhardt. Embedding trees in the rationals. *Proc. Nat. Acad. Sci. U.S.A.*, 67:1748–1753, 1970.

[Baumgartner *et al.*, 1993] J. Baumgartner, A. Hajnal, and S. Todorcevic. Extensions of the Erdős-Rado theorem. In *Finite and Infinite Combinatorics in Sets and Logic (Banff, AB, 1991)*, pages 1–17. Kluwer Acad. Publ., Dordrecht, 1993.

[Baumgartner, 1970] J. Baumgartner. Decompositions and embeddings of trees. *Notices Amer. Math. Soc.*, 18:967, 1970. Abstract, preliminary report.

[Baumgartner, 1973] J. Baumgartner. All $\aleph_1$-dense sets of reals can be isomorphic. *Fund. Math.*, 79(2):101–106, 1973.

[Baumgartner, 1974] J. Baumgartner. A short proof of Hindman's theorem. *J. Combinatorial Theory Ser. A*, 17:384–386, 1974.

[Baumgartner, 1975a] J. Baumgartner. Canonical partition relations. *J. Symbolic Logic*, 40(4):541–554, 1975.

[Baumgartner, 1975b] J. Baumgartner. Partition relations for uncountable ordinals. *Israel J. Math.*, 21(4):296–307, 1975.

[Baumgartner, 1975c] J. Baumgartner. Topological properties of Specker types. *Notices Amer. Math. Soc.*, 22:A–219, 1975. Abstract.

[Baumgartner, 1976a] J. Baumgartner. Almost-disjoint sets, the dense set problem and the partition calculus. *Ann. Math. Logic*, 9(4):401–439, 1976.

[Baumgartner, 1976b] J. Baumgartner. A new class of order types. *Ann. Math. Logic*, 9(3):187–222, 1976.

[Baumgartner, 1980] J. Baumgartner. Chains and antichains in $\mathcal{P}(\omega)$. *J. Symbolic Logic*, 45(1):85–92, 1980.

[Baumgartner, 1982] J. Baumgartner. Order types of real numbers and other uncountable orderings. In *Ordered Sets (Banff, Alta., 1981)*, volume 83 of *NATO Adv. Study Inst. Ser. C: Math. Phys. Sci.*, pages 239–277. Reidel, Dordrecht, 1982.

[Baumgartner, 1983a] J. Baumgartner. Iterated forcing. In *Surveys in set theory*, volume 87 of *London Math. Soc. Lecture Note Ser.*, pages 1–59. Cambridge Univ. Press, Cambridge, 1983.

[Baumgartner, 1983b] J. Baumgartner. Iterated forcing. In A. Mathias, editor, *Surveys in Set Theory*, pages 1–59. London Math. Soc., London, 1983. London Math. Soc. Lecture Note Series 87.

[Baumgartner, 1984] J. Baumgartner. Applications of the Proper Forcing Axiom. In *Handbook of Set-Theoretic Topology*, pages 913–959. North-Holland, Amsterdam, 1984.

[Baumgartner, 1985] J. Baumgartner. Bases for Aronszajn trees. *Tsukuba J. Math.*, 9(1):31–40, 1985.

[Baumgartner, 1989] J. Baumgartner. Remarks on partition ordinals. In *Set Theory and its Applications (Toronto, ON, 1987)*, volume 1401 of *Lecture Notes in Math.*, pages 5–17. Springer, Berlin, 1989.

[Baumgartner, 1997] J. Baumgartner. In Memoriam: Paul Erdős 1913-1996. *B. Symbolic Logic*, 3(1):70–71, 1997.

[Beckmann *et al.*, 2008] A. Beckmann, M. Goldstern, and N. Preining. Continuous Fraïssé Conjecture. *Order*, 25(4):281–298, 2008.

[Ben-David and Shelah, 1986] S. Ben-David and S. Shelah. Souslin trees and successors of singular cardinals. *Ann. Pure Appl. Logic*, 30(3):207–217, 1986.

[Bernstein, 1901] F. Bernstein. *Untersuchungen aus der Mengenlehre.* PhD thesis, inaugural dissertation, Georg-August-Universität Göttingen, printed in Halle, 1901. Reprinted in *Mathematsche Annalen*, vol. 61 (1905), pp. 111-155.

[Bernstein, 1908] F. Bernstein. Zur Theorie der trigonometrischen reihen. *Berichte über die Verhandlungen der Königlich Sächsischen Gesellschaft der Wissenschaften zu Leipzig, Mathematische-Physische Klasse*, 60:325–338, 1908.

[Bing, 1951] R. H. Bing. Metrization of topological spaces. *Canadian J. Math.*, 3:175–186, 1951.

[Birkhoff, 1940] G. Birkhoff. *Lattice Theory.* American Mathematical Society, New York, 1940.

[Blass, 1972] A. Blass. Weak partition relations. *Proc. Amer. Math. Soc.*, 35:249–253, 1972.

[Blass, 1981] A. Blass. A partition theorem for perfect sets. *Proc. Amer. Math. Soc.*, 82(2):271–277, 1981.

[Blass, 2009] A. Blass. Homogeneous sets from several ultrafilters. *Topology Appl.*, 156(16):2581–2594, 2009.

[Bloch, 1953] G. Bloch. Sur les ensembles stationnaires de nombres ordinaux et les suites distinguées de fonctions régressives. *C. R. Acad. Sci. Paris*, 236:265–268, 1953.

[Blumberg, 1920] H. Blumberg. Hausdorff's Grundzüge der Mengenlehre. *Bull. Amer. Math. Soc.*, 3(3):116–129, 1920.

[Bogart *et al.*, 1990] K. Bogart, C. Greene, and J. Kung. The impact of the chain decomposition theorem on classical combinatorics. In *The Dilworth Theorems: Selected Papers of Robert P. Dilworth*, pages 19–29. Birkhäuser, Boston, 1990.

[Bollobás, 1997] B. Bollobás. Paul Erdős – life and work. In *The Mathematics of Paul Erdős I*, pages 1–41. Springer, Berlin, 1997.

[Bonnet and Pouzet, 1969] R. Bonnet and M. Pouzet. Extension et stratification d'ensembles dispersés. *C. R. Acad. Sci. Paris Sér. A-B*, 268:A1512–A1515, 1969.

[Bonnet and Pouzet, 1982] R. Bonnet and M. Pouzet. Linear extensions of ordered sets. In *Ordered sets (Banff, Alta., 1981)*, volume 83 of *NATO Adv. Study Inst. Ser. C: Math. Phys. Sci.*, pages 125–170. Reidel, Dordrecht, 1982.

[Booth, 1970] D. Booth. Ultrafilters on a countable set. *Ann. Math. Logic*, 2(1):1–24, 1970.

[Bourbaki, 1939] N. Bourbaki. *Éléments de mathématique. Part I. Les structures fondamentales de l'analyse. Livre I. Théorie des ensembles (Fascicule de résultats).* Actual. Sci. Ind., no. 846. Hermann & Cie., Paris, 1939.

[Brendle, 2006] J. Brendle. Cardinal invariants of the continuum and combinatorics on uncountable cardinals. *Ann. Pure Appl. Logic*, 144(1-3):43–72, 2006.

[Brown *et al.*, 1977] L. G. Brown, R. G. Douglas, and P. A. Fillmore. Extensions of C^*-algebras and K-homology. *Ann. of Math. (2)*, 105(2):265–324, 1977.

[Brualdi and Scrimger, 1968] R. Brualdi and E. Scrimger. Exchange systems, matchings, and transversals. *J. Combinatorial Theory*, 5:244–257, 1968.

[Bry, 1981] F. Bry. Note on a theorem of J. Folkman on transversals of infinite families with finitely many infinite members. *J. Combin. Theory Ser. B*, 30(1):100–102, 1981.

[Bukovský, 1966] L. Bukovský. Consistency theorems connected with some combinatorial problems. *Comment. Math. Univ. Carolinae*, 7:495–499, 1966.

[Cantor, 1883] G. Cantor. Über unendliche, lineare Punktmannichfaltigkeiten. *Math. Ann.*, 21:545–591, 1883.

[Carlson and Simpson, 1984] T. Carlson and S. Simpson. A dual form of Ramsey's theorem. *Adv. in Math.*, 53(3):265–290, 1984.

[Carlson and Simpson, 1990] T. Carlson and S. Simpson. Topological Ramsey theory. In *Mathematics of Ramsey Theory*, volume 5 of *Algorithms Combin.*, pages 172–183. Springer, Berlin, 1990.

[Carlson *et al.*, 2006] T. Carlson, N. Hindman, and D. Strauss. An infinitary extension of the Graham-Rothschild parameter sets theorem. *Trans. Amer. Math. Soc.*, 358(7):3239–3262 (electronic), 2006.

[Carlson, 1984] T. Carlson. The pin-up conjecture. In *Axiomatic Set Theory (Boulder, Colo., 1983)*, pages 41–62. Amer. Math. Soc., Providence, RI, 1984.

[Carlson, 1987] T. Carlson. An infinitary version of the Graham-Leeb-Rothschild theorem. *J. Combin. Theory Ser. A*, 44(1):22–33, 1987.

[Carlson, 1988] T. Carlson. Some unifying principles in Ramsey theory. *Discrete Math.*, 68(2-3):117–169, 1988.

[Carruth, 1942] P. Carruth. Arithmetic of ordinals with applications to the theory of ordered Abelian groups. *Bull. Amer. Math. Soc.*, 48:262–271, 1942.

[Cayley, 1857] A. Cayley. On the theory of analytical forms called Trees. *Phil. Magazine*, 13:172–176, 1857.

[Cayley, 1875] A. Cayley. On the analytic forms called Trees, with applications to the theory of chemical combinations. *Brit. Assoc. Report*, pages 257–305, 1875.

[Cayley, 1881] A. Cayley. On the analytic forms called Trees. *Amer. J. Math.*, 4:266–268, 1881.

[Chang and Keisler, 1973] C. C. Chang and H. J. Keisler. *Model Theory*. North-Holland Publishing Co., Amsterdam, 1973. Studies in Logic and the Foundations of Mathematics, Vol. 73; Second edition published in 1977; Third edition in 1990 with additional material.

[Chang, 1965] C. C. Chang. A note on the two cardinal problem. *Proc. Amer. Math. Soc.*, 16:1148–1155, 1965.

[Chang, 1972] C. C. Chang. A partition theorem for the complete graph on ω^ω. *J. Combinatorial Theory Ser. A*, 12(3):396–452, 1972.

[Chang, 1998] C. C. Chang. The writing of the MV-algebras. *Studia Logica*, 61(1):3–6, 1998. Many-valued logics.

[Choksi *et al.*, 1989] J. R. Choksi, S. J. Eigen, J. C. Oxtoby, and V. S. Prasad. The work of Dorothy Maharam on measure theory, ergodic theory and category algebras. In *Measure and Measurable Dynamics (Rochester, NY, 1987)*, volume 94 of *Contemp. Math.*, pages 57–71. Amer. Math. Soc., Providence, RI, 1989.

[Choodnovsky, 1975] G. Choodnovsky. Combinatorial properties of compact cardinals. In *Infinite and Finite Sets (Colloq., Keszthely, 1973; dedicated to P. Erdős on his 60th birthday), Vol. I*, pages 109–130, Colloq. Math. Soc. János Bolyai, Vol. 10. North-Holland, Amsterdam, 1975.

[Ciesielski and Rassias, 2009] K. Ciesielski and T. Rassias. On Stan Ulam and his mathematics. *Aust. J. Math. Anal. Appl.*, 6(1):Art. 1, 9, 2009.

[Cohen, 1954] L. Cohen. October meeting in New York. *Bull. Amer. Math. Soc.*, 60(1):20–37, 1954. Report on the October 22–24, 1953 meeting.

[Cohn, 2000] P. N. Cohn. Obituary: Arthur Harold Stone (1916-2000). *Bulletin of the London Mathematical Society*, 34:613–618, 2000.

[Comfort and Negrepontis, 1982] W. W. Comfort and S. Negrepontis. *Chain Conditions in Topology*, volume 79 of *Cambridge Tracts in Mathematics*. Cambridge University Press, Cambridge, 1982.

[Comfort, 1977] W. W. Comfort. Ultrafilters: some old and some new results. *Bull. Amer. Math. Soc.*, 83(4):417–455, 1977.

[Comfort, 1997] W. W. Comfort. An interview with Arthur Stone. *Topological Commentary*, 2(1):1–10, 1997. Posted to http://at.yorku.ca/t/o/p/c/16.htm on February 3, 1997; reproduced from the Handbook of the History of General Topology.

[Comfort, 2005] W. Comfort. Honoring Neil Hindman. *Sci. Math. Jpn.*, 62(2):317–320, 2005.

[Cooke, 1993] R. Cooke. The Luzin archive. *Istor.-Mat. Issled.*, 34:246–255, 1993. In Russian.

[Cooke, 2010] R. Cooke. Naming infinity [book review]. *Math. Intelligencer*, 32(1):59–64, 2010.

[Crossley, 1975] J. Crossley. Reminiscences of logicians. In *Algebra and Logic (Fourteenth Summer Res. Inst., Austral. Math. Soc., Monash Univ., Clayton, 1974)*, volume 450 of *Lecture Notes in Mathematics*, pages 1–62. Springer-Verlag, Berlin, 1975. Reported by J. N. Crossley, With contributions by C. C. Chang, John Crossley, Jerry Keisler, Steve Kleene, Mike Morley, Vivienne Morley, Andrzej Mostowski, Anil Nerode, Gerald Sacks, Peter Hilton and David Lucy.

[Cuesta Dutari, 1943] N. Cuesta Dutari. Teoria decimal de los tipos de orden. *Revista Mat. Hisp.-Amer. (4)*, 3:186–205, 242–268, 1943. Decimal theory of the order types.

[Cummings and Foreman, 1998] J. Cummings and M. Foreman. The tree property. *Adv. Math.*, 133(1):1–32, 1998.

[Cummings *et al.*, 2001] J. Cummings, M. Foreman, and M. Magidor. Squares, scales and stationary reflection. *J. Math. Log.*, 1(1):35–98, 2001.

[Cummings, 1997] J. Cummings. Souslin trees which are hard to specialise. *Proc. Amer. Math. Soc.*, 125(8):2435–2441, 1997.

[Cummings, 2005] J. Cummings. Notes on singular cardinal combinatorics. *Notre Dame J. Formal Logic*, 46(3):251–282 (electronic), 2005.

[Darby and Laver, 1998] C. Darby and R. Laver. Countable length Ramsey games. In *Set Theory (Curaçao, 1995; Barcelona, 1996)*, pages 41–46. Kluwer Acad. Publ., Dordrecht, 1998.

[Darby, 1999] C. Darby. Negative partition relations for ordinals ω^{ω^α}. *J. Combin. Theory Ser. B*, 76:205–222, 1999. Notes circulated in 1995.

[de Jongh and Parikh, 1977] D. de Jongh and R. Parikh. Well-partial orderings and hierarchies. *Nederl. Akad. Wetensch. Proc. Ser. A* **80**=*Indag. Math.*, 39(3):195–207, 1977.

[Dedekind, 1968] R. Dedekind. *Gesammelte mathematische Werke. Bände I–III.* Herausgegeben von Robert Fricke, Emmy Noether und Öystein Ore. Chelsea Publishing Co., New York, 1968.

[Delsarte, 2000] J. Delsarte. Compte rendu de la réunion Bourbaki du 14 janvier 1935. *Gaz. Math.*, 84:16–18, 2000.

[Demidov and Levshin, 1999] S. S. Demidov and B. V. Levshin, editors. *Delo akademika Nikolaya Nikolaevicha Luzina.* Russkiĭ Khristianskiĭ Gumanitarnyĭ Institut, St. Petersburg, 1999.

[Denjoy, 1946] A. Denjoy. *L'énumération transfinie, Livre I: La notion de rang.* Gauthier-Villars, Paris, 1946. Bibliography compiled by G. Choquet; see pages XXVII-XXVIII for material on Suslin's problem.

[Denjoy, 1953] A. Denjoy. L'ordination des ensembles. *C. R. Acad. Sci. Paris*, 236:1393–1396, 1953.

[Deuber, 1989] Walter A. Deuber. Developments based on Rado's dissertation "Studien zur Kombinatorik". In *Surveys in Combinatorics, 1989 (Norwich, 1989)*, volume 141 of *London Math. Soc. Lecture Note Ser.*, pages 52–74. Cambridge Univ. Press, Cambridge, 1989.

[Devlin and Johnsbråten, 1974] K. Devlin and H. Johnsbråten. *The Souslin Problem*, volume 405 of *Lecture Notes in Mathematics*. Springer-Verlag, Berlin, 1974.

[Devlin and Shelah, 1978] K. Devlin and S. Shelah. A weak version of $\diamondsuit$ which follows from $2^{\aleph_0} < 2^{\aleph_1}$. *Israel J. Math.*, 29(2-3):239–247, 1978.

[Devlin and Shelah, 1979a] K. Devlin and S. Shelah. A note on the normal Moore space conjecture. *Canad. J. Math.*, 31(2):241–251, 1979.

[Devlin and Shelah, 1979b] K. Devlin and S. Shelah. Souslin properties and tree topologies. *Proc. London Math. Soc. (3)*, 39(2):237–252, 1979.

[Devlin, 1972] K. Devlin. Note on a theorem of J. Baumgartner. *Fund. Math.*, 76(3):255–260, 1972.

[Devlin, 1973] K. Devlin. *Aspects of Constructibility.* Lecture Notes in Mathematics, Vol. 354. Springer-Verlag, Berlin, 1973.

[Devlin, 1974] K. Devlin. Order types, trees, and a problem of Erdős and Hajnal. *Period. Math. Hungar.*, 5:153–160, 1974.

[Devlin, 1978] K. Devlin. $\aleph_1$-trees. *Ann. Math. Logic*, 13(3):267–330, 1978.

[Devlin, 1979a] K. Devlin. *Some Partition Theorems and Ultrafilters on ω.* PhD thesis, Dartmouth College, 1979. James E. Baumgartner, advisor.

[Devlin, 1979b] K. J. Devlin. Variations on $\diamondsuit$. *J. Symbolic Logic*, 44(1):51–58, 1979.

[Devlin, 1983] K. Devlin. A new construction of a Kurepa tree with no Aronszajn subtree. *Fund. Math.*, 118(2):123–127, 1983.

[Di Prisco and Todorcevic, 1999] C. Di Prisco and S. Todorcevic. A cardinal defined by a polarized partition relation. *Israel J. Math.*, 109:41–52, 1999.

[Dilworth, 1950] R. Dilworth. A decomposition theorem for partially ordered sets. *Ann. of Math. (2)*, 51:161–166, 1950.

[Dilworth, 1960] R. Dilworth. Some combinatorial problems on partially ordered sets. In *Proc. Sympos. Appl. Math., Vol. 10*, pages 85–90. American Mathematical Society, Providence, R. I., 1960.

[Donder, 1982] H. Donder. Prikry's principle for ω_2 and $0^{\#}$. Cited by [Kanamori, 1986], 1982.

[Doner *et al.*, 1978] J. Doner, A. Mostowski, and A. Tarski. The elementary theory of well-ordering—a metamathematical study. In *Logic Colloquium '77 (Proc. Conf. Wrocław, 1977)*, pages 1–54. North-Holland, Amsterdam, 1978.

[Dow, 1989] A. Dow. Review of five articles by Todorčević. *J. Symbolic Logic*, 54(2):653–658, 1989.

[Dow, 1992] A. Dow. Set theory in topology. In *Recent Progress in General Topology (Prague, 1991)*, pages 167–197. North-Holland, Amsterdam, 1992.

[Dowker, 1951] C. H. Dowker. On countably paracompact spaces. *Canadian J. Math.*, 3:219–224, 1951.

[Drake, 1974] F. Drake. *Set Theory: An Introduction to Large Cardinals*, volume 76 of *Studies in Logic and the Foundations of Mathematics.* North-Holland Publishing Company, Inc., Amsterdam-New York, 1974.

[Ducasse and Curry, 1962] C. J. Ducasse and H. Curry. Early history of the Association of Symbolic Logic. *J. Symbolic Logic*, 27(3):255–258, 1962.

[Duffus, 2003] D. Duffus. Obituary: Ivan Rival. *Order*, 20(2):173–183, 2003.

[Dushnik and Miller, 1940] B. Dushnik and E. Miller. Concerning similarity transformations of linearly ordered sets. *Bull. Amer. Math. Soc.*, 46:322–326, 1940.

[Dushnik and Miller, 1941] B. Dushnik and E. Miller. Partially ordered sets. *Amer. J. of Math.*, 63:600–610, 1941.

[Dushnik, 1931] B. Dushnik. A note on transfinite ordinals. *Bull. Amer. Math. Soc.*, 37(12):860–862, 1931.

[Džamonja and Shelah, 1999] M. Džamonja and S. Shelah. Similar but not the same: various versions of ♣ do not coincide. *J. Symbolic Logic*, 64(1):180–198, 1999.

[Džamonja and Thompson, 2005] M. Džamonja and K. Thompson. Universality results for well-founded posets. *Sarajevo J. Math.*, 1(14)(2):147–160, 2005.

[Džamonja and Thompson, 2006] M. Džamonja and K. Thompson. A poset hierarchy. *Cent. Eur. J. Math.*, 4(2):225–241 (electronic), 2006.

[Džamonja *et al.*, 2009] M. Džamonja, J. Larson, and W. Mitchell. A partition theorem for a large dense linear order. *Israel J. Math.*, 171:237–284, 2009.

[Editors, 1977] The Editors. Géza Fodor (1927–1977). *Acta Sci. Math. (Szeged)*, 39(3-4):203–204, 1977.

[Egorov, 1911] D. Egorov. Sur les suites de fonctions mesurables. *C. R. Acad. Sci. Paris*, 152:244–246, 1911.

[Ehrenfeucht *et al.*, 2008] A. Ehrenfeucht, V. W. Marek, and M. Srebrny, editors. *Andrzej Mostowski and Foundational Studies.* IOS Press, Amsterdam, 2008.

[Ehrenfeucht, 1957] A. Ehrenfeucht. Application of games to some problems of mathematical logic. *Bull. Acad. Polon. Sci. Cl. III.*, 5:35–37, IV, 1957.

[Ehrenfeucht, 1959] A. Ehrenfeucht. Decidability of the theory of the linear ordering relation. *Notices Amer. Math. Soc.*, 6(3):268, 1959. Abstract 556-38.

[Ehrenfeucht, 1961] A. Ehrenfeucht. An application of games to the completeness problem for formalized theories. *Fund. Math.*, 49:129–141, 1961. Received January 1, 1960.

[Eisworth, 2009] T. Eisworth. A note on strong negative partition relations. *Fund. Math.*, 202(2):97–123, 2009.

[Eisworth, 2010] T. Eisworth. Successors of singular cardinals. In *Handbook of Set Theory, II*, chapter 15, pages 1229–1350. Springer, Berlin, 2010.

[Ellentuck, 1974] E. Ellentuck. A new proof that analytic sets are Ramsey. *J. Symbolic Logic*, 39:163–165, 1974.

[Erdős and Hajnal, 1958] P. Erdős and A. Hajnal. On the structure of set-mappings. *Acta Math. Acad. Sci. Hungar*, 9:111–131, 1958.

[Erdős and Hajnal, 1962a] P. Erdős and A. Hajnal. On a classification of denumerable order types and an application to the partition calculus. *Fund. Math.*, 51:117–129, 1962.

[Erdős and Hajnal, 1962b] P. Erdős and A. Hajnal. Some remarks concerning our paper "On the structure of set-mappings". Non-existence of a two-valued σ-measure for the first uncountable inaccessible cardinal. *Acta Math. Acad. Sci. Hungar.*, 13:223–226, 1962.

[Erdős and Hajnal, 1966] P. Erdős and A. Hajnal. On a problem of B. Jónsson. *Bull. Acad. Polon. Sci. Sér. Sci. Math. Astronom. Phys.*, 14:19–23, 1966.

[Erdős and Hajnal, 1971a] P. Erdős and A. Hajnal. Ordinary partition relations for ordinal numbers. *Period. Math. Hungar.*, 1(3):171–185, 1971.

[Erdős and Hajnal, 1971b] P. Erdős and A. Hajnal. Unsolved problems in set theory. In *Axiomatic Set Theory (Proc. Sympos. Pure Math., UCLA, 1967)*, volume 13, pages 17–48. Amer. Math. Soc., Providence, R.I., 1971.

[Erdős and Hajnal, 1974] P. Erdős and A. Hajnal. Solved and unsolved problems in set theory. In *Proceedings of the Tarski Symposium, (Proc. Sympos. Pure Math., UC Berkeley, CA, 1971)*, volume 25, pages 261–265. Amer. Math. Soc., Providence, R.I., 1974.

[Erdős and Milner, 1972] P. Erdős and E. Milner. A theorem in the partition calculus. *Canad. Math. Bull.*, 15:501–505, 1972.

[Erdős and Rado, 1950] P. Erdős and R. Rado. A combinatorial theorem. *J. London Math. Soc. (2)*, 25:249–255, 1950.

[Erdős and Rado, 1952a] P. Erdős and R. Rado. Combinatorial theorems on classifications of subsets of a given set. *Proc. London Math. Soc. (3)*, 2:417–439, 1952.

[Erdős and Rado, 1952b] P. Erdős and R. Rado. Solution to problem 4358. *Amer. Math. Monthly*, 59(4):255–257, 1952.

[Erdős and Rado, 1953] P. Erdős and R. Rado. A problem on ordered sets. *J. London Math. Soc. (2)*, 28(4):426–438, 1953.

[Erdős and Rado, 1956] P. Erdős and R. Rado. A partition calculus in set theory. *Bull. Amer. Math. Soc.*, 62:427–489, 1956.

[Erdős and Rado, 1959] P. Erdős and R. Rado. A theorem on partial well-ordering of sets of vectors. *J. London Math. Soc.*, 34:222–224, 1959.

[Erdős and Rado, 1960] P. Erdős and R. Rado. Intersection theorems for systems of sets. *J. London Math. Soc.*, 35:85–90, 1960.

[Erdős and Szekeres, 1935] P. Erdős and G. Szekeres. A combinatorial problem in geometry. *Compositio Math.*, 2:463–470, 1935.

[Erdős and Tarski, 1943] P. Erdős and A. Tarski. On families of mutually exclusive sets. *Ann. of Math. (2)*, 44(2):315–329, April 1943.

[Erdős and Tarski, 1961] P. Erdős and A. Tarski. On some problems involving inaccessible cardinals. In *Essays on the Foundations of Mathematics*, pages 50–82. Magnes Press, Hebrew Univ., Jerusalem, 1961.

[Erdős *et al.*, 1965] P. Erdős, A. Hajnal, and R. Rado. Partition relations for cardinal numbers. *Acta Math. Acad. Sci. Hungar.*, 16:93–196, 1965.

[Erdős *et al.*, 1968] P. Erdős, A. Hajnal, and E. Milner. On sets of almost disjoint subsets of a set. *Acta Math. Acad. Sci. Hungar*, 19:209–218, 1968.

[Erdős *et al.*, 1984] P. Erdős, A. Hajnal, A. Máté, and R. Rado. *Combinatorial Set Theory: Partition Relations for Cardinals.* North-Holland Publishing Co., Amsterdam, 1984.

[Erdős *et al.*, 1997] P. Erdős, S. Jackson, and R. Mauldin. On infinite partitions of lines and space. *Fund. Math.*, 152(1):75–95, 1997.

[Erdős, 1942] P. Erdős. Some set-theoretical properties of graphs. *Revista de la Universidad Nacional de Tucumán, Serie A, Matemática y Fisica Téorica*, 3:363–367, 1942.

[Erdős, 1949] P. Erdős. Advanced problem 4358. *Amer. Math. Monthly*, 56:480, 1949.

[Erdős, 1950] P. Erdős. Some remarks on set theory. *Proc. Amer. Math. Soc.*, 1:127–141, 1950.

[Erdős, 1981] P. Erdős. My Scottish Book "problems". In *The Scottish Book: Mathematics from the Scottish Café*, pages 35–43. Birkhäuser, Boston, 1981.

[Erdős, 1987] P. Erdős. Some problems on finite and infinite graphs. In *Logic and Combinatorics (Arcata, Calif., 1985)*, pages 223–228. Amer. Math. Soc., Providence, R.I., 1987.

[Erdős, 1996] P. Erdős. On some of my favorite theorems. In D. Miklós, V. Sós, and T. Szőnyi, editors, *Combinatorics, Paul Erdős is Eighty. Vol. 2*, volume 2 of *Bolyai Society Mathematical Studies*, pages 97–132. János Bolyai Mathematical Society, Budapest, 1996. Papers from the International Conference on Combinatorics held at Keszthely, July 19–24, 1993.

[Everett and Whaples, 1949] C. Everett and G. Whaples. Representations of sequences of sets. *Amer. J. Math.*, 71:287–293, 1949.

[Farah, 2000] I. Farah. Analytic quotients: theory of liftings for quotients over analytic ideals on the integers. *Mem. Amer. Math. Soc.*, 148(702):xvi+177, 2000.

[Farah, 2011] I. Farah. All automorphisms of the Calkin algebra are inner. *Ann. of Math.* (2), 173(2): 619–661, 2011.

[Feferman, 2004] S. Feferman. Tarski's conception of logic. *Ann. Pure Appl. Logic*, 126(1):5–13, 2004.

[Felscher, 1962] W. Felscher. Doppelte Hülleninduktion und ein Satz von Hessenberg und Bourbaki. *Arch. Math.*, 13:160–165, 1962.

[Fitzpatrick, 1997] B. Fitzpatrick, Jr. Some aspects of the work and influence of R. L. Moore. In *Handbook of the History of General Topology, Vol. 1*, pages 41–61. Kluwer Acad. Publ., Dordrecht, 1997.

[Fleissner, 1975] W. Fleissner. When is Jones' Space Normal? *Proc. Amer. Math. Soc.*, 50:375–378, 1975.

[Fleissner, 1984] W. Fleissner. The normal Moore space conjecture and large cardinals. In *Handbook of Set-Theoretic Topology*, pages 733–760. North-Holland, Amsterdam, 1984.

[Fodor, 1955] G. Fodor. Generalization of a theorem of Alexandroff and Urysohn. *Acta Sci. Math. Szeged*, 16:204–206, 1955.

[Fodor, 1956] G. Fodor. Eine Bemerkung zur Theorie der regressiven Funktionen. *Acta Sci. Math. Szeged*, 17:139–142, 1956.

[Folkman, 1970] J. Folkman. Transversals of infinite families with finitely many infinite members. *J. Combinatorial Theory*, 9:200–220, 1970.

[Foreman and Hajnal, 2003] M. Foreman and A. Hajnal. A partition relation for successors of large cardinals. *Math. Ann.*, 325(3):583–623, 2003.

[Foreman *et al.*, 1988] M. Foreman, M. Magidor, and S. Shelah. Martin's Maximum, saturated ideals, and nonregular ultrafilters. I. *Ann. of Math. (2)*, 127(1):1–47, 1988.

[Foreman *et al.*, 2001] M. Foreman, M. Magidor, and R. Schindler. The consistency strength of successive cardinals with the tree property. *J. Symbolic Logic*, 66(4):1837–1847, 2001.

[Foreman, 1998a] M. Foreman. An $\aleph_1$-dense ideal on $\aleph_2$. *Israel J. Math.*, 108:253–290, 1998.

[Foreman, 1998b] M. Foreman. Generic large cardinals: new axioms for mathematics? In *Proceedings of the International Congress of Mathematicians, Vol. II (Berlin, 1998)*, pages 11–21 (electronic), 1998. Extra Vol. II.

[Foreman, 2010] M. Foreman. Ideals and Generic Elementary Embeddings. In *Handbook of Set Theory, II*, chapter 13, pages 885–1147. Springer, Berlin, 2010.

[Fraïssé, 1948] R. Fraïssé. Sur la comparaison des types d'ordres. *C. R. Acad. Sci. Paris*, 226:1330–1331, April 1948.

[Fraïssé, 1953] R. Fraïssé. *Sur quelques classifications des systèmes de relations*. PhD thesis, Université de Paris, 1953.

[Fraïssé, 1954] R. Fraïssé. Sur quelques classifications des systèmes de relations. *C.R. Acad. Sci. Paris Sér. A–B*, 1:35–182, 1954.

[Fraïssé, 1986] R. Fraïssé. *Theory of Relations*, volume 118 of *Studies in Logic and the Foundations of Mathematics*. North-Holland Publishing Co., Amsterdam, 1986. Translated from the French by P. Clote.

[Fraïssé, 2000] R. Fraïssé. *Theory of Relations. Revised edition with an Appendix by Norbert Sauer, rev. ed.* Elsevier, Amsterdam; London, 2000. Studies in Logic and the Foundations of Mathematics; vol 145.

[Franchella, 1997] M. Franchella. On the origins of Dénes König's infinity lemma. *Arch. Hist. Exact Sci.*, 51(1):3–27, 1997.

[Fremlin, 1989] D. Fremlin. Measure algebras. In *Handbook of Boolean Algebras, Vol. 3*, pages 877–980. North-Holland, Amsterdam, 1989.

[Frewer, 1981] M. Frewer. Felix Bernstein. *Jahresber. Deutsch. Math.-Verein.*, 83(2):84–95, 1981.

[Frič and Kent, 2000] R. Frič and D. Kent. In memoriam of Professor Josef Novák. *Czechoslovak Math. J.*, 50(125)(1):221–223, 2000.

[Friedman, 1975] H. Friedman. One hundred and two problems in mathematical logic. *J. Symbolic Logic*, 40:113–129, 1975.

[Frolík and Koutník, 1985] Z. Frolík and V. Koutník. The eightieth birthday of Professor Josef Novák. *Czechoslovak Math. J.*, 35(110)(2):338–344, 1985.

[Frolík and Zítek, 1975] Z. Frolík and F. Zítek. The seventieth anniversary of Professor Josef Novák. *Czechoslovak Math. J.*, 25(100):337–342, 1975.

[Furstenberg and Weiss, 1978] H. Furstenberg and B. Weiss. Topological dynamics and combinatorial number theory. *J. Analyse Math.*, 34:61–85 (1979), 1978.

[Furstenberg, 1981] H. Furstenberg. *Recurrence in Ergodic Theory and Combinatorial Number Theory*. Princeton University Press, Princeton, N.J., 1981. M. B. Porter Lectures.

[Gaifman and Specker, 1964] H. Gaifman and E. Specker. Isomorphism types of trees. *Proc. Amer. Math. Soc.*, 15:1–7, 1964.

[Gallai, 1967] T. Gallai. Transitiv orientierbare Graphen. *Acta Math. Acad. Sci. Hungar*, 18:25–66, 1967.

[Galvin and Hajnal, 1975] F. Galvin and A. Hajnal. Inequalities for cardinal powers. *Ann. of Math. (2)*, 101:491–498, 1975.

[Galvin and Larson, 1975] F. Galvin and J. Larson. Pinning countable ordinals. *Fund. Math.*, 82:357–361, 1975. Collection of articles dedicated to Andrzej Mostowski on his sixtieth birthday, VIII.

[Galvin and Prikry, 1973] F. Galvin and K. Prikry. Borel sets and Ramsey's theorem. *J. Symbolic Logic*, 38:193–198, 1973.

[Galvin and Prikry, 1976] F. Galvin and K. Prikry. Infinitary Jonsson algebras and partition relations. *Algebra Universalis*, 6(3):367–376, 1976.

[Galvin and Shelah, 1973] F. Galvin and S. Shelah. Some counterexamples in the partition calculus. *J. Combin Theory Ser. A*, 15:167–174, 1973.

[Galvin *et al.*, 1995] F. Galvin, A. Hajnal, and P. Komjáth. Edge decompositions of graphs with no large independent sets. *Publ. Inst. Math. (Beograd) (N.S.)*, 57(71):71–80, 1995. Đuro Kurepa memorial volume.

[Galvin, 1968a] F. Galvin. A generalization of Ramsey's theorem. *Notices Amer. Math. Soc.*, 15, 1968. Abstract 68T-368, received January 26, 1968.

[Galvin, 1968b] F. Galvin. Partition theorems for the real line (abstract). *Notices Amer. Math. Soc.*, 15, 1968. Erratum 16 (1969), 1095.

[Galvin, 1971] F. Galvin. Lectures by Fred Galvin. UCLA Seminar Notes taken April 27, 1971 by Marcia Brown, 1971.

[Galvin, 1975] F. Galvin. On a partition theorem of Baumgartner and Hajnal. In *Infinite and Finite Sets (Colloq., Keszthely, 1973; dedicated to P. Erdős on his 60th birthday), Vol. II*, pages 711–729. Colloq. Math. Soc. János Bolyai, Vol. 10. North-Holland, Amsterdam, 1975.

[Galvin, 1980] F. Galvin. Chain conditions and products. *Fund. Math.*, 108(1):33–48, 1980.

[Galvin, 1994] F. Galvin. A proof of Dilworth's chain decomposition theorem. *Amer. Math. Monthly*, 101(4):352–353, 1994.

[Galvin, 2010] F. Galvin. Email to Larson dated November 8. With a copy to Prikry, 2010.

[Gandy and Sacks, 1967] R. Gandy and G. Sacks. A minimal hyperdegree. *Fund. Math.*, 61:215–223, 1967.

[Ghouila-Houri, 1962] A. Ghouila-Houri. Caractérisation des graphes non orientés dont on peut orienter les arêtes de manière à obtenir le graphe d'une relation d'ordre. *C. R. Acad. Sci. Paris*, 254:1370–1371, 1962.

[Gilmore and Hoffman, 1964] P. Gilmore and A. Hoffman. A characterization of comparability graphs and of interval graphs. *Canad. J. Math.*, 16:539–548, 1964.

[Ginsburg, 1953a] S. Ginsburg. *Order Types and Similarity Transformations.* PhD thesis, University of Michigan, 1953. Ben Dushnik, advisor.

[Ginsburg, 1953b] S. Ginsburg. Some remarks on order types and decompositions of sets. *Trans. Amer. Math. Soc.*, 74:514–535, 1953.

[Ginsburg, 1955] S. Ginsburg. Order Types and Similarity Transformations. *Trans. Amer. Math. Soc.*, 79:341–361, 1955.

[Gleyzal, 1940] A. Gleyzal. Order types and structure of orders. *Trans. Amer. Math. Soc.*, 48:451–466, 1940.

[Gödel, 1940] K. Gödel. *The Consistency of the Axiom of Choice and of the Generalized Continuum Hypothesis with the Axioms of Set Theory.* Annals of Mathematics Studies, no. 3. Princeton University Press, Princeton, N. J., 1940.

[Gottschalk, 1951] W. Gottschalk. Choice functions and Tychonoff's theorem. *Proc. Amer. Math. Soc.*, 2:172, 1951.

[Gowers *et al.*, 2008] T. Gowers, J. Barrow-Green, and I. Leader, editors. *The Princeton Companion to Mathematics.* Princeton University Press, Princeton, NJ, 2008.

[Gowers, 2003] W. Gowers. Ramsey methods in Banach spaces. In *Handbook of the Geometry of Banach Spaces, Vol. 2*, pages 1071–1097. North-Holland, Amsterdam, 2003.

[Graham and Kantor, 2006] L. Graham and J. Kantor. A comparison of two cultural approaches to mathematics: France and Russia, 1890–1930. *Isis*, 97(1):56–74, 2006.

[Graham and Kantor, 2007] L. Graham and J. Kantor. Religious heresy and mathematical creativity in Russia. *Math. Intelligencer*, 29(4):17–22, 2007.

[Graham and Kantor, 2009] L. Graham and J. Kantor. *Naming Infinity.* The Belknap Press of Harvard, Cambridge, MA, 2009. A true story of religious mysticism and mathematical creativity.

[Graham and Nešetřil, 1997] R. Graham and J. Nešetřil, editors. *The Mathematics of Paul Erdős. I*, volume 13 of *Algorithms and Combinatorics.* Springer-Verlag, Berlin, 1997.

[Graham and Nešetřil, 2002] R. Graham and J. Nešetřil. Ramsey Theory in the work of Paul Erdős. In G. Halász, L. Lovász, M. Simonovits, and V. Sós, editors, *The Mathematics of Paul Erdős II*, volume 11 of *Bolyai Society Mathematical Studies*, pages 339–365. János Bolyai Mathematical Society and Springer-Verlag, Budapest, 2002.

[Graham and Rothschild, 1971] R. Graham and B. Rothschild. Ramsey's theorem for n-parameter sets. *Trans. Amer. Math. Soc.*, 159:257–292, 1971.

[Graham and Rothschild, 1974] R. Graham and B. Rothschild. Some recent developments in Ramsey theory. In *Combinatorics (Proc. Advanced Study Inst., Breukelen, 1974), Part 2: Graph theory; foundations, partitions and combinatorial geometry*, pages 61–76. Math. Centre Tracts, No. 56. Math. Centrum, Amsterdam, 1974.

[Graham *et al.*, 1972] R. Graham, K. Leeb, and B. Rothschild. Ramsey's theorem for a class of categories. *Advances in Math.*, 8:417–433, 1972.

[Graham *et al.*, 1990] R. Graham, B. Rothschild, and J. Spencer. *Ramsey Theory*. John Wiley & Sons Inc., New York, second edition, 1990.

[Gregory, 1976] J. Gregory. Higher Souslin trees and the generalized continuum hypothesis. *J. Symbolic Logic*, 41(3):663–671, 1976.

[Gruenhage and Mashburn, 1999] G. Gruenhage and J. Mashburn. On the decomposition of order-separable posets of countable width into chains. *Order*, 16(1):171–177, 1999.

[Haddad and Sabbagh, 1969a] L. Haddad and G. Sabbagh. Calcul de certains nombres de Ramsey généralisés. *C. R. Acad. Sci. Paris Sér. A-B*, 268:A1233–A1234, 1969.

[Haddad and Sabbagh, 1969b] L. Haddad and G. Sabbagh. Nouveaux résultats sur les nombres de Ramsey généralisés. *C. R. Acad. Sci. Paris Sér. A-B*, 268:A1516–A1518, 1969.

[Haddad and Sabbagh, 1969c] L. Haddad and G. Sabbagh. Sur une extension des nombres de Ramsey aux ordinaux. *C. R. Acad. Sci. Paris Sér. A-B*, 268:A1165–A1167, 1969.

[Hájek *et al.*, 2008] P. Hájek, V. W. Marek, and P. Vopěnka. Mostowski and Czech-Polish cooperation in mathematical logic. In *Andrzej Mostowski and Foundational Studies*, pages 403–405. IOS Press, Amsterdam, 2008.

[Hájek, 1971] P. Hájek. Sets, semisets, models. In *Axiomatic Set Theory (Proc. Sympos. Pure Math., Vol. XIII, Part I, Univ. California, Los Angeles, Calif., 1967)*, pages 67–81. Amer. Math. Soc., Providence, R.I., 1971.

[Hajnal and Hamburger, 1999] A. Hajnal and P. Hamburger. *Set Theory*, volume 48 of *London Mathematical Society Student Texts*. Cambridge University Press, Cambridge, 1999. Translated from the 1983 Hungarian original, *Halmazeimélet*, by Attila Máté.

[Hajnal and Juhász, 1967] A. Hajnal and I. Juhász. Discrete subspaces of topological spaces. *Nederl. Akad. Wetensch. Proc. Ser. A 70=Indag. Math.*, 29:343–356, 1967.

[Hajnal and Juhász, 1968] A. Hajnal and I. Juhász. On hereditarily α-Lindelöf and hereditarily α-separable spaces. *Ann. Univ. Sci. Budapest. Eötvös Sect. Math.*, 11:115–124, 1968.

[Hajnal and Juhász, 1969] A. Hajnal and I. Juhász. Discrete subspaces of topological spaces. II. *Nederl. Akad. Wetensch. Proc. Ser. A 72=Indag. Math.*, 31:18–30, 1969.

[Hajnal and Juhász, 1974] A. Hajnal and I. Juhász. On hereditarily α-Lindelöf and α-separable spaces. II. *Fund. Math.*, 81(2):147–158, 1974. Collection of articles dedicated to Andrzej Mostowski on the occasion of his sixtieth birthday, II.

[Hajnal and Komjáth, 1997] A. Hajnal and P. Komjáth. A strongly non-Ramsey order type. *Combinatorica*, 17(3):363–367, 1997.

[Hajnal and Larson, 2010] A. Hajnal and J. Larson. Partition relations. In *Handbook of Set Theory, I*, chapter 2, pages 129–213. Springer, Berlin, 2010.

[Hajnal *et al.*, 1987] A. Hajnal, A. Kanamori, and S. Shelah. Regressive partition relations for infinite cardinals. *Trans. Amer. Math. Soc.*, 299(1):145–154, 1987.

[Hajnal, 1960] A. Hajnal. Some results and problems in set theory. *Acta Math. Hungar.*, 11:227–298, 1960.

[Hajnal, 1961] A. Hajnal. Proof of a conjecture of S. Ruziewicz. *Fund. Math.*, 50:123–128, 1961.

[Hajnal, 1964] A. Hajnal. Remarks on the theorem of W. P. Hanf. *Fund. Math.*, 54:109–113, 1964.

[Hajnal, 1971] A. Hajnal. A negative partition relation. *Proc. Nat. Acad. Sci. U.S.A.*, 68:142–144, 1971.

[Hajnal, 1997] A. Hajnal. Paul Erdős' set theory. In R. Graham and J. Nešetřil, editors, *The Mathematics of Paul Erdős, II*, volume 14 of *Algorithms and Combinatorics*, pages 352–393. Springer, Berlin, 1997.

[Hales and Jewett, 1963] A. Hales and R. Jewett. Regularity and positional games. *Trans. Amer. Math. Soc.*, 106:222–229, 1963.

[Hall Jr., 1948] M. Hall Jr. Distinct representatives of subsets. *Bull. Amer. Math. Soc.*, 54:922–926, 1948.

[Hall, 1935] P. Hall. On representatives of subsets. *J. London Math. Soc.*, 10:26–30, 1935.

[Halmos and Vaughan, 1950] P. Halmos and H. Vaughan. The marriage problem. *Amer. J. Math.*, 72:214–215, 1950.

[Halpern and Läuchli, 1966] J. Halpern and H. Läuchli. A partition theorem. *Trans. Amer. Math. Soc.*, 124:360–367, 1966.

[Halpern and Lévy, 1964] J. Halpern and A. Lévy. The Ordering Theorem does not imply the Axiom of Choice. *Notices Amer. Math. Soc.*, 11, 1964. Abstract.

[Halpern and Lévy, 1971] J. Halpern and A. Lévy. The Boolean prime ideal theorem does not imply the axiom of choice. In *Axiomatic Set Theory (Proc. Sympos. Pure Math., Vol. XIII, Part I, Univ. California, Los Angeles, Calif., 1967)*, pages 83–134. Amer. Math. Soc., Providence, R.I., 1971.

[Halpern, 1962] J. Halpern. *Contributions to the Study of Independence of the Axiom of Choice.* PhD thesis, University of California, Berkeley, 1962.

[Halpern, 1964] J. Halpern. The independence of the axiom of choice from the Boolean prime ideal theorem. *Fund. Math.*, 55:57–66, 1964.

[Hanazawa, 1982] Masazumi Hanazawa. On Aronszajn trees with a non-Souslin base. *Tsukuba J. Math.*, 6(2):177–185, 1982.

[Hanf, 1964] W. Hanf. On a problem of Erdős and Tarski. *Fund. Math.*, 53:325–334, 1964.

[Harary, 1983] F. Harary. A tribute to F. P. Ramsey. *J. Graph Theory*, 7:1–7, 1983.

[Harzheim, 2008] E. Harzheim. A construction of subsets of the reals which have a similarity decomposition. *Order*, 25(2):79–83, 2008.

[Hausdorff, 1901] F. Hausdorff. Über eine gewisse Art geordneter Mengen. *Berichte über die Verhandlungen der Königlich Säsischen Gesellschaft der Wissenschaften zu Leipzig, Mathematisch-Physiche Klasse*, 53:460–475, 1901.

[Hausdorff, 1906] F. Hausdorff. Untersuchungen über Ordnugstypen I, II, III. *Berichte über die Verhandlungen der Königlich Säsischen Gesellschaft der Wissenschaften zu Leipzig, Mathematisch-Physiche Klasse*, 58:106–169, 1906.

[Hausdorff, 1907] F. Hausdorff. Untersuchungen über Ordnugstypen IV, V. *Berichte über die Verlhandlungen der Königlich Säsischen Gesellschaft der Wissenschaften zu Leipzig, Mathematisch-Physiche Klasse*, 59:84–159, 1907.

[Hausdorff, 1908] F. Hausdorff. Grundzüge einer Theorie der geordnete Mengenlehre. *Math. Ann.*, 65:435–505, 1908.

[Hausdorff, 1909] F. Hausdorff. Die graduierung nach dem endverlauf. *Abhandlungen der Königlich Säsischen Gesellschaft der Wissenschaften zu Leipzig, Mathematisch-Physiche Klasse*, 31:295–334, 1909.

[Hausdorff, 1914] F. Hausdorff. *Grunzüge der Mengenlehre.* Verlag Veit & Co, Leipzig, Germany, 1914. Reprinted by Chelsea Publishing Company, 1949, 1965.

[Hausdorff, 1916] F. Hausdorff. Die mätigkeit der Borelischen Mengen. *Math. Ann.*, 77:430–437, 1916.

[Hausdorff, 1936] F. Hausdorff. Summen von $\aleph_1$ mengen. *Fund. Math.*, 26:241–255, 1936.

[Hausdorff, 1957] F. Hausdorff. *Mengenlehre.* Chelsea Publishing Company, New York, 1957. third, revised edition of [Hausdorff, 1914]. Translated by John R. Auman as *Set Theory.*

[Hausdorff, 2002] F. Hausdorff. *Felix Hausdorff—gesammelte Werke. Band II.* Springer-Verlag, Berlin, 2002. "Grundzüge der Mengenlehre". ["Foundations of set theory"], Edited and with commentary by E. Brieskorn, S. D. Chatterji, M. Epple, U. Felgner, H. Herrlich, M. Hušek, V. Kanovei, P. Koepke, G. Preuß, W. Purkert and E. Scholz.

[Hausdorff, 2005] F. Hausdorff. *Hausdorff on Ordered Sets*, volume 25 of *History of Mathematics.* American Mathematical Society, Providence, RI, 2005. Translated from the German, edited and with commentary by J. M. Plotkin.

[Heath, 1964] R. W. Heath. Screenability, pointwise paracompactness, and metrization of Moore spaces. *Canad. J. Math.*, 16:763–770, 1964.

[Helson, 1949] H. Helson. On a problem of Sikorski. *Colloquium Math.*, 2:7–8, 1949.

[Hessenberg, 1906] G. Hessenberg. *Grundbegriffe der Mengenlehre.* Vandenhoek & Rupert, Göttingen, 1906. see also Abhandlungen der Fries'schen Schule, N.S. 1 (1906), 4.Heft, 487-706.

[Higman, 1952] G. Higman. Ordering by divisibility in abstract algebras. *Proc. London Math. Soc. (3)*, 2:326–336, 1952.

[Hilbert, 1892] D. Hilbert. Über die irreducibilitat Ganzer rationaler functionen mit Ganzzahligen coefficienten. *J. Reine Angew. Math.*, 110:104–129, 1892.

[Hilbert, 1900] D. Hilbert. Mathematische Probleme. Vortrag, gehalten auf dem internationalem Mathematiker-Kongress zu Paris. 1900. *Nachrichten von der Königlichen Gesellschaft der Wissenschaften zu Göttingen*, pages 253–297, 1900. Translated in the *Bulletin of the American Mathematical Society* 8 (1902), 437-479.

[Hildebrandt, 1929] T. H. Hildebrandt. Letter to R. L. Moore dated October 15, 1929. From the *R. L. Moore Collection, Archives of American Mathematics*, Center for American History, The University of Texas at Austin, 1929.

[Hindman and McCutcheon, 2002] N. Hindman and R. McCutcheon. One sided ideals and Carlson's theorem. *Proc. Amer. Math. Soc.*, 130(9):2559–2567 (electronic), 2002.

[Hindman and Strauss, 1998] N. Hindman and D. Strauss. *Algebra in the Stone-Čech Compactification*, volume 27 of *de Gruyter Expositions in Mathematics*. Walter de Gruyter & Co., Berlin, 1998. Theory and applications.

[Hindman, 1974] N. Hindman. Finite sums from sequences within cells of a partition of N. *J. Combinatorial Theory Ser. A*, 17:1–11, 1974.

[Hindman, 2005] N. Hindman. Algebra in the Stone-Čech compactification and its applications to Ramsey theory. *Sci. Math. Jpn.*, 62(2):321–329, 2005.

[Hindman, 2006] N. Hindman. The mathematics of Bruce Rothschild. *J. Combin. Theory Ser. A*, 113(1):2–20, 2006.

[Holz *et al.*, 1999] M. Holz, K. Steffens, and E. Weitz. *Introduction to Cardinal Arithmetic*. Birkhäuser Advanced Texts: Basler Lehrbücher. [Birkhäuser Advanced Texts: Basel Textbooks], Basel, 1999.

[Hrušák, 2001] M. Hrušák. Life in the Sacks model. *Acta Univ. Carolin. Math. Phys.*, 42(2):43–58, 2001. 29th Winter School on Abstract Analysis (Lhota nad Rohanovem/Zahrádky u České Lípy, 2001).

[Huntington, 1905a] E. Huntington. The continuum as a type of order: an exposition of the modern theory. *Ann. of Math. (2)*, 6(4):151–184, 1905.

[Huntington, 1905b] E. Huntington. The continuum as a type of order: an exposition of the modern theory. *Ann. of Math. (2)*, 7(1):15–43, 1905.

[Huntington, 1905c] E. Huntington. *The Continuum as a Type of Order; An Exposition of the Modern Theory, with an Appendix on the Transfinite Numbers*. Harvard University Press, Cambridge, MA, 1905.

[Huntington, 1905d] E. Huntington. A set of postulates for real algebra, comprising postulates for a one dimensional continuum and the theory of groups. *Trans. Amer. Math. Soc.*, 6:17–41, 1905.

[Huntington, 1917] E. Huntington. *The Continuum and Other Types of Serial Order: With an Introduction to Cantor's Transfinite Numbers*. Harvard University Press, second edition, 1917. Dover edition, 1955.

[Huuskonen, 1995] T. Huuskonen. Observations about Scott and Karp trees. *Ann. Pure Appl. Logic*, 76(3):201–230, 1995.

[Igoshin, 1996] V. I. Igoshin. A short biography of Mikhail Yakovlevich Suslin. *Uspekhi Mat. Nauk*, 51(3(309)):3–16, 1996. Translation in *Russian Math. Surveys*, 51(3):371–383.

[Jech, 1967] T. Jech. Non-provability of Souslin's hypothesis. *Comment. Math. Univ. Carolinae*, 8(2):291–305, 1967.

[Jech, 1971] T. Jech. Trees. *J. Symbolic Logic*, 36:1–14, 1971.

[Jech, 1972] T. Jech. Automorphisms of ω_1-trees. *Trans. Amer. Math. Soc.*, 173:57–70, 1972.

[Jech, 1995] T. Jech. Singular cardinals and the pcf theory. *Bull. Symbolic Logic*, 1(4):408–424, 1995.

[Jech, 2003] T. Jech. *Set Theory*. Springer Monographs in Mathematics. Springer-Verlag, Berlin, 2003. The third millennium edition, revised and expanded, 4th printing, 2006.

[Jensen, 1968] R. Jensen. Souslin's hypothesis is incompatible with $V = L$. *Notices Amer. Math. Soc.*, 15(6), 1968. Abstract 68T-E31 received August 20, 1968.

[Jensen, 1969a] R. Jensen. Automorphism properties of Soulin continua. *Notices Amer. Math. Soc.*, 16(3), 1969. Abstract 69T-E24 received December 2, 1968.

[Jensen, 1969b] R. Jensen. Some combinatorial principles in V and in L. Handwritten notes posted electronically on Jensen's web site. From the cover page: These notes presenting results of myself and Kunen were written at the Rockefeller University in 1969. Much of the work can be found in Frank Drake's book on large cardinals. Ronald Jensen 2002, 1969.

[Jensen, 1972] R. Jensen. The fine structure of the constructible hierarchy. *Ann. Math. Logic*, 4:229–308; erratum, ibid. 4 (1972), 443, 1972. With a section by Jack Silver.

[Johnston, 1956] J. Johnston. Universal infinite partially ordered sets. *Proc. Amer. Math. Soc.*, 7:507–514, 1956.

[Jones, 1933] F. B. Jones. Concerning normal and completely normal spaces. *Bull. Amer. Math. Soc.*, 39(9), 1933. Abstract 39-9-253-t (paper presented at the October 28, 1933 meeting of the AMS at Columbia University; and listed in the report on that meeting on page 828 of vol. 39, issue 11).

[Jones, 1937] F. B. Jones. Concerning normal and completely normal spaces. *Bull. Amer. Math. Soc.*, 43(10):671–677, 1937.

[Jones, 1952] F. B. Jones. On the separation of the set of pairs of a set. *J. Elisha Mitchell Sci. Soc.*, 68:44–45, 1952.

[Jones, 1953] F. B. Jones. On certain well-ordered monotone collections of sets. *J. Elisha Mitchell Sci. Soc.*, 69:30–34, 1953.

[Jones, 1954] F. B. Jones. On a property related to separability in metric spaces. *J. Elisha Mitchell Sci. Soc.*, 70:30–33, 1954.

[Jones, 1980] F. B. Jones. Metrization, non-metrization and a bit of history. In *Topology Conference, 1979 (Greensboro, N.C., 1979)*, pages 1–4. Guilford College, Greensboro, N.C., 1980.

[Jones, 1993] F. B. Jones. Some glimpses of the early years. In *Papers on General Topology and Applications (Madison, WI, 1991)*, volume 704 of *Ann. New York Acad. Sci.*, pages xi–xxii. New York Acad. Sci., New York, 1993.

[Jones, 1997] F. B. Jones. The beginning of topology in the United States and the Moore school. In *Handbook of the History of General Topology, Vol. 1*, pages 97–103. Kluwer Acad. Publ., Dordrecht, 1997.

[Jones, 1999] A. Jones. *Some Results in the Partition Calculus.* PhD thesis, Dartmouth College, 1999.

[Jones, 2000a] A. Jones. A polarized partition relation using elementary substructures. *J. Symbolic Logic*, 65(4):1491–1498, 2000.

[Jones, 2000b] A. Jones. A short proof of a partition relation for triples. *Electronic Journal of Combinatorics*, 7(1):Research Paper 24, 9 pp. (electronic), 2000.

[Jónsson, 1956] B. Jónsson. Universal relational systems. *Math. Scand.*, 4:193–208, 1956.

[Juhász and Weiss, 1978] I. Juhász and W. Weiss. On a problem of Sikorski. *Fund. Math.*, 100(3):223–227, 1978.

[Juhász, 1980] I. Juhász. A survey of S- and L-spaces. In *Topology, Vol. II (Proc. Fourth Colloq., Budapest, 1978)*, volume 23 of *Colloq. Math. Soc. János Bolyai*, pages 675–688, Amsterdam, 1980. North-Holland.

[Jullien, 1969] P. Jullien. *Contribution á l'étude des types d'ordre dispersés.* PhD thesis, Marseilles, 1969.

[Kanamori, 1982] A. Kanamori. On Silver's and related principles. In *Logic Colloquium '80 (Prague, 1980)*, pages 153–172. North-Holland, Amsterdam, 1982.

[Kanamori, 1986] A. Kanamori. Partition relations for successor cardinals. *Adv. in Math.*, 59(2):152–169, 1986.

[Kanamori, 1994] A. Kanamori. *The Higher Infinite.* Springer-Verlag, Berlin, 1994. Large cardinals in set theory from their beginnings.

[Kanamori, 2004] A. Kanamori. Zermelo and set theory. *Bull. Symbolic Logic*, 10(4):487–553, 2004.

[Kanamori, 2009] A. Kanamori. *The Higher Infinite.* Springer-Verlag, Berlin, second, corrected edition edition, 2009. Large cardinals in set theory from their beginnings.

[Kanamori, 2011a] A. Kanamori. Historical remarks on Suslin's Problem. In J. Kennedy and R. Kossak, editors, *Set Theory, Arithmetic and Foundations of Mathematics: Theorems, Philosophies*, volume 36 of *Lecture Notes in Logic*, page in press. Assoc. Symbol. Logic, 2011.

[Kanamori, 2011b] A. Kanamori. Kunen and set theory. To appear in *Topology and its Applications*, 2011.

[Kaplan, 1971] W. Kaplan. Mathematics at the University of Michigan. In P. Duran, editor, *A Century of Mathematics in America, (Univ. California, Los Angeles, Calif., 1967)*, volume 3, Part I, pages 179–134. Amer. Math. Soc., Providence, R.I., 1971.

[Keisler and Rowbottom, 1965] H. J. Keisler and F. Rowbottom. Constructible sets and weakly compact cardinals. *Notices Amer. Math. Soc.*, 12:373–374, 1965. Abstract 65T-186, received February 8, 1965.

[Keisler and Tarski, 1964] H. J. Keisler and A. Tarski. From accessible to inaccessible cardinals. Results holding for all accessible cardinal numbers and the problem of their extension to inaccessible ones. *Fund. Math.*, 53:225–308, 1964.

[Keisler, 1976] H. J. Keisler. *Elementary Calculus: An Infinitesimal Approach.* Prindle, Weber and Schmidt, 1976.

[Kelley, 1955] J. Kelley. *General Topology.* D. Van Nostrand Company, Inc., 1955.

[Kelley, 1959] J. Kelley. Measures on Boolean algebras. *Pacific J. Math.*, 9:1165–1177, 1959.

[Kelley, 1989] J. Kelley. Once over lightly. In *A Century of Mathematics in America, Part III*, volume 3, pages 471–493. American Mathematical Society, Providence, RI, 1989.

[Kleene and Post, 1954] S. Kleene and E. Post. The upper semi-lattice of degrees of recursive unsolvability. *Ann. of Math. (2)*, 59:379–407, 1954.

[Kleinberg, 1970] E. Kleinberg. Strong partition properties for infinite cardinals. *J. Symbolic Logic*, 35:410–428, 1970.

[Kleinberg, 1977] E. Kleinberg. *Infinitary Combinatorics and the Axiom of Determinateness*, volume 612 of *Lecture Notes in Mathematics.* Springer-Verlag, Berlin, 1977.

[Knaster, 1945] B. Knaster. Sur une propriété caractéristique de l'ensemble des nombres réels. *Rec. Math. [Mat. Sbornik] N. S.*, 16(58):281–290, 1945.

[Koepke, 1983] P. Koepke. *A Theory of Short Core Models and Some Applications.* PhD thesis, Albert-Ludwigs-Universität Freiburg im Breisgau, 1983.

[Koepke, 1984] P. Koepke. The consistency strength of the free-subset property for ω_ω. *J. Symbolic Logic*, 49(4):1198–1204, 1984.

[Kojman and Shelah, 1992] M. Kojman and S. Shelah. Nonexistence of universal orders in many cardinals. *J. Symbolic Logic*, 57(3):875–891, 1992.

[Kojman and Shelah, 1993] M. Kojman and S. Shelah. μ-complete Souslin trees on μ^+. *Arch. Math. Logic*, 32(3):195–201, 1993.

[Kojman, 1998] M. Kojman. Representing embeddability as set inclusion. *J. London Math. Soc. (2)*, 58(2):257–270, 1998.

[Kolata, 1981] G. Kolata. Ronald L. Graham. In *Science Year, 1981: the World Book Annual Science Supplement*, pages 372–387. World Book, Inc., 1981.

[Komjáth and Shelah, 1995] P. Komjáth and S. Shelah. Universal graphs without large cliques. *J. Combin. Theory Ser. B*, 63(1):125–135, 1995.

[Komjáth and Shelah, 2000] P. Komjáth and S. Shelah. Two consistency results on set mappings. *J. Symbolic Logic*, 65(1):333–338, 2000.

[Komjáth, 1982/86] P. Komjáth. The mathematical work of Géza Fodor. *Mat. Lapok*, 33(4):235–242 (1987), 1982/86.

[Komjáth, 1998] P. Komjáth. Distinguishing two partition properties of ω_1. *Fund. Math.*, 155(1):95–99, 1998.

[Komjáth, 1999] P. Komjáth. Some remarks on the partition calculus of ordinals. *J. Symbolic Logic*, 64(2):436–442, 1999.

[Kőnig and Valkó, 1926] D. Kőnig and S. Valkó. Über mehrdeutige Abbildungen von Mengen. *Math. Ann.*, 95(1):135–138, 1926.

[Kőnig, 1926] D. Kőnig. Sur les correspondances multivoques. *Fund. Math.*, 8:114–134, 1926.

[Kőnig, 1927] D. Kőnig. Über eine schlussweise aus dem endlichen ins unendlichen. *Acta Sci. Math. (Szeged)*, 3:121–130, 1927.

[Kőnig, 1931] D. Kőnig. Grápjok és mátrixok (graphs and matrices). *Mat. Fiz. Lapok*, 38:116–119, 1931. In Hungarian with German summary.

[Kőnig, 1936] D. Kőnig. *Theorie der endlichen und unendlichen Graphen: Kombinatorische Topolgie der Streckenkomplexe.* Akademischen Verlagsgesellschaft, Leipzig, 1936. (Reprinted: Chelsea, New York, 1950, ix+258.).

[Kőnig, 1990] D. Kőnig. *Theory of Finite and Infinite Graphs.* Birkhäuser Boston, Inc., Boston, MA, 1990. Translated from the German by Richard McCoart; with a commentary by W. T. Tutte and biographical sketch by T. Gallai.

[Koppelberg, 1989] S. Koppelberg. *Handbook of Boolean Algebras. Vol. 1.* North-Holland Publishing Co., Amsterdam, 1989.

[Kozen, 1988] D. Kozen. A finite model theorem for the propositional μ-calculus. *Studia Logica*, 47(3):233–241, 1988.

[Kruskal, 1954] J. Kruskal. *Theory of Well-Quasi-Ordering.* PhD thesis, Princeton University, 1954. Roger Lyndon, advisor.

[Kruskal, 1960] J. Kruskal. Well-quasi-ordering, the tree theorem, and Vazsonyi's conjecture. *Trans. Amer. Math. Soc.*, 95:210–225, 1960.

[Kunen and Tall, 2000] K. Kunen and F. Tall. The real line in elementary submodels of set theory. *J. Symbolic Logic*, 65(2):683–691, 2000.

[Kunen and Vaughan, 1984] K. Kunen and J. Vaughan, editors. *Handbook of Set-Theoretic Topology*. North-Holland Publishing Co., Amsterdam, 1984.

[Kunen, 1968] K. Kunen. *Inaccessibility Properties of Cardinals*. PhD thesis, Stanford University, 1968.

[Kunen, 1971] K. Kunen. A partition theorem. *Notices Amer. Math. Soc.*, 19(2):Abstract #71T–E17, 1971.

[Kunen, 1977] K. Kunen. Combinatorics. In *Handbook of Mathematical Logic, Part B*, pages 371–401. North-Holland, Amsterdam, 1977.

[Kunen, 1978] K. Kunen. Saturated ideals. *J. Symbolic Logic*, 43(1):65–76, 1978.

[Kunen, 1980] K. Kunen. *Set Theory*. North-Holland Publishing Co., Amsterdam, 1980. An introduction to independence proofs.

[Kunen, 1983] K. Kunen. *Set Theory*. North-Holland Publishing Co., Amsterdam, 1983. An introduction to independence proofs, Reprint of the 1980 original.

[Kung *et al.*, 2009] J. Kung, G.-C. Rota, and C. Yan. *Combinatorics: The Rota Way*. Cambridge Mathematical Library. Cambridge University Press, Cambridge, 2009.

[Kuratowski, 1922] K. Kuratowski. Une méthode d'élimination des nombres transfinis des raisonnements mathématiques. *Fund. Math.*, 3:76–107, 1922.

[Kuratowski, 1972] K. Kuratowski. Wacław Sierpiński (1882–1969). *Acta Arith.*, 21:1–5. (1 plate, errata insert), 1972.

[Kuratowski, 1980] K. Kuratowski. *A Half Century of Polish Mathematics*, volume 108 of *International Series in Pure and Applied Mathematics*. Pergamon Press Inc., Elmsford, N.Y., 1980. Remembrances and reflections, translated from the Polish by Andrzej Kirkor, with a preface by S. Ulam.

[Kurepa, 1934a] Đ. Kurepa. Sur le continu linéaire. *C. R. Acad. Sci. Paris*, 198:703–707, 1934.

[Kurepa, 1934b] Đ. Kurepa. Sur les ensembles ordonnés. *C. R. Acad. Sci. Paris*, 198:882–885, 1934.

[Kurepa, 1934c] Đ. Kurepa. Tableaux ramifiés d'ensembles. *C. R. Acad. Sci. Paris*, 199:112–114, 1934.

[Kurepa, 1934d] Đ. Kurepa. Tableaux ramifiés d'ensembles, Espaces pseudo-distanciés. *C. R. Acad. Sci. Paris*, 198:1563–1565, 1934.

[Kurepa, 1935] Đ. Kurepa. Ensembles ordonnées et ramifiés. *Publ. Math. Univ. Belgrade*, 4:1–138, 1935. A35.

[Kurepa, 1937a] Đ. Kurepa. Ensembles linéaires et une classe de tableaux ramifiés (Tableaux ramifiés de M. Aronszajn). *Publ. Inst. Math. (Beograd)*, 6:129–160, 1937.

[Kurepa, 1937b] Đ. Kurepa. L'hypothèse du continu et les ensembles partiellement ordonnés. *C. R. Acad. Sci. Paris*, 205:1196–1198, December 1937.

[Kurepa, 1937c] Đ. Kurepa. Transformations monotones des ensembles partiellement ordonnés. *C. R. Acad. Sci. Paris*, 205:1033–1035, November 1937.

[Kurepa, 1939] Đ. Kurepa. Sur la puissance des ensembles partillement ordonnés. *C. R. Soc. Sci. Varsovie, Cl. Math.*, 32:61–67, 1939. Sometimes the journal is listed in Polish: Sprawozdania Towarzystwo Naukowe Warszawa Mat.-Fiz. as in the Math Review by Bagemihl.

[Kurepa, 1940a] Đ. Kurepa. Transformations monotones des ensembles partiellement ordonnés. *Revista Ci., Lima*, 42:827–846, 1940.

[Kurepa, 1940b] Đ. Kurepa. Une propriété des familles d'ensembles bien ordonnées linéares. *Studia Math.*, 9:22–35, 1940.

[Kurepa, 1941] Đ. Kurepa. Transformations monotones des ensembles partiellement ordonnés. *Revista Ci., Lima*, 43:483–500, 1941.

[Kurepa, 1942] Đ. Kurepa. À propos d'une généralisation de la notion d'ensembles bien ordonnés. *Acta Math.*, 75(1:2):139–150, December 1942.

[Kurepa, 1948a] Đ. Kurepa. L'hypothèse du continu et le problème de Souslin. *Acad. Serbe Sci. Publ. Inst. Math.*, 2:26–36, 1948.

[Kurepa, 1948b] Đ. Kurepa. Sur les ensembles ordonnés dénombrables. *Hrvatsko Prirodoslovno Društvo. Glasnik Mat.-Fiz. Astr. Ser. II.*, 3:145–151, 1948.

[Kurepa, 1950] Đ. Kurepa. La condition de Souslin et une propriété caractéristique des nombres réels. *C. R. Acad. Sci. Paris*, 231:1113–1114, November 1950.

[Kurepa, 1952] Đ. Kurepa. Sur une propriété caractéristique du continu linéaire et le problème de Suslin. *Acad. Serbe Sci. Publ. Inst. Math.*, 4:97–108, 1952.

[Kurepa, 1953] Đ. Kurepa. On reflexive symmetric relations and graphs. *Slovenska Akad. Znan. Umet. Razred Mat. Fiz. Tehn. Vede. Ser. A.*, 4(4):65–92, 1953.

[Kurepa, 1954] Đ. Kurepa. On symmetrical binary relations. *Conseil Acad. RFP Yougoslavie Bull. Scientifique*, 2(1), 1954.

[Kurepa, 1959a] Đ. Kurepa. On the cardinal number of ordered sets and of symmetrical structures in dependence on the cardinal numbers of its chains and antichains. *Glasnik Mat.-Fiz. Astronom. Društvo Mat. Fiz. Hrvatske Ser. II*, 14:183–203, 1959.

[Kurepa, 1959b] Đ. Kurepa. Sur la puissance des ensembles partillement ordonnés. *Glasnik Mat.-Fiz. Astronom. (Ser. II)*, 14:205–211, 1959. (reprint of C39, [Kurepa, 1939]) Reproduced from C.R.Soc. Sci. Varsovie Cl. Math. 32, 1939, 61-67.

[Kurepa, 1968] Đ. Kurepa. On *A*-trees. *Publ. Inst. Math. (Beograd) (N.S.)*, 8 (22):153–161, 1968.

[Kurepa, 1969] Đ. Kurepa. Around the general Suslin problem. In *Proc. Internat. Sympos. on Topology and its Applications (Herceg-Novi, 1968)*, pages 239–245. Savez Društava Mat. Fiz. i Astronom., Belgrade, 1969.

[Kurepa, 1996] Đ. Kurepa. *Selected Papers of Đuro Kurepa.* Matematički Institut SANU, Belgrade, 1996. Edited and with commentaries by Aleksandar Ivić, Zlatko Mamuzić, Žarko Mijajlović and Stevo Todorčević.

[Kuzawa, 1970] M. Kuzawa. Fundamenta Mathematicae: An examination of its founding and significance. *Amer. Math. Monthly*, 77(5):485–492, 1970.

[la Vallée Poussin, 1916] C. de la Vallée Poussin. *Intégrales de Lebesgue, fonctions d'ensemble, classes de Baire.* Gauthier-Villars, Paris, 1916. Second edition, 1934, 193 pp.

[Larson, 1972] J. Larson. *Some Arrow Relations.* PhD thesis, University of California, Berkeley, 1972.

[Larson, 1973] J. Larson. A short proof of a partition theorem for the ordinal ω^ω. *Ann. Math. Logic*, 6:129–145, 1973.

[Larson, 1975] J. Larson. Partition theorems for certain ordinal products. In *Infinite and Finite Sets (Colloq., Keszthely, 1973; dedicated to P. Erdős on his 60th birthday), Vol. II*, pages 1017–1024. Colloq. Math. Soc. János Bolyai, Vol. 10. North-Holland, Amsterdam, 1975.

[Larson, 1980] J. Larson. A counterexample in the partition calculus for an uncountable ordinal. *Israel J. Math.*, 36(3-4):287–299, 1980.

[Larson, 1998] J. Larson. An ordinal partition from a scale. In *Set Theory (Curaçao, 1995; Barcelona, 1996)*, pages 109–125. Kluwer Acad. Publ., Dordrecht, 1998.

[Larson, 2000] J. Larson. An ordinal partition avoiding pentagrams. *J. Symbolic Logic*, 65(3):969–978, 2000.

[Larson, 2008] J. Larson. Notes of private conversation with R. Jensen and W. Mitchell, September 2008.

[Läuchli and Leonard, 1966] H. Läuchli and J. Leonard. On the elementary theory of linear order. *Fund. Math.*, 59:109–116, 1966.

[Laver and Shelah, 1981] R. Laver and S. Shelah. The $\aleph_2$-Souslin hypothesis. *Trans. Amer. Math. Soc.*, 264(2):411–417, 1981.

[Laver, 1969] R. Laver. *Order Types and Well-Quasi-Orderings.* PhD thesis, University of California, Berkeley, 1969. Ralph McKenzie, Advisor.

[Laver, 1971] R. Laver. On Fraïssé's order type conjecture. *Ann. of Math. (2)*, 93:89–111, 1971.

[Laver, 1973] R. Laver. An order type decomposition theorem. *Ann. of Math. (2)*, 98:96–119, 1973.

[Laver, 1975] R. Laver. Partition relations for uncountable cardinals $\leq 2^{\aleph_0}$. In *Infinite and Finite Sets (Colloq., Keszthely, 1973; dedicated to P. Erdős on his 60th birthday), Vol. II*, pages 1029–1042. Colloq. Math. Soc. Janós Bolyai, Vol. 10. North-Holland, Amsterdam, 1975.

[Laver, 1976] R. Laver. Well-quasi-orderings and sets of finite sequences. *Math. Proc. Cambridge Philos. Soc.*, 79(1):1–10, 1976.

[Laver, 1978a] R. Laver. Better-quasi-orderings and a class of trees. In *Studies in foundations and combinatorics*, volume 1 of *Adv. in Math. Suppl. Stud.*, pages 31–48. Academic Press, New York, London, 1978.

[Laver, 1978b] R. Laver. A saturation property on ideals. *Compositio Math.*, 36(3):233–242, 1978.

[Laver, 1979] R. Laver. Linear orders in $(\omega)^\omega$ under eventual dominance. In *Logic Colloquium '78 (Mons, 1978)*, volume 97 of *Stud. Logic Foundations Math.*, pages 299–302. North-Holland, Amsterdam, 1979.

[Laver, 1982] R. Laver. An $(\aleph_2, \aleph_2, \aleph_0)$-saturated ideal on ω_1. In *Logic Colloquium '80 (Prague, 1980)*, pages 173–180. North-Holland, Amsterdam, 1982.

[Laver, 1984] R. Laver. Products of infinitely many perfect trees. *J. London Math. Soc. (2)*, 29(3):385–396, 1984.

[Laver, 1987] R. Laver. Random reals and Souslin trees. *Proc. Amer. Math. Soc.*, 100(3):531–534, 1987.

[Lebesgue, 1905] H. Lebesgue. Sur les fonctions représentables analytiquement. *J. Math. Pures Appl. (6)*, 1(1):139–216, 1905. Downloaded from http://portail.mathdoc.fr/JMPA/.

[Leeb, 1973] K. Leeb. Vorlesungen über Pascaltheorie. *Arbeitsberichte des Inst. für Math. Maschinene and Daterveravbeitung*, 6(7), 1973.

[Lefschetz, 1930] S. Lefschetz. *Topology*, volume 12 of *American Mathematical Society Colloquium Publications*. American Mathematical Society, New York, 1930.

[Lévy and Solovay, 1967] A. Lévy and R. Solovay. Measurable cardinals and the continuum hypothesis. *Israel J. Math.*, 5:234–248, 1967.

[Losada and Todorcevic, 2000] M. Losada and S. Todorcevic. Chains and antichains in Boolean algebras. *Fund. Math.*, 163(1):55–76, 2000.

[Louveau *et al.*, 1993] A. Louveau, S. Shelah, and B. Veličković. Borel partitions of infinite subtrees of a perfect tree. *Ann. Pure Appl. Logic*, 63(3):271–281, 1993.

[Louveau, 1974] A. Louveau. Une démonstration topologique de théorèmes de Silver et Mathias. *Bull. Sci. Math. (2)*, 98(2):97–102, 1974.

[Lusin and Sierpiński, 1918] N. Lusin and W. Sierpiński. Sur quelques propriétés des ensembles. *Bull. Internat. Acad. Sci. Cracovie Ser. A*, 4:35–48, 1918.

[Lusin and Sierpiński, 1923] N. Lusin and W. Sierpiński. Sur une ensemble non measurable b. *J. Math. Pures Appl. (9)*, 12:53–72, 1923.

[Lusin, 1914] N. Lusin. Sur un problème de M. Baire. *C. R. Acad. Sci. Paris*, 158:1268–1261, 1914.

[Lusin, 1917] N. Lusin. Sur la classification de M. Baire. *C. R. Acad. Sci. Paris*, 164:91–94, 1917.

[Lusin, 1934] N. Lusin. *Sur les suites stationnaires.* Number 149 in Actualités Scientifiques et Industrielles. Hermann et Cie., Paris, 1934. French translation of Trudy Fiz.-Mat. Inst. Mat. Otdel. 5:125–147, 1934.

[Magidor and Shelah, 1996] M. Magidor and S. Shelah. The tree property at successors of singular cardinals. *Arch. Math. Logic*, 35(5-6):385–404, 1996.

[Magill, 1996] K. Magill. An interview with John Isbell. *Topological Commentary*, 1(2), 1996. Posted to http://at.yorku.ca/t/o/p/c/06.htm on May 20, 1996; last revised May 10, 1996.

[Maharam, 1947] D. Maharam. An algebraic characterization of measure algebras. *Ann. of Math. (2)*, 48:154–167, 1947.

[Maharam, 1948] D. Maharam. Set functions and Souslin's Hypothesis. *Bull. Amer. Math. Soc.*, 54:587–590, 1948.

[Malitz, 1968] J. Malitz. The Hanf number for complete $L_{\omega_1,\omega}$-sentences. In *The Syntax and Semantics of Infinitary Languages*, volume 72 of *Lecture Notes in Mathematics*, pages 166–181. Springer-Verlag, Berlin, 1968.

[Marczewski, 1930] E. Marczewski. Sur l'extension de l'ordre partiel. *Fund. Math.*, 16(2):386–389, 1930. Published under the name Szpilrajn.

[Marczewski, 1945] E. Marczewski. Sur deux proprétés des classes d'ensembles. *Fund. Math.*, 33:303–307, 1945. Published under the name Szpilrajn-Marczewski.

[Marczewski, 1947] E. Marczewski. Séparabilité et multiplication cartésienne des espaces topologiques. *Fund. Math.*, 34:127–143, 1947.

[Mardešić, 1999] S. Mardešić. Absolute neighborhood retracts and shape theory. In *History of Topology*, pages 241–269. North-Holland, Amsterdam, 1999.

[Martin and Solovay, 1970] D. A. Martin and R. Solovay. Internal Cohen extensions. *Ann. Math. Logic*, 2(2):143–178, 1970.

[Matet, 1993] P. Matet. Happy families and completely Ramsey sets. *Arch. Math. Logic*, 32(3):151–171, 1993.

[Mathias, 1968] A. Mathias. On a generalization of Ramsey's theorem. *Notices Amer. Math. Soc.*, 15, 1968. Abstract 68T-E19, received July 18, 1968.

[Mathias, 1972] A. Mathias. Solution of problems of Choquet and Puritz. In *Conference in Mathematical Logic—London '70 (Bedford Coll., London, 1970)*, pages 204–210. Lecture Notes in Math., Vol. 255. Springer, Berlin, 1972.

[Mathias, 1973] A. Mathias. On sequences generic in the sense of Prikry. *J. Austral. Math. Soc.*, 15:409–414, 1973.

[Mathias, 1977] A. Mathias. Happy families. *Ann. Math. Logic*, 12(1):59–111, 1977.

[Mathias, 1979] A. Mathias. Surrealist landscape with figures (a survey of recent results in set theory). *Period. Math. Hungar.*, 10(2-3):109–175, 1979.

[Mathias, 1997] A. Mathias. Letter to Odell on the historical remarks concerning Ellentuck and others in notes for his Mons lectures. Odell delivered a series of lectures in 1997 in Mons at the conference *Analyse et Logique*, October 1997.

[Mauldin, 1981] R. Mauldin, editor. *The Scottish Book*. Birkhäuser Boston, Mass., 1981. Mathematics from the Scottish Café, including selected papers presented at the Scottish Book Conference held at North Texas State University, Denton, Tex., May 1979.

[McAuley and Rao, 1981] L. F. McAuley and M. M. Rao, editors. *General Topology and Modern Analysis*, New York, 1981. Academic Press Inc. [Harcourt Brace Jovanovich Publishers]. In honor of the retirement of Professor F. Burton Jones.

[Menger, 1927] K. Menger. Zur allgemeinen Kurventheorie. *Fund. Math.*, 10:96–115, 1927.

[Meyers, 1991] D. Meyers. Hanf's eulogy. *Rev. Modern Logic*, 1(4):355–357, 1991.

[Michael, 1956] E. Michael. Continuous selections. I. *Ann. of Math. (2)*, 63:361–382, 1956.

[Michael, 1962] E. Michael. A note on intersections. *Proc. Amer. Math. Soc.*, 13:281–283, 1962.

[Miller, 1943] E. Miller. A note on Souslin's problem. *Amer. J. Math.*, 65(4):673–678, 1943.

[Miller, 1989] A. Miller. Infinite combinatorics and definability. *Ann. Pure Appl. Logic*, 41(2):179–203, 1989.

[Milliken, 1975a] K. Milliken. Ramsey's theorem with sums or unions. *J. Combinatorial Theory Ser. A*, 18:276–290, 1975.

[Milliken, 1975b] K. Milliken. *Some Results in Ramsey Theory.* PhD thesis, University of California, Los Angeles, 1975.

[Milliken, 1979] K. Milliken. A Ramsey theorem for trees. *J. Combin. Theory Ser. A*, 26(3):215–237, 1979.

[Milliken, 1981] K. Milliken. A partition theorem for the infinite subtrees of a tree. *Trans. Amer. Math. Soc.*, 263(1):137–148, 1981.

[Milner and Prikry, 1986] E. Milner and K. Prikry. A partition theorem for triples. *Proc. Amer. Math. Soc.*, 97(3):488–494, 1986.

[Milner and Prikry, 1991] E. Milner and K. Prikry. A partition relation for triples using a model of Todorčević. *Discrete Math.*, 95(1-3):183–191, 1991. Delivered at a conference *Directions in infinite graph theory and combinatorics* held in Cambridge, England in 1989.

[Milner and Rado, 1965] E. Milner and R. Rado. The pigeon-hole principle for ordinal numbers. *Proc. London Math. Soc. (3)*, 15:750–768, 1965.

[Milner and Shelah, 1975] E. Milner and S. Shelah. Some theorems on transversals. In *Infinite and Finite Sets (Colloq., Keszthely, 1973; dedicated to P. Erdős on his 60th birthday), Vol. III*, pages 1115–1126. North-Holland, Amsterdam, 1975.

[Milner, 1969] E. Milner. Partition relations for ordinal numbers. *Canad. J. Math.*, 21:317–334, 1969.

[Milner, 1971] E. Milner. A finite algorithm for the partition calculus. In *Proceedings of the Twenty-Fifth Summer Meeting of the Canadian Mathematical Congress (Lakehead Univ., Thunder Bay, Ont., 1971)*, pages 117–128. Lakehead Univ., Thunder Bay, Ont., 1971.

[Milner, 1975] E. Milner. Transversal theory. In *Proceedings of the International Congress of Mathematicians (Vancouver, B. C., 1974), Vol. 1*, pages 155–169. Canad. Math. Congress, Montreal, Que., 1975.

[Milner, 1982] E. Milner. On the decomposition of partially ordered sets into directed sets. In *Proceedings of the International Mathematical Conference, Singapore 1981 (Singapore, 1981)*, volume 74 of *North-Holland Math. Stud.*, pages 85–90. North-Holland, Amsterdam, 1982.

[Milner, 1985] E. Milner. Basic wqo- and bqo-theory. In *Graphs and Order (Proceedings of the NATO Advanced Study Institute held in Banff, Alta., May 18–31, 1984)*, volume 147 of *NATO Adv. Sci. Inst. Ser. C Math. Phys. Sci.*, pages 487–502. Reidel, Dordrecht, 1985.

[Milner, 1994] E. Milner. The use of elementary substructures in combinatorics. *Discrete Math.*, 136(1-3):243–252, 1994. Delivered at the conference *Trends in discrete mathematics* held October 29–31, 1992 in Bielefeld.

[Mirsky, 1971] L. Mirsky. *Transversal Theory. An Account of Some Aspects of Combinatorial Mathematics.* Mathematics in Science and Engineering, Vol. 75. Academic Press, New York, 1971.

[Mitchell, 1972] W. Mitchell. Aronszajn trees and the independence of the transfer property. *Ann. Math. Logic*, 5:21–46, 1972.

[Miyamoto, 1988] T. Miyamoto. *Some Results in Forcing.* PhD thesis, Dartmouth College, 1988. Advisor J. E. Baumgartner.

[Monk and Bonnet, 1989a] J. D. Monk and R. Bonnet, editors. *Handbook of Boolean Algebras. Vol. 2.* North-Holland Publishing Co., Amsterdam, 1989.

[Monk and Bonnet, 1989b] J. D. Monk and R. Bonnet, editors. *Handbook of Boolean Algebras. Vol. 3.* North-Holland Publishing Co., Amsterdam, 1989.

[Monk and Scott, 1964] D. Monk and D. Scott. Additions to some results of Erdős and Tarski. *Fund. Math.*, 53:335–343, 1964.

[Montalbán, 2007] A. Montalbán. On the equimorphism types of linear orderings. *Bull. Symbolic Logic*, 13(1):71–99, 2007.

[Moore, 1916] R. L. Moore. On the foundations of plane analysis situs. *Trans. Amer. Math. Soc.*, 17(2):131–164, 1916.

[Moore, 1927] R. L. Moore. Abstract sets and the foundations of analysis situs. *Bull. Amer. Math. Soc.*, 33(2), 1927. Abstract 23 (paper presented at the December 28—30, 1926 Annual meeting of the AMS).

[Moore, 1932] R. L. Moore. *Foundations of Point Set Theory*, volume 13 of *American Mathematical Society Colloquium Publications.* American Mathematical Society, New York, 1932.

[Moore, 1935] R. L. Moore. A set of axioms for plane analysis situs. *Fund. Math.*, 25:13–28, 1935.

[Moore, 1942] R. L. Moore. Concerning separability. *Proc. Nat. Acad. Sci. U. S. A.*, 28:56–58, 1942. [Fitzpatrick, 1997, 48] says a space homeomorphic to Moore's road space as described in Moore's letter to Hildebrandt appears in this paper.

[Moore, 1988] G. Moore. The origins of forcing. In *Logic Colloquium '86 (Hull 1986)*, volume 124 of *Studies in Logic and the Foundations of Mathematics*, pages 143–173. North Holland, Amsterdam, 1988. Frank R. Drake and John K. Truss, editors.

[Moore, 2000] J. Moore. Continuous colorings associated with certain characteristics of the continuum. *Discrete Math.*, 214(1-3):263–273, 2000.

[Moore, 2002] J. Moore. Open colorings, the continuum and the second uncountable cardinal. *Proc. Amer. Math. Soc.*, 130(9):2753–2759 (electronic), 2002.

[Moore, 2006a] J. Moore. A five element basis for the uncountable linear orders. *Ann. of Math. (2)*, 163(2):669–688, 2006.

[Moore, 2006b] J. Moore. A solution to the L Space Problem. *J. Amer. Math. Soc.*, 19(3):717–736 (electronic), 2006.

[Moore, 2008a] G. Moore. The emergence of open sets, closed sets, and limit points in analysis and topology. *Historia Math.*, 35(3):220–241, 2008.

[Moore, 2008b] J. Moore. Structural analysis of Aronszajn trees. In *Logic Colloquium 2005*, volume 28 of *Lect. Notes Log.*, pages 85–106. Assoc. Symbol. Logic, Urbana, IL, 2008.

[Morley and Vaught, 1962] M. Morley and R. Vaught. Homogeneous universal models. *Math. Scand.*, 11:37–57, 1962.

[Moschovakis, 2009] Y. N. Moschovakis. *Descriptive Set Theory*, volume 155 of *Mathematical Surveys and Monographs.* American Mathematical Society, Providence, RI, second edition, 2009.

[Mostowski and Tarski, 1949] A. Mostowski and A. Tarski. Arithmetical classes and types of well-ordered systems (abstract). *Bull. Amer. Math. Soc.*, 55(65), 1949. Abstract 5–1–78; erratum: ibid. 55:1192, 1949.

[Mostowski, 1938] A. Mostowski. Über gewisse universelle relationen. *Ann. Soc. Polonaise de Math.*, 17:117–118, 1938.

[Mostowski, 1939] A. Mostowski. Über die unabhängigkeit des wohlordnugssatzes vom ordnugsprinzip. *Fund. Math.*, 32:201–252, 1939.

[Motzkin, 1967] T. Motzkin. Cooperative classes of finite sets in one and more dimensions. *J. Combinatorial Theory*, 3:244–251, 1967.

[Mycielski, 1964] J. Mycielski. Independent sets in topological algebras. *Fund. Math.*, 55:139–147, 1964.

[Mycielski, 1967] J. Mycielski. Algebraic independence and measure. *Fund. Math.*, 61:165–169, 1967.

[Nash-Williams, 1963] C. Nash-Williams. On well-quasi-ordering finite trees. *Proc. Cambridge Philos. Soc.*, 59:833–835, 1963.

[Nash-Williams, 1965a] C. Nash-Williams. On well-quasi-ordering infinite trees. *Proc. Cambridge Philos. Soc.*, 61:697–720, 1965.

[Nash-Williams, 1965b] C. Nash-Williams. On well-quasi-ordering transfinite sequences. *Proc. Camb. Phil. Soc.*, 61:33–39, 1965.

[Nash-Williams, 1978] C. Nash-Williams. Another criterion for marriage in denumerable societies. *Ann. Discrete Math.*, 3:165–179, 1978. Advances in graph theory (Cambridge Combinatorial Conf., Trinity College, Cambridge, 1977).

[Nash-Williams, 2000] C. Nash-Williams. Obituary: Eric Charles Milner. *Bull. London Math. Soc.*, 32(1):91–104, 2000.

[Nathanson, 2008] M. Nathanson. Tennenbaum at Penn and Rochester. *Integers*, 8(2):2–5, 2008.

[Nerode, 2004] A. Nerode. Preface. *Ann. Pure Appl. Logic*, 126(1):1–2, 2004. This preface is for a volume which contains a collection of papers presented at the *Alfred Tarski Centenary Conference*, held in Warsaw, May 27-June 1, 2001.

[Nešetřil and Rödl, 1990] J. Nešetřil and V. Rödl, editors. *Mathematics of Ramsey Theory*, volume 5 of *Algorithms and Combinatorics*. Springer-Verlag, Berlin, 1990.

[Neumer, 1951] W. Neumer. Verallgemeinerung eines Satzes von Alexandroff und Urysohn. *Math. Z.*, 54:254–261, 1951.

[Neumer, 1958] W. Neumer. Kritische Zahlen und bestimmt divergente transfinite Funktionen. *Math. Z.*, 70:190–192, 1958.

[Nosal, 1974] E. Nosal. On a partition relation for ordinal numbers. *J. London Math. Soc. (2)*, 8:306–310, 1974.

[Nosal, 1979] E. Nosal. Partition relation for denumerable ordinals. *J. Combin. Theory Ser. B*, 27:190–197, 1979.

[Novák, 1950] J. Novák. A paradoxical theorem. *Fund. Math.*, 37:77–83, 1950.

[Novák, 1952] J. Novák. On partition of an ordered continuum. *Fund. Math.*, 39:53–64 (1953), 1952.

[Nyikos, 1980a] P. Nyikos. F. Burton Jones's contributions to the normal Moore space problem. In *Topology Conference, 1979 (Greensboro, N.C., 1979)*, pages 27–38. Guilford College, Greensboro, N.C., 1980.

[Nyikos, 1980b] P. Nyikos. A provisional solution to the normal Moore space problem. *Proc. Amer. Math. Soc.*, 78(3):429–435, 1980.

[Nyikos, 2001] P. Nyikos. A history of the normal Moore space problem. In *Handbook of the History of General Topology, Vol. 3*, volume 3 of *Hist. Topol.*, pages 1179–1212. Kluwer Acad. Publ., Dordrecht, 2001.

[Ochan, 1941] Y. S. Ochan. Sur une question liée au problème de Souslin. *Bull. Acad. Sci. URSS. Sér. Math. [Izvestia Akad. Nauk SSSR]*, 5:423–426, 1941. Appeared in Mathematical Reviews under the name G. Otchan.

[Ochan, 1967] Y. S. Ochan. *Sbornik zadach po metodam matematicheskoi fiziki.* Izdat. "Vysš. Škola", Moscow, 1967. Appeared in Mathematical Reviews under the name Ju. S. Očan.

[Ohkuma, 1960] T. Ohkuma. Comparability between ramified sets. *Proc. Japan Acad.*, 36:383–388, 1960.

[Ohkuma, 1961] T. Ohkuma. Comparability between ramified sets. *Osaka Math. J.*, 13:75–123, 1961.

[Ore, 1942] O. Ore. Theory of equivalence relations. *Duke Math. J.*, 9:573–627, 1942.

[Ostaszewski, 1976] A. Ostaszewski. On countably compact, perfectly normal spaces. *J. London Math. Soc. (2)*, 14(3):505–516, 1976.

[Oxtoby, 1989] J. Oxtoby. Biographical note. In *Measure and Measurable Dynamics (Rochester, NY, 1987)*, volume 94 of *Contemp. Math.*, pages xvii–xviii. Amer. Math. Soc., Providence, RI, 1989. Biographical Note (Dorothy Maharam).

[Paris and Harrington, 1977] J. Paris and L. Harrington. A mathematical incompleteness in Peano arithmetic. In *Handbook of Mathematical Logic, Part D*, pages 1133–1142. North-Holland, Amsterdam, 1977.

[Parker, 2005] J. Parker. *R. L. Moore: Mathematician and Teacher.* MAA Spectrum. Mathematical Association of America, Washington, DC, 2005.

[Paul, 1997] S. Paul. *Die Moskauer mathematische Schule um N. N. Lusin*, volume 11 of *Berliner Studien zur Wissenschaftsphilosophie & Humanontogenetik [Berlin Studies on the Philosophy of Science and Human Ontogenetics].* B. Kleine Verlag GmbH, Bielefeld, 1997. Entstehungs- und Entwicklungsgeschichte. Arbeitsprinzipien. Zerfall. Unter besonderer Berücksichtigung der Kampagne gegen Lusin im Sommer 1936. [Origin and evolution. Working principles. Disintegration. With special consideration of the campaign against Luzin in the summer of 1936].

[Pearl, 2007] E. Pearl, editor. *Open Problems in Topology. II.* Elsevier B. V., Amsterdam, 2007.

[Perles, 1963] M. Perles. On Dilworth's theorem in the infinite case. *Israel J. Math.*, 1:108–109, 1963.

[Phillips and Weaver, 2007] N. C. Phillips and N. Weaver. The Calkin algebra has outer automorphisms. *Duke Math. J.*, 139(1):185–202, 2007.

[Plotkin, 1993] J. Plotkin. Who put the "back" in back-and-forth? In *Logical Methods (Ithaca, NY, 1992)*, volume 12 of *Progr. Comput. Sci. Appl. Logic*, pages 705–712. Birkhäuser Boston, Boston, MA, 1993.

[Pouzet and Sauer, 2006] M. Pouzet and N. Sauer. From well-quasi-ordered sets to better-quasi-ordered sets. *Electron. J. Combin.*, 13(1):Research Paper 101, 27 pp. (electronic), 2006.

[Pouzet, 1992] M. Pouzet. Obituary: Ernest Corominas. *Order*, 9(1):1–3, 1992.

[Prikry, 1968] K. Prikry. *Changing Measurable into Accessible Cardinals.* PhD thesis, University of California, Berkeley, 1968. Jack Silver, advisor.

[Prikry, 1972] K. Prikry. On a set-theoretic partition problem. *Duke Math. J.*, 39:77–83, 1972.

[Prikry, 2010a] K. Prikry. Email to Galvin and Larson. Dated November 10, 2010., 2010.

[Prikry, 2010b] K. Prikry. Email to Larson. Dated October 28, 2010., 2010.

[Prömel and Voigt, 1983] H. Prömel and B. Voigt. Canonical partition theorems for parameter sets. *J. Combin. Theory Ser. A*, 35(3):309–327, 1983.

[Prömel and Voigt, 1985a] H. Prömel and B. Voigt. Baire sets of k-parameter words are Ramsey. *Trans. Amer. Math. Soc.*, 291(1):189–201, 1985.

[Prömel and Voigt, 1985b] H. Prömel and B. Voigt. Canonical forms of Borel-measurable mappings $\Delta\colon [\omega]^\omega \to \mathbb{R}$. *J. Combin. Theory Ser. A*, 40(2):409–417, 1985.

[Prömel and Voigt, 1990] H. Prömel and B. Voigt. Graham-Rothschild parameter sets. In *Mathematics of Ramsey Theory*, volume 5 of *Algorithms Combin.*, pages 113–149. Springer, Berlin, 1990.

[Prömel, 2005] H. Prömel. Complete disorder is impossible: the mathematical work of Walter Deuber. *Combin. Probab. Comput.*, 14(1):3–16, 2005.

[Rado, 1931] R. Rado. *Studien zur Kombinatorik.* PhD thesis, Friedrich-Wilhelms-Universität zu Berlin, 1931.

[Rado, 1933] R. Rado. Studien zur kombinatorik. *Mathematische Zeitschrift*, 36:424–480, 1933. Eingeganen am 15. Februar 1932.

[Rado, 1943] R. Rado. Note on combinatorial analysis. *Proc. London Math. Soc. (2)*, 48:122–160, 1943.

[Rado, 1949a] R. Rado. Axiomatic treatment of rank in infinite sets. *Canadian J. Math.*, 1:337–343, 1949.

[Rado, 1949b] R. Rado. Covering theorems for ordered sets. *Proc. London Math. Soc. (2)*, 50:509–535, 1949.

[Rado, 1954a] R. Rado. Direct decomposition of partitions. *J. London Math. Soc.*, 29:71–83, 1954.

[Rado, 1954b] R. Rado. Partial well-ordering of sets of vectors. *Mathematika*, 1:89–95, 1954.

[Rado, 1964] R. Rado. Universal graphs and universal functions. *Acta Arith.*, 9:331–340, 1964.

[Rado, 1967] R. Rado. Note on the transfinite case of Hall's theorem on representatives. *J. London Math. Soc.*, 42:321–324, 1967.

[Rado, 1970] R. Rado. Some partition theorems. In *Combinatorial Theory and its Applications, III (Proc. Colloq., Balatonfüred, 1969)*, pages 929–936. North-Holland, Amsterdam, 1970.

[Rado, 1981] R. Rado. Theorems on intervals of ordered sets. *Discrete Math.*, 35:199–201, 1981.

[Rado, 1986] R. Rado. Note on canonical partitions. *Bull. London Math. Soc.*, 18(2):123–126, 1986.

[Ramsey, 1930] F. Ramsey. On a problem of formal logic. *Proc. London Math. Soc. (2)*, 30(1):264–286, 1930. Reprinted in Graham, Rothschild and Spencer [1990, 19–21].

[Rebholz, 1974] J. Rebholz. Consequences of morass and diamond. *Ann. Math. Logic*, 7:361–385, 1974.

[Ricabarra, 1956] R. Ricabarra. Partitions in sets of ordinal numbers. *Rev. Mat. Cuyana*, 2:1–27, 1956. The volume appeared in 1958.

[Ricabarra, 1958] R. Ricabarra. *Conjuntos ordenados y ramificados (Contribución al estudio del problema de Suslin)*. Instituto de Matematica, Universidad Nacional del Sur, Bahía Blanca, 1958.

[Richards, 1971] C. Richards. Presentation ceremony to Professor Richard Rado at the University of Reading, May 7, 1971. *Bull. Inst. Math. Appl.*, 7:237–240, 1971.

[Riguet, 1948] J. Riguet. Relations binaires, fermetures, correspondances de Galois. *Bull. Soc. Math. France*, 76:114–155, 1948.

[Robinson, 1941] R. Robinson. On the simultaneous approximation of two real numbers. *Bull. Amer. Math. Soc.*, 47:512–513, 1941.

[Rogers, 1998] C. A. Rogers. Obituary: Richard Rado. *Bull. London Math. Soc.*, 30(2):185–195, 1998.

[Rogers, 1999] J. Rogers, Jr. F. Burton Jones (1910–1999)—an appreciation. In *Proceedings of the 1999 Topology and Dynamics Conference (Salt Lake City, UT)*, volume 24, pages 2–14, 1999.

[Roitman, 1978] J. Roitman. A reformulation of S and L. *Proc. Amer. Math. Soc.*, 69(2):344–348, 1978.

[Roitman, 1984] J. Roitman. Basic S and L. In *Handbook of Set-Theoretic Topology*, pages 295–326. North-Holland, Amsterdam, 1984.

[Rosenstein, 1982] J. Rosenstein. *Linear orderings*, volume 98 of *Pure and Applied Mathematics*. Academic Press Inc. [Harcourt Brace Jovanovich Publishers], New York, 1982.

[Rosłanowski and Shelah, 2001] A. Rosłanowski and S. Shelah. Historic forcing for depth. *Colloq. Math.*, 89(1):99–115, 2001.

[Rothberger, 1948] F. Rothberger. On some problems of Hausdorff and of Sierpiński. *Fund. Math.*, 35:29–46, 1948.

[Rotman, 1970] B. Rotman. A mapping theorem for countable well-ordered sets. *Proc. London Math. Soc.*, 15:750–768, 1970.

[Rotman, 1971] B. Rotman. Remarks on some theorems of Rado on universal graphs. *J. London Math. Soc. (2)*, 4:123–126, 1971.

[Rowbottom, 1964] F. Rowbottom. *Large Cardinals and Small Constructible Sets*. PhD thesis, University of Wisconsin, Madison, 1964.

[Rudin, 1950] M. E. Rudin. Concerning abstract spaces. *Duke Math. J.*, 17:317–327, 1950. Published under her maiden name, Estill.

[Rudin, 1951] M. E. Rudin. Separation in non-separable spaces. *Duke Math. J.*, 18:623–629, 1951. Published under her maiden name Estill.

[Rudin, 1952] M. E. Rudin. Concerning a problem of Souslin's. *Duke Math. J.*, 19:629–639, 1952. Published under her maiden name, Estill.

[Rudin, 1953] W. Rudin. *Principles of Mathematical Analysis*. McGraw-Hill Book Company, Inc., New York-Toronto-London, 1953.

[Rudin, 1955] M. E. Rudin. Countable paracompactness and Souslin's problem. *Canad. J. Math.*, 7:543–547, 1955.

[Rudin, 1969] M. E. Rudin. Souslin's conjecture. *Amer. Math. Monthly*, 76:1113–1119, 1969.

[Rudin, 1971] M. E. Rudin. A normal space X for which $X \times I$ is not normal. *Bull. Amer. Math. Soc.*, 77:246, 1971.

[Rudin, 1972] M. E. Rudin. A normal hereditarily separable non-Lindelöf space. *Illinois J. Math.*, 16:621–626, 1972.

[Rudin, 1975a] M. E. Rudin. *Lectures on Set Theoretic Topology*. Published for the Conference Board of the Mathematical Sciences by the American Mathematical Society, Providence, R.I., 1975. Expository lectures from the CBMS Regional Conference held at the University of Wyoming, Laramie, Wyo., August 12–16, 1974, Conference Board of the Mathematical Sciences Regional Conference Series in Mathematics, No. 23.

[Rudin, 1975b] M. E. Rudin. The metrizability of normal Moore spaces. In *Studies in Topology (Proc. Conf., Univ. North Carolina, Charlotte, N.C., 1974; dedicated to Math. Sect. Polish Acad. Sci.)*, pages 507–516. Academic Press, New York, 1975.

[Rudin, 1980] M. E. Rudin. S and L spaces. In *Surveys in General Topology*, pages 431–444. Academic Press, New York, 1980.

[Rudin, 1997] M. E. Rudin. The early work of F. B. Jones. In *Handbook of the History of General Topology, Vol. 1*, pages 85–96. Kluwer Acad. Publ., Dordrecht, 1997.

[Rudin, 2008] M. E. Rudin. A mathematical friendship. *Topology Appl.*, 155(8):752–753, 2008.

[Russell, 1920] B. Russell. *Introduction to Mathematical Philosophy*. G. Allen & Unwin, second edition, 1920.

[Sacks, 1971] G. Sacks. Forcing with perfect closed sets. In *Axiomatic Set Theory (Proc. Sympos. Pure Math., Vol. XIII, Part I, Univ. California, Los Angeles, Calif., 1967, Dana Scott, editior)*, pages 331–355. Amer. Math. Soc., Providence, R.I., 1971.

[Sacks, 1999] G. Sacks. Recursion theoretic memories 1954–1978. In *Models and Computability (Leeds, 1997)*, volume 259 of *London Math. Soc. Lecture Note Ser.*, pages 367–376. Cambridge Univ. Press, Cambridge, 1999.

[Sanders, 1968] J. Sanders. *A Generalization of a Theorem of Schur*. PhD thesis, Yale University, New Haven, Conn., 1968. Öystein Ore, advisor.

[Šanin, 1946] N. Šanin. A theorem from the general theory of sets. *C. R. (Doklady) Acad. Sci. URSS (N.S.)*, 53:399–400, 1946.

[Scanlan, 1999] M. Scanlan. Edward Vermilye Huntington. *American National Biography*, 11:534–356, 1999.

[Scheepers, 1993] M. Scheepers. Gaps in ω^ω. In *Set Theory of the Reals (Ramat Gan, 1991)*, volume 6 of *Israel Math. Conf. Proc.*, pages 439–561. Bar-Ilan Univ., Ramat Gan, 1993.

[Schipperus, 1999] R. Schipperus. *Countable Partition Ordinals*. PhD thesis, University of Colorado, Boulder, CO 80302, 1999.

[Schipperus, 2010] R. Schipperus. Countable partition ordinals. *Annals of Pure and Applied Logic*, 161:1195–1215, 2010.

[Schlindwein, 1989] C. Schlindwein. Special nonspecial $\aleph_1$-trees. In *Set Theory and its Applications (Toronto, ON, 1987)*, volume 1401 of *Lecture Notes in Mathematics*, pages 160–166. Springer-Verlag, Berlin, 1989.

[Schur, 1917] I. Schur. Über die Kongruenz $x^m + y^m = z^m (\text{mod. } p)$. *Jahresbericht der Deutschen Mathematiker-Vereinigung*, 25:114–117, 1917.

[Scott, 1971] D. Scott, editor. *Axiomatic Set Theory, Part I*. American Mathematical Society, Providence, R.I., 1971.

[Severini, 1910] C. Severini. Sulle Successioni di Funzioni Ortogonali. *Atti dell'Accademia Gioenia, serie 5*, 3(5):1–7, 1910.

[Shelah and Stanley, 1982] S. Shelah and L. Stanley. *S*-forcing. I. A "black-box" theorem for morasses, with applications to super-Souslin trees. *Israel J. Math.*, 43(3):185–224, 1982. Corrigendum: 53:304–314, 1986.

[Shelah and Stanley, 1987] S. Shelah and L. Stanley. A theorem and some consistency results in partition calculus. *Ann. Pure Appl. Logic*, 36(2):119–152, 1987.

[Shelah and Stanley, 1988] S. Shelah and L. Stanley. Weakly compact cardinals and nonspecial Aronszajn trees. *Proc. Amer. Math. Soc.*, 104(3):887–897, 1988.

[Shelah and Steprāns, 1987] S. Shelah and J. Steprāns. Extraspecial *p*-groups. *Ann. Pure Appl. Logic*, 34(1):87–97, 1987.

[Shelah and Steprāns, 1988] S. Shelah and J. Steprāns. A Banach space on which there are few operators. *Proc. Amer. Math. Soc.*, 104(1):101–105, 1988.

[Shelah, 1973] S. Shelah. Notes on combinatorial set theory. *Israel J. Math.*, 14:262–277, 1973.

[Shelah, 1975] S. Shelah. Notes on partition calculus. In *Infinite and Finite Sets (Colloq., Keszthely, 1973; dedicated to P. Erdős on his 60th birthday), Vol. III*, pages 1257–1276. North-Holland, Amsterdam, 1975.

[Shelah, 1976] S. Shelah. Decomposing uncountable squares to countably many chains. *J. Combinatorial Theory Ser. A*, 21(1):110–114, 1976.

[Shelah, 1977] S. Shelah. Whitehead groups may be not free, even assuming CH. I. *Israel J. Math.*, 28(3):193–204, 1977.

[Shelah, 1978a] S. Shelah. *Classification Theory and the Number of Nonisomorphic Models*, volume 92 of *Studies in Logic and the Foundations of Mathematics*. North-Holland Publishing Co., Amsterdam, 1978.

[Shelah, 1978b] S. Shelah. Jonsson algebras in successor cardinals. *Israel J. Math.*, 30(1-2):57–64, 1978.

[Shelah, 1980a] S. Shelah. Independence results. *J. Symbolic Logic*, 45(3):563–573, 1980.

[Shelah, 1980b] S. Shelah. A note on cardinal exponentiation. *J. Symbolic Logic*, 45(1):56–66, 1980.

[Shelah, 1980c] S. Shelah. Whitehead groups may not be free, even assuming CH. II. *Israel J. Math.*, 35(4):257–285, 1980.

[Shelah, 1982a] S. Shelah. Better quasi-orders for uncountable cardinals. *Israel J. Math.*, 42:177–226, 1982.

[Shelah, 1982b] S. Shelah. *Proper Forcing*, volume 940 of *Lecture Notes in Mathematics*. Springer-Verlag, Berlin, 1982.

[Shelah, 1984a] S. Shelah. Can you take Solovay's inaccessible away? *Israel J. Math.*, 48(1):1–47, 1984.

[Shelah, 1984b] S. Shelah. On universal graphs without instances of CH. *Ann. Pure Appl. Logic*, 26(1):75–87, 1984.

[Shelah, 1985] S. Shelah. More on the weak diamond. *Ann. Pure Appl. Logic*, 28(3):315–318, 1985.

[Shelah, 1988a] S. Shelah. Successors of singulars, cofinalities of reduced products of cardinals and productivity of chain conditions. *Israel J. Math.*, 62(2):213–256, 1988.

[Shelah, 1988b] S. Shelah. Was Sierpiński right? I. *Israel J. Math.*, 62:335–380, 1988. Sh: 276.

[Shelah, 1989] S. Shelah. Consistency of positive partition theorems for graphs and models. In *Set Theory and its Applications (Toronto, ON, 1987)*, volume 1401 of *Lecture Notes in Math.*, pages 167–193. Springer, Berlin, 1989.

[Shelah, 1990a] S. Shelah. Strong negative partition above the continuum. *J. Symbolic Logic*, 55(1):21–31, 1990.

[Shelah, 1990b] S. Shelah. Universal graphs without instances of CH: revisited. *Israel J. Math.*, 70(1):69–81, 1990.

[Shelah, 1994] S. Shelah. *Cardinal Arithmetic*, volume 29 of *Oxford Logic Guides*. The Clarendon Press Oxford University Press, New York, 1994. Oxford Science Publications.

[Shelah, 1998] S. Shelah. *Proper and Improper Forcing*. Perspectives in Mathematical Logic. Springer-Verlag, Berlin, second edition, 1998.

[Shelah, 2000a] S. Shelah. The generalized continuum hypothesis revisited. *Israel J. Math.*, 116:285–321, 2000.

[Shelah, 2000b] S. Shelah. On what I do not understand (and have something to say). I. *Fund. Math.*, 166(1-2):1–82, 2000. Saharon Shelah's anniversary issue.

[Shelah, 2010] S. Shelah. Diamonds. *Proc. Amer. Math. Soc.*, 138(6):2151–2161, 2010.

[Shepherdson, 1951a] J. C. Shepherdson. Well-ordered sub-series of general series. *Proc. London Math. Soc. (3)*, 1:291–307, 1951.

[Shepherdson, 1951b] J. C. Shepherdson. Inner models for set theory. I. *J. Symbolic Logic*, 16:161–190, 1951.

[Shepherdson, 1952] J. C. Shepherdson. Inner models for set theory. II. *J. Symbolic Logic*, 17:225–237, 1952.

[Shepherdson, 1953] J. C. Shepherdson. Inner models for set theory. III. *J. Symbolic Logic*, 18:145–167, 1953.

[Shepherdson, 1954] J. C. Shepherdson. On two problems of Kurepa. *Pacific J. Math.*, 4:301–304, 1954.

[Shields, 1987] A. Shields. Luzin and Egorov. *Math. Intelligencer*, 9(4):24–27, 1987.

[Shore, 1974] R. Shore. Square bracket partition relations in *L*. *Fund. Math.*, 84(2):101–106, 1974.

[Sierpiński, 1912] W. Sierpiński. *Zarys Teorii Mnogości*. Wydawnictwo Kasy im. Mianowskiego, Warsaw, 1912. An Outline of Set Theory. This book was based on 1909 lectures, 1st edition 1912, 2nd edition 1923, 3rd edition in 1928.

[Sierpiński, 1919] W. Sierpiński. Sur une théoréme équivalent á l'hypothése du continu. *Bull. Internat. Acad. Sci. Cracovie Ser. A*, pages 1–3, 1919.

[Sierpiński, 1924] W. Sierpiński. Sur l'hypothèse du continu. *Fund. Math.*, 5:177–187, 1924.

[Sierpiński, 1928a] W. Sierpiński. *Leçons sur les nombres transfinis*. Gauthiers-Villars, Paris, 1928. 512.8 S5721 at UF.

[Sierpiński, 1928b] W. Sierpiński. Sur une décomposition d'ensembles. *Monatsh. Math. Phys.*, 35(1):239–242, 1928.

[Sierpiński, 1931] W. Sierpiński. Sur les familles croissantes de sous-ensembles d'un ensemble dénombrable. *L'Enseignement Mathématique*, 30:240–242, 1931.

[Sierpiński, 1933] W. Sierpiński. Sur une problème de la théorie des relations. *Annali Scuola Norm. Super. Pisa Ser. 2*, 2(1):285–287, 1933.

[Sierpiński, 1934] W. Sierpiński. *Hypothèse du Continu.* Seminarjum Matematyczne Uniwersytetu Warszawskiego, Warsaw, 1934. Monografje Matematyczne, Tome 4.

[Sierpiński, 1937] W. Sierpiński. Sur les suites transfinies finalement disjointes. *Fund. Math.*, 28:115–119, 1937.

[Sierpiński, 1946] W. Sierpiński. Sur les types d'ordre de puissance du continu. *Revista Ci., Lima*, 48:305–307, 1946.

[Sierpiński, 1948] W. Sierpiński. Sur un problème de la théorie générale des ensembles equivalent un problème de Souslin. *Fund. Math.*, 35:165–174, 1948.

[Sierpiński, 1950] W. Sierpiński. Sur les types d'ordre des ensembles linéaires. *Fund. Math.*, 37:253–264, 1950.

[Sierpiński, 1958] W. Sierpiński. *Cardinal and Ordinal numbers.* Polska Akademia Nauk, Monografie Matematyczne. Tom 34. Państwowe Wydawnictwo Naukowe, Warsaw, 1958.

[Sierpiński, 1965] W. Sierpiński. *Cardinal and Ordinal Numbers.* PWN-Polish Scientific Publishers, Warszawa, 1965. Second edition revised.

[Sikorski, 1950] R. Sikorski. Remarks on some topological spaces of high power. *Fund. Math.*, 37:125–136, 1950.

[Silver, 1966] J. Silver. *Some Applications of Model Theory in Set Theory.* PhD thesis, University of California, Berkeley, 1966.

[Silver, 1970a] J. Silver. Every analytic set is Ramsey. *J. Symbolic Logic*, 35:60–64, 1970.

[Silver, 1970b] J. Silver. A large cardinal in the constructible universe. *Fund. Math.*, 69:93–100, 1970.

[Silver, 1971a] J. Silver. The independence of Kurepa's conjecture and two-cardinal conjectures in model theory. In *Axiomatic Set Theory (Proc. Sympos. Pure Math., Vol. XIII, Part I, Univ. California, Los Angeles, Calif., 1967)*, pages 383–390. Amer. Math. Soc., Providence, R.I., 1971.

[Silver, 1971b] J. Silver. Some applications of model theory in set theory. *Ann. Math. Logic*, 3(1):45–110, 1971.

[Silver, 1975] J. Silver. On the singular cardinals problem. In *Proceedings of the International Congress of Mathematicians (Vancouver, B. C., 1974), Vol. 1*, pages 265–268. Canad. Math. Congress, Montreal, Que., 1975.

[Simpson, 1970] S. Simpson. Model-theoretic proof of a partition theorem (abstract). *Notices Amer. Math. Soc.*, 18, 1970. Abstract.

[Skolem, 1920] T. Skolem. Logisch-kombinatorische Untersuchungen über die Erfülbarkeit oder Beweisbarkeit mathematischer Sätze nebst einem Theorem über dichte Mengen. *Krist. Vid. Selsk. Skr. I*, 4:1–36, 1920. published in 1922; reprinted in van Heijenoort, 1967, third printing 1977.

[Skolem, 1933] T. Skolem. Ein kombinatorischer Satz mit Anwendung auf ein logisches Entscheidungsproblem. *Fund. Math.*, 20:1–36, 1933.

[Skolem, 1970] T. Skolem. *Selected Works in Logic.* Edited by J. Fenstad. Universitetsforlaget, 1970.

[Slomson, 1972] A. B. Slomson. Generalized quantifiers and well orderings. *Arch. Math. Logik. Grundlagenforsch.*, 15:57–73, 1972.

[Soare, 1968a] R. Soare. Sets with no subsets of higher arithmetic degree. *Notices Amer. Math. Soc.*, 15, 1968. Abstract 68T-478, received April 1, 1968.

[Soare, 1968b] R. Soare. Sets with no subsets of higher degree, preliminary report. *Notices Amer. Math. Soc.*, 15, 1968. Abstract 656-37, received February 8, 1968.

[Soare, 1969] R. Soare. Sets with no subsets of higher degree. *J. Symbolic Logic*, 34:53–56, 1969.

[Soifer, 2009] A. Soifer. *The Mathematical Coloring Book.* Springer, New York, 2009. Mathematics of coloring and the colorful life of its creators, With forewords by Branko Grünbaum, Peter D. Johnson, Jr. and Cecil Rousseau.

[Solovay and Tennenbaum, 1971] R. Solovay and S. Tennenbaum. Iterated Cohen extensions and Souslin's problem. *Ann. of Math. (2)*, 94:201–245, 1971.

[Solovay, 1964] R. Solovay. The measure problem. *J. Symbolic Logic*, 29:227–228, 1964. Abstract for the Association of Symbolic Logic meeting of July 13–17, 1964 held at the University of Bristol; received July 6, 1964.

[Solovay, 1965] R. Solovay. The measure problem. *Notices Amer. Math. Soc.*, 12:217, 1965. Abstract 65T-62, received October 26, 1964.

[Solovay, 1967] R. Solovay. A nonconstructible Δ^1_3 set of integers. *Trans. Amer. Math. Soc.*, 127:50–75, 1967.

[Solovay, 1970] R. Solovay. A model of set-theory in which every set of reals is Lebesgue measurable. *Ann. of Math. (2)*, 92:1–56, 1970.

[Solovay, 1971] R. Solovay. Real-valued measurable cardinals. In *Axiomatic Set Theory (Proc. Sympos. Pure Math., Vol. XIII, Part I, Univ. California, Los Angeles, Calif., 1967)*, pages 397–428. Amer. Math. Soc., Providence, R.I., 1971.

[Solovay, 2010a] R. Solovay. 1963–66. Memo of November 13, 2010 sent to J. Larson, November 2010.

[Solovay, 2010b] R. Solovay. My cv, etc. Memo of November 6, 2010 sent to J. Larson, November 2010.

[Sós, 2002] V. Sós. Turbulent years: Erdős in his correspondence with Tur'an. In Simonovits Lalász, Lovász and Sós, editors, *The Mathematics of Paul Erdős I*, volume 11 of *Bolyai Society Mathematical Studies*, pages 85–146. János Bolyai Mathematical Society and Springer-Verlag, Hungary, 2002.

[Specker, 1949] E. Specker. Sur un problème de Sikorski. *Colloquium Math.*, 2:9–12, 1949.

[Specker, 1957] E. Specker. Teilmengen von Mengen mit Relationen. *Comment. Math. Helv.*, 31:302–314, 1957.

[Specker, 1990] E. Specker. *Selecta.* Birkhäuser Verlag, Basel, 1990. Edited and with a preface by Gerhard Jäger, Hans Läuchli, Bruno Scarpellini and Volker Strassen, With a biography of Ernst Specker by Jonas Meon.

[Spector, 1956] C. Spector. On degrees of recursive unsolvability. *Ann. of Math. (2)*, 64:581–592, 1956.

[Spencer, 1973] J. Spencer. *Paul Erdős: The Art of Counting.* The MIT Press, Cambridge, MA, 1973.

[Spencer, 1983] J. Spencer. Ramsey theory and Ramsey theoreticians. *J. Graph Theory*, 7:15–23, 1983.

[Staar, 1979] G. Staar. Élete a matematika (his life in mathematics). *Természet Világa*, 3, 1979.

[Stanley *et al.*, 1991] L. Stanley, D. Velleman, and C. Morgan. $\omega_3\omega_1 \to (\omega_3\omega_1, 3)^2$ requires an inaccessible. *Proc. Amer. Math. Soc.*, 111(4):1105–1118, 1991.

[Stern, 1978] J. Stern. A Ramsey theorem for trees, with an application to Banach spaces. *Israel J. Math.*, 29(2-3):179–188, 1978.

[Stewart, 1966] D. Stewart. (title unknown). Master's thesis, University of Bristol, 1966. Master of Science Thesis.

[Stoll, 1949] R. Stoll. Equivalence relations in algebraic systems. *Amer. Math. Monthly*, 56:372–377, 1949.

[Strauss, 1984] D. Strauss. Obituary: Clifford Hugh Dowker. *Bull. London Math. Soc.*, 16(5):535–541, 1984.

[Suslin, 1917] M. Suslin. Sur une définition des ensembles mesurables B sans nombres transfinis. *C. R. Acad. Sci. Paris*, 164:88–91, 1917.

[Suslin, 1920] M. Suslin. Problème 3. *Fund. Math.*, 1, 1920.

[Szeptycki, 1980] P. Szeptycki. Introduction to this issue. *Rocky Mountain J. Math.*, 10(1):1–6, 1980.

[Szeptycki, 1983] P. Szeptycki. Nachman Aronszajn (1907–1980). *Wiadom. Mat.*, 25(1):89–96, 1983.

[Tall, 1969] F. Tall. *Set-Theoretic Consistency Results and Topological Theorems Concerning the Normal Moore Space Conjecture and Related Problems.* PhD thesis, University of Wisconsin, Madison, 1969. Advisor Mary Ellen Rudin.

[Tall, 1977] F. Tall. Set-theoretic consistency results and topological theorems concerning the Normal Moore Space Conjecture and related problems. *Dissertationes Math. (Rozprawy Mat.)*, 148:53, 1977.

[Tall, 1993] F. Tall, editor. *The Work of Mary Ellen Rudin*, volume 705 of *Annals of the New York Academy of Sciences.* New York Academy of Sciences, New York, 1993. Papers from the Summer Conference on General Topology and Applications in honor of Mary Ellen Rudin held in Madison, Wisconsin, June 26–29, 1991.

[Tarkowski, 1960] S. Tarkowski. On the comparability of dendrites. *Bull. Acad. Polon. Sci. Sér. Sci. Math. Astronom. Phys.*, 8:39–41, 1960.

[Tarski, 1928] A. Tarski. Sur la décomposition des ensembles en sous-ensembles presque disjoints. *Fund. Math.*, 12:188–205, 1928.

[Tarski, 1929] A. Tarski. Sur la décomposition des ensembles en sous-ensembles presque disjoints. *Fund. Math.*, 14:205–215, 1929.

[Taylor, 1975] A. Taylor. *Some Results in Partition Theory.* PhD thesis, Dartmouth College, Hanover, N.H., 1975. Advisor J. E. Baumgartner; 113 pages.

[Taylor, 1976] A. Taylor. A canonical partition relation for finite subsets of ω. *J. Combinatorial Theory Ser. A*, 21(2):137–146, 1976.

[Taylor, 1978] A. Taylor. Partitions of pairs of reals. *Fund. Math.*, 99(1):51–59, 1978.

[Taylor, 1981] A. Taylor. Diamond principles, ideals and the normal Moore space problem. *Canad. J. Math.*, 33(2):282–296, 1981.

[Taylor, 1982] A. Taylor. A note on van der Waerden's theorem. *J. Combin. Theory Ser. A*, 33(2):215–219, 1982.

[Tennenbaum, 1959] S. Tennenbaum. Non-archimedian models for arithmetic. *Notices Amer. Math. Soc.*, 6, 1959. Abstract.

[Tennenbaum, 1968] S. Tennenbaum. Souslin's problem. *Proc. Nat. Acad. Sci. U.S.A.*, 59:60–63, 1968.

[Todorcevic and Farah, 1995] S. Todorcevic and I. Farah. *Some Applications of the Method of Forcing.* Yenisei Series in Pure and Applied Mathematics. Yenisei, Moscow, 1995.

[Todorcevic and Väänänen, 1999] S. Todorcevic and J. Väänänen. Trees and Ehrenfeucht-Fraïssé games. *Ann. Pure Appl. Logic*, 100(1-3):69–97, 1999.

[Todorcevic and Veličković, 1987] S. Todorcevic and B. Veličković. Martin's Axiom and partitions. *Compositio Math.*, 63(3):391–408, 1987.

[Todorčević, 1979a] S. Todorčević. *Rezultati i dokazi nezavisnosti u kombinatornoj teoriji skupova.* PhD thesis, University of Belgrade, 1979. Information from Mathematics Genealogy Project.

[Todorcevic, 1979b] S. Todorcevic. Rigid Aronszajn trees. *Publ. Inst. Math. (Beograd)*, 27(41):259–265, 1979. The SANU site lists it as appearing in 1980.

[Todorcevic, 1981a] S. Todorcevic. On the S-space problem. *Amer. Math. Soc. Abstracts*, 2(4), 1981. Abstract 81T-04-301.

[Todorcevic, 1981b] S. Todorcevic. A partition relation for partially ordered sets. *Amer. Math. Soc. Abstracts*, 2(3), 1981. Abstract 81T-03-227.

[Todorcevic, 1981c] S. Todorcevic. Some consequences of $\mathrm{MA} + \neg\mathrm{wKH}$. *Topology Appl.*, 12(2):187–202, 1981.

[Todorcevic, 1981d] S. Todorcevic. Trees, subtrees and order types. *Ann. Math. Logic*, 20(3):233–268, 1981.

[Todorcevic, 1983a] S. Todorcevic. Forcing positive partition relations. *Trans. Amer. Math. Soc.*, 280(2):703–720, 1983.

[Todorcevic, 1983b] S. Todorcevic. On a conjecture of R. Rado. *J. London Math. Soc. (2)*, 27(1):1–8, 1983.

[Todorcevic, 1984a] S. Todorcevic. A combinatorial property of sets of irrationals. Preprint, July 1984.

[Todorcevic, 1984b] S. Todorcevic. A note on the proper forcing axiom. In *Axiomatic set theory (Boulder, Colo., 1983)*, volume 31 of *Contemp. Math.*, pages 209–218. Amer. Math. Soc., Providence, RI, 1984.

[Todorcevic, 1984c] S. Todorcevic. Trees and linearly ordered sets. In *Handbook of Set-Theoretic Topology*, pages 235–293. North-Holland, Amsterdam, 1984.

[Todorcevic, 1985a] S. Todorcevic. Partition relations for partially ordered sets. *Acta Math.*, 155(1-2):1–25, 1985.

[Todorcevic, 1985b] S. Todorcevic. Remarks on chain conditions in products. *Compositio Math.*, 55(3):295–302, 1985.

[Todorcevic, 1986a] S. Todorcevic. Reals and positive partition relations. In *Logic, Methodology and Philosophy of Science, VII (Salzburg, 1983)*, pages 159–169. North-Holland, Amsterdam, 1986.

[Todorcevic, 1986b] S. Todorcevic. Remarks on cellularity in products. *Compositio Math.*, 57(3):357–372, 1986.

[Todorcevic, 1987] S. Todorcevic. Partitioning pairs of countable ordinals. *Acta Math.*, 159(3-4):261–294, 1987.

[Todorcevic, 1988a] S. Todorcevic. A chain decomposition theorem. *J. Combin. Theory Ser. A*, 48(1):65–76, 1988.

[Todorcevic, 1988b] S. Todorcevic. Oscillations of real numbers. In *Logic Colloquium '86 (Hull, 1986)*, pages 325–331. North-Holland, Amsterdam, 1988.

[Todorcevic, 1989] S. Todorcevic. *Partition Problems in Topology*, volume 84 of *Contemporary Mathematics*. American Mathematical Society, Providence, RI, 1989.

[Todorcevic, 1994] S. Todorcevic. Some partitions of three-dimensional combinatorial cubes. *J. Combin. Theory Ser. A*, 68(2):410–437, 1994.

[Todorcevic, 1998a] S. Todorcevic. Basis problems in combinatorial set theory. In *Proceedings of the International Congress of Mathematicians, Vol. II (Berlin, 1998)*, pages 43–52 (electronic), 1998. Extra Vol. II.

[Todorcevic, 1998b] S. Todorcevic. Countable chain condition in partition calculus. *Discrete Math.*, 188(1-3):205–223, 1998.

[Todorcevic, 2000] S. Todorcevic. A dichotomy for P-ideals of countable sets. *Fund. Math.*, 166(3):251–267, 2000.

[Todorcevic, 2007a] S. Todorcevic. Lipschitz maps on trees. *J. Inst. Math. Jussieu*, 6(3):527–556, 2007.

[Todorcevic, 2007b] S. Todorcevic. *Walks on Ordinals and their Characteristics*, volume 263 of *Progress in Mathematics*. Birkhäuser Verlag, Basel, 2007.

[Todorcevic, 2010] S. Todorcevic. *Introduction to Ramsey Spaces*, volume 174 of *Annals of Mathematics Studies*. Princeton University Press, Princeton, NJ, 2010.

[Transue, 1969] W. Transue, editor. *Proceedings of the Auburn Topology Conference*. Auburn University, Auburn, Ala., 1969. Held at Auburn University, Auburn, Ala., March 13–15, 1969, Dedicated to F. Burton Jones on the occasion of his 60th birthday.

[Turán, 1941] P. Turán. Eine Extremalaufgabe aus der Graphentheorie. *Mat. Fiz. Lapok*, 48:436–452, 1941.

[Ulam, 1930] S. Ulam. Zur masstheorie in der allgemeinen mengenlehre. *Fund. Math.*, 16(1):140–150, 1930.

[Ulam, 1974] S. Ulam. *Sets, Numbers, and Universes: Selected Works*. The MIT Press, Cambridge, Mass.-London, 1974. Edited by W. A. Beyer, J. Mycielski and G.-C. Rota, Mathematicians of Our Time, Vol. 9.

[Ulam, 1976] S. Ulam. *Adventures of a Mathematician*. Charles Scribner's Sons, New York, 1976.

[Urysohn, 1924] P. Urysohn. Un théorème sur la puissance des ensembles ordonnés. *Fund. Math.*, 5:14–19, 1924.

[van der Waerden, 1927] B. van der Waerden. Beweis einer baudetschen vermutung. *Nieuw Archief voor Wiskunde*, 19:212–216, 1927.

[van der Waerden, 1930] B. van der Waerden. *Moderne Algebra. Teil 1*, volume 33 of *Der Grundlehren der mathematischen Wissenschaften*. Julius Springer, Berlin, 1930.

[van der Waerden, 1937] B. van der Waerden. *Moderne Algebra I*. Julius Springer, Berlin, second edition, 1937. First edition copyrighted in 1931.

[van der Waerden, 1953] B. van der Waerden. *Modern Algebra*, volume I. Frederick Ungar Publishing Co., second edition, 1953. English translation by Fred Blum of the second edition from 1937.

[van der Waerden, 1954] B. van der Waerden. Einfall und überlegung in der Mathematik. III. *Elemente der Math.*, 9:49–56, 1954.

[van der Waerden, 1971] B. van der Waerden. How the proof of Baudet's conjecture was found. In *Studies in Pure Mathematics (Presented to Richard Rado)*, pages 251–260. Academic Press, London, 1971.

[van der Waerden, 1998] B. van der Waerden. Wie der Beweis der Vermutung von Baudet gefunden wurde. *Elem. Math.*, 53(4):139–148, 1998.

[van Mill and Reed, 1990] J. van Mill and G. Reed, editors. *Open Problems in Topology*. North-Holland Publishing Co., Amsterdam, 1990.

[Vaughan, 1980] J. E. Vaughan. Some cardinals related to c and topology. In *Topology Conference, 1979 (Greensboro, N.C., 1979)*, pages 39–49. Guilford College, Greensboro, N.C., 1980.

[Vaught, 1952] R. Vaught. On the equivalence of the axiom of choice and a maximal principle. *Bull. Amer. Math. Soc.*, 58, 1952. Abstract 87t (presented by title) from the December 1, 1951 meeting of the American Mathematical Society in Pasadena, California.

[Vaught, 1963] R. Vaught. Models of complete theories. *Bull. Amer. Math. Soc.*, 69:299–313, 1963.

[Vaught, 1997] R. Vaught. On the work of Andrzej Ehrenfeucht in model theory. In *Structures in logic and computer science*, volume 1261 of *Lecture Notes in Computer Science*, pages 1–13. Springer-Verlag, Berlin, 1997.

[Veblen, 1906] O. Veblen. Book Review: The Continuum as a Type of Order: An Exposition of the Modern Theory. With an Appendix on the Transfinite Numbers. *Bull. Amer. Math. Soc.*, 12(6):302–305, 1906.

[Velickovic and Woodin, 1998] B. Velickovic and W. H. Woodin. Complexity of reals in inner models of set theory. *Ann. Pure Appl. Logic*, 92(3):283–295, 1998.

[Velleman, 1984a] D. Velleman. ω-Morasses, and a weak form of Martin's Axiom provable in ZFC. *Trans. Amer. Math. Soc.*, 285(2):617–627, 1984.

[Velleman, 1984b] D. Velleman. Simplified morasses. *J. Symbolic Logic*, 49(1):257–271, 1984.

[Velleman, 1984c] D. Velleman. Souslin trees constructed from morasses. In *Axiomatic Set Theory (Boulder, Colo., 1983)*, volume 31 of *Contemp. Math.*, pages 219–241. Amer. Math. Soc., Providence, RI, 1984.

[Velleman, 1990] D. Velleman. Partitioning pairs of countable sets of ordinals. *J. Symbolic Logic*, 55(3):1019–1021, 1990.

[Vopěnka, 1965] P. Vopěnka. On $m1x$-model of set theory. *Bull. Acad. Polon. Sci. Sér. Sci. Math. Astronom. Phys.*, 13:267–272, 1965.

[Vuksanovic, 2002] V. Vuksanovic. A combinatorial proof of a partition relation for $[\mathbb{Q}]^n$. *Proc. Amer. Math. Soc.*, 130(10):2857–2864, 2002.

[Watson, 1993] S. Watson. Mary Ellen Rudin's early work on Suslin spaces. In *The Work of Mary Ellen Rudin (Madison, WI, 1991)*, volume 705 of *Ann. New York Acad. Sci.*, pages 168–182. New York Acad. Sci., New York, 1993.

[Welsh, 2003] D. Welsh. Crispin St. J. A. Nash-Williams (1932–2001). *Bull. London Math. Soc.*, 35(6):829–844, 2003.

[Williams, 1977] N. Williams. *Combinatorial Set Theory.* North Holland, Amsterdam, 1977.

[Wilson, 1991] R. Wilson. Mathematical review of "theory of finite and infinite graphs" authored by Dénes Kőnig. *Mathematical Reviews of the American Mathematical Society*, 91f:01026, 1991. MR1035708.

[Wolfsdorf, 1980] K. Wolfsdorf. Der Beweis eines Satzes von G. Chudnovsky. *Arch. Math. Logik Grundlag.*, 20:161–171, 1980.

[Wolfsdorf, 1988] K. Wolfsdorf. Das Theorem von Hales-Jewett und Mögliche Verallgemeinerungen. *European J. Combin.*, 9(1):91–95, 1988.

[Wolk, 1962] E. Wolk. The comparability graph of a tree. *Proc. Amer. Math. Soc.*, 13:789–795, 1962.

[Wolk, 1965] E. Wolk. A note on "The comparability graph of a tree". *Proc. Amer. Math. Soc.*, 16:17–20, 1965.

[Woodall, 1972] D. Woodall. Two results on infinite transversals. In *Combinatorics (Proc. Conf. Combinatorial Math., Math. Inst., Oxford, 1972)*, pages 341–350. Inst. Math. Appl., Southend, 1972.

[Woodin, 1999] W. H. Woodin. *The Axiom of Determinacy, Forcing Axioms, and the Nonstationary Ideal*, volume 1 of *de Gruyter Series in Logic and its Applications*. Walter de Gruyter & Co., Berlin, 1999.

[Yesenin-Volpin, 1954] A. S. Yesenin-Volpin. The unprovability of Suslin's hypothesis without the axiom of choice in the system of axioms of Bernays-Mostowski. *Doklady Akad. Nauk SSSR (N.S.)*, 96:9–12, 1954.

[Zarankiewicz, 1947] K. Zarankiewicz. Sur les relations synétriques dans l'ensemble fini. *Colloquium Math.*, 1:10–14, 1947.

[Zermelo, 1904] E. Zermelo. Beweis, daß jede Menge wohlgeordnet werden kann. *Math. Ann.*, 59(4):514–516, 1904.

[Zitarelli, 2004] D. Zitarelli. The origin and early impact of the Moore method. *Amer. Math. Monthly*, 111(6):465–486, 2004.

LARGE CARDINALS WITH FORCING

Akihiro Kanamori

This chapter describes, following the historical development, the investigation of large cardinal hypotheses using the method of forcing. Large cardinal hypotheses, also regarded as strong axioms of infinity, have stimulated a vast mainstream of modern set theory, and William Mitchell's chapter in this volume deals with their investigation through inner models, Menachem Kojman's chapter with their involvement in the study of singular cardinals, and Paul Larson's chapter with the accentuated direction of determinacy hypotheses. With the subject seen as a larger, integrated whole, we in this chapter incorporate aspects of these others while pursuing the directions involving forcing.

The author's other chapter in this volume, "Set Theory from Cantor to Cohen" (henceforth referred to as CC for convenience), had presented the historical development of set theory through to the creation of the method of forcing. Also, the author's book, *The Higher Infinite* [2003], provided the theory of large cardinals, including determinacy, up to a first plateau of combinatorial sophistication. In this chapter we first pick up the historical threads to large cardinals in that other chapter and pursue the developments involving forcing, thereby providing a complementary overlay to the book. There is no attempt at being comprehensive; rather we describe the earlier pivotal results and take up the more persistent themes, those that have mainly driven the subject.

1 TO THE MODERN THEORY

1.1 Beginnings

In this first section, we recall the early large cardinal landmarks, as pointed out in the author's other chapter CC, and then continue through to the first substantive elaborations of the theory in the 1960s. CC describes and develops the basics of set theory and should serve as a general reference for concepts, notations, and publications not again given here, and [Kanamori, 2003] frames a larger mathematical context. The emergence of large cardinal hypotheses dates well back to the beginnings of set theory; in our quick recapitulation, we proceed anachronistically in the modern terms of transfinite ordinals and cardinals as formulated within set theory in the 1930s work of John von Neumann.

Felix Hausdorff, the first developer of the transfinite after Cantor, carried out a substantive, far-reaching investigation of uncountable ordertypes and cardinals in

Handbook of the History of Logic: Sets and Extensions in the Twentieth Century.
Volume editor: Akihiro Kanamori. General editors: Dov M. Gabbay, Paul Thagard and John Woods.

the first decade of the 20th Century. He brought in the basic concept of cofinality (the *cofinality* $\mathrm{cf}(\kappa)$ of a cardinal κ is the least cardinal λ such that κ is the union of λ sets each of cardinality less than κ) and the distinction between *singular* ($\mathrm{cf}(\kappa) < \kappa$) and *regular* ($\mathrm{cf}(\kappa) = \kappa$) cardinals. And for the first time, he broached a large cardinal concept, an uncountable, regular limit cardinal, now also known as a *weakly inaccessible* cardinal, as a natural closure point for cardinal limit processes. In penetrating work early in the next decade, Paul Mahlo considered hierarchies of such cardinals based on fixed-point phenomena and used for the first time the concept of stationary set. For a cardinal κ, $C \subseteq \kappa$ is *closed unbounded* (in κ) if it is *closed*, i.e. if $\alpha < \kappa$ and $\bigcup(C \cap \alpha) = \alpha$ then $\alpha \in C$, and *unbounded* i.e. $\bigcup C = \kappa$; and $S \subseteq \kappa$ is *stationary* (in κ) if $C \cap S \neq \emptyset$ for any closed unbounded $C \subseteq \kappa$. A cardinal κ is now called *weakly Mahlo* if $\{\alpha < \kappa \mid \alpha \text{ is regular}\}$ is stationary in κ. Such a cardinal is weakly inaccessible, a limit of such, and so forth, and thus is conceptually transcendent of such cardinals.[1]

Ernst Zermelo in a remarkable 1930 paper offered his final axiomatization of set theory. Modern ZFC is recognizable, in that he incorporated the axioms of Replacement and Foundation, thereby promoting a view of the set-theoretic universe as stratified in a cumulative hierarchy $V = \bigcup V_\alpha$ indexed by ordinals forming a "regular class", where

$$V_0 = \emptyset;\ \ V_{\alpha+1} = \mathcal{P}(V_\alpha);\ \text{ and } V_\delta = \textstyle\bigcup_{\alpha<\delta} V_\alpha \text{ for limit ordinals } \delta\,.$$

Zermelo provided natural models of his axioms, each a V_κ where κ, when uncountable, is a *(strongly) inaccessible* cardinal (regular and *strong limit*, i.e. if $\lambda < \kappa$ then $2^\lambda < \kappa$). Zermelo in particular focused attention on strong limit cardinals, and e.g. one has correspondingly that κ is *(strongly) Mahlo* if κ is inaccessible and weakly Mahlo, when κ is also a strongly limit of inaccessible cardinals, and so forth.

The journal volume containing Zermelo's paper also contained Stanisław Ulam's seminal paper on measurable cardinals, which would become focal among large cardinals. A *filter over* a set Z is a family of subsets of Z closed under the taking of supersets and of intersections. (Usually excluded from consideration as trivial are the *principal filters*, those of form $\{X \subseteq Z \mid A \subseteq X\}$ for some set $A \subseteq Z$.) An *ultrafilter U over Z* is a maximal filter over Z, i.e. for any $X \subseteq Z$, either $X \in U$ or else $Z - X \in U$. For a cardinal λ, a filter is *λ-complete* if it is closed under the taking of intersections of fewer than λ members. Finally, an uncountable cardinal κ is *measurable iff* there is a κ-complete ultrafilter over κ. Ulam was motivated by measure-theoretic considerations, and he viewed his work as about $\{0,1\}$-valued measures, the measure 1 sets being the sets in the ultrafilter. Ulam established that *measurable cardinals are inaccessible*, and this initial airing generated a question that was to keep the spark of large cardinals alive for the next three decades: *Can the least inaccessible cardinal be measurable?*[2]

[1] This foregoing paragraph elaborates the beginning of CC §2.4. CC cites publications, and as mentioned earlier, it should serve as a reference for this until we get well into our second section.

[2] The foregoing two paragraphs elaborate parts of CC §3.2.

A decade later, Tarski provided a systematic development of these concepts in terms of ideals. An *ideal over* a set Z is a family of subsets of Z closed under the taking of subsets and of unions. This is the "dual" notion to filters; if I is an ideal (resp. filter) over Z, then $\breve{I} = \{Z - X \mid X \in I\}$ is its dual filter (resp. ideal). An ideal is λ-*complete* if its dual filter is. A more familiar conceptualization in mathematics, Tarski investigated the general notion of ideal on a Boolean algebra in place of the power set algebra $P(Z)$. For an ideal $I \subseteq P(Z)$, the I-*measure zero* (negligible) sets are the members of I, the I-*positive measure* (non-negligible) sets are the members of $P(Z) - I$, and the I-*measure one* (all but negligible) sets are the members of the dual filter. For a regular cardinal $\lambda > \omega$, a basic example is the *nonstationary ideal* NS_λ, consisting of the nonstationary subsets of λ, an ideal which is λ-complete, and the corresponding dual filter, the *closed unbounded filter* $\mathcal{C}_\lambda$.

At the very least as counterpoint to large cardinals, attention must be paid to how set theory was launched on an independent course as a distinctive field of mathematics by Kurt Gödel's 1930s construction of L, leading to the relative consistency of the Axiom of Choice and the Generalized Continuum Hypothesis. To formulate L in the modern way as an archetypal construction, we affirm notation and concepts in connection with Tarski's later, systematic definition of satisfaction-in-a-structure.

For a first-order language, a structure N interpreting that language (i.e. a specification of a domain of discourse as well as interpretations of the function and predicate symbols), a formula $\varphi(v_1, v_2, \ldots, v_n)$ of the language with the (free) variables as displayed, and $a_1, a_2, \ldots, a_n$ in the domain of N,

$$N \models \varphi[a_1, a_2, \ldots, a_n]$$

asserts that the formula φ is satisfied in N according to Tarski's recursive definition when v_i is interpreted as a_i. A subset y of the domain of N is *first-order definable over N iff* there is a formula $\psi(v_1, v_2, \ldots, v_{n+1})$ and $a_1, a_2, \ldots, a_n$ in the domain of N such that

$$y = \{z \in N \mid N \models \psi[a_1, a_2, \ldots, a_n, z]\}.$$

(The first-order definability of k-ary relations is analogously formulated with v_{n+1} replaced by k variables.)

Through Tarski's recursive definition and an "arithmetization of syntax" whereby formulas are systematically coded by natural numbers, the satisfaction relation $N \models \varphi[a_1, a_2, \ldots, a_n]$ for *sets* N is definable in ZF set theory. On the other hand, by Tarski's result on the "undefinability of truth", the satisfaction relation for V itself is not first-order definable in V.

For any set x let $\mathrm{def}(x)$ denote the collection of subsets of x first-order definable over the structure $\langle x, \in \rangle$ with domain x and the membership relation restricted to it.

Then define:

$$L_0 = \emptyset;\ L_{\alpha+1} = \mathrm{def}(L_\alpha),\ L_\delta = \bigcup\{L_\alpha \mid \alpha < \delta\} \text{ for limit ordinals } \delta;$$

and the *constructible universe*

$$L = \bigcup_\alpha L_\alpha \, .$$

With the formalizability of def(x), one can formalize the Axiom of Constructibility $V = L$, i.e. $\forall x(x \in L)$. To have a larger context, we affirm the following for a class X: For a set-theoretic formula φ, φ^X denotes φ with its quantifiers restricted to X and this extends to set-theoretic terms t (like $\bigcup x$, $P(x)$, and so forth) through their definitions to yield t^X. X is an *inner model iff* X is a transitive class containing all the ordinals such that φ^X is a theorem of ZF for every axiom φ of ZF. What Gödel did was to show in ZF that L is an inner model which satisfies AC and GCH. He thus established a relative consistency which can be formalized as an assertion: Con(ZF) implies Con(ZFC + GCH). Gödel also showed that for any inner model M of ZF that $L \subseteq M$, so that L is the canonically minimal inner model; it would be to counterpoint this minimality with "maximality" considerations that would further stimulate the investigation of large cardinals.

In later years Gödel speculated about the possibility of deciding propositions like CH outright with large cardinal hypotheses based on the heuristics of *reflection*, and later, *generalization*. In a 1946 address he suggested the consideration of "stronger and stronger axioms of infinity" and reflection down from V: "Any proof of a set-theoretic theorem in the next higher system above set theory (i.e. any proof involving the concept of truth ...) is replaceable by a proof from such an axiom of infinity." In a 1947 expository article on the Continuum Problem Gödel presumed that CH would be shown independent from ZF and speculated more concretely about possibilities with large cardinals. He argued that the axioms of set theory do not "form a system closed in itself" and so the "very concept of set on which they are based suggests their extension by new axioms that assert the existence of still further iterations of the operation of 'set of'." In an unpublished footnote toward a 1966 revision of the article, Gödel acknowledged "extremely strong axioms of infinity of an entirely new kind", generalizations of properties of ω "supported by strong arguments from analogy." These heuristics would surface anew in the 1960s, when the theory of large cardinals developed a self-fueling momentum of its own, stimulated by the emergence of forcing and inner models.[3]

From the 1930s, the direct investigation of the transfinite as extension of number was advanced by the emergence of *infinite combinatorics*. The investigation of partition relations and trees would be particularly relevant for large cardinals. In modern terms, for ordinals α, β, and δ and $n \in \omega$ the *partition property*

$$\beta \longrightarrow (\alpha)^n_\delta$$

is the assertion that for any partition $f\colon [\beta]^n \to \delta$ of the n-element subsets of β into δ cells, there is an $H \subseteq \beta$ of order type α *homogeneous* for the partition, i.e. all the n-element subsets of H lie in the same cell. Frank Ramsey in 1930 established the infinite Ramsey Theorem: $\omega \longrightarrow (\omega)^n_i$ for every $n, i \in \omega$.

[3]The foregoing paragraphs on Gödel and his work draw on CC §3.4, but also elaborate first-order definability.

A *tree* is a partially ordered set T such that the predecessors of any element are well-ordered. The *αth level* of T consists of those elements whose predecessors have order-type α, and the *height* of T is the least α such that the αth level of T is empty. Generally, a *chain* in a partially ordered set is a linearly ordered subset, and an *antichain* is a subset consisting of pairwise incompatible elements. A *cofinal branch* of T is a chain with elements at every non-empty level of T. Finally, for a cardinal κ, a *κ-tree* is a tree of height κ each of whose levels has cardinality less than κ, and κ has the *tree property iff* every κ-tree has a cofinal branch. Kőnig's Lemma, of 1927, is the assertion that *ω has the tree property.*

Djuro Kurepa, in his 1935 thesis, carried out the first systematic study of transfinite trees. In modern terms, an *Aronszajn tree* is an ω_1-tree without a cofinal branch, i.e. a counterexample to the tree property for ω_1. Kurepa gave Nachman Aronszajn's proof that *there is an Aronszajn tree.* A *Suslin* tree is an ω_1-tree with no uncountable chains or antichains. Kurepa reduced a question of Mikhail Suslin to Suslin's Hypothesis, that *there are no Suslin trees.* Finally, a *Kurepa* tree is an ω_1-tree with at least ω_2 cofinal branches, and Kurepa's Hypothesis, the assertion that there are such trees.

Paul Erdős was the prominent figure of a strong Hungarian tradition in combinatorics, and he and his collaborators simply viewed the transfinite as a rich source of combinatorially interesting problems. One of the early advances was the 1943 Erdős-Tarski paper, which concluded with an intriguing list of six combinatorial problems, the positive solution to any, as it was to turn out, amounting to the existence of a large cardinal. In a footnote various implications were noted, one of them being essentially that *for inaccessible κ, the tree property for κ implies $\kappa \longrightarrow (\kappa)^2_2$*, generalizing Ramsey's $\omega \longrightarrow (\omega)^2_2$ and making explicit the Kőnig Lemma property needed.[4]

1.2 Model-theoretic methods

The investigation of large cardinal hypotheses was crucially and essentially informed by the infusion of model-theoretic methods into set theory. Model theory was developed most concertedly by Tarski and his students in the 1950s and 1960s at the University of California at Berkeley. They carefully laid out satisfaction-in-a-structure; *theories* (deductively closed collections of sentences) and their *models*; algebratization with *Skolem functions* and *hulls*; and *elementary substructures and embeddings.* For structures $\mathcal{A}$ and $\mathcal{B}$ for the same language, $j: \mathcal{A} \to \mathcal{B}$ is an *elementary embedding* if for any $a_1, \ldots, a_n$ from the domain of $\mathcal{A}$, $\langle a_1, \ldots, a_n \rangle$ satisfies in $\mathcal{A}$ the same formulas that $\langle j(a_1), \ldots, j(a_n) \rangle$ does in $\mathcal{B}$; and when j is the identity, $\mathcal{A}$ is an *elementary substructure* of $\mathcal{B}$, denoted $\mathcal{A} \prec \mathcal{B}$. The construction of models freely used transfinite methods and soon led to new questions in set theory, but also set theory was to be decisively advanced by the infusion of model-theoretic methods. In particular, Gödel's pivotal work with L years earlier would be recast in systematic terms and generalized.

[4]The foregoing four paragraphs draw on CC §3.5.

A precursory result was a 1949 generalization by Andrzej Mostowski of the Mirimanoff-von Neumann result that every well-ordered set is order-isomorphic to exactly one ordinal with membership. A set X is *transitive* if $\bigcup X \subseteq X$, i.e. X is "closed under membership". A binary relation R on a set X is *extensional* if distinct members of X have distinct R-predecessors, and *well-founded* if every non-empty $Y \subseteq X$ has an R-minimal element (or, assuming AC, there is no infinite R-descending sequence). *If R is an extensional, well-founded relation on a set X, then there is a unique transitive set T and an isomorphism of the structure $\langle X, R\rangle$ onto $\langle T, \in\rangle$, i.e. a bijection $\pi\colon X \to T$ such that for any $x, y \in X$, $x\,R\,y$ iff $\pi(x) \in \pi(y)$.* $\langle T, \in\rangle$ is the *transitive collapse* of X, and π the *collapsing isomorphism.*

András Hajnal and Azriel Levy, in their theses of the mid-1950s, developed generalizations of L that were to become basic in a richer setting. For a set A, Hajnal formulated the *constructible closure $L(A)$ of A*, i.e. the smallest inner model M such that $A \in M$, and Levy formulated the *class $L[A]$ of sets constructible relative to A*, i.e. the smallest inner model M such that for every $x \in M$, $A \cap x \in M$. To formulate $L(A)$, define: $L_0(A) =$ the smallest transitive set $\supseteq \{A\}$ (to ensure that the resulting class is transitive); $L_{\alpha+1}(A) = \mathrm{def}(L_\alpha(A))$; $L_\delta(A) = \bigcup_{\alpha<\delta} L_\alpha(A)$ for limit $\delta > 0$; and finally $L(A) = \bigcup_\alpha L_\alpha(A)$. To formulate $L[A]$, first let $\mathrm{def}^A(x)$ denote the collection of subsets of x first-order definable over $\langle x, \in, A\cap x\rangle$, i.e. $A \cap x$ is now interpreting a predicate that can be used in the definitions. Then define: $L_0[A] = \emptyset$; $L_{\alpha+1}[A] = \mathrm{def}^A(L_\alpha[A])$; $L_\delta[A] = \bigcup_{\alpha<\delta} L_\alpha[A]$ for limit $\delta > 0$; and finally $L[A] = \bigcup_\alpha L_\alpha[A]$. With the "trace" $\overline{A} = A \cap L[A]$ one has $L_\alpha[\overline{A}] = L_\alpha[A]$ for every α and so $L[\overline{A}] = L[A]$.

$L(A)$ realizes the algebraic idea of building up a model starting from a set of generators, and $L[A]$ the logical idea of building up a model using A construed as a predicate. $L(A)$ may not satisfy AC since it may not have a well-ordering of A, yet $L[A]$ always satisfies that axiom. This distinction was only to surface later, as both Hajnal and Levy took A to be a set of ordinals, when $L(A) = L[A]$, and used these models to establish conditional independence results of the sort: if the failure of CH is consistent, then so is that failure together with $2^\lambda = \lambda^+$ for sufficiently large cardinals λ. In the coming expansion of the 1960s, both Hajnal and Levy would be otherwise engaged, with Hajnal becoming a major combinatorial set theorist and collaborator with Erdős, and Levy, a pioneer in the investigation of independence results.

The set-theoretic generalization of first-order logic allowing transfinitely indexed logical operations clarified the size of measurable cardinals. Extending familiarity by abstracting to a new domain, Tarski in 1962 formulated the *strongly compact* and *weakly compact* cardinals by ascribing natural generalizations of the key compactness property of first-order logic to the corresponding infinitary languages. These cardinals had figured in the 1943 Erdős-Tarski paper in equivalent combinatorial formulations that were later seen to imply that *a strongly compact cardinal is measurable,* and *a measurable cardinal is weakly compact.* Tarski's student Hanf then established, using the satisfaction relation for infinitary languages, that *there are many inaccessible cardinals (and Mahlo cardinals) below a weakly compact cardinal. A fortiori, the least inaccessible cardinal is not measurable.* This breakthrough was the first result about the size of measurable cardinals since Ulam's original 1930 paper and was greeted as a spectacular success for metamathematical methods. Hanf's work radically altered size intuitions about problems coming to be understood in terms of large cardinals and ushered in model-theoretic methods into the study of large cardinals beyond the Mahlo cardinals. Weak compactness was soon seen to have a variety of characterizations, most notably, *κ is weakly compact iff $\kappa \to (\kappa)^2_2$ iff $\kappa \to (\kappa)^n_\lambda$ for every $n \in \omega$ and $\lambda < \kappa$ iff κ is inaccessible and has the tree property,* and this was an early, significant articulation of the large cardinal extension of context for effecting known proof ideas and methods.[5]

The concurrent emergence of the *ultraproduct construction* in model theory set the stage for the development of the modern theory of large cardinals. The construction was brought to the forefront by Tarski and his students after Jerzy Łoś's 1955 adumbration of its fundamental theorem, and the new method of constructing concrete models brought set theory and model theory even closer together in a surge of results and a lasting interest in ultrafilters.

The ultraproduct construction was driven by the algebraic idea of making a structure out of a direct product of structures as modulated (or "reduced") by a filter. The particular case when all the structures are the same, the *ultrapower*, was itself seen to be substantive. To briefly describe a focal case for set theory, let N be a set, construed as a structure with $\in$, and U an ultrafilter over a set Z. On ${}^Z N$, the set of functions from Z to N, define

$$f =_U g \quad \textit{iff} \quad \{i \in Z \mid f(i) = g(i)\} \in U\ .$$

The filter properties of U imply that $=_U$ is an equivalence relation on ${}^Z N$, so with $(f)_U$ denoting the corresponding equivalence class of f, set ${}^Z N/U = \{(f)_U \mid f \in {}^Z N\}$. Next, the filter properties of U show that a binary relation E_U on ${}^Z N/U$ can be unambiguously defined by

$$(f)_U \ E_U \ (g)_U \quad \textit{iff} \quad \{i \in Z \mid f(i) \in g(i)\} \in U\ .$$

$=_U$ is thus a *congruence* relation, one that preserves the underlying structure; this sort of preservation is crucial in ultraproduct and classical, antecedent constructions with filters. The *ultrapower* of N by U is then defined to be the structure $\langle {}^Z N/U, E_U\rangle$. The crux of the construction is the fundamental *Łoś's Theorem: For a formula $\varphi(v_1, \ldots, v_n)$ and $f_1, \ldots, f_n \in {}^Z N$,*

$$\langle {}^Z N/U, E_U\rangle \models \varphi[(f_1)_U, \ldots, (f_n)_U] \quad \textit{iff} \quad \{i \in Z \mid \mathcal{N} \models \varphi[f_1(i), \ldots, f_n(i)]\} \in U\ .$$

Satisfaction in the ultrapower is thus reduced to satisfaction on a large set of coordinates, large in the sense of U. The proof is by induction on the complexity

[5]The foregoing paragraphs of this subsection draw on CC §3.6.

of φ using the filter properties of U, the ultrafilter property for the negation step, and AC for the existential quantifier step.

E_U is an extensional relation, and crucially, well-founded when U is $\aleph_1$-complete. In that case by Mostowski's theorem there is a collapsing isomorphism π of the ultrapower onto the transitive collapse $\langle M, \in\rangle$. Moreover, if for $x \in N$, c_x is the constant function: $N \to \{x\}$ and $j_U \colon N \to M$ is defined by $j_U(x) = \pi((c_x)_U)$, then j_U is an elementary embedding, i.e. for any formula $\varphi(v_1, \ldots, v_n)$ and $a_1, \ldots, a_n \in N$,

$$\langle N, \in\rangle \models \varphi[a_1, \ldots, a_n] \ \ \textit{iff} \ \ \langle M, \in\rangle \models \varphi[j_U(a_1), \ldots, j_U(a_n)]$$

by Łoś's Theorem. When we have well-foundedness, the ultrapower is identified with its transitive collapse and denoted $\mathrm{Ult}(N, U)$.

All of the foregoing is applicable, and will be applied, with proper classes N, as long as we replace the equivalence class $(f)_U$ by sets

$$(f)_U^0 = \{g \in (f)_U \mid g \text{ has minimal rank }\}$$

("Scott's trick"), and take Łoś's Theorem as a schema for formulas.

The model theorist H. Jerome Keisler established penetrating connections between combinatorial properties of ultrafilters and of their ultraproducts, and in particular took the ultrapower of a measurable cardinal κ by a κ-complete ultrafilter over κ to provide a new proof of Hanf's result that there are many large cardinals below a measurable cardinal. With Ulam's concept shown in a new light as providing well-founded ultrapowers, Dana Scott then struck on the idea of taking the ultrapower of the entire universe V by a κ-complete ultrafilter over a measurable κ, exploiting the resulting well-foundedness to get an elementary embedding $j \colon V \to \mathrm{Ult}(V, U)$. Importantly, κ is the *critical point*, i.e. $j(\alpha) = \alpha$ for every $\alpha < \kappa$ yet $\kappa < j(\kappa)$: Taking e.g. the identity function id : $\kappa \to \kappa$, $\{\xi < \kappa \mid \alpha < \xi < \kappa\} \in U$ for every $\alpha < \kappa$, so that $\kappa \leq \pi((\mathrm{id})_U) < j(\kappa)$ by Łoś's Theorem. If $V = L$, then $\mathrm{Ult}(V, U) = L$ by the definability properties of L, but this confronts $\kappa < j(\kappa)$, e.g. if κ were the least measurable cardinal. (One could also appeal to the general fact that $U \notin \mathrm{Ult}(V, U)$; that one "loses" the ultrafilter when taking the ultrapower would become an important theme in later work.) With this Scott established that *if there is a measurable cardinal, then* $V \neq L$. Large cardinal assumptions thus assumed a new significance as a means for "maximizing" possibilities away from Gödel's canonical, delimitative construction.

The ultrapower construction provided one direction of a new characterization that established a central structural role for measurable cardinals: *There is an elementary embedding* $j \colon V \to M$ *for some* M *with critical point* δ *iff* δ *is a measurable cardinal.* Keisler provided the converse direction: With j as hypothesized, $U_j \subseteq P(\delta)$ defined "canonically" by $X \in U_j$ *iff* $\delta \in j(X)$ is a δ-complete ultrafilter over δ. Generating ultrafilters thus via "ideal" elements would become integral to the theory of ultrafilters and large cardinals.

This characterization, when viewed with the focus on elementary embeddings, raises a point that will be even more germane, and thus will be emphasized later, in connection with strong hypotheses. That a $j\colon V \to M$ is elementary is not formalizable in set theory because of the appeal to the satisfaction relation for V, let alone the assertion that there is such a class j. Thus the "characterization" is really one of giving a formalization, one that provides operative sense through the ultrapower construction. Ulam's original concept was thus made intrinsic to set theory with the categorical imperative of elementary embeddings. In any event ZFC is never actually transcended in consistency results; one can always work in a sufficiently large V_α.

In Scott's $j\colon V \to M = \mathrm{Ult}(V, U)$ the concreteness of the ultrapower construction delivered ${}^\kappa M \subseteq M$, i.e. M is closed under the taking of arbitrary (in V) κ-sequences, so that in particular $V_{\kappa+1} \cap M = V_{\kappa+1}$. Through this agreement strong reflection conclusions can be drawn. U is *normal iff* $\pi((\mathrm{id})_U) = \kappa$, i.e. the identity function is a "least non-constant" function, a property that can be easily arranged. For such U, since κ is inaccessible, it is so in M and hence by Łoś's Theorem $\{\xi < \kappa \mid \xi \text{ is inaccessible}\} \in U$—the inaccessible cardinals below κ have measure one. An analogous argument applies to any $V_{\kappa+1}$ property of κ like weak compactness, and so, as would typify large cardinal hypotheses, measurability articulates its own sense of reflective transcendence over "smaller" large cardinals.

Normality went on to become staple to the investigation of ideals and large cardinals. Formulated for an ideal I over a cardinal λ, I is *normal iff* whenever a function f is regressive on an $S \in P(\lambda) - I$, there is a $T \in P(S) - I$ on which f is constant. The 1956 Fodor's Lemma is the assertion that the nonstationary ideal NS_λ is normal for regular $\lambda > \omega$, and a multitude of "smallness" properties other than nonstationarity has been seen to lead to normal ideals.

Through model-theoretic methods set theory was brought to the point of entertaining elementary embeddings into well-founded models. It was soon to be transfigured by a new means for getting well-founded *extensions* of well-founded models.[6]

2 THE ADVENT OF FORCING

2.1 Cohen

Paul Cohen (1934-2007) in April 1963 established the independence of AC from ZF and the independence of CH from ZFC. That is, Cohen established that Con(ZF) implies Con(ZF + $\neg$AC) and that Con(ZFC) implies Con(ZFC + $\neg$CH). Already prominent as an analyst, Cohen had ventured into set theory with fresh eyes and an open-mindedness about possibilities. These results solved two central problems of set theory. But beyond that, Cohen's proofs were the inaugural examples of a new technique, *forcing*, which was to become a remarkably general and flexible method for extending models of set theory. Forcing has strong intuitive underpinnings and reinforces the notion of set as given by the first-order ZF axioms with prominent

[6]The foregoing paragraphs on the ultraproduct construction considerably elaborates the latter part of CC §3.6, by setting out the basics of the construction.

uses of Replacement and Foundation. If Gödel's construction of L had launched set theory as a distinctive field of mathematics, then Cohen's method of forcing began its transformation into a modern, sophisticated one.

Cohen's approach was to start with a model M of ZF and adjoin a set G that witnesses some desired new property. This would have to be done in a minimal fashion in order that the resulting extension $M[G]$ also model ZF, and so Cohen devised special conditions on both M and G. Cohen's key idea was to tie G closely to M through a partially ordered system of sets in M called *conditions* that would approximate G. While G may not be a member of M, G is to be a subset of some $Y \in M$ (with $Y = \omega$ a basic case), and these conditions would "force" assertions about the eventual $M[G]$, e.g. by deciding some of the membership questions, whether $x \in G$ or not, for $x \in Y$. The assertions are to be just those expressible in a ramified language with terms developed in M, and Cohen developed a corresponding *forcing relation* $p \Vdash \varphi$, "p forces φ", between conditions p and formulas φ, a relation with properties reflecting his approximation idea. For example, if $p \Vdash \varphi$ and $p \Vdash \psi$, then $p \Vdash \varphi \wedge \psi$. The conditions are ordered according to the constraints they impose on the eventual G, so that if $p \Vdash \varphi$, and q is a stronger condition, then $q \Vdash \varphi$. It was crucial to Cohen's approach that the forcing relation, like the ramified language, be definable in M.

The final ingredient which gives this whole scaffolding life is the incorporation of a certain kind of set G. Stepping out of M and actually assuming that it is countable, Cohen enumerated the formulas of the ramified language in a countable sequence and required that G be completely determined by a sequence of stronger and stronger conditions $p_0, p_1, p_2, \ldots$ such that for every formula φ of the ramified language exactly one of φ or $\neg\varphi$ is forced by some p_n. Such a G is called a *generic* set. Cohen was able to show that the resulting $M[G]$ does indeed satisfy the axioms of ZF: Every assertion about $M[G]$ is already forced by some condition; the forcing relation is definable in M; and so the ZF axioms holding in M, most crucially Replacement and Foundation, can be applied to the ramified terms and language to derive corresponding forcing assertions about the ZF axioms holding in $M[G]$.

Cohen first described the case when $G \subseteq \omega$ and the conditions p are functions from some finite subset of ω into $\{0, 1\}$ and $p \Vdash \dot{n} \in \dot{G}$ if $p(n) = 1$ and $p \Vdash \dot{n} \notin \dot{G}$ if $p(n) = 0$. Today, a G so adjoined to M is called a *Cohen real over* M. If subsets of ω are identified with reals as traditionally construed, that G is generic can be extrinsically characterized by saying that G meets every open dense set of reals lying in M. Generally, a $G \subseteq \kappa$ analogously adjoined with conditions of cardinality less than κ is called a *Cohen subset of* κ. Cohen established the independence of CH by adjoining a set which in effect is a sequence of many Cohen reals. It was crucial that the cardinals in the ground model and generic extension coincide, and Cohen to this end drew out and relied on the important *countable chain condition* (c.c.c.): Any antichain, i.e. collection of mutually incompatible conditions, is countable (where two forcing conditions said to be *incompatible* if they have no common, stronger condition).

Several features of Cohen's arguments would quickly be reformulated, reorganized, and generalized, but the thrust of his approach through definability and genericity would remain. Cohen's great achievement lies in devising a concrete procedure for extending well-founded models of set theory in a minimal fashion to well-founded models of set theory with new properties but without altering the ordinals.

Just in the first weeks after Cohen's discovery, Solomon Feferman, who had been extensively consulted by Cohen as he was coming up with forcing, established further independences elaborating $\neg$AC and about definability. Levy soon joined in this work and pursued both directions, formulating the "Levy collapse" of an inaccessible cardinal, and Stanley Tennenbaum established the failure of Suslin's Hypothesis by generically adjoining a Suslin tree with finite conditions.

The Levy collapse would become of central importance for large cardinals. Cohen had shown how to "collapse" a cardinal: Given a regular cardinal κ and $\alpha > \kappa$ in a ground model M, a generic surjection G of α onto κ can be adjoined using as conditions approximations of cardinality less than κ. In the Levy collapse of an inaccessible cardinal $\lambda > \kappa$, such collapsing functions are collectively adjoined for every $\alpha < \lambda$, with the inaccessibility of λ ensuring that it becomes the successor cardinal to κ in the generic extension. The Levy collapse of an inaccessible cardinal to make it ω_1 or ω_2, the first or second uncountable cardinal of the extension, would become basic to subsequent developments.

The extent and breadth of the expansion of set theory henceforth described dwarfs all that has been described before, both in terms of the numbers of people involved and the results established. With clear intimations of a new and concrete way of building models, set theorists rushed in and, with forcing becoming method, were soon establishing a cornucopia of relative consistency results, truths in a wider sense, with some illuminating classical problems of mathematics. ZFC became quite unlike Euclidean geometry and much like group theory, with a wide range of models being investigated for their own sake.[7]

2.2 Solovay and forcing

Robert Solovay above all epitomized this period of sudden expansion in set theory with his mathematical sophistication and central results about and with forcing, and in the areas of large cardinals and descriptive set theory. Following initial graduate study in differential topology, Solovay turned to set theory after hearing a May 1963 lecture by Cohen. Just weeks after, Solovay elaborated the independence of CH by characterizing the possibilities for the size of 2^κ for regular κ and made the first exploration of a range of cardinals. Building on this William Easton in late 1963 established the definitive result for powers of regular cardinals: *Suppose that* GCH *holds and F is a class function from the class of regular cardinals to cardinals such that for regular $\kappa \leq \lambda$, $F(\kappa) \leq F(\lambda)$ and the cofinality* $\mathrm{cf}(F(\kappa)) > \kappa$. *Then there is a (class) forcing extension preserving cofinalities in which $2^\kappa = F(\kappa)$*

[7]The foregoing subsection draws on CC §4.1, but also elaborates on the Levy collapse.

for every regular κ. Thus, as Solovay had seen locally, the *only* restriction beyond monotonicity on the power function for regular cardinals is that given by a well-known constraint, the classical Zermelo-Kőnig inequality that $\mathrm{cf}(2^\kappa) > \kappa$ for any cardinal κ. Easton's result enriched the theory of forcing with the introduction of proper classes of forcing conditions, the basic idea of a product analysis, and the now familiar concept of *Easton support.* The result focused interest on the possibilities for powers of *singular* cardinals and the Singular Cardinals Hypothesis (SCH), which asserts that 2^κ for singular κ is the least possible with respect to the powers 2^μ for $\mu < \kappa$ as given by monotonicity and the Zermelo-Kőnig inequality. This requires in particular that for singular strong limit cardinals κ, $2^\kappa = \kappa^+$. Easton's models satisfy SCH, and the *Singular Cardinals Problem*, to determine the range of possibilities for powers of singular cardinals, would become a major stimulus for the further development of set theory much as the Continuum Problem had been for its early development.

In the Spring of 1964 Solovay established a result remarkable for its mathematical depth and revelatory of what standard of argument was possible with forcing: *If there is an inaccessible cardinal, then in a* ZF *inner model of a forcing extension the Principle of Dependent Choices* (DC) *holds and every set of reals is Lebesgue measurable, has the Baire property, and has the perfect set property.* Solovay's inner model is precluded from having a well-ordering of the reals, but DC is a choice principle implying the regularity of ω_1 and sufficient for the formalization of the traditional theory of measure and category on the real numbers. Thus, Solovay's work vindicated the early descriptive set theorists in the sense that the regularity properties can consistently hold for all sets of reals in a *bona fide* model for the classical mathematical analysis of the reals. To prove his result Solovay applied the Levy collapse of an inaccessible cardinal to make it ω_1. For the Lebesgue measurability he introduced a new kind of forcing beyond Cohen's direct ways of adjoining new sets of ordinals or collapsing cardinals, that of adding a *random real*, given by forcing with the Borel sets of positive measure as conditions and p stronger than q when $p - q$ is null. In contrast to Cohen reals, a random real meets every measure one subset of the unit interval lying in the ground model. Solovay's work not only opened the door to a wealth of different forcing arguments, but to this day his original definability arguments remain vital to descriptive set theory.

The perfect set property, central to Cantor's direct approach to the Continuum Problem through definability, led to the first acknowledged instance of a new phenomenon in set theory: the derivation of *equi-consistency* results between large cardinal hypotheses and combinatorial propositions about low levels of the cumulative hierarchy. Forcing showed just how relative the Cantorian concept of cardinality is, since bijective functions could be adjoined to models of set theory and powers like $2^{\aleph_0}$ can be made arbitrarily large with relatively little disturbance. For instance, large cardinals were found to satisfy substantial propositions even after they are Levy collapsed to ω_1 or ω_2. Conversely, such propositions were found to entail large cardinal propositions in an L-like inner model, mostly pointedly the very same initial large cardinal hypothesis. Thus, for some large cardinal property

$\varphi(\kappa)$ and proposition ψ, there is a direction $\text{Con}(\exists\kappa\varphi(\kappa)) \to \text{Con}(\psi)$ established by a collapsing forcing argument, and a converse direction $\text{Con}(\psi) \to \text{Con}(\exists\kappa\varphi(\kappa))$ established by witnessing $\varphi(\kappa)$ in an inner model.

Solovay's result provided the forcing direction from an inaccessible cardinal to the proposition that every set of reals has the perfect set property and ω_1 is regular. But Ernst Specker in 1957 had in effect established that if this obtains, then ω_1 (of V) is inaccessible in L. Thus, Solovay's use of an inaccessible cardinal was actually necessary, and its collapse to ω_1 complemented Specker's observation. The emergence of such equi-consistency results is a subtle realization of earlier hopes of Gödel for deciding propositions via large cardinals. Forcing, however, quickly led to the conclusion that there could be no direct implication for CH itself: Levy and Solovay, also in 1964, established that measurable cardinals neither imply nor refute CH, with an argument generalizable to other inaccessible large cardinals. Rather, CH and many other propositions would be reckoned with in terms of consistency, the methods of forcing and inner models being the operative modes of argument.

Building on his Lebesgue measurability result Solovay in 1965 reactivated the classical descriptive set theory program of investigating the extent of the regularity properties (in the presence of AC) by providing characterizations in terms of forcing and definability concepts for the $\mathbf{\Sigma}^1_2$ sets, the level at which Gödel through his work on L had in fact established from $V = L$ the failure of the properties. This led to the consistency relative to ZFC of the Lebesgue measurability of all $\mathbf{\Sigma}^1_2$ sets. Also, the characterizations showed that the regularity properties for $\mathbf{\Sigma}^1_2$ sets follow from existence of a measurable cardinal. Thus, although measurable cardinals do not decide CH, they do establish the perfect set property for $\mathbf{\Sigma}^1_2$ sets so that "CH holds for the $\mathbf{\Sigma}^1_2$ sets".[8] [9]

In a separate initiative, Solovay [1966; 1971] established the equi-consistency of the existence of a measurable cardinal and the "real-valued" measurability of $2^{\aleph_0}$, i.e. that there is a (countably additive) measure extending Lebesgue measure to all sets of reals. For the forcing direction, Solovay starting with a measurable cardinal adjoined random reals and applied the Radon-Nikodym Theorem of analysis, and for the converse direction, he starting with a real-valued measure enlisted the inner model constructed relative to the ideal of measure zero sets. This consistency result provided context for an extended investigation of the possibilities for the continuum as structured by such a measure. Through this work the concept of *saturated ideal*, first studied by Tarski, was brought to prominence as a generalization of having a measurable cardinal applicable to the low levels of the cumulative hierarchy. For an ideal over a cardinal κ, I is *λ-saturated iff* for any $\{X_\alpha \mid \alpha < \lambda\} \subseteq P(\kappa) - I$ there are $\beta < \gamma < \lambda$ such that $X_\beta \cap X_\gamma \in P(\kappa) - I$ (i.e. the

[8] A coda after many years: Although Solovay's use of an inaccessible cardinal for universal Lebesgue measurability seemed *ad hoc* at the time, in 1979 Saharon Shelah [1984] established in a *tour de force* that if ZF + DC and all Σ^1_3 sets of reals are Lebesgue measurable, then ω_1 is inaccessible in L.

[9] The foregoing paragraphs of §2.2 draws on CC §4.1 and §4.2. We now proceed anew, citing publications, for the rest of this chapter.

corresponding Boolean algebra has no antichains of cardinality λ). The ideal of measure zero sets is $\aleph_1$-saturated, and Solovay showed that if I is any κ-complete λ-saturated ideal over κ for some $\lambda < \kappa$, then $L[I] \models$ "κ is measurable".

Solovay's work also brought to the foreground the concept of *generic ultrapower* and *generic elementary embedding.* For an ideal I over κ, forcing with the members of $P(\kappa) - I$ as conditions and p stronger than q when $p - q \in I$ engenders an ultrafilter on the ground model $P(\kappa)$. With this one can construct an ultrapower of the ground model in the generic extension and a corresponding elementary embedding. It turns out that the κ^+-saturation of the ideal ensures that this generic ultrapower is well-founded. Thus, a synthesis of forcing and ultrapowers is effected, and this raised enticing possibilities for having such large cardinal-type structure low in the cumulative hierarchy.

The development of the theory of forcing went hand in hand with this procession of central results. Solovay had first generalized forcing to arbitrary partial orders of conditions, proceeding in terms of incompatible members and dense sets and Levy's concept of *generic filter.* In his work on the Baire property for his 1964 model, Solovay came to the idea of assigning values to formulas from a complete Boolean algebra. Loosely speaking, the value would be the supremum of all the conditions forcing it. Working independently, Solovay and Scott developed the idea of recasting forcing entirely in terms of *Boolean-valued models.* This approach showed how to replace Cohen's ramified languages by a more direct induction on rank and how to avoid his dependence on a countable model. Boolean-valued functions play the role of sets, and formulas involving these functions are assigned Boolean-values by recursion respecting logical connectives and quantifiers. By establishing in ZFC that e.g. there is a complete Boolean algebra assigning the formula expressing $\neg$CH Boolean value one, a semantic construction was replaced by a syntactic one that directly secured relative consistency.

Still, the view of forcing as a way of actually extending models held the reservoir of sense and the promise of discovery, and after Shoenfield [1971] popularized an approach to the forcing relation that captured the gist of the Boolean-valued approach, forcing has been generally cast as a matter of partial orders and generic filters. Boolean algebras would nonetheless underscore and enhance the setting: partial orders are to have a maximum element 1; one is attuned to the *separativity* of partial orders, the property that unique Boolean completions; Boolean-values are used when illuminating; and embedding results for forcing partial orders are cast, as most algebraically informative, in terms of Boolean algebras.

By the 1970s there would be a further assimilation of both the syntactic and semantic approaches, in that generic extensions would be "taken" of V. In this the current approach, a partial order $\langle P, < \rangle$ of conditions is specified to a purpose, with $p < q$ for p being stronger than q. A class V^P of P-names defined recursively is used in forcing assertions, with a canonical name $\check{x}$ corresponding to $x \in V$. A $D \subseteq P$ is *dense* if for any $p \in P$ there is a $d \in D$ with $d \leq p$. An $F \subseteq P$ is a *filter* if (i) if $p \in F$ and $p \leq q$, then $q \in F$, and (ii) if $p_1, p_2 \in F$ then there is an $r \in F$ with $r \leq p_1$ and $r \leq p_2$. Finally, $G \subseteq P$ is a *V-generic filter* if G is a filter such that for every dense $D \subseteq P$, $G \cap D \neq \emptyset$. One posits such a G and takes a generic extension $V[G]$, its properties argued for on the basis of combinatorial properties of P. For inner or transitive set models M, one proceeds analogously to define *M-generic filters* meeting every dense set belonging to M and takes generic extensions $M[G]$.

In this one goes against the sense of V as *the* universe of all sets and Tarski's "undefinability of truth", but actually V has become *schematic* for a ground model. Generic extensions of inner models M are taken with M-generic G, and moreover, successive iterated extensions are taken, exacerbating any preoccupation with a single universe of sets. As the techniques of forcing were advanced, the methodology was itself soon to be woven into set theory as part of its postulations.

Solovay and Tennenbaum earlier in 1965 (cf. their [1971]) had established the consistency of Suslin's Hypothesis, that there are no Suslin trees, illuminating a classical question from 1920; they had made ground-breaking use of iterated forcing to "kill" Suslin trees by forcing with them as partial orders and shooting cofinal branches through them. D. Anthony Martin pointed out that the Solovay-Tennenbaum argument actually established the consistency of a closure of forcing extensions of a certain kind, an instrumental "axiom" now known as Martin's Axiom (MA): *For any countable-chain-condition partial order P and collection $\mathcal{D}$ of fewer than $2^{\aleph_0}$ dense subsets of P, there is a filter $G \subseteq P$ meeting every member of $\mathcal{D}$.* Thus method became axiom, and many consistency results could now be simply stated as direct consequences of a single umbrella proposition. CH technically implies MA, but the Solovay-Tennenbaum argument established the consistency of MA with the continuum being arbitrarily large.

While classical results with CH had worked on an $\aleph_0 / \aleph_1$ dichotomy, MA established a $<2^{\aleph_0} / 2^{\aleph_0}$ dichotomy. For example, [Martin and Solovay, 1970] established that MA implies that the union of fewer than $2^{\aleph_0}$ Lebesgue measure zero sets is again Lebesgue measure zero. Sierpiński in 1925 had established that every $\mathbf{\Sigma}^1_2$ set of reals is the union of $\aleph_1$ Borel sets. Hence, MA and $2^{\aleph_0} > \aleph_1$ implies that every $\mathbf{\Sigma}^1_2$ set of reals is Lebesgue measurable. Many further results plied the $<2^{\aleph_0} / 2^{\aleph_0}$ dichotomy to show that under MA inductive arguments can be carried out in $2^{\aleph_0}$ steps that previously succeeded under CH in $\aleph_1$ steps. The Continuum Problem was newly illuminated as a matter of method, by showing that CH as a construction principle could be generalized to $2^{\aleph_0}$ being arbitrarily large.

Glancing across the wider landscape, forcing provided new and diverse ways of adjoin generic reals and other sets, and these led to new elucidations, for example about cardinal characteristics, or invariants, of the continuum and combinatorial structures and objects, like ultrafilters over ω. The work on Suslin's Hypothesis in hand and with the possibilities afforded by Martin's Axiom, the investigation of general topological notions gathered steam. With Mary Ellen Rudin and her students at Wisconsin breaking the ground, new questions were raised for general topological spaces about separation properties, compactness-type covering properties, separability and metrizability, and corresponding cardinal characteristics.

2.3 $0^\#$ and $L[U]$

The infusion of forcing into set theory induced a broad context extending beyond its applications and sustained by model-theoretic methods, a context which included central developments about large cardinals having their source in Scott's 1961 result that measurable cardinals contradict $V = L$. Haim Gaifman [1964] invented *iterated ultrapowers* and established seminal results about and with the technique, results which most immediately stimulated definitive work in the formative theses of Silver and Kunen.

Jack Silver in his 1966 Berkeley thesis (cf. his [1971c]) provided a structured sense of transcendence over L in terms of the existence of a special set of natural numbers $0^\#$ ("zero sharp") which refined an earlier formulation of Gaifman and was quickly investigated by Solovay [1967] in terms of definability. Mostowski and Andrzej Ehrenfeucht in their [1956] had developed theories whose models have *indiscernibles*, implicitly ordered members of the domain all of whose n-tuples satisfy the same formulas. They had applied Ramsey's Theorem in compactness arguments to get models generated by indiscernibles, models consequently having many automorphisms. Silver applied partition properties satisfied by measurable cardinals to produce indiscernibles *within* given structures, particularly in the initial segment $\langle L_{\omega_1}, \in\rangle$ of the constructible universe. With definability and Skolem hull arguments, Silver was able to isolate a canonical collection of sentences to be satisfied by indiscernibles, a theory whose models cohere to get L itself as generated by canonical ordinal indiscernibles—a dramatic accentuation of the original Gödel generation of L. $0^\#$ is that theory coded as a real, and as Solovay emphasized, $0^\#$ is the only possible real to satisfy a certain Π^1_2 relation, one whose complexity arises from its asserting that to every countable well-ordering there corresponds a well-founded model of the coded theory. The canonical class, closed and unbounded, of ordinal indiscernibles is often called the *Silver indiscernibles*. Having these indiscernibles substantiates $V \neq L$ in drastic ways: Each indiscernible ι has various large cardinal properties and satisfies $L_\iota \prec L$, so that by a straightforward argument the satisfaction relation for L is definable from $0^\#$. The theory of $0^\#$ was seen to relativize, and for reals $a \in {}^\omega\omega$ the analogous $a^\#$ for the inner model $L[a]$ would play focal roles in descriptive set theory as based on definability.

Kunen's main large cardinal results emanating from his 1968 Stanford thesis (cf. [1970]) would be the definitive structure results for inner models of measurability. For U a normal κ-complete ultrafilter over a measurable cardinal κ, the inner model $L[U]$ of sets constructible relative to U is easily seen with $\overline{U} = U \cap L[U]$ to satisfy $L[U] \models$ "$\overline{U}$ is a normal κ-complete ultrafilter". With no presumption that κ is measurable (in V) and taking $U \in L[U]$ from the beginning, call $\langle L[U], \in, U\rangle$ a *κ-model iff* $\langle L[U], \in, U\rangle \models$ "U is a normal κ-complete ultrafilter over κ". Solovay observed that in a κ-model, the GCH holds above κ by a version of Gödel's argument for L and that κ is the only measurable cardinal by a version of Scott's argument. Silver [Silver, 1971a] then established that the full GCH holds, thereby establishing the relative consistency of GCH and measurability; Silver's proof turned on a local structure $L_\alpha[U]$ being *acceptable* in the later parlance of inner model theory.

Kunen made Gaifman's technique of iterated ultrapowers integral to the subject of inner models of measurability. For a κ-model $\langle L[U], \in, U\rangle$, the ultrapower of $L[U]$ by U with corresponding elementary embedding j provides a $j(\kappa)$-model $\langle L[j(U)], \in, j(U)\rangle$, and this process can be repeated. At limit stages, one can take the direct limit of models, which when well-founded can be identified with the transitive collapse. Indeed, by Gaifman's work these iterated ultrapowers are always well-founded, i.e. κ-models are *iterable.* Kunen showed that the λth iterate of a κ-model for any regular $\lambda > \kappa^+$ is of form $\langle L[\mathcal{C}_\lambda], \in, \mathcal{C}_\lambda \cap L[\mathcal{C}_\lambda]\rangle$, where $\mathcal{C}_\lambda$ again is the closed unbounded filter over λ, so that remarkably, constructing relative to a filter definable in set theory leads to an inner model of measurability. With this, there can be *comparison* of κ-models and κ'-models by iterating them up to a sufficiently large λ. This comparison possibility led to the structure results: (1) *for any κ-model and κ'-model with $\kappa < \kappa'$, the latter is an iterated ultrapower of the former,* and (2) *for any κ, there is at most one κ-model.* It then followed that if κ is measurable and U_1 and U_2 are any κ-complete ultrafilters over κ, then $L[U_1] = L[U_2]$. These various results argued forcefully for the coherence and consistency of the concept of measurability. As for the methods employed, in the subsequent development of inner model theory iterability and comparison would remain as basic features of the subject.

Kunen importantly emphasized that iterated ultrapowers can be taken of an inner model M with respect to an ultrafilter U even if $U \notin M$, as long U is an *M-ultrafilter*, i.e. U in addition to having M related ultrafilter properties also satisfies an "amenability" condition for M. A crucial dividend was a characterization of the existence of $0^\#$ that secured its central importance in inner model theory. With $0^\#$, any increasing shift of the Silver indiscernibles provides an elementary embedding $j\colon L \to L$. Kunen established conversely that such an embedding generates indiscernibles, so that *$0^\#$ exists iff there is a (non-identity) elementary embedding $j\colon L \to L$.* Starting with such an embedding Kunen defined the corresponding ultrafilter U over the critical point and showed that U is an L-ultrafilter with which the iterated ultrapowers of L are well-founded. The successive images of the critical point were seen to be indiscernibles for L, giving $0^\#$. As inner model theory was to develop, this sharp analysis would become schematic: the "sharp" of an inner model M would encapsulate transcendence over M, and the *non-rigidity* of M, that there is a (non-identity) elementary embedding $j\colon M \to M$, would provide equivalent structural sense.

2.4 Jensen and constructibility

These various results were set against a backdrop of an increasing articulation of Gödel's original notion of constructibility. Levy [1965] had put forward the appropriate hierarchy for the first-order formulas of set theory: A formula is Σ_0 and Π_0 if it is bounded, i.e. having only quantifiers expressible in terms of $\forall v \in w$

and $\exists v \in w$, and recursively, a formula is Σ_{n+1} if it is of the form $\exists v_1 \ldots \exists v_k \varphi$ where φ is Π_n and is Π_{n+1} if it is of the form $\forall v_1 \ldots \forall v_k \varphi$ where φ is Σ_n. Two basic points about discounting bounded quantifiers are that Σ_0 formulas are absolute for transitive structures, i.e. they hold in such structures just in case they hold in V, and that if φ is Σ_n (resp. Π_n) then $\exists v \in w\varphi$ and $\forall v \in x\varphi$ are equivalent in ZFC to Σ_n (resp. Π_n) formulas by uses of Replacement.

Gödel's original GCH result with L was newly seen in light of the structured context for definability. For N and M construed as structures with $\in$, $j\colon N \to M$ is a *Σ_n-elementary embedding iff* for any Σ_n $\varphi(v_1, \ldots, v_k)$ and $x_1, \ldots, x_k \in N$, $N \models \varphi[x_1, \ldots, x_k]$ *iff* $M \models \varphi[j(x_1), \ldots, j(x_k)]$. N is a *Σ_n-elementary substructure* of M, denoted $N \prec_n M$, *iff* the identity map is Σ_n-elementary. Analysis of the satisfaction relation established that being an L_α is a Σ_1 property, and this led to the Condensation Lemma:

> If α is a limit ordinal and $N \prec_1 L_\alpha$,
> then the transitive collapse of N is L_β for some $\beta \leq \alpha$.

Operatively, one applies this lemma with Skolem's algebraic approach to logic by taking N to be a *Σ_1-Skolem hull* in L_α: For any Σ_0 formula $\varphi(v_1, \ldots, v_n, v_{n+1})$ and $x_1, \ldots, x_n \in L_\alpha$, if $\langle L_\alpha, \in \rangle \models \varphi[x_1, \ldots x_n, y]$ for $y \in L_\alpha$, let $f_\varphi(x_1, \ldots, x_n)$ be such a y. Then let N be the algebraic closure of some subset of L_α under these Skolem functions. The road from the Condensation Lemma to Gödel's Fundamental Theorem for the consistency of GCH is short. Generally, the lemma articulates a crucial hierarchical cohesion, and its various emanations would become fundamental to all inner model theory.

The consummate master of constructibility was to be Ronald Jensen, whose first systematic analysis transformed the subject with the introduction of the *fine structure theory* for L. Jensen's work is distinguished by the persistent pursuit of internal logical structure, the sophistication of the local apparatus developed, and a series of remarkable successes with reverberations throughout the whole expanse of set theory. Jensen [1968] made an initial breakthrough by showing that $V = L$ *implies the failure of Suslin's Hypothesis*, i.e. (there is a Suslin tree)^L, applying L for the first time after Gödel to establish a relative consistency result about a classical proposition. The initial breach had been when Tennenbaum in 1963 (cf. his [1968]) had adjoined a Suslin tree with forcing with finite conditions, and then Thomas Jech [1967] had provided another forcing argument using countable conditions; Jensen at first pitched his construction in the guise of a forcing argument, one in fact like Tennenbaum's. This is the paradigmatic case of what would become a recurring phenomenon: A combinatorial existence assertion is first shown to be relatively consistent with ZFC using forcing, and then that assertion is shown to hold in L, the minimal inner model.

The lack of cofinal branches in Suslin trees is complemented by their abundance in Kurepa trees. An early, eye-catching forcing result was due to Silver [1971b]: *If there is an inaccessible cardinal, then in the extension by the Levy collapse to make it ω_2, Kurepa's Hypothesis fails, i.e. there are no Kurepa trees.* Notably, in a 1966 masters' thesis D. Stewart [1966] showed how to adjoin a Kurepa tree by forcing. Inspired by Jensen's Suslin tree construction, the ubiquitous Solovay established: (there is a Kurepa tree)^L. Moreover, it was noted that if ω_2 (of V) is not inaccessible in L, then Solovay's argument still works to establish that there is a Kurepa tree. Thus, another equi-consistency result illuminated a classical issue: *The failure of Kurepa's Hypothesis is equi-consistent with having an inaccessible cardinal.*

Jensen, and others, isolated the combinatorial features of L that enabled these Suslin and Kurepa constructions, and in 1969 he together with Kunen worked out a larger theory. The focus was mainly on two combinatorial principles for a regular cardinal κ, $\diamondsuit_\kappa$ ("diamond") and a strengthening, $\diamondsuit^+_\kappa$ ("diamond plus"). Stating the first,

$\diamondsuit_\kappa$ There is a sequence $\langle S_\alpha \mid \alpha < \kappa \rangle$ with $S_\alpha \subseteq \alpha$ such that for any $X \subseteq \kappa$, $\{\alpha < \kappa \mid X \cap \alpha = S_\alpha\}$ is stationary in κ .

Just $\diamondsuit$ is implicitly $\diamondsuit_{\omega_1}$. $\diamondsuit_\kappa$ implies $\bigcup_{\alpha<\kappa} \kappa^{|\alpha|} = \kappa$ (so that $\diamondsuit$ implies CH) as every bounded subset of κ occurs in a $\diamondsuit_\kappa$ sequence. Indeed, a $\diamondsuit_\kappa$ sequence is an enumeration of the bounded subsets of κ that can accommodate every $X \subseteq \kappa$ in anticipatory constructions where $X \cap \alpha$ appearing in the enumeration for many α's suffices. Within a few years $\diamondsuit$ would be on par with CH as a construction principle with wide applications in topology, algebra, and analysis.[10]

Jensen abstracted his Suslin tree result to: (1) *if* $V = L$, *then* $\diamondsuit_\kappa$ *holds for every regular* $\kappa > \omega$, and (2) *if* $\diamondsuit_{\omega_1}$ *holds, then there is a Suslin tree.* Solovay's result was abstracted to higher, κ*-Kurepa trees*, κ-trees with at least κ^+ cofinal branches, in terms of a new cardinal concept, *ineffability*, arrived at independently by Jensen and Kunen: *If* $V = L$ *and* $\kappa > \omega$ *is regular, then* $\diamondsuit^+_\kappa$ *holds iff* κ *is not ineffable.* Ineffable cardinals, stronger than weakly compact cardinals, would soon be seen to have a range of involvements and an elegant theory. As for higher Suslin trees, how they might be constructed in L initially motivated a new combinatorial principle, one that soon figured in a sophisticated forcing argument.

The crowning achievement of the investigation of Suslin's Hypothesis was its joint consistency with CH, Con(ZFC) $\rightarrow$ Con(ZFC + CH + SH), established by Jensen (cf. [Devlin and Johnsbraten, 1974]). In the Solovay-Tennenbaum consistency proof for SH, cofinal branches had been adjoined iteratively to Suslin trees as they arose and direct limits were taken at limit stages, a limiting process that conspired to adjoin new reals so that CH fails. Jensen, with considerable virtuosity for the time, devised a way to kill Suslin trees less directly and effected the iteration according to a curtailed tree-like indexing—so that no new reals are ever adjoined. That indexing is captured by the $\kappa = \omega_1$ case of the combinatorial principle $\square_\kappa$ ("square"):

$\square_\kappa$ There is a sequence $\langle C_\alpha \mid \alpha \text{ a limit ordinal } < \kappa^+ \rangle$ such that for $\alpha < \kappa^+$:

[10]Another coda of Shelah's after many years: In 2007 he established that for successors $\lambda^+ > \omega_1$, $2^\lambda = \lambda^+$ actually implies $\diamondsuit_{\lambda^+}$, so that the two are equivalent.

(a) $C_\alpha \subseteq \alpha$ is closed unbounded in α,
(b) for β a limit point of C_α, $C_\alpha \cap \beta = C_\beta$, and
(c) for $\omega \leq \mathrm{cf}(\alpha) < \kappa$, the order-type of C_α is less than κ.

$\square_\omega$ is immediate, as witnessed by any ladder system, i.e. a sequence $\langle C_\alpha \mid \alpha$ a limit ordinal $< \omega_1\rangle$ such that C_α is of order-type ω and cofinal in α. $\square_\kappa$ for $\kappa > \omega$ brings out the tension between the desired (b) and the needed (c). As such, $\square_\kappa$ came to guide many a construction of length κ^+ based on components of cardinality $< \kappa$.

Jensen showed that $\square_\kappa$ can be adjoined by straightforward forcing with initial approximations. Continuing his work on constructibility, Jensen soon established: *If $V = L$, then $\square_\kappa$ holds for every κ.* As for higher Suslin trees, a *κ-Suslin tree* is expectedly a κ-tree with no chains or antichains of cardinality κ. Jensen's work on these trees having actually motivated his formulation of $\square_\kappa$, he established, generalizing his result for $\kappa = \omega_1$: (1) *for any κ, $\diamondsuit_{\kappa^+}$ and $\square_\kappa$ imply that there is a κ^+-Suslin tree,* and, for limit cardinals κ, the characterization (2) *there is a κ-Suslin tree iff κ is not weakly compact.* It is a notable happenstance that Suslin's early, 1920 speculation would have such extended ramifications in modern set theory.

Jensen's results that $\square_\kappa$ holds in L and (2) above were the initial applications of his fine structure theory. Unlike Gödel who had focused with L on relative consistency, Jensen regarded the investigation of how the constructible hierarchy grows by examining its behavior at arbitrary levels as of basic and intrinsic interest. And with his fine structure theory Jensen developed a considerable and intricate machinery for this investigation. A pivotal question became: when does an ordinal α first get "singularized", i.e. what is the least β such that there is in $L_{\beta+1}$ an unbounded subset of α of smaller order-type, and what Σ_n definitional complexity does this set have? One is struck by the contrast between Jensen's attention to such local questions as this one, at the heart of his proof of $\square_\kappa$, and how his analysis could lead to major large-scale results of manifest significance.

Jensen's CH + SH consistency result established the consistency of CH together with the failure of $\diamondsuit_{\omega_1}$. What about the possible failure of $\square_{\omega_1}$? Solovay observed: *If a Mahlo cardinal is Levy collapsed to make it ω_2, then in the extension $\square_{\omega_1}$ fails.* Moreover, Jensen's argument for $\square_\kappa$ works to show that if κ^+ is not Mahlo in L, then $\square_\kappa$ holds. Thus, among the increasing number of results complementing the historical emergence of large cardinal hypotheses was yet another equi-consistency result: *The failure of $\square_{\omega_1}$ is equi-consistent to having a Mahlo cardinal.*

3 STRONG HYPOTHESES

3.1 Large large cardinals

With elementary embedding having emerged as a systemic concept in set theory, Solovay and William Reinhardt at Berkeley in the late 1960s formulated interrelated large cardinal hypotheses stronger than measurability. Reinhardt conceived *extendibility*, and he and Solovay independently, *supercompactness*. (See the later [Solovay *et al.*, 1978] for an exposition.) A cardinal κ is *γ-supercompact iff* there is an elementary embedding $j\colon V \to M$ for some inner model M, with critical point κ and $\gamma < j(\kappa)$ such that ${}^{\gamma}M \subseteq M$, i.e. M is closed under the taking of arbitrary γ-sequences. κ is *supercompact iff* κ is γ-supercompact for every γ. Evidently, the heuristics of generalization and reflection were at work here, as κ is measurable *iff* κ is κ-supercompact, and stronger closure properties imposed on the target model M ensure stronger reflection properties. For example, if κ is 2^{κ}-supercompact with witnessing $j\colon V \to M$, then $M \models$ "κ is measurable", since ${}^{2^{\kappa}}M \subseteq M$ implies that every ultrafilter over κ is in M, and so if $U_j \subseteq P(\kappa)$ is defined canonically from j by $X \in U_j$ *iff* $\kappa \in j(X)$, then $\{\xi < \kappa \mid \xi \text{ is measurable}\} \in U_j$ by Łoś's Theorem. Supercompactness was initially viewed as an ostensible strengthening of Tarski's strong compactness in that, with the focus on elementary embedding, reflection properties were directly incorporated. Whether strong compactness is actually equivalent to supercompactness became a new, "identity crisis" issue.

Reinhardt entertained a *prima facie* extension of these ideas, that there is a (non-identity) elementary embedding $j\colon V \to V$. With suspicions soon raised, Kunen [1971a] dramatically established that this is inconsistent with ZFC by applying an Erdős-Hajnal partition relation result, a combinatorial contingency making prominent use of the Axiom of Choice. This contingency pointed out a specific lack of closure of the target model: *For any elementary embedding $j\colon V \to M$ with critical point κ, let λ be the supremum of $\kappa < j(\kappa) < j^2(\kappa) < \ldots$. Then, $V_{\lambda+1} \not\subseteq M$.* This lack of closure has essentially stood as the weakest known to this day.

A net of hypotheses consistency-wise stronger than supercompactness was soon cast across the conceptual space delimited by Kunen's inconsistency. For $n \in \omega$, κ is *n-huge iff* there is an elementary embedding $j\colon V \to M$, for some inner model M, with critical point κ such that ${}^{j^n(\kappa)}M \subseteq M$. As had been first formulated by Kunen, κ is *huge iff* κ is 1-huge. If κ is huge, then $V_\kappa \models$ "there are many supercompact cardinals". Thematically close to Kunen's inconsistency were several hypotheses articulated for further investigation, e.g. there is a (non-identity) elementary embedding $j\colon V_\lambda \to V_\lambda$ for some λ.

The appearance of proper classes in these various formulations raises issues about legitimacy. By Tarski's "undefinability of truth", the satisfaction relation for V is not definable in V, and the elementary embedding characterization of measurability already suffers from this shortcoming. However, the γ-supercompactness of κ can be analogously formulated in terms of the existence of a "normal" ultrafilter over the set $\mathcal{P}_\kappa\gamma = [\gamma]^{<\kappa} = \{x \subseteq \gamma \mid |x| < \kappa\}$. Similarly, n-hugeness can also be recast. As for Kunen's inconsistency, his argument can be regarded as establishing: *There is no (non-identity) elementary embedding $j\colon V_{\lambda+2} \to V_{\lambda+2}$ for any λ.*

The details on γ-supercompactness drew out new, generalizing concepts for filters (and so, for ideals). Suppose that Z is a set and F a filter over $P(Z)$ (so $F \subseteq P(P(Z)))$. Then F is *fine iff* for any $a \in Z$, $\{x \in P(Z) \mid a \in x\} \in F$. F is *normal iff* whenever f is a function satisfying $\{x \in P(Z) \mid f(x) \in x\} \in F$, i.e. f is a choice function on a set in F, there is an $a \in Z$ such that $\{x \in P(Z) \mid f(x) = a\} \in F$, i.e. f is constant on a set in F. When Z is a cardinal κ and $\kappa = \{x \in P(\kappa) \mid x \in \kappa\} \in F$, then this new normality reduces to the previous concept. With an analogous reduction to filters over $\mathcal{P}_\kappa\gamma = [\gamma]^{<\kappa}$, we have the formulation: κ is γ-supercompact *iff* there is a κ-complete, fine, normal ultrafilter over $\mathcal{P}_\kappa\gamma$. Also, Tarski's concept of a cardinal κ being strongly compact can be formulated as being γ-compact for every $\gamma \geq \kappa$, where κ is *γ-compact iff* there is a κ-complete, fine ultrafilter over $\mathcal{P}_\kappa\gamma$. These formulations inspired a substantial combinatorial investigation of filters over sets $\mathcal{P}_\kappa\gamma$, and a general, structural approach to filters over sets $P(Z)$.

Whether it is in these large cardinal hypotheses or the transition from V to $V[G]$ in forcing, the appeal to the satisfaction relation for V is liberal and unabashed in modern set-theoretic practice. Yet ZFC remains parsimoniously the official theory and this carries with it the necessary burden of formalization. On the other hand, it is the formalization that henceforth carries the operative sense; for example, the ultrafilter characterization of γ-supercompactness delivers through the concreteness of the ultrapower construction critical properties that become part of the concept in its use. It has become commonplace in modern set theory that informal assertions and schematic procedures often convey an incipient intentional sense, but formalization refines that sense with workable structural articulations.

3.2 Determinacy

Before proceeding with the historical development of the ZFC theory of large cardinals, we draw back to tuck in an account of a development from a different quarter, a development that at first seemed novel and orthogonal but eventually became woven into the larger fabric of the investigation of large cardinals as strong propositions. The investigation of the determinacy of infinite games is the most distinctive and intriguing development of modern set theory, and the correlations eventually achieved with large cardinals the most remarkable and synthetic.[11] For a set X and $A \subseteq {}^\omega X$, let $G_X(A)$ denote the following "infinite two-person game with perfect information": There are two players, I and II. I initially chooses an $x(0) \in X$; then II chooses an $x(1) \in X$; then I chooses an $x(2) \in X$; then II chooses an $x(3) \in X$; and so forth:

$$\begin{array}{llllll} I\ : & x(0) & & x(2) & & \ldots \\ II: & & x(1) & & x(3) & \ldots \end{array}$$

Each player before making each of his moves is privy to the sequence of previous moves ("perfect information"); and the players together specify an $x \in {}^\omega X$. I wins $G_X(A)$ if $x \in A$ (the "payoff" set), and otherwise II wins. A *strategy* is a function

[11]See Paul Larson's chapter for more details on determinacy.

that tells a player what move to make given the sequence of previous moves. A *winning strategy* is a strategy such that if a player plays according to it he always wins no matter what his opponent plays. A is *determined* if either I or II has a winning strategy in $G_X(A)$.

David Gale and James Stewart [1953] initiated the study of these games and observed that *if* $A \subseteq {}^{\omega}X$ *is open (in the product topology) then* A *is determined.* The simple argument turned on how membership is secured at a finite stage, and a basic stratagem in the further investigations of determinacy would be the reduction to such "open games". Focusing on the basic case $X = \omega$ and noting that a strategy then can itself be construed as a real, Gale and Stewart showed by diagonalizing through all strategies that *assuming* AC *there is an undetermined* $A \subseteq {}^{\omega}\omega$. Determinacy itself would come to be regarded as a regularity property, but there were basic difficulties from the beginning. Gale and Stewart asked whether all Borel sets of reals are determined, and in the decade that followed only sets very low in the Borel hierarchy were shown to be determined.

Infinitely long games involving reals had been considered as early as in the 1920s by mathematicians of the Polish school. With renewed interest in the subject in the 1950s, and with determinacy increasingly seen to be potent in its consequences, Jan Mycielski and Hugo Steinhaus [1962] proposed the following axiom, now known as the *Axiom of Determinacy* (AD):

Every $A \subseteq {}^{\omega}\omega$ is determined.

With AD contradicting AC they proposed from the beginning that in the ZFC context the axiom should hold in some inner model. Solovay pointed out the natural candidate $L(\mathbb{R})$, the constructible closure of the reals $\mathbb{R} = {}^{\omega}\omega$, observing that if AD holds then $\mathrm{AD}^{L(\mathbb{R})}$, i.e. AD holds in $L(\mathbb{R})$. Further restricted hypotheses would soon be applied to the tasks at hand: Projective Determinacy (PD) asserts that every projective $A \subseteq {}^{\omega}\omega$ is determined; $\mathbf{\Sigma}^1_n$-determinacy, that every $\mathbf{\Sigma}^1_n$ set A is determined; and so forth.

By 1964, games to specific purposes had been devised to show that for $A \subseteq {}^{\omega}\omega$ there is a closely related $B \subseteq {}^{\omega}\omega$ (a continuous preimage) so that if B is determined, then A is Lebesgue measurable, and similarly for the Baire property and the perfect set property. Moreover, AD does imply a limited choice principle, that *every countable set consisting of sets of reals has a choice function.* Thus, the groundwork was laid for the reign of AD in $L(\mathbb{R})$ to enforce the regularity properties for all sets of reals there as well as a local choice principle, and unfettered uses of AC relegated to the universe at large.

In 1967 two results drew determinacy to the foreground of set theory, one about the transfinite and the other about definable sets of reals. Solovay established that AD *implies that* ω_1 *is measurable*, injecting emerging large cardinal techniques into a novel setting without AC. David Blackwell [1967] provided a new proof via the determinacy of open games of a classical result of Kuratowski that *the* $\mathbf{\Pi}^1_1$ *sets have the reduction property,* one of several properties investigated in descriptive set theory for the projective sets. These results stimulated interest because of their immediacy and new approach to proof, that of devising a game and appealing to the existence of winning strategies to deduce a dichotomy. Martin in particular saw the potentialities at hand and soon made incisive contributions to investigations with and of determinacy. He initially made a simple but crucial observation based on the construal of strategies as reals that would have myriad applications: *Under* AD *the filter over the Turing degrees generated by the cones is an ultrafilter.*

Martin and Yiannis Moschovakis independently in 1968 extended the reduction property through the projective hierarchy by playing games and assuming PD, realizing a methodological goal of the classical descriptive set theorists by carrying out an inductive propagation. This was Martin's initial application of his ultrafilter on Turing cones, and the idea of ranking ordinal-valued functions via ultrafilters, so crucial in later arguments, first occurred here.

Already in 1964 Moschovakis had abstracted a property stronger and more intrinsic than reduction, the prewellordering property, from the classical analysis of $\mathbf{\Pi}^1_1$ sets. A relation $\preceq$ is a *prewellordering* if it is a well-ordering except possibly that there could be distinct x and y such that $x \preceq y$ and $y \preceq x$. While a well-ordering of a set A corresponds to a bijection of A into an ordinal, a prewellordering corresponds to a surjection onto an ordinal—a stratification of A into well-ordered layers. A class $\mathbf{\Gamma}$ of sets of reals has the *prewellordering property* if for any $A \in \mathbf{\Gamma}$ there is a prewellordering of A such that both it and its complement are in $\mathbf{\Gamma}$ in a strong sense. This property supplanted the reduction property in the Martin-Moschovakis First Periodicity Theorem, which implied that under PD the prewellordering property holds periodically for the projective classes: $\mathbf{\Pi}^1_1$, $\mathbf{\Sigma}^1_2$, $\mathbf{\Pi}^1_3$, $\mathbf{\Sigma}^1_4, \ldots$.

As for Solovay's result, he in fact established that *under* AD *the closed unbounded filter* $\mathcal{C}_{\omega_1}$ *is an ultrafilter* by using a game played with countable ordinals and simulating it with reals. Martin provided an alternate proof using his ultrafilter on Turing cones, and then Solovay in 1968 used Martin's result to establish that *under* AD ω_2 *is measurable.* With an apparent trend set, quite unexpected was the next advance. Martin in 1970 established that under AD the ω_n's for $3 \leq n < \omega$ are all singular with cofinality ω_2! This was a by-product of an incisive analysis of Martin's of $\mathbf{\Sigma}^1_3$ sets under AD.

Martin [1970] also reactivated the earlier project of securing more and more determinacy by establishing that if there is a measurable cardinal, then $\mathbf{\Pi}^1_1$-determinacy holds, or in refined terms, *if* $a^{\#}$ *exists, then* $\Pi^1_1(a)$*-determinacy holds.* The proof featured a remarkably simple reduction to an open game, based on indiscernibles and homogeneity properties, of form $G_X(A)$ for a set X of ordinals. This ground-breaking proof served both to make plausible the possibility of getting PD from large cardinals as well as getting $\mathbf{\Delta}^1_1$-determinacy, Borel Determinacy, in ZFC—both directions to be met with complete success in later years.

The advances in the investigation of definable sets of reals with determinacy would be in terms of their analysis as projections of trees. For purposes of descriptive set theory, T is a *tree on* $\omega \times \kappa$ *iff* (a) T consists of pairs $\langle s, t\rangle$ where s is a finite sequence drawn from ω and t is a finite sequence drawn from κ of the same length, and (b) if $\langle s,t\rangle \in T$, s' is an initial segment of s and t' is a initial segment of t of the same length, then also $\langle s',t'\rangle \in T$. For such T, $[T]$ consists of pairs $\langle f,g\rangle$ corresponding to infinite branches, i.e. f and g are ω-sequences such that for any finite initial segment s of f and finite initial segment t of g of the same length, $\langle s,t\rangle \in T$. In modern terms, $A \subseteq {}^{\omega}\omega$ is *κ-Suslin iff* there is a tree on $\omega \times \kappa$ such that $A = p[T] = \{f \mid \exists g(\langle f,g\rangle \in [T])\}$. $[T]$ is a closed set in the space of $\langle f,g\rangle$'s where $f\colon \omega \to \omega$ and $g\colon \omega \to \kappa$, and so otherwise complicated sets of reals, if shown to be κ-Suslin, are newly comprehended as projections of closed sets. The analytic ($\mathbf{\Sigma}^1_1$) sets are exactly the ω-Suslin sets. Shoenfield [1961] had established that *every $\mathbf{\Sigma}^1_2$ set is ω_1-Suslin.*

To address the classical issue of uniformization, Moschovakis [1971] abstracted from the proof of the classical, Kondô $\mathbf{\Pi}^1_1$ Uniformization Theorem what would become the central structural concept in the investigation of definable sets of reals with determinacy. A *scale* on a set $A \subseteq {}^{\omega}\omega$ is an ω-sequence of ordinal-valued functions on A satisfying convergence and continuity properties, and a class $\mathbf{\Gamma}$ of sets of reals has the *scale property* if for any $A \in \mathbf{\Gamma}$ there is a scale on A whose corresponding graph relations are in $\mathbf{\Gamma}$ in a strong sense. Having a scale on A corresponds to having $A = p[T]$ for a tree T in such a way that, importantly, from A is definable a member of A through a minimization process ("choosing the honest leftmost branch").

In the early 1970s Moschovakis, Martin, and Alexander Kechris proceeded with scales to provide a detailed analysis of the projective sets under PD in terms of Borel sets and as projections of trees, based on the *projective ordinals* $\boldsymbol{\delta}^1_n$ = the supremum of the lengths of the $\mathbf{\Delta}^1_n$ prewellorderings. For example, the $\mathbf{\Sigma}^1_{2n+2}$ sets are exactly the $\boldsymbol{\delta}^1_{2n+1}$-Suslin sets. The projective ordinals themselves were subjected to considerable scrutiny, with penetrating work of Kunen particularly advancing the theory, and were found to be measurable and to satisfy strong partition properties. However, where exactly the $\boldsymbol{\delta}^1_n$ for $n \geq 5$ are in the aleph hierarchy would remain a mystery until the latter 1980s, when Steve Jackson [1988] in a *tour de force* settled the question with a deep analysis of the ultrafilters and partition properties involved. As an otherwise complete structure theory for projective sets was being worked out into the 1970s, Martin in 1974 (cf. [1975]) returned to a bedrock issue for the regularity properties and established in ZFC that $\mathbf{\Delta}^1_1$-determinacy, Borel Determinacy, holds.

3.3 Elaborations

Although large large cardinals were developed particularly to investigate the possibilities for elementary embeddings and were quickly seen to have a simple but elegant basic theory, what really intimated their potentialities were new forcing results in the 1970s and 1980s, especially from supercompactness, that established new relative consistencies, even of assertions low in the cumulative hierarchy. The earliest, orienting result along these lines addressed the Singular Cardinals Problem. The "Prikry-Silver" result provided the first instance of a failure of the

Singular Cardinal Hypothesis by drawing together two results of independent significance, themselves crucial as methodological advances.

Karel Prikry in his 1968 Berkeley thesis (cf. his [1970]) had set out a simple but elegant notion of forcing that changed the cofinality of a measurable cardinal without collapsing any cardinals. With U a normal κ-complete ultrafilter over κ, *Prikry forcing* for U has as conditions $\langle p, A\rangle$ where p is a finite subset of κ and $A \in U$. For conditions $\langle p, A\rangle$ and $\langle q, B\rangle$, the first is stronger than the second if $p \supseteq q$ and $\alpha \in p - q$ implies $\alpha > \max(q)$ and $A \cup (p - q) \subseteq B$. A condition thus specifies a finite initial part of a new ω-cofinalizing subset of κ, and further members are to be added on top from a set large in the sense of being in U. Applying a partition relation available for normal ultrafilters, Prikry established a focal property, that for any condition $\langle p, A\rangle$ and forcing statement, there is a $B \subseteq A$ such that $B \in U$ and $\langle p, B\rangle$ decides the statement, i.e. extending p is unnecessary. Hence, e.g. the κ-completeness of U implies that V_κ remains unchanged in the forcing extension yet the cofinality of κ now becomes ω.

Prikry forcing may at first have seemed a curious possibility for singularization. However, that a Prikry generic sequence also generates the corresponding U in simple fashion and also results from indiscernibles made them a central feature of measurability. Prikry forcing would be generalized in various directions and for a variety of purposes. With the capabilities made available for changing cofinalities, equi-consistency connections would eventually be established between large cardinals on the one hand and formulations in connection with the Singular Cardinals Problem on the other.

Silver in 1971 first established the relative consistency of having a measurable cardinal κ satisfying $2^\kappa > \kappa^+$, a proposition that Kunen [1971b] had shown to be substantially stronger than just having measurability. Forcing over the model constructed by Silver with Prikry forcing yielded the first counterexample to the Singular Cardinals Hypothesis by providing a singular strong limit cardinal κ satisfying $2^\kappa > \kappa^+$.

To establish his result, Silver provided a technique for extending elementary embeddings into generic extensions and thereby preserving large cardinal properties. To get at what is at issue, suppose that $j\colon V \to M$ is an elementary embedding, P is a notion of forcing, and G is V-generic for P. To extend (or "lift") j to an elementary embedding for $V[G]$, the natural scheme would be to get an M-generic G' for $j(P)$ and extend j to an elementary embedding from $V[G]$ into $M[G']$. But for this to work with the forcing terms, it would be necessary to enforce

$$(*) \qquad \forall p \in G\,(j(p) \in G')\,.$$

For getting a measurable cardinal κ satisfying $2^\kappa = \kappa^{++}$, Silver started with an elementary embedding as above with critical point κ and devised a P for adjoining κ^{++} Cohen subsets of κ. In order to establish a close connection between P and $j(P)$ toward securing $(*)$, he took P to be a uniform iteration of forcings to adjoin λ^{++} Cohen subsets of λ for every inaccessible cardinal λ up to and including κ. Then with the shift from κ to $j(\kappa)$, $j(P)$ can be considered a two-stage iteration of P followed by a further iteration Q. Now with G V-generic for P, G is also M-generic for P, and in $M[G]$ one should devise an H $M[G]$-generic for Q such that the combined generic $G' = G * H$ satisfies $(*)$.

But how is this to be arranged? Silver was able to control the $j(p)$'s for $p \in G$ by a single, *(strong) master condition* $q \in Q$, and build in $V[G]$ an H $M[G]$-generic over Q with $q \in H$ to satisfy $(*)$. For getting both q and H, he needed that M be closed under arbitrary κ^{++}-sequences. Thus he established: *If κ is κ^{++}-supercompact, then there is a forcing extension in which κ is measurable and $2^\kappa = \kappa^{++}$.*[12] Silver's preparatory "reversed Easton" forcing with Easton support and master condition constructions of generic filters would become staple ingredients for the generic extension of elementary embeddings.

What about the use of very strong hypotheses in consistency results? A signal, 1972 result of Kunen (cf. [1978]) brought into play the strongest hypothesis to date for establishing a consistency result about the low levels of the cumulative hierarchy. Kunen [1970] had established that having a κ-complete κ^+-saturated ideal over a successor cardinal κ has consistency strength substantially stronger than having a measurable cardinal. Kunen now showed: *If κ is huge, then there is forcing extension in which $\kappa = \omega_1$ and there is an $\aleph_1$-complete $\aleph_2$-saturated ideal over ω_1.* With a $j \colon V \to M$ with critical point κ, $\lambda = j(\kappa)$, and ${}^\lambda M \subseteq M$ as given by the hugeness of κ, Kunen collapsed κ to ω_1 and followed it was a collapse of λ to ω_2 in such a way so as to be able to define a saturated ideal. Crucially, the first collapse was a "universal" collapse P iteratively constructed so that the second collapse can be absorbed into $j(P)$ in a way consistent with j applied to P, and this required ${}^\lambda M \subseteq M$. Hence, a sufficient algebraic argument was contingent on a closure property for an elementary embedding, one plucked from the emerging large cardinal hierarchy. In the years to come, Kunen's argument would be elaborated and emended to become the main technique for getting various sorts of saturated ideals over accessible cardinals. As for the proposition that there is an $\aleph_1$-complete $\aleph_2$-saturated ideal over ω_1 itself, Kunen's result set an initial high bar for the stalking of its consistency strength, but definitive work of the 1980s would show that far less than hugeness suffices.

William Mitchell would become a major developer of inner models for large cardinals, the first after Kunen. His Berkeley thesis (cf. [1972]), however, made important strides in a different direction. A κ-tree is *special* if it is the union of fewer than κ antichains. Aronszajn's original construction of an Aronszajn tree actually provided a special tree. Analogously, a κ-Aronszajn tree is a κ-tree with no cofinal branches, i.e. a counterexample to the tree property for κ, and a generalization of Aronszajn's construction shows that if κ is regular and $2^\alpha \leq \kappa$ for every $\alpha < \kappa$, then there is a special κ^+-Aronszajn tree. Mitchell showed: *If there is a Mahlo cardinal, then there is a forcing extension making it ω_2 in which there are no special ω_2-Aronszajn trees.* The forcing added Cohen reals to

[12] To mention an elegant coda, work of Woodin and Gitik in the 1980s showed that having a measurable cardinal satisfying $2^\kappa > \kappa^+$ is equi-consistent with the proposition $\exists\kappa(o(\kappa) = \kappa^{++})$, formulated below in the text.

make the eventual $2^{\aleph_0}$ necessarily ω_2, and had sophisticated features that could count as "countable support iteration" and "termspace forcing", important in later iterated forcing. Mitchell also observed that if there are no special ω_2-Aronszajn trees, then ω_2 is Mahlo in L. Hence we have the equi-consistency: *Having no special ω_2-Aronszajn trees is equi-consistent to having a Mahlo cardinal.* Recall that a characterization of a cardinal κ being weakly compact is being inaccessible and having the tree property (i.e. having no κ-Aronszajn trees at all). Mitchell's advisor Silver applied Mitchell's forcing argument and made a corresponding inner model observation to establish an equi-consistency result for the tree property at an accessible cardinal: *ω_2 having the tree property is equi-consistent to having a weakly compact cardinal.*

After a brief excursion into category theory, Mitchell [1974] soon provided the first substantive extension of Kunen's inner model results and brought to prominence a new, delimitative, large cardinal hypothesis. For normal κ-complete ultrafilters U and U' over κ, define the *Mitchell order* $U' \lhd U$ *iff* $U' \in \mathrm{Ult}(V, U)$, i.e. there is an $f\colon \kappa \to V$ representing U' in the ultrapower, so that $\{\alpha < \kappa \mid f(\alpha)$ is a normal α-complete ultrafilter over $\alpha\} \in U$ and κ is already a limit or measurable cardinals. $U \lhd U$ always fails, and generally, $\lhd$ is a well-founded relation by a version of Scott's argument that measurable cardinals contradict $V = L$. Consequently, to each U can be recursively assigned a rank $o(U) = \sup\{o(U') + 1 \mid U' \lhd U\}$, and to a cardinal κ, the supremum $o(\kappa) = \sup\{o(U) + 1 \mid U$ is a normal κ-complete ultrafilter over $\kappa\}$. By a cardinality argument, if $2^\kappa = \kappa^+$ then $o(\kappa) \leq \kappa^{++}$.

The hypothesis $o(\kappa) = \delta$ provided an "order" of measurability calibrated by δ, with larger δ corresponding to stronger assumptions on κ. For the investigation of these orders, Mitchell devised the concept of a *coherent* sequence of ultrafilters ("measures") and was able to establish canonicity results for inner models $L[\mathcal{U}] \models$ "$\mathcal{U}$ is a coherent sequence of ultrafilters". A coherent sequence $\mathcal{U}$ is a doubly indexed system of normal α-complete ultrafilters $\mathcal{U}(\alpha, \beta)$ over α such that $\mathcal{U}(\kappa, \beta) \lhd \mathcal{U}(\kappa, \beta')$ for $\beta < \beta'$ at the κth level, and the earlier levels contain just enough ultrafilters necessary to represent these $\lhd$ relationships in the respective ultrapowers. (Technically, if $j\colon V \to \mathrm{Ult}(V, \mathcal{U}(\kappa, \beta'))$, then $j(\mathcal{U}){\restriction}\{(\alpha, \beta) \mid \alpha \leq \kappa\} = \mathcal{U}{\restriction}\{(\alpha, \beta) \mid \alpha < \kappa \vee (\alpha = \kappa \wedge \beta < \beta')\}$, i.e. $j(\mathcal{U})$ through κ is exactly $\mathcal{U}$ "below" (κ, β').)

Mitchell first affirmed that these $L[\mathcal{U}]$'s are *iterable* in that arbitrary iterated ultrapowers via ultrafilters in $\mathcal{U}$ and its successive images are always well-founded. He then effected a comparison: *Any $L[\mathcal{U}_1]$ and $L[\mathcal{U}_2]$ have respective iterated ultrapowers $L[\mathcal{W}_1]$ and $L[\mathcal{W}_2]$ such that $\mathcal{W}_1$ is an initial segment of $\mathcal{W}_2$ or vice versa.* This he achieved through a process of *coiteration* of least differences: At each stage, one finds the lexicographically least coordinate at which the current iterated ultrapowers of $L[\mathcal{U}_1]$ and $L[\mathcal{U}_2]$ differ and takes the respective ultrapowers by the differing ultrafilters; the difference is eliminated as ultrafilters never occur in their ultrapowers. With this coiteration, Mitchell established that in $L[\mathcal{U}]$ the only normal α-complete ultrafilters over α for any α are those that occur in $\mathcal{U}$ and other propositions like GCH that showed these models to be L-like. Coiteration would henceforth be embedded in inner model theory, and with his models $L[\mathcal{U}]$ modeling $o(\kappa) = \delta$ for $\delta < \kappa^{++L[\mathcal{U}]}$, $\exists\kappa(o(\kappa) = \kappa^{++})$ would become the delimitative proposition of his analysis.

Much of the early formative work on strong large cardinal hypotheses and their integration into modern set theory through consistency results was carried out by Menachem Magidor, whose subsequent, broad-ranging initiatives have considerably advanced the entire subject. After completing his Hebrew University thesis in 1972 on supercompact cardinals, Magidor in the 1970s established a series of penetrating forcing consistency results involving strong hypotheses. In 1972-3 (cf. [1976]) he illuminated the "identity crisis" issue of whether supercompactness and strong compactness are distinct concepts by establishing: (1) *It is consistent that the least supercompact cardinal is also the least strong compact cardinal*, and (2) *It is consistent that the least strong compact cardinal is the least measurable cardinal (and so, much smaller than the least supercompact cardinal).* The proofs showed how changing many cofinalities with Prikry forcing to destroy measurable cardinals can be integrated into arguments about extending elementary embeddings.

In 1973 Magidor (cf. [1977a]) showed how with strong hypotheses the least limit cardinal $\aleph_\omega$ can violate the Singular Cardinals Hypothesis: *If κ is supercompact, there is a forcing extension in which κ is $\aleph_\omega$ as a strong limit cardinal yet $2^{\aleph_\omega} > \aleph_{\omega+1}$.* The strong elementary embedding hypothesis allowed for an elaborated Prikry forcing interspersed with Levy collapses. With this first net thrown out, it would be one of the great elaborations of method that equi-consistency results would eventually be achieved with weaker hypotheses for the violation of SCH, and already at $\aleph_\omega$.

3.4 Silver's theorem and covering

In mid-1974 Silver [1975] established that *if κ is a singular cardinal with* $\mathrm{cf}(\kappa) > \omega$ *and $2^\lambda = \lambda^+$ for $\lambda < \kappa$, then $2^\kappa = \kappa^+$.* This was a dramatic event and would stimulate dramatic developments. There had been precious little in the way of results provable in ZFC about cardinal arithmetic, and in the early ruminations about the Singular Cardinals Problem it was quite unforeseen that the power of a singular cardinal can be so constrained. An analogous preservation result had been observed by Scott for measurable cardinals, and telling was that Silver used large-cardinal ideas connected with generic ultrapowers.

Silver's result spurred broad-ranging investigations both into the combinatorics and avenue of proof and into larger, structural implications. The basis of his argument was a ranking of ordinal-valued functions on $\mathrm{cf}(\kappa)$. Let $\langle\gamma_\alpha \mid \alpha < \mathrm{cf}(\kappa)\rangle$ be a sequence of ordinals unbounded in κ and for $\alpha < \mathrm{cf}(\kappa)$ let $\tau_\alpha\colon P(\gamma_\alpha) \to 2^{\gamma_\alpha}$ be a bijection. For $X \subseteq \kappa$ let f_X on $\mathrm{cf}(\kappa)$ be defined by: $f_X(\alpha) = \tau_\alpha(X \cap \gamma_\alpha)$, noting that $X_1 \neq X_2$ implies f_{X_1} and f_{X_2} differ for sufficiently large α. Then 2^κ is mirrored through these eventually different functions, which one can work to order according to an ideal over the uncountable $\mathrm{cf}(\kappa)$. The combinatorial possibilities of such rankings led to a series of limitative results on the powers of singular cardinals of uncountable cofinality, starting with the results of Fred Galvin and Hajnal [1975], of which the paradigmatic example is that if $\aleph_{\omega_1}$ is a strong limit cardinal, then $2^{\aleph_{\omega_1}} < \aleph_{(2^{\aleph_1})^+}$.

In the wake of Silver's proof, Thomas Jech and Prikry [1976] defined a κ-complete ideal over κ to be *precipitous iff* the corresponding generic ultrapower à la Solovay is well-founded. They thus put the focus on a structural property of saturated ideals that Silver had simulated to such good effect. Jech and Prikry pointed out that a proof of Kunen's for saturated ideals using iterated ultrapowers can be tailored to show: *If there is a precipitous ideal over κ, then κ is measurable in an inner model.* Then Mitchell observed (cf. [Jech *et al.*, 1980]): *If a measurable cardinal is Levy collapsed to ω_1, then a normal ultrafilter generates a precipitous ideal over ω_1 in the extension.* Hence, a first equi-consistency result was achieved for measurability and ω_1 in ZFC. With combinatorial characterizations of precipitousness soon in place, well-foundedness as thus modulated by forcing became a basic ingredient in a large-scale investigation of strong properties tailored to ideals and generic elementary embeddings.

The most dramatic and penetrating development from Silver's Theorem was Jensen's work on "covering" for L and its first extensions, the most prominent advances of the 1970s in set theory. Jensen had found Silver's result a "shocking discovery", and was stimulated to intense activity. By the end of 1974 he had made prodigious progress, solving the Singular Cardinals Problem in the absence of $0^\#$ in three manuscripts, "Marginalia to a Theorem of Silver" and its two sequels. The culminating result featured an elegant and focal formulation of intuitive immediacy, the Covering Theorem (or "Lemma") for L: *If $0^\#$ does not exist, then for any uncountable set X of ordinals there is a $Y \in L$ with $|Y| = |X|$ such that $Y \supseteq X$.* (The first published account was given in [Devlin and Jensen, 1975]; without the "uncountable" there would be a counterexample using "Namba forcing".) This covering property expresses a global affinity between V and L, and its contrapositive provides a surprisingly simple condition sufficient for the existence of $0^\#$ and the ensuing indiscernible generation of L. As such, Jensen's theorem would find wide applications for implicating $0^\#$ and would provide a new initiative in inner model theory for encompassing stronger hypotheses.

The Covering Theorem gave the essence of Jensen's argument that in the absence of $0^\#$ the Singular Cardinals Hypotheses holds: Suppose that κ is singular and for reckoning with the powers of smaller cardinals consider $\lambda = \sup\{2^\mu \mid \mu < \kappa\}$. If there is a $\nu < \kappa$ such that $\lambda = 2^\nu$, then the functions f_X defined as above adapted to the present situation satisfy $f_X\colon \mathrm{cf}(\kappa) \to 2^\nu$, and so $\lambda \leq 2^\kappa \leq (2^\nu)^{\mathrm{cf}(\kappa)} \leq \lambda$. If on the other hand λ is the strict supremum of increasing 2^μ's, then $\mathrm{cf}(\lambda) = \mathrm{cf}(\kappa)$ and so the Zermelo-Kőnig inequality would dictate the least possibility for 2^κ to be λ^+. However, if for any $X \subseteq \kappa$ the range of f_X is covered by a $Y \subseteq \lambda$ with $Y \in L$ of cardinality $\mathrm{cf}(\kappa) \cdot \aleph_1$, then: there are $2^{\mathrm{cf}(\kappa)\cdot\aleph_1}$ subsets of each such Y and by the GCH in L, at most $|\lambda^{+L}|$ such Y. Hence, we would have $2^\kappa \le 2^{\mathrm{cf}(\kappa)\cdot\aleph_1} \cdot |\lambda^{+L}| \le \lambda^+$.

The Covering Theorem also provided another dividend that would grow in separate significance as having the *weak covering property*: *Assume that* $0^\#$ *does not exist. If* κ *is singular, then* $\kappa^{+L} = \kappa^+$. If to the contrary $\kappa^{+L} < \kappa^+$, then $\mathrm{cf}(\kappa^{+L}) < \kappa$. Let $X \subseteq \kappa^{+L}$ be unbounded so that $|X| < \kappa$ and let $Y \in L$ cover X with $|Y| = |X| \cdot \aleph_1$. But then, the order-type of Y would be less than κ, contradicting the regularity of κ^{+L} in L.

A crucial consequence of weak covering is that *in the absence of* $0^\#$, $\square_\kappa$ *holds for singular* κ, since a $\square_\kappa$ sequence in the sense of L is then a $\square_\kappa$ sequence in V. The weak covering property would itself become pivotal in the study of inner models corresponding to stronger and stronger hypotheses, and the failure of $\square_\kappa$ for *singular* κ would become a delimitative proposition. Solovay (cf. [Solovay *et al.*, 1978]) had already established an upper bound on consistency by showing in the early 1970s that *if* κ *is* λ^+*-supercompact and* $\lambda \ge \kappa$, *then* $\square_\lambda$ *fails.*

Jensen's ingenious proof of the Covering Theorem for L proceeded by taking a counterexample X to covering with $\tau = \sup(X)$ and $|X|$ minimal; getting a certain Σ_1-elementary $j\colon L_\gamma \to L_\tau$ which contains X in its range through a Skolem hull construction so that $|\gamma| = |X|$ and, as X cannot be covered, γ is a cardinal in L; and extending j to an elementary embedding from L into L, so that $0^\#$ exists. The procedure for extending j up to some large L_δ was to consider a directed system of embeddings of structures generated by $\xi \cup p$ for some $\xi < \gamma$ and p a finite subset of L_δ, the transitized components of the system all being members of L_γ as γ is a cardinal in L, and to consider the corresponding directed system consisting of the j images. The choice of γ insured that the new directed system is also well-founded, and so isomorphic to some L_ζ.

How can the proof of the Covering Theorem be adapted to establish a stronger result? The only possibility was to consider a larger inner model M and to establish that M has the *covering property*: for any uncountable set X of ordinals there is a $Y \in M$ with $|Y| = |X|$ such that $Y \supseteq X$. In groundbreaking work for inner model theory, Solovay in the early 1970s had developed a fine structure theory for inner models of measurability. Whilst a research student at Oxford University Anthony Dodd worked through this theory, and in early 1976 he and Jensen laid out the main ideas for extending the Covering Theorem to a new inner model, now known as the *Dodd-Jensen core model*, denoted K^{DJ}. (See their [1981] and the full exposition [Dodd, 1982].)

If $\langle L[U], \in, U\rangle$ is an inner model of measurability, say the κ-model, then there is a generic extension in which covering fails: If G is Prikry generic for U over $L[U]$, then G cannot be covered by any set in $L[U]$ of cardinality less than κ. Drawing back, there remains the possibility of "iterating out" the measurable cardinal: If $\langle L[U], \in, U\rangle$ is the κ-model, then $\langle L[W], \in, W\rangle$ is the λ-model for some $\lambda > \kappa$ exactly when it is an iterate of $\langle L[U], \in, U\rangle$, in which case $L[W] \subseteq L[U]$, $V_\kappa \cap L[U] = V_\kappa \cap L[W]$, and $U \notin L[W]$. Thus, if $\langle L[U_\alpha] \mid \alpha \in \mathrm{On}\rangle$ enumerates the inner models of measurability, then starting with any one of them, the process of iterating it through the ordinals converges to a proper class $\bigcap_\alpha L[U_\alpha]$ which has no inner models of measurability, with the stabilizing feature that for any γ, $V_\gamma \cap \bigcap_\alpha L[U_\alpha] = V_\gamma \cap L[U_\beta]$ for sufficiently large β. Assuming that there are inner models of measurability, K^{DJ} is in fact characterizable as this residue class. Aspiring to this, but without making any such assumption, Dodd and Jensen provided a formulation of K^{DJ} in ZFC.

K^{DJ} was the first inner model of ZFC since Gödel's L developed using distinctly new generating principles. Dodd and Jensen's approach was to take K^{DJ} as the union of L together with "mice". Loosely speaking, a *mouse* is a set $L_\alpha[U]$ such that

$$\langle L_\alpha[U], \in, U\rangle \models U \text{ is a normal ultrafilter over } \kappa$$

satisfying: (i) there is a subset of κ in $L_{\alpha+1}[U] - L_\alpha[U]$, so that U is on the verge of not being an ultrafilter; (ii) $\langle L_\alpha[U], \in, U\rangle$ is iterable in that all the iterated ultrapowers are well-founded, and (iii) fine structure conditions about a projectum below κ leading to (i). Mice can be compared by taking iterated ultrapowers, so that there is a natural prewellordering of mice, and moreover, crucial elements about L can be lifted to the new situation because there is a generalization of condensation: Σ_1-elementary substructures of mice, when transitized, are again mice. This led to $K^{\mathrm{DJ}} \models$ GCH, and that K^{DJ} in the sense of K^{DJ} is again K^{DJ}.

Mice generate indiscernibles through iteration, and so if $0^{\#}$ does not exist, then $K^{\mathrm{DJ}} = L$; if $0^{\#}$ exists but $0^{\#\#}$ does not, then $K^{\mathrm{DJ}} = L[0^{\#}]$; and this continues through the transfinite by coding sequences of sharps. On the other hand, K^{DJ} has no simple constructive analysis from below and is rather like a maximal inner model on the brink of measurability: Its own "sharp", that there is an elementary embedding $j\colon K^{\mathrm{DJ}} \to K^{\mathrm{DJ}}$, is equivalent to the existence of an inner model of measurability. Indeed, this was Dodd and Jensen's primary motivation for the formulation of K^{DJ}. They used it in place of the elementary embedding characterization of the existence of $0^{\#}$, together with the L-like properties of K^{DJ}, to establish (cf. [1982a]) the *Covering Theorem for* K^{DJ}: *If there is no inner model of measurability, then* K^{DJ} *has the covering property.* This has the attendant consequences for the Singular Cardinals Problem. Moreover, Dodd and Jensen were able to establish a covering result for inner models of measurability that accommodates Prikry forcing. Solovay had devised a set of integers $0^{\dagger}$ ("zero dagger"), analogous to $0^{\#}$, such that $0^{\dagger}$ exists exactly when for some κ-model $L[U]$ there is an elementary embedding $j\colon L[U] \to L[U]$ with critical point above κ. Dodd and Jensen (cf. [1982b]) established: *If* $0^{\dagger}$ *does not exist yet there is an inner model of measurability, then for the* κ*-model* $L[U]$ *with* κ *least, either (a)* $L[U]$ *has the covering property, or (b) there is a Prikry generic* G *for* U *over* $L[U]$ *such that* $L[U][G]$ *has the covering property.* Prikry forcing provides the only counterexample to covering! Hence, the inner models thus far considered were also "core models", models on the brink so that the lack of covering leads to the next large cardinal hypothesis.

In the light of the Dodd-Jensen work, Mitchell in the later 1970s (cf. [1984; 1985]) developed the core model $K[\mathcal{U}]$ for coherent sequences $\mathcal{U}$ of ultrafilters, which corresponds to his $L[\mathcal{U}]$ as K^{DJ} does to $L[U]$. The mice are now sets of form $J_\alpha[W]$ with iterability and fine structure properties, where W is an ultrafilter sequence with $\mathcal{U}$ as an initial segment. Under the assumption that there is no inner model satisfying $\exists\kappa(o(\kappa) = \kappa^{++})$, Mitchell established the weak covering property for $K[\mathcal{U}]$, i.e. that $(\kappa^+)^{K[\mathcal{U}]} = \kappa^+$ for singular κ. With this he showed that several propositions have at least the consistency strength of $\exists\kappa(o(\kappa) = \kappa^{++})$. One such proposition was that there is an $\aleph_1$-complete $\aleph_2$-saturated ideal over ω_1, establishing a new lower bound in consistency strength for Kunen's consistency result from a huge cardinal. Mitchell [1985] established, generalizing the Dodd-Jensen result with Prikry generic sets, a full covering theorem for $K[\mathcal{U}]$ cast in terms of coherent systems of indiscernibles and drew further conclusions about singular cardinals.

3.5 Forcing consistency results through the 1970s

Through the 1970s a wide range of variegated forcing consistency results were established at a new level of sophistication that clarified relationships among combinatorial propositions and principles and often drew in large cardinal hypotheses and stimulated the development of method, especially in iterated forcing. A conspicuous series of results resolved questions of larger mathematics (Whitehead's Problem, Borel's Conjecture, Kaplansky's Conjecture, the Normal Moore Space Problem) in terms of relative consistency and set-theoretic principles, newly affirming the efficacy and adjudicatory character of set theory. In what follows, we mention these results in turn as we continue to pursue the larger longitudinal themes, necessarily saying less and less about matters of increasing complexity.

In 1974 Magidor [1978] made a basic contribution to the theory of changing cofinalities, the first after Prikry. Magidor established: *If a measurable cardinal κ is of Mitchell order $o(\kappa) \geq \lambda$ for a regular $\lambda < \kappa$, then there is a forcing extension preserving cardinals in which* $\mathrm{cf}(\kappa) = \lambda$. Generalizing Prikry forcing, Magidor's conditions consisted of a finite sequence of ordinals and a sequence of sets drawn from normal ultrafilters in the Mitchell order, the sets providing for the possible ways of filling out the sequence. Like Prikry's forcing, Magidor's may at first have seemed a curious possibility for a new singularization. However, one of the discernments of Mitchell's core model for coherent sequences of measures is that, remarkably [1987]: *If a regular cardinal κ in V satisfies $\omega < \mathrm{cf}(\kappa) < \kappa$ in a generic extension, then V has an inner model in which $o(\kappa)$ is at least that cofinality.* Thus, the capability of changing cofinalities was exactly gauged; "Prikry-Magidor" generic sets as sequences of indiscernibles would become a basic component of Mitchell's covering work.

The most salient result of Magidor's of this period were one that provided counterweight to Silver's result for the least limit cardinal $\aleph_\omega$. Magidor [1977b] showed: *If κ is a huge cardinal, then there is a forcing extension in which $\kappa = \aleph_\omega$, $2^{\aleph_n} = \aleph_{n+1}$ for $n \in \omega$, yet $2^{\aleph_\omega} > \aleph_{\omega+1}$.* This was the first time after Kunen's saturated ideal result (cf. [1978]) that the strength of a huge cardinal was invoked; with it, the forcing conditions from Magidor's earlier SCH argument [1977a] could be made more liberal in order to ensure that GCH held below $\aleph_\omega$.

With respect to the Jech-Prikry-Mitchell equi-consistency of measurability and precipitousness, Magidor (cf. [Jech *et al.*, 1980]) showed that absorptive properties of the Levy collapse of a measurable cardinal to ω_1 can be exploited by subsequently "shooting" closed unbounded subsets of ω_1 through stationary sets to get: *If there is a measurable cardinal κ, then there is a forcing extension in which $\kappa = \omega_1$ and NS_{ω_1} is precipitous.* Thus a basic, definable ideal can be precipitous, and this naturally became a principal point of departure for the investigation of ideals.

The move of Saharon Shelah into set theory in the early 1970s brought in a new and exciting sense of personal initiative that swelled into an enhanced purposiveness across the subject, both through his solutions of major outstanding problems as well as through his development of new structural frameworks. A phenomenal mathematician, Shelah from his 1969 Hebrew University thesis on has worked in model theory and eventually infused it with a transformative, abstract classification theory for models. In both model theory and set theory he has remained eminent and has produced results at a furious pace, with nearly 1000 items currently in his bibliography (his papers are currently archived at http://shelah.logic.at/).

In set theory Shelah was initially stimulated by specific problems. He typically makes a direct, frontal attack, bringing to bear extraordinary powers of concentration, a remarkable ability for sustained effort, an enormous arsenal of accumulated techniques, and a fine, quick memory. When he is successful on the larger problems, it is often as if a resilient, broad-based edifice has been erected, the traditional serial constraints loosened in favor of a wide, fluid flow of ideas and the final result almost incidental to the larger structure. *What* has been achieved is more than a just succinctly stated theorem but rather the erection of a whole network of robust arguments.

Shelah's written accounts have acquired a certain notoriety that in large part has to do with his insistence that his edifices be regarded as autonomous conceptual constructions. Their life is to be captured in the most general forms, and this entails the introduction of many parameters. Often, the network of arguments is articulated by complicated combinatorial principles and transient hypotheses, and the forward directions of the flow are rendered as elaborate transfinite inductions carrying along many side conditions. The ostensible goal of the construction, that succinctly stated result that is to encapsulate it, is often lost in a swirl of conclusions.

Shelah's first and very conspicuous advance in set theory was his 1973, definitive results on Whitehead's Problem in abelian group theory: Is every Whitehead group, i.e. an abelian group G satisfying $\mathrm{Ext}^1(G, \mathbb{Z}) = 0$, free? Shelah [1974], established that $V = L$ implies that this is so, and that Martin's Axiom implies that there is a counterexample. Shelah thus established for the first time that a strong purely algebraic statement is undecidable in ZFC. With his L result specifically based on diamond-type principles, Shelah brought them into prominence with his further work on them, which were his first incursions into iterated forcing. By the late 1970s his increasing understanding of and work in iterated forcing would put a firm spine on much of the variegated forcing arguments about the continuum.

With an innovative argument pivotal for iterated forcing, Richard Laver [1976] established the consistency of Borel's Conjecture: *Every set of reals of strong measure zero is countable.* CH had provided a counterexample, and Laver established the consistency with $2^{\aleph_0} = \aleph_2$. His argument featured the adjunction of what are now called *Laver reals* in the first clearly set out *countable support iteration*, i.e. an iteration with non-trivial local conditions allowed only at countably many coordinates. The earlier Solovay-Tennenbaum argument for the consistency of MA had relied on finite support, and the Mitchell argument about Aronszajn trees, an involved countable support with a "termspace" forcing. Laver's work showed that countable support iteration is both manageable and efficacious for preserving certain framing properties of the continuum to establish the consistency of propositions with $2^{\aleph_0} = \aleph_2$. Interestingly, a trade-off would develop however: while finite support iterations put all cardinals $\geq \aleph_2$ on an equal footing with respect to the continuum, countable support iterations restricted the continuum to be at most $\aleph_2$. With a range of new generic reals coming into play with the widening investigation of the continuum, James Baumgartner [1983] formulated a property common to the corresponding partial orders, *Axiom A*, which in particular ensured the preservation of ω_1. He showed that the countable support iteration of Axiom A forcings is Axiom A, thereby uniformizing the iterative adjunction of the known generic reals.

All this would retrospectively have a precursory air, as Shelah soon established a general, subsuming framework. Analyzing Jensen's consistency argument for SH + CH and coming to grips with forcing names in iterated forcing, Shelah came to the concept of *proper forcing* as a general property that preserves ω_1 and is preserved in countable support iterations. The instrumental formulation of properness is given in an appropriately broad setting:

First, for a regular cardinal λ, let $H(\lambda) = \{x \mid |\text{tc}(\{x\})| < \lambda\}$, the sets hereditarily of cardinality less than λ. The $H(\lambda)$'s provide another cumulative hierarchy for V, one stratified into layers that each satisfy Replacement. Whereas the V_α's for limit α satisfy every ZFC axiom except possibly Replacement, the $H(\lambda)$'s satisfy every ZFC axiom except possibly Power Set. A partial order $\langle P, < \rangle$ is *proper* if for any regular $\lambda > 2^{|P|}$ and countable $M \prec H(\lambda)$ with $P \in M$, every $p \in P \cap M$ has a $q \leq p$ such that $q \Vdash \dot{G} \cap M$ is M-generic. (Here, $\dot{G}$ a canonical name for a generic filter with respect to P, and q forcing this genericity assertion has various combinatorial equivalents.)

A general articulation of how all countable approximations are to have generic filters has been achieved, and its presentation under countable support iterations exhibited the efficacy of this remarkable move to a new plateau. Shelah soon devised variants and augmentations, and in a timely monograph *Proper Forcing* [1982] revamped forcing for combinatorics and the continuum with systemic proofs of new and old results. As later presented in Shelah's mature book [1998] proper forcing has become a staple part of the methods of modern set theory, with its applications wide-ranging and the development of its extended theory a fount of research.

In light of Shelah's work and Martin's Axiom, Baumgartner in the early 1980s established the consistency of a new encompassing forcing axiom, the *Proper Forcing Axiom* (PFA): *For any proper partial order P and collection $\mathcal{D}$ of $\aleph_1$ dense subsets of P, there is a filter $G \subseteq P$ meeting every member of $\mathcal{D}$.* Unlike MA, the consistency of PFA required large cardinal strength and moreover could not be achieved by iteratively taking care of the partial orders at issue, as new proper partial orders occur arbitrarily high in the cumulative hierarchy. Baumgartner established: *If there is a supercompact cardinal κ, then there is a forcing extension in which $\kappa = \omega_2$ and* PFA *holds.* In an early appeal to the full global reflection properties available at a supercompact cardinal Baumgartner iteratively took care of the emerging proper partial orders along a special diamond-like sequence that anticipates all possibilities. Laver [1978] had formulated such a sequence, the "Laver diamond", toward establishing what has become a useful result for forcing theory; in a forcing extension he made a supercompact cardinal "indestructible" by any further forcing from a substantial, useful class of forcings. PFA became a widely applied forcing axiom, showcasing Shelah's concept, but beyond that, it would itself become a pivotal hypothesis in the large cardinal context.

4 NEW EXPANSION

4.1 Into the 1980s

The 1980s featured a new and elaborating expansion in set theory significantly beyond the successes, already remarkable, of the previous decade. There were new methods and results of course, but more than that there were successful *maximizations* in several directions—definitive and evidently optimal results—and successful *articulations* at the interstices—new concepts and refinements that filled out the earlier explorations. Our narrative now becomes even more episodic in increasingly inverse relation to the broad-ranging and penetrating developments.

In 1977 Lon Radin toward his Berkeley thesis developed (cf. [1982]) an ultimate generalization of the Prikry and Magidor forcings for changing cofinalities, a generalization which could in fact adjoin a closed unbounded subset, consisting of formerly regular cardinals, to a large cardinal κ while maintaining its regularity and further substantive properties. As graduate students at Berkeley, Hugh Woodin and Matthew Foreman saw the possibilities abounding in Radin forcing. While an undergraduate at Caltech, Woodin did groundbreaking work (cf. [Dales and Woodin, 1987]), built on by Solovay, on the consistency of Kaplansky's Conjecture (Is every homomorphism on the Banach algebra of continuous functions on the unit interval continuous?) and now with Radin forcing in hand Woodin would produce his first series of remarkable results. By 1979 Foreman and Woodin (cf. their [1991]) had the essentials for establishing: *If there is a supercompact cardinal κ, then there is forcing extension in which V_κ as a model of* ZFC *satisfies* *that* GCH *fails everywhere, i.e.* $2^\lambda > \lambda$ *for every* λ. This conspicuously subsumed the early Magidor result getting $\aleph_\omega$ a strong limit yet $2^{\aleph_\omega} > \aleph_{\omega+1}$ and put Radin forcing on the map for establishing global consistency results.

Shelah soon established two re-orienting results about powers of singular cardinals. Having come somewhat late into the game after Silver's Theorem, Shelah had nonetheless extended some of the limitative results about such powers, even to singular κ such that $\aleph_\kappa = \kappa$. Shelah [1983] subsequently established: *If there is a supercompact cardinal* κ *and* α *is a countable ordinal, then there is a forcing extension in which* κ *is* $\aleph_\omega$ *as a strong limit cardinal yet* $2^{\aleph_\omega} = \aleph_{\alpha+1}$. He thus extended Magidor's result by showing that the power of $\aleph_\omega$ can be made arbitrarily large below $\aleph_{\omega_1}$. In 1980 Shelah established (cf. his book [1982]) the general result that *for any limit ordinal* δ, $\aleph_\delta^{\mathrm{cf}(\delta)} < \aleph_{(|\delta|^{\mathrm{cf}(\delta)})^+}$, *so that in particular if* $\aleph_\omega$ *is a strong limit cardinal, then* $2^{\aleph_\omega} < \aleph_{(2^{\aleph_0})^+}$. Shelah thus brought countable cofinality, the one cofinality unattended to by Silver's Theorem, into the scheme of things by establishing a bound in ZFC analogous to the Galvin-Hajnal bound for uncountable cofinalities. Shelah's argument, based on the possible cofinalities of "reduced products" of a cofinal subset of $\aleph_\delta$, would evolve into a generally applicable method by the late 1980's, the remarkable *pcf theory*.

Mitchell [1978] made a new breakthrough for the inner model theory of large large cardinals by developing such a model for "hypermeasurable cardinals", e.g. a measurable cardinal κ such that for some normal ultrafilter U over κ, $P(P(\kappa)) \subseteq \mathrm{Ult}(V, U)$, so that every ultrafilter over κ is in the ultrapower. This at least captured a substantial consequence of the 2^κ-supercompactness of κ, and so engendered the hope of getting L-like inner models for such strong hypotheses. Supercompactness, while increasingly relied on in relative consistency results owing to its reflection properties, was out of reach, but the Mitchell result suggested an appropriate weakening: A cardinal κ is *α-strong iff* there is an elementary embedding $j\colon V \to M$ for some inner model M, with critical point κ and $\alpha < j(\kappa)$ such that $V_\alpha \subseteq M$. (One can alternately require that the αth iterated power set $P^\alpha(\kappa)$ be a subset of M, which varies the definition only for small α like $\alpha = 2$ but makes the definition more germane for them.) κ is *strong iff* it is α-strong for every α.

Dodd and Jensen (cf. [Dodd, 1982]) soon simplified Mitchell's presentation in what turned out to be a basic methodological advance for the development of inner model theory. While introducing certain redundancies, they formulated a general way of analyzing an elementary embedding in terms of *extenders*. The idea, anticipated in Jensen's proof of the Covering Theorem, is that elementary embeddings between inner models can be approximated arbitrarily closely as direct limits of ultrapowers with concrete features reminiscent of iterated ultrapowers.

Suppose that N and M are inner models of ZFC, $j\colon N \to M$ is elementary with a critical point κ, and $\beta > \kappa$. Let $\zeta \geq \kappa$ be the least ordinal satisfying $\beta \leq j(\zeta)$; the simple ("short") case is $\zeta = \kappa$, and the general case is for the study of stronger hypotheses. For each finite subset a of β, define E_a by:

$$X \in E_a \ \textit{iff} \ X \in P([\zeta]^{|a|}) \cap N \ \wedge \ a \in j(X) .$$

This is another version of the idea of generating ultrafilters from embeddings. E_a may not be in N, but $\langle N, \in, E_a \rangle \models$ "E_a is a κ-complete ultrafilter over $[\zeta]^{|a|}$". The (κ,β)-*extender derived from* j is $E = \langle E_a \mid a \text{ is a finite subset of } \beta \rangle$.

For each finite subset a of β, $\mathrm{Ult}(N, E_a)$ is seen to be elementarily embeddable into M, so that in particular $\mathrm{Ult}(N, E_a)$ is well-founded and hence identified with its transitive collapse, say M_a. Next, for $a \subseteq b$ both finite subsets of β, corresponding to how members of a sit in b there is a natural elementary embedding $i_{ab}\colon M_a \to M_b$. Finally,

$$\langle \langle M_a \mid a \text{ is a finite subset of } \beta \rangle, \langle i_{ab} \mid a \subseteq b \rangle \rangle$$

is seen to be a directed system of structures with commutative embeddings, so stipulate that $\langle M_E, \in_E \rangle$ is the direct limit, and let $j_E\colon N \to M_E$ be the corresponding elementary embedding. We thus have the *extender ultrapower* of N by E as a direct limit of ultrapowers. The crucial point is that the direct limit construction ensures that j_E and M_E approximate j and M "up to β", e.g. if $|V_\alpha|^M \leq \beta$, then $|V_\alpha|^M = |V_\alpha|^{M_E}$, i.e. the cumulative hierarchies of M and M_E agree up to α. Having formulated extenders derived from an embedding, a (κ,β)-*extender* is a sequence $E = \langle E_a \mid a \text{ is a finite subset of } \beta \rangle$ that satisfies various abstracted properties that enable the above construction.

In a manuscript circulated in 1980, Dodd and Jensen worked out inner models for strong cardinals. Building on the previous work of Mitchell, Dodd and Jensen formulated *coherent sequences of extenders*, built inner models relative to such, and established GCH in these models. The arguments were based on extending the established techniques of securing iterability and comparison through coiteration. The GCH result was significant as precursory for the further developments in inner model theory based on "iteration trees". Thus, with extenders the inner model theory was carried forward to encompass strong cardinals, newly arguing for the coherence and consistency of the concept. There would however be little further progress until 1985, for the aspiration to encompass stronger hypotheses had to overcome the problem of "overlapping extenders", having to carry out comparison through coiteration for local structures built on (κ_1,β_1)-extenders and (κ_2,β_2)-extenders with $\kappa_1 \leq \kappa_2 < \beta_1$. The difficulty here is one of "moving generators": if an extender ultrapower is taken with a (κ_1,β_1)-extender and then with a (κ_2,β_2)-extender, then $\kappa_2 < \beta_1$ implies that the generating features of the first extender ultrapower has been shifted by the second ultrapower and so one can no longer keep track of that ultrapower in the coiteration process. In any event, a crucial inheritance from this earlier work was the *Dodd-Jensen Lemma* about the minimality of iterations copied across embeddings, which would become crucial for all further work in inner model theory.

Through the 1970s and into the 1980s, those in the burgeoning field of *set-theoretic topology* continued the topological investigation of the continuum and general, transfinite spaces, and large cardinals figured more and more as strong propositions were studied. One prominent problem was whether every normal Moore space was metrizable. Notably, William Fleissner [1982] showed that were that the case there would be an inner model with a measurable cardinal. With large cardinals intimated, Peter Nyikos [1980] then established that a certain instrumental "axoim" implies that every normal Moore space is metrizable, an axiom which Kunen had shown holds if κ random reals are adjoined to a strong compact cardinal κ. Hence, *if there is a strong compact cardinal, then in a forcing extension every normal Moore space is metrizable.* With this as one striking example, large cardinals, and even the Proper Forcing Axiom, began to play substantial roles in new relative consistency results in this area feature general topological articulations. The rich and wide-ranging *Handbook of Set-Theoretic Topology* [Kunen and Vaughn, 1984] summed up the progress, and its many articles set the tone for further work.

Conspicuous in combinatorics and topology would be the work of Stevo Todorcevic. Starting with his doctoral work with Kurepa, in 1979 he carried out an incisive analysis of uncountable trees—Suslin, Aronszajn, Kurepa trees and variants—and their linearizations and isomorphism types. In 1983 he [1984] dramatically reoriented the sense of strength for the Proper Forcing Axiom by showing that PFA *implies that* $\square_\kappa$ *fails for every* $\kappa > \omega$. PFA had previously been shown consistent relative to the existence of a supercompact cardinal. With the failure of $\square_\kappa$ for singular κ having been seen as having quite substantial consistency strength, PFA was itself seen for the first time as a very strong proposition. Todorcevic would go from strength to strength, making substantial contributions to the theory of partition relations, eventually establishing definitive results about ω_1 as the archetypal uncountable order-structure (cf. his book [2007]).

Starting in 1980 Foreman made penetrating inroads into the possibilities for very strong propositions holding low in the cumulative hierarchy based on the workings of generic elementary embeddings. Extending Kunen's work and deploying Silver's master condition idea, Foreman [1982] initially used 2-huge cardinals to get model-theoretic transfer principles to hold and [1983] saturated ideals to exist among the range of $\aleph_n$'s. He would soon focus on generic elementary embeddings and corresponding ideals themselves, even making them postulational for set theory.

In a major 1984 collaboration in Jerusalem, Foreman, Magidor, and Shelah (cf. [1984]) established penetrating results that led to a new understanding of strong propositions and the possibilities with forcing. The focus was on a new, maximal forcing axiom: A partial order P *preserves stationary subsets of* ω_1 *iff* stationary subsets of ω_1 remain stationary in any forcing extension by P, and with this we have *Martin's Maximum* (MM): *For any* P *preserving stationary subsets of* ω_1 *and collection* $\mathcal{D}$ *of* $\aleph_1$ *dense subsets of* P*, there is a filter* $G \subseteq P$ *meeting every member of* $\mathcal{D}$. This subsumes PFA and is a maximally strong forcing axiom in that there is a P which does not preserve stationary subsets of ω_1 for which the conclusion fails. Foreman, Magidor, and Shelah established: *If there is a supercompact cardinal* κ*, then there is a forcing extension in which* $\kappa = \omega_2$ *and*

MM *holds.*

Shelah had considered a weakening of properness called *semiproperness*, a notion for forcing that could well render uncountable cofinalities countable. To iterate such forcings, it had to be faced that the countable cofinality of limit stages cannot be ascertained in advance, and so he developed *revised countable support iteration* (RCS) based on names for the limit stage indexing. Foreman, Magidor, and Shelah actually carried out Baumgartner's PFA consistency proof for semiproper forcings with RCS iteration to establish the consistency of the analogous *Semiproper Forcing Axiom* (SPFA). Their main advance was that, although a partial order that preserves stationary subsets of ω_1 is not necessarily semiproper, it is in this supercompact collapsing context.[13]

Foreman, Magidor, and Shelah then established the relative consistency of several propositions by deriving them directly from MM. One such proposition was that NS_{ω_1} is $\aleph_2$-saturated. Hence, not only was the upper bound for the consistency strength of having an $\aleph_1$-complete $\aleph_2$-saturated ideal over ω_1 considerably reduced from Kunen's huge cardinal, but for the first time the consistency of NS_{ω_1} itself being $\aleph_2$-saturated was established relative to large cardinals. Another formative result was simply that MM actually implies that $2^{\aleph_0} = \aleph_2$, starting a train of thought about forcing axioms actually determining the size of the continuum.[14]

With their work Foreman, Magidor, and Shelah had overturned a long-held view about the scaling down of large cardinal properties. In the first flush of new hypotheses and propositions, Kunen had naturally enough collapsed a large cardinal to ω_1 in order to transmute strong properties of the cardinal into an $\aleph_1$-complete $\aleph_2$-saturated ideal over ω_1, and this sort of direct connection had become the rule. The new discovery was that a collapse of a large cardinal to ω_2 instead can provide enough structure to secure such an ideal. In fact, Foreman, Magidor, and Shelah showed that even the usual Levy collapse of a supercompact cardinal to ω_2 engenders an $\aleph_1$-complete $\aleph_2$-saturated ideal over ω_1. In terms of method, the central point is that the existence of sufficiently large cardinals implies the existence of substantial generic elementary embeddings with small critical points like ω_1. Woodin's later strengthenings and elaborations of these results would have far-reaching consequences.

4.2 Reflecting stationary sets

Here we work in a discussion of stationary set reflection. This serves mainly as an illustrative example of a theme that interweaves through the mainstream of large cardinals, but could well be considered a topic sufficiently prominent to warrant separate description, particularly because of the way it displays aspects of many emerging issues. Another theme that would have served as well is Chang's

[13]Eventually, Shelah [1987] did establish that MM and SPFA are equivalent.

[14]It would be by different and elegant means that Todorcevic would show in 1990 (cf. [Bekkali, 1991]) that PFA already implies that $2^{\aleph_0} = \aleph_2$.

Conjecture.[15]

For $S \subseteq \kappa$ stationary in κ, S *reflects iff* there an $\alpha < \kappa$ such that $S \cap \alpha$ is stationary in α. In these terms, we let $\text{Refl}(S)$ be the assertion that every stationary subset of S reflects. That $S \subseteq \kappa$ is stationary in κ is a basic "second-order" property of $\langle \kappa, S \rangle$, and with reflection having become an important heuristic in set theory, stationary set reflection commended itself as a specific, second-order reflection possibility for investigation. It is a direct consequence of a characterizing reflection property of weakly compact cardinals that *if κ is weakly compact, then* $\text{Refl}(\kappa)$. Jensen [1972] raised interest in stationary set reflection by getting the converse in L as an initial dividend of his fine structure theory: *If $V = L$, then κ is weakly compact iff* $\text{Refl}(\kappa)$. On the other hand, Kunen in 1972 (cf. [1978]) quickly complemented Jensen's result by showing: *If κ is weakly compact, then there a cardinal-preserving forcing extension in which κ is inaccessible but not weakly compact, yet* $\text{Refl}(\kappa)$ *holds.* Kunen introduced a way of forcing a κ-Suslin tree through a large cardinal κ, thereby precluding its weak compactness, in such a way that the forcing combined by the further forcing for shooting a cofinal branch through the adjoined tree is essentially the same as having just added one Cohen subset of κ in the first place. For the actual model, Kunen made an initial application of Silver's "reversed Easton" technique (cf. 3.3) by adjoining for every inaccessible $\lambda < \kappa$ a Cohen subset of λ, and then followed this with his way of adjoining a κ-Suslin tree. Shooting a cofinal branch through the tree would then resurrect the large cardinal properties of κ because of the reversed Easton preparation, and Kunen exploited this promixity to show that $\text{Refl}(\kappa)$ already holds in his model.

What about $\text{Refl}(\kappa)$ for successor cardinals κ? For regular $\lambda < \kappa$, define the paradigmatic stationary sets

$$S^\kappa_\lambda = \{\alpha < \kappa \mid \text{cf}(\alpha) = \lambda\}.$$

Then for regular λ, clearly $S^{\lambda^+}_\lambda$ cannot reflect. Beyond this initial restriction, forcing and L expectedly work against stationary set reflection: For any regular $\kappa \geq \omega_2$, a non-reflecting stationary subset of S^κ_ω can be adjoined by straightforward forcing with initial segments. The Jensen combinatorial principle $\Box_\lambda$, holding in L, directly implies that $\text{Refl}(S)$ fails for every stationary $S \subseteq \lambda^+$.

Large cardinals provided the counterweight. Reminiscent of the Mitchell-Silver result on the tree property (cf. 3.3) but more immediate in its proof, Baumgartner [1976] observed how the collapse of a weakly compact cardinal preserves a significant attribute: *If a weakly compact cardinal is Levy collapsed to make it ω_2, then* $\text{Refl}(S^{\omega_2}_\omega)$ *holds in the extension.* $\text{Refl}(S^{\omega_2}_\omega)$ implies the failure of $\Box_{\omega_1}$, which by Jensen's fine structure argument for the principle holding in L implies that ω_2 must be Mahlo there. Thus, as for many propositions in modern set theory,

[15] For some representative articles, [Silver, 1971b] established the consistency of Chang's Conjecture relative to a large cardinal; Donder (cf. [Donder *et al.*, 1981]) established the converse with the core model K^{DJ}; [Donder and Koepke, 1983] investigated higher versions; and [Levinski *et al.*, 1990] established the relative consistency of the version at $\aleph_{\omega+1}$.

the consistency strength of $\text{Refl}(S^{\omega_2}_\omega)$ was initially bracketed, here between having a weakly compact cardinal and having a Mahlo cardinal. The expectation was nonetheless for weak compactness. This was especially so as Magidor (cf. [1982]) showed that *if for any pair of sets* $S_1, S_2 \subseteq \omega_2$ *both stationary in* ω_2 *there is an* $\alpha < \omega_2$ *such that both* $S_1 \cap \alpha$ *and* $S_2 \cap \alpha$ *are stationary in* α*, then* ω_2 *is weakly compact in* L. This "pair reflection" actually held in the Levy collapse of a weakly compact cardinal to make it ω_2 by checking Baumgartner's argument, and so at least there was this equi-consistency.

It was somewhat surprising then, when in the late 1970s, Leo Harrington and Shelah (cf. [1985]) established: *If a Mahlo cardinal, after first arranging* CH*, is Levy collapsed to make it* ω_2*, then* $\text{Refl}(S^{\omega_2}_\omega)$ *holds in the extension.* Their result involved a refined auxiliary forcing argument, which revealed a sense in which a concrete second-order reflection requires only a first-order overlay as given by a Mahlo cardinal. Remarkably, reflection and pair reflection were exactly and separately gauged in consistency strength!

Shelah also noted a generalization of Baumgartner's result, one exhibiting the global reflection possibilities provided by supercompactness: *If a supercompact cardinal is Levy collapsed to make it* ω_2*, then in the extension* $\text{Refl}(S^\kappa_\omega)$ *holds for every regular* $\kappa \geq \omega_2$.

As in many situations, ω_2 typified an accessible cardinal in the investigation of the consistency strength of stationary set reflection, as described above, but with ω_2 paradigmatic of successors of *regular* cardinals. For successors of *singular* cardinals the emerging inner model theory showed that the failure of $\square_\kappa$ for singular κ has strong consistency strength, and so also $\text{Refl}(\kappa^+)$. With $\omega_{\omega+1}$ typifying a successor of a singular cardinal, Magidor [1982] established: *If infinitely many supercompact cardinals are iteratively Levy collapsed to become the* ω_n*'s, then in the extension* $\text{Refl}(\omega_{\omega+1})$ *holds.* The intricate proof involved the lifting of various supercompact embeddings to the extension and then a careful analysis of how stationary reflection via these embeddings at intermediate stages is preserved to the end. This result, like others of Magidor's from this period, featured the liberal use of strong hypotheses providing strong elementary embeddings. In this case however, the hypothesis of the theorem has not as yet been significantly reduced in subsequent work for getting $\text{Refl}(\omega_{\omega+1})$. As with several strong propositions, it has been sufficient that a beachhead was secured, and a stalking for equi-consistency has not proceeded—unlike for the Singular Cardinals Problem (cf. 4.4) with its immediacy as a pressing issue for set theory. Be that as it may, Magidor's result was a seminal one for the investigation of the combinatorics and consistency of properties at successors of singular cardinals, a subject which has intimately involved strong large cardinal hypotheses through forcing.[16]

The 1984 Foreman-Magidor-Shelah collaboration led to the formulation of a strong stationary set reflection principle. The sets $\mathcal{P}_\kappa X = \{x \subseteq X \mid |x| < \kappa\}$ having emerged in the ultrafilter formulations of large large cardinals (cf. 3.1), the concepts of closed unbounded and stationary have natural generalizations to

[16]See Eisworth's chapter in [Foreman and Kanamori, 2010].

this index set: $C \subseteq \mathcal{P}_\kappa X$ is *closed unbounded* if it is *closed*, i.e. under the taking of $\subseteq$-chains of length less than κ, and unbounded, i.e. for any $y \in \mathcal{P}_\kappa X$ there is an $x \supseteq y$ such that $x \in C$. $S \subseteq \mathcal{P}_\kappa X$ is *stationary* if $C \cap S \neq \emptyset$ for any closed unbounded $C \subseteq \mathcal{P}_\kappa X$. Foreman, Magidor, and Shelah showed that their Martin's Maximum implies the following Reflection Principle: *For any regular $\lambda \geq \omega_2$, if $S \subseteq \mathcal{P}_{\omega_1}\lambda$ is stationary, then there is an $X \subseteq \lambda$ with $|X| = \aleph_1$ and $X \supseteq \omega_1$ such that $S \cap \mathcal{P}_{\omega_1}X$ is stationary in $\mathcal{P}_{\omega_1}X$.* Remarkably, the main-cited consequences of MM already follow from this reflection principle, except that it only implies that NS_{ω_1} is precipitous. Todorcevic (cf. [Bekkali, 1991]) formulated a strong reflection principle, one that also follows from MM but moreover has as a consequence that NS_{ω_1} is $\aleph_2$-saturated. Qi Feng and Jech [1998] also came up with a stronger reflection principle, but they soon showed it to be equivalent to Todorcevic's. Thus, stationary set reflection itself garnered a historically focal goal of large cardinal hypotheses.

4.3 Consistency of determinacy

The developments of the 1980s which are the most far-reaching and presentable as sustained narrative have to do with the stalking of the consistency of determinacy. By the late 1970s a more or less complete structure theory for the projective sets was in place, a resilient edifice founded on determinacy with both strong buttresses and fine details. In 1976 the researchers had started the Cabal Seminar in the Los Angeles area, and in a few years, with John Steel and Woodin having joined the ranks, attention began to shift to sets of reals beyond the projective sets, to inner models, and to questions of overall consistency. Most of the work before the crowning achievements of the later 1980s appears in the several proceedings of the Cabal Seminar, [Kechris and Moschovakis, 1978; Kechris *et al.*, 1981; Kechris *et al.*, 1983; Kechris *et al.*, 1988].

The question of the overall consistency of determinacy came increasingly to the fore. Is AD consistent relative to some large cardinal hypothesis? Or, with its strong consequences, can AD subsume large cardinals in some substantial way or be somehow orthogonal? A decade after his initial result that the existence of a measurable cardinal implies $\mathbf{\Pi}^1_1$-determinacy, Martin [1980] applied its homogeneity idea with a "Martin-Solovay" tree representation for $\mathbf{\Pi}^1_2$ sets, together with algebraic properties of elementary embeddings posited close to Kunen's large cardinal inconsistency, to establish $\mathbf{\Pi}^1_2$-determinacy. A direction was set but generality only came in 1984, when Woodin showed that an even stronger large cardinal hypothesis implies $\mathrm{AD}^{L(\mathbb{R})}$. So, a mooring was secured for AD after all in the large cardinal hierarchy. With Woodin's hypothesis apparently too remote it would now be a question of scaling it down according to the methods becoming available for proofs of determinacy, perhaps even achieving an equi-consistency result.

The rich 1984 Foreman-Magidor-Shelah work would have crucial consequences for the stalking of consistency also for determinacy. Shelah carried out a version of their collapsing argument that does not add any new reals but nonetheless gets an $\aleph_1$-complete $\aleph_2$-saturated ideal over ω_1. Woodin then pointed out that with no new reals adjoined the generic elementary embedding induced by such an ideal can be used to establish that the ground model $L(\mathbb{R})$ reals are actually Lebesgue measurable. Thus Shelah and Woodin (cf. their [1990]) had established an outright result: *If there is a supercompact cardinal, then every set of reals in* $L(\mathbb{R})$ *is Lebesgue measurable.* This result not only portended the possibility of getting $\mathrm{AD}^{L(\mathbb{R})}$ from a supercompact cardinal, but through the specifics of the argument stimulated the reducing of the hypothesis. While Woodin was visiting Jerusalem in June 1984, he came up with what is now known as a *Woodin cardinal.* The hypothesis was then reduced as follows: *If there are infinitely many Woodin cardinals with a measurable cardinal above them, then every set of reals in* $L(\mathbb{R})$ *is Lebesgue measurable.* An early suggestion of optimality of hypothesis was that if the "infinitely" is replaced by "n" for some $n \in \omega$, then one can conclude that every $\mathbf{\Sigma}^1_{n+2}$ set of reals is Lebesgue measurable. The measurable cardinal hovering above would be a recurring theme, the purpose loosely speaking to maintain a stable environment with the existence of sharps.

Especially because of its subsequent centrality, it is incumbent to give an operative definition of Woodin cardinal: For a set A, κ is *α-A-strong iff* there is an elementary embedding $j\colon V \to M$ witnessing that κ is α-strong which moreover preserves A: $A \cap V_\alpha = j(A) \cap V_\alpha$. A cardinal δ is *Woodin iff* for any $A \subseteq V_\delta$, there is a $\kappa < \delta$ which is α-A-strong for every $\alpha < \delta$.

A Woodin cardinal, evidently a technical, consistency-wise strengthening of a strong cardinal, is an important example of concept formation through method. The initial air of contrivance gives way to seeing that Woodin cardinal seemed to encapsulate just wanted is needed to carry out the argument for Lebesgue measurability. That argument having been based on first collapsing a large cardinal to get a saturated ideal and then applying the corresponding generic elementary embedding, Woodin later in 1984 stalked the essence of method and formulated *stationary tower forcing* (cf. [1990]). An outgrowth of the Foreman-Magidor-Shelah work, this notion of forcing streamlines their forcing arguments to show that a Woodin cardinal suffices to get a generic elementary embedding $j\colon V \to M$ with critical point ω_1 and ${}^\omega M \subseteq M$. With a new, minimizing large cardinal concept isolated, there would now be dramatic new developments both in determinacy and inner model theory. One important scaling down result was the early 1985 result of Shelah (cf. [1987a]): *If* κ *is Woodin, then in a forcing extension* $\kappa = \omega_1$ *and* NS_{ω_1} *is* $\aleph_2$*-saturated.* The large cardinal strength now seemed minimal for getting such an ideal, and there was anticipation of achieving an equi-consistency.

Steel in notes of Spring 1985 developed an inner model for a weak version of Woodin cardinal. While inner models for strong cardinals had only required linear iterations for comparison, the new possibility of overlapping extenders and moving generators had led Mitchell in 1979 to develop *iteration trees* of iterated ultrapowers for searching for possible well-founded limits of models along branches. A particularly simple example of an iteration tree is an *alternating chain*, a tree consisting of two ω-length branches with each model in the tree an extender ultrapower of the one preceding it on its branch, via an extender taken from a corresponding model in the other branch. Initially, Steel tried to avoid alternating chains, but the Foreman-Magidor-Shelah work showed that for dealing with Woodin cardinals they would be a necessary part. Their use soon led to a major breakthrough in the investigation of determinacy.

In the Fall of 1985 Martin and Steel (cf. [1988; 1989]) showed that Woodin cardinals imply the existence of alternating chains in which both branches have well-founded direct limits, and used this to establish: *If there are infinitely many Woodin cardinals, then* PD *holds.* This was a culmination of method in several respects. In the earlier Martin results getting $\mathbf{\Pi}^1_1$-Determinacy and $\mathbf{\Pi}^1_2$-Determinacy, trees on $\omega \times \kappa$ for some cardinal κ had been used, to each node of which were attached ultrafilters in a coherent way that governed extensions. Kechris and Martin isolated the relevant concept of *homogeneous tree*, the point being that sets of reals which are the projections $p[T]$ of such trees T—the *homogeneously Suslin sets*—are determined. With PD, the scale property had been propagated through the projective hierarchy. Now with Woodin cardinals, having representations via homogeneous trees was propagated, getting determinacy itself. In particular, Martin and Steel established: *If $n \in \omega$ and there are n Woodin cardinals with a measurable cardinal above them, then $\mathbf{\Pi}^1_{n+1}$-determinacy holds.*

Within weeks after the Martin-Steel breakthrough, Woodin used it together with stationary towers to investigate tree representations in $L(\mathbb{R})$ to establish: *If there are infinitely Woodin cardinals with a measurable cardinal above them, then* $\mathrm{AD}^{L(\mathbb{R})}$ *holds.* With the consistency strength of AD having been gauged by this result, Woodin soon established the crowning equi-consistency result: *The Axiom of Determinacy is equi-consistent with the existence of infinitely many Woodin cardinals.*[17]

This was a remarkable achievement of the concerted effort to establish the consistency strength of AD along the large cardinal hierarchy. But even this would just be a beginning for Woodin, who would go from strength to strength in establish many structural results involving AD and stronger principles, to become preeminent with Shelah in set theory.

4.4 Into the 1990s

In the later 1980s set theory continued to expand apace in various directions, and we conclude our historical survey by mentioning here a few of the most prominent developments, each of a different character but all being decisive advances.

In inner model theory, Stewart Baldwin in 1987 made a suggestion, one which Mitchell then newly forwarded, which led to a crucial methodological advance. Up to then, the extender models $L[\vec{E}]$ constructed relative to a coherent sequence of extenders $\vec{E}$ had each extender E in the sequence "measure" all the subsets in $L[\vec{E}]$ of the critical point. The Baldwin-Mitchell idea was to construct only with

[17]See the Neeman and Koellner-Woodin chapters of [Foreman and Kanamori, 2010] for proofs of each direction benefitting from hindsight.

"partial" extenders E which if indexed at γ only measures the sets in $L_\gamma[\vec{E}\restriction\gamma]$. This together with a previous Mitchell strategy of carrying out the comparison process using finely calibrated partial ultrapowers ("dropping to a mouse") led to a comparison process internal to $L[\vec{E}]$ based on the use of fine structure. The infusion of fine structure made the development of the new extender models more complex, but with this came the important dividends of a more uniform presentation, a much stronger condensation, and a more systematic comparison process. During 1987-9, Mitchell and Steel (cf. their [1994]) worked out the details and showed that if there is a Woodin cardinal then there is an inner model $L[\vec{E}]$, L-like in satisfying GCH and so forth, in which there is a Woodin cardinal. The process involved the correlating of iteration trees for $L[\vec{E}]$ with iteration trees in V. A canonical, fine structural inner model of a Woodin cardinal newly argued for the consistency of the concept, as well as provided a great deal of understanding about it as set in a finely tuned, layer-by-layer hierarchy.

What about a core model "up to" a Woodin cardinal, in analogy to K^{DJ} for $L[U]$? In 1990, Steel (cf. [1996]) solved the "core model iterability problem" by showing that large cardinals in V are not necessary for showing that certain models $L[\vec{E}]$ have sufficient iterability properties. With this, he constructed a new core model, first building a "background certified" K^c based on extenders in V and then the "true" core model K. Steel was thus able to extend the previous work of Mitchell on the core model $K[\mathcal{U}]$ up to $\exists\kappa(o(\kappa) = \kappa^{++})$ to establish e.g.: *If there is an $\aleph_1$-complete $\aleph_2$-saturated ideal over ω_1 and a measurable cardinal, then there is an inner model with a Woodin cardinal.* Thus, Shelah's 1985 forcing result and Steel's, except for the artifact of "the measurable cardinal above", had calibrated an important consistency strength, and what had become a central goal of forcing and inner model theory was handily achieved.

In the early 1990s, Steel, Mitchell, and Ernest Schimmerling (cf. their [1997]) pushed the Jensen covering argument over the hurdles of the new fine structural Steel core model K to establish a covering lemma up to a Woodin cardinal. Schimmerling [1994] both established combinatorial principles in K as well established new consistency strengths, e.g. PFA *implies that there is an inner model with a Woodin cardinal.*

The later 1980s featured a distinctive development that led to a new conceptual framework of applicability to singular cardinals, new incisive results in cardinal arithmetic, and a re-orienting of set theory to new possibilities for outright theorems of ZFC. Starting in late 1987 Shelah returned to the work on bounds for powers of singular cardinals and drew out an extensive underlying structure of possible cofinalities of reduced products, soon codified as *pcf theory*. With this emerged new work in *singular cardinal combinatorics*, with Shelah himself initially providing applications to model theory, partition relations, Jónsson algebras, Boolean algebras, and cardinal arithmetic. This last was epitomized by a dramatic result that exhibited how the newly seen structural constraints impose a tight bound: *If δ is a limit ordinal with $|\delta|^{\mathrm{cf}(\delta)} < \aleph_\delta$ then $\aleph_\delta^{\mathrm{cf}(\delta)} < \aleph_{(|\delta|^{+4})}$, so that in particular if $\aleph_\omega$ is a strong limit cardinal, then $2^{\aleph_\omega} < \aleph_{\omega_4}$.* This substantively betters the former bound $\aleph_{(2^{\aleph_0})^+}$ of his 1980 result (cf. [1982]), and complements his [1983] consistency result from that time that values for $2^{\aleph_\omega}$ up to $\aleph_{\omega_1}$ are possible. Whether the gap betweeen $\aleph_{\omega_1}$ and $\aleph_{\omega_4}$ can be closed is a prominent open problem.

Suppose that A is an infinite set of cardinals and F is a filter over A. The *product* ΠA consists of functions f with domain A such that $f(a) \in a$ for every $a \in A$. For $f, g \in \Pi A$, the relation $=_F$ defined by $f =_F g$ *iff* $\{a \in A \mid f(a) = g(a)\} \in F$ is an equivalence relation on ΠA, and the *reduced product* $\Pi A/F$ consists of the equivalence classes. We can impose order, officially on $\Pi A/F$ but still working with functions themselves, by: $f <_F g$ *iff* $\{a \in A \mid f(a) < g(a)\} \in F$.

Shelah's new theory took as central the investigation of the possible cofinalities function:

$$\mathrm{pcf}(A) = \{\mathrm{cf}(\Pi A/D) \mid D \text{ is an ultrafilter over } A\}$$

as calibrated by the ideals

$$J_{<\lambda}[A] = \{b \subseteq A \mid \mathrm{cf}(\Pi A/D) < \lambda \text{ whenever } D \text{ is an ultrafilter over } A \text{ such that } b \in D\}.$$

These concepts had appeared before in Shelah's work, notably in his 1980 result $\aleph_\delta^{\mathrm{cf}(\delta)} < \aleph_{(|\delta|^{\mathrm{cf}(\delta)})^+}$, but now they became autonomous and were propelled forward by the discovery of unexpectedly rich structure.

With an eye to substantive cofinal subsets A of a singular cardinal, the abiding assumption was that A is a set of regular cardinals satisfying $|A| < \min(A)$. With this one gets that for any ultrafilter D over A, $\mathrm{cf}(\Pi A/D) < \lambda$ *iff* $D \cap J_{<\lambda}[A] \neq \emptyset$, and further, that $\mathrm{pcf}(A)$ has a maximum element. At the heart is the striking result that $J_{<\lambda^+}[A]$ is generated by $J_{<\lambda}[A]$ together with a single set $B_\lambda \subseteq A$. Shelah in fact got "nice" generators B_λ derived from imposing the structure of elementary substructures of a sufficiently large $H(\Psi)$. A bulwark of the theory is then that with $\mu = \sup A$ and $F = \{X \subseteq \mathrm{cf}(\mu) \mid |\mathrm{cf}(\mu) - X| < \mathrm{cf}(\mu)\}$ the filter of co-bounded subsets of $\mathrm{cf}\mu$, it is a theorem of ZFC that $\Pi A/F$ has a linearly ordered, cofinal sequence of length μ^+—a *scale* for μ. The careful control on the possible cofinalities led as well, when A consists of all the regular cardinals in an interval of cardinals, to $|\mathrm{pcf}(A)| \leq |A|^{+++}$, and in particular to the $\aleph_{\omega_4}$ bound mentioned above.

Shelah's work on pcf theory to 1993 appeared in his book *Cardinal Arithmetic* [1994], and since then he has further developed the theory and provided wide-ranging applications. Through its applicability pcf theory has to a significant extent been woven into modern set theory as part of the ZFC facts of singular cardinal combinatorics.

The Singular Cardinal Hypothesis (SCH) and the train of results starting with the Prikry-Silver result of the early 1970s were to be decisively informed by results of Moti Gitik. Gitik's work exhibits a steady engagement with central and difficult issues of set theory and a masterful virtuosity in the application of sophisticated techniques over a broad range. Gitik [1980] had established through an iterated

Prikry forcing the conspicuous singularization result that: *If there is a proper class of strongly compact cardinals, then in a* ZF *inner model of a class forcing extension every infinite cardinal has cofinality* ω. Mentioned earlier was the mid-1970s result that that NS_{ω_1} is precipitous is equi-consistent with having a measurable cardinal. Gitik [1984] established: *The precipitousness of* NS_{ω_2} *is equi-consistent with having a measurable cardinal* κ *such that* $o(\kappa) = 2$ *in the Mitchell order.* The difficult, forcing direction required considerable ingenuity because of inherent technical obstructions.[18]

Turning to the work on SCH, in 1988 Woodin dramatically weakened the large cardinal hypothesis needed to get a measurable cardinal κ satisfying $2^\kappa > \kappa^+$, using a technique of "altering generic objects". With the subsequent use of Prikry forcing leading to the failure of SCH, a proposition technically strengthening measurability now sufficed for the failure of SCH. He also showed that one can in fact get Magidor's conclusion that $\aleph_\omega$ could be the least cardinal at which GCH fails. Soon afterwards Gitik established both directions of an equi-consistency: First, he established that one can get the consistency of Woodin's proposition from just $\exists\kappa(o(\kappa) = \kappa^{++})$. Then, he applied a result from Shelah's pcf theory to Mitchell's $K[\mathcal{U}]$ analysis to establish, bettering a previous result of Mitchell, that $\exists\kappa(o(\kappa) = \kappa^{++})$ is actually necessary to get the failure of SCH. Hence, *The failure of* SCH *is equi-consistent with* $\exists\kappa(o(\kappa) = \kappa^{++})$.

Woodin's model in which GCH first fails at $\aleph_\omega$ required a delicate construction to arrange GCH below and an ingenious idea to get $2^{\aleph_\omega} = \aleph_{\omega+2}$. How about getting $2^{\aleph_\omega} > \aleph_{\omega+2}$? In a signal advance of method, Gitik and Magidor in 1989 provided a new technique to handle the general Singular Cardinals Problem with appropriately optimal hypotheses. The Prikry-Silver two-stage approach, first making 2^κ large and then singularizing κ without adding any new bounded subsets or collapsing cardinals, had been the basic model for attacking the Singular Cardinals Problem. Gitik and Magidor showed how to add many subsets to a large cardinal κ while *simultaneously* singularizing it without adding any new bounded subsets or collapsing cardinals. Thus, it became much easier to arrange any particular continuum function behavior below κ, like achieving GCH below, while at the same time making 2^κ large. Moreover, the new method smacked of naturalness and optimality.

The new Gitik-Magidor idea was to add many new Prikry ω-sequences corresponding to κ-complete ultrafilters over κ while maintaining the basic properties of Prikry forcing. There is an evident danger that if these Prikry sequences are too independent, information can be read from them that corresponds to new reals being adjoined. The solution was to start from a sufficient strong large cardinal hypothesis and develop an extender-based Prikry forcing structured on a "nice system" of ultrafilters $\langle U_\alpha \mid \alpha < \lambda\rangle$, a system such that for many $\alpha \leq \beta < \lambda$ there is a ground model function $f \colon \kappa \to \kappa$ such that: For all $X \subseteq \kappa$, $X \in U_\alpha$ *iff* $f^{-1}(X) \in U_\beta$. (Having such a projection function is the classical way of connecting two ultra-

[18]Gitik eventually established [1995; 1997] definitive equi-consistency results for having NS_κ precipitous for $\kappa > \omega^2$ of analogous sort involving the Mitchell order.

filters together, and one writes that $U_\alpha \leq_{\mathrm{RK}} U_\beta$ under the Rudin-Keisler partial order.) By this means one has the possibility of adding new subsets of κ, corresponding to different Prikry sequences, which are still dependent on each other so that no new bounded subsets need necessarily be added in the process. Gitik and Magidor worked out how their new approach leads to what turns out to be optimal or near optimal consistency results, and incorporating collapsing maps as in previous arguments of Magidor and Shelah, they got models in which GCH holds below $\aleph_\omega$ yet $2^{\aleph_\omega} = \aleph_{\alpha+1}$ for any prescribed countable ordinal α.

In subsequent work Gitik, together with Magidor, Mitchell, and others, have considerably advanced the investigation of powers of singular cardinals. Equi-consistency results have been achieved for large powers of singular cardinals along the Mitchell order and with α-strong cardinals, and uncountable cofinalities have been encompassed, the investigation ongoing and with dramatic successes.

4.5 Ideals

In this last section we take a quick look at later developments involving ideals. In the 1990s ideals low in the cumulative hierarchy yet having strong consistency strength became more and more focal as being of intrinsic interest, to the extent that postulations about them were put forward as themselves large cardinal axioms as modulated by forcing.

The continuing investigation of determinacy led to the study of a very strong property of ideals. An ideal I over a cardinal κ is *λ-dense iff* there is a family $D \subseteq P(\kappa) - I$ with $|D| = \lambda$ such that for any $X \in P(\kappa)$, there is a $Y \in D$ such that $Y - X \in I$. This is a natural notion of density for the algebra $P(\kappa) - I$. Clearly, if I is λ-dense and $\lambda < \mu$, then I is μ-saturated. Woodin drew attention to $\aleph_1$-complete $\aleph_1$-dense ideals over ω_1 by establishing the consistency of their existence first from strong determinacy hypotheses and, by a notably simple argument, from a large cardinal hypothesis just weaker than having a huge cardinal. Stimulating interest on how strong ideals over ω_1 may affect CH, Shelah [1986] showed by an elegant argument that ZFC imposes an intriguing limitation: *If* $2^{\aleph_0} < 2^{\aleph_1}$ *(e.g. if* CH *holds), then* NS_{ω_1} *is not* $\aleph_1$*-dense.*

Pushing the connections between ideals and the consistency of determinacy, Woodin established the following, using results of Steel about $L(\mathbb{R})$: *If there is an* $\aleph_1$*-complete,* $\aleph_1$*-dense ideal over* ω_1*, then* $\mathrm{AD}^{L(\mathbb{R})}$ *holds.* Then in late 1992, Woodin was able to establish the following converse for consistency strength: *Assume* $\mathrm{AD}^{L(\mathbb{R})}$*; then there is a forcing extension of* $L(\mathbb{R})$ *satisfying* ZFC + NS_{ω_1} *is* $\aleph_1$*-dense.* Quite remarkably, AD is equi-consistent with NS_{ω_1} being $\aleph_1$-dense! This encapsulation made ideal hypotheses central to modern set theory.

Is NS_{ω_1} being $\aleph_2$-saturated consistent with CH? This was a prominent question for some time, with the consistency results for $\aleph_2$-saturation all achieved with $\neg$CH and Shelah's actually deriving $\neg$CH from $\aleph_1$-denseness. Again it was Woodin; he established in descriptive set-theoretic terms: *If* NS_{ω_1} *is* $\aleph_2$*-saturated and there is a measurable cardinal, then* $\boldsymbol{\delta}^1_2 = \omega_2$. Since then there are, hierarchically, prewellorderings of length at least ω_2, CH fails in a strong, definable sense. Pursuing this work, Woodin in the mid-1990s (cf. [1999]) developed a vast machinery for forcing over models of determinacy, his $\mathbb{P}_{\max}$ extensions, that get at canonical maximal models of ZFC + $\neg$CH. This in turn led (cf. [2001a; 2001b]) to his Ω-logic, and a possible, structured solution to the Continuum Problem itself. With this we have peered ahead a long distance in a vast, remarkably rich direction; we return to the bedrock of large cardinals and forcing to draw this chapter to a close.

Foreman (cf. his chapter in [Foreman and Kanamori, 2010]) has provided a systematic approach to ideals and generic elementary embeddings, to the point of postulating axioms as was first done on his [1986]. To set things out once again, what is at play is the basic synthesis of forcing and ultrapowers whereby one starts with an ideal I over a cardinal κ; forces with $P(\kappa) - I$ where p is stronger than q if $p - q \in I$; thus produces an ultrafilter over the ground model $P(\kappa)$; and then gets a "generic" elementary embedding of the ground model into the corresponding ultrapower. With the possibilities of ideals occurring low in the cumulative hierarchy, so that large cardinal ideas can be brought to bear on classical problems of set theory, an enormous subject has grown.

At the broadest level are the "three parameters" describing the strength of a generic elementary embedding $j \colon V \to M$: how j moves the ordinals; how large and closed M is; and the nature of the forcing that provided j. This last is the new parameter at play beyond the "conventional" large cardinal hypotheses. Ideals through their forcing properties thus assuming a crucial role, another guiding theme is the distinction between "natural" ideals that have intrinsic definitions and ideals "induced" by elementary embeddings. Strong ideal assumptions gain an autonomy as "generic large cardinals" in their own right, and there is a further delineation according to *consequences* of generic large cardinals and *consistency results* about them.

Of consequences, we mentioned Shelah's and Woodin's $\neg$CH results above, and there are implications for GCH; SCH; graphs and groups; stationary set reflection; Suslin and Kurepa trees; partition properties; and descriptive set theory. Moreover, there are remarkable limitative results on ideals properties e.g. the Gitik-Shelah result (cf. their [1997]) that *if κ is regular and $\delta^+ < \kappa$, then the ideal generated by* NS_κ *and* $\{\alpha < \kappa \mid \mathrm{cf}(\alpha) = \delta\}$ *is not κ^+-saturated,* and their result (cf. their [1989]) that *there is no $\aleph_1$-complete $\aleph_0$-dense nowhere prime ideal.*[19]

Of consistency results, beyond the several we have already mentioned there is Foreman's relative consistency of having an $\aleph_1$-complete $\aleph_1$-dense (uniform) ideal over ω_2 from a strong hypothesis about two coordinated huge-type cardinals. In general terms, Foreman has shown how the "conventional" large cardinals and the "generic" large cardinals interleave in consistency strength, sometimes with equi-consistencies at play. At this general confluence of the methods of forcing and ultrapowers, there is a wealth of possibilities.

[19]An ideal is *nowhere prime* if it restricted to no set becomes a maximal ideal, dual to an ultrafilter.

BIBLIOGRAPHY

[Baumgartner, 1976] J. E. Baumgartner. A new class of order types. *Annals of Mathematical Logic*, 9:187–222, 1976.

[Baumgartner, 1983] J. E. Baumgartner. Iterated forcing. In A. R. D. Mathias, editor, *Surveys in Set Theory*, volume 87 of *London Mathematical Society Lecture Note Series*, pages 1–59. Cambridge University Press, 1983.

[Bekkali, 1991] M. Bekkali. *Topics in Set Theory*, volume 1476 of *Lecture Notes in Mathematics*. Springer-Verlag, 1991.

[Blackwell, 1967] D. Blackwell. Infinite games and analytic sets. *Proceedings of the National Academy of Sciences USA*, 58:1836–1837, 1967.

[Dales and Woodin, 1987] H. G. Dales and W. H. Woodin. *An Introduction to Independence for Analysts*, volume 115 of *London Mathematical Society Lecture Note Series*. Cambridge University Press, 1987.

[Devlin and Jensen, 1975] K. J. Devlin and R. B. Jensen. Marginalia to a theorem of Silver. In G. H. Müller, A. Oberschelp, and K. Potthoff, editors, *ISILC Logic Conference, Kiel, 1974*, volume 499 of *Lecture Notes in Mathematics*, pages 115–142. Springer, 1975.

[Devlin and Johnsbraten, 1974] K. J. Devlin and H. Johnsbraten. *The Souslin Problem*, volume 405 of *Lecture Notes in Mathematics*. Springer-Verlag, 1974.

[Dodd and Jensen, 1981] A. J. Dodd and R. B. Jensen. The core model. *Annals of Mathematical Logic*, 20:43–75, 1981.

[Dodd and Jensen, 1982a] A. J. Dodd and R. B. Jensen. The covering lemma for K. *Annals of Mathematical Logic*, 22:1–30, 1982.

[Dodd and Jensen, 1982b] A. J. Dodd and R. B. Jensen. The covering lemma for $L[U]$. *Annals of Mathematical Logic*, 22:127–135, 1982.

[Dodd, 1982] A. J. Dodd. *The Core Model*, volume 61 of *London Mathematical Society Lecture Note Series*. Cambridge University Press, 1982.

[Donder and Koepke, 1983] H.-D. Donder and P. G. Koepke. On the consistency strength of 'accessible' Jonsson cardinals and of the weak Chang conjecture. *Annals of Pure and Applied Logic*, 25:233–261, 1983.

[Donder *et al.*, 1981] H.-D. Donder, R. B. Jensen, and B. J. Koppelberg. Some applications of the core model. In *Set theory and model theory*, volume 872 of *Lecture Notes in Mathematics*. Springer-Verlag, 1981.

[Ehrenfeucht and Mostowski, 1956] A. Ehrenfeucht and A. M. Mostowski. Models of axiomatic theories admitting automorphisms. *Fundamenta Mathematicae*, 43:50–68, 1956.

[Feng and Jech, 1998] Q. Feng and T. J. Jech. Projective stationary sets and a strong reflection principle. *Journal of the London Mathematical Society*, 58:271–283, 1998.

[Fleissner, 1982] W. G. Fleissner. If all normal moore spaces are metrizable, then there is an inner model with a measurable cardinal. *Transactions of the American Mathematical Society*, 273:365–373, 1982.

[Foreman and Kanamori, 2010] M. Foreman and A. Kanamori, editors. *Handbook of Set Theory*. Springer, 2010.

[Foreman and Woodin, 1991] M. Foreman and W. H. Woodin. The generalized continuum hypothesis can fail everywhere. *Annals of Mathematics*, 133:1–35, 1991.

[Foreman *et al.*, 1984] M. Foreman, M. Magidor, and S. Shelah. Martin's Maximum, saturated ideals, and nonregular ultrafilters I. *Annals of Mathematics*, 127:1–47, 1984.

[Foreman, 1982] M. Foreman. Large cardinals and strong model theoretic-transfer properties. *Transactions of the American Mathematical Society*, 272:427–463, 1982.

[Foreman, 1983] M. Foreman. More saturated ideals. In A. S. Kechris, D. A. Martin, and Y. N. Moschovakis, editors, *Cabal Seminar 79-81. Proceedings of the Caltech-UCLA Logic Seminar 1979-81*, volume 1019 of *Lecture Notes in Mathematics*, pages 1–27. Springer-Verlag, 1983.

[Foreman, 1986] M. Foreman. Potent axioms. *Transactions of the American Mathematical Society*, 294:1–28, 1986.

[Gaifman, 1964] H. Gaifman. Measurable cardinals and constructible sets (abstract). *Notices of the American Mathematical Society*, 11:771, 1964.

[Gale and Stewart, 1953] D. Gale and F. M. Stewart. Infinite games with perfect information. In *Contributions to the Theory of Games, vol. 2*, Annals of Mathematical Studies #28, pages 245–266. Princeton University Press, 1953.

[Galvin and Hajnal, 1975] F. Galvin and A. Hajnal. Inequalities for cardinal powers. *Annals of Mathematics*, 101:491–498, 1975.

[Gitik and Shelah, 1989] M. Gitik and S. Shelah. Forcings with ideals and simple forcing notions. *Israel Journal of Mathematics*, 68:129–160, 1989.

[Gitik and Shelah, 1997] M. Gitik and S. Shelah. Less saturated ideals. *Proceedings of the American Mathematical Society*, 125:1523–1530, 1997.

[Gitik, 1980] M. Gitik. All cardinals can be singular. *Israel Journal of Mathematics*, 35:61–88, 1980.

[Gitik, 1984] M. Gitik. The nonstationary ideal on $\aleph_2$. *Israel Journal of Mathematics*, 48:257–288, 1984.

[Gitik, 1995] M. Gitik. Some results on the nonstationary ideal. *Israel Journal of Mathematics*, 92:61–112, 1995.

[Gitik, 1997] M. Gitik. Some results on the nonstationary ideal. II. *Israel Journal of Mathematics*, 99:175–188, 1997.

[Harrington and Shelah, 1985] L. A. Harrington and S. Shelah. Some exact equiconsistency results in set theory. *Notre Dame Journal of Formal Logic*, 26:178–188, 1985.

[Jackson, 1988] S. Jackson. AD and the projective ordinals. In A. S. Kechris, D. A. Martin, and J. R. Steel, editors, *Cabal Seminar 81-85. Proceedings, Caltech-UCLA Logic Seminar 1981-1985*, volume 1333 of *Lecture Notes in Mathematics*, pages 117–220. Springer-Verlag, 1988.

[Jech and Prikry, 1976] T. J. Jech and K. L. Prikry. On ideals of sets and the power set operation. *Bulletin of the American Mathematical Society*, 82:593–595, 1976.

[Jech *et al.*, 1980] T. J. Jech, M. Magidor, W. J. Mitchell, and K. L. Prikry. Precipitous ideals. *The Journal of Symbolic Logic*, 45:1–8, 1980.

[Jech, 1967] T. J. Jech. Non-provability of Souslin's hypothesis. *Commentationes Mathematicae Universitatis Carolinae*, 8:291–305, 1967.

[Jensen, 1968] R. B. Jensen. Souslin's hypothesis is incompatible with $V = L$ (abstract). *Notices of the American Mathematical Society*, 1968.

[Jensen, 1972] R. B. Jensen. The fine structure of the constructible hierarchy. *Annals of Mathematical Logic*, pages 229–308, 1972.

[Kanamori, 2003] A. Kanamori. *The Higher Infinite. Large Cardinals in Set Theory from their Beginnings.* Springer, 2003. Second edition.

[Kechris and Moschovakis, 1978] A. S. Kechris and Y. N. Moschovakis, editors. *Cabal Seminar 76-77. Proceedings of the Caltech-UCLA Logic Seminar 1976-77*, volume 689 of *Lecture Notes in Mathematics.* Springer-Verlag, 1978.

[Kechris *et al.*, 1981] A. S. Kechris, D. A. Martin, and Y. N. Moschovakis, editors. *Cabal Seminar 77-79. Proceedings of the Caltech-UCLA Logic Seminar 1977-79*, volume 839 of *Lecture Notes in Mathematics.* Springer-Verlag, 1981.

[Kechris *et al.*, 1983] A. S. Kechris, D. A. Martin, and Y. N. Moschovakis, editors. *Cabal Seminar 79-81. Proceedings of the Caltech-UCLA Logic Seminar 1979-81*, volume 1019 of *Lecture Notes in Mathematics.* Springer-Verlag, 1983.

[Kechris *et al.*, 1988] A. S. Kechris, D. A. Martin, and J. R. Steel, editors. *Cabal Seminar 81-85. Proceedings of the Caltech-UCLA Logic Seminar 1981-85*, volume 1333 of *Lecture Notes in Mathematics.* Springer-Verlag, 1988.

[Kunen and Vaughn, 1984] K. Kunen and J. E. Vaughn. *Handbook of Set-Theoretic Topology.* North-Holland, 1984.

[Kunen, 1970] K. Kunen. Some applications of iterated ultrapowers in set theory. *Annals of Mathematical Logic*, 1:179–227, 1970.

[Kunen, 1971a] K. Kunen. Elementary embeddings and infinitary embeddings. *The Journal of Symbolic Logic*, 36:407–413, 1971.

[Kunen, 1971b] K. Kunen. On the GCH at measurable cardinals. In R. O. Gandy, editor, *Logic Colloquium '69*, pages 107–110. North-Holland, 1971.

[Kunen, 1978] K. Kunen. Saturated ideals. *The Journal of Symbolic Logic*, 43:65–76, 1978.

[Laver, 1976] R. Laver. On the consistency of Borel's Conjecture. *Acta Mathematica*, 137:151–169, 1976.

[Laver, 1978] R. Laver. Making the supercompactness of κ indestructible under κ-directed closed forcing. *Israel Journal of Mathematics*, 29:385–388, 1978.

[Levinski *et al.*, 1990] J.-P. Levinski, M. Magidor, and S. Shelah. Chang's Conjecture for $\aleph_\omega$. *Israel Journal of Mathematics*, 69:161–172, 1990.

[Levy, 1965] A. Levy. A hierarchy of formulas in set theory. *Memoirs of the American Mathematical Society*, 57, 1965.

[Magidor, 1976] M. Magidor. How large is the first strongly compact cardinal? or: A study on identity crises. *Annals of Mathematical Logic*, 10:33–57, 1976.

[Magidor, 1977a] M. Magidor. On the singular cardinals problem. I. *Israel Journal of Mathematics*, 28:1–31, 1977.

[Magidor, 1977b] M. Magidor. On the singular cardinals problem. II. *Annals of Mathematics*, 106:514–547, 1977.

[Magidor, 1978] M. Magidor. Changing cofinality of cardinals. *Fundamenta Mathematicae*, 99:61–71, 1978.

[Magidor, 1982] M. Magidor. Reflecting stationary sets. *The Journal of Symbolic Logic*, 47:755–771, 1982.

[Martin and Solovay, 1970] D. A. Martin and R. M. Solovay. Internal Cohen extensions. *Annals of Mathematical Logic*, 2:1443–178, 1970.

[Martin and Steel, 1988] D. A. Martin and J. R. Steel. Projective determinacy. *Proceedings of the National Academy of Sciences USA*, 85:6582–6586, 1988.

[Martin and Steel, 1989] D. A. Martin and J. R. Steel. A proof of Projective Determinacy. *Journal of the American Mathematical Society*, 2:1–73, 1989.

[Martin, 1970] D. A. Martin. Measurable cardinals and analytic games. *Fundamenta Mathematicae*, 66:287–291, 1970.

[Martin, 1975] D. A. Martin. Borel determinacy. *Annals of Mathematics*, 102:363–371, 1975.

[Martin, 1980] D. A. Martin. Infinite games. In O. Lehto, editor, *Proceedings of the International Congress of Mathematicians, Helsinki 1978*, pages 269–273. Academia Scientiarum Fennica, 1980.

[Mitchell and Steel, 1994] W. J. Mitchell and J. R. Steel. *Fine Structure and Iteration Trees*, volume 3 of *Lecture Notes in Logic*. Springer, 1994.

[Mitchell *et al.*, 1997] W. J. Mitchell, E. Schimmerling, and J. R. Steel. The covering lemma up to a Woodin cardinal. *Annals of Pure and Applied Logic*, 84:219–255, 1997.

[Mitchell, 1972] W. J. Mitchell. Aronszajn trees and the independence of the transfer property. *Annals of Mathematical Logic*, 5:21–46, 1972.

[Mitchell, 1974] W. J. Mitchell. Sets constructible from sequences of ultrafilters. *The Journal of Symbolic Logic*, 39:57–66, 1974.

[Mitchell, 1978] W. J. Mitchell. Hypermeasurable cardinals. In M. Boffa, D. van Dalen, and K. McAloon, editors, *Logic Colloquium '78*, Studies in Logic and the Foundations of Mathematics, pages 303–316. North-Holland, 1978.

[Mitchell, 1984] W. J. Mitchell. The core model for sequences of measures. I. *Mathematical Proceedings of the Cambridge Philosophical Society*, 95:229–260, 1984.

[Mitchell, 1985] W. J. Mitchell. The core model for sequences of measures. II. Available at http://www.math.ufl.edu/ wjm/papers, 1985.

[Mitchell, 1987] W. J. Mitchell. Applications of the covering lemma for sequences of measures. *Transactions of the American Mathematical Society*, 299:41–58, 1987.

[Moschovakis, 1971] Y. N. Moschovakis. Uniformization in a playful universe. *Bulletin of the American Mathematical Society*, 77:731–736, 1971.

[Mycielski and Steinhaus, 1962] J. Mycielski and H. Steinhaus. A mathematical axiom contradicting the axiom of choice. *Bulletin de l'Académe Polonaise des Sciences, Série des Sciences Mathématiques, Astronomiques et Physiques*, 10:1–3, 1962.

[Nyikos, 1980] P. J. Nyikos. A provisional solution to the normalMoore space problem. *Proceedings of the American Mathematical Society*, 78:429–435, 1980.

[Prikry, 1970] K. L. Prikry. Changing measurable cardinals into accessible cardinals. *Dissertationes Mathematicae (Rozprawy Matematyczne)*, 68:5–52, 1970.

[Radin, 1982] L. B. Radin. Adding closed cofinal sequences to large cardinals. *Annals of Mathematical Logic*, 22:243–261, 1982.

[Schimmerling, 1994] E. Schimmerling. Combinatorial principles in the core model for one Woodin cardinal. *Annals of Pure and Applied Logic*, 74:153–201, 1994.

[Shelah and Woodin, 1990] S. Shelah and W. H. Woodin. Large cardinals imply that every reasonably definable set of reals is Lesbegue measurable. *Israel Journal of Mathematics*, 70:381–394, 1990.

[Shelah, 1974] S. Shelah. Infinite abelian groups, Whitehead problem and some constructions. *Israel Journal of Mathematics*, 18:243–256, 1974.

[Shelah, 1982] S. Shelah. *Proper Forcing*, volume 940 of *Lecture Notes in Mathematics*. Springer-Verlag, 1982.

[Shelah, 1983] S. Shelah. The singular cardinals problem: Independence results. In A. R. D. Mathias, editor, *Surveys in Set Theory*, volume 87 of *London Mathematical Society Lecture Note Series*, pages 116–134. Cambridge University Press, 1983.

[Shelah, 1984] S. Shelah. Can you take Solovay's inaccessible away? *Israel Journal of Mathematics*, 48:1–47, 1984.

[Shelah, 1986] S. Shelah. *Around Classification Theory of Models*, volume 1182 of *Lecture Notes in Mathematics*. Springer-Verlag, 1986.

[Shelah, 1987] S. Shelah. Semiproper Forcing Axiom implies Martin's Maximum but not PFA^+. *The Journal of Symbolic Logic*, 52:360–367, 1987.

[Shelah, 1987a] S. Shelah. Iterated forcing and normal ideals on ω_1. *Israel Journal of Mathematics*, 60:345–380, 1987.

[Shelah, 1994] S. Shelah. *Cardinal Arithmetic*, volume 29 of *Oxford Logic Guides*. Oxford University Press, 1994.

[Shelah, 1998] S. Shelah. *Proper and Improper Forcing*, volume 29 of *Perspectives in Mathematical Logic*. Springer-Verlag, 1998.

[Shoenfield, 1961] J. R. Shoenfield. The problem of predicativity. In Y. Bar-Hillel, E. I. J. Poznanski, M. O. Rabin, and A. Robinson, editors, *Essays on the Foundations of Mathematics*, pages 132–139. Magnes Press, 1961.

[Shoenfield, 1971] J. R. Shoenfield. Unramified forcing. In D. S. Scott, editor, *Axiomatic Set Theory*, volume 13(1) of *Proceedings of Symposia in Pure Mathematics*, pages 357–381. American Mathematical Society, Providence, 1971.

[Silver, 1971a] J. H. Silver. The consistency of the GCH with the existence of a measurable cardinal. In Dana S. Scott, editor, *Axiomatic Set Theory*, Proceedings of Symposia in Pure Mathematics, pages 391–395. American Mathematical Society, Providence, 1971.

[Silver, 1971b] J. H. Silver. The independence of Kurepa's conjecture and two-cardinal conjectures in model theory. In D. S. Scott, editor, *Axiomatic Set Theory*, volume 13(1) of *Proceedings of Symposia in Pure Mathematics*, pages 383–390. American Mathematical Society, Providence, 1971.

[Silver, 1971c] J. H. Silver. Some applications of model theory in set theory. *Annals of Mathematical Logic*, 3:45–110, 1971.

[Silver, 1975] J. H. Silver. On the singular cardinals problem. In *Proceedings of the International Congress of Mathematicians, Vancouver, B. C., 1974*, volume 1, pages 265–268. Canadian Mathematical Congress, 1975.

[Solovay and Tennenbaum, 1971] R. M. Solovay and S. Tennenbaum. Internal Cohen extensions and Souslin's problem. *Annals of Mathematics*, 94:201–245, 1971.

[Solovay *et al.*, 1978] R. M. Solovay, W. N. Reinhardt, and A. Kanamori. Strong axioms of infinity and elementary embeddings. *Annals of Mathematical Logic*, 13:73–116, 1978.

[Solovay, 1966] R. M. Solovay. Real-valued measurable cardinals (abstract). *Notices of the American Mathematical Society*, 13:721, 1966.

[Solovay, 1967] R. M. Solovay. A nonconstructible Δ^1_3 set of integers. *Transactions of the American Mathematical Society*, 127:50–75, 1967.

[Solovay, 1971] R. M. Solovay. Real-valued measurable cardinals. In Dana S. Scott, editor, *Axiomatic Set Theory*, volume 13(1) of *Proceedings of Symposia in Pure Mathematics*. American Mathematical Society, Providence, 1971.

[Steel, 1996] J. R. Steel. *The Core Model Iterability Problem*. Springer, 1996.

[Stewart, 1966] D. H. Stewart. Master's thesis, University of Bristol, 1966.

[Tennenbaum, 1968] S. Tennenbaum. Souslin's problem. *Proceedings of the National Academy of Sciences USA*, 59:60–64, 1968.

[Todorcevic, 1984] S. Todorcevic. A note on the Proper Forcing Axiom. In James E. Baumgartner, D. Anthony Martin, and Saharon Shelah, editors, *Axiomatic Set Theory*, volume 31 of *Contemporary Mathematics*, pages 209–218. American Mathematical Society, 1984.

[Todorcevic, 2007] S. Todorcevic. *Walks on Ordinals and their Characteristics*, volume 263 of *Progress in Mathematics*. Birkhäuser, 2007.

[Woodin, 1990] W. H. Woodin. Supercompact cardinals, sets of reals, and weakly homogeneous trees. *Proceedings of the National Academy of Sciences USA*, 85:6587–6591, 1990.

[Woodin, 1999] W. H. Woodin. *The Axiom of Determinacy , Forcing Axioms and the Non-stationary Ideal*, volume 1 of *de Gruyter Series in Logic and Its Applications.* de Gruyter, 1999.
[Woodin, 2001a] W. H. Woodin. The Continuum Hypothesis. I. *Notices of the American Mathematical Society*, 48:567–576, 2001.
[Woodin, 2001b] W. H. Woodin. The Continuum Hypothesis. II. *Notices of the American Mathematical Society*, 48:681–690, 2001.

INNER MODELS FOR LARGE CARDINALS

William J. Mitchell

One new idea leads to another, that to a third, and so on through a course of time until someone, with whom no one of these is original, combines all together, and produces what is justly called a new invention.[1]

1 INTRODUCTION

This chapter will discuss the development of inner models for large cardinals during the last part of the last century, beginning at the end of Kanamori's chapter [2011b], which is to say at about 1965. There are two major themes in this development: The first of these is the expansion of Gödel's class L of constructible sets to obtain L-like models of ZFC which are capable of containing any larger cardinals existing in the universe, and the second is the development of core model theory, which provides the possibility of proving that the universe is well approximated by such a model.

The main part of the chapter is accordingly divided into two themes. The first is in Section 2, which describes the process of incorporating large cardinal properties into the constructibility paradigm, and the discovery of strong uniqueness properties of the resulting models, allowing the class of large cardinals to be seen as a well-ordered extension of the ordinals. The second theme is begun in Section 3, which describes the evolution of combinatorial principles in L, leading through the development of fine structure to the discovery of the Covering Lemma. The generalization of this development into a core model theory for larger cardinals is described in Section 4. The two themes merge in Section 5, which describes the extension of the models to the level of Woodin cardinals, where the models involved rely on fine structure and other techniques of core model theory, even when the maximality implied by the core model is not assumed.

Section 1 provides an introduction, which discusses the basic concepts of constructibility, on the one hand, and of large cardinals, on the other, as they were understood prior to the start of the main narrative. Both of these topics are covered more fully in other chapters of this volume: constructibility in [Kanamori, 2011b] (also see [Kanamori, 2007]) and large cardinals in [Kanamori, 2011a].

[1] Thomas Jefferson, quoted in Bedini, "Godfather of American Invention," p. 82.

Handbook of the History of Logic: Sets and Extensions in the Twentieth Century.
Volume editor: Akihiro Kanamori. General editors: Dov M. Gabbay, Paul Thagard and John Woods.

1.1 *Constructibility*

According to Kanamori [2011b], "An *inner model* is a transitive class containing all the ordinals such that, with membership and quantification restricted to it, the class satisfies each axiom of ZF." If we restrict consideration to definable classes, we have three principal examples. The first example is the universe V which, although not generally thought of as an inner model, does satisfy the definition; and it also provides, via the cumulative hierarchy defined by von Neumann in 1929, the beginnings of the structural analysis which will be seen in other inner models to be discussed.

The second example of an inner model is the class of *hereditarily ordinal definable* sets (HOD) which was introduced independently by several people to give an easier proof of the consistency of the Axiom of Choice. The model HOD has no intrinsic structure: the cumulative hierarchy in V is needed to see that the class HOD is even definable, and moreover the formula defining it is not absolute, so that the model $\mathrm{HOD}^{(\mathrm{HOD})}$, that is, the model HOD as defined inside HOD, may well be a proper subclass of HOD. Its primary use has been to produce, with the help of forcing, models which do not satisfy the Axiom of Choice. However, a precursor to the core model and covering lemma can be seen in Vopěnka's theorem [Vopěnka and Hájek, 1972] stating that any set of ordinals is in a generic extension of HOD. The theme of HOD as a core model has recently been investigated: see, for example, [Steel, 1995].

The third example of an inner model, and the prototype of the models with which this chapter is concerned, is Gödel's class L of constructible sets, introduced by Gödel [1938; 1939; 1940]. The model L is defined using a hierarchy like that of the first example, but the hierarchy incorporates definability by modifying the successor step to $L_{\alpha+1} = \operatorname{def}(L_\alpha)$ where $\operatorname{def}(X)$ is the set of subsets of X which are first-order definable over the model $(X, \in)$, using parameters from X. In this case the definability is restricted to the emerging class being defined, and hence the definition of the class L is independent of the universe in which it is being defined. Thus the class L is minimal in the sense that it is contained in any model of ZF which contains all of the ordinals. A simple induction shows that it satisfies the Axiom of Choice and that $|L_\alpha| = |\alpha|$ for all infinite ordinals α. To prove the Generalized Continuum Hypothesis (GCH) it then suffices to show that $\mathcal{P}(\kappa) \subseteq L_{\kappa^+}$ for every cardinal κ. This is a consequence of what has become known as the principle of *condensation:*

LEMMA 1 (Condensation.) *Suppose that α is an infinite ordinal and $X \prec L_\alpha$. Then $X \cong L_{\bar{\alpha}}$ for some ordinal $\bar{\alpha} \leq \alpha$.*

This principle is one of the major themes in the developments described in this chapter.

Relative constructibility

Relative constructibility was introduced independently by Hajnal [1956; 1961] and Levy [1957; 1960], who gave definitions which were equivalent in their original applications, but are quite different in general. Hajnal modified the definition of L to define $L(A)$, where A is an arbitrary transitive set, by setting $L_0(A) = A$. Thus $L(A)$ is the smallest model of ZF which contains $A \cup \{A\}$. The model $L(A)$ need not satisfy the Axiom of Choice, and indeed is most commonly used in applications where the Axiom of Choice does not hold.

Levy's definition of $L[A]$, on the other hand, retained from L the definition of $L_0[A] = \emptyset$, and instead modified the successor step, defining $L_{\alpha+1}[A]$ to be the set of subsets of $L_\alpha[A]$ which are definable in $L_\alpha[A]$ using parameters from $L_\alpha[A]$ together with the predicate A. Thus $L[A]$ is the smallest model M of ZF which has $A \cap M$ as a member, and $L[A]$ always satisfies the Axiom of Choice; however A is not, in general, a subset of $L[A]$. The model $L[A]$ can be viewed as the smallest model which contains the structure of the set A, and since this chapter is concerned with models including large cardinal structure, essentially all models discussed will be of this type.

In both cases the definability of the model requires the use of A as a parameter. For this reason the condensation principle, Lemma 1, is severely weakened. It is, in general, valid for $L(A)$ only in the case that $A \subseteq X$. For $L[A]$ there is the general statement:

LEMMA 2. *Suppose $\pi\colon M \cong X \prec L_\alpha[A]$ is the inverse of the the transitive collapse of X. Then $M = L_{\bar\alpha}[\bar A]$ for some ordinal $\bar\alpha \leq \alpha$, where $\bar A = \pi^{-1}[A \cap X]$.*

In particular, if $\bar A = A \cap L_{\bar\alpha}[\bar A]$ then $X \cong L_{\bar\alpha}[A]$.

Levy originally applied this condensation principle in cases in which the set A is contained in a transitive subset of X, so that $A = \bar A$. Using this he could prove, for example, that if $A \subseteq \kappa$ then $2^\lambda = \lambda^+$ in $L[A]$ for all $\lambda \geq \kappa$. The successful application of models of the form $L[A]$ to large cardinal theory has come from choosing the set A to encode the large cardinal structure, and then using that structure to see that $\bar A = A \cap L_{\bar\alpha}[\bar A]$ holds even in cases that π is not the identity on A.

In the model $L[A]$ the set A is being regarded as providing structure. If U is a ultrafilter on a cardinal κ, for example, then the model $L(U \cup \{U\})$ would include all subsets of κ, whether or not they have any relation to measurability. The model $L[U]$, on the other hand, contains only those subsets of κ which are required by the possibility of using U as a predicate. As will be seen, these are exactly the sets which are required by the existence of *any* measure on κ.

1.2 Large cardinals

Of the large cardinal properties known at the start of this history, two will dominate it: measurability and supercompactness.

Measurable cardinals were introduced by Ulam [1930], but we will use the later formulation in terms of elementary embeddings and normal ultrafilters: A cardinal κ is *measurable* if there is an elementary embedding $i\colon V \to M$ into a well-founded model M with *critical point* κ, that is, such that $i(\kappa) > \kappa$, but $i(\nu) = \nu$ for all $\nu < \kappa$. The normal ultrafilter, or *measure*, associated with this embedding is $U = \{\, x \subseteq \kappa : \kappa \in i(x) \,\}$. The measure U in turn leads to an embedding via the ultrapower construction, $i^U : V \to \mathrm{Ult}(V, U)$.

The concept of a measurable cardinal will serve as both a goal and a benchmark. The first part of the story will largely concern the effect which the presence of a measurable cardinal has on the constructible universe L, and the understanding of the minimal model $L[U]$ containing a measurable cardinal. The rest of the story will concern the generalization of this model and associated techniques to accommodate larger cardinals.

A number of large cardinals weaker than a measurable cardinal were known at the start of this history: these include inaccessible, Mahlo, and weakly compact cardinals. The most important for our purposes are Ramsey cardinals, which satisfy the partition relation $\kappa \to (\kappa)_2^{<\omega}$ and their generalization the α-Erdős cardinals, which satisfy the partition relation $\kappa \to (\alpha)_2^{<\omega}$.

Supercompact cardinals were introduced, at the start of the period of this history, by Reinhardt and Solovay although they were only published later in [Solovay *et al.*, 1978]: a cardinal κ is λ-*supercompact* if there is an elementary embedding $i\colon V \to M$ with critical point κ such that ${}^{\lambda}M \subseteq M$, and is *supercompact* if it is λ-supercompact for all λ. In contrast to measurable cardinals, it enters this history only as a goal: as this is being written, there is still no fully developed L-like model for even a κ^+-supercompact cardinal.

Given this fact, it is natural to ask about cardinal properties intermediate between measurable cardinals; at the start of this history, however there were, with one possible exception, no such cardinals. The possible exception is a *strongly compact cardinal.* This notion was defined by Tarski [1962] using infinitary logic but characterized by Reinhardt and Solovay as being supercompactness without normality: that is, with the condition ${}^{\lambda}M \subseteq M$ replaced by the covering property $\forall x \in [M]^{\lambda}\, \exists y \in M\ (x \subseteq y \wedge |y| < j(\kappa))$. It is conjectured that having a strongly compact cardinal is equiconsistent with having a supercompact cardinal, and proving this has been described as the holy grail problem of inner model theory.

A number of intermediate large cardinal properties have since been discovered, and will be described in due course. Most of these are modifications either of measurability or of supercompactness, but one exception is worth mentioning here: *Woodin cardinals*, discovered in 1984, have come to have an independent importance comparable to that of measurable and supercompact cardinals.

2 DEVELOPMENT OF INNER MODELS WITHOUT FINE STRUCTURE

2.1 *Large cardinals in L*

The study of L-like models for large cardinals naturally starts with L itself. Gödel [1938] observed that any inaccessible or Mahlo cardinal in V has the same property in L, simply because these properties are downward absolute to any inner model. After the discovery of weakly compact cardinals it was quickly realized that they too are downward absolute to L, by a more complicated argument. The same was true of several related stronger properties.

The study of inner models for larger cardinals can be said to have started with Dana Scott's [1961] proof that there cannot be any measurable cardinals in L, by observing that otherwise there would be an elementary embedding $i^U\colon L \to M = \operatorname{Ult}(L, U)$ in M such that $i^U(\kappa) > \kappa$, where κ is the least measurable cardinal in L. This is impossible because M, as an inner model of the minimal model L, must be equal to L; but M satisfies the sentence asserting that $i(\kappa)$ is the least measurable cardinal.

The response to Scott's observation took two directions. The first direction, which concentrated on the model L in order to further characterize the consequences of the existence in V of a measurable cardinal, led to Silver's discovery of $0^{\#}$. The other direction developed the model $L[U]$ as the minimal model, analogous to L, which contains a measurable cardinal. These two approaches use many of the same techniques, and they eventually merged in Dodd and Jensen's construction of the core model.

2.2 *What a measurable cardinal says about L*

Scott's result was followed up in the model L by Gaifman [1974] and Rowbottom [1971]. Gaifman, working in 1963, realized that the ultrapower embedding used by Scott could be iterated, and in the next year he realized that this iteration could be carried out using a countable elementary substructure of an appropriate V_λ. This showed, for example, that every cardinal of V is inaccessible in L. Rowbottom obtained similar results using partition properties. He reproved an earlier result of Erdős and Hajnal that any measurable cardinal is a Ramsey cardinal: *If U is a measure on κ and $f\colon [\kappa]^{<\omega} \to \lambda$ for $\lambda < \kappa$ then there is a set $A \in U$ such that f is constant on $[A]^n$ for each $n \in \omega$.* He then used this to show that κ is what came to be known as a *Rowbottom cardinal,* which is a cardinal κ with the following property: *Suppose that $(\kappa, A, \dots)$ is any structure with universe κ, that ρ is a cardinal less than κ, and $A \subseteq \kappa$ with $\rho < |A| < \kappa$. Then there is a set $X \subseteq \kappa$ with $\rho \subseteq X$ and $|X \cap A| = \rho$* such that $(X, A \cap X, \dots) \prec (\kappa, A, \dots)$. He then used condensation to show that this implies that every successor cardinal of V is inaccessible in L.

This analysis of the structure of L in the presence of a measurable cardinal took its final form with Silver's discovery of $0^{\#}$. He observed that starting from a Ramsey cardinal, or even an ω_1-Erdős cardinal, one could obtain an uncountable set of indiscernibles for L, that is, a set of ordinals A such that any two finite increasing sequences of members of A of the same length satisfy the same formulas in L. Using the condensation property of L, together with model-theoretic methods of Ehrenfeucht and Mostowski, he then showed that any such set could be extended to a proper class I of indiscernibles with the property that every set $x \in L$ is definable in L using only parameters from I. Furthermore, he noted that if the class I of indiscernibles is chosen so that its ωth member is as small as possible, then the class I would be closed and unbounded, and hence uniquely determined. Finally he defined the set $0^{\#} \subseteq \omega$ to be

$$0^{\#} = \{ \ulcorner \phi(x_0, \dots, x_{n-1}) \urcorner : L \models \phi(c_0, \dots, c_{n-1}) \} \tag{1}$$

and showed that $0^{\#}$ can be used to construct, and can be constructed from, the class I of indiscernibles. This class has since become known as the class of *Silver indiscernibles.*

Solovay independently discovered the set $0^{\#}$, and observed that the singleton $\{0^{\#}\}$ is Π^1_2-definable over the reals, and thus $0^{\#}$ is a Δ^1_3 subset of ω which is not constructible.

The existence of $0^{\#}$ has three important consequences. (i) As mentioned above, the set $\{0^{\#}\}$ is definable over the real numbers, by an absolute Π^1_2 formula which describes the process of obtaining I from $0^{\#}$. (ii) It takes the results of Rowbottom and Gaifman to a natural conclusion: if α is any member of I, and in particular any uncountable cardinal of V, then α is, in L, inaccessible and weakly compact—indeed α has every large cardinal property which can hold in L. (iii) It implies that there is a nontrivial elementary embedding $L \to L$. Indeed every increasing embedding from I into I can be extended to an elementary embedding from L into L, and every elementary embedding $i\colon L \to L$ is determined by its restriction to the class I. This analysis was rounded out by Kunen, who proved that the existence $0^{\#}$ follows from the existence of any nontrivial embedding $i\colon L \to L$.

Other work of Silver reinforced the view that the existence of $0^{\#}$ is the weakest large cardinal property not compatible with L. He noted [Silver, 1970] that although the existence of $0^{\#}$, and hence $V \neq L$, follows from the existence of an ω_1-Erdős cardinal, any cardinal which is α-Erdős for any ordinal — or even all ordinals — less than ω_1^L has the same property in L.

It is evident that the same analysis which gave $0^{\#}$ can be carried out for any set $a \subseteq \omega$, by defining $a^{\#}$ to be the theory of a closed unbounded class of indiscernibles for the structure $(L[a], \in, a)$. In particular the sharp construction can be iterated: setting $0^{\alpha\#} = \langle 0^{\nu\#} : \nu < \alpha \rangle^{\#}$ for any countable ordinal α. This construction gives the beginning of a hierarchy of inner models, $L[\langle 0^{\alpha\#} : \alpha < \theta \rangle]$. The case when θ is equal to ω_1 in the resulting model gives a minimal model for the statement that $a^{\#}$ exists for all reals a, which is the correct hypothesis for many results giving consequences for the continuum of the existence of a measurable cardinal. For example, results of Martin and Solovay [1969], including their $\mathbf{\Sigma}^1_3$-absoluteness result, follow from this hypothesis.

2.3 $L[U]$

The other direction, defining a model analogous to L with a measurable cardinal, was initiated by Solovay. If U is a measure on a cardinal κ then it is easy to see that $U \cap L[U] \in L[U]$, and that $U \cap L[U]$ is a measure on κ. In addition, the same arguments as were used for L can be used to show that $L[U]$ satisfies the axioms of ZFC.

Condensation, however, is more problematic. It was straightforward to show that it holds for any set $X \prec L[U]$ such that $\kappa \subseteq X$, and it follows that $L[U] \models 2^\lambda = \lambda^+$ for all $\lambda \geq \kappa$. Solovay also showed that κ is the only measurable cardinal in $L[U]$.

Condensation does not hold for arbitrary sets $X \prec L_\alpha[U]$; however [Silver, 1971] extended Gödel's proof that $2^\lambda = \lambda^+$ in L to cardinals $\lambda < \kappa$ in $L[U]$ by using Rowbottom's partition theorem (cf. Section 2.1) to find, for each $x \subseteq \lambda$ in $L[U]$, a set $X \prec L_a[U]$ with $X \cap \kappa \in U$ and $\lambda \cup \{x\} \subseteq X$ but $|\mathcal{P}(\lambda) \cap X| = \lambda$. Then condensation does hold for such X: if $\pi \colon L_{\bar{\alpha}}[\bar{U}] \cong X$ is the transitive collapse then $\{\lambda \in \kappa : \pi(\lambda) = \lambda\} \in U$ and hence $\bar{U} = \pi^{-1}[U \cap X] = U \cap L_{\bar{\alpha}}[U]$. This is precisely what is needed to show that $L[U] \models 2^\lambda = \lambda^+$.

The final step towards establishing $L[U]$ as a close analog of the model L was taken by Kunen in his thesis [1968; 1970], which adapted from Gaifman the use of iterated ultrapowers. The thesis began by refining Gaifman's theory of iterated ultrapowers. Kunen defined an M-ultrafilter on a transitive model M of set theory to be an ultrafilter U, not necessarily a member of M, on $\mathcal{P}(\kappa) \cap M$ such that $X \cap U \in M$ whenever $X \in M$ and $|X|^M = \kappa$, and he showed that this property sufficed for the construction of an iterated ultrapower $\mathrm{Ult}_\alpha(M, U)$.

He then turned to the model $L[U]$, showing that *if the models $L[U]$ and $L[U']$ satisfy that U and U', respectively, are measures on the same cardinal κ, then $L[U] = L[U']$ and $U \cap L[U] = U' \cap L[U]$.*

It follows that $U \cap L[U]$ is the only measure in $L[U]$, and that $L[U]$ is a definable subset of any class model of ZFC + "κ is a measurable cardinal".

Kunen's proof used several techniques which became important to the later development of inner model theory. First, he took from Gaifman the idea of iterated ultrapowers: if M is a model and U a measure on κ in M, then the iterated ultrapowers $i^U_\alpha \colon M \to \mathrm{Ult}_\alpha(M, U)$ are defined by setting $\mathrm{Ult}_0(M, U) = M$, $\mathrm{Ult}_{\alpha+1}(M, U) = \mathrm{Ult}(\mathrm{Ult}_\alpha(M, U), i_\alpha(U))$, and taking the direct limit when α is a limit ordinal. He observed that if $L[U]$ and $L[U']$ are as assumed, then $\mathrm{Ult}_{\kappa^{++}}(L[U], U) = \mathrm{Ult}_{\kappa^{++}}(L[U'], U')$, since both models are equal to $L[\mathcal{C}]$ where $\mathcal{C}$ is the closed unbounded filter on κ^{++}. Further, he showed that if Γ is any proper subclass of the image $\{ i^U_{\kappa^{++}}(\alpha) : \alpha \text{ an ordinal} \}$ in $L[\mathcal{C}]$ such that $\kappa \subseteq \Gamma$ then $L[U]$ is isomorphic to the Skolem hull of Γ in $L[\mathcal{C}]$, and similarly for $L[U']$. Finally, he noted that the intersection Γ of the ranges of $i^U_{\lambda^{++}}$ and $i^{U'}_{\lambda^{++}}$ is a proper class, so that each of $L[U]$ and $L[U']$ is the Skolem hull of Γ in $L[\mathcal{C}]$. It follows that $L[U] = L[U']$.

Kunen also applied these arguments to the case when $L[U]$ and $L[U']$ satisfy that U and U', respectively, are measures on different cardinals κ and κ'. He showed that in this case, if $\kappa < \kappa'$, then $L[U'] = \mathrm{Ult}_\alpha(L[U], U)$ for some ordinal α. Thus an appropriate analog to L for the theory ZFC + "there is a measurable cardinal" is the model $L[U]$ satisfying that U is a measure on some κ, where U is chosen so that κ is as small as possible.

Techniques such as these were also applied in situations in which no actual measurable cardinal was assumed. Several examples of this were due to Kunen, but perhaps the most important example is Solovay's proof [1971] that the existence of a real-valued measurable cardinal is equiconsistent with that of a (two-valued) measurable cardinal. One direction of this involved showing that if μ is a real-valued measure on κ and F is the filter of sets $x \subseteq \kappa$ such that $\mu(x) = 1$, then $F \cap L[F]$ is an ultrafilter on κ in $L[F]$.

2.4 More measurable cardinals

The definition of the model $L[U]$ can easily be extended to models of the form $L[\mathcal{U}]$, where $\mathcal{U} = \langle U_\nu : \nu < \alpha \rangle$ is a sequence of measures on an increasing sequence of cardinals $\langle \kappa_\nu : \nu < \alpha \rangle$. Solovay defined $0^\dagger$ by using indiscernibles for $L[U]$ in the same way in which $0^\#$ is defined using indiscernibles for L. He briefly hoped that models with more than one measurable cardinal would lead to further progress in descriptive set theory, but quickly realized that a model of the form $L[U_0, \ldots, U_{n-1}]$ with n measurable cardinals has essentially the same properties as does $L[U]$. Silver [1971] stated that the techniques he used for $L[U]$ can be extended to a model of the form $L[\mathcal{U}]$, provided that α is countable. Kunen similarly pointed out that the same techniques he used for $L[U]$ worked for models $L[\mathcal{U}]$ where $\mathcal{U}$ is a sequence of arbitrary length α, so long as α was smaller than the first measurable cardinal κ_0 of the sequence $\mathcal{U}$. In [Kunen, 1970] he used such models to show that one could obtain models with arbitrarily many measurable cardinals from a strongly compact cardinal κ, and in [Kunen, 1971] he reached the same conclusion from the assumption that the GCH fails at a measurable cardinal κ, or that every κ-complete filter can be extended to an ultrafilter.

Kunen's hypothesis, that there were fewer than measurably many measurable cardinals, was not by any means a rigid limit on the effectiveness of his techniques; however longer sequences of measures did entail additional complications, and a new framework was needed to deal with these complications in a general way. This framework came out of consideration of a much less important question: how many different measures could a single cardinal carry? Kunen showed that the model $L[U]$ had only one measure, and Kunen and Paris [1971] had presented a model in which there are 2^{2^κ} measures, the maximum possible. However no model was known with any intermediate number of measures, and they ended their paper with the following question: "Can the number of normal ultrafilters on a measurable cardinal be intermediate between 1 and 2^{2^κ}? Can this number be 2?"

Kunen [1970] gave an inductive proof that every measurable cardinal has a measure concentrating on smaller nonmeasurable cardinals: Suppose that for each measurable cardinal $\lambda < \kappa$ we have such a measure U_λ on λ, and let U be a measure on κ which does concentrate on smaller measurable cardinals. Then there is a second measure on κ, concentrating on nonmeasurable cardinals, which can be defined by setting $U_\kappa = \{x \subseteq \kappa : \{\lambda < \kappa : x \cap \lambda \in U_\lambda\} \in U\}$, or equivalently $U_\kappa = i(\langle U_\lambda : \lambda < \kappa \rangle)_\kappa$. Seeing this Mitchell, then finishing up his graduate work at Berkeley, asked whether there could be a model in which these are the only two measures on κ? It should be noted that the model $L[U, U_\kappa]$ would not do: this would be equal to $L[U]$, since the measures U and U_κ would agree on every set constructed using them as predicates. Two years later, driving back to Berkeley for the summer after a digression into category theory, he realized how to construct such a model. In [Mitchell, 1974] he defined an order on the ultrafilters on κ, by saying that $U \triangleleft U'$ if $U \in \mathrm{Ult}(V, U')$. Thus, for example, $U_\kappa \triangleleft U$ in Kunen's example discussed above. By an argument like that which Scott had used to prove that there are no measurable cardinals in L, Mitchell showed that the partial order $\triangleleft$ is well-founded, and this fact allowed a definition of the order of a measure, $o(U) = \sup\{\, o(U') + 1 : U' \triangleleft U \,\}$, and of a cardinal, $o(\kappa) = \sup\{\, o(U) + 1 :$ U is a measure on $\kappa\,\}$. He then defined a *coherent sequence of ultrafilters* to be a function $\mathcal{U}$, with domain of the form $\{\, (\kappa, \beta) : \beta < o^{\mathcal{U}}(\kappa) \,\}$ for some function $o^{\mathcal{U}}$, which witnesses that $o(\mathcal{U}(\kappa, \beta)) = \beta$ in the sense that $i^{\mathcal{U}(\kappa,\beta)}(\mathcal{U})(\kappa, \alpha) = \mathcal{U}(\kappa, \alpha)$ for all $\alpha < \beta < o^{\mathcal{U}}(\kappa)$, and he showed that if $\mathcal{U}$ is a coherent sequence then many of the basic properties of $L[U]$ can be extended to $L[\mathcal{U}]$. In particular, $L[\mathcal{U}]$ is a model of ZFC + GCH, and the only normal measures in $L[\mathcal{U}]$ are the sets $\mathcal{U}(\kappa, \beta) \cap L[\mathcal{U}]$ with $\beta < o^{\mathcal{U}}(\kappa)$. Thus, provided suitable coherent sequences could be found, this construction provided models with any desired number δ of normal measures on a cardinal κ, with $0 \leq \delta \leq \kappa^{++}$. Since the models satisfied GCH, this was the maximum possible number of measures.

For any finite n it was easy, if given a cardinal κ with $o(\kappa) \geq n$, to find a coherent sequence $\mathcal{U}$ with $o^{\mathcal{U}}(\kappa) = n$. This gave models which exactly n measures, for any finite n. Finding coherent sequences with $o^{\mathcal{U}}(\kappa)$ infinite was more difficult. Mitchell used what he later termed a μ-measurable cardinal to obtain sequences with $o(\kappa) = \lambda$ for any $\lambda \leq \kappa^{++}$. It is a commentary on the growth of our understanding of large cardinals that the smallest cardinal property generally known at the time which implied the existence of a μ-measurable cardinal — which is barely stronger than $\exists\kappa\, o(\kappa) = \kappa^{++}$ — was supercompactness. Later it was determined that the μ-measurable cardinal is not necessary. It was shown, first by using core model theory but later, in [Mitchell, 1983], with more elementary methods: *If $o(\kappa) = \delta$ then there is an inner model $L[\mathcal{U}]$ in which either $o^{\mathcal{U}}(\kappa) = \delta$ or else $o^{\mathcal{U}}(\kappa) = \kappa^{++L[\mathcal{U}]} < \delta$.* This was achieved by using a sequence $\mathcal{U}$ which might not be fully coherent in V: instead the measures on the sequence $\mathcal{U}$ were only required to have the correct order. In $L[\mathcal{U}]$ the sequence becomes fully coherent, because no set is constructed in $L[\mathcal{U}]$ which would witness a failure of coherence. This concept was independently discovered by A. Dodd and has come to be used as an important tool, referred to as *Doddages*, in core model theory.

Iterating the least difference

The essential tool in the analysis of $L[\mathcal{U}]$ was a generalization of Kunen's use of iterated ultrapowers which has come to be known as the process of *iterating the least difference.* This process is central to the development of inner model theory, and it seems appropriate to describe it a bit more fully so as to have a framework for discussing how its use has influenced the development of later inner models. To this end we will give a brief sketch of the proof of a simple application: *every measure in $L[\mathcal{U}]$ is a member of the sequence $\mathcal{U}$.*

Supposing the contrary, we may assume that $\mathcal{U}$ is the shortest sequence for which the assertion to be proved is false, and κ is the smallest cardinal for which there is a measure W in $L[\mathcal{U}]$ which is not on the sequence $\mathcal{U}$.

The proof uses a comparison of $L[\mathcal{U}]$ with the ultrapower $\mathrm{Ult}(L[\mathcal{U}], W)$ of that model by the supposed extraneous ultrafilter W. The comparison is made by iterated ultrapowers of the two models,

$$\begin{array}{llcl} i_{0,\theta}\colon & L[\mathcal{U}] & \Longleftarrow\!\Longrightarrow & M_0 \xrightarrow{i_{01}} M_1 \xrightarrow{i_{12}} \cdots M_\theta \\ & \Big\downarrow{\scriptstyle i^W} & & \\ j_{0,\theta}\colon & \mathrm{Ult}(L[\mathcal{U}], W) & = & N_0 \xrightarrow{j_{01}} M_1 \xrightarrow{j_{12}} \cdots N_\theta \end{array} \tag{2}$$

The term "iterating the least difference" describes the way the iterated ultrapowers $i_{0,\theta}$ and $j_{0,\theta}$ are defined: at stage ν we have $M_\nu = L[i_{0\nu}(\mathcal{U})]$ and $N_\nu = L[j_{0\nu} \circ i^W(\mathcal{U})]$. The next models $M_{\nu+1}$ and $N_{\nu+1}$ in the sequence are defined by simply identifying the first ultrafilter occurring in one of the sequences $i_{0\nu}(\mathcal{U})$ and $j_{0\nu} \circ i^U(\mathcal{U})$ which does not occur in the other, and taking the ultrapower of that model by that ultrafilter. A crucial fact, the proof of which we skip over, is that this process always terminates, either with the final models M_θ and N_θ being equal, or else with one being a proper initial segment of the other. In this case the models must be equal, since each satisfies the formula stating that its sequence of measures is the shortest sequence for which the statement being proved is false.

Now we can conclude the proof of the statement: we want to show that the ultrafilter W, supposedly not on the sequence $\mathcal{U}$, is in fact equal to the first ultrafilter used in the iterated ultrapower $i_{0\theta}$. It can be shown that this ultrafilter is on κ; write it as $U = \mathcal{U}(\kappa, \beta)$, so $i_{01} = i^U$. Now suppose to the contrary that $U \neq W$, and let x be the first subset of κ on which W and U disagree. The argument is concluded by showing that

$$x \in U \iff \kappa \in i^U(x) = i_{01}(x) \iff \kappa \in i_{0\theta}(x) \tag{3}$$

$$x \in W \iff \kappa \in i^W(x) \iff \kappa \in j_{0\theta} \circ i^W(x) \tag{4}$$

and noting that the whole construction, and in particular the set x, is definable in $L[\mathcal{U}]$, so that $i_{0\theta}(x) = j_{0,\theta} \circ i^W(x)$. Hence $W = U$.

This concludes the proof of the assertion that there are no measures in $L[\mathcal{U}]$ other than those on the sequence $\mathcal{U}$. We now isolate, for future reference, three factors which are necessary to the success of this argument:

Iterability. All iterated ultrapowers of $L[\mathcal{U}]$, and in particular the final models M_θ and N_θ in the comparison, are well-founded.

Equal comparison. The final models of the iteration are equal: neither of M_θ or N_θ is a proper initial segment of the other.

No moving generators. None of the critical points of ultrafilters used in the iterated ultrapowers of diagram (2) are moved by later ultrapowers in the same iteration.

Two of these terms, *iterability* and *moving generators*, are standard. For the models currently under consideration, iterability can be insured by using countably complete ultrafilters: either, as in this case, countably complete in $L[\mathcal{U}]$ or, in the case of a proof where the iteration is not definable in $L[\mathcal{U}]$, countably complete in V. Later, for larger cardinals, other methods became necessary. At the time of writing, iterability is still the principal limiting factor for the provable existence of inner models.

In the proof of $x \in U \iff x \in W$ sketched above, the lack of *moving generators* was used only with respect to the critical point κ of the embeddings i^W and $i_{0,1} = i^U$. It is also needed, with respect to all critical points, in the omitted proof that the comparison terminates. It was not recognized as a difficulty at this point in the history, since it is ensured by the coherence of the sequence $\mathcal{U}$ of measures. Later, in the 1980s when models for cardinals beyond a strong cardinal are developed (as we shall describe), it will be seen to have (in a more general form) a critical influence: it will require a major change in the structure of the iterated ultrapowers used for this method of comparison by iterating the least difference. (The term "moving generators" comes from the extenders used in such models.)

The definability technique used in the proof above to ensure equal comparison was implicitly used by Kunen, but was isolated by Mitchell [1974] and given the name *ϕ-minimality.* An important theme in the development of core models has been the introduction of new techniques to ensure equal comparison.

3 FINE STRUCTURE AND THE COVERING LEMMA IN L

Most of Jensen's work was originally distributed as handwritten, mimeographed notes made shortly after the results were obtained. These notes, though usually undated, provide a valuable source from which to follow his evolving views.

The first suggestion of fine structure came in his Habilitationsschrift [Jensen, 1967]. The introduction states that the main problem is to determine, given a transitive set M, for which ordinals α is the structure $L_\alpha(M)$ closed under given logical operations, and states that some of this is related to interests of his advisor, G. Hasenjaeger. The first theorem stated asserts that $\mathcal{P}(M) \cap L_{\alpha+1}(M) \subseteq L_\alpha(M)$ *if and only if there is no $L_\alpha(M)$-definable map from M onto L_α*, which recognizably presages (in contrapositive form) the basic theorems of fine structure.

An appendix specifically gives the Σ_1 case: *every subset of M which is Σ_1-definable over $L_\alpha(M)$ is a member of $L_\alpha(M)$ if and only if there is no map from M onto $L_\alpha(M)$ which is Σ_1-definable over $L_\alpha(M)$.* Jensen has stated that the name "fine structure" came from Gandy's comments on the Habilitationsschrift. Jensen has regretted the name, as it also names an unfashionable area of algebraic geometry.

In [1968?] Jensen proved that there is a Suslin tree in L. The proof was based on Tennenbaum's [1968] forcing construction of a Suslin tree: Jensen observed that in the tree T given by Tennenbaum's construction, the nodes T_α of T at level $\alpha < \omega_1$ were generic branches through the first α levels of T, and noted that it would be enough if they were generic only over the countable model L_β, where $\beta \geq \alpha$ is least such that $\mathcal{P}(\omega) \cap L_{\beta+1} \not\subseteq L_\beta$. Shortly after that, Solovay used a similar method to prove that there is a Kurepa tree in L, and several people then independently discovered the combinatorial principle $\diamondsuit$, and proved that $\diamondsuit$ is true in L and allows the construction of a Suslin tree. In [Jensen, 1969a], written at the Rockefeller University in 1969, Jensen discussed results of Kunen and himself concerning $\diamondsuit$ as well as the stronger principle $\diamondsuit^+$ which is needed to prove that there is a Kürepa tree.

In the notes [Jensen, 1970], expanded from lectures at Kiev, he presented a number of results relating to $\diamondsuit$ and trees, including generalizations of $\diamondsuit$ (which is $\diamondsuit_{\omega_1}$) and Suslin trees to larger cardinals. At the end he considered the question of a Suslin tree on an inaccessible cardinal κ in L. Now if κ is weakly compact then there is no Aronszajn tree on κ, and hence no Suslin tree, so his argument needed another constraint. The construction of a ω_1-Suslin tree T proceeded by recursion on the level α of T, and for a limit ordinal α the definition of the αth level of the tree had two parts: first showing that the tree of height α constructed so far had branches, and then selecting countably many of these branches to continue the tree. To define a κ-Suslin tree for larger κ, these two parts had conflicting requirements: there must be a stationary set $E \subseteq \kappa$ of ordinals at which fewer than κ many nodes would be selected to continue the tree, but the proof that there are branches at each level α requires a closed unbounded subset $C_\alpha \subseteq \alpha$ disjoint from E along which a branch can be defined by recursion. This was not a problem in defining a κ^+-Suslin tree for a regular cardinal κ, as one could take E to be the set of ordinals of cofinality κ and pick sets C_a of order type at most κ, so that the limit points of C_α all have cofinality less than κ. In [Jensen, 1970], Jensen described a new combinatorial principle which he called K^*_κ, and showed that K^*_κ + $\diamondsuit_\kappa$ implies that there is a κ-Suslin tree. The principle K^*_κ asserts that there are sets E and C_ν as needed, and Jensen states as a theorem that K^*_κ holds in L for every cardinal κ which is inaccessible, but not weakly compact. He omits the proof: "Der Beweis hiervon ist leider sehr lang and wird deshalb nicht gebracht".

The principle K^*_κ is essentially a square principle, but omits the vital coherence condition: $C_\beta = C_\alpha \cap \beta$ whenever β is a limit point of C_α. This coherence condition appears at the end of Jensen's proof [1969?b] of the consistency of the Suslin Hypothesis with CH, in an addendum in which Jensen corrects an error, pointed out by Solovay, in section I of the note. The correction, which Jensen characterizes as "restoring the 'unsimplified' proof", uses a principle which he calls +, and which we can recognize as essentially $\square_{\omega_1}$. In this note Jensen uses a preliminary forcing extension to make the principle + true.

In [Jensen, 1972], Jensen finally gave a complete exposition of the fine structure of L, ending with the definition of $\square_\kappa$ and of global square, and completing the proof that in L there is a κ-Suslin tree for every regular cardinal $\kappa > \omega$ which is not weakly compact. In this result the case in which κ was the successor of a singular cardinal λ was new, relying on the principle $\square_\lambda$. This principle was also critical in a final section written by Silver which proved that L satisfies the transfer property $(\kappa, \delta) \to (\lambda^+, \lambda)$ for any cardinals $\kappa < \delta$ and λ. This notation means that any theory with a model of type (κ, δ) also has a model of type (λ^+, λ), where a model $(M, U, \cdots)$ is said to have type (κ, λ) if M has size κ and the unary predicate U has size λ.

Details of the fine structure construction are far out of the scope of this paper, but it will be useful to have a simplified (and in some aspects inaccurate) description. We begin with an innovation which, though inessential and technical, was important to the later study of L and of fine structure models in general: the introduction of the J_α hierarchy in place of the usual L_α hierarchy. The J_α hierarchy was based on the set of rudimentary functions, defined by Gandy in [1974]. The two hierarchies are essentially equivalent (whence the characterization of the innovation as 'inessential'); however the new sequence allowed a much cleaner analysis, largely because every finite sequence of members of J_α is a member of J_α. The fact that the parameters, which were finite sequences of ordinals, could not be treated as single objects in L_α had greatly complicated the fine structure analysis of the L_α hierarchy. The J_α hierarchy has been widely used since, and is universally used in any inner model theory involving fine structure. We will use the J_α hierarchy henceforth; however a reader who prefers to read J_α as L_α will lose little or nothing in accuracy or comprehension.

Jensen took from Kripke and Platek the term *projectum*. He defined the projectum of J_α to be the least ordinal $\rho \leq \alpha$ such that there is a subset of ρ which is in $J_{\alpha+1} \setminus J_\alpha$, and for $n \in \omega$ he defined the Σ_n *projectum* of J_α to be the least ordinal ρ^α_n such that such a subset is Σ_n-definable over J_α with parameters from J_α. He answered the question left open in [Jensen, 1967] by showing that there is, for each ordinal α and each $n \in \omega$, a Σ_n-definable map from ρ^α_n onto J_α; however what is more important is the construction by which the map is obtained. He defined by recursion on $n \in \omega$ the Σ_n-core $\mathbb{C}_n(J_\alpha)$ as a structure $(J_{\rho^\alpha_n}, A^\alpha_n)$ in which A^α_n coded the Σ_n theory of J_α, with parameters from ρ^α_n. In the recursion step, he used the method from his Habilitationsschrift to define a $\Sigma_1^{\mathbb{C}_n(J_\alpha)}$-definable map from the Σ_1-projectum of $\mathbb{C}_n(J_\alpha)$, which is the Σ_{n+1} projectum ρ^α_{n+1} of J_α. He then defined $\mathbb{C}_{n+1}(J_\alpha) = (J_{\rho^\alpha_{n+1}}, A^\alpha_{n+1})$ where A^α_{n+1} encoded in a natural way the Σ_1 theory of $\mathbb{C}_n(J_\alpha)$ and hence, by extension, the Σ_{n+1} theory of J_α.

The construction was entirely canonical. In particular, given $\mathbb{C}_n(J_\alpha)$ one could recover J_α. In addition—and this is a critical point—it is preserved by embeddings: he showed that any substructure $X \prec_{\Sigma_1} \mathbb{C}_n(J_\alpha)$ of the Σ_n-code of J_α is isomorphic to the Σ_n code of some $J_{\bar\alpha}$, and furthermore the collapse map $\pi\colon \mathbb{C}_n(J_{\bar\alpha}) \cong X \prec_{\Sigma_1} \mathbb{C}_n(J_\alpha)$ can be extended to a Σ_{n+1} elementary map from $J_{\bar\alpha} \to J_\alpha$. We will refer to this property as *downward extension of embeddings*, but it should be noted that it can be regarded as an enhancement of the Condensation Lemma 1: it says that condensation applies at each of the infinitely many levels of the Levy hierarchy between J_α and $J_{\alpha+1}$.

The importance of this may be seen in the proof of the principle $\square_\kappa$, which states that there is a sequence of closed unbounded subsets of C_α, defined for limit ordinals $\alpha < \kappa^+$, such that the order type of C_α is at most κ, and if β is a limit point of C_α then $C_\beta = C_\alpha \cap \beta$. Jensen's construction has a number of cases, but the interesting case is that in which $\kappa < \alpha < \kappa^+$ and α has uncountable cofinality. Jensen considers the first ordinal γ such that α is singular in $J_{\alpha+1}$. Then the least witness $\sigma\colon \xi \to \alpha$ to this singularity is Σ_{n+1}-definable over J_γ for some $n \in \omega$. By careful use of the Σ_n code of J_γ, Jensen finds a tower of Σ_{n+1} elementary substructures $X_\nu \prec_{\Sigma_{n+1}} J_\gamma$ with the following property: For limit $\nu < \xi$ set $\alpha_\nu = \sup \sigma``\nu$ and let $\pi_\nu\colon X_\nu \cong J_{\gamma_\nu}$ be the collapse map. Then π_ν preserves the definition of the map σ, so that $\sigma\restriction\nu$ is the least witness to the singularity of α_ν, just as σ was for α. In particular, if σ is used to define C_α and β is any limit member of C_α, then $\beta = \alpha_\nu$ for some $\nu < \xi$, so C_β is defined using $\sigma\restriction\nu$ in the same way as the definition of C_α used σ. Thus $C_\beta = C_\alpha \cap \beta$.

The property $\square_\kappa$ has given rise to an extensive literature. In addition to its consequences, a number of weaker principles have been considered, starting with weak square, $\square^*_\kappa$, which was defined by Jensen by allowing κ many club subsets of each $\alpha < \kappa^+$ instead of only one. In particular, some of the scale and club guessing principles which Shelah has introduced in connection with his pcf theory can be regarded as weak versions of $\square_\kappa$. Little of this theory is relevant, however, to this history: since square holds in the inner models we are discussing, the study of weaker properties is moot. The reader may see the chapter [Kojman, 2011], covering singular cardinal combinatorics, for more about these properties.

Following the completion of [Jensen, 1972], Jensen went on to prove that the gap-1 transfer property is true in L. The gap-1 transfer property states that if $\alpha > \beta^+$ are cardinals then any theory which has a model of type (α, β) also has models of type (κ^{++}, κ) for any cardinal κ. For this he defined a new combinatorial structure, which he called a *gap-1 morass*, and used the fine structure to show that such a morasses exists in L on every cardinal. The gap-1 ω_2-morass, for example, expressed J_{ω_2} as an intricate direct limit of countable structures. This structure permitted a structure of size ω_2 to be defined as a limit of recursively defined countable models, in somewhat the same way as Chang's original proof that $(\alpha^+, \alpha) \to (\omega_1, \omega)$ constructed a model of size ω_1 as a limit of countable models.

From here, Jensen went on to define gap-n morasses for each $n \in \omega$, and he used these to prove transfer theorems with larger gaps: *If* $V = L$ *and* $n \in \omega$ *then* $(\lambda^{+n}, \lambda) \to (\kappa^{+n}, \kappa)$ for any infinite cardinals λ and κ. Part of the motivation here was in the hope that this would lead to a solution to the leading open problem of set theory of the time: the Singular Cardinal Hypothesis (SCH). The hope was this construction would lead to a forcing notion over L which would give a model in which $2^{\omega_n} = \omega_{n+1}$ for all $n < \omega$, but $2^{\omega_\omega} > \omega_{\omega+1}$. The actual outcome was quite different.

Before continuing, it is worth noting that while the combinatorial principles described above had their origin in the constructible universe L, they have largely moved away from constructivity and inner model theory in general — indeed a primary part of the motivation for their definition was to abstract away from L the combinatorial properties used in applications. This is not true of the next topic: the covering lemma remains tightly tied to the inner model program and has remained at its center.

3.1 The Singular Cardinal Hypothesis and the covering lemma

The origins of the Singular Cardinal Hypothesis are described in chapter [Kanamori, 2011b]. Shortly after Cohen's proof of the independence of the Continuum Hypothesis, Easton [1970] completely settled the general question of the size of the power sets of regular cardinals; however this left the size of the power set of a singular cardinal λ completely open. The Singular Cardinals Hypothesis, SCH, is the assertion that $\lambda^{\mathrm{cf}(\lambda)} = \max(\lambda^+, 2^{\mathrm{cf}(\lambda)})$ for every singular cardinal λ. For the case when λ is a strong limit cardinal, this can be simplified to $2^\lambda = \lambda^+$. Easton's model satisfies SCH, but Silver showed that a model in which SCH failed could be constructed by starting with a κ^{++} supercompact cardinal. It was commonly believed that no such hypothesis was necessary: there should be an Easton-like theorem for singular cardinals with no large cardinal hypothesis.

While trying to prove such a theorem for singular cardinals, Silver in 1974 (cf. [1975]) discovered a new restriction to the size of the power set of a singular cardinal: *Suppose that λ is a singular cardinal of uncountable cofinality, and $2^\gamma = \gamma^+$ for stationarily many $\gamma < \lambda$. Then $2^\lambda = \lambda^+$.* In particular, if λ is the first singular cardinal at which SCH fails, then λ has cofinality ω.

Silver's proof started by using forcing to obtain a generic ultrafilter U on $\mathcal{P}^V(\mathrm{cf}(\lambda))$ extending the closed unbounded filter, and then considered the embedding $i^U : V \to M = \mathrm{Ult}(V, U)$. The model M is not well-founded; however if we set $\bar{\lambda} = \sup\{\, i(\alpha) : \alpha < \lambda \,\}$ then the ordering in M of the set of ordinals below $\bar{\lambda}$ is λ-like,[2] and that of $(\bar{\lambda}^+)^M$ is λ^+-like. Also $M \models 2^{\bar{\lambda}} = \bar{\lambda}^+$ by the Łos theorem, and from these facts he deduced that $2^\lambda = \lambda^+$ in V.

Only a few weeks later, Jensen distributed a handwritten note entitled *Marginalia to a Theorem of Silver* [Jensen, 1974b], which gave the first statement of the covering lemma. In spite of the name of the note, the proof of the covering lemma is not closely related to the proof of Silver's theorem: the only thing they have in common is that both involve embeddings on a singular cardinal. A close reading of

[2] A linear ordering $(R, \leq)$ is said to be λ-*like*, where λ is a cardinal, if R has size λ but each proper initial segment has size less than λ.

the main theorem of [Jensen, 1974b] gives a hint as to how thinking about Silver's result led Jensen to the covering lemma.

The main theorem of [Jensen, 1974b] is in three parts, all of which concern a singular cardinal β of uncountable cofinality. The first part, Theorem I(i), is essentially a straightforward application of Silver's method, though Jensen (independently of Baumgartner and Prikry) eliminated the use of forcing from the proof: *Assume that $A \subseteq \beta$ is such that*[3] $L_\beta[A] = H_\beta$ *and* $2^\lambda = \lambda^+$ *on a stationary subset of* β, *and that* $\mathrm{cf}^L(\beta) = \mathrm{cf}(\beta)$ *and* $(\beta^+)^L = \beta^+$. *Then* $\mathcal{P}(\beta) \subseteq L[A]$. The second part, Theorem I(ii), applies Theorem I(i) in the special case $A = \emptyset$. The hypothesis that $2^\lambda = \lambda^+$ for all $\lambda < \beta$ follows automatically from the hypothesis that $H_\beta = L_\beta$, but there remain the awkward extra assumptions $\mathrm{cf}^L(\beta) = \mathrm{cf}(\beta)$ and $(\beta^+)^L = \beta^+$. Jensen used his newly developed fine structure to show that these assumptions also follow automatically. Consider, for example, the statement that $\mathrm{cf}^L(\beta) = \mathrm{cf}(\beta)$. Jensen used the singularity of β to obtain an elementary substructure X of L_β of size $\mathrm{cf}(\beta) < \beta$ such that $X \cap \beta$ is cofinal in β. By the condensation property, there is an isomorphism $\pi\colon L_{\bar\beta} \cong X \prec L_\beta$ for some $\bar\beta < \beta$. Since $H_\beta = L_\beta$, the preimage $\bar\sigma$ of σ is in L. In particular, $\bar\beta$ is singular in L. Let $\bar\alpha \geq \bar\beta$ be least such that $\bar\beta$ is singular in $L_{\bar\alpha+1}$.

To complete the proof, Jensen introduced two innovations. First, he generalized the ultrapower construction to extend the embedding π to an embedding $\tilde\pi\colon L_{\bar\alpha} \to L_\alpha$ for some $\alpha \geq \beta$. This generalization later gave rise to the notion of an *extender*, which has become central to the theory of inner models for large cardinals. Second, an *upward extension of embeddings* principle, a counterpart to the downward extension of embeddings principle used for the $\square$ principles, enabled this extension to be further extended to preserve the definition over $L_{\bar\alpha}$ of a witness to the singularity of $\bar\beta$. It followed that the same definition, applied to L_α, gave a witness to the singularity of β: thus β was singular in L.

In Theorem I(iii) Jensen let this argument stand on its own, and replaced the hypothesis that $L_\beta = H_\beta$ with the much weaker assumption that $0^\#$ does not exist. The result is recognizably the covering lemma (for uncountable cofinality): *if β is a singular cardinal with* $\mathrm{cf}(\beta) > \omega$, *and if* $\mathrm{cf}^L(\beta) = \beta$ *or* $(\beta^+)^L < \beta^+$, *then* $0^\#$ *exists.*

The argument again uses the embedding π as an extender: If $\bar\sigma$ is not singular in L then π can be extended to a nontrivial map $\tilde\pi\colon L \to L$, which by Kunen's theorem implies that $0^\#$ exists.

The restriction in [Jensen, 1974b] to cardinals β of uncountable cofinality came from the need to show that the target of the extended embedding $\tilde\pi$, which was obtained by the use of π as an extender, was well-founded. The main theorem of the second note, [Jensen, 1974c], removes this restriction: *Assume* $\neg 0^\#$. *Then* $\mathrm{cf}^L(\lambda) \geq \omega_2 \implies \mathrm{cf}(\lambda) = \mathrm{cf}^L(\lambda)$. The proof replaces the previous use of the assumption that $\mathrm{cf}(\lambda) > \omega$ with the facts that λ^+ and ω_1 each have uncountable cofinality. The use of ω_1 in the proof dictates the restriction $\mathrm{cf}^L(\lambda) \geq \omega_2$. As he

[3]Here H_β is the set of all sets which are hereditarily of size less than β, that is, sets whose transitive closure has size less than β.

remarks, it would be "reasonable but wrong to suppose the theorem to hold for $\mathrm{cf}(\lambda) \geq \omega_1$" because of the counterexample in [Namba, 1971].

The third and final note in this series, [Jensen, 1974a], gives the final form to the statement of the covering lemma: *"Theorem: Assume $\neg 0^{\#}$. Let X be an uncountable set of ordinals. Then there is a set $Y \in L$ such that $|X| = |Y|$ and $X \subseteq Y$."* He lists as consequences all of the results in the two earlier notes.

As we have seen, the fine structure was a basic component of Jensen's proof of the covering lemma; however some have considered its use to be a flaw in the theory. Silver has given a proof which avoids fine structure by using what he called "machines"; these machines were perhaps simpler than the fine structure but were, at least in this author's opinion, less intuitive. In any case they have not been generalized to the larger theory described in the rest of this chapter. Another proof, due to Magidor [1990], replaces the fine structure with the naïve Skolem function for Σ_n sets: using, for a Σ_n relation R over L_α, the function $h(x) = \mu y\ R(x, y)$. This function is not Σ_n-definable over L_α, but surprisingly the covering lemma can be proved by replacing the use of the Σ_n code $\mathbb{C}_n(J_\alpha)$ with the closure of J_α under the function h. Magidor has had some success in generalizing this method to core models with sequences of measurable cardinals, but unfortunately it does not appear to be suitable even for the full theory of such sequences, let alone the larger core models we describe in section 5.3.

The main result of Magidor's paper [1990] cited above gave an interesting alternate conclusion from the proof of the covering lemma: *If $0^{\#}$ does not exist then every primitively closed set of ordinals is a countable union of sets in L.* Another interesting development is the strong covering theorem, an observation contained in an unpublished note of Carlson: *If $0^{\#}$ does not exists, then for any pair of uncountable cardinals $\delta < \kappa$, with δ regular, there is a set $C \subseteq [\kappa]^{\delta}$ in L which is unbounded and closed under increasing unions of sequences of length δ.*

4 THE CORE MODEL

4.1 Up to one measurable cardinal

It now seemed that, with the new machinery available, the truth in L of any question could be decided, and attention turned to larger models. An obvious choice was $L[U]$, and Solovay extended Jensen's fine structure to $L[U]$ and verified that $\diamondsuit_\kappa$ and $\square_\kappa$ held there for all κ. However, while the model $L[U]$ is, in many respects, a suitable generalization of L, it has number of defects centering on the fact that its construction requires a known ultrafilter U. In some applications this problem was overcome, notably in the use by Solovay and Kunen of models $L[F]$, where F was the cardinal filter or the closed unbounded filter, under hypotheses which implied that F became an ultrafilter in $L[F]$. The usefulness of such strategies was limited, however. There is, for example, no evident filter associated with the failure of Singular Cardinal Hypothesis. Furthermore, the model $L[U]$ gives little insight into large cardinal properties lying between L and $L[U]$: it does not for example, provide a minimal model for a Ramsey cardinal.

Jensen and Anthony Dodd started with Solovay's fine structure for $L[U]$, but took it in a differerent direction. Instead of assuming that there is a model $L[U]$ with a measurable cardinal, they used approximations to such a model, which they called *mice*. There have been many conjectures as to where the name "mouse" came from but, according to Jensen, he chose the term simply because he felt that the concept was important and he wanted to give it a name which had not been used elsewhere in mathematics. A mouse is a structure $M = J_\alpha[U]$ which has three properties: (i) it satisfies the statement that U is a measure on some κ, (ii) all of its iterated ultrapowers are well-founded, and (iii) it admits a fine structure analysis, with projectum ρ^M smaller than the critical point of the ultrapower. The paradigmatic example of a mouse comes from $0^\#$. If U is a measure on κ, then $J_{\kappa+1}[U]$ is equal to $J_{\kappa+1}$, but $0^\#$ is Σ_1-definable in the structure $(J_{\kappa+1}[U], \in, U)$: the Gödel number $\ulcorner\phi(c_0, \ldots, c_n)\urcorner$ is in $0^\#$ if and only if

$$\exists X \in U^n \cap J_{\kappa+1}[U]\ \forall(\nu_0, \ldots, \nu_{n-1}) \in X\ J_{\kappa+1}[U] \models \phi(\nu_0, \ldots, \nu_{n-1}).$$

This fact is rather awkward for the fine structure of $L[U]$, taken by itself: what is "fine" about a subset of ω which does not appear until after a measurable cardinal? Taking the transitive collapse of the Σ_1-Skolem hull of $J_{\kappa+1}[U]$, however, gives a countable structure $M = J_{\bar{\kappa}+1}[\bar{U}]$, over which $0^\#$ is similarly definable. Thus M and $0^\#$ are equiconstructible.

Another way to view the equivalence between M and $0^\#$ is to take an iterated ultrapower of M of length Ω, the class of ordinals. In the resulting class model, the measurable cardinal is Ω and the initial segment below Ω is isomorphic to L, with the critical points of the embedding being the Silver indiscernibles. This can be compared with [Gaifman, 1974], in which he started with a countable substructure of an appropriate V_τ containing a measurable cardinal, and obtained a countable model X whose iterations generated a closed unbounded class of inaccessibles of L containing the cardinals of V. The $0^\#$ mouse M is the ultimate refinement of Gaifman's model X, and is the weakest Dodd-Jensen mouse.

It is important to realize that the measure U in a mouse $M = J_\alpha[U]$ is not a measure in V; indeed the fine structure analysis ensures that $J_{\alpha+1} \models |\alpha| = \rho^M < \text{crit}(U)$, so that U is not a measure in any model larger than M. Thus if $M = J_\alpha[U]$ and $J_{\alpha'}[U']$ are two mice then neither is an initial part of the other, as always happens at two levels J_α and $J_{\alpha'}$ of the L hierarchy. However the mice M and M' can be compared by use of iterated ultrapowers, as Kunen did with $L[U]$ and $L[U']$, because the fact that a mouse M has a fine structure, and that its projectum ρ is smaller than its critical point, ensures that the iterated ultrapowers will preserve the definition of the new subset of its projectum ρ. Thus the fine structure of a mouse can be used to ensure the criterion of equal comparison from section 2.4, similarly to the way ϕ-minimality was used at that point.

The use of comparison by iterated ultrapowers gave a well-ordering of the mice, defined by setting $M \lhd M'$ if M is a member of an iterated ultrapower of M'. This opened the way for Dodd and Jensen to define the core model as $L[\mathcal{M}]$, where $\mathcal{M}$

is the class of all mice. Although they called this model K, we will refer to it as K^{DJ} to distinguish it from later generalizations. Using properties of mice, they showed that K^{DJ} had most of the basic properties of L: It is a model of ZFC, it has a condensation property and hence satisfies GCH, and it satisfies $\diamondsuit_\kappa$ and $\square_\kappa$ for all cardinals κ. One important difference was in the complexity of its definition: while a model $M = J_\alpha$ need only be well-founded to be an initial segment of L, a model $M = J_\alpha[U]$ must have all of its iterated ultrapowers well-founded. This meant that for $x \subseteq \omega$, the formula "$x \in K^{\mathrm{DJ}}$" is Σ^1_3 while the sentence "$x \in L$" is Σ^1_2, and thus the canonical well-ordering of the reals of K is Δ^1_3 as in $L[U]$, instead of Δ^1_2 as in L.

Dodd and Jensen showed that K^{DJ} bridged the gap between L and $L[U]$ in the sense that we would have $K^{\mathrm{DJ}} = L$ in the event $0^\#$ did not exist, while if there were a model $L[U]$ with a measure U on κ, then K^{DJ} was equal to the initial ordinal length segment of $\mathrm{Ult}_{\mathrm{ON}}(L[U], U)$ or, equivalently, to $\bigcap_{\alpha \in \mathrm{ON}} \mathrm{Ult}_\alpha(L[U], U)$. In particular, $V_\kappa^{K^{\mathrm{DJ}}} = V_\kappa^{L[U]}$.

Finally, they proved a pair of covering lemmas. The covering lemma stated for K^{DJ} in [Dodd and Jensen, 1982a] assumed that there is no inner model with a measurable cardinal and had the same statement as that for L: *If there is no inner model with a measurable cardinal, then for any uncountable set x of ordinals there is a set $y \supseteq x$ in K^{DJ} of the same cardinality.* The following paper, [Dodd and Jensen, 1982b], assumed that a model $L[U]$ with a measurable cardinal existed, but $0^\dagger$, the sharp of $L[U]$, did not. Here the conclusion bifurcated. As Prikry had shown in his thesis [1970], there is a forcing notion which, given a measurable cardinal κ, adds an ω sequence C of ordinals, cofinal in κ, without collapsing any cardinals. However the necessary modification to the covering lemma was minimal. With the term "covering property" having the expected meaning from above: *Suppose that U is a measure in $L[U]$, the critical point κ is as small as possible, and $0^\dagger$ does not exist. Then either $L[U]$ has the covering property, or else there is a Prikry sequence C such that the model $L[U, C]$ has the covering property.* Furthermore the model $L[U, C]$ is unique, though the sequence C is unique only up to finite changes.

The first publication of this work was in a handwritten note, [Dodd and Jensen, 1976] titled *A Modest Remark.* The title referred not to the covering lemma, but to an alleged theorem claiming that the existence of a measurable cardinal was inconsistent with ZF. The error was quickly discovered independently by several parties in widespread locations: Jensen had incorrectly assumed that a cardinal κ of cofinality ω could be collapsed onto ω_1 without also collapsing κ^+. Jensen had, as he admitted, written and hurriedly distributed the note without consulting its putative coauthor Dodd. In spite of the error, this note provided a valuable early exposition of Dodd and Jensen's theory of the core model. The formal publication of the theory came in three papers: [Dodd and Jensen, 1981] gave the basic construction of K^{DJ} and the two papers [Dodd and Jensen, 1982a] and [Dodd and Jensen, 1982b] gave the proof of the covering lemma for K^{DJ} and $L[U]$ respectively. In addition Dodd wrote a book, [Dodd, 1982], giving an exposition of K^{DJ} and looking towards extensions of this model.

They showed that, just as the existence of a nontrivial embedding $i\colon L \to L$ implies that $0^{\#}$ exists, the existence of a nontrivial embedding $i\colon K^{\mathrm{DJ}} \to K^{\mathrm{DJ}}$ implies that there is a model $L[U]$ with a measurable cardinal. Where the critical point of an embedding $i\colon L \to L$ is always a Silver indiscernible, however, Jensen showed in [2001] (see also [Mitchell, 2010b, Theorem 3.43]) the critical point of an embedding $i\colon K^{\mathrm{DJ}} \to K^{\mathrm{DJ}}$ may be smaller than the measurable cardinal in $L[U]$.

The most important immediate consequence of the Dodd-Jensen core model theory is that a failure of the Singular Cardinal Hypothesis entails that $0^{\dagger}$, the sharp for a model $L[U]$ with a measure, exists. The proof of this implication was the same as the proof using L: the three coveriing lemmas of Jensen and Dodd imply that if $0^{\dagger}$ does not exist, then one of the models K^{DJ}, $L[U]$ or $L[U, C]$ has the covering property. Since all of these models satisfy GCH, this implies that SCH holds in V.

An important later consequence was Jensen's extension in [Donder *et al.*, 1981] of Shoenfield's theorem [Shoenfield, 1961] that any model M of set theory containing ω_1 is absolute for $\mathbf{\Sigma}^1_2$ formulas. Jensen's theorem stated: *Assume $a^{\#}$ exists for every real a, and that there is no inner model with a measurable cardinal. Then K^{DJ} is absolute for $\mathbf{\Sigma}^1_3$ formulas.* His proof actually showed that the same is true of any model $M \supseteq K^{\mathrm{DJ}}$.

Another consequence of the Dodd-Jensen covering lemma was a generalization, due to Mitchell, of Kunen's theorem [1970] that all Jónsson cardinals in $L[U]$ are Ramsey: Mitchell [1979b] showed that all Jónsson cardinals in V are Ramsey in K^{DJ}. It followed that a minimal model for a Ramsey cardinal could be constructed by taking the least initial segment of K^{DJ} which satisfies that there is a Ramsey cardinal. This was generalized by Jensen in [Donder *et al.*, 1981] to show that any δ-Erdős, and even any δ-Jónsson cardinal, is δ-Erdős in K^{DJ}. Thus the core model provides minimal models for these cardinals.

Before going on, a word about terminology is in order. At this point we have three covering lemmas for three different models: for the model L if $0^{\#}$ does not exist, for the Dodd-Jensen core model K^{DJ} if $0^{\#}$ exists but $L[U]$ does not, and for $L[U]$ if it exists but $0^{\dagger}$ does not. In the current terminology any of these models would, under the appropriate hypothesis, be called the *core model* and written as K. Much of the rest of this chapter is concerned with the continuing effort to extend this theory to encompass larger cardinals.

It will be useful here to consider some of the properties of the core model as described so far in the article in order to get some sense of what can be expected of larger core models, both those in the remainder of this history and those to come in the future. One such property is that the model is structured in a hierarchy, with each step in the hierarchy containing only sets which self-evidently must be in any model containing the previous steps of the hierarchy together with a measurable cardinal. Thus the result is guaranteed to be the smallest possible model of set theory containing all of the ordinals and every fragment of a measurable cardinal. Following this intuition, the model which is today called the *core model*, and is denoted by the symbol K, can be defined — as a goal, though not mathematically — as *the minimal model of* ZFC *containing all of the large cardinal structure existing in the universe* (cf. [Mitchell, 2010a], section 5).

Some of the other properties of the core model K, whether L, K^{DJ} or $L[U]$, described so far provide evidence that the core model constructions described thus far satisfy this definition, provided that $0^\dagger$ does not exist. The core model is rigid, in the sense that there is no nonidentity embedding $i\colon K \to K$, although this sense of rigidity is weakened in the case of $L[U]$ by the existence of embeddings $i\colon L[U] \to M$ given by iterated ultrapowers. The core model is absolute: in the case $K = L$ or $K = K^{\mathrm{DJ}}$ this is true in the strong sense that $K^M = K \cap M$ for any uncountable model M of set theory, but again the statement in the case $K = L[U]$ is weaker as K^M may be an iterated ultrapower of $L[U]$.

Finally, the covering lemma shows that the model K is close to V; again, the statement of the covering lemma is significantly weakened in the case $K = L[U]$.

4.2 More measures

The obvious next step was to incorporate the new core model theory into Mitchell's model $L[\mathcal{U}]$ for a sequence of measures, thereby extending the core model construction to include any number of measures. Mitchell had been attempting to do so even before Dodd and Jensen completed their work on K^{DJ}. Now, with notes available on the Dodd-Jensen techniques, he was able to proceed.

Mitchell called his model $K(\mathcal{F})$; however we will write $K[\mathcal{F}]$, using square brackets to agree with the standard notation for relative constructibility. The sequence $\mathcal{F}$ was a sequence of filters in V, which was intended to become a sequence of ultrafilters in $K[\mathcal{F}]$; however it is noteworthy that neither the definition of $K[\mathcal{F}]$ nor the fine structure analysis made any use of the structure of $\mathcal{F}$: what is defined is really a relativized core model, built above an arbitrary set $\mathcal{F}$.[4]

The model $K[\mathcal{F}]$ was defined to be $L[\mathcal{F}, \mathcal{M}]$, where $\mathcal{M}$ was the class of structures called *$\mathcal{F}$-mice*. The part of this work which was published in [Mitchell, 1984a] ended at the point where fine structure was to be introduced. It proved that the model $K[\mathcal{F}]$ was the union of the class of $\mathcal{F}$-mice, and that if $\mathcal{F}$ was *strong* — that is, it was a coherent iterable sequence of measures in $K[\mathcal{F}]$ — then $K[\mathcal{F}]$ was a model of ZFC + GCH + the set of reals has a Δ^1_3 well-ordering + every normal ultrafilter is a member the sequence $\mathcal{F}$.

The rest of this paper was never published, though it had some distribution as the manuscript [Mitchell, 1985]. Section 4, which analyzed the fine structure of the model, took up more than half of this manuscript, 65 out of about 110 double-spaced pages. Much of this was a straightforward extension of the Dodd-Jensen fine structure of $L[U]$, as extended to sequences of measures by using Mitchell's techniques from $L[\mathcal{U}]$. An $\mathcal{F}$-mouse M was a structure $J_\alpha[\mathcal{G}]$, where $\mathcal{G}$ has $\mathcal{F}$ as an initial segment and, above that, a coherent sequence of measures corresponding to

[4]Of course this statement assumes (as will we, unless stated otherwise) that $\mathcal{F}$ is not a proper class.

the single measure in the Dodd-Jensen mice. Some technical adaptations were, of course, needed to accommodate new sets being directly defined from the members of the sequence, and from the fact of working above the arbitrary set $\mathcal{F}$. The major complication involved the criterion of equal comparison: a new complement was needed to the use of mice, as described in section 2.4. We will return to this point when we look at the Dodd-Jensen lemma, which is a more general alternative to the technical trick used in this paper.

Only at this point was it possible to prove the crucial observation that if $\mathcal{F}$ is strong then $K[\mathcal{F}\restriction\alpha] \subseteq K[\mathcal{F}]$, and in particular that the definition of $K[\mathcal{F}]$ in the case that $\mathcal{F}$ is a proper class, $K[\mathcal{F}] = \bigcup_{\alpha\in\mathrm{ON}} K[\mathcal{F}\restriction\alpha]$, makes sense.

Perhaps the most important change from Dodd and Jensen's analysis of K^{DJ} is that the fine structure analysis took on a new role: it enabled a sequence $\mathcal{F}$ to be defined by recursion. More specifically, the main theorem of section 4 was: *Suppose $\mathcal{F}$ is a sequence of filters such that for each (α,β) in the domain of $\mathcal{F}$, the filter $\mathcal{F}(\alpha,\beta)$ (i) is a normal ultrafilter of order β on $K[\mathcal{F}\restriction(\alpha,\beta)]$ and (ii) is countably complete. Then $\mathcal{F}$ is strong in the sense that each filter $\mathcal{F}(\alpha,\beta)$ in the sequence $\mathcal{F}$ is an ultrafilter in $K[\mathcal{F}]$.* This was used to define a preliminary version of the core model, which is now known as K^{c}, the *countably complete core model*, by setting K^{c} equal to the model $K[\mathcal{F}]$ where $\mathcal{F}$ is the maximal sequence of countably complete measures, defined recursively by adding a filter $\mathcal{F}(\alpha,\beta)$ whenever it satisfies the criterion above.

A full covering lemma, in the style of that in the case that the core model is equal to L or K^{DJ}, or even $L[U]$, is not possible for these models. Mitchell instead used what the reader may recognize as Jensen's first version of the covering lemma in [Jensen, 1974b]: he said that an inner model M satisfies the *weak covering lemma* if, for every sufficiently large strong limit cardinal λ, the successor $(\lambda^+)^M$ of λ as evaluated in M is the same as the successor λ^+ as calculated in V. Mitchell used recursion to define what came to be known as the countably complete (later, with Steel's work, countably certified) core model $K^{\mathrm{c}} = K[\mathcal{F}]$, by requiring that each filter $\mathcal{F}(\alpha,\beta)$ on the sequence $\mathcal{F}$ be a countably complete $K[\mathcal{F}\restriction(\alpha,\beta)]$-ultrafilter. The countable completeness ensured that all iterated ultrapowers were well-founded, and this in turn allowed the use of fine structure to show that each filter $\mathcal{F}(\alpha,\beta)$ was still an ultrafilter in the full model K^{c}. Mitchell's proof that K^{c} has the weak covering property was essentially the same as Dodd and Jensen's proof that K^{DJ} has the covering property, here applied to $\kappa = (\lambda^+)^{K^{\mathrm{c}}}$ which has, if it is smaller than λ^+ of V, cofinality less than λ.

One important consequence of the weak covering lemma is that it provides a new technique for satisfying the criterion of equal comparison: a model $K[\mathcal{F}]$ which satisfies the weak covering property is *universal* in the sense that when $K[\mathcal{F}]$ is compared, using iterated ultrapowers, with any other model $K[\mathcal{F}']$ then the model $K[\mathcal{F}]$ will compare at least as long as $K[\mathcal{F}']$: the final model in the iteration of $K[\mathcal{F}']$ will always be an initial segment of the final model in the iteration of $K[\mathcal{F}]$. In particular, if $K[\mathcal{F}']$ is also universal then the final models of the two iterations will be equal.

The weak covering lemma not only showed that K^{c} is universal; it also provided a criterion for the transitive collapse of an elementary submodel $X \prec K^{\mathrm{c}}$ to be universal. This criterion is a counterpart to the particular case of condensation for L that any elementary subset of L which contains a proper class of ordinals is isomorphic to L. For $K[\mathcal{F}]$ it was not enough to assume that X is a proper class: Mitchell defined X to be *thick* if, for a stationary class of singular cardinals λ the order type of $X \cap \lambda^+$ is equal to λ^+. It follows from the weak covering lemma that not only is K^{c} universal, but also the transitive collapse of any thick class $X \prec K^{\mathrm{c}}$ is universal. This fact was used in Section 7 to obtain the true core model, which was written $K[\mathcal{F}_{\mathrm{M}}]$ but today would be written simply as K. The sequence $\mathcal{F}_{\mathrm{M}}$ was defined recursively in the same way as $\mathcal{F}^{\mathrm{c}}$, but with countable completeness replaced with the requirement that $\mathrm{Ult}(K[\mathcal{G}], \mathcal{F}(\alpha, \beta))$ be well-founded whenever $\mathcal{G}$ is a strong sequence which extends $\mathcal{F}{\restriction}(\alpha, \beta)$. This fact was used to show that $K[\mathcal{F}_{\mathrm{M}}]$ is iterable by showing that it is an elementary submodel of K^{c}.

Consequences of the covering lemma for sequences of measures

The unpublished article [Mitchell, 1985] concluded with a number of applications of the weak covering lemma in K: *Any of the following assertions imply the existence of an inner model of $\exists\kappa\ o(\kappa) = \kappa^{++}$:* (i) *$\kappa$ and κ^+ are both weakly compact,* (ii) *κ is κ^+ strongly compact,* (iii) *κ is measurable and $\kappa^+ > (\kappa^+)^{K[\mathcal{F}]}$,* (iv) *$\kappa$ is measurable and $2^\kappa > \kappa^+$,* (v) *there is a κ-complete ultrafilter U on κ such that $i^U(\kappa) = \kappa^+$ (in particular, the Axiom of Determinacy holds),* (vi) *every κ-complete filter can be extended to a κ-complete ultrafilter, or* (vii) *there is a κ^+-saturated ideal on a successor cardinal κ.*

These results relied on consequences of the weak covering lemma, and in particular on one which did not originally appear, at least in its strongest form, in [Mitchell, 1985], but was added as an addendum dated July 1985 after Steel pointed out its omission: *If there is no inner model of $\exists\kappa\, o(\kappa) = \kappa^{++}$ then every elementary embedding $i\colon K[\mathcal{F}_{\mathrm{M}}] \to N$ into a well-founded model N is an iterated ultrapower by measures in $\mathcal{F}_{\mathrm{M}}$.*

The arguments from K used in this proof seemed much stronger than the core model K itself, and it was expected that, once larger core models had been constructed, the same arguments would yield a stronger conclusion. This turned out to be true for all but clause (iii): A closer at this look led Mitchell to make the suggestion that $\exists\kappa\, o(\kappa) = \kappa^{++}$ was the exact large cardinal strength of the failure of GCH at a measurable cardinal; this suggestion was later confirmed by work of Woodin (using a slightly stronger assumption) and Gitik [1989].

Of the the topics one might have expected to see in [Mitchell, 1985], two were omitted. One was the weak covering lemma for the case of cardinals κ which are not countably closed, that is, such that $\lambda^\omega \geq \kappa$ for some $\lambda < \kappa$, and the other was the Singular Cardinal Hypothesis, or, indeed, any version of the covering lemma beyond the weak covering lemma. The first of these was the less consequential, but took the longest to resolve: the proof for κ which are not ω closed does not require any techniques except those in [Mitchell, 1985], together with those used for L in [Jensen, 1974c]; however no published, or even manuscript, exposition of the extension of the covering lemma for sequences of measures to the non-closed case appeared until [Mitchell and Schimmerling, 1995] (for a much more powerful version of the core model).

Applying this model to the SCH problem was much more troublesome. The most complicated case of the Dodd-Jensen covering lemma is that in which $0^\dagger$ does not exist and there is a model $L[U]$ with a measurable cardinal, along with a Prikry sequence C for that measure. Then the model $L[U, C]$ provides covering sets. The set C is not quite unique — this is a change from previous versions of the covering lemma — but the model $L[U, C]$ is unique. Mitchell showed in [Mitchell, 1984b] that a straightforward generalization of this is false in general. If λ is an inaccessible limit of a set I of measurable cardinals in V, then there is a generic extension $V[G]$ in which each $\kappa \in I$ is made singular by a Prikry sequence over V, but there is no sequence of Prikry sequences which is *uniform*: that is, for any sequence $\langle C_\kappa : \kappa \in I \rangle$ of Prikry sequences there is a sequence of sets $\langle A_\kappa : \kappa \in I \rangle$ in the ground model such that each set A_κ is a member of the measure on κ, but nevertheless $\{ \kappa \in I : C_\kappa \not\subseteq^* A_\kappa \}$ is unbounded in κ.

In this model, there is a uniform sequence of Prikry sequences for each subset of I which is bounded in κ, and this proved to be the general situation. Mitchell defined a notion of a *system of indiscernibles* which generalized, in a natural way, the notion of a Prikry sequence, and he proved a version of the covering lemma stating that for any uncountable set X in V, there is a system $\mathcal{C}$ of indiscernibles for K such that X is contained in a set $Y \in K[\mathcal{C}]$ of the same cardinality. However the system $\mathcal{C}$ is, in contrast to the case in the Dodd-Jensen covering lemma for $L[U, C]$, essentially local: it depends on the set X, and its size is at most that of X. Furthermore, while there is a sense in which the choice of $\mathcal{C}$ is, on its domain, unique up to finite variations, this is not enough to directly generalize the Jensen proof that the nonexistence of $0^\#$ implies SCH.

This version of the covering lemma was published in [Mitchell, 1987], which included several of its consquences. The simplest of these gave a converse to Magidor's construction in [Magidor, 1978]: *Suppose that there is no inner model satisfying $\exists\kappa\, o(\kappa) = \kappa^{++}$, and that λ is a singular cardinal of uncountable cofinality δ which is regular in K. Then $o(\lambda) \geq \delta$ in K.* An additional result in [Mitchell, 1987] showed that if there is a model of ZF in which the closed unbounded ultrafilter on ω_1 is an ultrafilter then κ is a (weak) repeat point, a result which Mitchell later showed (in unpublished work) is best possible. A final result considered the model $L[i]$, constructed from an embedding $i \colon V \to M$: it was shown that if there is no inner model of $\exists\kappa\, o(\kappa) = \kappa^{++}$ then the model $L[i]$ contains an iterate of the core model K, and is equal to an iterate of K if the embedding is an ultrapower embedding $i \colon V \to \mathrm{Ult}(V, U)$.

Several succeeding papers yielded further consequences: notably [Mitchell, 1991], which characterized the ways (assuming no inner model of $\exists\kappa\, o(\kappa) = \kappa^{++}$) in which a cardinal could become singular. Not until [Mitchell, 1992b] did he publish the partial results he obtained on the strength of SCH, and by this time [Gitik, 1991] had already appeared, giving the final solution to this question: *If* SCH *fails, then there is an inner model satisfying* $\exists\kappa\, o(\kappa) = \kappa^{++}$. Gitik's proof of this result required deep techniques from Shelah's pcf theory in addition to Mitchell's results. As related above, he had already strengthened a result of Woodin to prove that the existence of such a model is sufficient to obtain a forcing extension in which κ is measurable and $2^\kappa = \kappa^{++}$, and hence one in which SCH failed.

At about the same time, Mitchell [1992a] generalized Jensen's absoluteness result mentioned earlier: *Suppose that there is no inner model of* $\exists\kappa\, o(\kappa) = \kappa^{++}$, $a^\#$ *exists for every real* a, *and* M *is a model containing* K, *or an iterated ultrapower of* K. *Then every* $\mathbf{\Sigma}^1_3$ *formula true in* V *is true in* M. But, as we shall see, Steel shortly afterward had a stronger result, which was actually published earlier.

5 EXTENDER MODELS: WOODIN CARDINALS AND BEYOND

5.1 Moving beyond measurable cardinals

With the core model understood up to $\exists\kappa\, o(\kappa) = \kappa^{++}$, both Mitchell and the team of Dodd and Jensen turned their attention to larger cardinals. The smallest of these was a μ-measurable cardinal, defined in [Mitchell, 1979a] to be a cardinal κ such that there exists an embedding $i\colon V \to M$ with $\kappa = \operatorname{crit}(i)$ such that the induced ultrafilter $\mu = \{\, x \subseteq \kappa : \kappa \in i(x)\,\}$ is a member of M. Such an embedding cannot be obtained from an ultrapower using a normal ultrafilter, but it can be obtained by using a nonnormal ultrafilter on V_κ, namely $W = \{\, x \subseteq V_\kappa : \mu \in i(x)\,\}$. The next natural step beyond a μ-measurable cardinal is a $\mathcal{P}^2(\kappa)$ strong cardinal, that is, a cardinal κ with an embedding $i\colon V \to M$ such that $\mathcal{P}^2(\kappa) \subseteq M$. Such an embedding cannot be represented by a ultrafilter on V_κ, since any such ultrafilter would be a member of M. Thus significant progress beyond $\exists\kappa\, o(\kappa) = \kappa^{++}$ depended on the invention of a more flexible representation for large cardinal embeddings.

Such a representation was discovered by Mitchell [1979a] and subsequently simplified by Jensen. Of these two, it was Jensen's formulation which has become standard. This notion, which Jensen called an *extender*, is the same construction as was used to extend the collapse embedding in the proof of the covering lemma. An embedding $i\colon V \to M$, with critical point κ is said to be an *ultrapower by a* (κ, λ)*-extender* if $\operatorname{crit}(i) = \kappa$ and $M = \{\, i(f)(a) : a \in [\lambda]^{<\omega}\,\}$. In this case the (κ, λ)-extender E such that $M = \operatorname{Ult}(V, E)$ can be represented as a sequence of ultrafilters $\langle\, E_a : a \in [\lambda]^{<\omega}\,\rangle$, with $E_a = \{\, x \in \kappa^{|a|} : a \in i(x)\,\}$. The embedding i is then a direct limit of the ultrapower embeddings $i_a\colon V \to \operatorname{Ult}(V, E_a)$, with the image of $[f]_{E_a}$ in $\operatorname{Ult}(V, E)$ being written as $[f]_{E,a}$.

Mitchell's construction in [Mitchell, 1979a] was effectively the same, but complicated by using, instead of all ordinals less than λ, only the *generators* of the embedding i, that is, the set of ordinals $\alpha < \lambda$ such that $\alpha \neq i(f(a))$ for any $a \in [\alpha]^{<\omega}$. This decision to separate out the generators proved to be unnecessary and undesirable: for example, the question of which ordinals are generators depends on the power set of κ, and hence may differ between an embedding $i\colon V \to M$ and the restriction of that embedding to an inner model.

Jensen's notion of an extender had a side effect of eliminating a significant complication in Mitchell's models $L[\mathcal{U}]$: the need for coherence functions. If $\alpha < \beta < o(\kappa)$ then the requirement that $\mathcal{U}(\kappa,\alpha) \triangleleft \mathcal{U}(\kappa,\beta)$ entails that there is a function $f \in {}^{\kappa}\kappa$ such that $\alpha = [f]_{\mathcal{U}(\alpha,\beta)}$. Mitchell's core model construction, for example, required special care to ensure that no measure $\mathcal{U}(\kappa,\beta)$ was used in the construction until all of its coherence functions had appeared. If this same model is recast by representing the measures as extenders then this becomes unnecessary: the measure $U = \mathcal{U}(\kappa,\beta)$ has the single generator κ and is most conservatively represented as a $(\kappa, \kappa+1)$ extender; however it can be equivalently represented as a (κ,λ) extender for any $\lambda > \kappa$. If $\lambda \geq \beta$ then, since $\alpha = i(\mathbf{id})(\alpha)$, where **id** is the identity function, we can write $\alpha = [\mathbf{id}]_{\mathcal{U}(\kappa,\beta),\{\alpha\}}$ for all $\alpha < \beta$. Thus the identity function is the only coherence function needed.

In addition to Mitchell, Dodd and Baldwin both started using extenders to construct models with larger cardinals. All three claimed, at least briefly, to be able to deal with cardinals up to a *superstrong cardinal*, that is, a cardinal κ with an embedding $i\colon V \to M$ with $V_{i(\kappa)} \subseteq M$; however their actual success was much more modest.

The extension of the models was straightforward up to a strong cardinal, that is, a cardinal κ such that for each $\lambda > \kappa$ there is an elementary embedding $i\colon V \to M$ with $V_\lambda \subseteq M$. This, however, was only a short step towards a superstrong cardinal, and cardinals larger than this presented a new difficulty. Recall that one of the three criteria given in Section 2.4 for the success of a comparison by iterating the least differences was that generators not be moved. The use of the term "moving generators" at that point in the history was somewhat anachronistic: since the only generator of a normal ultrafilter U is its critical point, the coherence of the sequence of ultrafilters was enough to ensure that no generators were moved. For a (κ,λ)-extender E, in contrast, every ordinal in the interval $[\kappa,\lambda)$ is a generator, so that the principle of no moving generators would seem to require that if E is used in an iteration, then the rest of the iteration cannot involve any (κ',λ')-extender E' with $\kappa' < \lambda$. This difficulty begins to occur as soon as there is a cardinal which is strong past the next measurable cardinal, that is, a cardinal κ having an embedding $i\colon V \to M$ with critical point κ such that some cardinal $\lambda > \kappa$ has a measure U such that $U \in M$. While the problem of moving generators can easily be dealt with at this level, a general solution turned out to be much more elusive.

A secondary problem at this level involves the indexing of the sequence. The indexing $\mathcal{U}(\kappa,\beta)$ used by Mitchell relied on the fact that given two measures on different cardinals, the one on the larger cardinal is stronger. Again, this is always true up to a strong cardinal, where a cardinal κ is *strong* if for any λ there is an elementary embedding $i\colon V \to M$ with critical point κ such that $V_\lambda \subseteq M$. However, if κ is a strong cardinal, and there is also an extender E' with critical point $\kappa' > \kappa$, then there is an extender E on κ such that E' is in a member of the codomain $\mathrm{Ult}(V, E)$ of the embedding i^E, so that $E' \lhd E$.

Of the three working on the problem at this time, Baldwin [1986] was the only one whose work was published, and the one whose work went furthest in terms of the large cardinal properties accommodated. In order to deal with the problem of moving generators, he relied on a different, carefully defined indexing of the extenders, designed so that, although the iteration did have moving generators, there was guaranteed to be a closed unbounded set of stages of the iteration where the generators of the extender used were not later moved. This was sufficient to ensure the success of the comparison — at least for the cardinals discussed in the paper, which included hierarchies of what he called β-hypermeasurable and β-hyperhypermeasurable cardinals. The apparent problem with Baldwin's idea was that there did not seem to be a clear scheme for finding a suitable indexing as one moved to larger cardinals, and indeed Steel has asserted that Baldwin's paper came close to the limit of what can be achieved with such techniques.

The model described in Dodd's notes [1984?] allowed for a class of strong cardinals, weaker than the cardinals in Baldwin's model but enough to require consideration of moving generators. Like Baldwin's, it relied on a reindexing of the sequence and on finding a closed unbounded set of stages of the iteration generators which were not moved, and it shared the shortcoming of Baldwin's techniques. However Dodd's notes are important for introducing two new techniques into the comparison process which became important for later developments. One of these was the Dodd-Jensen lemma which, as was pointed out earlier, serves as an important complement to the use of mice in the verification that the criterion of equal comparison is satisfied. The lemma states that an iterated ultrapower is minimal among all Σ_0-elementary embeddings mapping into the same final model; more specifically: *Suppose $i\colon M \to M'$ is an iterated ultrapower of the mouse M, and $\sigma\colon M \to M'$ is a Σ_0-elementary embedding. Then* $\mathrm{range}(\sigma)$ *is cofinal in M', and $\sigma(\alpha) \geq i(\alpha)$ for all ordinals α in M.* A typical application occurs in the fine structure analysis: in this case $\sigma = j \circ \pi$, where $\pi\colon M \to J_\alpha[\mathcal{E}]$ arises from the transitive collapse of a Skolem hull in $J_\alpha[\mathcal{E}]$, and j and i are iterated ultrapowers on M and $J_\alpha[\mathcal{E}]$ respectively arising from the comparison of these models. The model M can be treated as a mouse, and since its projectum is not moved by the iterated ultrapower i on M this implies that the last model M' of this iteration is not a proper initial segment of the last model of the iteration j of $J_\alpha[\mathcal{E}]$. It is not immediately clear, however, that the embedding j does not move the projectum, and hence the mouse criterion cannot be used to rule out the possibility that the last model of the iteration j is a proper initial segment of M'. The Dodd-Jensen lemma does rule this out, since i is an iterated ultrapower and hence minimal.

There is some question about the origin of the Dodd-Jensen lemma. The author's memory is that it came as a result of a message from Dodd pointing out a hole in the author's proposals. The author, after some thought, filled the hole, and Dodd and Jensen turned his argument into a general lemma. The author remembers doubting the general statement and being startled to be told that the proof was his own. However Jensen has no memory of this, and it is likely that he and Dodd independently discovered the proof before the author's letter was received. However the author is certainly responsible for the current name of the lemma.

The other new technique in [Dodd, 1984?], was the use of a larger model to back up an iteration. As used by Dodd, this replaces the use of regarding M as a mouse in the argument described in the second paragraph above. Supposing, for the sake of contradiction, the iterated ultrapower of the collapsed model M involves an extender E which moves the projectum of M, he instead applies the extender E to the full model $J_\alpha[\mathcal{E}]$, and institutes a comparison between $J_\alpha[\mathcal{E}]$ and $\mathrm{Ult}(J_\alpha[\mathcal{E}], E)$. Since both sides of this comparison may be regarded as iterated ultrapowers of $J_\alpha[\mathcal{E}]$, the Dodd-Jensen lemma may be applied to both sides, leading to the conclusion that the two embeddings are equal. It follows that E was a member of $J_\alpha[\mathcal{E}]$, which contradicts the assumption that it was necessary to use it in the original comparison. This technique was later developed into the structure which Steel called a *phalanx.*

Overlapping extenders and iteration trees

At a talk at the 1979 Association of Symbolic Logic meeting in Biloxi, Mississippi, the author confidently asserted that an inner model for a supercompact cardinal was at hand. This confidence was largely based on a alternative technique of iteration. This alternative, which seems to be forced if one wishes to exclude moving generators, is to generalize the technique of iteration. The iterated ultrapowers up to this point had been linear: they looked like

$$L[\mathcal{E}] = M_0 \xrightarrow{i_{0,1}} M_1 \xrightarrow{i_{1,2}} \cdots M_\omega \xrightarrow{i_{\omega,\omega+1}} M_{\omega+1} \xrightarrow{i_{\omega+1,\omega+2}} \cdots M_\theta \tag{5}$$

where θ is the length of the iteration. In this construction, each model $M_{\alpha+1}$ is the ultrapower of the previous model: $M_{\alpha+1} = \mathrm{Ult}(M_\alpha, E_\alpha)$ where E_α is the least extender in the sequence $i_{0,\alpha}(\mathcal{E})$ of $M_\alpha = L[i_{0,\alpha}(\mathcal{E})]$ which is not in the corresponding sequence on the other side of the comparison. In the new construction, the extender $\mathcal{E}_\gamma$ is chosen in the same way, but $M_{\alpha+1}$ may be an ultrapower by an earlier model in the sequence: $M_{\alpha+1} = \mathrm{Ult}(M_{\alpha^*}, E_\alpha)$ where α^* is the least ordinal such that i^{E_α} moves generators of the extender E_{α^*} used to define M_{α^*+1}. This gives a tree ordering, defined by taking the ordinal α^* to be the immediate predecessor of $\alpha+1$. If M_γ and $M_{\gamma'}$ are two models of the iteration, then the iteration provides an embedding $i_{\gamma,\gamma'}\colon M_\gamma \to M_{\gamma'}$ between them if and only if γ precedes γ' in the tree. Any branch of this tree looks like a linear iteration, except that the extenders used to construct the embeddings along the branch may be taken from models in other branches, rather than from the models to which they are applied.

The resulting *iteration trees* satisfy the criterion of no moving generators, and they can be used in much the same way as linear iterations. In particular, a comparison between two models $L[\mathcal{E}]$ and $L[\mathcal{E}']$, using the process of iteration by the least difference but organized as a tree iteration, will terminate if it can be carried out far enough.

The use of nonlinear iterations eliminates moving generators, but introduces a new difficulty: that of proving iterability, that is to say, verifying that the comparison process can be carried out "far enough". This has proved to be far more difficult than for linear iterations, for which iterability can be ensured simply by using countably complete extenders. There was some question even at successor stages: Since the extender E_α used to define $M_{\alpha+1}$ is not a member of M_{α^*}, why does $\mathrm{Ult}(M_{\alpha^*}, E_\alpha)$ even exist, and if it does, why is it well-founded? The primary difficulty, however comes a limit stages. Suppose that the iteration has been carried out for ν stages, where ν is a limit ordinal. If b is any branch of the tree $(\nu, \prec)$ then $i_{\alpha,\alpha'}$ is defined for any $\alpha < \alpha'$ in b, so we can define M_b to be the direct limit of the system $(\langle M_\alpha : \alpha \in b \rangle, \langle i_{\alpha,\alpha'} : \alpha \prec \alpha' \in b \rangle)$. If there is a branch b which is cofinal in the tree such that M_b is well-founded, then the model M_b is a possible choice for M_ν; however it is not evident that there will be any such branch. Indeed it is not clear that the tree even has a cofinal branch b, let alone a well-founded one; indeed, one could imagine a worst case in which $\alpha^* = 0$ for all $\alpha > 0$, so that the tree has length ω, but every branch has length 2.

The author recalls attempting, while on the airplane home from Biloxi, to substantiate the claim which he had made of being on the verge of a model for a supercompact cardinal, and being shocked on convincing himself that iterability necessarily fails in the region of a supercompact cardinal. This conclusion was later confirmed by Neeman's discovery [Neeman, 2004] of an infinite iteration tree with no branches of length 4. This was a serious setback, since there seemed to be no interesting cardinals between measurability and supercompactness, and furthermore the problem of iterability seemed to be very difficult even below a supercompact cardinal.

Woodin cardinals

Discouraged, Mitchell essentially gave up work on such inner models; however his Biloxi talk did have one important effect: it attracted Steel to work on the topic. Steel was largely ignorant of inner models, and did not understand the later parts of the talk; however he was excited at the idea of models incorporating hypermeasures. He had regarded the coherent sequence framework as being only a technical elaboration of Kunen's work, but realized that models capable of incorporating embeddings with all of $\mathcal{P}^2(\kappa)$ in the ultrapower gave a much more general instance of the "more of V contained in M" hierarchy of large cardinals. With Martin's help, he studied the available literature on inner models. During this early period, although he recognized the seriousness of the problem of moving generators, Steel was inhibited from using nonlinear iterations because he realized that, for countable models, the definition of iterability for such iterations would require a Π^1_3 formula for its expression. This would give a Δ^1_3 well-ordering of the reals; however the state of knowledge about the Axiom of Determinacy at that time had led to a near universal belief that the well-ordering of the reals would not become that complicated until well past a supercompact cardinal.

Little further progress was made for the first five years. The breakthrough came from work of Foreman, Magidor and Shelah in the spring of 1984 which led to [Foreman *et al.*, 1988]. This did not immediately involve inner models at all; indeed it contradicted some of the aims of the inner model program. One of the results in this paper was that if it is consistent that there is a supercompact cardinal, then it is consistent that the filter of nonstationary sets on ω_1 is saturated. This in turn implies that, in a generic extension, there is an embedding $i\colon V \to M$ which, if it were in V rather than in a generic extension, would be much stronger than a supercompact embedding. Such an embedding contradicted an implicit premise of the inner model program, that any such embedding would extend a large cardinal embedding existing in the ground model.

Shelah and Woodin, in [Shelah and Woodin, 1990] and [Woodin, 1988], built on the techniques and results of [Foreman *et al.*, 1988] to prove: *If there is a supercompact cardinal, then in the model* $L(\mathbb{R})$ (i) *all sets of reals are Lebesgue measurable,* (ii) *all sets of reals have the property of Baire, and* (iii) *the partition property* $\omega \to (\omega)^\omega_2$ *holds.* As a consistency result this was not strong: all three are true in Solovay's model [1970] arising from the Levy collapse of an inaccessible cardinal to become ω_1 (with (iii) having been added to Solovay's results by Mathias [1977]). However the fact that, given a supercompact cardinal, these consequences were outright true in $L(\mathbb{R})$ was striking, especially in view of the fact that all are consequences of the Axiom of Determinacy.

Attempts to weaken the hypothesis ultimately led to Woodin's definition of what came to be called a *Woodin cardinal.* An inaccessible cardinal δ is said to be Woodin if *for all functions* $f\colon \delta \to \delta$ *there is an embedding* $i\colon V \to M$ *with critical point* $\kappa < \delta$ *such that* $V_{i(f)(\kappa)} \subseteq M$. The last theorem in [Woodin, 1988] stated that *if there is a model with a Woodin cardinal, together with a sharp for that model, then every* Σ^1_3 *set of reals is Lebesgue measurable.*

An important tool in dealing with Woodin cardinals was a modification, invented by Woodin, of the forcing of [Foreman *et al.*, 1988]. The new forcing, which he called the *stationary tower*, came in two forms. The first, which was used for results such as that stated in the last paragraph, collapsed cardinals so that the Woodin cardinal in the ground model became ω_1 in the generic extension. The other preserved the Woodin cardinal and defined, in the generic extension $V[G]$, an elementary embedding $i^G\colon V \to M$ with the remarkable property that $(V_\delta)^M = (V_\delta)^{V[G]}$. This second form ruled out a core model K with the properties which had been expected on the basis of earlier work. The forcing collapses many successors of singular cardinals, so that a core model K either would not have the weak covering property, or else $K^{V[G]} = i^G(K) \neq (K)^V$, in which case the core model would be changed by set forcing.

The introduction of Woodin cardinals spurred the development of inner models in two ways: first, it gave as a goal an important new large cardinal property which was more accessible than supercompactness, and, second, it helped to explain some of the difficulties which had been encountered. About this time, Steel began working with Martin, and they proceeded to develop a theory of iteration trees. The first result of this collaboration was published in [Martin and Steel, 1988] and [Martin and Steel, 1989], which proved: *If there is a model with n Woodin cardinals and a measurable cardinal above them, then all Π^1_{n+1} sets are determined.* Thus, they obtained projective determinacy from a model with infinitely many Woodin cardinals. Later they used a result of Woodin to prove that if there is a model with ω many Woodin cardinals with a measurable cardinal above them, then the Axiom of Determinacy holds in $L(\mathbb{R})$.

The proof of Π^1_{n+1} determinacy involved first developing the theory of iteration trees far enough to show that no countable iteration tree using countably complete extenders could have ill-founded limits on all branches with the ill-foundedness witnessed by a continuous function on the nodes of the iteration tree. Thus, by constructing a tree having such a witness to ill-foundedness on all but one of its branches, they could ensure that the remaining branch had a well-founded limit. They then used the Woodin cardinal to find a special type of iteration tree, which they called an *alternating chain*, consisting of just two branches, each of length ω. They constructed the tree so that the well-foundedness of one of the branches implied the existence of a strategy, and they ensured that the limit on that branch was well-founded by constructing the other branch to be ill-founded.

In the search for inner models for a Woodin cardinal, one key fact was observed by Martin and Steel: *Suppose that M is an extender model, and $\mathcal{T}$ is an iteration tree on M having two distinct branches b and c with well-founded limits M_b and M_c. Then there is model with a Woodin cardinal.* In fact if δ is the supremum of the lengths of the extenders used in $\mathcal{T}$, then δ is Woodin with respect to any function $f \in M_b \cap M_c$. In particular, if M is a class model then δ is Woodin in $L[\mathcal{E}]$, where $\mathcal{E} = \mathcal{E}^{M_b}\restriction\delta = \mathcal{E}^{M_c}\restriction\delta$.

This had two important consequences. One was obvious: in order to reach a minimal model with a Woodin cardinal, it would be enough to define a sequence $\mathcal{E}$ by recursion in such a way that, so long as it did not reach a Woodin cardinal, every iteration tree had at least one branch with a well-founded limit. The other consequence was an absoluteness result, which enchanced the importance of countable iteration trees: it was not possible, using a homogeneous forcing such as a Levy collapse, to add a new well-founded branch to an iteration tree which did not reach a Woodin cardinal.

Their strategy for finding well-founded branches was to assume that a tree had no well-founded branches, take an elementary substructure to obtain a countable model having such an ill-founded tree, and then define maps from models in this iteration tree into V which commuted with the tree embeddings, thus obtaining a contradiction since the union of the range of the maps would be an ill-founded collection of sets in V. Using this strategy, they succeeding in proving in [Martin and Steel, 1994] that *if there are n Woodin cardinals, for $n < \omega$, then there is a proper class model M_n satisfying* ZFC + *there are n Woodin cardinals* + $\mathbb{R}$ *has a Σ^1_{n+2} well-order.* This result was essentially best possible in the sense that by their previous result the existence of n Woodin cardinals plus a measurable cardinal above them implies that Π^1_{n+1} determinacy holds, and this in turn implies that there is no Δ^1_{n+2} well-ordering of the reals.

This Martin-Steel model does have a serious weakness, however. The models introduced in this paper have the form $L[\mathcal{E}]$, where $\mathcal{E}$ is a sequence of extenders. The comparison process for these models, however, does not use ultrapowers of the models themselves, but instead relies on ultrapowers in a larger model constructed from a sequence $\mathcal{F}$ of *background extenders.* For each extender E in the sequence $\mathcal{E}$ there is a background extender F in the sequence $\mathcal{F}$. The extender F extends E, however it is substantially stronger than E: it is an extender in V, not just $L[\mathcal{E}]$, and if E is a (κ, λ)-extender then F is satisfies $V_{\lambda+2} \subseteq \mathrm{Ult}(V, F)$. Because of this use of external iterations, many of the structural properties of their model were unclear. For example, they were not able to show that GCH holds in their model, although they did show, using a result of Jensen, that it satisfies CH.

A second forcing notion discovered by Woodin sheds further light on the question of iterability. The extender algebra, which according to Woodin was probably discovered in July of 1986, is related to the forcing of Vopěnka [Vopěnka and Hájek, 1972], by which any set is generic over HOD. Let M be a model of set theory with a Woodin cardinal δ such that the Woodinness of δ is witnessed by extenders in M which are iterable in the universe V (though perhaps not in M). The extender algebra is a forcing of size δ, defined in M, with the property that, for any set x in V, there is an iterated ultrapower $i\colon M \to M^*$ of M such that x is M^*-generic over $i(P)$. Thus, for example, if there is a supercompact cardinal in V then there is, for each λ, a generic extension of M which has a λ-supercompact cardinal — even if the model M has no more than one Woodin cardinal. The apparent contradiction is avoided because the model M does not know the strategy for iterating itself, and hence the argument cannot be carried out inside M. Indeed this forcing can be used to show that there is no fully iterable model of the form $L[A]$, with A a set, having a Woodin cardinal: any such model would have, in a generic extension, a model $L[A']$ having the same property with A' contained in the smallest measurable cardinal of $L[A]$. In fact Woodin has shown that if there is a iterable model with a Woodin cardinal, then $x^{\#}$ exists for every set x.

This does not exclude the possibility that there is a a strategy in V for iterating the model M, but it does imply that an iteration of the model M will not, in general, be fully internal to M even if it is internal in the sense that all extenders involved are members of the appropriate model of the iteration tree, and even if (as in the case of the genericity iteration of the last paragraph) the iteration is defined inside M. Of course this is not always such a difficulty as it may appear, since in most cases an iteration tree is used which is not defined inside M. The iteration trees used to compare two models M and N, for example, require knowledge of both models and hence cannot be carried out inside of either.

5.2 Fine structural extender models

Encouraged by the work of Martin and Steel, and prodded by work of Baldwin, Mitchell returned to work on inner models with extenders in 1986. Since he intended that the end result was a core model, he assumed from the start that the models would include mice as well as extenders.

A significant innovation was suggested by a note of Baldwin and greatly simplified the presentation of the model. Recall that Mitchell's original core model $K(\mathcal{F})$ had the form $L[\mathcal{F}, \mathcal{M}]$ where $\mathcal{F}$ was the sequence of ultrafilters of the model and $\mathcal{M}$ was a class of mice. This was awkward at the time, and promised to become more so in the newer models, as the mice would be expected to mimic the full core model and hence would need to themselves contain mice. In the new presentation the model had the form $L[\mathcal{E}]$, being constructed entirely from extenders $\mathcal{E}_\gamma$ in the sequence $\mathcal{E}$. However these extenders were not necessarily extenders on the full model $L[\mathcal{E}]$; instead $\mathcal{E}_\gamma$ was an extender on the model $J_\gamma[\mathcal{E}\restriction\gamma]$ constructed up to that point. The result was that the mice in the model $L[\mathcal{E}]$ were just the initial segments $M = J_\alpha[\mathcal{E}]$ of $L[\mathcal{E}]$. As a consequence, the instances of the Condensation Property, Lemma 1, which are needed for the analysis of fine structure would hold automatically.

In 1987 Mitchell visited Steel at UCLA, and described the idea of using partial extenders. Steel had, before this time, also been thinking about a fine structural model. After discussing the new idea with Jensen, who was visiting at the same time, he was excited to realize that it did, in fact, cut through many of the complications involved in the fine structure.

Although much hard work remained, the way was now open to precisely characterizing the structure of the extender models as they are known today. However there remained the iterability problem, that is, the problem of showing that such models could be constructed so that there was a strategy for choosing the branches in iteration trees which maintained the well-foundedness of the iterates. For this, Steel started with the techniques which he and Martin had developed for their previous inner model. The use of external iterations in that model had been dictated by the need for agreement between the models in the tree; under the new structure this agreement was ensured by the fine structure, so the iterations would be internal in the sense that all of the extenders used were taken from the models in the tree. However, he retained the use of background extenders. The construction proceeded through a delicate recursion, which defined a sequence of models M_α. To define $M_{\alpha+1}$, the model M_α would first be extended, either by constructing one step further or, if a suitably backgrounded extender was available, by adding that extender to the sequence. If the resulting model N_α did not contain any bounded subsets of M_α which were not already in M_α, then $M_{\alpha+1}$ would be equal to N_α. Otherwise the fine structural collapse would be used to obtain the core structure of N_α, and $M_{\alpha+1}$ would be that core structure. In this case, any extenders in N_α which had critical points greater than its projectum would become partial extenders in $M_{\alpha+1}$. This process converged, and yielded a model $L[\mathcal{E}]$ which satisfied the desired properties, inluding GCH and a Δ^1_3 well ordering of the reals, and which would have a Woodin cardinal assuming that there was one in V. This work was ultimately published as the book [Mitchell and Steel, 1994].

Steel later extended this work in [Steel, 1993] to models with arbitrarily many Woodin cardinals, and Neeman [2002] has defined such extender models slightly beyond a Woodin limit of Woodin cardinals. This is the best result known giving, from the assumption of a large cardinal, an iterable extender model $L[\mathcal{E}]$ for that (or even a smaller) cardinal. However if one is willing to add appropriate iterability hypotheses to the large cardinal assumptions then this construction can be carried out as far as, and even somewhat beyond, a superstrong cardinal, that is, a cardinal κ such that there is an embedding $i\colon V \to M$ with critical point κ such that $V_{j(\kappa)} \subseteq \kappa$. Doing so relies on a modification, due to Jensen and using a suggestion of Sy Friedman, of the construction of $L[\mathcal{E}]$. Suppose that E is a (κ, λ)-extender to be added to the sequence $\mathcal{E}$, where E cannot be represented as an extender of length less than λ. The construction of [Mitchell and Steel, 1994] adds E as $\mathcal{E}_\gamma$ where $\gamma = \lambda^+$ in $L[\mathcal{E}{\restriction}\lambda^+]$, a value for γ which was used because it is the least choice such that the mapping $E \mapsto \gamma$ is one-to-one. The Jensen construction sets $\gamma = i^E(\kappa^+)$, which avoids some anomalies in the construction. This approach was introduced in [Jensen, 1997], and a detailed exposition is given in the last part of the book [Zeman, 2002].

Most applications of the models $L[\mathcal{E}]$ rely on core model techniques, which will be discussed in the next subsection; however one striking application can be described here. The discussion of fine structure in this history started with the development of Jensen's definition of the principle $\square_\kappa$ and the proof that it holds in L. This result had been extended a number of times to larger models: by Solovay to $L[U]$, to K^{DJ} in [Welch, 1979], and for arbitrary sequences with measures of order 0 in [Wylie, 1989]. Schimmerling [1995] extended this to a model with $\exists\kappa\, o(\kappa) = \kappa^{++}$ and proved weaker square principles for larger models. Finally Schimmerling and Zeman, working first independently and then together, completed this program. Jensen had defined a *subcompact cardinal* to be a cardinal κ such that for each $B \subseteq \kappa^+$ there is a $\mu < \kappa$, $A \subseteq \mu^+$ and an elementary embedding $j\colon (H_{\mu^+}, \in, A) \to (H_{\kappa^+}, \in, B)$. This is a property somewhat stronger than superstrong, but weaker then κ^+-supercompact, and Jensen had proved that $\square_\kappa$ fails for any subcompact cardinal. Schimmerling and Zeeman [2001; 2004] proved a converse for extender models: *the property $\square_\kappa$ holds in any extender model $L[\mathcal{E}]$ for any cardinal κ which is not subcompact.*

5.3 $L[\mathcal{E}]$ as a core model

The models presented in [Mitchell and Steel, 1994] gave the form for a core model. Like the core models previously developed by Dodd and Jensen and by Mitchell, and unlike $L[U]$, they were built up recursively from below using mice, and so could present an apparently minimal model for any cardinal smaller than a Woodin cardinal. Furthermore all of the previously described core models could be presented as a model in the new form $L[\mathcal{E}]$. In the case of the Mitchell's core model for sequences of measures, and arguably even for the Dodd-Jensen core model K^{DJ}, doing so yielded a great gain in clarity. However, [Mitchell and Steel, 1994] did not present a core model construction: they relied on the Martin-Steel technique for the proof of iterability, and that required starting with extenders in the real world, called background extenders. Indeed, the background extenders had to be stronger than the extenders in the model $L[\mathcal{E}]$. Thus, like the model $L[U]$, they would not give a model with a Woodin cardinal unless a Woodin cardinal was already known to exist in the universe. A core model construction, in contrast, should expose the latent large cardinal structure even if the large cardinal has been destroyed by, for example, a forcing extension collapsing cardinals.

Before proceeding, it will be useful to compare the effect of Woodin cardinals in an inner model with that of measurable cardinals. The presence of measurable cardinals changed the theory in two major respects: (i) models were required to be iterable, rather than merely well-founded, and (ii) the covering lemma had to be adapted to allow for the possible presence of Prikry sequences.

In the case of measurable cardinals, iterability can be ensured by using only countably complete measures. Furthermore, if M is a model with only measurable cardinals for which all countable iterations are well-founded, then iteration of M of any length are well-founded. Hence any model M containing ω_1 is iterable in V if and only if $M \models$ "I am iterable" — indeed any iterated ultrapower in V can be embedded into an iterated ultrapower which is defined inside M.

None of these observations are true for extenders. Two important conjectures assert that the first is true for countable iteration trees: If $\mathcal{T}$ is a countable iteration tree, using only extenders which are countably complete in the model in which they appear, then the *Cofinal Branches Hypothesis* (CBH) asserts that $\mathcal{T}$ has a cofinal branch with a well-founded limit, and the *Unique Branches Hypothesis* (UBH) asserts that it has at most one such branch. Surprisingly, and fortunately, $\omega_1 + 1$-iterability, that is, iterability for trees of length at most ω_1, is sufficient for much of inner model theory.

This increased complexity of the concept of iterability does not actually begin at a Woodin cardinal; in fact it is a consideration in any model with overlapping extenders. However it is mitigated, short of a Woodin cardinal, by the Martin-Steel result in [Martin and Steel, 1994] that any iteration tree with two well-founded branches of length δ induces a model in which δ is Woodin. This means that, except at a Woodin cardinal, the property of an iteration tree having a well-founded branch is not changed by homogeneous forcing, such as a Levy collapse.

Clause (ii) above, the effect of Prikry forcing on the covering lemma at a measurable cardinal, is paralleled by a much larger effect at a Woodin cardinal: as was pointed out earlier, the stationary tower forcing witnesses that a model with a Woodin cardinal δ may not have even the weak covering property below δ.

This goal of eliminating the need for background extenders in V was reached, or nearly reached, by Steel [1996] in 1990. The key observation was that a much weakened background extender could be used: an extender on a cardinal κ would only need background extenders for κ-sized restrictions of the extender, and such background extenders could be obtained by assuming the presence of a single measurable cardinal in V. By adapting the recursive construction of [Mitchell and Steel, 1994] to this weaker notion of background extenders, he defined a model of height Ω which he called K^{c}, corresponding to Mitchell's countably closed core model K^{c}; in this case Steel took the superscript 'c' to mean "certified'.' He proved, assuming there is no inner model with a Woodin cardinal, that this model satisfies what he called a "cheapo" covering lemma: if μ is the given measure on κ, then K^{c} is a model of height κ such that the set of cardinals $\lambda < \kappa$ such that $\lambda^{+} = (\lambda^{+})^{K^{c}}$ is a member of μ. This version of the weak covering theorem implies that K^{c} is universal (as a model of height Ω), and enabled him to define the true core model K as a elementary submodel of K^{c}. As with Mitchell's core model for sequences of measures, this proof gave an alternate direct definition of K, as a model $L[\mathcal{E}]$ where the members $\mathcal{E}_{\gamma}$ of the sequence are chosen by induction on γ. However the characterization of which extenders to add to the sequence so as to ensure the final model is iterable is somewhat more delicate than in the case of Mitchell's model.

Among the applications given in [2002], Steel showed that if there is a measurable cardinal Ω then the existence of a model with a Woodin cardinal follows from the existence of a presaturated ideal on ω_1, and the existence of a tree on V_{Ω} with two cofinal well-founded branches implies the existence of a model with two Woodin cardinals. In other applications, the assumption of a measurable cardinal could be dispensed with because it was implied by the hypothesis, either in V or in an inner model. Thus the existence of a model with a Woodin cardinal follows from Martin's Maximum, from the assumption that every set of reals is weakly homogeneous, or (using an unpublished result of Woodin) from the existence of a strongly compact cardinal. Steel also reproved Woodin's result that a model with a Woodin cardinal follows from Δ^1_2 determinacy plus the assumption that $a^{\#}$ exists for all reals a, and in 1993 he extended a result of Jensen by showing that if there are two measurable cardinals in V, and there no Woodin cardinals in K^{c}, then K^{c} is Σ^1_3 correct.

The results of [Steel, 1996] raised several obvious questions. Is the measurable cardinal really necessary? Does Steel's core model satisfy the weak covering property? Is there a core model for more than a single Woodin cardinal?

The first question was still unanswered at the end of the century. Although it was easy to see that the full strength of a measurable cardinal was not necessary, it did not seem possible to fully eliminate the need for some large cardinal in V. This has continued to be an active field of research during the first decade of the current century, and it seems likely that this problem has been solved by work of Jensen and Steel.

The second question was answered by Mitchell, Schimmerling and Steel in [Mitchell *et al.*, 1997], who showed that if there is no model with a Woodin cardinal, then $(\kappa^{+})^{K} = \kappa^{+}$ for all countably closed singular cardinals. The assumption of countable closure was eliminated shortly afterward by Mitchell and

Schimmerling [1995]. Another answer was given in [Schimmerling and Woodin, 2001]: *Suppose that $L[\mathcal{E}]$ is an iterable extender model with no measurable cardinals and no extenders of superstrong type. Then either a sharp exists for $L[\mathcal{E}]$, or else* $\mathrm{cf}((\kappa^+)^{L[\mathcal{E}]}) \geq |\kappa|$ *whenever $|\kappa|$ is a countably closed cardinal.* Note that although $L[\mathcal{E}]$ does not have any measurable cardinals, it may have submodels $L[\mathcal{E}\restriction\gamma]$ with Woodin or stronger cardinals.

Steel tackled the third question, of getting larger core models, in [Steel, 2002]. He succeeded in defining a model K^c as in [Steel, 1996], having the 'cheapo' covering lemma, so long as there is no model of a non-tame mouse, that is, no mouse having a Woodin cardinal δ and also an extender $\mathcal{E}_\gamma$ with $\mathrm{crit}(\mathcal{E}_\gamma) < \delta \leq \gamma$. This construction of K^c required only $\omega_1 + 1$-iterability, that is, every suitable iteration tree of length at most ω_1 has a branch. In order to obtain the model K, however, he needed to show that the model was $\Omega + 1$-iterable, that is, that any iteration tree of height at most the measurable cardinal Ω had a well-founded branch. He was successful enough in this to show that K exists, provided that there is no model with infinitely many Woodin cardinals, and used this to show that the existence of an Ω-iterable premouse of height Ω with infinitely many Woodin cardinals is equivalent to any of the following three statements: (i) the first-order theory of $L(\mathbb{R})$ is unchanged by small forcing, (ii) the Axiom of Determinacy Is true in $L(\mathbb{R})$ after any small forcing, and (iii) no small forcing extension adds an uncountable sequence of distinct reals in $L(\mathbb{R})$. This was a result that had been reached independently, and slightly different methods, by Woodin. Steel also showed in [Steel, 2002] that a failure of the Unique Branch Hypothesis for trees of arbitrary length, that is, the existence of a nonoverlapping iteration tree $\mathcal{T}$ with two distinct well-founded branches, implies that there is an inner model with ω many Woodin cardinals.

One of the obstacles to extending the construction K^c beyond the restriction to tame mice in [Steel, 2002] was removed by [Neeman and Steel, 1999]. The proof of the Dodd-Jensen lemma, which is heavily used in establishing the basic fine structure, depends on the fact that iteration trees have at most one well-founded branch, and this is not known to be true for the trees needed for larger models. In this paper Neeman and Steel gave a variant of the Dodd-Jensen lemma, which they called the *weak Dodd-Jensen lemma*, which does not require unique branching but which is strong enough for the construction of K^c. With this result, iterability is the only gap that remains to be filled in order to construct K^c for any cardinal up to a superstrong cardinal. Andretta, Neeman and Steel [2001] used the weak Dodd-Jensen lemma to show that K^c exists provided that there is no non-domestic mouse, that is, that there is no sharp for a model with a proper class of Woodin cardinals and a proper class of strong cardinals. This is currently the best result known on the existence of K^c.

6 EPILOGUE

The end of the twentieth century is the cutoff date for this volume, but it is far from being a natural end point for the subject of this chapter. The temptation to try to continue the exposition to the present, however, is countered by the observation that the present is unlikely to be a better ending point, and that stopping at the end of the century yields the benefit of a few years perspective. However it seems desirable to mention three recent advances which promise to be major subjects of investigation in the future.

The first of these is *core model induction,* which was introduced by Woodin as a new technique for obtaining models higher in the hierarchy of Woodin cardinals. An interesting aspect of this technique is the use, in addition to the ordinary mice described earlier in this history, of *hybrid mice*, which are mice which include, in addition to extenders, information about a strategy for iterating those extenders. This recognizes the fact that, at and above a Woodin cardinal, iterability is an important part of the large cardinal structure. An exposition of the core model induction is given in [Schimmerling and Steel, 2006].

The other two advances give at least partial answers to questions raised earlier in this history. One of these questions concerns the elimination of the measurable cardinal, which was needed in [Steel, 1996] to obtain a core model up to a Woodin cardinal. After progress was made by several people, the problem appears to have been solved by the use of a new technique of Jensen, which he calls *stacks of mice.* The technique was introduced in a paper by Jensen, Schimmerling, Schindler and Steel, [Jensen *et al.*, 2009], which applies the technique to obtain a model with a nondomestic mouse from any of several hypotheses, including the failure of both $\square_\kappa$ and $\square(\kappa)$ at a countably closed cardinal $\kappa \geq \omega_3$. The use of stacks to construct the core model K up to a Woodin cardinal, without the use of a measurable cardinal, is due to Jensen and Steel and is currently unpublished.

The other major advance is due to Woodin, who has discovered a framework which gives inner models for essentially every large cardinal property which has been considered, including properties involving nontrivial embeddings $i\colon V_\lambda \to V_\lambda$. He avoids a problem, which Steel had called *moving spaces* and which seemed to show that an extender model $L[\mathcal{E}]$ could not contain cardinals larger than a supercompact cardinal, by defining a modified extender model $L[\mathcal{E}]$ in which the critical points of the extenders in the sequence $\mathcal{E}$ are bounded, and then showing that the desired embeddings with larger critical point are constructed from $\mathcal{E}$. This work is as yet unpublished.

BIBLIOGRAPHY

[Andretta *et al.*, 2001] A. Andretta, I. Neeman, and J. Steel. The domestic levels of K^c are iterable. *Israel Journal of Mathematics*, 125:157–201, 2001.

[Baldwin, 1986] S. Baldwin. Between strong and superstrong. *The Journal of Symbolic Logic*, 51(3):547–559, 1986.

[Dodd and Jensen, 1976] A. Dodd and R. Jensen. A modest remark. Handwritten manuscript, 1976. Available in [Jensen, 2006].

[Dodd and Jensen, 1981] A. Dodd and R. Jensen. The core model. *Annals of Mathematical Logic*, 20(1):43–75, 1981.

[Dodd and Jensen, 1982a] A. Dodd and R. Jensen. The covering lemma for K. *Annals of Mathematical Logic*, 22(1):1–30, 1982.

[Dodd and Jensen, 1982b] A. Dodd and R. Jensen. The covering lemma for $L[U]$. *Annals of Mathematical Logic*, 22(2):127–135, 1982.

[Dodd, 1982] A. Dodd. *The core model*, volume 61 of *London Mathematical Society Lecture Note Series*. Cambridge University Press, Cambridge, 1982.

[Dodd, 1984?] A. Dodd. Notes on strong cardinals. Handwritten notes, as transcribed in TEX by Apollo Hogan, 1984?

[Donder *et al.*, 1981] H. Donder, R. Jensen, and B. Koppelberg. Some applications of the core model. In *Set theory and Model Theory (Bonn, 1979)*, volume 872 of *Lecture Notes in Mathematics*, pages 55–97. Springer, Berlin, 1981.

[Easton, 1970] W. Easton. Powers of regular cardinals. *Annals of Mathematical Logic*, 1:139–178, 1970.

[Foreman and Kanamori, 2010] M. Foreman and A. Kanamori, editors. *Handbook of Set Theory*. Springer, 2010.

[Foreman *et al.*, 1988] M. Foreman, M. Magidor, and S. Shelah. Martin's Maximum, saturated ideals, and nonregular ultrafilters. I. *Annals of Mathematics*, 127(1):1–47, 1988.

[Gaifman, 1974] H. Gaifman. Elementary embeddings of models of set-theory and certain subtheories. In *Axiomatic Set Theory (Proc. Sympos. Pure Math., Vol. XIII, Part II, Univ. California, Los Angeles, Calif., 1967)*, pages 33–101. Amer. Math. Soc., Providence R.I., 1974.

[Gandy, 1974] R. Gandy. Set-theoretic functions for elementary syntax. In *Axiomatic Set Theory (Proc. Sympos. Pure Math., Vol. XIII, Part II, Univ. California, Los Angeles, Calif., 1967)*, pages 103–126. Amer. Math. Soc., Providence, R.I., 1974.

[Gitik, 1989] M. Gitik. The negation of the singular cardinal hypothesis from $o(\kappa) = \kappa^{++}$. *Annals of Pure and Applied Logic*, 43(3):209–234, 1989.

[Gitik, 1991] M. Gitik. The strength of the failure of the singular cardinal hypothesis. *Annals of Pure and Applied Logic*, 51(3):215–240, 1991.

[Gödel, 1938] K. Gödel. The consistency of the axiom of choice and of the generalized continuum hypothesis. *Proceedings of the National Academy of Sciences U.S.A.*, 24:556–557, 1938.

[Gödel, 1939] K. Gödel. Consistency-proof for the generalized continuum-hypothesis. *Proceedings of the National Academy of Sciences U.S.A.*, 25:220–224, 1939.

[Gödel, 1940] K. Gödel. *The Consistency of the Continuum Hypothesis*. Annals of Mathematics Studies, no. 3. Princeton University Press, Princeton, N. J., 1940.

[Hajnal, 1956] A. Hajnal. On a consistency theorem connected with the generalized continuum problem. *Zeitschrift für Mathematische Logic und Grundlagen der Mathematik*, 2:131–136, 1956.

[Hajnal, 1961] A. Hajnal. On a consistency theorem connected with the generalized continuum problem. *Acta Mathematica Academiae Scientiarum Hungaricae*, 12:321–376, 1961.

[Jensen *et al.*, 2009] R. Jensen, E. Schimmerling, R. Schindler, and J. Steel. Stacking mice. *The Journal of Symbolic Logic*, 74(1):315–315, March 2009.

[Jensen, 1967] R. Jensen. Stufen der konstrktiblen hierarchie. Habilitationsschrift, June 1967. Available on [Jensen, 2006].

[Jensen, 1968?] R. Jensen. Souslin's hypothesis is incompatible with $V = L$, 1968. Available on [Jensen, 2006].

[Jensen, 1969a] R. Jensen. Some combinatorial properties of L and V. Handwritten manuscript, 1969. Available on [Jensen, 2006] or at `http://www.mathematik.hu-berlin.de/~raesch/org/jensen.html`.

[Jensen, 1969?b] R. Jensen. Souslin's hypothesis is compatible with CH. Handwritten manuscript, 1969? Available on [Jensen, 2006].

[Jensen, 1970] R. Jensen. Kombinatorische eigenschaften der konstruktiblen hierarchie. Handwritten manuscript, 1970. Available on [Jensen, 2006].

[Jensen, 1972] R. Jensen. The fine structure of the constructible hierarchy. *Annals of Mathematical Logic*, 4:229–308; erratum, ibid. 4 (1972), 443, 1972. With a section by Jack Silver.

[Jensen, 1974a] R. Jensen. Marginalia III. Handwritten manuscript, 1974. Available on [Jensen, 2006].

[Jensen, 1974b] R. Jensen. Marginalia to a theorem of Silver. Handwritten manuscript, 1974. Available on [Jensen, 2006].

[Jensen, 1974c] R. Jensen. More marginalia. Handwritten manuscript, 1974. Available on [Jensen, 2006].

[Jensen, 1997] R. Jensen. A new fine structure for higher core models, August 1997. Available on [Jensen, 2006]; also available at `http://www.mathematik.hu-berlin.de/~raesch/org/jensen.html`.

[Jensen, 2001] R. Jensen. On some problems of Mitchell, Welch and Vickers, 2001. Available on [Jensen, 2006] or at `http://www.mathematik.hu-berlin.de/~raesch/org/jensen.html`.

[Jensen, 2006] R. Jensen. CDROM with compilation of Jensen notes. Notes stamped "Seminar Für Logik und Grundlagenforshung Universität Bonn", 2006.

[Kanamori, 2007] A. Kanamori. Gödel and set theory. *The Bulletin of Symbolic Logic*, 13(2):153–188, 2007.

[Kanamori, 2011a] A. Kanamori. Large cardinals with Forcing. In Kanamori *et al.* [2011, chapter 4].

[Kanamori, 2011b] A. Kanamori. Set theory from Cantor to Cohen. In Kanamori *et al.* [2011, chapter 1].

[Kanamori *et al.*, 2011] A. Kanamori, D. Gabbay, and J. Woods, editors. *Sets and Extensions in the Twentieth Century*, volume 6, *Handboo for the History of Logic*. Elsevier, 2011.

[Kojman, 2011] M. Kojman. Singular cardinals: from Hausdorff's gaps to Shelah's pcf theory. In Kanamori *et al.* [2011, chapter 7].

[Kunen and Paris, 1971] K. Kunen and J. Paris. Boolean extensions and measurable cardinals. *Annals of Mathematical Logic*, 2(4):359–377, 1971.

[Kunen, 1968] K. Kunen. *Inaccessibility properties of cardinals*. PhD thesis, Stanford University, 1968.

[Kunen, 1970] K. Kunen. Some applications of iterated ultrapowers in set theory. *Annals of Mathematical Logic*, 1:179–227, 1970.

[Kunen, 1971] K. Kunen. On the GCH at measurable cardinals. In *Logic Colloquium '69 (Proc. Summer School and Colloq., Manchester, 1969)*, pages 107–110. North-Holland, Amsterdam, 1971.

[Levy, 1957] A. Levy. Indépendance conditionnelle de $V = L$ et d'axiomes qui se rattachent au système de M. Gödel. *Comptes Rendus Hebdomadaires des Séances de l'Academie des Sciences, Paris*, 245:1582–1583, 1957.

[Levy, 1960] A. Levy. A generalization of Gödel's notion of constructibility. *The Journal of Symbolic Logic*, 25:147–155, 1960.

[Magidor, 1978] M. Magidor. Changing cofinality of cardinals. *Fundamenta Mathematicae*, 99(1):61–71, 1978.

[Magidor, 1990] M. Magidor. Representing sets of ordinals as countable unions of sets in the core model. *Transactions of the American Mathematical Society*, 317(1):91–126, 1990.

[Martin and Solovay, 1969] D. A. Martin and R. Solovay. A basis theorem for $\mathbf{\Sigma}^1_3$ sets of reals. *Annals of Mathematics*, 89:138–159, 1969.

[Martin and Steel, 1988] D. A. Martin and J. Steel. Projective determinacy. *Proceedings of the National Academy of Sciences U.S.A.*, 85(18):6582–6586, 1988.

[Martin and Steel, 1989] D. A. Martin and J. Steel. A proof of projective determinacy. *Journal of the American Mathematical Society*, 2(1):71–125, 1989.

[Martin and Steel, 1994] D. A. Martin and J. Steel. Iteration trees. *Journal of the American Mathematical Society*, 7(1):1–73, 1994.

[Mathias, 1977] A. Mathias. Happy families. *Annals of Mathematical Logic*, 12(1):59–111, 1977.

[Mitchell and Schimmerling, 1995] W. Mitchell and E. Schimmerling. Weak covering without countable closure. *Mathematical Research Letters*, 2(5):595–609, 1995.

[Mitchell and Steel, 1994] W. Mitchell and J. Steel. *Fine structure and iteration trees*, volume 3 of *Lecture Notes in Logic*. Springer-Verlag, Berlin, 1994.

[Mitchell *et al.*, 1997] W. Mitchell, E. Schimmerling, and J. Steel. The covering lemma up to a Woodin cardinal. *Annals of Pure and Applied Logic*, 84(2):219–255, 1997.

[Mitchell, 1974] W. Mitchell. Sets constructible from sequences of ultrafilters. *The Journal of Symbolic Logic*, 39:57–66, 1974.

[Mitchell, 1979a] W. Mitchell. Hypermeasurable cardinals. In *Logic Colloquium '78 (Mons, 1978)*, volume 97 of *Studies in Logic and the Foundations of Mathematics*, pages 303–316. North-Holland, Amsterdam, 1979.

[Mitchell, 1979b] W. Mitchell. Ramsey cardinals and constructibility. *The Journal of Symbolic Logic*, 44(2):260–266, 1979.

[Mitchell, 1983] W. Mitchell. Sets constructed from sequences of measures: revisited. *The Journal of Symbolic Logic*, 48(3):600–609, 1983.

[Mitchell, 1984a] W. Mitchell. The core model for sequences of measures. I. *Mathematical Proceedings of the Cambridge Philosophical Society*, 95(2):229–260, 1984.

[Mitchell, 1984b] W. Mitchell. Indiscernibles, skies, and ideals. In *Axiomatic Set Theory (Boulder, Colo., 1983)*, volume 31 of *Contemporary Mathematics*, pages 161–182. American Mathematical Society, Providence, RI, 1984.

[Mitchell, 1985] W. Mitchell. The core model for sequences of measures, II. A scanned typescript is available at `http://www.math.ufl.edu/~wjm/papers/`, 1985.

[Mitchell, 1987] W. Mitchell. Applications of the covering lemma for sequences of measures. *Transactions of the American Mathematical Society*, 299(1):41–58, 1987.

[Mitchell, 1991] W. Mitchell. Definable singularity. *Transactions of the American Mathematical Society*, 327(1):407–426, 1991.

[Mitchell, 1992a] W. Mitchell. $\mathbf{\Sigma}^1_3$-absoluteness for sequences of measures. In *Set theory of the Continuum (Berkeley, CA, 1989)*, volume 26 of *Mathematical Sciences Research Institute Publications*, pages 311–355. Springer, New York, 1992.

[Mitchell, 1992b] W. Mitchell. On the singular cardinal hypothesis. *Transactions of the American Mathematical Society*, 329(2):507–530, 1992.

[Mitchell, 2010a] W. Mitchell. Beginning inner model theory. In Foreman and Kanamori [2010].

[Mitchell, 2010b] W. Mitchell. The covering lemma. In Foreman and Kanamori [2010].

[Namba, 1971] K. Namba. Independence proof of (ω, ω_α)-distributive law in complete Boolean algebras. *Commentarii Mathematici Universitatis Sancti Pauli*, 19:1–12, 1971.

[Neeman and Steel, 1999] I. Neeman and J. Steel. A weak Dodd-Jensen lemma. *The Journal of Symbolic Logic*, 64(3):1285–1294, 1999.

[Neeman, 2002] I. Neeman. Inner models in the region of a Woodin limit of Woodin cardinals. *Annals of Pure and Applied Logic*, 116(1-3):67–155, 2002.

[Neeman, 2004] I. Neeman. The Mitchell order below rank-to-rank. *The Journal of Symbolic Logic*, 69(4):1143–1162, 2004.

[Prikry, 1970] K. Prikry. Changing measurable into accessible cardinals. *Dissertationes Mathematicae (Rozprawy Matematyczne)*, 68:55, 1970.

[Rowbottom, 1971] F. Rowbottom. Some strong axioms of infinity incompatible with the axiom of constructibility. *Annals of Mathematical Logic*, 3(1):1–44, 1971.

[Schimmerling and Steel, 2006] E. Schimmerling and J. Steel. The strength of AD. Available at `http://wwwmath.uni-muenster.de/logik/Personen/rds/AD_skript.pdf`, July 2006.

[Schimmerling and Woodin, 2001] E. Schimmerling and W. H. Woodin. The Jensen covering property. *The Journal of Symbolic Logic*, 66(4):1505–1523, 2001.

[Schimmerling and Zeman, 2001] E. Schimmerling and M. Zeman. Square in core models. *The Bulletin of Symbolic Logic*, 7(3):305–314, 2001.

[Schimmerling and Zeman, 2004] E. Schimmerling and M. Zeman. Characterization of $\square_\kappa$ in core models. *Journal of Mathematical Logic*, 4(1):1–72, 2004.

[Schimmerling, 1995] E. Schimmerling. Combinatorial principles in the core model for one Woodin cardinal. *Annals of Pure and Applied Logic*, 74(2):153–201, 1995.

[Scott, 1961] D. Scott. Measurable cardinals and constructible sets. *Bulletin de l'Academie Polonaise des Sciences, Série des Sciences Mathématiques, Astronomiques et Physiques*, 9:521–524, 1961.

[Shelah and Woodin, 1990] S. Shelah and W. H. Woodin. Large cardinals imply that every reasonably definable set of reals is Lebesgue measurable. *Israel Journal of Mathematics*, 70(3):381–394, 1990.

[Shoenfield, 1961] J. Shoenfield. The problem of predicativity. In *Essays on the Foundations of Mathematics*, pages 132–139. Magnes Press, Hebrew Univ., Jerusalem, 1961.

[Silver, 1970] J. Silver. A large cardinal in the constructible universe. *Fundamenta Mathematicae*, 69:93–100, 1970.

[Silver, 1971] J. Silver. The consistency of the GCH with the existence of a measurable cardinal. In *Axiomatic Set Theory (Proc. Sympos. Pure Math., Vol. XIII, Part I, Univ. California, Los Angeles, Calif., 1967)*, pages 391–395. American Mathematical Society, Providence, R.I., 1971.

[Silver, 1975] J. Silver. On the singular cardinals problem. In *Proceedings of the International Congress of Mathematicians (Vancouver, B. C., 1974), Vol. 1*, pages 265–268. Canad. Math. Congress, Montreal, Que., 1975.

[Solovay *et al.*, 1978] R. Solovay, W. Reinhardt, and A. Kanamori. Strong axioms of infinity and elementary embeddings. *Annals of Mathematical Logic*, 13(1):73–116, 1978.

[Solovay, 1970] R. Solovay. A model of set theory in which every set of reals is Lebesgue measurable. *Annals of Mathematics*, 92:1–56, 1970.

[Solovay, 1971] R. Solovay. Real-valued measurable cardinals. In *Axiomatic Set Theory (Proc. Sympos. Pure Math., Vol. XIII, Part I, Univ. California, Los Angeles, Calif., 1967)*, pages 397–428. Amer. Math. Soc., Providence, R.I., 1971.

[Steel, 1993] J. Steel. Inner models with many Woodin cardinals. *Annals of Pure and Applied Logic*, 65(2):185–209, 1993.

[Steel, 1995] J. Steel. $\mathrm{HOD}^{L(\mathbf{R})}$ is a core model below Θ. *The Bulletin of Symbolic Logic*, 1(1):75–84, 1995.

[Steel, 1996] J. Steel. *The core model iterability problem*, volume 8 of *Lecture Notes in Logic*. Springer-Verlag, Berlin, 1996.

[Steel, 2002] J. Steel. Core models with more Woodin cardinals. *The Journal of Symbolic Logic*, 67(3):1197–1226, 2002.

[Tarski, 1962] A. Tarski. Some problems and results relevant to the foundations of set theory. In *Logic, Methodology and Philosophy of Science (Proc. 1960 Internat. Congr.)*, pages 125–135. Stanford Univ. Press, Stanford, Calif., 1962.

[Tennenbaum, 1968] S. Tennenbaum. Souslin's problem. *Proceedings of the National Academy of Sciences U.S.A.*, 59:60–63, 1968.

[Ulam, 1930] S. Ulam. Zur Masstheorie in der allgemeinen Megenlehre. *Fundamenta Mathematicae*, 16:140–150, 1930.

[Vopěnka and Hájek, 1972] P. Vopěnka and P. Hájek. *The Theory of Semisets*. Academia (Publishing House of the Czechoslovak Academy of Sciences), Prague, 1972.

[Welch, 1979] P. Welch. *Combinatorial Principles In the Core Model*. PhD thesis, University of Oxford, 1979.

[Woodin, 1988] W. H. Woodin. Supercompact cardinals, sets of reals, and weakly homogeneous trees. *Proceedings of the National Academy of Sciences U.S.A.*, 85(18):6587–6591, 1988.

[Wylie, 1989] D. Wylie. *Condensation and square in a higher core model*. PhD thesis, Massachusetts Institute of Technology, 1989.

[Zeman, 2002] M. Zeman. *Inner models and large cardinals*, volume 5 of *de Gruyter Series in Logic and its Applications*. Walter de Gruyter & Co., Berlin, 2002.

A BRIEF HISTORY OF DETERMINACY

Paul B. Larson

1 INTRODUCTION

Determinacy axioms are statements to the effect that certain games are *determined*, in that each player in the game has an optimal strategy.[1] The commonly accepted axioms for mathematics, the Zermelo-Fraenkel axioms with the Axiom of Choice (ZFC; see [Jech, 2003; Kunen, 1983]), imply the determinacy of many games that people actually play. This applies in particular to many *games of perfect information*, games in which the players alternate moves which are known to both players, and the outcome of the game depends only on this list of moves, and not on chance or other external factors. Games of perfect information which must end in finitely many moves are determined. This follows from the work of Ernst Zermelo [1913], Dénes Kőnig [1927] and László Kálmar [1928], and also from the independent work of John von Neumann and Oskar Morgenstern (in their 1944 book, reprinted as [2004]).

As pointed out by Stanisław Ulam [1960], determinacy for games of perfect information of a fixed finite length is essentially a theorem of logic. If we let $x_1,y_1,x_2,y_2,\ldots,x_n,y_n$ be variables standing for the moves made by players I (who plays $x_1,\ldots,x_n$) and II (who plays $y_1,\ldots,y_n$), and A (consisting of sequences of length $2n$) is the set of runs of the game for which I wins, the statement

$$\exists x_1 \forall y_1 \ldots \exists x_n \forall y_n \langle x_1, y_1, \ldots, x_n, y_n \rangle \in A$$

essentially asserts that the first player has a winning strategy in the game, and its negation,

$$\forall x_1 \exists y_1 \ldots \forall x_n \exists y_n \langle x_1, y_1, \ldots, x_n, y_n \rangle \notin A$$

essentially asserts that the second player has a winning strategy.[2]

We let ω denote the set of natural numbers $0, 1, 2, \ldots$; for brevity we will often refer to the members of this set as "integers". Given sets X and Y, ${}^X Y$ denotes

[1] That is, either one of the players has a strategy that guarantees a win, or both players have strategies which guarantee at least a tie. Some of the games considered in this chapter allow ties, but most do not.

[2] If there exists a way of choosing a member from each nonempty set of moves of the game, then these statements are actually equivalent to the assertions that the corresponding strategies exist. Otherwise, in the absence of the Axiom of Choice the statements above can hold without the corresponding strategy existing.

Handbook of the History of Logic: Sets and Extensions in the Twentieth Century.
Volume editor: Akihiro Kanamori. General editors: Dov M. Gabbay, Paul Thagard and John Woods.

the set of functions from X to Y. The *Baire space* is the space ${}^{\omega}\omega$, with the product topology. The Baire space is homeomorphic to the space of irrational real numbers (see [Moschovakis, 2009, p. 9], for instance), and we will often refer to its members as "reals" (though in various contexts the Cantor space ${}^{\omega}2$, the set of subsets of ω ($\mathcal{P}(\omega)$) and the set of infinite subsets of ω ($[\omega]^{\omega}$) are all referred to as "the reals").

Given $A \subseteq {}^{\omega}\omega$, we let $G_{\omega}(A)$ denote the game of perfect information of length ω in which the two players collaborate to define an element f of ${}^{\omega}\omega$ (with I choosing $f(0)$, II choosing $f(1)$, I choosing $f(2)$, and so on), with I winning a run of the game if and only if f is an element of A. A game of this type is called an *integer game*, and the set A is called the *payoff set*. A *strategy* in such a game for player I (II) is a function Σ with domain the set of sequences of integers of even (odd) length such that for each $a \in \mathrm{dom}(\Sigma)$, $\Sigma(a)$ is in ω. A run of the game (partial or complete) is said to be *according to* a strategy Σ for player I (II) if every initial segment of the run of odd (nonzero even) length is of the form $a^\frown\langle\Sigma(a)\rangle$ for some sequence a. A strategy Σ for player I (II) is a *winning strategy* if every complete run of the game according to Σ is in (out of) A. We say that a set $A \subseteq {}^{\omega}\omega$ is *determined* (or the corresponding game $G_{\omega}(A)$ is determined) if there exists a winning strategy for one of the players. These notions generalize naturally for games in which players play objects other than integers (for instance, *real games*, in which they play elements of ${}^{\omega}\omega$) or games which run for more than ω many rounds (in which case player I typically plays at limit stages).

The study of determinacy axioms concerns games whose determinacy is neither proved nor refuted by the Zermelo-Fraenkel axioms ZF (without the Axiom of Choice). Typically such games are infinite. Axioms stating that infinite games of various types are determined were studied by Stanisław Mazur, Stefan Banach and Ulam in the late 1920's and early 1930's; were reintroduced by David Gale and Frank Stewart [1953] in the 1950's and again by Jan Mycielski and Hugo Steinhaus [1962] in the early 1960's; gained interest with the work of David Blackwell [1967] and Robert Solovay in the late 1960's; and attained increasing importance in the 1970's and 1980's, finally coming to a central position in contemporary set theory.

Mycielski and Steinhaus introduced the Axiom of Determinacy (AD), which asserts the determinacy of $G_{\omega}(A)$ for all $A \subseteq {}^{\omega}\omega$. Work of Banach in the 1930's shows that AD implies that all sets of reals satisfy the property of Baire. In the 1960's, Mycielski and Świerczkowski proved that AD implies that all sets of reals are Lebesgue measurable, and Mycielski showed that AD implies countable choice for reals. Together, these results show that determinacy provides a natural context for certain areas of mathematics, notably analysis, free of the paradoxes induced by the Axiom of Choice.

Unaware of the work of Banach, Gale and Stewart [1953] had shown that the AD contradicts ZFC. However, the proof used a wellordering of the reals given by the Axiom of Choice, and therefore did not give a nondetermined game of this type with definable payoff set. Starting with Banach's work, many simply definable payoff sets were shown to induce determined games, culminating in D.

Anthony Martin's celebrated 1974 result [1975] that all games with Borel payoff set are determined. This result came after Martin had used measurable cardinals to prove the determinacy of games whose payoff set is an analytic sets of reals.

The study of determinacy gained interest from two theorems in 1967, the first due to Solovay and the second to Blackwell. Solovay proved that under AD, the first uncountable cardinal ω_1 is a measurable cardinal, setting off a study of strong Ramsey properties on the ordinals implied by determinacy axioms. Blackwell used open determinacy (proved by Gale and Stewart) to reprove a classical theorem of Kazimierz Kuratowski. This also led to the application, by John Addison, Martin, Yiannis Moschovakis and others, of stronger determinacy axioms to produce structural properties for definable sets of reals. These axioms included the determinacy of $\undertilde{\Delta}^1_n$ sets of reals, for $n \geq 2$, statements which would not be proved consistent relative to large cardinals until the 1980's.

The large cardinal hierarchy was developed over the same period, and came to be seen as a method for calibrating consistency strength. In the 1970's, various special cases of $\undertilde{\Delta}^1_2$ determinacy were located on this scale, in terms of the large cardinals needed to prove them. Determining the consistency (relative to large cardinals) of forms of determinacy at the level of $\undertilde{\Delta}^1_2$ and beyond would take the introduction of new large cardinal concepts. Martin (in 1978) and W. Hugh Woodin (in 1984) would prove $\undertilde{\Pi}^1_2$-determinacy and $\mathrm{AD}^{L(\mathbb{R})}$ respectively, using hypotheses near the very top of the large cardinal hierarchy. In a dramatic development, the hypotheses for these results would be significantly reduced through work of Woodin, Martin and John Steel. The initial impetus for this development was a seminal result of Matthew Foreman, Menachem Magidor and Saharon Shelah which showed, assuming the existence of a supercompact cardinal, that there exists a generic elementary embedding with well-founded range and critical point ω_1. Combined with work of Woodin, this yielded the Lebesgue measurability of all sets in the inner model $L(\mathbb{R})$ from this hypothesis. Shelah and Woodin would reduce the hypothesis for this result further, to the assumption that there exist infinitely many Woodin cardinals below a measurable cardinal.

Woodin cardinals would turn out to be the central large cardinal concept for the study of determinacy. Through the study of tree representations for sets of reals, Martin and Steel would show that $\undertilde{\Pi}^1_{n+1}$-determinacy follows from the existence of n Woodin cardinals below a measurable cardinal, and that this hypothesis was not sufficient to prove stronger determinacy results for the projective hierarchy. Woodin would then show that the existence of infinitely many Woodin cardinals below a measurable cardinal implies $\mathrm{AD}^{L(\mathbb{R})}$, and he would locate the exact consistency strengths of $\undertilde{\Delta}^1_2$-determinacy and $\mathrm{AD}^{L(\mathbb{R})}$ at one Woodin cardinal and ω Woodin cardinals respectively.

In the aftermath of these results, many new directions were developed, and we give only the briefest indication here. Using techniques from inner model theory, tight bounds were given for establishing the exact consistency strength of many determinacy hypotheses. Using similar techniques, it has been shown that almost every natural statement (i.e., not invented specifically to be a counterexample) implies directly those determinacy hypotheses of lesser consistency strength. For instance, by Gödel's Second Incompleteness Theorem, ZFC cannot prove that the AD holds in $L(\mathbb{R})$, as the latter implies the consistency of the former. Empirically, however, every natural extension of ZFC without this limitation (i.e., not proved consistent by AD) does appear to imply that AD holds in $L(\mathbb{R})$. This sort of phenomenon is taken by some as evidence that the statement that AD holds in $L(\mathbb{R})$, and other determinacy axioms, should be counted among the true statements extending ZFC.

The history presented here relies heavily on those given by Jackson [2010], Kanamori [1995; 2003], Moschovakis [2009], Neeman [2004] and Steel [2008a]. As the title suggests, this is a selective and abbreviated account of the history of determinacy. We have omitted many interesting topics, including, for instance, Blackwell games [Blackwell, 1969; Martin, 1998; Martin *et al.*, 2003] and proving determinacy in second-order arithmetic [Louveau and Saint-Raymond, 1987; Louveau and Saint-Raymond, 1988; Koellner and Woodin, 2010].

2 EARLY DEVELOPMENTS

The first published paper in mathematical game theory appears to be Zermelo's paper [1913] on chess. Although he noted that his arguments apply to all games of reason not involving chance, Zermelo worked under two additional chess-specific assumptions. The first was that the game in question has only finitely many states, and the second was that an infinite run of the game was to be considered a draw. Zermelo specified a condition which is equivalent to having a winning strategy in such a game guaranteeing a win within a fixed number of moves, as well as another condition equivalent to having a strategy guaranteeing that one will not lose within a given fixed number of moves. His analysis implicitly introduced the notions of *game tree*, *subtree* of a game tree, and *quasi-strategy*.[3]

The paper states indirectly, but does not quite prove, or even define, the statement that in any game of perfect information with finitely many possible positions such that infinite runs of the game are draws, either one player has a strategy that guarantees a win, or both players have strategies that guarantee at least a draw. A special case of this fact is determinacy for games of perfect information of a fixed finite length, which is sometimes called Zermelo's Theorem.

Kőnig [1927] applied the fundamental fact now known as *Kőnig's Lemma* to the study of games, among other topics. While Kőnig's formulation was somewhat different, his Lemma is equivalent to the assertion that every infinite finitely branching tree with a single root has an infinite path (a path can be found by iteratively choosing any successor node such that the tree above that node is infinite). Extending Zermelo's analysis to games in which infinitely many positions are possible while retaining the condition that each player has only finitely many options

[3]As defined above, a strategy for a given player specifies a move in each relevant position; a quasi-strategy merely specifies a set of acceptable moves. The distinction is important when the Axiom of Choice fails, but is less important in the context of Zermelo's paper.

at each point, Kőnig used the statement above to prove that in such a game, if one player has a strategy (from a given point in the game) guaranteeing a win, then he can guarantee victory within a fixed number of moves. The application of Kőnig's Lemma to the study of games was suggested by von Neumann.

Kálmar [1928] took the analysis a step further by proving Zermelo's Theorem for games with infinitely many possible moves in each round. His arguments proceeded by assigning transfinite ordinals to nodes in the game tree, a method which remains an important tool in modern set theory. Kálmar explicitly introduced the notion of a winning strategy for a game, though his strategies were also quasi-strategies as above. In his analysis, Kálmar introduced a number of other important technical notions, including the notion of a *subgame* (essentially a subtree of the original game tree), and classifying strategies into those which depend only on the current position in the game and those which use the history of the game so far.[4]

Games of perfect information for which the set of infinite runs is divided into winning sets for each player appear in a question by Mazur in the Scottish Book, answered by Banach in an entry dated August 4, 1935 (see [Mauldin, 1981, p. 113]). Following up on Mazur's question (still in the Book), Ulam asked about games where two players collaborate to build an infinite sequence of 0's and 1's by alternately deciding each member of the sequence, with the winner determined by whether the infinite sequence constructed falls inside some predetermined set E. Essentially raising the issue of determinacy for arbitrary $G_\omega(E)$, Ulam asked: for which sets E does the first player (alternately, the second player) have a winning strategy? (Section 2.1 below has more on the Banach-Mazur game.)

Games of perfect information were formally defined in 1944 by von Neumann and Morgenstern [2004]. Their book also contains a proof that games of perfect information of a fixed finite length are determined (page 123).

Infinite games of perfect information were reintroduced by Gale and Stewart [1953], who were unaware of the work of Mazur, Banach and Ulam (Gale, personal communication). They showed that a nondetermined game can be constructed using the Axiom of Choice (more specifically, from a wellordering of the set of real numbers).[5] They also noted that the proof from the Axiom of Choice does not give a definable undetermined game, and raised the issue of whether determinacy might hold for all games with a suitably definable payoff set. Towards this end, they introduced a topological classification of infinite games of perfect information, defining a game (or the set of runs of the game which are winning for the first player) to be *open* if all winning runs for the first player are won at some finite

[4]See [Schwalbe and Walker, 2001] for much more on these papers of Zermelo, Kőnig and Kálmar.

[5]Given a set Y, we let AC_Y denote the statement that whenever $\{X_a \mid a \in Y\}$ is a collection of nonempty sets, there is a function f with domain Y such that $f(a) \in X_a$ for all $a \in Y$. Zermelo's *Axiom of Choice* [1904] is equivalent to the statement that AC_Y holds for all sets Y. A linear ordering $\leq$ of a set X is a *wellordering* if every nonempty subset of X has a $\leq$-least element. The Axiom of Choice is equivalent to the statement that there exist wellorderings of every set. Kőnig's Lemma is a weak form of the Axiom of Choice and cannot be proved in ZF (see [Lévy, 1979, Exercise IX.2.18]).

stage (i.e., if, whenever $\langle x_0, x_1, x_2, \ldots \rangle$ is a winning run of the game for the first player, there is some n such that the first player wins all runs of the game extending $\langle x_0, \ldots, x_n \rangle$). Using this framework, they proved a number of fundamental facts, including the determinacy of all games whose payoff set is a Boolean combination of open sets (i.e., in the class generated from the open sets by the operations of finite union, finite intersection and complementation). The determinacy of open games would become the basis for proofs of many of the strongest determinacy hypotheses. Gale and Stewart also asked a number of important questions, including the question of whether all Borel games are determined (to be answered positively by Martin [1975] in 1974).[6] Classifying games by the definability of their payoffs sets would be an essential tool in the study of determinacy.

2.1 *Regularity properties*

Early motivation for the study of determinacy was given by its implications for regularity properties for sets of reals. In particular, determinacy of certain games of perfect information was shown to imply that every set of reals has the property of Baire and the perfect set property, and is Lebesgue measurable.[7] These three facts themselves each contradict the Axiom of Choice. We will refer to Lebesgue measurability, the property of Baire and the perfect set property as the *regularity properties*, the fact that there are other regularity properties notwithstanding.

Question 43 of the Scottish Book, posed by Mazur, asks about games where two players alternately select the members of a shrinking sequence of intervals of real numbers, with the first player the winner if the intersection of the sequence intersects a set given in advance. Banach posted an answer in 1935, showing that such games are determined if and only if the given set is either meager (in which case the second player wins) or comeager relative to some interval (in which case the first player wins), i.e., if and only if the given set has the Baire property (see [Oxtoby, 1980, pp. 27-30], [Kanamori, 2003, pp. 373-374]). This game has come to be known as the Banach-Mazur game. Using an enumeration of the rationals, one can code intervals with rational endpoints with integers, getting a game on integers.

Morton Davis [1964] studied a game, suggested by Dubins, where the first player plays arbitrarily long finite strings of 0's and 1's and the second player plays individual 0's and 1's, with the payoff set a subset of the set of infinite binary sequences as before. Davis proved that the first player has a winning strategy in

[6]The *Borel* sets are the members of the smallest class containing the open sets and closed under the operations of complementation and countable union. The collection of Borel sets is generated in ω_1 many stages from these two operations. A natural process assigns a measure to each Borel set (see, for instance, [Halmos, 1950]).

[7]A set of reals X has the *property of Baire* if $X \triangle O$ is meager for some open set O, where the *symmetric difference* $A \triangle B$ of two sets A and B is the set $(A \setminus B) \cup (B \setminus A)$, and $A \setminus B = \{x \in A \mid x \notin B\}$. A set of reals X has the *perfect set property* if it is countable or contains a perfect set (an uncountable closed set without isolated points). A set of reals X is *Lebesgue measurable* if there is a Borel set B such that $X \triangle B$ is a subset of a Borel measure 0 set. See [Oxtoby, 1980].

such a game if and only if the payoff set contains a perfect set, and the second player has a winning strategy if and only if the payoff set is finite or countably infinite. The determinacy of all such games then implies that every uncountable set of reals contains a perfect set (asymmetric games of this type can be coded by integer games of perfect information). It follows that under AD there is no set of reals whose cardinality falls strictly between $\aleph_0$ and $2^{\aleph_0}$.[8]

Mycielski and Stanisław Świerczkowski [1964] showed that the determinacy of certain integer games of perfect information implies that every subset of the real line is Lebesgue measurable. Simpler proofs of this fact were later given by Leo Harrington (see [Kanamori, 2003, pp. 375-377]) and Martin [2003].

By way of contrast, an argument of Vitali [1905] shows that under ZFC there are sets of reals which are not Lebesgue measurable. Banach and Tarski ([1924], see also [Wagon, 1993]), building on work of Hausdorff [1914], showed that under ZFC the unit ball can be partitioned into five pieces which can be rearranged to make two copies of the same sphere, again violating Lebesgue measurability as well as physical intuition. As with the undetermined game given by Gale and Stewart, the constructions of Vitali and Banach-Tarski use the Axiom of Choice and do not give definable examples of nonmeasurable sets. Via the Mycielski-Świerczkowski theorem, determinacy results would rule out the existence of definable examples, for various notions of definability.

2.2 Definability

As discussed above, ZFC implies that open sets are determined, and implies also that there exists a nondetermined set. The study of determinacy was to merge naturally with the study of sets of reals in terms of their definability (i.e., descriptive set theory), which can be taken as a measure of their complexity. In this section we briefly introduce some important definability classes for sets of reals. Standard references include [Moschovakis, 1980; Kechris, 1995]. While we do mention some important results in this section, much of the section can be skipped on a first reading and used for later reference.

A *Polish space* is a topological space which is separable and completely metrizable. Common examples include the integers ω, the reals $\mathbb{R}$, the open interval $(0,1)$, the Baire space ${}^{\omega}\omega$, the Cantor space ${}^{\omega}2$ and their finite and countable products. Uncountable Polish spaces without isolated points are a natural setting for studying definable sets of reals. For the most part we will concentrate on the Baire space and its finite powers.

Following notation introduced by Addison [1958],[9] open subsets of a Polish space are called $\utilde{\Sigma}^0_1$, complements of $\utilde{\Sigma}^0_n$sets are $\utilde{\Pi}^0_n$, and countable unions of $\utilde{\Pi}^0_n$

[8]i.e., for every set X, if there exist injections $f\colon \omega \to X$ and $g\colon X \to 2^{\omega}$, then either X is countable or there exists a bijection between X and 2^{ω}.

[9]The papers [Addison, 1958] and [Addison, 1959] appear in the same volume of *Fundamenta Mathematicae*. The front page of the volume gives the date 1958-1959. The individual papers have the dates 1958 and 1959 on them, respectively.

sets are $\underset{\sim}{\Sigma}^0_{n+1}$. More generally, given a positive $\alpha < \omega_1$, $\underset{\sim}{\Sigma}^0_\alpha$ consists of all countable unions of members of $\bigcup_{\beta<\alpha} \underset{\sim}{\Pi}^0_\beta$, and $\underset{\sim}{\Pi}^0_\alpha$ consists of all complements of members of $\underset{\sim}{\Sigma}^0_\alpha$. The Borel sets are the members of $\bigcup_{\alpha<\omega_1} \underset{\sim}{\Sigma}^0_\alpha$.

A *pointclass* is a collection of subsets of Polish spaces. Given a pointclass $\Gamma \subseteq \mathcal{P}({}^\omega\omega)$, we let $\mathrm{Det}(\Gamma)$ and Γ-*determinacy* each denote the statement that $G_\omega(A)$ is determined for all $A \in \Gamma$. Philip Wolfe [1955] proved $\underset{\sim}{\Sigma}^0_2$-determinacy in ZFC. Davis [1964] followed by proving $\underset{\sim}{\Pi}^0_3$-determinacy. Jeffrey Paris [1972] would prove $\underset{\sim}{\Sigma}^0_4$-determinacy. However, this result was proved after Martin had used a measurable cardinal to prove analytic determinacy (see Section 5.2).

Continuous images of $\underset{\sim}{\Pi}^0_1$ sets are said to be $\underset{\sim}{\Sigma}^1_1$, complements of $\underset{\sim}{\Sigma}^1_n$ sets are $\underset{\sim}{\Pi}^1_n$,and continuous images of $\underset{\sim}{\Pi}^1_n$ sets are $\underset{\sim}{\Sigma}^1_{n+1}$. For each $i \in \{0, 1\}$ and $n \in \omega$, the pointclass $\underset{\sim}{\Delta}^i_n$ is the intersection of $\underset{\sim}{\Sigma}^i_n$ and $\underset{\sim}{\Pi}^i_n$. The *boldface projective pointclasses* $\underset{\sim}{\Sigma}^1_n$, $\underset{\sim}{\Pi}^1_n$, $\underset{\sim}{\Delta}^1_n$ were introduced independently by Nikolai Luzin [1925c; 1925b; 1925a] and Wacław Sierpiński [1925]. The notion of a boldface pointclass in general (i.e., possibly non-projective) is used in various ways in the literature. We will say that a pointclass Γ is *boldface* (or *closed under continuous preimages* or *continuously closed*) if $f^{-1}[A] \in \Gamma$ for all $A \in \Gamma$ and all continuous functions f between Polish spaces (where A is a subset of the codomain). The classes $\underset{\sim}{\Sigma}^0_n$, $\underset{\sim}{\Pi}^0_n$, $\underset{\sim}{\Delta}^0_n$ are also boldface in this sense.

The pointclass $\underset{\sim}{\Sigma}^1_1$ is also known as the class of *analytic sets*, and was given an independent characterization by Mikhail Suslin [1917]: a set of reals A is analytic if and only if there exists a family of closed sets D_s (for each finite sequence s consisting of integers) such that A is the set of reals x for which there is an ω-sequence S of integers such that $x \in \bigcap_{n\in\omega} D_{S\restriction n}$.[10] Suslin showed that there exist non-Borel analytic sets, and that the Borel sets are exactly the $\underset{\sim}{\Delta}^1_1$ sets.

We let $\exists^0$ and $\exists^1$ denote existential quantification over the integers and reals, respectively, and $\forall^0$ and $\forall^1$ the analogous forms of universal quantification. Given a set $A \subseteq ({}^\omega\omega)^{k+1}$, for some positive integer k, $\exists^1 A$ is the set of $(x_1, \ldots, x_k) \in ({}^\omega\omega)^k$ such that for some $x \in {}^\omega\omega$, $(x, x_1, \ldots, x_k) \in A$, and $\forall^1 A$ is the set of $(x_1, \ldots, x_k) \in ({}^\omega\omega)^k$ such that for all $x \in {}^\omega\omega$, $(x, x_1, \ldots, x_k) \in A$. Given a pointclass Γ, $\exists^1\Gamma$ consists of $\exists^1 A$ for all $A \in \Gamma$, and $\forall^1\Gamma$ consists of $\forall^1 A$ for all $A \in \Gamma$. It follows easily that for each positive integer n, $\exists^1\underset{\sim}{\Pi}^1_n = \underset{\sim}{\Sigma}^1_{n+1}$ and $\forall^1\underset{\sim}{\Sigma}^1_n = \underset{\sim}{\Pi}^1_{n+1}$.

Given a pointclass Γ, $\neg\Gamma$ is the set of complements of members of Γ, and Δ_Γ is the pointclass $\Gamma \cap \neg\Gamma$; Γ is said to be *selfdual* if $\Delta_\Gamma = \Gamma$. A set $A \in \Gamma$ is said to be Γ-complete if every member of Γ is a continuous preimage of A. If Γ is closed under continuous preimages and Γ-determinacy holds, then $\neg\Gamma$-determinacy holds. Each of the regularity properties for a set of reals A are given by the determinacy of games with payoff set simply definable from A (indeed, continuous preimages of A), but not necessarily with payoff A itself. It follows that when Γ is a boldface pointclass, Γ-determinacy implies the regularity properties for sets of reals in Γ.

[10] For S a function with domain ω, and $n \in \omega$, $S\restriction n = \langle S(0), \ldots, S(n-1)\rangle$.

A simple application of Fubini's theorem shows that if Γ is a boldface pointclass and there exists in Γ a wellordering of a set of reals of positive Lebesgue measure, then there is a non-Lebesgue measurable set in Γ. Skipping ahead for a moment, in the early 1970's Alexander Kechris and Martin, using a technique of Solovay called *unfolding*, proved that for each integer n, $\utilde{\Pi}^1_n$-determinacy plus *countable choice for sets of reals*[11] implies that all $\utilde{\Sigma}^1_{n+1}$ sets of reals are Lebesgue measurable, have the Baire property and have the perfect set property (see [Kanamori, 2003, pp. 380-381]).

As developed by Stephen Kleene, the *effective* (or *lightface*) pointclasses Σ^0_n, Π^0_n, Δ^0_n [1943] and Σ^1_n, Π^1_n, Δ^1_n [1955c; 1955b; 1955a] are formed in the same way as their boldface counterparts, starting instead from Σ^0_1, the collection of open sets O such that the set of indices for basic open sets contained in O (under a certain natural enumeration of the basic open sets) is recursive (see [Moschovakis, 2009], for instance). Sets in Σ^0_1 are called *semirecursive*, and sets in Δ^0_1 are called *recursive*. Given $a \in {}^\omega\omega$, $\Sigma^0_1(a)$ is the collection of open sets O such that the set of indices for basic open sets contained in O is recursive in a, and the *relativized lightface projective pointclasses* $\Sigma^0_n(a)$, $\Pi^0_n(a)$, $\Delta^0_n(a)$, $\Sigma^1_n(a)$, $\Pi^1_n(a)$, $\Delta^1_n(a)$ are built from Σ^0_n in the manner above. It follows that each boldface pointclass is the union of the corresponding relativized lightface classes (relativizing over each member of ${}^\omega\omega$).

Following [Moschovakis, 2009], a pointclass is *adequate* if it contains all recursive sets and is closed under finite unions and intersections, bounded universal and existential integer quantification (see [Moschovakis, 2009, p. 119]) and preimages by recursive functions.[12] The arithmetic and projective pointclasses are adequate (see [Moschovakis, 2009, pp. 118-120]).

Given a Polish space X, an integer k, a set $A \subseteq X^{k+1}$ and $x \in X$, A_x is the set of $(x_1, \ldots, x_k)$ such that $(x, x_1, \ldots, x_k) \in A$. A set $A \subseteq X^{k+1}$ in a pointclass Γ is said to be *universal* for Γ if each subset of X^k in Γ has the form A_x for some $x \in X$. Pointclasses of the form Σ^1_n, Π^1_n have universal members. Those of the form Δ^1_n do not. Each member of each boldface pointclass is of the form A_x for A a member of the corresponding effective class. Conversely, as each member of each lightface projective pointclass listed above is definable, each member of each corresponding boldface pointclass is definable from a real number as a parameter.

A set of reals is said to be Σ^2_1 (Π^2_1) if is definable by a formula of the form $\exists X \subseteq \mathbb{R}\,\phi$ ($\forall X \subseteq \mathbb{R}\,\phi$), where all quantifiers in ϕ range over the reals or the integers.

In the *Lévy hierarchy* [Lévy, 1965b], a formula ϕ in the language of set theory is Δ_0 (equivalently Σ_0, Π_0) if all quantifiers appearing in ϕ are bounded (see [Jech, 2003, Chapter 13]); Σ_{n+1} if it has the form $\exists x\psi$ for some Π_n formula ψ; and Π_{n+1}

[11]The statement that whenever X_n ($n \in \omega$) are nonempty sets of reals, there is a function $f\colon \omega \to \mathbb{R}$ such that $f(n) \in X_n$ for each n. Countable choice for sets of reals is a consequence of AD, as shown by Mycielski [1963] (see Section 2.3).

[12]A function f from a Polish space X to a Polish space Y is said to be *recursive* if the set of pairs $x \in X$, $n \in \omega$ such that $f(x)$ is in the n-th basic open neighborhood of Y is semi-recursive.

if it has the form $\forall x\psi$ for some Σ_n formula ψ. A set is *Σ_n-definable* if it can be defined by a Σ_n formula (and similarly for Π_n). We say that a model M is *Γ-correct*, for a class of formulas Γ, if for all ϕ in Γ and $x \in M$, $M \models \phi(x)$ if and only if $V \models \phi(x)$. If M is a model of ZF, we say that a set in M is Σ_n^M if it is definable by a Σ_n formula relativized to M (and similarly for other classes of formulas).

Gödel's inner model L is the smallest transitive model of ZFC containing the ordinals. For any set A, Gödel's constructible universe L generalizes to two inner models $L(A)$ and $L[A]$, developed respectively by András Hajnal [1956; 1961] and Azriel Lévy [1957; 1960] (see [Jech, 2003, Chapter 13] or [Kanamori, 2003, p. 34]). Given a set A, $L(A)$ is the smallest transitive model of ZF containing A and the ordinals and $L[A]$ is the smallest transitive model of ZF containing the ordinals and closed under the function $X \mapsto A \cap X$. Alternately, $L(A)$ is constructed in the same manner as L, but introducing the members of the transitive closure of the set $\{A\}$ at the first level, and $L[A]$ is constructed as L, but by adding a predicate for membership in A to the language.[13] When A is contained in L, $L(A)$ and $L[A]$ are the same. While $L[A]$ is always a model of AC, $L(A)$ need not be. Indeed, $L(\mathbb{R})$ is a model of AD in the presence of suitably large cardinals, and is thus a natural example of a "smaller universum" as described in the quote from [Mycielski and Steinhaus, 1962] in Section 2.3.

Though it can be formulated in other ways, we will view the set $0^\#$ ("zero sharp") as the theory of a certain class of ordinals which are indiscernibles over the inner model L. This notion was independently isolated by Solovay [1967] and by Jack Silver in his 1966 Berkeley Ph.D. thesis (see [1971]). The existence of $0^\#$ cannot be proved in ZFC, as it serves as a sort of transcendence principle over L. For instance, if $0^\#$ exists then every uncountable cardinal of V is a strongly inaccessible cardinal in L.[14] For any set X there is an analogous notion of $X^\#$ ("X sharp") serving as a transcendence principle over $L(X)$ (see [Kanamori, 2003]).

2.3 The Axiom of Determinacy

The Axiom of Determinacy, the statement that all length ω integer games of perfect information are determined, was proposed by Mycielski and Steinhaus [1962].[15] In a passage that anticipated a commonly accepted view of determinacy, they wrote

> It is not the purpose of this paper to depreciate the classical mathematics with its fundamental "absolute" intuitions on the universum of sets (to which belongs the axiom of choice), but only to propose

[13] A set x is *transitive* if $z \in x$ whenever $y \in x$ and $z \in y$. The *transitive closure* of a set x is the smallest transitive set containing x.

[14] A cardinal κ is *strongly inaccessible* if it is uncountable, regular and a strong limit (i.e., $2^\gamma < \kappa$ for all $\gamma < \kappa$). If κ is a strongly inaccessible cardinal, then V_κ is a model of ZFC. Hence, by Gödel's Second Incompleteness Theorem, the existence of strongly inaccessible cardinals cannot be proved in ZFC. See [Jech, 2003] for the definition of V_α, for an ordinal α.

[15] We continue to use the now-standard abbreviation AD for the Axiom of Determinacy; it was called (A) in [Mycielski and Steinhaus, 1962].

> another theory which seems very interesting although its consistency is problematic. Our axiom can be considered as a restriction of the classical notion of a set leading to a smaller universum, say of determined sets, which reflect some physical intuitions which are not fulfilled by the classical sets ... Our axiom could be considered as an axiom added to the classical set theory claiming the existence of a class of sets satisfying (A) and the classical axioms (without the axiom of choice).

Mycielski and Steinhaus summarized the state of knowledge of determinacy at that time, including the fact that AD implies that all sets of reals are Lebesgue measurable and have the Baire property, and they noted that by results of Kurt Gödel and Addison [1958], there is in Gödel's constructible universe L (and thus consistently with ZFC) a Δ^1_2 wellordering of the reals, and thus a Δ^1_2 set which is not determined.

In his [1963], Mycielski proved several fundamental facts about determinacy, including the fact that AD implies countable choice for set of reals (he credits this result to Świerczkowski, Dana Scott and himself, independently). Thus, while AD contradicts the Axiom of Choice, it implies a form of Choice which suffices for many of its most important applications, including the countable additivity of Lebesgue measure. Via countable choice for sets of reals, AD implies that ω_1 is regular.[16] Mycielski also showed that AD implies that there is no uncountable wellordered sequence of reals. In conjunction with the perfect set property, this implies that under determinacy, ω_1^V is a strongly inaccessible cardinal in the inner model L (and even in $L[a]$ for any real number a), a fact which was to be greatly extended by Solovay, Martin and Woodin. Harrington [1978] would show that Π^1_1-determinacy implies that $0^\#$ exists, and thus that Π^1_1-determinacy is not provable in ZFC.

In the same paper, Mycielski showed that ZF implies the existence of an undetermined game of perfect information of length ω_1 where the players play countable ordinals instead of integers. An interesting aspect of the proof is that it does not give a specific undetermined game. As a slight variant on Mycielski's argument, consider the game in which the first player plays a countable ordinal α (and then makes no other moves for the rest of the game) and the second player plays a sequence of integers coding α, under some fixed coding of hereditarily countable sets by reals.[17] Since the first player cannot have a winning strategy in this game, determinacy for the game implies the existence of an injection from ω_1 into $\mathbb{R}$, which contradicts AD but is certainly by itself consistent with ZF, as it follows from ZFC. Later results of Woodin would show that, assuming the consistency

[16] The ordinal ω_1 is the first uncountable ordinal. A cardinal κ is *regular* if, for every $\gamma < \kappa$, every function $f\colon \gamma \to \kappa$ has range bounded in κ. Under ZFC, every successor cardinal is regular. Feferman and Lévy [1963] (see also [Howard and Rubin, 1998, pp. 153-154]) showed that the singularity of ω_1 is consistent with ZF. Moti Gitik [1980] showed that it is consistent with ZF (relative to large cardinals) that ω is the largest regular cardinal.

[17] The *hereditarily countable* sets are those sets whose transitive closures are countable. Such sets are naturally coded by sets of integers.

of certain large cardinal hypotheses, ZFC is consistent with the statement that every integer game of length ω_1 with payoff set definable from real and ordinal parameters is determined (see Section 6.2, and [Neeman, 2004, p. 298]). Mycielski noted that under AD there are no nonprincipal ultrafilters[18] on ω (this follows from Lebesgue measurability for all sets of reals plus a result of Sierpiński [1938] showing that nonprincipal ultrafilters on ω give rise to nonmeasurable sets of reals), which implies that every ultrafilter is countably complete (i.e., closed under countable intersections). Finally, in a footnote on the first page of the paper, Mycielski reiterated a point made in the passage quoted above from his paper with Steinhaus, suggesting that an inner model containing the reals could satisfy AD. In a followup paper, Mycielski [1966] presented a number of additional results, including the fact that there is a game in which the players play real numbers whose determinacy implies *uniformization* (see Section 3.2) for subsets of the plane, another weak form of the Axiom of Choice.

In 1964, a year after Paul Cohen's invention of forcing, Solovay [1970] proved that if there exists a strongly inaccessible cardinal, then in a forcing extension there exists an inner model containing the reals in which every set of reals satisfies the regularity properties from Section 2.1. Shelah [1984] later showed that a strongly inaccessible cardinal is necessary, in the sense that the Lebesgue measurability of all sets of reals implies that ω_1 is strongly inaccessible in all models of the form $L[a]$, for $a \subseteq \omega$. In the introduction to his paper, Solovay conjectured (correctly, as it turned out) that large cardinals would imply that AD holds in $L(\mathbb{R})$.

1967 saw two major results in the study of determinacy, one by Blackwell [1967] and the other by Solovay. Reversing chronological order by a few months, we discuss Blackwell's result and its consequences in the next section, and Solovay's in Section 4. These two sections correspond roughly to the work of the so-called Cabal Seminar, whose proceedings were published in four volumes covering the years 1976 to 1985 [Kechris and Moschovakis, 1978a; Kechris *et al.*, 1981b; Kechris *et al.*, 1983; Kechris *et al.*, 1988] (at this time, one volume of a planned reissue has appeared [Kechris *et al.*, 2008]).

3 REDUCTION AND SCALES

Blackwell [1967] used open determinacy to reprove a theorem of Kuratowski [1936] stating that the intersection of any two analytic sets A, B in a Polish space Y is also the intersection of two analytic sets A' and B' such that $A \subseteq A'$, $B \subseteq B'$ and $A' \cup B' = Y$.[19] Briefly, the argument is as follows. Since A and B are analytic, there exist continuous surjections $f\colon {}^{\omega}\omega \to A$ and $g\colon {}^{\omega}\omega \to B$. For

[18] An *ultrafilter* on a nonempty set X is a collection U of nonempty subsets of X which is closed under supersets and finite intersections, and which has the property that for every $A \subseteq X$, exactly one of A and $X \setminus A$ is in U. An ultrafilter is *nonprincipal* if it contains no finite sets. The existence of nonprincipal ultrafilters on ω follows from ZFC, but (as this result shows) requires the Axiom of Choice.

[19] Blackwell describes the discovery of his proof in [Albers and Alexanderson, 1985, p. 26].

each finite sequence $\langle n_0, \ldots, n_k\rangle$, let $\Omega(\langle n_0, \ldots, n_k\rangle)$ be the set of $x \in {}^\omega\omega$ with $\langle n_0, \ldots, n_k\rangle$ as an initial segment; let $R(\langle n_0, \ldots, n_k\rangle)$ be the closure (in Y) of the f-image of $\Omega(\langle n_0, \ldots, n_k\rangle)$; and let $S(\langle n_0, \ldots, n_k\rangle)$ be the closure of the g-image of $\Omega(\langle n_0, \ldots, n_k\rangle)$. Then for each $z \in Y$, let $G(z)$ be the game where players I and II build x and y in ${}^\omega\omega$, with I winning if for some integer k, $z \in R(x\restriction k) \setminus S(y\restriction k)$, II winning if for some integer k, $z \in S(y\restriction k) \setminus R(x\restriction(k+1))$, and the run of the game being a draw if neither of these happens. Roughly, each player is creating a real (x or y) to feed into his function, and trying to maintain for as long as possible that the corresponding output can be made arbitrarily close to the target real z; the loser is the first player to fail to maintain this condition. Let A' be the set of z for which player I has a strategy guaranteeing at least a draw, and let B' be the set of z for which player II has such a strategy. Then the determinacy of open games implies that ${}^\omega\omega = A' \cup B'$, and $A \subseteq A'$, $B \subseteq B'$ and $A' \cap B' = A \cap B$ follow from the fact that A is the range of f and B is the range of g. The sets A' and B' are analytic, as A' is a projection of the set of pairs (ϕ, z) such that ϕ is (a code for) a strategy for I in $G(z)$ guaranteeing at least a draw, which is Borel, and similarly for B'.[20]

3.1 *Reduction, separation, norms and prewellorderings*

In his [1936], Kuratowski defined the *reduction theorem* (now called the *reduction property*) for a pointclass Γ to be the statement that for any A, B in Γ there exist disjoint A', B' in Γ with $A' \subseteq A$, $B' \subseteq B$ and $A' \cup B' = A \cup B$. He showed in this paper that $\utilde{\Pi}^1_1$ and $\utilde{\Sigma}^1_2$ have the reduction property; Addison [1958] showed this for $\Pi^1_1(a)$ and $\Sigma^1_2(a)$, for each real number a. Blackwell's argument proves the reduction property for $\utilde{\Pi}^1_1$, working with the corresponding $\utilde{\Sigma}^1_1$ complements.

Kuratowski also defined the *first separation theorem* (now called the *separation property*) for a pointclass Γ to be the statement that for any disjoint A, B in Γ there exists C in Δ_Γ with $A \subseteq C$ and $B \cap C = \emptyset$. This property had been studied by Sierpiński [1924] and Luzin [1930a] for initial segments of the Borel hierarchy. Kuratowski also noted that the reduction property for a pointclass Γ implies the separation property for $\neg\Gamma$. Luzin [1927, pp. 51-55] proved that the pointclass $\utilde{\Sigma}^1_1$ satisfies the separation property, by showing that disjoint $\utilde{\Sigma}^1_1$ sets are contained in disjoint Borel sets. Petr Novikov [1935] showed that $\utilde{\Pi}^1_2$ satisfies the separation property and $\utilde{\Sigma}^1_2$ does not. Novikov [1935] (in the case of $\utilde{\Sigma}^1_2$ sets) and Addison [1958] showed that if Γ satisfies the reduction property and has a so-called *doubly universal* member, and Δ_Γ has no universal member, then Γ does not have the separation property, so $\neg\Gamma$ does not have the reduction property.[21]

[20] A *projection* of a set $A \subseteq ({}^\omega\omega)^k$ (for some integer $k \geq 2$) is a set of the form $\{(x_0, \ldots, x_{i-i}, x_{i+1}, \ldots, x_{k-1}) \mid \exists x_i (x_0, \ldots, x_{k-1}) \in A\}$, for some $i < k$.

[21] Members U,V of a pointclass Γ are *doubly universal* for Γ if for each pair A,B of members of Γ there exist an $x \in {}^\omega\omega$ such that $U_x = A$ and $V_x = B$. The non-selfdual projective pointclasses (e.g., $\Sigma^1_1(a)$, $\Pi^1_1(a)$, $\Sigma^1_2(a)$, $\Pi^1_2(a)$, ...) all have doubly universal members.

Addison [1958; 1959] showed that if all real numbers are constructible, then the reduction property holds for $\utilde{\Sigma}^1_k$, for all $k \geq 2$.

Inspired by Blackwell's argument, Addison and Martin independently proved that $\utilde{\Delta}^1_2$-determinacy implies that $\utilde{\Pi}^1_3$ has the reduction property. Since the pointclass $\utilde{\Sigma}^1_3$ has a doubly universal member, this shows that $\utilde{\Delta}^1_2$-determinacy implies the existence of a nonconstructible real. This fact also follows from Gödel's result (discussed in [Addison, 1959]) that the Lebesgue measurability of all Δ^1_2 sets implies the existence of a nonconstructible real. Determinacy would soon be shown to imply stronger structural properties for the projective pointclasses.

The key technical idea behind the (pre-determinacy) results listed above on separation and reduction for the first two levels of the projective hierarchy was the notion of *sieve* (in French, *crible*). This construction first appeared in a paper of Lebesgue [1905], in which he proved the existence of Lebesgue-measurable sets which are not Borel. In Lebesgue's presentation, a sieve is an association of a closed subset F_r of the unit interval $[0, 1]$ to each rational number r in this interval. The sieve then represents the set of $x \in [0, 1]$ such that $\{r \mid x \in F_r\}$ is wellordered, under the usual ordering of the rationals. Using this approach, Luzin and Sierpiński [1918; 1923] showed that $\utilde{\Sigma}^1_1$ sets and $\utilde{\Pi}^1_1$ sets are unions of $\aleph_1$ many Borel sets.

Much of the classical work of Luzin, Sierpiński, Kuratowski and Novikov mentioned here was redeveloped in the lightface context by Kleene [1943; 1955c; 1955b; 1955a], who was unaware of their previous work. The two theories were unified primarily by Addison (for example, [1959]). While Blackwell's argument generalizes throughout the projective hierarchy, Moschovakis ([1967; 1969; 1970; 1971], see also [2009, pp. 202-206]) developed via the effective theory a generalization of the Luzin-Sierpiński approach (decomposing a set of reals into a wellordered sequence of simpler sets) which could be similarly propagated. Moschovakis's goal was to find a uniform approach to the theory of Π^1_1 and Σ^1_2; he was unaware of either Kuratowski's work or determinacy (personal communication). He extracted the following notions, for a given pointclass Γ: a Γ-*norm* for a set A is a function $\rho\colon A \to \mathrm{On}$ for which there exist relations $R^+ \in \Gamma$ and $R^- \in \neg\Gamma$ such that for any $y \in A$,

$$x \in A \wedge \rho(x) \leq \rho(y) \leftrightarrow R^+(x, y) \leftrightarrow R^-(x, y);$$

a pointclass Γ is said to have the *prewellordering property* if every $A \in \Gamma$ has a Γ-norm.[22] The prewellordering property was first explicitly formulated by Moschovakis in 1964; the definition just given is a reformulation due to Kechris. Kuratowski [1936] and Addison [1958] had shown that a variant of the property implies the reduction property; the same holds for the prewellordering property as defined by Moschovakis. Moschovakis applied Novikov's arguments to show that

[22] A *prewellordering* is a binary relation which is wellfounded, transitive and total. A function ρ from a set X to the ordinals induces a prewellording $\preceq$ on X by setting $a \preceq b$ if and only if $\rho(a) \leq \rho(b)$. Conversely, a prewellordering $\preceq$ on a set X induces a function ρ from X to the ordinals, where for each $a \in X$, $\rho(a)$ (the $\preceq$-*rank* of a) is the least ordinal α such that $\rho(b) < \alpha$ for all $b \in X$ such that $b \preceq a$ and $a \not\preceq b$. The range of ρ is called the *length* of $\preceq$.

if Γ is a projective pointclass such that $\forall^1\Gamma \subseteq \Gamma$, and Γ has the prewellordering property, then so does the pointclass $\exists^1\Gamma$. Martin and Moschovakis independently completed the picture in 1968, proving what is now known as the First Periodicity Theorem.

THEOREM 1 (First Periodicity Theorem.) *Let Γ be an adequate pointclass and suppose that Δ_Γ-determinacy holds. Then for all $A \in \Gamma$, if A admits a Γ-norm, then $\forall^1 A$ admits a $\forall^1\exists^1\Gamma$-norm.*

COROLLARY 2 ([Addison and Moschovakis, 1968; Martin, 1968].) *Let Γ be an adequate pointclass closed under existential quantification over reals, and suppose that Δ_Γ-determinacy holds. If Γ satisfies the prewellordering property, then so does $\forall^1\Gamma$.*

Projective Determinacy (PD) is the statement that all projective sets of reals are determined. By the First Periodicity Theorem, under Projective Determinacy the following pointclasses have the prewellordering property, for any real a:

$$\Pi^1_1(a), \Sigma^1_2(a), \Pi^1_3(a), \Sigma^1_4(a), \Pi^1_5(a), \Sigma^1_6(a), \ldots$$

By contrast (see [Kanamori, 2003, pp. 409–410]), in L the pointclasses with the prewellordering property are

$$\Pi^1_1, \Sigma^1_2, \Sigma^1_3, \Sigma^1_4, \Sigma^1_5, \Sigma^1_6, \ldots$$

3.2 Scales

As noted above, the Axiom of Determinacy contradicts the Axiom of Choice, but it is consistent with, and even implies, certain weak forms of Choice. If X and Y are nonempty sets and A is a subset of the product $X \times Y$, a function f *uniformizes* A if the domain of f is the set of $x \in X$ such that there exists a $y \in Y$ with $(x, y) \in A$, and such that for each x in the domain of f, $(x, f(x)) \in A$. A consequence of the Axiom of Choice, *uniformization* is the statement that for every $A \subseteq \mathbb{R} \times \mathbb{R}$ there is a function f which uniformizes A. Uniformization is not implied by AD, as it fails in $L(\mathbb{R})$ whenever there are no uncountable wellordered sets of reals ([Solovay, 1978b]; see Section 3.3).

Uniformization was implicitly introduced by Jacques Hadamard [1905], when he pointed out that the Axiom of Choice should imply the existence of functions on the reals which disagree everywhere with every algebraic function over the integers. Luzin [1930b] explicitly introduced the notion of uniformization and showed that such functions exist. He also announced several results on uniformization, including the fact that all Borel sets (but not all $\undertilde{\Sigma}^1_1$ sets) can be uniformized by $\undertilde{\Pi}^1_1$ functions. The result on Borel sets was proved independently by Sierpiński. Novikov [1935] showed that every $\undertilde{\Sigma}^1_1$ set of pairs has a $\undertilde{\Sigma}^1_2$ uniformization.

A pointclass Γ is said to have the *uniformization property* if every set of pairs in Γ is uniformized by a function in Γ. Motokiti Kondo [1938] showed that the pointclasses $\undertilde{\Pi}^1_1$ and $\undertilde{\Sigma}^1_2$ have the uniformization property. The effective version of this result (i.e., for Π^1_1 and Σ^1_2) was proved by Addison. In some sense this is as far as one can go in ZFC : Lévy [1965a] would show that consistently there exist Π^1_2 sets that cannot be uniformized by any projective function.

After studying Kondo's proof, Moschovakis in 1971 isolated a property for sets of reals which induces uniformizations. Given a set A and an ordinal γ, a *scale* (or a γ-scale) on A into γ is a sequence of functions $\rho_n\colon A \to \gamma$ $(n \in \omega)$ such that whenever

- $\{x_i \mid i \in \omega\} \subseteq A$ and $\lim_{i\to\omega} x_i = x$, and
- the sequence $\langle \rho_n(x_i) : i \in \omega\rangle$ is eventually constant for each $n \in \omega$,

then $x \in A$ and, for every $n \in \omega$, $\rho_n(x)$ is less than or equal to the eventual value of $\langle \rho_n(x_i) : i \in \omega\rangle$. The scale is a Γ-*scale* if there exist $R^+ \in \Gamma$ and $R^- \in \neg\Gamma$ such that for all $y \in A$ and all $n \in \omega$,

$$x \in A \wedge \rho_n(x) \leq \rho_n(y) \leftrightarrow R^+(n,x,y) \leftrightarrow R^-(n,x,y).$$

A pointclass Γ has the *scale property* if every A in Γ has a Γ-scale. Moschovakis [1971] proved the following three theorems about the scale property.

THEOREM 3. *If Γ is an adequate pointclass, $A \in \Gamma$, and A admits a Γ-scale, then $\exists^1 A$ admits a $\exists^1\forall^1\Gamma$-scale.*

THEOREM 4 (Second Periodicity Theorem.) *Suppose that Γ is an adequate pointclass such that Δ_Γ-determinacy holds. Then for all $A \in \Gamma$, if A admits a Γ-scale, then $\forall^1 A$ admits a $\forall^1\exists^1\Gamma$-scale.*

THEOREM 5. *Suppose that Γ is an adequate pointclass which is closed under integer quantification. Suppose that Γ has the scale property, and that Δ_Γ-determinacy holds. Then Γ has the uniformization property.*

Kondo's proof of uniformization for $\underset{\sim}{\Pi}^1_1$ shows that $\Pi^1_1(a)$ has the scale property for every real a (see [Kanamori, 2003, p. 419]). It follows that under $\underset{\sim}{\Delta}^1_{2n}$-determinacy, $\underset{\sim}{\Pi}^1_{2n+1}$ and $\underset{\sim}{\Sigma}^1_{2n+2}$ have the scale property, and every $\underset{\sim}{\Pi}^1_{2n+1}$ relation on the reals can be uniformized by a $\underset{\sim}{\Pi}^1_{2n+1}$ relation (and similarly for $\underset{\sim}{\Sigma}^1_{2n+2}$). Furthermore, under Projective Determinacy, for any real a, the projective pointclasses with the scale property are the same as those with the prewellordering property: $\Pi^1_1(a)$, $\Sigma^1_2(a)$, $\Pi^1_3(a)$, $\Sigma^1_4(a)$, $\Pi^1_5(a)$, $\Sigma^1_6(a)$, etc..

A *tree* on a set X is a collection of finite sequences from X closed under initial segments. Given sets X and Z, a positive integer k and a tree T on $X^k \times Z$, the *projection* of T, $p[T]$, is the set of $x \in (X^\omega)^k$ such that for some $z \in Z^\omega$, $(x\restriction n, z\restriction n) \in T$ for all $n \in \omega$ (strictly speaking, this definition involves the identification of finite sequences of k-tuples with k-tuples of finite sequences). If one substitutes the Baire space ${}^\omega\omega$ for $\mathbb{R}$, Suslin's construction for analytic sets (see Section 2.2) essentially presents them as projections of trees on $\omega \times \omega$, modulo the representation of closed intervals. Many descriptive set theorists, starting perhaps with Luzin and Sierpiński [1923], used trees to represent sets of reals, except that they converted these trees to linear orders via what is now known as the Kleene-Brouwer ordering (after [Brouwer, 1924] and [Kleene, 1955c]). The explicit use of projections of trees as we have presented them here is due to Richard Mansfield [1970]. As pointed out in [Kechris and Moschovakis, 1978b], given an ordinal γ, a γ-scale for a subset A of the Baire space naturally gives rise to a tree on $\omega \times \gamma$ such that $p[T] = A$. Given a set Z, a subset of the Baire space is said to be Z-*Suslin* if it is the projection of a tree on $\omega \times Z$. Suslin's representation of analytic sets shows that a set is analytic if and only if it is ω-Suslin. Some authors use "Suslin" to mean "analytic". We will follow a different usage, however, and say that a subset of the Baire space is *Suslin* if is γ-Suslin for some ordinal γ.

Given a tree T on $\omega \times Z$ and a wellordering of Z, a member of $p[T]$ can be found by following the so-called *leftmost* infinite branch through T (similar to the proof of Kőnig's Lemma, one picks a path through the tree by taking the least next step which is the initial segment of an infinite path through the tree). In a similar manner, a tree on $(\omega \times \omega) \times \gamma$, for some ordinal γ, induces a uniformization of the projection of the tree.

3.3 The game quantifier

Given a Polish space X and a set $B \subseteq X \times {}^{\omega}\omega$, we let $⅁B$ denote the set of $x \in X$ such that I has a winning strategy in $G_\omega(B_x)$. If Γ is a pointclass, $⅁\Gamma$ is the class $\{⅁B \mid B \in \Gamma\}$. The following facts appear in [Moschovakis, 2009, pp. 245-246].

THEOREM 6. *If* Γ *is an adequate pointclass then the following hold.*

- $⅁\Gamma$ *is adequate and closed under* $\exists^0$ *and* $\forall^0$.
- $\exists^1\Gamma \subseteq ⅁\Gamma$ *and* $\forall^1\Gamma \subseteq ⅁\Gamma$.
- *If* $\mathrm{Det}(\Gamma)$ *holds, then* $⅁\Gamma \subseteq \forall^1\exists^1\Gamma$.

The First Periodicity Theorem can be stated more generally as the fact that if an adequate pointclass Γ has the prewellordering property, then so does $⅁\Gamma$, and the Second Periodicity Theorem can be similarly stated as saying that if an adequate pointclass Γ has the scale property, then so does $⅁\Gamma$ (see [Moschovakis, 2009, pp. 246,267]). The propagation of these properties through the projective pointclasses then follows from Theorem 6, given that they hold for $\undertilde{\Pi}^1_1$ (and its variants).

Modifying the notion of Γ-scale by dropping the requirement that $\rho_n(x)$ is less than or equal to the eventual value of $\langle \rho_n(x_i) : i \in \omega \rangle$, one gets the notion of Γ-*semiscale*. Moschovakis's *Third Periodicity Theorem* [1973] concerns the definability of winning strategies and is stated using the game quantifier and the notion of semiscale.

THEOREM 7 (Third Periodicity Theorem.) *Let* Γ *be an adequate pointclass, and suppose that* $\mathrm{Det}(\Gamma)$ *holds. Fix* $A \subseteq {}^{\omega}\omega$ *in* Γ, *and suppose that* A *admits a* Γ-

semiscale and that I has a winning strategy in the game $G_\omega(A)$. Then I has a winning strategy coded by a subset of ω in $\Game\Gamma$.

One consequence the Third Periodicity Theorem in conjunction with Theorem 6 is the following [Moschovakis, 1973]: for any $n \in \omega$, if $\utilde{\Sigma}^1_{2n}$-determinacy holds, $A \subseteq \omega^\omega$ is $\Sigma^1_{2n}(a)$ for some real a and I has a winning strategy in the game with payoff A, then I has a winning strategy coded by a subset of ω in $\Delta^1_{2n+1}(a)$.

Let $\Game^1$ denote the game quantifier for *real games*, games of length ω where the players alternate playing real numbers. Then $\Game^1\utilde{\Sigma}^0_1$ defines the *inductive* sets of reals.[23] Moschovakis [1978] showed that the inductive sets have the scale property. Moschovakis [1983] showed that, assuming the determinacy of all games with payoff in the class built from the inductive sets by the operations of projection and complementation, coinductive sets have scales in this class. Building on this work, Martin and Steel [1983] showed that the pointclass $\utilde{\Sigma}^2_1$ has the scale property in $L(\mathbb{R})$. Kechris and Solovay had shown that if there is no wellordering of the reals in $L(\mathbb{R})$, then there exists in $L(\mathbb{R})$ a set of reals that cannot be uniformized, the set of pairs (x, y) such that y is not ordinal definable from x (i.e., definable from x and some ordinals). This set is Π^2_1 in $L(\mathbb{R})$.

The Solovay Basis Theorem says that if $P(A)$ is a Σ^2_1 relation on subsets of ${}^\omega\omega$ and there exists a witness to $P(A)$ in $L(\mathbb{R})$, then there is a Δ^2_1 witness. This reflection result, along with the Martin-Steel theorem on scales in $L(\mathbb{R})$, compensates in many circumstances for the fact that not every set of reals has a scale in $L(\mathbb{R})$.

Steel [1983] applied Jensen's fine structure theory [1972] to the study of scales in $L(\mathbb{R})$, refining and unifying a great deal of work on scales and Suslin cardinals. Extending [Martin and Steel, 1983], he showed that for each positive ordinal α, determinacy for all sets of reals in $J_\alpha(\mathbb{R})$ implies that the pointclass $\Sigma_1^{J_\alpha(\mathbb{R})}$ has the scale property.

Martin [1983] showed how to propagate the scale property using the game quantifier for integer games of fixed countable length (this subsumes propagation by the quantifier $\Game^1$), and Steel [1988; 2008b] did the same for certain games of length ω_1.

3.4 Partially playful universes

The periodicity theorems showed that determinacy axioms imply structural properties for sets of reals beyond the classical regularity properties. It remained to show that these hypotheses were necessary. Towards this end, Moschovakis (see [Becker, 1978]) identified for each integer n (under the assumption of $\utilde{\Delta}^1_k$-determinacy, where k is the greatest even integer less than n) the smallest transitive Σ^1_n-correct model of ZF + Dependent Choices (DC) which contains all the

[23]Formally, this definition requires a definable association of ω-sequences of reals to individual reals. Alternately, a set of reals is inductive if it is in $\Sigma_1^{J_{\kappa_\mathbb{R}}(\mathbb{R})}$, where J refers to Ronald Jensen's constructibility hierarchy and $\kappa_\mathbb{R}$ is the least κ such that $J_\kappa(\mathbb{R})$ is a model of Kripke-Platek set theory.

ordinals (Joseph Shoenfield [1961] had shown that L is Σ^1_2-correct).[24] This model satisfies AC and $\underset{\sim}{\Delta}^1_k$-determinacy and has a Σ^1_{n+1} wellordering of the reals. In this model, Π^1_i has the scale property for all odd $i \leq n$, and Σ^1_i has the scale property for all other positive integers i.

Kechris and Moschovakis [1978b] introduced the models $L[T_{2n+1}]$, where T_{2n+1} denotes the tree for a Π^1_{2n+1}-scale for a complete Π^1_{2n+1} set. Moschovakis showed that $L[T_1] = L$, and conjectured that $L[T_{2n+1}]$ is independent of the choice of complete set and scale when for all n. This conjecture was proved by Howard Becker and Kechris in [1984].

Solovay [1969] showed that if $L \cap \mathbb{R}$ is countable, then it is the largest countable Σ^1_2 set of reals (i.e., a countable Σ^1_2 set which contains all other such sets). Kechris and Moschovakis [1972] showed that for each positive integer n, if $\mathrm{Det}(\Delta^1_{2n})$ holds then there exists a largest countable Σ^1_{2n+2} set. The largest countable Σ^1_{2n} set came to be called $\mathcal{C}_{2n}$. Kechris [1975] showed that under Projective Determinacy there is for each integer n a largest countable Π^1_{2n+1} set, which he also called $\mathcal{C}_{2n+1}$. The case $n = 0$ follows from ZF + DC and was shown independently by David Guaspari, Kechris and Gerald Sacks [1976]. Kechris also showed that under Projective Determinacy there are no largest countable Σ^1_{2n+1} or Π^1_{2n} sets. It follows that under Projective Determinacy the lightface projective pointclasses with a largest countable set are the same as those in the zig-zag pattern above for the prewellording property and the scale property. Harrington and Kechris [1981] showed (under the assumption that AD holds in $L(\mathbb{R})$) that the reals of each $L[T_{2n+1}]$ are exactly $\mathcal{C}_{2n+2}$, for all integers n (the case $n = 1$ was due to Kechris and Martin).

Kechris showed (assuming Projective Determinacy) that each model $L[\mathcal{C}_{2n}]$ satisfies $\mathrm{Det}(\underset{\sim}{\Delta}^1_{2n-1})$ but not $\mathrm{Det}(\underset{\sim}{\Sigma}^1_{2n-1})$, and has a Δ^1_{2n} wellordering of its reals. Martin would show that $\mathrm{Det}(\underset{\sim}{\Delta}^1_{2n})$ implies $\mathrm{Det}(\underset{\sim}{\Sigma}^1_{2n})$ for each positive integer n.

3.5 Wadge degrees

In 1968, William Wadge considered the following game, given two sets of reals A and B: I builds a real x, II builds a real y, and II wins if $x \in A \leftrightarrow y \in B$. Determinacy for this class of games is known as *Wadge determinacy.* Given two sets of reals A, B, we say that $A \leq_W B$ (A has *Wadge rank* less than or equal to B, or is *Wadge reducible* to B) if there is a continuous function f such that for all reals x, $x \in A$ if and only if $f(x) \in B$ (i.e., such that $A = f^{-1}[B]$). Wadge determinacy implies that for any two sets of reals A, B, either $A \leq_W B$ (in the case that II has a winning strategy) or $\omega^\omega \setminus B \leq_W A$ (in the case that I does), from which it follows that for any two pointclasses closed under continuous preimages,

[24] The Axiom of Dependent Choices (DC) is the statement that if R is a binary relation on a nonempty set X, and if for each $x \in X$ there is a $y \in X$ such that xRy, then there exists an infinite sequence $\langle x_i : i < \omega \rangle$ such that $x_i R x_{i+1}$ for all $i \in \omega$. This statement is a weakening of the Axiom of Choice, sufficient to prove Kőnig's Lemma, the regularity of ω_1 and the wellfoundedness of ultrapowers by countably complete ultrafilters. See [Jech, 2003].

either the two classes are dual (i.e., a pair of the form Γ, $\neg\Gamma$) or one is contained in the other. Wadge showed that $\leq_W$ is wellfounded on the Borel sets, and Martin, using an idea of Leonard Monk, using AD + DC, extended this to all sets of reals under AD + CD (see [Van Wesep, 1978b]).

Wadge determinacy and the wellfoundedness of the Wadge hierarchy divide $\mathcal{P}(\omega^\omega)$ into equivalence classes by Wadge reducibility and order these classes into a wellfounded hierarchy, where each level consists either of one selfdual equivalence class, or two non-selfdual classes, one consisting of all the complements of the members of the other. Wadge determinacy also implies that every non-selfdual adequate pointclass has a universal set (see [Van Wesep, 1978b, p. 162]).

The discovery of Wadge determinacy led to further progress on separation and reduction. Robert Van Wesep [1978a] proved that under AD, if Γ is a non-selfdual pointclass which is closed under continuous preimages, then Γ and $\neg\Gamma$ cannot both have the separation property. Kechris, Solovay and Steel [1981c] showed that under AD + DC, if $\Gamma \subseteq L(\mathbb{R})$ is nonselfdual boldface pointclass and Γ is closed under countable intersections and unions and either $\exists^1$ or $\forall^1$, but not complements, then either Γ or $\neg\Gamma$ has the prewellordering property. In 1981, Steel [1981] showed that under AD, if Γ is a nonselfdual pointclass closed under continuous preimages, then either Γ or $\neg\Gamma$ has the separation property, and if one assumes in addition that Δ_Γ is closed under finite unions, then either Γ or $\neg\Gamma$ has the reduction property.

4 PARTITION PROPERTIES AND THE PROJECTIVE ORDINALS

A cardinal κ is *measurable* if there is a nonprincipal κ-complete ultrafilter on κ, where *κ-completeness* means closure under intersections of fewer than κ many elements. In ZFC measurable cardinals are strongly inaccessible. In 1967, Solovay (see [Jech, 2003, p. 633] or [Kanamori, 2003, p. 348]) showed that AD implies that the club filter on ω_1 is an ultrafilter, which implies that ω_1 is a measurable cardinal.[25] Ulam had shown that under ZFC there are stationary, co-stationary subsets of ω_1; Solovay's result shows the opposite under AD. Solovay also showed that under AD every subset of ω_1 is constructible from a real (i.e., exists in $L[a]$ for some real number a). Since the measurability of ω_1 implies that the sharp of each real exists, this gives another proof that the club filter on ω_1 is an ultrafilter, since for any real a, if $a^\#$ exists, then every subset of ω_1 in $L[a]$ either contains or is disjoint from a tail of the a-indiscernibles below ω_1, which is a club set.

A *Turing degree* is a nonempty subset of $\mathcal{P}(\omega)$ closed under equicomputability. A *cone* of Turing degrees is the set of all degrees above (or computing) a given degree.[26] Martin [1968] showed that under AD the cone measure on Turing degrees is an ultrafilter, i.e., that every set of Turing degrees either contains or is disjoint

[25]A subset of an ordinal is *closed unbounded* (or *club*) if it is unbounded and closed in the order topology on the ordinals, and *stationary* if it intersects every club set. The *club filter* on an ordinal γ consists of all subsets of γ containing a club set.

[26]See [Soare, 1987; Cooper, 2004] for more on the Turing degrees, including a more precise statement of their definition.

from a cone. This important fact has a relatively short and simple proof: the two players collaborate to build a real, with the winner decided by whether the Turing degree of the real falls inside the payoff set; the cone above the degree of any real coding a winning strategy must contain or be disjoint from the payoff set. Martin used this result to find a simpler proof of the measurability of ω_1. Solovay followed by showing that ω_2 is measurable as well. *Turing determinacy* is the restriction of AD to payoff sets closed under Turing equivalence. This form of determinacy is easily seen to suffice for Martin's result. In the early 1980's, Woodin would show that, in $L(\mathbb{R})$, AD and Turing determinacy are equivalent.

Given an ordered set X and an ordinal β, $[X]^\beta$ denotes the set of subsets of X of ordertype β. Given ordinals, α, β, δ and γ, the expression $\alpha \longrightarrow (\beta)^\gamma_\delta$ denotes the statement that for every function $f\colon [\alpha]^\gamma \to \delta$, there exists an $X \in [\alpha]^\beta$ such that f is constant on $[X]^\gamma$. Frank Ramsey [1930] proved that $\omega \longrightarrow (\omega)^n_2$ holds for each positive $n \in \omega$ (this fact is known as *Ramsey's Theorem*). For infinitary partitions, Paul Erdős and András Hajnal [1966] showed (in ZFC) that for any infinite cardinal κ there is a function $f\colon [\kappa]^\omega \to \kappa$ such that for every $X \in [\kappa]^\kappa$, the range of $f \restriction X$ is all of κ.

In 1968, Adrian Mathias [1968; 1977] showed that $\omega \longrightarrow (\omega)^\omega_2$ holds in Solovay's model from [1970], in which all sets of reals satisfy the regularity properties. A set $Y \subseteq [\omega]^\omega$ is said to be *Ramsey* if there exists an $X \in [\omega]^\omega$ such that either $[X]^\omega \subseteq Y$ or $[X]^\omega \cap Y = \emptyset$. The statement $\omega \longrightarrow (\omega)^\omega_2$ is equivalent to the statement that every subset of $[\omega]^\omega$ is Ramsey. Prikry [1976] showed that under $\mathrm{AD}_\mathbb{R}$ (determinacy for games of perfect information of length ω for which the players play real numbers) every subset of $[\omega]^\omega$ is Ramsey. It follows from the main theorem of [Martin and Steel, 1983] that AD + $V{=}L(\mathbb{R})$ implies that every such set is Ramsey. Whether AD alone suffices is still an open question.

In late 1968, Martin (see [Kanamori, 2003, p. 392]) showed that AD implies $\omega_1 \longrightarrow (\omega_1)^\omega_2$ (this implies for instance that the club filter on ω_1 is an ultrafilter). Kenneth Kunen then showed that AD implies that ω_1 satisfies the weak partition property, where a cardinal κ satisfies the *weak partition property* if $\kappa \longrightarrow (\kappa)^\alpha_2$ holds for every $\alpha < \kappa$. Martin followed by showing that $\omega_1 \longrightarrow (\omega_1)^{\omega_1}_2$, again under AD. The proof actually shows $\omega_1 \longrightarrow (\omega_1)^{\omega_1}_{2^\omega}$ and $\omega_1 \longrightarrow (\omega_1)^{\omega_1}_\alpha$ for every countable ordinal α. Martin and Paris (in an unpublished note [March, 1971], see [Kechris, 1978]) showed that under AD + DC, ω_2 has the weak partition property.

Before continuing with this line of results, we briefly discuss the Coding Lemma and the projective ordinals.

4.1 Θ, the Coding Lemma and the projective ordinals

Following convention, we let Θ denote the least ordinal that is not a surjective image of $\mathbb{R}$. Under ZFC, $\Theta = \mathfrak{c}^+$, but under AD, Θ is a limit cardinal, as noted by Harvey Friedman (see [Kanamori, 2003, p. 398]). This fact follows from a theorem known as the *Coding Lemma*, due to Moschovakis [1970], extending earlier work of Friedman and Solovay.

Given a subset P of some Polish space, let $\utilde{\Sigma}^1_1(P)$ denote the pointclass of sets which are Σ^1_1-definable using P and individual reals as parameters.

THEOREM 8 (Coding Lemma.) *Assume* ZF + AD. *Let* $\preceq$ *be a prewellordering of a set of reals* X. *Let* ξ *be the length of* $\preceq$ *and let* A *be a subset of* ξ. *Then there exists a* $Y \subseteq X$ *in* $\utilde{\Sigma}^1_1(\preceq)$ *such that* A *is the set of* $\preceq$*-ranks of elements of* Y.

As an immediate consequence, under AD, if $\xi < \Theta$, then there is a surjection from $\mathbb{R}$ onto $\mathcal{P}(\xi)$ (furthermore, if $\alpha < \Theta^M$ for some wellfounded model M of ZF containing the reals, then such a surjection can be found in M). The proof of the Coding Lemma uses a version of Kleene's Recursion Theorem (first proved in [Kleene, 1938] for partial recursive functions on the integers), which can be stated as saying that given a suitable coding under which each real x codes a continuous partial function $\hat{x}$ (our notation) on the reals, for each two-variable continuous partial function g on the reals there is a real x such that $\hat{x}(w) = g(x, w)$ for all reals w.

If Γ is a pointclass, δ_Γ denotes the supremum of the lengths of the prewellorderings of the reals in Δ_Γ. The notation $\utilde{\delta}^1_n$ is used to denote $\delta_{\utilde{\Sigma}^1_n}$ (which is the same as $\delta_{\utilde{\Pi}^1_n}$). The *projective ordinals* are the ordinals $\utilde{\delta}^1_n$, for $n \in \omega \setminus \{0\}$. It follows from the results of [Luzin and Sierpiński, 1923] that $\utilde{\Sigma}^1_1$ prewellorderings of the reals have countable length, and therefore that the ordinal $\utilde{\delta}^1_1$ is equal to ω_1. Moschovakis [1970] showed (under AD, using the Coding Lemma) that for each $n \in \omega$, $\utilde{\delta}^1_{n+1}$ is a cardinal, and that $\utilde{\delta}^1_{2n+1}$ is regular and (using just PD) strictly less than $\utilde{\delta}^1_{2n+2}$. Martin showed (without AD) that $\utilde{\delta}^1_2 \leq \omega_2$ (see [Kechris and Moschovakis, 1978b]); together these results show that under AD, $\utilde{\delta}^1_2 = \omega_2$.

Kunen and Martin (see [Kechris and Moschovakis, 1978b]) independently established from ZF + DC that every wellfounded κ-Suslin prewellordering has length less than κ^+ (this fact is sometimes called the *Kunen-Martin Theorem*). Moschovakis ([1970]; see [Moschovakis, 2009, 4C.14]) showed (from PD) that any $\utilde{\Pi}^1_{2n+1}$-norm on a complete $\utilde{\Pi}^1_{2n+1}$ set has length $\utilde{\delta}^1_{2n+1}$ (this result also uses Kleene's Recursion Theorem). By the scale property for $\utilde{\Pi}^1_{2n+1}$ sets (under the assumption of DC + $\utilde{\Delta}^1_{2n}$-determinacy, given $n \in \omega$ [Moschovakis, 1971]), every $\utilde{\Pi}^1_{2n+1}$ set (and thus every $\utilde{\Sigma}^1_{2n+2}$ set) is $\utilde{\delta}^1_{2n+1}$-Suslin, and, since $\utilde{\delta}^1_{2n+1}$ is regular, every $\utilde{\Sigma}^1_{2n+1}$ set is λ-Suslin for some $\lambda < \utilde{\delta}^1_{2n+1}$. It follows that under the same hypothesis, $\utilde{\delta}^1_{2n+2} \leq (\utilde{\delta}^1_{2n+1})^+$, and under AD that $\utilde{\delta}^1_{2n+2} = (\utilde{\delta}^1_{2n+1})^+$ for each $n \in \omega$.

Kechris [1974] proved (assuming AD) that $\utilde{\delta}^1_{2n+1}$ is a successor cardinal (its predecessor is called λ_{2n+1}). It follows from his arguments, and those of the previous paragraph, that the pointclasses $\utilde{\Sigma}^1_{2n+2}$ and $\utilde{\Sigma}^1_{2n+1}$ are exactly the $\utilde{\delta}^1_{2n+1}$-Suslin and λ_{2n+1}-Suslin sets respectively.

Given an ordinal λ, the λ-Borel sets of reals are those in the smallest class containing the open sets and closed under complements and well-ordered unions of length less than λ. Martin showed that if κ is a cardinal of uncountable cofinality, then all κ-Suslin sets are κ^+-Borel. He also showed (using AD + DC, the Coding Lemma and Wadge determinacy) the $\underset{\sim}{\delta}^1_{2n+1}$-Borel sets are $\underset{\sim}{\Delta}^1_{2n+1}$, for each $n \in \omega$ (the reverse inclusion follows from the results of Moschovakis [1971] mentioned above). Using this fact, Kechris proved (again, under AD) that λ_{2n+1} has cofinality ω. It follows (under AD) that $\underset{\sim}{\delta}^1_{2n} < \underset{\sim}{\delta}^1_{2n+1}$ for each $n \in \omega$, so that under AD the sequence $\langle \underset{\sim}{\delta}^1_{n+1} : n \in \omega \rangle$ is a strictly increasing sequence of successor cardinals. Kunen [March, 1971] showed that $\underset{\sim}{\delta}^1_n$ is regular for each positive $n \in \omega$.

Solovay noted that under AD, Θ is the Θ-th cardinal, and that under the further assumption of $V{=}L(\mathbb{R})$, Θ is regular (see [Kanamori, 2003, p. 398]). He showed [1978b] that under DC, Θ has uncountable cofinality, and also that ZFC + $\mathrm{AD}_{\mathbb{R}}$ + $\mathrm{cf}(\Theta) > \omega$ proves the consistency of ZF + $\mathrm{AD}_{\mathbb{R}}$, so that by Gödel's Second Incompleteness Theorem, if ZF + $\mathrm{AD}_{\mathbb{R}}$ is consistent, then so is ZFC + $\mathrm{AD}_{\mathbb{R}}$ + $\mathrm{cf}(\Theta) = \omega$. Kechris [1984], using the proof of the Third Periodicity Theorem and work of Martin, Moschovakis and Steel on scales [Martin *et al.*, 1982], showed that DC follows from AD + $V{=}L(\mathbb{R})$. Woodin (see [Kechris, 1984]) strengthened Solovay's result that DC does not follow from AD by showing that, assuming AD + $V{=}L(\mathbb{R})$ there is an inner model of a forcing extension satisfying ZF + AD + $\neg\mathrm{AC}_\omega$ (DC directly implies AC_ω). Whether AD implies $\mathrm{DC}({}^\omega\omega)$ (DC for relations on ${}^\omega\omega$) is still open.

4.2 Partition properties and ultrafilters

Kunen in an unpublished note [September, 1971b] proved that $\underset{\sim}{\delta}^1_{2n} \to (\underset{\sim}{\delta}^1_{2n})^\lambda_2$ for all positive $n \in \omega$ and $\lambda < \omega_1$, under AD. He also showed [September, 1971a] (under the same hypothesis) that $\underset{\sim}{\delta}^1_{2n} \to (\underset{\sim}{\delta}^1_{2n})^{\underset{\sim}{\delta}^1_{2n}}_2$ is false. Martin, in another unpublished note from 1971, showed that $\underset{\sim}{\delta}^1_{2n+1} \to (\underset{\sim}{\delta}^1_{2n+1})^\lambda_2$ for all positive $n \in \omega$ and $\lambda < \omega_1$, under AD.

While Erdős and Hajnal [1958] had shown how to derive partition properties from measurable cardinals, Eugene Kleinberg proved the following result in the other direction, which shows (via $\lambda = \omega$) that $\underset{\sim}{\delta}^1_n$ is measurable for each positive $n \in \omega$.[27]

THEOREM 9 ([Kleinberg, 1970].) *If $\lambda < \kappa$, λ is regular, and $\kappa \to (\kappa)^{\lambda+\lambda}_2$ holds, then C^λ_κ is a normal ultrafilter over κ.*

In 1970, Kunen proved, using Martin's result on the cone measure on the Turing degrees, that under AD, any ω_1-complete filter on an ordinal $\lambda < \Theta$ can be extended to an ω_1-complete ultrafilter, and that every ultrafilter on an ordinal less than Θ is definable from ordinal parameters (see [Kanamori, 2003, pp. 399-400]). Solovay [1978b] proved that under $\mathrm{AD}_{\mathbb{R}}$, there is a normal ultrafilter on $\mathcal{P}_{\aleph_1}\mathbb{R}$: for each $A \subseteq \mathcal{P}_{\aleph_1}\mathbb{R}$, consider the game where I and II collaborate to build a sequence $\langle s_i : i < \omega \rangle$ consisting of finite sets of reals, and I wins if and only if

[27]We let C^λ_κ denote the filter generated by the set of λ-closed unbounded subsets of κ. A filter is *normal* if every regressive function on a set in the filter is constant on a set in the filter.

$\bigcup\{s_i : i \in \omega\} \in A$.[28] This implies (again, under $\mathrm{AD}_{\mathbb{R}}$) that for each ordinal $\gamma < \Theta$ there is a normal ultrafilter on $\mathcal{P}_{\aleph_1}\gamma$. It is not known whether AD suffices for this result, though Harrington and Kechris [1981] showed that if AD holds and γ is less than a Suslin cardinal, then there is a normal ultrafilter on $\mathcal{P}_{\aleph_1}\gamma$.[29] Extending work of Becker, Woodin [1983a] showed that there is just one such ultrafilter for each $\gamma < \Theta$, if either $\mathrm{AD}_{\mathbb{R}}$ holds or AD holds and γ is below a Suslin cardinal.

A cardinal κ is said to have the *strong partition property* if $\kappa \to (\kappa)^{\kappa}_{\mu}$ holds for every $\mu < \kappa$. As mentioned above, Martin showed that under AD, ω_1 has the strong partition property. In late 1977, Kechris adapted Martin's argument to show that under AD there exists a cardinal κ with the strong partition property such that the set of $\lambda < \kappa$ with the strong partition property is stationary below κ (see [Kanamori, 2003, p. 432]). Pushing this further, Kechris, Kleinberg, Moschovakis and Woodin [1981a] showed (using a uniform version of the Coding Lemma) that AD implies that unboundedly many cardinals below Θ have the strong partition property and are stationary limits of cardinals with the strong partition property. They also showed that whenever λ is an ordinal below a cardinal with the strong partition property, all λ-Suslin sets are determined. Using work of Steel [1983] and Martin [1983], Kechris and Woodin [1983] showed that in $L(\mathbb{R})$, AD is equivalent to the assertion that Θ is a limit of cardinals with the strong partition property, and also to the statement that all Suslin sets are determined. James Henle, Mathias and Woodin [1985] later showed that the first equivalence does not follow from ZF + DC, since the existence of a nonprincipal ultrafilter on ω is consistent with Θ being a limit of cardinals with the strong partition property.

A key step in the proof of the Kechris-Woodin theorem was a transfer theorem extending results of Harrington and Martin (discussed in Section 5.3). Harrington and Martin had shown from ZF + DC that, for each real a, $\Pi^1_1(a)$-determinacy is equivalent to determinacy for the larger class $\bigcup_{\beta<\omega^2} \beta\text{-}\Pi^1_1(a)$. Kechris and Woodin showed, from the same hypothesis, that for all positive integers k, $\utilde{\Delta}^1_{2k}$-determinacy is equivalent to $\Game^{(2k-1)} \bigcup_{\beta<\omega^2} \beta\text{-}\utilde{\Pi}^1_1$-determinacy, where $\Game^{(2k-1)}$ indicates an application of $2k-1$ many instances of the game quantifier $\Game$. By Theorem 6, this means that $\utilde{\Delta}^1_{2k}$-determinacy implies $\utilde{\Pi}^1_{2k}$-determinacy. Martin had proved the lightface version in 1973 (see [Kechris and Solovay, 1985]). Later results of Woodin and Itay Neeman [1995] would show that $\utilde{\Pi}^1_{n+1}$-determinacy is equivalent to $\Game^{(n)} \bigcup_{\beta<\omega^2} \beta\text{-}\utilde{\Pi}^1_1$-determinacy for all $n \in \omega$.

4.3 Cardinals, uniform indiscernibles and the projective ordinals

A cardinal κ is *Ramsey* if for every function $f\colon [\kappa]^{<\omega} \to \{0,1\}$ (where $[\kappa]^{<\omega}$ denotes the finite subsets of κ) there exists $A \in [\kappa]^{\kappa}$ such that for each $n \in \omega$,

---

[28] Given a cardinal κ and a set X, $\mathcal{P}_{\kappa}X$ denotes the collection of subsets of X of cardinality less than κ. An ultrafilter U on $\mathcal{P}_{\kappa}X$ is *normal* if for each $Y \in U$, if f is a regressive function on Y (i.e., if $\mathrm{dom}(f) = Y$ and $f(A) \in A$ for all nonempty $A \in Y$) then f is constant on a set in U.

[29] An ordinal (necessarily a cardinal) κ is said to be *Suslin* if there is a set of reals which is κ-Suslin but not λ-Suslin for any $\lambda < \kappa$.

$f \restriction [\kappa]^n$ is constant. Measurable cardinals are Ramsey, and if there exists a Ramsey cardinal then the sharp of each real number exists. Assuming the existence of a Ramsey cardinal, Martin and Solovay [1969] showed that nonempty $\underset{\sim}{\Sigma}^1_3$ subsets of the plane have $\underset{\sim}{\Delta}^1_4$ uniformizations. As mentioned above, Lévy [1965a] had shown that ZFC does not suffice for this result. Martin and Solovay used an analysis of sharps for reals, and modeled their argument after the proof of the Kondo-Addison theorem. Mansfield [1971] extended the Martin-Solovay analysis to show (using a measurable cardinal) that nonempty $\underset{\sim}{\Pi}^1_2$ sets are uniformized by $\underset{\sim}{\Pi}^1_3$ functions.

Given a positive ordinal α, u_α denotes the αth *uniform indiscernible*, the αth ordinal which is a Silver indiscernible for each real number. As bijections between ω and countable ordinals can be coded by reals, the first uniform indiscernible, u_1, is ω_1. It follows from the basic analysis of sharps that all uncountable cardinals are uniform indiscernibles, so $u_2 \leq \omega_2$. By applying the Kunen-Martin theorem inside models of the form $L[a]$, for a a real number, and applying the basic analysis of sharps, Martin showed that $\underset{\sim}{\delta}^1_2 = u_2$ if the sharp of every real exists (see [Kechris, 1978]). Recall that by the results of Section 4.1, $\underset{\sim}{\delta}^1_2 = \omega_2$, under AD.

Martin showed from ZF plus the assumption that the sharp of each real exists that every $\underset{\sim}{\Sigma}^1_3$ set is u_ω-Suslin, and from AD that $u_\omega = \omega_\omega$ (see [Kanamori, 2003, pp.203-204]). By the Kunen-Martin Theorem, then, AD implies that $\underset{\sim}{\delta}^1_3 \leq \omega_{\omega+1}$. Solovay had shown that if the sharp of every real exists, then $u_{\xi+1}$ has the same cofinality as u_2, for every positive ordinal ξ (see [Kechris, 1978]). Since $u_\omega = \omega_\omega$, it follows that each ω_n $(n \geq 2)$ is of the form u_{k+1} for some positive integer k, and thus that each such ω_n has cofinality ω_2. It follows that under AD + DC, $\underset{\sim}{\delta}^1_3 = \omega_{\omega+1}$, since $\underset{\sim}{\delta}^1_3$ is a regular cardinal, and therefore that $\underset{\sim}{\delta}^1_4 = \omega_{\omega+2}$. Kunen and Solovay would then show that $u_n = \omega_n$ for all n satisfying $1 \leq n \leq \omega$.

In 1971, Kunen reduced the computation of $\underset{\sim}{\delta}^1_5$ to the analysis of certain ultrapowers of $\underset{\sim}{\delta}^1_3$ (see [Kechris, 1978]; as part of his analysis, Kunen showed that $\underset{\sim}{\delta}^1_3$ has the weak partition property, see [Solovay, 1978a]). The completion of this project was to take another decade. In the early 1980's, Martin proved new results analyzing these ultrapowers, and Steve Jackson, using joint work with Martin, computed $\underset{\sim}{\delta}^1_5$. The following theorem [Jackson, 1988; Jackson, 1999] completes the calculation of the $\underset{\sim}{\delta}^1_n$'s.

THEOREM 10 (Jackson.) *Assume* AD. *Then for* $n \geq 1$, $\underset{\sim}{\delta}^1_{2n+1}$ *has the strong partition property and is equal to* $\omega_{w(2n-1)+1}$, *where* $w(1) = \omega$ *and* $w(m+1) = \omega^{w(m)}$ *in the sense of ordinal exponentiation.*

Jackson's proof of this theorem was over 100 pages long. Elements of his argument (as presented in [Jackson, 1999]) include the Kunen-Martin theorem, Kunen's $\underset{\sim}{\Delta}^1_3$ coding for subsets of ω_ω [Solovay, 1978a], Martin's theorem that $\underset{\sim}{\Delta}^1_{2n+1}$ is closed under intersections and unions of sequences of sets indexed by ordinals less than $\underset{\sim}{\delta}^1_3$, and so-called homogeneous trees, a notion which traces back to [Martin and Solovay, 1969] and a result of Martin discussed in the next section.

5 DETERMINACY AND LARGE CARDINALS

As discussed above, a strongly inaccessible cardinal is an uncountable regular cardinal which is closed under cardinal exponentiation. If κ is strongly inaccessible, then V_κ is a model of ZFC, so that the existence of strongly inaccessible cardinals is not a consequence of ZFC. While there is no technical definition of *large cardinal*, a typical large cardinal notion (in the context of the Axiom of Choice) specifies a type of strongly inaccessible cardinal. Examples of this type include Ramsey cardinals, measurable cardinals, Woodin cardinals and supercompact cardinals. The *large cardinal hierarchy* orders large cardinals by *consistency strength*. That is, large cardinal notion A is below large cardinal notion B in the hierarchy if the existence of cardinals of type B implies the consistency of cardinals of type A. It is a striking empirical fact that the large cardinal hierarchy is linear, modulo open questions (the examples just given were listed in increasing order, for instance). Even more striking is the fact that many set-theoretic statements having no ostensible relationship to large cardinals are equiconsistent with some large cardinal notion.[30]

By results of Mycielski (discussed in Section 2.3), AD implies that ω_1 is strongly inaccessible in L, which means that AD cannot be proved in ZFC. Moreover, Solovay's result that AD implies the measurability of ω_1 implies that under AD, ω_1 (as computed in the full universe) is a measurable cardinal in certain inner models of AC, such as HOD.[31] As we shall see in this section, the relationship between large cardinals and determinacy runs in both directions: various forms of determinacy imply the existence of models of ZFC containing large cardinals, and the existence of large cardinals can be used to prove the determinacy of certain definable sets of reals.

5.1 *Measurable cardinals*

Solovay [1969] showed in 1965 that if there exists a measurable cardinal then every uncountable $\utilde{\Sigma}^1_2$ set of reals contains a perfect set. This result was proved independently by Mansfield (see [Solovay, 1969]). Martin [1970] showed that in fact analytic determinacy follows from the existence of a Ramsey cardinal.

Roughly, the idea behind Martin's proof is that if A is the projection of a tree T on $\omega \times \omega$ and χ is a Ramsey cardinal, one can modify the original game for A to require the second player to play, in addition to his usual moves, a function $G^* \colon \omega^{<\omega} \to \chi$ witnessing (via the wellfoundedness of the ordinal χ) that the fragment of T corresponding to the real produced by the two players in their moves from the original game has no infinite branches, and thus that this real is not in the projection of T. This modified game is closed, and thus determined, by Gale-Stewart. If the second player has a winning strategy in the modified game,

[30] [Kanamori, 2003] is the standard reference for the large cardinal hierarchy.

[31] The inner model HOD (a model of ZFC) consists of all sets x such that every member of the transitive closure of $\{x\}$ is ordinal-definable (see [Jech, 2003, Chapter 13]).

then he has a winning strategy in the original game by ignoring his extra moves. In general there is no reason that a winning strategy for the first player in the modified game will induce a winning strategy for the original game. However, if χ is a Ramsey cardinal, then there is uncountable $X \subseteq \chi$ such that, as long as the range of G^* is contained in X, the first player's strategy does not depend on the extra moves for the second player. Using this fact, the first player can convert his winning strategy in the modified game into a winning strategy in the original game. The notion of a determined (often closed) auxiliary game and a method for transferring strategies from the auxiliary game to the original game is the basis of many determinacy proofs.

Martin later proved the following refinement.

THEOREM 11. *If the sharp of every real exists, then $\underset{\sim}{\Pi}^1_1$-determinacy holds.*

In the 1970's Kunen and Martin independently developed the notion of a *homogeneous* tree, following a line of ideas deriving from Martin's proof of $\underset{\sim}{\Pi}^1_1$-determinacy (see [Kechris, 1981]). Given a set Z and a cardinal κ, a tree on $\omega \times Z$ is said to be *κ-homogeneous* if for each $\sigma \in \omega^{<\omega}$ there is a κ-complete ultrafilter μ_σ on $Z^{|\sigma|}$ such that

- for each $\sigma \in \omega^{<\omega}$, $\{z \mid (\sigma, z) \in T\} \in \mu_\sigma$;
- $p[T]$ is the set of $x \in \omega^\omega$ such that the sequence $\langle \mu_{x \restriction i} : i \in \omega \rangle$ is countably complete.[32]

A tree is said to be *homogeneous* if it is $\aleph_1$-homogeneous. A set of reals is said to be *homogeneously Suslin* if it is the projection of a homogeneous tree. There are related notions of *weakly homogeneous tree* and *weakly homogeneously Suslin set* of reals, involving a more involved relationship with a set of ultrafilters. Though it was not the original definition, let us just say that a tree on a set of the form $\omega \times (\omega \times Z)$ is weakly homogeneous if and only if the corresponding tree on $(\omega \times \omega) \times Z$ is homogeneous, and note that a set of reals is weakly homogeneously Suslin if and only if it is the projection of a homogeneously Suslin set of pairs.

Martin's proof then shows the following.

THEOREM 12 (Martin.) *Homogeneously Suslin sets are determined.*

The unfolding argument mentioned in Section 2.2 then shows that weakly homogeneously Suslin sets satisfy the regularity properties.

In retrospect, Martin's proof of analytic determinacy can be broken into two parts, the fact that homogeneously Suslin sets are determined, and the fact that if there is a Ramsey cardinal then $\underset{\sim}{\Pi}^1_1$ sets are homogeneously Suslin.

The results of [Martin and Solovay, 1969] can similarly be reinterpreted. If $\underset{\sim}{\Pi}^1_1$ sets are homogeneously Suslin, then $\underset{\sim}{\Sigma}^1_2$ sets are weakly homogeneously Suslin. The Martin-Solovay construction can be seen as a method for taking a γ-weakly

[32]i.e., for each sequence $\langle A_i \mid i \in \omega \rangle$ such that each $A_i \in \mu_{x \restriction i}$ there exists a $t \in Z^\omega$ such that $t \restriction i \in A_i$ for each i.

homogeneous tree T (for some cardinal γ) and producing a tree S on $\omega \times \gamma'$, for some ordinal γ', projecting to the complement of the projection of T. From this follows that all $\underset{\sim}{\Pi}^1_2$ sets, and thus all $\underset{\sim}{\Sigma}^1_3$ sets, are projections of trees on the product of ω with some ordinal. More sophisticated arguments can be carried out from the existence of sharps, using the fact that sharps give ultrafilters over certain inner models.

5.2 Borel determinacy

In 1968, Friedman [1970] showed that the Replacement axiom is necessary to prove Borel determinacy, even for sets invariant under Turing degrees (he also showed that analytic determinacy cannot hold in a forcing extension of L). As refined by Martin, his results show (for each $\alpha < \omega_1$) that ZFC - Power Set - Replacement + "the αth iteration of the power set of ${}^\omega\omega$ exists" does not prove the determinacy of all $\underset{\sim}{\Sigma}^0_{1+\alpha+3}$ sets.

James Baumgartner mixed the method of Martin's $\underset{\sim}{\Pi}^1_1$-determinacy proof with Davis's $\underset{\sim}{\Sigma}^0_3$-determinacy proof to give a new proof of $\underset{\sim}{\Sigma}^0_3$-determinacy in ZFC. Using a similar approach, Martin proved Det($\underset{\sim}{\Sigma}^0_4$) from the existence of a weakly compact cardinal,[33] and then Paris [1972] proved it in ZFC (Paris noted at the end of his paper that his argument could be carried out without the power set axiom, assuming instead only that the ordinal ω_1 exists).

Andreas Blass [1975] and Mycielski (1967, unpublished) independently proved that $AD_{\mathbb{R}}$ is equivalent to determinacy for integer games of length ω^2. The key idea in Blass's proof was to reduce determinacy in the given game to determinacy in another, auxiliary, game in such a way that one player's moves in the auxiliary game correspond to fragments of his strategy in the original game. Martin [1975] used this basic idea to prove Borel determinacy in 1974 (the auxiliary game was in fact an open game). In his [1985], Martin gave a short, inductive, proof of Borel determinacy, and introduced the notion of *unraveling* a set of reals—roughly, finding an association of the set to a clopen set in a larger domain with a map sending strategies in one game to strategies in the other. In his [1990], Martin extended this method to games of length ω played on any (possibly uncountable) set, with Borel payoff (in the corresponding sense). Neeman [2000; 2006b] would unravel $\underset{\sim}{\Pi}^1_1$ sets from the assumption of a measurable cardinal κ of Mitchell rank κ^{++} (proved to be an optimal hypothesis by Steel [1982]; see [Jech, 2003, pp. 357-360] for the definition of Mitchell rank). Complementing Friedman's theorem, Martin proved that for each $\alpha < \omega_1$, the determinacy of each Boolean combination of $\underset{\sim}{\Sigma}^0_{\alpha+2}$ sets follows from ZF - Power Set - Replacement + Σ_1-Replacement + "the αth iteration of the power set of ${}^\omega\omega$ exists".

[33] A cardinal κ is *weakly compact* if $\kappa \rightarrow (\kappa)^2_2$. Weakly compact cardinals are below the existence of $0^\#$ and above strongly inaccessible cardinals in the consistency strength hierarchy (see [Kanamori, 2003, pp. 76,472]).

5.3 The $\underset{\sim}{\Pi}^1_1$ difference hierarchy

Given a countable ordinal α and a real a, a set of reals X is said to be α-$\Pi^1_1(a)$ if there is wellordering of ω of length α recursive in a with corresponding rank function $R\colon \omega \to \alpha$ and a $\Pi^1_1(a)$ subset A of $\omega \times {}^\omega\omega$ such that

- for all $n, m \in \omega$, if $R(n) < R(m)$ then
$$\{x \mid (m, x) \in A\} \subseteq \{x \mid (n, x) \in A\};$$
- X is the set of reals x for which the least ξ such that either $\xi = \alpha$ or $\xi < \alpha$ and $(R^{-1}(\xi), x) \notin A$ is odd.

This notation has its roots in [Hausdorff, 1908]. When a is itself recursive one writes α-Π^1_1. The union of the sets α-$\Pi^1_1(a)$ for all reals a is denoted α-$\underset{\sim}{\Pi}^1_1$. The union of the sets α-$\underset{\sim}{\Pi}^1_1$ for all $\alpha < \omega_1$ is denoted Diff($\underset{\sim}{\Pi}^1_1$). Note that Diff($\underset{\sim}{\Pi}^1_1$) is a proper subclass of $\underset{\sim}{\Delta}^1_2$.

Friedman [1971] extended Theorem 11 to show that Det(3-$\underset{\sim}{\Pi}^1_1$) follows from the existence of the sharp of every real. Martin in 1975 then extended this result to show that the existence of $0^\#$ is equivalent to Det($\bigcup_{\beta<\omega^2} \beta$-$\Pi^1_1$) (see [DuBose, 1990]). Harrington [1978] then proved the converse to Theorem 11 by showing that Det($\Pi^1_1(a)$) implies the existence of $a^\#$, for each real a.

For the purposes of the next theorem, say that a model has α measurable cardinals and indiscernibles if there exists a set of ordertype α consisting of measurable cardinals of the model, and there exist uncountably many ordinal indiscernibles of the model above the supremum of these measurable cardinals. Martin proved the following theorem after Harrington's result.

THEOREM 13. *For any real a and any ordinal α recursive in a, the following are equivalent.*

- Det($\bigcup_{\beta<\omega^2} (\omega^2 \cdot \alpha + \beta)$-$\Pi^1_1(a)$).
- Det($(\omega^2 \cdot \alpha + 1)$-$\Pi^1_1(a)$).
- *There is an inner model of* ZFC *containing a and having α many measurable cardinals and indiscernibles.*

Still, a large-cardinal consistency proof of Det($\underset{\sim}{\Delta}^1_2$), the hypothesis used by Addison and Martin in their extension of Blackwell's argument, remained beyond reach. John Green [1978] showed that Det(Δ^1_2) implies the existence of an inner model with a measurable cardinal of Mitchell rank 1.

5.4 Larger cardinals

In Section 4 we defined a measurable cardinal to be a cardinal κ such that there exists a nonprincipal κ-complete ultrafilter on κ. Equivalently, under the Axiom of

Choice, κ is measurable if and only if there is a nontrivial elementary embedding j from the full universe V into some inner model M whose critical point is κ, i.e., such that κ is the least ordinal not mapped to itself by j. Many large cardinal notions can be expressed both in terms of ultrafilters and in terms of embeddings, though in the Choiceless context (without the corresponding form of Łoś's Theorem, see [Jech, 2003, p. 159]) it is the definition in terms of ultrafilters which is relevant. For instance, a cardinal κ is *supercompact* if for each $\lambda > \kappa$ there exists a normal fine ultrafilter on $\mathcal{P}_\kappa\lambda$.[34] Under the Axiom of Choice, κ is supercompact if and only if for every $\lambda > \kappa$ there is an elementary embedding j from V into an inner model M such that the critical point of j is κ and M is closed under sequences of length λ. Every supercompact cardinal is a limit of measurable cardinals. An even larger large cardinal notion is the huge cardinal, where an uncountable cardinal κ is *huge* if for some cardinal $\lambda > \kappa$ there is a κ-complete normal fine ultrafilter on $[\lambda]^\kappa$ (where "normal" and "fine" are defined in analogy with the supercompact case, see [Kanamori, 2003, p. 331]). Under AC, κ is huge if and only if there is an elementary embedding $j\colon V \to M$ with critical point κ such that M is closed under sequences of length $j(\kappa)$. The existence of huge cardinals does not imply the existence of supercompact cardinals, but it does imply their consistency.

Kunen [1971] put a limit on the large cardinality hierarchy, showing in ZFC that there is no nontrivial elementary embedding from V into itself. A corollary of the proof is that for any elementary embedding j of V into any inner model M, if δ is the least ordinal above the critical point of j sent to itself by j, then $V_{\delta+1} \not\subseteq M$. In 1978, Martin [1980] proved $\utilde{\Pi}^1_2$-determinacy from hypothesis I2, which states that for some ordinal δ there is a nontrivial elementary embedding of V into an inner model M with critical point less than δ such that $V_\delta \subseteq M$ and $j(\delta) = \delta$. In 1984, Woodin proved $\mathrm{AD}^{L(\mathbb{R})}$ from I0, the statement that for some ordinal δ there is a nontrivial elementary embedding from $L(V_{\delta+1})$ into itself with critical point below δ, thus verifying Solovay's conjecture that $\mathrm{AD}^{L(\mathbb{R})}$ would follow from large cardinals. I0 is one of the strongest large cardinal hypotheses not known to be inconsistent. The inner model program at the time had produced models for many measurable cardinals, hypotheses far short of I2, and so there was little hope of showing that I2 and I0 were necessary for these results.

New large cardinal concepts would prove to be the missing ingredient. Given an ideal I on a set X, forcing with the Boolean algebra given by the power set of X modulo I gives a V-ultrafilter on the power set of X.[35] The ideal I is said to be *precipitous* if the ultrapower of V by this generic ultrafilter is wellfounded in all generic extensions. If the underlying set X is a cardinal κ, the ideal I is said to be *saturated* if the Boolean algebra $\mathcal{P}(\kappa)/I$ has no antichains of cardinality κ^+.[36]

[34]Given a cardinal κ and a set X, a collection U of subsets of $\mathcal{P}_\kappa X$ is *fine* if it contains the collection of supersets of each element of $\mathcal{P}_\kappa X$.

[35]An *ideal* is a collection of sets closed under subsets and finite unions. Given a model M and a set X in M, an *M-ultrafilter* is a subset of $\mathcal{P}(X) \cap M$ closed under supersets and finite intersections such that for every $A \subseteq X$ in M, exactly one of A and $X \setminus A$ is in U. Note that U does not need to be an element of M.

[36]An *antichain* in a partial order (or a Boolean algebra) is a set of pairwise incompatible

If κ is a regular cardinal, saturation of I implies precipitousness. Huge cardinals were invented by Kunen [1978], who used them to produce a saturated ideal on ω_1.

In early 1984, Matthew Foreman, Menachem Magidor and Shelah [1988] showed that if there exists a supercompact cardinal—a hypothesis much weaker than I0 or I2—then there is an ω_1-preserving forcing making the nonstationary ideal on ω_1 (NS_{ω_1}) saturated.

Foreman (see [1986]) and Magidor [1980] had earlier made a connection between generic elementary embeddings[37] and regularity properties for reals. Magidor [1980] in particular had shown that the Lebesgue measurability of Σ^1_3 sets followed from the existence of a generic elementary embedding with critical point ω_1 and wellfounded image model (the existence of such an embedding follows from the Foreman-Magidor-Shelah result mentioned above). Woodin noted that these arguments plus earlier work of his (see [1986]) could be used to extend this to Lebesgue measurability for all projective sets. Woodin also noted that arguments from [Foreman *et al.*, 1988] could be used to prove the Lebesgue measurability of all sets of reals in $L(\mathbb{R})$, if one could force to produce a saturated ideal on ω_1 without adding reals. Shelah then noted that techniques from [Shelah, 1998] could be modified to do just that. It followed then that the existence of a supercompact cardinal implies that all sets of reals in $L(\mathbb{R})$ are Lebesgue measurable.

Woodin and Shelah then addressed the problem of weakening the hypotheses needed for the Lebesgue measurability of all projective sets of reals. We follow the account in [Neeman, 2004]. Woodin noted that a superstrong cardinal sufficed. Shelah then isolated a weaker notion now known as a *Shelah cardinal*, and showed that the existence of $n+1$ Shelah cardinals implies that $\underset{\sim}{\Sigma}^1_{n+2}$ sets are Lebesgue measurable.

DEFINITION 14. A cardinal κ is a *Shelah cardinal* if for every $f\colon \kappa \to \kappa$ there is an elementary embedding $j\colon V \to N$ with critical point κ such that $V_{j(f)(\kappa)} \subseteq N$.

Woodin noted that by switching the quantifiers in Shelah's definition one obtained a weaker, still sufficient, hypothesis, now known as a Woodin cardinal.

DEFINITION 15. A cardinal δ is a *Woodin cardinal* if for each function $f\colon \delta \to \delta$ there exists an elementary embedding $j\colon V \to M$ with critical point $\kappa < \delta$ closed under f such that $V_{j(f)(\kappa)} \subseteq M$.

Woodin proved that the existence of n Woodin cardinals below a measurable cardinal implies the Lebesgue measurability of $\underset{\sim}{\Sigma}^1_{n+2}$ sets, the same amount of measurability that would follow from $\underset{\sim}{\Pi}^1_{n+1}$-determinacy. All of this work was done within a few weeks of the Foreman-Magidor-Shelah result on the saturation of NS_{ω_1}. In [Shelah and Woodin, 1990] the hypothesis for the statement that all sets of reals in $L(\mathbb{R})$ are Lebesgue measurable and have the property of Baire

elements. In the case of a Boolean algebra of the form $\mathcal{P}(\kappa)/I$, an antichain is a collection of subsets of κ not in I which pairwise have intersection in I.

[37] A *generic elementary embedding* is an elementary embedding of the universe V into some class model M which exists in a forcing extension of V.

was reduced to the existence of ordertype $\omega + 1$ many Woodin cardinals. The hypothesis was to be reduced even further.

Woodin extracted from the Foreman-Magidor-Shelah results a one-step forcing for producing generic elementary embeddings with critical point ω_1, and developed it into a general method, now known as the *stationary tower.* Using this he showed (by the fall of 1984, see his [1988]) that if there exists a supercompact cardinal, then every set of reals in $L(\mathbb{R})$ is weakly homogeneously Suslin.

Steel had been working on the problem of finding inner models for supercompact cardinals. Inspired by the results of Foreman, Magidor, Shelah and Woodin, he begin to work on producing models for Woodin cardinals, and had some partial results by the spring of 1985, producing inner models with certain weak variants of Woodin cardinals. These models were generated by sequences of *extenders*, directed systems of ultrafilters which collectively generate elementary embeddings whose images contain more of V than possible for embeddings generated by a single ultrafilter. Special cases of extenders had appeared in Jensen's proof of the Covering Lemma, and in [Mitchell, 1979]. Jensen formulated the general notion of extender, which first appeared in [Dodd, 1982]. Steel and Martin saw that the problem of building models with Woodin cardinals was linked to the problem of proving determinacy, and they sets their sight on this problem in the late spring of 1985.

One key combinatorial problem related to elementary embeddings is whether infinite iterations of these embeddings produce wellfounded models. Kunen [1970] had shown that the answer was positive for iterations derived from a single ultrafilter. With extenders the situation was more complicated, as the iterations did not need to be linear but could produce trees of models with no rule for finding a path through the tree leading to a wellfounded model (indeed, this nonlinearity was essential, since otherwise the models would have simply definable wellorderings of their reals). The simplest such tree, a so-called *alternating chain*, is countably infinite and consists of two infinite branches. Martin and Steel saw that the issue of wellfoundedness for the direct limits along the two branches was linked. This observation led to the following theorem, proved in August of 1985.

THEOREM 16 (Martin-Steel [1989].) *Suppose that λ is a Woodin cardinal and A is a λ^+-weakly homogeneously Suslin set of reals. Then for any $\gamma < \lambda$, ${}^{\omega}\omega \setminus A$ is γ-homogeneously Suslin.*

It follows from this and the fact that analytic sets are homogeneously Suslin in the presence of a measurable cardinal that if there exist n Woodin cardinals below a measurable cardinal, then $\underset{\sim}{\Pi}^1_{n+1}$ sets are determined, and that Projective Determinacy follows from the existence of infinitely many Woodin cardinals.

Combined with Woodin's application of the stationary tower mentioned above, the Martin-Steel theorem implied that $\mathrm{AD}^{L(\mathbb{R})}$ follows from the existence of a supercompact cardinal. By the end of 1985, Woodin had improved the hypothesis to the existence of infinitely many Woodin cardinals below a measurable cardinal (see [Larson, 2004]).

THEOREM 17 (Woodin.) *If there exist infinitely many Woodin cardinals below a measurable cardinal, then* AD *holds in* $L(\mathbb{R})$.

In the spring of 1986, Martin and Steel [1994] produced extender models with n Woodin cardinals and Δ^1_{n+2} wellorderings of the reals. Such a model necessarily has a Σ^1_{n+2} set which is not Lebesgue measurable, and fails to satisfy Π^1_{n+1}-determinacy.

Skipping ahead for a moment, let $(*)_n$ be the statement that for each real x there exists an iterable model M containing x and n Woodin cardinals plus the sharp of V^M_δ, for δ the largest of these Woodin cardinals. For odd n, the equivalence of $\undertilde{\Pi}^1_{n+1}$-determinacy and $(*)_n$ was proved by Woodin in 1989. That $(*)_n$ implies $\undertilde{\Pi}^1_{n+1}$-determinacy for all n was proved by Neeman [1995] in 1994. Roughly, Neeman's methods work by considering a modified game in which one player builds an iteration tree and makes moves in the image of the original game by the embeddings given by the tree. In 1995, Woodin proved that $\undertilde{\Pi}^1_{n+1}$-determinacy implies $(*)_n$ for even $n > 0$.

As discussed above, homogeneously Suslin and weakly homogeneously Suslin sets of reals played an important role in applications of large cardinals to regularity properties for sets of reals, as early as the 1969 results of Martin and Solovay. Qi Feng, Magidor and Woodin [1992] introduced a related tree representation property for sets of reals. Given a cardinal κ, a set $A \subseteq \omega^\omega$ is *κ-universally Baire* if there exist trees S, T such that $p[S] = A$ and S and T project to complements in every forcing extension by a partial order of cardinality less than or equal to κ.

Woodin (see [Kanamori, 2003; Larson, 2004]) showed that if δ is a Woodin cardinal, then δ-universally Baire sets of reals are $<\delta$-weakly homogeneously Suslin. It follows from the arguments of [Martin and Solovay, 1969] that if $A \subseteq \omega^\omega$ is κ^+-weakly homogeneously Suslin, then it is κ-universally Baire.

Feng, Magidor and Woodin showed that if there exist two Woodin cardinals, then every universally Baire set is determined. Neeman later improved the hypothesis to one Woodin cardinal. In addition to the following theorem, Feng, Magidor and Woodin showed that $\mathrm{Det}(\undertilde{\Pi}^1_1)$ is equivalent to the statement that every $\undertilde{\Sigma}^1_2$ set of reals is universally Baire.

THEOREM 18 (Feng, Magidor and Woodin [1992].) *Assume* $\mathrm{AD}^{L(\mathbb{R})}$. *Then the following are equivalent.*

- $\mathrm{AD}^{L(\mathbb{R})}$ *holds in every forcing extension.*
- *Every set of reals in* $L(\mathbb{R})$ *is universally Baire.*

Woodin's *Tree Production Lemma* is a powerful means for showing that sets of reals are universally Baire (see [Larson, 2004]). Woodin's proof of Theorem 17 proceeded by applying the lemma to the set $\mathbb{R}^\#$. Informally, the lemma can be interpreted as saying that a set of reals A is δ-universally Baire if for every real r generic for a partial order in V_δ, either r is in the image of A for every $\mathbb{Q}_{<\delta}$-embedding for which r is in the image model, or r is in the image of A for no such embedding.

The partial order $\mathrm{Col}(\omega, {<}\delta)$ consists of all finite partial functions p from $\omega \times \delta$ to δ, with the requirement that $p(n, \alpha) \in \alpha$ for all (n, α) in the domain of p ($\mathrm{Col}(\omega, {<}\delta)$ is ordered by inclusion). If δ is a regular cardinal, then δ is the ω_1 of any forcing extension by $\mathrm{Col}(\omega, {<}\delta)$.

THEOREM 19 (Tree Production Lemma.) *Suppose that δ is a Woodin cardinal. Let ϕ and ψ be binary formulas, and let x and y be arbitrary sets, and assume that the empty condition in the stationary tower $\mathbb{Q}_{<\delta}$ forces that for each real r,*

$$M \models \psi(r, j(y)) \Leftrightarrow V[r] \models \phi(r, x),$$

where $j\colon V \to M$ is the induced elementary embedding. Then there exist trees T,S such that $p[T] = \{r \mid \psi(r, y)\}$ and T and S project to complements after forcing with $\mathrm{Col}(\omega, {<}\delta)$.

Woodin followed this by determining the exact consistency strength of AD. The forward direction of the next theorem (proved in [Koellner and Woodin, 2010]) shows from ZF + AD that there exist infinitely many Woodin cardinals in an inner model of a forcing extension (HOD of the forcing extension with respect to certain parameters) of V. The proof built on a sequence of results, starting with Solovay's theorem that AD implies that ω_1 is a measurable cardinal, which, as mentioned above, also shows that ω_1 (as defined in V) is measurable in the inner model HOD. Becker (see [Becker and Moschovakis, 1981]) had shown that, under AD, ω_1^V is the least measurable in HOD. Becker, Martin, Moschovakis and Steel then showed that under AD + $V{=}L(\mathbb{R})$, $\underset{\sim}{\delta}^2_1$ is β-strong in HOD, where β is the least measurable cardinal greater than $\underset{\sim}{\delta}^2_1$ in HOD.[38] In the 1980's, Woodin showed under the same hypothesis that $\underset{\sim}{\delta}^2_1$ is β-strong in HOD for every $\beta < \Theta$ (and that $\underset{\sim}{\delta}^2_1$ is the least ordinal with this property), and that Θ is Woodin in HOD.

THEOREM 20 (Woodin.) *The following are equiconsistent.*

- ZF + AD
- *There exist infinitely many Woodin cardinals.*

The following theorem illustrates the reverse direction of the equiconsistency (see [Steel, 2009]). It can be seen as a special case of the Derived Model Theorem, discussed in Section 6.1.

THEOREM 21 (Woodin.) *Suppose that λ is a limit of Woodin cardinals, and $G \subseteq \mathrm{Col}(\omega, {<}\lambda)$ is V-generic filter. Let $\mathbb{R}^* = \bigcup\{\mathbb{R}^{V[G\restriction\alpha]} \mid \alpha < \lambda\}$. Then* AD *holds in $L(\mathbb{R}^*)$.*

The results of Section 5.3 illustrate the difficulties in proving the determinacy of Π^1_2 sets. Woodin resolved this problem in 1989. The forward direction of the following theorem is proved in [Koellner and Woodin, 2010]. The proof was inspired in part by a result of Kechris and Solovay [1985], saying that in models of

[38] A cardinal κ is *β-strong* if there is an elementary embedding $j\colon V \to M$ with critical point κ such that $V_\beta \subseteq M$, and *${<}\delta$-strong* if it is β-strong for all $\beta < \delta$.

the form $L[a]$ for $a \subseteq \omega$, Δ^1_2-determinacy implies the determinacy of all ordinal definable sets of reals. Standard arguments show that if Δ^1_2 determinacy holds, then it holds in $L[x]$ for some real x. Woodin showed that if V is $L[x]$ for some real x, and Δ^1_2-determinacy holds, then $\omega_2^{L[x]}$ is a Woodin cardinal in HOD. Recall (from the end of Section 4.2) that Δ^1_2-determinacy and Π^1_2-determinacy are equivalent, by a result of Martin.

THEOREM 22 (Woodin.) *The following are equiconsistent.*

- ZFC + Det(Δ^1_2).
- ZFC + *There exists a Woodin cardinal.*

The following theorem illustrates the reverse direction. Its proof can be found in [Neeman, 2010, p. 1926]. The partial order $\mathrm{Col}(\omega, \delta)$ consists of all finite partial functions from ω to δ, ordered by inclusion.

THEOREM 23 (Woodin.) *If δ is a Woodin cardinal and $G \subseteq \mathrm{Col}(\omega, \delta)$ is a V-generic filter, then Δ^1_2-determinacy holds in $V[G]$.*

6 LATER DEVELOPMENTS

In this final section we briefly review some of the developments that followed the results of the previous section. As discussed in the introduction to this chapter, the set of topics presented here is by no means complete. The first two subsections discuss forms of determinacy ostensibly stronger than AD, in models larger than $L(\mathbb{R})$. The next discusses applications of determinacy to the realm of AC, via producing models of AC by forcing over models of determinacy. In the last two subsections we present some results which derive forms of determinacy from their ostensibly weak consequences, or from statements having no obvious relationship to determinacy. Many of the results of the last two subsections are applications of the study of canonical inner models for large cardinals.

6.1 AD^+ and $\mathrm{AD}_{\mathbb{R}}$

Moschovakis [1981] proved that under AD, if λ is less than Θ, A is a set of functions from ω to λ and A is Suslin and co-Suslin, then the game $G_\omega(A)$ is determined, where here the players play elements of λ. Woodin formulated the following axiom as an attempt to capture the consequences of being an inner model of a model of AD with the same reals and Suslin representations for all sets of reals in the inner model. A set of reals A is said to be ∞-Borel if there is a set of ordinals S and binary formula ϕ such that $A = \{x \in \mathbb{R} \mid L[x, S] \models \phi(x, S)\}$. A Suslin representation is an example of a witness to a set of reals being ∞-Borel.

DEFINITION 24. AD^+ is the conjunction of the following statements.

- $\mathrm{DC}({}^\omega\omega)$

- Every set of reals is ∞-Borel.
- If $\lambda < \Theta$ and $\pi \colon \lambda^\omega \to \omega^\omega$ is a continuous function, then $\pi^{-1}[A]$ is determined for every $A \subseteq \omega^\omega$.

It is an open question whether AD implies AD$^+$, though it is known that AD$^+$ holds in all models of AD of the form $L(A, \mathbb{R})$, where A is a set of reals (some of the details of the argument showing this appear in [Jackson, 2010]). It is not known whether AD$_\mathbb{R}$ implies AD$^+$, though AD$^+$ does follow from AD$_\mathbb{R}$ + DC.

The following consequences of AD$^+$ were announced in [Woodin, 1999].

THEOREM 25 (ZF + DC($^\omega\omega$).) *If* AD$^+$ *holds and* $V = L(\mathcal{P}(\mathbb{R}))$, *then*

- *the pointclass* Σ^2_1 *has the scale property,*
- *every* Σ^2_1 *set of reals is the projection of a tree in* HOD,
- *every true* Σ_1*-sentence is witnessed by a* $\underset{\sim}{\Delta}^2_1$ *set of reals.*

Woodin's *Derived Model Theorem*, proved around 1986, gives a means of producing models of AD$^+$. The model $L(\mathbb{R}^*, Hom^*)$ in the following theorem is said to be a *derived model* (over the ground model). A tree T is said to be $<\lambda$-absolutely complemented if there is a tree S such that $p[T] = \mathbb{R} \setminus p[S]$ in all forcing extensions by partial orders of cardinality less than λ.

THEOREM 26 (Derived Model Theorem; Woodin (see [Steel, 2009]).) *Let* λ *be a limit of Woodin cardinals. Let* $G \subseteq \mathrm{Col}(\omega, <\lambda)$ *be a* V*-generic filter. Let*

- $\mathbb{R}^*$ *be* $\bigcup_{\alpha<\lambda} \mathbb{R}^{V[G\restriction\alpha]}$;
- Hom^* *be the collection of sets of the form* $p[T] \cap \mathbb{R}^*$, *for* T *a* $<\lambda$*-absolutely complemented tree in* V;
- Γ *be the collection of sets of reals* A *in* $V[G]$ *such that* $L(A, \mathbb{R}^*) \models \mathrm{AD}^+$.

Then

- $L(Hom^*, \mathbb{R}^*) \models \mathrm{AD}^+$.
- Hom^* *is the collection of Suslin, co-Suslin sets of reals in* $L(Hom^*, \mathbb{R}^*)$.
- $L(\Gamma, \mathbb{R}^*) \models \mathrm{AD}^+$.
- Hom^* *is the collection of Suslin, co-Suslin sets of reals in* $L(\Gamma, \mathbb{R}^*)$.

The Derived Model Theorem has a converse, also due to Woodin, which says that all models of AD$^+$ arise in this fashion.

THEOREM 27 (Woodin.) *Let* M *be a model of* AD$^+$, *and let* Γ *be the collection of sets of reals which are Suslin, co-Suslin in* M. *Then in a forcing extension of* M *there is an inner model* N *such that* $L(\mathbb{R}^*, \Gamma)$ *is a derived model over* N.

In unpublished work, Woodin has shown that over AD, $\mathrm{AD}_{\mathbb{R}}$ is equivalent to some of its ostensibly weak consequences (see [Woodin, 1999]). The implication from (4) to (1) in the following theorem is due independently to Martin. The implication from (1) to (4) relies heavily on work of Becker [1985].

THEOREM 28 (Woodin.) *Assume* ZF + DC. *Then the following are equivalent.*

1. $\mathrm{AD}_{\mathbb{R}}$
2. AD + *Uniformization*
3. AD + *Every set of reals has a scale*
4. AD + *Every set of reals is Suslin*

Woodin would also produce models of $\mathrm{AD}_{\mathbb{R}}$ from large cardinals.

THEOREM 29 (Woodin.) *Suppose that there exists a cardinal δ of cofinality ω which is a limit of Woodin cardinals and $<\delta$-strong cardinals. Then there is a forcing extension in which there is an inner model containing the reals and satisfying* $\mathrm{AD}_{\mathbb{R}}$.

6.2 Long games

As mentioned above, Blass [1975] and Mycielski showed that determinacy for games of length ω^2 is equivalent to $\mathrm{AD}_{\mathbb{R}}$. For each $n \in \omega$, determinacy for games of length $\omega + n$ is equivalent to AD (think of the game as being divided in two parts, where in the first part (of length ω) the players try to obtain a position from which they have a winning strategy in the second; the winning strategy in the second part can be coded by an integer, and thus uniformly chosen).

Martin and Woodin independently showed that $\mathrm{AD}_{\mathbb{R}}$ is equivalent to determinacy for games of length α for each countable $\alpha \geq \omega^2$. Determinacy for games of length $\omega \cdot 2$ easily gives uniformization. It follows from this and Theorem 28 that $\mathrm{AD}_{\mathbb{R}}$ is equivalent to determinacy for games of length α for each $\alpha \geq \omega \cdot 2$.

While AD does not imply uniformization, the Second Periodicity Theorem shows that PD implies the uniformization of projective sets. It follows that PD is equivalent to PD for games of length less than ω^2. As noted by Neeman [2005], the techniques from this proof can be used to prove the determinacy of games of length ω^2 with analytic payoff from $\mathrm{AD}^{L(\mathbb{R})}$ plus the existence of $\mathbb{R}^{\#}$.

Steel [1988] considered *continuously coded games*, games where each stage of the game is associated with an integer, and the game ends when an associated integer is repeated. Such a game must end after countably many rounds, but runs of the game can have any countable length. Steel proved that ZF + AD + DC + "every set of reals has a scale" + "ω_1 is $\mathcal{P}(\mathbb{R})$-supercompact" implies the determinacy of all continuously coded games.

Extending this result, Neeman [2004; 2005; 2006a] proved a number of determinacy results for games of variable countable length. The proofs of many of these results reduced the determinacy of long games to the iterability of models containing large cardinals.

We give one example. Given $C \subseteq \mathbb{R}^{<\omega_1}$, let $G_{\text{local}}(L, C)$ be the game where players I and II alternate playing natural numbers so as to define elements z_ξ of the Baire space. The game ends at the first γ such that γ is uncountable in $L[z_\xi : \xi < \gamma]$, with I winning if the sequence $\langle z_\xi : \xi < \gamma \rangle$ is in C. It follows from mild large cardinal assumptions (for instance, the existence of the sharp of every subset of ω_1) that γ must be countable.

The existence of M satisfying the hypotheses of the following theorem follows from the existence in V of a sharp for a Woodin limit of Woodin cardinals [Neeman, 2002]. Given a pointclass Γ, a set C consisting of countable sequences of reals is said to be Γ *in the codes* if the set of reals coding members of C (under a suitably definable coding) is in Γ.

THEOREM 30 (Neeman [2004].) *Suppose that there exists an iterable class model M with a cardinal θ such that θ is a Woodin limit of Woodin cardinals in M and countable in V. Then the games $G_{local}(L, C)$ are determined for all C which are $\Game_\omega(<\omega^2{-}\Pi^1_1)$ in the codes.*

Woodin used techniques from the proof of this theorem and his work on AD^+ to prove the following.

THEOREM 31 (Woodin.) *Suppose that there is a Woodin limit of Woodin cardinals and a proper class of inaccessible limits of Woodin cardinals above it. Let A be universally Baire and let Λ be the pointclass of recursive preimages of A. Let C be Λ in the codes. Then $G_{\text{local}}(L, C)$ is determined.*

Again using some of the techniques from the proof of Theorem 30, Woodin proved that if there exists an iterable class model M with a cardinal θ which is a Woodin limit of Woodin cardinals in M and countable in V, then it is consistent that all games on integers of length ω_1 with payoff definable from reals and ordinals are determined (see Exercise 7F.15 of [Neeman, 2004]). Larson and Shelah [2008] showed that it is possible to force that some such game is not determined.

We give one more result of Neeman, proving the determinacy of certain games of length ω_1. In Theorem 32 below, $\mathcal{L}^+$ is the language of set theory with one additional unary predicate. Given an integer k and a sequence $\bar{S}$ of stationary sets indexed by $[\omega_1]^{<k}$, $[\bar{S}]$ is the collection of increasing k-tuples $\langle \alpha_0, \ldots, \alpha_{k-1} \rangle$ from ω_1 such that each initial segment of length $j \leq k$ is in $S_{\langle \alpha_0, \ldots, \alpha_{j-l} \rangle}$. The game $G_{\omega_1,k}(\bar{S}, \phi)$ is a game of length ω_1 in which the players collaborate to build a function $f \colon \omega_1 \to \omega_1$. Then player I wins if there is a club C such that $\langle L_{\omega_1}, r \rangle \models \phi(\alpha_0, \ldots, \alpha_{k-1})$ for all $\langle \alpha_0, \ldots, \alpha_{k-1} \rangle \in [\bar{S}] \cap [C]^k$, and player II wins if there is a club C such that $\langle L_{\omega_1}, r \rangle \models \neg\phi(\alpha_0, \ldots, \alpha_{k-1})$ for all $\langle \alpha_0, \ldots, \alpha_{k-1} \rangle \in [\bar{S}] \cap [C]^k$. Though there can be runs of the game for which neither player wins, determinacy for this game in the sense of Theorem 32 refers to the existence of a strategy for one player or the other that guarantees victory.

The model 0^W is the minimal iterable fine structural inner model M which has a top extender predicate whose critical point is Woodin in M. The existence of such a model is not known to follow from large cardinals.

The last part of the conclusion of Theorem 32 extends a result of Martin, who showed that for any recursive enumeration $\langle B_i : i < \omega \rangle$ of the $<\omega^2$-Π^1_1 sets, the set of i such that I has a winning strategy in $G_\omega(B_i)$ is recursively isomorphic to $0^{\#}$.

THEOREM 32 (Neeman [2007a].) *Suppose that 0^W exists. Let $k < \omega$. Let $\bar{S}$ be a sequence of mutually disjoint stationary sets indexed by $[\omega_1]^{<k}$. Let ϕ be a $\mathcal{L}^+$ formula with k free variables. Then the game $G_{\omega_1,k}(\bar{S}, \phi)$ is determined. Furthermore, the winner of each such game depends only on ϕ and not on $\bar{S}$, and the set of ϕ for which the first player has a winning strategy is recursively equivalent to 0^W.*

If one allows the members of $\bar{S}$ all to be ω_1, then there are undetermined games of this type, as observed by Greg Hjorth (see [Neeman, 2007a]). If one allows the members of $\bar{S}$ all to be ω_1 and changes the winning condition for player II to be simply the negation of the winning condition for I, then one can force from a strongly inaccessible limit of measurable cardinals that some game of this type is not determined [Larson, 2005].

6.3 Forcing over models of determinacy

Steel and Van Wesep [1982] showed that by forcing over a model of $\mathrm{AD}_{\mathbb{R}}$ + "Θ is regular" (the hypothesis they used was actually weaker) one can produce a model of ZFC in which NS_{ω_1} is saturated and $\underset{\sim}{\delta}^1_2 = \omega_2$. This was the first consistency proof of either of these two statements with ZFC. Martin had conjectured that "$\forall n \in \omega\, \underset{\sim}{\delta}^1_n = \aleph_n$" is consistent with ZFC, and this verified the conjecture for the case $n = 2$. Woodin [1983b] subsequently reduced the hypothesis to AD.

Shelah [1998] later showed that it was possible to force the saturation of NS_{ω_1} from a Woodin cardinal. Woodin [1999] proved that the saturation of NS_{ω_1} plus the existence of a measurable cardinal implies that $\underset{\sim}{\delta}^1_2 = \omega_2$. Woodin then turned his proof into a general method for producing models of ZFC by forcing over models of determinacy. The most general form of this method, a partial order called $\mathbb{P}_{\max}$, consists roughly of a directed system containing all countable models of ZFC with a precipitous ideal on ω_1. In the presence of large cardinals, the resulting extension is maximal for Π_2 sentences in $H(\omega_2)$, even with predicates for NS_{ω_1} and each set of reals in $L(\mathbb{R})$. In this model, NS_{ω_1} is saturated and $\underset{\sim}{\delta}^1_2 = \omega_2$. There are many variants of $\mathbb{P}_{\max}$. One of these variants, called $\mathbb{Q}_{\max}$, produces a model in which NS_{ω_1} is $\aleph_1$-dense (i.e., $\mathcal{P}(\omega_1)/\mathrm{NS}_{\omega_1}$ has a dense subset of cardinality $\aleph_1$; this implies saturation), from the assumption that AD holds in $L(\mathbb{R})$. No other method is known for producing a model of ZFC in which NS_{ω_1} is $\aleph_1$-dense.

Steel [1995] showed that under AD, $\mathrm{HOD}^{L(\mathbb{R})}$ is a core model below Θ. Woodin then showed that the entire model $\mathrm{HOD}^{L(\mathbb{R})}$ can be viewed as a core model plus a fragment of its iteration strategy. Using this approach, Steel showed that for every regular $\kappa < \Theta$, the ω-club filter over κ is an ultrafilter in $L(\mathbb{R})$. Woodin used this to show that ω_1 is $<\Theta$-supercompact in $L(\mathbb{R})$, and showed that ω_1 is huge to κ for each measurable $\kappa < \Theta$, improving results of Becker. Neeman [2007b] used this approach to prove, for each $\lambda < \Theta$, the uniqueness of the normal ultrafilter on $\mathcal{P}_{\aleph_1}\lambda$ witnessing the λ-supercompactness of ω_1. Solovay had proved this using determinacy for games on λ, and Harrington and Kechris [1981] proved the determinacy of these games from AD for $\lambda < \underset{\sim}{\delta}^2_1$. Woodin proved the uniqueness of these ultrafilters witnessing supercompactness up to $\underset{\sim}{\delta}^2_1$.

Neeman and Woodin independently used this approach to show that, assuming AD + $V = L(\mathbb{R})$, one could force without adding reals to obtain ZFC + $\underset{\sim}{\delta}^1_n = \omega_2$, for any $n \geq 3$. It is still unknown whether $\underset{\sim}{\delta}^1_n$ can equal ω_n for any $n > 2$ (under ZFC).

6.4 *Determinacy from its consequences*

Woodin [1982] conjectured that Projective Determinacy follows from the statement that all projective sets are Lebesgue measurable, have the Baire property and can be uniformized by projective functions (all consequences of PD). This conjecture was refuted by Steel in 1997. If one requires the uniformization property for the scaled projective pointclasses, then the conjecture is still open. Woodin did prove the following version of the conjecture in the late 1990's. The reverse direction of the theorem used work of Steel in inner model theory.

THEOREM 33 (Woodin.) *Assume* ZF + DC + $V{=}L(\mathbb{R})$. *Then the following are equivalent.*

- AD
- *Every set of reals is Lebesgue measurable and has the property of Baire, and every Σ^2_1 subset of $^2({}^\omega\omega)$ can be uniformized.*

Woodin had proved another equivalence in the early 1980's.

THEOREM 34 (Woodin.) *Assume* ZF + DC + $V{=}L(\mathbb{R})$. *Then the following are equivalent.*

- AD
- *Turing determinacy.*

It is apparently an open question whether AD follows from ZF + DC + $V{=}L(\mathbb{R})$ plus either of (a) for every $\alpha < \Theta$ there is a surjection of ${}^\omega\omega$ onto $\mathcal{P}(\alpha)$; (b) Θ is inaccessible.

Steel [1996] showed that the reduction property for $\underset{\sim}{\Pi}^1_3$ plus the existence of sharps for reals implies that there are inner models with Woodin cardinals.

Determinacy would turn out to be necessary for some of its earliest applications. For instance, Steel [1996] showed that $\underset{\sim}{\Sigma}^1_3$-separation plus the existence of sharps for all reals implies $\underset{\sim}{\Delta}^1_2$-determinacy. Hjorth [1996] showed that $\underset{\sim}{\Pi}^1_2$-determinacy follows from Wadge determinacy for $\underset{\sim}{\Pi}^1_2$ sets. Earlier, Harrington had shown that, for each real x, $\underset{\sim}{\Pi}^1_1(x)$-Wadge determinacy implies that $x^{\#}$ exists.

6.5 Determinacy from other statements

Determinacy axioms such as PD and $\mathrm{AD}^{L(\mathbb{R})}$ imply the consistency of ZFC (plus certain large cardinal statements) and so cannot be proved in ZFC. Empirically, however, these statements appear to follow from every natural statement of sufficient consistency strength. This includes a number of statements ostensibly having little relation to determinacy. In this section we give a few examples of this phenomenon. Most of these arguments use inner model theory, and our presentation relies heavily on [Schimmerling, 2010].

The following theorem shows, among other things, that in the presence of large cardinals, even mere forcing-absoluteness for the theory of $L(\mathbb{R})$ implies $\mathrm{AD}^{L(\mathbb{R})}$. The theorem is due to Steel and Woodin independently (see [Steel, 2002]).

THEOREM 35. *Suppose that κ is a measurable cardinal. Then the following are equivalent.*

- *For all partial orders $\mathbb{P} \in V_\kappa$, the theory of $L(\mathbb{R})$ is not changed by forcing with $\mathbb{P}$.*
- *For all partial orders $\mathbb{P} \in V_\kappa$,* AD *holds in $L(\mathbb{R})$ after forcing with $\mathbb{P}$.*
- *For all partial orders $\mathbb{P} \in V_\kappa$, all sets of reals in $L(\mathbb{R})$ are Lebesgue measurable after forcing with $\mathbb{P}$.*
- *For all partial orders $\mathbb{P} \in V_\kappa$, there is no ω_1-sequence of reals in $L(\mathbb{R})$ after forcing with $\mathbb{P}$.*

As discussed in Section 6.3, Woodin showed using a variation of $\mathbb{P}_{\max}$ that over a model of AD one can force to produce a model of ZFC in which the nonstationary ideal on ω_1 is $\aleph_1$-dense. Woodin developed a technique known as the *core model induction*, an application of descriptive set theory and core model theory, and used it to show that the $\aleph_1$-density of NS_{ω_1} implies $\mathrm{AD}^{L(\mathbb{R})}$. It follows that the two statements are equiconsistent. Steel had previously shown, using core models, that Projective Determinacy follows from CH plus the existence of a homogeneous ideal on ω_1 (a weaker assumption that the $\aleph_1$-density of NS_{ω_1}, which is in turn inconsistent with CH, by a theorem of Shelah).

A sequence $C = \langle C_\alpha : \alpha < \lambda \rangle$ (for some ordinal λ) is called a *coherent sequence* (of length λ) if each C_β is a club subset of β, and $C_\alpha = \alpha \cap C_\beta$ whenever α is a limit point of C_β. A *thread* of such a coherent sequence C is a club set $D \subseteq \lambda$ such that $C_\alpha = \alpha \cap D$ for all limit points α of D. The principle $\square(\lambda)$ says that there is a coherent sequence of length λ with no thread. The principle $\square_\kappa$ says that there is a coherent sequence C of length κ^+ such that the ordertype of C_α is at most κ, for each limit $\alpha < \lambda$. The principles $\square_\kappa$ were isolated in the 1960's by Jensen [1972], who showed that $\square_\kappa$ holds in L for all infinite cardinals κ (see [Devlin, 1984, p. 141]). The principles $\square(\lambda)$ were later extracted from Jensen's work by Stevo Todorcevic.

Todorcevic [1984] showed that the Proper Forcing Axiom (PFA) implies that $\square(\kappa)$ fails at all cardinals κ of cofinality at least ω_2, from which it follows that $\square_\kappa$ fails for all uncountable cardinals. The failure of these square principles implies the failure of covering theorems for certain inner models, from which one can derive inner models with large cardinals. Using this general approach, Ernest Schimmerling [1995] proved that PFA implies $\utilde{\Delta}^1_2$-determinacy. Woodin extended this proof to show that PFA implies PD. Woodin also showed that PFA plus the existence of a strongly inaccessible cardinal implies $\mathrm{AD}^{L(\mathbb{R})}$, using the core model induction. Roughly, the idea is to inductively work through the Wadge degrees to build canonical inner models which are correct for each Wadge class. The induction works through the gap structure highlighted in [Steel, 1983].

Alessandro Andretta, Neeman and Steel [2001] showed that PFA plus the existence of a measurable cardinal implies the existence of a model of $\mathrm{AD}_\mathbb{R}$ containing all the reals and ordinals. Steel [2005] showed that if $\square_\kappa$ fails for a singular strong limit cardinal κ, then AD holds in $L(\mathbb{R})$.

The following theorem is due to Schimmerling [2007]. Steel later strengthened the conclusion to $\mathrm{AD}^{L(\mathbb{R})}$.

THEOREM 36. *If* $\kappa \geq \max\{\aleph_2, c\}$ *and* $\square(\kappa)$ *and* $\square_\kappa$ *fail, then Projective Determinacy holds.*

Todorcevic (see [Bekkali, 1991]) and Boban Veličković [1992] showed that PFA implies that $2^{\aleph_0} = 2^{\aleph_1} = \aleph_2$. This gives another route towards showing that PFA implies that the AD holds in $L(\mathbb{R})$.

Schimmerling and Martin Zeman used the core model induction to prove the following theorem [2001]. They had previously derived Projective Determinacy from the failure of a weaker version of $\square_\kappa$ at a weakly compact cardinal; Woodin had then derived $\mathrm{AD}^{L(\mathbb{R})}$ from the same hypothesis.

THEOREM 37. *If* κ *is a weakly compact cardinal and* $\square_\kappa$ *fails, then* AD *holds in* $L(\mathbb{R})$.

Steel [1996] showed that if NS_{ω_1} is saturated and there is a measurable cardinal, then $\utilde{\Delta}^1_2$-determinacy holds. The hypothesis of the measurable cardinal was later removed through work of Jensen and Steel. In 2000, Steel and Zoble derived $\mathrm{AD}^{L(\mathbb{R})}$ from another principle of Todorcevic, known as the *Strong Reflection Principle at* ω_2.

We conclude with one final example. As mentioned in Section 2.3, Gitik showed that if there is a proper class of strongly compact cardinals, then there is a model of ZF in which all infinite cardinals have cofinality ω. Using the core model induction, Daniel Busche and Ralf Schindler [2009] showed that this statement implies that AD holds in the $L(\mathbb{R})$ of a forcing extension of HOD, for some partial order in HOD.

ACKNOWLEDGEMENTS

The author is supported in part by NSF grant DMS-0801009.

Gunter Fuchs helped with some original sources in German. The author would like to thank Akihiro Kanamori, Alexander Kechris, Tony Martin, Jan Mycielski, Robert Solovay and John Steel for making many helpful suggestions.

BIBLIOGRAPHY

[Addison and Moschovakis, 1968] J. Addison and Y. Moschovakis. Some consequences of the axiom of definable determinateness. *Proceedings of the National Academy of Sciences U.S.A.*, 59:708–712, 1968.

[Addison, 1958] J. Addison. Some consequences of the axiom of constructibility. *Fundamenta Mathematicae*, 46:337–357, 1958.

[Addison, 1959] J. Addison. Separation principles in the hierarchies of classical and effective descriptive set theory. *Fundamenta Mathematicae*, 46:123–135, 1959.

[Albers and Alexanderson, 1985] D. Albers and G. Alexanderson, editors. *Mathematical People.* Birkhäuser Boston Inc., Boston, MA, 1985. Profiles and interviews.

[Andretta *et al.*, 2001] A. Andretta, I. Neeman, and J. Steel. The domestic levels of K^c are iterable. *Israel Journal of Mathematics*, 125:157–201, 2001.

[Banach and Tarski, 1924] S. Banach and A. Tarski. Sur la décomposition des ensembles de points en parties respectivement congruentes. *Fundamenta Mathematicae*, 6:244–277, 1924.

[Becker and Kechris, 1984] H. Becker and A. Kechris. Sets of ordinals constructible from trees and the third Victoria Delfino problem. In *Axiomatic Set Theory*, volume 31 of *Contemporary Mathematics*, pages 13–29. American Mathematical Society, Providence, 1984.

[Becker and Moschovakis, 1981] H. Becker and Y. Moschovakis. Measurable cardinals in playful models. In *Cabal Seminar 77–79 (Proc. Caltech-UCLA Logic Sem., 1977–79)*, volume 839 of *Lecture Notes in Mathematics*, pages 203–214. Springer, Berlin, 1981.

[Becker, 1978] H. Becker. Partially playful universes. In *Cabal Seminar 76–77 (Proc. Caltech-UCLA Logic Sem., 1976–77)*, volume 689 of *Lecture Notes in Mathematics*, pages 55–90. Springer, Berlin, 1978.

[Becker, 1985] H. Becker. A property equivalent to the existence of scales. *Transactions of the American Mathematical Society*, 287(2):591–612, 1985.

[Bekkali, 1991] M. Bekkali. *Topics in Set Theory*, volume 1476 of *Lecture Notes in Mathematics.* Springer-Verlag, Berlin, 1991. Lebesgue measurability, large cardinals, forcing axioms, rho-functions, Notes on lectures by Stevo Todorčević.

[Blackwell, 1967] D. Blackwell. Infinite games and analytic sets. *Proceedings of the National Academy of Sciences U.S.A.*, 58:1836–1837, 1967.

[Blackwell, 1969] D. Blackwell. Infinite G_δ-games with imperfect information. *Zastosowania Matematyki*, 10:99–101, 1969.

[Blass, 1975] A. Blass. Equivalence of two strong forms of determinacy. *Proceedings of the American Mathematical Society*, 52:373–376, 1975.

[Brouwer, 1924] L. Brouwer. Beweis dass jede volle Funktion gleichmässig stetig ist. *Koninklijke Nederlandse Akademie van Wetenschappen, Amsterdam*, 27:189–193, 1924.

[Busche and Schindler, 2009] D. Busche and R. Schindler. The strength of choiceless patterns of singular and weakly compact cardinals. *Annals of Pure and Applied Logic*, 159(1-2):198–248, 2009.

[Cooper, 2004] S. B. Cooper. *Computability Theory.* Chapman & Hall/CRC, Boca Raton, FL, 2004.

[Davis, 1964] M. Davis. Infinite games of perfect information. In *Advances in Game Theory*, pages 85–101. Princeton Univversity Press, Princeton, N.J., 1964.

[Devlin, 1984] K. J. Devlin. *Constructibility.* Perspectives in Mathematical Logic. Springer-Verlag, Berlin, 1984.

[Dodd, 1982] A. J. Dodd. *The Core Model*, volume 61 of *London Mathematical Society Lecture Note Series.* Cambridge University Press, Cambridge, 1982.

[DuBose, 1990] D. DuBose. The equivalence of determinancy and iterated sharps. *The Journal of Symbolic Logic*, 55(2):502–525, 1990.

[Erdős and Hajnal, 1958] P. Erdős and A. Hajnal. On the structure of set mappings. *Acta Mathematica Academiae Scientiarum Hungaricae*, 9:111–131, 1958.

[Erdős and Hajnal, 1966] P. Erdős and A. Hajnal. On a problem of B. Jónsson. *Bulletin d'Académie Polonaise des Sciences, Série des Sciences Mathématiques, Astronomiques et Physiques*, 14:19–23, 1966.

[Feferman and Lévy, 1963] S. Feferman and A. Lévy. Independence results in set theory by Cohen's method II. *Notices of the American Mathematical Society*, 10:593, 1963. Abstract.

[Feng *et al.*, 1992] Q. Feng, M. Magidor, and W. H. Woodin. Universally Baire sets of reals. In *Set Theory of the Continuum (Berkeley, CA, 1989)*, volume 26 of *Mathematical Sciences Research Institute Publications*, pages 203–242. Springer-Verlag, New York, 1992.

[Foreman *et al.*, 1988] M. Foreman, M. Magidor, and S. Shelah. Martin's Maximum, saturated ideals, and nonregular ultrafilters. I. *Annals of Mathematics*, 127(1):1–47, 1988.

[Foreman, 1986] M. Foreman. Potent axioms. *Transactions of the American Mathematical Society*, 294(1):1–28, 1986.

[Friedman, 1970] H. Friedman. Higher set theory and mathematical practice. *Annals of Mathematical Logic*, 2(3):325–357, 1970.

[Friedman, 1971] H. Friedman. Determinateness in the low projective hierarchy. *Fundamenta Mathematicae*, 72(1):79–95. (errata insert), 1971.

[Gale and Stewart, 1953] D. Gale and F. Stewart. Infinite games with perfect information. In *Contributions to the theory of games, vol. 2*, Annals of Mathematics Studies, no. 28, pages 245–266. Princeton University Press, Princeton, N. J., 1953.

[Gitik, 1980] M. Gitik. All uncountable cardinals can be singular. *Israel Journal of Mathematics*, 35(1-2):61–88, 1980.

[Green, 1978] J. Green. *Determinacy and the existence of large measurable cardinals.* PhD thesis, University of California at Berkeley, 1978.

[Hadamard, 1905] J. Hadamard. Cinq letters sur la théorie des ensembles. *Bulletin de la Societé mathématique de France*, 33:261–273, 1905.

[Hajnal, 1956] A. Hajnal. On a consistency theorem connected with the generalized continuum problem. *Zeitschrift für Mathematische Logik und Grundlagen der Mathematik*, 2:131–136, 1956.

[Hajnal, 1961] A. Hajnal. On a consistency theorem connected with the generalized continuum problem. *Acta Mathematica Academiae Scientiarum Hungaricae*, 12:321–376, 1961.

[Halmos, 1950] P. Halmos. *Measure Theory.* D. Van Nostrand Company, Inc., New York, N. Y., 1950.

[Harrington and Kechris, 1981] L. Harrington and A. Kechris. On the determinacy of games on ordinals. *Annals of Mathematical Logic*, 20:109–154, 1981.

[Harrington, 1978] L. Harrington. Analytic determinacy and $0^\sharp$. *The Journal of Symbolic Logic*, 43(4):685–693, 1978.

[Hausdorff, 1908] F. Hausdorff. Grundzüge einer Theorie der geordneten Mengen. *Mathematische Annalen*, 65:435–505, 1908.

[Hausdorff, 1914] F. Hausdorff. Bemerkung über den Inhalt von Punktmengen. *Mathematische Annalen*, 75:428–434, 1914.

[Henle *et al.*, 1985] J. Henle, A. Mathias, and W. H. Woodin. A barren extension. In *Methods in Mathematical Logic (Caracas, 1983)*, volume 1130 of *Lecture Notes in Mathematics*, pages 195–207. Springer, Berlin, 1985.

[Hjorth, 1996] G. Hjorth. $\mathbf{\Pi}^1_2$ Wadge degrees. *Annals of Pure and Applied Logic*, 77(1):53–74, 1996.

[Howard and Rubin, 1998] P. Howard and J. Rubin. *Consequences of the Axiom of Choice*, volume 59 of *Mathematical Surveys and Monographs.* American Mathematical Society, 1998.

[Jackson, 1988] S. Jackson. AD and the projective ordinals. In *Cabal Seminar 81–85*, volume 1333 of *Lecture Notes in Mathematics*, pages 117–220. Springer, Berlin, 1988.

[Jackson, 1999] S. Jackson. A computation of δ^1_5. *Memoirs of the American Mathematical Society*, 140(670):viii+94, 1999.

[Jackson, 2010] S. Jackson. Structural Consequences of AD. In M. Foreman and A. Kanamori, editors, *Handbook of Set Theory.* Springer, 2010.

[Jech, 2003] T. Jech. *Set Theory.* Springer Monographs in Mathematics. Springer-Verlag, Berlin, 2003. The third millennium edition, revised and expanded.

[Jensen, 1972] R. Jensen. The fine structure of the constructible hierarchy. *Annals of Mathematical Logic*, 4:229–308; erratum, ibid. 4 (1972), 443, 1972. With a section by Jack Silver.

[Kalmár, 1928] L. Kalmár. Zur theorie der abstrakten Spiele. *Acta Scientiarum Mathematicarum Szeged*, 4:65–85, 1928. 1928 -29.

[Kanamori, 1995] A. Kanamori. The emergence of descriptive set theory. In *From Dedekind to Gödel (Boston, MA, 1992)*, volume 251 of *Synthese Library*, pages 241–262. Kluwer Acad. Publ., Dordrecht, 1995.

[Kanamori, 2003] A. Kanamori. *The Higher Infinite.* Springer Monographs in Mathematics. Springer-Verlag, Berlin, second edition, 2003. Large cardinals in set theory from their beginnings.

[Kechris and Moschovakis, 1972] A. Kechris and Y. Moschovakis. Two theorems about projective sets. *Israel Journal of Mathematics*, 72:391–399, 1972.

[Kechris and Moschovakis, 1978a] A. Kechris and Y. Moschovakis, editors. *Cabal Seminar 76–77*, volume 689 of *Lecture Notes in Mathematics*, Berlin, 1978. Springer.

[Kechris and Moschovakis, 1978b] A. Kechris and Y. Moschovakis. Notes on the theory of scales. In *Cabal Seminar 76–77 (Proc. Caltech-UCLA Logic Sem., 1976–77)*, volume 689 of *Lecture Notes in Mathematics*, pages 1–54. Springer, Berlin, 1978. Reprinted in [Kechris *et al.*, 2008].

[Kechris and Solovay, 1985] A. Kechris and R. Solovay. On the relative consistency strength of determinacy hypotheses. *Transactions of the American Mathematical Society*, 290(1):179–211, 1985.

[Kechris and Woodin, 1983] A. Kechris and W. H. Woodin. Equivalence of partition properties and determinacy. *Proceedings of the National Academy of Sciences U.S.A.*, 80(6 i.):1783–1786, 1983.

[Kechris *et al.*, 1981a] A. Kechris, E. Kleinberg, Y. Moschovakis, and W. H. Woodin. The axiom of determinacy, strong partition properties and nonsingular measures. In *Cabal Seminar 77–79 (Proc. Caltech-UCLA Logic Sem., 1977–79)*, volume 839 of *Lecture Notes in Mathematics*, pages 75–99. Springer, Berlin, 1981.

[Kechris *et al.*, 1981b] A. Kechris, D. A. Martin, and Y. Moschovakis, editors. *Cabal Seminar 77–79*, volume 839 of *Lecture Notes in Mathematics*, Berlin, 1981. Springer.

[Kechris *et al.*, 1981c] A. Kechris, R. Solovay, and J. R. Steel. The axiom of determinacy and the prewellordering property. In *Cabal Seminar 77–79 (Proc. Caltech-UCLA Logic Sem., 1977–79)*, volume 839 of *Lecture Notes in Mathematics*, pages 101–125. Springer, Berlin, 1981.

[Kechris *et al.*, 1983] A. Kechris, D. A. Martin, and Y. Moschovakis, editors. *Cabal Seminar 79–81*, volume 1019 of *Lecture Notes in Mathematics*, Berlin, 1983. Springer-Verlag.

[Kechris *et al.*, 1988] A. Kechris, D. A. Martin, and J. Steel, editors. *Cabal Seminar 81–85*, volume 1333 of *Lecture Notes in Mathematics*, Berlin, 1988. Springer-Verlag.

[Kechris *et al.*, 2008] A. Kechris, B. Löwe, and J. Steel, editors. *Games, Scales, and Suslin cardinals. The Cabal Seminar. Vol. I*, volume 31 of *Lecture Notes in Logic*. Association for Symbolic Logic, Chicago, IL, 2008.

[Kechris, 1974] A. Kechris. On projective ordinals. *The Journal of Symbolic Logic*, 39:269–282, 1974.

[Kechris, 1975] A. Kechris. The theory of countable analytical sets. *Transactions of the American Mathematical Society*, 202:259–297, 1975.

[Kechris, 1978] A. Kechris. AD and projective ordinals. In *Cabal Seminar 76–77 (Proc. Caltech-UCLA Logic Sem., 1976–77)*, volume 689 of *Lecture Notes in Mathematics*, pages 91–132. Springer, Berlin, 1978.

[Kechris, 1981] A. Kechris. Homogeneous trees and projective scales. In *Cabal Seminar 77–79 (Proc. Caltech-UCLA Logic Sem., 1977–79)*, volume 839 of *Lecture Notes in Mathematics*, pages 33–73. Springer, Berlin, 1981.

[Kechris, 1984] A. Kechris. The axiom of determinacy implies dependent choices in $L(\mathbf{R})$. *The Journal of Symbolic Logic*, 49(1):161–173, 1984.

[Kechris, 1995] A. Kechris. *Classical Descriptive Set Theory*, volume 156 of *Graduate Texts in Mathematics*. Springer-Verlag, New York, 1995.

[Kőnig, 1927] D. Kőnig. Über eine Schlusswiese aus dem endlichen ins Unendliche. *Acta Scientiarum Mathematicarum Szeged*, 3:121–130, 1927.

[Kleene, 1938] S.. Kleene. On notation for ordinal numbers. *The Journal of Symbolic Logic*, 3:150–155, 1938.

[Kleene, 1943] S. Kleene. Recursive predicates and quantifiers. *Transactions of the American Mathematical Society*, 53:41–73, 1943.

[Kleene, 1955a] S. Kleene. Arithmetical predicates and function quantifiers. *Transactions of the American Mathematical Society*, 79:312–340, 1955.

[Kleene, 1955b] S. Kleene. Hierarchies of number-theoretic predicates. *Bulletin of the American Mathematical Society*, 61:193–213, 1955.

[Kleene, 1955c] S. Kleene. On the forms of the predicates in the theory of constructive ordinals. II. *American Journal of Mathematics*, 77:405–428, 1955.

[Kleinberg, 1970] E. Kleinberg. Strong partition properties for infinite cardinals. *The Journal of Symbolic Logic*, 35:410–428, 1970.

[Koellner and Woodin, 2010] P. Koellner and W. H. Woodin. Large cardinals from determinacy. In M. Foreman and A. Kanamori, editors, *Handbook of Set Theory*. Springer, 2010.

[Kondo, 1938] M. Kondo. Sur l'uniformization des complementaires analytiques et les ensembles projectifs des las second classe. *Japanese Journal of Mathematics*, 15:197–230, 1938.

[Kunen, 1970] K. Kunen. Some applications of iterated ultrapowers in set theory. *Annals of Mathematical Logic*, 1:179–227, 1970.

[Kunen, 1971] K. Kunen. Elementary embeddings and infinitary combinatorics. *The Journal of Symbolic Logic*, 36:407–413, 1971.

[Kunen, 1978] K. Kunen. Saturated ideals. *The Journal of Symbolic Logic*, 43(1):65–76, 1978.

[Kunen, 1983] K. Kunen. *Set Theory*, volume 102 of *Studies in Logic and the Foundations of Mathematics*. North-Holland Publishing Co., Amsterdam, 1983. An introduction to independence proofs, Reprint of the 1980 original.

[Kunen, March, 1971] K. Kunen. A remark on Moschovakis's uniformization theorem. Circulated note, March, 1971.

[Kunen, September, 1971a] K. Kunen. Some more singular cardinals. Circulated note, September, 1971.

[Kunen, September, 1971b] K. Kunen. Some singular cardinals. Circulated note, September, 1971.

[Kuratowski, 1936] K. Kuratowski. Sur les théorèmes de séparation dans la théorie des ensembles. *Fundamenta Mathematicae*, 26:183–191, 1936.

[Larson and Shelah, 2008] P. Larson and S. Shelah. The stationary set splitting game. *Mathematical Logic Quarterly*, 54(2):187–193, 2008.

[Larson, 2004] P. Larson. *The Stationary Tower*, volume 32 of *University Lecture Series*. American Mathematical Society, Providence, RI, 2004. Notes on a course by William Hugh Woodin.

[Larson, 2005] P. Larson. The canonical function game. *Archive for Mathematical Logic*, 44(7):817–827, 2005.

[Lebesgue, 1905] H. Lebesgue. Sur les fonctions représentables analytiquement. *Journal de Matheématiques Pures et Appliquées*, 1:139–216, 1905.

[Lévy, 1957] A. Lévy. Indépendance conditionnelle de $V = L$ et d'axiomes qui se rattachent au système de M. Gödel. *Comptes Rendus Hebdomadaires des Séances de l'Académie des Sciences, Paris*, 245:1582–1583, 1957.

[Lévy, 1960] A. Lévy. A generalization of Gödel's notion of constructibility. *The Journal of Symbolic Logic*, 25:147–155, 1960.

[Lévy, 1965a] A. Lévy. Definability in axiomatic set theory. I. In *Logic, Methodology and Philos. Sci. (Proc. 1964 Internat. Congr.)*, pages 127–151. North-Holland, Amsterdam, 1965.

[Lévy, 1965b] A. Lévy. A hierarchy of formulas in set theory. *Memoirs of the American Mathematical Society*, 57:76, 1965.

[Lévy, 1979] A. Lévy. *Basic Set Theory*. Springer-Verlag, Berlin, 1979.

[Louveau and Saint-Raymond, 1987] A. Louveau and J. Saint-Raymond. Borel classes and closed games: Wadge-type and Hurewicz-type results. *Transactions of the American Mathematical Society*, 304(2):431–467, 1987.

[Louveau and Saint-Raymond, 1988] A. Louveau and J. Saint-Raymond. The strength of Borel Wadge determinacy. In *Cabal Seminar 81–85*, volume 1333 of *Lecture Notes in Mathematics*, pages 1–30. Springer, Berlin, 1988.

[Luzin and Novikov, 1935] N. Luzin and P. Novikov. Choix effectif d'un point dans un complementaire analytique arbitraire, donne par un crible. *Fundamenta Mathematicae*, 25:559–560, 1935.

[Luzin and Sierpiński, 1918] N. Luzin and W. Sierpiński. Sur quelques propriétés des ensembles (A). *Bulletin de l'Académie des Sciences Cracovie, Classe des Sciences Mathématiques, Série A*, pages 35–48, 1918.

[Luzin and Sierpiński, 1923] N. Luzin and W. Sierpiński. Sur un ensemble non measurable B. *Journal de Mathématiques Pures et Appliqueées*, 2(9):53–72, 1923.

[Luzin, 1925a] N. Luzin. Les proprietes des ensembles projectifs. *Comptes Rendus Hebdomadaires des Séances de l'Académie des Sciences, Paris*, 180:1817–1819, 1925.

[Luzin, 1925b] N. Luzin. Sur les ensembles projectifs de M. Henri Lebesgue. *Comptes Rendus Hebdomadaires des Séances de l'Académie des Sciences, Paris*, 180:1318–1320, 1925.

[Luzin, 1925c] N. Luzin. Sur un problème de M. Emile Borel et les ensembles projectifs de M. Henri Lebesgue: les ensembles analytiques. *Comptes Rendus Hebdomadaires des Séances de l'Académie des Sciences, Paris*, 164:91–94, 1925.

[Luzin, 1927] N. Luzin. Sur les ensembles analytiques. *Fundamenta Mathematicae*, 10:1–95, 1927.

[Luzin, 1930a] N. Luzin. Analogies entre les ensembles mesurables B et les ensembles analytiques. *Fundamenta Mathematicae*, 16:48–76, 1930.

[Luzin, 1930b] N. Luzin. Sur le problème de M. J. Hadamard d'uniformisation des ensembles. *Comptes Rendus Hebdomadaires des Séances de l'Académie des Sciences, Paris*, 190:349–351, 1930.

[Magidor, 1980] M. Magidor. Precipitous ideals and $\mathbf{\Sigma}^1_4$ sets. *Israel Journal of Mathematics*, 35(1-2):109–134, 1980.

[Mansfield, 1970] R. Mansfield. Perfect subsets of definable sets of real numbers. *Pacific Journal of Mathematics*, 35:451–457, 1970.

[Mansfield, 1971] R. Mansfield. A Souslin operation for Π^1_2. *Israel Journal of Mathematics*, 9:367–379, 1971.

[Martin and Paris, March, 1971] D. A. Martin and J. Paris. AD $\to \exists$ exactly 2 normal measures on ω_2, March, 1971.

[Martin and Solovay, 1969] D. A. Martin and R. Solovay. A basis theorem for $\mathbf{\Sigma}^1_3$ sets of reals. *Annals of Mathematics*, 89:138–159, 1969.

[Martin and Steel, 1983] D. A. Martin and J. Steel. The extent of scales in $L(\mathbf{R})$. In *Cabal Seminar 79–81*, volume 1019 of *Lecture Notes in Mathematics*, pages 86–96. Springer, Berlin, 1983. Reprinted in [Kechris *et al.*, 2008].

[Martin and Steel, 1989] D. A. Martin and J. Steel. A proof of projective determinacy. *Journal of the American Mathematical Society*, 2(1):71–125, 1989.

[Martin and Steel, 1994] D. A. Martin and J. Steel. Iteration trees. *Journal of the American Mathematical Society*, 7(1):1–73, 1994.

[Martin *et al.*, 1982] D. A. Martin, Y. Moschovakis, and J. Steel. The extent of definable scales. *Bulletin of the American Mathematical Society*, 6:435–440, 1982.

[Martin *et al.*, 2003] D. A. Martin, I. Neeman, and M. Vervoort. The strength of Blackwell determinacy. *The Journal of Symbolic Logic*, 68(2):615–636, 2003.

[Martin, 1968] D. A. Martin. The axiom of determinateness and reduction principles in the analytical hierarchy. *Bulletin of the American Mathematical Society*, 74:687–689, 1968.

[Martin, 1970] D. A. Martin. Measurable cardinals and analytic games. *Fundamenta Mathematicae*, 66:440–442, 1970.

[Martin, 1975] D. A. Martin. Borel determinacy. *Annals of Mathematics*, 102(2):363–371, 1975.

[Martin, 1980] D. A. Martin. Infinite games. In *Proceedings of the International Congress of Mathematicians (Helsinki, 1978)*, pages 269–273, Helsinki, 1980. Acad. Sci. Fennica.

[Martin, 1983] D. A. Martin. The real game quantifier propagates scales. In *Cabal Seminar 79–81*, volume 1019 of *Lecture Notes in Mathematics*, pages 157–171. Springer, Berlin, 1983. Reprinted in [Kechris *et al.*, 2008].

[Martin, 1985] D. A. Martin. A purely inductive proof of Borel determinacy. In *Recursion Theory (Ithaca, N.Y., 1982)*, volume 42 of *Proceedings of Symposia in Pure Mathematics*, pages 303–308. Amer. Math. Soc., Providence, RI, 1985.

[Martin, 1990] D. A. Martin. An extension of Borel determinacy. *Annals of Pure and Applied Logic*, 49(3):279–293, 1990.

[Martin, 1998] D. A. Martin. The determinacy of Blackwell games. *The Journal of Symbolic Logic*, 63(4):1565–1581, 1998.

[Martin, 2003] D. A. Martin. A simple proof that determinacy implies Lebesgue measurability. *Rendiconti del Seminario Matematico Università e Politecnico de Torino*, 61(4):393–397, 2003.

[Mathias, 1968] A. Mathias. On a generalization of Ramsey's theorem. *Notices of the American Mathematical Society*, 15:931, 1968.

[Mathias, 1977] A. Mathias. Happy families. *Annals of Pure and Applied Logic*, 12(1):59–111, 1977.

[Mauldin, 1981] R. Mauldin, editor. *The Scottish Book*. Birkhäuser Boston, Mass., 1981. Mathematics from the Scottish Café, Including selected papers presented at the Scottish Book Conference held at North Texas State University, Denton, Tex., May 1979.

[Mitchell, 1979] W. Mitchell. Hypermeasurable cardinals. In *Logic Colloquium '78 (Mons, 1978)*, volume 97 of *Studies in Logic and the Foundations of Mathematics*, pages 303–316. North-Holland, Amsterdam, 1979.

[Moschovakis, 1967] Y. Moschovakis. Hyperanalytic predicates. *Transactions of the American Mathematical Society*, 129:248–282, 1967.

[Moschovakis, 1969] Y. Moschovakis. Abstract first order computability I and II. *Transactions of the American Mathematical Society*, 138:427–504, 1969.

[Moschovakis, 1970] Y. Moschovakis. Determinacy and prewellorderings of the continuum. In *Mathematical Logic and Foundations of Set Theory (Proc. Internat. Colloq., Jerusalem, 1968)*, pages 24–62. North-Holland, Amsterdam, 1970.

[Moschovakis, 1971] Y. Moschovakis. Uniformization in a playful universe. *Bulletin of the American Mathematical Society*, 77:731–736, 1971.

[Moschovakis, 1973] Y. Moschovakis. Analytical definability in a playful universe. In *Logic, Methodology and Philosophy of Science IV (Proc. Fourth Internat. Congr., Bucharest, 1971)*, pages 77–85. Studies in Logic and Foundations of Math., Vol. 74. North-Holland, Amsterdam, 1973.

[Moschovakis, 1978] Y. Moschovakis. Inductive scales on inductive sets. In *Cabal Seminar 76–77 (Proc. Caltech-UCLA Logic Sem., 1976–77)*, volume 689 of *Lecture Notes in Mathematics*, pages 185–192. Springer, Berlin, 1978. Reprinted in [Kechris *et al.*, 2008].

[Moschovakis, 1980] Y. Moschovakis. *Descriptive Set Theory*, volume 100 of *Studies in Logic and the Foundations of Mathematics*. North-Holland Publishing Co., Amsterdam, 1980.

[Moschovakis, 1981] Y. Moschovakis. Ordinal games and playful models. In *Cabal Seminar 77–79 (Proc. Caltech-UCLA Logic Sem., 1977–79)*, volume 839 of *Lecture Notes in Mathematics*, pages 169–201. Springer, Berlin, 1981.

[Moschovakis, 1983] Y. Moschovakis. Scales on coinductive sets. In *Cabal Seminar 79–81*, volume 1019 of *Lecture Notes in Mathematics*, pages 77–85. Springer, Berlin, 1983.

[Moschovakis, 2009] Y. Moschovakis. *Descriptive Set Theory*, volume 155 of *Mathematical Surveys and Monographs*. American Mathematical Society, Providence, RI, 2009.

[Mycielski and Steinhaus, 1962] J. Mycielski and H. Steinhaus. A mathematical axiom contradicting the axiom of choice. *Bulletin de l'Académie Polonaise des Sciences. Série des Sciences Mathématiques, Astronomiques et Physiques*, 10:1–3, 1962.

[Mycielski and Świerczkowski, 1964] J. Mycielski and S. Świerczkowski. On the Lebesgue measurability and the axiom of determinateness. *Fundamenta Mathematicae*, 54:67–71, 1964.

[Mycielski, 1963] J. Mycielski. On the axiom of determinateness. *Fundamenta Mathematicae*, 53:205–224, 1963. 1963 -64.

[Mycielski, 1966] J. Mycielski. On the axiom of determinateness. II. *Fundamenta Mathematicae*, 59:203–212, 1966.

[Neeman, 1995] I. Neeman. Optimal proofs of determinacy. *The Bulletin of Symbolic Logic*, 1:327–339, 1995.

[Neeman, 2000] I. Neeman. Unraveling $\mathbf{\Pi}^1_1$ sets. *Annals of Pure and Applied Logic*, 106(1-3):151–205, 2000.

[Neeman, 2002] I. Neeman. Inner models in the region of a Woodin limit of Woodin cardinals. *Annals of Pure and Applied Logic*, 116(1-3):67–155, 2002.

[Neeman, 2004] I. Neeman. *The Determinacy of Long Games*, volume 7 of *de Gruyter Series in Logic and its Applications*. Walter de Gruyter GmbH & Co. KG, Berlin, 2004.

[Neeman, 2005] I. Neeman. An introduction to proofs of determinacy of long games. In *Logic Colloquium '01*, volume 20 of *Lecture Notes in Logic*, pages 43–86. Assoc. Symbol. Logic, Urbana, IL, 2005.

[Neeman, 2006a] I. Neeman. Determinacy for games ending at the first admissible relative to the play. *The Journal of Symbolic Logic*, 71(2):425–459, 2006.

[Neeman, 2006b] I. Neeman. Unraveling Π^1_1 sets, revisited. *Israel Journal of Mathematics*, 152:181–203, 2006.

[Neeman, 2007a] I. Neeman. Games of length ω_1. *Journal of Mathematical Logic*, 7(1):83–124, 2007.

[Neeman, 2007b] I. Neeman. Inner models and ultrafilters in $L(\mathbb{R})$. *The Bulletin of Symbolic Logic*, 13(1):31–53, 2007.

[Neeman, 2010] I. Neeman. Determinacy from large cardinals. In M. Foreman and A. Kanamori, editors, *Handbook of Set Theory*. Springer, 2010.

[Novikov, 1935] P. Novikov. Sur la séparabilité des ensembles projectifs de seconde class. *Fundamenta Mathematicae*, 25:459–466, 1935.

[Oxtoby, 1980] J. Oxtoby. *Measure and Category*, volume 2 of *Graduate Texts in Mathematics*. Springer-Verlag, New York, second edition, 1980. A survey of the analogies between topological and measure spaces.

[Paris, 1972] J. Paris. ZF $\vdash \Sigma^0_4$ determinateness. *The Journal of Symbolic Logic*, 37:661–667, 1972.

[Prikry, 1976] K. Prikry. Determinateness and partitions. *Proceedings of the American Mathematical Society*, 54:303–306, 1976.

[Ramsey, 1930] F. Ramsey. On a problem of formal logic. *Proceedings of the London Mathematical Society*, 30(2):2–24, 1930.

[Sacks, 1976] G. Sacks. Countable admissible ordinals and hyperdegrees. *Advances in Mathematics*, 20(2):213–262, 1976.

[Schimmerling and Zeman, 2001] E. Schimmerling and M. Zeman. Square in core models. *The Bulletin of Symbolic Logic*, 7(3):305–314, 2001.

[Schimmerling, 1995] E. Schimmerling. Combinatorial principles in the core model for one Woodin cardinal. *Annals of Pure and Applied Logic*, 74(2):153–201, 1995.

[Schimmerling, 2007] E. Schimmerling. Coherent sequences and threads. *Advances in Mathematics*, 216(1):89–117, 2007.

[Schimmerling, 2010] E. Schimmerling. A core model toolbox and guide. In M. Foreman and A. Kanamori, editors, *Handbook of Set Theory*. Springer, 2010.

[Schwalbe and Walker, 2001] U. Schwalbe and P. Walker. Zermelo and the early history of game theory. *Games and Economic Behavior*, 34(1):123–137, 2001. With an appendix by Ernst Zermelo, translated from German by the authors.

[Shelah and Woodin, 1990] S. Shelah and W. H. Woodin. Large cardinals imply that every reasonably definable set of reals is Lebesgue measurable. *Israel Journal of Mathematics*, 70(3):381–394, 1990.

[Shelah, 1984] S. Shelah. Can you take Solovay's inaccessible away? *Israel Journal of Mathematics*, 48(1):1–47, 1984.

[Shelah, 1998] S. Shelah. *Proper and Improper Forcing*. Perspectives in Mathematical Logic. Springer-Verlag, Berlin, second edition, 1998.

[Shoenfield, 1961] J. Shoenfield. The problem of predicativity. In *Essays on the Foundations of Mathematics*, pages 132–139. Magnes Press, Hebrew University, Jerusalem, 1961.

[Sierpiński, 1924] W. Sierpiński. Sur une propriété des ensembles ambigus. *Fundamenta Mathematicae*, 6:1–5, 1924.

[Sierpiński, 1925] W. Sierpiński. Sur une class d'ensembles. *Fundamenta Mathematicae*, 7:237–243, 1925.

[Sierpiński, 1938] W. Sierpiński. Fonctions additives non complètement additives et fonctions non mesurables. *Fundamenta Mathematicae*, 30:96–99, 1938.

[Silver, 1971] J. Silver. Some applications of model theory in set theory. *Annals of Mathematical Logic*, 3(1):45–110, 1971.

[Soare, 1987] R. Soare. *Recursively Enumerable Sets and Degrees*. Perspectives in Mathematical Logic. Springer-Verlag, Berlin, 1987. A study of computable functions and computably generated sets.

[Solovay, 1967] R. Solovay. A nonconstructible Δ^1_3 set of integers. *Transactions of the American Mathematical Society*, 127:50–75, 1967.

[Solovay, 1969] R. Solovay. On the cardinality of Σ^1_2 sets of reals. In *Foundations of Mathematics (Symposium Commemorating Kurt Gödel, Columbus, Ohio, 1966)*, pages 58–73. Springer, New York, 1969.

[Solovay, 1970] R. Solovay. A model of set theory in which every set of reals is Lebesgue measurable. *Annals of Mathematics*, 92:1–56, 1970.

[Solovay, 1978a] R. Solovay. A Δ^1_3 coding of the subsets of ω_ω. In *Cabal Seminar 76–77 (Proc. Caltech-UCLA Logic Sem., 1976–77)*, volume 689 of *Lecture Notes in Mathematics*, pages 133–150. Springer, Berlin, 1978.

[Solovay, 1978b] R. Solovay. The independence of DC from AD. In *Cabal Seminar 76–77 (Proc. Caltech-UCLA Logic Sem., 1976–77)*, volume 689 of *Lecture Notes in Mathematics*, pages 171–183. Springer, Berlin, 1978.

[Steel and Van Wesep, 1982] J. Steel and R. Van Wesep. Two consequences of determinacy consistent with choice. *Transactions of the American Mathematical Society*, 272(1):67–85, 1982.

[Steel, 1981] J. Steel. Determinateness and the separation property. *The Journal of Symbolic Logic*, 46(1):41–44, 1981.

[Steel, 1982] J. Steel. Determinacy in the Mitchell models. *Annals of Mathematical Logic*, 22(2):109–125, 1982.

[Steel, 1983] J. Steel. Scales in $L(\mathbf{R})$. In *Cabal Seminar 79–81*, volume 1019 of *Lecture Notes in Mathematics*, pages 107–156. Springer, Berlin, 1983. Reprinted in [Kechris *et al.*, 2008].

[Steel, 1988] J. Steel. Long games. In *Cabal Seminar 81–85*, volume 1333 of *Lecture Notes in Mathematics*, pages 56–97. Springer, Berlin, 1988. Reprinted in [Kechris *et al.*, 2008].

[Steel, 1995] J. Steel. $\mathrm{HOD}^{L(\mathbf{R})}$ is a core model below Θ. *The Bulletin of Symbolic Logic*, 1(1):75–84, 1995.

[Steel, 1996] J. Steel. *The Core Model Iterability Problem*, volume 8 of *Lecture Notes in Logic*. Springer-Verlag, Berlin, 1996.

[Steel, 2002] J. Steel. Core models with more Woodin cardinals. *The Journal of Symbolic Logic*, 67(3):1197–1226, 2002.

[Steel, 2005] J. Steel. PFA implies $\mathrm{AD}^{L(\mathbb{R})}$. *The Journal of Symbolic Logic*, 70(4):1255–1296, 2005.

[Steel, 2008a] J. Steel. Games and scales. Introduction to Part I. In *Games, Scales, and Suslin cardinals. The Cabal Seminar. Vol. I*, volume 31 of *Lecture Notes in Logic*, pages 3–27. Assoc. Symbol. Logic, Chicago, IL, 2008.

[Steel, 2008b] J. Steel. The length-ω_1 open game quantifier propagates scales. In *Games, Scales, and Suslin cardinals. The Cabal Seminar. Vol. I*, volume 31 of *Lecture Notes in Logic*, pages 260–269. Assoc. Symbol. Logic, Chicago, IL, 2008.

[Steel, 2009] J. Steel. The derived model theorem. In *Logic Colloquium 2006*, Lecture Notes in Logic, pages 280–327. Association of Symbolic Logic, Chicago, IL, 2009.

[Suslin, 1917] M. Suslin. Sur une définition des ensembles measurable B sans nombres transfinis. *Comptes Rendus Hebdomadaires des Séances de l'Académie des Sciences, Paris*, 164:88–91, 1917.

[Todorcevic, 1984] S. Todorcevic. A note on the proper forcing axiom. In *Axiomatic Set Theory (Boulder, Colo., 1983)*, volume 31 of *Contemporary Mathematics*, pages 209–218. Amer. Math. Soc., Providence, RI, 1984.

[Ulam, 1960] S. Ulam. *A Collection of Mathematical Problems*. Interscience Tracts in Pure and Applied Mathematics, no. 8. Interscience Publishers, New York-London, 1960.

[Van Wesep, 1978a] R. Van Wesep. Separation principles and the axiom of determinateness. *The Journal of Symbolic Logic*, 43(1):77–81, 1978.

[Van Wesep, 1978b] R. Van Wesep. Wadge degrees and descriptive set theory. In *Cabal Seminar 76–77 (Proc. Caltech-UCLA Logic Sem., 1976–77)*, volume 689 of *Lecture Notes in Mathematics*, pages 151–170. Springer, Berlin, 1978.

[Veličković, 1992] B. Veličković. Forcing axioms and stationary sets. *Advances in Mathematics*, 94(2):256–284, 1992.

[Vitali, 1905] G. Vitali. Sul problema della misura dei gruppi di punti di una retta. *Tip. Gamberini e Parmeggiani*, pages 231–235, 1905.

[von Neumann and Morgenstern, 2004] J. von Neumann and O. Morgenstern. *Theory of Games and Economic Behavior*. Princeton University Press, Princeton, NJ, 2004. Reprint of the 1980 edition [Princeton Univ. Press, Princeton, NJ].

[Wagon, 1993] S. Wagon. *The Banach-Tarski Paradox*. Cambridge University Press, Cambridge, 1993. With a foreword by Jan Mycielski, Corrected reprint of the 1985 original.

[Wolfe, 1955] P. Wolfe. The strict determinateness of certain infinite games. *Pacific Journal of Mathematics*, 5:841–847, 1955.

[Woodin, 1982] W. H. Woodin. On the consistency strength of projective uniformization. In *Proceedings of the Herbrand Symposium (Marseilles, 1981)*, volume 107 of *Studies in Logic and the Foundations of Mathematics*, pages 365–384. North-Holland, Amsterdam, 1982.

[Woodin, 1983a] W. H. Woodin. AD and the uniqueness of the supercompact measures on $P_{\omega_1}(\lambda)$. In *Cabal Seminar 79–81*, volume 1019 of *Lecture Notes in Mathematics*, pages 67–71. Springer, Berlin, 1983.

[Woodin, 1983b] W. H. Woodin. Some consistency results in ZFC using AD. In *Cabal Seminar 79–81*, volume 1019 of *Lecture Notes in Math.*, pages 172–198. Springer, Berlin, 1983.

[Woodin, 1986] W. H. Woodin. Aspects of determinacy. In *Logic, Methodology and Philosophy of Science, VII (Salzburg, 1983)*, volume 114 of *Studies in Logic and the Foundations of Mathematics*, pages 171–181. North-Holland, Amsterdam, 1986.

[Woodin, 1988] W. H. Woodin. Supercompact cardinals, sets of reals, and weakly homogeneous trees. *Proceedings of the National Academy of Sciences of the United States of America*, 85(18):6587–6591, 1988.

[Woodin, 1999] W. H. Woodin. *The Axiom of Determinacy, Forcing Axioms, and the Nonstationary Ideal*, volume 1 of *de Gruyter Series in Logic and its Applications*. Walter de Gruyter & Co., Berlin, 1999.

[Zermelo, 1904] E. Zermelo. Beweis, dass jede menge wohlgeordnet werden kann. *Mathematische Annalen*, 59:514–516, 1904.

[Zermelo, 1913] E. Zermelo. Über eine Anwendung der Mengenlehre auf die Theorie des Schachspiels. In *Proceedings of the Fifth International Congress Mathematics*, pages 501–504, Cambridge, MA, 1913. Cambridge University Press.

SINGULAR CARDINALS: FROM HAUSDORFF'S GAPS TO SHELAH'S PCF THEORY

Menachem Kojman

The mathematical subject of singular cardinals is young and many of the mathematicians who made important contributions to it are still active. This makes writing a history of singular cardinals a somewhat riskier mission than writing the history of, say, Babylonian arithmetic.

Yet exactly the discussions with some of the people who created the 20th century history of singular cardinals made the writing of this article fascinating. I am indebted to Moti Gitik, Ronald Jensen, István Juhász, Menachem Magidor and Saharon Shelah for the time and effort they spent on helping me understand the development of the subject and for many illuminations they provided. A lot of what I thought about the history of singular cardinals had to change as a result of these discussions.

Special thanks are due to István Juhász, for his patient reading for me from the Russian text of Alexandroff and Urysohn's *Memoirs*, to Salma Kuhlmann, who directed me to the definition of singular cardinals in Hausdorff's writing, and to Stefan Geschke, who helped me with the German texts I needed to read and sometimes translate.

I am also indebted to the Hausdorff project in Bonn, for publishing a beautiful annotated volume of Hausdorff's monumental *Grundzüge der Mengenlehre* and for Springer Verlag, for rushing to me a free copy of this book; many important details about the early history of the subject were drawn from this volume.

The wonderful library and archive of the Institute Mittag-Leffler are a treasure for anyone interested in mathematics at the turn of the 20th century; a particularly pleasant duty for me is to thank the institute for hosting me during my visit in September of 2009, which allowed me to verify various details in the early research literature, as well as providing me the company of many set theorists and model theorists who are interested in the subject.

I an grateful to all colleagues who read early versions of this manuscript and who made many valuable comments. Aki Kanamori, who spent too many hours of his time reading and correcting my writing, deserves my deepest gratitude. Needless to say, all mistakes which remained in this text are solely mine.

Every mathematician has come across Hausdorff's name, typically in more than one area of mathematics. That was the case with me as well, but it was not until

Handbook of the History of Logic: Sets and Extensions in the Twentieth Century.
Volume editor: Akihiro Kanamori. General editors: Dov M. Gabbay, Paul Thagard and John Woods.

the writing of this chapter that I truly grasped the influence of Hausdorff's work on modern set theory and, reading his letters, had the chance to appreciate from afar his delicate sense of humour and modesty. This article is devoted to his memory.

1 INTRODUCTION

Singular cardinals appeared on the mathematical world stage two years before they were defined. At the third international congress of mathematicians in Heidelberg, on August 10, 1904, Julius König allegedly refuted Georg Cantor's two major conjectures simultaneously, by proving that the continuum could not be well-ordered in any of the order-types in Cantor's list of alephs.

König's proof went as follows: Suppose that the continuum is some $\aleph_\beta$. Look at the sequence $\aleph_{\beta+1}, \aleph_{\beta+2}, \ldots, \aleph_{\beta+n}, \ldots$ for all natural numbers n. Its limit $\aleph_{\beta+\omega}$ is equal to the sum $\sum_n \aleph_{\beta+n}$.

König next presented his newly proved special case of what is now known as "König's inequality", which states that

$$\prod \aleph_{\beta+n+1} > \sum \aleph_{\beta+n}, \tag{1}$$

to derive the inequality

$$(\aleph_{\beta+\omega})^{\aleph_0} > \aleph_{\beta+\omega}. \tag{2}$$

So far, König is absolutely correct. The general theorem now known as "König's inequality" was obtained by Zermelo and presented in Göttingen in the same year, but published only four years later [Zermelo, 1908]. Next, König quotes a theorem from Felix Bernstein's Göttingen dissertation: $(\aleph_\alpha)^{\aleph_0} = \aleph_\alpha \cdot 2^{\aleph_0}$ for every ordinal α. Substituting $\beta + \omega$ for α and using $2^{\aleph_0} = \aleph_\beta$ one gets

$$(\aleph_{\beta+\omega})^{\aleph_0} = \aleph_{\beta+\omega} \tag{3}$$

Now (2) and (3) contradict each other. Ergo: $2^{\aleph_0}$ is not an aleph!

This had a dramatic effect:

> Though mathematics rarely makes a ripple in the daily press today, the local papers then were full of reports describing König's sensational discovery.

This is from Cantor's biography by Dauben [1990, p.249], where Cantor's consternation with the situation is described in detail (p.248):

> As Schoenflies once put it, belief that the power of the continuum was equal to $\aleph_1$ was a basic dogma with Cantor. Not only did König's proof challenge this dogma, but it further implied that the continuum could not be well-ordered. This rendered doubtful another article of Cantorian faith—that *every* set could be well-ordered.

Cantor was immediately suspicious of König's proof, particularly of Bernstein's lemma. Making a pun on König's last name, "Cantor quipped that whatever had been done to produce the alleged proof, he suspected the king less than the king's ministers." (Dauben [1990, p.249])

Indeed, Bernstein's lemma was false for exactly the type of cardinals required in König's use of it. Felix Hausdorff, after coming back from Wengen, Switzerland, where he and Hilbert had spent a short vacation after the Heidelberg congress, poetically described König's and Bernstein's combined flaw in a letter to Hilbert in the following month (cf. [Hausdorff, 2002, p.11]):

> After the continuum problem plagued me in Wengen almost like an obsession, my first look here was naturally directed to Bernstein's dissertation. The bug lay exactly in the expected place, on page 50 [...] Bernstein's argument employs a recursion from $\aleph_\mu$ which does not hold for alephs that have no immediate predecessor, that is, exactly those alephs which König required.[1]

That is, Bernstein's flaw lies in the fact that exactly the limit cardinal for which König uses Bernstein's equality—$\aleph_{\beta+\omega}$—escapes the recursion Bernstein uses to prove the equality, as it has no immediate predecessor.

As [Hausdorff, 2002] mentions, it was for a while erroneously thought that Zermelo was the one who detected the error in König's false proof. The debate continues whether it was actually Hausdorff (cf. [Hausdorff, 2002, p.10]) or someone else. Recently Moore [2009, p.825] suggested that a letter by Otto Blumenthal to Émile Borel dated December 1, 1904, may indicate that "König himself was the first to realize that his proof was not valid, followed (independently of each other) by Cantor, Bernstein and Zermelo."

Singular cardinals could not have received a louder introduction than the one they received in the 1904 Heidelberg episode: at an international congress, with all other parallel sessions canceled by the organizers to allow everyone to attend König's sensational lecture, in the presence of Cantor and Hilbert, and in direct relation to the two most important problems in set theory of the time.

Furthermore, disregarding the circumstances and considering only the mathematical concepts, *infinite products of distinct regular cardinals*, which are the most important objects in the modern elementary theory of singular cardinals, were introduced in König's pioneering result. König showed that the product $\prod_n \aleph_{\beta+n}$ is embeddable in the power set of $\aleph_{\beta+\omega}$ and diagonalizable against the sum of cardinals $\sum_n \aleph_{\beta+n} = \aleph_{\beta+\omega}$. He takes Cantor's diagonal argument for the inequality $2^\kappa > \kappa$ an important step further and applies it in a setting with a singular cardinal. Using the fact that $\aleph_{\beta+\omega}$ is a countable sum, he can use diagonalization

[1]Nachdem das Continuumproblem mich in Wengen beinahe wie eine Monomanie geplagt hatte, galt hier mein erster Blick natürlich der Bernsteinschen Dissertation. Der Wurm sitzt genau an der vermutheten Stelle, S. 50 [...] Bernsteins Betrachtung giebt eine Recursion von $\aleph_{\mu+1}$ auf $\aleph_\mu$, versagt aber für solche $\aleph_\mu$, die keinen Vorgänger haben, also gerade für die Alephs, für die Herr J. König sie nothwendig braucht.

to show that the total number of countable subsets of $\aleph_{\beta+\omega}$ exceeds $\aleph_{\beta+\omega}$. Zermelo [1908] extended this result to general products and observed that it implies Cantor's theorem that the power set of any set had larger cardinality than the set itself. Thus, the König-Zermelo inequality held all the information that Cantorian diagonalization could give at the time about cardinal exponentiation, and surprisingly enough continued to do so for many more decades.

The importance of this interplay between infinite products and infinite sums, which König discovered and Zermelo phrased in full generality, goes well beyond cardinal arithmetical considerations—it is fundamental for every mathematical structure of singular cardinality. Even a superficial examination of König's argument shows than it actually proves more that it states — e.g., that for every collection of $\aleph_{\beta+\omega}$ small subsets of $\aleph_{\beta+\omega}$ there exists a countable subset with finite intersection with each set in the family. A central tool for the study of singular cardinals has been introduced by König already in 1904 in Heidelberg, but was not put to appropriate use at that time.

König's Heidelberg episode clearly brought out the issue that would accompany the development of singular cardinals theory for many years to come. Singular cardinals were regarded as a curiosity, an obstacle, a threat to Cantor's conjectures. They are the points in the list of alephs at which Bernstein's inductive argument and many other arguments that would be attempted in the years to come break down, exactly because of the fact that some of the small subsets of the cardinal are unbounded in it — a property described as "singularity" by König [1905], and later adapted by Hausdorff as the name for these cardinals.

It is instructive to compare König's diagonal argument to the Baire category theorem, proved just three years earlier: both are proved via variations on Cantor's diagonal argument for $2^{\aleph_0} > \aleph_0$, with the Baire argument using first category sets instead of singletons and the König argument using subsets of cardinality smaller than the singular cardinal instead of singletons. Both theorems had short, simple proofs that were accessible to everyone and both were immediately accepted as true. One can argue that both theorems had similar aesthetic appeal. Yet, the Baire theorem rapidly became a part of the general mathematical language—it is probably one of the most frequently used theorems in mathematics and has numerous applications and generalizations—and many other uses of Cantorian diagonalization in analysis followed the Baire theorem example. But hardly any research at all followed the König-Zermelo inequality for many decades. The next development in singular cardinal arithmetic in the direction that the inequality marked came only with Silver's theorem in 1974 (see below).

Why did not König himself, Cantor, or, more realistically, Zermelo, who published the general formulation of König's inequality, or Hausdorff, pursue König's discovery further? König, at 1904 only one year before his retirement, is determined to disprove Cantor's well-ordering conjecture as much as Cantor is determined to prove it. He would spend the last eight years of his life trying to refute the Axiom of Choice, which Zermelo phrased and used to prove the Well-Ordering theorem shortly after the Heidelberg congress.

Cantor, who had previously been on good terms with König and appreciated his work, was agitated and probably angry that König put him in such public discomfort in Heidelberg, not having told him beforehand about the contents of his talk. Neither of them pursues König's discovery further.

Zermelo may have been more interested in the Axiom of Choice and in axiomatization issues in general than in exploring singular cardinal arithmetic further.

Hausdorff, who was an ideal candidate for studying the products that appeared in König's proof, with which he was well acquainted, and who would soon launch an intensive study of the infinite product ω^ω—concludes his letter to Hilbert as follows (cf. [Hausdorff, 2002, p.11]):

> I had written to this effect to Herr König while still on the road and explained this as far as I could without using Bernstein's work, but so far received no reply. Now I am yet even more convinced that König's proof is wrong and I believe König's theorem to be the peak of impossibility. On the other hand, I am sure you too do not believe that Herr Cantor discovered during these last few weeks what he has been looking for in vain for the last 30 years. Thus, your problem No. 1 seems to remain after the Heidelberg congress exactly where you left it after the Paris congress.
>
> But perhaps, as I am writing this, one of the combating parties is already in possession of the truth. I am very curious to see the printed proceedings of the congress.[2]

In König's paper, published in 1905 in *Mathematische Annalen*, he indeed retracted his claim and pointed to Bernstein's mistake (see below for further discussion of this paper).

Once the doubts that König's announcement cast had been lifted, effort to solve Cantor's two big open problems resumed. What is important to observe is that apart from its potential hazard to Cantor's problem, Hausdorff says nothing about König's inequality. This mathematically valuable statement seems to have been welded to the wrong proof in which it played a part.

Would König's important discovery have received wider attention had it been presented on its own, and not as part of the false refutation of the continuum hypothesis? The circumstances so dictated that the mathematical contents of König's inequality was overshadowed by König's erroneous use of it. By the time König's congress paper appeared in 1905, all König claimed to have proved was

[2]Ich hatte in diesem Sinne, soweit Ich ohne Benutzung der Bernsteinschen Arbetin konnte, schon von unterwegs an Herrn König geschrieben, aber keine Antwort erhalten, bin also umso mehr geneigt, den König'schen Beweis für falsch und den König'schen Satz für den Gipfel des Unwahrscheinlichen zu Halten. Anderseits werden wohl auch Sie kaum den Eindruck gewonnen haben, dass Herr Cantor das, was er seit 30 Jahren vergeblich sucht, in den letzten Wochen gefunden haben sollte, und so scheint Ihr Problem Nr. 1 nach dem Heidelberger Congress genau dort to stehen, wo Sie es aur dem Pariser Congress verlassen haben.

Aber vielleicht ist, während ich dies schriebe, doch schon eine der streitenden Parteien im Besitze der Wahrheit. Ich bin sehr gespannt auf die gedruckten Verhandlungen des Congresses.

that the continuum could not be well-ordered in an order-type of the form $\aleph_{\beta+\omega}$. However, his argument essentially proved that the continuum could not be well-ordered in a type of countable cofinality; as "cofinality" had not been yet defined, he could not state this. (Hausdorff later attributed the full statement to König.) But this claim is incomparably less dramatic that his original statement that the continuum could not be well-ordered at all.

The name "singular cardinal" and the circumstances of its emergence as a name for non-regular limit cardinals requires some explanation. This name appears for the first time in König's [1905] following his congress lecture, in a passage in which he explains the mistake in his lecture:

> It follows that the assumption that the continuum is equivalent to a well-ordered set would be certainly false if Bernstein's theorem was correct n general. Unfortunately its proof has a significant gap since for $\aleph_\omega$ and each of the "singular" well-ordered sets considered above, the assumption that each countable subset is included in a [proper] initial segment is not correct anymore.
>
> I mention this mainly in order to retract the conclusion that I drew from Bernstein's theorem in my talk at the congress under the assumption that the theorem was correct.

From a distance of over a hundred years the scorn in König's articulation is still audible: if not for these "singular" order-types, as König puts it, his proof would still be correct. Later in the paper König brings up the possibility that the continuum is not well-orderable and (as a non-well-ordered cardinal) larger than every well-orderable set. In this case, he writes, Bernstein's theorem follows.

Technically, König's original use of "singular" means "an uncountable cardinal with unbounded countable subsets", that is, singular of countable cofinality. The name "singular" was adopted by Hausdorff to describe cardinals whose cofinality type was smaller than themselves, once cofinality had been defined in his 1906 paper on order-types, and became the standard name for these cardinals ever since. This name was far from neutral. Hausdorff's division of cardinals to "regular" and "singular" suggested that regular cardinals were the objects that deserved serious attention, and that the singulars were less important. This view was sustained for a surprisingly long time.

Zermelo [1904] introduced the Axiom of Choice and proved that every set could be well-ordered, and Hausdorff launched an intensive project of classifying order-types, perhaps hoping he could solve Cantor's continuum problem in this way. Hausdorff's first result in set theory (obtained independently by Bernstein) was that the number of countable order-types is equal to the continuum. Moore [1989] expresses the opinion that Hausdorff was doing this so that he could prove that the number of countable well-order-types, which Cantor had already shown to be $\aleph_1$, was equal to the total number of countable order-types, and in this way solve Cantor's continuum problem, but Steprans [2010] feels that Hausdorff was "primarily interested in the classification of linear order-types, but ... nevertheless recognized the significance of his work to the Continuum Problem."

Although Hausdorff did treat singular cardinals in his work, they were cast aside by him and the other researchers of the time in favor of the more burning issue of the continuum. The (consistent) possibilities that the continuum itself was a singular or a regular limit cardinal were not raised at that time at all. This may be partly due to the confusion between the $\aleph$ operation and the $\beth$ operation; the continuum is clearly not closed under the latter, therefore (if the two are the same, as GCH claims) it is not closed under the former. It took a stimulus from measure theory to bring the possibility of a limit continuum into the foreground (see Sect. 6.1).

Neither fashion nor prejudice could hide singular cardinals from the mathematical eye for long. The interrelation between singular cardinalities and basic properties of standard mathematical objects is so profound and varied that singular cardinals inevitably appeared in set-theoretic investigation in several mainstream branches of mathematics during the long period of time in which they were ignored by the organized set-theoretic research. Some of these investigations foundered for decades in the absence of the right set-theoretic tools. In Section 3 we shall look at some of these occurrences of singulars.

Set theory had to experience several remarkable metamathematical developments before set theorists developed an elementary theory of singular cardinals. It had to be realized first that there were no more rules for regular cardinal arithmetic than those given by the König-Zermelo inequality (and the trivial weak monotonicity). It was actually the singular cardinals that possessed a theory, and hence deserved a close study. A theory of singular cardinal arithmetic was finally developed without metamathematical means, via the algebraic study of small infinite products of different regular cardinals by Shelah in the 1980s and 1990s under the name of *pcf theory*. *A posteriori*, Shelah's theory shed an explanatory light on the instances in which singular cardinals appeared naturally in topology, algebra and elsewhere, and in many cases could be used to directly continue the early investigations of singulars.

A central ingredient in Shelah's pcf theory is the theory of exact upper bounds of increasing sequences of ordinal functions modulo some ideal on an infinite set. Given an ideal I of subsets over an infinite set A one defines a quasi-ordering on the class of functions from A to the ordinal numbers, On^A. For two such functions f and g, f is smaller modulo I than g if except for a set in I it holds for all $a \in A$ that $f(a) < g(a)$. This relation is denoted by $<_I$ and similarly $\leq_I$ and $=_I$ are defined. A typical example is the ideal of finite sets—in which case the relations $<_I$, $\leq_I$ and $=_I$ are customarily denoted by $<^*$, $\leq^*$ and $=^*$.

A central notion in pcf theory is that of a *scale*—a sequence of functions in an infinite product of regular cardinals which is linearly ordered and dominating for the relation $<_I$ for some ideal I. Shelah's results on scales are remarkably similar to what Hausdorff was looking for in his study of sequences of functions from natural numbers to natural numbers modulo the ideal of finite sets (cf. [Steprans,

2010]). Hausdorff knew that the existence of an ω_1-scale in $(\omega^\omega, <^*)$ followed from CH, but failed to prove the existence of such a scale without CH. He later managed to prove without CH (after having first proved with CH) the existence of an (ω_1, ω_1^*)-gap in $(\omega^\omega, <^*)$. A (κ, κ^*)-gap in a partially ordered set is a pair of sequences of length κ, the first increasing, the second decreasing, with each element in the first sequence being smaller than every element in the second, and with no member of the partially ordered set between the sequences.

The proof of the existence of an (ω_1, ω_1^*)-gap in $(\omega^\omega, <^*)$ was one of Hausdorff's most important results—partly so because Shelah's work in the 1990s showed that Hausdorff could not have proved more than that without using additional axioms. The irony is exquisite. Hausdorff, who discovered everything one could discover without metamathematics about gaps in $(\omega^\omega, <^*)$, is unaware that this is all that can be proved about gaps. Shelah, who knows everything there is to know about the independence phenomena surrounding the continuum proves that there are no ZFC gaps in the product $\prod_n \omega_n$ except the gaps Hausdorff had discovered in 1909, and proves the existence of an $(\aleph_{\omega+1})$-scale in a product of the form $\prod_{n\in B} \omega_n$ by totally elementary means. As Shelah writes in the introduction to his *Cardinal Arithmetic* [1994a], "Cantor should have no problems understanding and (so I feel) appreciating the theorems and even most proofs in this book."

Did Hausdorff miss the opportunity to discover pcf theory? Could he have extended his study of scales and gaps in ω^ω to scales and gaps in the (much better behaved) product $\prod_n \omega_n$? Perhaps so; there is no *mathematical* reason for not discovering at least some of the elementary theory of singular cardinals in Hausdorff's lifetime. We shall see in Section 3 that early investigations of structures related to singular cardinals came very close to discovering elements from pcf theory.

One should recall that the general notion of an ideal was not yet available to Hausdorff, that the Łoś ultraproduct theorem appears much later,[3] that a good part of the ZFC combinatorics Shelah employs was motivated by the combinatorial principles which Ronald Jensen discovered with fine structure theory in Gödel's constructible universe in the 1970s (like $\diamondsuit$, $\square$ and others), and that the most fundamental concept of "covering" was not arrived at in Hausdorff's lifetime. But it is still not hard to imagine Hausdorff discovering some of the basic facts about products of distinct regular cardinals. The sophistication and difficulty of what Hausdorff achieved in his 1909 construction of an (ω_1, ω_1^*)-gap certainly indicates that he would have discovered quite a bit of what there was to discover about infinite products of regular cardinals which arise in the study of singular cardinals, had he studied the appropriate products.[4]

I will risk stating that it was not mathematical ability that Hausdorff lacked;

[3] It is, however, remarkable that Hausdorff [1909] discovered a substantial part of the structure of a reduced product modulo the ideal of finite sets in his construction of an "arithmetic pantachie". See [Plotkin, 2005, p.264].

[4] I recently heard from Ronald Jensen the following words on this matter: "After Silver's result was obtained I found an elementary proof of it, which I wrote down and circulated. As I was writing the proof I thought, 'Hausdorff could have proved this theorem!' "

rather, it was the independence results which blew away the chaff from the wheat in cardinal arithmetic in the 1960s. History shows that singular cardinal arithmetic was pursued only after Cohen's, Solovay's and Easton's results made it absolutely clear that there was no elementary theory of regular cardinal arithmetic.[5]

What is more important, from the historical point of view, is that the impressive research about the structure of the product ω^ω which Hausdorff, Rothberger and later researchers have conducted, whose discoveries were later delineated by the huge body of complementary independence results that followed Cohen's discovery of forcing, created an intuition about the structure of ω^ω whose analog for the structure of the product $\prod_n \omega_n$ were totally misleading. The structural properties of each of these two infinite products are very different from those of the other and the independence phenomenon has a totally different effect in each (as will be described in Section 5). Thus, the influence of the understanding of $(\omega^\omega, <^*)$ on the understanding of $(\prod_n \omega_n, <^*)$ was more by contrast than by analogy.

Sociologically, rather than mathematically, before the method of forcing was discovered and used by Easton to show that apart from weak monotonicity and the König-Zermelo inequality there were no other rules governing regular cardinal exponentiation, attention was not diverted to singular cardinal exponentiation.

After the discovery of forcing and the development of large cardinals theory, the understanding of singular cardinal arithmetic progressed simultaneously in three tracks—consistency results were proved with large cardinal forcing, ZFC results were proved by a variety of methods, first non-elementary, then elementary, and consistency strength was achieved via the inner-model program.

Chronologically, the metamathematical investigations of singular cardinals preceded the development of their elementary theory. Still, in this history the elementary theory is described first. The reader should not forget that Prikry and Silver proved the consistency of the failure of the Singular Cardinal Hypothesis and Magidor proved the consistency of its failure at $\aleph_\omega$ *before* Silver's theorem was discovered (see Section 4.1 below)—this fact will be emphasized below.

Although the modern elementary theory of singular cardinals is accessible, in principle, to Hausdorff or to Cantor, the part played in its development by a variety of sophisticated metamathematical means is significant.

In retrospect, Cantor, König and Zermelo discovered by 1904 all the rules that could be discovered for *regular* cardinal exponentiation. The fact that no more new rules were discovered between 1904 and the 1970s was not because set theorists were less skilled in these decades, but because there are no other rules to discover. Had Hausdorff known that the König-Zermelo inequality was all that one could prove about regular cardinals, he and others who followed in his footsteps might have turned their attention to investigating singular cardinals.

[5]This dynamics of set-theoretic discoveries in the modern era is evident also in the working mathematician's private sphere, rather than that of the community at large. I heard from Shelah in the 1980s the view that it is good mathematical practice to try and prove by forcing the *negation* of your intended theorem before trying to prove it. He claimed that many of his ZFC proofs came after attempts to force the consistency of the negation had failed, leaving behind valuable clues about how the proof should go.

Among the duties that kept König busy in 1905 as a member of the Hungarian Academy of Science was his appointment to serve on the committee of the newly established Bolyai prize. In that year, to mark the centennial anniversary of János Bolyai, and also partly stirred by the new Swedish Nobel Prize and the new Russian Lubachevsky prize, the Hungarian Academy of Science established the Bolyai prize, to be awarded to an important treatise or book published in the ten years preceding the award.

Two candidates were considered in 1905 for the Bolyai prize: Henri Poincaré and David Hilbert. Poincaré won the 1905 prize, and Hilbert was re-nominated and won the next prize in 1910.[6] The prize regulations required that the prize committees each consist of two Hungarian academy members and two foreign academy members. The two foreign members in 1905 were the Frenchman Jean-Gaston Darboux and the German Felix Klein, and in 1910 they were Poincaré, the 1905 prize winner (who became a foreign academy member after his win) and the Swede Gösta Mittag-Leffler. In both the 1905 and 1910 Bolyai prize committees served the same two Hungarian academy members: Gusztáv Rados and Gyula (Julius) König.

How is all this related to singular cardinals? Well, it is quite surprising that the third winner of the Bolyai prize, in the year 2000, was Saharon Shelah, for his book *Cardinal Arithmetic* [1994a]. The Bolyai prize had been discontinued for almost a century, and was resumed in 2000. The first time it was awarded in the new era was for a monograph on singular cardinals.

The history of the Bolyai prize demarcates the story of singular cardinals in the 20th century. All the story unfolded below happens between the year in which the founder of the theory of infinite products of different regular cardinals served on the Bolyai prize committee and the year 2000, in which Shelah won the prize for his monograph on the study of singular cardinals via the algebraic theory of such products of sets of regular cardinals, called pcf theory.

The two *mathematical* events which demarcate the development of singular cardinals theory in the 20th century would have to be Hausdorff's famous construction of an (ω_1, ω_1^*)-gap inside $(\mathbb{R}^\omega, <^*)$[7] and Shelah's construction of an $\aleph_{\omega+1}$-scale in $(\prod_{n \in B} \omega_n, <^*)$ for some infinite set $B \subseteq \omega$. Hausdorff had previously hoped to construct an ω_1-scale and had also previously [1908] constructed such a gap with the aid of CH, and regarded his elimination of CH from the proof as a major

[6]The mathematical politics of that time is quite fascinating. It is perhaps worthwhile to mention in the context of this article that Poincaré opposed modern set theory, and declared that "Cantorism was dead", as Hausdorff states in a letter to Hilbert in 1907 (cf. [Plotkin, 2005]). Hilbert, in contrast to Poincaré, was the most prominent defender of set theory from the attacks on which were made, in Hausdorff's words, "by such medieval means!" ([Plotkin, 2005]).

[7]I choose Hausdorff's gap theorem and not König's inequality, because although the inequality introduced the products used in the study of singular cardinals, no development followed it for many decades. Hausdorff proved his gap theorem for sequences of real numbers—that is, he worked in the structure $(\mathbb{R}^\omega, <^*)$. It had soon become clear that one could work with $(\omega^\omega, <^*)$ or even $(2^\omega, \leq^*)$. The latter structure is indeed the one Hausdorff uses in publishing his second construction of a gap in [1936] (see also the English translation in [Plotkin, 2005]). A discussion of these issues in the context of set theory of the real numbers is found in [Steprans, 2010].

achievement by which "the first closer relationship between the continuum and the second number class [i.e. ω_1] would be established". We know in hindsight that this was all the relationship Hausdorff could hope to find between the continuum and ω_1. Shelah, on the other hand, had also first proved the existence of an $\aleph_{\omega+1}$-scale with the aid of an additional assumption (that $2^{\aleph_0} \leq \aleph_{\omega+1}$) but continued to discover that this was in fact a ZFC theorem. At that point, the structural analogy between $(\omega^\omega, <^*)$ and $(\prod_n \omega_n, <^*)$ broke down; the independence over ZFC of the existence of an ω_1-scale in the former versus the existence in ZFC of an $\aleph_{\omega+1}$-scale in a sub-product of the latter, made the difference between the theory of ω^ω and the theory of $\prod_n \omega_n$ manifest.

More than 80 years after their discovery, singular cardinals did receive their proper status. By the time Shelah published his book on singular cardinal arithmetic, it was sufficiently clear to everyone that there is no theory of regular cardinal arithmetic, that there was no need for Shelah to title his book "Singular Cardinal Arithmetic"; "Cardinal Arithmetic" was quite enough.[8] The absence of the word "singular" in Shelah's title symbolizes thus the singular cardinals' triumph as a central set-theoretic subject.

2 THE BEGINNING: HAUSDORFF'S WORK

Hausdorff, whose interest in set theory had begun shortly before the 1904 meeting, proceeded after 1904 to advance the theory of Cantor's alephs, and grounded them in the broader setting of linearly ordered sets. It is hard to exaggerate the importance of his 1906–1909 work on order-types. To this day, this work is fundamental.[9]

For our purposes, though, the most important part of this work is Hausdorff's [1906] definition, for the first time, of the *cofinality type* of a linearly ordered set, and the division of cardinals into *regular* cardinals—those whose cofinality types are equal to themselves—and *singular* cardinals, whose cofinality types are smaller than themselves. Here too Hausdorff is being politely aware of König's earlier imprecise use of the word "type" in saying that the continuum could not be well-ordered in any of the types $\aleph_{\beta+\omega}$. The right statement is, of course, that the continuum could not be well-ordered in any type of countable cofinality, which now after Hausdorff's definition is expressible precisely.

Hessenberg [1906] also defined cofinality independently of Hausdorff in the same year, in a text intended for general readership. Hessenberg uses a different terminology than Hausdorff. He talks about the "kernel" Kern(M) of a linearly ordered set M (a cofinal subset of M).

Hausdorff proved that all successor cardinals are regular, or, equivalently, that

[8]Shelah reports that he had considered also the names "Cofinality Arithmetic" or "The Skeleton of Set Theory" for a while.

[9]The whole corpus of Hausdorff's work on ordered sets is fortunately available in English translation in Plotkin's lovely book [2005].

all singular cardinals are limit. The modern eye may overlook the fact that Hausdorff did not hesitate to use Zermelo's newly formulated Axiom of Choice, which is today generally accepted as a true axiom, but which at that time was still in dispute. Hausdorff had to use the Axiom of Choice, as without the axiom it may consistently hold that *all* uncountable cardinals are singular, as proved by Moti Gitik [1980] from the assumption of the consistency of existence of a proper class of large cardinals.

Hausdorff was aware of the possibility of a *regular limit* cardinal. He writes in his *Gründzuge der Mengenlehre* about regular limit cardinals ([Hausdorff, 2002, p.131]):

> Wenn es also reguläre Anfangzahlen mit Limesindex gibt (und es is bisher nicht gelungen, in dieser Annahme einen Widerspruch zu entdecken), so is die kleinste unter ihnen von einer so exorbitanten Größe, daß sie für die üblichen Zwecke der Mengenlehre kaum jemals in Betracht kommen wird.

Hausdorff describes the regular limit cardinals "whose existence so far did not lead to any contradiction" as being "of such an exorbitant magnitude that they can hardly be taken into account for the usual purposes of set theory." It is completely clear that Hausdorff does not consider the possibility that the continuum itself may be a regular limit cardinal. He also does not make a distinction between a strong limit regular—a "strongly inaccessible", that is, a limit which is closed under exponentiation—and a regular limit, a limit cardinal whose cofinality is equal to itself.

It is very reasonable to speculate that Hausdorff did not consider the possibility of a singular continuum either. As König proved, the possibility that the continuum was singular of countable cofinality did not exist, but there still remained the (consistent) possibility that the continuum was singular of uncountable cofinality.

This speculation is consistent with Hausdorff's formulation of the Generalized Continuum Hypothesis (GCH) in his [1908], which was the fourth in a sequence of works on order-types. Hausdorff calls GCH *Cantor's Aleph Hypothesis* and formulates it as the statement that $\kappa^{<\kappa} = \kappa$ for every *regular* κ. He warns his readers not to forget that he only requires it for regular cardinals; we know that he himself was aware of the possibility of a regular limit cardinal. He does mention in a footnote that for a singular cardinal μ the equality $\mu = \mu^{<\mu}$ does not hold, and generously credits König's 1905 paper for this fact, although, strictly speaking, König stated this only for singular cardinals of type $\aleph_{\beta+\omega}$, that is, the ω-th successor of some cardinal, whereas the full generality of the statement for all singular cardinals follows from Zermelo's formulation.

It is not clear whether Hausdorff regarded GCH as a "dogma" in the sense that Schoenflies ascribed to Cantor's view of CH (see quotation above). It is quite certain that at the time Hausdorff defines singular cardinal he does not consider the possibility that CH could be false, and in particular does not consider the possibility of a singular continuum. Hausdorff's results in the 1909 paper on order-types illustrated the usefulness of GCH, whose consistency was established by Kurt Gödel in 1937, and it is possible that Hausdorff also regarded GCH as true in the same sense.[10]

Among the striking pioneering results Hausdorff [1908] proved is the result, assuming GCH and AC, that in every infinite cardinal a *universal linearly ordered set* exists.[11] The argument Hausdorff uses is a saturation argument. Model theorists would most likely not declare Hausdorff as the first model theorist, but Hausdorff in fact counts types and realizes them to obtain saturated order-types, many years before the formal birth of model theory. Saturation arguments were formulated and used by Vaught and Morley in the 1950s and are central in model theory, but they appear for the first time in Hausdorff's work on linearly ordered sets.

For limit cardinals Hausdorff introduces the first construction of a "special model", in model-theoretic terminology. In a previous paper he proved the uniqueness of saturated linearly ordered sets and their universality without deciding their existence, and in [1908] establishes the existence of such orderings in all infinite cardinalities from GCH. For this construction, it is important that the limit cardinal at hand is a *strong limit*, that is, closed under the $\beth$ function, which indeed follows from his assumption of GCH. In this way Hausdorff generalized Cantor's treatment of countable order-types and initiated the uniform treatment of the class of all infinite cardinals.

It is important to acknowledge Hausdorff's contribution here to establishing the study of the class of cardinal numbers as standard mathematical object, similar to the way all natural numbers are studied in number theory. In his work on order-types, Hausdorff for the first time proves a structure theorem, one could even say, a model-theoretic theorem, concerning *all* infinite cardinals, *and* for the first time phrases and applies GCH. Similar proofs are now standard in modern algebra, topology, Boolean algebra and so on.

Hausdorff declared the regular well-order-types and their inverses as being "fundamental building blocks", or "atoms" in the structural theory of linearly ordered sets (as, say, is demonstrated by his work on scattered order-types). This is certainly true, but may be misleading about the function cardinals serve in measuring properties of structures (e.g. the weight of a topological space, the cofinality of a *partially* ordered set, etc.). It is unclear to what extent this work of Hausdorff's influenced the general perception of singular cardinals, as most of his achievements here were not included in his famous and very influential book *Grundzüge der Mengelehre* (see below).

Another aspect of Hausdorff's use of CH is quite modern and merits attention. The Continuum Hypothesis is the equation $2^{\aleph_0} = \aleph_1$, which asserts that the cardinality of the real line ($2^{\aleph_0}$) is equal to that of the first uncountable ordinal. This

[10]The use of GCH for combinatorial arguments was made routine by Erdős, who elegantly used this assumption numerous times in his papers. Andras Hajnal coined the humorous acronym "ZFE", "Zermelo-Fraenkel-Erdős", for ZFC + GCH.

[11]A universal linearly ordered set of cardinality $\aleph_\alpha$ is a linearly ordered set of cardinality $\aleph_\alpha$ with the property that every linearly ordered set of cardinality $\aleph_\alpha$ order-embeds into it.

equation has since been used often for deriving consequences about the structure of the real line and its many related structures. The Polish school, mainly Sierpiński, excelled in this subject. It was much later that uses of CH (and more generally, of GCH) became popular for proving (combinatorial, model-theoretic, algebraic) theorems about the ordinals themselves in branches of modern mathematics in which the notion of an infinite cardinal played a central role, like infinite Abelian group, infinite combinatorics and model theory. Hausdorff's construction of a universal linearly ordered set of cardinality $\aleph_1$ is an early and rare example in which CH is used to prove a property of $\aleph_1$, rather than of $\mathbb{R}$.

Finally, another remarkable feature of Hausdorff's work with CH is that he obtained the first, and what would remain forever a rare instance of a "verifiable consequence" of CH. Let us recall Gödel's [1947] objection to CH as not having enough "verifiable consequences", that is, consequences which turn out to be ZFC theorems. Among the myriad consequences of CH listed in Sierpinski's book [1934], almost none proved to be a ZFC theorem.[12] Hausdorff first constructed an (ω_1, ω_1^*)-gap with CH, and then was able to prove the existence of such a gap in ZFC. It is now known that he could not hope to construct any other gap in ZFC! The relation between ω_1 and $2^{\aleph_0}$ stands, of course, in strong contrast to the relation between $(\aleph_\omega)^{\aleph_0}$ and $\aleph_{\omega+1}$, as in ZFC there exists an $\aleph_{\omega+1}$-scale inside $(\prod_n \omega_n)$ (see below for further discussion).

In 1914 Hausdorff published his very influential *Grundzüge der Mengenlehre*, which was dedicated to "the creator of set theory, Herr Georg Cantor". In this surprisingly concise volume Hausdorff presents the basics of modern set theory, Borel sets, metric space theory, measure theory and many other subjects that have become indispensable for modern mathematics. But no extensive treatment of singular cardinals is given. His theorem about universal order-types in all cardinals from GCH, for one example, stayed outside of this book (see [Plotkin, 2005] for further discussion of this issue). It could be that Hausdorff did not consider singular cardinals as sufficiently relevant to the central problem of set theory—Cantor's Continuum Hypothesis. Although Hausdorff himself defined the singular cardinals rigorously and made important contributions to their theory, his influential volume on set theory maintained the early approach to singular cardinals as a marginal subject.

3 EARLY OCCURRENCES OF SINGULAR CARDINALS IN MATHEMATICS

Cardinals are more than just abstract objects which one studies for their own sake. They are used to measure the size, and other properties, of mathematical structures. One expects that the arithmetical properties of the cardinality or other cardinal characteristics of a mathematical structure provide information about the structure itself. In the case of finite structures, this is hardly surprising. Recall,

[12] I know of exactly one: number 46, communicated to me by E. Grzegorek (cf. his [1985]).

for instance, that a finite group whose cardinality divisible by a prime p has p-subgroups or that if the cardinality of a finite group is a power of a prime then the group must have a non-trivial center—that is, the cardinality of a finite group provides information about the group's algebraic structure.

Long before the elementary arithmetic of singular cardinals was discovered, certain arithmetical properties of singular cardinals surfaced in various well-known classes of mathematical structures—topological spaces, Abelian groups and Boolean algebras are some of the examples—as holding significant structural information. All these occurrences of singular cardinals are understood better in retrospect from the point of view of the elementary theory of singular cardinals, but it is interesting to assess them in the context in which they emerged.

We shall review a limited number of such occurrences in chronological order, leaving out many others, perhaps just as interesting and important. It appears that a singular cardinal characteristic of a structure was often more informative than a regular one. In some of the examples we examine in this section one may feel that singular cardinals were treated with a certain resentment, even when they occurred quite naturally. An exception to that general impression is found in the work of the Hungarian school in infinite combinatorics—Erdős, Rado, Hajnal and others.

3.1 Singular cardinals in topology: the work of Alexandroff and Urysohn

Two of the many young mathematicians who read Hausdorff's newly published *Grundzüge* carefully were Pavel Alexandroff and Pavel Urysohn from the new Moscow School of mathematics.[13] Their interest in the notion of *topological compactness*, which was formulated by Fréchet and presented in Hausdorff's book, led them to reformulate this notion by quantifying over the class of infinite cardinals in their *Memoirs* [1929].

A point of complete accumulation of a subset $A \subseteq X$ is a point $x \in X$ each of whose open neighborhoods $u \ni x$ satisfies $|A| = |A \cap u|$. Alexandroff and Urysohn observed that for every topological space X the following three properties are equivalent, and called a space which satisfied them "bicompact":

A Every infinite subset has a point of complete accumulation.

B Every well-ordered decreasing intersection of nonempty closed sets has a nonempty intersection.

C Every open cover has a finite subcover.

[13]About the "Lusitania" group in Moscow, lead by Egorov and Luzin, and the development of descriptive set theory, see [Graham and Kantor, 2009]. As part of the Lusitania group's excessive interest in names, each member in the club was named by some aleph. "Recruits were called $\aleph_0$. ... Alexandroff and Urysohn soon achieved the high rank of $\aleph_5$. Luzin himself was given the name $\aleph_{17}$. Egorov was $\aleph_\omega$." [Graham and Kantor, 2009, p.116]

Condition C is nowadays called "compactness". This term was used then for what now is called "countable compactness" (the property that every infinite set has an accumulation point). Condition B is of course equivalent to the statement that every open cover that is linearly ordered by inclusion has a finite subcover.

A reasonable strategy for proving the equivalence was to try to prove it *separately* for each cardinal λ being the relevant parameter. However, this equivalence for a single infinite cardinal λ holds only if λ is a *regular* cardinal. To prove the equivalence between A, B, and C as stated, that is, *for all cardinals together*, they considered for each of the conditions A, B and C its relativization to the class of regular cardinals A^r, B^r and C^r and actually prove the equivalence of all six conditions.

Using their equivalence and the well-ordering of the cardinals, Alexandroff and Urysohn obtain the following elegant equivalent definition of topological compactness: a space X is compact if and only if for every *regular* cardinal λ, X satisfies the λ-complete accumulation point property. This result is very modern in flavor. In particular, the Alexandroff-Urysohn characterization of topological compactness in terms of complete accumulation points allowed them to ignore sets of points of singular cardinality: a space X is compact iff every infinite subset of X of regular cardinality has a complete accumulation point iff every infinite subset—of regular or of singular cardinality—has a complete accumulation point. This was actually convenient, as the standard operations one does with countable *sequences* in a topological space generalize to sequences of regular length, but not to sequences of singular length (e.g., that a sequence can be thinned out to either a constant or a 1-1 sequence of the same order-type). Thus, not having to deal with sequences of singular length was an advantage.

Thus, this set-theoretic characterization of compactness is a very interesting early example of "elimination of singulars" from the natural mathematical discourse. Alexandroff and Urysohn skip over the singular cardinals elegantly. An open cover of singular cardinality μ can be made, by unionizing, an increasing cover of regular cardinality $\mathrm{cf}(\mu)$. The complete accumulation point property for $\mathrm{cf}(\mu)$ implies that the unionized cover has a strictly smaller subcover, and hence also the original cover has a strictly smaller subcover. The equation $\mu = \sum_{\alpha<\mathrm{cf}(\mu)} \mu_\alpha$ is found useful for the calculus of accumulation points.

As Juhász and Szentmiklóssy point out in their [2009], the property of having complete accumulation points satisfies a sort of additivity: if $\mu = \sum_{i<\kappa} \lambda_i$ where κ is regular and $\langle \lambda_i : i < \kappa \rangle$ is an increasing sequence of regular cardinals, then whenever a space X satisfies the complete accumulation point condition for κ and for all λ_i then it satisfies it also for μ. There is no need to go through the language of open covers. Thus, if the accumulation point condition holds for all *regular* cardinals, it holds for all cardinals.

For a topological space X let $\mathrm{CAP}(X)$ be the class of all cardinals λ for which X satisfies the λ-complete accumulation point property. What Alexandroff and Urysohn discovered was that $\mathrm{CAP}(X)$ is "closed under singular sums", that is, if κ and each κ_α for $\alpha < \kappa$ belong to $\mathrm{CAP}(X)$ then also $\sum_{\alpha<\kappa} \kappa_\alpha$ belongs to $\mathrm{CAP}(X)$.

It is quite ironic that in their elimination of singular cardinals from the discussion of accumulation points Alexandroff and Urysohn indirectly prove an important topological property of singular cardinals that rests on their arithmetical structure. One can compare their discovery to Euler's discovery that if every prime number is a sum of four squares, then also every composite positive integer is. In this analogy the regular cardinals correspond to primes, and the singular cardinals to composites. However, Euler did not think (I hope) that composite numbers were of secondary importance! In fact, the desired theorem, eventually proved by Lagrange, was that *every* positive integer was a sum of four squares. It seems that this point of view is missing in the *Memoirs* about the class of cardinal numbers.

This example, discovered by Alexandroff and Urysohn, is the first in a line of many examples of structural properties indexed by a singular cardinal which are inherited from smaller regular cardinals. Cardinal exponentiation, the freeness of an Abelian group and several other topological properties demonstrate that the fact a singular cardinal is built from smaller ones is significant to many mathematical structures of such cardinality.

Next in their discussion of compactness Alexandroff and Urysohn consider a localization of their complete accumulation point condition. They consider *finally compact* spaces, that is, spaces in which the complete accumulation point condition holds for all regular cardinals in an end-segment of the cardinals, and pay special attention to the case in which all uncountable regular cardinals are considered. It is right here that their elimination of singular cardinals fights back. They are unable to prove the full correspondence they had before, and have to settle for a weaker implication. To the modern eye, the reason is clear: eliminating the regular cardinal $\aleph_0$ influences the topological behavior of subsets of cardinality μ for all μ with countable cofinality, not only $\aleph_0$.

However, if one held the view that only regular cardinals mattered—a view that the characterization of compactness supported—then it made sense to conjecture that if the CAP condition held for all regular uncountable cardinals then every open cover could be reduced to a countable subcover, that is, that this property of a space X was equivalent to the *Lindelöf property*.[14]

If, indeed, $\aleph_0$ is removed from the class of *all* infinite cardinals, that is, X is required to satisfy the λ-complete accumulation point property for *every* $\lambda > \aleph_0$, then indeed X is Lindelöf (as the implication from accumulation points to smaller subcovers does not require regularity). However, this property is too strong to *characterize* Lindelöfness, as there are Lindelöf spaces that do not satisfy the complete accumulation point for singular cardinals.[15] But as they required this only for *regular* $\lambda > \aleph_0$, Alexandroff and Urysohn were only able to prove that a space X which satisfies this requirement is *linearly Lindelöf*, that, is, each of its linearly ordered covers has a countable subcover.

The question of whether linearly Lindelöf and Lindelöf are the same remained

[14]A topological space X satisfies the Lindelöf property if each of its open covers has a countable subcover.

[15]Arhangelskii [2008] believes that this was known to Alexandroff and Urysohn.

unanswered for a long time. Alexandroff and Urysohn did not come up with an example of a space X which satisfied the complete accumulation point condition for every regular $\lambda > \aleph_0$ but failed to have the Lindelöf property. If such an example existed, then in it the iterated process of shrinking covers, which Alexandroff and Urysohn described, not to reach a subcover of cardinality $\aleph_0$, it would have to stop at a *singular* of countable cofinality. Thus, the topological weight[16] of such a space would have to be at least $\aleph_\omega$.

Such a non-Lindelöf space was indeed discovered in 1959 by the Russian mathematician A. Miščenko [1962]. His space was a subspace of the product $\prod_n(\omega_n+1)$ with the product topology. In other words, Miščenko used a singular cardinal to construct his example, in fact, a product of distinct regular cardinals. An even easier example is provided in unpublished work of Gruenhage and Buzyakova from the 1990s: Let X be the space of all functions in $2^{\aleph_\omega}$ which satisfy $|f^{-1}(1)| < \aleph_\omega$, with the topology induced from standard product topology on $2^{\aleph_\omega}$. This space is linearly Lindelöf but not Lindelöf.

For our purposes, this historic development illustrates simultaneously the naturalness and fundamentality of singular cardinals on the one hand and the resistance to acknowledge them on the other hand. Linearly Lindelöf spaces are investigated in general topology and their properties are compared with those of Lindelöf spaces. The main point is that a topologist who is interested in the Lindelöf property will inevitably be driven to consider spaces of singular weight—a fact that at first glance may not seem very natural.

Quite recently, Juhász and Szentmiklóssy [2009] proved the following result: if a space X satisfies the complete accumulation point condition for all regular *uncountable* cardinals and in addition for $\aleph_\omega$, the smallest singular cardinal, then X satisfies the condition for *all* $\lambda > \aleph_0$ and is hence Lindelöf. The only tool they use which Alexandroff and Urysohn did not have is Shelah's pcf scales theorem. This is a good example of how very early results about singular cardinals in topology were continued both naturally and by elementary means after a remarkably long pause.

3.2 More topology in products: Rudin's space

Miščenko's space motivated the famous construction of a Dowker space in ZFC by Mary Ellen Rudin [1971]. The problem of Dowker spaces began with a homotopy extension theorem due to Karol Borsuk [1936] which uses the following assumption about a normal space[17] X: the product $X \times [0,1]$ is also normal. At the time, Borsuk did not know whether this indeed was an additional assumption. Perhaps all normal spaces X satisfied that their product with the closed unit interval is also normal.

[16]The *weight* of a topological space X is the smallest cardinality of a topological base for X.

[17]A topological space X is *normal* if any two closed and disjoint subsets of X can be separated by two open sets.

In 1951 the British topologist Clifford Dowker [1951] gave an internal characterization of the normality of $X \times [0,1]$. For a normal X, $X \times [0,1]$ is normal if and only if X is countably paracompact.[18] The name "Dowker space" became a standard name for normal yet not countably paracompact spaces. The existence of such spaces was still unknown at the time Dowker obtained this characterization.

The first Dowker space in ZFC was constructed by Rudin in 1969 (cf. [1971]) and for over 20 years was the only known ZFC example. Rudin was motivated by Miščenko's construction of the subspace of $\prod_n(\omega_n + 1)$. Miščenko raised in his paper the question of whether his space was Dowker or not. Very soon it was observed by many that Miščenko's space was not normal, hence not a Dowker space.

Rudin made the following modifications to Miščenko's construction: she used only functions whose values are everywhere of uncountable cofinality, and she replaced the product topology by the box product topology.[19] Now she was able to prove normality, making essential use of the fact that her space was a *P-space*, that is, satisfied that any countable intersection of open sets was open. She also proved that her space was not countably paracompact by exposing a combinatorial property of unbounded subsets of her partially ordered space.

The cardinality of Rudin's Dowker space is $(\aleph_\omega)^{\aleph_0}$. For many years, Rudin herself promoted the "small Dowker space problem", that is, of finding a ZFC example of a Dowker space with "small" cardinal characteristics, like cardinality, weight and local character. She called her own space, of cardinality and weight $(\aleph_\omega)^{\aleph_0}$ and local character $\aleph_\omega$, "huge" and "ugly". When a ZFC Dowker space of cardinality $2^{\aleph_0}$ and weight $2^{2^{\aleph_0}}$ was constructed by Zoltan Balogh [1996] in a paper titled "A small Dowker space in ZFC" it was accepted by Rudin and many others in the set-theoretic topology community as a solution of the small Dowker space problem. Was the continuum provably smaller that $(\aleph_\omega)^{\aleph_0}$? Even after it was known that the continuum was not bounded in ZFC, and could be, for example, a regular limit cardinal, still the opinion prevailed that $2^{\aleph_0}$ was "small", while $\aleph_\omega$ was "large".

What is absolutely remarkable about Rudin's construction is that Rudin's clear resistance to singular cardinals did not stop her from implicitly discovering several important ingredients of the systematic theory of small products of regular cardinals. Looking carefully through her proof of the failure of countable paracompactness, one can trace elements of pcf theory appearing. In fact, neither normality nor the absence of countable paracompactness required the use of *all* functions in the product. The Rudin Dowker space contains a closed Dowker subspace of cardinality $\aleph_{\omega+1}$ [Kojman and Shelah, 1998], smaller in cardinality than all but the first ω members of the class of possible values of the continuum.

The interested reader may read more about singular cardinals in topology and

---

[18]A normal space X is *countably paracompact* if for every sequence $\{D_n\}_n$ of closed subsets of X which satisfies $\bigcap_n D_n = \emptyset$ there are open sets $U_n \supseteq D_n$ such that $\bigcap_n U_n = \emptyset$.

[19]The *box product topology* on a product $\prod_{i \in I} X_i$ is the topology generated by all sets of the form $\prod_{i \in I} u_i$ where $u_i \subseteq X_i$ is open.

in Boolean algebra in Juhász's book [1980] and in Monk's book [1996], both dealing with cardinal characteristics. Singular cardinals often play a different role in this setting than regular cardinals—we shall mention as a final example just one result, due to Erdős and Tarski [1943], that a singular cellularity[20] of a topological space or of a complete Boolean algebra is always attained.

3.3 The Czech school's investigations of the algebra $\mathcal{P}_{<\mu}(\mu)$

A cardinal number is also referred to in the set-theoretic literature since Hausdorff's time as an "initial ordinal", that is, an ordinal which is not equinumerous with any smaller ordinal. A cardinal is, then, an ordinal which possesses "new" properties, not possessed by smaller ordinals.

It makes sense to discover the *genuinely new* information stored in *the power set* $\mathcal{P}(\kappa)$ of a cardinal κ by modding-out all smaller subsets of κ and inspecting the the quotient Boolean algebra. Let $\mathcal{P}(\kappa)/{<}\kappa$ denote the *completion* of the quotient algebra of $\mathcal{P}(\kappa)$ over the ideal of all subsets of κ of cardinality smaller than κ. For the case $\kappa = \aleph_0$ this algebra corresponds to $\beta\omega \setminus \omega$, where $\beta\omega$ is the *Čech-Stone* compactification of ω. These Boolean algebras were addressed by the Czech school of set theory, and several interesting achievements were obtained, some of which are highly relevant to singular cardinals.

Among the cardinal characteristics studied in this setting one finds the *distributivity number* of $\mathcal{P}(\kappa)/{<}\kappa$, which, in forcing terminology, is the smallest cardinality of a new sequence of ordinals which is added by extending the universe of sets with the Boolean algebra as a forcing notion. In 1972 Bohuslav Balcar and Petr Vopěnka began the study of distributivity in quotient algebras $\mathcal{P}(\kappa)/{<}\kappa$ for infinite cardinals κ. Their first interesting discovery about singular cardinals was quit surprising then. While for regular cardinals κ the distributivity number of $\mathcal{P}(\kappa)/{<}\kappa$ was undecidable over ZFC, Balcar and Vopěnka [1972] proved that for *singular* cardinals μ the distributivity number of $\mathcal{P}(\mu)/{<}\mu$ is completely determined in ZFC by the cofinality of μ as follows: for μ of countable cofinality the distributivity number is ω_1 and for μ of uncountable cofinality it is ω. Additional parameters of distributivity in $\mathcal{P}(\mu)/{<}\mu$ for a singular μ were computed in a series of papers [Balcar and Franěk, 1987; Balcar and Simon, 1989; Balcar and Simon, 1988], usually under additional set-theoretic assumptions.

The Balcar-Vopěnka discovery was made already at a time when the intuition about cardinal characteristics of the continuum emerged that almost everything about them was independent of the axioms. Thus, it was remarkable that the naturally derived structures $\mathcal{P}(\mu)/{<}\mu$ are much more stable for singular cardinals than for regular cardinals.

In their [1987], Balcar and Franěk were able to show that under certain additional axioms, the quotient algebra $\mathcal{P}(\mu)/{<}\mu$ was actually isomorphic to a collapse

[20]The *cellularity* of a topological space is the supremum over all cardinalities of families of pairwise disjoint nonempty open sets of the space.

algebra $\mathrm{Col}(\omega_1, \mu^{\aleph_0})$:[21] in the case that μ had countable cofinality the algebra $\mathcal{P}(\mu)/{<\mu}$ was isomorphic to $\mathrm{Col}(\omega_1, \mu^{\aleph_0})$ and in the other case to $\mathrm{Col}(\omega, \mu^+)$.

The question may have arisen naturally whether these two algebras could be isomorphic just in ZFC, if not for an earlier independence result due to Baumgartner [1973]. By Baumgartner's result on almost disjoint families a negative *consistent* answer was given to the isomorphism question: it was consistent that these two algebras had different cellularities, or, in more detail, that $\mathcal{P}(\mu)/{<\mu}$ has strictly bigger cellularity than that of $\mathrm{Col}(\omega_1, \mu^{\aleph_0})$ for a singular μ of countable cofinality. The higher cellularity, one should point out, resulted from subsets of μ of regular cardinality larger than ω.

Balcar and Simon [1995] then conjectured that it would turn out in ZFC that $\mathrm{Col}(\omega_1, \mu^{\aleph_0})$ was isomorphic to a *complete subalgebra* of $\mathcal{P}(\mu)/{<\mu}$. This was proved by them with various additional axioms, and was finally settled just in ZFC by Kojman and Shelah [2001] with pcf theory.

As was the case with the topological properties of Rudin's space, after the discovery of pcf theory it turned out that some of the Boolean-algebraic properties of $\mathcal{P}(\mu)/{<\mu}$ for a singular μ are not that related to the usual power-set function $\kappa \mapsto 2^\kappa$ but rather closely related to the "binomial" function on singular cardinals (see Section 4 below).

3.4 The Erdős-Rado work in the partition calculus and the Erdős-Hechler work on MAD families over a singular

Paul Erdős was one of the first mathematicians to discover the charm of singular cardinals. The combinatorial arguments he employed in his considerations of singular cardinals were often very elegant. There are two other characteristics of Erdős' approach to infinite combinatorics. The first is his insistence *not* to exclude singular cardinals from the general formulation of a theorem. The other is his readiness to assume GCH for counting arguments.

Among the early results in partition calculus one finds the Erdős-Dushnik-Miller [1941] partition relation, valid for *every* infinite cardinal κ:

$$\kappa \longrightarrow (\kappa, \omega)^2.$$

This symbol means that for every partition of unordered pairs from κ to two parts, if there is no infinite subset of κ all of whose pairs lie in the first part, then there exists a subset of κ of cardinality κ all of whose pairs lie in the other part.

Dushnik and Miller [1941] were interested in this relation for the purpose of showing that any two enumerations of a set in the order-type of its cardinality agree on a large set (which follows from this relation trivially), and were able to prove it for *regular* cardinals. Erdős provided the elegant argument for the *singular* case.

[21]The collapse algebra $\mathrm{Col}(\omega_1, \kappa)$ is the unique complete Boolean algebra in which the partially ordered set of all countable partial functions from ω_1 to κ is densely embedded. This poset is the standard forcing notion for collapsing the cardinal κ to ω_1.

Erdős and Rado begin their monumental work *A Partition calculus in set theory* [1956] by quoting previous results. The very first is Ramsey's theorem. Then comes the relation by Dushnik and Miller above. The third is a negative[22] partition relation:

$$\aleph_{\omega_0} \not\to (\aleph_1, \aleph_{\omega_0})^2.$$

All along the way singular cardinals are treated in this work as equal to their fellow regular cardinals and many a time the combinatorial argument required for a singular case is actually prettier than in the regular case.

It is interesting to point out that a "polarized partition relation" which was proved for singular cardinals of countable cofinality in [Erdős *et al.*, 1965] and was left open for singular cardinals of uncountable cofinality (even with GCH) was finally proved by Shelah [1998] to follow from the assumption that the singular cardinal is a strong limit but its exponent is larger that its successor.[23]

Another beautiful example from the Hungarian school of infinite combinatorics is Hajnal's free set theorem. Erdős' review of Hajnal's paper tells the story completely:

> Let S be a set of power m. To every $X \in S$ there corresponds a subset $f(X)$ of S. We assume that there exists a cardinal number $n < m$ so that for every X, $f(X)$ has power less than m. A subset S' of S is called free if for every every $X, Y \in S'$, $X \notin f(Y)$ and $Y \notin f(X)$.
>
> Ruziewicz conjectured that there always exists a free subset of power m. The first question of this type is due to P. Turán and the first positive results are due to G. Grünwald and D. Lázár. If m is regular the conjecture was first proved by Sierpiński and if m is cofinal to ω it was proved by S. Piccard. Assuming the generalised continuum hypothesis the reviewer proved the conjecture in full generality. The author now proves the conjecture without any hypothesis in a surprisingly simple and ingenious way.

The theorem was first proved for regular cardinals, then for singular cardinals with cofinality ω, then for all cardinals ("in full generality") with GCH (by Erdős himself) and finally in ZFC for all cardinals. Obviously, it was the singular cardinals case that required more ingenuity than the regular cardinal case, and, also obviously, it was not called "the free set theorem" before *all* cardinals were covered in ZFC.

Erdős and Steven Hechler have studied maximal almost disjoint families over a singular cardinal in their [1975]. For a singular cardinal μ denote by $\mathrm{MAD}(\mu)$ the collection of cardinalities of *maximal almost disjoint* families over μ, that is, cardinalities of families $\mathcal{F} \subseteq \mathcal{P}(\mu)$ satisfying that each member of the family has

[22]The negation line on the arrow means that the partition relation expressed by the arrow notation does not hold.

[23]That is, from the *negation* of the Singular Cardinal Hypothesis, for which see the next section.

cardinality μ, the intersection of any two members in the family has cardinality smaller than μ and every set $A \subseteq \mu$ of cardinality μ has intersection of size μ with some member of the family. For obvious reasons, families whose cardinality is smaller than $\mathrm{cf}(\mu)$ are not considered.

Erdős and Hechler began their investigations with an observation that resembles very much the additivity of the complete accumulation point spectrum described in section 3.1 above, and with pretty much the "same" proof: they proved that $\mathrm{MAD}(\mu)$ is closed under singular suprema. That is, if $\{\lambda_i : i < \theta\} \subseteq \mathrm{MAD}(\mu)$ and $\lambda := \sum_{i<\theta} \lambda_i > \theta$, then $\lambda \in \mathrm{MAD}(\mu)$. Using this, they proved that for a singular μ it could be that $\mu \in \mathrm{MAD}(\mu)$—a relation which clearly cannot hold for a regular cardinal! The simplest case to state is that if $2^{\aleph_0} < \aleph_\omega$ then every cardinal in the interval $[2^{\aleph_0}, \aleph_\omega]$ belongs to $\mathrm{MAD}(\aleph_\omega)$.

Then Erdős and Hechler raised the question of whether their assumption was necessary, and in fact conjectured that $2^{\aleph_0} > \aleph_\omega$ together with Martin's Axiom would imply that $\aleph_\omega \notin \mathrm{MAD}(\aleph_\omega)$. Their conjecture was answered positively in [Kojman *et al.*, 2004], in which new ways to control $\mathrm{MAD}(\mu)$ for a singular μ were introduced using pcf theory. The Erdős-Hechler additivity method works plainly only up to μ itself; with pcf theory one can get almost-increasing *smooth* unions up to some cardinal larger than μ, and prove that the whole interval of cardinals from $\min \mathrm{MAD}(\mu)$ to that cardinal is contained in $\mathrm{MAD}(\mu)$.

3.5 Singular cardinal compactness and Whitehead's problem

The algebraic compactness that Shelah discovered and which shall be described now is a phenomenon which is unique to singular cardinals. Let us give a simple example: if an Abelian group G is of singular cardinality μ and G is *almost free*, that is, each of the subgroups of G of cardinality smaller than $|G|$ is free, then G itself is free. This implication does not hold, in general, for Abelian groups of regular cardinality. In the constructible universe, for example, this phenomenon of algebraic compactness does not hold for any regular cardinal other than the weakly compact cardinals, hence consistently does not hold for any regular cardinal at all. The implication "almost free" $\Longrightarrow$ "free" holds for all cardinals—regular or singular—above a compact cardinal.

The discovery of singular cardinal compactness was related to Shelah's solution of *Whitehead's problem* in infinite Abelian groups. Whitehead's problem in infinite Abelian group theory is the following: is every Whitehead group a free Abelian group? An Abelian group G is Whitehead if $\mathrm{Ext}(G, \mathbb{Z}) = 0$, that is, the functor Ext on G is trivial. Every free Abelian group is Whitehead, and Whitehead's problem, which stood open for many years, was whether the converse implication is also true.

Shelah constructed a non-free Whitehead group from $\mathrm{MA}(\aleph_1)$, and thus showed that the implication "Whitehead" $\Longrightarrow$ "free" could not be proved in ZFC (see the discussion in [Steprans, 2010] of the role of this development for the set theory of the continuum). Could this implication be *refuted* in ZFC? Since Shelah's counterexample utilized the additional axiom MA($\aleph_1$), it did not produce a refutation in ZFC.

Shelah proved that from Gödel's axiom of constructibility it followed that every Whitehead group is free, hence that Whitehead's implication cannot be refuted in ZFC. The proof was by induction on cardinality, utilizing the fact that the Whitehead property is hereditary. Given a Whitehead group of cardinality λ, each of its subgroups of smaller cardinality is Whitehead as a subgroup of a Whitehead group, and by the induction hypothesis is therefore free. For a regular cardinal λ, the constructibility axiom is used to assure that also G itself is free. But in the case λ is singular, Shelah discovered a very interesting fact: *every* Abelian group of singular cardinality all of whose subgroups of smaller cardinality are free, is itself free—just in ZFC.

A previous theorem by Paul Hill stated that a group of cardinality μ which is almost free is free whenever μ is singular of cofinality ω_0 or ω_1. It had also been known that every countable subgroup of $\mathbb{Z}^\omega$ is free, but that there are subgroups of $\mathbb{Z}^\omega$ of cardinality $\aleph_1$ which are not free (cf. [Eklof and Mekler, 2002]).

In fact Shelah has done much more. He phrased a simple set of axioms for an abstract notion of freeness, axioms which freeness in Abelian groups satisfies, and proved his *Singular Cardinals Compactness* theorem [Shelah, 1975]: Every structure of singular cardinality which is almost free is free. Many other natural problems fell under the scope of Shelah's set of axioms for freeness.

A reader who is interested in almost-freenes may want to look at the book [Eklof and Mekler, 2002] on this subject.

4 THE ARITHMETIC OF SINGULAR CARDINALS

Cantor and Hessenberg proved that for every infinite cardinal κ the equality $2^\kappa = \kappa^\kappa$, which implies trivially that whenever $\lambda \leq \kappa$ the exponentiation λ^κ is equal to 2^κ. Thus, exponentiations in which the base is not larger than the exponent, reduced to exponentiations with base 2.

The König-Zermelo inequality showed that for a singular μ, $\mu^{\mathrm{cf}(\mu)} > \mu$. Thus, there are nontrivial exponentiations in which the base is larger than the exponent. Zermelo proved in general that whenever $\lambda = \sum_{i<\theta} \lambda_i$ with the λ_i are increasing with i, the product $\prod_i \lambda_i$ has larger cardinality than λ.

Before continuing with the arithmetic of singular cardinals, it is important to note that the "usual" power set function $\kappa \mapsto 2^\kappa$ is not always sufficiently informative for understanding singular cardinals. Suppose that $\lambda = \sum_{i<\theta} \lambda_i$ where θ is regular and the λ_i are increasing. By the usual distributive law, known to Cantor, it follows that

$$2^{\sum_{i<\theta} \lambda_i} = \prod_{i<\theta} 2^{\lambda_i}. \tag{4}$$

In other words, the power set of a singular cardinal is itself representable as a product of the sort considered by König. However, the cardinalities of such products have lives of their own. While in the case of a product of *uncountably* many different cardinals there is some influence on the product from the exponentiations of its factors, in countable products this is more delicate. In particular, the first fixed point of the aleph function μ may be a strong limit and have an arbitrary large 2^μ.

The equation (4) above constituted an observation by Lev Bukovský [1965] and independently by Hechler (see [Jech, 2003]) that if in this case the function $\kappa \mapsto 2^\kappa$ is eventually constant below a singular, then its value persists at the singular itself.

Bukovský in fact proved that all exponentiations are reducible to the *gimmel function* given by $\kappa \mapsto \kappa^{\mathrm{cf}(\kappa)}$. For regular cardinals the gimmel function is exactly the power-set function, and for singular cardinals it is again related to König-type products. If one knows what are the values of 2^κ for regular κ and what are the values of $\mu^{\mathrm{cf}(\mu)}$ for singular μ, then one knows what is κ^λ for every κ and λ.

This indicates that for singular cardinals μ it is not the function $\mu \mapsto 2^\mu$ which matters, but rather $\mu \mapsto \mu^{\mathrm{cf}(\mu)}$. What can be said about this function? Well, the cardinal $\mu^{\mathrm{cf}(\mu)}$ is clearly the total number of functions from $\mathrm{cf}(\mu)$ to μ, which is equal to the total number of $\mathrm{cf}(\mu)$-subsets of μ, that is $|[\mu]^{\mathrm{cf}(\mu)}|$.

Let us now consider a different function, which Cantor neglected to introduce: the binomial $\binom{\mu}{\mathrm{cf}(\mu)}$ which we define as the *cofinality* of the partially ordered set $([\mu]^{\mathrm{cf}(\mu)}, \subseteq)$. This is the smallest cardinality of a collection of $\mathrm{cf}(\mu)$-subsets of μ which *covers* all of $[\mu]^{\mathrm{cf}(\mu)}$. It is justified to call this function "the binomial" because if you apply it to natural numbers $k \leq n$ it gives exactly the usual binomial (because two different k-subsets of n do not cover each other).

It is obvious that $\mu^{\mathrm{cf}(\mu)} = \max\{2^{\mathrm{cf}(\mu)}, \binom{\mu}{\mathrm{cf}(\mu)}\}$. Thus, one can use the binomial for singular cardinals rather than the gimmel function for determining all exponentiations.

König's original argument needs to be only slightly modified to prove, for a singular μ, that

$$\binom{\mu}{\mathrm{cf}(\mu)} > \mu;$$

in fact it gives more, namely that the almost disjointness number of $[\mu]^{\mathrm{cf}(\mu)}$ is larger than μ.

The Singular Cardinals Hypothesis (SCH) is, basically, the statement that exponentiation of singular cardinals assumes the smallest possible value. It was first phrased in a special way as a conditional:

$$\bigwedge_{\kappa<\mu} 2^\kappa < \mu \Longrightarrow 2^\mu = \mu^+.$$

We find it in this form in Azriel Levy's review of [Easton, 1970]:

> The corresponding question concerning the singular $\aleph_\alpha$'s is still open, and seems to be one of the most difficult open problems of set theory in the post-Cohen era. It is, e.g., unknown whether for all $n(n < \omega \rightarrow 2^{\aleph_n} = \aleph_{n+1})$ implies $2^{\aleph_\omega} = \aleph_{\omega+1}$ or not.

This formulation says nothing when μ is not a strong limit.

Then SCH was phrased in a more general way, which makes sense also in when μ is not a strong limit (see, e.g. [Jech, 2003, p.61]):

$$2^{\mathrm{cf}(\mu)} < \mu \Longrightarrow \mu^{\mathrm{cf}(\mu)} = \mu^+.$$

It can be put equivalently as $\mu^{\mathrm{cf}(\mu)} = \max\{\mu^+, 2^{\mathrm{cf}(\mu)}\}$, since for any cardinal $\kappa < \mu$ it holds that if $2^\kappa > \mu$ then $\mu^\kappa = 2^\kappa$. This formulation allows that $2^{(\mathrm{cf}(\mu))^+} > \mu$ while remaining meaningful. When $2^{\mathrm{cf}(\mu)} > \mu$, this formulation says nothing.

However, there is no need to make any conditions about $2^{\mathrm{cf}(\mu)}$ either: the Binomial Hypothesis (BH) is simply

$$\binom{\mu}{\mathrm{cf}(\mu)} = \mu^+.$$

This statement is meaningful also in the case $2^{\mathrm{cf}(\mu)}$ is larger than μ.

This separation of cardinal arithmetic to the exponent function on regular cardinals and to the binomial on singular cardinals enables the separate study of each of the functions independently of the other. Thus, for example, one may ask whether the Binomial Hypothesis for $\aleph_\omega$ holds or not regardless of what the continuum is. (One may observe that adding any number of Cohen reals, for example, has absolutely no effect on $\binom{\aleph_\omega}{\omega}$.)

The gimmel function was not dealt with before Bukovský [1965]. The function $\kappa \mapsto 2^\kappa$ for regular κ was completely understood with Easton's results, that followed Cohen and Solovay. Apart from the König-Zermelo inequality and weak monotonicity there are no rules governing $\kappa \mapsto 2^\kappa$ *on the regular cardinals.*

This created a drastically wrong intuition about the Singular Cardinals Hypothesis. Mathematicians in the 1970s expected that Easton's results would be extended to the singular cardinals. The truth turned out to be totally different. The following table, incorporating present knowledge, illustrates the vastly different behavior of $2^{\aleph_0}$ versus $\binom{\aleph_\omega}{\omega}$.

Properties \ Term	$2^{\aleph_0}$	$\binom{\aleph_\omega}{\omega}$
Possible values of the term	The class of all cardinals κ satisfying $\mathrm{cf}(\kappa) \neq \omega$	Only Regular Cardinals in the interval $[\aleph_{\omega+1}, \aleph_{\omega_4})$. Only regular cardinals below $\aleph_{\omega_1}$ are known to actually occur in models of ZFC.
Structural relations with least possible value	Hardly any structural relation between canonical structures on $2^{\aleph_0}$ and ω_1	Many structural relations between canonical structures on $(\aleph_\omega)^{\aleph_0}$ and $\aleph_{\omega_1}$
Consistency Strength of values other than the least possible one	All possible values are equiconsistent with ZFC	Any value other than $\aleph_{\omega+1}$ requires the consistency of a measurable κ of order at least κ^{++}
Implications of the negations of CH/BH	No information on ω_1 is available in ZFC from $\omega_1 < 2^{\aleph_0}$	$\aleph_{\omega+1} < \binom{\aleph_\omega}{\omega}$ has many implications for $\aleph_{\omega_1}$, some of which are strengthening of consequences of BH

The behavior of the power-set function at $\aleph_0$ is not unique to this cardinal—it is shared by every regular cardinal. There is no bound for 2^κ for a regular κ, and worse, by Easton's result. The binomial of singular cardinals behaves differently. On singular cardinals which are not fixed points of the aleph function there are bounds. At the least fixed point there is no bound. However, above the first cardinal which is compact for Abelian groups (that is, above it, every almost free Abelian group is free) BH holds [Shelah, 1994a]. One can say that some chaos is displayed in the behavior of the binomial, but only for a "small" part of the universe, and at a high consistency-strength cost.

It had already been amply confirmed from Sierpiński's days that CH, that is, the equation $2^{\aleph_0} = \aleph_1$, has many consequences for the structure of the continuum. Less evident is that there is information stored in this equation for $\aleph_1$. However, the negation of CH has very little information about what happens at that cardinality. If Martin's Axiom holds, that all cardinals which are smaller than the continuum "look like" $\aleph_0$. Nothing of the structure of the continuum is inherited by its subsets of cardinality $\aleph_1$. For example, if CH holds then by Hausdorff's work there is a universal linearly ordered set of cardinality $\aleph_1$, a fact which does not necessarily hold without CH. Another example is that by CH there are many pairwise incomparable separable, $\aleph_1$-dense order-types; but if the continuum is larger than $\aleph_1$ there may be only a single isomorphism type of such a set, by [Baumgartner, 1973]. Finally, the existence of an ω_1-scale in $(\omega^\omega, <^*)$, which follows from CH, is not a ZFC theorem, nor is the existence of scale of any length. All this background makes Hausdorff's fantastic discovery of a ZFC (ω_1, ω_1^*)-gap shine in solitude.

In strong contrast to this situation, much of what is true for $\aleph_{\omega+1}$ when BH holds remains true when BH fails. For example, the existence of an $\aleph_{\omega+1}$-scale in $(\prod_n \aleph_n, <^*)$, which is a triviality from BH, is actually a ZFC theorem, that is, holds whether BH holds or not. Furthermore, if BH fails, not only does an $\aleph_{\omega+1}$-scale exist, but actually one exists with additional, combinatorial properties which follow from $V = L$ but are not true for all scales in general—a scale which is called "very good", that is, in which a closed unbounded set of initial segments of cofinality ω_1 are equivalent to strictly increasing sequences (see [Shelah, 1994a, Chap. II]).

An even more intriguing example is the polarized partition relation for a singular with uncountable cofinality [Shelah, 1998], which at the moment is known to follow *only* from the assumption that the singular is a strong limit which violates SCH. As Shelah described it:

> This is a good example of a major thesis from [Shelah, 1994a]:
>
> **Thesis 1.2** Whereas CH and GCH are good (helpful, strategic) assumptions having many consequences and, say, $\neg$CH is not, the negation of GCH at singular cardinals (i.e. for μ strong limit singular $2^\mu > \mu^+$ or, really, the strong hypothesis: $\mathrm{cf}(\mu) < \mu \Longrightarrow \mathrm{pp}(\mu) > \mu^+$) is good (helpful, strategic) assumption.

The fact that the negation of BH is also useful allows one to prove theorems by dividing the proof to cases: one argument if BH holds and another if it fails. (Such, for example, was Shelah's first proof of the existence of an $\aleph_{\omega+1}$-scale.)

In summary, most of the intuition that was gathered about CH and the negation of CH in many decades simply shatters when approaching BH. BH is hard to violate, its negation has consequences, and there are deep, unbreakable connections between $\binom{\aleph_\omega}{\omega}$, $(\aleph_\omega)^{\aleph_0}$, and $\aleph_{\omega+1}$ with many consequences. Several mathematical structures constructed on $(\aleph_\omega)^{\aleph_0}$ reflect down to (absolute) objects of cardinality $\aleph_{\omega+1}$ because the machinery which dictates their properties is binomial rather than exponential. For example, a closed Dowker space of cardinality $\aleph_{\omega+1}$ was constructed inside Rudin's original space, whose cardinality is $(\aleph_\omega)^{\aleph_0}$ [Kojman and Shelah, 1998], with a pcf scale and without forming unrestricted infinite products, which appear in Rudin's construction.

4.1 Annus Mirabilis: the development of singular cardinal arithmetic in 1974

The beginning of modern singular cardinal arithmetic can certainly be dated to Silver's 1974 theorem about the exponentiation of singulars of uncountable cofinality. It is not only the theorem itself which marks the beginning of the modern era, but the realization that followed its publication, that there actually was a theory of singular cardinal arithmetic to be discovered.

By 1974 Solovay had already proved that SCH holds above a compact cardinal, and Magidor had proved that SCH may fail at $\aleph_\omega$, starting from a supercompact cardinal. But Magidor himself, like everyone else, thought that large cardinals were not necessary for such consistency results and that with time it would be discovered that also singular cardinal arithmetic did not really exist. Jensen was investigating his morasses with the intention of putting them to use to proving the consistency of the negation of SCH without appealing to the consistency of large cardinals.

Magidor arrived in Berkeley in 1974 shortly after Jensen left Berkeley for Bonn. Magidor, who was interested in ultrapowers of the universe of sets, was trying to establish that *non-regular ultrafilters* on ω_1 could not exist, by examining ultraproducts of the universe taken with such an ultrafilter. Rather than disproving the existence of such ultrafilters, his efforts led to a surprising discovery, that if there is such an ultrafilter and GCH held below $\aleph_{\omega_1}$ it would continue to hold at $\aleph_{\omega_1}$ itself. By modifying Magidor's proof, Silver proved in ZFC that GCH does not fail for the first time at a singular of uncountable cofinality.

Silver's own account of the discovery of his famous theorem is as follows:

> The immediate stimulus for this result was some work of Kanamori and Magidor[1] concerning nonregular ultrafilters on ω_1. The other principal influences were the result of Scott concerning GCH at measurable cardinals, Some work of Keisler on ultrapowers ..., the two-cardinal theory developed by several model-theorists, some work of Prikry and Silver on indecomposable ultrafilters ... as well as Cohen's and work on nonstandard models of set theory ... [24]

---

[24]The footnote below the line is Silver's and so is the formulation of the properties of the ultrafilter.

[1]The result of Magidor states, in particular: If there is a regular, nonuniform ultrafilter over ω_1 and $2^{\aleph_\alpha} = \aleph_{\alpha+1}$ for all $\alpha < \omega_1$ then $2^{\aleph_{\omega_1}} = \aleph_{\omega_1+1}$.

We shall not survey all of the developments that Silver lists in this passage; rather, the reader is invited to appreciate the enormous quantity and sophistication of techniques that went into stimulating what very soon turned out to be an elementary theorem.

Magidor had presented in a seminar the following proof: Suppose that there is a non-regular ultrafilter on ω_1. By a theorem of Kanamori [1976] who built on earlier work by Ketonen [1976], also a weakly normal such ultrafilter exists. Hence, in the ultraproduct of the universe of sets (which is not necessarily well founded) the constant functions with values ω_α for $\alpha < \omega_1$ are cofinal below the identity function.

Taking a non-wellfounded ultraproduct of the universe of sets, Magidor showed that if GCH was true below $\aleph_{\omega_1}$, then it could not fail at $\aleph_{\omega_1}$. Magidor was hoping to obtain a contradiction with these considerations.

Silver, who was present at the seminar, observed that one did not need to assume that there was such an ultrafilter; rather, the situation which Magidor obtained from this assumption could be forced, without any assumptions, by collapsing 2^{ω_1} to ω, and rather than a real ultrapower of the universe, a *generic* ultrapower of the ground universe sufficed.

The result was stunning:

THEOREM 1 (Silver.) *If* GCH *holds below* $\aleph_{\omega_1}$ *then it holds at* $\aleph_{\omega_1}$.

This result should be viewed in the light of the prevailing view in those days, that Easton's result would soon be generalized also to singular cardinals. Baumgartner and Prikry [1977], and independently Jensen, discovered elementary proofs of Silver's theory shortly after its discovery. Baumgartner and Prikry soon published a self-contained, purely combinatorial proof of Silver's theorem and Jensen circulated mimeographed copies of essentially the same elementary proof.

More mimeographed notes were being circulated from Bonn by Jensen, who was studying the consequences of Silver's result. After several preliminary partial results, Jensen realized that what he was doing in showing that L computes cofinality correctly amounted to showing that any subset from V was covered by a subset in L of the same cardinality plus $\aleph_1$. Jensen's covering lemma was discovered:

THEOREM 2 (Jensen's Covering Lemma.) *If* $0^\#$ *does not exist, then every set of ordinals* A *is covered by a set of ordinals* $A' \in L$ *such that* $|A'| \leq |A| + \aleph_1$.

$0^\#$ is a set of integers whose existence is implied by large cardinals and implies that L is very small compared to V (cf. Sect. 6.2). The lemma provides a surprisingly weak condition for the existence of $0^\#$.

The consequences of this lemma were dramatic. We quote from the introduction to the paper by Devlin and Jensen [1975]:

> Jensen's effort to produce a positive solution [getting SCH to fail by forcing] over L led to total failure. Silver's work then led him to consider the problem from a new perspective. He discovered the statement "$0^\#$ does not exist" ... implies a negative solution to all cases of the singular cardinal problem. But then there cannot be a positive forcing solution over L, since every generic extension of L by a set of conditions satisfies $\neg 0^\#$. ... Our main theorem says, in effect, that if $\neg 0^\#$ then the "essential structure" of cardinalities and cofinalities in L is retained in V.

Jensen continued his work and together with Dodd published in [1981] their work on the core model, from which it followed that for Magidor's result at least the assumption of a measurable cardinal was necessary.

Magidor and Solovay obtained a *bound* on the power of a strong limit singular cardinal of uncountable cofinality from the assumption of a Ramsey cardinal, but their proof was soon subsumed by the proof of Galvin and Hajnal [1975] of a bound in ZFC. Galvin and Hajnal proved, e.g., that if $\aleph_{\omega_1}$ is a strong limit cardinal then $2^{\aleph_{\omega_1}} < \aleph_{(2^{\aleph_1})^+}$. Jech and Prikry [1976] were able to extend such bounds to more cases, like fixed points of the aleph function, using the additional assumption of a precipitous ideal on ω_1, and Magidor [1977a] obtained the bound $(\aleph_{\omega_1})^{\aleph_1} < \aleph_{\omega_2}$ from Chang's conjecture.

Both Silver's theorem and Solovay's theorem would turn out to be special cases of more general pcf theorems. This is the time to describe this theory, to which we devote a separate section.

5 SHELAH'S PCF THEORY

Before describing the historical development of Shelah's pcf theory it is perhaps useful to describe the theory itself. The name "pcf" stands for "possible cofinalities". The basic discovery in Shelah's pcf theory is that products of small sets of regular cardinals have a basis for their set of *possible cofinalities.*

The notion of cofinality was introduced by Hausdorff for linearly ordered sets, as the smallest well-order-type of a *cofinal* subset of the linearly ordered set. With linearly ordered sets, "cofinal" and "unbounded" are the same. Not so with partially ordered sets, where these two parameters may differ.

In a partially ordered set with no finite unbounded set, the least cardinality of an unbounded subset is always a regular cardinal, which is called the *bounding number* of the poset. The cofinality, or *dominating number*, of a poset is the least cardinality of a cofinal, or *dominating*, set, and this time may be of singular cardinality. The bounding and dominating numbers are denoted by $\mathfrak{b}(P,\le)$ and $\mathfrak{d}(P,\le)$ respectively.

When in a poset it happens that the bounding number and the potentially larger dominating number are the same number, it is said that the poset has *true cofinality* that number, which is denoted by $\mathrm{tcf}(P,\le)$.

An infinite product of cardinals $\prod_{i\in A}\lambda_i$ consists of all functions $f : A \to \mathrm{On}$ which satisfy that $f(i) \in \lambda_i$ for all $i \in A$. There is a natural partial ordering on this product, namely, $f < g \iff f(i) < g(i)$ for all $i \in A$. As mentioned earlier, this partial ordering can be "modded out" using an ideal I on A to get $f <_I g$ iff $\{i \in I\colon g(i) \le f(i)\} \in I$. In the well-studied case $(\omega^\omega, <^*)$ where $<^*$ is $<_I$ for the ideal I of finite subsets of ω, the dominating number is uncountable, but undecidable in ZFC. It may be singular. The bounding number is also uncountable, not greater than the dominating number and undecidable in ZFC. Whether $\mathfrak{b}(\omega^\omega, <^*) = \mathfrak{d}(\omega^\omega, <^*)$ or not is also not decidable in ZFC. Lastly, it may be the case that for every ultrafilter U over ω, when modding out by (the dual of) U, hence making $(\omega^\omega, <_U)$ a linearly ordered set in which bounding and domination agree, this number is always strictly larger than $\mathfrak{b}(\omega^\omega, <^*)$.

This discussion may be technical, but it is important to mention these facts as the results which pcf theory obtained go against the intuition that may have been created by earlier results about ω^ω.

In the product $(\prod_n \omega_n, <^*)$, for example, the bounding number is equal to $\aleph_{\omega+1}$ in ZFC (that is, absolutely), the dominating number is *always* regular and for *every* ideal I over ω the bounding number $\mathfrak{b}(\prod_n \omega_n, <_I)$ is equal to the cofinality modulo some ultrafilter disjoint to I, actually to the true cofinality of some non-maximal ideal extending I. The exact details are less important than the general theme that in the product $\prod_n \omega_n$ there exists decidable structure, while in ω^ω there is none.

We quote now the "basis theorem" for all possible cofinalities of structures of the form $(\prod_n \omega_n, <_I)$ for some ideal I:

THEOREM 3 (Shelah's pcf theorem.) *Let A be a set of regular cardinals which satisfies $|A| < \min A$. Let* $\mathrm{pcf}(A)$ *be the set of all bounding numbers of all reduced products $(\prod A, <_I)$ modulo some proper ideal $I \subseteq \mathcal{P}(A)$. Then for every $\lambda \in$* $\mathrm{pcf}(A)$ *there is a set $B_\lambda \subseteq A$, called the* pcf generator *for λ, such that, letting $J_{<\lambda}$ be the ideal generated by $\{B_\theta : \theta \in \mathrm{pcf}(A)$ and $\theta < \lambda\}$:*

- $\lambda = \mathfrak{b}(\prod A, <_{J_{<\lambda}}) = \mathrm{tcf}(\prod B_\lambda, <_{J_{<\lambda}})$.
- *For every proper ideal $I \subseteq \mathcal{P}(A)$ the bounding number $\mathfrak{b}(\prod A, <_I)$ is the least λ for which $B_\lambda \notin I$ and $\mathfrak{d}(\prod A, <_I)$ is the last λ for which $B_\lambda \notin I$.*
- $\mathrm{tcf}(\prod A, <_I) = \lambda$ *iff $\lambda \in \mathrm{pcf}(A)$ is unique such that $B_\lambda \notin I$.*

Implicit in the theorem is that every bounding number is also a true cofinality. With this theorem it is easy to determine what is the bounding number modulo an ideal I and whether or not it is also a true cofinality: the bounding number of $(\prod A, <_I)$ is simply the least λ for which $B_\lambda \notin I$ and true cofinality exists if and only if for every $\theta \ne \lambda$ in $\mathrm{pcf}(A)$ it holds that $B_\theta \in I$.

Shelah's pcf theorem is an algebraic basis theorem for the space of all linearly ordered homeomorphic images of the partially ordered product $(\prod A, <)$. It is *not* something one expects to find if guided by the intuition accumulated in the area of independence results surrounding ω^ω, or the continuum in general.

The modern way to prove this theorem uses the density of exact sequences, that is, for every $<_I$-increasing sequence $\langle f_\alpha : \alpha < \lambda\rangle$ for regular λ there is a $<_I$-increasing sequence $\langle f'_\alpha : \alpha < \lambda\rangle$ such that $f_\alpha \leq f'_\alpha$ for all α and $\langle f'_\alpha : \alpha < \lambda\rangle$ has an exact upper bound $g : A \to \mathrm{On}$. The existence of the latter sequence is proved using some ZFC combinatorics, typically, a stationary set in an intrinsically important ideal $I[\lambda]$, with which an exact sequence is constructed so that two of the three possibilities in Shelah's Trichotomy Theorem are ruled out, leaving only the exact upper bound possibility.

The relation of pcf to arithmetic is as follows: for a singular μ let $\mathrm{pp}(\mu)$ be the supremum of all $\mathrm{tcf}(\prod A, <_I)$ for $A \subseteq \mu$ an unbounded set of regular cardinals of cardinality $\mathrm{cf}(\mu)$ and I an ideal over A which contains all bounded subsets of A. For many singular cardinals it is known that $\mathrm{pp}(\mu)$ is the binomial $\binom{\mu}{\mathrm{cf}(\mu)}$. If fact, the consistency of the opposite is not known.

In particular, if some end-segment A of all regular cardinals below a singular μ has order-type smaller than μ and hence without loss of generality satisfies $|A| < \min A$, then $\max \mathrm{pcf}(A) = \mathrm{cf}([\mu]^{\mathrm{cf}\,\mu}, \subseteq)$. In other words, the binomial function at μ is equal to $\max \mathrm{pcf}(A)$. For such μ,

$$\mu^{\mathrm{cf}(\mu)} = \max \mathrm{pcf}(A) + 2^{\mathrm{cf}(\mu)}.$$

This part of the theory is obtained with the "convexity" or "no hole" theorem, which states that if the set A is an interval of regular cardinals then also $\mathrm{pcf}(A)$ is an interval of regular cardinals.

Another important pcf theorem is that for regular uncountable products, e.g. $\prod_{\alpha<\omega_1} \aleph_{\alpha+1}$, true cofinality exists and is equal to the least possible value, that is, in the case quoted, $\mathrm{tcf}(\prod_{\alpha<\omega_1} \aleph_{\alpha+1}, <_{\mathrm{NS}}) = \aleph_{\omega_1+1}$.[25] This is the underlying basis for, e.g., Silver's theorem.

The pcf theorem is a ZFC theorem, but did not start out as a ZFC theorem. It was not discovered all at once. We describe now the gradual emergence of its components. Let us identify now the exact place where pcf theory was born: Shelah's [1978] construction of a Jonsson algebra on $\aleph_{\omega+1}$ from the assumption $2^{\aleph_0} \leq \aleph_{\omega+1}$.

A Jonsson algebra is an algebra of functions—an infinite set with countably many (finitary) operations—that has no proper subalgebra of the same cardinality as the whole algebra. There is a Jonsson algebra on ω and if there is a Jonsson algebra on a cardinal κ there is one on κ^+. Thus a straightforward induction gives that there are Jonsson algebras on $\aleph_n$ for all $n \in \omega$. What about $\aleph_\omega$? This is still an open problem! The next cardinal in line is, then, $\aleph_{\omega+1}$. Can one combine the Jonsson algebras on each $\aleph_n$ to create one on $\aleph_{\omega+1}$? From modern pcf theory the answer is obviously positive: if $\langle f_\alpha : \alpha < \aleph_{\omega+1}\rangle \subseteq \prod_{n\in B_{\aleph_{\omega+1}}} \omega_n$ is $<^*$-increasing and $<^*$-cofinal in the product then every subset of this scale of the same cardinality

[25] NS denotes the ideal of non-stationary subsets of ω_1.

as the whole scale achieves unbounded values in unboundedly many ω_n, and so a Jonsson algebra is easily created from such a scale.

Does such a scale exist? This is exactly the third in a list of three questions with which the paper [Shelah, 1978] begins, the first being whether a Jonsson algebra on $\aleph_{\omega+1}$ exists in ZFC.

Shelah defines a precursor of the pcf operation, $Psc_D(\overline{\lambda})$ where $\overline{\lambda} = \langle \lambda_i : i < \delta \rangle$ is a sequence of regular cardinals and D is a filter containing all co-bounded sets of δ. Now Psc, "possible scales", is the set of all cardinals λ obtainable as follows: for some subsequence of $\langle \lambda_i : i < \delta \rangle$, replace λ_i in the product by 1. This corresponds to changing the filter D to its restriction $D \restriction B$ to a subset $B \subseteq \delta$. Now the cardinal which is obtained is simply the bounding number of the product $(\prod_{i<\delta} \lambda_i, <_{D \restriction B})$. Since a single filter is considered, there are at most 2^δ localizations $D \restriction B$, hence the set of possible scales has cardinality at most 2^δ.

What Shelah does next is to prove the first version of the "no-hole" theorem, using the assumption that the continuum is small:

THEOREM 4. *Suppose* $\overline{\lambda} = \langle \lambda_i : i < \kappa \rangle$ *is increasing,* κ *regular, and* $\lambda_* = \sum_{i<\kappa} \lambda_i$. *If* $\lambda \in Psc_D \overline{\lambda}$, $\lambda_* < \mu < \lambda$, μ *regular,* D $\aleph_1$*-complete or* $2^\kappa < \mu$, *then* $\mu \in Psc_D \langle \lambda'_i : i < \kappa \rangle$ *for some* $\lambda'_i \leq \lambda_i$, $\langle \lambda'_i : i < \kappa \rangle$ *is not D-trivial.*

Thus he gets a maximal filter D on ω with $\mathrm{cf}(\prod_n \omega_n, <_D) = \aleph_{\omega+1}$, and uses this to get a Jonsson algebra.

We first find here the main themes of pcf theory, but should also appreciate, in retrospect, the way Shelah managed to circumvent the difficulty which may arise by choosing inadvertently a maximal filter which creates cofinality larger than $\aleph_{\omega+1}$. One cannot just prove that for an arbitrary maximal filter the cofinality shall be as desired.

The consideration of $D \restriction B$ already predicts the concept of a pcf generator. Also the concept of a least upper bound is present and least upper bounds are obtained by using the cardinal arithmetic assumptions.

It is quite fascinating to look at this early work and trace the exact point in history at which pcf theory emerged. It is also important to observe that the motivation for this development was a concrete combinatorial problem; not a systematic study of cardinal arithmetic.

In 1980 Shelah was working on stronger versions of Jensen's covering lemma. One of the discoveries made in this context led to the next important step in the development of pcf theory. Shelah discovered that the intersection of a sufficiently closed elementary submodel M of size $\aleph_1$ (for example) with the ordinals below $\aleph_\omega$, say, is *determined* by the model's *characteristic function* $\chi_M(n) := \sup M \cap \omega_n$. By carefully choosing particular pcf scales and employing a sophisticated argument, Shelah obtained control over characteristic functions of sufficiently closed submodels. Now Shelah could use pcf theory to bound $(\aleph_\omega)^{\aleph_0}$.

Thus we find the full scheme of the pcf proof for upper bounds for singular cardinal exponentiation appearing in Shelah's 1980 result:

$$\forall n(2^{\aleph_n} < \aleph_\omega) \Longrightarrow 2^{\aleph_\omega} < \aleph_{(2^{\aleph_0})^+}.$$

pcf theory of a set A was developed here with ultrafilters and with the additional assumption that $2^{|A|} < \min A$. That was a reasonable assumption in the context of the application, but Shelah was aware then of the importance of separating pcf theory from the exponentiation of regular cardinals.

This proof appeared in Shelah's book *Proper Forcing* [1982] right after the sections dealing with the strong covering lemma. The ideal $J_{<\lambda}(A)$ was defined as the collection of all $B \subseteq A$ such that for every ultrafilter U containing B the cofinality of the product mod U is smaller than λ. Also the notions of exact upper bound and true cofinality of an ideal appear.

Shelah obtained his result in 1980 while visiting Ohio State University, where Hugh Woodin had just presented a proof using large cardinals that $(\aleph_\omega)^{\aleph_0}$ could be arbitrarily large with $\aleph_\omega$ being strong limit. After leaving Ohio State University, Woodin discovered a gap in his proof, and was trying to fill it for a while until Shelah's proof of the bound was communicated to him.

Shelah presented his proof of the bound at the logic meeting in Patras, Greece, 1980. Woodin was not present in that meeting, but his earlier announcement created the suspicion (or the hope, for some) that large cardinals would soon be proved inconsistent. However, all large cardinals involved in this incident are still alive and well.

The remaining developments in pcf theory we shall refer to here appear in Shelah's book *Cardinal Arithmetic* [1994a]. In [1990], which became Chapter I of [1994a], Shelah replaces the assumption $2^{|A|} < \min A$ by $|A| < \min A$. The existence of a single generator for $J_{<\lambda^+}$ over $J_{<\lambda}$ for $\lambda \in \text{pcf}(A)$ is not yet proved. This is an important step forward in separating the exponent function from the binomial function.

A major advance is made in Chapter II of [1994a]; Shelah isolated the Trichotomy Theorem for sequences of ordinal functions which are increasing modulo some ideal. This discovery is presented as "Claim 1.2" on page 41 of the book, a presentation which does little justice to its formidable contents. Shelah's trichotomy says, roughly, that for *any* ideal I over an infinite set A and *any* $<_I$-increasing sequence of ordinal functions $\langle f_\alpha : \alpha < \lambda \rangle$ on A of regular length $\lambda > |A|^+$, either the sequence is equivalent modulo some larger ideal to a sequence inside the product of small sets (that is, in case A is countable, the sequence is inside a copy of ω^ω); or there is explicit evidence that the saturation of I exceeds λ; or there is an exact upper bound to the sequence modulo I. If the ideal I is maximal or if $\lambda > 2^{|A|}$, neither of the latter two alternatives can hold, and thus the sequence has an exact upper bound.

This is a remarkable discovery: the class of ordinal functions on a set A satisfies Dedekind completeness for all increasing sequences of length larger than $2^{|A|}$ when quasi-ordered modulo *any* ideal over A. This means that Hausdorff could not have discovered any gap other than the one he discovered in On^ω. Even if Hausdorff was allowed to choose any ideal he wanted over ω and search for gaps of any cardinal length λ in the class of *all* cardinal functions on ω, he would find no other gap in ZFC than the one he discovered: If CH holds, then every sequence of ordinal functions on ω of length $\geq \omega_2$ which is increasing modulo some ideal I over ω has an exact upper bound modulo the same ideal I. Put more bluntly: Hausdorff had discovered the single exceptional case in the whole multitude of $<_I$-increasing sequences of ordinal functions on ω for any ideal I!

Using his trichotomy, Shelah proceeds to prove one of the most useful pcf theorems, the existence of a μ^+-scale for every singular μ, i.e. a sequence which is increasing and cofinal of length μ^+ inside a product $(\prod_{i<\mathrm{cf}(\mu)} \kappa_i, <_{bd})$ for an increasing sequence $\langle \kappa_i : i < \mathrm{cf}(\mu) \rangle$ of regular cardinals with supremum μ. This had many applications, e.g. the existence of a Jonsson algebra on $\aleph_{\omega+1}$ in ZFC, as mentioned earlier.

The development here resembles the steps that Hausdorff had taken some eighty years earlier. Hausdorff first sought an ω_1-scale—which he found, with the aid of CH, but could not find in ZFC—and then moved to construct his gap. Shelah first got an $\aleph_{\omega+1}$-scale from $2^{\aleph_0} \leq \aleph_{\omega+1}$, and then proceeded to discover that such a scale existed in ZFC. One cannot help but wonder what would have happened if Hausdorff was driven to study the product $(\prod_n \omega_n, <^*)$.

To apply the strength of the Trichotomy Theorem inside the product $(\prod A, <_I)$ two ingredients were needed: club guessing and having stationary sets in an intrinsically important ideal $I[\lambda]$. Club guessing was discovered by Shelah in 1988 in his model-theoretic work—to prove that a non-superstable theory has many pairwise non-embeddable models at a singular cardinality. It is a prediction principle which is a formal weakening of Jensen's diamond principle, weaker in the sense that it predicts only closed unbounded subsets (rather than all subsets) of a cardinal with the prediction requirement weakened from equality to inclusion. Formally, if $\kappa < \lambda$ are regular cardinals, a club guessing sequence on $S^\lambda_\kappa := \{a < \lambda \colon \mathrm{cf}(\alpha) = \kappa\}$ is a sequence $\overline{C} = \langle c_\delta : \delta \in S^\lambda_\kappa \rangle$ such that for each $\delta \in S^\lambda_\kappa$, c_δ is a closed unbounded subset of δ of order-type κ and such that for every closed unbounded $E \subseteq \lambda$ there is a $\delta \in S^\lambda_\kappa$ such that $c_d \subseteq E$. The main gain in weakening Jensen's diamond principle in this way is, as Shelah proved, that if $\kappa^+ < \lambda$ and both κ and λ are regular, then a club-guessing sequence on S^λ_κ exists *in* ZFC.

Club guessing appears in [Shelah, 1994a], but the existence of a stationary set $S \subseteq S^\kappa_\lambda$ in $I[\lambda]$ for $\kappa^+ < \lambda = \mathrm{cf}(\lambda)$ does not appear in the book—although Shelah knew this for a long time before the book was printed. These two facts are related: the existence of a stationary set in $I[\lambda]$ is proved using club guessing. With these two facts one proves the density of exact sequences in products $(\prod A, <_I)$ and can prove with them the existence of pcf generators.

In Chapter VIII of [Shelah, 1994a], Shelah proves the existence of a single pcf generator without using the later approach. The mechanism of the proof of a stationary set in $I[\lambda]$ appears in disguise in the proof of the existence of a generator. The other property of pcf that is proved in the chapter is the *localization* theorem, which states that if $\lambda \in \mathrm{pcf}(B)$ where $B \subseteq \mathrm{pcf}(A)$ then there is a subset $C \subseteq B$ of cardinality at most $|A|$ such that $\lambda \in \mathrm{pcf}(C)$. In the very first section of the chapter Shelah proves a pcf generalization of the Galvin-Hajnal theorem for his pp function.

By the Fall of 1989 all ingredients were ready to the next great achievement of pcf theory—the $\aleph_{\omega_4}$ bound. Shelah was visiting Mathematical Sciences Research Institute in Berkeley, where, according to testimony in his own book, he was eager to contradict Leo Harrington's opinion that cardinal arithmetic was "dead" ([Shelah, 1994a, p.359]):

> I was interested in this particularly since 80, but the last impetus was my failure in Fall 89 to convince Harrington to retract his 86 statement: "cardinal arithmetic, this was a great problem, [but is now] essentially finished."

The set $\text{pcf}(\{\aleph_n : n \in \omega\})$ is an interval of regular cardinals on which the pcf operation defines a closure operation, hence a topology. This topology is countably-tight,[26] by the localization theorem. Furthermore, this topology is related to the order topology by the representation theorem of μ^+. Now a purely combinatorial argument proves that such a structure has cardinality smaller than ω_4.

This proves that $\max \text{pcf}(\{\aleph_n : n \in \omega\}) < \aleph_{\omega_4}$. By the equality of max pcf with the binomial it follows that

$$\text{cf}([\aleph_\omega]^\omega, \subseteq) < \aleph_{\omega_4}.$$

Thus, unlike the exponentiation $2^{\aleph_0}$ which has no bound in the list of alephs, the binomial of $\aleph_\omega$ is bounded by $\aleph_{\omega_4}$.

This gives immediately the following bounds:

$$2^{\aleph_0} < \aleph_\omega \Longrightarrow (\aleph_\omega)^{\aleph_0} < \aleph_{\omega_4};$$

$$\forall n(2^{\aleph_n} < \aleph_\omega) \Longrightarrow 2^{\aleph_\omega} < \aleph_{\omega_4}.$$

The significant improvement in bound $\aleph_{\omega_4}$ over Shelah's 1980 bound $\aleph_{(2^{\aleph_0})^+}$ is that the subscript ω_4 does not contain any exponentiation. This is perhaps Shelah's best known application of pcf theory.

As Shelah writes about the problem of what the actual bound is:

> So really our problem breaks to two:
>
> (a) Is it consistent to have such [a pcf] structure on specific $\gamma(*)$?
>
> (b) If there is such a structure can we then force a suitable cardinal arithmetic such that it is the structure corresponding to it?
>
> What is accomplished here ... is that in (a), for $\aleph_\omega$, necessarily $\gamma(*) < \aleph_4$. ... But (b) is widely open, and it does not seem unreasonable at all, to me at least.

[26] Countable tightness in a topological space X is the property that for every subset $A \subseteq X$ and a point x in the closure of A there is a countable $B \subseteq A$ with x in the closure of B.

The same proof works to bound the binomial of every singular which is not a fixed point of the aleph function. For the first fixed point the theory breaks down, as there is no interval of regular cardinals of the right form unbounded in this singular cardinal. As Shelah puts it:

> ...before 1980 we could have thought that for example for strong limit λ of cofinality $\aleph_0$ there are no restrictions on 2^λ ... we know now there are some but have less clear idea what to expect for fix points.

In the same chapter the relation between $\mathrm{pp}(\mu)$ and the binomial of μ is clarified for singular μ. Earlier, Shelah [1986] proved that if μ is strong limit singular of uncountable cofinality with no weakly inaccessible cardinal below it then 2^μ is smaller than the first weakly inaccessible cardinal.

In [1983] Shelah had already proved that for the first fixed point of order ω_1 of the aleph function of cofinality $\aleph_1$[27] this was the best upper bound. Also, for every limit δ of uncountable cofinality, for club many $\alpha < \delta$ of countable cofinality $\mathrm{pp}(\beth_\alpha)$ is equal to the binomial of $\beth_\alpha$. Finally, if $\mathrm{cf}(\delta) = \aleph_0$ and the binomial of $\beth_\delta$ is strictly larger than $\mathrm{pp}(\beth_\delta)$, then $\mathrm{pp}(\beth_\delta)$ must be very large: there is some fixed point μ of the aleph function of uncountable cofinality between $\beth_\delta$ and its pp.

A remarkable result of Moti Gitik [2005] showed that the least fixed point of the aleph function, which is the least cardinal for which pcf theory does not give a bound on its binomial, indeed has no bound on its binomial. This result required the full machinery of large cardinal forcing developed by Magidor and Gitik in over 20 years. This result of Gitik closed a circle that began in 1980 when Woodin thought he could prove that there was no bound on the power set of a strong limit $\aleph_\omega$. In exactly the same year Shelah discovered the $\aleph_{(2^{\aleph_0})^+}$ bound. Gitik proved for the first fixed point of the aleph function what Woodin tried to prove for $\aleph_\omega$.

In the very first section of [1994a] Shelah proves that the least μ for which $\mathrm{pp}(\mu) > \mu^+$ satisfies that $\mathrm{pp}(\mu)$ is equal to its binomial. This was the missing part for completing the work of Gitik and Mitchell on the consistency strength of the Singular Cardinal Hypothesis. Gitik and Mitchell proved that the negation of SCH is equiconsistent with a measurable κ of order κ^{++}.

Shelah continued to develop pcf theory in many directions. He [1997] investigated relaxations of the condition $|A| < \min A$ and considered the theory in models without AC. Research on pcf theory and on its application is conducted presently by set-theorists and by mathematicians from other fields.

It should be made clear that only a small part of pcf theory had been described here. Not even all of the material in *Cardinal Arithmetic* has been described, while the volume of results in pcf theory and its application has doubled since the appearance of the book. Shelah produces formidable results faster than what the author of this paper can read, and even a well-known text like the book contains

[27] A cardinal $\aleph_\alpha$ is a fixed point if $\aleph_\alpha = \alpha$. It is a fixed point of order two if it is a fixed point and is equal to its own index in the increasing enumeration of fixed points, and so forth.

too many results to digest in a life time. A reader who wishes to learn more about pcf theory should consult the *analytical guide* at the end of the book and Shelah's archive: http://shelah.logic.at.

5.1 *The revised* GCH *above* $\beth_\omega$

A result of special status is Shelah's [2000b]. In this work Shelah provides a new perspective on cardinal arithmetic by proving in ZFC that a natural cardinal arithmetic function asymptotically assumes the smallest possible value.

We may recall the value of $2^{\aleph_0}$ is undecidable in a formidable way: the axiom system ZFC allows $2^{\aleph_0}$ to be any cardinal $\aleph_\alpha$ for which $\mathrm{cf}(\aleph_\alpha) \neq \omega$. Thus CH is very far from being "true in ZFC". The situation with GCH on regular cardinals is not much better, as by Easton's result there are no restrictions on regular cardinal exponentiation other than König's lemma and weak monotonicity.

With full exponentiation being separated into the binomial on singular cardinals and exponentiation on regular cardinals, it made sense to clarify what rules governed the binomial. As long as the singular at hand is not a fixed point of countable cofinality of the aleph function, there is a bound (whether the "4" in the bound is best possible or not is a separate issue) and the binomial of the first fixed point of countable cofinality is *unbounded.*

In 1990, the British model theorist Wilfrid Hodges communicated to Shelah a problem posed by another British mathematician, Kishor Kale: Do the cardinalities of countably generated topologies satisfy the continuum hypothesis? Suppose that τ is a topology, not necessarily Hausdorff, nor even T_0, which has a countable basis. Does it follow that the number of open sets with respect to τ is either countable or equal to the continuum?

Shelah's answer was positive and appeared in [1993], but that was not the end of this investigation. He examined what happens if one replaces $\aleph_0$ by $\beth_\omega$. In [1994b] Shelah proved the analogous theorem for $\beth_\omega$: if a topology τ has a base of cardinality at most μ and for some strong limit cardinal λ of cofinality ω and $\lambda \leq \mu < 2^\lambda$, then the size of τ is either $\leq \mu$ or $\geq 2^\lambda$. Shelah's first proof used cardinal arithmetic assumptions that were shortly later replaced by the revised GCH theorem.

In [2000b] Shelah formulated a different binomial function: $\lambda^{[\kappa]}$ is the least cardinality of a family of κ-subsets of λ with the property that every κ-subset of λ is contained in the union of fewer that κ members of the family. Clearly, for every singular cardinal μ it holds that $\mu \leq \mu^{[\mathrm{cf}(\mu)]} \leq \binom{\mu}{\mathrm{cf}(\mu)}$. However, a totally new phenomenon emerged with this revised form of the binomial. This new function seemed to be able to absorb almost all independence fluctuations, and assumes asymptotically the smallest possible values. Shelah proved in ZFC that for every $\lambda \geq \beth_\omega$, for every sufficiently large $\kappa < \beth_\omega$ it holds that $\lambda^{[\kappa]} = \lambda$. This result is Shelah's "revised GCH in ZFC".

This theorem has applications (some of which are described below) and will have more applications in the future, but its primary significance is in changing the attitude towards the axiomatic system ZFC in the context of infinite cardinal arithmetic. Placing Easton's result that there are no rules of singular cardinal exponentiation at one extreme, this theorem secures the opposite extreme position. The revised GCH in ZFC demonstrates that a natural cardinal arithmetic function assumes the smallest possible value asymptotically.

The asymptotic minimality of the revised binomial is not an isolated cardinal arithmetic fact, but actually has applications which change the standard intuition about using cardinals in mathematics. Let us quote an example.

A Boolean algebra B is μ-linked if $B \setminus \{0\}$ is the union of at most μ sets, in each of which any two elements are compatible (have non-zero intersection); it is ccc, i.e. satisfies the countable chain condition, if in every uncountable subset of the algebra there are two elements with non-zero intersection. Hajnal, Juhász and Szentmiklóssy in their [1997] proved the following interesting structure theorem for Boolean algebras: If $\mu = \mu^{\aleph_0}$ then every ccc Boolean algebra B whose cardinality satisfies $|B| \leq \mu^{+\omega} < 2^{\mu}$ is μ-linked.

What about larger cardinalities below 2^{μ}? Well, the same authors proved that this was independent: for $|B| = \mu^{+\omega+1}$ the condition could hold or could fail. This may indicate that nothing more can be proved on the relation between chain condition and μ-linkedness, as Shelah writes ([2000a, Sect. 8]):

> This gives the impression of essentially closing the issue, and so I would have certainly thought some years ago,...

Now Shelah proves: If $\mu = \mu^{<\beth_\omega}$ and B is a ccc Boolean algebra whose cardinality satisfies $|B| < 2^{\mu}$ then B is μ-linked. Here the result is gotten for *all* cardinals below 2^{μ}, rather than only for the first $\omega + 1$ of them. The proof uses the revised GCH above $\beth_\omega$.

What has changed in Shelah's perspective? It is the awareness that things do not always become more complicated along the sequence of cardinals. Once Hajnal, Juhász and Szentmiklóssy pushed their proof up to $\mu^{+\omega}$ and showed that the next case is consistently false, then why should anyone bother to look higher? What was shown is that there is a qualitative difference between the condition $\mu = \mu^{\aleph_0}$ and the condition $\mu = \mu^{\beth_\omega}$.

This is a new sort of intuition about infinite cardinals. One may say, more mature, or less naïve. One finds such a point of view among the mathematicians who look at the natural numbers. There it is not expected that simple intuitions will guide the researcher correctly. Many years of number theoretic investigations have accustomed number theorists to expect surprises. With cardinal arithmetic this is only beginning.

Kurt Gödel [1947] introduced "verifiable consequences" as a means by which new axioms should be evaluated. A new axiom could help mathematicians easily prove theorems which may later be proved without it. Such is, for example, the Axiom of Choice. In the same paper Gödel uses his own criterion to reject CH. Among all consequences of CH, he explains, none turned out to be "really" true. That of course projected on Gödel's own constructibility axiom $V = L$, which implies CH. Later mathematical discoveries, like Scott's theorem (see below) also indicated that L may be very far from the "real" universe of sets.

However, as one moves up the list of cardinal numbers, the axiom $V = L$ seems to have more and more "verifiable consequences", if not literally, then at least metaphorically. Asymptotic regularity along the sequence of cardinal numbers emerged in many combinatorial discoveries of Shelah in the sense that combinatorial principles which hold in L by Jensen's work had weaker forms that are true in ZFC—from some point onwards.

Shelah's revised GCH in ZFC is the culmination of the "eventual regularity" theme, or, the verifiability of $V = L$ theme, which emerged from previous pcf discoveries. The standard GCH is of course not a ZFC theorem, and even the binomial hypothesis is not a ZFC theorem, though its violation requires consistency strength, but the revised GCH is a ZFC theorem—above $\beth_\omega$. A. Kanamori writes in his review of [Shelah, 2000b]:

> Bewildered by the sheer volume of [Shelah's] work, one can lose the sense for those few among the many achievements that might have an even higher claim to significance. The current paper under review, number 460, is such a paper, and although it was written in the early 1990s, it has not appeared in print until now. ...This is a remarkable ZFC result about the small size of covering families. Whether or not one regards the result as an appropriate "solution" to Hilbert's first problem, it is certainly a triumph for the author's pcf theory, as developed in his book.

In spite of its evident significance, Shelah's [2000b] claim that this result was a solution to Hilbert's first problem did not stir any serious discussion. On the one hand, this is certainly not Hilbert's original problem. On the other hand, after Easton's result, which Hilbert of course was not acquainted with, the problem, viewed broadly as "to discover the rules of cardinal exponentiation", needed to be reformulated. If one adopts Hilbert's problem to "is cardinal arithmetic simple" or even "do the arithmetic functions assume their least possible values" then this is indeed a partial solution: a natural cardinal arithmetical function assumes *asymptotically* its least possible values, in every universe of sets which satisfies ZFC.

As an application of his theorem, Shelah proved the following: for every $\lambda \geq \beth_\omega$, $2^\lambda = \lambda^+$ if and only if Jensen's diamond holds on λ^+. Kanamori addresses in his review this new connection between GCH and Jensen's diamond:

> The two principles had long been known to be distinct at small cardinals, but Shelah's result says that for sufficiently large λ, they are equivalent! There are further applications to topology and to model theory. Once again, the power of the author's pcf theory is demonstrated with surprising and elegantly concise results provable in ZFC.

Shelah's discovery of the revised GCH in ZFC enabled many new proofs by induction along the cardinals. Some of these inductions have to begin above $\beth_\omega$, though. A new type of intuition has formed: there is eventual regularity in ZFC. While in the small cardinals there may be a radically different behavior in V than the behavior in L, asymptotically V behaves in many ways like L.

Most of Shelah's diamond consequences were superseded by an even more surprising theorem of his, from 2007: for every uncountable λ, $2^\lambda = \lambda^+$ if and only if diamond holds on λ^+. Thus, $\aleph_0$ is the *only* infinite cardinal on which and instance of GCH may hold *without* being accompanied by Jensen's diamond on its successor.

This is, philosophically, an interesting example in the development of the subject, as the "simplest case" was again, as in other cases, not discovered first: Shelah first proved the equivalence of instances of GCH with Jensen's diamond for fairly large cardinals, and only later, and with a simpler proof, proved it for all successors of uncountable cardinals.

6 FROM MEASURABLE CARDINALS TO LARGE CARDINAL FORCING

Measurable cardinals are never singular! But measurable cardinals theory nevertheless induced several major developments in singular cardinal theory.

6.1 Ulam's discovery of measurable cardinals and a regular limit continuum

The discussion of Ulam's [1930] discoveries about measurable cardinals belongs naturally to the opening lines of a history of large cardinals. This chapter of modern set theory is, however, relevant to the history of singular cardinals.

The first relevance is that by proving that the least real-valued measurable cardinal is either at most the continuum or else a strong limit cardinal, Ulam directed attention for the first time to the possibility that the continuum may be a limit cardinal. Ulam discovered that in yet another branch of mathematics—measure theory—the relation between a cardinal and its cofinality is useful. Thus, the least cardinal which carries a real-valued measure cannot be a successor, and cannot be singular. This makes the least real-valued measurable cardinal a regular limit—one of those "exorbitantly big" cardinal numbers which Hausdorff predicted would not be related to any meaningful mathematical consideration. Furthermore, if the real-valued measure is atomless, the continuum is at least as large as the least real-valued measurable.

6.2 Scott's theorem

Set theory developed as a modern discipline from Gödel's work on the constructible universe. The role of formal logic and the concept of the universe of set theory became central, and the line of research of set theory as a *set of axioms* became a central branch of investigation.

Łoś's discovery of ultrapowers and Ulam's work on measurable cardinals were combined by Scott [1961] to introduce the construction of the ultraproduct embedding of the universe of set theory into a transitive class. Two theorems were proved by Scott: first, that if GCH fails at a measurable cardinal κ then it fails in many cardinals below κ; second, that if a measurable cardinals exists, then $V \neq L$.

This was the first time in which the way L sits inside V was investigated, and in which it was shown that L could be different from V—if large cardinals exist.

After Scott's theorem further work was done by Gaifman and by Rowbottom, who showed that if large cardinals exist not only did L differ from V but actually L was very small—-e.g. contains only countably many real numbers.

Silver's Ph.D. thesis completed this line of investigation by showing that a small L was *very* small—that assuming a set of integers $0^{\#}$ exists as given by large cardinals, all cardinals from V, for instance, are indiscernible, inaccessible and so on in L. In particular, singular cardinals in V are regular from the point of view of L. This meant that singular cardinals could be much larger in an inner model than what they seemed in V. This possibility was manifested much more sharply after Prikry showed how a measurable could be turned into a singular of countable cofinality by forcing while preserving all cardinals as cardinals; yet already at that stage of history, the importance of singular cardinals to the study of the relation between V and L became evident.

The role singular cardinals played in this branch of set theory became much more central when Jensen derived the first consequences from the assumption that $0^{\#}$ did not exist after Silver's theorem about singular cardinals. Jensen's covering lemma was discovered under direct influence of Silver's theorem, and implied that if $0^{\#}$ did not exist, that all singular cardinals in V are also singular in L. Thus, although their elementary theory was not yet explored in depth, singular cardinals served as a central tool for detecting the distance between L and V.

6.3 Generic sequences: Prikry's discovery

After the invention of forcing, measurable cardinals were found related to singular cardinal in a surprisingly intimate way.

In 1970 Karel Prikry made a formidable discovery: that an unbounded ω-sequence to a measurable cardinal κ could be introduced by a forcing notion, called ever since its discovery "Prikry forcing", without collapsing the cardinal, without introducing bounded subsets of κ and without changing the universe above κ. In short, Prikry's forcing does virtually nothing else other than changing the cofinality of κ to ω. In other words, the singular of cofinality ω is a strong limit regular—in fact measurable—cardinal in an inner model which is hidden below the full universe V by a thin crust.

The unbounded sequences in a singular cardinal received a totally different meaning after Prikry's discovery. Prikry sequences are still the basis to all modern large cardinal forcing constructions for increasing the power of a strong limit singular cardinal. Prikry himself observed that his forcing could provide a counterexample to the SCH if it could be arranged to have a measurable cardinal κ which violates GCH. Indeed, Silver managed to start with an even larger large cardinal—a supercompact—and by forcing turn it into a measurable κ with $2^\kappa > \kappa^+$, actually, with 2^κ arbitrarily large. Now following Silver's forcing by Prikry's, this κ became a strong limit singular of cofinality ω which satisfied $2^\kappa > \kappa^+$. It is important to observe that the singular κ was very high on the list of alephs. In retrospect, this made a difference, as for strong limit singular cardinals below the least fixed point of the aleph function it is, as we now know, impossible to make their power set arbitrarily large. The feeling at the time was that the use of a large cardinal to obtain a violation of SCH would turn out to be unnecessary.

The next major development was Magidor's technique of changing the cofinality of a large cardinal. Magidor combined a system of Levy collapses with Prikry's method of adding an ω-sequence to a large cardinal and obtained in [1977b] a model in which $\aleph_\omega$ was strong limit and $2^{\aleph_\omega}$ was larger than $\aleph_{\omega+1}$, as large as $\aleph_{\omega+\omega+1}$. Thus, the first singular cardinal—$\aleph_\omega$—was shown to consistently violate SCH. The difference between the violation of SCH at a "large" singular by combining Prikry and Silver forcing, described above, to Magidor's result is not merely quantitative. In retrospect we know that the structure of the product of the ω_n's imposes strong restrictions on the power set of $\aleph_\omega$, while at larger singular cardinals, e.g. fixed points of the aleph function, such limitations are not present. Magidor's forcing construction had to overcome these difficulties, before pcf theory was discovered.

In a subsequent paper [1977c] Magidor started from a larger large cardinal, managed to obtain an even more stunning result: $\aleph_\omega$ is the first cardinal violating GCH. This means that for every $n \in \omega$ it holds that $2^{\aleph_n} = \aleph_{n+1}$ but $2^{\aleph_\omega} = \aleph_{\omega+2}$ in Magidor's model.

Still, in 1973, Magidor—like everyone else working in this field —was convinced that large cardinals were not truly needed to force a failure of SCH and that eventually Easton's result would be extended to all cardinals by some forcing over, say, L.

Magidor was able to obtain violations of SCH in an arbitrary long initial segment of the ordinals with the techniques he introduced. After Radin's forcing [1982] was introduced, with which a club is added to a large cardinal, Foreman and Woodin [1991] were able to improve on Magidor's techniques to obtain a model in which GCH fails at every cardinal. Cummings [1992] used Radin forcing to obtain a model in which GCH holds at regular cardinals and fails at singular cardinals.

In [1983] Shelah built on Magidor's work to increase the lower bound on the power of a strong limit $\aleph_\omega$. For an arbitrary $\alpha < \omega_1$ Shelah proved the consistency of $2^{\aleph_\omega} = \aleph_{\alpha+1}$ while $2^{\aleph_n} < \aleph_\omega$ for all n.

An important breakthrough in Prikry-type forcing was obtained by Woodin in the early 1980s (unfortunately, unpublished). Woodin was able to replace the supercompact cardinal by a $(\kappa, \kappa^{++} + 1)$ extender in obtaining GCH up to $\aleph_\omega$ and a gap of 2 above. This was a major step nearing the exact consistency strength, which is a measurable κ of order κ^{++}.

Gitik and Magidor, who, as they write in the paper [1992], were "grateful for the opportunity to work together in Berkeley", introduced a new type of forcing to obtain GCH below $\aleph_\omega$ and $2^{\aleph_\omega} = \aleph_{\alpha+1}$ for an arbitrary countable ordinal α. They used a directed system of extenders related by projections so that a Prikry sequence of a higher one created the Prikry sequence of a lower one. The directedness secured the κ^{++}-chain condition for their forcing. They obtained the lower bound of Shelah, with GCH below $\aleph_\omega$, from a weaker large cardinal hypothesis.

In a series of papers Gitik investigated this type of forcing further. He managed to replace the Woodin $(\kappa, \kappa^{++}+1)$ extender by a (κ, κ^{++}) extender, and continued to tighten the lower bounds obtained by Mitchell on the consistency strength of the negation of SCH. By around 1990 Gitik and Mitchell (cf. [Gitik, 1989; Gitik, 1991]) completed the mission of computing the exact consistency strength of the failure of SCH at $\aleph_\omega$: a measurable κ of order κ^{++}.

With Gitik's proof that there was no bound on the binomial of the least fixed point of the aleph function, basically all cardinal arithmetic problems concerning $\kappa \mapsto 2^\kappa$ that were expressible before the discovery of pcf theory were solved.

7 THE 21ST CENTURY

Singular cardinals theory is very active in the present century. In 2004 a workshop titled "Cardinal arithmetic at work" was held in Jerusalem, where more than half of the results which were presented were about singular cardinals. In 2004 in Banff, a meeting was held which was solely devoted to "singular cardinals combinatorics".

Many important results have been obtained about singular cardinals in the last decade. Of these let us mention Matteo Viale's proof [2006] that the Proper Forcing Axiom implies SCH, which built on earlier work by Moore [2006]. Assaf Sharon's forcing construction of the failure of SCH with combinatorial conditions and finally Itai Neeman's proof [2009] of the consistency of the failure of SCH at a singular of countable cofinality μ together with the tree property at μ^+.

More work is being done by Shelah and others on pcf without AC. Applications of pcf theory to topology, measure theory and algebra continue to flow.

The mathematical subject of singular cardinals is most likely only in it beginning, and someone should certainly revise its history in the 22nd century.

8 SUMMARY AND CONCLUDING REMARKS

The understanding of singular cardinals developed in the 20th century in a chaotic way. The main feature of this development, as seen from the description above, is that existing techniques were not accompanied by suitable conjectures. In other words, not all that could be discovered with elementary methods was discovered; much of it was not even conjectured.

Singular cardinals appeared in many subjects, and their centrality to mathematics was well established. Within set theory itself, singular cardinals play a role in several contexts. First, they serve as gauges for the distance between L and V, and thus have metamathematical significance. Second, their arithmetic is where ZFC could restore some of its lost honor after it was discovered that it could prove next to nothing about *regular* cardinal arithmetic.

Finally, in tracing the major developments in the theory, one finds that concrete mathematical problems were several times the stimulus for important discoveries in set theory. This is true for set theory in general, of course, as Cantor discovered the uncountability of $\mathbb{R}$ in his investigations of uniqueness of trigonometric series, and is evident in the history of singular cardinals as well. Topological investigations related to homotopy led to the first constructions of spaces in the product $\prod_n \omega_n$ whose more advance properties touched pcf theory; the problem of a Jonsson algebra on $\aleph_{\omega+1}$ is where pcf theory is directly born; and Kishor Kale's innocent looking problem about the number of open sets in a countably generated topology let quite quickly to Shelah's discovery of the revised GCH in ZFC.

Many of the advanced techniques for treating singular cardinals are widespread. Large cardinal forcing, L-like combinatorics and pcf theory are being used daily by young set theorists, and perhaps soon by mathematicians from neighboring subjects. There is still much left to be explored.

Having traced the fascinating connection between Hausdorff's work and Shelah's work, it is fitting to conclude this article by the following mathematical problem which is related to the work of both mathematicians. This problem is a good benchmark for the existing techniques, and it illustrates that not all singular cardinals are alike.

Hausdorff proved that for every strong limit cardinal μ there is a universal linearly ordered set of cardinality μ. A partial converse was proved by Kojman and Shelah [1992] for an *initial segment* of the singular cardinals: For every singular μ below the least fixed point of second order, if μ is not strong limit, then a universal linearly ordered set of cardinality μ does not exist. (A fixed point is a cardinal $\mu = \aleph_\mu$ and a fixed point of the second order is a cardinal μ which is the μ-th fixed point.)

Does this converse hold for all cardinals? Does it fail at the least fixed point of second order? Does it fail higher?

BIBLIOGRAPHY

[Alexandrov and Urysohn, 1929] P. S. Alexandrov and P. S. Urysohn. Mémoire sur les espaces topologique compacts. *Verhandelingen der Koninklijke Nederlandse Akademie van Wetenschappen Tweed Reeks. Afdeling Natuurkunde*, 14:1–93, 1929.

[Arhangelskii, 2008] A. Arhangelskii. Homogeneity and complete accumulation points. *Topology Proceedings*, 32:239–243, 2008. Spring Topology and Dynamics Conference.

[Balcar and Franěk, 1987] B. Balcar and F. Franěk. Completion of factor algebras of ideals. *Proceedings of the American Mathematical Society*, 100(2):205–212, 1987.

[Balcar and Simon, 1988] B. Balcar and P. Simon. On collections of almost disjoint families. *Commentationes Mathematicae Universitatis Carolinae*, 29(4):631–646, 1988.

[Balcar and Simon, 1989] B. Balcar and P. Simon. Disjoint refinements. In J. D. Monk and R. Bonnet, editors, *Handbook of Boolean Algebra*, volume 2. North-Holland, Amsterdam, 1989.

[Balcar and Simon, 1995] B. Balcar and P. Simon. Baire number of the spaces of uniform ultrafilters. *Israel Journal of Mathematics*, 92:263–272, 1995.

[Balcar and Vopěnka, 1972] B. Balcar and P. Vopěnka. On systems of almost disjoint sets. *Bulletin de l'Académie Polonaise des Sciences, Série des Sciences Mathématiques, Astronomiques et Physiques*, 20:421–424, 1972.

[Balogh, 1996] Z. Balogh. A small Dowker space in ZFC. *Proceedings of the American Mathematical Society*, 124:2555–2560, 1996.

[Baumgartner and Prikry, 1977] J. Baumgartner and K. Prikry. Singular cardinals and the generalized continuum hypothesis. *The American Mathematical Monthly*, 84(2):108–113, 1977.

[Baumgartner, 1973] J. Baumgartner. All $\aleph_1$-dense sets can be isomorphic. *Fundamenta Mathematicae*, 79:101–106, 1973.

[Borsuk, 1936] K. Borsuk. Sur les prolongements des transformations continues. *Fundamenta Mathematicae*, 28:99–110, 1936.

[Bukovský, 1965] L. Bukovský. The continuum problem and powers of alephs. *Commentationes Mathematicae Universitatis Carolinae*, 6:181–197, 1965.

[Cummings, 1992] J. Cummings. A model in which GCH holds at successors but fails at limits. *Transactions of the American Mathematical Society*, 329:1–39, 1992.

[Dauben, 1990] J. W. Dauben. *Georg Cantor: His Mathematics and Philosophy of the Infinite.* Princeton University Press, Princeton, 1990.

[Devlin and Jensen, 1975] K. Devlin and R. Jensen. Marginalia to a theorem of Silver. In $\models$*ISILC Logic Conference (Proc. Internat. Summer Inst. and Logic Colloq., Kiel, 1974)*, volume 499 of *Lecture Notes in Mathematics*, pages 115–142. Springer-Verlag, Berlin, 1975.

[Dodd and Jensen, 1981] A. Dodd and R. Jensen. The core model. *Annals of Pure and Applied Logic*, 20:43–75, 1981.

[Dowker, 1951] C. H. Dowker. On countably paracompact spaces. *Canadian Journal of Mathematics*, 3:219–224, 1951.

[Dushnik and Miller, 1941] B. Dushnik and E. W. Miller. Partially ordered sets. *American Journal of Mathematics*, 63:600–610, 1941.

[Easton, 1970] W. B. Easton. Powers of regular cardinals. *Annals of Mathematical Logic*, 1:139–178, 1970.

[Eklof and Mekler, 2002] P. C. Eklof and A. H. Mekler. *Almost Free Modules. Set-Theoretic Methods*, volume 65 of *North-Holland Mathematical Library*. North-Holland, Amsterdam, 2002. Revised edition, xxii+597 pp.

[Erdős and Hechler, 1975] P. Erdős and S. Hechler. On maximal almost-disjoint families over singular cardinals. In A. Hajnal, R. Rado, and V. T. Sós, editors, *Infinite and Finite Sets (Keszthely, Hungary 1973)*, volume 10 of *Colloquia Mathematica Societatis János Bolyai*, pages 597–604. North-Holland, Amsterdam, 1975.

[Erdős and Rado, 1956] P. Erdős and R. Rado. A partition calculus in set theory. *Bulletin of the American Mathematical Society*, 62:427–489, 1956.

[Erdős and Tarski, 1943] P. Erdős and A. Tarski. On families of mutually exclusive sets. *Annals of Mathematics*, 44:315–329, 1943.

[Erdős *et al.*, 1965] P. Erdős, A. Hajnal, and R. Rado. Partition relations for cardinal numbers. *Acta Mathematica Academiae Scientiarum Hungaricae*, 16:93–196, 1965.

[Foreman and Woodin, 1991] M. Foreman and W. H. Woodin. The generalized continuum hypothesis can fail everywhere. *Annals of Mathematics*, 133:1–35, 1991.

[Galvin and Hajnal, 1975] F. Galvin and A. Hajnal. Inequalities for cardinal powers. *Annals of Mathematics*, 101:491–498, 1975.

[Gitik and Magidor, 1992] M. Gitik and M. Magidor. The Singular Cardinals Hypothesis revisited. In *Set Theory of the Continuum (Berkeley, CA, 1989)*, volume 26 of *Mathematical Sciences Research Institute Publicatons*, pages 243–279. Springer, New York, 1992.

[Gitik, 1980] M. Gitik. All uncountable cardinals can be singular. *Israel Journal of Mathematics*, 35:61–88, 1980.

[Gitik, 1989] M. Gitik. The negation of the singular cardinal hypothesis from $o(\kappa) = \kappa^{++}$. *Annals of Pure and Applied Logic*, 43:209–34, 1989.

[Gitik, 1991] M. Gitik. The strength of the failure of the singular cardinal hypothesis. *Annals of Pure and Applied Logic*, 51:215–240, 1991.

[Gitik, 2005] M. Gitik. No bound for the first fixed point. *Journal of Mathematical Logic*, 5:193–246, 2005.

[Gödel, 1947] K. Gödel. What is Cantor's continuum problem? *The American Mathematical Monthly*, 54:515–525, 1947.

[Graham and Kantor, 2009] L. Graham and J.-M. Kantor. *Naming Infinity: A True Story of Religious Mysticism and Mathematical Creativity*. The Belknap Press of Harvard Uiversity Press, 2009.

[Grzegorek, 1985] E. Grzegorek. Always of the first category sets (II). In Z. Frolik, V. Souček, and J. Vinárek, editors, *Proceedings of the 13th Winter School on Abstract Analysis, Section of Analysis*, pages 43–48. Circolo Matematico di Palermo, Palermo, 1985.

[Hajnal *et al.*, 1997] A. Hajnal, I. Juhász, and Z. Szentmiklóssy. On CCC Boolean algebras and partial orders. *Commentationes Mathematicae Universitatis Carolinae*, 38:537–544, 1997.

[Hausdorff, 1906] F. Hausdorff. Untersuchungen über ordnungstypen I,II,III. *Bericht über die Verhandlungen der Königlich Sächsischen Gesellschaft der Wissenschaften zu Leipzig, Mathematisch-Physische Klasse*, 58:106–169, 1906.

[Hausdorff, 1908] F. Hausdorff. Gründzuge einer Theorie der geordneten Mengen. *Mathematische Annalen*, 65:435–505, 1908.

[Hausdorff, 1909] F. Hausdorff. Die graduierung nach dem endverlauf. *Abhandlungen der Königlich Sächsischen Gesellschaft der Wissenschaften zu Leipzig, Mathematisch-Physische Klasse*, 31:295–334, 1909.

[Hausdorff, 1936] F. Hausdorff. Summen von $\aleph_1$ mengen. *Fundamenta Mathematicae*, 26:241–255, 1936.

[Hausdorff, 2002] F. Hausdorff. *Gesammelte Werke, Band II, Grundzüge der Mengenlehre*. Springer-Verlag, Berlin, 2002. Edited and with commentary by E. Brieskorn, S. D. Chatterji, M. Epple, U. Felgner, H. Herrlich, M. Husek, V. Kanovei, P. Koepke, G. Preuß, W. Purkert and E. Scholz, xviii+884 pp.

[Hessenberg, 1906] G. Hessenberg. *Grundbegriffe der Mengenlehre*. Vandenhoeck and Ruprecht, Göttingen, 1906.

[Jech and Prikry, 1976] T. Jech and K. Prikry. On ideals of sets and the power set operation. *Bulletin of the American Mathematical Society*, 82:593–595, 1976.

[Jech, 2003] T. Jech. *Set Theory*. Springer, Berlin, 2003. Third millennium edition, revised and expanded.

[Juhász and Szentmiklóssy, 2009] I. Juhász and Z. Szentmiklóssy. Interpolation of κ-compactness and PCF. *Commentationes Mathematicae Universitatis Carolinae*, 50:315–320, 2009.

[Juhász, 1980] I. Juhász. *Cardinal Functions in Topology—Ten Years Later*, volume 123 of *Mathematical Centre Tracts*. Mathematisch Centrum, Amsterdam, 1980. Second edition, iv+160 pp.

[Kanamori, 1976] A. Kanamori. Weakly normal filters and irregular ultrafilters. *Transactions of the American Mathematical Society*, 220:393–399, 1976.

[Ketonen, 1976] J. Ketonen. Nonregular ultrafilters and large cardinals. *Transactions of the American Mathematical Society*, 224:61–73, 1976.

[Kojman and Shelah, 1992] M. Kojman and S. Shelah. Non-existence of universal orders in many cardinals. *The Journal of Symbolic Logic*, 57:875–891, 1992.

[Kojman and Shelah, 1998] M. Kojman and S. Shelah. A ZFC Dowker space in $\aleph_{\omega+1}$: An application of PCF theory to topology. *Proceedings of the American Mathematical Society*, 126:2459–2465, 1998.

[Kojman and Shelah, 2001] M. Kojman and S. Shelah. Fallen cardinals. *Annals of Pure and Applied Logic*, 109:117–129, 2001.

[Kojman *et al.*, 2004] M. Kojman, W. Kubis, and S. Shelah. On two problems of Erdős and Hechler: new methods in singular madness. *Proceedings of the American Mathematical Society*, 132:3357–3365, 2004.

[König, 1905] J. König. Über die Grundlagen der Mengenlehre und das Kontinuumproblem. *Mathematische Annalen*, 61:156–160, 1905.

[Magidor, 1977a] M. Magidor. Chang's conjecture and powers of singular cardinals. *The Journal of Symbolic Logic*, 42:272–276, 1977.

[Magidor, 1977b] M. Magidor. On the singular cardinals problem. I. *Israel Journal of Mathematics*, 28:1–31, 1977.

[Magidor, 1977c] M. Magidor. On the singular cardinals problem. II. *Annals of Mathematics*, 106:517–547, 1977.

[Miščenko, 1962] A. Miščenko. On finally compact spaces. *Doklady Akademii Nauk SSSR*, 145:1224–1227, 1962.

[Monk, 1996] J. D. Monk. *Cardinal invariants on Boolean algebras*, volume 142 of *Progress in Mathematics*. Birkhauser Verlag, Basel, 1996. x+298 pp.

[Moore, 1989] G. H. Moore. Towards a history of Cantor's continuum problem. In *The history of Modern Mathematics: Ideas and Their Reception (Poughkeepsie 1989)*, volume 1, pages 79–121. Academic Press, Boston, 1989.

[Moore, 2006] J. T. Moore. The Proper Forcing Axiom, Prikry forcing, and the Singular Cardinals Hypothesis. *Annals of Pure and Applied Logic*, 140:128–132, 2006.

[Moore, 2009] G. H. Moore. Review of: Ernst Zermelo: An Approach to His Life and Work by Heinz-Dieter Ebbinghaus. *Notices of the American Mathematical Society*, 56:823–827, 2009.

[Neeman, 2009] I. Neeman. Aronszajn trees and failure of the singular cardinal hypothesis, 2009. Preprint,to appear in the Journal of Mathematical Logic.

[Plotkin, 2005] J. M. Plotkin, editor. *Hausdorff on Ordered Sets*, volume 35 of *History of Mathematics*. American Mathematical Society, 2005.

[Radin, 1982] L. B. Radin. Adding closed cofinal sequences to large cardinals. *Annals of Mathematical Logic*, 22:243–261, 1982.

[Rudin, 1971] M. E. Rudin. A normal space X for which $X \times I$ is not normal. *Fundamenta Mathematicae*, 73:179–186, 1971.

[Scott, 1961] D. S. Scott. Measurable cardinals and constructible sets. *Bulletin de l'Académie Polonaise des Sciences, Série des Sciences Mathématiques, Astronomiques et Physiques*, 9:521–524, 1961.

[Shelah, 1975] S. Shelah. A compactness theorem for singular cardinals, free algebras, Whitehead problem and transversals. *Israel Journal of Mathematics*, 21:319–349, 1975.

[Shelah, 1978] S. Shelah. Jonsson algebras in successor cardinals. *Israel Journal of Mathematics*, 30:57–64, 1978.

[Shelah, 1982] S. Shelah. *Proper Forcing*, volume 940 of *Lecture Notes in Mathematics*. Springer-Verlag, 1982.

[Shelah, 1983] S. Shelah. The singular cardinals problem: independence results. In A. R. D. Mathias, editor, *Surveys in Set Theory*, volume 87 of *London Mathematical Society Lecture Note Series*, pages 116–134. Cambridge University Press, Cambridge, 1983.

[Shelah, 1986] S. Shelah. On power of singular cardinals. *Notre Dame Journal of Formal Logic*, 27:263–299, 1986.

[Shelah, 1990] S. Shelah. Products of regular cardinals and cardinal invariants of products of boolean algebras. *Israel Journal of Mathematics*, 70:129–187, 1990.

[Shelah, 1993] S. Shelah. Number of open sets for a topology with a countable basis. *Israel Journal of Mathematics*, 83:369–374, 1993.

[Shelah, 1994a] S. Shelah. *Cardinal Arithmetic*, volume 29 of *Oxford Logic Guides*. Oxford University Press, New York, 1994.

[Shelah, 1994b] S. Shelah. Cardinalities of topologies with small base. *Annals Pure and Applied Logic*, 68, 1994.

[Shelah, 1997] S. Shelah. The pcf-theorem revisited. In R. L. Graham and J. Nesetril, editors, *The Mathematics of Paul Erdős, II*, pages 420–459. Springer, Berlin, 1997.

[Shelah, 1998] S. Shelah. A polarized partition relation and failure of GCH at singular strong limit. *Fundamenta Mathematicae*, 155:153–160, 1998.

[Shelah, 2000a] S. Shelah. Cellularity of free products of Boolean algebras (or topologies). *Fundamenta Mathematicae*, 166:153–208, 2000.

[Shelah, 2000b] S. Shelah. The Generalized Continuum Hypothesis revisited. *Israel Journal of Mathematics*, 116:285–321, 2000.

[Sierpiński, 1934] W. Sierpiński. *Hypothèse du Continu*, volume 4 of *Monografie Matematyczne*. Warsaw, 1934. Second, revised edition, Chelsea, New York 1956.

[Steprans, 2012] J. Steprans. History of the continuum in the 20th century, 2012. this volume.

[Ulam, 1930] S. Ulam. Zur Masstheorie in der allgemeinen Mengenlehre. *Fundamenta Mathematicae*, pages 140–150, 1930.

[Viale, 2006] M. Viale. The Proper Forcing Axiom and the Singular Cardinal Hypothesis. *The Journal of Symbolic Logic*, 71:473–479, 2006.

[Zermelo, 1904] E. Zermelo. Beweis, dass jede Menge wohlgordnet werden kann. *Mathematische Annalen*, 59:514–516, 1904.

[Zermelo, 1908] E. Zermelo. Untersuchungen über die Grundlagen der Mengenlehre. I. *Mathematische Annalen*, 65:261–281, 1908.

ALTERNATIVE SET THEORIES

M. Randall Holmes, Thomas Forster, and Thierry Libert

1 INTRODUCTION

The one thing that all alternative set theories have in common is the fact that they are alternatives to ZF or ZFC. They are not what the biologists call *monophyletic* nor do they form what the philosophers call a *natural kind*. Nor for that matter is there anything alternative about them in the vulgar sense: people who study alternative set theory are no more New Age or eco-friendly than any other kind of set theorist. What unites alternative set theories is what they are not. To this end the treatments of the alternative set theories will potter off into their little separate corners at the earliest opportunity.

The diversity of set-theoretic systems that have been considered since the discovery of the paradoxes is either unrecognized or disregarded. For one system has always overshadowed the others: Zermelo's and its variants, notably ZF, which is now the presumed referent of the phrase "axiomatic set theory". And this is perfectly understandable since—as well as providing a natural foundational framework for ordinary mathematics—no other system has proved to be more appropriate for investigating questions about infinity, which—after all—was the original motivation for Cantor's work.

Yet sets are *abstract* objects, and therefore subject to mathematical abstraction and generalization. Thus, as happened with the concept of number, it was to be expected that mathematical constructions, as algebraic, topological, or more generally categorical ones, would be used to extend the concept of set in one way or another. One often cites standard algebraic constructions in the theory of fields as inspiring the usual model-theoretic techniques for proving relative consistency and independence results in ZF. But a more significant example can be found in [Barwise and Moss, 1996], where the move from the integers to the rationals is used to motivate the enrichment of the ordinary well-founded universe of sets by so-called "hypersets", which provides solutions to certain reflexive set-theoretic equations. At least, the independence of the Axiom of Foundation and subsequently the consistency of anti-foundation axioms, discussed in [Aczel, 1988], have shown that the *iterative* conception of sets cannot be regarded as the sole motivation for ZF-like axiomatizations of set theory, and *a fortiori* as the sole method of avoidance of the paradoxes.[1]

[1]But see the discussion below of the realization of something like the cumulative hierarchy (and so of the iterative conception) in theories with anti-foundation axioms.

Handbook of the History of Logic: Sets and Extensions in the Twentieth Century.
Volume editor: Akihiro Kanamori. General editors: Dov M. Gabbay, Paul Thagard and John Woods.

Arguably, what characterizes ZF-like axiomatizations then is the principle of *specialized comprehension*, cited in [Church, 1974], according to which "we seek axioms which are special cases of the general comprehension axiom and which promise to maintain consistency while at the same time being adequate for a large variety of mathematical purposes". In the systems we are going to review the selection of admissible instances of comprehension is not guided by mathematical purposes, whatever they are, but is essentially of a *logico-syntactic* nature in order to block the paradoxes.

One famous example of such systems is Quine's "New Foundations" [Quine, 1937]. This belongs to the respectable category of *type-theoretic* approaches, which make use of syntactical criteria to prohibit circularity in definitions — as typically in the instance of comprehension leading to Russell's paradox, i.e., $\{x \mid \neg(x \in x)\}$. A section on type theory is provided to support the discussion of New Foundations.

Another logico-syntactic device which can be used in a type-free setting to eradicate Russell's paradox is to proscribe negation in formulas defining sets, or at least to tamper with its use. Systems following that prescription will be referred to as *positive set theory.* A section below gives the reader an overview of such systems, in connection with the general topic of *topological set theory.* Although the subject would also encompass systems based on certain non-classical logics, these are not going to be discussed in this chapter. A historical account of positive set theory from that perspective can be found in [Libert, 2004].

Nor are we going to discuss set theories motivated by the application of intuitionistic logic. This is not because we do not consider these theories interesting, but because they introduce a new level of complexity which would make this chapter much longer.

We will briefly cover simple type theory [Wang, 1970] and Zermelo set theory [Zermelo, 1908], which are not strictly speaking alternative set theories (type theory is not precisely set theory and Zermelo set theory is not precisely "alternative", as it is the original version of ZFC: but we do present a version of Zermelo set theory which qualifies as an alternative set theory). We will discuss theories with proper classes: VGB set theory [Gödel, 1940] and Kelley-Morse set theory [Kelley, 1976] are very close to ZFC; Ackermann set theory [Levy, 1959] appears quite distant in its motivation from ZFC but turns out to be almost the same theory. We then discuss a further class of theories which are quite close to ZFC, in which the Axiom of Extensionalityis weakened to allow atoms or in which the Axiom of Foundation is negated.

The next section takes us farther from the familiar. New Foundations [Quine, 1937] and the known-to-be consistent theories allied with it are motivated by simple type theory rather than Zermelo set theory and superficially look very different from the usual set theory. Some of the consistent fragments remain alien to ZFC on closer inspection, but the most mathematically fluent system of this kind, NFU (Jensen's modification of NF to allow urelements in [Jensen, 1969]) turns out to be quite closely related to the usual set theory in a way we will describe.

We then discuss positive set theories. The most mathematically fluent system of positive set theory is the system GPK^+_∞ defined by Olivier Esser [Esser, 1999], which is based on earlier work on positive comprehension and "hyperuniverses" in which Marco Forti is prominent (see [Forti and Honsell, 1996b]). This system is elegant, superficially different from standard set theory but underneath deeply related to a powerful extension of ZFC, and surprisingly strong: its model theory takes us into the domain of large cardinalsof moderate strength (this is also true of the higher reaches of the model theory of natural extensions of NFU, but in the case of Esser's theory one is involved with large cardinals from the outset). Esser's set theory and earlier related systems of Forti and others have topological motivation, and these theories might be called topological set theories.

We discuss a pair of set theories motivated by the surprising application of model theory to analysis subsumed under the term "nonstandard analysis". One of these, Nelson's IST [Nelson, 1977], is a deliberate adaptation of ZFC to support nonstandard analysis. The other, Vopěnka's "alternative set theory" or AST [Vopěnka, 1979], is more *sui generis*, and differs more fundamentally in its outlook from ordinary set theory: it is mathematically interesting though as a set theory it is rather weak.

Finally, we discuss two curiosities. The first one honestly exhibits a characteristic which has been ascribed by philosophers to all systems of set theory (incorrectly, we believe). The double extension set theory of Andrzej Kisielewicz [1998] is a fascinating and bizarre system of set theory which is frankly an *ad hoc* solution to the paradoxes. Appropriately, we do not yet know whether this system is consistent. It is at least as strong as ZFC, for reasons which seem to resemble those which make the Ackermann set theory as strong as ZFC.

The other curiosity is the theory obtained by adjoining the existence of an elementary embedding $j : V \to V$ to Zermelo set theory with the Rank Axiom, which has been studied by Paul Corazza [2000]. The existence of such an embedding is inconsistent with ZFC (though it is not known to be inconsistent with ZF), and this extension of Zermelo set theory is not a weak theory but is in fact one of the strongest set theories ever proposed.

Zermelo set theory (first proposed in [Zermelo, 1908]) was not in our view motivated by an *ad hoc* attempt to resolve the paradoxes. Zermelo appears to have catalogued constructions in set theory which had actual applications in mathematical reasoning and which were required to implement his proof that the Axiom of Choice implies the well-ordering theorem. The motivation of simple type theory seems to us to be similar (though more motivated by logical considerations than by mathematical practice): its history is enormously complicated (see [Wang, 1970]). New Foundations ([Quine, 1937], hereinafter NF) does seem to have started out as an *ad hoc* attempt to further simplify "simple type theory" to eliminate the annoyance of type indices: but the simplification which motivates New Foundations is mathematically very appealing and the fact that almost any reasonable weakening of NF seems to give a demonstrably consistent theory suggests that Quine was onto something. Positive set theory appears to have an elegant topological motivation. A final point is that most of these set theories, different as they appear to be from one another, seem to converge on the same picture of the world of set theory, though there are some interesting byways.

2 THE BASICS: TYPE THEORY AND THE ORIGINAL THEORY OF ZERMELO

In this section we describe two systems which are not "alternative": they are standard theories. These are simple type theory (in a streamlined form—not the complex theory of *Principia Mathematica* [Whitehead and Russell, 1910-13] but the much simpler system whose evolution is described in [Wang, 1970]) and the original set theory of Zermelo [1908], which differs in some ways from its final modern development.

2.1 Simple type theory

The simple type theory TST is a first-order many-sorted theory with equality and membership as primitive relations. The sorts, called "types" are indexed by the natural numbers $0, 1, 2 \ldots$. Atomic formulas are well-formed if they take one of the forms $x^n = y^n$ or $x^n \in y^{n+1}$, where the superscript indicates the type of the variable.

The intuition is that type 0 is inhabited by some objects of an unspecified nature called "individuals", type 1 is inhabited by sets of individuals, type 2 is inhabited by sets of sets of individuals, and so forth; in general type $n + 1$ is inhabited by sets of type n objects.

There are two axiom schemes which make up the core of the theory and two further axiom schemes which are usually assumed. We advocate the use of variables without explicit type superscripts in this system, in situations where the types assigned can be deduced from the context or in which any types for the variables compatible with the syntax will work, but we will sometimes give explicit type indices for clarity.

Axiom of Extensionality: $(\forall AB.A = B \leftrightarrow (\forall x.x \in A \leftrightarrow x \in B))$. Objects of any positive type are equal iff they have the same elements.

Axiom of Comprehension: For any formula ϕ in which A does not appear free, $(\exists A.(\forall x.x \in A \leftrightarrow \phi))$. Any condition on objects of a given type defines a set of the next higher type.

Definition: For any formula ϕ in which A^{n+1} does not appear free, $\{x^n \mid \phi\}^{n+1}$ is defined as the unique A^{n+1} such that $(\forall x^n.x^n \in A^{n+1} \leftrightarrow \phi)$ (which exists by Comprehension and is unique by Extensionality).

Note that in the statement of the axiom schemes we do not assign types explicitly to the variables: any assignment of types which works will give an instance of the scheme.

We define V^{n+1} as $\{x^n \mid x^n = x^n\}^{n+1}$ and $\emptyset^{n+1}$ as $\{x^n \mid x^n \neq x^n\}^{n+1}$. We define $\{x^n, y^n\}^{n+1}$ as $\{z^n \mid z^n = x^n \vee z^n = y^n\}^{n+1}$. Type indices may be omitted.

Ordered pairs can be defined using the usual Kuratowski definition $(x, y) = \{\{x\}, \{x, y\}\}$, and relations and functions can then be defined as usual. It is unfortunate that in the notation $x\,R\,y$ the type of R is 3 higher than that of x or y, and similarly the type of f is three higher than that of x in the notation $f(x)$; there is a way to change this displacement to the more natural 1 (by developing an ordered pair which is of the same type as its projections) but that development would take us a bit far afield.

We define 0 as $\{\emptyset\}$: note that we can assign any type ≥ 2 to this object, and in fact we define 0 in each type at or above 2. For each set A, define $A + 1$ as $\{a \cup \{x\} \mid a \in A \wedge x \notin a\}$. $A + 1$ is the set of all objects obtained by adding a single new element to an element of A. Now $0+1$ is the set of all one-element sets, which we call 1, and $1+1$ is the set of all two-element sets, which we call 2, and so forth. We define $\mathbb{N}$, the set of natural numbers, as the intersection of all sets which contain 0 and are closed under the extended "successor" operation. Notice that a version of each natural number is defined in each type with index ≥ 2, and a set $\mathbb{N}$ is defined in each type with index ≥ 3. The elements of the natural numbers are the finite sets: we define the set $\mathbb{F}$ of finite sets as $\bigcup \mathbb{N}$.

We can now assert the remaining axioms usually assumed in TST.

Axiom of Infinity: $V^{n+1} \notin \mathbb{F}^{n+2}$. The universe (of objects of a given type) is not a finite set.

Axiom of Choice: Any pairwise disjoint collection of nonempty sets (of a given type) has a choice set (of the next lower type).

This theory figures in our discussion here primarily as the basis for the development of New Foundations and related theories, but there are other interesting relationships between TST and other alternative set theories.

This theory has been described carelessly as the simple type theory of Russell, but this is historically inaccurate. Russell had something like this in mind in the Appendix to *Principles of Mathematics* [Russell, 1903] though even there this is not clear. In the formalization of *Principia Mathematica* (hereinafter PM, [Whitehead and Russell, 1910-13]) Whitehead and Russell were hampered by the lack of a set-theoretical definition of ordered pair. They knew that relations were sets of ordered pairs, but they had no set-based notion of ordered pair, so in fact in PM they defined ordered pairs as a sort of relation! The type system of PM contains types of n-ary relations over each finite sequence of previously defined types, which makes it quite complex: it is further complicated by predicativity restrictions which motivate the further subdivision of types into "orders". Ramsey [1926] pointed out that the Axiom of Reducibility renders the orders redundant, and so they should simply be dropped. Norbert Wiener [1914] gave the first set-theoretical definition of ordered pair, defining (x, y) as $\{\{\{x\}, \emptyset\}, \{\{y\}\}\}$, a definition which has certain formal merits. He pointed out that this made it possible to collapse the complex type system of relations of PM to a simple linear hierarchy of set types. Hao Wang [1970] has an interesting discussion of the history of simple type theory which suggests that this kind of theory was actually first described in detail around 1930.

Whitehead and Russell were aware of the ambiguity of the notation (what computer scientists call "polymorphism") which tempts us to omit explicit type indices: they called it "systematic ambiguity" in PM and took advantage of it in the same way to avoid giving explicit indications of type in his much more complex type system. (In fact, oddly, PM does not have a notation for types).

2.2 The original system of Zermelo

The axioms of Zermelo's original theory of [1908] are presented. The content is presented (this is not a literal translation). Familiarity with common set-theoretical notation is assumed.

Axiom of Extensionality: If every element of a set A is also an element of B and *vice versa*, then $A = B$.

Axiom of Elementary Sets: $\emptyset$ exists; for any object x, $\{x\}$ exists; for any objects x and y, $\{x, y\}$ exists.

Axiom of Separation: If $P(x)$ is a proposition with a definite truth-value for each $x \in M$, $\{x \in M \mid P(x)\}$ exists.

Power Set Axiom: For every set A, the power set $\mathcal{P}(A)$, the set of all subsets of A, exists.

Union Axiom: For any set A, the union $\bigcup A$, the set of all elements of elements of A, exists.

Axiom of Choice: Any pairwise disjoint collection of nonempty sets has a choice set.

Axiom of Infinity: There is a set Z such that $\emptyset \in Z$ and for each $x \in Z$, $\{x\} \in Z$.

The Axiom of Extensionality is phrased in a way which allows non-set elements of the domain of discourse (objects with no elements distinct from one another and from the empty set, which are often called *atoms* or *urelements*). In modern theories, it is usually assumed that every object is a set.

The Axiom of Elementary Sets is usually replaced with the Axiom of Pairing, which asserts the existence of $\{x, y\}$ for any objects x and y. The existence of $\emptyset$ follows from Separation and the existence of any specific set (such as the one given in Infinity). It is a more modern observation that the existence of $\{x\} = \{x, x\}$ is a special case of pairing.

Zermelo did not have the modern formulation of the conditions in the Axiom of Separation in terms of formulas of the first-order language of set theory, and apparently when he saw this formulation he did not like it: he believed that it was too restrictive. The true formulation of Zermelo's intentions may be second-order.

The Axiom of Choice takes a modern form. It is interesting that Zermelo proves his theorem that the well-ordering principle follows from choice in his [1908], without having the ability to express ordered pairs as sets (which was present in this theory of course, but first recognized by Norbert Wiener [1914]).

The Axiom of Infinity differs from its usual modern form which asserts the existence of a set which contains $\emptyset$ and is closed under the von Neumann successor operation $x \mapsto x \cup \{x\}$. A different definition of the natural numbers is the reason: Zermelo defines the natural numbers as the iterated singletons of the empty set. It is an interesting fact that though the mathematical effect of either axiom of infinity is the same, these axioms are not equivalent in the presence of the other axioms of Zermelo set theory: neither one implies the other. They have the same consistency strength: each theory can interpret the other. The two axioms of infinity are equivalent in the presence of the very powerful Axiom of Replacement; they are also equivalent in the presence of a weaker axiom which asserts that each object belongs to some rank of the cumulative hierarchy.

The Axiom of Foundation is absent from Zermelo's original theory though it often appears in modern lists of axioms for Zermelo set theory.

Axiom of Foundation: For any non-empty set x, there is a $y \in x$ such that $y \cap x = \emptyset$.

This axiom asserts that the membership relation restricted to any set is a well-founded relation. It rules out such things as self-membered sets. This is a consequence of the view that the world of set theory is constructed by the iterated application of the power set operation to an initial collection (the empty set or perhaps a set of atoms) through stages indexed by the ordinals. We will see below that variations of ZFC have been studied whose distinctive feature is that Foundation is omitted and replaced with principles which flatly contradict it.

This subsection introduces the original system of Zermelo, but does not aim to correct common usage: "Zermelo set theory" in the sequel will refer to the theory with the von Neumann form of the Axiom of Infinity and with the Axiom of Foundation.

For reference we include a statement of the final axiom scheme which turns Zermelo set theory (in either form) to the usual set theory ZFC:

Axiom of Replacement: If a is a set and ϕ is a formula such that we can prove $(\forall x.(\exists! y.\phi))$, then $\{y \mid (\exists x \in a.\phi)\}$ is a set.

2.3 The relationship between these systems. Mac Lane set theory

The simple theory of types and Zermelo set theory appear superficially to live at the same level of strength. Both of them can talk about an infinite set (type 0 for TST, the set provided by Infinity for Zermelo) and its iterated power sets. Each theory proves the existence of the cardinals $\aleph_n$ for each n but cannot prove the existence of $\aleph_\omega$. But Zermelo set theory is stronger: Kemeny proved in his PhD thesis [1950] (which was never published) that Zermelo set theory proves the consistency of TST.

The source of strength in Zermelo set theory is the ability to quantify over the entire universe in instances of Separation. The variant of Zermelo set theory in which Separation is restricted to Δ_0 formulas (those in which every quantifier is bounded in a set) [with the von Neumann form of Infinity and the Axiom of Foundation] is known as Mac Lane set theory (because it was advocated as a foundational system by Saunders Mac Lane [1986]), and Mac Lane set theory is demonstrably mutually interpretable with TST.

Adrian Mathias has written a beautiful survey of relations between Mac Lane set theory, Zermelo set theory, and some other systems, found in [Mathias, 2001a], [Mathias, 2001b].

2.4 Mac Lane or Zermelo set theory as an alternative set theory

As witnessed by the program of Mac Lane mentioned above, Zermelo set theory or variants of Zermelo set theory have been pressed into service themselves as alternative set theories, presumably by workers nervous about the high consistency strength of ZFC.

One is then faced with the problem that the implementations of cardinal and ordinal numbers traditional in ZFC do not work in Zermelo set theory. Cardinals are implemented as initial ordinals, so we need only consider the ordinals. And here we have a serious problem. Zermelo set theory with the modern form of Infinity proves the existence of the von Neumann ordinals $0, 1, 2, 3, \ldots \omega, \omega{+}1, \omega{+}2, \omega{+}3 \ldots$ but not of $\omega \cdot 2$. But these theories prove the existence of much longer well-orderings.

An elegant solution is obtained by adding the Axiom of Foundation (or restricting our attention to well-founded sets), and further stipulating that every set belongs to a rank of the cumulative hierarchy. It is then possible to define the cardinality of a set A as the set of all sets B which are equinumerous with A and of the lowest rank which contains such sets, and define the order type of a well-ordering $\leq$ as the set of all well-orderings similar to $\leq$ and of the lowest rank which contains such well-orderings (notice that we no longer identify cardinals with initial ordinals if we use these implementations). This is called "Scott's trick", introduced in [Scott, 1955].

It remains to explain how to stipulate that every set belongs to a rank of the cumulative hierarchy.

Definition: A set H is a *subhierarchy* iff H is well-ordered by inclusion, $\bigcup I \in H$ for each $I \subseteq H$, and for any $x \in H$ other than $\bigcup H$, the minimal element of H properly including x is $\mathcal{P}(x)$. A set r is a *rank* iff there is a subhierarchy H such that $r \in H$.

Axiom of Rank: For every set x there is a rank r such that $x \in r$.

Theorem: For any subhierarchies H_1 and H_2, either $H_1 \subseteq H_2$ or $H_2 \subseteq H_1$.

Proof: Omitted.

Corollary: Ranks are well-ordered by inclusion (all ranks, not just the ones in any given subhierarchy).

Definition: For any set x, the rank of x is the minimal rank which contains x as a subset.

Definition: For any set A, $|A| \equiv_{\texttt{def}} \{B \in r_A \mid B \sim A\}$ where r_A is the minimal rank which contains a set equinumerous with A. A *cardinal number* is a set $|A|$.

Definition: For any well-ordering W let r_W be the minimal rank which contains a well-ordering isomorphic to W. Define $ot(W)$ as $\{W' \in r_W \mid W \approx W'\}$. An *ordinal number* is a set of the form $ot(W)$.

Definition: For any ordinal α, V_α is the rank (if there is one) such that the order type of the inclusion order on the ranks properly included in α is α.

It is further interesting to observe that the following axioms suffice to present this extension of Zermelo (or Mac Lane) set theory.

Axiom of Extensionality: Sets with the same elements are the same.

Axiom of (Bounded) Separation: For any set A and (Δ_0) formula ϕ in which B is not free, $(\exists B.x \in B \leftrightarrow x \in A \wedge \phi)$.

Axiom of Power Set: For any set A, there is a set $\mathcal{P}(A)$ whose members are exactly the subsets of A.

Axiom of Rank: Every set belongs to some rank.

Axiom of Infinity: The von Neumann ordinal ω exists. [it is equivalent to assume the existence of the set of Zermelo natural numbers in this context.]

Axiom of Choice: Any pairwise disjoint collection of nonempty sets has a choice set.

The axioms of Pairing and Union in the original Zermelo axiom set turn out to be redundant. If x and y belong to ranks r_x and r_y, $\{x, y\} = \{z \in r_x \cup r_y \mid z = x \vee z = y\}$, where $r_x \cup r_y$ is a set because it is the larger of the two ranks. If x belongs to the rank r_x, $\bigcup x = \{y \in r_x \mid (\exists z.y \in z \wedge z \in x)\}$. The Axiom of Foundation follows quite directly from the Axiom of Rank as well.

A book-length development of set theory in this style is given by Michael Potter [2004]. The Axiom of Rank adds no essential strength to Zermelo or MacLane set theory: see [Mathias, 2001a].

For reference and comparison, we give the familiar definition of ordinal usual in ZFC, originally due to von Neumann.

Definition: A set A is *transitive* iff for all x, y, $x \in y \wedge y \in A$ implies $x \in A$. This admits the equivalent formulations $\bigcup A \subseteq A$ and $A \subseteq \mathcal{P}(A)$.

Definition: A (von Neumann) ordinal is a transitive set α which is strictly well-ordered by the membership relation.

Definition: The (von Neumann) order type of a well-ordering $\leq$ is the unique (von Neumann) ordinal α such that $\leq$ is isomorphic to the inclusion order on α.

The Axiom of Replacement suffices to prove that every well-ordering has a von Neumann order type. This is not a theorem of Zermelo set theory (or any of the related theories of similar strength that we have described).

3 THEORIES WITH CLASSES

3.1 General considerations

In any version of set theory, we find ourselves wanting to talk about collections which are not sets as if they were sets. It is easy to see that in a certain sense such talk is harmless: if (M, E) is a model of set theory, we could add $\mathcal{P}(M)$ (the true power set of M) as a new domain, and add the membership relation of elements of M in elements of $\mathcal{P}(M)$ to obtain a theory with signature

$$(M, \mathcal{P}(M), E, \in \cap (M \times \mathcal{P}(M))),$$

then finally identify each preimage of an element of M under E with that element of M and collapse E and $\in \cap (M \times \mathcal{P}(M)$ into a single relation appropriately. Of course this requires some strength in the metatheory, but not very much.

In such an extended theory, the domain of the original theory is definable as the collection of elements. In the family of theories with classes that we give first, general objects are called *classes* and objects which are elements of classes are called *sets*. The primitive relations of the theory are equality and membership as usual.

We give the following

Definition: $set(x) \equiv (\exists Y. x \in Y)$: a *set* is a class which is an element of some class.

Convention: We use upper case letters for general variables and lower case letters for variables restricted to the sets.

Any theory of this kind has two characteristic axioms as preamble.

Axiom of Extensionality: $(\forall AB.A = B \leftrightarrow (\forall x.x \in A \leftrightarrow x \in B))$

Axiom of {Predicative} Class Comprehension: Where ϕ is a formula {in which all quantifiers are restricted to the domain of sets} in which A does not appear free, $(\exists A.(\forall x.x \in A \leftrightarrow \phi))$. Note the use of the variable case convention here: if we were to assume that variables ranged over all classes, this would be $(\exists A.(\forall x.x \in A \leftrightarrow set(x) \wedge \phi))$.

Definition: The object A such that $(\forall x.x \in A \leftrightarrow \phi)$ (where A is not free in ϕ) is denoted by $\{x \mid \phi\}$.

The differences between predicative and impredicative class comprehension will be brought out in the next subsection. The motivating discussion above suggests that any set theory can be extended to a theory of classes with extensionality and full class comprehension. Thus impredicative class comprehension as such adds no additional strength to a set theory, but it does add considerable expressive power.

3.2 Von Neumann-Gödel-Bernays and Kelley-Morse set theory

Two theories with sets and classes are given here. The theory of sets in each extends ZFC. The first theory, called VGB (von Neumann-Gödel-Bernays set theory — [Gödel, 1940] is a reference) is a conservative extension of ZFC: it proves nothing about sets that ZFC does not. This theory was originally proposed by von Neumann [1925] but as a theory of functions rather than sets. We give our own presentation of a theory with the same mathematical force.

Notice that if x and y are sets, the existence of $\{x, y\}$, $\bigcup x$, $\mathcal{P}(x)$ (understood as the collection of sub*sets* of x) follows from Class Comprehension: the additional axioms corresponding to the axioms of Pairing, Union, and Power Set assert that these classes are sets.

Axiom of Extensionality: as above

Axiom of Predicative Class Comprehension: as above

Axiom of Separation: Every subclass of a set is a set.

Axiom of Power Set: for any set x, $\mathcal{P}(x)$ is a set.

Axiom of Set Union: for any set x, $\bigcup x$ is a set.

Axiom of Infinity: There is a set which contains $\emptyset$ and contains $x \cup \{x\}$ if it contains x, for any set x.

Axiom of Limitation of Size: A class C is not a set iff there is a class bijection from the class V of all sets to C.

Axiom of Foundation: For any non-empty set x, there is an element y of x such that $x \cap y = \emptyset$.

Many of the axioms of ZFC (or even of Zermelo set theory) may seem to be missing here, but this is not the case.

The Axiom of Infinity implies that there are at least three distinct sets (say the von Neumann numerals $0, 1, 2$) so there cannot be a bijection between a class $\{x, y\}$ and the universe, so $\{x, y\}$ must be a set so Pairing holds. Since Pairing holds, ordered pairs of sets are sets, and so every logical relation on sets is realized by a class of ordered pairs.

The Axiom of Choice (in a strong form) follows from Limitation of Size (surprise!). The class of von Neumann ordinals is not a set, so there is a class bijection from the class of von Neumann ordinals to the class of all sets, which determines a well-ordering of the universe in the obvious way. From any element of a collection of pairwise disjoint sets, select the least element with respect to this global well-ordering: collect these least elements to build a choice set.

The Axiom of Replacement follows from Limitation of Size. Suppose a is a set and $(\forall x \in a.(\exists! y.\phi))$. The class $B = \{y \mid (\exists x \in a.\phi)\}$ exists as a class by Class Comprehension. Note that $B = F``a$ where $F = \{(x, y) \mid \phi\}$. Define $F^{-1}(b)$ as $\{a \in A \mid F(a) = b\}$: notice that F^{-1} is an injection from B into $\mathcal{P}(a)$. $\{F^{-1}(b) \mid b \in B\}$ is a set because it is a subclass of $\mathcal{P}(a)$: if there were a class bijection from V to B, there would be a class bijection from V to this set, which contradicts Limitation of Size. Thus B is a set and Replacement holds.

If one wishes to avoid proving Choice, one can replace Limitation of Size with the following weaker axiom:

Weak Limitation of Size: For any set a and class B, if there is a bijection from a to B then B is a set.

It is straightforward to check that the derivation of Replacement above still succeeds. In the presence of the other axioms and weak Limitation of Size, we can deduce full Limitation of Size from a strong version of Choice (such as "there is a class well-ordering of the universe of sets") but not from weaker forms such as "every pairwise disjoint set of sets has a choice set".

The role of Limitation of Size in encapsulating the powerful axioms of Replacement and Choice is one of the striking features of VGB class theory. The other striking feature, found in all the original presentations, and obscured in our presentation so far, is that VGB class theory is finitely axiomatizable, because the axiom scheme of Predicative Class Comprehension can be replaced by finitely many of its instances.

We present a reduction of Predicative Class Comprehension to finitely many axioms. This proceeds by induction on the structure of formulas.

To handle classes in which the top level logical operator in the defining formula is propositional, note that

$$\{x \mid \phi \wedge \psi\} = \{x \mid \phi\} \cap \{x \mid \psi\}$$

and

$$\{x \mid \neg\phi\} = V - \{x \mid \phi\},$$

and every propositional operator can be defined in terms of conjunction and negation. So we stipulate that if A and B are classes, $A \cap B$ and $V \setminus A$ are classes.

We stipulate that for any sets x and y, $\{x, y\}$ is a set and so $(x, y) = \{\{x\}, \{x, y\}\}$ is a set. We define $(a_1, a_2, \ldots, a_n)$ as $(a_1, (a_2, \ldots, a_n))$. We stipulate that $(x_1) = x_1$: a 1-tuple is its sole item.

We define π_1 as $\{((x, y), x) \mid x = x\}$ and π_2 as $\{((x, y), y) \mid x = x\}$. We define $A|B$ for A and B classes as $\{(x, y) \mid (\exists z.(x, z) \in A \wedge (z, y) \in B)\}$. Note that if f and g are functions in the usual sense, $f \circ g = g|f$. We stipulate that π_1 is a class, π_2 is a class, and if A and B are classes, so is $A|B$. Note that each of the sets $\pi_i^n = \{((a_1, \ldots, a_i, \ldots, a_n), a_i) \mid a_1 = a_1\}$, the function sending an n-tuple to its ith element, is a class because constructible from the projection functions and relative product, except for $\pi_1^1 = \{(x, x) \mid x = x\}$, which we stipulate is a class.

For any class R we stipulate that $\text{dom}(R) \equiv \{x \mid (\exists y.(x, y) \in R)\}$ is a class and $R^{-1} \equiv \{(y, x) \mid (x, y) \in R\}$ is a class.

Define $R \otimes S$ as $(R|\pi_1^{-1}) \cap (S|\pi_2^{-1})$. Define $R_1 \otimes R_2 \otimes \ldots \otimes R_n$ as $R_1 \otimes (R_2 \otimes \ldots \otimes R_n)$ (same grouping as for n-tuples).

Our strategy is first to fix a finite sequence of variables $x_1, \ldots, x_n$ and show how to define $\{(x_1, \ldots, x_n) \mid \phi\}$ for any formula in which no variable other than the x_i's appears free.

We provide $\{(x_1, x_2) \mid x_1 = x_2\}$ and $\{(x_1, x_2) \mid x_1 \in x_2\}$ as basic classes (actually, we have already provided the former).

If we have already shown $X_\psi = \{(b_1, \ldots b_p) \mid \psi\}$ to be a class, where all free variables in ψ appear among the b_i's, then $\{(a_1, \ldots, a_n) \mid \psi(a_{s_1}, \ldots, a_{s_p})\}$ can be shown to exist using the constructions we have already defined and one additional construction. It is constructible as $\text{dom}((\pi_n^{s_1} \otimes \ldots \otimes \pi_n^{s_p}) \cap V \times X_\psi)$, so we stipulate that for any class X, $V \times X$ is a class (and for any classes X and Y, $X \times Y = (V \times X)^{-1} \cap (V \times Y)$ is thus a class).

If we have already shown $X_\phi = \{(a_1, \ldots, a_n) \mid \phi\}$ to be a class, we construct $\{(a_1, \ldots, a_n) \mid (\exists a_i.\phi)\}$ as $\text{dom}(\pi_n^1 \otimes \ldots \otimes \pi_n^{i-1} \otimes (V \times V) \otimes \pi_n^{i+1} \otimes \ldots \otimes \pi_n^n) \cap (V \times X_\phi)$. Of course there are typographical variations of this if $i = 1$, $i = n$.

Now we show how to construct an arbitrary $\{x \mid \phi\}$. Let $x, x_2, x_3, \ldots, x_n$ be all the variables free or bound in ϕ. Construct $\{(x, x_2, \ldots, x_n) \mid \phi\}$. For each of the x_i's which is free in ϕ, intersect this class with $(\pi_n^i)^{-1}``\{x_i\}$ ($R^{-1}``A \equiv \text{dom}(R^{-1} \cap V \times A)$; pairing implies the existence of singletons as sets and so as classes). Finally, the domain of this intersection is $\{x \mid \phi\}$.

We present the finite list of axioms which we have just shown to imply predicative class comprehension.

boolean intersection: For any classes A and B, $A \cap B = \{x \mid x \in A \wedge x \in B\}$ is a class.

complement: For any class A, $V \setminus A = \{x \mid x \notin A\}$ is a class.

pairing: For any sets x and y, $\{x, y\} = \{z \mid z = x \vee z = y\}$ is a set.

definition of ordered pair: (x, y) is defined as $\{\{x\}, \{x, y\}\}$.

projections: $\pi_1 = \{((x, y), x) \mid x = x\}$ is a class. $\pi_2 = \{((x, y), y) \mid x = x\}$ is a class.

relative product: For any classes R and S, $R|S = \{(x, y) \mid (\exists z.(x, z) \in R \wedge (z, y) \in S)\}$ is a class.

converse: For any class R, $R^{-1} = \{(y, x) \mid (x, y) \in R\}$ is a class.

domain: For any class R, $\mathrm{dom}(R) = \{x \mid (\exists y.(x, y) \in R)\}$ is a class.

primitive relations: $\{(x, x) \mid x = x\}$ and $\{(x, y) \mid x \in y\}$ are classes.

cartesian product: for any classes A and B, $A \times B = \{(a, b) \mid a \in A \wedge b \in B\}$ is a class (we only used $V \times A$ in the construction above, but one might as well provide the general construction).

Various finite sets of constructions have been given for this purpose. The primitives here are basic operations of the algebras of sets and relations, and the approach is inspired by the work of Givant and Tarski in [Tarski and Givant, 1987], in which they show that standard axioms for relation algebra augmented with basic properties of projection operations are sufficient to interpret first-order logic.

One can note here that our elimination of the Axiom of Pairing from the axioms of VGB is undone when we "unpack" the predicative comprehension scheme, but we are still correct that it is not needed in the presence of that scheme: what we have apparently revealed is that it is a component of that scheme. It is unique among those components in being a set existence principle rather than a class existence principle.

Kelley-Morse set theory (see the appendix to Kelley's [1976]) differs from VGB simply in having the full scheme of class comprehension instead of the predicative scheme. We can quantify over all classes in constructions of both classes and sets in Kelley-Morse set theory. We remarked above that adding classes which satisfy impredicative class comprehension to a set theory adds no real strength to the theory, but impredicative class comprehension can be used here to prove the existence of sets as well as classes, since any subclass of a set is a set and any class which can be placed in a class one-to-one correspondence with a set is a set.

Kelley-Morse set theory is stronger than VGB in a very marked way: Kelley-Morse set theory proves the consistency of VGB. Further, the class comprehension scheme of Kelley-Morse set theory cannot be finitely axiomatized.

The significant nature of the difference between VGB and Kelley-Morse set theory is not at first obvious. One way of seeing it is to observe that the classes of VGB can be identified with formulas of the language of ZFC; we will explain in what sense this is true and why it is not true of Kelley-Morse set theory.

If (M, E) is a model of ZFC, we show how to construct a model of VGB (with the weak version of Limitation of Size) which satisfies exactly the same theorems. A *preclass* is a formula with no variable other than x free in the language of ZFC augmented with constants for each element of M. We define $\phi \sim \psi$ as holding iff

$(\forall x.\phi \leftrightarrow \psi)$ holds in (M, E). We define a *class* as an equivalence class of preclasses under $\sim$. We define $[\phi]\, E^*\, [\psi]$ as holding iff $(\exists y.(\forall x.x \in y \leftrightarrow \phi) \wedge \psi[y/x])$. It is straightforward to establish that the structure (C, E^*), where C is the set of classes, satisfies VGB (with the weak form of Limitation of Size; if a global well-ordering of the universe is added, we can get full Limitation of Size) and moreover satisfies exactly the same sentences about sets as (M, E). This establishes that VGB (at least the version with weak Limitation of Size) proves nothing about sets that is not provable in ZFC.

We pause to observe that we can define ordered pairs of classes as classes. Let σ be the function which sends each natural number to its successor and fixes each other object. For classes A and B, define (A, B) as $\sigma``A \cup \{0\} \times \sigma``B \cup \{0\}$. From a cartesian product of nonempty classes one can of course extract the first and second projections. This means that we can represent formulas with arbitrary class parameters as classes in either VGB or Kelley-Morse set theory by a suitable coding scheme. What we can do in Kelley-Morse set theory that we cannot do in VGB is define satisfaction for formulas with no quantifiers over classes. In fact, we can define the structure defined in the previous paragraph over the universe of sets and code every element of it (set or preclass) as a set, so we obtain a class structure which is a model of VGB and we can prove the consistency of VGB. This shows that Kelley-Morse set theory is essentially stronger than VGB.

3.3 Ackermann set theory

The set theory of Ackermann (see [Levy, 1959]) has sets and classes but not in the same sense as VGB or Kelley-Morse. In the former theories, classes which are not sets are not elements; in Ackermann set theory some non-set classes are elements of other classes. The notion of set in Ackermann set theory is an independent primitive notion not definable in terms of the usual primitives of equality and membership: in fact, the whole thing works precisely because sethood is not definable.

Ackermann set theory is a first-order theory. General objects are called *classes*. Primitive relations are equality and membership and there is a primitive unary predicate of sethood.

Here are the five axioms of the theory.

Extensionality: Classes with the same elements are equal.

Class Comprehension: For any formula ϕ in which the variable A is not free, $(\exists A.(\forall x.x \in A \leftrightarrow set(x) \wedge \phi))$.

Elements: Any element of a set is a set.

Subclasses: Any subclass of a set is a set.

Set Comprehension: Let ϕ be a formula in which the variable A is not free, in which the predicate *set* does not occur, and in which all free variables other than x denote sets, and suppose $(\forall x.\phi \rightarrow set(x))$. Then $(\exists A.set(A) \wedge (\forall x.x \in A \leftrightarrow \phi))$.

We give the flavor of this theory by doing some basic proofs.

Universe: There is a class V which contains exactly the sets as elements (this is obvious from Class Comprehension).

Empty Set: $\emptyset$ is a set. The formula $x \neq x$ satisfies the conditions of Set Comprehension vacuously.

Pairing: Suppose a and b are sets. Then $\{a, b\}$ is a set.

Proof: The formula $x = a \vee x = b$, for any sets a and b, satisfies the conditions of Set Comprehension: it does not mention set, the parameters a and b are sets, and if it is true of x then x is a set.

Union: Suppose a is a set. Then $(\exists y.x \in y \wedge y \in a)$ satisfies the conditions of Set Comprehension. Note that if $x \in y$ and $y \in a$, then y must be a set by the Axiom of Elements (it belongs to the set a) and so x must also be a set (as it belongs to the set y).

Power Set: Suppose a is a set. Then $(\forall y.y \in x \rightarrow y \in a)$ satisfies the conditions of Set Comprehension. If x satisfies this condition it is a set by the Axiom of Subclasses.

Infinity: There is a set which contains $\emptyset$ as an element and is closed under the operation $x \mapsto x \cup \{x\}$.

Proof: "x belongs to every class C such that $\emptyset \in C$ and $(\forall X.X \in C \rightarrow X \cup \{X\} \in C)$" is a formula which satisfies the conditions of Set Comprehension. Note that $\emptyset$ is a set and for any set x, $x \cup \{x\}$ is a set (by applications of Pairing and Union proved above). This means that the class V of all sets is a class with the desired closure conditions. This verifies that if x satisfies this condition then $x \in V$ so x is a set! Note that the condition here quantifies over all *classes* (in fact we cannot quantify over all sets in an instance of Set Comprehension, because this would require us to mention the set predicate).

Separation: If a is a set, $\{x \in a \mid \phi\}$ is a set for any formula ϕ at all, by the Axiom of Subclasses.

Transitive Closure: The transitive closure $TC(A)$ of a set A, the intersection of all transitive sets which contain A as a subset, is a set.

Proof: Let A be a set. Consider the formula "x belongs to every class C such that $A \in C$ and for any $x \in C$, $\bigcup x \in C$". Note that V has the closure property in question, so this formula satisfied the requirements for Set Comprehension and defines a set. The union of this set is a transitive set containing A as a subset (and in fact is the minimal such set).

The proofs of Union and Power Set seem to be "set up" in advance by the axioms of Elements and Subclasses, but the fact that we can prove Infinity is a surprise.

At this point we have shown that Ackermann set theory has the strength of Zermelo set theory and so is adequate for classical mathematics.

We pause before considering how much stronger this theory might be to wonder what the motivation of the theory might be. Suppose that sets are actual (completed, definite) collections and classes are merely potential collections. The collection of all sets is a merely potential collection (however many actual sets we have constructed, we can construct some more). The class of all sets with a given property is a potential collection (as we built more and more sets we may add more objects to this class). An element of an actual collection is actual; a subclass of an actual collection is actual. The tricky part (as always) is getting one's mind around the interpretation of Set Comprehension. A collection defined mentioning only actual parameters (and in particular not mentioning the merely potential collection V of all sets) all of whose members turn out to be actual. . .is actual. This is not to our minds entirely convincing, but it may be suggestive.

We now develop an argument for the proposition that the well-founded sets of Ackermann set theory satisfy ZFC.

Definition: A *hierarchy* is a class which is well-ordered by inclusion, contains the unions of all its subclasses as elements, and in which the immediate successor of each element in the inclusion order (if it has one) is its power class. A *rank* is a class which belongs to a hierarchy. (Note that this definition is similar to the definition of *subhierarchy* in section 3.4, but note also that there is a distinction between set and class here which is not drawn in the earlier context).

Comments: A linear order is a well-ordering iff each subclass of its domain has a least element; since our class comprehension principles are apparently limited to classes of sets this might mean less than it appears to mean once we get past set ranks: we would not necessarily be able to carry out inductions on non-set ranks because not all definable subcollections of arbitrary classes are necessarily classes. The power class of a class is the class of all subclasses of that class, if such a class exists.

Theorem: For any two hierarchies H_1 and H_2 which are classes of sets, either $H_1 \subseteq H_2$ or $H_2 \subseteq H_1$.

Comment: We omit the proof of this result as we did with the similar result stated in section 3.4: we note that the restriction to hierarchies of sets is necessary because our comprehension principles are restricted.

Corollary: The set ranks are well-ordered by inclusion.

Definition: A *good rank* is a rank which is either included in or includes each other rank, and such that the ranks included in it are well-ordered by inclusion.

Theorem: The class of all ranked sets is a good rank.

Proof: Any hierarchy which contains all the set ranks will contain the class of all ranked sets as an element: then every element of such a hierarchy (and so every rank) will either be a set rank (and so be included in the class of all ranked sets) or include the class of all ranked sets. A hierarchy which does not contain all ranked sets will have its union a set rank, which is a subset of the class of all ranked sets.

Theorem: The class of all ranked sets is not the maximal good rank in the inclusion order.

Proof: If it were, it would be a set by set comprehension, as it would be definable without reference to the sethood predicate (as the maximal good rank in the inclusion order) and all of its elements are sets (since it is the class of all ranked sets). If the class of all ranked sets were a set, its power class would also be a set, so it would itself be a ranked set, and an element of itself. It is straightforward to establish by induction on inclusion that all set ranks are non-self-membered.

Corollary: There are non-sets which are elements: any rank which properly includes the class of all ranked sets has the class of all ranked sets itself as an element!

Theorem: Each of the axioms remains true if the term "ranked set" replaces the term "set".

Proof: Extensionality does not mention sethood at all. Any definable subclass of the ranked sets is a class by Class Comprehension, so the modified version of Class Comprehension holds. Any element of a ranked set is a ranked set: one proves by induction on the inclusion order that all set ranks are transitive. So the modified version of the Axiom of Elements holds. Any subclass of a ranked set is a ranked set: if x is included in a set rank r, the power set of x is included in the power set of r. So the modified version of the Axiom of Subclasses holds. A class x which is defined without reference to sethood, with no parameters but ranked set parameters, and all of whose members are ranked sets is a set by Set Comprehension. x is included as a subset in a rank (the class of all ranked sets will serve). The first rank which includes x must be a set, because we just defined it using only a set parameter, and all of its elements are sets. Thus the modified version of Set Comprehension holds.

Comment: We have just shown that we can without affecting the strength of the theory simply add the assertion that every set belongs to a rank as an axiom.

Theorem: Let $\phi(x, y)$ be a formula containing only ranked set parameters which does not mention the sethood predicate and which is functional (for every ranked set x there is exactly one ranked set y such that $\phi(x, y)$). Then for any ranked set a, $\{y \mid (\exists x \in a.\phi(x, y))\}$ is a set.

Proof: This set exists by set comprehension, by inspection of its definition. There is a rank which includes all of its elements, namely the class of all ranked sets, and the minimal rank in the inclusion order which includes this set is also a set by set comprehension, so it follows that this set is a ranked set.

Comment: This assertion looks like the Axiom of Replacement, but some delicacy is required to show that the Axiom of Replacement holds for all formulas. The difficulty is that the formula ϕ in an instance of Replacement as directly translated from ZFC may contain quantifiers over all sets, which would be quantifiers over the class of sets in Ackermann set theory, to which set comprehension could not be applied. Nonetheless Replacement does hold. The idea is to show that $(\exists x \in V.\phi(a_1, \ldots, a_n))$ can be translated to a formula not mentioning nonset parameters or the sethood predicate if ϕ is equivalent to a formula not mentioning nonset parameters or the sethood predicate. Define $r(a_1, \ldots, a_n)$ as the least rank in the inclusion order which includes an x such that $\phi(a_1, \ldots, a_n)$, if there is one, and otherwise the empty rank. Note that this rank will be a set in any case. $(\exists x \in V.\phi(a_1, \ldots, a_n))$ is equivalent to $(\exists x \in r(a_1, \ldots, a_n).\phi(a_1, \ldots, a_n))$. By this technique we can systematically eliminate quantifiers over the domain of all sets from formulas that meet the appropriate syntactical restrictions. We should now have given at least a strong indication why the final theorem holds. Full details of the equivalence of Ackermann and ZF are found in [Reinhardt, 1970].

Theorem: The interpretation of the Axiom of Replacement (and, as we have already seen, of all the other axioms of ZFC) holds on the domain of ranked sets of Ackermann set theory.

The consistency of Ackermann set theory also follows from the consistency of ZFC, in a way which reveals something about the nature of the domain of "sets" in Ackermann set theory.

Augment the language of ZFC with an additional symbol M. Provide axioms asserting that M is transitive and contains all subsets of its elements. For each formula ϕ and variables x, y add an axiom to the effect that $(\forall x.(\exists y.\phi)) \leftrightarrow (\forall x \in M.(\exists y \in M.\phi))$ (for all values of any parameters). Note that a corollary of this is that $(\forall x.\phi) \leftrightarrow (\forall x \in M.\phi)$ will be an axiom for any ϕ (use a dummy variable y that does not appear in ϕ). It is straightforward to show that any finite collection of such axioms is consistent with ZFC. If $(\forall x.(\exists y.\phi))$ holds, we arrange for M to be closed under a suitable operation (applying replacement); if $(\forall x.(\exists y.\phi))$ does not hold, we arrange for M to include a counterexample (if ϕ contains parameters, this is still a closure operation, as we need counterexamples for all appropriate values of the parameters).

This extension of ZFC has a model if ZFC has a model. We claim that this is a model of Ackermann set theory, with the sethood predicate taken as meaning "is an element of M". The only axiom which requires any attention is Set Comprehension. Let ϕ be a predicate not mentioning M with all parameters other than x belonging to M. We may suppose without loss of generality that there is only one parameter $a \in M$. So we express the predicate as $\phi(x, a)$ and further suppose that for any $a \in M$, $\phi(x, a)$ is in M. Note that in our original model of ZFC it will be the case for every $a \in M$ that $\{x \mid \phi(x, a)\}$ exists as a set in ZFC (it will be in the power set of M), and so for every $a \in M$ there is a (dummy) $b \in M$ such that $\{x \mid \phi(x, a)\}$ exists, and so for every a whatsoever, $\{x \mid \phi(x, a)\}$ exists. From this it follows by a more natural application of the special axioms for M that for each $a \in M$, $\{x \mid \phi(x, a)\} \in M$, and this establishes that the interpretation of Ackermann set theory's Set Comprehension Axiom holds in the extended model of ZFC.

It is worth noting that though externally M is a model of ZFC, the extended model of ZFC does not necessarily think that M is a model of ZFC (after all, the original model may not see any models of ZFC, in which case the extended model will not see any either). In Ackermann set theory, the class V of all sets cannot be shown to be a model of ZFC, though in some external sense it is, for similar reasons.

3.4 A pocket set theory

This final theory with classes is an expansion by one of the authors (Holmes) of a suggestion of Rudy Rucker. It is the weakest set theory discussed so far in this chapter. It is also very funny.

It is a folk observation that there are two sizes of infinite sets which occur in nature (that is, which occur naturally in mathematics outside of set theory). These are $\aleph_0$ and c, the cardinalities of the set of natural numbers and the set of real numbers respectively. The theory under consideration here asserts that these are the *only* infinite cardinalities (the alternative set theory of Vopěnka has this consequence). The very amusing aspect is that with carefully crafted axiomatics the assertion that these are the only two infinite cardinalities turns out to be almost the entire theory, and the theory is strong enough to support most classical mathematics.

Pocket set theory is a theory with sets and classes. We have the usual axioms and definitions: this is a first-order single sorted theory with equality and membership as primitive relations.

Extensionality: $(\forall AB.A = B \leftrightarrow (\forall x.x \in A \leftrightarrow x \in B))$

Definition: $set(x) \equiv (\exists Y.x \in Y)$

Class Comprehension: For any formula ϕ and variable A not free in ϕ, $(\exists A.(\forall x.x \in A \leftrightarrow set(x) \wedge \phi))$.

Definition: We define $\{x \mid \phi\}$ as the unique A (A not free in ϕ) such that $(\forall x.x \in A \leftrightarrow set(x) \wedge \phi)$.

Definition: For any sets a and b, we define $\{a, b\}$ as $\{x \mid x = a \vee x = b\}$. We define $\{a, a\}$ as $\{a\}$. We define (a, b) as $\{\{a\}, \{a, b\}\}$. Note that these are strictly definitions: we do not as yet know that there are any sets at all, nor have we assumed that an unordered pair of sets is itself a set.

Definition: A *relation* is a class of ordered pairs. For any relation R, we define R^{-1} as $\{(y, x) \mid (x, y) \in R\}$ (if this class exists, which requires that (y, x) be a set for each $(x, y) \in R$). A relation is a function f such that $(\forall xyz.(x, y) \in f \wedge (x, z) \in f \rightarrow y = z)$. A function f is a bijection iff f^{-1} exists and is a function. We define $\text{dom}(R)$ for any relation R as $\{x \mid (\exists y.(x, y) \in R)\}$. We say that f is a bijection from A to B if f is a bijection, $\text{dom}(f) = A$ and $\text{dom}(f^{-1}) = B$. Sets A and B are said to be the same size iff there is a bijection from A to B.

Now we give the axioms which provide the specific content of this theory (so far we have just given the generic theory of sets and classes and some definitions).

Definition: A class C is *infinite* iff there is a bijection from C to some proper subclass of C. A class C is *proper* if it is not a set.

Axiom of Infinite Sets: There is an infinite set. Any two infinite sets are the same size.

Axiom of Proper Classes: Any two proper classes are the same size, and any class the same size as a proper class is proper.

These axioms assert that there are just two infinite cardinalities for classes, the cardinality of the infinite sets and the cardinality of the proper classes.

This may not seem like a mathematically adequate set of axioms. But it is.

Definition: We fix an infinite set I. We define R as the class $\{x \mid x \notin x\}$, the Russell class, which is demonstrably proper.

Theorem: The empty class is a set.

Proof: Note that $\emptyset = \{x \mid x \neq x\}$ does exist as a class by Class Comprehension. Suppose that $\emptyset$ is not a set and so is proper. It follows that $\emptyset$ is the same size as R, so R is empty from which it follows that every set is self-membered. It follows further that $\{I\}$ is a set (because it is certainly not the same size as $\emptyset$) but also $\{I\} \notin \{I\}$, because $I \neq \{I\}$, because $\{I\}$ clearly is not the same size as any of its proper subclasses, so it cannot be the infinite I, so $\{I\} \in R$, which is a contradiction.

Theorem: For any set x, $\{x\}$ is a set.

Proof: Suppose otherwise. Then $\{x\}$ is proper and must be the same size as R, so R has just one element, which must be $\emptyset$ as obviously $\emptyset \in R$. From this it follows that $\{I, \emptyset\}$ is a set (because both I and $\emptyset$ are sets and there cannot be a bijection from $\{I, \emptyset\}$ to $\{x\}$), but neither I nor $\emptyset$ can have exactly two elements, so $\{I, \emptyset\} \in R$, which is a contradiction.

Theorem: For any sets x and y, $\{x, y\}$ is a set.

Proof: R contains $\emptyset$, $\{\emptyset\}$ and $\{\{\emptyset\}\}$, so $\{x, y\}$ cannot be the same size as R.

Corollary: (x, y) is a set for any sets x and y.

Theorem: For any formula $\phi(x, y)$, there is a class $\{(x, y) \mid \phi(x, y)\}$ which implements it. This makes it possible to demonstrate the existence of definable bijections as classes (note that until this point we have never appealed to the existence of specific bijections but only to the logical impossibility of the existence of bijections between concrete finite sets of specific sizes).

Definition: A von Neumann ordinal is a transitive set which is strictly well-ordered by the membership relation (where this means that every sub*class* of a von Neumann ordinal has a membership-minimal element).

Theorem: The class of all von Neumann ordinals is strictly well-ordered by the membership relation. The proof of this is standard, involved, and omitted.

Theorem: The class of all von Neumann ordinals is not a set.

Proof: This class is transitive and it is strictly well-ordered by membership. If it were a set it would be a von Neumann ordinal and so self-membered and so not strictly well-ordered by membership, which is a contradiction.

Theorem: If two classes are each the same size as a subclass of the other, then they are the same size. This is the Cantor-Schröder-Bernstein theorem and the standard proof is omitted.

Theorem: The Axiom of Choice holds.

Proof: By Cantor-Schröder-Bernstein, the universe V of all sets and R are the same size (R is the same size as $R \subseteq V$ and V is the same size as

$$\{\{\{x\}, \emptyset\} \mid x \in V\} \subseteq R.)$$

R is the same size as the class of von Neumann ordinals. Thus the universe is the same size as the class of von Neumann ordinals and a specific bijection from V to the ordinals gives an ordinal indexing of the universe which induces a well-ordering in the obvious way. A global well-ordering of the universe obviously implies Choice: choose the least element in the well-ordering of each element of a pairwise disjoint family of nonempty sets in order to get a choice set.

Theorem: Each set is the same size as a von Neumann ordinal.

Proof: Each set is the same size as a set of von Neumann ordinals, which can be mapped by an obvious recursion to an initial segment of the von Neumann ordinals, which is itself a von Neumann ordinal, which will be a set because it is in bijection with a set.

Theorem: There is an infinite von Neumann ordinal.

Proof: The von Neumann ordinal which is the same size as I is infinite.

Theorem: All infinite von Neumann ordinals are the same size as the first infinite von Neumann ordinal ω (that is, all infinite ordinals are countably infinite).

Proof: All infinite sets are the same size.

Theorem: The class of all subsets of ω is not a set.

Proof: The standard argument for Cantor's Theorem shows that ω is not the same size as the class of its subsets. Since the class of all subsets of ω is clearly infinite, it cannot be a set, and so must be a proper class.

Construction: The usual constructions of integers, rational numbers, and real numbers as Dedekind cuts in the rationals can be carried out. The set of reals is the same size as the class of all sets of natural numbers for standard reasons, and so is a proper class. Individual reals are countable collections of rationals, thus sets.

Theorem: Each infinite subclass of the class of real numbers is either the size of the class of real numbers or countable. $c = \aleph_1$.

Proof: The real numbers make up a proper class. An infinite subclass of the reals is either a set, in which case it is the same size as the infinite set ω, or it is a proper class, in which case it is the same size as the proper class of all reals. The real numbers can be placed in one-to-one correspondence with the countable ordinals, as both collections are proper classes.

Comment: The principle of Limitation of Size of the original class theories, which appears here as the Axiom of Proper Classes, in this context allows us to prove not only the "Axiom" of Choice (as in the original theories) but also the Continuum Hypothesis! This theory is much weaker than the theory of types (all objects it constructs appear by type 5 on the most uncharitable interpretation of the capabilities of TST), but every mathematical object needed in physics is constructible here. The collection of *all* functions from the reals to the reals is too large, but notice that the collection of continuous functions from the reals to the reals is of size c and can be represented in fairly natural ways, and in general the constructions actually needed in mathematical physics (or any mathematics short of set theory and shorn of excessive levels of abstraction) do not transcend the cardinality c. Points of Hilbert space are countable sequences of real numbers (thus sets) and continuous functions on Hilbert space are representable just as continuous functions on the reals are representable, and so forth.

4 THEORIES WITH ATOMS AND THEORIES WITH ANTI-FOUNDATION AXIOMS

4.1 Weak extensionality and ZFA

Most modern set theories have sets (or more generally classes) as the only objects. This is odd because originally sets were conceived as collections of other, previously given objects. It has seemed less odd (at least to those who have completed their indoctrination in foundations of mathematics) since we have adopted the view that all mathematical objects are sets. It also affords a logical simplification of the theories.

The usual form of the axiom of extensionality is

Axiom of Extensionality: $(\forall AB.A = B \leftrightarrow (\forall x.x \in A \leftrightarrow x \in B))$

A more natural axiom from a naïve standpoint would be

Weak Axiom of Extensionality: $(\forall AB.class(A) \wedge class(B) \rightarrow (A = B \leftrightarrow (\forall x.x \in A \leftrightarrow x \in B)))$

Here we restrict the scope of extensionality to classes (we say "class" here to avoid collision with the notion of "sethood" found in the theories with sets and classes above; below in NFU we will use "set" for this concept).

Further, it seems natural to assert that classes are the only things which have elements:

Axiom of Classhood: $(\forall Ax.x \in A \rightarrow class(A))$

The non-classes have no elements. These objects are generally called *atoms* or *urelements.* There is (in most theories) an elementless class $\emptyset$ as well (and it is the only one). This observation about $\emptyset$ indicates that, though we can formalize theories with weak extensionality using a primitive notion of classhood there is no need to do this.

Definition: $class(x) \equiv x = \emptyset \vee (\exists y.y \in x)$

This treatment applies equally well to theories with or without proper classes. If no distinction is drawn between sets and classes, then the predicate *class* could equally well be written *set.* In either case it is definable if we are willing to take $\emptyset$ as a primitive notion instead of the classhood (or sethood) predicate.

In most cases it is mathematically preferable to suppose that everything is a set and adopt the strong form of extensionality. In the context of New Foundations the adoption of strong extensionality has very strong consequences: the theory NFU with weak extensionality is known to be consistent where the original theory NF of Quine is not. This will be discussed below. In Kripke-Platek set theory (see [Barwise, 1975]) it is found useful to have a variant KPU with urelements: this we do not discuss at all here.

The theory (or theories) ZFA obtained by replacing Extensionality with Weak Extensionality in the usual set theory ZFC have a notable practical application which make them worthy of note as an alternative set theory. These theories support a relatively straightforward proof of the independence of the Axiom of Choice from the other axioms of set theory [Fraenkel, 1922]. It is trickier to prove the relative consistency of $\neg$AC with extensional ZF: this was not done until forcing was developed by Cohen (and the proof owes something to the technique based in ZFA).

We note that there are two different sorts of ZFA, depending on whether the atoms make up a set or a proper class. A very strong denial of Extensionality would stipulate that every set is the same size as some set of atoms; a weaker nonextensional theory would provide that the universe consisted of the union of all the iterated power sets of a set A of atoms.

If we fix an infinite set of atoms A, we can consider the action of permutations on A on all sets: if π is a permutation of A, there is a uniquely determined permutation π^* acting on all sets such that $\pi^*(a) = \pi(a)$ for $a \in A$ and $\pi^*(B) = \pi^* ``B$ for all sets B (the existence of this uniquely determined permutation depends on Foundation).

We can then say that a set B has support $S \subseteq A$ iff every permutation π^* fixing each element of $A \setminus S$ also fixes B. The punchline is that the class of all sets with finite support in A satisfies all the axioms of ZFA except Choice: for certainly A itself has finite support in A and no linear order of A has finite support in A (much less any well-ordering of A). The demonstration that the other axioms of ZFA hold in the domain of sets with finite support in A is technically involved and beyond the scope of this chapter.

While there might seem to be philosophical advantages to providing many non-classes with no elements (because it is a very sophisticated ontological perspective indeed that would lead us to decide that everything is a set), the mathematical advantage generally seems to be with stipulating that everything is a set, and where we do find mathematical applications for atoms they do not seem to have anything to do with the pre-set-theoretical reasons for supposing that there are such objects.

4.2 Aczel's AFA

Nowadays when people use the phrase "anti-foundation axiom" it is almost always to denote specifically the anti-foundation axiom of Forti and Honsell [Forti and Honsell, 1983] rather than any of the other anti-foundation axioms that have appeared from time to time. Broadly the common noun 'anti-foundation axiom' is used to refer to an axiom (scheme) which can be added to the axioms of ZF(C) once the Axiom of Foundation has been dropped. Thus anti-foundation axioms are alternative in the sense of being alternative to the axiom scheme of Foundation *once the other axioms of* ZF *have been agreed on.* This is a different sense of 'alternative' from the way in which NF or GPK^+_∞ are alternative.

Forti-Honsell's axiom (which is called AFA by Peter Aczel in his [Aczel, 1988]) is the most interesting and gives rise to the clearest mathematics, and this is probably because of all the anti-foundation axioms it is that one which most clearly arises from a sensible conception of what sets are. It arises from the idea that sets are things denoted by set pictures.

Definition: The *field* of a relation R is defined as the union of its domain and range. A pair (R, r) is a *set picture* iff R is a relation, r is an element of the field of R, and there is a map f with domain the field of R with the property that $f(x) = \{f(y) \mid y\,R\,x\}$ for all x in the field of R. We say further that (R, r) is a *set picture* of $f(r)$. There is no immediate reason that a set picture should be a set picture of just one set (but this is the case in the first theory we consider).

In the usual set theory ZFC (or for that matter in ZF: choice is not an issue), the question of what relations are set pictures is fully settled in a very simple way.

Theorem (ZFC): A pair (R, r) with r in the field of R is a set picture iff the relation R is well-founded. Further it is the picture of a uniquely determined set.

The Axiom of Foundation implies immediately that any set picture must be well-founded, and the Mostowski Collapsing Lemma establishes that any well-founded relation is actually a set picture of a uniquely determined set.

Now we change our attention to the theory ZFC– which is obtained by omitting the Axiom of Foundation from ZFC. It now becomes conceivable that more general relations R may be set pictures.

Consider the very simple example of a reflexive relation R with field $\{r\}$. If (R, r) is a set picture, there is a set which is its own sole element. Note that nothing tells us that there cannot be many such sets if the possibility of one is admitted: the same relation may be the picture of many different sets.

It is a theorem of ZFC– that for each well-founded set picture (R, r) the set of which it is a picture is uniquely determined, but this depends strongly on the well-foundedness of R.

The axiom of Forti and Honsell which is called AFA by Aczel asserts this:

Axiom of Anti-Foundation: Each pair (R, r) with R a relation and r in the field of R is a set picture and the set picture of a uniquely determined set.

There are two things going on here: that every pair (R, r) is a set picture says that the Axiom of Foundation fails in very elaborate ways (though not in all possible ways as we will see); that each relation R is a set picture in *only one way* amounts to a strong form of extensionality. There is a set which its own sole element under AFA, but also there is exactly one such set.

At least one of the authors (Holmes) can testify that the appearance of Aczel's book which popularized this anti-foundation axiom caused a deal of philosophical excitement, as it seemed that the possibility of self-membered sets somehow transcended the iterative conception of set. This excitement was ill-founded (the author begs forgiveness for the pun): the relation of ZFC– + AFA to ZFC is intimate, and even in its own terms it reveals the stamp of the iterative conception of set. The theory does have technical uses in avoiding certain inconveniences of the universe of well-founded sets, but its philosophical differences from ZFC have been overstated.

ZFC– + AFA is readily interpreted in ZFC. If we have two relations R and S in ZFC we say that a relation b from the field of R to the field of S is a *bisimulation* iff $x\,b\,y$ iff for each $z\,R\,x$ there is $w\,S\,y$ such that $z\,b\,w$ and for each $w\,S\,y$ there is $z\,R\,x$ such that $z\,b\,w$. The objects of our interpretation of ZFA will be pairs (R, r) where R is a relation and r is an element of the field of R. Two pairs (R, r) and (S, s) are equivalent iff there is a bisimulation from the field of R to the field of an end extension of S which relates r to s (an end extension of S is a relation including S under which elements of the field of S have the same preimages they have under S). We say that $(R, r)\,E\,(S, s)$ iff (R, r) is equivalent to some (S, t) where $t\,S\,s$. E is the membership relation of our interpretation of ZFC– + AFA. We will not present a proof that this works in the limited space of this chapter.

In terms of ZFC– + AFA, a version of the cumulative hierarchy is readily recovered in spite of the ill-founded nature of the theory. The stages of this hierarchy are indexed by the ordinals as is the usual hierarchy of ranks. $H_0 = \{0, 1\}$ (it will clear in a moment why we cannot start with an empty rank). Once H_α is defined, we define $H_{\alpha+1}$ as the collection of all sets which have set pictures R with field a subset of H_α. At limit stages, we take unions as in the usual hierarchy. Observe for example that H_1 contains three sets, the empty set and $\{\emptyset\}$ (pictured by the usual order relation on $\{0, 1\}$), and the solitary object which is its own sole element (pictured in any reflexive relation). It is straightforward to show that the power set of H_α is included in $H_{\alpha+2}$ (and in $H_{\alpha+1}$ if α is infinite) but of course each new level of the hierarchy includes new sets defined in non-well-founded ways. However, the novelty is quite constrained: H_α is the same size as V_α for each infinite ordinal α.

4.3 Boffa's axiom

An anti-foundation axiom with effects rather different those of AFA was proposed by Boffa [1972]. Aczel's axiom allows relatively few non-well-founded objects because of its strong extensionality attributes (for example, it allows only one self-singleton). Boffa's axiom allows (speaking roughly) as many non-well-founded sets as possible. For example, under ZFC– with Boffa's anti-foundation axiom there is a proper class of self-singletons.

Definition: A relation R is *extensional* iff each element x of the range of R is uniquely determined by its preimage $R^{-1}``\{x\}$.

Definition: Let R be an extensional relation. A *set-labelling* of R is an injection s with domain a subset of the field of R with the property that if x is in the domain of s then $R^{-1}``\{x\}$ is included in the domain of s and

$$s(x) = \{s(y) \mid y\,R\,x\}.$$

Boffa's anti-foundation axiom: Any set-labelling of an extensional relation R can be extended to a set-labelling of R (not necessarily unique) whose domain is the entire field of R.

So, for example, if we define a relation R on the domain $\mathbb{N}$ so that $m\,R\,n$ iff either $n = 0$ or $m = n > 0$, we find that under Aczel's axiom this is a set picture of the unique self-singleton, whereas under Boffa's axiom we can extend the empty set-labelling of this set to a total set-labelling of its range and discover that each positive integer is sent to a self-singleton and 0 is sent to a countably infinite set of self-singletons. The same technique can be used to get as many distinct self-singletons as we might want.

Boffa's axiom is similarly motivated by the idea of sets as derived from set pictures, but with leeway for many sets to be represented by the same picture, and ZFC– with Boffa's axiom admits a fairly straightforward relative consistency proof from ZFC.

5 NEW FOUNDATIONS AND RELATED SYSTEMS

In this section we review the set theory "New Foundations" (NF) proposed by W. v. O. Quine [1937] and related systems. This system is not so far known to be consistent, and it was shown by Specker [1953] that it disproves the Axiom of Choice (and so it proves Infinity). However, there are several known subsystems of NF which are known to be consistent (none of which reproduce Specker's disproof of Choice, though one of them does allow use of Specker's argument to prove Infinity, as we will see below).

5.1 Stratified comprehension

New Foundations is a variant of TST rather than of Zermelo set theory, in spite of the fact that it is an untyped set theory. The starting point for the development of NF is the observation that TST enjoys a great deal of what Russell called "systematic ambiguity", modern set theorists call "typical ambiguity", and computer scientists call “polymorphism”. For any sentence ϕ in the language of TST, define ϕ^+ as the sentence obtained by raising the type of each variable in ϕ by one. Note that for any axiom ϕ, ϕ^+ is also an axiom, and that for any rule of inference allowing ψ to be deduced from χ, ψ^+ can be deduced from χ^+ using the same rule. It follows from these considerations that for any formula ϕ, if ϕ is a theorem so is ϕ^+. If we associate with each variable x a specific variable x^+ of the next higher type, we can extend the definition of ϕ^+ to all formulas. Note that each object $\{x \mid \phi\}$ defined by a set abstract has a precise analogue $\{x^+ \mid \phi^+\}$ in the next higher type: for example the Frege natural number 3 defined in type 2 (the set of all type 1 sets with three elements) has a precise analogue in type 3 (the set of all type 2 sets with three type 1 elements) and indeed in each higher type.

Quine's suggestion is that this is an indication that the types need not actually be different. He proposed that all the types might be the same domain: the sentence ϕ says the same thing as the sentence ϕ^+ and the object $\{x \mid \phi\}$ is the same object as $\{x^+ \mid \phi^+\}$: in our example, there is exactly one Frege natural number 3 which is simply the set of all sets with three elements.

The axioms of Quine's theory “New Foundations” (so called from the name of the paper [Quine, 1937]) are the extensionality and comprehension axioms of TST with all indications of type removed. The details follow.

NF is a first-order single-sorted theory with equality and membership as primitive relations.

Axiom of Extensionality: $(\forall AB.A = B \leftrightarrow (\forall x.x \in A \leftrightarrow x \in B))$.

Axiom of Comprehension: For any formula ϕ in which A does not appear free, and which can be converted to a well-formed formula of TST by an assignment of types to variables, $(\exists A.(\forall x.x \in A \leftrightarrow \phi))$ is an axiom.

Definition: For any formula ϕ in which A does not appear free, $\{x \mid \phi\}$ is defined as the unique A (if there is one) such that $(\forall x.x \in A \leftrightarrow \phi)$ (this exists by Comprehension if ϕ can be obtained by dropping types from a formula of TST and is unique if it exists by Extensionality).

This presentation of the axioms should make it clear why dropping types from the axioms of TST does *not* give us the inconsistent Axiom of Comprehension of naïve set theory. The Russell class $\{x \mid x \notin x\}$ is not provided by the Axiom of Comprehension of NF because there is no way to assign types to the variables in $x \notin x$ which gives a well-formed formula of TST.

It is usual to rephrase the comprehension axiom of NF in a way which hides the apparent dependence on the language of another theory.

Definition: A formula ϕ is said to be *stratified* iff there is a function σ from the set of all variables to the natural numbers (equivalently, to the integers) such that for each atomic subformula $x = y$ of ϕ we have $\sigma(x) = \sigma(y)$ and for each atomic subformula $x \in y$ of ϕ we have $\sigma(x) + 1 = \sigma(y)$. Such a function σ is called a *stratification* of ϕ.

Axiom of Stratified Comprehension: For any stratified formula ϕ, $(\exists A.(\forall x.x \in A \leftrightarrow \phi))$ is an axiom.

A stratification of ϕ of course codes an assignment of types to the variables in ϕ which converts ϕ to a well-formed formula of TST.

The dependence of the axiomatization of NF on an understanding of TST really is just apparent as we said above. The axiom scheme of Stratified Comprehension is equivalent to a finite collection of its instances, and so can be given as a list of particular comprehension axioms with no reference to types or stratification at all. The original reference for this is [Hailperin, 1944], but the implementation given there is terrible. The construction of such a finite axiomatization is similar to our development of the finite axiomatization of predicative class comprehension given above.

5.2 New Foundations with urelements

In this section we will explore the development of set theory with stratified comprehension, but we will begin with a critique of the theory history presents to us.

Motivation of NFU

Quine [Quine, 1937] claimed that the choice of strong extensionality over weak extensionality is purely a matter of convenience. He suggested that any urelements (objects with no elements distinct from the empty set and from one another) could be reinterpreted as self-singletons $x = \{x\}$, thus restoring strong extensionality. In Zermelo-style set theory this suggestion makes sense. One can define a (proper class) map f which sends each urelement and iterated singleton of an urelement to its singleton and redefine membership to the relation $x \in_f y \equiv_{def} x \in f(y)$. This will preserve the axioms of Zermelo-style set theory and convert all urelements to self-singletons. But this procedure makes no sense in a set theory with stratified comprehension, because the map f has an unstratified definition, so sets defined in terms of $\in_f$ cannot be expected to exist in general.

This suggests that the theory we should be considering is the theory NFU whose primitive notions are equality, membership, and the empty set, and whose axioms are as follows:

Axiom of the Empty Set: $(\forall x.x \notin \emptyset)$

Axiom of Weak Extensionality: $(\forall ABx.x \in A \rightarrow (A = B \leftrightarrow (\forall y.y \in A \leftrightarrow y \in B)))$

Axiom of Stratified Comprehension: For any stratified formula ϕ, $(\exists A.(\forall x.x \in A \leftrightarrow \phi))$ is an axiom.

Definition: $set(x) \equiv_{def} x = \emptyset \vee (\exists y.y \in x)$. Notice that $(\forall AB.set(A) \wedge set(B) \rightarrow (A = B \leftrightarrow (\forall y.y \in A \leftrightarrow y \in B)))$

Definition: For any stratified formula ϕ, we define $\{x \mid \phi\}$ as the unique A such that $set(x)$ and for all x, $x \in A \leftrightarrow \phi$. This exists by stratified comprehension and is unique by weak extensionality and the definition of the empty set.

And indeed history suggests that this is the theory we should be thinking about. The consistency of NF remains an open question after more than 70 years, but R. B. Jensen proved the consistency of NFU on quite weak assumptions in [Jensen, 1969]. Most of the standard development of mathematics in NF (which looks quite odd from the Zermelo-trained standpoint) works perfectly well in NFU, which has well-understood models which can be examined to see what is going on.

For this reason, we will follow the plan of giving the general presentation of mathematics with stratified comprehension in NFU, then following with a section in which distinctive features of mathematics in NF (notably Specker's disproof of the Axiom of Choice in [Specker, 1953]) are described.

A model construction for NFU

We first give a model construction for NFU. This is not the original consistency proof due to Jensen, but a similarly motivated model construction due to Maurice Boffa [1988].

We consider a model of Mac Lane set theory without the Axiom of Infinity or the Axiom of Choice (though both could be adjoined and usually will be) with an external automorphism j which moves a rank V_α of the cumulative hierarchy downward. The existence of such models is a standard result of model theory. The domain of the model of NFU to be constructed is the extension of the nonstandard rank V_α. The membership relation of the model is the relation $x \in_{\text{NFU}} y \equiv_{def} j(x) \in y \wedge y \in V_{j(\alpha)+1}$. Each element x of $V_{j(\alpha)+1}$ is assigned the extension of $j^{-1}(x)$ in the original model (notice in particular that $V_{j(\alpha)}$ is assigned the extension of V_α, the entire universe of the model!), and each element of $V_\alpha \setminus V_{j(\alpha)+1}$ is treated as an urelement. This should make it clear that weak extensionality is satisfied. Of course the empty set of the model of NFU is the empty set of the original model of set theory.

We argue that stratified comprehension is satisfied in the model. Let ϕ be a stratified formula in the language of NFU with a stratification σ. Let N be a natural number larger than the value of σ at any variable in ϕ. This formula has a translation ϕ_1 into the language of the model of Mac Lane in which the model construction is carried out. Each reference to $\emptyset$ in NFU is replaced with a reference to $\emptyset$ in the language of Mac Lane set theory, each atomic formula $u = v$ is replaced with $u = v$, each atomic formula $u \in v$ is replaced with $j(u) \in v \wedge v \in V_{j(\alpha)+1}$, and each quantifier is bounded in V_α. It might seem that we just need to construct the set $\{x \in V_\alpha \mid \phi_1\}$ and note that $j(\{x \in V_\alpha \mid \phi_1\})$ is the set which we assign this extension in the model of NFU. That is what we do in the end, but it requires justification. The problem is that the formula ϕ_1 contains references to the automorphism j, and the Axiom of Separation of Mac Lane set theory does not apply to formulas containing j, unless all references to j are in parameters (terms not containing bound variables). We show how to effect a transformation of ϕ_1 to an equivalent formula in which j occurs only in parameters. We begin by replacing each atomic formula $u = v$ with $j^{N-\sigma(v)}(u) = j^{N-\sigma(v)}(v)$. This is equivalent because j is an automorphism. Notice that $j^{N-\sigma(v)}(u) = j^{N-\sigma(u)}(u)$ because σ is a stratification. We replace each formula $j(u) \in v$ with $j^{N-\sigma(v)}(j(u)) \in j^{N-\sigma(v)}(v)$. This is equivalent because j is an automorphism. Notice that $j^{N-\sigma(v)}(j(u)) = j^{N-\sigma(u)}(u)$ because σ is a stratification. We replace formulas of the shape $v \in V_{j(\alpha)+1}$ with the equivalent $j^{N-\sigma(v)}(v) \in j^{N-\sigma(v)}(V_{j(\alpha)+1})$. These substitutions produce a formula ϕ_2. In ϕ_2, every variable u occurs with exactly $N - \sigma(u)$ applications of j. An equivalent formula is produced by replacing each $j^{N-\sigma(u)}(u)$ such that u is bound by a quantifier (restricted to V_α) with the variable u, replacing the bounding set with $j^{N-\sigma(u)}(V_\alpha)$. These substitutions produce a formula ϕ_3. The variable x occurs in ϕ_3 only in the context $j^{N-\sigma(x)}(x)$. ϕ_4 is obtained by replacing this term with x. Note that ϕ_4 contains no occurrences of j except in constants. $j^{-(N-\sigma(x))}(\{x \in j^{N-\sigma(x)}(V_\alpha) \mid \phi_4\})$ is the set $\{x \in V_\alpha \mid \phi_1\}$ of the original model of Mac Lane set theory, and the image of this set under j is the set $\{x \mid \phi\}$ of the model of NFU.

If we let the ordinal α above be a nonstandard natural number, we obtain a model of NFU in which the universe is finite! This witnesses an error in [Quine, 1937]: Quine says there that the existence of $\emptyset$, $\{\emptyset\}$, $\{\{\emptyset\}\}$, etc. shows that Infinity holds in NF. Infinity *does* hold in NF but not for this reason. This sequence of sets witnesses the fact that any model of NFU (including a model of NFU + "the universe is finite") must actually be externally infinite, but it does not witness the Axiom of Infinity because there is no reason to expect that it is a set (its definition is not stratified). In a model of NFU + "the universe is finite" it is *not* a set. It is interesting to observe that in fact this collection *can* be a set but this would take us far afield.

If we assume that Infinity and Choice hold in the underlying model of Mac Lane, Infinity and Choice hold in the model of NFU. We can obtain strong axioms of infinity by assuming them in the underlying model of Mac Lane set theory, but more satisfying results can be achieved by making strong assumptions about the automorphism j (which are generally equivalent to strong axioms of infinity in Mac Lane or ZFC, but give nicer characteristics to the model of NFU).

Mathematics in NFU

In this section, we outline the development of the foundations of mathematics in NFU. What we do here will also work in NF, as long as we do not assume Choice. Infinity, which we must add as an assumption to NFU for the purposes of this development, is a theorem of NF, as we will see in the next section.

We develop the natural numbers (following Frege's definition) exactly as we did above in TST. We define 0 as $\{\emptyset\}$. For each set A, define $A+1$ as $\{a \cup \{x\} \mid a \in A \wedge x \notin a\}$. $A + 1$ is the set of all objects obtained by adding a single new element to an element of A. Now $0 + 1$ is the set of all one-element sets, which we call 1, and

$1+1$ is the set of all two-element sets, which we call 2, and so forth. We define $\mathbb{N}$, the set of natural numbers, as the intersection of all sets which contain 0 and are closed under the extended "successor" operation. The elements of the natural numbers are the finite sets: we define the set $\mathbb{F}$ of finite sets as $\bigcup \mathbb{N}$.

Notice that NFU witnesses the coherence of Frege's implementation of the natural number n as the set of all sets with n elements.

We are not quite done. There is a bad possibility which we have avoided discussing in the development above. One can prove by mathematical induction that no natural number contains a proper subset of one of its elements. This implies that if the universal set V is finite, the natural number which contains it as an element is $\{V\}$. It then follows that the successor of V is $\emptyset$, and of course the successor of $\emptyset$ is also $\emptyset$. This gives an exception to Peano's fourth axiom (all the others are easily seen to hold for this implementation of the natural numbers).

We rule out this badness by adopting the

Axiom of Infinity: $V \notin \mathbb{F}$.

We further usually adopt the

Axiom of Choice: Each pairwise disjoint collection of nonempty sets has a choice set.

though in this context it should be noted that the Axiom of Choice is false in NF.

We discuss the extension of the definition of stratification to support term constructions. Where ψ is a stratified formula, let $(\iota x.\psi)$ represent the unique object x such that ψ (if there is one) and the empty set otherwise. We define $(\exists! x.\psi)$ as $(\exists x.\psi) \wedge (\forall xy.\psi \wedge \psi[y/x] \rightarrow x = y)$ (y being a fresh variable). Notice that this is stratified iff ψ was stratified. Now note that $\phi[(\iota x.\psi)/x]$ is equivalent to $((\exists! x.\psi) \wedge \phi) \vee (\neg(\exists! x.\psi) \wedge \phi[\emptyset/x])$. The correct notion of stratification for general terms containing definite description terms $(\iota x.\psi)$ is as follows. A stratification of a formula ϕ is a function σ from terms to natural numbers (or integers) such that for any atomic subformula $t = u$ we have $\sigma(t) = \sigma(u)$, for any atomic subformula $t \in u$ we have $\sigma(t) + 1 = \sigma(u)$ [note that t and u may be complex terms rather than variables], and $\sigma((\iota x.\psi)) = \sigma(x)$: each definite description is assigned the same type as the variable bound in it. To see that this works, check that in the transformations above a stratification of a formula is the same as the stratification of the transformed formula. We do not intend to explicitly use definite description terms in what follows, but any defined term construction introduced can be thought of as being implemented in this way.

As we noted above under TST, the usual Kuratowski definition $(x, y) \equiv_{def} \{\{x\}, \{x, y\}\}$ of the ordered pair is inconvenient in type theory because the pair thus defined is two types higher than its projections.

If we assume Infinity, it is possible to define an ordered pair on $\mathcal{P}^2(V)$ (the collection of sets of sets) which is the same type as its projections. In NF, since

$\mathcal{P}^2(V) = V$, this is a definable type level ordered pair on the universe. In NFU, we develop a modified interpretation of NFU in which there is a type level pair on the universe. It should also be noticed that it is provable in NFU + Infinity + Choice that there is a type level pair (this follows from the theorem $\kappa^2 = \kappa$ for infinite cardinals κ, in the special case $\kappa = |V|$, but this line of development requires that a lot of mathematics be developed using the Kuratowski pair first).

Definition: We define $\sigma(x)$ as $x+1$ if x is a natural number and as x otherwise.

Definition: We define $\sigma_1(x)$ as $\{\sigma(y) \mid y \in x\}$ and $\sigma_2(x)$ as $\sigma_1(x) \cup \{0\}$, for any $x, y \in \mathcal{P}(V)$.

Observations: Notice that $\sigma_i(x) = \sigma_i(y) \rightarrow x = y$ for $i = 1, 2$, and note also that $\sigma_1(x) \neq \sigma_2(y)$ for any x, y.

Definition: We define σ_1“x as $\{\sigma_1(y) \mid y \in x\}$, σ_2“x as $\{\sigma_2(y) \mid y \in x\}$, and (x, y) (the Quine ordered pair of x and y) as σ_1“$x \cup \sigma_2$“y, for any $x, y \in \mathcal{P}^2(V)$.

Definition: For any $w \in \mathcal{P}^2(V)$, we define $\pi_1(w)$ as $\{x \mid \sigma_1(x) \in w\}$ and $\pi_2(w)$ as $\{x \mid \sigma_2(x) \in w\}$.

Observation: $\pi_1(x, y) = x$; $\pi_2(x, y) = y$; $(\pi_1(x), \pi_2(x)) = x$, for all $x, y \in \mathcal{P}^2(V)$.

This gives an ordered pair on sets of sets which is assigned the same value as its projections by any stratification (due to Quine [1945] and thus called the Quine ordered pair) . Now for a trick. We define a new interpretation of NFU whose domain is the extension of $\mathcal{P}^2(V)$ in the model we start with. The empty set of the new interpretation is the empty set of the old model. The equality relation of the new interpretation is the equality relation of the old interpretation restricted to the new domain. The membership relation $x \in_{new} y \equiv_{def} x \in y \wedge y \in \mathcal{P}^3(V)$. In other words, each set of sets of sets retains its original extension (so all subsets of the domain of the new interpretation are realized) and sets of sets which are not sets of sets of sets are treated as urelements. The verification that the new interpretation satisfies the axioms of NFU is straightforward. The type level ordered pair on sets of sets of the old interpretation serves as a type level pair in the new interpretation. If we replace natural numbers n with their restrictions $n \cap \mathcal{P}^2(V)$ to sets of sets in the definition of the Quine pair, and use the modified Quine pair to carry out the construction of the new interpretation, then the type level pair of the new interpretation will coincide with the original (unmodified) Quine pair of the new interpretation on sets of sets. The reason for this is that the restricted natural number $n \cap \mathcal{P}^2(V)$ becomes a natural number in the new interpretation.

The upshot of this discussion is that we might as well adjoin relations π_1 and π_2 with the same stratification conditions as equality to our language, define (x, y) as $(\iota z.z\pi_1 x \wedge z\pi_2 y)$, define $\pi_1(x)$ as $(\iota y.x\pi_1 y)$ and $\pi_2(x)$ as $(\iota y.x\pi_2 y)$, then adjoin an

Axiom of Ordered Pairs: $\pi_1(x, y) = x$; $\pi_2(x, y) = y$; $(\pi_1(x), \pi_2(x)) = x$.

We could but do not need to stipulate that the type level ordered pair coincides with the Quine pair on sets of sets. The introduction of type level ordered pairs by axiom in NFU is a proposal of one of the authors (Holmes) in his [Holmes, 1998].

The Axiom of Ordered Pairs is actually an inessential strengthening of the Axiom of Infinity (in the presence of Choice, they are equivalent); we have just shown that NFU + Infinity interprets NFU + Ordered Pairs. The function sending x to $(x, 0)$ has a stratified definition and maps the universe V to its proper subset $V \times \{0\}$ which is sufficient to show that Ordered Pairs implies Infinity.

Whether we introduce a primitive type level pair as we have just done or use the Kuratowski pair, the definitions of relations and functions and the basic properties of and operations on relations and functions are similar. There is a precise correspondence between the functions and relations implementable with a type level pair and those implementable with the Kuratowski pair. One has to bear in mind with the Kuratowski pair that the pair is two types higher than its projections: this causes some inconveniences. A fundamental difference between the Quine pair and the Kuratowski pair is that the projection operations of the Quine pair (or of a primitive type level pair) are set functions, while the projection functions of the Kuratowski pair are not. If $\langle x, y \rangle$ is the Kuratowski pair, there is a set function π_1^* such that $\pi_1^*(\langle x, y \rangle) = \{\{x\}\}$ but there is no set function π_1^* such that $\pi_1^*(\langle x, y \rangle) = x$. Notice that the double application of the singleton operation ensures that the relative types of the two occurrences of x are the same.

We define cardinal number. For any set A, we define the relation $A \sim B$ as holding iff there is a bijection f from A onto B. This is proved to be an equivalence relation in a standard way. We define $|A|$ as $\{B \mid B \sim A\}$. It is useful to observe that $|A|$ is exactly the same set whether we use the type level pair or the Kuratowski pair (because every function implemented with the type level pair has an exact analogue implemented with the Kuratowski pair and *vice versa*). $|A|$ is one type higher than A for purposes of stratification. Note that the natural numbers defined above are cardinal numbers of finite sets. This can be proved by mathematical induction. Objects which are $|A|$ for some set A are called *cardinal numbers*.

We define isomorphism types of relations. $R \approx S$ is defined as holding iff there is a bijection f from the field of R onto the field of S such that $x \, R \, y \leftrightarrow f(x) \, S \, f(y)$. Isomorphism is an equivalence relation on relations for standard reasons. The isomorphism type of a relation R is defined as $\{S \mid S \approx R\}$. The identity of the isomorphism type of R does depend on the pair used, and so does the type differential, because the isomorphism type is a set of functions and so a set of sets of ordered pairs. If we use the type level pair, the isomorphism type of R is one type higher than R for purposes of stratification; if we use the Kuratowski pair it is three types higher than R.

Isomorphism types of well-orderings are of special interest to us. If $\leq$ is a well-ordering, we define $ot(\leq)$ as the isomorphism type of $\leq$. An isomorphism type of a well-ordering is called the *order type* of a well-ordering, and an object which is the order type of some well-ordering is called an *ordinal number*.

These collections are very big. The analogous classes in Zermelo-style set theory are proper. Further, we have the universal set $V = \{x \mid x = x\}$, the very standard of bigness. We can contemplate $|V|$, the cardinality of the universe. We can prove that there is a natural well-ordering on the ordinal numbers in a quite standard way, then consider the order type Ω of the natural order on the ordinal numbers. We seem to be very close to paradox here, as $|V|$ and Ω are precisely the objects which instigate the classic paradoxes of Cantor and Burali-Forti!

However, we will now discover that NFU has its own characteristic way of averting paradox. We begin with the Cantor paradox of the largest cardinal. The Cantor theorem in TST takes the form $|\mathcal{P}_1(A)| < |\mathcal{P}(A)|$, where $\mathcal{P}_1(A)$ is the set of one-element subsets of A; the usual form $|A| < |\mathcal{P}(A)|$ would be ill-typed. The proof of the Cantor theorem of TST is inherited by NFU, so we have $|\mathcal{P}_1(A)| < |\mathcal{P}(A)|$ for any set A and so in particular $|\mathcal{P}_1(V)| < |\mathcal{P}(V)|$. Now of course we can add in $|\mathcal{P}_1(V)| < |\mathcal{P}(V)| \leq |V|$, so we get the odd result $|\mathcal{P}_1(V)| < |V|$. This is odd because we can "see" the one-to-one correspondence between these two sets defined by the singleton map. But there is no reason to believe that the singleton map is a set function, and we have just shown that it can't be.

Since we have a counterexample to the intuitively plausible proposition $|\mathcal{P}_1(A)| = |A|$ $(A = V)$ and since it is easy to show that $|A| = |B| \leftrightarrow |\mathcal{P}_1(A)| = |\mathcal{P}_1(B)|$, we define an operation on cardinals with defining equation $T(|A|) = |\mathcal{P}_1(A)|$. We have just shown that $T(|V|) < |V|$, so this is a nontrivial operation. There are sets A such that $|\mathcal{P}_1(A)| = |A|$: standard finite sets have this property, for example. Such sets are called *cantorian* sets and their cardinals are called cantorian cardinals. An even stronger property (motivated by the apparent witness our incorrect intuition that $|\mathcal{P}_1(A)| = |A|$ for all A, and also holding of concrete finite sets) is this: a set A (and its cardinality) is said to be *strongly cantorian* (s.c). iff $\iota\lceil A$, the restriction of the singleton map to A, is a set. Clearly strongly cantorian sets are cantorian; the converse is a strong axiom of infinity.

The Burali-Forti paradox is resolved in an even more interesting way. One can prove in a quite standard way that for any two well-orderings, either one is similar to an initial segment of the other, or the two are similar (the three alternatives being mutually exclusive). This determines a natural linear order on the ordinal numbers (*qua* isomorphism types of well-orderings) which is itself a well-ordering and so has an isomorphism type Ω. The paradox depends on the "obvious" observation that the order type of the restriction of this order to the ordinals less than α is α. The resolution depends on the discovery that this "obvious" observation is false.

For any relation R, we can define a relation $R^\iota = \{(\{x\}, \{y\}) \mid x\, R\, y\}$. If the isomorphism type of R is ρ, we define $T(\rho)$ as the isomorphism type of R^ι. It is straightforward to show that the definition of $T(\rho)$ does not depend on the choice of the relation R. Note that R^ι is one type higher than R for purposes of stratification and $T(\rho)$ is one type higher than ρ. We have just seen that the singleton map is not a function, so we have no reason to believe that $\rho = T(\rho)$

in general. A relation type ρ is cantorian iff $T(\rho) = \rho$ and strongly cantorian iff $T(\rho) = \rho$ is witnessed for each $R \in \rho$ by a set function equal to the restriction of the singleton map to the field of ρ.

Suppose that $\leq$ is a well-ordering belonging to an ordinal number α. For each x in the field of $\leq$, we can define α_x as the order type of the restriction of $\leq$ to $\{y \mid y \leq x\}$. If $\leq$ is also used for the order on ordinal numbers, we have $x \leq y \leftrightarrow \alpha_x \leq \alpha_y$. This would establish the desired isomorphism between the natural order on the ordinals less than α and the element $\leq$ of α if the map $x \mapsto \alpha_x$ were a function. But this cannot be shown to be the case, for α_x is two types higher than x for purposes of stratification (four types if we were using the Kuratowski pair). What *is* a function is $\{\{x\}\} \mapsto \alpha_x$; the application of the singleton operation repairs the failure of stratification. So what we can actually show is that $(\leq^{\iota})^{\iota}$ is similar to the natural order on the ordinals less than α, so the order type of the ordinals less than α is $T^2(\alpha)$.

Now the paradox resolves itself. For the order type of the ordinals less than Ω is $T^2(\Omega)$ by the preceding discussion, and it then follows that $T^2(\Omega) < \Omega$. This is not altogether appetizing, because it is easy to show that for any ordinals α and β, $T(\alpha) < T(\beta) \leftrightarrow \alpha < \beta$, so $T^2(\alpha) < T^2(\beta) \leftrightarrow \alpha < \beta$, so $\ldots T^6(\Omega) < T^4(\Omega) < T^2(\Omega) < \Omega$. This does not give an infinite descending sequence of ordinals in NFU, because T is not a function and the sequence just revealed cannot be shown to be a set (fortunately!). This does imply that any set model of NFU has ordinals whose elements are not well-orderings from an external standpoint. If our working set theory were NFU, this would not lead to contradiction, because though the domain V of the "model" of NFU in which we would work is a set, the membership relation $\in$ of the "model" is not a set.

Notice that the T operation appears to be an external endomorphism of the ordinals. In our model construction for NFU, we used an external automorphism of our model of set theory which moved ordinals. This similarity is not an accident. The T operation on isomorphism types is closely related to the automorphism j of the underlying model of set theory if we are working in a Boffa model (the same is true of the T operation on cardinals defined above).

The considerations so far give us some assurance that the original system NF does not easily fall prey to the paradoxes. We *know* that NFU does not (unless a quite weak fragment of the usual set theory is inconsistent).

We now turn our attention to the implementation of mathematics in NF(U). We have already described the implementations of natural numbers, ordered pairs, relations, functions, and cardinal and ordinal numbers. Operations on cardinal and ordinal numbers admit quite usual definitions (hereafter in this subsection we assume without comment that the pair is type level). For example, $|A| + |B| = |(A\times\{0\})\cup(B\times\{1\})|$ and $|A|\cdot|B| = |A\times B|$ are quite familiar-looking definitions of cardinal addition and multiplication. If we define B^A as the collection of functions from A to B, we might want to define $|B|^{|A|}$ as $|B^A|$, but we do not for reasons of relative type. The type of $|B^A|$ is one higher than that of A or B: we correct this by defining $|B|^{|A|}$ as $T^{-1}(|B^A|)$. This is not a total operation: there are cardinals κ such that there is no λ such that $T(\kappa) = \lambda$, such as $\kappa = |V|$. A special case of this definition is $2^{|A|} = T^{-1}(|\{0,1\}^A|) = T^{-1}(|\mathcal{P}(A)|)$ (the last equation uses the one-to-one correspondence between sets and their characteristic functions). The argument for Cantor's theorem above shows that $|A| < 2^{|A|}$ for all A for which $2^{|A|}$ is defined: we have $|\mathcal{P}_1(A)| < |\mathcal{P}(A)|$: applying T^{-1} to both sides of the inequality gives $|A| < T^{-1}(|\mathcal{P}(A)|)$, whenever the latter is defined, and so $|A| < 2^{|A|}$. It is straightforward to establish that the T operation distributes over all these operations and relations in the obvious sense, and so does T^{-1} where it is defined. Very similar considerations apply to operations on ordinal numbers, though all the usual operations on ordinals turn out to be total. One should note that the identification between cardinal numbers and initial ordinals does not hold here, so one needs to make use from time to time of explicit operations $card(\alpha)$ = the cardinality of the field of an element of the ordinal α and $init(\kappa)$ = the smallest order type of a well-ordering of a set of size κ.

All of the mathematical constructions we have done so far can be clarified by looking at how one would do the same mathematics in TST. The appearance of T and T^{-1} is unmysterious there: a set of size $T(\kappa)$ is "the same size" as a set of size κ but appears one type higher. A set of size $T^{-1}(\kappa)$ is a set the same size as a set of size κ but one type lower, and it is not surprising that this does not work for all κ because higher types are larger (being "power sets" of lower types).

Some mathematical constructions are awkward in NFU because they are awkward in TST. A general class of constructions for which this is true is "indexed families of sets". The difficulties are best illustrated by giving an example. We are going to develop the definition of the product of a family of cardinals, $\Pi_{i\in I}\kappa_i$. The product is the size of the generalized cartesian product of a collection of sets A_i each of cardinality κ_i. The generalized cartesian product $\times_{i\in I}A_i$ is the collection of all functions f such that $f(i) \in A_i$ for each $i \in I$. The problem that now arises is how to read the various indexed expressions here. The difficulty is that the sets A are one type lower than the cardinals κ, so if we read A_i as $A(i)$ (letting A be a function from I to sets) then we need to read κ_i as $\kappa(\{i\})$, so κ can be construed as a function from $\mathcal{P}_1(I)$ to cardinals. The set $\times_{i\in I}A_i$ can then be defined explicitly as the set of all functions f such that $f(i) \in \mathcal{P}_1(A(i))$ for each $i \in I$. We cannot have $f(i) \in A_i$ because once again the types are wrong: the type of i is the same as the type of $A(i)$ and so is one higher than the type of an element of A_i: we fix this by having f send each $A(i)$ to the singleton of one of its elements. We want $\Pi_{i\in I}\kappa_i$ to be of the same type as each of the cardinals κ_i. $\times_{i\in I}A_i$ is a set of functions which are each one type higher than elements of I, so it is itself two types higher than an element of I, and its cardinality is three types higher than that of an element of I. We want $\Pi_{i\in I}\kappa_i$ to be of the same type as the $\kappa(i)$'s. A $\kappa_i = \kappa(\{i\})$ is one type higher than that of an element of i. So the full definition of $\Pi_{i\in I}\kappa_i$ is $T^{-2}(|\times_{i\in I} A_i|)$, where $A(i) \in \kappa(\{i\})$ for each $i \in I$ (the existence of such an A and the well-definedness of this cardinal depends on Choice; this leaves open the possibility that products of indexed families of cardinals may be undefined because T^{-2} is partial, and indeed they may). There is no particular mystery to do with NF(U) in this construction: it is a construction in TST imported into NFU. The awkwardness could be reduced in special cases if we had index sets made up of self-singletons (if $i = \{i\}$ for each $i \in I$). One can prove the consistency of the existence of such index sets, but it is also provable that they are all small sets (such an index set would certainly be strongly cantorian, and it is consistent with NFU that every s.c. set is the same size as a set of self-singletons).

There are some mathematical techniques appropriate to NFU which take advantage of the fact that it is a type-free theory and cannot be replicated in TST. These generally require "reasonable" assumptions about the behavior of T operations on cardinals and ordinals which are not provable in NFU (and are in effect strong axioms of infinity).

We defined the notion of *strongly cantorian set* above. "Strongly cantorian" is a notion of smallness (every subset of a strongly cantorian set is strongly cantorian). A variable x restricted to a strongly cantorian set in a set definition does not need to have a type assigned to it for purposes of stratification. The reason is that its type can be freely manipulated: if $x \in A$ and A is s.c., then $x = \bigcup(\iota\lceil A(x)) = (\iota\lceil A)^{-1}(\{x\})$: in one of these expressions the type of x is raised by one and in one it is lowered by one, and these operations can be iterated.

The first special axiom we might consider is Rosser's Axiom of Counting (proposed by Rosser [1953]), which asserts that the set of natural numbers is strongly cantorian. Rosser's original formulation was the assertion that $\{1, \ldots, n\} \in n$ for each natural number n: the connection of this to our usual counting procedure should be clear and the theorem provable in NF(U) is $\{1, \ldots, n\} \in T^2(n)$ It is clear that Rosser's axiom implies and is implied by the assertion that each natural number n is cantorian ($T(n) = n$). It is only very slightly less obvious that this is equivalent to the assertion that the set of natural numbers is s.c. It is provable in NFU + Infinity that $\mathbb{N}$ is cantorian, and the natural inductively defined bijection from $\mathbb{N}$ to $\mathcal{P}_1(\mathbb{N})$ sends $T(n)$ to $\{n\}$ for each n, so will witness that $\mathbb{N}$ is s.c. if and only if every natural number is cantorian.

The practical effect of the use of Rosser's Axiom of Counting is that one does not need to assign types to variables with natural number values. In fact, the same becomes true of many familiar classes of mathematical objects, as the class of s.c. sets is closed under cartesian product, power set, and the formation of function spaces, so for example the set of real numbers or the set of points in Hilbert space will be s.c. sets.

The metamathematical effect of the Axiom of Counting is surprising. It does not actually add practical strength on the level of arithmetic (even arithmetic of higher order): we simply avoid notational complications. The truly substantial effects are in set theory. NFU + Infinity does not prove the existence of $\beth_\omega$; NFU + Counting proves the existence of $\beth_{init(\beth_n)}$ for each natural number n. NFU + Counting has models $\beth_{init(\beth_\omega)}$ is needed in the construction) and NF + Counting is known to prove the consistency of NF (such independence results are rare for NF).

The strong principle of mathematical induction (induction for unstratified as well as stratified formulas) is stronger than the Axiom of Counting (which it easily implies). Its precise strength is not known, but it is weaker than ZFC, as an ω-model of NFU can be constructed on fairly weak assumptions (existence of $\beth_{init(\beth_{\omega_1})}$ is enough). Again, its effects are not seen in higher-order arithmetic (which is adequately handled by stratified induction) but in set theory.

An assumption considered by C. Ward Henson [1973] (in a slightly more restricted form) is "every cantorian set is strongly cantorian", which we call CS (short for "the Axiom of Cantorian Sets"). Notice that Infinity and CS together imply Counting. The strength of NFU + Infinity + CS is rather shocking: it is equiconsistent with ZFC + "there is an n-Mahlo cardinal" for each concrete natural number n (see [Holmes, 2001]). Rieger-Bernays permutation methods can be used to convert a model of this theory to one in which the hereditarily strongly cantorian sets make up a model of ZFC with an n-Mahlo cardinal for each concrete n.

There are still stronger assumptions which have been investigated and shown to correspond in strength to further strong axioms of infinity, but this is enough to show the pattern: natural axioms formulated in NFU which regularize the behavior of strongly cantorian sets have surprisingly strong effects on the consistency strength of the extended version of NFU which results. We give one more example: the Axiom of Small Ordinals (introduced in [Holmes, 1998] and [Holmes, 2001]) in its weakest form asserts that any definable class of strongly cantorian ordinals is the intersection of the class of strongly cantorian ordinals (which is not a set) with some set. NFU + Infinity + Small Ordinals has the same strength as Kelley-Morse set theory + "the proper class ordinal is weakly compact" or ZFC - Power Set + "there is a weakly compact cardinal" (Solovay showed this in [Solovay, preprint]).

5.3 Peculiarities of NF

We now address the oddities of New Foundations itself. These are twofold. On the negative side, no consistency proof for NF is known, nor is there any proof that NF is any stronger than TST with the Axiom of Infinity (or equivalently Mac Lane set theory with Infinity) which is somewhat weaker than Zermelo set theory. Any suspicion that NF must be strong because it allows "big" sets such as the universe should be dispelled by the observation that NFU, which proves the existence of the same big sets for the same reasons, is weaker than Peano arithmetic, and NFU + Infinity (which is a more reasonable set theory) is exactly as strong as TST + Infinity or Mac Lane set theory with Infinity. There is a useful distinction to be drawn between "big" objects such as the universe V or the ordinal Ω and "large" objects such as inaccessible or measurable cardinals.

On the positive side, NF is known to prove the Axiom of Infinity and disprove the Axiom of Choice [Specker, 1953]. This means that NFU + Choice (which is known to be consistent) proves the existence of urelements! Specker gave separate proofs of these two results, but of course the proof that Choice is false also proves Infinity: if the universe were finite, it could be well-ordered, which would allow the proof of Choice in a familiar way (it is worth noting here that all the standard equivalences between forms of Choice hold in NF(U), as long as one is careful to state them in stratified forms).

We present Specker's proof that the Axiom of Choice is false.

Theorem: (NF, due to Specker). $\neg$AC.

Proof: Suppose otherwise. Then the natural order on the cardinal numbers would be a well-ordering (this is equivalent to *AC* for quite standard reasons).

We know from above that the exponential map $\kappa \mapsto 2^\kappa$ is partial. We define for each cardinal κ a set called the exp-closure of κ (this is a nonce notion): the exp-closure of κ is the smallest set which contains κ and contains 2^λ whenever it contains λ and 2^λ exists. We define the set SM (German *Speckermenge*, "the Specker set", a coinage of Thomas Forster) as the set of all cardinals whose exp-closure is finite. This set is nonempty: for example, $|V|$ belongs to SM.

Now comes a move which is peculiar to NF. We have $2^{|\mathcal{P}_1(V)|} = |\mathcal{P}(V)|$ by the definition of exponentiation. But this means that $2^{T(|V|)} = |V|$ (this requires $|\mathcal{P}(V)| = |V|$, which is a consequence of strong extensionality). Properties of the T operation then tell us that $2^{T^{n+1}(|V|)} = T^n(|V|)$ for each concrete natural number n, so all the cardinals $T^n(|V|)$ (which do not make up any kind of set sequence!) are in SM.

We argue that for any cardinal κ in SM, we also have $T^n\kappa \in SM$ and $T^{-1}(\kappa) \in SM$, if the latter cardinal exists. Suppose $\kappa \in SM$. We give the name exp to the map $\kappa \mapsto 2^\kappa$. If λ is an element of the exp-closure of κ (so of the form $exp^n(\kappa)$ for some natural number n) then $T(\lambda)$ is in the exp-closure of $T(\kappa)$, because it is equal to $exp^{T(n)}(T(\kappa))$ (this proof is considerably simpler if the Axiom of Counting is assumed, which would give $T(n) = n$!). Now consider the largest element ν of the exp-closure of κ. It must be greater than $T(|V|)$, or else its image under exp would be defined (notice that linear ordering of cardinals by the natural order is used here). $exp(T(\nu))$ will be defined, since $T(\nu) \leq T(|V|)$: since $\nu > T(|V|)$, we have $T(\nu) > T^2(|V|)$ and so $exp(T(\nu)) \geq exp(T^2(|V|)) = T(|V|)$. From this it follows that $exp^2(T(\nu))$ either fails to exist or is equal to $|V|$. So if the cardinality of the exp-closure of κ is n, the cardinality of the exp-closure of $T(\kappa)$ will be either $T(n)+1$ or $T(n)+2$, and in any case finite, so $T(\kappa) \in SM$. Now suppose that $\kappa = T(\lambda)$, and that $\lambda \notin SM$. This would imply that $exp^n(\lambda)$ was defined for every n, and it would follow that $T(exp^n(\lambda)) = exp^{T(n)}(\kappa)$ was defined for every n, and though the T operation might fail to be the identity on $\mathbb{N}$ it is onto: it follows that $\kappa \notin SM$ contrary to hypothesis. So if $\kappa \in SM$ and $T^{-1}(\kappa)$ exists, it follows that $T^{-1}(\kappa) \in SM$.

Because SM is a set of cardinals, it has a smallest element μ. The results above imply that $T(\mu) = \mu$. For otherwise we would have to have $T(\mu) > \mu$, whence $T^{-1}(\mu)$ would exist, belong to SM, and be less than μ, contrary to choice of μ (because the T operation preserves order on cardinals and its domain of definition is downward closed). Now let n be the finite number of elements in the exp-closure of μ. We see from the previous paragraph that the number of elements in the exp-closure of $T(\mu)$ must be $T(n)+1$ or $T(n)+2$, and of course this must be the same number. But $n = T(n) \bmod 3$ is easy to show, so $n = T(n)+1$ or $n = T(n)+2$ is impossible.

As we have already remarked, there is a nicer proof of this (in detail; the basic idea is the same) if the Axiom of Counting is assumed, found in [Holmes, 1998] (mod 3 arithmetic does not come into play), but we will shortly indicate reasons why we do not regard the assumption of Counting as harmless.

For any cardinal κ, we define the *Specker tree* of κ as the smallest set which contains κ and includes the preimage under exp of each of its elements. This is the set of all cardinals λ such that $exp^n(\lambda) = \kappa$ for some $n \in \mathbb{N}$. If the Axiom of Choice is assumed, it is straightforward to show that the Specker tree of any cardinal has finite depth (there is an n such that for no λ is $exp^n(\lambda) = \kappa$). It is a theorem of ZF that the Specker tree of any cardinal is well-founded, due to Forster [1976] (using Sierpinski's result that $\aleph(\kappa) < exp^3(\kappa)$, where $\aleph$ is the Hartogs aleph function: $\aleph(\kappa)$ is the first ordinal which is not the order type of a well-ordering of a subset of a set of size κ [this definition would need to be modified slightly to be stratified]). Thus in ZF the Specker tree of a cardinal must have an ordinal rank in an obvious sense, which if AC is assumed must be finite. It is an open question in ZF whether it is possible for a cardinal to have a Specker tree of infinite rank (and so have inverse images under exp of every index); it is a theorem of NF + Counting that the Specker tree of the cardinality of the universe has infinite rank: so here we have a combinatorial situation arising in what should not be a terribly strong extension of NF whose possibility has not been established in the usual Choice-free mathematics.

The only obvious mathematical advantage of NF over NFU which we see is that the Quine pair is definable in NF. There may be other elegant features of mathematical development in NF which follow from the absence of urelements. The use of Choice in mathematics is pervasive, which is a strong apparent *disadvantage* of NF over NFU. Now it might be said that the form of Choice most often used in classical mathematics is Dependent Choices, and so far as anyone knows DC is not inconsistent with NF. Unfortunately, no one has any idea how to provide a relative consistency proof for DC relative to NF.

The only method for relative consistency and independence proofs which has been used extensively with NF is the Rieger-Bernays permutation method, which can prove various amusing results, but is limited by the fact that stratified sentences (i.e, anything which makes sense in TST) are invariant in permutation models. Orey showed that NF + Counting is essentially stronger than NF using metamathematical techniques. It is possible to emulate work done for NFU to show that NF + Infinity + CS proves the relative consistency of the existence of n-Mahlos, but we do not know that NF + Counting does not prove the existence of

n-Mahlos. Similar considerations apply to other strong extensions of NFU whose consistency strength is known. Recently, one of the authors has shown that forcing arguments can be carried out in NF, but without some Choice forcing is not useful. A typical result which can be shown is that if NF + DC is consistent then NF + DC + "the continuum is well-ordered" is consistent.

The appeal that NF has over NFU for some is that it is a theory of pure sets. In NF (as in ZF) everything without exception is a set. This appeal should be resisted or at least viewed with care. Notice that in ZF the idea that something is a pure set can be expressed and used to define subsets using the Separation Axiom. In NFU, the predicate "is a pure set" has no stratified definition, and so cannot be used in a set definition (if there were such a definition, then the collection of pure sets in any model of NFU would be a model of NF, and this is not the case in known models). These considerations are intimately related to the fact that there are easy ways to interpret ZFA in ZF and *vice versa*, whereas all efforts to interpret NF in NFU have failed.

5.4 Extensional fragments of New Foundations

NFU, which is known to be consistent, differs from NF "only" in that extensionality is weakened to allow urelements; it has the same comprehension scheme. There are other fragments of NF which are known to be consistent which include the strong Axiom of Extensionality and perforce involve limitations on Stratified Comprehension. We describe two of them.

In the same year (1969) that Jensen showed the consistency of NFU, Grishin [1969] showed the consistency of NF_3, the fragment of New Foundations whose axioms are the strong Extensionality Axiom and all those comprehension axioms which admit a stratification whose range has no more than three elements: that is, all those axioms which can be typed in TST using types 0,1,2. He also showed that NF_4 is the same theory as NF, and that NF_3 + "$\{\{\{x\}, y\} \mid x \in y\}$ exists" is the same theory as NF.

To put this result in context, we cite a model-theoretic result of Specker. He showed that NF is equiconsistent with TST plus the scheme "$\phi \leftrightarrow \phi^+$" (for each formula ϕ). This scheme is more briefly called Amb (the ambiguity scheme). This result is a verification of Quine's motivation for NF. The technique of his proof generalizes to relate fragments of NF to fragments of TST. For example, NFU is equiconsistent with TSTU, the version of TST in which urelements are allowed in each positive type. Grishin's system NF_3 is equiconsistent with TST_3 + Amb, where TST_3 is the system of type theory with only three types (0,1,2), and the notation ϕ^+ of course only applies to formulas not mentioning type 2. It turns out that *all* infinite models of TST_3 satisfy Amb, so the consistency of NF_3 and indeed of powerful extensions of NF_3 is easy to establish, and it appears that NF_3 is a much more natural theory than NF. On the other hand, NF_3 is not an environment in which it is easy to do mathematics (essentially because TST_3 is rather restrictive). Henrard, in unpublished work, has shown that NF_3 does admit a theory of cardinal number. Pabion has shown that NF_3 + Infinity is precisely as strong as second-order arithmetic.

Marcel Crabbé [1982] defined a system NFP (for "predicative NF") and a stronger system NFI (which Holmes has called "mildly impredicative NF"). NFI consists of the strong Axiom of Extensionality and those axioms $\{x \mid \phi\}$ exists" which admit a stratification in which the type assigned to $\{x \mid \phi\}$ (one higher than the type assigned to x) is the highest type in the range. NFP further restricts to those comprehension axioms in which no variable of the type of $\{x \mid \phi\}$ is bound in ϕ. Crabbé showed that NFI (and so of course NFP) is consistent. NFI is exactly as strong as second order arithmetic and NFP is weaker than Peano arithmetic.

A striking feature of NFP is that one can make use of the Specker argument for the failure of Choice to prove Infinity. All the known consistent fragments (including NFP and NFI) are consistent with Choice. The argument goes like this. Use $\iota(x)$ as a nonce notation for $\{x\}$. For any formula ϕ, for sufficiently large n, "$\{\iota^n(x) \mid \phi\}$ exists" will be a comprehension axiom of NFP (make n large enough, and the variable representing $\iota^n(x)$ will have type higher than any of the variables in ϕ). From this it follows that NFP + Union = NF. Argue in NFP that Infinity holds as follows: if Union holds then Infinity follows by Specker's argument in NF; if Union does not hold, then Infinity holds because finite sets have unions. There is of course a bit of detail to this argument hidden in this sketch. NFP has interesting relationships to weak subsystems of arithmetic, and Holmes has shown that it is precisely equiconsistent with the ramified theory of types of Whitehead and Russell's *Principia Mathematica* without the Axiom of Reducibility (and it is much simpler!) NFI admits just enough impredicativity to implement impredicative arithmetic, and is as we observed above precisely as strong as second-order arithmetic.

5.5 Reflections on New Foundations and ZF

Perhaps the single most important point to make about the Quine systems is that—properly understood—they do not contradict ZF at all. This point is in any case important, being both fundamental and correct, but it is important also in the sense that there is a continuing and pressing need to make it. Let us explain.

ZF and ZF-like theories are the endeavour to axiomatise our understanding of the cumulative hierarchy of well-founded sets. NF casts its net a little wider, in that it is an attempt to not only do that but to reason about some other sets as well. It is possible to define a predicate "is a well-founded set" in NF, and this predicate[2] enables us to reason about well-founded sets in NF. *No known NF theorem about well-founded sets contradicts anything provable in ZF or any large cardinal upgrade thereof.* Models of strong extensions of NFU can be constructed in which the well-founded sets make up a model of (a strong extension of) ZFC.[3] If

[2]Which of course is unstratified; were it stratified, its extension would be a set and we would have a paradox.

[3]It is true that NF refutes the Axiom of Choice, but the only known failures of AC involve

you are interested in facts about well-founded sets then you will find nothing in NF to offend you. If you wish to be offended by NF then you will have to believe in the proposition that all sets are well-founded. Of course a lot of people do believe this proposition, but one of the most striking things about the set-theoretic literature is the complete absence of any arguments for it. Not only does one find any arguments, one is hard put to find even an acknowledgement that one might come in handy. The closest thing we can find is Boolos [1971] "There does not seem to be any argument that is guaranteed to persuade someone who really does not see the peculiarity of a set's belonging to itself, or to one of its members *etc.*, that these states of affairs are peculiar". NF's peculiar charm is that it attempts to axiomatise the behaviour of large collections while taking care not to say anything that will contradict what we know or believe about well-founded sets. It's the lover that the spouse can't object to. That being so, one would expect NF to be the toast of Society. But there are several respects in which NF gets what in popular parlance would be called a *bum rap.* Admittedly NF does not prove mathematical induction for unstratified properties of natural numbers, and it disappoints in other ways by somehow neglecting to establish that the sets that are conceptually small have all the nice properties that one might expect them to have. But this shows merely that we haven't yet determined what axioms should be added to NF; that's our fault not NF's.[4] I think the most illuminating parallel for NF in this context is with Zermelo set theory, Z. Z is clearly a theory about the cumulative hierarchy (or at least it has been so interpreted in retrospect)—and although no one is suggesting that it makes allegations about the cumulative hierarchy that are *false*—it does stand accused of not achieving absolutely everything that was asked of it. NF is an attempt to axiomatise not only the cumulative hierarchy but some sets beyond it. One shouldn't expect one's first attempt in this to be successful any more than one would expect that Z should be not only our first word on the axiomatisation of the cumulative hierarchy but our last as well.[5] Clearly we are going to have to add axioms to NF to ensure that whenever ZF $\vdash \phi$ then NF $\vdash \phi^{WF}$ (where ϕ^{WF} is the relativisation of ϕ to the well-founded sets); the fact that this task has not been completed is no cause to belittle the NF project: it simply means that there are axioms that remain to be found. One could of course simply add this aforementioned principle by main force as an axiom scheme, but the idiomatic and gentler way to achieve this result would be to add judiciously designed axioms concerning big sets. In NFU the Axiom of Cantorian Sets ensures that this is true in a certain permutation model. In [Forster, 2006] it is shown how axioms postulating good behaviour on the part of the T function systematically correspond to axioms giving the existence of well-founded sets of high rank. Clearly

the sets that in NF studies we call **big** (following a suggestion of Thomas Forster) as opposed to *large*, as in *large* cardinals in ZF. These are the collections like the universal set, and the set of all cardinals and the set of all ordinals: collections denoted by expressions which in ZF-like theories will pick out proper classes. Of course in NFU there is no such problem.

[4]For NFU, the question of what axioms to add has been extensively investigated, and these are probably similar to what would be added to NF.

[5]Why is it that I always find my spectacles in the *last* place I look for them not the *first*?

this is the way in which we have to go. The reason why nobody has explored this in the detail which will eventually be needed is that the whole project is stalled—and will remain stalled until nerves around the consistency question for NF have been calmed: adding new axioms to a theory not yet known to be consistent invites a wealth of disparaging metaphors—*running before one can walk* or *building castles in the air* ... (of course such remarks do not apply to NFU). However, any reason for disparaging the project-to-find-sensible-extensions-for-NF is a reason for trying to prove NF inconsistent[6] and *that* project seems to have no takers.

In coming to grips with NF the hard part is to fully understand stratification, to know how to recognise a stratified formula when you see one. There is an easy rule of thumb with formulas that are in primitive notation, for one can just ask oneself whether the formula could become a wff of type theory by adding type indices. It's harder when one has formulas no longer in primitive notation, and the reader encounters these difficulties very early on, since even something as basic as the ordered pair is not a set-theoretic notion. How does one determine whether or not a formula is stratified when it contains subformulas like $y = f(x)$? The technical/notational difficulty here lands on top of—as so often—a conceptual difficulty. The answer is that of course one has to fix an implementation of ordered pair and stick to it. Does that mean that—for formulas involving ordered pairs—whether or not the given formula is stratified depends on how one implements ordered pairs? The answer is 'yes' but the situation is not as grave as this suggests. Let us consider again the formula $y = f(x)$. This is of course a molecular formula, and how we stratify it will depend on what formula it turns out to be in primitive notation once we have settled on an implementation of ordered pairs. If we use Kuratowski ordered pairs then the formula we abbreviate to $y = f(x)$ is stratified with x and y having the same type, and that type is three types lower than the type of f. If we use Quine ordered pairs then the formula we abbreviate to $y = f(x)$ is stratified with x and y having the same type, and that type is one type lower than the type of f. There are yet other implementations of ordered pair under which the formula we abbreviate to $y = f(x)$ is stratified with x and y having the same type, and that type is one or more types lower than the type of f.

The point is that our choice among the possible implementations will affect the difference in level between x (and y) and f but will not change the formula from a stratified one to an unstratified one. This is subject to two important provisos:

1. We restrict ourselves to ordered pair implementations that ensure that in $x = \langle y, z \rangle$ y and z are given the same type.

2. We do not admit self-application: $(f(f))$.[7]

[6] Thus Forster, but Holmes comments that some (not he!) will raise an objection of the form "NF does not satisfy choice and so is irrelevant to mathematics as she is done", which does not necessarily involve doubts about the *consistency* of NF.

[7] There is a further important distinction between type differential 0 and all other cases: the projection maps are set functions, which allows further constructions which could be construed as subverting the type security of the pair, though they are mathematically convenient. For

Thes two provisos are of course related. The second will seem reasonable to anyone who thinks that mathematics is strongly typed. (The typing system in NF interacts quite well with the endogenous strong typing system of mathematics; this is a striking and deep fact that has not so far attracted the attention it should). If we consider expressions like $x = \langle x, y \rangle$ we see that their truth-value depends on how we implement ordered pairs. There is a noncontroversial sense (entirely transparent in the theoretical CS tradition) in which expressions of this kind are not part of mathematics—in contrast to expressions like $x = \langle y, z \rangle$ which are. The only formulas whose stratification status are implementation-sensitive in this way are formulas that are not in this sense part of mathematics.

The second one is a bit harder to understand: why should we not have an implementation that compels y and z to be given different types in a stratification of $x = \langle y, z \rangle$—or even make the whole formula unstratified?

If we make $x = \langle y, z \rangle$ into something unstratified then we cannot be sure that $X \times Y$ exists, nor that compositions of relations (that are sets) are sets; converses of relations might fail to exist; and we will not really be able to do any mathematics. After all, $X \times Y$ is $\{z \mid (\exists x \in X)(\exists y \in Y)(z = \langle x, y \rangle)\}$ and if $z = \langle x, y \rangle$ is not stratified then the set abstraction expression might not denote a set.

However, even if we muck things up only to the extent of allowing $x = \langle y, z \rangle$ to be stratified with y and z of different types then we will find not only that some compositions of relations (that are sets) are not sets but also that, for some big sets X, the identity relation 1_X is not a set. For example 1_V would not be a set. Let's look into this last point a bit more closely. Suppse "$x = \langle y, z \rangle$" is stratified but with y and z being given different types. Then $X \times Y$ is $\{z \mid (\exists x \in X)(\exists y \in Y)(z = \langle x, y \rangle)\}$ which this time is stratified, so $X \times Y$ is a set. However if $R \subseteq X \times Y$ and $S \subseteq Y \times Z$ then $R \circ S$ is

$$\{w \mid (\exists x \in X)(\exists y \in Y)(\exists z \in Z)(\langle x, y \rangle \in R \wedge \langle y, z \rangle \in S \wedge w = \langle x, z \rangle)\}.$$

This is not stratified. If the difference between the types of the two components of an ordered pair is n, then x and y have types differing by n, and y and z too have types differing by n, and x and z have types differing by $2n$, so although we can stratify $\langle x, y \rangle \in R \wedge \langle y, z \rangle \in S$ we will not be able to stratify $\langle x, y \rangle \in R \wedge \langle y, z \rangle \in S \wedge w = \langle x, z \rangle$.

The problem with 1_X arises because $(\exists x \in X)(y = \langle x, x \rangle)$ is not stratified, so its extension is not certain to be a set. By the same token no permutation of a set can be relied upon to be a set. The (graph of the) relation of equipollence might fail to be reflexive, or symmetrical, or transitive.

The conclusion is that if we want our implementation of mathematical concepts into set theory to be tractable from the NF point of view, and to deliver the routine banalities about relational algebra that we take for granted (existence of 1_X, existence of compositions of relations, existence of transitive closures of

---

example, in NFU the existence of a type-level pair implies Infinity. In either theory, the definition of cardinal multiplication is much simpler with a type level pair, but the definition clearly involves a subversion of the security of the cartesian product type.

relations whose graphs are sets, and so on) then we want a pairing/unpairing function that interprets $x = \langle y, z\rangle$ as a stratified formula with y and z having the same type. One such ordered pair is the Kuratowski ordered pair that we all know and love. As it happens, in NF we usually use the Quine ordered pair which we defined in section 6.2.3.

Does the difference between Quine pairs and Kuratowski pairs matter? Much less than you might think. In some deep sense it doesn't matter at all.[8] Both of them make the formula "$\langle x, y\rangle = z$" stratified and give the variables x and y the same type; z takes a higher type in most cases (never lower). *How* much higher depends on the version of ordered pair being used, but there are very few formulas that come out stratified on one version of ordered pair but unstratified on another, and they are all pathological in ways reminscent of the paradoxes. The best way to illustrate this is by considering ordinals in NF. (In NF we implement ordinals as isomorphism classes of well-orderings). For any ordinal α the order type of the set (and it is a set) of the ordinals below α is well-ordered. In ZF one can prove that the well-ordering of the ordinals below α is of length α. In NF one cannot prove this equation for arbitrary α since the formula in the set abstract whose extension is the graph of the requisite isomorphism is not stratified for any implementation of ordered pair. Now any well-ordering R of a set A to length α gives rise to a well-ordering of $\{\{\{a\}\} \mid a \in A\}$, and if instead one tries to prove (in NF) that the ordinals below α are isomorphic to the well-ordering of length α decorated with curly brackets, one finds that the very assertion that there is an isomorphism between these two well-orderings comes out stratified or unstratified depending on one's choice of implementation of ordered pair! This is because, in some sense, the applications of the pairing function are *two* deep in well-ordering of the ordinals below α, but only *one* deep in the well-ordering of the set of double singletons. If we use Quine ordered pairs, the assertion is stratified—and indeed provable. If one uses Kuratowski ordered pairs (or Wiener ordered pairs) then the assertion is unstratified and refutable. However if one uses Kuratowki ordered pairs there is instead the assertion that the ordinals below α are isomorphic to the obvious corresponding well-ordering of $\{\{\{\{\{a\}\}\}\} \mid a \in A\}$, which comes out stratified (and provable). In general for each implementation of ordered pair there is a depth of nesting of curly brackets which will make a version of this equality come out stratified and true. This does not work with deviant implementations of ordered pair under which "$\langle x, y\rangle = z$" is unstratified or even with those which are stratified but give the variables x and y different types.

Let us try to prove Cantor's theorem. The key step in showing there is no surjection $f : X \twoheadrightarrow \mathcal{P}(X)$ by *reductio ad absurdum* is the construction of the diagonal set $\{x \in X \mid x \notin f(x)\}$. The proof relies on this object being a set, which it will be if "$x \in X \wedge x \notin f(x) \wedge f : X \to \mathcal{P}(X)$" is stratified. This in turn depends on "$(\exists y)(y \in \mathcal{P}(X) \wedge \langle y, x\rangle \in f \wedge f : X \to \mathcal{P}(X))$" being stratified. And it *isn't* stratified, because "$\langle y, x\rangle \in f$" compels '$x$' and '$y$' to be given the same type, while "$f : X \to \mathcal{P}(X)$" will compel 'y' to be given a type one higher than

[8] as long as one does not care whether the projection maps are set functions.

'x'. This is because we have subformulas '$x \in X$' and '$y \subseteq x$'. Notice that we can draw this melancholy conclusion without knowing whether the type of 'f' is one higher than that type of its argument, or two, or three ....

However if we try instead to prove that $\{\{x\} \mid x \in X\}$ is not the same size as $\mathcal{P}(X)$ (the form of Cantor's theorem appropriate in TST) we find that the diagonal set is defined by a stratified condition and exists, so the proof succeeds. This tells us that we cannot prove that $|X| = |\{\{x\} \mid x \in X\}|$ for arbitrary X: graphs of restrictions of the singleton function tend not to exist. (If they did, we would be able to prove Cantor's theorem in full generality). This gives rise to an endomorphism T on cardinals, where $T(|X|) =: |\{\{x\} \mid x \in X\}|$ which we have seen earlier. T misbehaves in connection with *big* sets. If $|X| = |\{\{x\} \mid x \in X\}|$ we say that X is **cantorian**. If the singleton function restricted to X exists, we say that X is **strongly cantorian**. Sets whose sizes are concrete natural numbers are strongly cantorian. $\mathbb{N}$ (the set of Frege natural numbers) is cantorian, but the assertion that it is strongly cantorian (which is the same as the assertion that every inductively finite set is cantorian) implies the consistency of NF.

NF has various natural notions of 'small'. *Cantorian* and *strongly cantorian* are two obvious examples. *Strongly Cantorian* turns out to be mathematically more natural. It has nicer closure properties and does more to ensure nice behaviour on the part of its bearers than *cantorian* does. Another natural notion is *well-founded*. We have recently established [Bowler and Forster, 2009] that every well-founded set is smaller than $T^n(|V|)$ for every concrete n, so there are no *big* well-founded sets. *Well-Ordered* sounds like a notion of smallness too, but there is a theorem of Hartogs' that says that there are in some sense arbitrarily large well-ordered sets. Hartogs' theorem in the exact form in which it is usually stated in ZF is not provable in NF but the form of Hartogs' theorem appropriate to TST is provable. There is no analogue for well-ordered sets of the result alluded to above for well-founded sets: there are well-ordered sets whose size is $\not\leq T(|V|)$.

Inductive Definitions in NF

Mathematics is replete with "inductively" defined sets—sets defined by a recursion. Any set theory that claims to provide a foundation for mathematics must prove that these sets exist, and the more transparent the proof is the happier we will all be. The paradigmatic inductively defined set is $\mathbb{N}$, the set of natural numbers. It is the $\subseteq$-least set containing 0 and closed under successor. That is to say it is the intersection of all sets that contain 0 and are closed under successor. When can we prove the existence of intersections? Well (as long as Y is nonempty) $\bigcap Y$ is a set because it is a subset of $\bigcup Y$—and *that* will be a set as long as Y is, because we have the Axiom of Union. That is the situation in ZF. However we should steel ourselves for trouble because *prima facie* there is no reason to suppose that the collection of all sets-containing-0-and-closed-under-successor is a set. However if there is even one set X that contains 0 and is closed under successor then we know that the set $\mathbb{N}$ that we are trying to define will be a subset of X. If Y is a set-containing-0-and-closed-under-successor then $X \cap Y$ will be another such. In these circumstances the intersection of *all* sets-containing-0-and-closed-under-successor is going to be the same as the intersection of *all-subsets-of-X*-containing-0-and-closed-under-successor. This last object is certainly going to be a set because it is a subset of X.

The usual formulation of the Axiom of Infinity in ZF is geared precisely to the problem of proving that $\mathbb{N}$—indeed a particular implementation of $\mathbb{N}$—exists. It says that there is a set which contains $\emptyset$ and is closed under $\lambda x.x \cup \{x\}$. In the von Neumann implementation of arithmetic $\emptyset$ is the implementation of 0 and $\lambda x.x \cup \{x\}$ is the implementation of successor. Thus it hands us on a plate an X containing-0-and-closed-under-successor.

The reader may feel that this is rather *ad hoc*. It is, and in two ways. For one thing this form of the Axiom of Infinity is unnecessarily specific and informative. It would be equally satisfactory to take the axiom in a form that says merely that there is an infinite set (one the same size as a proper subset of itself). It is *ad hoc* also in the sense that it solves only one case of the problem of proving the existence/sethood of inductively defined collections. The general problem remains unsolved: in ZF we cannot in general define inductively defined sets "top-down" as the intersection of a suitably closed family of sets; we cannot rely on there automatically being a set that contains the founders and is closed under the operations in question.[9]

The predicament can be illustrated by considering two examples of inductively defined families. In any topological space, the family of Borel sets is the smallest collection of sets containing all open and closed sets and closed under countable unions and complementation. In this case we have a set—namely the power set of the space in question—that contains all open and closed sets and is closed under complementation and countable union, so we can obtain the family of Borel sets as a subset of it. In contrast, if we try to prove the sethood of the collection of hereditarily countable sets (which is the $\subseteq$-least set containing all its countable subsets) there is no obvious natural set, visible in the light of day, which contains all its countable subsets.[10] What can we do? We can approach the desired collection "from below" by starting with the set of all the founder objects (which in the case of HC is empty) and iteratively closing under the operations of interest. In the case of HC this means adding at each stage—to the stage-in-hand—all countable subsets of that stage. Then we hope to reach a fixed point. In fact we do reach one—and after fewer than ω_2 steps—but establishing this by purely combinatorial means requires considerable ingenuity.

Typically these iterative "from below" constructions are done by iteration along the ordinals. It doesn't much matter how we implement ordinals, and in principle

[9]More accurately: we cannot do it as easily. Replacement is a great help but the process is complex and a known problem for beginners. In Zermelo set theory it is in general impossible, even with foundation or the Axiom of Rank.

[10]It is true that V_{ω_1} can be proved to contain all its countable subsets if we use AC_ω but this is a red herring, since the existence of HC can be proved without using AC_ω.

any sufficiently long well-ordering will do. (This point is not generally appreciated. The fact that von Neumann ordinals from $\omega \cdot 2$ onwards cannot be proved to exist in Zermelo doesn't mean that we cannot in Zermelo prove termination of iterative processes of arbitrary countable length: we most certainly can). There's the rub: how do we know that there always *is* a sufficiently long well-ordering? That's where Hartogs' theorem comes in. Suppose our desired object D can be obtained as the union of a well-ordered collection of stages that we construct by recursion over a (any) well-ordering. Hartogs' theorem says that for any set X there is a well-ordering too big to be embeddable into X. Thus if our desired object D really does exist, then a well-ordering too big to be embeddable into D will be a well-ordering long enough for us to construct the sequence of stages by recursion over it. Thus Hartogs' theorem tells us that if a recursive definition crashes, it won't be because we have run out of ordinals.

Of course we should not expect to be able to prove the sethood of arbitrary inductively defined collections: some such collections are paradoxical. But what one would like is to have smooth proofs of the existence/sethood of the nonparadoxical collections, and this we do not really have. (A good illustration of this is the difficulty—alluded to above—that we have in proving that the collection of hereditarily countable sets is a set).

In NF the existence of big sets restores the possibility of direct top-down definitions of inductively defined sets: any inductively defined set that can be defined at all can be given a direct "top-down" definition. (This is for the gratifyingly simple reason that—whatever your founders and operations—the universal set contains all founders and is closed under all operations, so when we take the intersection of the set of all sets containing the founders and closed under the operations we are not taking the intersection of the empty set). Thus we obtain the effect of Hartogs' theorem without actually having the theorem itself (as it is usually phrased in ZF).

That is not to say that every inductively defined collection is a set. An inductively defined collection (least thing containing this and closed under that) will be a set as long as the property of being-a-set-that-contains-this-and-is-closed-under-that is stratified. If that is a stratified property then the collection of all sets that contain-this-and-are-closed-under-that is a set and its intersection (which is the inductively defined collection we want) will be a set. This proviso is important, because we need somehow to block the existence of arbitrary inductively defined collections. The trouble is that this stratification proviso seems to block too much. The property of containing-all-your-countable-subsets is not stratified, so we cannot use NF axioms to prove the sethood of the intersection of all sets with it. The existence of HC (the set of well-founded hereditarily countable sets) is not provable in NF if NF is consistent. One should emphasise here that NF does not appear to have anything like a proof that HC cannot be a set. There is a suite of sensible-looking axioms that one can add to NF to get the well-founded sets to exhibit all the behaviour that ZF says they should. These axioms take the form of saying that the collections of cantorian ("nice") cardinals and ordinals are closed under certain natural operations.

The stratification constraint means that we have to be very careful when implementing these operations. An important example is cardinal exponentiation. Since (as we have seen) A and $\{\{y\} \mid y \in A\}$ are not reliably the same size, the two definitions of 2^α—(i) as $|\mathcal{P}(A)|$ where $|A| = \alpha$ or (ii) as $|\mathcal{P}(A)|$ where $|\{\{y\} \mid y \in A\}| = \alpha$—are not equivalent as they are in ZF. It makes a difference which one we choose. If we choose the second it turns out (easy to check) that the property of containing a given cardinal α and being closed under exponentiation is stratified. This means that the collection $\{\alpha, 2^\alpha, 2^{2^\alpha} \ldots\}$ is a set for all α. Were we to use the other definition it wouldn't be.

However, although such inductive constructions as can be executed at all can be executed in the direct top-down fashion, it is still possible to import ordinals into a description of this activity. Suppose our inductive construction starts from a set X with a stratified definition (so it is $\{x \mid \phi\}$ for some stratified formula ϕ with one free variable) and we want to obtain the least superset of X closed under some infinitary homogeneous operation. Examples would be: union of countable subsets; or $F(X) =: \{y \mid (\exists f : y \twoheadrightarrow X)($ is countable-to-one$\}$. F is the operation taking us from each stage to the next stage. The collection of F-stages is the least set containing X, and closed under F and unions of chains. It is of course a set, and it is—for the usual reasons—well-ordered by $\subseteq$. Therefore one can associate an ordinal with every F-stage. (As usual there are several ways of doing it: (i) the set of stages and the set of ordinals are alike well-ordered so there is a canonical map between them; (ii) each stage bounds an initial segment which has an ordinal for its length. (ii) is guaranteed to work even though (i) isn't). Notice that in this treatment we do not use Hartogs' theorem.

Now we are in a position to find an echo of the ZF way of doing things. The closure ordinal α is in a weak sense well-behaved. Let f be the map that sends the ordinal α to the αth stage in the construction. f has a stratified definition without parameters, so the expression

$$(\forall\alpha)(\forall\beta)(f(\alpha) = f(\beta) \longleftrightarrow f(T(\alpha)) = f(T(\beta)))$$

is stratified (fully stratified: it has no parameters and only the one free variable) and so it can be proved by induction on ordinals. This means that if α is the closure ordinal (that is to say, the least β such that $f(\beta) = f(\beta + 1)$) then so is $T(\alpha)$.

It would close the circle very nicely if we knew that every closure ordinal of a stratified recursion were strongly cantorian, but we see no proof. Perhaps it's a very strong assumption.

Armed as we now are with a clearer understanding of recursive definitions in NF, we can return to an earlier topic. There does seem to be a faint chance that the big sets could yet tell us something about well-founded sets. If we define exponentiation of cardinals in NF so that 2^α is $|\mathcal{P}(A)|$ where α is $|\{\{x\} : x \in A\}|$— as we have in fact resolved to do—we find that for some cardinals α the cardinal 2^α cannot be defined. A brief look at the definition will reassure us that if α is a cardinal of the form $|\{\{x\} \mid x \in A\}|$ then 2^α is defined, but that 2^α is not defined if $\alpha \not\leq T(|V|)$. So, for a cardinal α, it may well happen that the sequence

$$\alpha, 2^\alpha, 2^{2^\alpha} \ldots$$

is finite. We now know that this object will always be a set in NF, since the property of containing α and being closed under exponentiation is stratified. In the NF literature this set is called $\phi(\alpha)$. Now let A be suitable big set (this is only ever going to give us interesting results if A is big) and α its cardinal and consider the sequence

$$|\phi(\alpha)|, |\phi(T(\alpha))|, |\phi(T^2(\alpha))| \ldots$$

of natural numbers. We can prove that $|\phi(T(\alpha))| \geq T(|\phi(\alpha)|)+1$ always; we know that for some α it can happen that for n sufficiently large $|\phi(T^n(\alpha))|$ is $\aleph_0$ but it is known that, for at least some big α (such as $|V|$) $|\phi(T^n(\alpha))|$ is finite for at least all concrete n. If $|\phi(T^n(\alpha))|$ is finite for all natural numbers n of the model then the model has a class of natural numbers definable with a big cardinal α as a parameter. *Prima facie* it is a proper class rather than a set because its definition is highly unstratified, but—again *prima facie*—there is no obvious reason why the axiom saying that all such classes are sets should be inconsistent. Nor is there any obvious reason why sets defined in this way with big parameters should be definable without them, so we just might have here a way in which big sets can tell us something about rather more familiar and mundane sets like $\mathbb{N}$ and $\Re$.

6 POSITIVE SET THEORY

6.1 *Positive set theory from the Fregean notion of set*

The Fregean notion of set is about sets as *extensions* of propositional expressions, as presented in Frege's *Grundgesetze der Arithmetik*. More properly, given that Frege takes the notion of function as primitive in developing his system, and particularly in his analysis of propositional expressions, the Fregean notion of set is about sets as extensions of propositional *functions*. Anyway, whatever it is exactly, in its original presentation, it is inconsistent.

This was also true of the first systems proposed subsequently by the founders of the λ-calculus and related combinatory logics, which, like the Fregean paradise of type-free functions, were aimed at incorporating propositional notions, and which were thus exposed to paradoxes akin to Russell's, such as Curry's paradox.[11]

Nevertheless, logicians have always been trying to get back into the Fregean paradise in one way or another. And naturally, since the pure functional part of Frege's system, which is essentially the pure λ-calculus, was proved to be consistent, some have been tempted to locate the flaw in Frege's interpretation of the propositional component.

[11] We would here refer the reader to [Seldin, 2009] for a precise account of the history of the λ-calculus and combinatory logic.

Such a way out is taken in [Aczel, 1980] and followed through in [Flagg and Myhill, 1987a; Flagg and Myhill, 1987b; Flagg and Myhill, 1989] where consistent systems of illative λ-calculus are proposed. As noticed in [Flagg and Myhill, 1987a], that line of research is in fact closely related to much earlier work by Fitch, as [Fitch, 1948] and papers therein cited. Additionally, a related approach that makes explicit reference to Fitch's work as source of inspiration is [Scott, 1975].

All of these proposals apply the 'method of iterated truth definitions', which has been used successfully (in one variant or another) for consistency proofs of various systems related to positive set theory, as in [Gilmore, 1974], and whose origin can thus be traced back to Fitch. However, we shall not elaborate on them herein as the logics of the corresponding systems are essentially non-classical (excluded middle fails). That is the price of *unrestricted* comprehension, after all, to which one should add the tax of intensionality here, for another drawback common to those approaches is the loss of the basic principle of extensionality for sets — which is inherent in the Fregean notion of set.

But was Frege's original interpretation of the propositional component flawed? In fact, no, as lately shown in [Libert, 2008b], provided one is willing to restrict comprehension, of course. And the restriction just consists in proscribing the use in abstraction of those logical connectives that invalidate the fixed-point theorem, such as negation and implication, which respectively give rise to Russell's and Curry's paradoxes. Drastically, but remarkably enough, this indeed suffices to block the paradoxes. The fragment of Frege's system so obtained is presented as a system of illative λ-calculus in [Libert, 2008b]. Its pure set-theoretic part, which is extensional, will provide us with our first example of system of (pure) positive set theory. Here is its axiomatization.

Consider the language of set theory given by the following abstract syntax:

$$\begin{array}{rrcl}
\mathrm{var_a}: & \upsilon & ::= & x, y, z, \ldots \\
\mathrm{term_a}: & \tau & ::= & \upsilon \mid \{\upsilon \mid \varphi\} \\
\mathrm{form_a}: & \varphi & ::= & \bot \mid \top \mid \tau_1 \in \tau_2 \mid \\
 & & & \varphi_1 \wedge \varphi_2 \mid \varphi_1 \vee \varphi_2 \mid \forall \upsilon \varphi \mid \exists \upsilon \varphi \mid \\
 & & & \neg\varphi \mid \varphi_1 \to \varphi_2 \mid \varphi_1 \leftrightarrow \varphi_2
\end{array}$$

Note that the equality relation is not taken as primitive here. As usual, in any extensional framework, $x = y$ can be defined by $\forall z(z \in x \leftrightarrow z \in y)$, provided the following formulation of the Axiom of Extensionality is adopted:

$$\mathrm{ext}: \quad \forall z(z \in x \leftrightarrow z \in y) \to \forall z(x \in z \leftrightarrow y \in z)$$

Positive formulas, which we denote by adding the superscript '+' to metavariables, are those formulas obtained without using the last line of the formation rules. Naturally, Positive Comprehension in this language is defined as the Axiom of Comprehension restricted to the set $\mathrm{form}_\mathrm{a}^+$ of positive formulas, i.e.:

$$\{\mathrm{form}_\mathrm{a}^+\}: \quad \forall x(x \in \{x \mid \varphi^+\} \leftrightarrow \varphi^+)$$

And the first system of positive set theory on which we shall now focus is:

$$\mathrm{PST_a} :\equiv \mathrm{ext} + \{\mathrm{form}_\mathrm{a}^+\}$$

The consistency of $\mathrm{PST_a}$ was established in [Hinnion and Libert, 2003], without any reference to Frege's system. Note that the consistency of $\{\mathrm{form}_\mathrm{a}^+\}$ can easily be proved by a *term model* construction, using the method of iterated truth definitions, as in [Gilmore, 1974]; but that *extensional* term models can indeed be obtained by that method requires a subtle analysis of it, as shown in [Hinnion and Libert, 2003]. A historical account of consistency problems for positive comprehension principles can be found in [Libert, 2004], where these are traced to Skolem's late work, e.g. [Skolem, 1960] which already provides some insight into another consistency proof of $\mathrm{PST_a}$, as follows.

The appropriate way of looking at sets described by $\mathrm{PST_a}$ is as extensions of propositional *functions*, indeed. This was emphasized in [Libert, 2008a] where natural *topological* models are considered. And these are obtained in exactly the same way as the topological models of the pure λ-calculus: they come up from solutions to the reflexive equation $U \cong [U \to 2]$ in the category of Scott spaces, where $[U \to 2]$ is the set of Scott-continuous functions from U to the Sierpinski space 2 of truth values[12] — whereas we recall that Scott topological models of the pure λ-calculus appear as solutions to $U \cong [U \to U]$. Connections with the λ-calculus being disclosed, it was not long before $\mathrm{PST_a}$ could finally be related in [Libert, 2008b] to some fragment of Frege's system.

The condition on a propositional function for its extension to exist here is *Scott-continuity*, which can thus be regarded as a safety property for avoidance of the paradoxes. Given that $[U \to 2]$ is naturally isomorphic to the space of Scott *open* subsets of U, one may equally look at $\mathrm{PST_a}$ as the pure set-theoretic system associated with the limitation of comprehension to so-called *observable* predicates. Moreover, the following first-order axiom scheme, whose semantical interpretation is clear, is then satisfied in any topological model of $\mathrm{PST_a}$ described above:

$$(\Box): \quad \exists y \forall z (z \subseteq y \leftrightarrow \forall x (x \in z \to \varphi))$$

So a natural extension of $\mathrm{PST_a}$ as formal system is $\mathrm{PST_a^\Box} :\equiv \mathrm{PST_a} + (\Box)$, which is our first example of a system related to topological set theory.

Any extension of $\mathrm{PST_a}$ will, however, be characterized by a rather peculiar feature: the non-existence of singletons. For the reader may have noticed that the extensional equality relation is not positive, so it cannot be used in comprehension. And this seems to be unavoidable, as the consideration of $\{x \mid \{y \mid x \in x\} = \{y \mid \bot\}\}$ will show.[13] But this can be related to another feature of the formalization of $\mathrm{PST_a}$: the use in the object language of the abstraction operator '$\{\cdot \mid -\}$', a remnant of the development of that system from systems with functional abstraction (λ-calculi).

---

[12] I.e., $2 := \{1, 0\}$, with 1 as the true, 0 as the false, and where $\{1\}$ is open, but not $\{0\}$.

[13] Let $R := \{x \mid \{y \mid x \in x\} = \{y \mid \bot\}\}$. We have $R \in R$ iff $\{y \mid R \in R\} = \{y \mid \bot\}$ iff $R \in R \leftrightarrow \bot$ iff $\neg(R \in R)$.

Although it is related to the Fregean notion of set, such a system as $\mathrm{PST}_{\mathrm{a}}^{\Box}$ is likely to disconcert anyone who attempts to resurrect Frege's logicist program through it. Anyway, as we shall now see, there are more *natural* versions of positive set theory which are more appropriate for mathematical purposes.

6.2 Positive set theory seen from the Cantorian point of view

The Cantorian notion of set is sets-as-used-in-mathematics. Despite the emergence of paradoxes in its naïve formulation, the general understanding was that the mathematical notion of set should be coherent. This notion is now usually regarded as being based on an intuitively safe conception of sets as abstract collections, which we now describe.

In any instance of the abstraction process of set formation, it seems natural to assume that we must have available all the objects that are to be members of the set to be constructed, and that this always provides us with a *new* abstract object, which therefore cannot be involved in that particular instance of application of the abstraction process, nor in any previous one, but can be involved in subsequent ones.

That view is supported by the observation that sets occurring in every classical field of mathematics are indeed obtainable by *iterated* application of that "well-founded" abstraction process from some given collection of primitive objects (e.g. numbers). This gives rise to the intuitive *cumulative hierarchy* picture: a universe of sets built up in stages, which appears as the core of any set-theoretic universe described by what we called 'ZF-like axiomatizations' — the underlying principle of specialized comprehension essentially guaranteeing that that construction can be carried out.

Let us now have a closer look at sets that can be so obtained *ex nihilo* in a finite number of steps (this will avoid questions about infinity). The collection of those sets, each of which is obviously finite, is denoted by V_ω. It is called the universe of *hereditarily finite well-founded sets*, which is the first limit step of von Neumann's implementation of the cumulative hierarchy, i.e.:

$$V_0 := \emptyset, \qquad V_{n+1} := \mathcal{P}(V_n), \qquad V_\omega := \bigcup\{V_n \mid n < \omega\},$$

where $\mathcal{P}(\cdot)$ is the powerset operation available in any ZF-like axiomatization.

As widely known, the set-theoretic structure $\langle V_\omega; \in\rangle$ is an inner model of ZF minus infinity. Note also that $\langle V_\omega; \in\rangle$ is somehow captured by the family of (finite) set-theoretic structures $\langle V_n; \in\rangle$, $n > 0$, seeing that $x \in y$ for $x, y \in V_\omega$ iff there is some $n > 0$ such that $x, y \in V_n$ and $x \in y$. In categorical terms, we say that $\langle V_\omega; \in\rangle$ is the *inductive* (or *direct*) limit of the $\langle V_n; \in\rangle$'s. Let us formalize things carefully, if perhaps pedantically, in order to make clearer the comparison with the construction we are going to consider next.

For each $n > 0$, let $i_n : V_n \hookrightarrow V_{n+1} : x \mapsto x$ be the canonical inclusion map, and let $\langle V_n; \in_n^i\rangle$ be the set-theoretic structure defined by $x \in_n^i y$ iff $x \in i_n(y)$, that is, $x \in y$. So $\langle V_n, \in_n^i\rangle$ is nothing but $\langle V_n; \in\rangle$. We note that each i_n

is obviously an injective $\in$-morphism of $\langle V_n; \in^i_n \rangle$ into $\langle V_{n+1}; \in^i_{n+1} \rangle$, i.e., $x \in^i_n y$ implies $i_n(x) \in^i_{n+1} i_n(y)$. Thus we have properly defined a *direct* system $(\langle V_n; \in^i_n \rangle, i_n)_{0<n<\omega}$ of set-theoretic structures, whose so-called *direct* or *inductive* limit is just $\langle V_\omega; \in \rangle$. Indeed, $\varinjlim_i V_n = \bigcup \{V_n \mid 0 < n < \omega\} = V_\omega$, by definition, and $x \in^i_\omega y$ iff there exists $n > 0$ such that $x \in^i_n y$, i.e., iff $x \in y$.

Given that categorical construction, and the familiar set-theoretic structure that comes with it, it is natural to look at the *dual* one, as follows.

Let $p_1 : V_2 \twoheadrightarrow V_1 : x \mapsto \emptyset$, and then let $p_{n+1} : V_{n+2} \twoheadrightarrow V_{n+1}$, $n > 0$, be inductively defined by $p_{n+1}(x) := \{p_n(y) \mid y \in x\}$. It is easily seen by induction that $p_n \circ i_n = \mathrm{id}_{V_n}$, i.e., the restriction of p_n on V_n is id_{V_n}, so p_n is indeed a surjection. Now, for each $n > 1$, let $\langle V_n; \in^p_n \rangle$ be the set-theoretic structure defined by $x \in^p_n y$ iff $p_{n-1}(x) \in y$; we conveniently define $\in^p_1$ as $V_1 \times V_1$. It is easy to prove that each p_n is a surjective $\in$-morphism of $\langle V_{n+1}; \in^p_{n+1} \rangle$ onto $\langle V_n; \in^p_n \rangle$, i.e., $x \in^p_{n+1} y$ implies $p_n(x) \in^p_n p_n(y)$. Thus we have defined an *inverse* system $(\langle V_n; \in^p_n \rangle, p_n)_{0<n<\omega}$ of set-theoretic structures, which is naturally associated with the direct system above. Let $\langle N_\omega; \in^p_n \rangle$ be the so-called *inverse* or *projective* limit of that system, i.e.,

$$N_\omega := \varprojlim_p V_n = \{\xi \in \prod \{V_n \mid 0 < n < \omega\} \mid \forall n > 0, p_n(\xi_{n+1}) = \xi_n\},$$

and $\xi \in^p_\omega \zeta$ iff for all $n > 0$, $\xi_n \in^p_n \zeta_n$. What is that toy, yet natural, set-theoretic structure a model of?

The first thing to note is that $\langle N_\omega; \in^p_\omega \rangle$ is extensional. A proof of this would make use of the following 'extension lemma': given $x \in V_\omega, \zeta \in N_\omega$, if $x \in^p_n \zeta_n$ for some $n > 0$, there exists $\xi \in N_\omega$ such that $\xi_n = x$ and $\xi \in^p_\omega \zeta$. We would also invite the reader to convince himself that $\langle N_\omega; \in^p_\omega \rangle$ is an *end-extension* of $\langle V_\omega; \in \rangle$. More precisely, for any $x \in V_\omega$, let $\rho(x)$ be the rank of x, that is, the least $n < \omega$ such that $x \in V_{n+1}$. Then, one can prove that the map $e : V_\omega \to N_\omega$ defined by $e(x) := (\ldots, p_{\rho(x)-1}(p_{\rho(x)}(x)), p_{\rho(x)}(x), x, x, \ldots)$, is an embedding of $\langle V_\omega; \in \rangle$ as *initial part* of $\langle N_\omega; \in^p_\omega \rangle$, i.e.: e is an injective $\in$-morphism and if $\xi \in^p_\omega e(y)$ for some $y \in V_\omega$, there exists $x \in V_\omega$ such that $\xi = e(x)$ and $x \in y$. Thereupon we note that $\langle N_\omega; \in^p_\omega \rangle$ is really a proper extension of $\langle V_\omega; \in \rangle$. For instance, it is easy to see that $p_n(V_n) = V_{n-1}$ for all $n > 0$, so $\nu := (V_{n-1})_{0<n<\omega} \in N_\omega$, and then that for all ξ in N_ω, $\xi \in^p_\omega \nu$. Therefore, there is a universal set in $\langle N_\omega; \in^p_\omega \rangle$, namely ν, whereas there is no such set in $\langle V_\omega; \in \rangle$. There are by far many other unusual sets in $\langle N_\omega; \in^p_\omega \rangle$, simply because N_ω has the power of the continuum, whereas V_ω is countable!

As a matter of fact, N_ω could have been presented as a *topological* completion of V_ω, in the same way as the real numbers can be presented as the Cauchy completion of the rational numbers. The notion of convergence in N_ω is defined so that for all ξ in N_ω, $\xi = \lim_{n\to\infty} e(\xi_n)$. In other words, sets in N_ω can be described as limits of hereditarily finite well-founded sets. Other presentations of N_ω as a mathematical structure can be found in [Abramsky, 1988], where $\langle N_\omega; \in^p_\omega \rangle$ is depicted as the universe of *finitary non-well-founded sets*.

One precise characterization of N_ω is as the solution to a reflexive equation of the form $U \cong \mathcal{P}_{cl}(U)$ in the category of compact metric spaces, where $\mathcal{P}_{cl}(U)$ is the set of *closed* subsets of U. This reminds us of the topological models of $\mathrm{PST_a}$ discussed earlier, which appear as solutions to the equation $U \cong \mathcal{P}_{op}(U)$ in the category of Scott spaces, where $\mathcal{P}_{op}(U)$ is the set of *open* subsets. So this suggests that set-theoretic universes as $\langle N_\omega; \in^p_\omega \rangle$ could be related to positive set theory as well. Actually, such set-theoretic structures turn out to be the topological models associated with another natural version of positive set theory, which can be axiomatized as follows.

Consider the language of set theory whose formulation rules are:

$$\begin{array}{lll} \mathrm{var_b}: & \upsilon & ::= \quad x, y, z, \ldots \\ \mathrm{form_b}: & \varphi & ::= \quad \bot \mid \top \mid \upsilon_1 \in \upsilon_2 \mid \\ & & \qquad \varphi_1 \wedge \varphi_2 \mid \varphi_1 \vee \varphi_2 \mid (\forall \upsilon_1 \in \upsilon_2)\varphi \mid (\exists \upsilon_1 \in \upsilon_2)\varphi \mid \\ & & \qquad \neg\varphi \mid \varphi_1 \rightarrow \varphi_2 \mid \varphi_1 \leftrightarrow \varphi_2 \end{array}$$

Again, we call *positive* those formulas obtained without using the last line of the formation rules, which we indicate by the superscript '+' on metavariables. We stress the presence of bounded quantifiers as primitive *positive* operators here, but the absence of the abstractor '$\{\cdot \mid -\}$' as primitive term constructor. Positive comprehension, that is, the restriction of the Axiom of Comprehension to the set $\mathrm{form_b^+}$ of positive formulas, is then expressed in this language by:

$$\{\mathrm{form_b^+}\}: \quad \exists y \forall x (x \in y \leftrightarrow \varphi^+)$$

And the corresponding system of positive set theory in which we are interested now is:

$$\mathrm{PST_b} :\equiv \mathrm{ext} + \{\mathrm{form_b^+}\}$$

Note immediately that the equality relation, which is not primitive here, can however be defined by a *positive* formula, i.e.,

$$x = y :\equiv (\forall z \in y) z \in x \wedge (\forall z \in x) z \in y.$$

So singletons are innocent definable objects in $\mathrm{PST_b}$, in contrast with $\mathrm{PST_a}$. Thanks to bounded quantification, other set-theoretic operations that were not available in $\mathrm{PST_a}$ are now legitimate, as notably the power-set operation, $\mathcal{P}(y) := \{x \mid (\forall z \in x) z \in y\}$, which thus can be defined positively according to this criterion. Note also that the usual 'unbounded' quantifiers can be recovered from the primitive bounded ones, by defining $\forall \upsilon \varphi$ as $(\forall \upsilon \in \mathrm{V})\varphi$ and $\exists \upsilon \varphi$ as $(\exists \upsilon \in \mathrm{V})\varphi$, where V is the universal set, whose existence follows from $\mathrm{PST_b}$.

As said above, the consistency of $\mathrm{PST_b}$ can be established by means of N_ω, and more generally of topological models that appear as solutions to the reflexive equation $U \cong \mathcal{P}_{cl}(U)$ in some category of compact Hausdorff spaces. These topological models clearly satisfy the *dual* of the axiom scheme ($\Box$) in Section 2, namely:

$$(\Diamond): \quad \exists y \forall z (y \subseteq z \leftrightarrow \forall x (\varphi \rightarrow x \in z))$$

So the natural extension of PST_b as formal system is $\text{PST}_\text{b}^\Diamond :\equiv \text{PST}_\text{b} + (\Diamond)$.

Interestingly, it was shown in [Esser, 1997][14] that ZF minus infinity can be interpreted within $\text{PST}_\text{b}^\Diamond$, which definitely distinguishes that system from $\text{PST}_\text{a}^\Box$. In fact, ZF (with infinity) is naturally interpretable in $\text{PST}_\text{b}^\Diamond$ augmented with von Neumann's Axiom of Infinity.[15] The resulting system, called GPK_∞^+ in [Esser, 1997], is then proved to be mutually interpretable with a large cardinal extension of the Kelley-Morse set theory. So there is undoubtedly a natural version of positive set theory which is at least as suitable as ZF for mathematical purposes.

The relative consistency of GPK_∞^+ in [Esser, 1997] was essentially established in [Forti and Hinnion, 1989], which provides a general construction of a hierarchy of N_α's, such that for any weakly compact cardinal κ, N_κ is a topological model of GPK_∞^+ containing V_κ (the von Neumann hierarchy up to κ). Such structures, subsequently called *hyperuniverses*, have been extensively studied since then, e.g. in [Forti and Honsell, 1996a; Forti and Honsell, 1996b]. As is explained in [Forti and Hinnion, 1989], they were independently investigated in [Weydert, 1989] for the purpose of topological set theory, to which subject we shall now turn. Note incidentally that the constructions given in [Forti and Hinnion, 1989] and [Weydert, 1989] were both inspired by the pioneering work of Malitz [1976], which can also be classed as a contribution to topological set theory.

6.3 Topological set theory

We have met two variants of positive set theory, namely PST_a and PST_b, which can respectively be extended by the (first-order) *topological* axioms schemes $(\Box)$ and $(\Diamond)$.[16] These were suggested by the corresponding topological models, whose interest is manifest: in such topological set-theoretic structures, even though not all definable subsets of the domain can be abstracted (because of Russell's paradox), each of these can at least be optimally *approximated* by the largest [resp. the smallest] set contained in [resp. containing] it. So each of the axiom schemes $(\Box)$ and $(\Diamond)$ provides on its own an approximate version of the naïve comprehension scheme.

Such an idea of approximation of naïve comprehension — the *essence* of topological set theory — originated in a paper by Skala [1974], which was subsequently refined by Manakos [1984]. Although Skala's paper is cited in the references of [Weydert, 1989] and [Malitz, 1976], and so might have been a source of inspiration in those works, the proposal she made is clearly distinguishable from the others in that her system is based upon *both* $(\Box)$ and $(\Diamond)$, *together with* an axiom for the

[14] Esser's thesis was subsequently published in [Esser, 2004]; see also [Esser, 1999] for a related paper.

[15] In the form: 'there exists an infinite von Neumann ordinal'.

[16] By the way, we mention that PST_a could equally be extended by $(\Diamond)$, by duality, that is, $\text{PST}_\text{a}^\Diamond$ is consistent (see [Libert, 2008a] for details). On the other hand, $\text{PST}_\text{b}^\Box$ is inconsistent merely because $(\Box)$ is incompatible with the existence of all singletons (see [Libert and Esser, 2005]).

existence of complements, which we denote by $(\mathcal{C})$.[17] So it is not at all a system of positive set theory! Besides, Skala's proposal is purely axiomatic, with no attempt at giving any topological characterization of the models. As shown in [Libert and Esser, 2005], the model theory of Skala's system is quite simple though.

The natural models of $(\Box)+(\Diamond)+(\mathcal{C})$ appear as solutions to a reflexive equation of the form $U \simeq \mathcal{P}_{clop}(U)$ in the category of quasi-discrete spaces, where $\mathcal{P}_{clop}(U)$ is the set of *clopen* subsets of U, which precisely coincides for quasi-discrete spaces with both $\mathcal{P}_{cl}(U)$ and $\mathcal{P}_{op}(U)$ – the reason why both $(\Box)$ and $(\Diamond)$ hold in these models. Thereupon we recall that $\mathcal{P}_{clop}(U)$ is a complete Boolean algebra, so there is actually no restriction in Skala's set theory on the use of logical connectives and quantifiers in comprehension, which is in contrast with the situation in positive set theory. The restriction, for there should be one, is relegated to the atomic level and is rather of a syntactical nature here, as follows.

Consider the simple language of set theory whose formation rules are:

$$\begin{array}{rlll} \mathrm{var_c}: & \upsilon & ::= & x, y, z, \ldots \\ \mathrm{form_c}: & \varphi & ::= & \bot \mid \top \mid \upsilon_1 \in \upsilon_2 \mid \\ & & & \varphi_1 \wedge \varphi_2 \mid \varphi_1 \vee \varphi_2 \mid \forall \upsilon \varphi \mid \exists \upsilon \varphi \mid \\ & & & \neg \varphi \mid \varphi_1 \rightarrow \varphi_2 \mid \varphi_1 \leftrightarrow \varphi_2 \end{array}$$

Let $\varphi[\upsilon[$ denote any formula of $\mathrm{form_c}$ in which the variable υ does not appear on the *right-hand* side of $\in$, and let $\mathrm{form_c}[\upsilon[$ be the set of those formulas. Given a concrete variable, say x, we then naturally define the restriction of the axiom scheme of Comprehension to $\mathrm{form_c}[x[$ by:

$$\{\mathrm{form_c}[x[\}: \quad \exists y \forall x (x \in y \leftrightarrow \varphi[x[)$$

It is proved in [Libert and Esser, 2005] that ext + $(\Box)$ + $(\Diamond)$ + $(\mathcal{C})$ is equivalent to ext + $\{\mathrm{form_c}[x[\}$. So Skala's set theory can equally be presented as a type-free system based on a very simple syntactic criterion for avoidance of the paradoxes, which, in view of Russell's paradox, remains the only option if one wants a *Boolean* set-theoretic universe in a type-free setting.[18] But this has drastic consequences.

Thus we observe that $x = y$, which was defined by $\forall z(z \in x \leftrightarrow z \in y)$, is neither in $\mathrm{form_c}[x[$ nor in $\mathrm{form_c}[y[$. Hence, the extensional equality relation cannot be used in instances of comprehension. However, what is called the *indiscernibility relation*, which is defined by $x \doteq y :\equiv \forall z(x \in z \leftrightarrow y \in z)$, can be used instead. Furthermore, *quasi-singletons*, i.e., sets of the form $\{x \mid x \doteq y\}$, appear as atomic objects in Skala's set theory, for any set is provably equal to the union of the quasi-singletons of its members. That the naïve conception of set can somehow be salvaged when indiscernibility plays the role of equality may not be devoid of philosophical interest. This was taken seriously in [Apostoli and Kanda, to

[17] Manakos [1984] dropped this last assumption in order to consider other possible extensions of the basic system $(\Box)$ + $(\Diamond)$; see [Libert and Esser, 2005] for details.

[18] Clearly, the restriction on the abstracted variable to appear on *left-hand* side of $\in$, which is the other option, would seem even less sensible.

appear; Apostoli and Kanda, 2000], where elaborated quasi-discrete solutions to $U \simeq \mathcal{P}_{clop}(U)$ are constructed, without any reference to Skala set theory, but in connection with *rough set theory.*

On the other hand, the non-definability of the usual operation of forming singletons does not entail their non-existence. As a matter of fact, it is proved in [Libert and Esser, 2005] that *any* given set-theoretic universe in which all singletons exist (e.g. a model of ZF) can consistently be embedded as the class of objects admitting singletons in some model of Skala's set theory. In other words, although it is intrinsically weak, Skala's system is very adaptable.

In any case, such an axiomatic set theory should not be regarded as (and then compared to) a *classical* axiom system, that is, one for which we have in mind some *standard* interpretation — which is the case of ZF for instance. Rather, it appears as a *general* axiom system, as the one of the theory of Boolean algebras, to which it is closely related. After all, thinking of sets as extensions of predicates, one might expect that some set-theoretic universe or another could reflect the Boolean structure of classical logic.

Hopefully the various set-theoretic systems presented in this survey will have at least convinced the reader of what mathematicians have discovered through their investigations, namely that set theory is definitely *not* unique.

6.4 A development of mathematics in GPK^+_∞

In this section we give a brief self-contained presentation of Oliver Esser's system GPK^+_∞ of [Esser, 1999], the most mathematically effective system of positive set theory.

GPK^+_∞ is a first-order theory with equality and membership as primitive relations.

Extensionality: $(\forall AB.(\forall x.x \in A \leftrightarrow x \in B) \rightarrow A = B)$

Definition: Let the class of (bounded) positive formulas be the smallest class containing the formula $x \neq x$ (useful because uniformly false), all atomic formulas, and closed under conjunction, disjunction, bounded universal quantification $(\forall x \in A.\phi)$ and existential quantification $(\exists x \in A.\phi)$.

Positive Comprehension: For any positive formula ϕ in which A does not appear, $(\exists A.(\forall x.x \in A \leftrightarrow \phi))$.

Closure: For any formula ϕ, there is a set C such that $(\forall D.C \subseteq D \leftrightarrow (\forall x.\phi \rightarrow x \in D))$ (a set minimal in the inclusion order among all sets which include the possibly proper class $\{x \mid \phi\}$). This is what was referred to above as $(\diamond)$ and it should be noted that this is equivalent to the assertion that any definable class of sets has set intersection.

Definition: The von Neumann ordinal ω is the intersection of all sets which contain $\emptyset$ and are closed under the von Neumann successor operation $x \mapsto x \cup \{x\}$. By Closure, ω is a set.

Definition: A set x is *isolated* iff $\{x\}^c = \{y \mid y \neq x\}$ is a set.

Infinity: Each element of ω is isolated. (We have stated this axiom in a form specifically designed to make our brief development more efficient).

We start working toward an interpretation of ZF in this theory.

Observation: $\{x, y\}$ is a set ($\{z \mid z = x \vee z = y\}$) Moreover $z = \{x, y\}$ is equivalent to a positive formula $x \in z \wedge y \in z \wedge (\forall w \in z. z = x \vee z = y)$. It is important in a development of this theory to know what term constructions can appear freely in positive formulas, as we have just shown the unordered pair (and so the usual ordered pair) can. Similarly, $x \cup y$ and $x \cap y$ can be mentioned freely in positive formulas. For any set x, $\bigcup x = \{y \mid \exists z \in x. y \in z\}$ and $\mathcal{P}(x) = \{y \mid \forall z \in y. z \in x\}$. Note that the axioms of Pairing, Union, and Power Set of the usual set theory thus hold. $y = \bigcup x$ is known to be equivalent to a positive formula; $y = \mathcal{P}(x)$ is not known to expressible in positive form. We cannot expect the Axiom of Separation to hold, as the formula ϕ determining an instance of the axiom is not necessarily positive.

Observation: Let A be a set all of whose elements are isolated. Then $\{x \in A \mid \phi\}$ exists for *every* formula ϕ by an application of Closure (take the intersection of all the sets $A \setminus \{y\}$ such that $y \in A$ and $\neg\phi[y/x]$.

Similarly, if $\phi(x, y)$ is a functional formula $((\forall x.(\exists y!.\phi(x, y))))$ and A is a set all of whose elements are isolated, then $\{y \mid (\exists x \in A.\phi(x, y))\}$ is a set. Let F^* be the closure of the class of all (x, y) such that $\phi(x, y)$. There is a set $\{(x, y) \mid x \in A \wedge (x, y) \in F^*\}$. Let $F(x)$ denote the unique y such that $\phi(x, y)$. For each $x \in A$, there is a set $\{(u, v) \mid x \neq u \vee v = F(x)\}$. From the existence of these sets, it follows that the intersection of all sets containing the class $\{y \mid (\exists x \in A.\phi(x, y))\}$ is exactly that class, and so the class is a set.

Observation: If all elements of A are isolated, then A is isolated. The function d such that $d(x) = V \setminus \{x\}$ for each $x \in A$ is a set by the second half of the previous observation. We want to show that A^c is a set. $x \neq A$ is equivalent to $(\exists y \in A. y \neq x) \vee (\exists y \in x. y \notin A)$. This is in turn equivalent to $(\exists y \in A. y \in d(x)) \vee (\exists y \in x.(\forall z \in A. y \in d(z)))$. Finally, $u \in d(v)$ is equivalent to $(u, v) \in d$, and we know that pairs can be mentioned freely in positive formulas. Since $x \neq A$ is equivalent to a positive formula, A^c is a set as desired.

The observations above are sufficient to suggest that the class of hereditarily isolated sets satisfies the axioms of ZF, though some additional details would need to be attended to in a full presentation.

One can then develop an adequate theory of proper classes of hereditarily isolated sets (closures of arbitrary classes of hereditarily isolated sets). This theory will satisfy Kelley-Morse set theory, with one further surprise: one can prove that the proper class ordinal in the interpreted Kelley-Morse set theory is weakly compact.

Conversely, there is a construction of a model of GPK^+_∞ in ZFC with a weakly compact, and in fact one can show that GPK^+_∞ is precisely as strong as Kelley-Morse set theory with the proper class ordinal weakly compact. This is considerably stronger than ZFC.

An interesting weaker theory is obtained if one drops the Axiom of Infinity (which was referred to above as $\mathrm{PST}^\diamond_\mathrm{b}$, after the description of its natural model N_ω). This allows ω to have a non-isolated element, namely ω itself (ω becomes the topological limit point of the finite von Neumann ordinals). This theory has the same strength as second-order arithmetic.

7 SYSTEMS MOTIVATED BY NONSTANDARD ANALYSIS

In this section we discuss systems of set theory which can be regarded as motivated by the nonstandard analysis of Abraham Robinson [1966]. These theories can be characterized by the presence of what Vopěnka calls *proper semisets*: subcollections of sets which are proper classes. Of course, any set theory with a universal set must have proper semisets (the Russell class being a subclass of the universe) but in the theories here we postulate that "small" familiar sets such as $\mathbb{N}$ have proper subclasses.

7.1 Nonstandard analysis

Consider the first-order theory of the real numbers (the precise details of this theory are not important here). The collection of sentences $\{c > 0, c < 1, c < \frac{1}{2}, \ldots, c < \frac{1}{n}, \ldots\}$, where c is a new constant, is consistent (any finite subcollection is readily satisfied by taking c to be positive and sufficiently small), so by the compactness theorem the theory of the real numbers has models in which this collection of sentences is satisfied.

In such a model, we can think of c as a positive infinitesimal, for c is smaller than any of the familiar real numbers, because it is smaller than all the reciprocals of the familiar natural numbers. But note that of course there are "natural numbers" N of the model such that $\frac{1}{N} < c$.

If we define an "infinitesimal" as any real (of our nonstandard model) which is smaller than all standard $\frac{1}{n}$ in absolute value (a notion which cannot be defined in terms of the theory of the reals with which we started, because the notion of a standard natural number does not make sense there), then we can attempt the program of 17th century analysis.

For example, if f is a function from the reals to the reals definable in our original theory of real numbers, with a derivative $f'(x)$, it is straightforward to show that $\frac{f(x+dx)-f(x)}{dx}$ differs infinitesimally from $f'(x)$. So we can *define* $f'(x)$ for each standard real x as the unique standard real y which differs infinitesimally from

$\frac{f(x+dx)-f(x)}{dx}$ for each infinitesimal dx. A very appealing development of analysis in this style is possible.

Robinson presented nonstandard analysis in the framework of model theory. In his approach, one considers a model $\mathbb{R}^*$ of the theory of the real numbers, containing all the usual reals (which are called the "standard" reals) and having the property that any real of the model which is bounded in absolute value by a standard real differs infinitesimally from one of the standard reals.

There is an alternative approach, which is to "augment" our set theory with axioms which ensure that infinite sets such as $\mathbb{N}$ or $\mathbb{R}$ have the properties we expect and at the same time have proper subclasses (proper semisets) which can play the role of the collections of "standard" reals or natural numbers in the discussion above.

7.2 *Nelson's* IST

Without further ado, we present Nelson's IST (internal set theory, defined in [Nelson, 1977]) which is motivated exactly by the desire to streamline the development of nonstandard analysis.

IST is a first-order theory with the familiar predicates of equality and membership and an additional predicate $st(x)$ of "standardness". It is convenient to abbreviate $(\forall x.st(x) \rightarrow \phi)$ as $(\forall^{st} x.\phi)$ and $(\exists x.st(x) \wedge \phi)$ as $(\exists^{st} x.\phi)$.

The axioms of IST are of two sorts. First we have the axioms of standard set theory. Nelson uses the axioms of ZFC, but one could equally well use a weaker theory such as Mac Lane set theory. The axioms of Separation and Replacement of IST are modified in that the formulas mentioned in the schemes may not mention the predicate *st*: sets may not be defined using the predicate *st*.

There are three additional axioms in Nelson's theory which directly support the program of standard analysis.

Idealization: Let $\phi(x, y)$ be a formula not mentioning *st* or the variable f (it may have nonstandard parameters). "for every standard finite set f, $(\exists x.(\forall y \in f.\phi(x, y)))$" is equivalent to "$(\exists x.(\forall^{st} y.\phi(x, y)))$".

Corollary: If ϕ is taken to be the formula "x is finite and $y \in x$", we observe that the first formula is true and by Idealization equivalent to the second, and so we assert that there is a finite set which contains every standard object.

Standardization: For every standard set A, and for any formula ϕ whatsoever (this formula may mention *st* and nonstandard parameters freely), $(\exists^{st} B.(\forall^{st} x.x \in B \leftrightarrow x \in A \wedge \phi))$: the standard elements x of B are exactly the standard elements x of A such that ϕ.

Transfer: For any formula ϕ not mentioning the predicate *st* and containing only standard parameters, $(\forall x.\phi) \leftrightarrow (\forall^{st} x.\phi)$.

The consistency of these axioms is readily established by standard techniques of model theory. The advantage for the nonstandard analyst is that they can then use IST as their set-theoretic framework and not have to directly discuss nonstandard models of (say) the reals at all.

For the set theorist, it is interesting that the axioms of IST are not recognizably about analysis: they are set-theoretical in character, though certainly what they assert is rather odd.

We review how these axioms support the program of nonstandard analysis. Suppose that the reals are implemented as left sets of Dedekind cuts in the rationals (which can in turn be supposed implemented in set theory in a quuite usual way).

Idealization gives us a positive infinitesimal real number. We define an infinitesimal as a real number which is less in absolute value than any standard real number. Obviously 0 is infinitesimal. But we can define a positive infinitesimal: let ϵ be the intersection of all positive reals (considered as left sets of Dedekind cuts in the rationals) which belong to a specific finite set F containing all standard sets as elements (the existence of such a set being a corollary of Idealization). Clearly ϵ is a positive real (it is the minimum of a finite set of reals) but it is also less than every positive standard real (because all of them belong to F).

Standardization gives us a unique standard real number infinitesimally close to any real number which is bounded in absolute value by some standard real. Let r be a real number which is bounded above in absolute value by some standard real. By Standardization, there is a standard set r^* which contains the same standard elements (r being a subset of the standard set of rationals). It is straightforward to use Transfer to verify that r^* is also a real number (since all standard elements of r^* are rational numbers, all elements of r^* without exception are rational numbers; since the standard elements of r^* make up a downward closed subset of the standard reals without a largest element, r^* itself is a downward closed subset of the reals without a largest element: it is important here than r is bounded above in absolute value by some standard real, as otherwise r^* could be either empty or the entire set of rationals and so of course not a real). If r^* differed from r by a non-infinitesimal amount, there would be a standard real in their symmetric difference, contrary to the choice of r^*.

Transfer (and the restriction of Separation and Replacement to formulas not mentioning the new predicate *st*) ensures that the underlying set theory (and so the implementation of the reals in the underlying set theory) works exactly as expected, in spite of the obvious weirdness caused by the presence of proper subclasses in every infinite set (strictly speaking, we do not have a theory of classes in IST, but we will in the theory considered next, and it would be straightforward to add a theory of classes to IST itself)

7.3 Vopěnka's alternative set theory

The alternative set theory of Vopěnka [1979] is also usable for nonstandard analysis, but it is conceptually more different from the usual set theories. It is also much weaker, but mathematically rather interesting.

Vopěnka's theory has sets and classes. We have the usual axioms of Extensionality and Class Comprehension. We follow the convention that capital letters represent classes and lower-case letters represent sets.

Definition: "x is a set" means $(\exists A.x \in A)$ as usual.

Extensionality: $(\forall AB.(\forall x.x \in A \leftrightarrow x \in B) \rightarrow A = B)$. Classes with the same elements are the same.

Class Comprehension: For any formula ϕ in which A is not free, the universal closure of $(\exists A.(\forall x.x \in A \leftrightarrow \phi))$ is an axiom.

The axioms for sets are quite surprising. The theory asserted of sets is equivalent to ZFC with the Axiom of Infinity replaced by its negation, though the form presented has independent interest as a way to present the theory of hereditarily finite sets.

Empty set: The empty class $\emptyset$ (which exists by class comprehension) is a set.

Successor: For any sets x and y, $x \cup \{y\} = \{z \mid z \in x \vee z = y\}$, which exists by class comprehension, is a set.

Induction: For any formula ϕ in which all parameters are sets and all quantifiers are bounded in the class V of all sets, $\phi[\emptyset/z] \wedge (\forall xy \in V.\phi[x/z] \wedge \phi[y/z] \rightarrow \phi[x \cup \{y\}/z]) \rightarrow (\forall z \in V.\phi)$. This axiom says in effect that all sets are built up by the iterated application of the successor operation starting with the empty set.

The Axiom of Foundation for sets is usually listed as an axiom but follows from Induction.

Now come the peculiarities obviously related to nonstandard analysis.

Definition: A *semiset* is a subclass of a set. A *proper semiset* is a semiset which is not a set.

Semisets: There is a proper semiset.

Definition: A set is *finite* iff all of its subclasses are sets.

Definition: An order of type ω is a well-ordering whose domain is not finite but all initial segments of which have finite domain. A class is countable if there is a class bijection between it and the domain of an order of type ω. The class of finite von Neumann ordinals is provably an order of type ω.

Prolongation: Each countable function F can be extended to a set function.

The Axiom of Semisets has the effect of the Idealization Axiom in IST (it ensures that there are nonstandard sets, in this case nonstandard hereditarily finite sets). The notion of *finite* here can be understood as (roughly) capturing the notion of a set of standard finite size. The function of the Axiom of Prolongation is similar to that of standardization in IST. It is a distinct oddity of the alternative set theory as an implementation of nonstandard analysis that there is no analogue of the Axiom of Transfer.

We review the construction of number systems. There are two flavors of natural numbers: the system of all the von Neumann ordinals (nonstandard naturals) and the system of finite von Neumann ordinals (standard naturals). The real numbers can be defined as Dedekind cuts in the *standard* rationals (ratios of finite natural numbers and their additive inverses): these are not sets, but prolongation can be used to show that any real can be approximated infinitesimally by nonstandard rationals, and an interesting nonstandard analysis (not really the same as Robinson's) can be developed.

There are two more axioms in Vopěnka's system.

Vopěnka considers representations of superclasses of classes using relations on sets. A class relation R on a class A is said to code the superclass of inverse images of elements of A under R. A class relation R on a class A is said to extensionally code this superclass if distinct elements of A have distinct preimages. He "tidies up" the theory of such codings by adopting the very technical

Axiom of Extensional Coding: Every collection of classes which is codable is extensionally codable.

The final axiom tidies up the theory of infinite cardinality.

Axiom of Cardinalities: Any two uncountable classes are the same size (i.e, there is a class bijection between them).

Extensional Coding and Cardinalities together can prove Choice (any class can be well-ordered). Since Choice easily implies Extensional Coding, the former axiom might seem more natural. As in pocket set theory, there are only two infinite cardinalities, $\aleph_0$ and c. $\aleph_0$ is here the cardinality of the collection of standard natural numbers, and c is the cardinality of the collection of all natural numbers (?!).

A model of the alternative set theory in the usual set theory is a nonstandard model of V_ω of size ω_1 in which every countable external function extends to a function in the model. It might be best to suppose that this model is constructed inside L (the constructible universe) so that the Axiom of Cardinalities will be satisfied. The Axiom of Extensional Coding follows from Choice in the ambient set theory.

8 CURIOSITIES

In this section we introduce two eccentric theories. One of them has the property which is usually ascribed to New Foundations (we believe not entirely fairly) of being motivated by a syntactical trick without any semantic motivation. The double extension set theory of Andrzej Kisielewicz definitely has this property (and it is not known whether it is consistent).

The other is one of the strongest set theories ever proposed. It is known that it is inconsistent with ZFC that there is a proper elementary embedding from the universe to itself. But this proof does not work in Zermelo set theory: Zermelo set theory with an axiom asserting that there is an elementary embedding from V to V is of the order of consistency strength of the strongest extensions of ZFC that have been proposed.

8.1 Double extension set theory

Andrzej Kisielewicz has proposed in two papers (describing three systems: the not-known-to-be-inconsistent one is described in [Kisielewicz, 1998])) that the paradoxes of set theory can be averted by providing two different membership relations, and allowing extensions for each relation to be defined using formulas in the other relation. His strongest systems have been shown to be inconsistent by Holmes [2004]. Holmes [2005] also showed that the ordinals in his weaker system have startling properties.

We describe the version of Kisielwicz's theory which is not known to be inconsistent. This is a first-order theory with equality and two primitive relations $\in$ and ϵ, both of which are to be thought of as membership. For any formula ϕ, the formula ϕ^* is obtained by replacing $\in$ with ϵ and *vice versa* in ϕ.

Definition: A set x is said to be *regular* iff $(\forall y, y \in x \leftrightarrow y \,\epsilon\, x)$ (i.e., if it has the same extension with respect to both membership relations).

Axiom of Comprehension: For each formula ϕ which does not mention ϵ, in which any parameters are regular, and in which the variable A is not free, $(\exists A.(\forall x.x \,\epsilon\, A \leftrightarrow \phi) \wedge (\forall x.x \in A \leftrightarrow \phi^*))$. The object A is denoted by $\{x \mid \phi\}$.

Axiom of Mixed Extensionality: $(\forall AB.(\forall x.x \in A \leftrightarrow x \,\epsilon\, B) \rightarrow A = B)$ Note that any sets shown to be equal by mixed extensionality are necessarily regular.

Definition: We say that x is *partially contained* in y iff $(\forall z.z \in x \rightarrow z \in y) \vee (\forall z.z \,\epsilon\, x \rightarrow z \,\epsilon\, y)$. We say that a set *has regular elements* iff all of its elements in either sense are regular.

Axiom of Regular Containment: Any set partially contained in a set with regular elements is regular.

This set theory avoids Russell's paradox in the following curious way. There is a set $R = \{x \mid x \in x\}$ by Comprehension, but we have to recall what this means: $x \epsilon R \leftrightarrow x \in x$ and $x \in R \leftrightarrow x \epsilon x$. Thus we have $R \in R \leftrightarrow \neg R \epsilon R$: we conclude that R is not regular.

This theory is at least as strong as ZF (the class of hereditarily regular sets satisfies ZF), but the presentation of the proof is hampered by the extremely counterintuitive character of the reasoning involved: Holmes has commented that computer proof checking really recommends itself when reasoning with this system, because of the curious interchanges of the two membership relations in applications of comprehension. An important observation is that the proof that double extension set theory is strong turns out to be driven by a not immediately obvious formal resemblance of this theory to the Ackermann set theory discussed above.

The most recent result about this theory (whose consistency remains an entirely open question) is the result of Holmes that the formal symmetry between the two membership relations is broken: the sequence of von Neumann ordinals in terms of one of the relations is a proper initial segment of the sequence of von Neumann ordinals defined in terms of the other.

This theory would have some real interest if it could be showed to be consistent with respect to ZFC or some generally accepted extension thereof, because it is remarkably economical in its notions and axioms. But it is in practical terms rather hard to reason in.

8.2 Zermelo set theory with an elementary embedding

It is a theorem of ZFC due to Kunen [1971] that there is no nontrivial elementary embedding from V to V. The status of ZF with an axiom scheme asserting the existence of such an embedding is unclear (this extension of ZF would be extremely strong).

Kunen's proof depends on examination of the limit of the $j^n(\kappa)$'s, where j is the elementary embedding and κ is the first ordinal moved by j. This suggests that it might be profitable to consider Zermelo set theory with an elementary embedding (which would have a model if it were consistent with ZFC that there is an elementary embedding from a limit rank onto itself, which is among the strongest "axioms of infinity" ever proposed). The Axiom of Rank is useful here, as one wants an elementary embedding which moves a rank, not one which merely moves an ordinal (the latter being not obviously strong in Zermelo set theory).

We present a set of axioms for a first-order theory with primitives equality, membership, and a function j. An equivalent theory was introduced by Paul Corazza in private discussions with one of the authors (Holmes): we present it in a slightly different way. The theory presented in [Corazza, 2000] appears to be essentially the same, but the fact that it is naturally viewed as a version of Zermelo set theory rather than ZFC seems not to be mentioned.

Extensionality: $(\forall AB.(\forall x.x \in A \leftrightarrow x \in B) \rightarrow A = B)$

Separation: For any formula ϕ in which the variable A is not free, $(\forall B.\exists A.(\forall x.x \in A \leftrightarrow x \in B \wedge \phi))$. A is called $\{x \in B \mid \phi\}$ as usual.

Power Set: $(\forall x.(\exists y.(\forall z.z \in y \leftrightarrow z \subseteq x)))$

Rank: Every set is included in a rank (referring back to the definition of hierarchy and rank in our discussion of Zermelo set theory and Mac Lane set theory above).

Elementarity: For any formula ϕ in which j is not mentioned and in which no variable other than x appears free, $(\forall x.\phi \leftrightarrow \phi[j(x)/x])$.

Nontriviality: There is a rank r such that $j(r) \neq r$.

Choice: Every nonempty collection of disjoint sets has a choice set.

We have enjoyed stating these axioms minimally. It should be noted that Pairing and Union can be deduced from the axioms given and the Axiom of Rank just as in Zermelo or Mac Lane set theory with Rank. Infinity is not needed because it is straightforward to show that a rank moved by j must be infinite.

Further, it is straightforward to prove that the index κ of the first rank moved by j is inaccessible, Mahlo, weakly compact, measurable, etc. It follows further that the Axiom of Replacement holds for all formulas not mentioning j: the apparent weakness of this system is illusory. It could further be augmented with classes (with the embedding j extending to classes, and classes freely definable using j just as j can be freely used in instances of Separation in the version given here).

9 CONCLUSIONS

We believe that there are two conclusions to be drawn from this survey, one positive and one negative, for a program of alternative set theory.

The positive result is that there are mathematically fluent and useful systems of alternative set theory. The most unequivocal examples of this are quite close to standard set theory: Nelson's IST and the theory ZFC– + AFA are modifications of ZFC specifically designed to facilitate certain kinds of mathematics, and they are fit for their respective purposes. Further, there are at least two systems which at least appear to be conceptually fundamentally different from ZFC in which it is clear that mathematics can be done. These are the positive set theory GPK^+_∞ of Esser and reasonably strong extensions of NFU (practical considerations suggest that at least the axioms of Choice and Counting would be desirable).

The negative result (which on reflection might not be too negative) is that no fundamentally different picture of the mathematical universe emerges from the alternative theories. Of course certain advocates of some of these theories might dispute this. Our exact claim is that all these theories are mutually interpretable in well-understood ways (except where one theory is much stronger than another: we can interpret the alternative set theory in Mac Lane set theory but of course not *vice versa*). There is no mathematics which can be done in one theory but not in another (except where very low consistency strength is a barrier, as in the case of pocket set theory or the alternative set theory). We do *not* agree with the claim made by some that a theory such as NFU can only be understood to be consistent via its interpretation in ZFC (and so that it is not an independent approach to mathematics); we have shown elsewhere that a self-contained motivation of and development of NFU up to the consistency proof can be made entirely within the simple theory of types, and motivated in a way which has nothing to do with ZFC.

The one possible exception is NF itself, simply because we do not know what a model of NF looks like: there might possibly be some fundamental idea quite alien to the ZF-iste view of things found in a construction of a model of NF (though we do not expect this). The situation in NFU + Infinity + Choice is well-understood: the advocate of this theory will discover in the theory of isomorphism classes of well-founded extensional relations a structure precisely analogous to an initial segment of the cumulative hierarchy of the usual set theory. In GPK^+_∞, the ranks of the cumulative hierarchy indexed by isolated ordinals are there to be examined, but there is more stuff not in the hierarchy.

There is no justification for a revolution here: the current hegemony of ZFC as the set theory in which mathematics is done in practice does not hamper the progress of mathematics. The most that can be said is that it might benefit a mathematician interested in foundations to know that things *could* be done differently, and study of one or more of these systems would serve that purpose.

ACKNOWLEDGEMENTS

Part of this chapter was written in January 2008 while Thierry Libert was visiting Thomas Forster at the DPMMS in Cambridge, with the support of the Belgian 'Fonds National de la Recherche Scientifique'.

BIBLIOGRAPHY

[Abramsky, 1988] S. Abramsky. A Cook's Tour of the Finitary Non-Well-Founded Sets. In *We Will Show Them: Essays in honour of Dov Gabbay*, S. Artemov, H. Barringer, A. d'Avila Garcez, L. C. Lamb, and J. Woods, eds., vol. 1, pp. 1–18, College Publications, 2005.

[Aczel, 1980] P. Aczel. On the notion of proposition, truth and set. In *The Kleene Symposium*, North-Holland, Amsterdam, 1980.

[Aczel, 1988] P. Aczel. *Non-well-founded Sets*, CSLI Lecture Notes no. 14, Stanford, 1988.

[Apostoli and Kanda, to appear] P. Apostoli and A. Kanda. Parts of the continuum: towards a modern ontology of science. To appear in *The Poznan Studies in the Philosophy of Science and the Humanities*, ed. L. Nowak.

[Apostoli and Kanda, 2000] P. Apostoli and A. Kanda. Approximation spaces of type-free sets. In *Lecture Notes in Artificial Intelligence*, J. G. Carbonell and J. Siekmann, eds., vol. 2005, pp. 98–105, Springer-Verlag, 2000.

[Barwise and Moss, 1996] J. Barwise and L. Moss. *Vicious Circles*, CSLI Lecture Notes, no. 60, Stanford, 1996.

[Boffa, 1972] M. Boffa. Forcing et Negation de l'axiome de Fondement, *Mem. Acad. Sci. Belg.*, 40:fasc. 7, 1972.

[Boffa, 1988] M. Boffa. ZFJ and the consistency problem for NF, *Jahrbuch der Kurt Gödel Gesellschaft*, pp. 102-106, Wien, 1988.

[Boolos, 1971] G. Boolos, The iterative conception of set, *Journal of Philosophy*, 68:215–31, 1971.

[Bowler and Forster, 2009] N. Bowler and T. Forster. Normal subgroups of infinite symmetric groups with an application to stratified set theory, *The Journal of Symbolic Logic*, 74:17–26, 2009.

[Church, 1974] A. Church. Set theory with a universal set. In *Proceedings of the Tarski symposium*, L. Henkin ed., *Proceedings of Symposia in Pure Mathematics*, vol. XXV, American Mathematical Society, pp. 297–308, 1974.

[Corazza, 2000] P. Corazza. The Wholeness Axiom and Laver Sequences, *Annals of Pure and Applied Logic*, 105:157–260, 2000.

[Crabbé, 1982] M. Crabbé. On the consistency of an impredicative subsystem of Quine's NF, *The Journal of Symbolic Logic*, 47:131–36, 1982.

[Esser, 1997] O. Esser. Interprétations mutuelles entre une théorie positive des ensembles et une extension de la théorie de Kelley-Morse. Ph. D. Thesis, Université Libre de Bruxelles, 1997.

[Esser, 1999] O. Esser. On the consistency of a positive theory, *Mathematical Logic Quarterly*, 45:105–116, 1999.

[Esser, 2004] O. Esser. *Une théorie positive des ensembles*, Cahiers du centre de logique, no. 13, Centre national de recherches en logique, Academia-Bruylant, Louvain-La-Neuve, 2004.

[Fitch, 1948] F. B. Fitch. *The Journal of Symbolic Logic*, 13:95–106, 1948.

[Flagg and Myhill, 1987a] R. Flagg and J. Myhill. Implication and analysis in classical Frege structures, *Annals of Pure and Applied Logic*, 34:33–85, 1987.

[Flagg and Myhill, 1987b] R. Flagg and J. Myhill. κ-continuous lattices and comprehension principles for Frege structures, *Annals of Pure and Applied Logic*, 36: 1–16, 1987.

[Flagg and Myhill, 1989] R. Flagg and J. Myhill. A type-free system extending ZFC, *Annals of Pure and Applied Logic*, 43: 79–97, 1989.

[Fraenkel, 1922] A. Fraenkel. Über den Begriff 'definit' und die Unabhängigkeit des Auswahlaxioms, Sitzungsberichte der Preussischen Akademie derWissenschaften, Physikalisch-mathematische Klasse, pp. 253–257, 1922. Translated in [van Heijenoort, 1967, 284–289].

[Forster, 1976] T. E. Forster. *N.F.*, Ph.D. thesis, University of Cambridge, 1976.

[Forster, 2006] T. E. Forster. Permutations and wellfoundedness: the true meaning of the bizarre arithmetic of Quine's NF. *The Journal of Symbolic Logic*, 71:227–240, 2006.

[Forti and Hinnion, 1989] M. Forti and R. Hinnion. The consistency problem for positive comprehension principles, *The Journal of Symbolic Logic*, 54:1401–1418, 1989.

[Forti and Honsell, 1996a] M. Forti and F. Honsell. Choice principles in Hyperuniverses, *Annals of Pure and Applied Logic*, 77:35–52, 1996.

[Forti and Honsell, 1996b] M. Forti and F. Honsell. A general construction of hyperuniverses, *Theoretical Computer Science*, 156:203–215, 1996.

[Forti and Honsell, 1983] M. Forti and F. Honsell, Set theory with free construction principles, *Annali Scuola normale superiore, Pisa, classe di scienze*, 10:493–522, 1983.

[Gilmore, 1974] P. Gilmore. The consistency of partial set theory without extensionality. In *Axiomatic Set Theory*, Proceedings of Symposia in Pure Mathematics, vol. 13, American Mathematical Society, pp. 147–153, Providence, 1974.

[Gödel, 1940] K. Gödel. *The Consistency of the Axiom of Choice and of the Generalized Continuum Hypothesis with the Axioms of Set Theory*, Princeton University Press, Princeton, 1940.

[Grishin, 1969] V. N. Grishin. Consistency of a fragment of Quine's NF system, *Sov. Math. Dokl.*, 10:1387–1390, 1969.

[Hailperin, 1944] T. Hailperin. A set of axioms for logic, *The Journal of Symbolic Logic*, 9:1–19, 1944.

[Henson, 1973] C. W. Henson. Type-raising operations in NF, *The Journal of Symbolic Logic*, 38:59–68, 1973.

[Hinnion and Libert, 2003] R. Hinnion and T. Libert. Positive abstraction and extensionality, *The Journal of Symbolic Logic*, 68:828–836, 2003.

[Holmes, 2001] M. R. Holmes. Strong axioms of infinity in NFU, *The Journal of Symbolic Logic*, 66:87–116, 2001. (Brief notice of errata with corrections appeared later in 2001.)

[Holmes, 1998] M. R. Holmes. *Elementary Set Theory with a Universal Set*, volume 10 of the Cahiers du Centre de logique, Academia, Louvain-la-Neuve (Belgium), 1998. A revised and corrected version is available online.

[Holmes, 2004] M. R. Holmes. Paradoxes in double extension set theories, *Studia Logica*, 77:41–57, 2004.

[Holmes, 2005] M. R. Holmes. The structure of the ordinals and the interpretation of ZF in double extension set theory, *Studia Logica*, 79:357–372, 2005.

[Jensen, 1969] R. B. Jensen. On the consistency of a slight (?) modification of Quine's 'New Foundations', *Synthese*, 19:250–263, 1969.

[Kisielewicz, 1998] A. Kisielewicz. A very strong set theory?, *Studia Logica* 61:171–178, 1998.

[Kelley, 1976] J. Kelley. *General Topology*, appendix, Springer-Verlag 1976.

[Kemeny, 1950] J. G. Kemeny. *Type theory vs. set theory* (abstract). *The Journal of Symbolic Logic*, 15:78, 1950.

[Barwise, 1975] J. Barwise. *Admissible Sets and Structures*, Springer-Verlag, 1975.

[Levy, 1959] A. Levy. On Ackermann's set theory. *The Journal of Symbolic Logic*, 24:154–166, 1959.

[Libert, 2004] T. Libert. Semantics for Naïve Set Theory in Many-valued Logics: Technique and Historical Account. In *The Age of Alternative Logics*, J. van Benthem and G, Heinzmann eds, Kluwer Academic, 2004.

[Libert, 2008a] T. Libert. Positive abstraction and extensionality revisited, *Log. Anal.*, 51:149–165, 2008.

[Libert, 2008b] T. Libert. Positive Frege and its Scott-style semantics, *Mathematical Logic Quarterly*, 54:410–434, 2008.

[Libert and Esser, 2005] T. Libert and O. Esser. On topological set theory, *Mathematical Logic Quarterly*, 51:263–273, 2005.

[Mac Lane, 1986] S. Mac Lane. *Mathematics, Form and Function*, Springer-Verlag 1986.

[Malitz, 1976] R. J. Malitz. Set Theory in which the Axiom of Foundation Fails. Ph. D. thesis, University of California, Los Angeles, 1976.

[Manakos, 1984] Manakos, J., On Skala's set theory, *Zeitschrift für mathematische Logik und Grundlagen der Mathematik*, 30:541–546, 1984.

[Mathias, 2001a] A. R. D. Mathias. The strength of Mac Lane set theory, *Annals of Pure and Applied Logic*, 110:107–234, 2001.

[Mathias, 2001b] A. R. D. Mathias. Slim models of Zermelo set theory, *The Journal of Symbolic Logic*, 66:487–496, 2001.

[Nelson, 1977] E. Nelson. Internal set theory, a new approach to nonstandard analysis, *Bulletin of the American Mathematical Society*, 83:1165–1198, 1977.

[Potter, 2004] M. Potter. *Set Theory and its Philosophy*, Oxford University Press, 2004.

[Quine, 1937] W. V. O. Quine. New Foundations for Mathematical Logic, *American Mathematical Monthly*, 44:70–80, 1937.

[Quine, 1945] W. V. O. Quine. On ordered pairs, *The Journal of Symbolic Logic*, 10:95–96, 1945.

[Ramsey, 1926] F. Ramsey. The foundations of mathematics, *Proceedings of the London Mathematical Society 2*, 25:338–384, 1926.

[Reinhardt, 1970] W. Reinhardt. Ackermann's set theory equals ZF, *Annals of Mathematical Logic*, 2:189–249, 1970.

[Kunen, 1971] K. Kunen. Elementary embeddings and infinitary combinatorics, *The Journal of Symbolic Logic*, 407–413, 1971.

[Robinson, 1966] A. Robinson. *Nonstandard Analysis*, North-Holland, London, 1966.

[Rosser, 1953] J. B. Rosser. *Logic for Mathematicians*, second edition, Chelsea, New York, 1973.

[Russell, 1903] B. Russell, *The Principles of Mathematics*, appendix, W. W. Norton Company, New York, 1996.

[Scott, 1955] D. Scott. Definitions by abstraction in axiomatic set theory, *Bulletin of the American Mathematical Society*, 61:442, 1955.

[Scott, 1975] D. Scott. Combinators and classes. In *λ-Calculus and computer science*, Lecture Notes in Computer Science, vol. 37, C. Böhm, ed., pp. 1–26, Springer, Berlin, 1975.

[Seldin, 2009] J. Seldin. Logic from Church to Curry. Iin *Logic from Russell to Church*, Dov M. Gabbay and John Woods, eds., pp. 819–874. Volume 5 of the *Handbook of the History of Logic*. Amsterdam: North-Holland, 2009.

[Skala, 1974] H. Skala. An alternative way of avoiding the set-theoretical paradoxes, *Zeitschrift für mathematische Logik und Grundlagen der Mathematik*, 20:233–237, 1974.

[Skolem, 1960] T. Skolem. Investigations on a comprehension axiom without negation in the defining propositional functions, *Notre Dame Journal of Formal Logic*, 1:13–22. 1960.

[Solovay, preprint] R. Solovay. *The consistency strength of NFUB*, preprint, available through logic e-prints on the WWW.

[Specker, 1953] E. P. Specker. The axiom of choice in Quine's 'New Foundations for Mathematical Logic', *Proceedings of the National Academy of Sciences of the USA*, 39:972-975, 1953.

[Tarski and Givant, 1987] A. Tarski and S. Givant, *A Formalization of Set Theory without Variables*, American Mathematical Society, Providence, 1987.

[van Heijenoort, 1967] J. van Heijenoort, (ed). *From Frege to Gdel: A Source Book in Mathematical Logic, 1879-1931*, Jean van Heijenoort, ed., Harvard University Press, 1967.

[von Neumann, 1925] von Neumann, John, An Axiomatization of Set Theory, (appeared 1925): English translation in [van Heijenoort, 1967].

[Vopěnka, 1979] P. Vopěnka. *Mathematics in the alternative set theory*, Teubner-Verlag, Leipzig, 1979.

[Wang, 1970] H. Wang. *Logic, Computers, and Sets*, p. 406, Chelsea, 1970.

[Weydert, 1989] E. Weydert. How to Approximate the Naïve Comprehension Scheme Inside of [*sic*] Classical Logic. Ph. D. thesis, Bonner Mathematische Schriften Nr.194, Bonn 1989.

[Whitehead and Russell, 1910-13] A. N. Whitehead and B. Russell, *Principia Mathematica*, 3 vols, Cambridge University Press, 1910, 1912, and 1913. The abridgment is generally easier to find: Whitehead, Alfred North, and Russell, Bertrand, *Principia Mathematica (to *56)*, Cambridge University Press, 1967.

[Wiener, 1914] N. Wiener. A simplification of the logic of relations, *Proceedings of the Cambridge Philosophical Society*, 17:387–390, 1914.

[Zermelo, 1908] E. Zermelo. Untersuchen über die Grundlagen der Mengenlehre I, *Mathematische Annalen*, 65:261–281, 1908.

TYPES, SETS, AND CATEGORIES

John L. Bell

This essay is an attempt to sketch the evolution of type theory from its beginnings early in the last century to the present day. Central to the development of the type concept has been its close relationship with set theory to begin with and later its even more intimate relationship with category theory. Since it is effectively impossible to describe these relationships (especially in regard to the latter) with any pretensions to completeness within the space of a comparatively short article, I have elected to offer detailed technical presentations of just a few important instances.

1 THE ORIGINS OF TYPE THEORY

The roots of type theory lie in set theory, to be precise, in Bertrand Russell's efforts to resolve the paradoxes besetting set theory at the end of the 19th century. In analyzing these paradoxes Russell had come to find the set, or class, concept itself philosophically perplexing, and the theory of types can be seen as the outcome of his struggle to resolve these perplexities. But at first he seems to have regarded type theory as little more than a *faute de mieux.*

In June 1901 Russell learned of the set-theoretic paradox, known to Cantor by 1899, that issued from applying the latter's theorem concerning power sets to the class V of all sets. According to that theorem the class of all subclasses of V would, impossibly, have to possess a cardinality exceeding that of V. This stimulated Russell to formulate the paradox that came to bear his name (although it was independently discovered by Zermelo at about the same time) concerning the class of all classes not containing themselves (or predicates not predicable of themselves). Realizing that his paradox applied also to the logical system previously elaborated by Frege in his *Grundgesetze*, Russell communicated it to Frege in June 1902, occasioning, in the latter's own words, the "greatest surprise, and I would almost say, consternation, since it has shaken the basis on which I intended to build arithmetic". Frege responded by attempting to patch up the postulate of his system (Basic Law V) subject to the paradox, fashioning a repair that Russell was initially inclined to endorse (see the final note to Appendix A of *The Principles of Mathematics*)[1]). Frege's repair in fact proved inadequate; but in any case Russell had already begun to search for a solution to the paradox as it applied to classes. This led to his proposing the theory, or doctrine, of types.

[1]Russell [1903, 1].

Handbook of the History of Logic: Sets and Extensions in the Twentieth Century.
Volume editor: Akihiro Kanamori. General editors: Dov M. Gabbay, Paul Thagard and John Woods.

It is of interest to see how Russell's doctrine of types emerges from his analysis of classes in the *Principles.* He puts forward the following definition of class (p. 515):

> A class is an object uniquely determined by a propositional function, and determined equally by any equivalent propositional function.

Here Russell is, in essence, adopting an extensional view of classes.[2] He then considers the following possible characterizations of the class notion:

> A class may be identified with (α) the predicate, (β) the class concept, (γ) the concept of the class, (δ) Frege's range, (ε) the numerical conjunction of the terms of the class, (ζ) the whole composed of the terms of the class.

To understand what Russell intends by the first three of these, let us follow him in considering the term *man.* Here (α) then corresponds to the predicate ...*is human*, (β) to the function of *man* in propositions such as *Socrates is a man*, and (γ) to the concept of the *class of men.* Clearly each of these characterizations is intensional in nature, so that none of them can be taken as a satisfactory determination of the notion of the essentially extensional notion of class.

Under (δ), "Frege's range" refers to the device introduced by Frege in the *Grundgesetze* by which each function is assigned an object called its *range* in such a way that two functions are assigned the same range precisely when they have identical values for all values of their arguments. This is Basic Law V, which Russell had already shown to be inconsistent. In view of this Russell is led to observe "(δ) suffers from a doubt as to their being such an entity, and also from the fact that, if ranges are terms, the contradiction is inevitable." Once again, then, an unsatisfactory determination.

Under (ε) we have the *class-as-many.* Russell regards this notion as "logically unobjectionable" since he thinks that the existence of classes-as-many is guaranteed ontologically, and so at the very least classes-as-many can always be assumed to exist without falling into contradiction. But classes-as-many suffer from the drawback that they are not single entities, except when the class has only one member.

Finally, category (ζ) is the *class-as-one.* The problem here is that Russell's paradox shows that classes-as–one do not always exist.

[2]But interestingly Russell earlier remarks (p. 513) that he was led to an extensional view of classes "against [his] inclination", and only because of "the necessity of discovering some entity determinate for a given propositional function, and the same for any equivalent propositional function." This would seem to be essentially the same reason Frege was led to introduce ranges, or, more pertinently, their special case extensions of concepts. Russell goes on to say that the only entities he has been able to discover which meet the requirements are, first, the class as one of all objects making the given propositional function true, and, second, the class (also as one) of all propositional functions equivalent to the given function. One notes that the first expedient involves a type reduction, and the second a type augmentation.

Russell is accordingly left with no satisfactory way of defining classes. Yet, as he says, "without a single object to represent an extension, Mathematics crumbles." In the end he is forced to adopt (ε), the class-as-many, as the definition of a class. This requires abandoning the doctrine that the subject of a proposition is always a single term; instead, it can be essentially many terms. Once this is accepted, one sees that the plural nature of classes does not prevent them from being "counted as though each were a genuine unity".

But in that case, Russell observes, it then becomes necessary "to distinguish (1) terms, (2) classes, (3) classes of classes, and so on *ad infinitum*". Moreover, these collections will be disjoint, and to able to assert $x \in u$ requires that the collection to which x belongs should be of a degree one lower than that to which u belongs. This expedient leads to a resolution of the paradox, since $x \in x$ has now been rendered a meaningless proposition. The hierarchy of collections (1), (2), (3), ... is the germ of the doctrine of types.

In Appendix B to the *Principles*, entitled *The Doctrine of Types*, Russell states that he is putting forward the doctrine "tentatively, as affording a possible solution of the contradiction"; but it "requires, in all probability, to be transformed into some subtler shape before it can answer all difficulties." He proceeds to explain exactly what he means by a type:

> Every propositional function $\varphi(x)$ ... has, in addition to its range of truth, a range of significance, i.e. a range within which x must lie if $\varphi(x)$ is to be a proposition at all, whether true or false. This is the first point in the theory of types; the second point is that ranges of significance form types, i.e. if x belongs to the range of significance of $\varphi(x)$, then there is a class of objects, the type of x, all of which must also belong to the range of significance of $\varphi(x)$, however φ may be varied; and the range of significance is always either a single type or a sum of several whole types.

As observed above, Russell believed the doctrine of types to be adequate for resolving the paradox for classes. But according to his definition of type, propositions themselves can serve as the constituting objects for types. Russell proceeds to formulate a paradox for propositions similar to that for classes, which he does not think is resolved by the doctrine of types as he has formulated it. This paradox may be stated in the following way. For any class P of propositions, write $\sqcap P$ for the proposition *every proposition in P is true*. Russell makes the key (but as we shall see below, questionable) assumption that, for classes P, Q of propositions,

$$P \neq Q \Rightarrow \sqcap P \neq \sqcap Q, \tag{*}$$

or equivalently (in classical logic, at least)

$$\sqcap P = \sqcap Q \Rightarrow P = Q. \tag{**}$$

Now let Q be the class of propositions p for which there is a class P of propositions such that $p = \sqcap P$ and $p \notin P$. If one now writes q for $\sqcap Q$, then, using (**), $q \in Q \Leftrightarrow q \notin Q$. Faced with this conclusion Russell then remarks

> It is possible of course to hold that propositions themselves are of various types and that their logical products must have propositions of only one type as factors. But this suggestion seems harsh and highly artificial.

A number of years were to pass before Russell came finally to adopt this "harsh and highly artificial" suggestion in the form of the ramified theory of types.

Russell remarks that if the identity relation is replaced by the relation of *logical equivalence,* then there is no reason to uphold implication (*), and the contradiction disappears. But this escape is blocked, according to Russell, because

> it is quite self-evident that equivalent propositional functions are often not identical.

At this point, then, Russell could see no way out of the contradiction.

Now in deriving the contradiction it is (*) (or (**)) that is the culprit. Writing **Prop** for the class of all propositions, each asserts that the map $P \mapsto \sqcap P$ is one-one from the class of subclasses of **Prop** to **Prop**. And this, as is now well-known (and likely also then to Russell himself), violates Cantor's theorem. In Boolos [1997] it is shown how to prove Cantor's theorem through the provision of explicit counterexamples.[3] We will adapt his argument to produce a counterexample to (**). By Boolos's argument there exists a subclass M of **Prop** and a strict well-ordering $\prec$ of M such that (1) $\sqcap M \in M$ and, (2) for $p \in M$, $p = \sqcap\{q \in \mathbf{Prop} : q \prec p\}$.[4] Now let $N = \{q : q \prec \sqcap M\}$. Then $\sqcap M \notin N$ and so it follows from (1) that $M \neq N$. But, by (1) and (2), $\sqcap M = \sqcap N$. The classes of propositions M and N thus together constitute a counterexample to (**).

It is tempting to speculate that, had Russell known of this argument, he might have been led to abandon (*) and so perhaps have been more inclined to accept the adequacy of his early theory of types.

Since Russell was by no means convinced that his initial theory of types would prove adequate for resolving the logical paradoxes that had arisen in the foundations of mathematics, he developed a number of other theories in attempting to meet the challenge. But he soon became aware of the inadequacies of these and returned once more to the theory of types. The refined theory of types which resulted from his labours was first outlined in a paper published in 1908: *Mathematical logic as based on the theory of types.*[5] Russell prefaces his paper by remarking that while the theory he is about to present "recommends itself ... in the first instance, by the ability to solve certain contradictions," it "seems not wholly dependent on this indirect recommendation." For it has also "a certain consonance with common sense which makes it inherently credible." It is the the-

[3]Boolos's argument is itself an adaptation of the first argument Zermelo used to derive the well-ordering theorem from the axiom of choice.

[4]If we express the assertion $q \prec p$ by saying that q is *below* p, then (2) may be expressed as: for any proposition $p \in M, p$ is identical with the proposition *every proposition below* p *is true.*

[5]Russell [1908].

ory's latter feature which, in part, led Russell to put it forward as a foundation for mathematical logic — and so also for mathematics.

Russell offers an account of the various contradictions with which he is concerned, finding that they all have in common "the assumption of a totality such that, if it were legitimate, it would at once be enlarged by new members defined in terms of itself". This leads to the enunciation of the rule which he calls the *vicious circle principle*, to wit:

> Whatever involves all of a collection must not be one of a collection, or, conversely: If, provided a certain collection had a total, it would have members only definable in terms of that total, then the said collection has no total.

Russell recognizes that this principle, being purely negative, "suffices to show that many theories are wrong, but it does not show how the errors are to be rectified." Russell sees a positive theory emerging from an analysis of the use of the term "all", and in particular from drawing a careful distinction between it and the term "any". While restrictions must, on pain of contradiction, be imposed on the use of "all", no such restrictions need be placed on "any". Thus "all" corresponds to the bound variable of universal quantification, which ranges over a type — the *type* of that variable — while "any" corresponds to a free variable for which the name of anything can be substituted, irrespective of type.

Russell then proceeds to introduce a theory of types for propositions — an expedient which, it will be recalled, he had shunned in 1903 as being "harsh and artificial". The entities populating Russell's typed logical universe are in fact of three kinds: *individuals, propositions,* and *(propositional) functions.* He defines a *type* as "the range of significance of a propositional function, that is, as the collection of arguments for which the said function has values." Individuals are decreed to occupy the first, or lowest type. Types are assigned to propositions through the device of *orders.* Thus first-order propositions are those whose quantified variables (if any) are individuals: these form the second type. Second-order propositions or functions are then those whose quantified variables are of at most second type, and so on.

From the hierarchy of propositions Russell derives a hierarchy of propositional functions. Now the hierarchy of propositions may be considered one-dimensional in that just a single number is required to specify its position in the hierarchy. The hierarchy of propositional functions, however, is multidimensional since the type of a function involves not just its order as a proposition, but also the orders of its arguments, or free variables. It is this feature of Russell's theory of types which led to its being called the *ramified* theory of types.

As Quine[6] has pointed out, the logical universe presented by Russell has a number of ambiguous features. For instance, in Russell's formulation, individuals are objects, but propositions and propositional functions are notations, dependent

[6] In his commentary on Russell's paper in van Heijenoort [1967].

on the vicissitudes of presentation. That being the case, it is far from clear whether Russell is "assigning types to his objects or to his notations."

Russell next addresses the question of whether his theory is adequate for the development of mathematics. He gives examples from elementary mathematics to show that the answer is no unless some method is found "of reducing the order of a propositional function without affecting the truth or falsehood of its values". To achieve this he introduces the *Axiom of Reducibility,* which asserts that every function is coextensive with a predicative function (in the same arguments). Here by a *predicative* function is meant one in which the types of quantified variables do not exceed the types of the arguments. (All others are termed *impredicative.*) Russell justifies this postulate on the grounds that it has all the useful consequences flowing from the class concept, without having to go so far as to admit the actual presence of classes.

As Quine[7] observes, this expedient is "oddly devious". If every propositional function is coextensive with a predicative one, then attention could have been confined to these latter from the beginning. In other words, the "ramification" of types is rendered superfluous; the types of propositional functions could have been made dependent solely on their arguments. Indeed, "The axiom of reducibility is self-effacing: if it is true, the ramification it was meant to cope with was pointless to begin with."[8]

In the same year (1908) that Russell published his paper on logical types, Ernst Zermelo published his paper *Investigations in the foundations of set theory I.*[9] In it he presents the first axiomatization of set theory. While both Russell's and Zermelo's theories were intended to furnish a foundation for mathematics — a framework, that is, within which the usual development of mathematics could proceed free of contradiction — there are a number of differences between the two approaches. Russell had always maintained that mathematics is, at bottom, logic, and his system naturally reflects that conviction. For Russell, once the correct logical framework had been found, mathematics would then follow; hence his concern with the logical paradoxes, and his consequent elaboration of the theory of types as a strategy for circumventing them. Once this had been taken care of, the development of mathematics could proceed along the lines mapped out by the great 19th century German mathematicians Weierstrass, Dedekind, and Cantor, whose work had so impressed Russell.

In formulating his axiom system for set theory Zermelo was animated by rather different concerns. In 1904 he published his proof that every set could be well-ordered. The proof excited considerable controversy, because of its use of the axiom of choice and the vagueness of the set-theoretic background. This led Zermelo to publish a new, improved proof (again based on the axiom of choice, but in a new formulation) of the theorem in 1908, which was itself to be underpinned by the sharpened, axiomatized notion of set presented in *Investigations.* While designed

[7] *Op. cit.*
[8] *Op. cit.*
[9] Zermelo [1908].

to avoid the set-theoretic paradoxes, the primary purpose of Zermelo's system was to provide the working mathematician with a reliable foundational toolkit.

While for Russell sets or classes were essentially logical entities, Zermelo treated them as mathematical objects subject to certain axiomatic conditions. The most interesting of these latter is the third, the *Axiom of Separation.* This axiom is the counterpart, for sets considered as mathematical objects, of the principle governing their introduction as logical entities, namely, as extensions of propositional functions. Zermelo formulates his axiom by introducing the notion of *definite property*, and then stipulating that a definite property separates a subset from a previously given set. Sadly, Zermelo's notion of definite property is itself far from being definite, since in its formulation Zermelo invokes "the universally valid laws of logic", but neglects to specify what these are. This shortcoming was later rectified, in various ways, by Weyl, Fraenkel, Skolem and von Neumann. Zermelo's system, as modified and extended by the latter three as well as Gödel and Bernays, in due course became the standard foundation of mathematics, a role it still plays today.

Type theory has, by contrast, undergone a more tortuous development and has met with a more complex reception. Its initial phase culminated in Whitehead and Russell's monumental trilogy *Principia Mathematica*[10] of 1910–13. This is a fully elaborated treatment of the framework introduced by Russell in his 1908 paper. It includes both an axiom of infinity and an axiom of reducibility.

Classes play a very minor role in *Principia.* By the time *Principia* was written, Russell had come to regard classes as dubious logical entities and so sought to avoid having to postulate their existence outright. Whitehead and Russell see the introduction of classes as arising from the necessity to deal with *extensions*:

> It is an old dispute whether formal logic should concern itself mainly with intensions or with extensions. In general, logicians whose training was mainly philosophical have decided for intensions, while those whose training was mainly mathematical have decided for extensions. The facts seem to be that, while mathematical logic requires extensions, philosophical logic refuses to supply anything except intensions. Our theory of classes recognizes and reconciles these two apparently opposite facts, by showing that an extension (which is the same as a class) is an incomplete symbol, whose use always acquires its meaning through a reference to an intension.

In order to be able to handle extensions, and yet at the same time avoiding any untoward ontological commitments, in *Principia* classes are introduced, then, as *incomplete symbols*, that is, symbols whose "*uses* are defined, but [which] themselves are not assumed to mean anything at all". Thus, "classes...are merely symbolic or linguistic conveniences, not genuine objects as their members are if they are individuals". In other words, classes are mere shadows.

[10]Whitehead and Russell [1910–13].

The reconditeness[11] of *Principia* and its focus on philosophical niceties far removed from the concerns of working mathematicians hardly commended it to them as a practical foundation for their subject. Zermelo's axiomatic approach to classes, by contrast, seemed simpler and closer to the actual practice of mathematics.

Nevertheless, the concept of the ramified hierarchy has been an important, if usually unremarked, influence in mathematical logic. For instance, in the 1930s Gödel discovered that, if the ramification of properties over the natural numbers is extended into the transfinite, the process breaks down at level ω_1, the first uncountable ordinal. (This may be considered a provable form of the Axiom of Reducibility.) This led him to introduce the hierarchy of constructible sets which he employed to prove the relative consistency of the continuum hypothesis. Here we have a significant, if indirect, application of type-theoretic ideas to set theory.

2 CRITIQUING RAMIFIED TYPES

A number of mathematicians/logicians, notably H. Weyl, L. Chwistek and F. P. Ramsey, attempted to remedy what they had identified as technical deficiencies in *Principia.* In his *Das Kontinuum* of 1918 Weyl presents a predicative formulation of mathematical analysis — not, as Russell and Whitehead had attempted, by introducing a hierarchy of logically ramified types, which Weyl seems to have regarded as too complicated — but rather by confining the comprehension principle to formulas whose bound variables range over just the initial given entities (numbers). Thus he restricts analysis to what can be done in terms of natural numbers with the aid of three basic logical operations, together with the operation of substitution and the process of "iteration", i.e., primitive recursion. Weyl recognized that the effect of this restriction would be to render unprovable many of the central results of classical analysis[12] — e.g., Dirichlet's principle that any bounded set of real numbers has a least upper bound — but he was prepared to accept this as part of the price that must be paid for the security of mathematics.

In 1925 Chwistek formulated what he called the theory of *constructive types.*[13] This is essentially the system of *Principia* purged of all existential propositions,

---

[11]In this connection it is worth quoting the following extract from a review of *Principia* in a 1911 number of the London magazine *The Spectator:*

> It is easy to picture the dismay of the innocent person who out of curiosity looks into the later part of the book. He would come upon whole pages without a single word of *English below* the headline; he would see, instead, scattered in wild profusion, disconnected Greek and Roman letters of every size interspersed with brackets and dots and inverted commas, with arrows and exclamation marks standing on their heads, and with even more fantastic signs for which he would with difficulty so much as find names.

[12]This would also have been the case for a ramified type theory lacking the Axiom of Reducibility.

[13]Chwistek [1925].

including the Axiom of Reducibility. It has the property that each of its symbols can be obtained by the application of a finite number of operations applied to a set of initial symbols. Chwistek's system is in fact no stronger than Weyl's.

The year 1926 saw the publication of Ramsey's important paper *The Foundations of Mathematics*.[14] The core of Ramsey's paper is a critique of the logical system of *Principia Mathematica*, together with a proposal for setting it right. Ramsey identifies three "great defects in *Principia*". The first, that in it only definable classes are admitted, in violation of "the extensional attitude of modern mathematics, an essential part of which is "the possibility of indefinable classes and relations in extension." He continues

> The mistake is made not by having a primitive proposition asserting that all classes are definable, but by giving a definition of class which applies only to definable classes, so that all mathematical propositions about some or all classes are misinterpreted. This misinterpretation is not merely objectionable on its own account in a general way, but is especially pernicious in connection with the Multiplicative Axiom, which is a tautology when properly interpreted, but when misinterpreted after the fashion of *Principia Mathematica* becomes a significant empirical proposition, which there is no reason to suppose true.

By the multiplicative axiom Ramsey means the version of the axiom of choice asserting that, for any nonempty family $K$of disjoint nonempty sets, there is a set having exactly one member in common with each set in K. Ramsey says that according to the extensional view he takes of classes, "the Multiplicative Axiom seems... the most evident tautology." It only becomes really doubtful when "the class whose existence it asserts must be one definable by a propositional function of the sort which occurs in *Principia*". Ramsey shows that the multiplicative axiom (more precisely, its logical equivalent that the cardinalities of any pair of sets are comparable) can in fact be falsified in a *Principia*-like system in which there are very few atomic propositional functions.

Ramsey's chief objection against the theory of types presented in *Principia* is the fact that, in order to furnish an adequate foundation for mathematics, it requires the introduction of the "illegitimate" Axiom of Reducibility. Ramsey attributes this to the fact that the authors of *Principia* have failed to observe that the contradictions whose avoidance is one of the work's chief purposes actually fall into two "fundamentally distinct" groups. The first of these, of which Russell's paradox is representative, "consists of contradictions which, were no provision made against them, would occur in a logical or mathematical system itself." They "involve only logical or mathematical terms such as class and number, and show that there must be something wrong with our logic or mathematics." On the other hand the contradictions in the second group, of which the Liar Paradox and the "heterological" paradox are representative, "are not purely logical, and cannot be stated in logical terms alone; for they all contain some reference to thought,

[14] Ramsey [1926].

language, or symbolism, which are not formal but empirical terms." So "they may be due not to faulty logic or mathematics, but to faulty ideas concerning thought and language."

Ramsey points out that the theory of types really consists of two distinct parts "directed respectively against the two groups of contradictions", unified only by "being both deduced in a somewhat sloppy way from the 'vicious circle principle'." But it is essential, he says, to consider them separately. The contradictions in the first group can be eliminated by noting that a propositional function cannot meaningfully take itself as argument, which leads to a division of "functions and classes into a hierarchy of types according to their possible arguments." As a result, there are functions of individuals, functions of functions of individuals, etc. And then "the assertion that a class is a member of itself is neither true nor false, but meaningless." Ramsey regards this part of the theory of types — the part which would later evolve into the simple theory of types — as "unquestionably correct."

As for the second part of the theory, which is designed to contend with the second group of contradictions, it "requires further distinctions between the different functions which take the same arguments." These distinctions are reflected in the orders of *Principia*, the classification of functions according to the level of their bound variables. Ramsey agrees that the contradictions in the second group are circumvented by the introduction of orders, but this resolution "lands us in an almost equally serious difficulty, for it invalidates many important mathematical arguments which appear to contain exactly the same fallacy as the contradictions." To overcome this deficiency, the authors of *Principia* introduce the Axiom of Reducibility, a postulate which in Ramsey's view "there is no reason to suppose true", whose truth, indeed, would be merely "a happy accident and not a logical necessity." "Such an axiom," Ramsey asserts, "has no place in mathematics, and anything which cannot be proved without using it cannot be regarded as proved at all."

Ramsey then considers the question of why the Axiom of Reducibility fails to reproduce the contradictions (those in the second group) the ramified theory of types was explicitly designed to avoid. For in asserting that any function is coextensive with a predicative function the Axiom of Reducibility "may appear to lose again whatever was gained by making the distinction." Ramsey says that the reason why these contradictions fail to reappear is their "peculiar nature". For they "are not purely mathematical, but all involve the ideas of thought or meaning", in a word, they are *intensional.* If any purely mathematical contradictions were to arise through the conflating of arbitrary with predicative functions, or from the identification of intensionally different, but extensionally equivalent functions, then these contradictions would, says Ramsey, be "reinstated by the Axiom of Reducibility, owing to the extensional nature of mathematics, in which [extensionally] equivalent functions are interchangeable." The fact that no such contradictions have been shown to arise underscores the intensional nature of the contradictions in the second group, making it "even more probable that they have a psychological or epistemological, and not a purely logical or mathematical solution." Ramsey's conclusion is that "there is something wrong with the account of the matter given in *Principia.*"

Ramsey's attitude towards the foundations of mathematics and logic was robustly extensional, and accordingly realist. This realist attitude insulated him from the worries about impredicative definition that had exercised Russell (and Weyl). In Ramsey's view, the distinction of orders of functions is just a complication imposed by the structure of our language and not, unlike the hierarchy of types, something inherent in the way things truly are. Ramsey's suggestion in *The Foundations of Mathematics* for repairing the defects in *Principia* reflects his realist orientation. In essence, his proposal is to render the whole apparatus of orders superfluous through the simple expedient of eliminating quantifiers in definitions. So a universal quantifier is then regarded as indicating a conjunction, and an existential quantifier a disjunction, of the collection of propositions obtained from a given propositional function by the substitution of a name of any one of the members of the collection of objects, itself conceived as existing independently of any definition we might envisage, which constitutes the propositional function's range of significance.[15] It does not matter whether it may be impossible in practice to write out the resulting expressions in full. Once the apparatus of orders is abandoned, the simple theory of types remains, and Ramsey was convinced that this would provide an adequate foundation for mathematics.

3 CHURCH'S VERSION OF THE SIMPLE THEORY OF TYPES

When the ramified structure of the type hierarchy is abandoned, what remains is, as we have seen, a *simple* theory of types, similar to that which Russell had originally proposed for classes-as-many.

Neither Chwistek nor Ramsey produced a detailed formulation of the simple theory of types, although it seems clear that either could have achieved this had he so wished. In fact the first fully worked out formulation is that of Carnap [1929]; later formulations include Tarski [1931] and Gödel [1931].

The form of the simple theory of types which has proved definitive was put forward by Church [1940]. Church's system is based on functions instead of relations or classes, and incorporates certain features of the λ-calculus which had been previously developed by him. Church's system has had a profound influence on computer science, particularly in the areas of automated theorem proving, computational logic, and formal methods. Today the field of type theory (in computer science at least) is largely the study of Church-style systems based on functions with λ-notation.

A straightforward version of Church's simple type theory is the following system

[15]On this reckoning, then, the statement ***Citizen Kane*** *has all the qualities that make a great film* would be taken as an abbreviation for something like ***Citizen Kane*** *is a film, brilliantly directed, superbly photographed, outstandingly performed, excellently scripted, etc.*

T.[16]

T is equipped with the following *types* and *terms*.

Types: $\mathbf{I}$, the type of individuals
$\mathbf{\Omega}$, the type of propositions or truth values
A collection of base types
Function types $(\mathbf{A} \to \mathbf{B})$, for any types $\mathbf{A}, \mathbf{B}$ (we omit the parentheses whenever possible)

Terms: Terms are introduced according to the following rules, so that each term t is assigned a type $\mathbf{A}$, written t: $\mathbf{A}$.
Variables x_A, y_A, z_A, ... of each type $\mathbf{A}$ (we omit the subscript whenever possible)
Individuals a, b, c, ... of type $\mathbf{I}$
Function application. If t: $\mathbf{A}$ and f: $\mathbf{A} \to \mathbf{B}$, then $f(t) : \mathbf{B}$
Function abstraction. If t: $\mathbf{B}$, then $(\lambda x_A.t) : \mathbf{A} \to \mathbf{B}$
Equality. If $t : \mathbf{A}, u : \mathbf{A}$, then $t = u : \mathbf{\Omega}$.

Terms of type $\mathbf{\Omega}$ are called *propositions* or *formulas.*

Notice that T lacks logical operators. It was observed by Henkin [1963] that they can in fact be defined in a system like T. Let ω, ω' be variables of type $\mathbf{\Omega}$, and let α, β be propositions. Then we write

$\top$	for	$(\lambda\omega.\omega) = (\lambda\omega.\omega)$
$\bot$	for	$(\lambda\omega.\top) = (\lambda\omega.\omega)$
$\neg\alpha$	for	$\alpha = \bot$
$\alpha \Leftrightarrow \beta$	for	$\alpha = \beta$
$\alpha \wedge \beta$	for	$\lambda f.f(\top)(\top) = \lambda f.f(\alpha)(\beta)$, with $f : \mathbf{\Omega} \to (\mathbf{\Omega} \to \mathbf{\Omega})$
$\alpha \Rightarrow \beta$	for	$\alpha \wedge \beta \Leftrightarrow \alpha$
$\forall x.\alpha$	for	$\lambda x.\alpha = \lambda x.\top$
$\alpha \vee \beta$	for	$\forall\omega[[(\alpha \Rightarrow \omega) \wedge (\beta \Rightarrow \omega)] \Rightarrow \omega]$
$\exists x.\alpha$	for	$\forall\omega[[\forall x(\alpha \Rightarrow \omega) \Rightarrow \omega]$

A simple type theory like T, employing just function application, function abstraction, and equality, can serve as a foundation for mathematics in the sense that the usual mathematical structures, e.g. the natural numbers and the reals, admit straightforward formulations within it. And with appropriate axioms (including an axiom of infinity to the effect that the type of individuals is infinite), it can be shown that simple type theory has the same proof-theoretic strength as bounded Zermelo set theory.

Church's formulation of the simple theory of types is evidently far from being a *faute de mieux*; it is, rather, the theory that naturally ensues from an analysis of the function concept. This is well summed up by Robin Gandy:

[16]Based on the presentation by Farmer [2006]; here we give a version free of the background assumption of classical logic.

> The simple theory of types provides a straightforward, reasonably secure foundation for the greater part of classical mathematics. This is why a number of authors (Carnap, Gödel, Tarski, Church, Turing) gave a precise formulation of it, and used it as a basis for metamathematical investigations. The theory is straightforward because it embodies two principles which (at least before the advent of modern abstract concepts) were part of the mathematicians' normal code of practice. Namely that a variable always has a precisely delimited range, and that a distinction must always be made between a function a and its arguments. [Gandy, 1977, p. 173]

4 TYPES VS. SETS

What is the difference between a type and a set? A first thought, naturally, is that types resemble the syntactic or grammatical categories forming the basis of language in that the specification of types furnish the conditions for expressions to be well-formed, or meaningful. Everybody knows (instinctively at least) that the linguistic phrase "___*sees*___" becomes grammatical only when the first blank is filled in with a noun or a noun phrase and the second with a noun, a noun phrase, or an adverb.

The typing concept also arises in concrete situations. Consider, for example, an automotive toolkit. Here one is provided with nuts, bolts and wrench bits to attach nuts to bolts. This gives three types: **N, B,** and **S.** Then given the instruction *Use ___ to attach ___ to ___* , it is implicit that the blanks be filled in with (names of) components of types **S, N,** and **B** respectively.

This case points up nicely the difference between types and sets. For example, we can consider the concrete *set* or *collection* of, say, nuts, of a given toolkit. If (as is so often the case) a few of these go missing, the *set* of nuts is changed, but not the associated *type*. The set, that is, the concrete embodiment of the type, is subject to variation, but the type is not.

Here is another illustration. We can take two toolkits and *amalgamate* them. Now while it would be possible to do this simply by jumbling their contents together — that is, to take their set-theoretic union — , in order for the result to remain a toolkit in the specified sense *typing must be respected*: the nuts, bolts and wrench bits of each toolkit must be conjoined only with the nuts, bolts and wrench bits, respectively, of the other.

Sets and types, while different, are nevertheless clearly related. Indeed, since each type gives rise to the set of entities of that type, a set may be regarded as a *type considered in extension.* Clearly different types can give rise to the same set: for example those old chestnuts the types "rational featherless biped" and "human being" are distinct, but their associated sets are (terrestrially, at least) identical.

Now sets may also be regarded as *properties* considered in extension. This, conjoined with the fact that sets may be regarded as types considered in extension, might suggest that types are just properties. But that would be a mistake. For while the domain of properties is closed under the usual logical operations, these operations are not (in general) defined for types at all. For example, we may consider "natural number" and "circle" as types, but we do not similarly admit "natural number or circle" as defining a type.

The realm of types is, in fact, now customarily taken to be closed not under logical operations but under what may be termed *mathematical* operations. These may include the Cartesian product $\times$, the exponential or function type operation $\rightarrow$, and the power type operation $\mathbf{P}$. Here, for types $\mathbf{A}$, $\mathbf{B}$, $\mathbf{A} \times \mathbf{B}$ is the common type of pairs (a, b) with a of type $\mathbf{A}$ and b of type $\mathbf{B}$, $\mathbf{PA}$ is the type of sets of entities of type $\mathbf{A}$, and $\mathbf{A} \rightarrow \mathbf{B}$ is the type of correspondences (functions) between entities of types $\mathbf{A}$, $\mathbf{B}$, respectively.

The idea that a set may be regarded as a type considered in extension suggests a way of representing (simple) type theory in set theory, viz., just interpret types as sets, and function types $\mathbf{A} \rightarrow \mathbf{B}$ as the set of functions from (the set corresponding to) $\mathbf{A}$ to (the set corresponding to) $\mathbf{B}$.

The representation of set theory within type theory is an altogether subtler affair. The problem here is to represent within type theory the membership relation, which is defined for all pairs of sets. An elegant way of doing this has been elaborated by Miquel [2001] using the notions of pointed graph and bisimilarity.

A *pointed graph* is a triple $\mathfrak{A} = (A, R, a)$ with A a set, R a binary relation on A, and $a \in A$. A *bisimilarity* between two pointed graphs $\mathfrak{A} = (A, R, a)$ and $\mathfrak{B} = (B, S, b)$ is a relation $\approx$ between A and B satisfying the conditions:

- $\forall x x' \in A \forall y \in B[x' R x \wedge x \approx y \Rightarrow \exists y' \in B(y' S y \wedge x' \approx y')]$
- $\forall y y' \in B \forall x \in A[y' S y \wedge x \approx y \Rightarrow \exists x' \in A(x' R x \wedge x' \approx y)]$
- $a \approx b$

Two pointed graphs $\mathfrak{A}, \mathfrak{B}$ are *bisimilar*, written $\mathfrak{A} \sim \mathfrak{B}$, if there is a bisimilarity between them. It is readily checked that bisimilarity is an equivalence relation.

We shall require the *transitive closure* $\tau(A)$ of a set A: this is the set whose elements are A itself, the members of A, the members of members of A, etc. Now let $\mathfrak{A}^*$ be the pointed graph $(\tau(A), \in |\tau(A), A)$. A straightforward induction on rank argument shows that, for any sets A, B,

$$\mathfrak{A}^* \sim \mathfrak{B}^* \Leftrightarrow A = B.$$

It follows that

$$A \in B \Leftrightarrow \exists X(X \in B \wedge \mathfrak{X}^* \sim \mathfrak{A}^*). \qquad (*)$$

This last equivalence,[17] which expresses set membership in terms of bisimilarity of pointed graphs — *shifted bisimilarity* — furnishes the basis for the representation

[17] While this equivalence was actually proved only for *well-founded* sets, its purpose is to suggest the form of the type-theoretic representation of the membership relationship for *arbitrary* (i.e., not necessarily well-founded) sets.

of sets as types. The idea is to interpret sets as pointed graphs, and then to observe that pointed graphs can be represented as types.

In type-theoretic terms, a pointed graph may be defined as a triple $(\mathbf{A}, R, a)$ with $\mathbf{A}$ a type, R a term of type $\mathbf{A} \to (\mathbf{A} \to \mathbf{\Omega})$, and a a term of type $\mathbf{A}$. The conditions for a bisimilarity between pointed graphs given above — and so also the representation of set membership in terms of shifted bisimilarity — are then readily translated into the language of type theory. This yields a faithful interpretation of set theory[18] in a certain version of type theory[19] under which sets correspond to pointed graphs and the membership relation corresponds to shifted bisimilarity.

5 CATEGORIES AND TYPES

When Eilenberg and Mac Lane first introduced (in 1945) the concepts of category and functor their concerns were far removed from the theory of types. Yet category theory and type theory have proved to have deep connections. In fact categories can *themselves* be viewed as type theories of a certain kind; this fact alone indicates that type theory is much more closely related to category theory than it is to set theory. In thinking of a category as a type theory, the objects of a category are regarded as *types* (or sorts) and the arrows as *mappings* between the corresponding types. Roughly speaking, a category may be thought of a type theory shorn of its syntax. In the 1970s Lambek[20] established that, viewed in this way, cartesian closed categories correspond to the typed λ-calculus. Later Seely [1984] proved that locally Cartesian closed categories correspond to Martin-Löf, or predicative, type theories. Lambek and Dana Scott independently observed that *C-monoids*, i.e., categories with products and exponentials and a single, nonterminal object correspond to the *untyped* λ-calculus. The analogy between type theories and categories has since led to what Jacobs [1999] terms a "type-theoretic boom", with much input from, and applications to, computer science.

5.1 Cartesian closed categories and the typed λ-calculus

As already mentioned, Lambek has shown that the typed λ-calculus corresponds to a cartesian closed category. Here we give a brief sketch of how this correspondence is established.

A *typed λ-calculus* is a theory formulated within a certain kind of formal language which we shall call a *λ-language*. A language Λ of this kind is equipped with the following *basic symbols*:

- a *type symbol* $\mathbf{1}$
- *ground types* $\mathbf{A}, \mathbf{B}, \mathbf{C}, \ldots$ (possibly none of these)

[18] I.e., intuitionistic Zermelo (or even Zermelo-Fraenkel) set theory.
[19] To be precise, polymorphic dependent type theory (see below) with a type of propositions.
[20] See the reference entries under Lambek and Lambek and Scott.

- *function symbols* **f**, **g**, **h**, ... (possibly none of these)
- *variables* $x_A, y_A, z_A, \ldots$ of each type **A**, where a *type* is as defined below
- unique entity ★ of type **1**.

The *types* of Λ are defined recursively as follows:

- **1** and each ground type are types
- $\mathbf{A}_1 \times \ldots \times \mathbf{A}_n$ is a type whenever $\mathbf{A}_1, \ldots, \mathbf{A}_n$ are, where, if $n = 1$, $\mathbf{A}_1 \times \ldots \times \mathbf{A}_n$ is $\mathbf{A}_1$, while if $n = 0$, $\mathbf{A}_1 \times \ldots \mathbf{A}_n$ is **1** (*product types*)
- $\mathbf{B}^A$ is a type whenever $\mathbf{A}, \mathbf{B}$ are.

Each function symbol **f** is assigned a *signature* of the form $\mathbf{A} \to \mathbf{B}$ where $\mathbf{A}, \mathbf{B}$ are types. We normally write $\mathbf{f} : \mathbf{A} \to \mathbf{B}$ to indicate this.

Terms of Λ and their associated *types* are defined recursively as follows. We write $\tau : \mathbf{A}$ to indicate that the term τ has type **A**.

Term: **type**	Proviso
$\bigstar : \mathbf{1}$	
$x_A : \mathbf{A}$ (we shall usually omit the subscript)	
$\mathbf{f}(\tau) : \mathbf{b}$	$\mathbf{f} : \mathbf{A} \to \mathbf{B} \quad \tau : \mathbf{A}$
$\langle \tau_1, \ldots, \tau_n \rangle : \mathbf{A}_1 \times \ldots \times \mathbf{A}_n$, where $\langle \tau_1, \ldots, \tau_n \rangle$ is τ_1 if $n = 1$, and ★ if $n = 0$.	$\tau_1 : \mathbf{A}_1, \ldots, \tau_n : \mathbf{A}_n$
$(\tau)_i : \mathbf{A}_i$ where $(\tau)_i$ is τ if $n = 1$	$\tau : \mathbf{A}_1 \times \ldots \times \mathbf{A}_n, 1 \leq i \leq n$
$\tau'\sigma : \mathbf{B}$	$\tau : \mathbf{B}^A \quad \sigma : \mathbf{A}$
$\lambda_x \tau : \mathbf{B}^A$	$x : \mathbf{A} \quad \tau : \mathbf{B}$

We write $\tau(x/\sigma)$ or $\tau(\sigma)$ for the result of substituting σ at each free occurrence of x in τ, where an occurrence of x is *free* if it does not appear within a term of the form $\lambda_x \rho$. A term σ is *substitutable* for a variable x in a term τ if no variable free in σ becomes bound when so substituted.

An *equation* in Λ is an expression of the form $\sigma = \tau$, where σ and τ are terms of the same type. A *theory* in Λ (or simply a λ-*theory*) is a set T of equations satisfying the following conditions:

- $\sigma = \sigma \in T \quad \sigma = \tau \in T \Rightarrow \tau = \sigma \in T$
  $\sigma = \tau \in T \;\&\; \tau = \upsilon \in T \Rightarrow \sigma = \upsilon \in T$
- $\sigma = \upsilon \in T \Rightarrow \tau'\sigma = \tau'\upsilon \in T \quad \sigma = \tau \in T \Rightarrow \lambda_x \sigma = \lambda_x \tau \in T$
- T contains all equations of the following forms:
  $\tau = \bigstar$ for $\tau : \mathbf{1}$

  $$(\langle \tau_1, \ldots, \tau_n \rangle)_i = \tau_i \quad \tau = \langle (\tau)_1, \ldots, (\tau)_n \rangle$$

$(\lambda_x \tau)'\sigma = \tau(x/\sigma)$ for all σ substitutable for x in τ

$\lambda_x(\tau' x) = \tau$ for all $\tau : \mathbf{B}^A$, provided $x : \mathbf{A}$ is not free in τ.

Now λ-theories comprise the objects of a category $\lambda - \mathcal{Th}$ whose arrows are the so-called *translations* between such theories. Let us define a *translation* $\mathbf{K}$ of a λ-language $\mathcal{L}$ into a λ-language $\mathcal{L}'$ to be a map which assigns to each type $\mathbf{A}$ of $\mathcal{L}$ a type $\mathbf{KA}$ of $\mathcal{L}'$ and to each function symbol $\mathbf{f} : \mathbf{A} \to \mathbf{B}$ a function symbol of $\mathcal{L}'$ of signature $\mathbf{KA} \to \mathbf{KB}$ in such a way that

$$\mathbf{K1} = 1,\ \mathbf{K}(\mathbf{A}_1 \times \ldots \times \mathbf{A}_n) = \mathbf{KA}_1 \times \ldots \mathbf{KA}_n,\ \mathbf{K}(\mathbf{B}^A) = \mathbf{KB}^{\mathbf{KA}}.$$

Any translation $\mathbf{K} : \mathcal{L} \to \mathcal{L}'$ may be extended to the terms of $\mathcal{L}$ in the evident recursive way — i.e., by defining $\mathbf{K}x_A = x_{\mathbf{K}A}, \mathbf{K}\bigstar = \bigstar, \mathbf{K}(\mathbf{f}(\tau)) = \mathbf{Kf}(\mathbf{K}_\tau), \mathbf{K}(\lambda_x \tau) = \lambda_{\mathbf{K}x}\mathbf{K}\tau$, etc. — so that if $\tau : \mathbf{A}$, then $\mathbf{K}\tau : \mathbf{KA}$. We shall sometimes write α_K for $\mathbf{K}\alpha$. If T, T' are λ-theories in $\mathcal{L}, \mathcal{L}'$ respectively, a translation $\mathbf{K} : \mathcal{L} \to \mathcal{L}'$ is a *translation of* T *into* T', and is written $\mathbf{K} : T \to T'$ if, for any equation $\sigma = \tau$ in T, the equation $\mathbf{L}\sigma = \mathbf{K}\tau$ is in T'.

As we shall see, λ-theories correspond to Cartesian closed categories in such a way as to make the category λ-$\mathcal{Th}$ equivalent to the category $\mathcal{Cart}$ of Cartesian closed categories. To establish this, we first introduce the notion of an *interpretation* I of a λ-language $\mathcal{L}$ in a cartesian closed category $\mathcal{C}$. This is defined to be an assignment

- to each type $\mathbf{A}$, a $\mathcal{C}$-object $\mathbf{A}_I$ such that:
  $(\mathbf{A}_1 \times \ldots \times \mathbf{A}_n)_I = (\mathbf{A}_1)_I \times \ldots \times (\mathbf{A}_n)_I$,
  $(\mathbf{B}^A)_I = (\mathbf{B}_I)^{A_I}$
  $\mathbf{1}_I = 1$, the terminal object of $\mathcal{C}$,
- to each function symbol $\mathbf{f} : \mathbf{A} \to \mathbf{B}$, a $\mathcal{C}$-arrow $\mathbf{f}_I : A_I \to B_I$.

We shall sometimes write $A_\mathcal{C}$ or just A for $\mathbf{A}_I$.

We extend I to terms of $\mathcal{L}$ as follows. If $\tau : \mathbf{B}$, write x for $(x_1, \ldots, x_n)$, any sequence of variables containing all variables of τ (and call such sequences *adequate* for τ). Define the $\mathcal{C}$-arrow.

$$[\![\tau]\!]_x : A_1 \times \ldots A_n \to B$$

recursively as follows:

$$\begin{aligned}
&[\![\bigstar]\!]_x = A_1 \times \ldots \times A_n \longrightarrow 1\\
&[\![x_i]\!]_x = \pi_i : A_1 \times \ldots \times A_n \longrightarrow A_i\\
&[\![\mathbf{f}(\tau)]\!]_x = \mathbf{f}_I \circ [\![\tau]\!]_x\\
&[\![\langle \tau_1, \ldots, \tau_n \rangle]\!]_x = \langle [\![\tau_1]\!]_x, \ldots, [\![\tau_n]\!]_x \rangle\\
&[\![(\tau)_i]\!]_x = \pi_i \circ [\![\tau]\!]_x\\
&[\![\tau'\sigma]\!]_x = ev \circ \langle [\![\tau]\!]_x, [\![\sigma]\!]_x \rangle\\
&[\![\lambda_y \tau]\!]_x = \widehat{[\![\tau]\!]_{xy}}
\end{aligned}$$

Here $\widehat{\ }$ denotes exponential transpose and *ev* the appropriate evaluation arrow in the definition of the exponential.[21]

Next, one shows that any λ-theory T determines a Cartesian closed category $\mathcal{C}(T)$. The objects of $\mathcal{C}(T)$ are taken to be the types of the language of T. The arrows of $\mathcal{C}(T)$ are pairs of the form (x, τ), where τ is a term with no free variables other than x, two such pairs (x, τ) and (y, σ) being identified whenever the equation $\tau = \sigma$ is a member of T. The identity arrow on a type $\mathbf{A}$ is the pair (x_A, x_A). The composite of $(x, \tau) : \mathbf{A} \to \mathbf{B}$ and $(y, \sigma) : \mathbf{B} \to \mathbf{C}$ is given by the pair $(x, \sigma(y/\tau))$. It is now readily checked that $\mathcal{C}(T)$ is a Cartesian closed category with terminal object the type $\mathbf{1}$ and in which products and exponentials are given by the analogous operations on types. Moreover,

$$
\begin{gathered}
1_A = (x_A, \bigstar) \\
\pi_i = (z, (z)_i) \text{ (with } z : \mathbf{A}_1 \times .. \times \mathbf{A}_n) \\
\langle (z, \sigma), (z, \tau) \rangle = (z, \langle \sigma, \tau \rangle) \\
\widehat{(z, \tau)} = (x, \lambda_y \tau(\langle x, y \rangle)) \text{ with } z : \mathbf{A} \times \mathbf{B}, x : \mathbf{A}, y : \mathbf{B} \\
ev_{C,A} = (y, (y)'_2(y)_1) \text{ with } y : \mathbf{C}^A \times \mathbf{A}.
\end{gathered}
$$

There is a canonical interpretation of $\mathcal{L}$ in $\mathcal{C}(T)$ — which we denote by the same expression $\mathcal{C}(T)$ — given by

$$\mathbf{A}_{\mathcal{C}(T)} = \mathbf{A} \qquad \mathbf{f}_{\mathcal{C}(T)} = (x, \mathbf{f}(x)).$$

A straightforward induction shows that, for any term τ with free variables $\mathbf{x} = (x_1, ..., x_n)$ of types $\mathbf{A}_1, ..., \mathbf{A}_n$, we have

$$[\![\tau]\!]_x = (z, \tau)$$

where z is a variable of type $\mathbf{A}_1 \times ... \times \mathbf{A}_n$.

Inversely, each Cartesian closed category $\mathcal{E}$ determines a λ-theory $\mathcal{T}\!h(\mathcal{E})$ in the following way. First, one defines the *internal language* $\mathcal{L}_{\mathcal{E}}$ of $\mathcal{E}$: the ground type symbols of $\mathcal{L}_{\mathcal{E}}$ are taken to match the objects of $\mathcal{E}$ other than 1, that is, for each $\mathcal{E}$-object A other than 1 we assume given a ground type $\mathbf{A}$. Next, we with each type symbol $\mathbf{A}$ of $\mathcal{L}_{\mathcal{E}}$ we associate the $\mathcal{E}$-object $\mathbf{A}_{\mathcal{E}}$ defined by

$$\mathbf{A}_{\mathcal{E}} = A \text{ for ground types } \mathbf{A}, \ (\mathbf{A} \times \mathbf{B})_{\mathcal{E}} = \mathbf{A}_{\mathcal{E}} \times \mathbf{B}_{\mathcal{E}}, \ (\mathbf{B}^A)_{\mathcal{E}} = \mathbf{B}_{\mathcal{E}}^{A}\mathcal{E}.$$

As function symbols in $\mathcal{L}_E$ we take triples of the form $(f, \mathbf{A}, \mathbf{B}) = \mathbf{f}$ with $f : \mathbf{A}_{\mathcal{E}} \to \mathbf{B}_{\mathcal{E}}$ in $\mathcal{E}$. The signature of $\mathbf{f}$ is $\mathbf{A} \to \mathbf{B}$.

The *natural interpretation* — denoted by $\mathcal{E}$ — of $\mathcal{L}_{\mathcal{E}}$ in $\mathcal{E}$ is determined by the assignments

$$\mathbf{A}_{\mathcal{E}} = A \text{ for ground types } \mathbf{A}, \ (f, \ \mathbf{A}, \ \mathbf{B}) = f.$$

[21] For these notions see the Appendix.

The λ-*theory* $\mathcal{T}h(\mathcal{E})$ *of* $\mathcal{E}$ is now defined in the following way. Given a pair of terms σ, τ of the same type in $\mathcal{L}_{\mathcal{E}}$, write $\mathbf{x}$ for the sequence of variables occurring free in either σ or τ. The theory $\mathcal{T}h(\mathcal{E})$ consists of all equations $\sigma = \tau$ for which the arrows $([\![\sigma]\!]_{\mathbf{x}})_{\mathcal{E}}$ and $([\![\tau]\!]_{\mathbf{x}})_{\mathcal{E}}$ coincide.

It can then be shown that the canonical functor $F : \mathcal{E} \to \mathcal{C}(\mathcal{T}(\mathcal{E}))$ defined by

$$FA = \mathbf{A} \text{ for each } \mathcal{E}\text{-object } A$$
$$Ff = (x, \mathbf{f}(x)) \text{ for each } \mathcal{E}\text{-arrow } f : A \to B$$

is an *isomorphism of categories.*

Similarly, for each λ-theory T there is a canonical translation $G : T \to \mathcal{T}h(\mathcal{C}(T))$ given by

$$G\mathbf{A} = \mathbf{A} \text{ for each type } \mathbf{A}$$
$$G\mathbf{f} = ((x, \mathbf{f}(x), \mathbf{A}, \mathbf{B})) \text{ for } \mathbf{f} : \mathbf{A} \to \mathbf{B}.$$

This translation is clearly an isomorphism in the category λ-$\mathcal{T}h$. Now let $\mathcal{C}art$ be the category whose objects are Cartesian closed categories and whose arrows are functors between these preserving terminal objects, products, and exponentials. The mappings $\mathcal{C}$ and $\mathcal{T}h$ act as functors λ-$\mathcal{T}h \to \mathcal{C}art$ and $\mathcal{C}art \to \lambda$-$\mathcal{T}h$ respectively; in view of the fact that, for any objects $\mathcal{E}$ of $\mathcal{C}art$, T of λ-$\mathcal{T}h$, $\mathcal{C}(\mathrm{Th}(\mathcal{E})) \cong \mathcal{E}$ and $\mathcal{T}h(\mathcal{C}(T)) \cong T$, these functors are equivalences. Accordingly λ-$\mathcal{T}h$ and $\mathcal{C}art$ are equivalent categories. This is one exact sense in which a formal theory is completely representable in categorical terms.

5.2 Logical languages and local set theories

What sort of category corresponds to the simple theory of types[22] itself? The answer to this question — a(n *elementary*) *topos* — was provided in 1969 by Lawvere and Tierney,[23] although it was not at first grasped that their work in fact could in fact be seen as answering that question. However, it soon became clear that toposes are associated with a natural version of simple type theory, based on intuitionistic logic, which we shall call *intuitionistic type theory.* The system to be described here, *local set theory,* is a modification, due to Zangwill [1977], of that of Joyal and Boileau, later published as their [1981]. It combines the most convenient features of simple type theory and set theory. A full account is given in Bell [1988].

A local set theory is a type-theoretic system built on the same primitive symbols $=, \in, \{ : \}$ as classical set theory, in which the set-theoretic operations of forming products and powers of types can be performed, and which in addition contains a "truth value" type acting as the range of values of "propositional functions"

[22]It should be stressed that here by the "simple theory of types" is meant the theory of types in Church's sense which includes a type of propositions. Without a type of propositions the simple theory of types is essentially just the typed λ-calculus.

[23]See Lawvere [1971, 1972].

on types. A local set theory is determined by specifying a collection of *axioms* formulated within a *local language* defined as follows.

A *local language* $\mathcal{L}$ has the following *basic symbols:*

- **1** (*unit type*) $\boldsymbol{\Omega}$ (*truth value type*)
- **S, T, U,...** (*ground types*: possibly none of these)
- **f, g, h,...** (*function symbols*: possibly none of these)
- $x_A, y_A, z_A, \ldots$ (*variables of each type* **A**, where a *type* is as defined below)
- ★ (*unique entity of type* **1**)

The *types* of $\mathcal{L}$ are defined recursively as follows:

- **1**, $\boldsymbol{\Omega}$ are types
- any ground type is a type
- $\mathbf{A}_1 \times \ldots \times \mathbf{A}_n$ is a type whenever $\mathbf{A}_1, \ldots, \mathbf{A}_n$ are, where, if $n = 1$, $\mathbf{A}_1 \times \ldots \times \mathbf{A}_n$ is $\mathbf{A}_1$, while if $n = 0$, $\mathbf{A}_1 \times \ldots \times \mathbf{A}_n$ is **1** (*product types*)
- **PA** is a type whenever **A** is (*power types*)

Each function symbol **f** is assigned a *signature* of the form $\mathbf{A} \to \mathbf{B}$, where $\mathbf{A}, \mathbf{B}$ are types; this is indicated by writing $\mathbf{f} : \mathbf{A} \to \mathbf{B}$.

Terms of $\mathcal{L}$ and their associated *types* are defined recursively as follows. We write $\tau : \mathbf{A}$ to indicate that the term τ has type **A**.

Term: **type**	Proviso
$\bigstar : \mathbf{1}$	
$x_A : \mathbf{A}$	
$\mathbf{f}(\tau) : \mathbf{B}$	$\mathbf{f}: \mathbf{A} \to \mathbf{B} \quad \tau : \mathbf{A}$
$\langle \tau_1, ..., \tau_n \rangle : \mathbf{A}_1 \times ... \times A_n$, where $\langle \tau_1, ..., \tau_n \rangle$ is τ_1 if $n = 1$, and ★ if $n = 0$.	$\tau_1 : \mathbf{A}_1, ..., \tau_n: \mathbf{A}_n$
$(\tau)_i$: $\mathbf{A}_i$ where $(\tau)_i$ is τ if n= 1	$\tau : \mathbf{A}_1 \times ... \times \mathbf{A}_n$, $1 \leq i \leq n$
$\{x_A\text{: } \alpha\}$: **PA**	$\alpha : \Omega$
$\sigma = \tau$: Ω	σ, τ of same type
$\sigma \in \tau$	$\sigma : \mathbf{A}$, τ: **PA** for some type **A**

Terms of type $\boldsymbol{\Omega}$ are called *formulas, propositions,* or *truth values.* Notational conventions we shall adopt include:

$\omega, \omega', \omega''$	variables of type $\mathbf{\Omega}$
α, β, γ	formulas
$x, y, z, \ldots$	$x_A, y_A, z_A \ldots$
$\tau(x/\sigma)$ or $\tau(\sigma)$	result of substituting σ at each free occurrence of x in τ: an occurrence of x is *free* if it does not appear within $\{x : \alpha\}$
$\alpha \leftrightarrow \beta$	$\alpha = \beta$
$\Gamma : \alpha$	sequent notation: Γ a finite set of formulas
$: \alpha$	$\emptyset : \alpha$

A term is *closed* if it contains no free variables; a closed term of type $\mathbf{\Omega}$ is called a *sentence.*

The *basic axioms* for $\mathcal{L}$ are as follows:

Unity: $x_1 = \bigstar$

Equality: $x = y, \alpha(z/x) : \alpha(z/y)$ (x, y free for z in α)

Products: $(\langle x_1, \ldots, x_n \rangle)_i = x_i$

$$x = \langle (x)_1, \ldots, (x)_n \rangle$$

Comprehension: $x \in \{x : \alpha\} \leftrightarrow \alpha$

The *rules of inference* for $\mathcal{L}$ are:

Thinning: $\dfrac{\Gamma : \alpha}{\beta, \Gamma : \alpha}$

Restricted Cut: $\dfrac{\Gamma : \alpha \quad \alpha, \Gamma : \beta}{\Gamma : \beta}$ (any free variable of α free in Γ or β)

Substitution: $\dfrac{\Gamma : \alpha}{\Gamma(x/\tau) : \alpha(x/\tau)}$ (τ free for x in Γ and α)

Extensionality: $\dfrac{\Gamma : x \in \sigma \leftrightarrow x \in \tau}{\Gamma : \sigma = \tau}$ (x not free in Γ, σ, τ)

Equivalence: $\dfrac{\alpha, \Gamma : \beta \qquad \beta, \Gamma : \alpha}{\Gamma : \alpha \leftrightarrow \beta}$

These axioms and rules of inference yield a system of *natural deduction* in $\mathcal{L}$. If S *is* any collection of sequents in $\mathcal{L}$, we say that the sequent $\Gamma : \alpha$ is *deducible from* S, and write $\Gamma \vdash_S \alpha$ provided there is a deduction of $\Gamma : \alpha$ using the basic axioms, the sequents in S, and the rules of inference. We shall also write $\Gamma \vdash \alpha$ for $\Gamma \vdash_\emptyset \alpha$ and $\vdash_S \alpha$ for $\emptyset \vdash_S \alpha$.

A *local set theory* in $\mathcal{L}$ is a collection S of sequents closed under deducibility from S. Any collection of sequents S *generates* the local set theory $S*$ comprising all the sequents deducible from S. The local set theory in $\mathcal{L}$ generated by $\emptyset$ is called *pure* local set theory in $\mathcal{L}$.

5.3 Logic in a local set theory

The *logical operations* in $\mathcal{L}$ are defined as follows:

Logical Operation	**Definition**
$\top$ (true)	$\bigstar = \bigstar$
$\alpha \wedge \beta$	$\langle \alpha, \beta \rangle = \langle \top, \top \rangle$
$\alpha \rightarrow \beta$	$(\alpha \wedge \beta) \leftrightarrow \alpha$
$\forall x \alpha$	$\{x : \alpha\} = \{x : \top\}$
$\perp$ (false)	$\forall \omega . \omega$
$\neg \alpha$	$\alpha \rightarrow \perp$
$\alpha \vee \beta$	$\forall \omega [(\alpha \rightarrow \omega \wedge \beta \rightarrow \omega) \rightarrow \omega]$
$\exists x \alpha$	$\forall \omega [\forall x (\alpha \rightarrow \omega) \rightarrow \omega]$

We also write $x \neq y$ for $\neg(x = y), x \notin y$ for $\neg(x \in y)$, and $\exists ! x \alpha$ for $\exists x[\alpha \wedge \forall y(\alpha(x/y) \rightarrow x = y)]$.

It can now be shown that the logical operations on formulas just defined satisfy the axioms and rules of free intuitionistic logic. We present just a few of the relevant derivations.

We write

$$\frac{\Gamma_1 : \alpha_1, \ldots, \Gamma_n : \alpha_n}{\Delta : \beta}$$

for deducibility of $\Delta : \beta$ from $\Gamma_1 : \alpha_1, \ldots, \Gamma_n : \alpha_n$.

$$\vdash x = x.$$

Derivation: $: (x)_1 = x.$

$$\alpha \vdash \alpha$$

Derivation:

$$\dfrac{\dfrac{\omega, \omega = \omega : \omega \qquad \dfrac{}{: \omega : \omega}}{\omega : \omega}}{\alpha : \alpha}$$

$$\frac{\Gamma : a \quad \Gamma : \beta}{\Gamma : \alpha \wedge \beta}$$

Derivation:

$$\dfrac{\dfrac{}{\beta : \beta = \top} \qquad \dfrac{\dfrac{}{\alpha : \alpha = \top} \qquad \dfrac{}{\alpha = \top, \beta = \top : \alpha \wedge \beta}}{\alpha, \beta = \top : \alpha \wedge \beta}}{\dfrac{\Gamma : \alpha \qquad \Gamma, \beta : \alpha \wedge \beta}{\Gamma : \alpha \wedge \beta}}$$

$$\dfrac{\Gamma : \alpha \leftrightarrow \beta}{\gamma : \{x : \alpha\} = \{x : \beta\}}$$

Deriviation:

$$\dfrac{\dfrac{\dfrac{\Gamma : \alpha \leftrightarrow \beta}{\dfrac{/}{\alpha, \Gamma : \beta} \qquad \dfrac{\backslash}{\beta, \Gamma : \alpha}}}{x \in \{x : \alpha\}, \Gamma : x \in \{x : \beta\} \quad x \in \{x : \beta\}, \Gamma : x \in \{x : \alpha\}}}{\dfrac{\Gamma : x \in \{x : \alpha\} \leftrightarrow x \in \{x : \beta\}}{\Gamma : \{x : a\} = \{x : \beta}}$$

In any local set theory the following *Modified Cut Rule* holds:

(i) $$\dfrac{\Gamma : \alpha \qquad \alpha, \Gamma : \beta}{\exists x_1(x_1 = x_1), \ldots, \exists x_n(x_n = x_n), \Gamma : \beta}$$

where $x_1, \ldots, x_n$ are the free variables of α not occurring freely in Γ or β.

(ii) $$\dfrac{\Gamma : \alpha \qquad \alpha, \Gamma : \beta}{\Gamma : \beta}$$

provided that, whenever $\mathbf{A}$ is the type of a free variable of α with no free occurrences in Γ or β, there is a closed[24] term of type $\mathbf{A}$.

5.4 Set theory in a local language

We can now introduce the concept of *set* in a local language. A *set-like* term is a term of power type; a *closed* set-like term is called an $(\mathcal{L}-)$ *set*. We shall use upper case italic letters $X, Y, Z, \ldots$ for sets, as well as standard abbreviations such as $\forall x \in X.\alpha$ for $\forall x(x \in X \rightarrow \alpha)$. The set theoretic *operations* and *relations* are defined as follows. Note that in the definitions of $\subseteq$, $\cap$, and $\cup$, X and Y *must be of the same type*:

[24] A term is *closed* if it contains no free variables.

Operation	Definition
$\{x \in X : \alpha\}$	$\{x : x \in X \wedge \alpha\}$
$X \subseteq Y$	$\forall x \in X. x \in Y$
$X \cap Y$	$\{x : x \in X \wedge x \in Y\}$
$X \cup Y$	$\{x : x \in X \vee x \in Y\}$
$x \notin X$	$\neg(x \in X)$
U_A or A	$\{x_A : \top\}$
$\emptyset_A$ or $\emptyset$	$\{x_A : \bot\}$
$E - X$	$\{x : x \in E \wedge x \notin X\}$
PX	$\{u : u \subseteq X\}$
$\bigcap U$ (U : **PPA**)	$\{x : \forall u \in U. x \in u\}$
$\bigcup U$ (U : **PPA**)	$\{x : \exists u \in U. x \in u\}$
$\bigcap_{i \in I} X_i$	$\{x : \forall i \in I. x \in X_i\}$
$\bigcup_{i \in I} X_i$	$\{x : \exists i \in I. x \in X_i\}$
$\{\tau_1, \ldots, \tau_n\}$	$\{x : x = \tau_1 \vee \ldots \vee x = \tau_n\}$
$\{\tau : \alpha\}$	$\{z : \exists x_1 \ldots \exists x_n (z = \tau \wedge \alpha)\}$
$X \times Y$	$\{\langle x, y\rangle : x \in X \wedge y \in Y\}$
$X + Y$	$\{\langle \{x\}, \emptyset\rangle : x \in X\} \cup \{\langle \emptyset, \{y\}\rangle : y \in Y\}$
$Fun(X, Y)$	$\{u : u \subseteq X \times Y \wedge \forall x \in X \exists! y \in Y. \langle x, y\rangle \in u\}$

The standard facts concerning the set-theoretic operations and relations now follow as straightforward consequences of their definitions.

Given a term τ such that

$$\langle x_1, ..., x_n\rangle \in X \vdash_S \tau \in Y$$

we write $(\langle x_1, ..., x_n\rangle \mapsto \tau)$ or simply $\mathbf{x} \mapsto \tau$ for

$$\{\langle x_1, ..., x_n\rangle, \tau : \langle x_1, ..., x_n\rangle \in X\}.$$

Clearly we have

$$\vdash_S (\langle x_1, ..., x_n\rangle \mapsto \tau)) \in Fun(X, Y),$$

and so we may think of $(\langle x_1, ..., x_n\rangle \mapsto \tau)$ as the function from X to Y determined by τ.

We now show that each local set theory determines a topos. Let S be a local set theory in a local language $\mathcal{L}$. Define the relation $\approx_S$ on the collection of all $\mathcal{L}$-sets by

$$X \approx_S Y \text{ iff } \vdash_S X = Y.$$

This is an equivalence relation. An *S-set* is an equivalence class $[X]_S$ — which we normally identify with X — of $\mathcal{L}$-sets under the relation $\approx_S$. An *S-map* f: $X \to Y$ is a triple (f, X, Y) — normally identified with f — of S-sets such that $\vdash_S f \in Y^X$. X and Y are, respectively, the *domain* dom(f) and the *codomain* cod(f) of f. It is now readily shown that the collection of all S–sets and maps forms a category $\mathcal{C}(S)$, the *category of S-sets,* in which the composite of two maps $f : X \to Y$and $g : Y \to Z$ is given by

$$g \circ f = \{\langle x, z\rangle : \exists y(\langle x, y\rangle \in f \wedge \langle y, z\rangle \in g)\}.$$

In fact, $\mathcal{C}(S)$ is a *topos*, the *topos of sets determined by* S. It has terminal object $U_{\mathbf{1}}$, the product of two objects (S-sets) X, Y is the S-set $X \times Y$, with projections given by

$$\pi_1 = (\langle x, y\rangle \mapsto x) : X \times Y \to X, \quad \pi_2 = (\langle x, y\rangle \mapsto x) : X \times Y \to Y,$$

its truth-value object is $(\mathbf{\Omega}, t)$, where $t : 1 \to \mathbf{\Omega}$ is the S-map $\{\langle *, \top\rangle\}$, and the power object of an object X is $(PX,\ e_X)$, where $e_X : X \times PX \to \mathbf{\Omega}$ is the S-map $\langle x, z\rangle \mapsto x \in z$. All this is proved in much the same way as for classical set theory.

It can also be shown that properties of maps in $\mathcal{C}(S)$ are characterized in just the same way as in ordinary set theory, e.g. a map is monic iff it is one-to-one, epic iff it is onto, and an isomorphism iff it is both. This fact will be used implicitly in the sequel.

5.5 Interpreting a local language in a topos: the soundness and completeness theorems

Let $\mathcal{L}$ be a local language and $\mathcal{E}$ a topos. An *interpretation* I of $\mathcal{L}$ in $\mathcal{E}$ is an assignment:

- to each type $\mathbf{A}$, of an $\mathcal{E}$-object $\mathbf{A}_I$ such that:
  $(\mathbf{A}_1 \times \ldots \times \mathbf{A}_n)_I = (\mathbf{A}_1)_I \times \ldots \times (\mathbf{A}_n)_I$.
  $(\mathbf{PA})_I = P\mathbf{A}_I$,
  $\mathbf{1}_I = 1$, the terminal object of $\mathcal{E}$,
  $\mathbf{\Omega}_I = \mathbf{\Omega}$, the truth-value object of $\mathcal{E}$.
- to each function symbol $\mathbf{f} : \mathbf{A} \to \mathbf{B}$, an $\mathcal{E}$-arrow $\mathbf{f}_I : \mathbf{A}_I \to \mathbf{B}_I$.

We shall sometimes write $A_{\mathcal{E}}$ or just A for $\mathbf{A}_I$.

We extend I to terms of $\mathcal{L}$ as follows. If $\tau : \mathbf{B}$, write x for $(x_1, \ldots, x_n)$, any sequence of variables containing all variables of τ (and call such sequences *adequate* for τ). Define the $\mathcal{E}$-arrow

$$[\![\tau]\!]_x : A_1 \times \ldots \times A_n \longrightarrow B$$

recursively as follows:

$$\begin{aligned}
&[\![\bigstar]\!]_x = A_1 \times \ldots \times \longrightarrow 1\\
&[\![x_i]\!]_x = \pi_i : A_1 \times \ldots \times A_n \longrightarrow A_i\\
&[\![\mathbf{f}(\tau)]\!]_x = \mathbf{f}_1 \circ [\![\tau]\!]_x\\
&[\![\langle\tau_1, \ldots, \tau_n\rangle]\!]_x = \langle[\![\tau_1]\!]_x, \ldots, [\![\tau_n]\!]_x\rangle\\
&[\![(\tau)_i]\!]_x = \pi_i \circ [\![\tau]\!]_x\\
&[\![\{y : \alpha\}]\!]_x = ([\![\alpha(y/u)]\!]_{ux} \circ \ \mathit{can})\hat{}
\end{aligned}$$

where in this last clause u differs from $x_1, \ldots, x_n$, is free for y in α, y is of type $\mathbf{C}$, (so that B is of type $\mathbf{PC}$), *can* is the canonical isomorphism $C \times (A_1 \times ... \times A_n) \cong C \times A_1 \times ... \times A_n$, and $\hat{f}$ is as defined for power objects.[25] To understand why, consider the diagrams

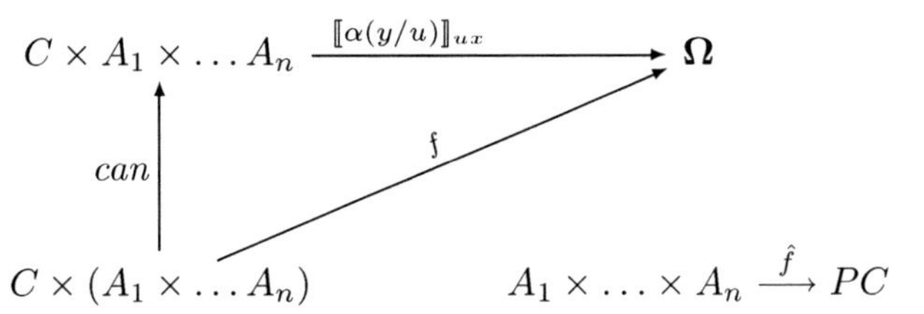

In set theory, $\hat{f}(a_1, \ldots, a_n) = \{y \in C : \alpha(y, a_1, ..., a_n)\}$, so we take $[\![\{y : \alpha\}]\!]_x$ to be $\hat{f}$.

Finally,

$$\begin{array}{lll} [\![\sigma = \tau]\!]_x = & eq_c \circ [\![\langle \sigma, \tau \rangle]\!]_x & \text{(with } \sigma, \tau : \mathbf{c}) \\ [\![\sigma \in \tau]\!]_x = & e_c \circ [\![\langle \sigma, \tau \rangle]\!]_x & \text{(with } \sigma : \mathbf{C}, \tau : \mathbf{PC} \text{ and where } e_c \text{ is as defined for} \\ & & \text{power objects)} \end{array}$$

If $\tau : \mathbf{B}$ is closed, then $\mathbf{x}$ may be taken to be the empty sequence $\emptyset$. In this case we write $[\![\tau]\!]$ for $\langle \tau \rangle_\emptyset$; this is an arrow $1 \to B$.

We note that

$$[\![\top]\!]_x = [\![\bigstar = \bigstar]\!]_x = eq \circ \langle [\![\bigstar]\!]_x, [\![\bigstar]\!]_x \rangle = T.$$

For any finite set $\Gamma = \{\alpha_1, \ldots, \alpha_m\}$ of formulas write

$$[\![\Gamma]\!]_{I,x} \text{ for } [\![\alpha_1]\!]_{I,x} \wedge \ldots \wedge [\![\alpha_m]\!]_{I,x} \text{ if } m \geq 1 \text{ or } T \text{ if } m = 0.$$

Given a formula α, let $\mathbf{x} = (x_1, ..., x_n)$ list all the free variables in $\Gamma \cup \{\alpha\}$; write

$$\Gamma \vDash_1 \alpha \text{ for } [\![\Gamma]\!]_{I,\mathbf{x}} \leq [\![\alpha]\!]_{I,\mathbf{x}}.$$

$\Gamma \vDash_I \alpha$ is read: "$\Gamma : \alpha$ is *valid* under the interpretation I in $\mathcal{E}$." If S is a local set theory, we say that I is a *model* of S if every sequent of S is valid under I. Notice that

$$\vDash_I \beta \text{ iff } [\![\beta]\!]_{\mathbf{x}} = T.$$

So if I is an interpretation in a *degenerate* topos, i.e., a topos possessing just one object up to isomorphism, then $\vDash_i \alpha$ for all α.

We write:

$$\begin{array}{lll} \Gamma \vDash \alpha & \text{for} & \Gamma \vDash_I \alpha \text{ for every } I \\ \Gamma \vDash_S \alpha & \text{for} & \Gamma \vDash_I \alpha \text{ for every model } I \text{ of } S. \end{array}$$

[25]See Appendix for definitions of all categorical notions used in the sequel.

It can be shown (laboriously) that the basic axioms and rules of inference of any local set theory are valid under every interpretation. This yields the

Soundness Theorem.

$$\Gamma \vdash \alpha \Rightarrow \Gamma \vDash \alpha \qquad \Gamma \vdash_S \alpha \Rightarrow \Gamma \vDash_S \alpha.$$

A local set theory S is said to be *consistent* if it is not the case that $\vdash_S \bot$. The soundness theorem yields the

Corollary. *Any pure local set theory is consistent.*

Proof. Set up an interpretation I of $\mathcal{L}$ in the topos $\mathcal{F}inset$ of finite sets as follows: $\mathbf{1}_I = 1, \mathbf{\Omega}_I = \{0,1\} = 2$, for any ground type $\mathbf{A}$, $\mathbf{A}_I$ is any nonempty finite set. Extend I to arbitrary types in the obvious way. Finally $\mathbf{f}_I : \mathbf{A}_I \rightarrow \mathbf{B}_I$ is to be any map from $\mathbf{A}_I$ to $\mathbf{B}_I$.

If $\vdash \bot$, then $\vdash \alpha$, so $\vDash_I \alpha$ for any formula α. In particular $\vDash_I u = v$, where u, v are variables of type $\mathbf{P1}$. Hence $[\![u]\!]_{I,uv} = [\![v]\!]_{I,uv}$, that is, the two projections $P1 \times P1 \rightarrow P1$would have to be identical, a contradiction. ■

Given a local set theory S in a language $\mathcal{L}$, define the *canonical interpretation* $C(S)$ of $\mathcal{L}$ in $\mathcal{C}(S)$ by:

$$\mathbf{A}_{C(S)} = U_A \ \mathbf{f}_{C(S)} = (x \mapsto \mathbf{f}(x)) : U_A \rightarrow U_B \text{ for } \mathbf{f} : \mathbf{A} \rightarrow \mathbf{B}.$$

A straightforward induction establishes

$$[\![\tau]\!]_{C(S),\mathbf{x}}, = (\mathbf{x} \mapsto \tau).$$

This yields

$$\Gamma \vDash_{C(S)} \alpha \Leftrightarrow \Gamma \vdash_S \alpha. \qquad (*)$$

For:

$$\begin{aligned} \vDash_{C(S)} \alpha &\Leftrightarrow [\![\alpha]\!]_{C(S)\mathbf{x}} = T \\ &\Leftrightarrow (\mathbf{x} \mapsto \alpha) = (\mathbf{x} \mapsto \top) \\ &\Leftrightarrow \vdash_S \alpha = \top \\ &\Leftrightarrow \vdash_S \alpha. \end{aligned}$$

Since $\Gamma \vdash_S \alpha \Leftrightarrow \vdash_S \gamma \rightarrow \alpha$, where γ is the conjunction of all the formulas in Γ, the special case yields the general one.

Equivalence (*) may be read as asserting that $\mathcal{C}(S)$ is a *canonical model* of S. This fact yields the

Completeness Theorem.

$$\Gamma \vDash \alpha \Rightarrow \Gamma \vdash \alpha \qquad \Gamma \vDash_S \alpha \Rightarrow \Gamma \vdash_S \alpha$$

Proof. We know that $C(S)$ is a model of S. Therefore, using (*),

$$\Gamma \vDash_S \alpha \Rightarrow \Gamma \vdash_{C(S)} \alpha \Rightarrow \Gamma \vdash_S \alpha.$$

■

5.6 *Every topos is linguistic: the equivalence theorem*

A topos of the form $\mathcal{C}(S)$ is called a *linguistic* topos. It can be shown that every topos is, in a certain sense, equivalent to a linguistic one. This fact was apparently established independently by a number of mathematicians: it appears for example, in Fourman [1977], Zangwill [1977], and Boileau and Joyal [1981].

Given a topos $\mathcal{E}$, we exhibit a theory $Th(\mathcal{E})$ and an equivalence $\mathcal{E} \simeq \mathcal{C}(Th(\mathcal{E}))$.

The local language $\mathcal{L}_E$ associated with $\mathcal{E}$ — also called the *internal language* of $\mathcal{E}$ — is defined as follows. The ground type symbols of $\mathcal{L}_\mathcal{E}$ are taken to match the objects of $\mathcal{E}$ other than its terminal and truth-value objects, that is, for each $\mathcal{E}$-object A (other than 1, $\mathbf{\Omega}$) we assume given a ground type $\mathbf{A}$ in $\mathcal{L}_\mathcal{E}$. Next, we define for each type symbol $\mathbf{A}$ an $\mathcal{E}$-object $\mathbf{A}_\mathcal{E}$ by

$$\begin{aligned}&\mathbf{A}_\mathcal{E} = \mathbf{A} \text{ for ground types } \mathbf{A},\\ &(\mathbf{A} \times \mathbf{B})_\mathcal{E} = \mathbf{A}_\mathcal{E} \times \mathbf{B}_\mathcal{E}{}^{26}\\ &(\mathbf{PA})_\mathcal{E} = P(\mathbf{A}_\mathcal{E}).\end{aligned}$$

The function symbols of $\mathcal{L}$ are triples $(f, \mathbf{A}, \mathbf{B}) = \mathbf{f}$ with $f : \mathbf{A}_\mathcal{E} \to \mathbf{B}_\mathcal{E}$ in $\mathcal{E}$. The signature of $\mathbf{f}$ is $\mathbf{A} \to \mathbf{B}$.[27]

The *natural interpretation* — denoted by $\mathcal{E}$ — of $\mathcal{L}_E$ in $\mathcal{E}$ is determined by the assignments

$$\mathbf{A}_\mathcal{E} = \mathbf{A} \text{ for each ground type } \mathbf{A} \qquad (f, \mathbf{A}, \mathbf{B})_\mathcal{E} = \mathbf{f}.$$

The local set theory $Th(\mathcal{E})$, the *theory of* $\mathcal{E}$, is the theory in $\mathcal{L}_\mathcal{E}$ generated by the collection of all sequents $\Gamma : \alpha$ such that $\Gamma \models_\mathcal{E} \alpha$ under the natural interpretation of $\mathcal{L}_\mathcal{E}$ in $\mathcal{E}$. Then we have

$$\Gamma \vdash_{Th(\mathcal{E})} \alpha \Leftrightarrow \Gamma \models_\mathcal{E} \alpha.$$

For if $\Gamma \vdash_{Th(\mathcal{E})} \alpha$ then by Soundness $\Gamma \models_{Th(\mathcal{E})} \alpha$, i.e., $\Gamma : \alpha$ is valid in every model of $Th(\mathcal{E})$. But by definition $\mathcal{E}$ is a model of $Th(\mathcal{E})$.

It can be shown that the canonical functor $F : \mathcal{E} \to C(Th(\mathcal{E}))$ defined by

$$\begin{aligned}FA &= U_\mathbf{A} \text{ for each } \mathcal{E}\text{-object } A\\ Ff &= (x \mapsto \mathbf{f}(x) : U_\mathbf{A} \to U_\mathbf{B} \text{ for each } \mathcal{E}\text{-arrow } f : A \to B\end{aligned}$$

is an equivalence of categories. This is the **Equivalence Theorem.**

Finally, we state two more facts about $Th(\mathcal{E})$.

A local set theory S in a language $\mathcal{L}$ is said to be

[26]Note that, if we write C for $A \times B$, then while $\mathbf{C}$ is a ground type, $\mathbf{A} \times \mathbf{B}$ is a product type. Nevertheless $\mathbf{C}_\mathcal{E} = (\mathbf{A} \times \mathbf{B})_\mathcal{E}$.

[27]Note the following: if $f : A \times B \to D$, in $\mathcal{E}$, then, writing C for $A \times B$ as in the footnote above, $(f, \mathbf{C}, \mathbf{D})$ and $(f, \mathbf{A} \times \mathbf{B}, \mathbf{D})$ are both function symbols of $\mathcal{L}_\mathcal{E}$ associated with f. But the former has signature $\mathbf{C} \to \mathbf{D}$, while the latter has the different signature $\mathbf{A} \times \mathbf{B} \to \mathbf{D}$.

- *well-termed* if whenever $\vdash_S \exists! x\alpha$, there is a term τ of $\mathcal{L}$ whose free variables are those of α with x deleted such that $\vdash_S \alpha(x/\tau)$;

- *well-typed* if for any S-set X there is a type symbol $\mathbf{A}$ of $\mathcal{L}$ such that $U_A \cong X$ in $\mathcal{C}(S)$.

Then, *for any topos $\mathcal{E}$, $Th(\mathcal{E})$ is well-termed and well-typed.*

5.7 Translations of local set theories

A *translation* $\mathbf{K} : \mathcal{L} \to \mathcal{L}'$ of a local language $\mathcal{L}$ into a local language $\mathcal{L}'$ is a map which assigns to each type $\mathbf{A}$ of $\mathcal{L}$ a type $\mathbf{KA}$ of $\mathcal{L}'$ and to each function symbol $\mathbf{f} : \mathbf{A} \to \mathbf{B}$ of $\mathcal{L}$ a function symbol $\mathbf{Kf}$: $\mathbf{KA} \to \mathbf{KB}$ of $\mathcal{L}'$ in such a way that

$$\mathbf{K1} = \mathbf{1},\ \mathbf{K\Omega} = \mathbf{\Omega},\ \mathbf{K}(\mathbf{A})_1 \times \ldots \times \mathbf{A}_n) = \mathbf{KA}_1 \times \ldots \times \mathbf{KA}_n,\ \mathbf{K}(\mathbf{PA}) = \mathbf{PKA}.$$

Any translation $\mathbf{K} : \mathcal{L} \to \mathcal{L}'$ may be extended to the terms of $\mathcal{L}$ in the evident recursive way — i.e., by defining $\mathbf{K}\bigstar = \bigstar$, $\mathbf{K}(\mathbf{f}(\tau)) = \mathbf{Kf}(\mathbf{k}\tau)$, $\mathbf{K}(\sigma \in \tau) = \mathbf{K}\sigma \in \mathbf{K}\tau$, etc — so that if $\tau : \mathbf{A}$, then $\mathbf{K}\tau : \mathbf{KA}$. We shall sometimes write α_K for $\mathbf{K}\alpha$.

If S, S' are local set theories in $\mathcal{L}, \mathcal{L}'$ respectively, a translation $\mathbf{K} : \mathcal{L} \to \mathcal{L}'$ is a *translation of* S into S', and is written $\mathbf{K} : S \to S'$ if, for any sequent $\Gamma : \alpha$ of $\mathcal{L}$,

$$\Gamma \vdash_S \alpha \Rightarrow \mathbf{K}\Gamma \vdash_{S'} \mathbf{K}\alpha, \tag{*}$$

where if $\Gamma = \{\alpha_1, ..., \alpha_n\}, \mathbf{K}\Gamma = \{\mathbf{K}\alpha_1, ..., \mathbf{K}\alpha_n\}$. If the reverse implication to (*) also holds, $\mathbf{K}$ is called a *conservative translation* of S into S'. If S' is an extension of S and the identity translation of S into S' is conservative, S' is called a *conservative extension of* S.

There is a natural correspondence between models of S in a topos $\mathcal{E}$ and translations of S into $Th(\mathcal{E})$: in particular the *identity translation* $Th(\mathcal{E}) \to Th(\mathcal{E})$ corresponds to the *natural interpretation* of $Th(\mathcal{E})$ in $\mathcal{E}$.

Now let $\mathcal{E}, \mathcal{E}'$ be toposes with specified terminal objects, products, projection arrows, truth-value objects, power objects and evaluation arrows: a functor $F : \mathcal{E} \to \mathcal{E}'$ which preserves all these is called a *logical functor*. It is easily seen that the canonical functor $\mathcal{E} \to \mathcal{C}(Th(\mathcal{E}))$ is logical.

If $\mathbf{K} : S \to S'$ is a translation, then for terms σ, τ of $\mathcal{L}, \vdash_S \sigma = \tau$ implies $\vdash_{S'} \mathbf{K}\sigma = \mathbf{K}\tau$, so that $\mathbf{K}$ induces a map C_K from the class of S-sets to the class of S'-sets via

$$C_{\mathbf{K}}([\sigma]_S) = [\mathbf{K}\sigma]_{S}.$$

$C_{\mathbf{K}}$ is actually a *logical functor* $C(S) \to C(S')$. Writing $\mathcal{L}$oc for the category of local set theories and translations, and $\mathcal{T}$op for the category of toposes and logical functors, C is a functor $\mathcal{L}\text{oc} \to \mathcal{T}\text{op}$. And reciprocally any logical functor $F : \mathcal{E} \to \mathcal{E}'$ induces a translation $Th(F) : Th(\mathcal{E}) \to Th(\mathcal{E}')$ in the natural way, so yielding a functor *Th*: $\mathcal{L}\text{oc} \to \mathcal{T}\text{op}$. C and *Th* are "almost" inverse, making $\mathcal{L}$oc and $\mathcal{T}$op "almost" equivalent.

Given a local set theory S in a language $\mathcal{L}$, define a translation $\mathbf{K} : \mathcal{L} \to \mathcal{L}_{\mathcal{L}(S)}$ by

$$\mathbf{KA} = U_A,\ \mathbf{Kf} = (f,\ \mathbf{A},\ \mathbf{B}) \text{ if } \mathbf{f} : \mathbf{A} \to \mathbf{B}.$$

An easy induction on the formation of terms shows that, for any term τ of $\mathcal{L}$, $[\![\tau]\!]_{\mathcal{C}(S)},\ x = [\![\mathbf{K}\tau]\!]_{\mathcal{C}(S)},\ \mathbf{K}_x$. It follows from this that $\mathbf{K}$ is a *conservative* translation of S into $Th(\mathcal{C}(S))$. For $\Gamma \vdash_S \alpha \Leftrightarrow \Gamma \vdash_{\mathcal{C}(S)} \alpha \Leftrightarrow \mathbf{K}\Gamma \vdash_{\mathcal{C}(S)} \mathbf{K}\alpha \Leftrightarrow \mathbf{K}\Gamma \vdash_{Th(\mathcal{C}(S))} \mathbf{K}\alpha$.

Accordingly *any local set theory can be conservatively embedded in one which is well-termed and well-typed.*

A particularly important instance of translation is the *adjunction of indeterminates* to a local set theory. Let us define a *constant* of type $\mathbf{A}$ in a local language $\mathcal{L}$ to be a term of the form $\mathbf{f}(\bigstar)$, where $\mathbf{f} : 1 \to \mathbf{A}$. Write $\mathcal{L}(\mathbf{c})$ for the language obtained from $\mathcal{L}$ by adding a new function symbol $\mathbf{c} : 1 \to \mathbf{A}$ and write c for $\mathbf{c}(\bigstar)$. Given a local set theory S in $\mathcal{L}$, and a formula α of $\mathcal{L}$ with exactly one free variable x of type $\mathbf{A}$, write $S(\alpha)$ for the theory in $\mathcal{L}(\mathbf{c})$ generated by S together with all sequents of the form: $\beta(x/c)$ where $\alpha \vdash_S \beta$. Since clearly $\vdash_{S(\alpha)} \alpha(x/c)$, it follows that $\vdash_{S(\alpha)} \exists x\alpha$. Also, an analysis of derivations establishes that proofs in $S(\alpha)$ are related to proofs in S by the condition:

$$\Gamma(x/c) \vdash_{S(\alpha)} \gamma(x/c) \Leftrightarrow \alpha, \Gamma \vdash_S \gamma. \tag{$*$}$$

for any sequent $\Gamma : \gamma$ of $\mathcal{L}$.

It is not hard to show that:

$$\vdash_S \exists x\alpha \quad \Leftrightarrow \quad S(\alpha) \text{ is a conservative extension of } S$$
$$\vdash_S \neg\exists x\alpha \quad \Leftrightarrow \quad S(\alpha) \text{ is inconsistent.}$$

In $S(\alpha), c$ behaves as an *indeterminate* of sort α in the sense that it can be arbitrarily assigned any value satisfying α. To be precise, the following can be proved:

Let S^* be a local set theory in a local language $\mathcal{L}^*$ and let $\mathbf{K} : S \to S^*$. Then for any constant c^* of $\mathcal{L}^*$ of type $\mathbf{KA}$ such tha $\vdash_{S^*} \alpha_{\mathbf{K}}(c^*)$, there is a unique translation $\mathbf{K}^* : S(\alpha) \to S$ extending $\mathbf{K}$ such that $\mathbf{K}^*(c) = c^*$.

If I is an S-set and α the formula $x \in I$, we write S_I for $S(\alpha)$ and call it the theory obtained from $S$by *adjoining an indeterminate element of I*. It follows from (*) above that, for any formula γ of $\mathcal{L}$ in which i is free for x,

$$\vdash_{S(I)} \gamma(x/c) \Leftrightarrow \vdash_S \forall i \in I\gamma(x/i).$$

If α is the formula $x = x$ with $x : \mathbf{A}$, then $S(\alpha)$ is written $S(\mathbf{A})$ and called the theory obtained from S by *adjoining an indeterminate of type* $\mathbf{A}$. In particular, let S_0 be the pure local set theory in the local language $\mathcal{L}_0$ with no ground types or function symbols. Evidently S_0 is an *initial object* in the category $\mathcal{L}$oc: there is a unique translation of $\mathcal{L}_0$ into any given local set theory S. (Similarly, the topos $C(S_0)$ is an initial object in the category top.) Now consider the theory $S_0(\mathbf{A})$, where $\mathbf{A}$ is a type symbol of $\mathcal{L}_0$: $\mathbf{A}$ may be considered a type symbol of *any* local language $\mathcal{L}$. If d is a constant of type $\mathbf{A}$ in $\mathcal{L}$, and S a local set theory in $\mathcal{L}$, there is then a unique translation $\mathbf{K} : S_0(\mathbf{A}) \to S$ mapping c to d. So $S_0(\mathbf{A})$ may be considered *the universal theory of an indeterminate of type* $\mathbf{A}$.

5.8 Classicality and the choice principle

Let S be a local set theory in a language $\mathcal{L}$. We make the following

Definitions.

- S is *classical* if $\vdash_S \forall\omega(\omega \vee \neg\omega)$. This is the full law of excluded middle for S.

- S is *sententially classical* if $\vdash_S \alpha \vee \neg\alpha$ for any sentence α. This is a weakened form of the law of excluded middle.

- For each S−set X : $\mathbf{PA}$ let $\Delta(X)$ be the set of closed terms τ such that $\vdash_S \tau \in X$. X is *standard* if for any formula α with at most the variable $x : \mathbf{A}$ free the following is valid:

$$\frac{\vdash_S \alpha(x/\tau) \text{ for all } \tau \text{ in } \Delta(X)}{\vdash_S \forall x \in X\alpha}$$

  S is *standard* if every S-set is so.

- S is *witnessed* if for any type symbol $\mathbf{A}$ of $\mathcal{L}$ and any formula α with at most the variable $x : \mathbf{A}$ free the following is valid:

$$\frac{\vdash_S \exists x\alpha}{\vdash_S \alpha(x/\tau) \text{ for some closed term } \tau : \mathbf{A}}$$

- S is *choice* if, for any S-sets X, Y and any formula α with at most the variables x, y free the following rule (the *choice rule*) is valid:

$$\frac{\vdash_S \forall x \in X \exists y \in Y\alpha(x, y)}{\vdash_S \forall x \in X\alpha(x, fx) \text{ for some } f : X \to Y}$$

- S is *internally choice* if under the conditions of the previous definition

$$\forall x \in X \exists y \in Y\alpha(x, y) \vdash_S \exists f \in \mathit{Fun}(X, Y)\forall x \in X \exists y \in Y[\alpha(x, y) \wedge \langle x, y\rangle \in f].$$

- An S-set X is *discrete* if

$$\vdash_S \forall x \in X\forall y \in X.x = y \vee x \neq y.$$

- A *complement* for an S-set $X : \mathbf{PA}$ is an S-set $Y : \mathbf{PA}$ such that

$$\vdash_S X \cup Y = A \wedge X \cap Y = \emptyset.$$

An S-set that has a complement is said to be *complemented.*

Now for some facts concerning these notions. In formulating our arguments we shall assume that our background metatheory is constructive, in that no use of the metalogical law of excluded middle will be made.

1. *Any of the following conditions is equivalent to the classicality of S:*

   (i) $\vdash_S \mathbf{\Omega} = \{\top, \bot\}$

   (ii) $\vdash_S \neg\neg\omega \to \omega$

   (iii) *any S-set is complemented,*

   (iv) *any S-set is discrete,*

   (v) $\mathbf{\Omega}$ *is discrete,*

   (vi) $\vdash_S 2 = \{0, 1\}$ *is well-ordered under the usual ordering,*

Proof. (iii) If S is classical, clearly $\{x : x \notin X\}$ is a complement for X. Conversely, if $\{\top\}$ has a complement U, then

$$\vdash_S \omega \in U \to \neg(\omega = \top) \to \neg\omega \to \omega = \bot.$$

Hence $\vdash_S U = \{\bot\}$, whence $\vdash_S \mathbf{\Omega} = \{\top\} \cup U = \{\top, \bot\}$.
(v) If $\mathbf{\Omega}$ is discrete, then $\vdash_S \omega = \top \vee \neg(\omega = \top)$, so $\vdash_S \omega \vee \neg\omega$.
(vi) If S is classical, then 2 is trivially well-ordered under the usual well-ordering. Conversely, if 2 is well-ordered, take any formula α, and define $X = \{x \in 2 : x = 1 \vee \alpha\}$. Then X has a least element, a, say. Clearly $\vdash_S a = 0 \leftrightarrow \alpha$, so, since $\vdash_S a = 0 \vee a = 1$, we get $\vdash_S a = 1 \leftrightarrow \neg\alpha$, and hence $\vdash_S \alpha \vee \neg\alpha$. ■

2. *For well-termed S, S choice $\Leftrightarrow$ S internally choice and witnessed.*

Proof. Suppose S is choice. If $\vdash_S \exists x\alpha$, let $u : 1$ and define $\beta(u, x) \equiv \alpha(x)$. Then $\vdash_S \forall u \in 1\exists x \in X\beta(u, x)$. Now choice yields an S-map $f : 1 \to X$ such that $\vdash_S \forall u \in 1\beta(u, f(u))$ i.e., $\vdash_S \beta(\bigstar, f\bigstar)$ or $\vdash_S \alpha(f\bigstar)$. By well-termedness, $f\bigstar$ may be taken to be a closed term τ, and we then have $\vdash_S \alpha(\tau)$. So S is witnessed.

To derive internal choice from choice, we argue as follows: let

$$X^* = \{x \in X : \exists y \in Y\alpha(x, y)\}.$$

Then $\vdash_S \forall x \in X^*\exists y \in Y\alpha(x, y)$. Accordingly choice yields a map $f : X^* \to Y$ such that $\vdash_S \forall x \in X^*\alpha(x, fx)$, i.e. $\vdash_S \forall x \in X^*\exists y \in Y[\langle x, y\rangle \in f \wedge \alpha(x, y)]$. Now

$$\forall x \in X\exists y \in Y\alpha(x, y)] \vdash_S X = X^* \vdash_S f \in \mathit{Fun}(X, Y)$$

so

$$\forall x \in X \exists y \in Y \alpha(x,y)] \vdash_S \forall x \in X \exists y \in Y[\langle x,y\rangle \in f \wedge \alpha(x,y)].$$

Hence

$$\forall x \in X \exists y \in Y \alpha(x,y) \vdash_S \exists f \in \mathit{Fun}(X,Y) \forall x \in X \exists y \in Y[\alpha(x,y) \wedge \langle x,y\rangle \in f],$$

as required. The converse is easy. ■

3. *If S is well-termed and well-typed, then S is choice $\Leftrightarrow S_X (= S(X))$ is witnessed for every S-set X.*

Proof. Suppose S is choice and $\vdash_{S(X)} \exists y \alpha(y)$. We may assume that X is of the form U_A, in which case α is of the form $\beta(x/c, y)$ with $x : \mathbf{A}$. From $\vdash_{S(X)} \exists y \beta(x/c, y)$ we infer $\vdash_S \forall x \exists y \beta(x/c, y)$. So using choice in S and the well-termedness of S we obtain a term $\tau(x)$ such that $\vdash_S \forall x \beta(x, \tau(x))$. Hence $\vdash_{S(X)} \beta(c, \tau(c))$, i.e., $\vdash_{S(X)} \alpha(\tau(c))$. Therefore S_X is witnessed.

Conversely, suppose S_X is witnessed for every S-set X, and that $\vdash_S \forall x \in X \exists y \in Y \alpha(x,y)]$. Then $\vdash_{S(X)} \exists y \in Y \alpha(c,y)]$, so there is a closed $\mathcal{L}_X$-term τ such that $\vdash_{S(X)} \tau \in Y \wedge \alpha(c, \tau)$. But τ is $\sigma(x/c)$ for some $\mathcal{L}$-term $\sigma(x)$. Thus $\vdash_{S(X)} \sigma(c) \in Y \wedge \alpha(c, \sigma(c))$, whence $\vdash_S \forall x \in X[\sigma(x) \in Y \wedge \alpha(x, \sigma)]$. Defining $f = (x \mapsto \sigma) : X \to Y$ then gives $\vdash_S \forall x \in X \alpha(x, fx)]$ as required. ■

4. **Diaconescu's Theorem**. *S choice $\Rightarrow$ S classical.*

   This is the version, for local set theories, of Diaconescu's [1975] result[28] that any topos in which the axiom of choice holds is Boolean.

Proof. Step 1. *S choice $\Rightarrow$ S_I choice for any S-set I.*
Proof of step 1. Suppose that S is choice, and

$$\vdash_{S(I)} \forall x \in X(c) \exists y \in Y(c) \alpha(x,y,c).$$

Then

$$\vdash_S \forall x \in X(i) \exists y \in Y(i) \alpha(x,y,i).$$

Define

$$X^* = \{\langle x,i\rangle : x \in X(i) \wedge i \in I\}, Y^* = \bigcup_{i \in I} Y(i).$$
$$\beta(u,y) \equiv \exists x \in X(i) \exists i \in I[u = \langle x,i\rangle \wedge \alpha(x,y,i) \wedge y \in Y(i)].$$

Then

$$\vdash_S \forall u \in X^* \exists y \in Y^* \beta(u,y).$$

So choice yields $f^* : X^* \to Y^*$ such that

$$\vdash_S \forall u \in X^* \beta(u, f^* u).$$

[28] First presented in purely logical form in Goodman and Myhill [1978].

i.e.,

$$\vdash_S \forall i \in I \forall x \in X(i)[\alpha(x, f^*(\langle x, i\rangle, i) \wedge f^*(\langle x, i\rangle) \in Y(i)],$$

whence

$$\vdash_S \forall x \in X(c)[\alpha(x, f^*(\langle x, c\rangle, c) \wedge f^*(\langle x, c\rangle) \in Y(c)],$$

Now define $f = (x \mapsto f^*(\langle x, c\rangle))$. Then $f : X(c) \to Y(c)$ in S_I and

$$\vdash_{S(I)} \forall x \in X(c)\alpha(x, fx, c).$$

This completes the proof of step 1.

Step 2. *S choice $\Rightarrow$ S sententially classical.*
Proof of step 2. Define $2 = \{0, 1\}$ and let $X = \{u \subseteq 2 : \exists y.y \in u\}$. Then

$$\vdash_S \forall u \in X \exists y \in 2.y \in u.$$

So by choice there is $f : X \to 2$ such that

$$\vdash_S \forall u \in X.fu \in u.$$

Now let α be any sentence; define

$$U = \{x \in 2 : x = 0 \vee \alpha\}, V = \{x \in 2 : x = 1 \vee \alpha\},$$

Then $\vdash_S U \in X \wedge V \in X$, so, writing $a = fU, b = fV$, we have

$$\vdash_S [a = 0 \vee \alpha] \wedge [b = 1 \vee \alpha],$$

whence

$$\vdash_S [a = 0 \wedge b = 1] \vee \alpha,$$

so that

$$\vdash_S a \neq b \vee \alpha. \qquad (*)$$

But $\alpha \vdash_S U = V \vdash_S a = b$, so that $a \neq b \vdash_S \neg\alpha$. It follows from this and (*) that

$$\vdash_S \alpha \vee \neg\alpha,$$

as claimed. This establishes step 2.

Moral of step 2: if pair sets have choice functions, then logic is classical.

Step 3 (obvious). *S classical $\Leftrightarrow$ S_Ω sententially classical.*
Finally, to prove Diaconescu's theorem, we observe that
S choice $\Rightarrow$ S_Ω choice $\Rightarrow$ S_Ω sententially classical $\Rightarrow$ S classical. ■

5.9 *Characterisation of* $\mathcal{S}et$

We remind the reader that we are assuming that our background metatheory is constructive. For definiteness we will take that metatheory to be intuitionistic set theory **IST.** Now consider the category $\mathcal{S}et$ of sets in **IST**. Its objects are all sets and its arrows all maps between sets. $\mathcal{S}et$ is a topos in with truth-value object $P1$. We seek a characterization of $\mathcal{S}et$ in terms of its associated local set theory, that is, *a characterization of the category of sets in type-theoretic terms.* This is to be carried out in a constructive manner.

To achieve this, we first define a local set theory S to be *full* if for each set I in the background metatheory there is a type symbol $\widehat{I}$ of the language $\mathcal{L}$ of S together with a collection $\{\widehat{i} : i \in I\}$ of closed terms each of type $\widehat{I}$ satisfying the following:

(i) $\vdash_S \widehat{i} = \widehat{j} \ \Rightarrow\ i = j$.

(ii) For any I-indexed family $\{\tau_i : i \in I\}$ of closed terms of common type $\mathbf{A}$, there is a term $\tau(x) : \mathbf{A}, x : \widehat{I}$ such that

$$\vdash_S \tau_i = \tau(\widehat{i}) \text{ for all } i \in I,$$

and, for any term $\sigma(x) : \mathbf{A}, x : \widehat{I}$, if $\vdash_S \tau_i = \sigma(\widehat{i})$ for all $i \in I$, then

$$\vdash_S \tau = \sigma.$$

We establish the *generalization principle for hatted type symbols: The following is valid for any formula* $\alpha(x)$ *with* $x : \widehat{I}$:

$$\frac{\vdash_S \alpha(\widehat{i}) \text{ for all } i \in I}{\vdash_S \forall x \alpha}$$

and similarly for more free variables. In particular, $\widehat{I}$ is standard.

Proof. Assume the premises. Then for any $i \in I$ we have $\vdash_S \alpha(\widehat{i}) = \top$ and it follows from the uniqueness condition that $\vdash_S \alpha(x) = \top$, whence have $\vdash_S \forall x \alpha$. ■

Note the following: (i) if S is well-termed and well-typed, then $\mathcal{C}(S)$ is full iff the *I-indexed sum of* 1 exists in $\mathcal{C}(S)$ for any (intuitive) set I. Here the I-indexed sum of an object X in a category $\mathcal{C}$ is an object $I \otimes X$ together with arrows $\sigma_i : X \to I \otimes X (i \in I)$ such that, for any arrows $f_i : X \to A$, there is a unique arrow $f : I \otimes X \to A$ such that $f_i = f \circ \sigma_i$ for all $i \in I$. In $\mathcal{S}et$, $I \otimes X$ is the disjoint union of I copies of X, so that, in particular, $I \otimes 1$ is essentially I itself.

(ii) S is standard iff 1 is a *generator* in $\mathcal{C}(S)$. Here an object A of a category $\mathcal{C}$ is a generator if, for any pair of $\mathcal{C}$-arrows $f, g : X \to Y$, $f = g$ whenever $f \circ h = g \circ h$ for all arrows $h : A \to X$.

Conditions (i) and (ii) are both satisfied by $\mathcal{S}et$.

We can now prove the

Theorem. *The following conditions on a well-termed, well-typed consistent local set theory S are equivalent:*

(i) *S is full and standard,*

(ii) $\mathcal{C}(S) \simeq \mathcal{S}et$.

Proof. Implication (ii) $\Rightarrow$ (i) has been discussed above. Now assume that S is full. Since S is well-termed, for any S-map $f : X \to Y$ we can write $f(\tau)$ for each closed term τ such that $\vdash_S \tau \in X$.

We define functors $\Delta : \mathcal{C}(S) \to \mathcal{S}et$, $\hat{} : \mathcal{S}et \to \mathcal{C}(S)$, which, under the specified conditions, we show to define an equivalence.

First, $\Delta(X)$ is the set of closed terms τ such that $\vdash_S \tau \in X$, where we identify σ, τ if $\vdash_S \sigma = \tau$. Given $f : X \to Y$, we define $\Delta(f)$ to be the map $(\tau \mapsto f(\tau)) : \Delta(X) \to \Delta(Y)$.

Next, given I in $\mathcal{S}et$, we define $\widehat{I}$ to be the S-set $U_{\widehat{I}}$. Given $f : I \to J$, there is a term $f(x) : \widehat{J}$ with $x : \widehat{I}$ such that that $\vdash_S \widehat{f}(\widehat{i}) = \widehat{f(i)}$ for all $i \in I$. We define $\widehat{f} : \widehat{I} \to \widehat{J}$ to be the S-map $x \mapsto \widehat{f}(x)$.

For any set I and any S-set X, we have natural maps $\eta_I : I \to \Delta(\widehat{I})$ and $\varepsilon : \widehat{\Delta(X)} \to X$ defined as follows:

$$\eta_I(i) = \widehat{i} \text{ for } i \in I; \vdash_S \varepsilon(\widehat{\tau}) = \tau \text{ for all } \tau \in \Delta(X).$$

Clearly η is monic. The same is true of ε since for $\sigma, \tau \in \Delta(X)$,

$$\vdash_S \varepsilon(\widehat{\sigma}) = \varepsilon(\widehat{\tau}) \to \sigma = \tau,$$

whence

$$\vdash_S \forall x \forall y [\varepsilon(x) = \varepsilon(y) \to x = y]$$

by generalization for hatted type symbols.

Now suppose that S is *standard*. We claim that then ε is epic and hence an isomorphism. For we have, for all $\tau \in \Delta(X), \vdash_S \varepsilon(\widehat{\tau}) = \tau$, whence $\vdash_S \exists y \varepsilon(y) = \tau$. Since X is standard, we infer that

$$\vdash_S \forall x \in X \exists y \varepsilon(y) = x,$$

so that ε is onto, hence epic.

Using the fact that ε is an isomorphism we can now show that η is epic, and hence also an isomorphism. To do this we require the readily established fact that, for $f : I \to J, \widehat{f} : \widehat{I} \to \widehat{J}$ is a map in S, and if $\widehat{f}$ is epic, then so is f.

Now consider $\widehat{\eta} : \widehat{I} \to \widehat{\Delta\widehat{I}}$. We note that

$$\varepsilon \circ \widehat{\eta} = 1_{\widehat{I}}. \qquad (*)$$

For if $i \in I$, then

$$\vdash_S \varepsilon(\widehat{\eta}(\widehat{i})) = \varepsilon(\widehat{\eta i}) = \eta i = \widehat{i}$$

It follows by generalization that

$$\vdash_S \forall x \in \widehat{I}\ \varepsilon(\eta x) = x,$$

whence (*).

Since ε is an isomorphism, it follows easily from (*) that $\widehat{\eta}$ is an isomorphism, hence also epic. Accordingly η is itself epic, and hence also an isomorphism.

We conclude that $(\Delta, \hat{\ })$ define an equivalence between $\mathcal{C}(S) \to \mathcal{S}et$, as required. ■

6 NEW FORMS OF TYPE THEORY AND THE DOCTRINE OF "PROPOSITIONS AS TYPES"

Type theory took a remarkable turn in the 1980s with the emergence of the *propositions as types* doctrine. Underlying this doctrine is the idealist notion, traceable to Kant, and central to Brouwerian intuitionism, that the meaning of a proposition does not derive from an absolute standard of truth external to the mind, but resides rather in the evidence for its assertability in the form of a mental construction or proof. Thus the central thesis of the "propositions-as-types" doctrine is that each proposition is to be *identified* with the type, set, or assemblage of its proofs.[29] As a result, such proof types, or sets of proofs, have to be accounted the *only* types, or sets. Strikingly, then, in the "propositions as types" doctrine, a type, or set, simply *is* the type, or set, of proofs of a proposition, and, reciprocally, a proposition *is* just the type, or set, of its proofs. These are truly radical identifications.

In a simple type theory or a local set theory, each type is independent of other types and is thus, so to speak, absolute or static; this holds in particular of the type of propositions or truth values. Now formulas or propositional functions in general manifest variation, since their values vary over, or depend on, the domain(s) of their free variables. Because of this they cannot be accurately represented as static types. This limitation makes it impossible for a simple type theory to realize faithfully the "propositions as types" doctrine. In order to achieve this it is necessary to develop a theory of "variable" or *dependent* types, wherein types can depend on, or "vary over" other types. In a dependent type theory, type symbols may take the form $\mathbf{B}(x)$, with x a variable of a given type $\mathbf{A} : \mathbf{B}(x)$ is then a type dependent on or varying over the type $\mathbf{A}$.

[29]This idea was advanced by Curry and Feys [1958] and later by Howard [1980]. As the *Curry-Howard correspondence* it has come to play an important role in theoretical computer science.

Such a theory — *constructive dependent type theory* (CDTT) — was introduced[30] by Per Martin-Löf [1975; 1982; 1984]. His theory, which has subsequently undergone much development, is also (as its name indicates) the first truly constructive theory of types, in the sense of being both predicative (so in particular it lacks a type of propositions) and based on intuitionistic logic. In introducing it Martin-Löf's purpose was to provide, as he put it in [1975] "a full scale system for formalizing intuitionistic mathematics as developed, for example, in the book by Bishop".[31] Martin-Löf's system provides a complete embodiment of the "propositions-as-types" doctrine.[32] Here is Martin-Löf himself on the latter in [1975]:

> Every mathematical object is of a certain kind or type. Better, a mathematical object is always given together with its type, that is it is not just an object: it is an object of a certain type. ... A type is defined by prescribing what we have to do in order to construct an object of that type... Put differently, a type is well-defined if we understand...what it means to be an object of that type. ... Note that it is required, neither that we should be able to generate somehow all the objects of a given type, nor that we should so to say know all of them individually. It is only a question of understanding what it means to be an arbitrary object of the type in question.
>
> A proposition is defined by prescribing how we are allowed to prove it, and a proposition holds or is true intuitionistically if there is a proof of it. ... Conversely, each type determines a proposition, namely, the proposition that the type in question is nonempty. This is the proposition which we prove by exhibiting an object of the type in question. On this analysis, there appears to be no fundamental difference between propositions and types. Rather, the difference is one of point of view: in the case of a proposition, we are not so much interested in what its proofs are as in whether it has a proof, that is, whether it is true or false, whereas, in the case of a type, we are of course interested in what its objects are and not only in whether it is empty or nonempty.

A key element in Martin-Löf's formulation of type theory is the distinction, which goes back to Frege, between *propositions* and *judgments.* Propositions (which, as we have seen, in Martin-Löf's systems are identified with types) are syntactical objects on which mathematical operations can be performed and which

[30]Dependent types were actually first studied in the late 1960s by de Bruijn and his colleagues at the University of Eindhoven in connection with the AUTOMATH project. CDTT has been employed as a basis for various computational devices employed for the verification of mathematical theories and of software and hardware systems in computer science.

[31]I.e. Bishop [1967].

[32]Martin-Löf's original calculus contained a type of all types. This assumption was shown to be inconsistent by Girard [1972]. Martin-Löf accordingly dropped this assumption in later versions of his theory.

bear certain formal relationships to other syntactical objects called proofs. Propositions and proofs are, so to speak, *objective* constituents of the system. Judgments, on the other hand, typically involve the *idealist* notion of "understanding" or "grasping the meaning of". Thus, for example, while $2 + 2 = 4$ is a proposition, "$2 + 2 = 4$ is a proposition" and "$2 + 2 = 4$ is a true proposition" are judgments.

Martin-Löf also follows Frege in taking the rules of inference of logic to concern judgments rather than propositions. Thus, for example, the correct form of the rule of $\rightarrow$-elimination is not

$$\frac{A \quad A \rightarrow B}{B}$$

but

$$\frac{A \text{ true} \quad A \rightarrow B \text{ true}}{B \text{ true}}.$$

That is, the rule does not say that the proposition B follows from the propositions A and $A \rightarrow B$, but that the *truth* of the proposition B follows from the *truth* of the proposition A conjoined with that of $A \rightarrow B$. In general, judgments may be characterized as expressions which appear at the conclusions of rules of inference.

Another important respect in which Martin-Löf follows Frege is in his insistence that judgments and formal rules be accompanied by full explanations of their *meaning*. (This is to be contrasted with the usual model-theoretic semantics which is really nothing more than a translation of one object-language into another.) In particular, the judgment A *is a proposition* may be made only when one knows what a (canonical) proof of A is, and the judgment A *is a true proposition* only when one knows how to find such a proof. Judgments, and the notion of truth, are thus seen to be mind-dependent.

Martin-Löf's various systems abound in subtle distinctions. For example, in addition to the distinction between proposition and judgment, there is a parallel distinction between *type* (or set) and *category*[33] (or species). In order to be able to judge that $A$is a category one must be able to tell what kind of objects fall under it, and when they are equal. To be in a position to make the further judgment that a category is a type, or set, one must be able to specify what its "canonical" or typical, elements are. In judging something to be a set, one must possess sufficient information concerning the its elements to enable quantification over it to make sense. Thus, for example, the natural numbers form a set $\mathbf{N}$, with canonical elements given by: 0 is a canonical element of $\mathbf{N}$, and if n is a canonical element of $\mathbf{N}$, then $n + 1$ is a canonical element of $\mathbf{N}$. On the other hand the collection of subsets of $\mathbf{N}$ forms a category, but not a set.

The "propositions-as-types" doctrine (which for convenience we abbreviate to PAT) gives rise to a correspondence between logical operators and operations on (dependent) types . Let us follow Tait's [1994] exposition of the idea in set-theoretic terms. To begin with, consider two propositions/types/sets $\mathbf{A}$ and $\mathbf{B}$.

[33]In this usage, of course, to be distinguished from the term as employed in its mathematical sense throughout the present article.

What should be required of a proof f of the implication $\mathbf{A} \to \mathbf{B}$? Just that, given any proof x of $\mathbf{A}$, f should yield a proof of $\mathbf{B}$, that is, f should be a function from $\mathbf{A}$ to $\mathbf{B}$. In other words, the proposition $A \to B$ is just the type of functions from $\mathbf{A}$ to $\mathbf{B}$:

$$\mathbf{A} \to \mathbf{B} = \mathbf{B}^A$$

Similarly, all that should be required of a proof c of the conjunction $\mathbf{A} \wedge \mathbf{B}$ is that it should yield proofs x and y of $\mathbf{A}$ and $\mathbf{B}$, respectively. From this point of view $\mathbf{A} \wedge \mathbf{B}$ is accordingly just the type $\mathbf{A} \times \mathbf{B}$ of all pairs (x, y), with x of type $\mathbf{A}$ (we write this as $x : \mathbf{A}$) and $y : \mathbf{B}$.

A proof of the disjunction $\mathbf{A} \vee \mathbf{B}$ is either a proof of $\mathbf{A}$ or a proof of $\mathbf{B}$ together with the information as to which of $\mathbf{A}$ or $\mathbf{B}$ it is a proof. That is, if we introduce the type $\mathbf{2}$ with the two distinct elements 0 and 1, a proof of $\mathbf{A} \vee \mathbf{B}$ may be identified as a pair (c, n) in which either c is a proof of $\mathbf{A}$ and n is 0, or c is a proof of $\mathbf{B}$ and n is 1. This means that $\mathbf{A} \vee \mathbf{B}$ should be construed as the disjoint union $\mathbf{A} + \mathbf{B}$ of $\mathbf{A}$ and $\mathbf{B}$.

The true proposition $\top$ may be identified with the one element type $\mathbf{1} = \{0\} : 0$ thus counts as the unique proof of $\top$. The false proposition $\bot$ is taken to be a proposition which lacks a proof altogether: accordingly $\bot$ is identified with the empty set $\emptyset$. The negation $\neg\mathbf{A}$ of a proposition $\mathbf{A}$ is defined as $\mathbf{A} \to \bot$, which therefore becomes identified with the set $\mathbf{A}^{\emptyset}$.

As we have already said, a proposition A is deemed to be true if it (i.e, the associated type $\mathbf{A}$) has an element, that is, if there is a function $\mathbf{1} \to \mathbf{A}$.Accordingly the *law of excluded middle* for a proposition A becomes the assertion that there is a function $\mathbf{1} \to \mathbf{A} + \emptyset^A$.

If a and b are objects of type $\mathbf{A}$, we introduce the *identity proposition* or *type* $a =_A b$ expressing that a and b are identical objects of type A. This proposition is true, that is, the associated type has an element, if and only if a and b are identical. In that case $\mathrm{id}(a)$ will denote an object of type a.

In PAT one must, as in Martin-Löf's system, distinguish sharply between *propositions*, which have proofs, and *judgments*, which do not. For example $0 =_2 0$ is a proposition, while "0 is of type $\mathbf{2}$" is a judgment. Rather than being true or false, a judgment is either assertable, or nonsensical.

In order to deal with the quantifiers we require operations defined on families of types, that is, types $\Phi(x)$ depending on objects x of some type $\mathbf{A}$. By analogy with the case $\mathbf{A} \to \mathbf{B}$, a proof f of the proposition $\forall x : \mathbf{A}\Phi(x)$, that is, an object of type $\forall x : \mathbf{A}\Phi(x)$, should associate with each $x : \mathbf{A}$ a proof of $\Phi(x)$. So f is just a function with domain A such that, for each $x : \mathbf{A}$, fx is of type $\Phi(x)$. That is, $\forall x : \mathbf{A}\Phi(x)$ is the *product* $\Pi x : \mathbf{A}\Phi(x)$ of the $\Phi(x)$s. We use the λ-notation in writing f as $\lambda x fx$.

A proof of the proposition $\exists x : \mathbf{A}\Phi(x)$, that is, an object of type $\exists x : \mathbf{A}\Phi(x)$, should determine an object $x : \mathbf{A}$ and a proof y of $\Phi(x)$, and *vice-versa*. So a proof of this proposition is just a pair (x, y) with $x : \mathbf{A}$ and $y : \Phi(x)$. Therefore $\exists x : \mathbf{A}\Phi(x)$ is the *disjoint union*, or *coproduct* $\coprod x : \mathbf{A}\Phi(x)$ of the $\Phi(x)$s.

To translate all this into the language of CDTT,[34] one uses the following concordance:

Logical Operation	Set-theoretic Operation	Type-theoretic Operation
$\wedge$	$\times$	$\times$
$\vee$	$+$	two-term dependent sum
$\rightarrow$	set exponentiation	type exponentiation
$\forall x$	Cartesian product $\prod_{i \in I}$	dependent product $\Pi x{:}\mathbf{A}$
$\exists x$	disjoint sum $\coprod_{i \in I}$	dependent sum $\coprod x{:}\mathbf{A}$

Of especial interest is the status of the *axiom of choice* in this framework. Again following Tait, we introduce the functions σ, π, π' of types $\forall x : \mathbf{A}(\Phi(x) \rightarrow \exists x : \mathbf{A}\Phi(x)), \exists x : \mathbf{A}\Phi(x) \rightarrow \mathbf{A}$, and $\forall y : (\exists x\Phi(x)).\Phi(\pi(y))$ as follows. If $b : \mathbf{A}$ and $c : \Phi(b)$, then σbc is (b, c). If $d : \exists x : \mathbf{A}\Phi(x)$, then d is of the form (b, c) and in that case $\pi(d) = b$ and $\pi'(d) = c$. These yield the equations

$$\pi(\sigma bc) = b\pi'(\sigma bc) = c\sigma(\pi d)(\pi\prime d) = d.$$

We may take the axiom of choice as the proposition

$$\forall x : \mathbf{A}\exists y : \mathbf{B}\Phi(x, y)) \rightarrow \exists f : B^A\forall x : \mathbf{A}\Phi(x, fx)). \tag{AC}$$

Remarkably, AC is correct under PAT, that is, *provable* in CDTT , as the following argument shows. Let u be a proof of the antecedent $\forall x : \mathbf{A}\exists y : \mathbf{B}\Phi(x, y))$. Then, for any $x : \mathbf{A}, \pi(ux)$ is of type $\mathbf{B}$ and $\pi'(ux)$ is a proof of $\Phi(x, \pi ux)$. So $s(u) = \lambda x.\pi(ux)$ is of type $\mathbf{B}^A$ and $t(u) = \lambda x.\pi'(ux)$ is a proof of $\forall x : \mathbf{A}\Phi(x, s(u)x)$. Accordingly $\lambda u.\sigma s(u)t(u)$ is a proof of $\forall x : \mathbf{A}\exists y : \mathbf{B}\Phi(x, y)) \rightarrow \exists x : \mathbf{B}^A\forall x : \mathbf{A}\Phi(x, fx))$.

Put informally, what this shows is that in CDTT the consequent of AC means *nothing more than its antecedent.* Indeed, in many versions of constructive mathematics the assertability of an alternation of quantifiers $\forall x\exists yR(x, y)$ means *precisely* that one is given a function f for which $R(x, fx)$ holds for all x.

We note that in ordinary set theory this argument establishes the *isomorphism* of the sets $\prod x : A \coprod y : \mathbf{B}\Phi(x, y))$ and $\coprod f : \mathbf{B}^A \prod x : \mathbf{A}\Phi(x, fx))$, but not the validity of the usual axiom of choice. In set theory AC is not represented by this isomorphism, but is rather (equivalent to) the equality in which $\prod$ is replaced by $\bigcap$ and $\coprod$ by $\bigcup$, namely

$$\bigcap_{x \in A} \bigcup_{y \in B} \Phi(x, y) = \bigcup_{f \in B^A} \bigcap_{x \in A} \Phi(x, fx).$$

[34]For a complete specification of the operations and rules of CDTT, see Chapter 10 of Jacobs [1999] or Gambino and Aczel [2005].

While in PAT, AC is provable, and so *a fortiori* has no "untoward" logical consequences, in intuitionistic set theory, or in the internal language of a topos this is far from being the case, for, as Diaconescu's theorem shows, in the latter AC implies the law of excluded middle. In other words, AC interpreted à la "propositions as types" is tautologous,[35] while construed set-, or topos-theoretically it is anything but, since so construed its affirmation yields classical logic. This prompts the question: what modification needs to be made to the "propositions-as-types" paradigm so as to yield the topos-theoretic interpretation of AC? An illuminating answer to this question has been given by Maietti [2005] through the use of so-called *monotypes* (or mono-objects), that is, (dependent) types containing at most one entity or having at most one proof. In $\mathcal{S}et$, mono objects are *singletons*, that is, sets containing at most one element.

Monotypes correspond to monic maps. This can be illustrated concretely by considering the categories $\mathcal{I}ndset$ of *indexed* sets and $\mathcal{S}et^{\rightarrow}$ of *bivariant* sets. The objects of $\mathcal{I}ndset$ are indexed sets of the form $M = \{\langle i, M_i\rangle : i \in I\}$ and those of $\mathcal{S}et^{\rightarrow}$ maps $A \rightarrow B$ in $\mathcal{S}et$, with appropriately defined arrows in each case. It can be shown that these two categories are equivalent. If we think of (the objects of) $\mathcal{S}et$ as representing simple or static types, then (the objects of) $\mathcal{I}ndset$, and hence also of $\mathcal{S}et^{\rightarrow}$, represent dependent or variable types. It is easily seen that a monotype, or object, in $\mathcal{I}ndset$, is precisely an object M for which each M_i has at most one element. Moreover, under the equivalence between $\mathcal{I}ndset$ and $\mathcal{S}et^{\rightarrow}$, such an object corresponds to a monic map- object in $\mathcal{S}et^{\rightarrow}$.

Now consider $\mathcal{S}et^{\rightarrow}$ as a topos. Under the topos-theoretic interpretation in $\mathcal{S}et^{\rightarrow}$, formulas correspond to monic arrows, which in turn correspond to mono-objects in $\mathcal{I}ndset$. Carrying this over entirely to $\mathcal{I}ndset$ yields the sought modification of the "propositions-as-types" paradigm to bring it into line with the topos-theoretic interpretation of formulas, namely, to take formulas or propositions to correspond to *mono*-objects, rather than to *arbitrary* objects. Let us call this the "formulas-as-monotypes" interpretation.

Finally let us reconsider AC under the "formulas-as-monotypes" interpretation within $\mathcal{S}et$. It will be convenient to rephrase AC as the assertion

$$\forall i \in I \exists j \in J M_{ij} \leftrightarrow \exists f \in J^I \forall i \in I M_{if(i)} \qquad (*)$$

where $\langle M_{ij} : i \in I, j \in J\rangle$ is any doubly indexed family of propositions (or sets). In the "propositions as types" interpretation, where (*) corresponds to the existence of an isomorphism between $\prod_{i\in I}\coprod_{j\in J} M_{ij}$and $\coprod_{f\in J^I}\prod_{i\in I} M_{if(i)}$. On the other hand, AC

[35] Precisely as Ramsey (*v. supra*) asserted, but in this case for quite different reasons. Ramsey construed, and accepted the truth of, the axiom of choice as asserting the objective existence of choice functions, given extensionally and so independently of the manner in which they might be described. But the intensional nature of constructive mathematics, and, in particular, of CDTT decrees that nothing is given completely independently of its description. This leads to a strong construal of the quantifiers which, as we have observed, trivializes the axiom of choice by rendering the antecedent of the implication constituting it essentially equivalent to the consequent. It is remarkable that the axiom of choice can be considered tautological both from an extensional and from an intensional point of view.

interpreted in the usual way, that is, using the rules of topos semantics, can be presented in the form of the distributive law

$$\bigcap_{i\in I}\bigcup_{j\in J} M_{ij} = \bigcup_{f\in J^I}\bigcap_{i\in I} M_{if(i)}. \tag{**}$$

In the "propositions-as-types" interpretation (as applied to $\mathcal{S}et$), the universal quantifier $\forall i \in I$ corresponds to the product $\prod_{i\in I}$ and the existential quantifier $\exists i \in I$ to the coproduct, or disjoint sum, $\coprod_{i\in I}$. Now in the "formulas-as-monotypes" interpretation, wherein formulas correspond to singletons, $\forall i \in I$ continues to correspond to $\prod_{i\in I}$, since the product of singletons is still a singleton. But the interpretation of $\exists i \in I$ is changed. In fact, the interpretation of $\exists i \in IM_i$ (with each M_i a singleton) now becomes $[\coprod_{i\in I} M_i]$, where for each set $X, [X] = \{u : u = 0 \wedge \exists x.x \in X\}$ is the *canonical singleton* associated with X.

It follows that, under the "formulas-as-monotypes" interpretation, the proposition $\forall i \in I\exists j \in JM_{ij}$ is interpreted as the singleton

$$\prod_{i\in I}[\coprod_{j\in J} M_{ij}] \tag{1}$$

and the proposition $\exists f \in J^I\forall i \in IM_{if(i)}$ as the singleton

$$[\coprod_{f\in J^I}\prod_{i\in I} M_{if(i)}]. \tag{2}$$

Under the "formulas-as-monotypes" interpretation AC would be construed as asserting the existence of an isomorphism between (1) and (2).

Now it is readily seen that to give an element of (1) amounts to no more than affirming that, for every $i \in I$, $\bigcup_{j\in J} M_{ij}$ is nonempty. But to give an element of (2) amounts to specifying maps $f \in J^I$ and g with domain I such that $\forall i \in Ig(i) \in M_{if(i)}$. It follows that to assert the existence of an isomorphism between (1) and (2), that is, to assert AC under the "formulas-as-monotypes" interpretation, is tantamount to asserting AC in the form (**), so leading in turn to classical logic. This is in sharp contrast with AC under the "propositions-as-types" interpretation, where, as we have seen, its assertion is automatically correct and so has no nonconstructive consequences.

In the "propositions as types" interpretation each logical operation corresponds to a categorical operation: $\wedge$ to $\times$, $\vee$ to $+$, $\rightarrow$ to exponentiation, $\perp$ to an initial element 0, $\neg$ to exponentiation by 0, $\forall$ to product and $\exists$ to coproduct. This suggests that CDTT should be interpretable in suitable categories, just as local set theories are interpretable in toposes. For some time it was conjectured that the appropriate categories in this respect were the so-called *locally Cartesian closed* categories:

these are finitely complete categories $\mathcal{E}$ such that, for each object A of $\mathcal{E}$, the "slice" category $\mathcal{E}/A$ is Cartesian closed. (Any topos is locally Cartesian closed.) It was already known that many of the mathematical constructions within a topos could be carried out within a locally Cartesian closed category. In such a category, the notion of "variable set", for example an "A-indexed set" , is represented by an arrow $B \to A$ of $\mathcal{E}$, that is, by an object of $\mathcal{E}/A$. In the interpretation of CDTT that was explicitly worked out by Seely [1984], types are interpreted as objects, and terms as arrows, of a locally Cartesian closed category $\mathcal{E}$. As Seely points out, this interpretation engenders a systematic ambiguity among the notions of type, predicate and term, and between object and proof: indeed a term of type A is an arrow into A, which is in turn a predicate over A, and an arrow $1 \to A$ may be regarded either as an object of type A or as a proof of the proposition A. (This ambiguity is, of course, shared by the "propositions as types" interpretation.) In precise analogy with the correspondence between local set theories and toposes, Seely establishes a correspondence between theories formulated in Martin-Löf's system — constructive dependent type theories — and locally Cartesian closed categories. The relationship between type theories and categories is investigated in a general setting in Jacobs [1999].

It is natural to ask what form of *set theory* is interpretable in CDTT. This has been worked out by Peter Aczel[36] and has become come known as *constructive set theory* (CST). This is a system of intuitionistic set theory in which the set-theoretic operations in reflect those on types in CDTT. n particular CST will admit products and exponentials of sets. But, given the predicative nature of CDTT, CST must also be predicative. This means, in particular, that in CST there can be no set of propositions, and no power sets. A particularly illuminating form of CST[37] employs the distinction between sets and classes, as in Gödel-Bernays set theory. In this version of CST, any predicate determines a class, but only certain classes are sets. Here a class may be thought of as something like a species or category in Martin-Löf's sense, that is, a range of objects all of which share a given property, and whose identity relations are fully determinate, but whose "extent" is not sufficiently fixed to admit quantification over it. A set, on the other hand, is a class with a fully determinate extent, hence supporting quantification. In CST, the class PA of all subsets — the power class — of a set A is almost never a set (the only exception being the power class of the empty set). On the other hand the exponential A^B of two sets A, B — the class of maps from B to A — is always a set, so in particular 2^A is always a set. In classical set theory 2^A is essentially the power set of A, which cannot be the case in CST. This means that in CST 2 cannot represent, as it does in classical set theory, the set of propositions or truth values. In classical set theory 2 is P1, the power set of a one-element set, but in CST P1, while of course a class, is not a set. It can be shown that classical set theory is equivalent to CST augmented by the law of excluded middle and the assertion that P1 is a set.

[36] See bibliographic entries under Aczel, Aczel and Gambino, and Aczel and Rathjen.

[37] See Aczel and Rathjen [2201].

Finally, a word on other forms of type theory that have emerged in the past few decades. *Polymorphic* type theory admits the presence of *type variables*, so that a type variable **X** may occur inside a type **A(X)**. The assignment of "types" to *type* variables (or terms) is achieved by the introduction of a new order of entities called *kinds*. Just as each ordinary term is assigned a type, so each type term is assigned a kind. In particular, there is a kind called **Type** which represents the "kind of all types".[38] Any type **A** then satisfies the judgment **A: Type**, and **Type** itself satisfies the judgment **Type: Kind.** In addition to **Type**, the presence of arbitrarily arbitrary "subkinds" of **Type** may be assumed. It is also assumed that **Kind** is closed under the operations corresponding to cartesian product and exponentiation, and that **Type** is closed under these latter and those corresponding to product and sum over type variables of an arbitrarily given kind.

Polymorphic and dependent type theories have been combined in various ways. One form is the *Calculus of Constructions* of Coquand and Huet [1988]. This has been extended to what Jacobs [1999] calls *Full Higher Order Dependent Type Theory*, or the *full theory* for short. In dependent type theory types can "depend on" types; in polymorphic type theory types can "depend on" kinds. In the full theory, the dependence is extended to the remaining possibilities: those of kinds on types and kinds on kinds. An account of this, and other amalgamations of polymorphic and dependent type theories, may be found in Jacobs [1999].

APPENDIX: BASIC CONCEPTS OF CATEGORY THEORY

A *category* $\mathcal{C}$ is determined by first specifying two classes $Ob(\mathcal{C}), Arr(\mathcal{C})$ — the collections of $\mathcal{C}$-*objects* and $\mathcal{C}$-*arrows*. These collections are subject to the following axioms:

- Each $\mathcal{C}$-arrow f is assigned a pair of $\mathcal{C}$-objects $\text{dom}(f)$, $\text{cod}(f)$ called the *domain* and *codomain* of f, respectively. To indicate the fact that $\mathcal{C}$-objects X and Y are respectively the domain and codomain of f we write $f : X \to Y$ or $X \xrightarrow{f} Y$. The collection of $\mathcal{C}$-arrows with domain X and codomain Y is written $\mathcal{C}(X, Y)$.
- Each $\mathcal{C}$-object X is assigned a $\mathcal{C}$-arrow $1_X : X \to X$ called the *identity arrow* on X.
- Each pair f, g of $\mathcal{C}$-arrows such that $\text{cod}(f) = \text{dom}(g)$ is assigned an arrow $g \circ f$: $\text{dom}(f) \to \text{cod}(g)$ called the *composite* of f and g. Thus if $f : X \to Y$ and $g : Y \to Z$ then $g \circ f : X \to Z$. We also write $X \xrightarrow{f} Y \xrightarrow{g} Z$ for $g \circ f$. Arrows f, g satisfying $\text{cod}(f) = \text{dom}(g)$ are called *composable.*
- *Associativity law.* For composable arrows (f, g) and (g, h), we have $h \circ (g \circ f) = h \circ (g \circ f)$.

[38]This is, of course, not the same as asserting that there is a *type* of all types, for to do so would be to rekindle Girard's paradox (see note 25 above).

- *Identity law.* For any arrow $f : X \to Y$, we have $f \circ 1_X = f = 1_Y \circ f$.

As a basic example of a category, we have the category $\mathcal{S}et$ of sets whose objects are all sets and whose arrows are all maps between sets (strictly, triples (f, A, B) with domain$(f) = A$ and range$(f) \subseteq B$.) Other examples of categories are the category of groups, with objects all groups and arrows all group homomorphisms and the category of topological spaces with objects all topological spaces and arrows all continuous maps. Categories with just one object may be identified with *monoids*, that is, algebraic structures with an associative multiplication and an identity element.

A *subcategory* $\mathcal{C}$ of a category $\mathcal{D}$ is any category whose class of objects and arrows is included in the class of objects and arrows of $\mathcal{D}$, respectively, and which is closed under domain, codomain, identities, and composition. If, further, for any objects C, C' of $\mathcal{C}$, we have $\mathcal{C}(C, C') = \mathcal{D}(C, C')$, we shall say that $\mathcal{C}$ is a *full* subcategory of $\mathcal{D}$.

Basic category-theoretic definitions

Commutative diagram (in category)	Diagram of objects and arrows such that the arrow obtained by composing the arrows of any connected path depends only on the endpoints of the path.
Initial object	Object 0 such that, for any object X, there is a unique arrow $0 \to X$ (e.g., $\emptyset$ in $\mathcal{S}et$)
Terminal object	Object 1 such that, for any object X, there is a unique arrow $X \to 1$ (e.g. any singleton in $\mathcal{S}et$)
Element of an object X	Arrow $1 \to X$
Monic arrow $X \rightarrowtail Y$	Arrow $f : X \to Y$ such that, for any arrows $g, h : Z \to X, f \circ g = f \circ h \Rightarrow g = h$ (in $\mathcal{S}et$, *one-one* map)
Epic arrow $X \twoheadrightarrow Y$	Arrow $f : X \to Y$ such that, for any arrows $g, h : Y \to Z, g \circ f = h \circ f \Rightarrow g = h$ (in $\mathcal{S}et$, *onto* map)
Isomorphism $X \cong Y$	Arrow $f : X \to Y$ for which there is $g : Y \to X$ such that $g \circ f = 1_X, f \circ g = 1_Y$

Product of objects X, Y

Object $X \times Y$ with arrows (projections) $X \xleftarrow{\pi_1} X \times Y \xrightarrow{\pi_2} Y$ such that any diagram

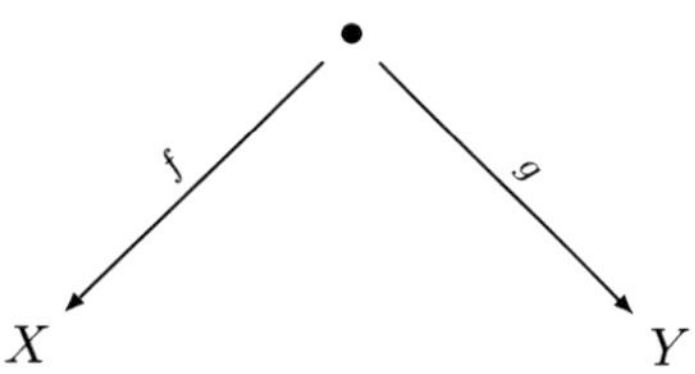

can be uniquely completed to a commutative diagram

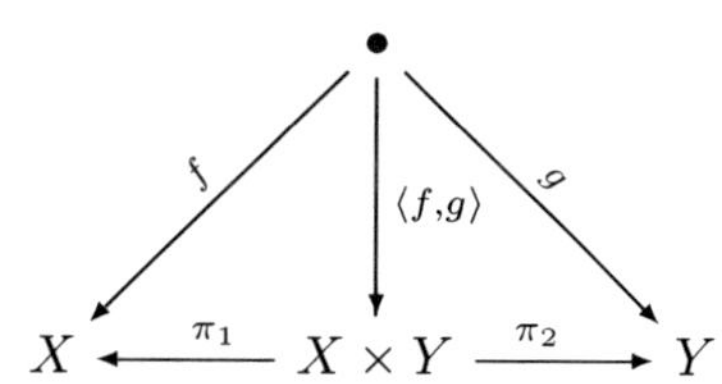

Product of arrows $f_1 : X_1 \to Y_1, f_2 : X_2 \to Y_2$

Unique arrow $f_1 \times f_2 : X_1 \times X_2 \to Y_1 \times Y_2$ making the diagram

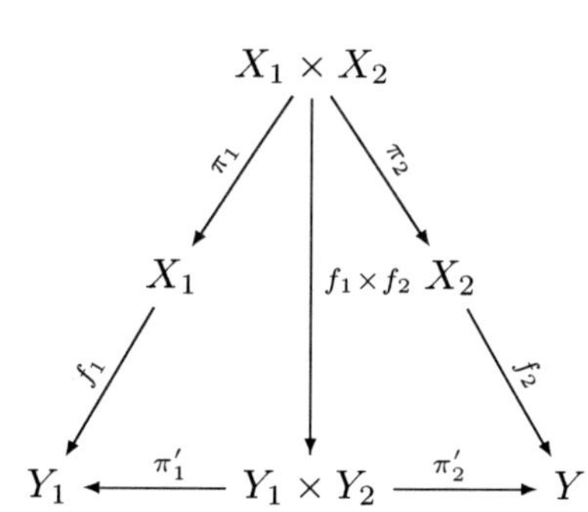

commute. I.e., $f_1 \times f_2 = \langle f_1 \circ \pi_1, f_2 \circ \pi_2 \rangle$.

Diagonal arrow on object X

Unique arrow $\delta_X : X \to X \times X$ making the diagram

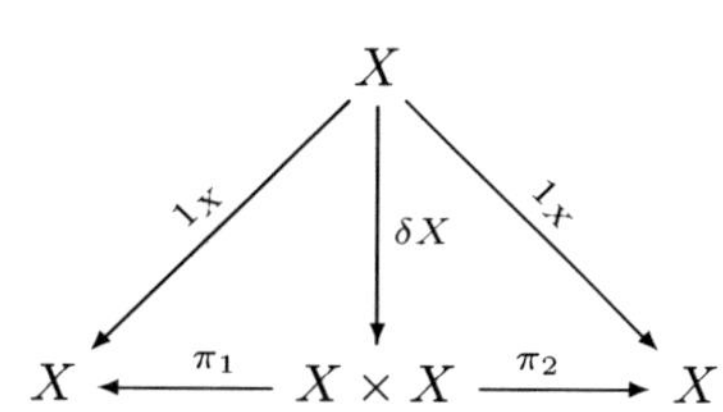

commute. I.e., $\delta_X = \langle 1_X, 1_X \rangle$.

Coproduct of objects X, Y — Object $X+Y$ together with a pair of arrows $X \xrightarrow{\sigma_1} X+Y \xleftarrow{\sigma_2} Y$ such that for any pair of arrows $X \xrightarrow{f} A \xleftarrow{g} Y$, there is a unique arrow $X+Y \xrightarrow{h} A$ such that the diagram

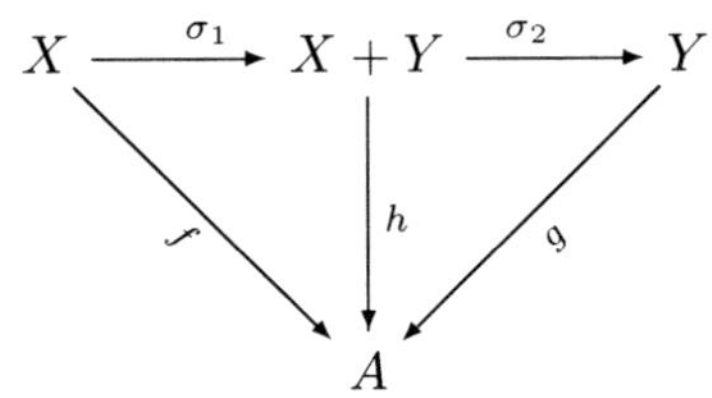

commutes.

Pullback diagram or *square* — Commutative diagram of the form

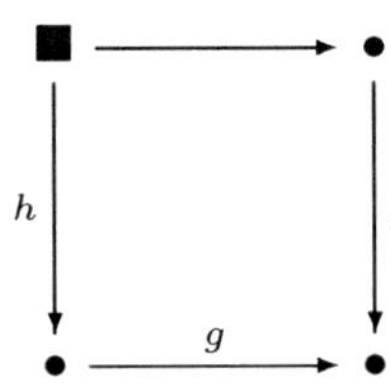

such that for any commutative diagram

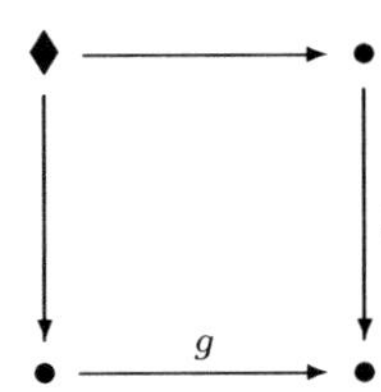

there is a unique $\blacklozenge \xrightarrow{!} \blacksquare$ such that

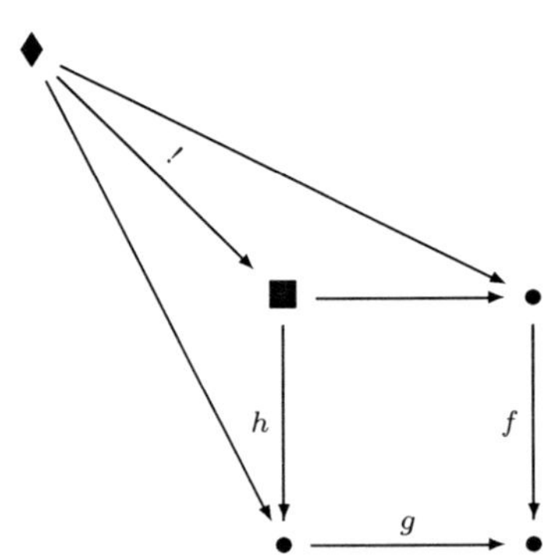

commutes. h is called the *pullback* of f (along g).

Equalizer of pair of arrows $\bullet \overset{f}{\underset{g}{\rightrightarrows}} \blacksquare$ — Arrow $\blacklozenge \xrightarrow{e} \bullet$ such that $f \circ e = g \circ e$ and, for any arrow $\blacktriangle \xrightarrow{e'} \bullet$ such that $f \circ e' = g \circ e'$ there is a unique $\blacktriangle \xrightarrow{u} \blacklozenge$ such that

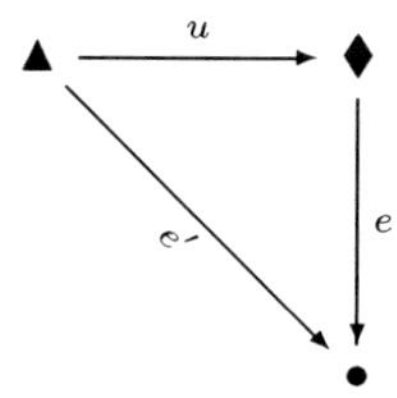

commutes.

Subobject of an object X — Pair (m, Y), with m a monic arrow $Y \rightarrowtail X$

Inclusion of subobjects — For $m : Y \rightarrowtail X, n : Z \rightarrowtail X, m \subseteq n$ iff there is $f : Y \to Z$ such that $n \circ f = m$.

Truth value object or *subobject* classifier — Object $\mathbf{\Omega}$ together with arrow $t : 1 \to \mathbf{\Omega}$ such that every monic $m : \bullet \rightarrowtail \blacklozenge$ (i.e., subobject of $\blacklozenge$) can be uniquely extended to a pullback diagram of the form

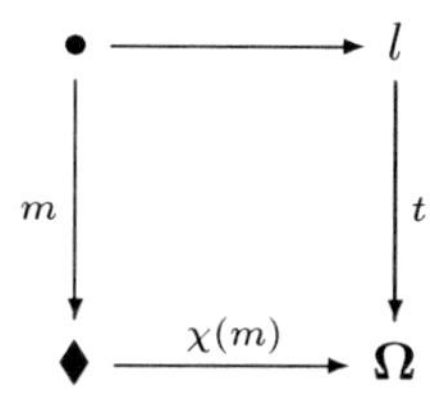

$\chi(m)$ is the *characteristic arrow* of m.
Conversely, any diagram of the form $\blacklozenge \xrightarrow{u} \mathbf{\Omega} \xleftarrow{t} 1$ must have a pullback.
The pullback of u will be written $\overline{u}$. It is necessarily monic.

The next few definitions assume that the ambient category has a terminal object, products and a subobject classifier. It can be shown that in any such category, every monic arrow has a pullback.

Partial ordering $\leq$ *on arrows to* $\mathbf{\Omega}$. For $u, v : A \to \mathbf{\Omega}, u \leq v$ iff $\overline{u} \subseteq \overline{v}$.

Intersection of subobjects For monics m, n with common codomain A, form the pullback square

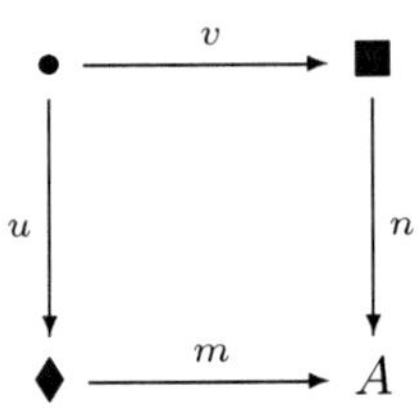

Then the map $m \circ u (= n \circ v)$ is the intersection $m \cap n$ of m and n.

Meet of two maps to $\mathbf{\Omega}$. For $u, v : A \to \mathbf{\Omega}, u \wedge v = \chi(\overline{u} \cap \overline{v})$.

Equality arrow $A \times A \to \mathbf{\Omega}$ $eq_A = \chi(\delta_A)$

Power object of an object X. An object PX together with an arrow ("evaluation") $e_X : X \times PX \to \mathbf{\Omega}$ such that, for any $f : X \times PX \to \mathbf{\Omega}$, there is a unique arrow $\widehat{f} : Y \to PX$ such that

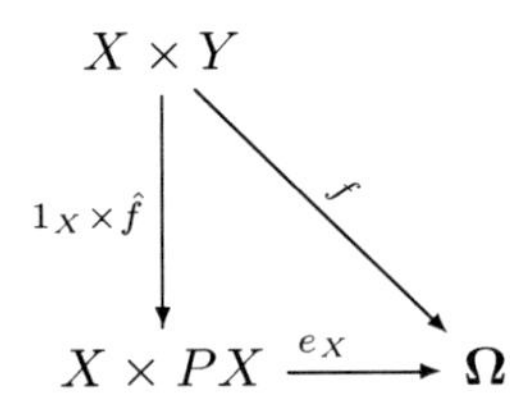

commutes. (In $\mathcal{S}et$, PX is the power set of X and e_X the characteristic function of the membership relation between X and PX.)

Exponential object of objects Y, X An object Y^X, together with an arrow ev: $X \times Y^X \to Y$ such that, for any arrow $f : X \times Z \to Y$ there is a unique arrow $\widehat{f} : Z \to Y^X$ — the *exponential transpose* of f — such that the diagram

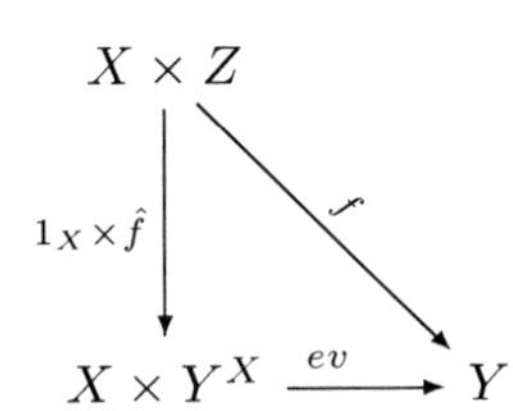

commutes. In $\mathcal{S}et$, Y^X is the set of all maps $X \to Y$ and ev is the map that sends (x, f) to $f(x)$.

Product of indexed set $\{A_i : i \in I\}$ *of objects* — Object $\prod_{i\in I} A_i$ together with arrows $\prod_{i\in I} A_i \xrightarrow{\pi_i} A_i (i \in I)$ such that, for any arrows $f_i : B \to A_i (i \in I)$ there is a *unique* arrow $h : B \longrightarrow \prod_{i\in I} A_i$ such that, for each $i \in I$, the diagram

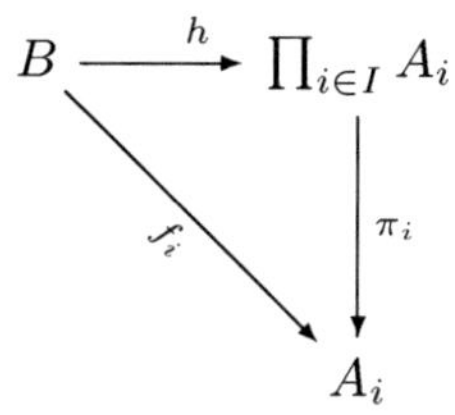

commutes.

Coproduct of indexed set $\{A_i : i \in I\}$ *of objects* — Object $\coprod_{i\in I} A_i$ together with arrows $A_i \xrightarrow{\sigma_i} \coprod_{i\in I} A_i$ $(i \in I)$ such that, for any arrows $f_i : A_i \to B(i \in I)$ there is a *unique* arrow $h : \coprod_{i\in I} A_i \longrightarrow B$ such that, for each $i \in I$, the diagram

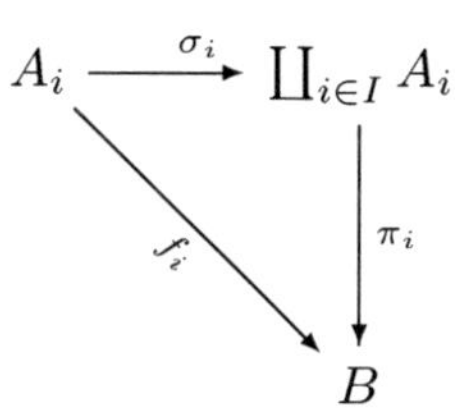

commutes.

A category is *Cartesian closed* if it has a terminal object, as well as products and exponentials of arbitrary pairs of its objects. It is *finitely complete* if it has a terminal object, products of arbitrary pairs of its objects, and equalizers. A *topos* is a category possessing a terminal object, products, a truth-value object, and power objects. It can be shown that every topos is cartesian closed and finitely complete (so that this notion of topos is equivalent to that originally given by Lawvere and Tierney). The category $\mathcal{S}et$ of sets is a topos.

More on products in a category. A *product* of objects $A_1, ..., A_n$ in a category $\mathcal{C}$ is an object $A_1 \times ... \times A_n$ together with arrows $\pi_i : A_1 \times ... \times A_n \to A_i$ for $i = 1, ..., n$, such that, for any arrows $f_i : B \to A_i, i = 1, ..., n$, there is a unique arrow, denoted by $\langle f_1, ..., f_n \rangle : B \to A_1 \times ... \times A_n$ such that $\pi_i \circ \langle f_1, ..., f_n \rangle = f_i, i = 1, ..., n$. Note that, when $n = 0$, $A_1 \times ... \times A_n$ is the terminal object 1. The category is said to *have finite products* if $A_1 \times ... \times A_n$ exists for all $A_1, ..., A_n$. If $\mathcal{C}$ has binary products, it has finite products, since we may take $A_1 \times ... \times A_n$ to be $A_1 \times (A_2 \times (... \times A_n)...)$. It is easily seen that the product operation is, up to isomorphism, commutative and associative. The relevant isomorphisms are called *canonical* isomorphisms.

A *functor* $F : \mathcal{C} \to \mathcal{D}$ between two categories $\mathcal{C}$ and $\mathcal{D}$ is a map that "preserves commutative diagrams", that is, assigns to each $\mathcal{C}$-object A a $\mathcal{D}$-object FA and to each $\mathcal{C}$-arrow $f : A \to B$ a $\mathcal{D}$-arrow $Ff : FA \to FB$ in such a way that:

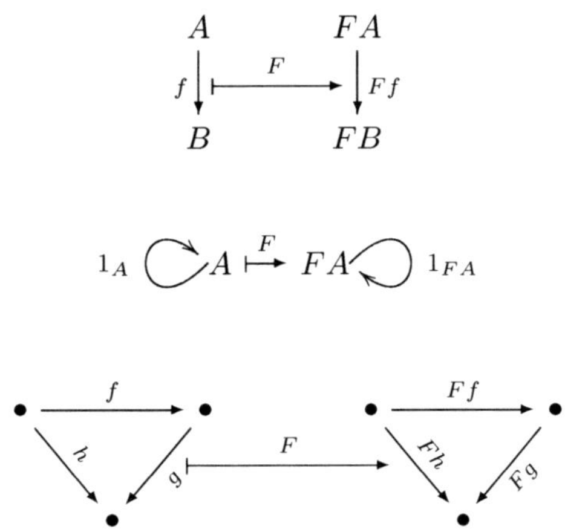

commutes commutes

A functor $F : \mathcal{C} \to \mathcal{D}$ is an *equivalence* if it is "an isomorphism up to isomorphism", that is, if it is

- *faithful*: $Ff = Fg \Rightarrow f = g$.
- *full*: for any $h : FA \to FB$ there is an $f : A \to B$ such that $h = Ff$.
- *dense*: for any $\mathcal{D}$-object B there is a $\mathcal{C}$-object A such that $B \cong FA$.

Two categories are *equivalent*, written $\simeq$, if there is an equivalence between them. Equivalence is the appropriate notion of "identity of form" for categories.

Given functors $F, G : \mathcal{C} \to \mathcal{D}$, a *natural transformation* between F and G is a map η from the objects of C to the arrows of D satisfying the following conditions.

- For each object A of $\mathcal{C}, \eta A$ is an arrow $FA \to GA$ in $\mathcal{D}$
- For each arrow $f : A \to A'$ in $\mathcal{C}$,

$$\begin{array}{ccc} FA & \xrightarrow{\eta A} & GA \\ {\scriptstyle Ff}\downarrow & & \downarrow{\scriptstyle Gf} \\ FA' & \xrightarrow{\eta A'} & GA' \end{array}$$

Finally, two functors $F : \mathcal{C} \to \mathcal{D}$ and $G : \mathcal{D} \to \mathcal{C}$ and are said to be *adjoint* to one another if, for any objects A of $\mathcal{C}$, B of $\mathcal{D}$, there is a "natural" bijection between arrows $A \to GB$ in $\mathcal{C}$ and arrows $FA \to B$ in $\mathcal{D}$. To be precise, for each such pair A, B we must be given a bijection $\varphi_{AB} : \mathcal{C}(A, GB) \to \mathcal{D}(FA, B)$ satisfying the "naturality" conditions

- for each $f : A \to A'$ and $h : A' \to GB, \varphi_{AB}(h \circ f) = \varphi_{A'B}(h) \circ Ff$
- for each $g : B \to B'$ and $h : A \to GB', \varphi_{AB}(Gg \circ h) = g \circ \varphi_{AB'}(h)$.

Under these conditions F is said to be *left adjoint* to G, and G *right adjoint* to F.

BIBLIOGRAPHY

[Aczel, 1978] P. Aczel. The type-theoretic interpretation of constructive set theory. In A. ManIntyre, L. Pacholski, and J. Paris. eds., *Logic Colloquium '77.*, pp. 55–66. North-Holland, 1978.

[Aczel, 1982] P. Aczel. The type-theoretic interpretation of constructive set theory: choice principles. In A. S. Troelstra and D. van Dalen, eds., *The L.E.J. Brouwer Centenary Symposium,* pp. 1–40. North-Holland, 1982.

[Aczel, 1986] P. Aczel. The type-theoretic interpretation of constructive set theory: inductive definitions. In R. Barcan Marcus, G.J.W. Dorn and P. Weingartner, eds., *Logic, methodology and Philosophy of Science VII,* pp. 17–49. North-Holland, 1986.

[Aczel and Gambino, 2002] P. Aczel and N. Gambino. Collection principles in dependent type theory. In P. Callaghan, Z. Luo, J. McKinna and R. Pollack, eds. *Types for Proofs and Programs,* vol. 2277 of *Lecture Notes on Computer Science,* pp. 1–23. Springer, 2002.

[Aczel and Gambino, 2005] P. Aczel and N. Gambino. The generalized type-theoretic interpretation of constructive set theory, 2005. Manuscript available on first author's webpage `www.cs.man.ac.uk/~petera/papers`

[Aczel and Rathjen, 2001] P. Aczel and M. Rathjen. Notes on Constructive Set Theory. Technical Report 40, Mittag-Leffler Institute, The Swedish Royal Academy of Sciences, 2001. Available on first author's webpage `www.cs.man.ac.uk/~petera/papers`

[Bell, 1988] J. L. Bell. *Toposes and Local Set Theories: An Introduction.* Oxford Logic Guides 14. Oxford: Clarendon Press, 1988.

[Bishop, 1967] E. Bishop. *Foundations of Constructive Analysis.* McGraw-Hill, 1967.

[Boileau, 1975] A. Boileau. *Types vs. Topos.* Thesis, Univesité de Montreal, 1975.

[Boileau and Joyal, 1981] A. Boileau and A. Joyal. La logique de topos. *The Journal of Symbolic Logic*, 46:6–16, 1981.

[Boolos, 1997] G. Boolos. Constructing Cantorian counterexamples. *Journal of Philosophical Logic* 26:37–39, 1997.

[Carnap, 1929] R. Carnap. *Abriss der Logistik.* Vienna, 1929.

[Church, 1940] A. Church. A formulation of the simple theory of types. *The Journal of Symbolic Logic*, 1:56–68, 1940.

[Chwistek, 1924] L. Chwistek. *Theory of Constructive Types.* Cracow, 1924.

[Coquand, 2006] T. Coquand. Type Theory. *Stanford Encyclopedia of Philosophy*, 2006

[Coquand and Huet, 1988] T. Coquand and G. Huet. The calculus of constructions. *Information and Computation*, 76:95–120, 1988.

[Crole, 1993] R. Crole. *Categories for Types.* Cambridge University Press, 1993.

[Curry and Feys, 1958] H. B. Curry and R. Feys. *Combinatory Logic.* North Holland, 1958.

[Diaconescu, 1975] R. Diaconescu. Axiom of choice and complementation. *Proceedings of the American Mathematical Society* 51, 176–8, 1975.

[Eilenberg and Mac Lane, 1945] S. Eilenberg and S. Mac Lane. General theory of natural equivalences. *Transactions of the American Mathematical Society*, 58:231–294, 1945. Reprinted in Eilenberg, S. and S. Mac Lane, *Eilenberg-Mac Lane: Collected Works.* New York: Academic Press, 1986.

[Farmer, 2006] W. Farmer. The Seven Virtues of Simple Type Theory, 2006. Manuscript available online at `imps.mcmaster.ca/doc/seven-virtues.pdf`

[Fourman, 1974] M. P. Fourman. *Connections between category theory and logic.* D. Phil. Thesis, Oxford University, 1974.

[Fourman, 1977] M. P. Fourman. The logic of topoi. In J. Barwise (ed.) *Handbook of Mathematical Logic.* Amsterdam: North-Holland, pp. 1053–90, 1977.

[Fourman *et al.*, 1979] M. P. Fourman, C. J. Mulvey, and D. S. Scott, eds. *Applications of Sheaves. Proc. L.M.S. Durham Symposium* 1977. Springer Lecture Notes in Mathematics 753, 1979.

[Gandy, 1977] R. O. Gandy. The simple theory of types. In R. Gandy, M. Hyland, eds., Logic Colloquium 76, North-Holland, Amsterdam, pp. 173–181, 1977.

[Girard, 1972] J.-Y. Girard. *Interprétation fonctionelle élimination des coupures dans l'arithmétique d'ordre supérieure.* Ph.D. thesis, Université Paris VII, 1972.

[Gödel, 1931] K. Gödel. Über formal untenscheidbare Sätze der Principia Mathematica und verwandter Systeme I. *Monatsh. Math. Phys.*, 38:349–360, 1931. Reprinted in English translation in *Godel, Collected Works vol. I.,* S. Feferman et al., eds. Oxford University Press, 1986.

[Goodman and Myhill, 1978] N. Goodman and J. Myhill. Choice implies excluded middle. *Z. Math Logik Grundlag. Math.*, 24:461, 1978.

[Henkin, 1963] L. Henkin. A theory of propositional types. *Fund. Math.*, 52:323–344, 1963.

[Howard, 1980] W. A. Howard. The formulae-as-types notion of construction. In J. R. Hindlet and J. P. Seldin (eds.), *To H. B. Curry: Essays on Combinatorial Logic. Lambda Calculus and Formalism,* pp. 479-490. New York and London: Academic Press, 1980.

[Jacobs, 1999] B. Jacobs. *Categorical Logic and Type Theory.* Amsterdam: Elsevier, 1999.

[Johnstone, 1977] P. T. Johnstone. *Topos Theory.* London: Academic Press, 1977.

[Johnstone, 2002] P. T. Johnstone. *Sketches of an Elephant: A Topos Theory Compendium, Vols. I and II.* Oxford Logic Guides Vols. 43 and 44, Oxford: Clarendon Press, 2002.

[Lambek, 1974] J. Lambek. Functional completeness of Cartesian categories. *Annals of Mathematical Logic*, 6:259–92, 1974.

[Lambek, 1980a] J. Lambek. From λ-calculus to Cartesian closed categories. In J.P. Seldin and J.R. Hindley, eds., *To H.B. Curry: Essays on Combinatory Logic, Lambda Calculus and Formalism.* London, Academic Press, pp. 375–402, 1980.

[Lambek, 1980b] J. Lambek. From types to sets. *Adv. In Math.*, 36:113–164, 1980.

[Lambek and Scott, 1980] J. Lambek and P. J. Scott. Intuitionist type theory and the free topos. *Journal Pure and Applied Algebra* 19:215–257, 1980.

[Lambek and Scott, 1981] J. Lambek and P. J. Scott. Intuitionist type theory and foundations. *Joural of Philosophical Logic*, 10:101–115, 1981.

[Lambek and Scott, 1986] J. Lambek and P. J. Scott. *Introduction to Higher-Order Categorical Logic.* Cambridge: Cambridge University Press, 1986.

[Lawvere, 1971] F. W. Lawvere. Quantifiers and sheaves. In *Actes du Congrès Intern. Des Math. Nice 1970,* tome I. Paris: Gauthier-Villars, pp. 329–34, 1971.

[Lawvere, 1972] F. W. Lawvere. Introduction to *Toposes, Algebraic Geometry and Logic.* Springer Lecture notes in Math. 274, pp. 1–12, 1972.

[Maietti, 2005] M. E. Maietti. Modular correspondence between dependent type theories and categories including pretopoi and topoi. *Math. Struct. Comp. Sci.*, 15:1089–1145, 2005.

[Maietti and Valentini, 1999] M. E. Maietti and S. Valentini. Can you add power-sets to Martin-Löf's intuitionistic set theory? *Mathematical Logic Quarterly*, 45:521–532, 1999

[Martin-Löf, 1975] P. Martin-Löf. An Intuitionistic theory of types; predicative part. In H. E. Rose and J. C. Shepherdson (eds.), *Logic Colloquium '73,* pp. 73-118. Amsterdam: North-Holland, 1975.

[Martin-Löf, 1982] P. Martin-Löf. Constructive mathematics and computer programming. In L. C. Cohen, J. Los, H. Pfeiffer, and K.P. Podewski (eds.), *Logic, Methodology and Philosophy of Science VI,* pp. 153–179. Amsterdam: North-Holland, 1982.

[Martin-Löf, 1984] P. Martin-Löf. *Intuitionistic Type Theory.* Naples: Bibliopolis, 1984.

[Miquel, 2001] A. Miquel. A strongly normalising Curry-Howard correspondence for IZF set theory. *Computer Science Logic,* Lecture Notes in Computer Science 2803, pp. 441–54. Springer, 2001.

[Nordström *et al.*, 1990] B. Nordström, K. Petersson, and J. M. Smith. *Programming in Martin-Löf's Type Theory.* Oxford University Press, 1990.

[Ramsey, 1926] F. P. Ramsey. The foundations of mathematics. *Proceedings of the London Mathematical Society*, 25:338–84, 1926.

[Russell, 1903] B. Russell. *The Principles of Mathematics.* Cambridge University Press, 1903.

[Russell, 1908] B. Russell. Mathematical logic as based on the theory of types. *Am. J. Math.*, 30:222–262, 1908. Reprinted in [van Heijenoort, 1967, 150–182].

[Seely, 1984] R. A. G. Seely. Locally Cartesian closed categories and type theory. *Math. Proc. Camb. Phil. Soc.*, 95:33–48, 1984.

[Seely, 1987] R. A. G. Seely. Categorical semantics for higher order polymorphic lambda calculus. *The Journal of Symbolic Logic*, 52:969–989, 1987.

[Tait, 1994] W. W. Tait. The law of excluded middle and the axiom of choice. In *Mathematics and Mind*, A. George, ed., pp. 45–70. New York: Oxford University Press, 1994.

[Tarski, 1931] A. Tarski. Sur les ensembles definissable de nombres réels I. *Fundamenta Mathematicae*, 17:210–29, 1931.

[van Heijenoort, 1967] J. van Heijenoort, ed. *From Frege to Gödel: A Source Book in Mathematical Logic, 1879-1931.* Harvard University Press, 1967.

[Weyl, 1918] H. Weyl. *Das Kontinuum.* Leipzig: Veit, 1918. English translation by S. Pollard and T. Bole as *The Continuum: A Critical Examination of the Foundation of Analysis,* tr. Kirksville, Mo.: Thomas Jefferson University Press, 1987.

[Whitehead and Russell, 1910–13] A. N. Whitehead and B. Russell. *Principia Mathematica.* 3 vols., Cambridge University Press, 1910–13.

[Zangwill, 1977] J. Zangwill. *Local Set Theory and Topoi.* M. Sc. thesis, Bristol University, 1977.

[Zermelo, 1908] E. Zermelo. Untersuchungen über die Grundlagen der Mengenlehre I. *Matematische Annalen*, 59:261–81, 1908. Reprinted in [van Heijenoort, 1967, 199–215].

THE HISTORY OF CATEGORICAL LOGIC: 1963–1977

Jean-Pierre Marquis and Gonzalo E. Reyes

Categorical logic, as its name indicates, is logic in the setting of category theory. But this description does not say much. Most readers would probably find more instructive to learn that categorical logic *is* algebraic logic, pure and simple. It is logic in an algebraic dressing. Just as algebraic logic encodes propositional logic in its different guises (classical, intuitionistic, etc.) by their Lindenbaum-Tarski algebras (Boolean algebras, Heyting algebras and so on), categorical logic encodes first-order and higher-order logics (classical, intuitionistic, etc.) by categories with additional properties and structure (Boolean categories, Heyting categories and so on). Thus, from the purely technical point of view, categorical logic constitutes a generalization of the algebraic encoding of propositional logic to first-order, higher-order and other logics. Furthermore, we shall present and discuss arguments (given by the main actors) to show that this encoding constitutes the correct generalization of the well-known algebraic encoding of propositional logics by the Lindenbaum-Tarski algebras.

The proper algebraic structures are not only categories, but also morphisms between categories, mainly functors and more specially adjoint functors. A key example is provided by the striking fact that quantifiers, which were the stumbling block to the proper algebraic generalization of propositional logic, can be seen to be adjoint functors and thus entirely within the categorical framework. As is usually the case when algebraic techniques are imported and developed within a field, e.g. geometry and topology, vast generalizations and unification become possible. Furthermore, unexpected concepts and results show up along the way, often allowing a better understanding of known concepts and results.

Categorical logic is not merely a convenient tool or a powerful framework. Again, as is usually the case when algebraic techniques are imported and used in a field, the very nature of the field has to be thought over. Furthermore, various results shed a new light on what was assumed to be obvious or, what turns out to be often the same on careful analysis, totally obscure. Thus, categorical logic is philosophically relevant in more than one way. The way it encodes logical concepts and operations reveals important, even essential, aspects and properties of these concepts and operations. Again, as soon as quantifiers are seen as adjoint functors, the traditional question of the nature of variables in logic receives a satisfactory analysis.

Handbook of the History of Logic: Sets and Extensions in the Twentieth Century.
Volume editor: Akihiro Kanamori. General editors: Dov M. Gabbay, Paul Thagard and John Woods.

Furthermore, many results obtained via categorical techniques have clear and essential philosophical implications. The systematic development of higher-order logic, type theory under a different name, and various completeness theorems are the most obvious candidates. But there is much more. Many important questions concerning the foundations of mathematics and the very nature of mathematical knowledge are inescapable. In particular, issues related to abstraction and the nature of mathematical objects emerges naturally from categorical logic.

This paper covers the period that can be qualified as the birth and the constitution of categorical logic, that is the time span between 1963 and 1977. No one will deny that categorical logic started with Bill Lawvere's Ph.D. thesis written in 1963 under S. Eilenberg's supervision and widely circulated afterwards. (It is now available on-line on the TAC web-site.) In his thesis, Lawvere offered a categorical version of *algebraic* theories. He also suggested that the category of categories could be taken as a foundation for mathematics and that sets could be analyzed in a categorical manner. In the years that followed, Lawvere tried to extend his analysis and sketched a categorical version of first-order theories under the name of *elementary theories*. Then, in 1969, in collaboration with Myles Tierney, Lawvere introduced the notion of an elementary topos, making an explicit connection with higher-order logic and type theories. Both Lawvere and Tierney were aiming at an elementary, that is first-order, axiomatic presentation of what are now called *Grothendieck toposes*, a special type of categories introduced by Alexandre Grothendieck in the context of algebraic geometry and sheaf theory. Soon after, connections with intuitionistic analysis, recursive functions, completeness theorems for various logical systems, differential geometry, constructive mathematics were made. We have decided to end our coverage in 1977 for the following reasons. First, we had to stop somewhere, otherwise we would have to write a book. Second, and this is a more serious reason, three independent events in 1977 mark more or less a turning point in the history of categorical logic. First, the book *First-Order Categorical Logic*, by Makkai and Reyes appears, a book that more or less codifies the work done by the Montreal school in the period 1970-1974 and now constitutes the core of categorical first-order logic. Second, the same year witnesses the publication of Johnstone's *Topos Theory*, the first systematic and comprehensive presentation of topos theory as it was known in 1974-75. (The reader should compare this edition with Johnstone's recent *Sketches of an Elephant*, a comprehensive reference on topos theory in three volumes.) Third, 1977 was also the year of the Durham meeting on applications of sheaf theory to logic, algebra and analysis, whose proceedings were published in 1979. We submit that around the end of the nineteen seventies, categorical logic was on firm ground and could be developed in various directions, which is precisely what happened, from theoretical computer science, modal logic and other areas.

The usual warnings, caveat and apologies are now necessary. It is impossible to cover, even in a long article such as this one and for such a short time period, all events involved in the history of categorical logic. This paper is but a first attempt at a more precise and detailed history of a complicated and fascinating period in the history of ideas. We hope that it will stimulate more work on the topic. We hasten to add that it also reflects our interests and (hopefully not too limited) knowledge of the field. It is our hope that it will nonetheless be useful to logicians and philosophers alike. We sincerely apologize to mathematicians, logicians and philosophers whose names ought to have appeared in this history but have not because of our ignorance.

1 THE BIRTH OF CATEGORY THEORY AND ITS EARLY DEVELOPMENTS

Category theory as a discipline in itself and was born in the context of algebraic topology in the nineteen forties. We will briefly sketch the history of category theory before the advent of categorical logic and rehearse the fundamental notions of the theory required for the exposition of the following sections.

1.1 Category theory: its origins

We will here only rehearse the ingredients required for the history of categorical logic. The reader is referred to [Landry and Marquis, 2005], [Marquis, 2006] and [Krömer, 2007] for more details.

Category theory made its official public appearance in 1945 in the paper entitled "General Theory of Natural Equivalences" written by Samuel Eilenberg and Saunders Mac Lane. This "off beat" and "far out" paper, as Mac Lane came to qualify it later [Mac Lane, 2002, 130], was meant to provide an autonomous framework for the concept of *natural transformation*, a concept whose generality, pervasiveness and usefulness had become clear to both of them during their collaboration on the clarification of an unsuspected link between group extensions and homology groups. Such a general, pervasive and conceptually useful notion seemed to deserve a precise, rigorous, systematic and abstract treatment.

Eilenberg and Mac Lane decided to devise an *axiomatic* framework in which the notion of natural transformation would receive an entirely general and autonomous definition. This is where *categories* came in. Informally, a natural transformation is a family of maps that provides a systematic "translation" or a "deformation" between two systems of interrelated entities within a given framework. But in order to give a precise definition of natural transformations, one needs to clarify the *systematic* nature of these deformations, that is, one has to specify *what* these deformations depend upon and *how* they depend upon it. Eilenberg and Mac Lane introduced what they called *functors* — the term was borrowed from Carnap — so that one could say between *what* the natural transformations were acting: a natural transformation is a family of maps between functors. Clearly, one has to define the notion of a functor: the concept of *category* was tailored for that purpose. The *systematic* nature of natural transformations was also made clear by categories themselves. Thus categories were introduced in 1945 and, as Mac

Lane reported (see [Mac Lane, 2002]), Eilenberg believed that their paper would be the only paper written on "pure" category theory.

As we have already mentioned, Eilenberg and Mac Lane gave a purely *axiomatic* definition of category in their original paper. It is worth mentioning that they explicitly avoided using a set-theoretical terminology and notation in the axioms themselves. Here is how their definition unfolds (only with a slightly different notation):

A *category* $\mathcal{C}$ is an aggregate of abstract elements X, called the *objects* of the category, and abstract elements f, called *mappings* of the category. Certain pairs of mappings f, g of $\mathcal{C}$ determine uniquely a product mapping $g \circ f$, satisfying the axioms C_1, C_2, C_3 below. Corresponding to each object X of $\mathcal{C}$, there is a unique mapping, denoted by 1_X satisfying the axioms C_4 and C_5. The axioms are:

C_1 The triple product $h \circ (g \circ f)$ is defined if and only if $(h \circ g) \circ f$ is defined. When either is defined, the associative law

$$h \circ (g \circ f) = (h \circ g) \circ f$$

holds. This triple product will be written as $h \circ g \circ f$.

C_2 The triple product $h \circ g \circ f$ is defined whenever both products $h \circ g$ and $g \circ f$ are defined. A mapping 1 of $\mathcal{C}$ will be called an *identity* of $\mathcal{C}$ if and only if the existence of any product $1 \circ f$ and $g \circ 1$ implies that $1 \circ f = f$ and $g \circ 1 = g$.

C_3 For each mapping f of $\mathcal{C}$ there is at least one identity 1_r such that $f \circ 1_r$ is defined, and at least one identity 1_l such that $1_l \circ g$ is defined.

C_4 The mapping 1_X corresponding to each object X is an identity.

C_5 For each identity 1 of $\mathcal{C}$ there is a unique object X of $\mathcal{C}$ such that $1_X = 1$.

The last two axioms "assert that the rule $X \longrightarrow 1_X$ provides a one-to-one correspondence between the set of all objects of the category and the set of all its identities. It is thus clear that the objects play a secondary role, and could be entirely omitted from the definition of a category. However, the manipulation of the applications would be slightly less convenient were this done." [Eilenberg and Mac Lane, 1945, 238] Thus, from a theoretical point of view, a category is determined by its mappings, but from a practical point of view, it is convenient to distinguish the objects from the mappings. Eilenberg and Mac Lane then state as a lemma that each mapping f has a unique domain (source) X and a unique codomain (target or range) Y and write $f : X \longrightarrow Y$.

Eilenberg and Mac Lane proceed to define *equivalences in* a category, nowadays called *isomorphisms*, thus: a mapping f is an *isomorphism* if it has an inverse, i.e. if there is a mapping g such that $g \circ f$ and $f \circ g$ are defined and are identities. Two objects X_1 and X_2 are said to be *isomorphic* if there is an isomorphism between them.

Eilenberg and Mac Lane gave four basic examples of categories: the category *Set* of sets with functions between them, the category *Top* of topological spaces with continuous functions, the category *TopGrp* of topological groups with continuous homomorphisms and the category *Ban* of Banach spaces with linear transformations with norm at most 1. This is a surprisingly short list of examples. They give more examples by defining the notion of a subcategory in the obvious fashion. Thus, they point out that given a category $\mathcal{C}$, the subcategory composed of the same objects as $\mathcal{C}$ but with mappings only the isomorphisms is a category, nowadays called a *groupoid*. The category of finite sets is also mentioned as well as other subcategories of the category of sets, e.g. for a fixed cardinal k, there is a category of all sets of power less than k together with all the mappings. By restricting the mappings between sets to be onto or injective, one obtains different subcategories of sets. Similarly, if one restricts the continuous maps to open maps between topological spaces, then one obtains a different subcategory of topological spaces. In §11 of their paper, Eilenberg and Mac Lane observe that any group G can be thought of as a category: it has only one object and its mappings are the elements of the group. They also point out in §20 that any preorder P can be viewed as a category.

It should be emphasized how truly secondary categories were for Eilenberg and Mac Lane at that point. In that respect, categories had an ambiguous status. It is clear that categories are *conceptually* required for the systematic and rigorous definition of natural transformations, but at the same time, they cannot be legitimate mathematical entities unless certain precautions are taken with respect to their size. Eilenberg and Mac Lane explicitly recognized this fact in §6 where they discuss foundational issues related to categories, e.g. the category of all sets is not a set, thus not a legitimate entity from the standard set-theoretical point of view.

> It should be observed first that the whole concept of a category is essentially an auxiliary one; our basic concepts are essentially those of a *functor* and of a natural transformation (...). The idea of a category is required only by the precept that every function should have a definite class as domain and a definite class as range, for the categories are provided as the domains and ranges of functors. Thus one could drop the category concept altogether and adopt an even more intuitive standpoint, in which a functor such as "Hom" is not defined over the category of "all" groups, but for each particular pair of groups which may be given. The standpoint would suffice for the applications, inasmuch as none of our developments will involve elaborate constructions on the categories themselves. [Eilenberg and Mac Lane, 1945, 247]

Although the definition of category was conceptually necessary, categories themselves were not doing any mathematical work. They were simply a way of systematizing the required data. However, it did not take long, approximately ten to fifteen years, before elaborate constructions on categories themselves became essential and thus the question concerning their nature became more pressing and we indeed see, for instance, Mac Lane, going back to the problem of the foundations of categories in the late fifties and early sixties. (See [Mac Lane, 1961].)

Eilenberg and Mac Lane's definition of a functor is given for n arguments. We will give the definition of a functor with one argument.

A *functor* F between categories $\mathcal{C}$ and $\mathcal{D}$ is a pair of functions, an *object-function* which associates to each object X of $\mathcal{C}$ an object $Z = F(X)$ in $\mathcal{D}$ and a *mapping function* which associates to each mapping f of $\mathcal{C}$ a mapping $h = F(f)$, such that

$$1. \quad F(1_X) = 1_{F(X)}$$

$$2. \quad F(g \circ f) = F(g) \circ F(f)$$

Such a functor is said to be *covariant*. Whenever a functor satisfies the equality

$$2^o. \quad F(g \circ f) = F(f) \circ F(g)$$

instead of 2, it is said to be *contravariant*.

Functors with the same domain category and the same codomain category can be connected to one another systematically or "naturally". This is precisely what the notion of natural transformation captures. Here is Eilenberg and Mac Lane's definition, restricted to functors $F, G : \mathcal{C} \longrightarrow \mathcal{D}$ in one argument.

A *natural transformation* $\tau : F \longrightarrow G$ between functors $F, G : \mathcal{C} \longrightarrow \mathcal{D}$ is a function that associates to each object X of $\mathcal{C}$ a mapping $\tau_X : F(X) \longrightarrow G(X)$ of $\mathcal{D}$ such that for any mapping $f : X \longrightarrow Y$, the following diagram commutes

$$\begin{array}{ccc} F(X) & \xrightarrow{\tau_X} & G(X) \\ {\scriptstyle F(f)}\downarrow & & \downarrow{\scriptstyle G(f)} \\ F(Y) & \xrightarrow[\tau_Y]{} & G(Y) \end{array}$$

that is, $G(f) \circ \tau_X = \tau_Y \circ F(f)$.

If each τ_X is an isomorphism, then τ is said to be a *natural isomorphism* (Eilenberg and Mac Lane said *natural equivalence*).

Given functors and natural transformations, it is possible to define categories of functors: its objects are functors $F : \mathcal{C} \longrightarrow \mathcal{D}$ and its mappings are natural transformations $\tau : F \longrightarrow G$. Eilenberg and Mac Lane find categories of functors "useful chiefly in simplifying the statements and proofs of various facts about functors" [Eilenberg and Mac Lane, 1942, 250] and not in themselves. This is one of the key elements that was about to change drastically in the following years.

Eilenberg and Mac Lane defined two other important notions in their original paper: the dual $\mathcal{C}^o$ of a category $\mathcal{C}$ in §13 and limits and colimits for directed sets in §21 and §22. Dual categories play an important conceptual role in category theory and categorical logic. Given a category $\mathcal{C}$, the dual category $\mathcal{C}^o$ has as its objects those of $\mathcal{C}$; the mappings f^o of $\mathcal{C}^o$ are in one-to-one correspondence $f \rightleftarrows f^o$

with the mappings of $\mathcal{C}$. If $f : X \longrightarrow Y$ is in $\mathcal{C}$, then $f^o : Y \longrightarrow X$ is in $\mathcal{C}^o$. The composition law is defined by the equation $f^o \circ g^o = (g \circ f)^o$, whenever $g \circ f$ is defined in $\mathcal{C}$.

Before we move on, let us now quickly underline what one does *not* find in Eilenberg and Mac Lane's paper. First, although the notion of a subcategory is clearly defined in the paper, properties of the inclusion functor, or for that matter, basic properties of functors in general, for instance being faithful, full and essentially surjective, are not identified.

Although Eilenberg and Mac Lane did define the notion of isomorphism of categories, they did *not* define the notion of *equivalence* of categories. The distinction between the two concepts might seem to be formally subtle, but it is crucial in the applications of category theory. The notion of isomorphism between categories is just the same as the notion of isomorphism between objects in a category: two categories $\mathcal{C}$ and $\mathcal{D}$ are said to be *isomorphic* if there is an isomorphism between them, that is if there are functors $F : \mathcal{C} \longrightarrow \mathcal{D}$ and $G : \mathcal{D} \longrightarrow \mathcal{C}$ such that $G \circ F = 1_{\mathcal{C}}$ and $F \circ G = 1_{\mathcal{D}}$, where $1_{\mathcal{C}}$ and $1_{\mathcal{D}}$ denote the obvious identity functors. Two categories $\mathcal{C}$ and $\mathcal{D}$ are said to be *equivalent* if there is an equivalence between them, that is if there are functors $F : \mathcal{C} \longrightarrow \mathcal{D}$ and $G : \mathcal{D} \longrightarrow \mathcal{C}$ and natural isomorphisms $\tau : G \circ F \xrightarrow{\sim} 1_{\mathcal{C}}$ and $\rho : F \circ G \xrightarrow{\sim} 1_{\mathcal{D}}$. Thus, in the case of an equivalence, composing the functors F and G does not yield the identity functors, but there are *systematic translations*, namely natural isomorphisms, of the compositions to the identity functors. From the point of view of category theory, the notion of equivalence of categories is fundamental.

Although Eilenberg and Mac Lane introduced functor categories, they do not mention the possibility of a category of categories nor do they notice that natural transformations compose in two different ways. Of course, they did not need these concepts and therefore did not have to consider them at all. Interestingly enough, *all* these notions — functors with specific properties, equivalence of categories and, *in a certain sense*, the category of categories — will play a crucial role in the development of categorical logic in the nineteen sixties. But it can certainly be said without hesitation that the construction that will occupy the center stage of the development of categorical logic is the construction of functor categories.

1.2 Category theory from 1945 until 1963

We will now sketch the development of category theory from 1945 until 1963, underlying the points that will prove to be indispensable for the development of categorical logic.

Although, Eilenberg and Mac Lane introduced and defined the basic concepts of category theory, we believe that it is reasonable to claim that they did not introduce category *theory* as such. Category *theory* started in the late fifties and early sixties. For the *theory* to get off the ground, properties of categories and functors had to be introduced and used systematically. Those arose naturally in specific applications.

Eilenberg and Steenrod quickly applied category theory to algebraic topology. Their book *Foundations of Algebraic Topology* was published in 1952, but it circulated in the form of notes well before that date. It was extremely influential in many different ways. First, because it provided a systematic presentation of algebraic topology and clarified immensely how algebraic topology had to be organized and developed. Second, many students learned algebraic topology from their book and thus assimilated basic categorical concepts along the way. Eilenberg and Steenrod did not develop category theory itself, all the definitions are taken directly from Eilenberg and Mac Lane's paper, but they did use it in such a way that *diagrams* became a fundamental tool in the proofs of various results. Third, homology and cohomology theories were now *functors* and comparing theories amounted to looking at natural transformations between them. Thus, some mathematical objects were now best thought of as being functors between categories.

The same remarks apply to the equally influential book by Cartan and Eilenberg *Homological Algebra*, published in 1956. This book more or less created the subject of homological algebra. It relied heavily on the language of categories and in the use of diagrams in proofs.

Both books brought a shift of focus with respect to the original paper by Eilenberg and Mac Lane. First, it can be said that functors were moving to the front stage or were at the very least just as important as natural transformations. Second, both presented an obvious problem that was about to become a fundamental and general heuristic principle: to find the appropriate category to define and develop a certain aspect of mathematics. In the first case, it was the appropriate setting to express clearly and precisely the duality between homology and cohomology theories. In the second case, it was the search for the appropriate setting for the notion of derived functor. This last heuristic principle led to a second fundamental shift: categories could now be considered in themselves, not only as domains and codomains of functors, but as formal contexts with specific categorical properties in which one could identify, define and develop a specific portion of mathematics.

Thus, these last two problems led to the use of categories within an axiomatic framework. More precisely, it became imperative to define certain concepts, develop specific theories by stipulating that a category satisfied appropriate categorical properties. [Buchsbaum, 1955], [Grothendieck, 1957] and [Heller, 1958] defined the notion of an Abelian category in which a large portion of homological algebra can be carried out from a purely abstract or formal point of view. Furthermore, functor categories became a prominent tool in various fields: algebraic topology, especially homotopy theory, homological algebra, algebraic geometry and more and more constructions on categories were required and seen to be useful. Finally, in 1958, Daniel Kan introduced the notion of adjoint functor, arguably the core concept of the whole theory and which plays a key role in categorical logic as well.

Informally and as a rough heuristic guide, adjoint functors can be thought of as conceptual inverses. The formal definition goes as follows. Two functors

$F : \mathcal{C} \longrightarrow \mathcal{D}$ and $G : \mathcal{D} \longrightarrow \mathcal{C}$ together with natural transformations $\eta : 1_{\mathcal{C}} \longrightarrow G \circ F$ and $\xi : F \circ G \longrightarrow 1_{\mathcal{D}}$ determine *an adjunction* if $G\xi \circ \eta G = 1_G$ and $\xi F \circ F\eta = 1_F$, that is the following triangles commute:

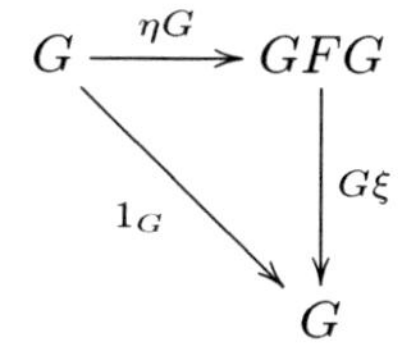

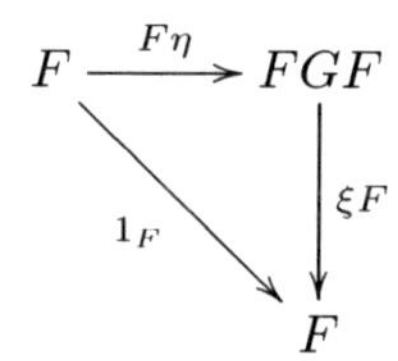

In the foregoing adjunction the functor F is said to be a *left adjoint* to G, denoted by $F \dashv G$, and G is said to be a *right adjoint* to F. As is usual in category theory, adjoints are determined up to a unique isomorphism, that is if $F \dashv G$ and $F \dashv G'$, then there is a unique natural isomorphism $G \longrightarrow G'$.

When Kan introduced the concept of adjoint functor in 1958, he immediately saw the generality of the concept, its usefulness and power in unifying various and apparently different results. Many important theorems and almost all fundamental mathematical constructions can be cast in the frame of adjoint functors. It is also a surprise to see that many fundamental constructions simply appear as adjoints to basic functors. (See [Mac Lane, 1998]; [Adamek *et al.*, 1990]; [Taylor, 1999] for many examples.)

By the time Grothendieck and Kan wrote their seminal papers, in 1955-56, constructions on categories, and in particular the construction and use of functor categories, had become pivotal. Their use of functor categories may not be totally unrelated to the fact that both mathematicians were coming from mathematical fields, namely functional analysis and homotopy theory respectively, in which functional spaces and their properties played a key role.

Adjoint functors and functor categories occupied right from the start a central role in categorical logic. The idea that logical operations, all logical operations, should appear as adjoints to basic functors was one of Lawvere's convictions and motivation.

2 LAUNCHING THE PROGRAM: 1963–1969

When they introduced the theory of categories in 1945, Eilenberg and Mac Lane suggested the possibility of "functorizing" the study of general algebraic systems. The author has carried out the first steps of this program, making extensive use of the theory of adjoint functors, as introduced by Kan and refined by Freyd [Lawvere, 1963, 869].

2.1 Basic principles

We now turn to logic proper. It is unquestionable that one person sparked the whole program of thinking about logic and the foundations of mathematics *in general* in a categorical spirit: F. William Lawvere. He did so in his very first work, namely his PhD thesis defended at Columbia under Eilenberg's supervision in 1963. It can certainly be said that the thesis already contains all the basic ideas that have guided Lawvere throughout his career and that have influenced the categorical community greatly. These ideas are in a nutshell:

1. To use the category of categories as a framework for mathematics, i.e. the category of categories should be the foundations of mathematics;
2. Every aspect of mathematics should be representable in one way or another in that framework; in other words, categories constitute the background to mathematical thinking in the sense that, in this framework, essential features of that thinking are revealed;
3. Mathematical objects and mathematical constructions should be thought of as *functors* in that framework;
4. In particular, sets always appear in a category, there are no such thing as sets by themselves, in fact there is no such thing as a mathematical concept by itself;
5. But sets form categories and the latter categories play a key role in the category of categories, i.e. in mathematics;
6. Adjoint functors occupy a key position in mathematics and in the development of mathematics; one of the guiding principles of the development of mathematics should be "look for adjoints to given functors"; in that way foundational studies are directly linked to mathematical practice and the distinction between foundational studies and mathematical studies is a matter of degree and direction, it is not a qualitative distinction;
7. As the foregoing quote clearly indicates, Lawvere is going back to the claim made by Eilenberg and Mac Lane that the "invariant character of a mathematical discipline can be formulated in these terms" [i.e. in terms of functoriality], [Eilenberg and Mac Lane, 1945, 145]. But now, in order to reveal this invariant character, extensive use of adjoint functors is made.
8. The invariant content of a mathematical theory is the "objective" content of that theory; this is expressed at various moments throughout his publications. To wit:

> As posets often need to be deepened to categories to *accurately reflect the content of thought*, so should inverses, in the sense of group theory, often be replaced by adjoints. Adjoints retain the virtue of being uniquely determined reversal attempts, and very often exist when inverses do not. [Lawvere, 1994, 47] (our emphasis)

9. Not only sets should be treated in a categorical framework, but also logical aspects of the foundations of mathematics should be treated categorically, in as much as they have an objective content. In particular, the logical and the foundational are directly revealed by adjoint functors.

As we have said, these ideas, as well as others, are more or less implicit in Lawvere's thesis. Lawvere's goal is general in the sense that it aims at incorporating the whole of mathematics. Furthermore, Lawvere's *usage* of categories reflects a change in their *status* among category theorists in the sixties and seventies. Lawvere recognizes explicitly that categories defined axiomatically constitute autonomous *kinds* or *types* and are, as such, independent of any underlying set-theoretical structures and structure preserving functions. Furthermore, categories become polymorphic: in addition to their usual role, they become the algebraic descriptions of formal systems and, as such, can be thought of as formal systems; but they also provide the underlying framework for semantics and, as such, can be thought of as universe of interpretations.

A summary of the main results of the thesis was communicated to the *Proceedings of the National Academy of Sciences USA* by Mac Lane and published in 1963. Essentially the same summary was presented at Berkeley in 1963 at a symposium on model theory and later published in 1965 in the proceedings of the meeting. Mac Lane also communicated Lawvere's axiomatization of the elementary theory of the category of sets in the same *Proceedings* in 1964. That axiomatization was not in the thesis as such. The following year, Lawvere presented an explicit axiomatization of the category of categories, published again in the *Proceedings* in 1966. Another paper, published in 1968, gives an account of the main elements of the thesis together with some new extensions.

In the early and mid-sixties, a certain methodological shift in attention can be detected in the work done by many category theorists. Following Kan and Grothendieck, certain mathematical theories are developed within categories, e.g. homotopy theory, algebra, and the development of these theories is done using the properties of categories more directly. (To mention but a few cases, all published in 1963: Bénabou with his categories with multiplication, C. Ehresmann with his structured categories and Eckmann and Hilton with their group-like structures in categories.) At the same time, adjoints receive more attention and are used more systematically, in particular to define and characterize various structures. Examples are once again provided by Bénabou's work on categories with multiplication, but others, slightly different examples can be found in Eilenberg and Moore and Kleisli on triples, as well as Eilenberg and Kelly with the notion of *closed category*. (See [Bénabou, 1963; Eilenberg and Kelly, 1966; Kleisli, 1965]). Needless to say, Grothendieck's work in algebraic geometry, based on the use of sheaf theory, is extraordinarily influential, but it is not in pure category theory.

Going back to his original program of clarifying the conceptual content of semantics, Lawvere realized that certain types of categories can be *defined* purely by stipulating that certain adjoint functors to *given elementary* functors exist. The definitions can be given by this data and nothing else. In a loose sense, defining a category via the existence of adjoints amounts to the claim that certain basic conceptual operations can be represented in that category. This in itself would probably not be of foundational relevance, were it not for the fact that the categories so defined correspond in a precise technical sense to *logical* concepts and theories. Thus the existence of certain adjoints to specific elementary functors amounts to a specification of logical structures and resources. With these ideas and results in his pocket, Lawvere could see that a program of "functorizing" the study of mathematical concepts in general could be formulated.

The presentation of these fundamental facts and the program that ensued were made at various conferences in the mid-sixties, published in the form of abstracts in 1966 and a series of papers published in 1969 and 1970. Among the latter, the paper entitled *Adjoints in Foundations* [Lawvere, 1969a] deserves special attention for its general philosophical orientation. It contains the seeds of a categorical program in logic and the foundations of mathematics. Two other papers of that period also contain important parts of that program: the paper [Lawvere, 1969b] on diagonal arguments presented in 1968 and published in 1969 and the paper [Lawvere, 1970b] on quantifiers and the comprehension schema as adjoints also presented in 1968 and published in 1970. The two abstracts [Lawvere, 1966; Lawvere, 1967] are also revealing and influential, for they concentrate on first-order logic. Finally, the discovery of the notion of elementary topos in collaboration with Tierney in 1969/70 provided the general framework in which the whole program could be cast and opened vast and rich possibilities that were unforeseen [Lawvere, 1970a].

We will now look more carefully at the details of this program. We will start with Lawvere's study of algebraic categories, look briefly at the elementary theory of the category of sets and then move to the so-called elementary theories. We will ignore Lawvere's axiomatization of the category of categories, since it did not have a direct impact on the development of categorical logic. Lawvere's work on the category of categories and the category of sets did not have the same fate as his work on universal algebra. Despite the fact that Lawvere's work on the category of categories suffered from a slight technical flaw, both it and his work on the category of sets were essentially metamathematical and category theory was not yet seen as a potentially useful framework for the latter. Studies on the category of categories that followed Lawvere's pioneering work were mathematically motivated and we speculate that no one saw what to do with the category of sets. It simply did not have a clear function. His work on algebraic theories, however, inspired much of what was to follow in logic, including Lawvere's own work, and it still constitutes the starting point of what are now called "doctrines", a term suggested to Lawvere by John Beck, in categorical logic and the categorical approach to universal algebra.

2.2 Lawvere's thesis: 1963

> Lawvere's imaginative thesis at Columbia University, 1963 contained his categorical description of algebraic theories, his proposal to treat sets without elements and a number of other ideas. I was stunned when I first saw it; in the spring of 1963, Sammy and I happened to get on the same airplane from Washington to New York. He handed me the just completed thesis, told me that I was the *reader*, and went to sleep. I didn't. ([Mac Lane, 1988, 346].)

> Essentially, algebraic theories are an invariant notion of which the usual formalism with operations and equations may be regarded as "presentation". [Lawvere, 1963, ii]

The main concept of Lawvere's thesis is the notion of *algebraic category.* The main result of the thesis is a *categorical* characterization of algebraic categories. Together with algebraic categories, Lawvere also introduced *algebraic theories* and *algebraic functors.* The three notions are intimately connected to one another. As Lawvere pointed out himself, there is a strong analogy between the way his work is developed and the theory of sheaves that had been just introduced at that time: in the same way that Grothendieck had provided an abstract characterization of categories of sheaves on topological spaces, Lawvere's goal was to characterize algebraic categories in a similar manner. The main tool of the thesis and what provides the key to the connections between these notions are adjoint functors. Thus, they constitute the methodological core of the thesis and of the whole approach. The framework is presented as a new foundation for universal algebra. In the very first chapter of the thesis, thus the underlying context of the work, there is a sketch of a first-order theory of the category of categories. Within that context sets are defined in categorical terms, the notion of equivalence of categories is given, as well as the category of small categories, the category of large categories, the category of finite sets, the category of small sets, and the category of large sets. A categorical version of the Peano postulates is also given.[1] But the bulk of that chapter is occupied by the presentation and development of the notions of adjoint functors and limits. In the second chapter, algebraic theories are introduced and the category of algebraic theories is defined and various properties of the category are proved, e.g. the existence of an adjoint that corresponds to the existence of free algebraic theories. Chapter three deals with algebraic categories (and here Lawvere explicitly exploits the analogy with sheaves). The notions of algebraic

[1]These axioms are sometimes called the Peano-Lawvere axioms. See for instance [Birkhoff and Mac Lane, 1967, 67].

semantics and algebraic structure are defined and a categorical characterization of algebraic categories is given. Chapter IV deals with algebraic functors and their adjoints. Finally, in chapter V, particular cases and extensions are considered.

Let us look at the central concepts and results of the thesis and see how the invariant content of universal algebra is analyzed.

First comes the notion of an algebraic theory. A group is usually thought of as a set together with some specified operations, e.g. multiplication, inverse and unit, satisfying certain identities, e.g. associativity, unit law and inverses. Formally, this is encoded by a signature, e.g. $(\times, (-)^{-1}, e)$ or some other, with the standard axioms. But it is clear that the signature can in general vary and so does the choice of axioms. These choices determine, though, the *same* theory, in the sense that all the definable operations and all the theorems are the same.[2] Lawvere's idea is to define a category which will encode all the information at once, thus independently of the choice of signature and axioms and call *that* category the theory. Thus, the latter category would in some sense be the objective encoding of the theory, independent of any presentation of it. This is in stark contrast with the traditional logical approach in which a theory T depends directly on a choice of primitive symbols, its signature, and a choice of axioms. Strictly speaking, changing the axioms changes the theory, although, for instance we still talk about *the* theory of groups.

Here is the general definition:[3]

Definition: an *algebraic theory* is a (small) category $\mathcal{A}$ such that:

(i) the objects of $\mathcal{A}$ is a denumerable set $\{A^0, A^1, A^2, \ldots, A^n, \ldots\}$ of distinct objects;

(ii) each object A^n is the product of A^1 with itself n times; thus, for each n, the projection maps $\pi_i^{(n)} : A^n \longrightarrow A$, for $i = 0, 1, \ldots, n-1$, exist;

(iii) for any n morphism $A^m \xrightarrow{\vartheta_i} A$, for $i = 0, 1, \ldots, n-1$, in $\mathcal{A}$, there exists exactly one morphism $A^m \xrightarrow{\langle \vartheta_0, \vartheta_1, \ldots, \vartheta_{n-1} \rangle} A^n$ such that $\pi_i^{(n)} \circ \langle \vartheta_0, \vartheta_1, \ldots, \vartheta_{n-1} \rangle = \vartheta_i$, for $i = 0, 1, \ldots, n-1$.

The arbitrary morphisms $A^n \xrightarrow{\eta} A$ are the *n-ary operations* of $\mathcal{A}$.

Thus, the theory, thought of as a category, contains all the possible operations systematically.

The underlying motivation is very simple and makes perfect sense once the notion of an *algebra of type* $\mathcal{A}$, also called an $\mathcal{A}$*-algebra*, has been given: it is

[2]The second author recalls that in one of his courses in the late 60's at Berkeley, Tarski told the class that there wasn't such a thing as *the* theory of groups, but there were infinitely many theories of groups.

[3]We are here presenting the direct definition given by Lawvere in his published papers. In his thesis, Lawvere defines the *category* of algebraic theories as a subcategory of the category of finite sets, which is itself a category in the category of categories. Pareigis' presentation is, in this respect, more faithful to Lawvere's original work. See [Pareigis, 1970].

simply a product preserving functor from an algebraic theory to the category of sets, $F : \mathcal{A} \longrightarrow Set$. Thus, $F(A)$ picks a set, and $F(A^n)$ is simply an n-fold product of $F(A)$, i.e. $F(A) \times \cdots \times F(A)$ n-times. An operation $A^n \xrightarrow{\eta} A$ becomes a standard set-theoretical operation $F(A) \times \cdots \times F(A) \xrightarrow{F(\eta)} F(A)$. Notice that an algebra of type $\mathcal{A}$ *is* a *functor*. It is also called a *model* of the theory $\mathcal{A}$. Thus, in particular, if $\mathcal{A}$ is the theory of groups, then each and every group *is* a functor.

Two categories can now be defined: 1. The category $\Im$ of algebraic theories whose objects are algebraic theories, morphisms are functors preserving products and taking 1 to 1; 2. The functor category $Set^{(\mathcal{A})}$ of all *product preserving* functors $\mathcal{A} \longrightarrow Set$, which can be thought of as the category of models of the theory $\mathcal{A}$. The latter category is called an *algebraic category.*

A morphism $f : \mathcal{A} \longrightarrow \mathcal{B}$ of algebraic theories induces a functor $Set^{(f)} : Set^{(\mathcal{B})} \longrightarrow Set^{(\mathcal{A})}$ by composition. The latter functor is called an *algebraic functor.* Furthermore, for any algebraic category $Set^{(\mathcal{A})}$, there is an obvious forgetful functor $U_{\mathcal{A}} : Set^{(\mathcal{A})} \longrightarrow Set$ which sends to each object $F(A^n)$ the underlying set and to each morphism the underlying set map. Notice that the forgetful functor is an algebraic functor.

Algebraic functors lead to another construction, named *algebraic semantics*: it assigns to each algebraic theory $\mathcal{A}$, the forgetful functor $U_{\mathcal{A}}$ and to each morphism $f : \mathcal{A} \longrightarrow \mathcal{B}$ of algebraic theories the algebraic functor $Set^{(f)} : Set^{(\mathcal{B})} \longrightarrow Set^{(\mathcal{A})}$. This is in fact itself a functor, sometimes called the *semantic functor*,

$$\mathrm{S} : \Im^{op} \longrightarrow \mathrm{K}$$

where K is the category of algebraic categories.

With these definitions, Lawvere's main results are:

1. Every algebraic functor has a (left) adjoint. This is the conceptual and unified formulation of various constructions in universal algebra, e.g. free algebras, tensor algebras, monoid rings, etc.

2. Algebraic semantics has a (left) adjoint, which can be called *algebraic structure.* This means that it is possible to recover an algebraic theory from the semantics, i.e. from the category of models.

3. The categorical characterization of algebraic categories: if $\mathcal{C}$ is a category with finite limits, has an abstractly finite regular projective generator G and every precongruence in $\mathcal{C}$ is a congruence, then there is an algebraic theory $\mathcal{A}$ and an equivalence $\Phi : \mathcal{C} \longrightarrow Set^{(\mathcal{A})}$. (There is no need to specify what the second and third conditions mean here. They are technical conditions that we do not have to look into.)

These results were not only remarkable for what they accomplished, but also for the research avenues they opened. These three points in themselves will become guidelines in categorical logic. There were some obvious generalizations that were taken up rapidly by others and not so obvious generalizations that had to wait for other concepts to be fully worked out.

1. The obvious generalization was to consider infinitary operations and related work in universal algebra. This was done quickly by Linton. (See [Linton, 1966b], [Linton, 1966a].)

2. One of the great advantages of the categorical language is that it is possible to replace the category *Set* of sets by an arbitrary category $\mathcal{C}$ with appropriate properties. [Eilenberg and Moore, 1965] and [Barr and Beck, 1966] used triples (monads) to extend Lawvere's work over an arbitraty base category. [Bénabou, 1968] considered the case of many-sorted theories. The identification of the properties of $\mathcal{C}$ required to do the work, expressed in categorical terms, leads to a *classification* of logical categories in *categorical* terms. The category *Set* of sets becomes a special, but very important, case of a type of category defined abstractly. Lawvere has given a characterization of algebraic categories. Further work lead to characterizations of similar categories, i.e. categorical characterization of semantic frameworks.

3. An algebraic theory as defined by Lawvere can be thought of as a data type.[4] Lawvere's work shows how syntactical information of a specific kind can be encoded by categories. The search for a proper generalization to cover all types of logical theories, not only the algebraic or equational case, is irresistible. More specifically, the task is to find a general procedure to move from a theory written in a given formal system to a category that would be the invariant formulation of the latter. The notion of algebraic theory was specifically tailored for algebraic structures and it is not clear how one can go from there to other cases, e.g. cases with quantifiers and relations. In particular, Lawvere considered single-sorted theories and a generalization to many-sorted seems natural, although in traditional logical presentations, we are used to the single-sorted case.

4. Once an element of one or all the previous points have been settled, the next task consists in looking at the various adjoint situations and see what one can obtain from them. For instance, it appears clearly from Lawvere's work that the adjoint situation is a special case of an algebraic duality and its importance is due to the fact that it is the very first case of such a duality where the category of sets appears as the dualizing object.

Mathematicians and logicians took up these tasks soon after Lawvere's thesis. Important results were obtained in the late sixties by Lawvere himself, but also by Lambek, Freyd, Linton, Isbell and by Gabriel and Ulmer. (See [Gabriel and Ulmer, 1971].) Parallel and independent developments were obtained by Ehresman and

[4]This is more than a metaphor. There is indeed a formal connection with data bases. See for instance [Plotkin, 2000].

his students, most notably Bénabou, also during that period. Ehresman's motivation was different and was mainly oriented towards the foundations of differential geometry, but it lead him to the notion of sketch which was developed by his students C. Lair and R. Guitart in the seventies and independently rediscovered by M. Makkai and R. Paré in the eighties. (See [Makkai and Paré, 1989].)

2.3 The elementary theory of the category of sets

Let us now briefly turn to Lawvere's axiomatization of the category of sets.

We should emphasize right from the start the fact that Lawvere never tried to *eliminate* sets altogether from the foundational landscape, but rather he tried to provide a *categorical* analysis of sets, that is a characterization of the category of sets. Mac Lane, for one, at first thought that the idea was absurd.

> ... and then [Lawvere] conceived the idea of giving a direct axiomatic description of the category of all categories. In particular, he proposed to do set theory without using the elements of a set. His attempt to explain this idea to Eilenberg did not succeed; I happened to be spending a semester in New York (at the Rockefeller University), so Sammy asked me to listen to Lawvere's idea. I did listen, and at the end I told him "Bill, you can't do that. Elements are absolutely essential to set theory." After that year, Lawvere went to California. [Mac Lane, 1988, 342]

Lawvere's basic insight is that, even for sets, one can know them by looking at the invariant content of the category of sets.

> Thus we seem to have partially demonstrated that even in foundations, not Substance but invariant Form is the carrier of the relevant mathematical information. [Lawvere, 1964, 1506]

> Some years ago I began an introductory course on Set Theory by attempting to explain the *invariant content* of the category of sets, for which I had formulated an axiomatic description. [Lawvere, 1994, 5] (our emphasis)

Let us briefly look at this invariant content, at least as Lawvere saw it in the early sixties.

Lawvere assumes the standard axioms for a category and then postulates that the category of sets has a terminal object 1, an initial object 0, binary products and coproducts, equalizers and coequalizers, thus all finite limits and colimits (axiom 1). He also assumes that it is Cartesian closed, i.e. that the object B^A, together with the known morphism and universal property, exist for any A and B (axiom 2). These two axioms are "structural" and are satisfied by many categories. Thus, as such they do not characterize a category of sets.

The next axiom is known as the Peano-Lawvere characterization of the natural numbers. It is therefore an axiom of infinity, but it contains more when it is assumed with the previous axioms.

Axiom 3: There is an object N together with morphisms $1 \xrightarrow{z} \mathrm{N} \xrightarrow{s} \mathrm{N}$ such that given any object X together with mappings $1 \xrightarrow{x_0} \mathrm{X} \xrightarrow{s} \mathrm{X}$, there is a unique morphism $\mathrm{N} \xrightarrow{x} \mathrm{X}$ such that $x_0 = z \circ x$ and $x \circ t = s \circ x$.

This is now the standard characterization of the natural number system by a universal property. A theorem on primitive recursion follows from it and the preceding axioms. However, Lawvere points out that it is satisfied, as well as the first two axioms, by the category **1**, i.e. the category with one object and one morphism. Thus, more is needed, both structurally and in terms of existence.

The next axiom is false in the category of categories, for it says that the terminal object 1 is a *generator*, i.e. if the morphisms $A \underset{g}{\overset{f}{\rightrightarrows}} B$ are different, then there is a morphism $x : 1 \longrightarrow A$ such that $f \circ x \neq g \circ x$. In more colloquial language, this axiom states that if f and g are different, then there is an element x, or a point, in A such that f and g are different on that element.

The fifth axiom is the axiom of choice. The last three axioms are *not* expressed in a categorical fashion. Surprisingly perhaps, Lawvere reintroduced the $\in$ notation in his axiomatization. Since there is a one-to-one correspondence between the morphisms $x : 1 \longrightarrow A$ and the elements $x \in A$ in the universe of sets, it seems reasonable to say that x is an *element* of A if and only if $x : 1 \longrightarrow A$. The remaining axioms are then expressed as follows:

Axiom 6: If A is not an initial object, then A has elements.

Axiom 7: An element of a sum is a member of one of the injections.

Axiom 8: There exists an object with more than one element.

Lawvere underlined the fact that the first seven axioms are satisfied by the category **1** with one morphism, thus the need for the eighth axiom. One could certainly object to the introduction of the $\in$ relation and the notion of element in a categorical framework and claim that this part of the axiomatization does not capture the invariant form of the category of sets.

Be that as it may, the claim that the invariant form of the concept of set is captured by the axioms is substantiated by a metatheorem and its corollary. Indeed, the metatheorem asserts that any two categories satisfying the eight axioms are equivalent, thus in particular any (complete) category satisfying the eight axioms is equivalent to the category of sets. Lawvere is in a position to state precisely what it means to be "the" category of sets: the properties are (mainly) expressed in categorical terms and the invariance amounts to the claim that any other category satisfying these properties is equivalent to it.

Lawvere's work did not open the door to further investigation along similar lines. The category of sets was not taken as a foundational framework. It was not studied and explored. Although there is no clear explanation of this fact, Johnstone suggested that the category of sets is simply too "rigid":

> In retrospect, the answer is that Lawvere's axioms were too specialized: the category of sets is an extremely useful object to have as a foundation for mathematics, but as a subject of axiomatic study it is not (*pace* the activity of Martin, Solovay *et al.*!) tremendously interesting — it is too "rigid" to have any internal structure. [Johnstone, 1977, xiii]

It is not exactly clear why it is such "an extremely useful object to have as a foundation for mathematics" if nothing is developed in it and no one proceeds to do more research within it. It seems that one of the most common reactions at the time was that ETCS was merely a translation in categorical terms of the standard axioms of set theory and that, as such, did not provide any genuinely new insight into the nature of sets (in contrast with the analysis provided with the axioms of elementary toposes, which was about to appear). Sociological and historical factors must probably be invoked also. Lawvere's paper appeared in 1964, very shortly after Cohen's proof of the independence of the continuum hypothesis was published. (See [Cohen, 1963].) Cohen's result, and probably most notably his method of forcing, deservedly attracted much attention. There is very little doubt that one of the key advantages of topos theory over ETCS is precisely that the former bridges the gap between a categorical description of sets and the method of forcing, whereas the latter is a category of sets satisfying the axiom of choice. It should be recorded that Miles Tierney's proof of the independence of the continuum hypothesis in a topos theoretical setting is done with respect to Lawvere's ETCS. The latter is taken to be the categorical expression of ZF set theory, reinforcing the impression that ETCS is nothing more than a translation.

Although the ETCS did not attract much attention, the general program was launched and it did not take long for Lawvere and others to see how logic could be and perhaps should be developed in a categorical setting.

2.4 Categorical logic: the program

> Categorical logic, in a very broad sense, can be seen to derive from the completeness and exactness properties of the category of sets, in a manner paralleling the earlier development of Abelian categories. [Makkai and Paré, 1989, 5]

Lawvere pursued his work in the categorification of logic and presented two papers at the Meeting of the Association of Symbolic Logic, one in 1965 and the other in 1966. Abstracts of these talks were published in *The Journal of Symbolic Logic* in the following year. The first one was received the 17 August 1965 and the second almost a year later, the 12 August 1966. They were published a year later. (See [Lawvere, 1966; Lawvere, 1967].) The second note offers a simplification of the first attempt. It is entirely clear that Lawvere is trying to extend his results obtained for algebraic theories to first-order theories in general. In these two notes, three elements stand out. First, it is the first time in print that the existential quantifier is presented as an adjoint. Second, every first-order theory with equality corresponds to a category with certain properties that Lawvere calls an *elementary theory.* In particular, certain elements of how semantics of first order languages ought to be done in a category are suggested. Third, in the second abstract, Lawvere sketches a categorical proof of the completeness theorem that amounts to the existence of specific adjoint functors. These abstracts were in fact influential and did serve as guides in the late sixties and early seventies for a categorical analysis of first-order logic. Here are the essential elements.

We will here present the notions as they are defined in 1966, for it is the latter that were referred to by various logicians who followed Lawvere's footsteps.

An *elementary theory* $\mathbb{T}$ is a small category with finite products, including the empty product 1, satisfying three conditions:

1. There is a distinguished object B together with two morphisms $1 \underset{\bot}{\overset{\top}{\rightrightarrows}} B$ such that for all objects X, the morphisms $\langle 1_X, \top_X \rangle : X \longrightarrow X \times B$ and $\langle 1_X, \bot_X \rangle : X \longrightarrow X \times B$, where $\top_X : X \longrightarrow 1 \xrightarrow{\top} B$ and $\bot_X : X \longrightarrow 1 \xrightarrow{\bot} B$, constitute a categorical coproduct of X with itself, that is $X \times B = X + X$; it follows that for each object X of $\mathbb{T}$, the functor $Hom(X, B)$[5] is in fact a Boolean algebra;

2. For any $f : X \longrightarrow Y, \varphi : X \longrightarrow B$, there is a morphism $\exists_f[\varphi] : Y \longrightarrow B$ such that for all $\psi : Y \longrightarrow B, \exists_f[\varphi] \leq_Y \psi$ if and only if $\varphi \leq_X \psi \circ f$;

3. There is a second distinguished object A of $\mathbb{T}$ such that every object X of $\mathbb{T}$ has a unique representation of the form $A^n \times B^k$, where n, k are finite non negative integers.

Morphisms $A^n \longrightarrow A$ are thought of as n-ary terms and morphisms $A^n \longrightarrow B$ are thought of as n-ary formulas. In particular, $Hom_{\mathbb{T}}(1, B)$ is the set of sentences of $\mathbb{T}$. A *model* of $\mathbb{T}$ is any product preserving functor $M : \mathbb{T} \longrightarrow Set$ which takes B to a two-element sets and which takes each $\exists_f[\varphi]$ to the image of φ under f, that is if, as above $f : X \longrightarrow Y$ and $\varphi : X \longrightarrow B$, then identifying $M(B)$ with the two-elements set **2**, we have that an element $y \in M(Y)$ is in $M(\exists_f[\varphi])^{-1}(1)$ if and only if *there exists* $x \in M(\varphi)^{-1}(1)$ such that $M(f)(x) = y$. The set $M(A)$ is called the *universe* of the model. Morphisms of models are natural transformations, thus yielding a category $Mod_{Set}(\mathbb{T})$ of set-models. Notice that the category Set satisfies all the axioms, with the only exception that it is not small. A morphism of theories $I : \mathbb{T} \longrightarrow \mathbb{T}'$ is a product preserving functor that preserves quantification. Lawvere

[5] $Hom(X, Y)$ denotes the collection of morphisms from X to Y in the category.

notices that there is an induced functor $I^* : Mod_{Set}(\mathbb{T}') \longrightarrow Mod_{Set}(\mathbb{T})$ which preserves universes.

In his first abstract, Lawvere extends the claim we find in his thesis to elementary theories as follows. Let $\mathfrak{T}$ be the category of elementary theories with morphisms of theories $I : \mathbb{T} \longrightarrow \mathbb{T}'$ and $\mathfrak{M}$ the category of set-models. Then, as indicated in the foregoing paragraph, there is a functor $\Phi : \mathfrak{T}^{op} \longrightarrow \mathfrak{M}$, called *elementary semantics* and Lawvere claims that it has an adjoint, called *elementary structure*. No details are given.

The main claim of the second abstract is a completeness theorem. The precise statement of the theorem is not given in the abstract. Lawvere enumerates three conditions that constitute the core of the *proof* of the theorem: it sketches how, given an elementary theory $\mathbb{T}$, a model M of $\mathbb{T}$ can be constructed.

As we will see, the notion of an elementary theory served as a guide in the search for the correct categorical characterization of first-order logic. Also, the description of the quantifiers as adjoints to substitution, more generally the description of the logical operations as adjoints and the basic element of the semantics left a definitive impact. Interestingly enough, Lawvere himself suggests, in his first abstract, that a given theory $\mathbb{T}$, as a category, could be thought of as the *"Sinn"* of a theory presented in a first order language, and the category $Mod_{Set}(\mathbb{T})$ could be thought as the *"Bedeutung"* of the theory. Whether this can be defended and how it compares to other analyses of sense and reference, for instance Frege's, remains to be clarified and as far as we know, no one took it up afterwards. (But look at Bell's *Epilogue* in his textbook on topos theory, i.e. [Bell, 1988, chapter 8]. Awodey's paper on structuralism is also relevant here, [Awodey, 1996].)

Here is how quantifiers can be thought of as adjoints to substitution. Recall that if $f : X \longrightarrow Y$ is a function between sets and $B \subseteq Y$, then the pullback of B along f is the subset of X defined by $f^*(B) = \{x \in X : f(x) \in B\}$. By considering B as a predicate of Y, say $B(y)$, the pullback may be considered as the predicate of X obtained from $B(y)$ by substituting $f(x)$ for y. Lawvere's remark was that the pull-back formation

$$f^* : \wp(Y) \longrightarrow \wp(X)$$

which is a functor between posets has a left adjoint and a right adjoint

$$\exists_f \dashv f^* \dashv \forall_f$$

Indeed, one can easily check that

$$\begin{aligned} \exists_f(A) &= \{y \in B : \exists_x f(x) = y \text{ and } x \in A\} \text{ and} \\ \forall_f(A) &= \{y \in B : \forall_x f(x) = y \text{ implies } x \in A\}. \end{aligned}$$

This was a key observation that convinced many mathematicians that this was the right analysis of quantifiers. They arise naturally as adjoints to an elementary operation, namely substitution, which appears as *the* basic operation of first-order logic, contrary to the classical view which defines this operation by recursion, as a derived one. By the way, this recursive definition is not universally valid. In particular, it gives wrong results in some non-classical logics (e.g. co-Heyting). Again, apart from the intrinsic interest of the conceptual analysis it provides, this point of view generalizes to categories other than the category of sets. Once the *categorical* properties of sets used in the characterization have been identified and defined, it is possible to transpose the analysis in new contexts, i.e. in other categories.

After 1966, Lawvere's own published work extends the connections between categories and logic. Three papers published in 1969 and 1970 are extremely important since they contain the seeds and the statement of a vast foundational program which was taken up and is still alive. One of the surprising features of these papers is that none of them contain a bibliography. There is no reference to other papers, books or earlier related work, which we believe can be taken as an sign of the originality of the work presented. Here are the basic methodological elements underlying this program.

1. The use of adjoint functors is emphasized both in practice and also from a more general point of view. In his paper *Diagonal arguments and Cartesian closed categories*, Lawvere defines a CC category as a category $\mathcal{C}$ equipped with three kinds of right adjoints:

1. a right adjoint 1 to the unique $\mathcal{C} \longrightarrow 1$;
2. a right adjoint $(- \times -)$ to the diagonal functor $\Delta : \mathcal{C} \longrightarrow \mathcal{C} \times \mathcal{C}$;
3. and for each object A in $\mathcal{C}$, a right adjoint $(-)^A$ to the functor $A \times - : \mathcal{C} \longrightarrow \mathcal{C}$, [Lawvere, 1969b]].

Before that paper, adjoints were used by Lawvere to show that any category satisfying certain specifications was equivalent to a fixed category of interest or to establish certain properties of given functors in a context. In this Lawvere was following the examples of Abelian categories and sheaves. As we have seen, the axioms for the category of sets did not mention adjoints explicitly and the notion of elementary theories stipulated the existence of two distinguished *objects* of a category. Right from the beginning, Lawvere emphasized the fact that Cartesian closed categories are algebraic versions of type theories. Notice two important facts in the foregoing characterization of a CC category: first, the first two adjoints are adjoints to basic functors or structural functors and as such arise "naturally" and express fundamental connections; second, the third adjoint depends on the second and, in this way, there is a hierarchy or an order in the way the fundamental operations are defined or related to one another. This is a key element of the analysis of logical operations in a categorical setting.

2. In connection with the role of adjoint functors, there is an *explicit* recognition of the levels of abstraction introduced by category theory, more specifically the fact that category theory now allows for purely abstract characterizations of mathematical domains.

> More recently, the search for universals has also taken a conceptual turn in the form of Category Theory, which began with viewing as a new mathematical object the totality of all morphisms of the mathematical objects of a given species A, and then recognizing that these new mathematical objects all belong to a common non-trivial species C *which is independent* of A. [Lawvere, 1969a, 281]. (our emphasis)

What is described by Lawvere should be clear : in the beginning, mathematicians started with already defined mathematical objects and structure-preserving functions and moved to a new object, namely the *category* of these objects of species A. But it was soon realized that such a category participated in a different species, namely a category of type C, which can be described *independently* of A. The first historical example of this phenomenon was provided by Abelian categories: one started with Abelian groups (or modules over a commutative ring), the latter constituting the objects of species A, moved to the category of Abelian groups (or modules over a commutative ring) and then to an Abelian category, which is a category of type C. From a conceptual and as well as a foundational point of view, the crucial step is that the latter can be described independently of the former. Thus, although a category of type C arises from given structures, and in the last century, that meant more often than not, structured sets with structure-preserving functions between them, once the abstract description has been given, it is possible to develop mathematics directly from categories of type C. Lawvere makes a bold generalization: he sees this case as a *general* phenomenon, even as a framework that should guide the development and analysis of mathematics. Of course, Abelian categories did not constitute his only example, e.g. algebraic theories and categories, Cartesian closed categories and hyperdoctrines, etc.

3. Lawvere uses Cartesian closed categories to present an analysis of well-known diagonal arguments, i.e. those of Cantor, Russell, Gödel and Tarski. The motivation is similar to the one indicated in the foregoing paragraph: these diagonal arguments are similar and thus seem to form a *species* of argument. Lawvere hopes to be able to disclose the common abstract structure underlying them. This abstract structure takes the form of a fixed point theorem based on the properties of Cartesian closed categories. (For an excellent presentation of these results together with work inspired by Lawvere along these lines, see [Yanofsky, 2003].) In the process of his analysis, Lawvere introduces an object of truth-values **2**, for this object appears in one way or another in all the arguments. This object, which will be called *the object of truth-values* and had already been introduced in the context of elementary theories, as well as the Cartesian closed structure, will become pivotal in his characterization of the notion of elementary topos in 1969. However, the object of truth-values will constitute an *obstacle* for the characterization of first-order logic.

4. In all three papers, Lawvere suggests extensions of his earlier work on algebraic theories to theories written in higher-order type theories and, as a special case, first-order theories. We should point out immediately that he was no longer

alone in looking for connections between categories and logic. Lambek's work on categorical analysis of deductive systems has to be mentioned at this stage. (See [Lambek, 1968a; Lambek, 1969; Lambek, 1972].) We will get back to Lambek's work later.) Lawvere's extensions are based on the following fundamental facts:

1.1. As we have already seen, the logical quantifiers can be presented as adjoint functors to the simple and fundamental operation of substitution (the analysis is extended in [Lawvere, 1969c; Lawvere, 1970b]);

1.2. The comprehension principle can be presented as an adjoint functor in a proper context [Lawvere, 1969a; Lawvere, 1970b];

1.3. Lawvere sketches how one can construct a category from a given theory formalized in higher-order logic [Lawvere, 1969b; Lawvere, 1970b].

We have to underline the fact once more that a categorical analysis of logical systems not only provided a novel and unifying understanding of logical operations and systems, but by the same token, it initiated a shift in the status of categories *themselves*. It is now possible to *identify* a type of category with a type of deductive system. The claim that category theory can be seen as a language can now be made more precise: category theory can be seen as a *formal* language for mathematics. However, it should also be emphasized that in 1969, these were all programmatic claims.

5. All these constructions are incorporated in a general framework that constitutes nothing less than the scaffolding of categorical logic as it developed afterwards.[6] This framework is presented in very broad strokes at the beginning and the end of the papers entitled "Adjointness in Foundations" and is itself of considerable philosophical interest.

Lawvere identified two aspects, which he qualifies as being "dual", since they appear to obey some sort of general duality inherent to mathematics, namely the Formal and the Conceptual. The Formal is more or less identified with the manipulation of symbols, either in deduction or calculations, whereas the Conceptual is identified with the content of these symbols, the subject matter of the Formal or what they refer to. Thus, at first sight, Lawvere's terminology coincides with the classical distinction between the syntax and the semantics of formal languages. However, Lawvere has the actual practice of mathematics in mind and therefore does not equate his distinction with the fundamental metamathematical distinction. In fact, he sees foundational research as being part of mathematics: "Being itself part of Mathematics, Foundations also partakes of the Formal-Conceptual duality." [Lawvere, 1969a, 281]. Thus, the syntax of a logical system is part of the Formal, whereas the semantics is part of the Conceptual. However, his presentation of the semantics is somewhat odd: "Naturally the formal tendency in

[6] At least before categorical logic became a standard tool in theoretical computer science. But even in this context, one can see the influence of Lawvere's suggestions.

Foundations can also deal with the conceptual aspect of mathematics, as when the semantics of a formalized theory $\mathbb{T}$ is viewed itself as another formalized theory $\mathbb{T}'$, or in a somewhat different way, as in attempts to formalize the study of the category of categories." [Lawvere, 1969a, 281]. Category theory is clearly put on the conceptual side of mathematics and, in fact, one can see that Lawvere sees his work on the foundations of universal algebra and the subsequent work on Cartesian closed categories and extensions thereof as being part of the *conceptualization* of the formal aspect of mathematics.[7] Indeed, he claims explicitly that "Foundations may conceptualize the formal aspect of mathematics, leading to Boolean algebras, cylindric and polyadic algebras, and certain of the structures discussed below." [Lawvere, 1969a, 282]. At the center of this conceptualization appear adjoint functors.

Adjoints are present in Foundations in two senses.

1. Lawvere introduces Cartesian closed categories and what he calls hyperdoctrines in that paper. The main property of these two concepts is that they are *entirely* given by adjoint functors. However, as we have already mentioned, Cartesian closed categories and hyperdoctrines correspond in a precise technical sense to logical frameworks. Thus, both Cartesian closed categories and hyperdoctrines are categorical codifications of logical structures, the algebraic counterpart of these structures. Adjoint functors are used to *define* the conceptual content of foundations;

2. However, adjoint functors also play a more general role. The Formal and the Conceptual mentioned above should be, but we are clearly at a programmatic stage here, related by adjoint functors. Lawvere is in fact more precise in the way these adjoints should show up and here we see Lawvere generalizing the work contained in his thesis. First, Lawvere suggests that one should consider *categories* of models of a theory, thus framing model theory in the context of category theory.[8] More specifically, a model, in the standard model theoretic sense, can de described as a *functor* from a category $\mathbb{T}$ to the category Set of sets. The category of such models is then a subcategory of the functor category $Set^{\mathbb{T}}$, or also written $\text{Mod}(\mathbb{T}, Set)$ or, $Mod_{Set}(\mathbb{T})$ as we have already done. Second, the category $\mathbb{T}$ is the categorical encoding of a given formal theory. In Lawvere's own term "The *invariant* notion of theory here appropriate has, in all cases considered by the author, been expressed most naturally by identifying a theory T itself with a category of a certain sort" [Lawvere, 1969a]. We emphasize the fact that Lawvere is looking, once again, for an *invariant* notion of a theory and that this invariant notion is provided by a category. Third, and adjoint functors enter

[7]Indeed, in the nineties Lawevere published, in collaboration with Steve Schanuel, a book entitled "Conceptual Mathematics" (see [Lawvere and Schanuel, 1997]).

[8]We have to point out that Freyd had made explicit connections between category theory and model theory before. See [Freyd, 1965].

the picture explicitly at this stage, in principle there ought to be an adjoint pair of functors encapsulating the general duality expressed above

$$\text{Theories}^{op} \underset{\text{structure}}{\overset{\text{semantics}}{\rightleftarrows}} \text{Conceptual.}$$

The conceptual is identified with the category of models of $\mathbb{T}$.[9] However, the formal is *not* identified with the invariant formulation of the theory, since clearly there are aspects of the formal, e.g. specific rules of computation or derivation, that are inherent to a formal framework. Therefore, Lawvere suggests that there is a further adjoint situation, left unspecified this time

$$\text{Formal} \rightleftarrows \text{Theories}$$

which describes "the presentation of the invariant theories by means of the formalized languages appropriate to the species" [Lawvere, 1969a, 295]. Since adjoint functors compose, one should get a family of adjoint functors

$$\text{Formal}^{op} \rightleftarrows \text{Conceptual}$$

which we started with.

As such, this description is completely programmatic. It is clearly a bold generalization of Lawvere's thesis. It is given as a completely general framework for foundational research. It is taken as being faithful to the essential elements of mathematical knowledge. Logicians who were about to enter the scene picked up that program and started to develop it systematically. It lead to what were called afterwards "categorical doctrines" and is presented as such by Kock and Reyes in their survey paper (see [Kock and Reyes, 1977]). Thus, we strongly disagree with the claim made by Corry:

> Lawvere himself proposed in an article of 1969 to connect the concept of duality, and other categorical concepts, with the epistemological issues related to the philosophy of mathematics. In order to do that, he identified two "dual aspects" of mathematical knowledge — the conceptual and the formal aspects — which appear in many domains of mathematics. .... Now Lawvere proposed to dedicate efforts to develop the second aspect, the conceptual one, embodied in category theory. This proposal, however, remained at the programmatic level and no one seems to have developed it further. [Corry, 1996, 388].

It *is* true that different mathematicians may have interpreted the formal and the conceptual according to their own convictions, but the mathematical content of Lawvere's proposal, including the sketch of the *mathematical* duality involved,

---

[9] We are not entirely faithful to Lawvere here. He takes the conceptual to be the functor category $\text{Mod}(\mathcal{C}at, [\mathcal{S}et^{\mathbb{T}}])$, which is of course slightly more general than what we have been describing.

lead very quickly to a host of important results. The fact is, no one had to quote Lawvere explicitly or say that their work was part of that program — and in fact, no one did quote Lawvere's paper —, for in a sense the program was already implicit in the *manner* Lawvere had set up his own work on universal algebra and that work, as we have seen, called for various generalizations and expansions that were taken up by various mathematicians.

6. One has to contrast categorical logic with other attempts at developing an algebraic framework for logic around the same time, e.g. Halmos's polyadic algebras and Tarski et. al.'s cylindrical algebras. The advantages of the categorical approach over the latter have been explicitly argued by Joyal and Reyes in the mid-seventies:

> 1. the concept of category is used in all branches of mathematics, whereas the structure of polyadic algebra is exotic. Thus, we may hope to extend the application of logic in mathematics
> 2. certain categories, used in different fields of mathematics are in fact theories and is useful [sic!] to consider them as such
> 3. as we shall see in this paper, constructions which are actually used in model theory are specializations of general categorical constructions (hence logic is no exception to the generalized use of categories in mathematics). [Joyal and Reyes, 1978, 5]

These considerations boil down to one fundamental fact: whereas cylindric algebras and polyadic algebras are isolated in the conceptual realm of mathematics, categories are omnipresent. The heuristic gain of using categories is therefore clear and powerful.

There are indications that the gain of categorical logic is more than heuristic. It is well known that the algebraic expression of propositional logic is given by lattice theory and in the latter adjoint functors are usually called Galois connections. Disagreements appear when higher-order operations like quantifiers are considered, in other words, disagreements appear as to how to generalize the algebraic framework to higher-order logical operations. The fact that categories are a generalization of posets suggests that they might yield the correct generalization. Indeed:

i. Concepts and results about propositional logic are special cases of concepts and results in the categorical setting;

ii. It is possible to extend naturally *proofs* in the propositional setting to the categorical setting;

iii. It is possible to obtain *new* results in the categorical setting;

iv. It is possible to make contacts with other areas of mathematics either by using results of different fields in the new context or by applying the new results in different fields.

It is a remarkable fact that the categorical machinery introduced for algebraic topology, homological algebra, homotopy theory and algebraic geometry constitutes at the same time the proper setting for an algebraic analysis of logic.

Thus, towards the end of the sixties and early seventies, the following 'dictionary' was being elaborated:

Logic	*Category Theory*
(many-sorted) theory	category
Sort	Object (of a category)
(sorted) formula	Sub-object
(sorted) term	Morphism
Interpretation	Functor
Set-theoretical model	Functor into Sets
Homomorphism	Natural transformation

In this table, category refers to categories with finite limits (pull-backs and terminal object) with possibly further structure satisfying some exactness conditions. This extra structure is determined by the choice of the logical connectives considered, as we will see.

3 1969–1970 ELEMENTARY TOPOS THEORY

From the logical point of view, Lawvere's thesis and subsequent work had introduced a starting point in two senses. First, it is the first systematic attempt to give a categorical version of logical concepts and second, the theories at first captured by Lawvere were purely equational theories. When Lawvere and Tierney introduced the concept of elementary topos, they introduced, in some sense, the other end of the spectrum, namely higher order theories or type theories. Of course, the relationships between elementary toposes and type theories still had to be clarified fully, but it was clear to everyone that there were intermediate cases to consider and characterize, or as they became called at that time, there were 'doctrines' to be defined categorically, in particular, the doctrine of first-order logic.

3.1 Elementary topos theory: 1969–1970

> The unity of opposites in the title is essentially that between logic and geometry, and there are compelling reasons for maintaining that geometry is the leading aspect. At the same time, in the present joint work with Myles Tierney there are important influences in the other direction... [Lawvere, 1970a, 329]

The program of investigating the connections between algebraic geometry and "intuitionistic" logic under the guidance of the form of objective dialiectics known as category theory was discussed and moved forward at a conference in January 1971 at Halifax, ... (...) Briefly we may say that the notion of topos summarizes in objective categorical form the essence of "higher-order logic" (we will explain below how the logical operators become morphisms in a topos) with no axiom of extensionality. This amounts to a natural and useful generalization of set theory to the consideration of "sets which internally develop". In a basic example of algebraic geometry, the development may be viewed as taking place along a parameter which varies over "rings of definition"; in a basic example from intuitionistic logic, the parameter is interpreted as varying over "stages of knowledge". [Lawvere, 1972, 1–3].

As the first sentence of the first foregoing quote indicates, in the early seventies, Lawvere insisted again and again on the dual nature of toposes: a geometric aspect and a logical aspect. As the quote suggests clearly, the geometric aspect was thought to be more important than the logical aspect at the beginning and that, to a certain extent, it subsumed the logical aspect. Even to this day, Lawvere presents the situation in similar terms:

In spite of its geometric origin, topos theory has in recent years sometimes been perceived as a branch of logic, partly because of the contributions to the clarification of logic and set theory which it has made possible. However, the orientation of many topos theorists could perhaps be more accurately summarized by the observation that what is usually called mathematical logic can be viewed as a branch of algebraic geometry, and it is useful to make this branch explicit in itself. [Lawvere, 2000, 717].

Going back to the first quote above, an explicit ideology associated with category theory can be discerned. Indeed, Lawvere himself talks about "unity of opposites" and of the "form of objective dialectics". This paper, published in the *Actes du Congrès International des Mathématiciens 1970*, was the first public presentation of elementary topos theory. In the bibliography, for this paper contains a short and cryptic bibliography, the first reference is to Mao Tsetung and his work entitled *On contradiction — Where do correct ideas come from?* Notice also that the title of the paper — *Quantifiers and Sheaves* — does not refer directly to toposes. This was probably not the original or the intended title of the paper, for as we have seen the first sentence of the paper refers to "the unity of opposites in the title". It is worth quoting a short passage of the first paragraph, for it weaves all these elements together in an interesting manner.

We first sum up the *principal contradictions* of the Grothendieck-Giraud-Verdier theory of topos in terms of four or five adjoint functors, significantly generalizing the theory to free it of reliance on an *external*

> notion of infinite limit (*in particular enabling one to claim that in a sense logic is a special case of geometry*). (our emphasis)

Notice that the axiomatization of the notion of elementary topos is not only seen as a significant generalization of the notion introduced by Grothendieck, but also as a way of giving the concept an autonomy, as a way of liberating the notion of topos from its reliance on an external notion.

For Lawvere, then, category theory was the objective form of dialectical materialism and adjoint functors were the exact formulation of principal contradictions. Lawvere felt that the logical and the geometric aspects of toposes were its main contradictions. Geometry was considered to be progressive and Logic was, at least for some time and in some circles, considered to be reactionary. Thus, many category theorists came to the conclusion that it was preferable to avoid the logical aspect altogether, by subsuming it under the geometrical aspect. This was certainly not Lawvere's own proposal if only because it would not be faithful to a dialectical outlook, it became natural to think that way in the early seventies. In particular, using logical techniques to solve a problem was sometimes considered to be reactionary or fascist. It is even said that set theory was considered to be essentially bourgeois since it is founded on the relationship of belonging. This background ideology might partly explain the difficult relationships that category theorists developed with the other parts of the mathematical and logical communities. Political ideas are rarely thought of as being inherent to mathematical theories.

Nowadays, people recognize that two faces are simply not enough to capture toposes, that toposes are multifaceted objects with many complementary aspects. (See, for instance, the introduction in [Johnstone, 2002].) The geometric aspect of toposes comes directly from their origins: Grothendieck defined toposes in the early sixties in the context of algebraic geometry via the use of sheaves. Grothendieck presented them as a generalization of the concept of topological spaces and, in fact, as potentially being the real object of study of topology. Grothendieck had already noted that a topos inherited many of the properties of the category of sets and, as such, could be considered to be a generalization of the latter. Thus, a topos was at the same time a generalization of the idea of a topological space and of the idea of a category of sets. However, Grothendieck and his students did not see or perhaps pay attention to its logical aspect.

Lawvere suggested thinking of a topos as a universe of variable sets, among which the standard sets are conceived as being without variation, or constant. He later suggested thinking of the standard sets as being abstract. (See [Lawvere, 1975a; Lawvere, 1976; Lawvere, 1994; Lawvere, 2003].) Once this viewpoint is adopted, one starts looking at the elementary properties of a topos as a universe of variable sets. As we have seen, Lawvere had already tried to axiomatize the universe of sets, or of constant sets. He had a larger class of universes to capture.

> When the main contradictions of a thing have been found, the scientific procedure is to summarize them in slogans which one then constantly

> uses as an ideological weapon for the further development and transformation of the thing. Doing this for "set theory" requires taking account of the experience that the main pairs of opposing tendencies in mathematics take the form of adjoint functors, and frees us of the mathematically irrelevant traces ($\in$) left behind by the process of accumulating ($\cup$) the power set (P) at each stage of a metaphysical "construction". Further, experience with sheaves, permutation representations, algebraic spaces, etc., shows that a "set theory" for geometry should apply not only to *abstract* sets divorced from time, space, ring of definition, etc., but also to more general sets which do in fact develop along such parameters. [Lawvere, 1970a, 329].

The hope was that these "geometric sets", or sets for geometry, would provide an adequate foundation for analysis, in particular functional analysis, as is illustrated by the very last sentence of the same paper: "In any topos satisfying (ω) [an additional condition satisfied by the topos of the so-called evolutive sets] each definition of the real numbers yields a definite object, but it is not yet known what theorems of analysis can be proved about it." [Lawvere, 1970a, 334].

Already in 1967, Lawvere was moving towards the notion of an elementary topos (as can be witnessed in his 1969 paper on diagonal arguments). As he himself made clear, his motivation was coming from continuum mechanics, whereas Myles Tierney's motivation was coming from sheaf theory.

> What was the impetus which demanded the simplification and generalization of Grothendieck's concept of topos, if indeed the applications to logic and set theory were not decisive? Tierney had wanted sheaf theory to be axiomatized for efficient use in algebraic topology. My own motivation came from my earlier study of physics. The foundation of the continuum physics of general materials, in the spirit of Truesdell, Noll, and others, involves powerful and clear physical ideas which unfortunately have been submerged under a mathematical apparatus including not only Cauchy sequences and countably additive measures, but also ad hoc choices of charts for manifolds and of inverse limits of Sobolev Hilbert spaces, to get at the simple nuclear spaces of intensively and extensively variable quantities. But, as Fichera lamented, all this apparatus may well be helpful in the solution of certain problems, but can the problems themselves and the needed axioms be stated in a direct and clear manner? And might this not lead to a simpler, equally rigorous account? [Lawvere, 2000, 726].

Thus in 1969, in collaboration with Myles Tierney, the correct axiomatization of the notion came about. Although the axioms given by Lawvere in 1970 at the Nice Congress are redundant, for the assumption that a topos has all finite colimits will be shown quickly afterwards to be a consequence of the other axioms, it contains the key ingredients of the characterization. In a nutshell, an elementary topos is a Cartesian closed category $\mathcal{E}$ with a subobject classifier. Formally, the definition is as follows.

DEFINITION
An *elementary topos* is a category $\mathcal{E}$ such that

1. it has pullbacks;
2. it has a terminal object 1;
3. the functor $X \times - : \mathcal{E} \longrightarrow \mathcal{E}$ has a right adjoint, denoted by $(-)^X : \mathcal{E} \longrightarrow \mathcal{E}$;
4. it has an object Ω, called the *subobject classifier*, together with a monic arrow $\top : 1 \rightarrowtail \Omega$ such that for any monic $m : A \rightarrowtail X$ there is a unique morphism $\phi : X \longrightarrow \Omega$ in $\mathcal{E}$ for which the following square is a pullback:

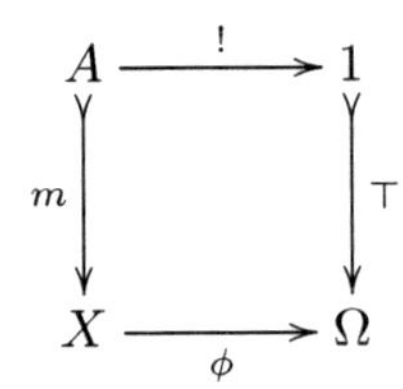

The morphism $\phi : X \longrightarrow \Omega$ is called the *characteristic morphism of* A and we will sometimes denote it by ϕ_A.

The third axiom could be replaced by the following:

(3′) it has, for each object X, an object PX and an arrow $\in_X : X \times PX \longrightarrow \Omega$ such that for every arrow $f : X \times Y \longrightarrow \Omega$ there is a unique arrow $g : Y \longrightarrow PX$ for which the following diagram commutes:

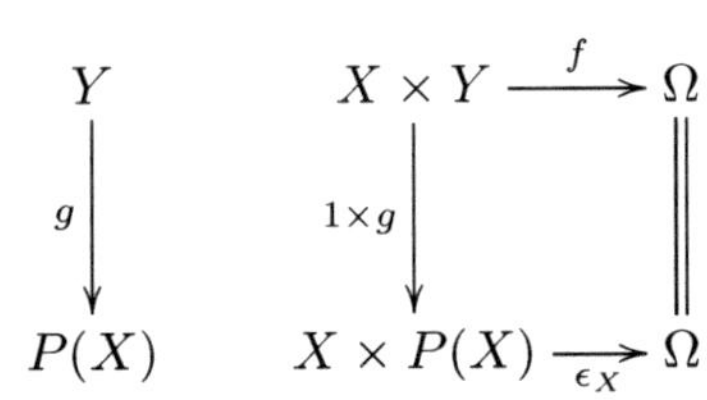

Notice that the definition can be given entirely in terms of the existence of adjoint functors to given functors. The notion of being a Cartesian closed category, as we have already mentioned, was introduced earlier by Lawvere in this spirit. The existence of the subobject classifer Ω can also be presented via the existence of an adjoint functor to a given functor. Since the adjoints whose existence is required by the axioms are adjoints to "structural" functors whose nature is fundamental or elementary, depending on the point of view, the notion of elementary topos ought to be seen as arising naturally from elementary constraints. We have to emphasize this fact once again: in a categorical set up, certain notions arise naturally from elementary functors in a hierarchical manner. They are not ad hoc nor do they depend on complicated contraptions. This is taken by many as being an indication of the fundamental character of the analysis provided by categorical logic and of the notions themselves.

Examples of toposes abound. The category of sets, as defined by the axioms of ZFC for instance, is a topos. Given a set X, the functor $X \times -$ is defined in the obvious way and its right adjoint is given by the set $X^Y = \{f : Y \longrightarrow X\}$. The subobject classifier Ω can be taken to be the set $2 = \{0, 1\}$ together with the map $\top : 1 \longrightarrow 2$, defined by $\top(*) = 1$. It should be observed that these sets are *canonical* choices and that from a purely categorical point of view, any isomorphic set with the appropriate function would do just as well. Thus, there is nothing special about the set $2 = \{0, 1\}$, any two elements set would do just as well (with an appropriate function picking out an element into it).

The other examples of toposes are more involved, but there is a class of examples that have the same general form: given a topos $\mathcal{E}$ and a small category $\mathcal{C}$ ("small" in a sense that can be made precise with respect to the topos $\mathcal{E}$) the functor category $\mathcal{E}^{\mathcal{C}^{op}}$ is a topos. Notice that it is a functor category. Specific important cases of this construction are the following. Let $\mathcal{E}$ be the category *Set* of sets and $\mathcal{C}$ be the lattice of open sets $\mathcal{O}(X)$ of a topological space X. The functor category $Set^{\mathcal{O}(X)^{op}}$ is a topos, called the topos of *presheaves* over X. There is an important subcategory of $Set^{\mathcal{O}(X)^{op}}$, denoted by $Sh(X)$, the category of *sheaves* over X and it can also be shown to be a topos (we will come back to sheaves in a very short while). A related class of toposes, especially important from the point of view of the history of logic, are those of the form $Set^{P^{op}}$, where P is a poset. Lawvere and Tierney saw that the usual models of intuitionistic logic and mathematics known at that time as well as the forcing method of Cohen were specific cases of these toposes. Finally, let $\mathcal{C}$ be a group $\mathcal{G}$ or, more generally a groupoid, then the functor category $Set^{\mathcal{G}^{op}}$, also denoted by $B\mathcal{G}$ and called the *category of representations of* $\mathcal{G}$, can also be shown to be a topos. These examples were all mentioned by Lawvere in his conference at the International Congress of Mathematicians held in Nice in 1970 (see [Lawvere, 1970a]).

Lawvere emphasized right from the start the logical structure of toposes. First, for any category $\mathcal{C}$ with finite limits, it is possible to define an *internal lattice object* L of $\mathcal{C}$ as an object of $\mathcal{C}$ together with two morphisms making certain diagrams commute. (See, for details, [Mac Lane and Moerdijk, 1994, IV.8].) If, furthermore, there is an additional binary morphism $\Rightarrow: L \times L \longrightarrow L$ satisfying the usual identities, then L is an *internal Heyting algebra*. It can be shown that for any object X of a topos $\mathcal{E}$, the power object PX is an internal Heyting algebra. In particular, it follows from the definition that $\Omega \approx P1$ and thus that the subobject classifier Ω is also an internal Heyting algebra. (See [Mac Lane and Moerdijk, 1994, 201] for details.) Thus all the propositional operations are definable in a topos $\mathcal{E}$ as morphisms of the subobject classifier: $1 \xrightarrow{\top} \Omega$, $\Omega \xrightarrow{\neg} \Omega$, $\bot = \neg \circ \top : 1 \rightarrow \Omega$, $\Omega \times \Omega \xrightarrow{\wedge} \Omega$, $\Omega \times \Omega \xrightarrow{\Rightarrow} \Omega$, $\Omega \times \Omega \xrightarrow{\vee} \Omega$. Furthermore, for any morphism $f : X \longrightarrow Y$ of $\mathcal{E}$, there is a morphism $Pf : PY \longrightarrow PX$ and the latter has a left adjoint

$\exists_f : PX \longrightarrow PY$ and a right adjoint $\forall_f : PX \longrightarrow PY$ or, equivalently, for any morphism $f : X \longrightarrow Y$ of $\mathcal{E}$, there are morphisms $\Omega^X \xrightarrow{\forall_f} \Omega^Y$ and $\Omega^X \xrightarrow{\exists_f} \Omega^Y$. Hence, quantifiers are also definable in any elementary topos. Needless to say, the logic of an arbitrary topos $\mathcal{E}$ is intuitionistic.

In certain specific toposes $\mathcal{E}$, the power object $P\Omega$ can be shown to be a Boolean algebra. In this case, we say that $\mathcal{E}$ is *Boolean*. Lawvere and Tierney noted that this condition is equivalent to the following:

1. the negation operator $\Omega \xrightarrow{\neg} \Omega$ satisfies $\neg \circ \neg = 1_\Omega$;
2. for every object X of $\mathcal{E}$, the Heyting algebra PX is a Boolean algebra;
3. every suboject is complemented, i.e. for every AX in $\mathcal{E}$, $A \vee \neg A = X$;
4. the morphisms $1 \xrightarrow{\top} \Omega$ and $1 \xrightarrow{\bot} \Omega$ induce an isomorphism $1 + 1 \simeq \Omega$, which means that the subobject classifier is two-valued.

In 1971, Radu Diaconescu gave a sufficient condition for a topos $\mathcal{E}$ to be Boolean. We say that a topos satisfies the axiom of choice (AC) if every epimorphism $p : X \longrightarrow Y$ has a section $s : Y \longrightarrow X$, that is $p \circ s = 1_Y$. It can easily be seen that this is only a different way of formulating the usual axiom of choice expressed in term of the existence of a choice function defined on a family of disjoint sets. Diaconescu showed that any topos $\mathcal{E}$ satisfying AC is necessarily Boolean. It was quickly observed afterwards that this statement has an "internal" version, in the sense that this property can be expressed solely in terms of the internal language of the topos. (We will come back to the internal language later.) A topos $\mathcal{E}$ satisfies the *internal axiom of choice* (IAC) if for any epimorphism $p : X \longrightarrow Y$ and for any object Z of $\mathcal{E}$, the morphism $X^Z \longrightarrow Y^Z$ is also an epimorphism. Any topos $\mathcal{E}$ satisfying AC also satisfies IAC, but the converse is false. It was shown that any topos $\mathcal{E}$ satisfying IAC is necessarily Boolean.

Given the foregoing definition of an elementary topos, it is natural to define a morphism of toposes as a functor preserving all the specified structure. Lawvere and Tierney called these morphisms *logical morphisms*. More specifically, a functor $T : \mathcal{E} \longrightarrow \mathcal{E}'$ which preserves, up to isomorphism, all finite limits, the subobject classifier and the exponential is a *logical morphism*. Such morphisms preserve the logical structure of a topos "on the nose".

However, Lawvere and Tierney considered and used primarily a different class of morphisms between toposes, those coming from the geometric side of toposes and already defined and heavily used by the Grothendieck school. The definition, directly lifted from sheaf theory, goes thusly: a *geometric morphism* $f : \mathcal{F} \longrightarrow \mathcal{E}$ between toposes is a pair of functors $f^* : \mathcal{E} \longrightarrow \mathcal{F}$ and $f_* : \mathcal{F} \longrightarrow \mathcal{E}$ such that $f^* \dashv f_*$ and f^* is left exact, that is it preserves all finite limits. The functor $f^* : \mathcal{E} \longrightarrow \mathcal{F}$ is called the *inverse image part* of the geometric morphism and the functor $f_* : \mathcal{F} \longrightarrow \mathcal{E}$ is called the *direct image part* of the geometric morphism.

Lawvere and Tierney then claimed that any geometric morphism $f : \mathcal{F} \longrightarrow \mathcal{E}$ between toposes can be factored into

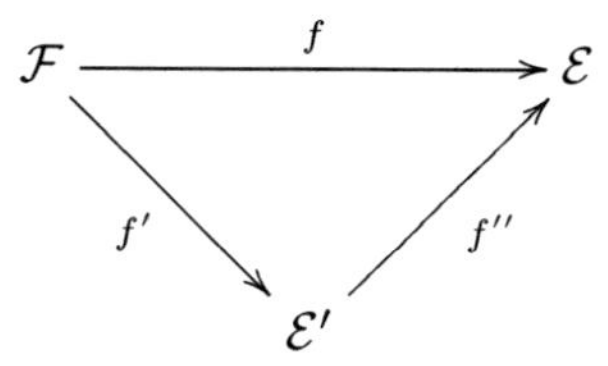

where f' and f'' are geometric morphisms such that f''_* is full and faithful while f'^* reflects isomorphisms. Lawvere and Tierney observed that the geometric morphism f'' is entirely determined by a single morphism $j : \Omega \longrightarrow \Omega$ in $\mathcal{E}$. This morphism, in turn, determines what Lawvere and Tierney called a *Grothendieck topology*, now also known as a *Lawvere-Tierney topology*. The formal definition is as follows.

DEFINITION
Let $\mathcal{E}$ be a topos and Ω be its subobject classifier. A *Lawvere-Tierney topology on* $\mathcal{E}$ is a morphism $j : \Omega \longrightarrow \Omega$ in $\mathcal{E}$ satisfying the following three properties:

(i) $j \circ \top = \top$;

(ii) $j \circ j = j$;

(iii) $j \circ \wedge = \wedge \circ (j \times j)$.

This definition was central to Lawvere and Tierney since it allowed them to make the connection to sheaf theory and Grothendieck toposes and, thus, putting the geometry at the forefront of the theory. For, one would not be able to see from the axioms of a topos *alone* that they *have* this essentially geometric character. (In the same way, perhaps, that one would not be able to see the *geometric* content of Lie groups from their axioms alone.) After all, the properties stated in the axioms are very much *set*-like or *logical*: taking products and finite limits in general, having the possibility to classify subobjects and forming function spaces (or power sets, depending on the choice of axioms). The definition of a Lawvere–Tierney topology does not look geometrical either. Since the subobject classifier Ω can be thought of as an object of truth-values, the morphism j can be viewed in a loose sense as a modal operator. However, the main examples of such topologies are given by sheaves of sets over a site, the latter being a category equipped with a Grothendieck topology which, in turn, is a categorical generalization of the notion of topology. It is precisely this connection between logic and algebraic geometry that Lawvere and Tierney were trying to capture: to give an elementary characterization of toposes in terms of adjoint functors that would allow them to recover Grothendieck's toposes. But not all examples of such topologies are geometrical in that sense. A more logical example is the double negation topology: as we have mentioned, any topos $\mathcal{E}$ has a negation operator $\Omega \xrightarrow{\neg} \Omega$ and it is easy to verify that the double negation $\Omega \xrightarrow{\neg\circ\neg} \Omega$ is a Lawvere-Tierney topology.

Before we give the definition of a j-sheaf, recall that a sheaf is, informally, a tool that allows one to look at a concept of local character from a global point of view. In other words, an object is a sheaf if it is possible to extend a local notion to the whole space in which it is defined.

The formal connection with sheaves comes in two steps. First, the morphism j determines a closure operator on the subobjects of each object X of $\mathcal{E}$. Given a subobject $A \longrightarrow X$, its *j-closure* $\bar{A}$ is given by the following pullback:

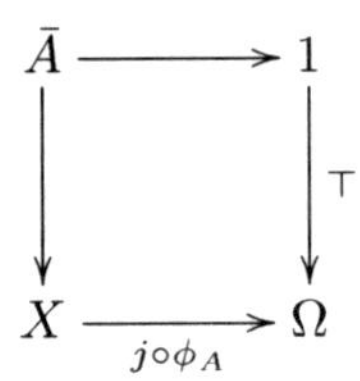

It can be shown that j is a Lawvere–Tierney topology if and only if the closure operator $A \mapsto \bar{A}$ has the following properties for all subobjects A, B of X : $A \wedge \bar{A} = A, \bar{\bar{A}} = \bar{A}, \overline{A \wedge B} = \overline{A} \wedge \overline{B}$.

Second, the closure operator allows us to characterize 'concepts of a local character' as follows. Given a monomorphism $i : A \longrightarrow X$, we say that i is a *dense monomorphism* if $\bar{A} = X$. Whenever such a dense monomorphism exists, the subobject A is said to be *dense in* X. A dense subobject represents a concept of local character. A j-sheaf is an object F of $\mathcal{E}$ such that any representation of a concept of local character in it can be extended to the whole space in which the concept lies. Formally, we say that an object F of $\mathcal{E}$ is a *j-sheaf* (for the Lawvere–Tierney topology) if for every dense monomorphism $i : A \longrightarrow X$ in $\mathcal{E}$ and every morphism $f : A \longrightarrow F$, there is a unique morphism $g : X \longrightarrow F$ making the following diagram commute:

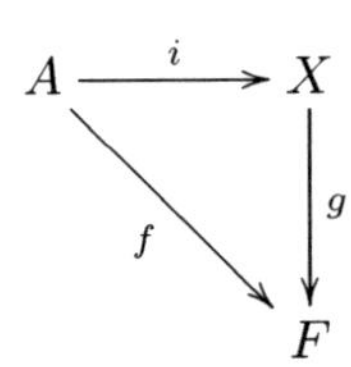

The full subcategory $\mathrm{Sh}_j\mathcal{E}$ of j-sheaves of $\mathcal{E}$ can be shown to be a topos. As Lawvere and Tierney have indicated, every Grothendieck topos arises in this way. Moreover, there is a *geometric morphism* $\mathrm{Sh}_j\mathcal{E} \longrightarrow \mathcal{E}$.

At that point, Lawvere indicates in his paper how Cohen's proof of the independence of the continuum hypothesis from the axioms of ZF could be translated into the language of topos theory. It was left to Tierney to present the full proof of this translation in 1972. (See [Tierney, 1972].)

This was the first public and printed presentation of elementary topos theory. The study and presentation of elementary toposes were taken up quickly afterwards by various mathematicians: Peter Freyd, Jean Bénabou, Gavin Wraith, Anders Kock, André Joyal, Chris Mikkelsen.

It is interesting to see what the first volume published in 1972 and dedicated to toposes, algebraic geometry and logic contains. (See [Lawvere, 1972].) Apart from Lawvere's introduction, which goes over the basic definition and concepts we have mentioned and Tierney's paper on the independence of the continuum hypothesis in the topos theoretical framework, there is nothing about the interactions between toposes, algebraic geometry and logic! The remaining papers are in order of appearance: Giraud on the classifying topos (*not* in the logical sense of that expression), Lambek on deductive systems and categories, Goodman and Myhill on a formalization of Bishop's constructive mathematics, Scott's paper on continuous lattices, Bucur on the applications of formalism of duality in algebraic geometry and Illusie on cotangent complex and deformations of torsors and group schemes. All these papers touch on issues that are clearly relevant and will even become important in their own right, e.g. classifying toposes or domain theory, but they do not constitute direct contributions to the subject at hand. This was about to change rapidly.

4 FOCUSING ON FIRST-ORDER LOGIC

> Part I of this volume consists of three of the first papers on functorial model theory, developing concretely the approach to algebraic logic according to which a "theory" (understood in a sense invariant with respect to various "presentations" by means of particular atomic formulas and particular axioms) is actually a category T having certain properties P and a model of T is any set-valued P-preserving functor. As a rough general principle, one could choose for P any collection of categorical properties which the category of sets satisfies, the choice determining the "doctrine" of theories of kind P, which is thus a (non-full) subcategory of the category of small categories. For example, the doctrine of universal algebra thus springs from the fact that the category of sets has the property P of having finite Cartesian products, while the doctrine of higher-order logic springs from the property of being a topos. [Lawvere, 1975a, 3].

Lawvere had emphasized right from the start the idea of replacing the category of sets by an arbitrary topos to develop mathematics, in particular analysis. The first publication on the application of topos theoretical methods to analysis appeared in 1974, but the ideas had been presented already in 1972. (See [Kock and Mikkelsen, 1974].) At the same time, there was a lot of work to do. Very quickly, presentations of topos theory became available. (See [Freyd, 1972; Kock and Wraith, 1971; Wraith, 1975] (presented in 1973)].) Since the axioms of topos theory are very much set-like, early works were concerned with clarifying the relationships between toposes and set theory. Of course, in this case, it is necessary to add axioms to the theory, e.g. an axiom of infinity and an axiom to guarantee that the toposes considered were Boolean. Thus, J. C. Cole and W. Mitchell both explored categories of sets and models of set theory independently and about the same time (see [Cole, 1973], in fact his PhD thesis defended at the University of Sussex in 1971; [Mitchell, 1972] (also presented in 1971)) and Osius followed closely. (Osius presented his results in 1973; they were published in 1974 and 1975.) As we will see, Mitchell introduced a method that had a direct impact on the development of categorical logic. Around the same time, Marta Bunge gave a topos theoretical proof of the independence of Souslin hypothesis in set theory and Kock, Lecouturier and Mikkelsen investigated the concepts of finiteness in toposes. (See [Bunge, 1974] and [Kock *et al.*, 1975].)

As far as *logic*, and in particular first-order logic, is concerned, the development of its history is extremely difficult to document accurately. When Gray wrote his history of sheaf theory, he already lamented on the situation:

> Perhaps the main aspect which is difficult to document in published works is the connection with logic. The best sources are Lawvere [1975] and [1976], together with Reyes [1974; 1975], and [1976], Lambek [unpublished when Gray wrote] and Makkai and Reyes [1976] and [1977], and also the articles in this volume. Unfortunately one of the most influential figures in this development, A. Joyal, has thus far not given us a written record of his work; however see [Labelle, 1971]. Besides topics discussed at the present meeting, future developments seem to be going in the directions of Bénabou [1975] ..., and recent unpublished work of Cole and Tierney on pseudolimits in the category of topoi. [Gray, 1979, 63].

This was presented in 1977 and published in 1979! When one consults the bibliographical sources mentioned in this paragrah, one discovers how poor and inaccessible these sources are in general. Lawvere's papers are important in themselves and do refer to Joyal, Bénabou, Reyes and others, but they do not allow one to reconstruct precisely who did what when and while [Reyes, 1974] and [Reyes, 1977] were easily available, [Reyes, 1975/1976] was not. Labelle [1971] was just as difficult to obtain, although Labelle's long exposition was influential among students at Montreal in the early seventies. Bénabou's paper was about fibered categories and, although relevant, could certainly not constitute an introduction to, or a survey of, his work in the area. Makkai and Reyes papers published in 1976 already build up on previous material, only sketch proofs and they certainly do not constitute and entry point to the field. The fact is, two of the main actors, namely Jean Bénabou in Paris and André Joyal in Montréal, never even wanted to publish their results. It seems that they simply did not believe in publishing at that time. Recall that we are in the late sixties and early seventies and events like *Mai 68* in Paris and the *Crise d'octobre* in 1970 in Québec were still very much in the air. Bénabou and Joyal did present their results during colloquia, seminars and at informal gatherings and thus people learned and knew what they had done, but the details of these contributions and the manner in which they were done are not available. What we do have is a list of talks given by Bénabou and Joyal during this period, collected from various bibliographical sources, mostly theses written during that period. First, here is the list of talks given by Bénabou and his students.

Bénabou and Celeyrette	*Généralités sur les topos de Lawvere et Tierney*	Paris	1971
Bénabou	*Structures syntaxiques*	Montreal	April 1973
	Catégories et logiques faibles	Oberwolfach	April 1973
Michel Coste	*Logique du premier ordre dans les topos élémentaires*	Paris	1973
	Logique d'ordre supérieur dans les topos élémentaires	Paris	November 1973

Here is the list of Joyal's talks:

Joyal	*Dogmes Booléens*	Montreal	4 talks, December 1969, January 1970
	Topos	Montreal	24 April 1971
	Topos et Logique	Montreal	5 October and 27 October 1971
	Théorèmes de descente finitaire; théorème de complétude	Montreal	Summer 1972
	Indépendance de l'hypothèse du continu	Montreal	7 November 1972
	Une méthode uniforme pour démontrer les théorèmes de Gödel, Kripke, Freyd, Mitchell, Barr	Montreal	14 November 1972
	Homotopie dans les topos (le théorème de Giraud)	Montreal	28 November 1972
	Qu'est-ce qu'une théorie équationnelle générale?	Montreal	January 1973
	Modèle universel d'une théorie équationnelle générale	Montreal	January 1973

	Théorie catégorique des fonctions récursives	Montreal	February 1973
	Forcing=Interprétation dans un Topos (I et II)	Montreal	23 February and 2 March 1973
	Univers récursifs	Montreal	March 1973
	Théorie formelle des topos et langage d'ordre supérieur	Montreal	2 April 1974
	Algèbre constructive	Montreal	15 November 1974
	Algèbre constructive et théorie des faisceaux	Montreal	2 December and 9 December 1974
	Théorème de Chevalley-Tarski	Montreal	10 July 1975

The lists are not complete. Notice that these talks, except for the first one in Joyal's list, were given after the introduction of elementary topos theory. There was work done before, as we will see. Some of this work, for instance Joyal's recursive universes (also called 'arithmetic universes'), have simply vanished for a long period of time (they are now coming back in circulation). They are mentioned *en passant* in some of Lawvere's papers, but there is no available systematic exposition of what they were and how they were developed.

There was a particularly important period in April 1973. Lawvere was in Montreal for that month, away from Perugia. Dana Schlomiuk had invited many category theorists for a month during which there were many seminars and discussions. It was an intense informal gathering and not an organized, planned and disciplined meeting. Another decisive event for the dissemination of categorical logic was the meeting of the summer of 1974, called *Séminaire de mathématiques supérieures*, organized by Shuichi Takahashi. During this meeting, André Joyal gave a series of talks on all the topics mentioned above, Gonzalo Reyes gave a series of talks on what was to become his collaborative book with Mihaly Makkai, Jean Bénabou gave a series of talks on fibered categories. Freyd presented tau-categories, what were to become *allegories* much later. (See [Freyd and Scedrov, 1990].) Eilenberg started a "bilingual' talk: one sentence in french, one sentence in english and alternating between the languages until the audience pleaded him to stick to one language. Also in the audience, a contingent of Italians coming with Lawvere: Meloni, Carboni, Riccioli (Feit). As far as we know, there is no official record of that meeting either.

The first *published* results appeared in 1973, although the work was done in 1970: one of Lawvere's student, Hugo Volger, presented a categorical version of a first-order logic and of the completeness theorem by following the leads given earlier by Lawvere himself, e.g. [Lawvere, 1966; Lawvere, 1967]; also in 1971,

Orville Keane, in his PhD thesis (under Freyd's supervision at the University of Pennsylvania), gave a categorical characterization of universal Horn theories. (See [Keane, 1975].)

4.1 Volger's work

It is instructive to look more closely at Volger's work, for although he obtained significant results, the community did not adopt them. Many category theorists felt that the notions defined were not the "right" notions, mostly because his results were not easily generalizable to non-Boolean cases. Volger modified the definitions given in the first paper "because my friends in Aarhus insisted that the construction should work in this [e.g. non-Boolean] more general case" [Volger, 1975b, 87]. Alas for Volger, the modified definitions suffered the same fate as the original ones.

In his first paper, Volger introduced three types of categories: logical categories, elementary theories and semantical categories. Elementary theories were meant to be the categorical encoding of first-order theories with equality, as Lawvere had already presented them. Logical categories were introduced to allow a generalization to higher-order logic and, finally, semantical categories were suppose to play the role of categories of sets, but the notion is weaker than the concept of topos. We will restrict ourselves to the notion of logical categories, for it contains the seeds of the demise of these notions and even of Volger's subsequent attempt to rescue them.

Volger's definition is based on Lawvere's notion of an elementary theory. However, Volger added two further conditions, for his "proof of the completeness theorem requires the addition of two new conditions to the original definition of elementary theories." [Volger, 1975a, 52]. A category $\mathcal{C}$ is said to be *logical* if it satisfies the following conditions:

i. $\mathcal{C}$ has finite products (in particular a terminal object 1);

ii. $\mathcal{C}$ has an object Ω which is a Boolean algebra with mappings $1 \underset{\bot}{\overset{\top}{\rightrightarrows}} \Omega$;

iii. For every $f : X \longrightarrow Y$ in $\mathcal{C}$, there is a functor $\exists_f : Hom_{\mathcal{C}}(X, \Omega) \longrightarrow Hom_{\mathcal{C}}(Y, \Omega)$, left adjoint to the functor $Hom(f, \Omega) : Hom(Y, \Omega) \longrightarrow Hom(X, \Omega)$;

iv. Given the following pullback

$$\begin{array}{ccc} X & \xrightarrow{(X,f)} & X \times Y \\ {\scriptstyle f}\downarrow & & \downarrow{\scriptstyle f \times Y} \\ Y & \xrightarrow[\Delta_Y]{} & X \times Y \end{array}$$

then we have that $\exists_{(X,f)}[1_Y f] = \exists_{\Delta_Y}[1_Y](f \times Y)$, where $1_Y = \top \circ !_Y$ and given the other pullback

$$\begin{array}{ccc} X \times Z & \xrightarrow{\pi_Z} & Z \\ {\scriptstyle 1_X \times g}\downarrow & & \downarrow{\scriptstyle g} \\ X \times Y & \xrightarrow[\pi_Y]{} & Y \end{array}$$

then we have that $\exists_{\pi_Y}[\varphi]\, g = \exists_{\pi_Z}[\varphi\,(X \times g)]$.

v. a) Let $e_Y = \exists_{\Delta_Y}[1_Y]$ denote the *equality on* Y. If $e_Y(f_1, f_2) = 1_X$, then $f_1 = f_2$;

b) $e_\Omega = \Leftrightarrow$, where $\Leftrightarrow$ denotes the bi-implication.

This is Volger's working definition, for all the results in his paper are for logical categories. Notice that the distinguished object A, what was meant to become the universe of interpretation, has now disappeared. The only distinguished object left is the Boolean algebra Ω. Notice that the category $\mathcal{S}$ of sets is a logical category but that an arbitrary topos $\mathcal{E}$ is not (which is somewhat disappointing given the inherent logical structure of an arbitrary topos). A functor $F : \mathcal{C} \longrightarrow \mathcal{D}$ between logical categories is said to be *logical* if it preserves finite products, the Boolean algebra object Ω together with 0, 1, $\neg$, $\wedge$ and quantification. A *Set-model* of a logical category $\mathcal{C}$ is a logical functor $M : \mathcal{C} \longrightarrow Set$ and a natural transformation $\alpha : M \longrightarrow N$ between two *Set*-models is called a *$\mathcal{C}$-embedding*.

However, Volger also gives a slightly modified definition of elementary theories. An *elementary theory* $\mathbb{T}$ is a category such that

i) $\mathbb{T}$ has two basic objects A and Ω such that every object X different from Ω has a specified representation as a finite power A^n of A, and $Hom_{\mathbb{T}}(\Omega, X)$ is empty.

ii) $Hom_{\mathbb{T}}(A^n, \Omega)$ is a Boolean algebra for every object A^n and $Hom_{\mathbb{T}}(f, \Omega)$ is a Boolean homomorphism for every $f : A^n \longrightarrow A^m$.

iii) For every $f : A^n \longrightarrow A^m$ in $\mathbb{T}$, there exists an existential quantifier $\exists_f$ which satisfies the conditions iii), iv) and v) of the definition of a logical category.

Although all the results are given for logical categories, Volger remarks that all the proofs given still hold, provided that appropriate but slight modifications are made. The most important result, naturally, is the completeness theorem for logical categories, formulated as follows:

THEOREM.
Let $\mathcal{C}$ be a small logical category such that $\bot_X \neq \top_X$ and $\exists_{!_X}[\top_X] = \top$ for every object X of $\mathcal{C}$.

1. For every pair of morphisms $f, g : X \longrightarrow Y$ there exists a *Set*-model M such that $M(f) \neq M(g)$
2. There exists a *Set*-model M such that card$(M(X))$ $\leq$card$(\mathcal{C})$ for every object X of $\mathcal{C}$.

The proof is a categorical adaptation of Henkin's completeness proof to a categorical framework along the lines indicated by Lawvere in his second abstract. The basic idea is to construct an extension $I : \mathcal{C} \longrightarrow \mathcal{C}'$ for the given logical category $\mathcal{C}$ for which $Hom_{\mathcal{C}'}(1, -) : \mathcal{C}' \longrightarrow Set$ has a model, in fact a canonical model. One obtains the required model by composition. It is shown that a logical category $\mathcal{C}$ has a canonical model if and only if 1) it is *maximally consistent*, that is $Hom_{\mathcal{C}}(1, \Omega) = \{0, 1\}$ and 2) it is *rich*, that is for every $\varphi : X \longrightarrow \Omega$ such that $\exists_{!X}\,[\varphi] = \top$, there exists $k : 1 \longrightarrow X$ with $\varphi \circ k = \top$.

As we have mentioned, Volger's proposal, although technically irreproachable, was not received favorably. The main problem was that it was too contrived, perhaps too *ad hoc*. While Volger was following Lawvere's path and suggestions, André Joyal together with Gonzalo Reyes were looking at categorical logic from a slightly different perspective.

4.2 The Montreal school

Category theory had a distinctive status in Montreal in the late sixties and early seventies. Joachim Lambek, who graduated from McGill under Zassenhaus in 1950, was already a distinguished algebraist at McGill in the sixties. He had spent his 1965–66 sabbatical year in Zurich at the ETH, where Beno Eckmann had asked him to give a graduate seminar in category theory, a subject that he was learning at the time. In his audience, he found Fritz Ulmer, John Beck, John Gray and Bill Lawvere. Also showing up were Eckmann himself, Peter Hilton, Pierre André and Paul Bernays. His seminar led to his Springer Lecture Notes *Completion of Categories* published in 1966. It is also during period that Lambek wrote his first paper on category theory, a generalization of Tarski's fixpoint theorem to categories. (See [Lambek, 1968b].) Before coming back to McGill, he invited one of the post-doctoral students at the ETH, Marta Bunge, who came in the fall of 1966 and was soon to join faculty. She had officially worked under Freyd's supervision but she had in fact been supervised both by Freyd and Lawvere and done category theory right from the start. Michael Barr, who started as an homological algebraist under the supervision of David Harrison at the University of Pennsylvania, had spent two years at Columbia as a post-doctoral fellow where he met Eilenberg before moving to Urbana where he found John Beck, John Gray, Alex Heller and Max Kelly and then spent the fall of 1967 in Zurich with Eckmann. He joined the McGill faculty in the fall of 1968. The logician Mihàly Makkai, who contributed rapidly to the development of categorical logic, arrived in Montreal in the fall of 1973 at the *Centre de recherches mathématiques* of the Université de Montréal and joined the McGill mathematics faculty in the fall of 1974.

At the Université de Montréal, one finds an unusual, interesting and perhaps unique combination of algebraic logic and category theory that might explain in part why Lawvere's ideas quickly found a good reception. Maurice Labbé, a student of Church from Princeton in the early fifties, acted as chair of the department during most of the sixties. Another student from Princeton, also supervised by Church and who graduated in 1959, Aubert Daigneault, took interest in Lawvere's approach right from the beginning. Léon LeBlanc, who graduated in 1960 from Chicago under Halmos' supervision in algebraic logic, was there in the early sixties until his untimely death in 1968. Thus, in the sixties, Daigneault and Leblanc were active in algebraic logic. Gonzalo Reyes, after completing his thesis in model theory at Berkeley under Craig's supervision, was hired by the Université de Montréal in 1967, in replacement of Léon Leblanc. In the late sixties, there was also an important number of category theorists. Jean Maranda, who graduated from McGill in 1953 under Zassenhaus, was closely following developments in category theory and had published some papers on the subject in the sixties, but died in 1971. (See [Maranda, 1962; Maranda, 1965; Maranda, 1966].) Two students who had written Ph.D. theses on category theory under Lambek's supervision were also hired in the sixties: first, Pierre Berthiaume in 1964 and then Dana Schlomiuk in 1967. When Reyes arrived, André Joyal was a graduate student at the same institution. Joyal arrived in 1963 as an undergraduate student and did his masters between 1967 and 1969, under Q.I. Rahman's supervision. Joyal joined the mathematics faculty of the *Université du Québec à Montréal* (UQAM) in 1970. Also important for his influence on colleagues was Shuichi Takahashi who was interested in category theory, topos theory and logic, among other things, already in the sixties. (See, for instance [Takahashi, 1965; Takahashi, 1969; Takahashi, 1974].) There were also a number of post-doctoral students, for instance Michel Jean who had also graduated from Berkeley under Henkin and Craig's supervision and Diana Dubrovsky, a student of Yiannis Moschovakis coming from UCLA, participating in the seminars.

There was an intense activity in Montreal in the seventies. McGill University, the Université de Montréal and the Université du Québec à Montréal, had their own seminars in category theory and there was also a Montreal Category Theory research seminar held at that time during weekends. Freyd, Lawvere, Beck, Tierney and, sometimes, even Eilenberg would show up at the latter seminar. Often, when no one was ready to speak, Joyal would go to the blackboard and present new ideas, new results and new theories. Although a student of Adrian Mathias at Cambridge, Robert Seely spent the year 1974–1975 in Montréal and interacted with Joyal, Reyes, Makkai, Lambek and their students. He came back to Montréal in 1977. Philip Scott, who had written his thesis under the supervision of Denis Higgs at the University of Waterloo, came to McGill in the fall of 1976 as a post-doctoral student. Reflecting all this activity, eight theses, Master's and PhDs, were written on categorical logic between 1973 and 1977 at the Université de Montréal. Thus, categorical logic was a very active field in Montreal from the early seventies until about 1977, the year of the publication of Makkai and Reyes book on categorical logic and of the Durham meeting on applications of sheaves. Afterwards, for reasons that are not entirely clear, there are very few theses submitted in categorical logic. André Joyal concentrated his efforts on the conceptual foundations of combinatorics, leading to his beautiful and influential work on species of structures. (See [Joyal, 1981].) Gonzalo Reyes, in collaboration with Anders Kock, started to work actively in synthetic differential geometry. (See, for instance, [Kock and Reyes, 1979].) Logic moved from the French speaking part of Montreal to the English speaking universities, mainly McGill University where Makkai, Lambek, Scott and Seely continued to do original and important work. (Scott left for Ottawa in 1982.) Reyes came back to logic in the late eighties and nineties in collaboration with the psychologist John Macnamara and Joyal also returned to foundational issues in the nineties, mainly in collaboration with Ieke Moerdijk [Macnamara and Reyes, 1994]; [Joyal and Moerdijk, 1997]. However, in contrast with the situation that prevailed in the seventies, few graduate students followed.

The influence of Grothendieck's work on logical research done during the seventies has to be underlined, and this, for two completely different reasons. The first reason is purely conceptual: many of the logical concepts developed were aimed at finding a purely logical translation of concepts of algebraic geometry. Thus, for instance, we read in the introduction of Makkai and Reyes:

> Finally, our treatment of categorical logic is geared towards establishing a link with Grothendieck's theory of (Grothendieck) topoi as it is exposed in SGA4. One of our main points is that some of the fundamental properties of notions in this theory (notably the notions of *topos*, *coherence* of, and in, topoi and *pretopos*) are purely logical.... It is a very interesting fact that notions originally developed for the purposes of (abstract) algebraic geometry turn out to be intimately related to logic and model theory. Compared to other existing versions of algebraic logic, categorical logic has the distinction of being concerned with objects that appear in mathematical practice. [Makkai and Reyes, 1977, 3].

Thus, Grothendieck's work was seen as central to development of algebraic geometry (and coincidentally to homological algebra, sheaf theory and category theory itself) and was part of the development of mathematics. Key notions of categorical logic were directly related to these developments, in fact they were seen as a conceptual clarification of some of it, e.g. coherent topos or the notion of a site, and thus justified almost as such.

The second reason is sociological but just as important: while category theorists were developing topos theory and categorical logic with toposes in mind, in France, Grothendieck had quit the mathematical scene altogether in 1970 and his successors were in fact playing down the importance of some of his mathematical ideas, in particular the concept of (Grothendieck) topos. Whereas the notion of schemes, Grothendieck's analysis of the foundations of algebraic geometry, slowly but surely came to constitute the bedrock of algebraic geometry — as witnessed by [Shafarevich, 1974; Hartshorne, 1977] and more recently [Eisenbud and Harris, 2000] —, toposes were more or less banished from the mathematical establishment, even from algebraic geometry. Indeed, they are not even defined in Shafarevich and Hartshorne. Thus, although categorical logic was directly linked to important ideas of Grothendieck's program, it was not linked to the ideas that the mathematical community was judging as being central. Hence, in a sense, the attempt to build bridges between logic and algebraic geometry did not bear its fruits, although mostly it seems, for sociological reasons.

4.3 The background

Let us come back to the late sixites and early seventies. Among Joyal's multiple interests in late sixties was Algebraic Geometry that he was studying by himself in the formidable treatise EGA, *Éléments de géométrie algébrique*, of Grothendieck and Dieudonné and SGA, *Séminaire de Géométrique Algébrique*, in particular SGA4 on topos theory, which was held in 1963/64 and circulated in limited form before its publication by Springer in the *Lecture Notes*. Sheaf theory and more generally category theory were omnipresent in these treatises. Reyes learned category theory including early work of Lawvere (along with some algebraic geometry) from Joyal, who in turn learned model theory *à la* Berkeley, in particular some of the results and ideas included in his PhD thesis, from Reyes. There are some indirect indications of the work done in 1969 and 1970, before the introduction of the notion of elementary topos hit the community.

We have already indicated that in 1968, Aubert Daigneault had started to look for relationships between Lawvere's elementary theories and polyadic and cylindric algebras. He considered both approaches in a categorical framework and proved some equivalences the category of elementary theories and the category of polyadic algebras. As far as we know, Daigneault's paper is the only published paper whose purpose is to investigate properties of Lawvere's elementary theories. Daigneault's motivation, in this paper, was explicitly "to call attention to Lawvere's important contribution to Algebraic Logic" [Daigneault, 1969/1970, 307]. Thus, Lawvere's contribution is ackowledged as being a contribution to *algebraic logic*. It is worth mentioning that Volger refers explicitly to Daigneault's paper in the introduction of his article on logical categories. In fact, Daigneault's proof of the equivalence between elemetary theories and polyadic algebras is taken by Volger as evidence in favor of the algebraic character of Lawvere's work.

> This concept of an elementary theory may also be viewed as an algebraization of first-order logic by categorical means in the following sense. The elementary theory and the structure preserving functors between them correspond to polyadic algebras and homomorphisms between them. ... The connections between elementary theories and polyadic algebras have been studied by Daigneault.... [Volger, 1975a, 51–52].

Daigneault uses Lawevere's second characterization of elementary theories to construct the relevant categories and the equivalence between them. Recall that in Lawvere's analysis, Boolean algebras play a key role and the existential quantifier is part of the data. Daigneault reformulates Lawvere's completeness theorem (and uses it) as follows:

THEOREM.
For any elementary theory $\mathbb{T}$ and $\varphi : 1 \longrightarrow B$, where B is the distinguished Boolean algebra (Volger's Ω) in $\mathbb{T}$ such that $\varphi \neq \bot$, there exists a model M of $\mathbb{T}$ such that $M(\varphi) = \top$.

Daigneault also gives the representation theorem for polyadic algebras. Let X be a non-empty set and I a denumerably infinite set (taken to represent individual variables). Then it is noted that the Boolean algebras of all functions $X^I \longrightarrow 2$ is a polyadic algebra C_X. The representation theorem is then:

Theorem: For any polyadic algebra P and any $q \in P$ such that $q \neq 0$ (the smallest element of P), there exists a non empty set X and a 2-valued representation $f : P \longrightarrow C_X$ such that $f(q) = 1$.

Although Daigneault uses Lawvere's theorem in his paper, he points out that Lawvere's completeness theorem could be reduced to the representation theorem for polyadic algebras and ultimately to the ordinary completeness theorem for polyadic algebras. Thus, already in 1968, one can see the connections between the various theorems, a theme that will become central in the early 1970's.

A second paper of Daigneault deserves to be looked at carefully, for it contains a key component of what is about to come. It is his paper entitled *Injective Envelopes* [1969], published in the *American Mathematical Monthly.* In this paper, Daigneault gives conditions that are sufficient to ensure the existence of injective envelopes in a category. The existence of enough injectives played a key role in the development of Abelian categories ten years earlier. In his paper, Daigneault gives a list of six equivalent conditions on a category $\mathcal{C}$ that are sufficient in order that, for any object of $\mathcal{C}$, there exists an injective essential extension. The first of these conditions is that for each morphism $f : X \longrightarrow Y$, there exists an epi-mono factorization, i.e. there is an epimorphism $g : X \longrightarrow Z$ and a monomorphism $h : Z \longrightarrow Y$ such that $f = h \circ g$. This condition is satisfied by any Abelian category. But it is satisfied in some non-additive categories as well. In particular, Daigneault considers the category of Boolean algebras with unit preserving homomorphism and shows that it satisfies the given conditions. (It should be mentioned at this point that the Stone duality theorem, and thus the category *Stone* of Stone spaces, is used to prove that surjective homomorphisms are indeed epimorphisms.) It is also shown that the injectives are precisely the complete Boolean algebras. Although Daigneault's paper makes no connection with Lawvere's elementary theories nor to polyadic algebras, it is known that the concept of an extension of a map played a role in the latter. The proof that complete Boolean algebras are injectives is attributed to Vincent Papillon, Reyes' first graduate student.

Vincent Papillon Master's thesis [1969] is entitled *Quelques aspects de l'enveloppe injective d'une algèbre de Boole.* The first chapter of the thesis starts off by presenting the results of the foregoing paper. In the remaining two chapters of the thesis, Papillon gives two different constructions of the injective envelope of a Boolean algebra: one follows Reyes's alternative definition of a complete ring of quotients of a ring A and the other follows Joyal's suggestion of using the Stone space associated to the Boolean algebra A. The thesis was submitted in April 1969.

We mention Daigneault's work and Papillon's thesis for two reasons. First, it shows that certain categories, which were not necessarily Abelian, satisfying certain simple exactness conditions like the existence of images, had interesting consequences, e.g. the existence of injective envelopes. Furthermore, the conditions revealed here are directly linked to the category of Boolean algebras, certainly relevant for classical propositional logic. Indeed, if one were to start with classical propositional theories, construct their Lindenbaum–Tarski algebras and consider the resulting category of such algebras, one would obtain a category satisfying these exactness conditions. Second, as Papillon's thesis shows clearly, Joyal and Reyes were probably aware of Lawvere's elementary theories and they were immersed in category theory, Boolean algebras and Stone spaces.

Indeed, in August of 1969, the *Notices of the American Mathematical Society* received an abstract from Joyal entitled *Boolean algebras as functors.* A γ-complete Boolan algebra, for γ an infinite regular cardinal, is a Booleal algebra closed under sups of power less than γ. A γ-field is a Boolean algebra isomorphic to a γ-complete field of sets. A γ-representable Boolean algebra, is a Boolean algebra isomorphic to a quotient of a γ-field by a γ-ideal. Let Set_γ denote the full subcategory of the category Set of sets, generated by sets of power less than γ. The functor category Set^{Set_γ} where the functors $Set_\gamma \longrightarrow Set$ commute with limits is equivalent to the category of γ-fields with γ-isomorphisms. When limits are restricted to diagrams of power less than γ, then there is an equivalence between the functor category Set^{Set_γ} and the category of γ-representable Boolean algebras (and γ-homomorphisms). Joyal obtains a similar equivalence between the category of *complete* Boolean algebras with a functor category Set^{Set_γ}, thus showing that Boolean algebras can be described as functors. A particular case of this result is attributed to Lawvere. What is interesting here is the *kind* of result Joyal is after: to be able to represent certain fundamental mathematical concepts as *functors.* As we have seen in the foregoing sections, this was very much in the spirit of Lawvere's work.

Approximately one year later, on November 4 1970, the same *Notices* receive five abstracts from Joyal. Three of them are put in the section of algebra and two of them are placed in the section on logic and foundations. The first three are entitled *Spectral Spaces and Distributive Lattices*, *Spectral Spaces II* and *Cohomology of Spectral Spaces* respectively. They are directly related to Grothendieck's work in algebraic geometry, in particular to results about schemes and sheaves. The two abstracts on logic and foundations are entitled *Polyadic Spaces and Elementary Theories* and *Functors which preserve elementary operations.* Thus, Joyal indeed knew about Lawvere's elementary theories from early on and was pursuing their study. Despite the fact that they were inserted in different sections by the editors of the *Notices* and that their titles suggest that they are about different topics altogether, they are in fact related to one another, illustrating how, already before the advent of elementary toposes, Joyal was connecting algebraic geometry and logic together. Furthermore, it is clear that Joyal's earlier work on Boolean algebras and Stone spaces played a role in his subsequent work.

Let us start with the two abstracts on logic and foundations. The first abstract is about polyadic *spaces*, not algebras, and elementary theories. Let Set_0 be the category of *finite* sets. A *polyadic space* is a functor $E : Set_0^{op} \longrightarrow Stone$, where $Stone$ is the category of Stone spaces and continuous maps such that

i) For each $f : n \longrightarrow m$ in Set_0, $E(f) : E(m) \longrightarrow E(n)$ is open and

ii) E transforms push-out squares in Set_0 into quasi pull-back squares in $Stone$.

A *morphism* $\alpha : E \longrightarrow E'$ of polyadic spaces is a natural transformation such that for any $f : n \longrightarrow m$ in Set_0, the square

$$\begin{array}{ccc} E(m) & \xrightarrow{E(f)} & E(n) \\ {\scriptstyle \alpha_m}\downarrow & & \downarrow{\scriptstyle \alpha_n} \\ E'(m) & \xrightarrow[E'(f)]{} & E'(n) \end{array}$$

is a quasi pull-back. (Recall, that given a category $\mathcal{C}$ with pullbacks, a square

$$\begin{array}{ccc} P & \longrightarrow & Y \\ \downarrow & & \downarrow \\ X & \longrightarrow & Z \end{array}$$

if the morphism $P \longrightarrow X \times_Z Y$ is an epimorphism.)

Given a finite set X, Joyal defines a polyadic space $\tilde{X}$ by $\tilde{X}(n) = \beta(X^n)$, the Stone-Čech compactification of X^n, and a *model* of a polyadic space E(based on X) to be a morphism $\widetilde{X} \longrightarrow E$. He then claims that it is possible to associate canonically a polyadic space to an elementary theory and vice-versa. No details are given. It is asserted that under this correspondence, the two concepts of model coincide and that classical theorems of logic are interpreted and proven. We are not told which classical theorems of logic are interpreted and proven. Finally, and this is an important remark, it is claimed, without any detail, that polyadic spaces arise naturally in algebraic geometry.

The second abstract does not mention polyadic spaces. But it contains a fundamental result that will turn out to be central later on. Consider the category

$\mathcal{E} = Set^{Set}$ with functors $Set \longrightarrow Set$ preserving finite limits and finite (disjoint) sums. It is claimed that these functors are precisely those preserving elementary operations of first-order logic. We are not told precisely what the latter operations are, except for projection and complement. But this is not the result reported, it is simply mentioned in the opening sentence of the abstract. The statement of the main result requires additional data. Let $\mathcal{U}$ denote the category whose objects are pairs $(I, \mathfrak{F})$, with $\mathfrak{F}$ an ultrafilter over the set I and whose morphisms are mappings preserving ultrafilters. Given two objects $(I, \mathfrak{F})$ and $(J, \mathfrak{G})$, define an equivalence relation on morphisms $(I, \mathfrak{F}) \longrightarrow (J, \mathfrak{G})$ by $f \equiv g$ if and only if $\exists F \in \mathfrak{F}$ such that $f|F = g|F$. Let $\mathcal{U}P$, called the category of *ultrapoints*, be the quotient category $\mathcal{U}/\equiv$. The main theorem is that there is an equivalence between $\mathcal{E}^{op}$ and the category of pro-objects of $\mathcal{U}P$ of ultrapoints. This is in itself an interesting duality theorem, but Joyal points out immediately afterwards that "the straightforward proof of this theorem can be adapted to prove a theorem of Keisler on elementary extensions of complete structures" [Joyal, 1971d, 967]. The theorem refered to here appeared in 1960 and states that an elementary class K is closed with respect to colimits if and only if K is characterized by a set of sentences constructed from finite conjunctions, finite disjunctions and existential quantifiers. We will give a more precise statement of the theorem later. Keisler's proof is model theoretical whereas Joyal's result stated here is purely categorical. The reference to the category of pro-objects is revealing, for the latter construction was introduced by Grothendieck and Deligne in SGA4, thus showing that Joyal was already well versed in Grothendieck toposes and related constructions.

This work was seen by Joyal and Reyes as being part of the search for an algebraic approach to logic. Indeed, in his paper published in 1972, Reyes used Joyal's work on Boolean algebras and talks about "Joyal's functorial approach to algebraic logic", which refers to the first abstract quoted above. In the bibliography of the same paper, Reyes announced a paper in preparation entitled "Éléments de logique algébrique", which never saw the day.

Let us now turn to the three abstracts in algebraic geometry. A *spectral space* X is a topological space satisfying two conditions: i) the quasi-compact open subsets of X form a base closed under finite (possibly empty) intersections; ii) every irreducible closed subset F possesses one and only one generic point x, i.e. $\overline{\{x\}} = F$. A morphism of spectral spaces is a quasi-compact mapping. A theorem asserting that a topological space X is a spectral space if and only if X is a limit of finite T_0-spaces is stated. The main theorem is, once more, an equivalence of categories between the opposite of the category $DLat$ of distributive lattices and the category S_p of spectral spaces. Nowadays, the so-called category of coherent locales is used instead of the category of spectral spaces to prove the duality theorem. (See [Johnstone, 1982, 65].) The link to Stone's duality theorem is immediate. Furthermore, as Joyal mentions himself, the underlying space of a quasi-compact quasi-separated scheme is a spectral space, making explicit once more the connection with algebraic geometry *à la* Grothendieck, this time with EGA.

The second abstract is exactly in the same vein: the main theorem asserts the existence of a right adjoint to the inclusion functor $Stone \longrightarrow S_p$. Furthermore, there is a covariant full embedding from the category of spectral spaces and the category of *ordered* Stone spaces. Finally, Joyal claims that a Stone space is ordered if and only if it is a limit of finite partially ordered sets. (See [Johnstone, 1982, 75] for details.)

Only the last abstract is directly related to algebraic geometry: in it, Joyal generalizes results of Grothendieck on the cohomology of quasi-compact schemes to spectral spaces.

What these abstracts show clearly is that Joyal was juggling with three conceptual realms during this period: algebraic geometry, including Grothendieck toposes, Stone-type dualities and Lawvere's elementary theories in the spirit of algebraic logic. Thus, when Lawvere and Tierney introduced the concept of elementary topos, Joyal already had results that could easily be seen to be directly related to that development. The notion of elementary topos simply provided the general framework to develop these ideas systematically.

Joyal, but not Reyes, was present at the meeting held at Dalhousie between January 16 and January 19, 1971. The goal of this meeting, according to Lawvere, was to investigate

> ... the connecctions between algebraic geometry and " intuitionistic " logic under the guidance of the form of objective dialectics known as category theory....
>
> Our own hopes in the success of the above general program were strengthened by initial progress in carrying out a more special program which will be outlined in the introduction. This is the development on the basis of elementary (first-order) axioms of a theory of " toposes " just good enough to be applicable not only to sheaf theory, algebraic spaces, global spectrum, etc. as originally envisaged by Grothendieck, Giraud, Verdier, and Hakim but also to Kripke semantics, abstract proof theory, and the Cohen-Scott-Solovay method for obtaining independence results in set theory. [Lawvere, 1972, 1].

According to Lawvere, seventy mathematicians participated in this meeting, among others Jean Bénabou, Marta Bunge, Jim Lambek, William Mitchell, Dana Schlomiuk, Dana Scott, Hugo Volger, to mention but those who are directly involved in our story. In the concept of elementary topos, geometry and logic are woven together in one structure, thus providing a natural setting for logic and algebraic geometry.

Let us now look carefully at the characterization of first-order logic developed by Joyal and Reyes.

4.4 *From logical theories to categories*

Joyal and Reyes realized that Volger's mistake had been to include the object Ω in the data of a logical category. They came to the conclusion that this object does not belong to first-order logic, neither in its syntax nor in its semantics. The key methodological insight was to start from a logical theory and build a category from it. The beauty of this approach is that in this way, one introduces, simultaneously, a hierarchy of logical theories together with a hierarchy of categories. Furthermore, the very first type of category thus obtained turned out to be an important type of abstract categories that had just been introduced and developed by Michael Barr for entirely different purposes, namely *regular categories*. (See [Barr *et al.*, 1971].) In addition, Joyal and Reyes made connections with elementary toposes and Joyal gave a uniform categorical treatment of the various completeness theorems. In fact, Lawvere reported that much himself in the introduction of the volume where Volger's papers were published. Indeed, in the very first page of the introduction, we read that the "much-researched intermediate doctrine of (classical) first-order logic corresponds to the fact P that the category of sets has finite limits, complements of subsets, and images of mappings (related by the condition of being a "regular" category...)." [Lawvere, 1975a, 3]. Volger himself acknowledges Joyal's work and remarks that Joyal has shown that his semantical categories are regular categories. (See [Volger, 1975a, 82].)

Joyal and Reyes started a joint paper that was finally published under Reyes's name in 1974, but a preliminary version of it was already circulating in 1972. (See [Reyes, 1974].) This is one of the very few published articles that present Joyal's work during that period. In the introduction, Reyes explicitly acknowledges Joyal's contribution. However, there is more literature that gives a portrait of the situation: as we have already mentioned, eight theses, Master's and PhDs, were written at the Université de Montréal on categorical logic between 1973 and 1977. Here is the complete list with the authors, subject, supervisors, degree and year:

Name	Title	Supervisor	Degree and year
Jean Dionne	Des théories élémentaires aux categories conceptuelles	Gonzalo Reyes	MSc July 1973
André Boileau	Les multiples splendeurs du forcing	Gonzalo Reyes	MSc December 1973
Christiane Rousseau	Triples, localization et topos	Pierre Berthiaume	MSc July 1974
Roch Ouellet	Axiomatization de la logique interne du premier ordre des topos, version inclusive et multisorte	Aubert Daigneault	PhD August 1974

Name	Title	Supervisor	Degree and year
Monique Robitaille-Giguère	Modèles d'une catégorie logique dans des topos de préfaisceaux et d'ensembles de Heyting	Gonzalo Reyes	MSc January 1975
Wedad Antonius	Théories cohérentes et prétopos	Gonzalo Reyes	MSc June 1975
André Boileau	Types versus Topos	Gonzalo Reyes	PhD July 1976
Christiane Rousseau	Théorie des topos et analyse complexe	Dana Schlomiuk	PhD April 1977

We assume that the work presented in Dionne, Ouellet, Robitaille-Giguère, Antonius and Boileau's Ph.D. thesis *roughly* reflects the chronological development and assimilation of the work done by Joyal and Reyes, although with a time shift. Also relevant are Reyes' later publications in 1977 and 1978 as well as Makkai and Reyes's papers published in 1976 which builds on the work done by Reyes and Joyal. Boileau's thesis is particularly interesting since it gives a general portrait of the work done in categorical logic at the time. Indeed, he refers to almost all the results presented in the theses by his colleagues.

Joyal and Reyes's strategy was to *start with* a first-order theory $\mathbb{T}$, *construct* a *small* category $\mathcal{C}_{\mathbb{T}}$ from it and then characterize in an abstract manner the type of category thus obtained. One of the goals was then to work with categorical methods to obtain significant results about the theory $\mathbb{T}$ or the underlying logic. There was an additional philosophical motivation. The very first formal step in the construction is to identify formulas of a theory $\mathbb{T}$ in a given language relative to their logical strength. More specifically, given a language $\mathcal{L}$ and a theory $\mathbb{T}$ in $\mathcal{L}$, one defines an equivalence relation between formulas of $\mathcal{L}$ by

$$\varphi \sim \psi \text{ if and only if } \mathbb{T} \vdash (\forall x_1) \cdots (\forall x_n)(\varphi \leftrightarrow \psi)$$

where $x_1, \ldots, x_n$ are the free variables of φ and ψ. The objects of the constructed category are the equivalence classes determined by the equivalence relation. One quickly recognizes that this construction is the categorical extension of the usual Lindenbaum–Tarski construction for propositional logic. One of the key insights here, which builds on Lawvere's work on quantifiers, is the fact that the Lindenbaum-Tarski construction can be lifted in a natural way so that when applied to first-order logic, it yields a category. Since categories are, among other things, generalizations of partial orders, this constitutes evidence that categorical logic provides a natural setting for the algebraic treatment of logic. From the point of view of Joyal, Reyes and other categorical logicians of that period, one of the underlying motivations was that logically equivalent formulas "mean" the same thing, that is we are dealing with *concepts* instead of their specific linguistic expressions. For that reason, these categories were labeled *conceptual categories* or *categories of concepts* (or sometimes *syntactical categories*). Morphisms between concepts are also equivalence classes of functional relations definable in the language.

As Jean Dionne underlies in the introduction of his thesis, it was believed that by replacing the theory $\mathbb{T}$ by the category $\mathcal{C}_{\mathbb{T}}$, the object of study would be at the same time simpler and more complete. The latter fact amounts to the claim that categories allow for more relations between concepts than the usual set-theoretical inclusion. Furthermore, models of theories can now be taken in various categories, in particular in toposes. The very last sentence of Dionne's thesis summarizes the belief that a categorical presentation $\mathcal{C}_{\mathbb{T}}$ of a theory $\mathbb{T}$ in a given formal system allows logicians to concentrate on the conceptual aspects of a theory and less on the syntactical details of a chosen presentation:

> Our method is therefore general enough to apply to a large variety of languages and we can conclude that it constitutes the true link between logic in the traditional sense and categorical logic : with the advantage that we can now work without using the axioms, rules of inference, formal demonstrations and the rest. We can even forget about variables, which were always inconvenient. Moving to categories, we can do logic in an " abstract " fashion in terms of objects, arrows, functors and diagrams, which is, among all the advantages mentioned in this work, not the least, since the logician is naturally more interested by concepts than by the indices to give to variables. [Dionne, 1973, 110] (our translation)

Thus, in the early seventies, the belief was that categorical methods were more objective or invariant than logical methods and that they would fruitfully replace logical methods. As far as we can tell, it was in April 1973, while he was in Montreal, that Bénabou showed that in some cases logical methods, via the so-called internal language, were more simple and direct than categorical methods to prove categorical properties of categories, in particular of certain toposes.

As we have already said, Joyal and Reyes' approach lead to a hierarchy of conceptual categories, the first of which resulting from the weakest first-order logical language yielding a category in the foregoing construction: regular categories. It was also quickly conjectured that the converse would be true, i.e. that one could start from a regular category $\mathcal{C}$ and obtain a theory $\mathbb{T}_{\mathcal{C}}$ in the appropriate language. Furthermore, it was common at that point to consider logic in a many-sorted language. (Indeed, Lawvere, Lambek, Bénabou and as we will see Mitchell had already indicated that many-sorted languages were natural in a categorical context.) For the presentation of many-sorted languages, Dionne refers to Feferman's lecture notes in proof theory published a few years earlier. (See [Feferman, 1968].) The same reference is found in [Makkai and Reyes, 1976b; Reyes, 1977; Makkai and Reyes, 1977]. We will here essentially follow [Boileau, 1976], for he gives a general and unified presentation that, as we have indicated, constitutes in our opinion a comprehensive synthesis of the work done by Joyal and Reyes between 1971 and 1973 approximately. (For a slightly different presentation and terminology, see [Johnstone, 2002].)

A (first-order) *primitive language* $\mathcal{L}_0$ is given by:

1. A set $\mathcal{S}$ of sorts. The set $\mathcal{T}$ of *types* is build from the set of sorts as follows:
   (a) if $S \in \mathcal{S}$, then $S \in \mathcal{T}$;
   (b) if $S_1, ..., S_n \in \mathcal{T}$, then $S_1 \times \cdots \times S_n \in \mathcal{T}$, where n can be 0; whenever it is 0, we denote it by the symbol 1; (notice that $S_1 \times \cdots \times S_n$ is merely a formal expression, not a product);
   (c) nothing else is a type.
2. A set $\mathcal{F}$ of *function symbols*, to each function symbol f, we associate its *type* $S_1 \times \cdots \times S_n \times S$ (with the last sort having a distinguished status); we write $f : S_1 \times \cdots \times S_n \longrightarrow S$ to indicate the type of f. Whenever n is 0, f is a *constant* of type S and we write $1 \xrightarrow{f} X$.
3. A set $\mathcal{R}$ of *relation symbols*, each such relation symbol R has a *type* $S_1 \times \cdots \times S_n$; we write $R \rightarrowtail S_1 \times \cdots \times S_n$ to indicate that R has type $S_1 \times \cdots \times S_n$.

Terms and formulas are defined by recursion as follows:

Definition of terms:

1. Every variable of type S is a term of type S;
2. If $f : S_1 \times \cdots \times S_n \longrightarrow S$ is a function symbol and $t_1, \ldots, t_n$ are terms of type $S_1, \ldots, S_n$ respectively, then $f(t_1, \ldots, t_n)$ is a term of type S; (if f is a constant, we simply write f);

Definition of formulas:

1. If t_1 and t_2 are terms of type S, then $t_1 = t_2$ is a formula;
2. If $R \rightarrowtail S_1 \times \cdots \times S_n$ is a relation symbol and $t_1, \ldots, t_n$ are terms of type $S_1, \ldots, S_n$, then $R(t_1, \ldots, t_n)$ is a formula;
3. $\top$ is a formula ('true' is a formula);

The notions of free variable, bound variable, closed formula and closed terms are defined in the usual manner. This completes the definition of $\mathcal{L}_0$. There are no logical connectives so far. But there are logical rules associated to it, as we will see.

However, interpreting logical systems in a categorical framework forced categorical logicians to reconsider basic and simple assumptions in a different light. Indeed, in a set-theoretical setting, sorts are assumed to be non-empty. In a categorical setting, the situation is different. For one thing, in certain toposes, in particular toposes of sheaves, there are objects without elements that are nonetheless inhabited. To illustrate the situation, consider the simple case of the topos Set^2 whose objects are pairs $\langle X, Y\rangle$ of sets and a morphism $\langle f, g\rangle : \langle X, Y\rangle \longrightarrow \langle V, W\rangle$ is given by a pair of functions $f : X \longrightarrow V$ and $g : Y \longrightarrow W$. Consider now any object of the form $\langle\emptyset, X\rangle$ where X is any non-empty set. Such an object is not empty (nor is it an initial object in the language of categories), but it is not like an arbitrary object $\langle X, Y\rangle$ either. From the categorical point of view, an element x of a set X corresponds to a morphism $1 \xrightarrow{x} X$ from the terminal object 1, that is any singleton set, to the set X in the category Set of sets. It is therefore reasonable to generalize this fact to categories with a terminal object (or finite limits in general): a (global) element of an object X in a category $\mathcal{C}$ is a morphism $1 \xrightarrow{x} X$. In the topos Set^2, the terminal object is the pair $\langle 1, 1\rangle$ and there is no morphism $\langle 1, 1\rangle \longrightarrow \langle\emptyset, X\rangle$. In this sense, we can say that the latter object does not have any (global) element, although it is not empty. In his introduction to the volume on model theory and toposes, Lawvere underlines this fact and comments rather harshly on the classical solution:

> Since a variable set may be partly empty and partly non-empty, the traditional model-theoretic banishment of empty models cannot be maintained, bringing to light a certain difficulty which the banishment obscured. Some claim that this difficulty is the "fact" that "entailment is not transitive", contrary to mathematical experience. However, the actual "difficulty" is that the traditional logical way of dealing with variables is inappropriate and hence should be abandoned. This traditional method (which by the way is probably one of the reasons why most mathematicians feel that a logical presentation of a theory is an absurd machine strangely unrelated to the theory or its subject matter) consists of declaring that there is one set I of variables on which all finitary relations depend, albeit vacuously on most of them; ... [Lawvere, 1975b, 5]

Various people, among others Mostowski, Hailperin in the fifties, and Bénabou and Joyal in 1973, had observed the failure of the transitivity of entailment with empty sorts. (See [Mostowski, 1951; Hailperin, 1953].) A simple case is as follows: the following two sequents are always valid, no matter what:

$$\forall x P(x) \vdash P(x)$$

$$P(x) \vdash \exists x P(x)$$

Thus, by transitivity of entailment, we obtain:

$$\forall x P(x) \vdash \exists x P(x)$$

which *is* an invalid entailment when interpreted in an empty sort. Mostowski suggested to simply dropping the transitivity of entailment. Hailperin gave an axiomatization in which entailments are restricted to sentences. Ideally, one would keep the transitivity of entailments and give a presentation of a logical system for arbitraty formulas.

As far as we can tell, the current solution to the problem was first presented in print in Roch Ouellet's PhD thesis [1974] defended at the Université de Montréal in 1974. (See also his [1981].) He explicitly attributes the key observation leading to the solution to Joyal. The solution consists in introducing the notion of the *support of a sequent* or what is now called a *context*: a *context* is a finite list $\vec{x} = x_1, \ldots, x_n$ of distinct variables. Notice that a context can be empty, i.e. n can be 0. The *type* of a context $\vec{x}$ is the string of (not necessarily distinct) sorts of the variables appearing in it. A *term-in-context* is an expression of the form $\vec{x} \cdot t$ where t is a term and $\vec{x}$ is a context containing all the free variables of t. Similarly, a *formula-in-context* is an expression of the form $\vec{x} \cdot \phi$, where ϕ is a formula and $\vec{x}$ is a context containing all the free variables occurring in ϕ.

The notion of logical consequence is then defined with respect to a context $\vec{x}$. A *sequent* is a sequence of symbols of the form $(\phi \vdash_{\vec{x}} \psi)$, where ϕ and ψ are formulas and $\vec{x}$ is a context containing all the free variables of ϕ and ψ. It is now easy to see that the foregoing failure of transitivity is eliminated, since the sequent of the conclusion, namely $\forall x P(x) \vdash \exists x P(x)$, has an empty context whereas the premises do not. Hence, if we require, as we will do, that entailments have to be done over uniform contexts or that a change of context has to obey certain restrictions, the conclusion will not follow. It should be emphasized that, as is done by Ouellet himself in his thesis, the process of finding a solution was guided all along by categorical semantics. Thus, in this case, the correct syntactical constraint was derived from categorical constraints.

Recall that in a sequent calculus, a deductive rule has the form $\frac{\Gamma}{\alpha}(R)$, where Γ is a (possibly empty) list of sequents and α is a sequent. The horizontal bar means that the rule R allows the move from Γ to α. The intended meaning is that whenever the sequents of Γ are valid, then α is valid. When Γ is empty, then in this case α is a logical axiom. Here are the axioms and rules of $\mathcal{L}_0$, which are in fact the rules of *equational logic*:

1. Structural rules:

   1.1. $\phi \vdash_{\vec{x}} \phi$ (identity axiom)

   1.2. $\dfrac{\phi \vdash_{\vec{x}} \psi \quad \psi \vdash_{\vec{x}} \theta}{\phi \vdash_{\vec{x}} \theta}$ (cut)

   1.3. $\frac{\phi \vdash_{\vec{x}} \psi}{\phi[\vec{s}/\vec{x}] \vdash_{\vec{y}} \psi[\vec{s}/\vec{x}]}$ (substitution) where $\vec{s} = s_1, \ldots, s_n$ is a list of (not necessarily distinct) terms of the same length and type as the context $\vec{x}$, $\phi\,[\vec{s}/\vec{x}]$ denotes the usual operation of simultaneous substitution with the usual proviso and $\vec{y}$ is any string of variables including all the variables occurring in the terms $\vec{s}$.

2. Logical rule:

2.1. $\phi \vdash_{\vec{x}} \top$;

3. Rule for equality:

3.1. $\top \vdash_{\vec{x}} x = x$.

The next step consists in introducing logical connectives in a certain order. The first extension of $\mathcal{L}_0$ is denoted by $\mathcal{L}_1$ by Boileau, but we will denote it by $\mathcal{L}_{reg}$ for *regular logic.*

Definition of formulas of $\mathcal{L}_{reg}$: formulas of $\mathcal{L}_{reg}$ consist of formulas of $\mathcal{L}_0$ together with the following additional clauses:

4. If ϕ and ψ are formulas, then $(\phi \wedge \psi)$ is a formula;

5. If ϕ is a formula, then $\exists x\phi$ is formula (where x is variable of some type; we sometimes write $(\exists x : S)\phi$ to indicate the type of x).

Axioms and rules of $\mathcal{L}_{reg}$: the axioms and rules of $\mathcal{L}_{reg}$ are those of $\mathcal{L}_0$ together with:

2. Logical rules:

3. $\phi \wedge \psi \vdash_{\vec{x}} \phi \quad \phi \wedge \psi \vdash_{\vec{x}} \psi \quad \dfrac{\phi \vdash_{\vec{x}} \psi \quad \phi \vdash_{\vec{x}} \theta}{\phi \vdash_{\vec{x}} \psi \wedge \theta}$;

4. $\dfrac{\phi \vdash_{\vec{x},y} \psi}{\exists y\phi \vdash_{\vec{x}} \psi}$ where y is not free in ψ and where a double line indicates that we can go either way;

5. $\phi \wedge \exists y\psi \vdash_{\vec{x}} \exists y\, (\varphi \wedge \psi)$ where y is not free in ϕ.

6. Rule for equality:

3.2. $((x_1 = y_1) \wedge \ldots \wedge (x_n = y_n)) \wedge \phi \vdash_{\vec{z}} \phi\, [\vec{y}/\vec{x}]$ where $\vec{z}$ is a context containing $\vec{x}$, $\vec{y}$ and the free variables of ϕ.

A *regular theory* $\mathbb{T}$ is a set of regular sequents, i.e. sequents in which the formulas are all regular formulas. The elements of $\mathbb{T}$ are called the *axioms* of $\mathbb{T}$. There are no examples of regular theories in [Dionne, 1973], nor in [Reyes, 1974]. Of course, there are numerous examples of regular *categories.*

The next extension quickly became the center of attention because of its metalogical properties. It is the language $\mathcal{L}_{coh}$, labeled $\mathcal{L}_2$ by Boileau, of *coherent logic.*

Definition of formulas of $\mathcal{L}_{coh}$: formulas of $\mathcal{L}_{coh}$ consist of formulas of $\mathcal{L}_{reg}$ together with the additional clauses:

6. $\perp$ is a formula ('false' is a formula);

7. If ϕ and ψ are formulas, then $(\phi \vee \psi)$ is a formula.

Axioms and rules of $\mathcal{L}_{coh}$ are those of $\mathcal{L}_{reg}$ together with:

2. 2.1. $\bot \vdash_{\vec{x}} \phi$;

   2.2. $\phi \vdash_{\vec{x}} \phi \vee \psi \quad \psi \vdash_{\vec{x}} \phi \vee \psi \quad \dfrac{\phi \vdash_{\vec{x}} \theta \quad \psi \vdash_{\vec{x}} \theta}{\phi \vee \psi \vdash_{\vec{x}} \theta}$;

   2.3. $\phi \wedge (\psi \vee \theta) \vdash_{\vec{x}} (\phi \wedge \psi) \vee (\phi \wedge \theta)$.

A *coherent theory* $\mathbb{T}$ is a set of coherent sequents. Although the language of coherent logic might seem to be rather weak at first, it was quickly seen that many mathematical theories are in fact axiomatized in that language. Thus, every equational theory is coherent, e.g. the theory of groups, the theory of rings, Boolean algebras, etc., provided one writes the axioms appropriately. The theory of fields of characteristic p, $p \geq 0$, is coherent, as well as the theory of algebraically closed fields. It was proved that any classical theory can be translated in the language of coherent logic provided that the latter is enriched with sufficiently many relational symbols. The latter claim is a corollary of the proof of the completeness theorem for coherent theories attributed to Makkai by Antonius. (See [Antonius, 1975, 35–36].) Coherent logic was seen has having an interesting position with respect to intuitionistic logic and classical logic. In the words of Reyes:

> If the reader looks at the formal system for coherent logic ..., he will notice that all axioms as well as rules of inference are intuitionistically valid (...) as well as classically valid. The distinction between the intuitionistic and classical interpretations of logical operations become irrelevant for this "absolute" logic.
>
> Furthermore, this logic besides being a part [sic], may be considered as a generalization of classical logic. Indeed, any classical theory may be rendered coherent by extending the language [Antonius]. The idea is trivial and may be seen from the following example : take
>
> $$\exists x \exists y (x \neq y)$$
>
> as our theory. Adding a new binary relation symbol D (to be thought of as $\neq$), the desired coherent theory is
>
> $$\begin{gathered} \top \vdash \exists x \exists y D(x,y) \\ \top \vdash x = y \vee D(x,y) \\ x = y \wedge D(x,y) \vdash \bot \end{gathered}$$
>
> The point is : the models which respect the coherent logic of the new theory are the usual classical models (respecting the full logic) of the old one. [Reyes, 1977, 23].

(Reyes did not specify the support of the sequents.) As we can see from the last sentence, the charms of coherent logic were particularly obvious from the model-theoretic point of view. But we should point out that the insistence on the formal system itself is something that came during the academic year 1973–74, for absolutely nothing is said about it before that date. As we have already pointed out, they were merely used as a springboard in order to rise to the categorical level.

The next two levels are the intuitionistic and the classical levels.

Definition of formulas of $\mathcal{L}_{int}$ (Boileau's $\mathcal{L}_3$): the formulas of $\mathcal{L}_{int}$ are those of $\mathcal{L}_{coh}$ together with:

8. If ϕ and ψ are formulas, then $(\phi \Rightarrow \psi)$ is a formula;

9. If ϕ is a formula, then $\forall x\phi$ is formula (or $(\forall x : S)\phi$).

Axioms and rules of $\mathcal{L}_{int}$ are those of $\mathcal{L}_{coh}$ together with:

2.8 $\dfrac{\phi \wedge \psi \vdash_{\vec{x}} \theta}{\psi \vdash_{\vec{x}} \phi \Rightarrow \theta}$

2.9 $\dfrac{\phi \vdash_{\vec{x},y} \psi}{\phi \vdash_{\vec{x}} \forall y\psi}$ where y is not free in ϕ.

Intuitionistic theories are defined in the obvious manner.

Finally, we come to classical first-order logic $\mathcal{L}_{Bool}$ (Boileau's $\mathcal{L}_4$). We now allow formulas of the form $\neg\phi$, which can be defined as $\phi \Rightarrow \perp$.

Axioms and rules of $\mathcal{L}_{Bool}$ are those of $\mathcal{L}_{int}$ together with:

2.10 $\dfrac{\phi \wedge \psi \vdash_{\vec{x}} \perp}{\psi \vdash_{\vec{x}} \neg\phi}$; a special case of the rule 2.8;

2.11 $\top \vdash_{\vec{x}} \phi \vee \neg\phi$.

A *proof* in these systems is defined in the usual fashion. We will write $\mathbb{T} \vdash: \alpha$, where α is a sequent $\phi \vdash_{\vec{x}} \psi$, to indicate that there is a proof of α from the axioms of $\mathbb{T}$. Whenever we will have to specify the underlying logic, we will write for instance $\mathbb{T} \underset{coh}{\vdash:} \alpha$, meaning that the proof is in coherent logic.

Dionne [1973] describes these languages for one-sorted systems and sketches the extension to many-sorted systems in the conclusion of his thesis. Furthermore, coherent logic does not receive a name: it is simply called regular logic with stable sup (for reasons that will appear clear once we look at categories corresponding to this language). This is also how they are called in Reyes [1974]. (Remember that the latter was already circulating in 1972.) This does suggest that in 1972 and 1973, coherent logic had not quite acquired its status or at least no one thought that it deserved to receive a name.

There is an obvious partial order between the foregoing logical systems: define $\mathcal{L}_i \leq \mathcal{L}_k$ if and only if $\mathcal{L}_i \subseteq \mathcal{L}_k$ and for any set of sequents $\mathbb{T} \cup \alpha$ of $\mathcal{L}_i$, we have $\mathbb{T} \underset{\mathcal{L}_i}{\vdash:} \alpha$ if and only if $\mathbb{T} \underset{\mathcal{L}_k}{\vdash:} \alpha$. It can be proved that resulting order relation is:

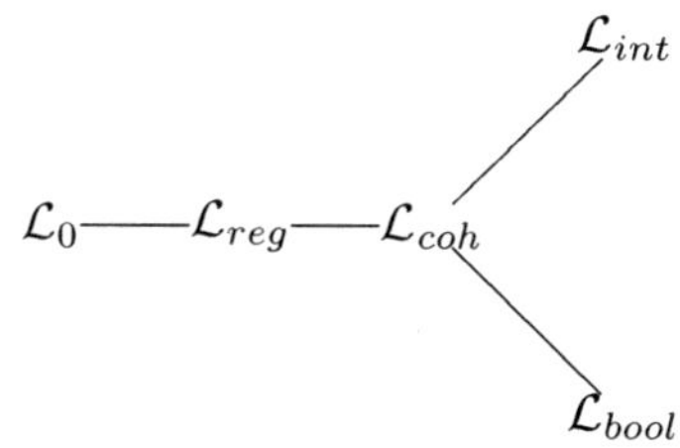

From a given theory $\mathbb{T}$ in a formal system $\mathcal{L}_i$, $0 \leq i \leq 4$, a category of concepts $\mathcal{C}_{\mathbb{T}}$ can be built. The construction of $\mathcal{C}_{\mathbb{T}}$ always follows the same pattern. (It should be pointed out that Makkai and Reyes introduced a slightly different construction later on, although the result is equivalent. See [Antonius, 1975, 14–15]. In their book, Makkai and Reyes sketch a method that uses a completeness theorem proved earlier but claim that the construction is described in great detail in Dionne's thesis, whereas the latter proceeds in a purely syntactic manner. See [Makkai and Reyes, 1977, 241–242].) As we have already indicated, the objects of $\mathcal{C}_{\mathbb{T}}$ are equivalence classes of formulas, where the equivalence relation is defined by:

$$\phi(\vec{x}) \sim \psi(\vec{x}) \text{ if and only if } \phi(\vec{x}) \vdash_{\vec{x}} \psi(\vec{x}) \text{ and } \psi(\vec{x}) \vdash_{\vec{x}} \phi(\vec{x}) \text{ are derivable sequents of } \mathbb{T}.$$

The equivalence class of $\phi(\vec{x})$ is denoted by $[\phi(\vec{x})]$ and it is sometimes called a *formal set.*

A morphism of formal sets is a *formal function.* Informally, a formal function is given by a formula that defines, in each model of the theory $\mathbb{T}$, an actual morphism between the interpreted formal sets. In particular, when the theory is interpreted in the category of sets, it defines a set-theoretical function between sets. Specifically, let $[\phi(\vec{x})]$ and $[\psi(\vec{y})]$ be two formal sets such that the contexts $\vec{x}$ and $\vec{y}$ are disjoint and let ρ be a formula with free variables in the context $\vec{x};\vec{y}$. Then we say that $\rho(\vec{x},\vec{y})$ defines a formal function from $[\phi(\vec{x})]$ to $[\psi(\vec{y})]$, denoted by $\langle \vec{x} \mapsto \vec{y} : \rho \rangle : [\phi(\vec{x})] \longrightarrow [\psi(\vec{y})]$, if

1. $\rho(\vec{x},\vec{y}) \vdash_{\vec{x},\vec{y}} (\varphi(\vec{x}) \wedge \psi(\vec{y}))$
2. $\rho(\vec{x},\vec{y}) \wedge \rho(\vec{x},\vec{z}) \vdash_{\vec{x},\vec{y},\vec{z}} \vec{y} = \vec{z}$
3. $\varphi(\vec{x}) \vdash_{\vec{x}} \exists \vec{y} \rho(\vec{x},\vec{y}))$.

To finish the definition of morphisms, we identify formal functions that are provably equivalent: that is, whenever we have

$$\rho(\vec{x},\vec{y}) \vdash_{\vec{x},\vec{y}} o(\vec{x},\vec{y}) \text{ and } o(\vec{x},\vec{y}) \vdash_{\vec{x},\vec{y}} \rho(\vec{x},\vec{y})$$

where $(\vec{x},\vec{y},o)$ defines a function from $[\phi(\vec{x})]$ to $[\psi(\vec{y})]$, then $(\vec{x},\vec{y},\rho)$ and $(\vec{x},\vec{y},o)$ belong to the same equivalence class $\langle \vec{x} \mapsto \vec{y} : \rho \rangle : [\phi(\vec{x})] \longrightarrow [\psi(\vec{y})]$. In a nutshell, the morphisms of $\mathcal{C}_{\mathbb{T}}$ are equivalence classes of formulas that are provably equivalent functional relations of $\mathbb{T}$. Composition of morphisms is defined by following the same general strategy. Given formal sets and formal functions

$$[\phi(\vec{x})] \xrightarrow{\langle \vec{x} \mapsto \vec{y} : \rho \rangle} [\psi(\vec{y})] \xrightarrow{\langle \vec{y} \mapsto \vec{z} : o \rangle} [\theta(\vec{z})]$$

the composition $\langle \vec{y} \mapsto \vec{z} : o \rangle \circ \langle \vec{x} \mapsto \vec{y} : \rho \rangle : [\phi(\vec{x})] \longrightarrow [\theta(\vec{z})]$is given by the formal function $\langle \vec{x} \mapsto \vec{z} : \kappa \rangle$ where $\kappa = \exists \vec{y}(\rho \wedge o)$. The verifications that composition is associative and that the identity morphisms exist and satisfy the required properties are extremely tedious. The moral is that one obtains a genuine *small* category $\mathcal{C}_{\mathbb{T}}$ for each theory $\mathbb{T}$ in a given language. The results can be summarized with the following table:

If $\mathbb{T}$ is in	**then $\mathcal{C}_{\mathbb{T}}$ is a**
$\mathcal{L}_{reg}$	Regular category
$\mathcal{L}_{coh}$	Coherent category
$\mathcal{L}_{int}$	Heyting category
$\mathcal{L}_{Bool}$	Boolean category

The foregoing construction can be seen to be a direct generalization of the standard Lindenbaum–Tarski algebra for a propositional theory. Indeed, it can be seen that if $\mathbb{T}$ is a *propositional* coherent theory, then $\mathcal{C}_{\mathbb{T}}$ is a preorder, in fact a distributive lattice. In the general case of a coherent theory $\mathbb{T}$, the standard Lindenbaum-Tarski algebra is simply a part of $\mathcal{C}_{\mathbb{T}}$, that is the part restricted to sentences, e.g. formulas in the empty context. This provides further evidence to the claim that categorical logic is algebraic logic.

Notice, and this is an important *philosophical* point, that a category of concepts $\mathcal{C}_{\mathbb{T}}$ is *not* a category of structured sets and structure preserving functions: morphisms are *equivalence classes* of morphisms. Thus, one of the main tools of categorical logic is the construction of categories that are in a sense very different from the categories arising in the practice of mathematics, e.g. the category of groups or the category of vector spaces.

Whereas Dionne's thesis consists in a detailed description of the constructions of conceptual categories from theories and a proof, in each case, that the resulting category has the appropriate properties, Reyes [1974] starts right from the start with a description of the categories. Of course, as we have already emphasized, within the categorical community, it was assumed that replacing a theory $\mathbb{T}$ by its category of concepts $\mathcal{C}_{\mathbb{T}}$ was not only mathematically fruitful but also philosophically motivated. One of the goals was to give a purely categorical characterization of these categories and then work with these instead of the syntactical presentations of the theories.

4.5 Algebraic logic: from regular to Boolean categories

Regular categories are categories in which any morphism factors uniquely as a (regular) epimorphism followed by a monomorphism. In contrast with Volger's logical categories and elementary theories, any topos $\mathcal{E}$ turns out to be a regular category and, furthermore, first-order logic becomes completely integrated in the overall framework at the center of which sits the notion of topos.

In order to state the definition of a regular category, we need to recall a few simple notions. The pullback of a pair of equal morphisms $X \xrightarrow{f} Y \xleftarrow{f} X$, when it exists, is called the *kernel pair* of f. Given two parallel morphisms $X \underset{g}{\overset{f}{\rightrightarrows}} Y$, a *coequalizer* of $\langle f, g\rangle$ is a morphism $q : Y \longrightarrow E$ such that $q \circ f = q \circ g$ and for any morphism $h : Y \longrightarrow Z$ such that $h \circ f = h \circ g$, there is a unique morphism $h' : E \longrightarrow Z$ such that $h' \circ q = h$. A morphism $f : X \longrightarrow Y$ is said to be a *regular epimorphism* when it is the coequalizer of a pair of arrows.

Definition: a category $\mathcal{C}$ is said to be *regular* if it satisfies the following properties:

1. It has all finite limits;
2. Coequalisers of kernel pairs exist;
3. Regular epimorphisms are stable under pullback.

The last condition simply means that if $f : X \longrightarrow Y$ is a regular epimorphism and $g : Z \longrightarrow Y$ is a morphism of $\mathcal{C}$, then the morphism $g' : X \times_Y Z \longrightarrow X$ obtained by pulling back f along g as in the following diagram

$$\begin{array}{ccc} X \times_Y Z & \xrightarrow{f'} & Z \\ {\scriptstyle g'}\downarrow & & \downarrow{\scriptstyle g} \\ X & \xrightarrow[f]{} & Y \end{array}$$

is also a regular epimorphism.

The category *Set* of sets is a regular category and, in fact, so is any elementary topos $\mathcal{E}$.

This is not the definition given by Reyes in his [1974], but it is equivalent to it. In Reyes' paper, a regular category is a category with finite limits, finite sups of subobjects of a given object and images, which we will define. As we have already indicated, one of the main properties of a regular category is that any morphism $f : X \longrightarrow Y$ can be factored uniquely, up to isomorphism,

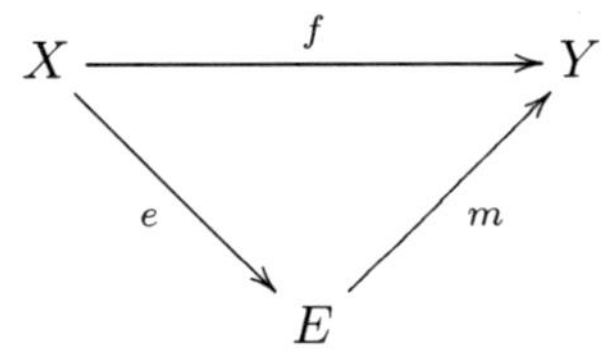

where $e : X \twoheadrightarrow E$ is a regular epimorphism and $m : E \longrightarrow Y$ is a monomorphism. The latter monomorphism is called the *direct image* of f and is denoted $\mathrm{Im}(f)$. $\mathrm{Im}(f)$ is defined only up to a unique isomorphism, but as such it determines a unique subobject of Y, which we denote $\exists_f(X)$. The choice of terminology is not a coincidence. Given any momomorphism $i : A \rightarrowtail X$, we can define $\exists_f(A) = \mathrm{Im}(f \circ i)$ and in this way $\exists_f$ is a well defined morphism from the subobjects of X to the subobjects of Y and it can be shown that, in fact, for any arrow $f : X \longrightarrow Y$ of a regular category $\mathcal{C}$, $\exists_f \dashv f^*$. Properties of the existential quantifier then follow immediately. In particular, what is now called the *Frobenius identity* holds in any regular category $\mathcal{C}$: for any morphism $f : X \longrightarrow Y$ and monomorphisms $i_A : A \rightarrowtail X$, $i_B : B \rightarrowtail Y$, $\exists_f(A \wedge f^*B) = \exists_f A \wedge B$, where both sides of the equality are subobjects of Y.

Notice how the latter characterization weaves together Lawvere's characterization of elementary theories and Daigneault's characterization of the condition for the existence of injective envelopes.

It is natural to consider *regular functors* between regular categories: a functor $F : \mathcal{C} \longrightarrow \mathcal{D}$ between regular categories is said to be *regular* if it preserves finite limits and coequalizers of kernel pairs. We can therefore define the category *RegCat* with objects (small) regular categories and with morphisms regular functors. (We should point out that it was immediately clear that it is a 2-category.)

Joyal and Reyes then introduced *regular categories with stable* $\vee$, now called *coherent categories*, but they were also called *logical categories* from 1974 until 1977 inclusively. It should be emphasized that all these categories are regular categories with additional *properties* and not additional *structure*.

A category $\mathcal{C}$ is said to be a *coherent category* if:

1. It is a regular category;
2. For each object X of $\mathcal{C}$, $\mathrm{Sub}(X)$ has finite sups;
3. The inverse image morphism $f^* : \mathrm{Sub}(Y) \longrightarrow \mathrm{Sub}(X)$ preserves finite sups, i.e. finite sups are stable under pullbacks.

A functor $F : \mathcal{C} \longrightarrow \mathcal{D}$ between coherent categories is said to be *coherent* if it is regular and if it preserves finite sups. Again, we can consider the category *Coh* of coherent categories and coherent functors between them.

A category $\mathcal{C}$ is said to be a *Heyting category* if:

1. It is a regular category;

2. For every morphism $f : X \longrightarrow Y$, the inverse image morphism $f^* : \mathrm{Sub}(Y) \longrightarrow \mathrm{Sub}(X)$ has a right adjoint $\forall_f : \mathrm{Sub}(X) \longrightarrow \mathrm{Sub}(Y)$;

3. For every object X of $\mathcal{C}$, $\mathrm{Sub}(X)$ is a lattice with smallest element.

It should be noted that the existence of the right adjoint $\forall_f$, for every f, implies the existence of the operation $\Rightarrow$ of implication in every lattice of subobjects. A *Heyting functor* between Heyting categories is a regular functor preserving $\forall_f$ and the lattice operations (in fact, supremums and implication suffice). The category of Heyting categories and Heyting functors will be denoted by $Heyt$.

A category $\mathcal{C}$ is said to be a *Boolean category* if

1. It is a Heyting category;

2. For every object X of $\mathcal{C}$, $\mathrm{Sub}(X)$ is a Boolean algebra.

A *Boolean functor* is a Heyting functor preserving all the lattice operations. The category of Boolean categories and Boolean functors will be denoted by $Bool$.

Thus, a theory $\mathbb{T}$ in a language $\mathcal{L}_i$ is now replaced by an i-category $\mathcal{C}_{\mathbb{T}}$. The goal is now to use concepts and methods of category theory to obtain results about logic. As usual, in a categorical framework, one first investigates the existence of adjoint functors to naturally given functors. Thus, there are obvious forgetful functors (in fact 2-functors):

$$Bool \longrightarrow Heyt \longrightarrow Coh \longrightarrow Reg.$$

Two questions arise immediately: 1. Are there adjoints to these natural forgetful functors? 2. How are these categories related to toposes? In other words, what are the functors between these categories and the category of toposes?

These questions are in part answered in [Reyes, 1974]. Let us now summarize the content of that paper, since it reflects in part the knowledge of the field as of 1971-1972 approximately. The first section, written essentially by Joyal, defines the various foregoing categories and proves some of their important properties. The second section, written by Reyes, gives a definition of Grothendieck toposes, proves that every Grothendieck topos is a Heyting category and defines the category $\mathbb{T}\mathrm{op}$ of toposes with *geometric* morphisms as functors. Section three of the paper investigates the links between Grothendieck toposes and logic, in particular the existence of (left-)adjoints to forgetful functors from $\mathbb{T}\mathrm{op}$ to other logical categories. Reyes attributes to Joyal the observation that there is a left adjoint $Sh(-)$: $Coh \longrightarrow \mathbb{T}\mathrm{op}$ to the forgetful functor $\mathbb{T}\mathrm{op} \longrightarrow Coh$ defined by: for each coherent category $\mathcal{C}$, $Sh(\mathcal{C})$ is the topos of sheaves for the so-called finite cover topology and for any *regular* functor $\mathcal{C} \xrightarrow{u} \mathcal{D}$, we get a geometric morphism of toposes $\widehat{u}$ given by the pair of adjoints $Sh(\mathcal{C}) \underset{u_*}{\overset{u^*}{\rightleftarrows}} Sh(\mathcal{D})$. Furthermore, and this is the key observation that will lead to important concepts and results, although it is not emphasized nor put in perspective in the paper, the functor $Sh(-)$ satisfies the following universal property: for every *coherent* functor $u : \mathcal{C} \longrightarrow \mathcal{E}$ into a topos, there is a unique (up to isomorphism) morphism of topos $Sh(-) \underset{u_*}{\overset{u^*}{\rightleftarrows}} \mathcal{E}$ such that the triangle

$$\begin{array}{ccc} \mathcal{C} & \xrightarrow{h} & Sh(\mathcal{C}) \\ & \searrow^{u} & \downarrow{u^*} \\ & & \mathcal{E} \end{array}$$

commutes, where the functor $h : \mathcal{C} \longrightarrow Sh(\mathcal{C})$ is the standard Yoneda embedding. As we will see, this result introduces without naming it what will be called the *classifying topos* of $\mathcal{C}$.

It is then shown in the next paragraph that there is no such left adjoint between the category of toposes and the category of Heyting categories. There is however a left adjoint to the forgetful functor from the category of Boolean toposes and the category of Boolean categories, provided that the functors in the latter category, that is the Boolean functors, preserve arbitrary sups.

In the last section of the paper, Reyes presents Joyal's completeness theorem for Boolean categories and explores links with models of non-standard analysis. We will only comment the completeness theorem.

Here is the formulation given in the paper: Let $\mathcal{C}$ be a small Boolean category. If X is distinct from the smallest object of $\wp(X)$, then there is a Boolean model $M : \mathcal{C} \longrightarrow Set$ such that $M(X) \neq \emptyset$. Three points have to be underlined. First, notice the similarity between this formulation and Daigneault's formulation of the completeness theorem and, thus, with Lawvere's formulation. Second, the proof of the completeness theorem given in the paper relies on what is essentially the same construction as the one stated in the second abstract on logic and foundations in the *Notices*, although there is no mention of pro-objects in the 1974 paper. This strongly suggests that Joyal had a proof of the completeness theorem by purely categorical means already in the fall of 1970. Third, although it is not worded in this manner in the paper, the theorem amounts to the following *representation* theorem: let $\mathcal{C}$ be a (small) coherent category. Then there is a *set* I and a *conservative* coherent functor $M : \mathcal{C} \longrightarrow Set^I$. Being "conservative" means, in this context, that M *reflects* isomorphisms, that is if $M(f)$ is an isomorphism in Set^I, then f is an isomorphism in $\mathcal{C}$. Notice here the similarity with the representation theorem for polyadic algebras, as formulated by Daigneault. It is striking that the set X in the latter formulation is replaced by the category Set of sets in this new formulation. Although the theorem is strictly speaking a representation theorem, it *implies* the completeness theorem, since the coherent category $\mathcal{C}$ is a conceptual category, i.e. a theory. Furthermore, since coherent functors between Boolean categories are in fact Boolean, a fact used explicitly by Joyal in his proof, the representation theorem for Boolean categories is a special case the theorem for coherent categories. Thus, one obtains directly the completeness theorem for classical first-order logic.

When compared to Volger's attempt at characterizing the doctrine of first-order logic, two elements standout in Joyal and Reyes's work. First, the elegance and simplicity of the hierarchy of categories defined, starting with regular categories, reveals a global conceptual coherence, no pun intended, of logical categories. Second, as we have already pointed out, regular categories also have an important status in category theory in general and thus their classification establishes important links between logical notions and categorical notions. Third, the approach seems to be at the right level of generality and flexibility, in contrast with Volger's work. In particular, it did not presuppose the existence of a subobject classifier, or an object of truth-value and still fitted in perfectly with toposes. The key was to find a way to articulate together fragments of first-order logic that would allow a categorical passage from one type of category to another and in such a way that one would not have to import additional structure, in the way Volger had done. It was certainly seen as an important gain that only properties were added to move from one level to the next. Finally, the connections with toposes made it possible to construct a dictionary between algebraic geometry and logic and one could hope that results and methods from both fields could interact and yield new and useful insights. Thus, a central piece of the puzzle of algebraic logic was now in place. The other parts were being added quickly. One of the pieces, already hinted at in Dionne's thesis, was to find a way to construct a theory $\mathbb{T}_{\mathcal{C}}$ from a category $\mathcal{C}$ in such a way that there is a canonical interpretation of $\mathbb{T}_{\mathcal{C}}$ in $\mathcal{C}$. But for the latter, one needs to have a general notion of interpretation of a theory in a category, which is the second piece. Once these have been defined appropriately, they fit together smoothly and the general conceptual picture emerging is of great beauty.

4.6 Constructing theories from categories

As early as 1971, the idea of interpreting a logical theory in a topos appeared in William Mitchell's work on the relation between Boolean toposes and set theory. In his paper, communicated by Mac Lane in October 1971 and published in 1972, he explicitly introduced a language $\mathcal{L}(\mathcal{E})$ for a topos $\mathcal{E}$ and two types of interpretations of $\mathcal{L}(\mathcal{E})$: an *external interpretation* and an *internal interpretation*. The internal interpretation would soon become the internal *language* of a topos and it would be used like any other deductive system to prove results *in* that specific theory, whereas the external interpretation would amount to using certain toposes as *models* of certain specified theories of sets. Both methods evolved quickly, but it seems clear that Bénabou, in his talk at Oberwolfach and Joyal in his talk in Montreal, both in early 1973, found a way to define the internal language $\mathcal{L}(\mathcal{C})$ for a weaker category $\mathcal{C}$. Makkai and Reyes quickly used these methods in 1973 and 1974. Their first joint papers were presented in November 1974 and published in 1976.

Before we look at the internal language of a category $\mathcal{C}$, we have to consider the notion of an interpretation M of a many-sorted language $\mathcal{L}$ in a category $\mathcal{C}$ with finite limits. The basic idea was to lift to categories the traditional Tarskian semantics as defined with sets. The translation requires, as usual, that clauses stated in terms of elements and subsets be reformulated in terms of morphisms and subobjects.

Assume $\mathcal{C}$ is a category with finite limits and $\mathcal{L}_0$ is defined as above. A $\mathcal{C}$-*interpretation* M of $\mathcal{L}_0$ consists of:

1. an object MS of $\mathcal{C}$ for each sort S;
2. a morphism $Mf : MS_1 \times \cdots \times MS_n \longrightarrow MS$ for each function symbol $f : S_1 \times \cdots \times S_n \longrightarrow S$; in particular, when $n = 0$ a morphism $1 \xrightarrow{f} MS$;
3. a subobject $MR \rightarrowtail MS_1 \times \cdots \times MS_n$ for each relation symbol $R \rightarrowtail S_1 \times \cdots \times S_n$.

Notice immediately that when $\mathcal{C}$ is the category Set of sets, then a Set-interpretation is an ordinary many-sorted structure of similarity type $\mathcal{L}_0$. Homomorphisms $h : M \longrightarrow N$ between interpretations are defined in the obvious fashion, i.e. they preserve the interpretations of functions and relations. As usual, interpretations are extended to all terms and formulas of the language $\mathcal{L}_i$, where the value of i, $0 \leq i \leq 4$, will be clear from the context below. Let $\vec{x} = x_1, \ldots, x_n$ be a context with types $S_1, \ldots, S_n$ respectively, t a term of type S and $\vec{x} \cdot t$ a term-in-context over $\mathcal{L}_i$. To every such term-in-context $\vec{x} \cdot t$, we assign a morphism $[\![\vec{x} \cdot t]\!]_M : MS_1 \times \cdots \times MS_n \longrightarrow MS$ by recursion as follows:

1. If t is a variable, then it is a variable x_i of type S_i, and thus $[\![\vec{x} \cdot t]\!]_M = 1_{S_i} \circ \pi_i$ (we are being pedantic: a variable is interpreted as the appropriate identity morphism on the relevant type, the projection morphism π_i is there to get rid of the 'dummy' variables in the context);
2. If t is $f(t_1, \ldots, t_m)$, t_i terms of type T_i, then by induction we have that $[\![\vec{x} \cdot t_i]\!]_M : MS_1 \times \cdots \times MS_n \longrightarrow MT_i$ for each t_i, and $[\![\vec{x} \cdot f(t_1, \ldots, t_m)]\!]_M$ is the composite $MS_1 \times \cdots \times MS_n \xrightarrow{\langle [\![\vec{x} \cdot t_1]\!]_M, \ldots, [\![\vec{x} \cdot t_m]\!]_M \rangle} MT_1 \times \cdots \times MT_m \xrightarrow{Mf} MS$;
3. In particular, when t is a constant f, then $[\![\vec{x} \cdot t]\!]_M$ is the composite morphism $MS_1 \times \cdots \times MS_n \longrightarrow 1 \xrightarrow{Mf} MS$.

A formula-in-context $\vec{x} \cdot \phi$ is interpreted as a subobject $[\![\vec{x} \cdot \phi]\!]_M \rightarrowtail MS_1 \times \cdots \times MS_n$ as follows:

1. If ϕ is $(t_1 = t_2)$, and t_1 and t_2 are of type T, then $[\![\vec{x} \cdot \phi]\!]_M$ is the equalizer of $MS_1 \times \cdots \times MS_n \underset{[\![\vec{x} \cdot t_2]\!]}{\overset{[\![\vec{x} \cdot t_1]\!]}{\rightrightarrows}} MT$.

2. If ϕ is $R(t_1, \ldots, t_m)$ with $t_1, \ldots, t_m$ of types $T_1, \ldots, T_m$ respectively, $[\![\vec{x} \cdot \phi]\!]_M$ is the pullback

$$\begin{array}{ccc}
[\![\vec{x} \cdot \phi]\!] & \longrightarrow & MR \\
\downarrow & & \downarrow \\
MS_1 \times \cdots \times MS_n & \xrightarrow[<\langle [\![\vec{x} \cdot t_1]\!]_M, \ldots, [\![\vec{x} \cdot t_m]\!]_M \rangle>]{} & MT_1 \times \cdots \times MT_m
\end{array}$$

3. If ϕ is $\top$, then $[\![\vec{x} \cdot \phi]\!]_M$ is $MS_1 \times \cdots \times MS_n$ or equivalently, the top element of $\mathrm{Sub}(MS_1 \times \cdots \times MS_n)$;

4. If ϕ is $(\psi \wedge \theta)$, then $[\![\vec{x} \cdot \phi]\!]_M$ is the pullback (in this case, since we are dealing with subobjects, it *is* the intersection or the inf)

$$\begin{array}{ccc}
[\![\vec{x} \cdot \phi]\!]_M & \rightarrowtail & [\![\vec{x} \cdot \psi]\!]_M \\
\downarrow & & \downarrow \\
[\![\vec{x} \cdot \theta]\!]_M & \rightarrowtail & MS_1 \times \cdots \times MS_n
\end{array}$$

5. If ϕ is $\perp$ and $\mathcal{C}$ is a coherent category, then $[\![\vec{x} \cdot \phi]\!]_M$ is the bottom element of $\mathrm{Sub}(MS_1 \times \cdots \times MS_n)$;

6. If ϕ is $(\psi \vee \theta)$ and $\mathcal{C}$ is a coherent category, then $[\![\vec{x} \cdot \phi]\!]_M$ is the union of the subobjects $[\![\vec{x} \cdot \psi]\!]_M$ and $[\![\vec{x} \cdot \theta]\!]_M$;

7. If ϕ is $(\psi \Rightarrow \theta)$ and $\mathcal{C}$ is a Heyting category, then $[\![\vec{x} \cdot \phi]\!]_M$ is the implication $[\![\vec{x} \cdot \psi]\!]_M \Rightarrow [\![\vec{x} \cdot \theta]\!]_M$ is the Heyting algebra $\mathrm{Sub}(MS_1 \times \cdots \times MS_n)$;

8. If ϕ is $\neg\psi$ (which is the same as $(\psi \Rightarrow \perp)$), and $\mathcal{C}$ is a Heyting category, then $[\![\vec{x} \cdot \phi]\!]_M$ is the Heyting negation $\neg[\![\vec{x} \cdot \psi]\!]_M$;

9. If ϕ is $(\exists y)\psi$, where y is a sort T, and $\mathcal{C}$ is a regular category, then $[\![\vec{x} \cdot \phi]\!]_M = \exists_\pi [\![\vec{x}, y \cdot \psi]\!]_M$ which is the image of the composite

$$[\![\vec{x}, y \cdot \psi]\!]_M \rightarrowtail MS_1 \times \cdots \times MS_n \times MT \xrightarrow{\pi\langle 1, \ldots, n\rangle} MS_1 \times \cdots \times MS_n$$

where '$\vec{x} \cdot y$' denotes the context resulting from the juxtaposition of y to the list $\vec{x}$;

10. If ϕ is $(\forall y)\psi$, where y is a sort T, and $\mathcal{C}$ is a Heyting category, then $[\![\vec{x} \cdot \phi]\!]_M = \forall_\pi [\![\vec{x}, y \cdot \psi]\!]_M$ where π is the projection

$$MS_1 \times \cdots \times MS_n \times MT \xrightarrow{\pi\langle 1,\ldots,n\rangle} MS_1 \times \cdots \times MS_n.$$

Again, when $\mathcal{C}$ is the category *Set* of sets, then all these definitions translate into Tarski's usual semantics for a many-sorted first-order language. With all these definitions at hand, we define the notion of satisfaction and model as follows.

Let M be a $\mathcal{C}$-interpretation. We say that M is a *model* of a sequent $\sigma = (\phi \vdash_{\vec{x}} \psi)$ or that σ is *satisfied in* M, we write as usual $M \models_{\mathcal{C}} \sigma$ or simply $M \models \sigma$, if $[\![\vec{x} \cdot \phi]\!]_M \leq [\![\vec{x} \cdot \psi]\!]_M$ in $\mathrm{Sub}(MS_1 \times \cdots \times MS_n)$. We say that M is a *model of a theory* $\mathbb{T}$ if all the axioms of $\mathbb{T}$ are satisfied in M.

We denote the *category of models of* $\mathbb{T}$ *in* $\mathcal{C}$ by $\mathrm{Mod}_{\mathcal{C}}(\mathbb{T})$. An important feature of this framework is that it is possible to transfer models of $\mathbb{T}$ in $\mathcal{C}$ in an appropriate category $\mathcal{D}$ along appropriate functors. More specifically, if $\mathbb{T}$ is a (regular, coherent, etc.) theory, then any (regular, coherent, etc.) functor $F : \mathcal{C} \longrightarrow \mathcal{D}$ induces a functor $\mathrm{Mod}(F) : \mathrm{Mod}_{\mathcal{C}}(\mathbb{T}) \longrightarrow \mathrm{Mod}_{\mathcal{D}}(\mathbb{T})$ in the obvious way.

Let $\mathcal{C}$ be a category with finite limits. Then the *internal language* $\mathcal{L}_{\mathcal{C}}$ of $\mathcal{C}$ is given as follows: for each object X of $\mathcal{C}$, there is a sort $\ulcorner X \urcorner$, the name of X; for each morphism $f : X_1 \times \cdots \times X_n \longrightarrow Y$ in $\mathcal{C}$, a function symbol $\ulcorner f \urcorner : \ulcorner X_1 \urcorner \times \cdots \times \ulcorner X_n \urcorner \longrightarrow \ulcorner Y \urcorner$ and for each subobject $R \rightarrowtail X_1 \times \cdots \times X_n$, a relation symbol $\ulcorner R \urcorner \rightarrowtail \ulcorner X_1 \urcorner \times \cdots \times \ulcorner X_n \urcorner$. (This is called the *extended internal language* in [Makkai and Reyes, 1977].) In other words, the language $\mathcal{L}_{\mathcal{C}}$ is nothing less than $\mathcal{C}$. There is a *canonical interpretation* of $\mathcal{L}_{\mathcal{C}}$ in $\mathcal{C}$, which is nothing less than the identity interpretation: each sort $\ulcorner X \urcorner$ is sent to the object X named by it, each function symbol $\ulcorner f \urcorner : \ulcorner X_1 \urcorner \times \cdots \times \ulcorner X_n \urcorner \longrightarrow \ulcorner Y \urcorner$ is sent to the morphism $f : X_1 \times \cdots \times X_n \longrightarrow Y$ it denotes and each relation symbol $\ulcorner R \urcorner \rightarrowtail \ulcorner X_1 \urcorner \times \cdots \times \ulcorner X_n \urcorner$ is sent to the subobject $R \rightarrowtail X_1 \times \cdots \times X_n$ it refers to. A sequent σ over $\mathcal{L}_{\mathcal{C}}$ will therefore have a canonical interpretation in $\mathcal{C}$. Whenever the sequent σ is verified in $\mathcal{C}$, we write $\mathcal{C} \models \sigma$. Thus, diagrams of $\mathcal{C}$ can be replaced by sequents over $\mathcal{L}_{\mathcal{C}}$. The *theory* $\mathbb{T}_{\mathcal{C}}$ *of* $\mathcal{C}$ is given by the collection of sequents of the internal language that are verified in $\mathcal{C}$ by the canonical interpretation. This is of course the purely model-theoretical point of view of a theory. The point of this apparently tautologous game is that it is possible to use the appropriate deductive system introduced above to prove properties of $\mathcal{C}$. More specifically, we can reproduce the foregoing system of theories as follows. Let $\mathcal{C}$ be a category. Then, the internal language $\mathcal{L}_{\mathcal{C}}$ is a special case of a primitive language $\mathcal{L}_0$. We can construct the theory $\mathbb{T}_{\mathcal{C}}$, denoted by $\mathbb{T}_0(\mathcal{C})$ by Boileau, by taking the axioms

i) $\top \vdash f(a) = a$ whenever $f : A \longrightarrow A$ is an identity morphism in $\mathcal{C}$;

ii) $\top \vdash g(f(a)) = h(a)$ whenever

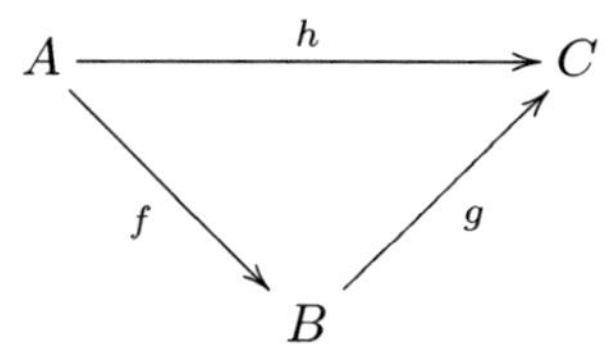

is a commutative triangle in $\mathcal{C}$.

If $\mathcal{C}$ is a regular category, then, to the axioms of $\mathbb{T}_0(\mathcal{C})$, the axioms

iii) $\top \vdash \exists a(a = a)$ and $\top \vdash a_1 = a_2$ whenever A is a terminal object of $\mathcal{C}$;

iv) $\top \vdash \exists a(h(a) = b \wedge k(a) = c)$ and $h(a_1) = h(a_2) \wedge k(a_1) = k(a_2) \vdash a_1 = a_2$ whenever

$$C \xleftarrow{k} A \xrightarrow{h} B$$

is a product diagram in $\mathcal{C}$;

v) $f(b) = g(b) \vdash \exists a(h(a) = b)$, $\exists a(h(a) = b) \vdash f(b) = g(b)$ and $h(a_1) = h(a_2) \vdash a_1 = a_2$ whenever

$$A \xrightarrow{h} B \underset{g}{\overset{f}{\rightrightarrows}} C$$

is an equalizer diagram in $\mathcal{C}$

to obtain the theory $\mathbb{T}_1(\mathcal{C})$. By adding the appropriate axioms, we associate a specific theory $\mathbb{T}_i(\mathcal{C})$ to coherent, Heyting and Boolean categories.

Using the appropriate rules, together with the axioms of $\mathbb{T}_{\mathcal{C}}$, one can deduce formally results about $\mathcal{C}$. Proceeding in this manner, we obtain the following table:

If $\mathcal{C}$ is a	Then $\mathcal{L}_{\mathcal{C}}$ is	And $\mathbb{T}_{\mathcal{C}}$ is
Category	$\mathcal{L}_0(\mathcal{C})$	$\mathbb{T}_0(\mathcal{C})$
Regular category	$\mathcal{L}_1(\mathcal{C})$	$\mathbb{T}_1(\mathcal{C})$
Coherent category	$\mathcal{L}_2(\mathcal{C})$	$\mathbb{T}_2(\mathcal{C})$
Heyting category	$\mathcal{L}_3(\mathcal{C})$	$\mathbb{T}_3(\mathcal{C})$
Boolean category	$\mathcal{L}_4(\mathcal{C})$	$\mathbb{T}_4(\mathcal{C})$

It is straightforward to prove a soundness theorem for the various logical systems involved. More specifically, given a (regular, coherent, Heyting, Boolean) theory, M a model of $\mathbb{T}$ in a (regular, coherent, Heyting, Boolean) category $\mathcal{C}$, if σ is a sequent (in the appropriate fragment of the first-order language $\mathcal{L}$) which is provable in $\mathbb{T}$, then $M \models_{\mathcal{C}} \sigma$.

4.7 Building bridges

The resulting situation was somewhat surprising and to some, perhaps, uncomfortable. Whereas we saw Dionne constructing categories from logical theories in order to apply categorical methods and obtain in this manner results about the original theories, it became possible to use logical methods to obtain results about categories. In a sense, the frontier between logical theories as categories and categories as logical theories is thereby vanishing. When Mihály Makkai arrived in Montreal in the fall of 1973, he solved some of the problems considered by Joyal and Reyes by using purely logical methods. These results lead to two joint papers with Reyes submitted in the fall of 1974. The first paragraph of their second paper is revealing in this respect:

> Here, just as in [9] [their previous paper], our method consists in translating any given problem for categories into a purely logical problem via the mechanism described in Sec. 1 in [9], and then using familiar methods developed in logic. (...)
>
> Our point of view here is that categories provide an algebraic formulation of logic, in the spirit of equating theories with categories and models with certain ("logical") functors. We contend that much of logic (model theory) can and should be expressed this way. On the other hand, we freely use methods tied to the concrete "presentations", i.e., to primitive symbols, formulas, etc., usually employed in model theory; in fact, this is the novelty of our approach from the point of view of the category theorist. There is an obvious analogy to abstract groups versus groups defined by generators and relations. Just as in groups, the abstract formulation (categories) may (and does) point up examples of "theories" in branches of mathematics quite distant from logic. [Makkai and Reyes, 1976a, 385].)

Thus, we see again the claim that a category is the abstract algebraic expression of theory presented in a specific language with specific axioms. The main advantage underlined here, however, is that the abstract expression opens up unforeseen connections with other branches of mathematics. But the tension was certainly still present. In a paper written in 1975-76 while he was on sabbatical from Montreal, Reyes opens with the following claims:

> The goal of topos theory is to develop a language and an efficient method for the study of *concepts* of local character (as well as constructions on such concepts) that are found in diverse branches of mathematics : topology, algebraic geometry, analytic geometry, ...
>
> To the *geometric* aspect (or topological), which is the dominant aspect, another is dialectically opposed : the *logical* aspect.([Reyes, 1978, 156].)(our translation)

We are back to the early rhetoric with the emphasis, once again, on the geometric aspect, although at this stage it is clear that there is some sort of dictionary between the two. It is more as if the geometric and the logical are two faces of the same coin (but in fact, there are other faces as well, so it is more a cube than a coin).

But developing the logical machinery required to present Makkai and Reyes' results has its costs. Presenting all the preliminary details and the proofs contained in the first paper published in 1976 will constitute the bulk of the chapters 2, 3, 4, 5 and 6 (out of 9) of their book published in 1977. Chapters 7, 8 and 9 present the proofs of the second paper.

Here are the key results presented in the two papers presented in 1974 by Makkai and Reyes.

Having presented the foregoing notion of interpretation, of the internal language and the theory $\mathbb{T}_{\mathcal{C}}$ of a category $\mathcal{C}$, Makkai and Reyes proved the following completeness theorems for categories.

Theorem: i) let $\mathcal{C}$ be a small category with finite limits and with finite stable sups (i.e. each finite family of subobjects of an object has a stable sup). Then there is a complete Boolean algebra $\mathcal{B}$ and a functor

$$M : \mathcal{C} \longrightarrow Sh(\mathcal{B})$$

such that M preserves all finite limits, (finite or infinite) stable sups, stable images (finite or infinite), distributive infs, and distributive $\forall_f(C)$'s, and such that M is conservative.

ii) If $\mathcal{C}$ has finite limits only, the same conclusion holds without the clauses for infs and $\forall_f$'s.

The proof rests entirely on *logical* methods. In the paper, it is barely sketched. It proceeds by replacing $\mathcal{C}$ by $\mathbb{T}_{\mathcal{C}}$ and then using properties of the canonical interpretation adapting completeness results in the literature, e.g. Mansfield 1972 and an unpublished manuscript of Higgs. We should emphasize at this stage the importance of Higgs's (still) unpublished manuscript, widely read and extremely influential at the time, on a categorical version of Boolean-valued models.

The theorem has, among others, the following corollaries:

Corollary 1 (an improvement of Barr's theorem): every Grothendieck topos $\mathcal{E}$ has a surjective Boolean point, i.e. there is a complete Boolean algebra $\mathcal{B}$ and a geometric morphism $p : Sh(\mathcal{B}) \longrightarrow \mathcal{E}$ of toposes such that $p^* : \mathcal{E} \longrightarrow Sh(\mathcal{B})$ is faithful. Moreover, and this is the improvement, $\mathcal{B}$ and p can be chosen so that p^* preserves all (finite or infinite) distributive infs and all distributive $\forall_f(C)$'s in $\mathcal{E}$.

Corollary 2: (i) Let $\mathcal{C}$ be a coherent category. Then there is a set Y and a functor M

$$M : \mathcal{C} \longrightarrow Sh(\wp(Y)),$$

where $\wp(Y)$ denotes the Boolean algebra of subsets of Y, such that M is as in the foregoing theorem except that the sups and infs to be preserved by M are only the finite stable (distributive) ones.

The main theorem implies, together with a compactness argument, Deligne's theorem: every coherent topos $\mathcal{E}$ has a surjective Boolean point $p : Sh(\wp(Y)) \longrightarrow \mathcal{E}$ for some set Y. In chapter six of their book, the formulation of the theorem takes the more eloquent form: if $\mathcal{E}$ is a coherent topos, then there is a conservative model $M : \mathcal{E} \longrightarrow Set^I$ into a Boolean topos of the form Set^I, with I a set. In the third chapter, the theorem is stated for coherent categories as follows:

Let $\mathcal{C}$ be a (small) coherent category. Then:

1. If A and B are two subobjects of X in $\mathcal{C}$ such that $A \nleq B$, then there is a coherent functor $M : \mathcal{C} \longrightarrow Set$ such that $M(A) \nleq M(B)$ as subobjects of $M(X)$.

2. There is a set I and a faithful coherent functor $M : \mathcal{C} \longrightarrow Set^I$.

The foregoing results follow from the main theorem in which one considers *distributive infs* and *distributive* $\forall$'s. When one turns to models that preserve *all* infs and *all* $\forall$'s, then one obtains similar results for intuitionistic logic. The results become:

Theorem: For every Grothendieck topos $\mathcal{E}$ there is a complete Heyting algebra H and a surjective H-valued point $p : Sh(H) \longrightarrow \mathcal{E}$ such that $p^* : \mathcal{E} \longrightarrow Sh(H)$ preserves all (finite or infinite) infs and all $\forall_f(C)$'s in $\mathcal{E}$.

Theorem: If the topos $\mathcal{E}$ has a surjective Boolean point of the form

$$Sh(\wp(Y)) \longrightarrow \mathcal{E}$$

for a set Y, then the topos $Sh(H)$ can be taken to be the category of sheaves over a topological space.

These *are* completeness results for intuitionistic logic. Again the proofs use purely logical methods. (See chapter 6.3 of [Makkai and Reyes, 1977].) In this setting, Joyal's completeness theorem, closely related to Kripke's completeness theorem for intuitionistic logic, takes the following form:

Theorem (Joyal, as we saw, probably in 1970): Let $\mathcal{C}$ be a coherent category and let $Mod_{Set}(\mathcal{C})$ the category of *coherent* functors $\mathcal{C} \longrightarrow Set$, thus the category of set-models of $\mathcal{C}$. There is a small full subcategory P of $Mod_{Set}(\mathcal{C})$ such that the evaluation functor $ev : \mathcal{C} \longrightarrow Set^P$ is conservative and preserves all finite limits, stable finite sups, stable images and stable $\forall_f(C)$ existing in $\mathcal{C}$.

Some important remarks are in order. First, it is now possible to see that completeness theorems for certain logical theories are *equivalent* to representation theorems for categories. Thus, for instance, the completeness theorem for first-order classical theories is equivalent to the representation theorem for *coherent* categories. As we have already pointed out, the key construction for the implication from the representation theorem to the completeness theorem is the construction of the category of concepts of a theory. To prove the other direction, one uses the internal theory $\mathbb{T}_{\mathcal{C}}$ of a small coherent category $\mathcal{C}$. Thus, one can say that the completeness theorem and the representation theorem are translation-equivalent to one another. Second, Joyal's theorem has a remarkable feature: the link between a theory and its category of models is provided by a *canonically defined functor*, the evaluation functor. Given two categories $\mathcal{C}$ and $\mathcal{D}$, the evaluation functor $ev : \mathcal{C} \longrightarrow \mathcal{D}^{(\mathcal{D}^{\mathcal{C}})}$ is *always* definable by $ev(X, F) = F(X)$ (and similarly for morphisms). Furthermore, the fact that, whenever $\mathcal{C}$ and $\mathcal{D}$ are coherent categories, ev is a coherent functor follows by abstract general nonsense. The specific observation is that it is also conservative.

In the early sixties, Lubkin-Freyd-Mitchell-Héron proved an important representation theorem for Abelian categories. The latter are categories satisfying certain abstract properties sufficient for the development of a large part of homological algebra. It is fair to say that Abelian categories together with the representation theorem just mentioned occupied a central position in the development of category theory in the sixties. Not only were they extremely powerful in their applications, but they also served as a model of the power of category theory itself. It should be mentioned at this point that the category *Set* is *not* an Abelian category. The typical example of an Abelian category is the category of Abelian groups or a category of modules over a (commutative) ring. Michael Barr, then at McGill University, was looking for non-additive versions of the Abelian notion as well as the corresponding representation theorem. As we have already mentioned, he succeeded in this attempt and introduced *regular* and *exact* categories in 1970 together with a representation theorem for exact categories, from which the known representation theorem for Abelian categories can be deduced. In contrast with the situation found with Abelian categories, the category *Set* is regular and exact. A fascinating fact is that a category is Abelian if and only if it is both additive and exact. It should be pointed out immediately that the notion of an exact category is simply a strengthening of the notion of a regular category and that only the latter property was used in Barr's proof of the representation theorem. Thus the category *Set* is regular (as well as coherent, Heyting, Boolean and a topos.) Barr's representation theorem for regular categories turns out to be equivalent to a completeness theorem for coherent logic when it is looked at from the proper angle.

First, the property of being conservative is crucial and captures a form of completeness. To understand this, it is imperative to see what the property of reflecting isomorphisms means in this particular case. The category *Set* is coherent and thus so is the category $Set^{Mod_{Set}(\mathcal{C})}$. Thus, the latter category inherits all its coherent properties from the category *Set*. Furthermore, the property of reflecting isomorphisms of the functor $e : \mathcal{C} \longrightarrow Set^{Mod_{Set}(\mathcal{C})}$ implies that whatever coherent property the category $Set^{Mod_{Set}(\mathcal{C})}$ has, so does the category $\mathcal{C}$. Hence the property of reflecting isomorphisms implies that any coherent category $\mathcal{C}$ has all coherent properties that *Set* has.

Second, the latter claim has to be compared with the Stone representation theorem for distributive lattices and Boolean algebras. The usual formulation of the Stone representation theorem is that any distributive lattice has an embedding into a power-set algebra. This statement can be shown to be equivalent to the claim that for any distributive lattice D, there is a 2-valued homomorphism $f : D \longrightarrow 2$

that reflects the order, i.e. if $f(x) \leq f(y)$, then $x \leq y$. This last statement can be interpreted as saying that any distributive lattice shares all the universal Horn properties of the 2-element lattice. Thus, moving from the propositional case to the first-order case requires that we replace the 2-element lattice by the category *Set*.

Finally, let us see how the representation theorem is equivalent to a completeness theorem. Assume the representation theorem for coherent categories and let $\mathcal{C}$ be the conceptual category $[\mathbb{T}]_{\text{coh}}$ for some coherent theory $\mathbb{T}$. Then the representation theorem yields automatically the completeness theorem. For the other direction, assume we have a coherent category $\mathcal{C}$ and let $\mathbb{T}_{\mathcal{C}}$ be its internal theory. Applying the classical completeness theorem for $\mathbb{T}_{\mathcal{C}}$, we get the representation theorem for $\mathcal{C}$, since the coherent functors $\mathcal{C} \longrightarrow Set$ are identical to the models of the internal theory $\mathbb{T}_{\mathcal{C}}$.

Let us come back to the evaluation functor $e : \mathcal{C} \longrightarrow Set^{Mod_{Set}(\mathcal{C})}$. Joyal showed that the functor e preserves all the existing Heyting structure that happens to exist in $Set^{Mod_{Set}(\mathcal{C})}$. In other words, if $\mathcal{C}$ is a Heyting category, then e is automatically a conservative Heyting functor. Hence, a representation theorem for Heyting categories is obtained for free: every small Heyting category has a conservative Heyting functor to a Heyting category of the form Set^P, where P is a small category (it has to be extracted from the large category $Mod_{Set}(\mathcal{C})$). Kripke's completeness theorem for first-order intuitionistic logic can be deduced from Joyal's theorem.

Let us close this section by coming back to Barr's representation theorem for regular categories. Joyal showed that Barr's theorem is equivalent to the claim that the evaluation functor $e : \mathcal{C} \longrightarrow Set^{Mod_{Set}(\mathcal{C})}$ is full and faithful for any small regular category $\mathcal{C}$ and where $Mod_{Set}(\mathcal{C})$ is the category of regular functors $\mathcal{C} \longrightarrow Set$. This is of course equivalent to a completeness result for first-order logic.

The second paper published in 1976 is important not only for the results it contains, but also for the *question* it asks. Indeed, one can argue that only a categorical formulation of logic allows the question to be stated precisely. Thus, in this case, it is not only the fact that we obtain a new result, but the very statement could not have been formulated without the categorical framework.

Since Lawvere's thesis, it was common to consider functors $I : \mathbb{T}_1 \longrightarrow \mathbb{T}_2$ between theories and the induced functors between models of theories

$$I^* : Mod(\mathbb{T}_2) \longrightarrow Mod(\mathbb{T}_1).$$

One of the questions that can be asked in this context is this: how are properties of I^* reflected in I? If the "logical" properties of I^* are indeed reflected in I, then certainly this is a kind of completeness. Since, T_1 and $\mathbb{T}_2$ are thought of as categories of concepts, it becomes natural to talk about a form of conceptual completeness whenever one can recover the properties of I from those of I^*. Let us now see how these are stated in Makkai and Reyes's paper.

First, they fix the categories they are working with: $\mathbb{T}_1$ and $\mathbb{T}_2$ are assumed to be small coherent categories and $I : \mathbb{T}_1 \longrightarrow \mathbb{T}_2$ is assumed to be a coherent functor. Furthermore, the category $Mod(\mathbb{T})$ is the category of all coherent functors

$F : \mathbb{T} \longrightarrow Set$, it is a full subcategory of $Set^{\mathbb{T}}$. Thus we have the usual induced functor $I^* : Mod(\mathbb{T}_2) \longrightarrow Mod(\mathbb{T}_1)$. The following results, although interesting in themselves, are used to prove what will be call the conceptual completeness theorem. The first theorem is an application of the completeness theorem obtained in their first paper:

Theorem 1A: Assume I^* is surjective, that is for every model M in $Mod(\mathbb{T}_1)$ there is a model N in $Mod(\mathbb{T}_2)$ such that $I^*(N) \simeq M$. Then I is injective on subobjects, i.e. if $X_1 \rightarrowtail X$ and $X_2 \rightarrowtail X$ in $\mathbb{T}_1$ and $I(X_1) \leq I(X_2)$ (in the ordering of subobjects of $I(X)$), then $X_1 \leq X_2$. In particular, I is faithful.

Theorem 1B (Beth definability theorem): Assume that $\mathbb{T}_1$ and $\mathbb{T}_2$ are Boolean categories. Assume also that if $f : I^*(N_1) \longrightarrow I^*(N_2)$ is an isomorphism, then there is a morphism $g : N_1 \longrightarrow N_2$ such that $I^*(g) = f$. Then I is full with respect to subobjects, i.e. if $Y \rightarrowtail I(X)$ in $\mathbb{T}_2$, then there is $X_1 \rightarrowtail X$ in $\mathbb{T}_1$ such that $Y \simeq I(X_1)$.

Theorem 1.1: if I^* is full, then I is full with respect to subobjects. If in addition I^* is surjective, then I is full.

Whereas theorem 1B is closely related to the Beth definability theorem, theorem 1.1 comes naturally in a categorical framework and does not require that the theories be Boolean. (See [Makkai and Reyes, 1977, 195–197] to see how theorem 1B is related to the classical Beth definability theorem.)

The conceptual completeness theorem is formulated for *pretoposes*, a notion introduced by Grothendieck in the sixties. A *pretopos* is a small coherent category having finite disjoint sums (coproducts), and effective equivalence relations.

Theorem (conceptual completeness): Assume $\mathbb{T}_1$ is a pretopos, $\mathbb{T}_2$ is coherent and $I : \mathbb{T}_1 \longrightarrow \mathbb{T}_2$ is a coherent functor. Then, if I^* is an equivalence of categories, then so is I.

Two facets of the result have to be emphasized. First, the conceptual completeness theorem provides a characterization of pretoposes among coherent categories. Indeed, the result fails whenever $\mathbb{T}_1$ is not a pretopos. Second, the theorem can be read as follows: suppose that $I : \mathrm{T}_1 \longrightarrow \mathbb{T}_2$ is thought of an interpretation and that $\mathbb{T}_2$ is considered to be an extension of $\mathbb{T}_1$. One could say that I is *strongly conservative* whenever I^* is an equivalence. Then the theorem asserts that pretoposes and only pretoposes do not have proper strongly extensions. Still put differently, one could say that although $\mathbb{T}_2$ is an extension of $\mathbb{T}_1$, it has the same category of models. The result shows, and this was in fact part of the motivation underlying the original conjecture formulated by Reyes, that finite disjoint sums of formulas and quotients of a formula by a provable (in the theory) equivalence relation are coherent logical operations (although not representable in ordinary logic), *but there are no others*, whence the conceptual *completeness*. It should be emphasized, once more, that all the proofs rely on logical methods. Andrew Pitts gave purely categorical proofs of these results in the eighties. (See [Pitts, 1987; Pitts, 1989].)

The next theorem is Keisler's theorem, but now stated explicitly for coherent logic.

Theorem 2.1. Let K be an elementary class, i.e. defined by any set of sentences in $\mathcal{L}_2$. Then the following are equivalent:

1. K is the class of models of a coherent theory over $\mathcal{L}_2$.
2. Cat_K, the category generated by K, i.e. whose objects are the many-sorted structures of similarity type $\mathcal{L}_2$ and homomorphisms between them, is the dual of the category of points of some coherent topos.
3. K contains the colimit of any directed family taken from Cat_K.

It is pointed out that the equivalence between (1) and (3) was proven by Keisler. It is not mentioned that the equivalence with (2), although formulated differently and using different means, was announced by Joyal in 1970. Makkai and Reyes' proof relies on the notion of *classifying topos*.

4.8 Classifying topos and generic model of a theory

The notion of classifying topos has its own complex history. The first appearance of the concept is usually attributed to Monique Hakim in the sixties, but her work was published only in 1972. (See [Hakim, 1972].) The *term* explicitly appears in Giraud's contribution to the first meeting dedicated to elementary toposes in Dalhousie in the winter of 1971. In this paper, Giraud presents three different constructions of what he calls the classifying topos of a stack $\mathcal{C}$ and states a universal property characterizing classifying toposes in the bicategory of toposes. Needless to say, in this form, the notion has no explicit connection with logic. As we have already mentioned, an implicit version of the notion also appeared in Reyes 1974, this time having a direct connection to logic. Thus, the notion was implicitly used by Joyal and Reyes already in 1971-1972. Myles Tierney and Jean Bénabou also discovered a version of the notion in the spring of 1975, as reported by Tierney himself in 1976. (See [Tierney, 1976, 211, 216 and 217]. See also [Bénabou, 1975], where the notion is mentioned but not defined.) Johnstone and Wraith introduced particular cases of the notion independently also in 1974.

To introduce the concept of classifying topos, we digress and recall some important facts of ring theory. Up until the middle of the 19th century, even people like Gauss, Galois and Abel took for granted that any polynomial with coefficients in a ring A had a zero *somewhere*, in an extension of A, although this had not been proved. Kronecker, following the lead of Cauchy, pointed out that the ring $A[X]/(p)$ contains the 'generic' zero $G = X + (p)$. Furthermore, this solution is universal in the sense that (taking $A = \mathbb{Z}$ to simplify), $G^* : RING(\mathbb{Z}[X]/(p), R) \simeq Zero_R(p)$, where $G^*(\varphi)$ is the obvious zero of p in R obtained from φ, namely $\varphi(G)$. Clearly, all of this can be generalized to ideals of polynomials, rather than single polynomials.

This result may be interpreted as showing that although the 'domain' A may fail to have a zero of a proper ideal $I \subset A[X_1, X_2, \ldots, X_n]$ there is an extension of this 'domain' namely $A[X_1, X_2, \ldots, X_n]/I$ which results from A by adding the generic zero G of I.

These ideas underline a far reaching analogy stressed by Joyal between rings and toposes and which can be summarized in the following table:

Ring theory	*Categorical logic*
Ring	Topos
Finitely presented ring	Coherent topos
$\mathbb{Z}$	*Set*
Ideal	Theory
Zero	Model
Proper ideal	Consistent theory
Generic zero	Generic model

Notice, once more, the algebraic inspiration underlying the analogy. The category *Set* plays a role analogous to the ring of integers $\mathbb{Z}$ in ring theory. The various constructions to be developed in topos theory are analogous to fundamental operations of ring theory, especially constructing from the data generic zero's to polynomials.

The classifying topos provided the context in which the notion of generic model could be defined and developed. Joyal and Reyes showed in the early seventies that for any coherent theory $\mathbb{T}$, there is a topos $\mathcal{B}(\mathbb{T})$ and a generic model G of $\mathbb{T}$ in $\mathcal{B}(\mathbb{T})$ satisfying the universal property: $G^* : \mathrm{TOP}(\mathcal{E}, \mathcal{B}(\mathbb{T})) \simeq \mathrm{Mod}_{\mathcal{E}}(\mathbb{T})$ where G^* is the functor which sends a geometric morphism (p^*, p_*) into a model $p^* \circ G$ of $\mathbb{T}$ in $\mathcal{E}$.

The topos $\mathcal{B}(\mathbb{T})$ is called the *classifying topos* of the theory $\mathbb{T}$ and it is a coherent topos in the sense of Grothendieck, that is roughly a topos defined by a topology having the property that every covering family is finite. The analogy with Kronecker's construction should be obvious. On the basis of this analogy we may view the classifying topos of $\mathbb{T}$ as the 'universe' $\mathcal{B}(\mathbb{T})$ which results from *Set* by forcing the existence of a generic model of $\mathbb{T}$.

One construction of $\mathcal{B}(\mathbb{T})$ proceeds as follows. First, given $\mathbb{T}$, construct the category of concepts $[\mathbb{T}]_{\mathrm{coh}}$. Define a Grothedieck topology J in $[\mathbb{T}]_{\mathrm{coh}}$ by taking all finite families $\{f_i : X_i \longrightarrow X : 1 \leq i \leq n\}$ such that the union of the images of the f_i is the whole of X. This yields a site $\left([\mathbb{T}]_{\mathrm{coh}}, J\right)$ and the topos of sheaves $\mathrm{Sh}\left([\mathbb{T}]_{\mathrm{coh}}, J\right)$. The latter is the classifying topos $\mathcal{B}(\mathbb{T})$ of $\mathbb{T}$. The Yoneda embedding $[\mathbb{T}]_{\mathrm{coh}} \longrightarrow \mathrm{Sh}\left([\mathbb{T}]_{\mathrm{coh}}, J\right)$ can be shown to be a coherent functor and it automatically yields the generic model G.

It turns out that every coherent topos is the classifying topos of a coherent theory (the details of this claim were worked out in Makkai & Reyes 1977). One of the first and important examples is the classifying topos of the theory of local rings, also called the Zariski topos. (This example goes back to Monique Hakim, although not from a logical point of view. See Hakim 1972.) Remember that the theory of local rings may be axiomatized by the following coherent theory. We add to the axioms for the theory of commutative rings with unity, the following two coherent axioms:

$$\begin{cases} \text{i. } 0 = 1 \vdash_{[]} \bot \\ \text{ii. } (\exists \mathrm{z})((x+y)z = 1) \vdash_{x,y} ((\exists z)(xz = 1) \vee (\exists z)(yz = 1)) \end{cases}$$

The topology of the site can be described explicitly and a description of the generic local ring can also be given explicitly.

A different but equally important example was given by Joyal. Joyal has shown that the topos of simplicial sets is the classifying topos of the theory of linear orders with different bottom and top elements. (See Mac Lane & Moerdijk 1994 for details.) Other examples have appeared in several investigations.

An important tool to find the coherent theory classified by a coherent topos is the conceptual completeness theorem mentioned above. In more detail, if $\mathbb{T}$ is a coherent theory which has a coherent model M in a coherent topos $\mathcal{E}$ such that there is an equivalence of categories given by the obvious functor $M^* : \mathrm{TOP}(\mathcal{E}, Set) \simeq \mathrm{Mod}_{Set}(\mathbb{T})$, then $\mathcal{E}$ is equivalent to $\mathcal{B}(\mathbb{T})$.

From the existence of the classifying topos and its 'converse', it follows that any coherent topos results from *Set* by adding a generic model of a suitable coherent theory in strict parallel to the fact that any finitely presented ring may be obtained as a quotient of a ring of polynomials with coefficients in $\mathbb{Z}$ divided by a suitable ideal.

The classifying topos theorem has an infinitary extension, as Makkai and Reyes have shown in their book. In fact, it turns out that every geometric theory has also a classifying topos and, furthermore, that every topos is the classifying topos of a geometric theory. Thus, every Grothendieck topos appears as the extension of *Set* obtained by adding a generic model of a suitable geometric theory; in complete analogy with the representation of an arbitrary ring as a ring of polynomials (in a possibly infinite number of indeterminates) modulo some ideal, showing how fruitful the analogy between topos theory and ring theory is. And it is far from clear that it has been fully exploited. For instance, is there any topos theoretical analog to Hilbert's celebrated Nullstellensatz?

All this work brought to the forefront the logical aspects of the work done by the Grothendieck School. For instance, as shown by Makkai and Reyes in their book, Deligne's theorem that coherent toposes have enough points turns out to be, modulo the theory of classifying toposes, equivalent to Gödel's completeness theorem for first order logic. Apart from Deligne's proof, there is a purely categorical proof given by Joyal that has strong similarities to Henkin's proof of the completeness theorem. Another result of Grothendieck about point of a topos being obtained as colimits of a set of points is a consequence of the downward Löwenheim-Skolem-Tarski theorem.

Here is the version presented by Makkai and Reyes in November 1974. Consider the functor category $\mathcal{E}^{\mathcal{E}_0}$ of all geometric functors $u^* : \mathcal{E}_0 \longrightarrow \mathcal{E}$, called the category of all $\mathcal{E}$-models of $\mathcal{E}_0$ and denoted by $Mod(\mathcal{E}_0, \mathcal{E})$. Let $\mathbb{T}$ be a coherent theory and $\mathcal{C}_{\mathbb{T}}$ its corresponding conceptual category. A $\mathcal{E}$-model of $\mathbb{T}$ in a topos $\mathcal{E}$ can be taken to be a coherent functor $M : \mathcal{C}_{\mathbb{T}} \longrightarrow \mathcal{E}$. Let $Mod(\mathbb{T}, \mathcal{E})$ denote the full subcategory of the functor category $\mathcal{E}^{\mathbb{T}}$ whose objects are the $\mathcal{E}$-models of $\mathbb{T}$. As usual, any $\mathcal{E}_0$-model $M_0 : \mathbb{T} \longrightarrow \mathcal{E}_0$ induces a functor $M_0^* : Mod(\mathcal{E}_0, \mathcal{E}) \longrightarrow \mathcal{E}^{\mathbb{T}}$. Now, $\mathcal{E}_0$ is a *classifying topos* of $\mathbb{T}$ with canonical model $M_0 : \mathbb{T} \longrightarrow \mathcal{E}_0$ if the induced functor $M_0^* : Mod(\mathcal{E}_0, \mathcal{E}) \longrightarrow Mod(\mathbb{T}, \mathcal{E})$ is an equivalence of categories, for any topos $\mathcal{E}$.

A slightly weaker definition of a classifying topos is given in their book: $\mathcal{E}_0$ is a *classifying topos* of $\mathbb{T}$ with canonical model $M_0 : \mathbb{T} \longrightarrow \mathcal{E}_0$ if for every $\mathcal{E}$-model $M : \mathbb{T} \longrightarrow \mathcal{E}$, for any topos $\mathcal{E}$, there is an $\mathcal{E}$-model of $\mathcal{E}_0$ and a unique (up to isomorphism) geometric functor $u^* : \mathcal{E}_0 \longrightarrow \mathcal{E}$ such that the diagram

$$\begin{array}{ccc} \mathbb{T} & \xrightarrow{M_0} & \mathcal{E}_0 \\ & {\scriptstyle M}\searrow & \downarrow{\scriptstyle u^*} \\ & & \mathcal{E} \end{array}$$

commutes.

The main fact is that every coherent theory $\mathbb{T}$ has a classifying topos $\mathcal{E}_0$ together with a canonical model, also called a *generic model* in their book, $M : \mathbb{T} \longrightarrow \mathcal{E}_0$. In the paper, it is stated that the classifying topos $\mathcal{E}_0$ can be constructed in a purely syntactical manner. The first part of the construction consists in constructing the small coherent category $\mathcal{C}_{\mathbb{T}}$ from $\mathbb{T}$. One then constructs, by formal means, a pretopos $\mathcal{E}_{\mathbb{T}}$ and a coherent functor $I : \mathcal{C}_{\mathbb{T}} \longrightarrow \mathcal{E}_{\mathbb{T}}$ satisfying the universal property for pretoposes. It can then be shown that the pretopos $\mathcal{E}_{\mathbb{T}}$ is in fact a topos.

Two other important facts about classifying toposes are stated. First, if $\mathcal{E}$ is a coherent topos, as defined by Grothendieck, then there is a coherent theory $\mathbb{T}$ such that $\mathcal{E}_{\mathbb{T}} \simeq \mathcal{E}$. In other words, every coherent topos is the classifying topos of some coherent theory. It is for this reason that Makkai and Reyes claimed that the notion of *coherent* topos is a logical notion. Second, let $\mathcal{E}$ be a coherent topos and $Coh(\mathcal{E})$ be the pretopos of coherent objects of $\mathcal{E}$. (See, for instance, Makkai & Reyes 1977, p. 276 for a definition of coherent objects.) Let M be an interpretation of $\mathcal{L}_2$ in $Coh(\mathcal{E})$. Suppose that M induces an equivalence $\widehat{M} : Mod(\mathcal{E}, \mathcal{S}) \longrightarrow Mod(\mathbb{T}, \mathcal{S})$. Then M is a $Coh(\mathcal{E})$-model of $\mathbb{T}$ and $\mathcal{E}$ is the classifying topos for $\mathbb{T}$ with canonical model M. This result is usually interpreted as saying that points are enough for classifying.

Makkai and Reyes, following suggestions given by Mulvey and Lawvere, used the above results to prove results of Hakim, which amounts to the claim that the so-called Zariski topos is the classifying topos for the coherent theory of nontrivial local rings and that the Etale topos is the classifying topos for the coherent theory of separably closed local rings.

The second paper does not end there. The last section mentions that some of the results obtained have an obvious generalization to models preserving infinite sups, a generalization to certain infinitary sentences. It is worth mentioning that when Makkai arrived in Montreal, he was already an expert in infinitary logic, more specifically model theory of formulas with denumerably long disjunctions and denumerably long conjunctions. Furthermore, Reyes' thesis was partly dedicated to infinitary languages.[10] However, the full presentation of the generalization to what is now called *geometric logic* will only be introduced in their book. Their paper contains only one illustration of the generalization, a special case of the conceptual completeness theorem that we will not state here. (For the statement of the theorem and the complete proof, see [Makkai and Reyes, 1977, 217–236].)

Although it was not mentioned explicitly in their papers, the completeness result for coherent logic, in fact as we will see for geometric logic, had an interesting and potentially very useful metalogical consequence. Here is how Reyes formulated it later:

Let $\mathbb{T}$ be a coherent theory in a language $\mathcal{L}$ and σ the universal closure of a formula $\varphi \Rightarrow \psi$, where φ and ψ are coherent. Then if $\mathbb{T} \models \sigma$ in the classical sense (every set-model of $\mathbb{T}$ is a set-model of σ), then for any Grothendieck topos $\mathcal{E}$, σ is valid in any $\mathcal{E}$-model of $\mathbb{T}$ [Reyes, 1978, 166, our translation].

The result means that to check that a *first-order coherent sequent* is valid intuitionistically, it is enough to check its validity in the topos of sets. Thus, the metatheorem allows direct and potentially interesting generalizations from a classical setting, where presumably the computations are easy and straightforward, to an arbitrary intuitionistic context. Reyes himself gives an application by proving that Swan's theorem is valid in any topos by giving a coherent formulation of it and verifying it in the topos *Set*. In fact, the metatheorem is true for *geometric logic* as well, to which we now turn.

4.9 Geometric logic

The notion of geometric logic appeared in print, although not as a full-blown formal system, in Reyes's paper on sheaves and logic in which it is presented as the logic of sheaves. Lawvere mentioned the idea in his conference in 1973 and in the published version of that conference. Underlying the idea of geometric logic is the concept of *geometric morphisms* between toposes introduced in the sixties by Grothendieck and was emphasized right from the start by Lawvere in 1970. In a nutshell, geometric logic is the logic preserved by a geometric morphism between toposes. Let us first give a precise definition of geometric logic.

Let $\mathcal{L}_0$ be as above. Then the class of *geometric formulas* $\mathcal{L}_{geo}$ over $\mathcal{L}_0$ is the class closed under those of coherent logic except that infinitary disjunctions $\vee_{i \in I} \varphi_i$ are now admitted. The infinitary rules are the obvious generalizations of the finite cases:

$\phi \vdash_{\vec{x}} \vee_{i \in I} \phi_i$ if $\phi \in \{\phi_i\}_{i \in I}$ and

[10] In a paper published in 1972, Reyes proved that every theory in $L_{\mu,\omega}$ is equivalent to a theory in $L_{\omega_1,\omega}$ in a specific technical sense that we won't clarify. Interestingly enough, Reyes used Joyal's Polyadic Spaces to prove his result. See [Reyes, 1972].

$$\frac{\phi \vdash_{\vec{x}} \theta \quad \bigvee_{i \in I} \phi_i \vdash_{\vec{x}} \theta}{\bigvee_{j \in J} \phi_j \vdash_{\vec{x}} \theta} \text{ where } \{\phi_j\}_{j \in J} = \{\phi\} \cup \{\phi_i\}_{i \in I}.$$

A sequent $\phi \vdash_{\vec{x}} \psi$ is *geometric* if both ϕ and ψ are geometric formulas. A *geometric theory* T is a theory in which all sequents are geometric. Two interesting examples of geometric theories are the theory of torsion Abelian groups and the theory of finite sets.

Many of the results obtained by Joyal, Reyes and Makkai generalize to the case of geometric logic. Certainly the most important result is that any geometric theory T has a classifying topos $\mathcal{E}_{\mathrm{T}}$ together with a generic model $M_0 : \mathrm{T} \longrightarrow \mathcal{E}_{\mathrm{T}}$. Furthermore, every topos $\mathcal{E}$ is the classifying topos of a geometric theory, that is for any topos $\mathcal{E}$, there is a geometric theory $\mathrm{T}_{\mathcal{E}}$ such that $\mathcal{E} \simeq \mathcal{E}_{\mathrm{T}_{\mathcal{E}}}$.

5 HIGHER-ORDER LOGIC AND TOPOSES

As we have already mentioned, the very axioms of elementary toposes point towards a certain conception of sets or more naturally a type theory. Accordingly, connections between previous set theories and elementary toposes were explored immediately showing that it is possible to construct models of a weak set theory from various extensions of elementary toposes. For that purpose and already in 1971, W. Mitchell, Cole and Osius developed an *internal language* for toposes. Here is how Osius characterizes the situation in 1973 at the *Colloque sur l'algèbre des catégories* in Amiens in France.

> Concerning the relationship between set theory and elementary topoi we consider two problems :
>
> 1. Find a general procedure of proving set-theoretical results internally in a topos (internal aspect).
> 2. Characterize certain topoi as " the category of sets " arising from certain types of models for set theory, thus generalizing the well-known results of Cole, W. Mitchell, Osius (external aspect). ([Osius, 1974b, 157])

In his lectures on elementary toposes given in the fall of 1973, Gavin Wraith writes:

> There are certain developments, due to J. Bénabou, which I should have liked to have included. Until recently, when one wished to carry out a construction in an elementary topos that was well enough understood in S, the category of sets and functions, one had to wrestle with pullback diagrams and the like. Bénabou's formal language permits one to dispense with these problems, and to proceed directly to the construction from its formal description. I believe that these methods must displace the older, clumsier ones. ([Wraith, 1975, 115–116])

Thus Bénabou had also developed a method that allowed the use of a language to prove results about toposes. It goes without saying that Wraith's opinion that this new method should replace the older, clumsier one was polemical at the time. But, indeed, it appeared quickly that the use of logical *methods* could be of real value in the development of topos theory. In some cases, the geometrical methods were convoluted, long and difficult to follow and their logical counterpart were direct, simple and almost immediate. Thus, there was a practical dimension to the problem that seemed to be worth developing fully. Since an elementary topos $\mathcal{E}$ can be thought of as a universe of sets, although not the usual sets of ZF in general, it should be possible to treat it as such, that is to use a topos $\mathcal{E}$ in the same way that one uses the category of sets. In particular, one would like to treat objects of $\mathcal{E}$ as sets with "elements" and write down definitions just as in the category of sets with the usual set-theoretical language.

As we have said, various people gave a positive answer to this question. But Bénabou and his student Michel Coste, Michael Fourman and André Joyal in collaboration with André Boileau, all in 1973-1974, are usually credited with the introduction of a full, rigorous description of an appropriate formal system. The striking fact is that the resulting higher-order type theory is purely algebraic.

5.1 Interpreting higher-order logic in toposes

We will first present Boileau and Joyal's approach, based on Joyal's talk given in April 1974, expanded and developed by Boileau in his Ph.D. thesis submitted in 1976 and then in a joint paper received by the editors of *The Journal of Symbolic Logic* in July 1977 but published only in 1981. (See [Boileau and Joyal, 1981].)

A (higher-order) *similarity type* or *signature* $\mathcal{T}$ is given by

1. A set $\mathcal{S}$ of sorts (which does not include the symbol Ω). The set $\mathcal{T}$ of *types* is build from the set of sorts as follows:
   (a) if $S \in \mathcal{S}$, then $S \in \mathcal{T}$;
   (b) if $S_1, ..., S_n \in \mathcal{T}$, then $\Omega(S_1, \ldots, S_n) \in \mathcal{T}$, where n can be 0;
   (c) nothing else is a type.

The type $\Omega(\ )$ is denoted by Ω.

1. A set $\mathcal{F}$ of *function symbols*, to each function symbol f, we associate its *type* $S_1 \times \cdots \times S_n \times S$ (with the last sort having a distinguished status); we write $f : S_1 \times \cdots \times S_n \longrightarrow S$ to indicate the type of f. Whenever n is 0, f is a *constant* of type S and we write $1 \xrightarrow{f} X$. Furthermore, a functional symbol $f : S_1 \times \cdots \times S_n \longrightarrow \Omega$ with codomain Ω will be called a *relational symbol.*

Given a signature $\mathcal{T}$, the *language* $\mathcal{L}(\mathcal{T})$ consists of the signature, for each type S, a countable list of variables. Terms and formulas are defined by recursion as follows:

Definition of terms:

1. every variable of type S is a term of type S;
2. if $f : S_1 \times \cdots \times S_n \longrightarrow S$ is a function symbol and $t_1, \ldots, t_n$ are terms of type $S_1, \ldots, S_n$ respectively, then $f(t_1, \ldots, t_n)$ is a term of type S; (if f is a constant, we simply write f);
3. if ϕ is a formula and if $x_1, \ldots, x_n$ is a list of distinct variables of type $S_1, \ldots, S_n$, then $\{(x_1, \ldots, x_n) : \phi\}$ is a term of type $\Omega(S_1, \ldots, S_n)$;

Definition of formulas:

1. if t_1 and t_2 are terms of type S, then $t_1 = t_2$ is a formula;
2. if $t_1, \ldots, t_n$ are terms of type $S_1, \ldots, S_n$ and if t is a term of type $\Omega(S_1, \ldots, S_n)$ then $(t_1, \ldots, t_n) \in t$ is a formula;
3. $\top$ is a formula ('true' is a formula);
4. if ϕ and ψ are formulas, then $(\phi \wedge \psi)$ is a formula;

The reader will have noticed how the higher-order structure emerges: it is the recursive nature of 1.b for types that leads to higher types.

The remaining connectives can be *defined* as follows:

D1. $\forall x \phi \equiv (\{x : \phi\} = \top)$;
D2. $(\phi \Rightarrow \psi) \equiv ((\phi \wedge \psi) = \phi)$;
D3. $(\phi \vee \psi) \equiv \forall w\,[((\phi \Rightarrow w) \wedge (\psi \Rightarrow w) \Rightarrow w) = \top]$
D4. $\perp \equiv \forall w\,(w = \top)$;
D5. $\exists x \phi \equiv \forall w\,[\forall x\,((\phi \Rightarrow w) \Rightarrow w) = \top]$.

Notice that the foregoing definitions are all equational, whence the purely algebraic character of logical operations in this framework. A *sequent* over a signature $\mathcal{T}$ is defined as above, that is as a sequence of symbols of the form $(\phi \vdash_{\vec{x}} \psi)$, where ϕ and ψ are formulas over $\mathcal{T}$ and $\vec{x}$ is a context containing all the free variables of ϕ and ψ. The deductive system is again a sequent calculus, only shorter since there are fewer cases to cover.

1. Axiom schemas:

A1. $\varphi \vdash_{\vec{x}} \varphi$.
A2. $\varphi \vdash_{\vec{x}} \top$.
A3. $\top \vdash_{\vec{x}} x = x$.
A4. $\varphi \wedge (x = t) \vdash_{\vec{x}} \varphi(t/x)$ provided t is free for x in φ.
A5. $\varphi \vdash_{\vec{y}} (x_1, \ldots, x_n) \in \{(x_1, \ldots, x_n) : \varphi\}$.
A6. $(x_1, \ldots, x_n) \in \{(x_1, \ldots, x_n) : \varphi\} \vdash_{\vec{y}} \varphi$.

2. Rules of inference:

R1. $\frac{\varphi \vdash_{\vec{x}} \psi}{\varphi \vdash_{\vec{x}, \vec{y}} \psi}$.

R2. $\dfrac{\varphi \vdash_{\vec{x}} \psi \quad \psi \vdash_{\vec{x}} \theta}{\varphi \vdash_{\vec{x}} \theta}$.

R3. $\dfrac{\varphi \vdash_{\vec{x}} \psi \quad \varphi \vdash_{\vec{x}} \theta}{\varphi \vdash_{\vec{x}} \psi \wedge \theta}$.

R4. $\frac{\varphi\vdash_{\vec{x}}\psi\wedge\theta}{\varphi\vdash_{\vec{x}}\psi}\ \frac{\varphi\vdash_{\vec{x}}\psi\wedge\theta}{\varphi\vdash_{\vec{x}}\theta}$.

R5. $\frac{\varphi\vdash_{\vec{x}}\psi}{\varphi(t/x)\vdash_{\vec{y}}\psi(t/x)}$

where t is a term of the same type as $\vec{x}$, and $\vec{y}$ is any string of variables including all the variables occurring in the term t.

R6. $\frac{\phi\wedge(x_1,\ldots,x_n)\in t_1\vdash_{\vec{x},\vec{y}}(x_1,\ldots,x_n)\in t_2\ \ \phi\wedge(x_1,\ldots,x_n)\in t_2\vdash_{\vec{x},\vec{y}}(x_1,\ldots,x_n)\in t_1}{\phi\vdash_{\vec{y}}t_1=t_2}$

where the distinct variables $x_1,\ldots,x_n$ are not free in ϕ, t_1, t_2.

Axioms 5 and 6 are sometimes called "comprehension axioms" and R6 is sometimes referred to as "extensionality".

This presentation is admittedly somewhat awkward. For instance, it is far from obvious that we are dealing with an intuitionistic logic. But it was tailored in such a way that its interpretation in a topos becomes almost direct and, furthermore, the categorical manipulations to be done afterwards are few and almost trivial. It was shown by Boileau to be equivalent to the following, more standard, intuitionistic type theory (see [Kleene, 1952] or [Hatcher, 1968]);

Axioms:

1. $A\supset(B\supset A)$
2. $(A\supset B)\supset((A\supset(B\supset C)\supset(A\supset C))$
3. $A\supset(A\vee B)$
4. $B\supset(A\vee B)$
5. $(A\supset C)\supset((B\supset C)\supset((A\vee B)\supset C))$
6. $A\supset(B\supset(A\wedge B))$
7. $(A\wedge B)\supset A$
8. $(A\wedge B)\supset B$
9. $(A\supset B)\supset((A\supset\neg B)\supset\neg A)$
10. $\neg A\supset(A\supset B)$
11. $\forall xA(x)\supset A(t)$ provided that x is free in $A(x)$ and t is free for x in $A(x)$.
12. $A(x)\supset\exists xA(x)$ provided that $x$is free in $A(x)$
13. $x=x$
14. $x_1=x_2\supset(A(x_1)\supset A(x_2))$ provided that x_2 is free for x_1 in $A(x_1)$
15. $QxA\equiv Qx(A\wedge x=x)$ where $Q\in\{\forall,\exists\}$
16. $\forall x_1\cdots\forall x_n((x_1,\ldots,x_n)\in X_1\equiv(x_1,\ldots,x_n)\in X_2)\supset X_1=X_2$
17. $A\equiv(x_1,\ldots,x_n)\in\{(x_1,\ldots,x_n):A\}$.

Rules of inference:

R1. $\frac{A\quad A\supset B}{B}$ provided that all free variables in A are free in B.

R2. $\frac{C\supset A}{C\supset\forall xA}$ provided that x is not free in C.

R3. $\frac{A\supset C}{\exists xA\supset C}$ provided that x is not free in C.

The first fourteen axioms are the standard axioms for first-order intuitionistic logic. Axiom 15 together with the restriction on the rule of modus ponens reflects the fact that there may be empty types. The higher-order axioms are also standard. (See also [Osius, 1975] for what amounts to an equivalent proof of the same claim and [Fourman, 1977]. We will come back to the later in a short while.)

The interpretation of a language $\mathcal{L}(\mathcal{T})$ in a topos $\mathcal{E}$ is an extension of the interpretation of first-order language in a category of the appropriate type. Thus, a *$\mathcal{T}$-structure* or an *interpretation* M in $\mathcal{E}$ consists of:

1. an object MS of $\mathcal{E}$ for each sort S of $\mathcal{T}$; this is extended to types by stipulating that $M(\Omega(S_1,\ldots,S_n)) = P(MS_1 \times \cdots \times MS_n)$ where $P(X)$ denotes the power-object of X in $\mathcal{E}$;
2. a morphism $Mf : MS_1 \times \cdots \times MS_n \longrightarrow MS$ for each function symbol $f : S_1 \times \cdots \times S_n \longrightarrow S$ of $\mathcal{T}$; in particular, when $n = 0$ a morphism $1 \xrightarrow{f} MS$.

Variables-in-context, functional symbols-in-context and conjunction-in-context are interpreted as in the foregoing section.

1. If t is $\{(y_1,\ldots,y_m) : \phi\}$, then $[\![\vec{x}\cdot t]\!]_M$ is the morphism

$$MS_1 \times \cdots \times MS_n \longrightarrow P(MS_1 \times \cdots \times MS_n);$$

1. If ϕ is $t_1 = t_2$, and t_1 and t_2 are of type T, then $[\![\vec{x}\cdot\phi]\!]_M$ is the composite

$MS_1\times\cdots\times MS_n \xrightarrow{\langle[\![\vec{x}\cdot t_1]\!],[\![\vec{x}\cdot t_2]\!]\rangle} MT\times MT \xrightarrow{\delta_{MT}} \Omega$, where the morphism $MT\times MT \xrightarrow{\delta_{MT}} \Omega$ is the characteristic morphism of the diagonal $MT \xrightarrow{\Delta_{MT}} MT \times MT$;

1. If ϕ is $(t_1,\ldots,t_m) \in t_{m+1}$, then $[\![\vec{x}\cdot\phi]\!]_M$ is the composite

$MS_1\times\cdots\times MS_n \xrightarrow{\langle[\![\vec{x}\cdot t_1]\!],\ldots,[\![\vec{x}\cdot t_m]\!]\rangle} (MT_1 \times \cdots \times MT_m)\times P(MT_1 \times \cdots \times MT_m) \xrightarrow{e} \Omega;$

1. If ϕ is $\top$, then $[\![\vec{x}\cdot\phi]\!]_M$ is the composite

$$MS_1 \times \cdots \times MS_n \longrightarrow 1 \xrightarrow{\top} \Omega.$$

As is obvious from the definition, every formula in context is interpreted as a morphism into the subobject classifier Ω.

Let M be a $\mathcal{T}$-structure in a topos $\mathcal{E}$. We say that M is a *model* of a sequent $\sigma = (\phi \vdash_{\vec{x}} \psi)$ or that σ is *satisfied in* M, we write as usual $M \models_{\mathcal{E}} \sigma$ or simply $M \models \sigma$, if $[\![\vec{x}\cdot\phi]\!]_M \leq [\![\vec{x}\cdot\psi]\!]_M$ in $P(MS_1\times\cdots\times MS_n)$. We say that M is a *model of a theory* $\mathbb{T}$ if all the axioms of $\mathbb{T}$ are satisfied in M.

As usual, a soundness theorem can now be proved: if $\mathbb{T} \vdash$: $(\varphi \vdash_{\vec{x}} \psi)$, then $\mathbb{T} \models (\varphi \vdash_{\vec{x}} \psi)$.

Given a theory $\mathbb{T}$ in the language $\mathcal{L}(\mathcal{T})$, it is once again possible to construct the conceptual category $\mathcal{C}_{\mathbb{T}}$ and show that it is in fact a topos, which we will denote by $\mathcal{E}_{\mathbb{T}}$. This yields the following:

Completeness Theorem: Let $\mathbb{T}$ be a higher-order type theory of similarity type $\mathcal{T}$. Then, $\mathbb{T}$ has a canonical model M_0, the canonical interpretation, in a topos $\mathcal{E}_{\mathbb{T}}$ such that $\mathbb{T} \vdash$: $(\varphi \vdash_{\vec{x}} \psi)$ if and only if $M_0 \models_{\mathcal{E}_{\mathbb{T}}} (\varphi \vdash_{\vec{x}} \psi)$.

Starting with an elementary topos $\mathcal{E}$, it is possible to define a higher-order signature $\mathcal{T}_{\mathcal{E}}$ from $\mathcal{E}$ by transposing the foregoing procedure appropriately and similarly the *internal language* $\mathcal{L}_{\mathcal{E}}$ and the type theory $\mathbb{T}_{\mathcal{E}}$ of $\mathcal{E}$ can be defined. The language $\mathcal{L}_{\mathcal{E}}$ is defined as follows:

1. To each object X of $\mathcal{E}$, we attribute a sort $\ulcorner X \urcorner$
2. To each morphism $f : X \longrightarrow Y$, we associate the function symbol $\ulcorner f \urcorner : \ulcorner X \urcorner \longrightarrow \ulcorner Y \urcorner$.
3. To each relation (or monomorphism) $\ulcorner R \urcorner \rightarrowtail \ulcorner X \urcorner$.

(There are slight technical difficulties one has to take care of here, in particular with products, but we will skip these formal details.) There is then a canonical interpretation $M : \mathcal{L}_{\mathcal{E}} \longrightarrow \mathcal{E}$ defined in the obvious way, i.e. $M(\ulcorner X \urcorner) = X$, $M(\ulcorner f \urcorner : \ulcorner X \urcorner \longrightarrow \ulcorner Y \urcorner) = f : X \longrightarrow Y$, $M(\ulcorner R \urcorner \rightarrowtail \ulcorner X \urcorner) = R \rightarrowtail X$. The theory $\mathbb{T}_{\mathcal{E}}$ is also defined in the obvious way, i.e. it is the collection of sequents $\varphi \vdash_{\vec{x}} \psi$ such that $M \models \varphi \vdash_{\vec{x}} \psi$. One could be a little more specific and describe $\mathbb{T}_{\mathcal{E}}$ with explicitly definable axioms in such a way that its theorems are precisely the valid sequents of the canonical interpretation M. For instance, one could take the collection of sequents of $\mathcal{L}_{\mathcal{E}}$ obtained from the categorical properties of $\mathcal{E}$ with a dictionary that would include, for instance:

$X \xrightarrow{f} X$ is the identity morphism in $\mathcal{E}$ is translated into $\top \vdash_x \ulcorner f \urcorner(x) = x$, and so on, as we have seen above.

One can then verify that indeed $M \models \mathbb{T}_{\mathcal{E}}$.

It follows from the foregoing procedure, that given any topos $\mathcal{E}$, one can show that there is a theory, the internal theory of $\mathcal{E}$, such that $\mathcal{E} \models \mathbb{T}_{\mathcal{E}}$.

Joyal proved in 1974 that every elementary topos $\mathcal{E}$ arises from a higher-order theory (this is called the "théorème d'engendrement" in Boileau & Joyal 1981). More specifically, the statement is that for any elementary topos $\mathcal{E}$, there is a higher-order theory, naturally $\mathbb{T}_{\mathcal{E}}$ such that $\mathcal{E}$ is equivalent (as a category) to $\mathcal{C}_{\mathbb{T}_{\mathcal{E}}}$, the category of concepts of $\mathbb{T}_{\mathcal{E}}$. In this sense, a topos is nothing but a higher-order type theory. One can therefore use $\mathbb{T}_{\mathcal{E}}$ to prove properties of $\mathcal{E}$ and *working internally* means working in $\mathbb{T}_{\mathcal{E}}$. For instance, Boileau shows that any topos $\mathcal{E}$ has finite colimits, exponentiation from $\mathbb{T}_{\mathcal{E}}$. He also shows Cantor's theorem, i.e. if in a topos $\mathcal{E}$, there is an object X and an epimorphism $X \twoheadrightarrow \Omega^X$, then $\mathcal{E}$ is trivial, that is $1 \simeq 0$, from $\mathbb{T}_{\mathcal{E}}$.

The reader will have noticed that we have not included an object $\mathbb{N}$ of natural numbers in our theory. This constitutes an extension of the fundamental theory. Other extensions include being Boolean, being two-valued, or satisfying the axiom of choice, which can be formulated in different ways.

We can now see the unity underlying Joyal's work from regular categories to elementary toposes. Indeed, we start with the language $\mathcal{L}_0$ presented in section 4.4 and define four new higher-order languages: $\mathcal{L}_5$, $\mathcal{L}_6$, $\mathcal{L}_7$ and $\mathcal{L}_8$ from $\mathcal{L}_0$ as follows. First, for $\mathcal{L}_5$. We close the set of sorts with the operation $\Omega(-,\ldots,-)$ to obtain the set of types. Terms and formulas are defined as before, with the symbols $\exists$, $\vee$, $\perp$, $\supset$, $\forall$ to form new formulas. The language $\mathcal{L}_6$ is defined like $\mathcal{L}_5$, except that we now allow the negation operator $\neg$ in the construction of formulas. $\mathcal{L}_7$ is constructed from $\mathcal{L}_0$ by adding a new sort $\mathbb{N}$ and two new functional symbols, 0 and s. Types and formulas are constructed as in $\mathcal{L}_5$. To construct $\mathcal{L}_8$, we proceed as with $\mathcal{L}_7$, but by including all types and formulas of $\mathcal{L}_6$ instead of $\mathcal{L}_5$. The logical systems corresponding to these languages are then defined in the expected manner, i.e. by adding the appropriate sequents at each stage. Recall that $\mathcal{L}_0$ is purely equational logic, $\mathcal{L}_1$ is regular logic, $\mathcal{L}_2$ is coherent logic, $\mathcal{L}_3$ is first order intuitionistic logic, $\mathcal{L}_4$ is first order classical logic. As we have seen, $\mathcal{L}_5$ is intuitionistic type theory, $\mathcal{L}_6$ is classical type theory, $\mathcal{L}_7$ is intuitionistic type theory with $\mathbb{N}$ and $\mathcal{L}_8$ classical type theory with $\mathbb{N}$. One can then prove that:

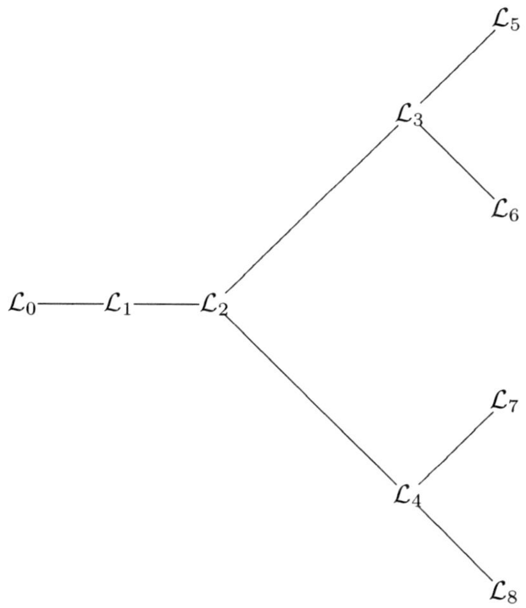

Needless to say, one can then move to conceptual categories and construct a diagram of functors between the respective categories thus obtained.

5.2 Fourman's approach

Let us now briefly consider Fourman's approach, since it differs from the foregoing in important and interesting ways.

Fourman starts with a primitive predicate, denoted by E, called the *existence predicate.* Thus, given a term t, one reads "Et" as "t exists". There is another primitive relation, namely a notion of *equivalence* between terms, i.e. $t \equiv s$. Equality of terms is then *defined* from the two primitive notions by $t = s \Leftrightarrow (t \equiv s) \wedge Et \wedge Es$. The definition of the formal system proceeds as follows. For convenience, another term is introduced, the *definite description* $Ix\varphi$, which is read "the unique x such that φ".

Definition: A *higher-order language* $\mathcal{L}$ is specified by the following data:

1. Two sets Sort and Const of sorts and constants.

2. A *power-type map* from $\bigcup_{n \in \omega} \text{Sort}^n \longrightarrow \text{Sort}$, written as $(S_1, \ldots, S_n) \mapsto [S_1, \ldots, S_n]$.

3. A map assigning a sort to each constant, $\# : \text{Const} \longrightarrow \text{Sort}$.

As in the foregoing case, for each sort S, there is a countable list of variables and each variable x has a sort $\#x$.

Definition of terms:

1. Every variable of sort S is a term of type S;

2. Every constant c of sort $\#c$ is a term;

3. If x is a variable of sort S and φ a formula, then $Ix\varphi$ is a term of sort S.

Definition of formulas:

1. If t is a term, then Et is a formula;

2. If t_1 and t_2 are terms of sort S, then $t_1 \equiv t_2$ is a formula;

3. If t is a term of sort $[S_1, \ldots, S_n]$ and $t_1, \ldots, t_n$ are of sorts $S_1, \ldots, S_n$ respectively, then $t\,(t_1, \ldots, t_n)$ is a formula;

4. If φ and ψ are formulas, then so are $(\varphi \wedge \psi)$ and $(\varphi \supset \psi)$;

5. If x is a variable and φ is a formula, then $\forall x \varphi$ is a formula.

The empty power sort $[]$ can be thought of as consisting of truth-values. Thus if t is a term of sort $[]$, then $t\,()$ is a formula and it asserts the proposition t.

The remaining connectives are also definable in this system, but we will skip the definitions.

1. Axioms

A1. $\varphi \supset (\psi \supset \varphi)$,
A2. $(\varphi \supset (\psi \supset \theta)) \supset ((\varphi \supset \psi) \supset (\varphi \supset \theta))$,
A3. $(\varphi \wedge \psi) \supset \varphi$
A4. $(\varphi \wedge \psi) \supset \psi$
A5. $(\varphi \supset (\psi \supset (\varphi \wedge \psi)))$
A6. $(\varphi\,[y/x] \wedge y \equiv z) \supset \varphi\,[z/x]$ (Substituvity of equivalents),
A7. $\forall x (x \equiv y \Leftrightarrow x \equiv z) \supset y \equiv z$ (Extensionality),
A8. $(\forall x \varphi \wedge Ex) \supset \varphi$ (Universal instantiation),
A9. $\forall y (y \equiv Ix\varphi \Leftrightarrow \forall x (\varphi \Leftrightarrow x \equiv y))$ (Description),
A10. $EIy\forall \vec{x} (\varphi \Leftrightarrow y(\vec{x}))$ (Comprehension)
A11. $y(\vec{x}) \supset Ey \wedge \bigwedge_{i=1}^{n} Ex_i$ (Predication).

2. Rules of inference:

R1. $\dfrac{\varphi \quad \varphi \supset \psi}{\psi}$,
R2. $\frac{\varphi}{\varphi[t/x]}$,

R3. $\frac{(\psi \wedge Ex) \supset \varphi}{\psi \supset \forall x \varphi}$ where x is not free in ψ.

The notion of derivability is defined as usual: $\Gamma \vdash \varphi$ means that φ is derivable from the set Γ of formulas. A theory T is a set of formulas such that if $\mathrm{T} \vdash \varphi$, then $\varphi \in \mathrm{T}$. Also, a theory T has an underlying language denoted by $\mathcal{L}_{\mathrm{T}}$.

The reader will have noticed that types haven't been mentioned. They are in fact defined: a *type* is a term of the form $Iy\forall x(\varphi \Leftrightarrow y(x))$. If x is of sort X, then this term is of sort $[X]$. It is abbreviated as $\{x : \varphi\}$. Thus, types are seen as "subsets" of sorts. A type is *definable* if and only if it is a closed term.

Fourman proceeds to prove that the conceptual category $\mathcal{C}_{\mathrm{T}}$ of a theory T in the higher-order language $\mathcal{L}_{\mathrm{T}}$ is in fact a topos, denoted as usual $\mathcal{E}_{\mathrm{T}}$. He explicitly attributes the construction of $\mathcal{C}_{\mathrm{T}}$ to Joyal. He then defines the interpretation of a language $\mathcal{L}$ in a topos $\mathcal{E}$ and proves soundness and completeness theorems. Finally, given a topos $\mathcal{E}$, he defines the internal language $\mathcal{L}_{\mathcal{E}}$ of $\mathcal{E}$ and its theory $\mathrm{T}_{\mathcal{E}}$ and proves that $\mathcal{E} \simeq \mathcal{E}_{\mathrm{T}_{\mathcal{E}}}$. Various extensions, including toposes with a natural number object, are then considered.

5.3 Revising logic: the debate

An interesting debate between Fourman and Boileau & Joyal followed. We will simply summarize the main points raised in the debate.

Here are Fourman's arguments in favor of his system:

1. Fourman's approach contains an *original* component, namely an existence predicate;

2. Because their approach lacks this predicate, Boileau and Joyal are forced to make awkward restrictions on the rule of modus ponens;

3. In the words of Fourman's himself: "we *emphatically* do not agree that because of this [i.e. the modifications to the rule of modus ponens] "the traditional logical way of dealing with variables ... should be abandoned". The traditional use of variables has much to commend it and there are far less drastic remedies to hand." [Fourman, 1977, 1054])

In a nutshell: Fourman insists on the original component of his system, which allows him to avoid revising the way variables are used in deductions. Still in other words, the basic principles of deduction stay the same and there is an original component to the theory.

It seems only fair to bring in Dana Scott's voice to the debate, since he was Fourman's thesis advisor and did write on related issues around the same time. Scott's position finds its justification in a certain interpretation of intuitionism.

> Standard formulations of intuitionistic logic, whether by logicians or by category theorists, generally do not take into account partially defined elements. (For a recent reference see [Makkai and Reyes, 1977, esp. pp. 144–163].) Perhaps there is a simple psychological reason : we dislike

> talking of those things not already proved to exist. Certainly we should not *assume* that things exist without making this assumption explicit. In classical logic the problem is not important, because it is always possible to split the definition (or theorem) into cases according as the object in question does or does not exist. In intuitionistic logic this way is not open to us, and the circumstance complicates many constructions, the theory of descriptions, for example. Many people I find do not agree with me, but I should like to advocate in a mild way in this paper what I consider a simple extension of the usual formulation of logic allowing reference to partial elements. (...)
>
> Technically the idea is to permit a wider interpretation of *free* variables. All bound variables retain their usual existential import (when we say something exists it does exist), but free variables behave in a more " schematic " way. Thus there will be no restrictions on the use of *modus ponens* or on the rule of *substitution* involving free variables and their occurrences. The laws of quantifiers require some modification, however, to make the existential assumptions explicit. The modification is very straightforward, and I shall argue that what has to be done is simply what is done naturally in making a *relativization* of quantifiers from a larger domain to a subdomain.
>
> (...)
>
> The idea of schematic free variables is not new for classical logic, and the literature on " free " logic (or logic without existence assumptions) is extensive. (...) All I have done in this essay is to make what seems to me to be the obvious carryover to intuitionistic logic, because I think it is necessary and convenient. For those who do not like this formulation, some comfort can be taken from the fact that in topos theory both kinds of systems are completely equivalent, and the domains of partial elements can be *defined* at higher types (...). However, in first-order logic something is lost in not allowing partial elements, as I shall argue along the way.
>
> (...)
>
> Is the existence predicate E an illusion? Was the equality predicate an illusion? No. We shall find in the section, with a full statement of the laws of equality, that E can always be defined in terms of quantification :... . However, both in conception and in the models of (intuitionistic) logic we have in mind, the existence predicate is more basic than equality and prior to it. [Scott, 1979, 660–662])

The psychological reason can probably be dismissed altogether. From an anthropological point of view, it is hard to sustain the claim that "we dislike talking of those things not already proved to exist". In fact, the contrary seems to be the case: as a whole, we seem to enjoy very much talking of those things whose actual existence is completely unknown. Of course, Scott is here talking about *mathematicians*, a different cultural group if there is one and he might be formulating a deontological maxim for mathematicians than anything else. Notice that Scott acknowledges the fact that as far as higher-order intuitionistic logic is concerned, that is topos theory, the various systems presented can be shown to be formally equivalent. We should emphasize, in the same vein, the fact that Fourman also "regards topos theory as the "algebraic" form of this higher-order intuitionistic logic" [Fourman, 1977, 1054]. First-order logic seems to be the real issue, although it is not so easy to separate the various logical systems in such a way.

Here is Boileau and Joyal's rebuttal.

1. First, as Fourman explicitly admits (see Fourman 1977, 1070), his system is an abstract codification of a category of H-sets, where H is a Heyting algebra, which was shown by Higgs to be equivalent to a topos of sheaves over the Heyting algebra of open sets of a topological space. Fourman is thus lead to consider a sheaf as a *set* with a complementary structure. Boileau and Joyal believe that the formal presentation should not present the objects as sets with a structure, but simply as sets with a specific logic.

2. The logical system developed by Boileau and Joyal is claimed to be more simple and natural than Fourman's system. One can work with the objects as if they were sets, provided that the logical rules, constructive in their workings if not in their spirit, are respected.

3. Fourman's system is perfectly legitimate for toposes, but the underlying intuition does not generalize to other cases. By adding the idea of the support of a sequent, one can treat logical systems in various categories.

4. The last point brings us to the underlying *unity* of Boileau and Joyal's approach. Although the notion of deduction is modified or restricted in contrast with the classical notion, one can treat various logical systems in a unified manner.

In a nutshell, Boileau and Joyal insists on the fact that Fourman's approach is somewhat *ad hoc* and that their approach is simpler and provides a unified framework to do and develop categorical logic in general. Furthermore, but this point is not underlined by the actors involved in the dispute, it is clear that Joyal's approach was right from the start motivated by the desire to minimize the distance between the syntax of categories and the syntax of logic. Evidence for this is easy to find: the choice of a Gentzen type presentation of logic, the choice of a multi-sorted system, the choice of rules which are simply rewriting of adjunctions between functors, and as we have seen, the higher-order type theory itself, which was chosen so that the interpretation of the connectives is simple and almost immediate in toposes. Stipulating the context of a deduction is nothing neither more nor less than the idea that a function *necessarily* has an underlying domain and codomain, although set theorists never conceived of functions that way. In contrast, Fourman and Scott are more preoccupied by purely *logical* issues and categories, in particular toposes, are seen as providing new metalogical resources in the study of intuitionistic and more generally constructive systems. An illustration of this position is found in their joint paper entitled *Sheaves and Logic,* also published in the proceedings of the Durham meeting and related to Scott's paper.

> Scott's paper in this volume describes a system of higher-order logic which may be interpreted in any topos. Here we describe the models of this logic given by sheaves over a complete Heyting algebra (cHa). These sheaf models subsume the more familiar Beth, Kripke and topological interpretations of intuitionistic logic, which correspond to interpretations in sheaves over the appropriate cHa of "truth values". They also provide a uniform way of extending these interpretations to higher-order logic and thus help to explain the models of analysis of Scott, Moschovakis and van Dalen. Once we go beyond first-order logic, these sheaf models are more general than Beth, Kripke or topological models. Models over a site (...) provide yet more generality. (...) Sites also arise naturally in first-order model theory once we take into account the comparisons between various models made possible by geometric morphisms. For a presentation of this theory see Makkai and Reyes 1977. However, we find a full-blown categorical presentation is often too abstract and results in very heavy machinery's being brought to bear on very simple problems. By restricting our attention to the special case of models over a cHa, we hope to make what is simple look simple. Models over cHa show clearly the link with traditional models for intuitionistic logic and are sufficient for many applications. ([Fourman and Scott, 1979, 303])

Thus, according to Fourman and Scott, logic has precedence over categorical ideas and methods. The latter are certainly useful, but there is no need to look for the most general result in that framework nor to find a unified framework that would be based on the categorical framework. Two specific rebuttals are in order. First, as Lawvere has emphasized in the eighties, one has to distinguish between *petit toposes* and *gros toposes.* Examples of petit toposes are precisely toposes of sheaves over a cHa (or a locale) and examples of gros toposes are toposes of all spaces of a certain type or the topos of simplicial sets. The point is that gros toposes have very different properties from petit toposes. Scott is here ignoring the class of gros toposes, which nonetheless are important not only in mathematics in general but for logic too. (See [Lawvere, 1986; Lawvere, 1989].) Second, for some applications, classifying toposes ("models over a site" in Scott's terminology) constitute a powerful tool. A classical example of this sort was given by Joyal in the seventies with his construction of the so-called formal reals for which local compactness fails. (See, [Ščedrov, 1984, section 4.2] and the whole paper for various examples or [Johnstone, 2002].)

6 THE METHOD OF FORCING IN TOPOSES: KRIPKE-JOYAL SEMANTICS

In the foregoing sections, the relationship between a topos $\mathcal{E}$ and its associated theory $\mathrm{T}_{\mathcal{E}}$ might seem to be, in some sense, too close for comfort. For $\mathcal{E}$ can be given in the form of $\mathrm{T}_{\mathcal{E}}$ and it is thus difficult to see how $\mathcal{E}$ could be useful in order to understand $\mathrm{T}_{\mathcal{E}}$. However, Grothendieck toposes were given and can be given independently of a theory and they are, in a precise technical sense that we will not clarify here, incompatible with *finitary* theories. Furthermore, Joyal quickly showed that Grothendieck toposes encompasses the usual semantics, at least for intuitionistic and classical logic: algebraic, topological and Kripke models. Joyal generalized Kripke's semantics for intuitionistic logic in the context of Grothendieck toposes. The interesting feature of the Kripke-Joyal semantics is that it can be manipulated classically. It should be emphasized immediately that Joyal's work was not a generalization for the sake of generalizing; it was seen and presented as an efficient *method* to prove results about toposes in the same way that Kripe semantics was used to prove results about intuitionistic logic.

Let $\mathcal{E}$ be an elementary topos, $\mathcal{L}$ an higher-order language and $M : \mathcal{L} \longrightarrow \mathcal{E}$ an interpretation of $\mathcal{L}$ in $\mathcal{E}$. Let $\mathcal{G}$ be a full sub-category of $\mathcal{E}$ such that

i) $\mathcal{G}$ is closed under subobjects, i.e. given $i : A \rightarrowtail X$ of $\mathcal{E}$ and X in $\mathcal{G}$, then i and therefore A is in $\mathcal{G}$;

ii) $|\mathcal{G}|$ is a set of generators for $\mathcal{E}$, that is if $X \underset{k}{\overset{h}{\rightrightarrows}} Y$ are such that $h \neq k$ in $\mathcal{E}$, then there is an object G in $\mathcal{G}$ and a morphism $A \xrightarrow{f} X$ such that $h \circ f \neq k \circ f$.

Definition (Kripke-Joyal semantics): Let G be an object of $\mathcal{G}$, $\varphi(x_1, \ldots, x_n)$ a formula of $\mathcal{L}$ and $G \xrightarrow{a_i} X_i$, $1 \leq i \leq n$, where $X_i = M$ ("the sort of x_i''). The morphism a_i is sometimes called a *generalized element* of X_i *at stage* G. We define the relation $G \Vdash \varphi(a_1, \ldots, a_n)$ and say that '$\varphi(a_1, \ldots, a_n)$ holds at stage G' or that 'G forces $\varphi(a_1, \ldots, a_n)$' as follows:

(1) $G \Vdash a_1 = a_2$ if and only if $a_1 = a_2 : G \longrightarrow X$;

(2) $G \Vdash a \in A$ if and only if the diagram

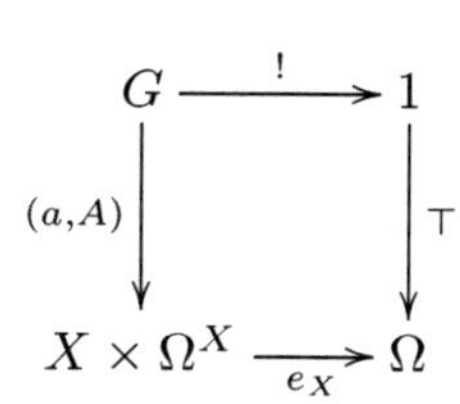

commutes;

(3) $G \Vdash R(a_1, \ldots, a_n)$ if and only if the triangle

$$\begin{array}{ccc} G & \xrightarrow{(a_1,\ldots,a_n)} & X_1 \times \cdots \times X_n \\ & \searrow & \uparrow \\ & & M(R(x_1,\ldots,x_n)) \end{array}$$

commutes;

(4) $G \Vdash \varphi(\vec{a}) \wedge \psi(\vec{b})$ if and only if $G \Vdash \varphi(\vec{a})$ and $G \Vdash \psi(\vec{b})$;

(5) $G \Vdash \varphi(a_1, \ldots, a_n) \vee \psi(b_1, \ldots, b_m)$ if and only if there are morphisms $t_1 : G_1 \longrightarrow G$ and $t_2 : G_2 \longrightarrow G$ in $\mathcal{G}$ such that $t_1 + t_2 : G_1 + G_2 \twoheadrightarrow G$ is an epimorphism and $G_1 \Vdash \varphi(a_1 t_1, \ldots, a_n t_1)$ and $G_2 \Vdash \psi(b_1 t_2, \ldots, b_m t_2)$;

(6) $G \Vdash \varphi(a_1, \ldots, a_n) \supset \psi(b_1, \ldots, b_m)$ if and only if for all $t : G' \longrightarrow G$ in $\mathcal{G}$, if $G' \Vdash \varphi(a_1 t, \ldots, a_n t)$, then $G' \Vdash \psi(b_1 t, \ldots, b_m t)$;

(7) $G \Vdash \neg\varphi(\vec{a})$ if and only if for all $t : G' \longrightarrow G$ in $\mathcal{G}$, if $G' \Vdash \varphi(a_1 t, \ldots, a_n t)$, then $G' \simeq 0$;

(8) $G \Vdash \forall z \varphi(z, a_1, \ldots, a_n)$ if and only if for all $t : G' \longrightarrow G$ in $\mathcal{G}$ and $b : G' \longrightarrow Z$, $G' \Vdash \varphi(b, a_1 t, \ldots, a_n t)$;

(9) $G \Vdash \exists z \varphi(z, a_1, \ldots, a_n)$ there is $\left\{G_i \xrightarrow{t_i} G : i \in I\right\}$ in $\mathcal{G}$ and there is $\left\{G_i \xrightarrow{b_i} Z : i \in I\right\}$ such that

- $(\forall i \in I)(G_i \Vdash \varphi(b_i, a_1 t_i, \ldots, a_n t_i)$
- $\forall X$ in $\mathcal{E} \forall h, k \in Hom_{\mathcal{E}}(G, X)(h \neq k \supset \exists i \in I(ht_i \neq kt_i))$.

The full subcategory $\mathcal{G}$ comes from the case of a Grothendieck topos generated by a site, whose image in the topos is $\mathcal{G}$. One can also consider special cases of interest. First, for an arbitrary topos $\mathcal{E}$, it is always possible to let $\mathcal{G} = \mathcal{E}$. Two other cases worth mentioning are: 1) $\mathcal{G} = Sub(1)$, which corresponds to toposes of sheaves over a topological space and yields a particularly simple formulation of the semantics and 2) the case when $\mathcal{G} = 1$, that is when the terminal object is a generator. In the latter case, the topos $\mathcal{E}$ is Boolean and it can be shown that the semantics "is" the classical set-theoretical semantics (this is not quite precise, but can be clarified easily).

One natural question to ask is how the internal semantics of a topos $\mathcal{E}$ relates to the Kripke-Joyal semantics. In fact, they coincide in the following sense:

Theorem: Let $\mathcal{L}'$ be the language build from $\mathcal{L}$ by adding to the latter new sorts, one for each G of $\mathcal{G}$ and new functional symbols, one for each $G \xrightarrow{a} M(S)$, where S is a sort of $\mathcal{L}$, and let M' be the obvious extension of M to $\mathcal{L}'$. Then:

(i) $G \Vdash \varphi(a_1, \ldots, a_n)$ if and only if $M' \models \top \vdash_{\vec{a}} \varphi(a_1(g), \ldots, a_n(g))$

(ii) If φ is a sentence of $\mathcal{L}$, then $M \models \top \vdash \varphi$ if and only if $\forall G$ in $\mathcal{G}(G \Vdash \varphi)$.

7 FIBRED CATEGORIES AND LOGIC

Although this development found its full expression in the late eighties and early nineties, it has to be mentioned since it has become an important part of categorical logic and it has its roots in work done in the late sixties and mid-seventies. We will concentrate on the roots themselves and only point to certain presentations of the contemporary literature.

As we have mentioned, Lawvere and Lambek were both in Zurich in 1965-66 and at that time Lawvere was developing his ideas on categorical logic while, as we have seen, Lambek was generalizing results about posets to categories. Lawvere asked Lambek whether he had noticed that Tarski's fixed point theorem for posets generalized to categories and Lambek answered by showing him his manuscript of *A fixpoint theorem for complete categories*, published in 1968. (A result that is now better known by theoretical computer scientists than by category theorists, although the distinction between these two groups is nowadays more and more difficult to make.) Lawvere, as we have also mentioned, returned to the issue of fixed point theorems in 1968 and published his *Diagonal Arguments and Cartesian Closed Categories*, in the proceedings of the meeting of the Batelle Institute in 1969. This paper not only contains the definition of a Cartesian Closed Category (CCC), but it also suggests explicitly that a CCC is an algebraic version of type theory, on the one hand, and, on the other hand, that a theory gives rise to a category with finite products, now called a Cartesian category. There is, furthermore, an indication that Lawvere saw a connection between CCCs and Church's λ-calculus. Although there is no explicit mention of this connection in the paper, there is a clear indication that Lawvere did indeed see a connection. It is presented right from the start, in the very definition of a Cartesian Closed category. The key ingredient is in the definition of what Lawvere called the λ-transform of a morphism. Let $\mathcal{C}$ be a CCC and $h : X \longrightarrow Y^A$ be a morphism. Then h is said to be the λ-transform of f if and only if the diagram

$$\begin{array}{ccc} A \times X & & \\ {\scriptstyle 1\times h}\downarrow & \searrow^{f} & \\ A \times Y^A & \xrightarrow[ev]{} & Y \end{array}$$

is commutative. In particular, when $X = 1$, the terminal object, it follows that every $f : A \longrightarrow Y$ gives rise to a unique morphism $f : 1 \longrightarrow Y^A$, called the 'name of f'. Lambek was to use this connection systematically later.

At the same conference, Lambek presented his second paper on the relationships between categories and deductive systems and certainly took note of this connection, since it comes back in his third paper on the topic. Although these papers were not influential at first, they contain ideas that were to become important later.

Already in the mid-fifties, Lambek, together with his colleague Findley, had noticed a similarity between certain results in the theory of bimodules and deductive systems. Lambek and Findley developed a notation for certain concepts and wrote a paper on the subject, entitled *Calculus of Bimodules*, that was never published, since the results contained therein were supposedly subsumed by Cartan and Eilenberg's work that was about to appear in their book on homological algebra. (There is, however, a published trace of their work in a M.Sc. thesis submitted under Maranda's supervision at the Université de Montréal, done by Vaillancourt. See [Vaillancourt, 1968].) Lambek used the same notation in his first investigation of linguistics in a paper published in 1958. (See [Lambek, 1958].) Once he had learned category theory, Lambek realized that they were in fact looking at certain morphisms in specific concrete categories and that the results could be reformulated and developed in a categorical context. Lambek published three papers on deductive systems and categories, the first one in 1968, the second one in 1969 and the third one in 1972. Although the three papers have the same title and seem to be forming a complete whole, they are in fact independent from one another. They are simply on the same theme and constitute various attempts at developing the same ideas, which is formulated at the beginning of the very first paper:

> Roughly speaking, the situation is this: While deductive systems may be viewed as free algebra on the category of pre-ordered sets, they may also be used to construct free algebras on the category of all categories in some universe. The crucial step in the argument is the construction of a category whose objects are terms of a given deductive system and whose maps are equivalence classes of proofs. ([Lambek, 1968a, 287].)

In modern terminology, the main problem could be formulated as follows: how to construct all arrows $A \longrightarrow B$ in the free biclosed monoidal category generated by a graph and to decide when two such arrows are equal. The first problem, namely the construction of the free category of the right sort, was solved by generalizing Gentzen's cut-elimination theorem, first to substructural logic and then to certain structured categories. The second problem, deciding when two arrows are equal, was attacked in full generality in the second paper *Deductive Systems and Categories II. Standard Constructions and Closed Categories.* It also introduces explicitly the notion of a free biclosed monoidal category. However, the paper contains a certain number of mistakes. The satisfactory solution to the second problem was finally presented in a paper published in 1993. (See [Lambek, 1993].)

In his third paper, *Deductive systems and categories III*, presented at Dalhousie in January 1972, Lambek starts with Cartesian closed categories and tries to establish an equivalence with combinatory logic. More specifically, he shows how to go from combinatory logic, defined in the paper as *ontologies*, to Cartesian closed categories, but he was unable to verify that one could go in the opposite direction. This problem was solved later, in fact in the early eighties. It is here that Lambek developed systematically the connection between CCCs and typed λ-calculi. Modifying slightly the foregoing notation (and the formal context somewhat, but we won't give the details), given a morphism $\varphi(x) : 1 \longrightarrow Y$, there exists a unique morphism $f : A \longrightarrow Y$ such that $fx = \varphi(x)$, where $x : 1 \longrightarrow A$. The name of f can in this case be written as $\lambda_{x \in A}\varphi(x)$ in the corresponding λ-calculus. In collaboration with Phil Scott, they showed that the category of typed λ-calculi (with translations as morphisms) and the category of Cartesian closed categories with weak natural number objects (and Cartesian closed functors preserving weak natural numbers objects on the nose) are equivalent. ([Lambek and Scott, 1986, 79, Theorem 11.3].)

However, the main moral of all these papers is already stated in the foregoing quote: it is the correspondence between propositions of a deductive system with the objects of a corresponding category and the (equivalence classes of) proofs of the deductive system and the morphisms of the corresponding category. This correspondence is now summarized by the slogan *propositions-as-objects* and *proofs-as-morphisms*. This fundamental correspondence constitutes the first element of three essential ideas at the heart of developments of categorical logic that took part from the mid to the late eighties. The other two ideas were, in the first place, Lawvere's analysis of the comprehension principle and Bénabou's work on fibred categories.

Later in 1968, at the conference on applications of categorical algebra that took place in New York, Lawvere extended the connections between CCCs and logic in the direction of higher-order logic and type theory. The goal was to provide an analysis of equality and the comprehension schema as adjoint functors to elementary functors, thus including them in the algebraic approach.

Lawvere's basic strategy was to organize the logical data in a more complicated structure, now called an *hyperdoctrine*, given by the following data. First, one starts with a category $\mathcal{T}$ of types assumed to be Cartesian closed. This is, in a sense, the base category with its underlying logic. The morphisms of $\mathcal{T}$ are thought of as terms. Then, the additional structure is given "over" the types, so to speak. For each type X, there is a Cartesian closed category $P(X)$, whose morphisms are "deductions over X", and which is thought of as the "attributes of type X". Furthermore, for every term $f : X \longrightarrow Y$, there is a functor $f^* : P(Y) \longrightarrow P(X)$, called by Lawvere "substitution along f", such that $(gf)^* = f^*g^*$ (strictly speaking the last equality should be an isomorphism, but Lawvere notes that equality holds in all the examples he considers). Finally, the operation of substitution has two adjoints, left and right, which are, respectively, the existential and the universal quantification along f. Again, Lawvere emphasizes the fact that the concept of hyperdoctrine is purely equational.

Lawvere then gives a series of examples, some of which are barely sketched. For instance, he suggests that any multi-sorted higher-order theory, intuitionistic or classical, yields an hyperdoctrine. The objects are the basic types of the theory, closed under the operations of product and exponentiation and the morphisms are equivalence classes of (tuples) of terms from the theory. The equivalence class is not defined nor the construction presented. Given an object X, $P(X)$ consists of all formulas of the theory whose free variables correspond to the type

X. Morphisms of $P(X)$ can be taken to be entailments and in this case $P(X)$ is a preordered set or "one may take suitable "homotopy classes" of deductions in the usual sense." [Lawvere, 1970b, 4] Needless to say, the last claim still has to be clarified. Lawvere himself cautiously writes that "one can write down an inductive definition of the "homotopy" relation, but the author does not understand well what results (some light is shed on this question by the work of Läuchli and Lambek cited above)." [Lawvere, 1970b, 4] Lambek's work referred to here is the work on coherence that was flawed. But Lawvere immediately moves on to add that "although such syntactically presented hyperdoctrines are quite important, it is fortunate for the intuition that there are also semantically-defined examples, as below." [Lawvere, 1970b, 4] And indeed, the examples involving sets are clear, interesting and support the basic intuition.

The simplest example, but perhaps not the most interesting, is the case when $\mathcal{T} = \mathit{Set}$, the category of (small) sets and mappings and, for each set X, $P(X) = 2^X$, the set of subsets of X, or equivalently the set of characteristic functions defined on X. In this case, f^* is simply the inverse image and the existential quantifier is the image. As such, this example is not surprising. More interesting perhaps is the case when $\mathcal{T}$ is the same but $P(X) = \mathit{Set}^X$. An attribute of X is then a family φ_x of sets indexed by the elements $x \in X$. A deduction $\varphi \longrightarrow \psi$ over X is any family of functions $\varphi_x \longrightarrow \psi_x$. In particular, $P(1)$ is the category Set of sets and can be thought of as the category of truth-values (an unusual many-valued logic). Lawvere conceives this example as "a kind of set-theoretical surrogate of proof theory" (Lawvere 1970, 4) and proceeds to speculate that "honest proof theory would presumably also yield a hyperdoctrine with nontrivial $P(X)$, but a syntactically-presented one." [Lawvere, 1970b, 4] It is precisely this insight that was about to become central to the developments in the eighties.

Going up on the ladder of abstraction, the next example consists in taking a category of small categories, which is Cartesian closed, and consider functor categories $P(\mathcal{C}) = \mathit{Set}^{\mathcal{C}}$ for the attributes over the type $\mathcal{C}$. Remember that Lawvere presented this material before the discovery of elementary toposes and he does not make an explicit connection with Grothendieck toposes. Needless to say, the connection has become important afterwards.

From a conceptual point of view, the important point is that hyperdoctrines are related to another categorical concept, although Lawvere did not exploit the connection. Indeed, he explicitly made a reference to fibred categories in the opening statement of his paper.

> The notion of hyperdoctrine was introduced (...) in an initial study of systems of categories connected by specific kinds of adjoints that arise in formal logic, proof theory, sheaf theory, and group-representation theory. .... Since then the author has noticed that yet another "logical operation", namely that which assigns to every formula φ its "extension" $\{x : \varphi(x)\}$ is characterized by adjointness, and that the "same" adjoint in a different hyperdoctrine leads to the notion of fibered category. [Lawvere, 1970b, 1]

What he says there is that the concept of hyperdoctrine leads to the notion of fibred category. Later in the same paper, Lawvere reports that "Gray, by introducing the appropriate notion of 2-dimensional adjointness, has shown that all the features of a hyperdoctrine, including our comprehension scheme, can be obtained by defining a type to be an arbitrary category and an attribute of type **B** to be any fibration over **B**." [Lawvere, 1970b, 3] Thus, from a purely conceptual point of view, one could start with fibrations and use the later framework to develop the connection between type theory and categorical logic. What was lacking at first, perhaps, was a clear motivation and the right tools to do so. As early as 1974, Jean Bénabou pushed the idea that fibrations should be taken as the foundation of category theory itself and laid the groundwork that was to lead, when combined with Lawvere's ideas on hyperdoctrines, the groundwork done in the seventies on first-order and higher-order logics and results by Lambek, Scott, Seely, Hyland, Pitts and others, to the developments of categorical logic in this direction in the eighties. (The reference on the subject is [Jacobs, 1999].)

Bénabou had already given a series of talks on fibrations and logic in the summer of 1974 in Montréal. In the fall of 1975, he presented in Paris two related papers. In the first one, he describes, explicitly in the context of 2-categories, the hierarchy of categorical doctrines, e.g. Cartesian, regular, coherent, etc., the notion of a generic model of a theory and results on finitary objects in a topos. The second paper is about small and locally small fibrations and ends with a proof of a classifying topos for geometric theories. (See [Bénabou, 1975].) Three aspects are striking in Bénabou's papers. First, the papers are thoroughly categorical: all definitions and methods are categorical. For instance, the various types of categorical doctrines are defined not by specifying a formal first-order language, but by giving properties on diagrams, thus in the spirit of Ehresmann and of what are now called sketches. Second, all the results are expressed in the context of 2-categories and 2-functors, and some in bicategories. Bénabou himself had introduced earlier a study of bicategories, an important generalization of category theory that was to come back on the scene in the late eighties and nineties. At the time, even some category theorists were reluctant to "higher-dimensional" categories (although 2-categories are strict and not as such higher-dimensional in the way bicategories are). Third, the use of fibrations for the conceptual clarification of fundamental issues related to categories. We underline these three aspects to emphasize the originality, coherence and rigor of Bénabou's work at the time. We will now summarize some of the results on fibrations contained in the papers.

Bénabou presupposes the definition of a fibration in his papers. First, we introduce a notational but very suggestive convention. In the context of fibred categories, we use the vertical notation $\begin{array}{c}\mathrm{E}\\ \downarrow p\\ \mathrm{B}\end{array}$ to denote a functor $p : \mathrm{E} \longrightarrow \mathrm{B}$. Given an object X in B, the *fibre* $p^{-1}(X)$ over X is the category whose objects are those objects A of E such that $p(A) = X$ and the morphisms are those $f : A \longrightarrow B$ in E such that $p(f) = 1_X$ in B. An object A in E such that $p(A) = X$ is said to be *above* X and, similarly, a morphism f in E such that $p(f) = u$ is said to be above u. The category B is called the *base category* and E the *total category*. This terminology goes back to the early 1940's when the notion of fibration was introduced in homotopy theory. Grothendieck extended it to categories in the early sixties.

Definition: Let $p : \mathrm{E} \longrightarrow \mathrm{B}$ be a functor.

1. A morphism $f : A \longrightarrow B$ in E is said to be *Cartesian over* $u : X \longrightarrow Y$ in B if $p(f) = u$ and if for every $g : C \longrightarrow B$ such that $p(g) = u \circ w$ for some $w : p(C) \longrightarrow X$, then there is a unique $h : C \longrightarrow A$ in E above w such that $f \circ h = g$. This situation is best depicted by the following diagram:

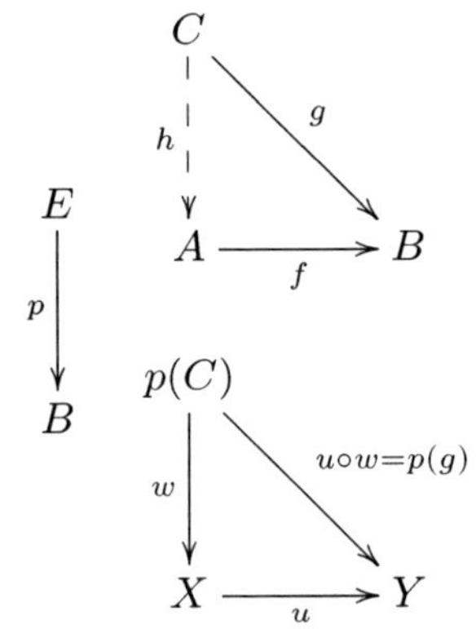

2. The functor $p : \mathrm{E} \longrightarrow \mathrm{B}$ is a *fibration* if for every B in E and $u : X \longrightarrow p(B)$ in B, there is a Cartesian morphism $f : A \longrightarrow B$ in E above u. A fibration is also called a *fibred category*.

Bénabou gave conditions for a fibration to be small or locally small. These are important conceptual clarifications. Up to that day, category theorists, and in particular those working in topos theory, were assuming an ambient universe that allowed one to assume what being small and locally small meant. Thus, a category of sets was more or less assumed for certain foundational purposes, in particular for issues of size. What Bénabou showed, and developed later in a paper published in 1985 is that the concept of fibration was precisely what was needed to give a conceptually sound analysis of these problems. In a fibration $p : \mathrm{E} \longrightarrow \mathrm{B}$, the base category B provides the "universe" for the total category E. Thus, being locally small in E means that the Hom-sets of E can be represented as objects of B. As we have said, the most interesting result with respect to categorical logic announced by Bénabou is the proof of the existence of classifying toposes from the fact that certain fibrations are small. See [Bénabou, 1985].

In the eighties, categorical logicians picked up on the notion of hyperdoctrines and the conceptual framework provided by Bénabou, sometimes in terms of indexed categories, as a semantical framework and developed the syntactical aspects of the work to obtain more specific type theories and completeness theorems for them in this framework. (See [Seely, 1983; Seely, 1984; Seely, 1987; Lamarche, 1991; Moggi, 1991] and, of course, [Jacobs, 1999].)

8 THE DURHAM MEETING

It seems more than appropriate to end with a brief survey of the Durham meeting that took place from July 9 to July 21 in the summer of 1977. It constitutes in our mind the endpoint of the first period initiated by Lawvere in 1963. The title of the meeting was *Applications of Sheaves* and the subtitle was *Applications of Sheaf Theory to Logic, Algebra, and Analysis.* It was edited by M.P. Fourman, C.J. Mulvey and D.S. Scott. Among the participants, we find P. Aczel, B. Banaschewski, M. Barr, J.M. Beck, J. Bénabou, M. Bunge, J.C. Cole, M. Coste, M.-F. Coste, R. Diaconescu, E.J. Dubuc, M.P. Fourman, P.J. Freyd, J.W. Gray, R.J. Grayson, J.M.E. Hyland, J. Isbell, P.T. Johnstone, A. Joyal, G.M. Kelly, J.F. Kennison, A. Kock, J. Lambek, R. Lavendhomme, W. Lawvere, M. Makkai, R.B. Mansfield, C.J. Mulvey, G.E. Reyes, D. Schlomiuk, D.S. Scott, R. Seely, G. Takeuti, M. Tierney, D. Van Dalen, H. Volger, G.C. Wraith and J.J. Zangwill. The list of talks is quite impressive but it does not correspond exactly to the list of papers published. For instance, G.E. Reyes gave two lectures on models in sheaves, but published a joint paper with Anders Kock on manifolds in formal differential geometry and a paper on Cramer's rule in the Zariski topos. Some of the talks were never transformed into papers. As usual, Bénabou and Joyal gave talks but did not submit papers. Bénabou talked about fibrations and Joyal gave a talk entitled *A topos as a space and a theory.* Makkai gave a talk on the syntactical constructions and basic properties of classifying toposes and Lambek gave a talk entitled *From λ-calculus to free topoi.* Peter Freyd presented results about complete higher-order theories. When one considers the published papers, one element stands: it is dominated by papers that are either about or in the context of sheaves over a topological space X, or equivalently H-sets, for H a complete Heyting algebra. Out of 33 papers (we exclude Gray's paper on the history of sheaf theory), at least 12 are clearly in the framework of a category of sheaves over a topological space X. The other striking feature is, as the title indicates, the search for concrete, precise *applications* of sheaf theory to various fields of mathematics. This second feature probably explains the first one. However, the underlying restriction, that is sheaves over a topological space or, more generally, sheaves over a complete Heyting algebra, certainly did not reflect the extent of the potential applications of the methods available. A more striking example, which we will simply mention, was given by Marie-Françoise Coste and Michel Coste shortly afterwards, using in an essential manner the classifying topos of a geometric theory. (See [Coste and Coste, 1979].) The search for applications continued for some time after and lead to interesting developments: models for synthetic differential geometry, the effective topos and Freyd's models for the independence of the axiom of choice, to mention but the most obvious. (See, for instance, [Kock, 1981; Moerdijk and Reyes, 1991] for synthetic differential geometry, [Hyland, 1982;

Hyland, 1991; van Oosten, 2008] for the effective topos and [Freyd, 1980; Blass and Scedrov, 1989] for Freyd's models.) But by the early eighties, it seems that the mathematical community was shunning away from abstract and general methods.

With hindsight, two elements stand out when we look at the early history of categorical logic. First, a categorical analysis of certain of the key concepts of logic and set theory *is* not only possible – an idea that was at first seen as being absurd since it was thought that logical and set theoretical concepts were too primitive for such an analysis – , but fruitful in as much as they naturally fall under the central notions of the theory, e.g. adjoint functors. What this analysis amounted to was an adequate *algebraic* treatment of these notions. Although the analysis was associated with a distinctive ideological component, strongly political in the late sixties and early seventies, according to which logic and geometry were opposites, the fabric weaving together these facets was nonetheless thoroughly *algebraic* and this was acknowledged by all the parties involved and seen as a potentially powerful gain. Second, the development of categorical logic was not, in contrast for instance with Grothendieck's program in algebraic geometry which was carried on with the Weil conjectures as the final target, fueled by specific problems or conjectures deemed to be important by the community of logicians or mathematicians. A large part of the work done consisted, at first, in finding what one would consider being an adequate categorical "translation" of concepts and results of logic. The surprise was to see that so many notions coming from algebraic geometry and algebraic topology were mathematically equivalent to logical notions when they were formulated adequately in a categorical framework. This in itself opened new avenues of research, new ways of thinking about certain problems and theories.

For those who were not charmed by the beauty and elegance of categorical manners, who were convinced that category theorists had put the cart before the horses or had fallen prey to a complicated and terrifying new gadget whose purpose was far from being clear, the standard objection was simply: can you prove something by these means that cannot be proved by other, meaning more "traditional", means? If what one does in categorical logic is simply the same but in a different, more complicated and ultimately irrelevant guise, why bother learning this general abstract nonsense in the first place? If one does not care about a grand unifying picture of mathematics and its foundations, of unsuspected links between domains resulting in a complex and rich network of abstract structures, then is there any genuine value in the categorical toolbox?

It should be pointed out immediately that these complaints strangely resemble what mathematicians often ask logicians in general. What is original here is that it is a complaint coming from mathematical logicians towards categorical logicians. If logicians can or could reply to mathematicians that they were interested in a specific *kind* of analysis of mathematical notions, namely a *foundational* analysis, categorical logicians had a hard time convincing logicians that their analyses are genuinely foundational. Notice, however, that many logicians, including Tarski for instance, do want to use logic as a tool for mathematics and not specifically as a *foundational* tool. Be that as it may, in our specific case, the culprit is certainly related to the *algebraic* nature of a categorical analysis and algebraic logic has been seen, at least since Frege, as being foundationally misguided since it implicitly subsumes logic as a part of mathematics, when Frege and his successors assumed – not without reasons, it should be added – that the opposite must be true and we are back, once again, to an ideological, or perhaps some would prefer to say *philosophical*, debate. One possible answer to logicians, but that we will develop elsewhere, is that categorical logic *does* indeed offer a genuine foundational analysis of various mathematical concepts, no matter how logicians define the notion of "foundational analysis". One can start with various formal systems, whether propositional, first-order or higher-order, coherent, constructive or classical, develop them systematically, define semantics for these formal systems and prove various completeness or other kinds of theorems. Moreover, as we have already mentioned, categorical logic provides bridges between constructive and classical approaches, geometry and logic, topology and logic, to name but the most obvious links. The fate of categorical logic is presently intimately tied to theoretical computer science, and to a certain extent to the foundations of homotopy theory and its place in mathematics and mathematical physics. No matter how categorical logic develops, no matter what route it travels along, its algebraic roots will always reveal themselves.

ACKNOWLEDGEMENTS

The authors would like to thank Michael Barr, Luc Bélair, André Boileau, Bill Boshuk, Marta Bunge, Bernard Hodgson, André Joyal, Joachim Lambek, Marie La Palme Reyes, Bill Lawvere, Mihály Makkai, Phil Scott and Robert Seely for their help. The first author would like to thank the SSHRC of Canada for its financial support while this research was done. Part of this work was done while the first author was on sabbatical at the Mathematics Research Center at Stanford.

BIBLIOGRAPHY

[Adamek *et al.*, 1990] J. Adamek, H. Herrlich, and G. E. Strecker. *Abstract and concrete categories.* John Wiley & Sons Inc., New York, 1990.

[Antonius, 1975] W. Antonius. *Théories cohérentes et prétopos.* PhD thesis, Université de Montréal, 1975.

[Awodey, 1996] S. Awodey. Structure in mathematics and logic: a categorical perspective. *Philos. Math.*, 4:209–237, 1996. 0031-8019.

[Barr and Beck, 1966] M. Barr and J. Beck. Acyclic models and triples. *Proc. Conf. Categorical Algebra (La Jolla, Calif., 1965)*, pp. 336–343, Springer, New York, 1966.

[Barr *et al.*, 1971] M. Barr, P. A. Grillet, and D. H. Van Osdol. *Exact Categories and Categories of Sheaves.* Springer-Verlag, Berlin, New York,, 1971.

[Bell, 1988] J. L. Bell. *Toposes and Local Set Theories*, Oxford Logic Guides, volume 14. The Clarendon Press Oxford University Press, 1988.

[Bénabou, 1963] J. Bénabou. Catégories avec multiplication. *C. R. Acad. Sci. Paris*, 256:1887–1890, 1963.

[Bénabou, 1968] J. Bénabou. Structures algébriques dans les catégories. *Cahiers Topologie Géom. Différentielle*, 10:1–126, 1968.

[Bénabou, 1975] J. Bénabou. Fibrations petites et localement petites. *C. R. Acad. Sci. Paris Sér. A-B*, 281:A897–A900, 1975.

[Bénabou, 1985] J. Bénabou. Fibered categories and the foundations of naive category theory. *J. Symbolic Logic*, 50:10–37, 1985.

[Birkhoff and Mac Lane, 1967] G. Birkhoff and S. Mac Lane. *Algebra*. Macmillan, London, 1967.

[Blass and Scedrov, 1989] A. Blass and A. Scedrov. Freyd's models for the independence of the axiom of choice. *Memoirs of the American Mathematical Society*, 79:1–134, 1989.

[Boileau and Joyal, 1981] A. Boileau and A. Joyal. La logique des topos. *J. Symbolic Logic*, 46:6–16, 1981.

[Boileau, 1976] A. Boileau. *Types versus topos*. PhD thesis, Montréal, 1976.

[Buchsbaum, 1955] D. A. Buchsbaum. Exact categories and duality. *Trans. Amer. Math. Soc.*, 80:1–34, 1955.

[Bunge, 1974] M. Bunge. Topos theory and Souslin's hypothesis. *Journal of Pure and Applied Algebra*, 4:159–187, 1974.

[Cartan and Eilenberg, 1956] H. P. Cartan and S. Eilenberg. *Homological algebra*. Princeton University Press, Princeton,, 1956.

[Cohen, 1963] P. J. Cohen. The independence of the continuum hypothesis. *Proceedings of the National Academy of Sciences USA*, 50:1143–1148, 1963.

[Cole, 1973] J. C. Cole. Categories of sets and models of set theory. In J. L. Bell, J. C. Cole, G. Priest, and A. Slomson, eds., *The Proceedings of the Bertrand Russell Memorial Conference (Uldum 1971)*, pp. 351–399. Bertrand Russell Memorial Conference, Leeds, 1973.

[Corry, 1996] L. Corry. *Modern Algebra and the Rise of Mathematical Structures*, volume 17 of *Science Networks. Historical Studies*. Birkhäuser, Basel, 1996.

[Coste and Coste, 1979] M.-F. Coste and M. Coste. Topologies for real algebraic geometry. In *Topos Theoretic Methods in Geometry*, volume 30, pp. 37–100. Aarhus University, Aarhus, 1979.

[Coste, 1974] M.-F. Coste. Construction d'un modèle booléen de la théorie des ensembles à partir d'un topos booléen. *C. R. Acad. Sci. Paris Sér. A*, 278:1073–1076, 1974.

[Daigneault, 1969] A. Daigneault. Injective envelopes. *American Mathematical Monthly*, 76:766–774, 1969.

[Daigneault, 1969/1970] A. Daigneault. Lawvere's elementary theories and polyadic and cylindric algebras. *Fundamenta Mathematicae*, 66:307–328, 1969/1970.

[Deligne, 1977] P. Deligne. *Cohomologie étale*. Lecture Notes in Mathematics, Vol. 569. Springer-Verlag, Berlin, 1977.

[Demazure and Grothendieck, 1962/1964a] A. Demazure and M. Grothendieck, eds. *Schémas en groupes. II: Groupes de type multiplicatif, et structure des schémas en groupes généraux*. Séminaire de Géométrie Algébrique du Bois Marie 1962/64 (SGA 3). Dirigé par M. Demazure et A. Grothendieck. Lecture Notes in Mathematics, Vol. 152. Springer-Verlag, Berlin, 1962/1964.

[Demazure and Grothendieck, 1962/1964b] A. Demazure and M. Grothendieck, eds. *Schémas en groupes. III: Structure des schémas en groupes réductifs*. Séminaire de Géométrie Algébrique du Bois Marie 1962/64 (SGA 3). Dirigé par M. Demazure et A. Grothendieck. Lecture Notes in Mathematics, Vol. 153. Springer-Verlag, Berlin, 1962/1964.

[Demazure and Grothendieck, 1970] A. Demazure and M. Grothendieck, eds. *Schémas en groupes. I: Propriétés générales des schémas en groupes*. Séminaire de Géométrie Algébrique du Bois Marie 1962/64 (SGA 3). Dirigé par M. Demazure et A. Grothendieck. Lecture Notes in Mathematics, Vol. 151. Springer-Verlag, Berlin, 1970.

[Dieudonné and Grothendieck, 1960] J. Dieudonné and A. Grothendieck. *Éléments de géométrie algébrique I*, volume 4. Publications Mathématiques de l'IHÉS, Paris, 1960.

[Dieudonné and Grothendieck, 1961a] J. Dieudonné and A. Grothendieck. *Éléments de géométrie algébrique II*, volume 8. Publications Mathématiques de l'IHÉS, Paris, 1961.

[Dieudonné and Grothendieck, 1961b] J. Dieudonné and A. Grothendieck. *Éléments de géométrie algébrique III. Première partie*, volume 11. Publications Mathématiques de l'IHÉS, Paris, 1961.

[Dieudonné and Grothendieck, 1963] J. Dieudonné and A. Grothendieck. *Éléments de géométrie algébrique III. Seconde partie*, volume 17. Publications Mathématiques de l'IHÉS, Paris, 1963.

[Dieudonné and Grothendieck, 1964] J. Dieudonné and A. Grothendieck. *Éléments de géométrie algébrique IV. Première partie*, volume 20. Publications Mathématiques de l'IHÉS, Paris, 1964.

[Dieudonné and Grothendieck, 1965] J. Dieudonné and A. Grothendieck. *Éléments de géométrie algébrique IV. Seconde partie*, volume 24. Publications Mathématiques de l'IHÉS, Paris, 1965.

[Dieudonné and Grothendieck, 1966] J. Dieudonné and A. Grothendieck. *Éléments de géométrie algébrique IV. Troisième partie*, volume 28. Publications Mathématiques de l'IHÉS, Paris, 1966.

[Dieudonné and Grothendieck, 1967] J. Dieudonné and A. Grothendieck. *Éléments de géométrie algébrique IV. Quatrième partie*, volume 32. Publications Mathématiques de l'IHÉS, Paris, 1967.

[Dionne, 1973] J. Dionne. *Des théories élémentaires aux catégories conceptuelles.* PhD thesis, Université de Montréal, 1973.

[Eilenberg and Kelly, 1966] S. Eilenberg and G. M. Kelly. Closed categories. *Proc. Conf. Categorical Algebra (La Jolla, Calif., 1965)*, pp. 421–562. Springer, New York, 1966.

[Eilenberg and Mac Lane, 1942] S. Eilenberg and S. Mac Lane. Natural isomorphisms in group theory. *Proeedings of the National Academy of Sciences USA*, 28:537–543, 1942.

[Eilenberg and Mac Lane, 1945] S. Eilenberg and S. Mac Lane. A general theory of natural equivalences. *Trans. Amer. Math. Soc.*, 58:231–294, 1945.

[Eilenberg and Moore, 1965] S. Eilenberg and J. C. Moore. Adjoint functors and triples. *Illinois J. Math.*, 9:381–398, 1965. 0019-2082.

[Eilenberg and Steenrod, 1952] S. Eilenberg and N. E. Steenrod. *Foundations of algebraic topology.* Princeton mathematical series, 15. Princeton University Press, Princeton, 1952.

[Eisenbud and Harris, 2000] D. Eisenbud and J. Harris. *The Geometry of Schemes*, volume 197 of *Graduate Texts in Mathematics.* Springer, New York, 2000.

[Feferman, 1968] S. Feferman. *Lectures on Proof Theory.* Proceedings of the Summer School in Logic (Leeds 1967). Springer Verlag, New York, 1968.

[Fourman, 1977] M. P. Fourman. The logic of topoi. In J. Barwise, ed., *Handbook of Mathematical Logic*, pp. 1053–1090. Elsevier, Amsterdam, 1977.

[Fourman and Scott, 1979] M. P. Fourman and D. S. Scott. Sheaves and logic. In *Applications of Sheaves (Proc. Res. Sympos. Appl. Sheaf Theory to Logic, Algebra and Anal., Univ. Durham, Durham, 1977)*, volume 753 of *Lecture Notes in Math.*, pp. 302–401. Springer, Berlin, 1979.

[Freyd and Scedrov, 1990] P. J. Freyd and A. Scedrov. *Categories, Allegories*, volume 39. North-Holland Publishing Co., Amsterdam, 1990.

[Freyd, 1965] P. Freyd. The theories of functors and models. *Theory of Models (Proc. 1963 Internat. Sympos. Berkeley)*, pp. 107–120. North-Holland, Amsterdam, 1965.

[Freyd, 1972] P. Freyd. Aspect of topoi. *Bulletin of the Autralian Mathematical Society*, 7:1–76, 1972.

[Freyd, 1980] P. Freyd. The axiom of choice. *Journal of Pure and Applied Algebra*, 19:103–125, 1980.

[Gabriel and Ulmer, 1971] P. Gabriel and F. Ulmer. *Lokal präsentierbare Kategorien.* Springer-Verlag, Berlin, New York,, 1971.

[Gray, 1979] J. W. Gray. Fragments of the history of sheaf theory. In M. P. Fourman, C. J. Mulvey, and D. S. Scott, eds., *Applications of Sheaves*, volume 753 of *Lecture Notes in Mathematics*, pp. 1–79. Springer, Berlin, 1979.

[Grothendieck, 1957] A. Grothendieck. Sur quelques points d'algèbre homologique. *Tôhoku Math. J. (2)*, 9:119–221, 1957.

[Grothendieck, 2003] A. Grothendieck. *Revêtements étales et groupe fondamental (SGA 1).* Documents Mathématiques (Paris) [Mathematical Documents (Paris)], 3. Société Mathématique de France, Paris, 2003.

[Hailperin, 1953] T. Hailperin. Quantification theory and empty individual-domains. *Journal of Symbolic Logic*, 18:197–200, 1953.

[Hakim, 1972] M. Hakim. *Topos Annelés et Schémas Relatifs*, volume 64 of *Ergebnisse des Mathematik und ihrer Grenzgebiete*. Springer Verlag, Berlin, 1972.

[Hartshorne, 1977] R. Hartshorne. *Algebraic Geometry*. Springer, New York, 1977.

[Hatcher, 1968] W. S. Hatcher. *Foundations of Mathematics*. W. B. Saunders Co., Philadelphia, 1968.

[Heller, 1958] A. Heller. Homological algebra in abelian categories. *Annals of Mathematics*, 68:484–525, 1958.

[Hyland, 1982] J. M. E. Hyland. The effective topos. In A.S. Troelstra and D. Van Dalen, eds., *The L.E.J. Brouwer Centerary Symposium*, volume 110 of *Studies in Logic and the Foundations of Mathematics*, pp. 165–216. North Holland, Amsterdam, 1982.

[Hyland, 1991] J. M. E. Hyland. First steps in synthetic domain theory. In A. Carboni, M.C. Pedicchio, and G. Rosolini, eds., *Category Theory (Como 1990)*, volume 1488, pp. 131–156. Springer-Verlag, Berlin, 1991.

[Jacobs, 1999] Bart Jacobs. *Categorical Logic and Type Theory*, volume 141. North-Holland Publishing Co., Amsterdam, 1999.

[Johnstone, 1977] P. T. Johnstone. *Topos Theory*. Academic Press [Harcourt Brace Jovanovich Publishers], London, 1977.

[Johnstone, 1982] P. T. Johnstone. *Stone Spaces*, volume 3 of *Cambridge Studies in Advanced Mathematics*. Cambridge University Press, Cambridge, 1982.

[Johnstone, 2002] P. T. Johnstone. *Sketches of an elephant: a topos theory compendium. Vol. 1*, volume 43. The Clarendon Press Oxford University Press, New York, 2002.

[Joyal and Moerdijk, 1997] A. Joyal and I. Moerdijk. *Algebraic Set Theory*, volume 220 of *London Mathematical Society Lecture Note Series*. Cambridge University Press, Cambridge, 1997.

[Joyal and Reyes, 1978] A. Joyal and G. E. Reyes. Forcing and generic models in categorical logic. Manuscript, 1978.

[Joyal, 1970] A. Joyal. Boolean algebras as functors. *Notices of the American Mathematical Society*, 17, 1970.

[Joyal, 1971a] A. Joyal. Cohomology of spectral spaces. *Notices of the American Mathematical Society*, 18:793, 1971.

[Joyal, 1971b] A. Joyal. Functors which preserve elementary operations. *Notices of the American Mathematical Society*, 18:967, 1971.

[Joyal, 1971c] A. Joyal. Polyadic spaces and elementary theories. *Notices of the American Mathematical Society*, 18:563, 1971.

[Joyal, 1971d] A. Joyal. Spectral spaces and distributive lattices. *Notices of the American Mathematical Society*, 18:393, 1971.

[Joyal, 1971e] A. Joyal. Spectral spaces ii. *Notices of the American Mathematical Society*, 18:618, 1971.

[Joyal, 1981] A. Joyal. A combinatorial theory of formal series. *Advances in Mathematics*, 42:1–82, 1981.

[Kan, 1958] D. M. Kan. Adjoint functors. *Trans. Amer. Math. Soc.*, 87:294–329, 1958. 0002-9947.

[Keane, 1975] O. Keane. Abstract Horn theories. In F. W. Lawvere, C. Maurer, and G. C. Wraith, eds., *Model Theory and Topoi*, volume 445, pp. 15–50. Springer-Verlag, Berlin, 1975.

[Kleene, 1952] S. C. Kleene. *Introduction to Metamathematics*. D. Van Nostrand Co., New York, 1952.

[Kleisli, 1965] H. Kleisli. Every standard construction is induced by a pair of adjoint functors. *Proc. Amer. Math. Soc.*, 16:544–546, 1965. 0002-9939.

[Kock and Mikkelsen, 1974] A. Kock and C. J. Mikkelsen. Topos-theoretic factorizations of non-standard extensions. In A. Hurd and P. Loeb, eds., *Victoria Symposium on Nonstandard Analysis 1972*, volume 369 of *Lecture Notes in Mathematics*, pp. 122–143. Springer, Berlin, 1974.

[Kock and Reyes, 1977] A. Kock and G. E. Reyes. Doctrines in categorical logic. In J. Barwise, ed., *Handbook of Mathematical Logic*, pp. 283–313. North Holland, Amsterdam, 1977.

[Kock and Reyes, 1979] A. Kock and G. E. Reyes. Manifolds in formal differential geometry. In M. P. Fourman, C. J. Mulvey, and D. S. Scott, eds., *Applications of Sheaves*, volume 753 of *Lecture Notes in Mathematics*, pp. 514–533. Springer, Berlin, 1979.

[Kock and Wraith, 1971] A. Kock and G. C. Wraith. *Elementary Toposes*, volume 30 of *Lecture Notes in Mathematics*. Matematisk Institut, Aarhus Universitet, Aarhus, 1971.

[Kock *et al.*, 1975] A. Kock, P. Lecouturier, and C. J. Mikkelsen. Some topos theoretic concepts of finiteness. In F. W. Lawvere, C. Maurer, and G. C. Wraith, eds., *Model Theory and Topoi*, volume 445 of *Lecture Notes in Mathematics*, pp. 209–283. Springer, Berlin, 1975.

[Kock, 1981] A. Kock. *Synthetic Differential Geometry*, volume 51 of *London Mathematical Society Lecture Note Series*. Cambridge University Press, Cambridge, 1981.

[Krömer, 2007] R. Krömer. *Tool and object*, volume 32 of *Science Networks. Historical Studies*. Birkhäuser Verlag, Basel, 2007.

[Lamarche, 1991] F. Lamarche. *Modelling Polymorphism with Categories*. PhD thesis, McGill University, 1991.

[Lambek and Scott, 1986] J. Lambek and P. J. Scott. *Introduction to Higher Order Categorical Logic*, volume 7 of *Cambridge Studies in Advanced Mathematics*. Cambridge University Press, Cambridge, 1986.

[Lambek, 1958] J. Lambek. The mathematics of sentence structure. *Amer. Math. Monthly*, 65:154–170, 1958.

[Lambek, 1968a] J. Lambek. Deductive systems and categories. I. syntactic calculus and residuated categories. *Mathematical Systems Theory*, 2:287–318, 1968.

[Lambek, 1968b] J. Lambek. A fixpoint theorem for complete categories. *Math. Z.*, 103:151–161, 1968.

[Lambek, 1969] J. Lambek. Deductive systems and categories II. standard constructions and closed categories. In P. J. Hilton, ed., *Category Theory, Homology Theory, and their Applications, I*, volume 86 of *Lecture Notes in Mathematics*, pp. 76–122, Springer-Verlag, Berlin, New York, 1969.

[Lambek, 1972] J. Lambek. Deductive systems and categories III. Cartesian closed categories, intuitionistic propositional calculus, and combinatory logic. In F. W. Lawvere, ed., *Toposes, Algebraic Geometry and Logic*, volume 274 of *Lecture Notes in Mathematics*, pp. 57–82, Springer-Verlag, Berlin, New York, 1972.

[Lambek, 1993] J. Lambek. From categorial grammar to bilinear logic. In *Substructural logics (Tübingen, 1990)*, volume 2 of *Stud. Logic Comput.*, pp. 207–237. Oxford Univ. Press, New York, 1993.

[Landry and Marquis, 2005] E. Landry and J.-P. Marquis. Categories in context: Historical, foundational, and philosophical. *Philos. Math.*, 12:1–43, 2005.

[Lawvere and Schanuel, 1997] F. W. Lawvere and S. H. Schanuel. *Conceptual mathematics*. Cambridge University Press, Cambridge, 1997. Reprint of the 1991 original.

[Lawvere, 1963] F. W. Lawvere. *Functorial Semantics of Algebraic Theories*. Ph.d., Columbia, 1963.

[Lawvere, 1964] F. W. Lawvere. An elementary theory of the category of sets. *Proc. Nat. Acad. Sci. USA*, 52:1506–1511, 1964.

[Lawvere, 1965] F. W. Lawvere. Algebraic theories, algebraic categories, and algebraic functors. *Theory of Models (Proc. 1963 Internat. Sympos. Berkeley)*, pp. 413–418. North-Holland, Amsterdam, 1965.

[Lawvere, 1966] F. W. Lawvere. Functorial semantics of elementary theories. *Journal of Symbolic Logic*, 31:294–295, 1966.

[Lawvere, 1967] F. W. Lawvere. Theories as categories and the completeness theorem. *Journal of Symbolic Logic*, 32:562, 1967.

[Lawvere, 1969a] F. W. Lawvere. Adjointness in foundations. *Dialectica*, 23:281–296, 1969.

[Lawvere, 1969b] F. W. Lawvere. Diagonal arguments and cartesian closed categories. In Peter Hilton, editor, *Category Theory, Homology Theory and their Applications II*, volume 92 of *Lecture Notes in Mathematics*, pp. 134–145. Springer-Verlag, Berlin, 1969.

[Lawvere, 1969c] F. W. Lawvere. Ordinal sums and equational doctrines. *Sem. on Triples and Categorical Homology Theory (ETH, Zürich, 1966/67)*, pp. 141–155, Springer, Berlin, 1969.

[Lawvere, 1970a] F. W. Lawvere. Quantifiers and sheaves. In M. Berger, J. Dieudonné, J. Leray, J.-L. Lions, P. Malliavin, and J. P. Serre, eds., *Actes du congrès international des mathématiciens*, pp. 329–334, Gauthier-Villars, Nice, 1970.

[Lawvere, 1970b] F. W. Lawvere. Equality in hyperdoctrines and comprehension schema as an adjoint functor. *Applications of Categorical Algebra (Proc. Sympos. Pure Math., Vol. XVII, New York, 1968)*, pp. 1–14, American Mathematical Society, Providence, RI, 1970.

[Lawvere, 1972] F. W. Lawvere. Introduction. *Toposes, algebraic geometry and logic (Conf., Dalhousie Univ., Halifax, N.S., 1971)*, pp. 1–12. Lecture Notes in Math., Vol. 274, Springer, Berlin, 1972.

[Lawvere, 1975a] F. W. Lawvere. Introduction. In F. W. Lawvere, C. Maurer, and G. C. Wraith, eds., *Model Theory and Topoi*, volume 445, pp. 3–14. Springer-Verlag, Berlin, 1975.

[Lawvere, 1975b] F. W. Lawvere. *Variable sets etendu and variable structure in topoi.* Department of Mathematics, Univeristy of Chicago, Chicago, Ill., 1975.

[Lawvere, 1976] F. W. Lawvere. Variable quantities and variable structures in topoi. *Algebra, topology, and category theory (a collection of papers in honor of Samuel Eilenberg)*, pp. 101–131, Academic Press, NY, 1976.

[Lawvere, 1986] F. W. Lawvere. Categories of spaces may not be generalized spaces as exemplified by directed graphs. *Rev. Colombiana Mat.*, 20:179–185, 1986.

[Lawvere, 1989] F. W. Lawvere. Qualitative distinctions between some toposes of generalized graphs. In *Categories in computer science and logic (Boulder, CO, 1987)*, volume 92 of *Contemp. Math.*, pp. 261–299. Amer. Math. Soc., Providence, RI, 1989.

[Lawvere, 1994] F. W. Lawvere. Tools for the advancement of objective logic: closed categories and toposes. *The logical foundations of cognition*, 4:43–56, 1994.

[Lawvere, 2000] F. W. Lawvere. Comments on the development of topos theory. In J.-P. Pier, ed., *Development of Mathematics: 1950-2000*, volume II, pp. 715–734. Birkhäuser, Basel, 2000.

[Lawvere, 2003] F. W. Lawvere. Foundations and applications: axiomatization and education. *Bull. Symbolic Logic*, 9(2):213–224, 2003.

[Linton, 1966a] F. E. J. Linton. Autonomous equational categories. *J. Math. Mech.*, 15:637–642, 1966.

[Linton, 1966b] F. E. J. Linton. Some aspects of equational categories. *Proc. Conf. Categorical Algebra (La Jolla, Calif., 1965)*, pp. 84–94, Springer, New York, 1966.

[Mac Lane and Moerdijk, 1994] S. Mac Lane and I. Moerdijk. *Sheaves in Geometry and Logic.* Springer-Verlag, New York, 1994. Corrected reprint of the 1992 edition.

[Mac Lane, 1961] S. Mac Lane. Locally small categories and the foundations of set theory. *Infinitistic Methods (Proc. Sympos. Foundations of Math., Warsaw, 1959)*, pp. 25–43, Pergamon, Oxford, 1961.

[Mac Lane, 1988] S. Mac Lane. Concepts and categories in perspective. In P. Duren, ed., *A Century of Mathematics in America, I*, volume I, pp. 323–365. American Mathematical Society, Providence, 1988.

[Mac Lane, 1998] S. Mac Lane. *Categories for the working mathematician*, volume 5 of *Graduate Texts in Mathematics.* Springer-Verlag, New York, 2 edition, 1998.

[Mac Lane, 2002] S. Mac Lane. Samuel Eilenberg and categories. *Journal of Pure and Applied Algebra*, 168:127–131, 2002.

[Macnamara and Reyes, 1994] J. Macnamara and G. E. Reyes, eds. *The Logical Foundations of Cognition*, volume 4 of *Vancouver Studies in Cognitive Science.* Oxford University Press, New York, 1994.

[Makkai and Paré, 1989] M. Makkai and R. Paré. *Accessible Categories: The Foundations of Categorical Model Theory*, volume 104. American Mathematical Society, Providence, RI, 1989.

[Makkai and Reyes, 1976a] M. Makkai and G. E. Reyes. Model-theoretical methods in the theory of elementary topoi and related categories. II. *Bulletin de l'Académie Polonaise des Sciences. Série des Sciences Mathématiques, Astronomiques et Physiques*, 24:385–392, 1976.

[Makkai and Reyes, 1976b] M. Makkai and G. E. Reyes. Model-theoretical methods in the theory of topoi and related categories I. *Bulletin de l'Académie Polonaise des Sciences. Série des Sciences Mathématiques, Astronomiques et Physiques*, 24:379–385, 1976.

[Makkai and Reyes, 1977] M. Makkai and G. E. Reyes. *First Order Categorical Logic*, volume 611 of *Lecture Notes in Mathematics.* Springer Verlag, New York, 1977.

[Maranda, 1962] J. M. Maranda. Some remarks on limits in categories. *Canad. Math. Bull.*, 5:133–146, 1962.

[Maranda, 1965] J. M. Maranda. Formal categories. *Canad. J. Math.*, 17:758–801, 1965.

[Maranda, 1966] J. M. Maranda. On fundamental constructions and adjoint functors. *Canad. Math. Bull.*, 9:581–591, 1966.

[Marquis, 2006] J.-P. Marquis. What is category theory? In Giandomenico Sica, editor, *What is Category Theory?*, pp. 221–255. Polimetrica, Monza, 2006.

[Mitchell, 1972] W. Mitchell. Boolean topoi and the theory of sets. *J. Pure Appl. Algebra*, 2:261–274, 1972.

[Moerdijk and Reyes, 1991] I. Moerdijk and G. E. Reyes. *Models of Smooth Infinitesimal Analysis*. Springer-Verlag, New York, 1991.

[Moggi, 1991] E. Moggi. A category-theoretic account of program modules. *Mathematical Structures in Computer Science*, 1:103–139, 1991.

[Mostowski, 1951] A. Mostowski. On the rules of proof in the pure functional calculus of the first order. *Journal of Symbolic Logic*, 16:107–111, 1951.

[Osius, 1974a] G. Osius. Categorical set theory: a characterization of the category of sets. *Journal of Pure and Applied Algebra*, 4:79–119, 1974.

[Osius, 1974b] G. Osius. The internal and external aspect of logic and set theory in elementary topoi. *Cahiers de topologie et géométrie différentielle*, 15:157–180, 1974.

[Osius, 1975] G. Osius. Logical and set theoretical tools in elementary topoi. In F. W. Lawvere, C. Maurer, and G. C. Wraith, eds., *Model Theory and Topoi*, volume 445 of *Lecture Notes in Mathematics*, pp. 297–346. Spinger, Berlin, 1975.

[Ouellet, 1974] R. Ouellet. *Axiomatization de la logique interne du premier ordre des topos, version inclusive et multisorte*. PhD thesis, Université de Montréal, 1974.

[Ouellet, 1981] R. Ouellet. Inclusive first-order logic. *Studia Logica*, 40:13–28, 1981.

[Papillon, 1969] V. Papillon. *Quelques Aspects de l'Enveloppe Injective d'une Algèbre de Boole*. PhD thesis, Université de Montréal, 1969.

[Pareigis, 1970] B. Pareigis. *Categories and Functors*. Academic Press, New York, 1970.

[Pitts, 1987] A. M. Pitts. Interpolation and conceptual completeness for pretoposes via category theory. In D. W. Kueker, E. G. K. Lopez-Escobar, and C. H. Smith, editors, *Mathematical Logic and Theoretical Computer Science*, volume 106 of *Lecture Notes in Pure and Applied Mathematics*, pp. 301–327. Dekker, New York, 1987.

[Pitts, 1989] A. M. Pitts. Conceptual completeness for first-order intuitionistic logic: an application of categorical logic. *Annals of Pure and Applied Logic*, 41:33–81, 1989.

[Plotkin, 2000] B. Plotkin. Algebra, categories and databases. In M. Hazewinkel, ed., *Handbook of Algebra*, volume 2, pp. 79–148. North-Holland, Amsterdam, 2000.

[Reyes, 1972] G. E. Reyes. Lw1w is enough: A reduction theorem for some infinitary languages. *Journal of Symbolic Logic*, 37:705–710, 1972.

[Reyes, 1974] G. E. Reyes. From sheaves to logic. In A. Daigneault, ed., *Studies in Algebraic Logic*, volume 9, pp. 143–204. The Mathematical Association of America, 1974.

[Reyes, 1975/1976] G. E. Reyes. Sheaves and concepts: a model-theoretic introduction to Grothendieck topoi. Technical Report 2, Aarhus University, 1975/1976.

[Reyes, 1977] G. E. Reyes. Sheaves and concepts: A model-theoretic interpretation of Grothendieck topoi. *Cahiers de topologie et géométrie différentielle*, XVIII:1–33, 1977.

[Reyes, 1978] G. E. Reyes. Théorie des modèles et faisceaux. *Adv. in Math.*, 30:156–170, 1978.

[Ščedrov, 1984] A. Ščedrov. Forcing and classifying topoi. *Mem. Amer. Math. Soc.*, 48:x+93, 1984.

[Scott, 1979] D. Scott. Identity and existence in intuitionistic logic. In *Applications of sheaves (Proc. Res. Sympos. Appl. Sheaf Theory to Logic, Algebra and Anal., Univ. Durham, Durham, 1977)*, volume 753 of *Lecture Notes in Math.*, pp. 660–696. Springer, Berlin, 1979.

[Seely, 1983] R. A. G. Seely. Hyperdoctrines, natural deduction and the beck condition. *Z. Math. Logik Grundlag. Math.*, 29:505–542, 1983.

[Seely, 1984] R. A. G. Seely. Locally cartesian closed categories and type theory. *Math. Proc. Cambridge Philos. Soc.*, 95:33–48, 1984.

[Seely, 1987] R. A. G. Seely. Categorical semantics for higher order polymorphic lambda calculus. *Journal of Symbolic Logic*, 52:969–989, 1987.

[Shafarevich, 1974] I. R. Shafarevich. *Basic Algebraic Geometry*. Springer, Berlin, 1974.

[Takahashi, 1965] S. Takahashi. A characterization of group rings as a special class of hopf algebras. *Canad. Math. Bull.*, 8:465–475, 1965.

[Takahashi, 1969] S. Takahashi. *Analysis in categories*. Queen's University, Kingston, Ont., 1969.

[Takahashi, 1974] S. Takahashi. *Méthodes logiques en géométrie diophantine*, volume 48 of *Séminaire de mathématiques supérieures*. Les Presses de l'Université de Montréal, Montréal, 1974.

[Taylor, 1999] P. Taylor. *Practical foundations of mathematics*, volume 59. Cambridge University Press, Cambridge, 1999.

[Tierney, 1972] M. Tierney. Sheaf theory and the continuum hypothesis. In *Toposes, algebraic geometry and logic (Conf., Dalhousie Univ., Halifax, N.S., 1971)*, pp. 13–42. Lecture Notes in Math., Vol. 274. Springer, Berlin, 1972.

[Tierney, 1976] M. Tierney. Forcing topologies and classifying topoi. In Alex Heller and Myles Tierney, editors, *Algebra, Topology and Category Theory (a collection of papers in honor of Samuel Eilenberg)*, pp. 211–219. Academic Press, New York, 1976.

[Vaillancourt, 1968] P. Vaillancourt. *Catégories et Systèmes Déductifs.* PhD thesis, Université de Montréal, 1968.

[van Oosten, 2008] J. van Oosten. *Realizability: an introduction to its categorical side*, volume 152 of *Studies in Logic and the Foundations of Mathematics.* Elsevier B. V., Amsterdam, 2008.

[Volger, 1975a] H. Volger. Completeness theorems for logical categories. In F. W. Lawvere, C. Maurer, and G. C. Wraith, eds., *Model Theory and Topoi*, volume 445, pp. 51–86. Springer-Verlag, Berlin, 1975.

[Volger, 1975b] H. Volger. Logical categories, semantical categories and topoi. In F. W. Lawvere, C. Maurer, and G. C. Wraith, eds., *Model Theory and Topoi*, volume 445, pp. 87–100. Springer-Verlag, Berlin, 1975.

[Wraith, 1975] G. C. Wraith. Lectures on elementary topoi. In F. W. Lawvere, C. Maurer, and G. C. Wraith, eds., *Model Theory and Topoi*, volume 445 of *Lecture Notes in Mathematics*, pp. 114–206. Springer, Berlin, 1975.

[Yanofsky, 2003] N. Yanofsky. A universal approach to self-referential paradoxes, incompleteness and fixed points. *Bulletin of Symbolic Logic*, 9:362–386, 2003.

RUSSELL'S ORDERS IN KRIPKE'S THEORY OF TRUTH AND COMPUTATIONAL TYPE THEORY

Fairouz Kamareddine, Twan Laan, and Robert Constable

ABSTRACT

In Russell's Ramified Theory of Types RTT as presented in *Principia Mathematica* by Whitehead and Russell [1910, 1927], two hierarchical concepts dominate: *orders* and *types*. The class of propositions over types is divided into different orders where a propositional function can only depend on objects of lower orders. The use of orders renders the logic part of RTT *predicative*. Ramsey [1926] and Hilbert and Ackermann [1928] considered the orders to be too restrictive and therefore removed them. This led to the development of Church's Simple Type Theory STT [1940] which uses types without orders. Since, numerous type systems abandoned the hierarchy of orders. For example, all the eight influential Pure Type Systems (PTSs) of the Barendregt cube [1992] avoid orders. Despite this lack of explicit use of orders in some modern type systems, orders still play an influential role in understanding hierarchy in modern type systems. In this chapter, we reflect on the use of orders in providing an adequate foundation for basic concepts in computer science and computational mathematics as expressed in Computational Type Theory CTT [Constable *et al.*, 1986; Allen *et al.*, 2006; Kamareddine and Laan, 2001] and in explaining the truth levels in Kripke's Theory of Truth KTT [Kripke, 1975; Kamareddine and Laan, 2001].

1 INTRODUCTION

The role of type theory in logic and mathematics has always been a restrictive one. In computer science it is used both to protect against errors and to code knowledge. The need for restrictions was realised at the beginning of the twentieth century, when Bertrand Russell showed that Frege's formalisation of logic was inconsistent.[1] Russell considered self-application to be the cause of the contradictions in Frege's *Grundgesetze der Arithmetik* [1892], and hence excluded all possibilities of self-application in his Ramified Theory of Types RTT [Russell, 1908; Whitehead and

[1]An English translation of Russell's letter to Frege in which this inconsistency is described can be found in [Heijenoort, 1967].

Handbook of the History of Logic: Sets and Extensions in the Twentieth Century.
Volume editor: Akihiro Kanamori. General editors: Dov M. Gabbay, Paul Thagard and John Woods.

Russell, 1910, 1927].[2] This was done through RTT's double hierarchy: one of types which can be seen as an elementary version of Church's well-known Simple Theory of Types STT [1940], and one of orders which can be compared with the truth hierarchy in Kripke's Theory of Truth KTT as we will see in this chapter (see also [Kamareddine and Laan, 1996]). Logic based on the double hierarchy of orders and types is usually called *predicative*. In 1925, Ramsey [1926] showed that, by making a distinction between language and meta-language, the orders can be removed from the system without re-introducing any known paradox. This led to many modern type systems being impredicative and to the pervasive belief that orders are too restrictive. For instance, the systems of the Barendregt cube [1992] that have the rule "$(\square, *)$" are all impredicative. Hence, a proof checker like Coq [The Coq Development Team, 2004], based on the Calculus of Constructions [Coquand and Huet, 1988], is itself founded on impredicative logic.

The difference between predicative and impredicative logic leads to fundamental differences in the development of mathematics in these logics. For instance, when constructing the real numbers out of the rationals (with Dedekind-cuts), the Theorem of the Least Upper Bound,[3] is not provable in predicative logic (see [Weyl, 1918]). This theorem is, however, a fundamental tool in real analysis as is illustrated in the work of Feferman (see for instance his [1964]). Nevertheless, mathematics with predicative logic is possible, and from a constructive point of view it is even attractive and a simpler basis for some fundamental notions in computer science. For instance, the proof assistant Nuprl [Constable *et al.*, 1986; Allen *et al.*, 2006] for CTT, a predicative type theory with orders, has been used to develop parts of constructive real analysis including the intermediate value theorem and other basic results [Chirimar and Howe, 1991]. It is likely that all of Bishop's book [1967] could be developed in this theory as well as in ITT, Martin-Löf's Intuitionistic Type Theory [1975; 1982; 1984].

Although the role of orders has been well understood in the proof theory community, the hierarchy of orders is less known than the hierarchy of types, as it became unpopular when Ramsey [1926] and Hilbert and Ackermann [1928] showed that one can avoid the paradoxes without this hierarchy. Despite this, orders help explain some hierarchies in modern systems. As a matter of fact, although Church followed Ramsey's [1926] simplification of RTT into STT, he still attempted to explain orders in his book [1941] and as late as in his [1976]. In this chapter we illustrate the use of orders to explain Kripke's hierarchy of truths and to reason about classifications of functions and objects in Nuprl, a modern system of computer mathematics.

Kripke's Hierarchy of Truths. As paradoxical sentences in natural language play a role similar to that of the paradoxes in logic and mathematics, type theory was used to eliminate the paradoxical sentences from natural language (see for

[2]Further details on the development of type theory can be found in [Kamareddine *et al.*, 2004].

[3]This theorem states that any non-empty set of real numbers with an upper bound has a least upper bound.

instance [Montague, 1973] and [Ranta, 1994]). Paradoxes moreover have been classified in two categories (see [Ramsey, 1926]): the logical and the semantical. The famous Russell's Paradox is logical whereas the famous Liar's Paradox is semantical. The semantical paradoxes usually involve the truth predicate T which gives the truth value of a proposition. Tarski [1936] shows that truth is undefinable and that having the truth predicate inside the language leads to contradictions. For this reason, he distinguishes between (object-) language and meta-language and allows the truth predicate only at the meta-level. Now, to talk about the truth of sentences in the meta-language, one needs a meta-meta-language and so on. Kripke [1975] considers Russell's Ramified Theory of Types and the theory of truth of Tarski to be too restrictive for a proper formalisation of natural language and presents a type-free theory where the truth predicate belongs to the language, in which nevertheless the known paradoxes do not occur. Kripke proposed to follow a certain hierarchy as with Russell but to take the fixed point of his hierarchy of languages to reach a language which has its own truth predicate with three values. We show in this chapter that this hierarchy can be elegantly explained via the notion of orders. We carefully compare Russell's RTT and Kripke's KTT. As Russell's system is said to be more restrictive than Kripke's, our comparison is carried out by coding Russell's expressions in Kripke's theory. The stronger restrictions in the Ramified Type Theory can be seen clearly: at several parts in the definition of the embedding the reader will notice that some type-theoretic properties of Russell's expressions are mentioned, but not used in this definition. We show that the embedding is conservative, i.e. that truth in Russell's theory and in Kripke's theory are the same, as far as formulas expressible in Russell's (more restrictive) system are concerned. In particular, we show that:

- At level n of KTT the truth or falsehood of all order-n propositions of RTT can be established.
- There are order-n propositions of RTT that get a truth value at an earlier stage in KTT.
- RTT is more restrictive than KTT: as some type restrictions are not needed in KTT, more formulas can be expressed in the latter.
- KTT can be regarded as the dual of STT where types have been removed and orders are maintained.

Computational Type Theory was created and implemented in 1984. The initial version is described in the 1986 book *Implementing Mathematics with the Nuprl Proof Development System* [Constable *et al.*, 1986] where the authors describe it on page 9 as "an expressive foundational theory for mathematics and computer science" which "unifies many diverse ideas from modern computing." Many of its fundamental concepts were also elaborated in a series of PhD theses and articles including [Allen, 1987; Howe, 1987; Howe, 1989; Howe, 1998; Harper, 1992; Mendler, 1987; Basin and Howe, 1991; Jackson, 1994; Jackson, 1995;

Murthy, 1991; Kopylov, 2003; Kreitz, 2003; Allen *et al.*, 2006]. CTT was also implemented in MetaPRL [Hickey *et al.*, 2003]. After 1995 Nuprl became a distributed system [Allen *et al.*, 2006; Kreitz, 2003] organized around a database architecture; in the Nuprl community this version is called Nuprl-5. Since 1984 considerable experience has been gained in using Nuprl and MetaPRL to formalize parts of mathematics, e.g. algebra [Jackson, 1995] and computing theory [Constable and Crary, 2001], to solve open problems [Howe, 1987; Murthy, 1991] in computer science and mathematics, and to apply the method of software design and implementation called *correct-by-construction* programming that involves extracting programs from constructive proofs of design specifications. Nuprl was also used to *verify systems* (among many examples are [Howe, 1996; Bickford, 2010]) and *optimize systems* [Liu *et al.*, 1999; Kreitz, 2004], for providing formal semantics for modern programming languages (including modules, dependent record, and objects) and for natural language, automating many tasks needed to verify and explain protocols, algorithms, and systems, and creating courseware that is grounded in fully formalized and computer checked explanations of key computing concepts (see booklets by Stuart Allen in the Math Library at www.nuprl.org). This experience sheds light on the viability of orders in a wide variety of useful work. In particular, we show here that:

- The PTS version of CTT, explicitly typed, shows the correctness of orders.
- RTT orders correspond to CTT orders which are *universes*.
- CTT admits an internal definition of proof valid for all orders.
- CTT is implicitly typed, and the power of polymorphism compensates for predicativity in defining basic notions from computing theory, including the class of all general recursive functions over any type.

Summary of the Chapter. In Section 2 we present a modern formalization of RTT and logical truth for RTT in Tarski's style. In Section 3 we briefly describe Kripke's Theory of Truth KTT. In Section 4 we describe RTT within KTT and investigate the similarity between the hierarchy of orders in RTT and the hierarchy of truths in KTT. We show that RTT is more restrictive with respect to self-reference than KTT and reflect on the use of orders versus types. In Section 5 we review a formal description of a part of the type system of Nuprl as a Pure Type System (PTS) [Terlouw, 1989]. The systems of the Barendregt cube are examples of PTSs. We also review how Nuprl in PTS style enables one to formalize the concept of order in Nuprl and to show its correctness. This order classifies types and terms of Nuprl into their relevant hierarchy of universes $\mathbb{U}_i$. In Section 6 we introduce CTT, an implicitly typed extensional type theory with orders whose logic is based on the propositions-as-types principle. This theory is implemented in the Nuprl proof assistant which has been in continuous use since 1984, and its wide variety of applications demonstrates the viability of CTT in logic, mathematics, and informatics. The applications have led to the addition of several new types and concepts to support computing theory and practice, including the definition of an internal proof type discussed here. These results and applications show the freedom and power allowed in a predicative theory with orders. We conclude in Section 7. Some parts of this chapter appear in greater detail in [Kamareddine *et al.*, 2004].

2 THE RAMIFIED THEORY OF TYPES RTT

The basic aim of RTT is to exclude the logical paradoxes from logic by eliminating all self-references. An extended philosophical motivation for RTT can be found in *Principia Mathematica* [Whitehead and Russell, 1910, 1927], pages 38–55. In this section we review a formal description of RTT that is both faithful to Russell's original informal presentation and compatible with the present formulations of type theories (see [Laan and Nederpelt, 1996; Kamareddine and Laan, 1996]).

In Subsection 2a we introduce propositional functions, the logical formulas of the "naïve" system of logic. In Subsection 2b we present a rule to assign a type to some of these propositional functions. The propositional functions that lead to the logical paradoxes are, of course, not typeable. In Subsection 2c substitution for RTT is discussed. This part is rather technical, but we need it in the proof of Lemma 5, which is essential in the proof of one of our fundamental results (Theorem 4.9). Lemma 5 helps us show that KTT can be regarded as a system based on RTT of which the types and not the orders have been removed.

2a Propositional functions

In this section we shall describe the set of propositions and propositional functions which Whitehead and Russell use in *Principia*. We give a modernised, formal definition which corresponds to the description in *Principia*.

At the basis of the system of our formalization there is

- an infinite set $\mathcal{A}$ of *individual-symbols*;
- an infinite set $\mathcal{V}$ of *variables*;
- an infinite set $\mathcal{R}$ of *relation-symbols* together with a map $\mathtt{a} : \mathcal{R} \to \mathbb{N}^+$ (indicating the *arity* of each relation-symbol).

0-ary relations are not explicitly used in *Principia* but could be added without problems. Since functions are relations in *Principia*, we will not introduce a special set of function symbols.

We assume that:

- $\{\mathtt{a}_1, \mathtt{a}_2, \ldots\} \subseteq \mathcal{A}$;
- $\{\mathtt{x}, \mathtt{x}_1, \mathtt{x}_2, \ldots, \mathtt{y}, \mathtt{y}_1, \ldots, \mathtt{z}, \mathtt{z}_1, \ldots\} \subseteq \mathcal{V}$;
- $\{\mathtt{R}, \mathtt{R}_1, \ldots, \mathtt{S}, \mathtt{S}_1, \ldots\} \subseteq \mathcal{R}$.

We will use the letters $x, y, z, x_1, \ldots$ as meta-variables over $\mathcal{V}$, and $R, R_1, \ldots$ as meta-variables over $\mathcal{R}$. Note that variables are written in `typewriter` style and that meta-variables are written in *italics*: $\mathtt{x}$ denotes one, *fixed* object in $\mathcal{V}$ whilst x denotes an *arbitrary* object of $\mathcal{V}$. We assume that there is an *order* (e.g. alphabetical) on the collection $\mathcal{V}$, and write $x < y$ if the variable x is ordered before the variable y. In particular, we assume that

$$\mathtt{x} < \mathtt{x}_1 < \ldots < \mathtt{y} < \mathtt{y}_1 < \ldots < \mathtt{z} < \mathtt{z}_1 < \ldots$$

We also have the logical symbols $\vee$, $\neg$ and $\forall$ in our alphabet, and the non-logical symbols: parentheses and the comma.

Definition 2.1. (Propositional Functions) We define a collection $\mathcal{F}$ of *propositional functions*, and for each element f of $\mathcal{F}$ we simultaneously define the collection $\text{FV}(f)$ of *free variables* of f:

1. If $R \in \mathcal{R}$ and $i_1, \ldots, i_{\mathsf{a}(R)} \in \mathcal{A} \cup \mathcal{V}$ then $R(i_1, \ldots, i_{\mathsf{a}(R)}) \in \mathcal{F}$.
   $\text{FV}(R(i_1, \ldots, i_{\mathsf{a}(R)})) \stackrel{\text{def}}{=} \{i_1, \ldots, i_{\mathsf{a}(R)}\} \cap \mathcal{V}$;

2. If $z \in \mathcal{V}$, $n \in \mathbb{N}$ and $k_1, \ldots, k_n \in \mathcal{A} \cup \mathcal{V} \cup \mathcal{F}$, then $z(k_1, \ldots, k_n) \in \mathcal{F}$.
   $\text{FV}(z(k_1, \ldots, k_n)) \stackrel{\text{def}}{=} \{z, k_1, \ldots, k_n\} \cap \mathcal{V}$.
   If $n = 0$, we write $z()$ so as to distinguish the propositional function $z()$ from the variable z;[4]

3. If $f, g \in \mathcal{F}$ then $f \vee g \in \mathcal{F}$ and $\neg f \in \mathcal{F}$. $\text{FV}(f \vee g) \stackrel{\text{def}}{=} \text{FV}(f) \cup \text{FV}(g)$; $\text{FV}(\neg f) \stackrel{\text{def}}{=} \text{FV}(f)$;

4. If $f \in \mathcal{F}$ and $x \in \text{FV}(f)$ then $\forall x[f] \in \mathcal{F}$. $\text{FV}(\forall x[f]) = \text{FV}(f) \setminus \{x\}$.

5. All propositional functions can be constructed by using the rules 1, 2, 3 and 4 above.

We use the letters f, g, h as meta-variables over $\mathcal{F}$.

Convention 2.2. (Variable Convention) We make the usual convention that a variable x in a propositional function f that is bound by the quantifier $\forall$ does not occur as a free variable in f. Moreover, different bound variables in f have different names.

A propositional function f must be seen as a proposition in which some parts (the free variables) have been left undetermined. It will turn into a proposition as soon as we assign values to all the free variables occurring in it. In this light, a *proposition* can be seen as a degenerate propositional function (with 0 free variables). It will be clear now what the intuition behind propositional function of the form $R(i_1, \ldots, i_{\mathsf{a}(R)})$, $f \vee g$, $\neg f$ and $\forall x[f]$ is. The intuition behind propositional

[4]It is important to note that a variable is not a propositional function. See for instance [Russell, 1903], Chapter VIII: "The variable", p. 94 of the 7th impression.

functions of the second kind is not so obvious. $z(k_1, \ldots, k_n)$ is a propositional function of *higher-order: z is a variable for a propositional function with n free variables; the argument list* $k_1, \ldots, k_n$ *indicates what should be substituted*[5] *for these free variables as soon as one assigns such a propositional function to z.*

Notice that there are propositional functions of the form $z(k_1, \ldots, k_n)$ (where $z \in \mathcal{V}$) but that expressions of the form $f(k_1, \ldots, k_n)$, where $f \in \mathcal{F}$, are not propositional functions. Even substituting f for z in $z(k_1, \ldots, k_n)$ does not lead to $f(k_1, \ldots, k_n)$, as the notion of substitution in RTT will appear to be quite different from the usual notion of substitution in first-order logic (see Subsection 2c for more details).

Example 2.3. Here are some higher-order propositional functions from ordinary mathematics.

- The propositional functions $\mathtt{z(x)}$ and $\mathtt{z(y)}$ in the definition of Leibniz equality:

$$\forall \mathtt{z}[\mathtt{z(x)} \leftrightarrow \mathtt{z(y)}]$$

- The propositional functions $\mathtt{z(0)}$, $\mathtt{z(x)}$ and $\mathtt{z(y)}$ in the formulation of complete induction:

$$[\mathtt{z(0)} \to (\forall \mathtt{x} \forall \mathtt{y}[\mathtt{z(x)} \to (\mathtt{S(x,y)} \to \mathtt{z(y)})])] \to \forall \mathtt{x}[\mathtt{z(x)}]$$

- $\mathtt{z()}$ in the formulation of the law of the excluded middle:

$$\forall \mathtt{z}[\mathtt{z()} \vee \neg \mathtt{z()}]$$

2b Ramified types

Not all propositional functions should be allowed in our language. For instance, the expression $\neg x(x)$ is a perfectly legal element of $\mathcal{F}$, nevertheless, it is the propositional function that makes it possible to derive Russell's Paradox. Therefore, types are introduced.

Definition 2.4. (Ramified Types)

1. $\iota^{\mathtt{0}}$ is a ramified type ($\mathtt{0}$ is called the *order* of this type);
2. If $t_1, \ldots, t_n$ are ramified types of orders $a_1, \ldots, a_n$ respectively, and $a > \max(a_1, \ldots, a_n)$, then $(t_1, \ldots, t_n)^a$ is a ramified type of order a;
3. All ramified types can be constructed using the rules 1 and 2.

[5]In *Principia*, it is not made clear *how* we should carry out such substitutions. We must depend on our intuition and on the way in which substitution is *used* in *Principia*. Nevertheless, it is hard and elaborate to give a proper definition of substitution. We present a short overview of this definition in Subsection 2c; for a motivation of this definition and its relation to β-reduction in the λ-calculus the reader should consult [Kamareddine *et al.*, 2004].

ι^0 represents the type of the individuals, and one can think of $(t_1, \ldots, t_n)^a$ as being the type of the propositional functions with n free variables, say $x_1, \ldots, x_n$, such that if we assign values k_1 of type t_1 to x_1, ..., k_n of type t_n to x_n, then we obtain a proposition. The type $()^a$ stands for the type of propositions of order a.

Russell strictly divides his propositional functions in orders. For instance, both $\forall \mathtt{p}[\mathtt{p}() \vee \neg \mathtt{p}()]$ and $\mathtt{R}(\mathtt{a})$ are propositions, but they are of different level: The earlier one presumes a full collection of propositions, hence (according to Russell) it cannot belong to the same collection of propositions as the propositions $\mathtt{p}$ over which it quantifies (among which $\mathtt{R}(\mathtt{a})$). This made Russell decide to let $\forall \mathtt{p}[\mathtt{p}() \vee \neg \mathtt{p}()]$ belong to a type of a *higher* order (level) than the order of $\mathtt{R}(\mathtt{a})$.

This can already be seen in the definition of ramified types: $(t_1, \ldots, t_n)^a$ can only be a type if a is strictly greater than each of the orders of the t_is.

Definition 2.5. Let $x_1, \ldots, x_n$ be a list of distinct variables, and $t_1, \ldots, t_n$ be a list of ramified types. We call $x_1{:}t_1, \ldots, x_n{:}t_n$ a *context* and call $\{x_1, \ldots, x_n\}$ its *domain.*

We write $\Gamma \vdash f : t$ to express that $f \in \mathcal{F}$ has type t in context Γ, and extend the variable convention to contexts: If x is bound in f, then x does not occur in the domain of Γ.

We use Γ, Δ to range over contexts and $t_1, t_2, \ldots$ to range over types.

We now present a set of typing rules for RTT. These rules are derived from and equivalent to the rules in [Laan, 1997], which are as close as possible to Russell's original ideas. We change our notation for propositional functions slightly: Instead of $\forall x[f]$ we write $\forall x{:}t[f]$, where t is some ramified type.

Definition 2.6. (Typing Rules for RTT)

- If $c \in \mathcal{A}$, then $\Gamma \vdash c : \iota^0$ for any context Γ;
- If $f \in \mathcal{F}$, and $x_1 < \ldots < x_n$ are the free variables of f, and $t_1, \ldots, t_n$ are types such that $x_i{:}t_i \in \Gamma$, then $\Gamma \vdash f : (t_1, \ldots, t_n)^a$ if and only if
  - If $f \equiv R(i_1, \ldots, i_{\mathtt{a}(R)})$ then $t_i = \iota^0$ for all i, and $a = 1$ (if $n > 0$) or $a = 0$ (if $n = 0$);
  - If $f \equiv z(k_1, \ldots, k_m)$ then there are $u_1, \ldots, u_m$ such that $z{:}(u_1, \ldots, u_m)^{a-1} \in \Gamma$, and $\Gamma \vdash k_i{:}u_i$ for all $k_i \in \mathcal{A} \cup \mathcal{F}$, and $k_i{:}u_i \in \Gamma$ for all $k_i \in \mathcal{V}$;
  - If $f \equiv f_1 \vee f_2$ then there are $u_1^{a_1}, u_2^{a_2}$ such that $\Gamma \vdash f_i : u_i^{a_i}$ and $a = \max(a_1, a_2)$;
    if $f \equiv \neg f'$ then $\Gamma \vdash f' : (t_1, \ldots, t_n)^a$.
  - If $f \equiv \forall x{:}t_0[f']$ then there is a j such that $\Gamma, x{:}t_0 \vdash f' : (t_1, \ldots, t_{j-1}, t_0, t_j, \ldots, t_n)^a$.

Example 2.7. $\neg \mathtt{x}(\mathtt{x})$ is not typeable in any context Γ.

Assume that we would have $\Gamma \vdash \neg\mathtt{x}(\mathtt{x}) : t$. Then t must be of the form $(u)^a$, with $\mathtt{x}{:}u \in \Gamma$, as $\neg\mathtt{x}(\mathtt{x})$ has one free variable. This implies $\Gamma \vdash \mathtt{x}(\mathtt{x}) : (u)^a$, hence by unicity of types below, $u \equiv (u')^{a-1}$, with $\mathtt{x} : u' \in \Gamma$. As Γ is a context, we have $u \equiv u'$, hence $u \equiv (u)^{a-1}$, which is impossible.

An important result (whose proof follows directly from the definition of $\Gamma \vdash f : t$) is the following:

Theorem 2.8. (Unicity of Types) If $\Gamma \vdash f : t$ and $\Gamma \vdash f : u$ then $t \equiv u$. □

2c *Substitution in* RTT

Substitution in RTT is not simply a syntactic operation of replacing a variable by an object, as is usual in first-order logic. This can be understood if we read the interpretation of the propositional function $z(k_1, \ldots, k_m)$. Substituting a propositional function f for the variable z should have as a result f, in which $k_1, \ldots, k_m$ are substituted for the free variables in f. So a substitution may result in a new substitution (and we may wonder whether this process will ever terminate). Below, we give a formal definition of substitution in RTT (needed in the proof of the Substitution Lemma 5).

Definition 2.9. Let $f \in \mathcal{F}$, $\Gamma \vdash f : t$, $k_1, \ldots, k_m \in \mathcal{A} \cup \mathcal{V} \cup \mathcal{F}$ and $x_1, \ldots, x_n \in \mathcal{V}$ such that

- If $k_i \in \mathcal{A}$ then $x_i{:}\iota^0 \in \Gamma$;
- If $k_i \in \mathcal{V}$ then there is a t such that both $k_i{:}t \in \Gamma$ and $x_i{:}t \in \Gamma$;
- If $k_i \in \mathcal{F}$ then there is a t such that $\Gamma \vdash k_i{:}t$ and $x_i{:}t \in \Gamma$.

We define $f[x_1, \ldots, x_m{:=}k_1, \ldots, k_m]$, the (simultaneous) substitution of $k_1, \ldots, k_m$ for $x_1, \ldots, x_m$ in f (shorthand $f[x_i{:=}k_i]$ if no confusion arises) by a double induction on the order and structure of f:

- $f \equiv R(i_1, \ldots, i_{\mathsf{a}(R)})$. Define $i'_j \stackrel{\text{def}}{=} \begin{cases} k_\ell & \text{if } i_j \equiv x_\ell \\ i_j & \text{if } i_j \notin \{x_1, \ldots, x_m\} \end{cases}$

  $f[x_i{:=}k_i] \stackrel{\text{def}}{=} R(i'_1, \ldots, i'_{\mathsf{a}(R)})$.
- $f \equiv z(h_1, \ldots, h_n)$. We distinguish two cases:
  1. $z \notin \{x_1, \ldots, x_m\}$. Define $h'_j \stackrel{\text{def}}{=} \begin{cases} k_\ell & \text{if } h_j \equiv x_\ell \\ h_j & \text{if } h_j \notin \{x_1, \ldots, x_m\} \end{cases}$

     $f[x_i{:=}k_i] \stackrel{\text{def}}{=} z(h'_1, \ldots, h'_n)$.
  2. $z \in \{x_1, \ldots, x_m\}$, assume $z \equiv x_p$.
     Define $h'_j \stackrel{\text{def}}{=} \begin{cases} k_\ell & \text{if } h_j \equiv x_\ell \\ h_j & \text{if } h_j \notin \{x_1, \ldots, x_m\} \end{cases}$
     Notice that, as z, x_p and k_p have the same type, k_p is a propositional function with n free variables, say $y_1 < \ldots < y_n$. Now: $f[x_i{:=}k_i] \stackrel{\text{def}}{=}$

$k_p[y_1,\ldots,y_n{:=}h'_1,\ldots,h'_n]$. Note that the object on the right is a correct substitution (with respect to the types of the y_j and the h'_j) and has already been defined, as k_p has the same order as z, which is exactly one less than the order of $z(h_1,\ldots,h_n)$.

- $f \equiv f_1 \vee f_2$. Then $f[x_i{:=}k_i] \stackrel{\text{def}}{=} f_1[x_i{:=}k_i] \vee f_2[x_i{:=}k_i]$.
- $f \equiv \neg f'$. Then $f[x_i{:=}k_i] \stackrel{\text{def}}{=} \neg f'[x_i{:=}k_i]$.
- $f \equiv \forall x{:}t[f']$. Then $f[x_i{:=}k_i] \stackrel{\text{def}}{=} \forall x{:}t[f'[x_i{:=}k_i]]$.
(We assume that $x \notin \{x_1,\ldots,x_m\}$.)

Substitution in RTT is quite different from usual notions of substitution in, for example, first-order logic or λ-calculus. For a good understanding of the rest of this chapter it is essential to see these differences.

There is no definition of substitution in *Principia Mathematica.* The above definition is based on what happens in *Principia* when a substitution seems to take place. The hardest part of the definition is a substitution of the form

$$z(h_1,\ldots,h_n)[x_1,\ldots,x_m{:=}k_1,\ldots,k_m]$$

where z is among the x_i: say, $z \equiv x_p$. We can assume that k_p is a propositional function with n free variables, say, $y_1 < \ldots < y_n$.

According to the definition, we first carry out the substitutions that have nothing to do with z (the definition of the h'_js). This part is similar to a usual first-order substitution. Now we must substitute k_p for z in $z(h'_1,\ldots,h'_n)$. The intuition on the propositional function $z(h'_1,\ldots,h'_n)$, that was explained at the end of Subsection 2a, prescribes that the arguments $h'_1,\ldots,h'_n$ must be substituted for the free variables $y_1,\ldots,y_n$ of k_p, as soon as k_p is substituted for z. This leads to a *new* substitution $k_p[y_1,\ldots,y_n{:=}h'_1,\ldots,h'_n]$. As the order of k_p is lower than the order of $z(h_1,\ldots,h_p)$, we may assume that the final result of this new substitution has already been defined.

To understand the notion better it may be helpful to treat the substitution

$$z(h_1,\ldots,h_n)[x_i{:=}k_i]$$

first as if it was a usual, first-order substitution, and write down $k_p(h'_1,\ldots,h'_n)$ as an informal, intermediate result. Then the substitution of the h'_j for the y_j in k_p can be seen as the contraction of the n β-redexes in $(\lambda y_1 \cdots y_n.k_p)h'_1 \cdots h'_n$. Notice, however, that $k_p(h'_1,\ldots,h'_n)$ is *not* a propositional function (see the explanation in Subsection 2a).

We give some examples of RTT-style substitutions in order to make the reader more familiar with this notion.

Example 2.10.

- $\mathtt{R(x_1,x_2)[x_1{:=}a_1] = R(a_1,x_2)}$. On the first-order level, RTT substitution is the same as in first-order logic.
- $\mathtt{z(R(x),y)[x{:=}a] = z(R(x),y)}$. Note that $\mathtt{x}$ is not a free variable of $\mathtt{z(R(x),y)}$! The substitution does not "continue" in the arguments of $\mathtt{z(R(x),y)}$: $\mathtt{z(R(x),y)[x{:=}a] \neq z(R(x)[x{:=}a],y)}$.
- $\mathtt{z(a)[z{:=}R(x)] = R(x)[x{:=}a] = R(a)}$.
- $\mathtt{z_1(R(x))[z_1{:=}z_2(a)] = z_2(a)[z_2{:=}R(x)] = R(x)[x{:=}a] = R(a)}$.
- $\mathtt{z_1(x_2,R(x_1))[x_2,z_1{:=}a,z_2(y)] = z_2(y)[y,z_2{:=}a,R(x_1)] = R(x_1)[x_1{:=}a] = R(a)}$. The reader might want to make some informal, intermediate steps in this substitution (as explained above): $\mathtt{z_1(x_2,R(x_1))[x_2,z_1{:=}a,z_2(y)]}$ first leads to $\mathtt{(z_2(y))(a,R(x_1))}$ as an intermediate result, and then to $\mathtt{z_2(y)[y,z_2{:=}a,R(x_1)]}$. Similarly, this new substitution first leads to $\mathtt{(R(x_1))(a)[z_2{:=}R(x_1)]}$ and then to $\mathtt{R(x_1)[x_1{:=}a]}$.

We will need the following results about substitutions.

LEMMA 1. *The order of f is greater than or equal to the order of the substitution $f[x_i{:=}k_i]$.* □

LEMMA 2. $\text{FV}(f[x_i{:=}g_i]) = (\text{FV}(f) \setminus \{x_1,\ldots,x_n\}) \cup \{g_i | g_i \in \mathcal{V}$ *and* $x_i \in \text{FV}(f)\}$. □

2d Logical truth for RTT in Tarski's style

With substitution properly defined, we can give a definition of logical truth in Tarski's style for RTT:

Definition 2.11. (Logical Truth for RTT) Let $f \in \mathcal{F}$ and assume $\text{FV}(f) = \emptyset$. We define $\text{RTT} \models f$:

- If $(a_1,\ldots,a_m) \in R$ then $\text{RTT} \models R(a_1,\ldots,a_m)$, for all individuals $a_1,\ldots,a_m$.
- If $\text{RTT} \models f_1$ or $\text{RTT} \models f_2$ then $\text{RTT} \models f_1 \vee f_2$.
- If not $\text{RTT} \models f$, then $\text{RTT} \models \neg f$.
- If $f \equiv \forall x{:}t[h]$ and for all g of type t, $\text{RTT} \models h[x{:=}g]$,[6] then $\text{RTT} \models \forall x{:}t[h]$.

Remark 2.12. At first sight, the reader might expect a clause for the case $f \equiv z(k_1,\ldots,k_m)$ in the above definition. However, since $\text{FV}(z(k_1,\ldots,k_m)) \supseteq \{z\}$, we have $\text{FV}(z(k_1,\ldots,k_m)) \neq \emptyset$. Propositional functions of the form $z(k_1,\ldots,k_m)$ only occur in the above definition in a form in which the variable z has been bound

[6] $\text{FV}(h[x{:=}g]) = \emptyset$ by Lemma 2.

by a quantifier. As was noted earlier (in Subsection 2a), expressions of the form $f(k_1, \ldots, k_n)$, where f is a propositional function, do not exist in RTT.

Remark 2.13. This definition of logical truth is quite informal. For example, the first clause "If $(a_1, \ldots, a_m) \in R$ then RTT $\models R(a_1, \ldots, a_m)$" requires the symbol R to be already fully interpreted and to denote a relation independently of any Tarskian assignment function. This is faithful to Russell, for whom the Tarskian notion of an uninterpreted formal language was quite alien.

3 KRIPKE'S THEORY OF TRUTH KTT

In this section, we describe Kripke's Theory of Truth KTT ([Kripke, 1975]). Kripke expresses higher-order formulas within a first-order language, using the fact that many interesting languages are rich enough to reflect their own syntax (for instance, via a Gödel numbering).

Let us assume a first-order language L, with variables ranging over a domain D, and primitive predicates interpreted by (totally defined) relations on D. Let us also assume two subsets S_1 and S_2 of D such that $S_1 \cap S_2 = \emptyset$. Kripke extends L to $L(S_1, S_2)$ by adding a monadic predicate $\mathtt{T}$. The main idea is to interpret $\mathtt{T}$ as a "truth predicate". S_1 contains the elements d of D for which $\mathtt{T}(d)$ holds (so it contains the (codes of) formulas which we consider to be "true"); S_2 contains those $d \in D$ for which $\neg\mathtt{T}(d)$ holds (hence it contains the (codes of) formulas which we consider to be "false"). We do not demand that $S_1 \cup S_2 = D$, hence $\mathtt{T}$ is a partial predicate over D.

Definition 3.1. (Logical Truth for KTT) Let L be a first-order language over a domain D with $\mathcal{R}$ as set of primitive predicates. Let f be a sentence in L. We define $L \models f$ as follows:[7]

f	$L \models f$	$L \models \neg f$
$R(d_1, \ldots, d_m)$	$(d_1, \ldots, d_m) \in R$	$(d_1, \ldots, d_m) \notin R$
$g_1 \wedge g_2$	$L \models g_1$ and $L \models g_2$	$L \models (\neg g_1) \vee (\neg g_2)$
$g_1 \vee g_2$	$L \models g_1$ or $L \models g_2$	$L \models (\neg g_1) \wedge (\neg g_2)$
$\forall x[g]$	$L \models g[x{:=}d]$ for all $d \in D$	$L \models \exists x[\neg g]$
$\exists x[g]$	$L \models g[x{:=}d]$ for some $d \in D$	$L \models \forall x[\neg g]$
$\neg\neg g$	$L \models g$	$L \models \neg g$

Here, $R \in \mathcal{R}$; $d, d_1, \ldots, d_m \in D$, and g, g_1, g_2 are formulas of L. Now let $S_1, S_2 \subseteq D$ such that $S_1 \cap S_2 = \emptyset$. KTT $\equiv L(S_1, S_2)$ is the first-order language over D with $\mathcal{R} \cup \{\mathtt{T}\}$ as the set of primitive predicates ($\mathtt{T} \notin \mathcal{R}$). We extend the definition of $L \models f$ to KTT $\models f$ by putting KTT $\models \mathtt{T}(d)$ iff $d \in S_1$ and KTT $\models \neg\mathtt{T}(d)$ iff $d \in S_2$.

[7]Notice that even though this definition is different from Tarski's definition, especially with respect to the definition of $L \models \neg f$, it is easy to prove the equivalence of both definitions. This is because all primitive predicates of L are totally defined. We took this definition however as we need to extend it for the partial predicate $\mathtt{T}$.

It is important (and easy) to notice that the extension of L to $L(S_1, S_2)$ is conservative:

LEMMA 3. *Let L be a first-order language over a domain D, let $S_1, S_2 \subseteq D$ such that $S_1 \cap S_2 = \emptyset$, and assume that f is a sentence in L. Then $L \models f$ if and only if $L(S_1, S_2) \models f$.* □

Now Kripke uses $\mathtt{T}$ as a predicate expressing truth by defining a hierarchy of languages. This hierarchy has much in common with Russell's hierarchy of orders. L was assumed to be able to reflect its own syntax, hence so is $L(S_1, S_2)$, for any S_1, S_2. Notice that the sentences of $L(S_1, S_2)$ do not depend on the sets S_1 and S_2, so we can take one Gödel numbering $\langle\rangle$, being a map from the formulas of $L(S_1, S_2)$ to D. The Kripke hierarchy of languages is defined by presenting a hierarchy of pairs of sets (S_1, S_2):

Definition 3.2. For any ordinal α we define a pair of sets $(S_{\alpha,1}, S_{\alpha,2})$ and a language KTT_α.

- $S_{0,1} \stackrel{\text{def}}{=} \emptyset$; $S_{0,2} \stackrel{\text{def}}{=} \emptyset$; $\text{KTT}_0 \stackrel{\text{def}}{=} L(S_{0,1}, S_{0,2})$.
- If $S_{\alpha,1}$, $S_{\alpha,2}$ and KTT_α have been defined, then we define:

$$\begin{array}{rcl} S_{\alpha+1,1} & \stackrel{\text{def}}{=} & \{\langle f\rangle \mid f \text{ is a sentence and } \text{KTT}_\alpha \models f\} \\ S_{\alpha+1,2} & \stackrel{\text{def}}{=} & \{\langle f\rangle \mid f \text{ is a sentence and } \text{KTT}_\alpha \models \neg f\} \cup \\ & & \{d \in D \mid d \not\equiv \langle f\rangle \text{ for all sentences } f \text{ of } \text{KTT}_\alpha\} \\ \text{KTT}_{\alpha+1} & \stackrel{\text{def}}{=} & L(S_{\alpha+1,1}, S_{\alpha+1,2}) \end{array}$$

- If α is a limit ordinal and $S_{\beta,1}$, $S_{\beta,2}$ and KTT_β have been defined for all $\beta < \alpha$, then

$$\begin{array}{rcl} S_{\alpha,i} & \stackrel{\text{def}}{=} & \bigcup_{\beta<\alpha} S_{\beta,i} \\ \text{KTT}_\alpha & \stackrel{\text{def}}{=} & L(S_{\alpha,1}, S_{\alpha,2}) \end{array}$$

LEMMA 4. **(Conservation of Knowledge)** *If $\alpha < \beta$ then $S_{\alpha,1} \subseteq S_{\beta,1}$ and $S_{\alpha,2} \subseteq S_{\beta,2}$.* □

We can see the construction of the languages KTT_α as a process of obtaining knowledge. At the initial stage, KTT_0, $\mathtt{T}(d)$ is not defined for any $d \in D$. There is no knowledge at all. Applying the definition of truth given for KTT_0, we obtain knowledge: Some sentences f can be judged true ($\text{KTT}_0 \models f$; we store the code of f in $S_{1,1}$), some other sentences g can be judged false ($\text{KTT}_0 \models \neg g$; the code of g is stored in $S_{1,2}$). It is not possible to judge all sentences. For instance, neither $\text{KTT}_0 \models \forall\mathtt{x}[\mathtt{T}(\mathtt{x}) \vee \neg\mathtt{T}(\mathtt{x})]$ nor $\text{KTT}_0 \models \neg\forall\mathtt{x}[\mathtt{T}(\mathtt{x}) \vee \neg\mathtt{T}(\mathtt{x})]$ hold, so $\langle\forall\mathtt{x}[\mathtt{T}(\mathtt{x}) \vee \neg\mathtt{T}(\mathtt{x})]\rangle$ belongs neither to $S_{1,1}$ nor to $S_{1,2}$.

The knowledge we obtained is expressed by the predicate T in KTT_1. In KTT_1 we know more about T than in KTT_0. Hence more sentences can be judged true or false; we store their codes in $S_{2,1}$ and $S_{2,2}$ respectively. The Lemma on Conservation of Knowledge 4 guarantees that this process only extends our knowledge, i.e.:

- Sentences that were judged to be true at level KTT_1 remain true at level KTT_2;
- Sentences that were judged to be false at level KTT_1 remain false at level KTT_2.

By iterating this process we arrive at the levels $\text{KTT}_3, \text{KTT}_4, \ldots, \text{KTT}_\omega, \text{KTT}_{\omega+1}, \ldots$. This limit does terminate however in that it has a fixed point.

4 RTT IN KTT

Both in RTT and in KTT we are confronted with a hierarchy. Russell constructs a hierarchy by dividing propositions and propositional functions into different orders, taking care that a propositional function f can only depend on objects of a lower order than the order of f.

Kripke does not make this distinction beforehand. He has only one truth-predicate (T), but decisions about truth of propositions are split into different levels: At the first level only decisions about propositions that do not involve T are made, at the second level decisions about propositions involving T for codes of first-level propositions are made, and so on.

In Subsection 4a we investigate the similarity between both hierarchies, by describing RTT within KTT. In Subsection 4b we investigate in which way RTT is more restrictive with respect to self-reference than KTT.

4a RTT *embedded in* KTT

To embed RTT in a first-order language L, we have to cope with two technical problems:

- We need to encode the notion of and the manipulation with (higher-order) propositional functions into a first-order language. The manipulation is especially important with respect to substitution, which in the higher-order situation is much more complicated than in the first-order case (cf. the definition of substitution 2.9).
- In Russell's theory, it is possible (and, due to the hierarchy of orders, in fact *only* possible) to quantify over only a subtype of all propositions. This makes it impossible to translate, for instance, the proposition $\forall \mathtt{p}{:}()^1[\mathtt{p}() \vee \neg\mathtt{p}()]$ directly by $\forall \mathtt{x}[\mathtt{T}(\mathtt{x}) \vee \neg\mathtt{T}(\mathtt{x})]$, as the quantifier in the latter also quantifies over (codes of) higher-order propositions.

As we do not want contexts to be involved in this coding, we assume that each variable in $\mathcal{V}$ has (implicitly) a superscript t, indicating its type. This makes it possible to do without contexts, as the types of the variables are now clear from the function in which they occur. For reasons of clarity, we will not write this superscript explicitly, as long as no confusion arises.

We propose the following solutions to the problems sketched above (we first give the definition and afterwards explain the thinking behind it):

Definition 4.1. Let KTT be the language L with domain $D = \mathcal{A}$, extended with for each ramified type t a monadic predicate $\mathtt{Typ}_t$, for each $n \in \mathbb{N}$ a $(n+1)$-ary function $\mathtt{App}_n$, and the monadic predicate $\mathtt{T}$ ($\mathtt{T}$ will play the same role as in Section 3). We code the typeable propositional functions f of $\mathcal{F}$ to formulas $\overline{f}$ in the language KTT. We do this by induction on the structure of f.

- If $f \equiv R(i_1, \ldots, i_{\mathtt{a}(R)})$, then f is present in the original language L and we take $\overline{f} \stackrel{\text{def}}{=} f$.
- If $f \equiv z(k_1, \ldots, k_m)$, write $K_i \equiv \langle\overline{k_i}\rangle$ for $k_i \in \mathcal{F}$, and $K_i \equiv k_i$ for $k_i \in \mathcal{A} \cup \mathcal{V}$. Define $\overline{f} \stackrel{\text{def}}{=} \mathtt{T}(\mathtt{App}_m(z, K_1, \ldots, K_m))$.
- If $f \equiv f_1 \vee f_2$, then $\overline{f} \stackrel{\text{def}}{=} \overline{f_1} \vee \overline{f_2}$.
- If $f \equiv \neg f'$, then $\overline{f} \stackrel{\text{def}}{=} \neg\overline{f'}$.
- If $f \equiv \forall x : u[f']$, then $\overline{f} \stackrel{\text{def}}{=} \forall x[\neg\mathtt{Typ}_u(x) \vee \overline{f'}]$.

We now give a formal interpretation to the newly introduced predicates $\mathtt{Typ}_t$ and $\mathtt{App}_n$.

Definition 4.2. For all ramified types $t \neq \iota^0$, let $\mathrm{Typ}_t \stackrel{\text{def}}{=} \{\langle\overline{f}\rangle \mid f \in \mathcal{F} \text{ and } f : t\}$ and $\mathrm{Typ}_{\iota^0} \stackrel{\text{def}}{=} D$.

Assume: $n \in \mathbb{N}$, $f \in \mathcal{F}$ is of type $(t_1, \ldots, t_n)$ and has free variables $x_1 < \ldots < x_n$. Assume also that for $i = 1, \ldots, n$, $k_i : t_i$ and either $d_i = k_i$ (if $t_i = \iota^0$) or $d_i = \langle\overline{k_i}\rangle$ (if $t_i \neq \iota^0$). We define:

$$\mathrm{App}_n(\langle\overline{f}\rangle, d_1, \ldots, d_n) \stackrel{\text{def}}{=} \langle\overline{f[x_1, \ldots, x_n{:=}k_1, \ldots, k_n]}\rangle.$$

From now on, we will interpret the function symbol $\mathtt{App}_n$ as the function App_n, and the relation symbol $\mathtt{Typ}_t$ as the relation Typ_t.

We make some remarks with respect to these definitions.

Remark 4.3. It is clear that the newly introduced functions App_n are used for carrying out substitutions, thus solving the first of the technical problems stated at the beginning of this subsection. The predicates Typ_t solve the second problem, as can be seen in the definition of $\overline{\forall x[f]}$.

Remark 4.4. Notice that we did not define the functions App_n on the full domain D^{n+1}. We could have done that, but will not need App_n on other elements of D^{n+1} than defined above.

Remark 4.5. At this point, our work is related to (but independent of) Paul Gilmore's work on NaDSet 1. NaDSet 1 is a theory of generalized abstraction which makes n-ary predication a primitive of the system, with the unary truth predicate being trivially definable upon this basis. For a useful connection between KTT and NaDSet 1, see [Feferman, 1984].

Remark 4.6. The extensions suggested above are of a mere technical character. Therefore, we think that we can still speak of an *embedding* of RTT within KTT.

Notation 4.7. To keep notations uniform, we sometimes want to speak about $\langle\overline{x}\rangle$ when we only intend to mention x, for $x \in \mathcal{V}$, and about $\langle\overline{a}\rangle$ when only meaning a, for $a \in \mathcal{A}$. Hence, we formally define: $\langle\overline{x}\rangle \stackrel{\text{def}}{=} x$ and $\langle\overline{a}\rangle \stackrel{\text{def}}{=} a$ for all $x \in \mathcal{V}$ and all $x \in \mathcal{A}$.

Below, we work in two systems: RTT and KTT. These systems have a different notion of substitution, though they use the same notation for expressing substitution. From the context, however, it will always be clear which kind of substitution is meant.

The language KTT above is similar to that presented in Section 3, and we construct KTT_α for each ordinal α as described in that section. We need the following lemma:

LEMMA 5. **(Substitution Lemma)** *Assume g is a propositional function of order m and $g[x{:=}k]$ is a proposition of order n. If $\text{KTT}_n \models \overline{g[x{:=}k]}$ then $\text{KTT}_m \models \overline{g}[x{:=}\langle\overline{k}\rangle]$.*

PROOF: We make the proof a little easier by proving that if: *If g is a propositional function of order m and $g[x_1,\ldots,x_p{:=}k_1,\ldots,k_p]$ is a proposition of order n, then 1 and 2 hold where*

1. $\text{KTT}_n \models \overline{g[x_1,\ldots,x_p{:=}k_1,\ldots,k_p]}$ implies
   $\text{KTT}_m \models \overline{g}[x_1,\ldots,x_p{:=}\langle\overline{k_1}\rangle,\ldots,\langle\overline{k_p}\rangle]$
2. $\text{KTT}_n \models \overline{\neg g[x_1,\ldots,x_p{:=}k_1,\ldots,k_p]}$ implies
   $\text{KTT}_m \models \overline{\neg g}[x_1,\ldots,x_p{:=}\langle\overline{k_1}\rangle,\ldots,\langle\overline{k_p}\rangle]$

We write $g[x_i{:=}k_i]$ as a shorthand for $g[x_1,\ldots,x_p{:=}k_1,\ldots,k_p]$ as long as no confusion arises, and use similar abbreviations for other substitutions. The proof is by induction on the structure of g.

- $g \equiv R(i_1,\ldots,i_{\mathsf{a}(R)})$. Then, by definition of $g[x_i{:=}k_i]$, $\overline{g[x_i{:=}k_i]} \equiv \overline{g}[x_i{:=}\langle\overline{k_i}\rangle]$. As $n \leq m$, the lemma follows by the Lemma on Conservation of Knowledge 4.

- $g \equiv z(h_1, \ldots, h_q)$. If $z \notin \{x_1, \ldots, x_p\}$ then again $\overline{g[x_i{:=}k_i]} \equiv \overline{g}[x_i{:=}\langle\overline{k_i}\rangle]$[8] and again the lemma follows from $n \leq m$ and the Lemma on Conservation of Knowledge 4.

  The interesting case is when $g \equiv z(h_1, \ldots, h_q)$ and $z \in \{x_1, \ldots, x_p\}$. To keep notations clear, we assume $p = 1$ and $z = x_1$. The reader may verify that the case $p > 1$ only complicates notation, not the proof. We only show 1 as 2 is similar. Assume $\text{KTT}_n \models \overline{g[x_i{:=}k_i]}$.

  As k_1 and z have the same type, k_1 has q free variables, say $y_1 < \ldots < y_q$, and by definition of substitution in RTT, $z(h_1, \ldots, h_q)[x_1{:=}k_1] \equiv k_1[y_i{:=}h_i]$. Notice that z and k_1 have the same order $(m-1)$, and that n, the order of $k_1[y_i{:=}h_i]$, is at most the order of k_1 (Lemma 1). This means: $n \leq m-1$. Using Lemma 4: $\text{KTT}_{m-1} \models \overline{k_1[y_i{:=}h_i]}$.

  By the definition of T we have: $\text{KTT}_m \models \mathtt{T}\left(\left\langle\overline{k_1[y_i{:=}h_i]}\right\rangle\right)$. We are now done because:

$$\begin{aligned} \overline{g}[x_1{:=}\langle\overline{k_1}\rangle] &\equiv \overline{z(h_1, \ldots, h_q)}[z{:=}\langle\overline{k_1}\rangle] \\ &\equiv \mathtt{T}(\mathtt{App}_q(z, \langle\overline{h_1}\rangle, \ldots, \langle\overline{h_q}\rangle))[z{:=}\langle\overline{k_1}\rangle] \\ &\equiv \mathtt{T}(\mathtt{App}_q(\langle\overline{k_1}\rangle, \langle\overline{h_1}\rangle, \ldots, \langle\overline{h_q}\rangle)) \\ &\equiv \mathtt{T}\left(\left\langle\overline{k_1[y_i{:=}h_i]}\right\rangle\right) \end{aligned}$$

- $g \equiv g_1 \vee g_2$. First, assume $\text{KTT}_n \models \overline{g[x_i{:=}k_i]}$. As $\overline{g[x_i{:=}k_i]} \equiv \overline{g_1[x_i{:=}k_i]} \vee \overline{g_2[x_i{:=}k_i]}$, there is j such that $\text{KTT}_n \models \overline{g_j[x_i{:=}k_i]}$. By the induction hypothesis, there is a j such that $\text{KTT}_m \models \overline{g_j}[x_i{:=}\langle\overline{k_i}\rangle]$, as the order of g_j is $\leq m$. Hence $\text{KTT}_m \models \overline{g_1}[x_i{:=}\langle\overline{k_i}\rangle] \vee \overline{g_2}[x_i{:=}\langle\overline{k_i}\rangle]$, so we are done.

  Now assume $\text{KTT}_n \models \overline{\neg g[x_i{:=}k_i]}$. Hence: $\text{KTT}_n \models \neg(\overline{g_1[x_i{:=}k_i]} \vee \overline{g_2[x_i{:=}k_i]})$, and $\text{KTT}_n \models \neg\overline{g_j[x_i{:=}k_i]}$ for $j = 1, 2$, and by the induction hypothesis, this means (again the order of the g_js are $\leq m$) $\text{KTT}_m \models \neg\overline{g_j}[x_i{:=}\langle\overline{k_i}\rangle]$ for $j = 1, 2$, hence $\text{KTT}_m \models \neg\overline{g_1}[x_i{:=}\langle\overline{k_i}\rangle] \wedge \neg\overline{g_2}[x_i{:=}\langle\overline{k_i}\rangle]$. So $\text{KTT}_m \models \neg(\overline{g_1}[x_i{:=}\langle\overline{k_i}\rangle] \vee \overline{g_2}[x_i{:=}\langle\overline{k_i}\rangle])$, and $\text{KTT}_m \models (\neg(g_1 \vee g_2))[x_i{:=}\langle\overline{k_i}\rangle]$.

- $g \equiv \neg g'$.

  If $\text{KTT}_n \models \overline{g[x_i{:=}k_i]}$ then use the induction hypothesis for g'.

  If $\text{KTT}_n \models \overline{\neg g[x_i{:=}k_i]}$ then $\text{KTT}_n \models \overline{g'[x_i{:=}k_i]}$, so by induction $\text{KTT}_m \models \overline{g'}[x_i{:=}\langle\overline{k_i}\rangle]$, so $\text{KTT}_m \models \overline{\neg\neg g'}[x_i{:=}\langle\overline{k_i}\rangle]$.

- $g \equiv \forall x{:}t[g']$.

  If $\text{KTT}_n \models \overline{g[x_i{:=}k_i]}$, then for all d such that $\text{Typ}_t(d)$ and $\text{KTT}_n \models \overline{g'[x_i{:=}k_i][x{:=}d]}$, hence for all these d, $\text{KTT}_m \models \overline{g'}[x_i{:=}\langle\overline{k_i}\rangle][x{:=}d]$, so $\text{KTT}_m \models \forall x[\overline{g'}[x_i{:=}\langle\overline{k_i}\rangle]]$, and $\text{KTT}_m \models \overline{g}[x_i{:=}\langle\overline{k_i}\rangle]$.

[8] This is because in this case, no higher-order substitutions occur, and the notion of RTT substitution coincides with ordinary, first-order substitution.

If $\text{KTT}_n \models \overline{\neg g[x_i{:=}k_i]}$ then there is a $d \in D$ such that $\text{Typ}_t(d)$ and $\text{KTT}_n \models \overline{\neg g'[x_i{:=}k_i]}[x{:=}d]$,
hence $\text{KTT}_m \models \overline{\neg g'}[x_i{:=}\langle\overline{k_i}\rangle][x{:=}d]$, and $\text{KTT}_m \models \exists x[\overline{\neg g'}[x_i{:=}\langle\overline{k_i}\rangle]]$.
Hence $\text{KTT}_m \models \exists x[\neg\overline{g'}[x_i{:=}\langle\overline{k_i}\rangle]]$ and $\text{KTT}_m \models \neg\forall x[\overline{g'}[x_i{:=}\langle\overline{k_i}\rangle]]$. □

Remark 4.8. We have actually proven a stronger fact: *Assume g is a propositional function of order m and $g[x{:=}k]$ is a proposition of order n. If $\text{KTT}_n \models \overline{g[x{:=}k]}$ then $\text{KTT}_p \models \overline{g}[x{:=}\langle\overline{k}\rangle]$, where $p = min(m, n+1)$.* This tells us more about the role of the predicate T: Although a substitution may lower the order of a propositional function by more than one, only one application of the predicate T is involved (hence only one level in the hierarchy of truths). However, in the theorem below we only need the (weaker) form in which we presented the Substitution Lemma originally.

Theorem 4.9. Let $f : ()^n \in \mathcal{F}$. Then: $\text{RTT} \models f$ if and only if $\text{KTT}_n \models \overline{f}$.

PROOF:

$\Leftarrow$ Due to the use of $\neg$ in the definition of $\text{KTT}_n \models \overline{f}$, we prove a little bit more:

- If $\text{RTT} \models f$ then $\text{KTT}_n \models \overline{f}$;
- If $\text{RTT} \models \neg f$ then $\text{KTT}_n \models \overline{\neg f}$.

These claims are proved simultaneously by induction on the definition of $\text{RTT} \models f$.

- $f \equiv R(d_1, \ldots, d_{\mathsf{a}(R)})$ for a $R \in \mathcal{R}$ and some $d_1, \ldots, d_{\mathsf{a}(R)} \in D$. Then $\overline{f} \equiv f$. As $\text{RTT} \models f$, we know that $(d_1, \ldots, d_{\mathsf{a}(R)}) \in R$, hence $\text{KTT}_n \models \overline{f}$. The proof is similar for $\neg f$.
- $f \equiv g_1 \vee g_2$. Then the orders of the g_is are either equal to, or smaller than n.
  First assume $\text{RTT} \models f$. Then we know that $\text{RTT} \models g_i$ for $i = 1$ or $i = 2$. By the induction hypothesis (and Conservation of Knowledge, if the order of g is $< n$), $\text{KTT}_n \models \overline{g_i}$, As $\overline{f} \equiv \overline{g_1} \vee \overline{g_2}$, $\text{KTT}_n \models \overline{f}$.
  Now assume $\text{RTT} \models \neg f$. Then it is not true that $\text{RTT} \models f$, so it is not true that $\text{RTT} \models g_i$ for $i = 1$ or $i = 2$. So $\text{RTT} \models \neg g_i$ for $i = 1, 2$. By the induction hypothesis (and, again, possibly Conservation of Knowledge), we have $\text{KTT}_n \models \overline{\neg g_i}$, hence, $\text{KTT}_n \models \neg\overline{g_i}$ for $i = 1, 2$.
  So $\text{KTT}_n \models \neg\overline{g_1} \wedge \neg\overline{g_2}$, and hence $\text{KTT}_n \models \overline{\neg f}$.
- $f \equiv \neg g$. If $\text{RTT} \models f$ then use the induction hypothesis on g to get $\text{KTT}_n \models \overline{\neg g}$, hence $\text{KTT}_n \models \overline{f}$.
  If $\text{RTT} \models \neg f$, then $\text{RTT} \models g$, so by induction $\text{KTT}_n \models \overline{g}$, so $\text{KTT}_n \models \neg\neg\overline{g}$, so $\text{KTT}_n \models \overline{\neg f}$.
- $f \equiv \forall x{:}t[g]$. Notice that g has order n.
  If $\text{RTT} \models f$ then for all $k{:}t$, $\text{RTT} \models g[x{:=}k]$. By the induction hypothesis, we know that for all $k : t$, $\text{KTT}_{m_k} \models \overline{g[x{:=}k]}$, where m_k is the order of $g[x{:=}k]$.

By the Substitution Lemma 5 we have: For all $k : t$, $\text{KTT}_n \models \overline{g}[x{:=}\langle\overline{k}\rangle]$. Hence, for all $d \in D$, $\text{KTT}_n \models \neg\text{Typ}_t(d) \vee \overline{g}[x{:=}d]$. Hence $\text{KTT}_n \models \overline{\forall x : t[g]}$.

The argument for $\text{RTT} \models \neg f$ is similar.

$\Rightarrow$ This is easy now. Assume, for the sake of the argument, not $\text{RTT} \models f$. Then $\text{RTT} \models \neg f$, hence $\text{KTT}_n \models \neg\overline{f}$ and $\text{KTT}_n \models \overline{f}$, which is a contradiction. □

This theorem clearly shows the relation between the orders in RTT and the levels of truth in KTT. The heart of the proof of Theorem 4.9 is in the proof of case $z(h_1, \ldots, h_q)$ of the Substitution Lemma 5. This is the only place in the proof where the properties of the predicate T are used. It is understandable that these properties must be used at exactly this place when we look at the definition of propositional functions and the typing rules for propositional functions. Exactly the possibility of constructing a propositional function of the form $z(h_1, \ldots, h_q)$ makes it possible to arrive at higher-order propositional functions and higher-order propositions. So exactly at this spot, Kripke's predicate T must appear, in order to raise one level in KTT as well.

Corollary 4.10. $\text{RTT} \models f$ if and only if $\text{KTT}_\omega \models \overline{f}$. □

We cannot improve the result of Theorem 4.9 in general: There are propositions f of order n in RTT whose code is provable at level KTT_n in KTT, but not at any lower level.

Theorem 4.11. Let $n > 0$, and f_n be the nth-order proposition $\forall \mathtt{p}{:}()^{n-1}[\mathtt{p}() \vee \neg\mathtt{p}()]$. Then:

$$\text{KTT}_m \models \overline{f_n} \text{ if and only if } m \geq n.$$

PROOF: $\Leftarrow$ follows from Theorem 4.9 and Lemma 4. $\Rightarrow$ is by induction on n. Observe that

$$\overline{f_n} \equiv \forall \mathtt{p}[\neg\mathtt{Typ}_{()^{n-1}}(\mathtt{p}) \vee (\mathtt{T}(\mathtt{App}_0(\mathtt{p})) \vee \neg\mathtt{T}(\mathtt{App}_0(\mathtt{p})))].$$

- $n = 1$. Let g be any proposition of order 0 in RTT. Then $\text{KTT}_0 \models \mathtt{Typ}_{()^0}(g)$ but as T is completely undefined at level 0, $\text{KTT}_0 \not\models \mathtt{T}(\mathtt{App}_0(g)) \vee \neg\mathtt{T}(\mathtt{App}_0(g))$. Hence, $\text{KTT}_0 \not\models \overline{f_1}$.
- Assume the theorem has been proved for all $n' < n$. Assume $m < n$ and $\text{KTT}_m \models \overline{f_n}$. By definition of $\models$, we have: $\text{KTT}_m \models \mathtt{T}(\mathtt{App}_0(\langle\overline{f_{n-1}}\rangle)) \vee \neg\mathtt{T}(\mathtt{App}_0(\langle\overline{f_{n-1}}\rangle))$, and for reasons of consistency: $\text{KTT}_m \models \mathtt{T}(\mathtt{App}_0(\langle\overline{f_{n-1}}\rangle))$, hence $\text{KTT}_m \models \mathtt{T}(\overline{f_{n-1}})$, so, by the definition of T: $\text{KTT}_{m-1} \models \overline{f_{n-1}}$, which contradicts the induction hypothesis, as $m - 1 < n - 1$. □

There are, however, propositions f of order n in RTT for which $\text{KTT}_m \models \overline{f}$ or $\text{KTT}_m \models \overline{\neg f}$ can already be established for $m < n$.

Example 4.12. Consider a proposition $g \equiv g_1 \vee g_2$ where g_1 is a true proposition of order m and g_2 is any proposition of order n. As g_1 is true in RTT, we have $\text{KTT}_m \models \overline{g_1}$, and therefore $\text{KTT}_m \models \overline{g}$.

4b The restrictiveness of Russell's theory

We illustrate the different approaches of Russell and Kripke by an example given by Kripke himself.

Example 4.13. Let D, $\mathcal{R}$, L, $S_{\alpha,i}$ and KTT_α be as in Section 3 where $\mathcal{R}$ contains two monadic predicates $\mathtt{V}$ and $\mathtt{W}$ which are collections of (codes of) utterances of persons $\mathtt{V}$ and $\mathtt{W}$. Now define

$$P \equiv \forall \mathtt{x}[\neg \mathtt{W}(\mathtt{x}) \vee \neg \mathtt{T}(\mathtt{x})]$$
$$Q \equiv \forall \mathtt{x}[\neg \mathtt{V}(\mathtt{x}) \vee \neg \mathtt{T}(\mathtt{x})]$$

(informally, P denotes: All utterances of $\mathtt{W}$ are false, and Q denotes: All utterances of $\mathtt{V}$ are false). Now distinguish two situations. In both situations, we want to know whether P and Q become true or false when passing through the hierarchy of languages KTT_0, KTT_1, ... Or, more formally, whether there is an α such that $\langle \mathtt{V} \rangle$ and $\langle \mathtt{W} \rangle$ belong to $S_{\alpha,1} \cup S_{\alpha,2}$.

1. $\mathtt{V} = \{\langle P \rangle\}$ and $\mathtt{W} = \{\langle Q \rangle\}$ (notice that $\mathtt{V}$ and $\mathtt{W}$ are just subsets of D). In this case, P is logically equivalent to $\neg\mathtt{T}(\langle Q \rangle)$ and Q is logically equivalent to $\neg\mathtt{T}(\langle P \rangle)$. As a consequence we have: if $\langle Q \rangle \in S_{\alpha,i}$ then $\langle P \rangle \in S_{\beta,3-i}$ for some $\beta < \alpha$, and if $\langle P \rangle \in S_{\alpha,i}$ then $\langle Q \rangle \in S_{\beta,3-i}$ for some $\beta < \alpha$. Hence $\langle P \rangle, \langle Q \rangle \notin S_{\alpha_i}$, for all α, i, so neither the truth of P nor the truth of Q will ever be established.

2. In the situation above, the only utterance of $\mathtt{V}$ was that anything said by $\mathtt{W}$ is false, and vice versa. In that case, it is also intuitively clear that it is impossible to say anything about the truth of P or Q. Now we change the situation. We assume that $\mathcal{R}$ also contains a third monadic predicate $\mathtt{R}$, and that $\mathtt{d}$ is an element of $\mathtt{R}$. We redefine $\mathtt{W}$:

   $$\mathtt{W} \stackrel{\text{def}}{=} \{\langle Q \rangle, \langle \mathtt{R}(\mathtt{d}) \rangle\}.$$

   This has drastic consequences. As $\text{KTT}_0 \models \mathtt{R}(\mathtt{d})$, $\langle \mathtt{R}(\mathtt{d}) \rangle \in S_{1,1}$, so $\text{KTT}_1 \models \mathtt{T}(\langle \mathtt{R}(\mathtt{d}) \rangle)$, hence $\text{KTT}_1 \models \neg P$. Therefore, $\langle P \rangle \in S_{2,2}$, so $\text{KTT}_2 \models \neg\mathtt{T}(\langle \mathtt{P} \rangle)$, hence:

   $$\text{KTT}_2 \models \neg P$$
   $$\text{KTT}_2 \models Q$$

   The fact that $\mathtt{W}$ utters a true sentence makes it possible to conclude at level 1 that P is false, irrespective of the fact that $\mathtt{W}$ has also uttered another sentence Q, of which we cannot establish the truth at level 1. The falsehood of P makes it possible to decide about Q at the next level, so the falsehood of P and the truth of Q could have been established at level 2.

In Russell's terminology it would not be possible to write expressions like P and Q at all: They are excluded beforehand, as P involves Q, therefore has to be of higher-order than Q, and Q involves P, therefore has to be of higher-order than P.

This indicates an important difference between RTT and KTT: Kripke allows much more expressions to be written down. In some situations these expressions will never obtain any truth-value (like P and Q in the first example), but in other situations (so: with other definitions of the primitive predicates) the same expressions will get a truth-value. Kripke concludes: *"it would be fruitless to look for an* intrinsic *criterion that will enable us to sieve out – as meaningless, or ill-formed – those sentences which lead to paradox"*.

Example 4.14. Another, more formal, example of a proposition f in KTT for which there is no $g \in \mathcal{F}$ with $\overline{g} \equiv f$ is the proposition $f \stackrel{\text{def}}{=} \forall \mathtt{x}[\mathtt{T}(\mathtt{x}) \vee \neg\mathtt{T}(\mathtt{x})]$:
Assume, for the sake of the argument, that $\overline{g} \equiv f$. Let m be the order of g. Then $\text{KTT}_m \models f$ or $\text{KTT}_m \models \neg f$. This implies $\text{KTT}_m \models \mathtt{T}(f_m) \vee \neg\mathtt{T}(f_m)$, where f_m is as in Theorem 4.11. By definition of T this means $\text{KTT}_{m-1} \models f_m$ or $\text{KTT}_{m-1} \models \neg f_m$, both contradicting Theorem 4.11.

4c Orders and types

RTT is based on a double hierarchy: One of types and one of orders. This double hierarchy is too restrictive. It is possible to develop Logic and Mathematics within RTT, but for instance the proof of the Supremum Theorem, which is fundamental in real analysis, cannot be given. The origin of the problem is the use of the so-called predicative and impredicative propositional functions.

Definition 4.15. Let $f \in \mathcal{F}$ be typeable in RTT. Assume f has free variables $x_1, \ldots, x_n$ of orders $m_1, \ldots, m_n$ respectively. f is called *predicative* if its order is equal to $\max(m_1, \ldots, m_n) + 1$; if its order is greater then f is called *impredicative.*

As the impredicative propositional functions cause problems, the "Axiom of Reducibility" is proposed in *Principia Mathematica.* This axiom is given by the following definition:

Definition 4.16. (Axiom of Reducibility)

For each $f \in \mathcal{F}$ there is a predicative *$g \in \mathcal{F}$ that is logically equivalent to f*

This axiom has been controversial from the moment it was introduced. Russell himself admits that *"it has a purely pragmatic justification: it leads to the desired results, and to no others. But clearly it is not the sort of axiom with which we can rest content."* Though serious efforts have been made to develop mathematics within RTT (for instance by Weyl [1918]), this has not become the usual practice. Ramsey [1926] shows that, by making distinction between language and meta-language, the orders can be removed from the system without re-introducing any known paradox. Hilbert and Ackermann [1928] present a similar idea. With this remark the type-theoretic fundamentals for the Simple Theory of Types STT, introduced by Church [1940], were laid, and orders have remained out of the important modern type systems up until now.

It is therefore interesting to notice the relation between orders in RTT and levels of truth in KTT, as formulated in Theorem 4.9. It shows that Kripke's system can be regarded as a system based on RTT, of which not the *orders*, but the *types* have been removed. In this way, KTT can be seen as a system that is dual to STT.

KTT, however, has a more subtle approach than many type theories as it does not exclude any, possibly "paradoxical", expression from the syntax, which is the usual type-theoretic approach. If an expression is paradoxical, it will not get a truth value at any level α of the hierarchy of Truths. Whether an expression is paradoxical or not does not only depend on its syntactic structure, but also on the domain D (see Example 4.13). So paradoxes are only excluded at the level of semantics.

The discussion above shows that the orders of RTT are not to be blamed for the restrictiveness of RTT. KTT is a system which contains orders but has only few restrictions towards self-application. It is the *combination* of orders and types that makes RTT restrictive.

5 THE NUPRL AND MARTIN-LÖF TYPE THEORIES

Martin-Löf's type theory ITT (see [Martin-Löf, 1975; Martin-Löf, 1982] and also [Nordström *et al.*, 1990]) was originally developed as a foundation of constructive mathematics, and the CTT relative of it was developed and implemented in Nuprl [Jackson, 1995; Constable *et al.*, 1986] to formalize and unify many basic concepts in computer science as well as constructive mathematics. A fundamental idea common to both theories is the interpretation of logic within type theory through the *propositions-as-types principle* (also called the Curry-Howard-de Bruijn isomorphism) where (roughly speaking), a proposition is interpreted as a type whose elements represent the *computational content of proofs* of the proposition. Hence, a false proposition is interpreted as the empty set and a true proposition is interpreted as a non-empty set. In order to prove that a proposition is true, we need to show that the proposition is inhabited. Nuprl can extract from proved propositions their computational content and treat *proofs as programs* [Bates and Constable, 1985].

This idea has proved extremely attractive from the computational point of view and has been exploited in many theorem provers (e.g., Alf, Agda, Nuprl, MetaPRL, and Coq). This idea was already exploited in de Bruijn's AUTOMATH which played an influential role in both provers Coq and Nuprl. In this section, we concentrate on a PTS-style variant of core Nuprl from 1984.[9]

[9]Since 1984 many new types and type constructors have been added to CTT; much of that work is summarized in [Allen *et al.*, 2006] and reported in several PhD theses available at www.nuprl.org.

5a A fragment of Nuprl in PTS style

We give a PTS style description of a variant of the earliest CTT type theory on which Nuprl was based (see [Jackson, 1995; Constable *et al.*, 1986]). We do not give a full presentation even of all of the early Nuprl type constructors, as we will only need parts of it. The description of the typing rules is given in a natural deduction style similar to that used in the Barendregt cube [1992], and Pure Type Systems [Terlouw, 1989]. This assumes an explicit typing of this subset of Nuprl's terms, but in Section 8 we will examine the actual *implicit typing* used in CTT.

For full details the reader is referred to [Kamareddine and Laan, 2001]. Here, we review the results that are relevant for our section on Computational Type Theory (see Section 6) and for the role of orders in achieving powerful type theory which remain computational.

In the next definition, $\mathbb{V}$, $\mathbb{Z}$ and $\mathbb{S} = \{*_1, *_2, \ldots\}$ represent the sets of variables, integers and sorts respectively. In an extension to the tradition of PTSs, the sort $*_a$ represents the propositions (and, more general, the types) of order $\leq a$. The sort $*_a$ corresponds to the Universe of Types $\mathbb{U}_a$ in [Jackson, 1995; Martin-Löf, 1982]. The symbol $\perp$ represents the undefined or a contradiction. Again as in PTSs, abstraction is represented by λ and Π. The remaining notions represent application, Cartesian products, pairing, and first and second projections.

Definition 5.1. The set of terms $\mathbb{T}$ is defined by the following abstract syntax:

$$\mathbb{T} ::= \mathbb{S} \mid \mathbb{V} \mid \perp \mid \mathbb{Z} \mid \mathbb{T}\mathbb{T} \mid \lambda\mathbb{V}{:}\mathbb{T}.\mathbb{T} \mid \Pi\mathbb{V}{:}\mathbb{T}.\mathbb{T} \mid \mathbb{T} \times \mathbb{T} \mid \langle \mathbb{T}, \mathbb{T} \rangle \mid \pi_1(\mathbb{T}) \mid \pi_2(\mathbb{T})$$

We let $\alpha, \beta, x, y, z, \ldots$ range over $\mathbb{V}$; $m, n, \ldots$ over $\mathbb{Z}$ and A, B, M, N, a, b over $\mathbb{T}$. When x does not occur free in B, we write $A \to B$ for $\Pi x{:}A.B$. Free and bound variables are defined as usual. $\mathrm{FV}(A)$ and $\mathrm{BV}(A)$ denote the set of free and bound variables of A. $A[x{:=}B]$ denotes the term in which all the free occurrences of x in A have been replaced by B. Syntactic equality of terms is taken modulo renaming of bound variables and as is standard, we adopt the Barendregt's convention where names of bound variables differ from the free ones in a term and where we use different bound names for different bound variables.

The next definition gives for each term A the order-free term $|A|$ (where all orders are collapsed into $*_1$), introduces the various notions of reduction on our Nuprl segment, and defines the typing relation of the PTS version of Nuprl.

Definition 5.2.

- For each term A we define a term $|A|$ (where we collapse orders) as follows:

$$\begin{array}{ll} |\, *_m \,| = *_1 & |\Pi x{:}A.B| = \Pi x{:}A.|B| \\ |x| = x & |A \times B| = |A| \times |B| \\ |\perp| = \perp & |\,\langle A, B \rangle\,| = \langle |A|, |B| \rangle \\ |\mathbb{Z}| = \mathbb{Z} & |\pi_1(M)| = \pi_1(|M|) \\ |MN| = |M||N| & |\pi_2(M)| = \pi_2(|M|) \\ |\lambda x{:}A.b| = \lambda x{:}A.|b| & \end{array}$$

- We take the axioms:

$$(\to_\beta) : (\lambda x{:}T.A)B \to_\beta A[x{:=}B]$$
$$(\to_\sigma) : \pi_1(\langle A, B\rangle) \to_\sigma A \text{ and } \pi_2(\langle A, B\rangle) \to_\sigma B.$$

- The reduction relations $\to_\beta$, $\to_\sigma$ and $\to_{\beta\sigma}$ generated respectively by the axioms $(\to_\beta)$, $(\to_\sigma)$ and their combination.
- The reflexive transitive closures $\twoheadrightarrow_\beta$, $\twoheadrightarrow_\sigma$ and $\twoheadrightarrow_{\beta\sigma}$ of $\to_\beta$, $\to_\sigma$ and $\to_{\beta\sigma}$.
- The symmetric closure $=_{\beta\sigma}$ of $\twoheadrightarrow_{\beta\sigma}$.
- *Contexts* as finite lists $x_1{:}A_1, \ldots, x_n{:}A_n$ of declarations $x_i{:}A_i$ with *domain* $\{x_1, \ldots, x_n\}$. Let Γ, Δ range over contexts and $\Gamma \subseteq \Delta$ if all declarations in Γ are also in Δ.
- A statement $\Gamma \vdash A : B$ is *derivable* if it can be deduced by repeated application of the rules given in Figure 1.
- Γ is called *legal* if there are A, B such that $\Gamma \vdash A : B$;
- A is called *legal* if there are Γ, B such that $\Gamma \vdash A : B$ or $\Gamma \vdash B : A$;
- A is called a Γ-*term* if there is a B such that $\Gamma \vdash A : B$ or $\Gamma \vdash B : A$;
- A is called a Γ-*type* if there is an n such that $\Gamma \vdash A : *_n$.

Note that the rule $(\subseteq)$ of Figure 1 simplifies the formulation without sacrificing expressivity. This can be seen through the simplicity of the rule (Π-form) without being restrictive.

The next theorem shows that important properties of PTSs also hold for the Nuprl formulation as a PTS except for the unicity of types. This is consistent with Nuprl and Martin-Löf's type theory where unicity of types does not hold. However, we have a weak unicity of types in the sense that if we collapse orders (using the collapse function $|.|$) then unicity of types will hold.

Theorem 5.3.

1. **(Church-Rosser Theorem for $\to_\beta$, $\to_\sigma$ and $\to_{\beta\sigma}$)** Let $r \in \{\beta, \sigma, \beta\sigma\}$. If $A \twoheadrightarrow_r B_1$ and $A \twoheadrightarrow_r B_2$ then there is a C such that $B_1 \twoheadrightarrow_r C$ and $B_2 \twoheadrightarrow_r C$.
2. **(Correctness of Types)** If $\Gamma \vdash A : B$ then there is an $n \geq 1$ such that $\Gamma \vdash B : *_n$.
3. **(Subject Reduction)** If $\Gamma \vdash A : B$ and $A \to_{\beta\sigma} A'$ then $\Gamma \vdash A' : B$.
4. **($\twoheadrightarrow_{\beta\sigma}$ Preserves Γ-terms)** If A is a Γ-term and $A \twoheadrightarrow_{\beta\sigma} A'$ then A' is a Γ-term.
5. **(Weak Unicity of Types)** If $\Gamma \vdash A : B_1$ and $\Gamma \vdash A : B_2$ then $|B_1| =_{\beta\sigma} |B_2|$.
6. If A is a Γ-term then there is a Γ-term B and an $n \geq 1$ such that $\Gamma \vdash A : B : *_n$.

$$
\begin{array}{lll}
\text{(Axioms)} & \begin{array}{lll} \vdash \bot : *_1 & \vdash *_n : *_{n+1} & (n \in \mathbb{N}) \\ \vdash \mathbb{Z} : *_1 & \vdash n : \mathbb{Z} & (n \in \mathbb{Z}) \end{array} & \\
\text{(Start)} & \dfrac{\Gamma \vdash A : *_n}{\Gamma, x{:}A \vdash x{:}A} & (x \text{ is } \Gamma\text{-fresh}) \\
\text{(Weak)} & \dfrac{\Gamma \vdash M : N \qquad \Gamma \vdash A : *_n}{\Gamma, x{:}A \vdash M : N} & (x \text{ is } \Gamma\text{-fresh}) \\
(\Pi\text{-form}) & \dfrac{\Gamma \vdash A : *_n \qquad \Gamma, x{:}A \vdash B : *_n}{\Gamma \vdash (\Pi x{:}A.B) : *_n} & \\
(\lambda) & \dfrac{\Gamma, x{:}A \vdash b : B \qquad \Gamma \vdash (\Pi x{:}A.B) : *_n}{\Gamma \vdash (\lambda x{:}A.b) : (\Pi x{:}A.B)} & \\
\text{(App)} & \dfrac{\Gamma \vdash M : (\Pi x{:}A.B) \qquad \Gamma \vdash N : A}{\Gamma \vdash MN : B[x{:=}N]} & \\
(\times\text{-form}) & \dfrac{\Gamma \vdash A : *_n \qquad \Gamma \vdash B : *_n}{\Gamma \vdash (A \times B) : *_n} & \\
\text{(Pairs)} & \dfrac{\Gamma \vdash a : A \qquad \Gamma \vdash b : B \qquad \Gamma \vdash (A \times B) : *_m}{\Gamma \vdash \langle a, b \rangle : (A \times B)} & \\
\text{(Left)} & \dfrac{\Gamma \vdash M : (A \times B)}{\Gamma \vdash \pi_1(M) : A} & \\
\text{(Right)} & \dfrac{\Gamma \vdash M : (A \times B)}{\Gamma \vdash \pi_2(M) : B} & \\
\text{(Conv)} & \dfrac{\Gamma \vdash M : A \qquad \Gamma \vdash B : *_n \qquad A =_{\beta\sigma} B}{\Gamma \vdash M : B} & \\
(\subseteq) & \dfrac{\Gamma \vdash A : *_n}{\Gamma \vdash A : *_{n+1}} &
\end{array}
$$

Figure 1. Rules of Nuprl as PTSs

5b *Orders in Nuprl*

[Kamareddine and Laan, 2001] adopts the position that the order of a term A is the smallest natural number n such that the type of A is of type $*_{n+1}$. By ($\subseteq$), this means that for any $m > n$, the type of A is also of type $*_m$. This captures the notion of orders à la Russell. Here is the formal definition of orders.

Definition 5.4.

- **(Γ-terms modulo A)** We define $[A]_\Gamma = \{A' \mid A' \text{ is } \Gamma\text{-term and } A =_{\beta\sigma} A'\}$.
- **(Order of a Term)** Assume A is a Γ-term. We define $\text{ord}_\Gamma(A)$, the *order* of A in Γ, as the smallest natural number a (i.e. $a \geq 0$) for which there are $A' \in [A]_\Gamma$ and B such that $\Gamma \vdash A' : B : *_{a+1}$.

The next lemma shows that any element $=_{\beta\sigma}$ to A has the same order as A (item 1). Even better, the lemma shows that there is an A' where $A \twoheadrightarrow_{\beta\sigma} A'$ rather than only $A =_{\beta\sigma} A'$ (item 2). Furthermore, If A itself is a type and n is the order of A, then not only the type of A is of type $*_{n+1}$, but also $A \twoheadrightarrow_{\beta\sigma} A'$ for some A' of type $*_n$ (see item 3). Moreover, $*_n$ can be regarded as the type of types of order $\leq n$ (item 4) and a term is always of a lower order than its type (item 5). Note however that if $\Gamma \vdash A : B$ then $\text{ord}_\Gamma(A) = \text{ord}_\Gamma(B) - 1$ may not hold: take $\Gamma = \emptyset$, $A \equiv *_1$ and $B \equiv *_3$. This is as expected because, by the inclusion rule ($\subseteq$), once A is of type $*_n$, it is of type $*_m$ for any $m \geq n$.

LEMMA 6. *Let A be a Γ-term and* $\text{ord}_\Gamma(A) = a$. *The following holds:*

1. *If $A' \in [A]_\Gamma$ then* $\text{ord}_\Gamma(A) = \text{ord}_\Gamma(A')$.
2. *There are A' and B such that $\Gamma \vdash A' : B : *_{a+1}$ and $A \twoheadrightarrow_{\beta\sigma} A'$.*
3. **(A type B reduces to a type B' of type $*_{\text{ord}(B)}$)** *Let B be a Γ-type and $b = \text{ord}_\Gamma(B)$. $\exists B'$ such that $\Gamma \vdash B' : *_b$ and $B \twoheadrightarrow_{\beta\sigma} B'$.*
4. **($*_a$ is the type of types of order $\leq a$)** *If P is a Γ-type in $\beta\sigma$-normal form, then $\Gamma \vdash P : *_a \Leftrightarrow \text{ord}_\Gamma(P) \leq a$.*
5. **(A term is of a lower order than its type)** *If $\Gamma \vdash A : B$ then* $\text{ord}_\Gamma(A) < \text{ord}_\Gamma(B)$.

[Kamareddine and Laan, 2001] provides a procedure that almost evaluates the order of Nuprl terms. In particular, it explains how the order of almost all complex terms (like $A \times B$) is evaluated in term of the orders of the components (A and B). The only case that fails is that of an application since one cannot evaluate the order of AB precisely in terms of the orders of A and B. Rather, in the case of an application AB, one can only establish that the order of AB is $\leq$ the order of A. More importantly also is the fact that a function can never take arguments of a higher-order than itself (Lemma 4).

LEMMA 7.

1. **(Order of Projections)** *For a* Γ*-term* $\langle A, B\rangle$, $\mathrm{ord}_\Gamma(\pi_1(\langle A, B\rangle)) = \mathrm{ord}_\Gamma(A)$ *and* $\mathrm{ord}_\Gamma(\pi_2(\langle A, B\rangle)) = \mathrm{ord}_\Gamma(B)$.

2. **(Orders of Constants and Sorts)** *Let* Γ *be a legal context. Then* $\mathrm{ord}_\Gamma(*_a) = a + 1$, $\mathrm{ord}_\Gamma(\bot) = 1$, $\mathrm{ord}_\Gamma(\mathbb{Z}) = 1$, *and* $\mathrm{ord}_\Gamma(n) = 0$.

3. *Let* C *be a* Γ*-term. The following holds:*

   (a) *If* $C \equiv x$ *where* $x{:}A \in \Gamma$ *then* $\mathrm{ord}_\Gamma(x) = \mathrm{ord}_\Gamma(A) - 1$.

   (b) *If* $C \equiv \Pi x{:}A.B$ *then* $\mathrm{ord}_\Gamma(\Pi x{:}A.B) = \max(\mathrm{ord}_\Gamma(A), \mathrm{ord}_{\Gamma,x:A}(B))$.

   (c) *If* $C \equiv \lambda x{:}A.b$ *then* $\mathrm{ord}_\Gamma(\lambda x{:}A.b) = \max(\mathrm{ord}_\Gamma(A) - 1, \mathrm{ord}_{\Gamma,x:A}(b))$.

   (d) *If* $C \equiv A \times B$ *or* $C \equiv \langle A, B\rangle$ *then* $\mathrm{ord}_\Gamma(C) = \max(\mathrm{ord}_\Gamma(A), \mathrm{ord}_\Gamma(B))$.

4. **(Order of an Application)** *If* $\Gamma \vdash M : \Pi x{:}P.Q$ *and* $\Gamma \vdash N : P$ *then* $\mathrm{ord}_\Gamma(N), \mathrm{ord}_\Gamma(MN) \leq \mathrm{ord}_\Gamma(M)$.

PROOF: For full details see [Kamareddine and Laan, 2001]. We only show the case of application: Let $m = \mathrm{ord}_\Gamma(M)$. $\exists M', R$ such that $\Gamma \vdash M' : R : *_{m+1}$ and $M \twoheadrightarrow_{\beta\sigma} M'$. By Subject Reduction, $\Gamma \vdash M' : \Pi x{:}P.Q$, so by Weak Unicity of Types, $|R| =_{\beta\sigma} |\Pi x{:}P.Q| \equiv \Pi x{:}P.|Q|$. By Church-Rosser $\exists R'$ such that $R \twoheadrightarrow_{\beta\sigma} R'$ and $\Pi x{:}P.|Q| \twoheadrightarrow_{\beta\sigma} |R'|$. Also, R' must be of the form $\Pi x{:}P'.Q'$, where $P \twoheadrightarrow_{\beta\sigma} P'$ and $|Q| \twoheadrightarrow_{\beta\sigma} |Q'|$. By Subject Reduction and Conversion, $\Gamma \vdash M' : (\Pi x{:}P'.Q') : *_{m+1}$. As m is minimal, $\mathrm{ord}_\Gamma(\Pi x{:}P'.Q') = m + 1$. Now, $m = \mathrm{ord}_\Gamma(M) = \mathrm{ord}_\Gamma(\Pi x{:}P'.Q') - 1 = \max(\mathrm{ord}_\Gamma(P') - 1, \mathrm{ord}_{\Gamma,x:P'}(Q') - 1) \geq \mathrm{ord}_\Gamma(P') - 1 = \mathrm{ord}_\Gamma(P) - 1 \geq \mathrm{ord}_\Gamma(N)$.

By conversion, $\Gamma \vdash N : P'$, so $\Gamma \vdash M'N : Q'[x{:=}N]$. As $MN =_{\beta\sigma} M'N$, we have $\mathrm{ord}_\Gamma(MN) = \mathrm{ord}_\Gamma(M'N) < \mathrm{ord}_\Gamma(Q'[x{:=}N]) \leq \mathrm{ord}_{\Gamma,x:P'}(Q') \leq \mathrm{ord}_\Gamma(\Pi x{:}P'.Q') = m + 1$, so $\mathrm{ord}_\Gamma(MN) \leq m$. □

This shows that a function can never take an argument of higher-order, and that the order of a term cannot increase when applying an argument to that term.

Finally, we state that our study of orders is different from the approach of [Harper and Pollack, 1991] whose main concerns were type checking and well-typedness in Coquand's CC^ω, extending CC^ω with anonymous universes to model Russell and Whitehead's *typical ambiguity* convention, and with definitions.

Another example that orders and universes play an influential role in powerful modern systems can be found in [Harper and Pollack, 1991].

6 COMPUTATIONAL TYPE THEORY CTT

In this section we look at Computational Type Theory as implemented in Nuprl and MetaPRL, the implicitly typed extensional theory CTT [Constable *et al.*, 1986;

Allen *et al.*, 2006] which supports *proofs as programs*. We will show how the implicitly typed theory offers expressive advantages that compensate for its predicative character. We point out how results from Section 5 apply to this theory. Implicit typing also supports polymorphism, and we show the advantages of polymorphism by defining subtyping as used in object oriented programming. Polymorphism is the basis for CTT definitions of dependent records, inheritance and objects [Kopylov, 2003]. This is another example of the fact that orders do not limit us in expressing *concepts that apply to types at all orders,* i.e. in all CTT universes. It is precisely such concepts that one would expect to be hard to express predicatively. For comparison with KTT we create an internal definition of proof using a syntactic fixed point and mention in that context how orders provide a stratified internal notion of truth.

We begin by presenting the intuitive ideas behind Computational Type Theory. This presentation will show that CTT is an *extensional type theory* in contrast to the *Calculus of Inductive Constructions* (CIC) [The Coq Development Team, 2004; Bertot and Castéran, 2004] which is considered an *Intensional Type Theory*. Extensionality is also compatible with orders even though it is a notion associated with set theory, another impredicative approach to mathematics that does not use orders. But extensionality comes at a cost, as it does in set theory, namely there is no type checking algorithm for membership in a type (nor in a set) because that question is undecidable for CTT and for standard set theories such as ZFC [Zermelo, 1908]. Computer scientists, unlike mathematicians, are accustomed to working in programming languages with type disciplines supported by strong static *type checkers*, such as Java, or *type inference algorithms* in the case of ML. Nevertheless, Nuprl has been used by computer scientists in many very practical ways, providing additional evidence that the concept of orders in an extensional theory is not overly restrictive.

6a Origins

Computational Type Theory, like all formal type theories, is related to RTT (which was never completely formalized). Moreover, CTT has been implemented as a software system, in both Nuprl and MetaPRL; of course RTT could not be implemented in 1927 for lack of computers and programming languages![10] However Church's simple types in his 1940 Simple Theory of Types STT [1940] are essentially the ramified types of RTT without orders. A variant of STT was implemented as HOL [Gordon and Melham, 1993]; it was not designed to formalize all classical mathematics, but rather *applied classical mathematics.* In contrast to RTT, STT is an elegantly small theory whose semantics can either be taken as intuitive or modeled in Zermelo set theory.

For Russell a type is the domain of significance of a *propositional function*, i.e. those objects that the proposition is about. In RTT that domain cannot include the proposition itself. Russell believed that it is not consistent to postulate a single

[10]Randel Holmes implemented this theory in 2000.

universal type of all meaningful objects on which propositions are defined because such a universal type would include the propositions being defined, creating a *vicious circle* in the definition. Instead, the universe must be divided into types. Computational type theory builds on these design decisions.

However, even before types are considered, a very key subsystem of CTT is defined, the *computation system* on terms. The computation system is defined using small step *operational semantics à la* Plotkin [Plotkin, 1981; Kahn, 1987]. The terms include all untyped lambda calculus expressions, and the operational semantics means that the term language can express all general computable functions and is thus Turing complete.[11] So the computation system includes all *computable functions* among its primitives, and to reason about one of them, it must have a domain which is a *type* of the theory. If $P(x)$ is a propositional function on a type T, then those objects for which it is true form a *class* denoted $\{x \mid P(x)\}$ in RTT and $\{x \mid T|P(x)\}$ in CTT. Classes are fundamental in RTT and are used to define the natural numbers; however, in CTT the integers are elements of a primitive type.

We have seen in Section 5 that CTT allows compound types built with type forming operations such as product, union, and function, and so forth starting with atomic types such as the integers. To characterize such operations on types as functions, the domain of the function would need to include types. If we allowed a type of all types, then we could do this, but as we have discussed, Russell pointed out obstacles to construction of such a type, and following RTT we do not allow such a type of all types. Instead we build large collections of types called *universes*, each closed under all the usual type constructors; the first universe $\mathbb{U}_1$ is formed by collecting all the types that do not require universes. The second universe $\mathbb{U}_2$ is built the same way except that we now add $\mathbb{U}_1$ as a new type allowed in the constructions, and allow $\mathbb{U}_1$ as an element of $\mathbb{U}_2$. In general, the nth universe $\mathbb{U}_n$ allows type expressions of universe level $i < n$, and each such $\mathbb{U}_i$ belongs to $\mathbb{U}_n$. This was presented in Section 5 by the axiom ($\subseteq$) where $*_n$ represents $\mathbb{U}_n$.

We see that the universe hierarchy is *cumulative*, so $\mathbb{U}_i \subseteq \mathbb{U}_{i+1}$. To express that an expression of the theory is a type, it is necessary that it belongs to a universe (see Definition 5.2). If an expression includes operators on $\mathbb{U}_i$, then it is in $\mathbb{U}_{i+1}$ as in this assertion about propositional functions, $\forall T : \mathbb{U}_i.\forall P : T \to \mathbb{U}_i.\forall x : T.\ P(x) \to P(x)$. The universe hierarchy corresponds to orders in RTT, and it renders the theory predicative.

Unlike in Section 5, in the extensional type theory, equality on the dependent function type and dependent product type is not syntactic equality but extensional equality. Thus $\Pi x{:}A.B = \Pi x{:}A'.B'$ iff $A = A'$ and for all $x{:}A$ we have $B(x) = B'(x)$.

6b *Philosophical issues*

Principia Mathematica was also designed to demonstrate that mathematics could be reduced to logic, say to truth values and higher-order functions over them,

[11] These are also called *general recursive functions* in much of the literature.

specifically to predicative logic which avoids the concept of all functions or all propositions. This is one reason that *Principia Mathematica* did not start with a primitive notion of natural numbers nor with specific individuals as in HOL; namely, Russell and Whitehead wanted to derive mathematical concepts such as natural numbers from purely logical concepts. Unfortunately, the core RTT foundation is inadequate for classical mathematics, even for number theory, without an *axiom of reducibility* (see Definition 4.16) among the typing rules, which is a way to introduce controlled impredicative notions, those whose definition quantifies over a collection including the concept being defined. The axiom of reducibility which we exhibited above, proved to be unpopular among mathematicians and other logicians (see [Kamareddine *et al.*, 2004]). Poincaré wrote harshly about it and about the cumbersome nature of RTT. There were no computer scientists around to make the case that machines could fill in vast amounts of detail and use editors to present a version that would read much more like ordinary mathematics than the raw formalism.

It is noteworthy that theories like CTT harmonize some of the philosophical disputes about the foundations of mathematics summarized well in Dummett's book on intuitionism [2000]. For example, serious use of implemented formal theories such as CTT and CIC illustrate David Hilbert's belief that mathematical truth can be reduced to mechanically checkable claims in a formal system, a belief called formalism. Founding CTT and CIC on the irreducible concepts of natural number and computation validates Brouwer's claim, to be discussed below, that number and computation are fundamental intuitions on which meaningful mathematics rests. CTT and CIC also follow Brouwer in equating mathematical truth with proof. Finally, the fact that among the primitive constructions of CTT are those that validate the laws of logic shows that logic is inseparable from mathematics, not quite in the way Russell imagined in his belief called logicism, but in a way that makes his notion of type fundamental to both logic and mathematics. Thus through the lens of CTT and CIC we recognize that each of the major philosophical schools contributed part of the underlying foundations of mathematics and computing. We see the interplay of these philosophical ideas in the next section. These points are discussed in the Scholarpedia article.[12]

6c Computation and data types

Building computation into the basic primitives

The small set of irreducible primitive notions in Church's Simple Type Theory STT does not include primitive concepts from computational mathematics, yet computational ideas have been an integral part of mathematics since the Greeks with their stress on ruler and compass constructions and since the Arabs with their stress on written algorithmic processes executed by humans; this is the tradition of *effective mathematics*.

[12]www.scholarpedia.org/article/Computational_type_theory

Algorithms and computation are central to intuitionism, a specific philosophy of mathematics originating circa 1907 with L. E. J. Brouwer (see [Heyting, 1975]) which accounts for computational concepts, especially computable functions. A key principle of this philosophy is that the primitive truths of mathematics are based on innate human *intuitions* about natural numbers and *constructions* on them, that logical reasoning is justified by the most general constructions, and that propositions are true only if we can prove them by these general constructions from the primitive truths. Thus intuitionistic mathematics includes among its primitive concepts effective computation on the natural numbers, and its logic is based on constructing evidence that can be grounded in computation. Brouwer developed this mathematics beyond law-like computation on numbers to include other mental constructions encompassing infinite sets and free choice sequences as objects of mathematics. He used free choice sequences to define real numbers (the continuum). Many of these concepts can be defined in CTT.

Constructive mathematics

What is nowadays called *constructive mathematics* is closely related to effective mathematics and intuitionistic mathematics. One of the seminal publications in (American) constructive mathematics is the book *Foundations of Constructive Analysis* by Errett Albert Bishop [1967]. In philosophical remarks in this book, Bishop disagreed with Brouwer about the continuum and defined real numbers using standard data types, in the way that they are now implemented as "infinite precision numbers" in some programming languages. For Bishop, a real number r is a computable function from the natural numbers to the rational numbers such that the absolute value of the difference between $r(n)$ and $r(m)$ is less or equal to $1/n + 1/m$. Many basic concepts and theorems from analysis have been implemented in CTT [Chirimar and Howe, 1991].

Bishop believed that constructive mathematics could be consistent with classical mathematics, something Brouwer denied. For example, in Brouwer's theory of real numbers, all functions from reals to reals are continuous (as is the case in recursive mathematics as well), in Bishop's account they are not, and indeed one can read Bishop's book as a valid account of the classical real numbers, but one that inherently provides information about how to compute with them. One of the goals of CTT is to formalize the concepts in Bishop's book and to *formalize all of mathematics constructively in such a way that the account can be understood classically yet there is enough information provided to enable practical computation.*[13]

[13]Thus far this approach has failed only when it comes to understanding the partial computable functions on a type in the sense that computer scientists prefer versus the sense in which mathematicians prefer. To the computer scientist, a partial function from A to B is a computing procedure that applies to elements of type A but might fail to terminate on some (or all) elements. To a mathematician, a partial function with this typing is a total function from a subset of A into B.

Like intuitionistic and constructive mathematics, Computational Type Theory axiomatizes digital computation on natural numbers and other recursive data types, but it does not adopt the *Church-Turing Thesis*; instead the theory relies on Bishop's insight that unless a person is interested in studying what cannot be computed, it is sufficient to specify a natural universal (Turing-complete) computing formalism and leave it open-ended allowing the possibility of extending the underlying programming framework yet preserving the mathematical theorems. Many logical questions can be explored starting from CTT because it can be extended in many directions, toward classical mathematics by adding the law of the excluded middle $P \vee \neg P$ (shown to be consistent with core CTT [Howe, 1996]), toward recursive mathematics by adding the Church-Turing Thesis [Beeson, 1985] as a logical principle, toward intuitionistic mathematics by axiomatizing free choice sequences, and towards *domain theory* [Scott, 1976] by typing partial functions in the sense of computer science as functions which may not converge on all elements of their domain type [Constable and Smith, 1993].

In CTT, the collection of all types is also left *open-ended*, and it is possible to define the standard data types such as numbers, tuples, arrays, lists, trees, etc. as well as the higher-order types such as functions from one type to another, infinite lists (streams), infinite trees, and so forth. The delicate point from a foundational point of view is whether to include the partial types such as all partially computable functions from a type to a type.[14]

In the extension of CTT with a logic of events [Bickford and Constable, 2008], say CTT-E, the programs include *distributed systems*, and proofs can be treated as processes; thus the programming logic applies to virtually all known important classes of programs. This connection to programming explains why CTT has been used extensively in important areas of computer science, namely formal methods and mathematical software engineering. CTT-E also connects to the logic of knowledge and the theory of agents in AI.

CTT is expressive enough to account for the idea of computational complexity of algorithms [Constable and Crary, 2001]; so it is possible to express the ideas of complexity theory such as the famous P=NP problem. It can also account for libraries of theorems and thus for systems [Allen *et al.*, 2006].

6d Elements of computational type theory

Terms and programs

In CTT all terms have a simple uniform syntax which is essentially an operator identifier followed by a list of subterms (possibly empty) each of which is in the

[14]Nuprl implements partial types using its concept of a *bar type* [Constable and Smith, 1993], and that extension provides a constructive domain theory. However, this extension of CTT is not consistent with classical mathematics, just as intuitionistic mathematics is not. The extension is not classically consistent because it is possible to show that the halting problem for partial functions on nontrivial types, say the natural numbers, is not decidable by *any function* of the theory; classically it is decidable by a noncomputable function.

scope of a list of binding variables (possibly empty), e.g. $op(\bar{x}_1.t_1; ...; \bar{x}_n.t_n)$. For example, the notation for functions has the form $\lambda(\bar{x}.b)$ where the body b can use the binding variables in the list $\bar{x}$. The identity function in this notation is $\lambda(x.x)$. Notice that this function is untyped and thus polymorphic, meaning that it belongs to many types, in particular to all types $A \to A$. This feature of starting with an untyped programming language that includes all algorithms (even nonterminating ones) distinguishes CTT and Martin-Löf's Intuitionistic Type Theory ITT [1975; 1982; 1984].[15] In CTT, even the famous Y combinator is allowed in assertions and used to define recursive functions. Thus as we mentioned above, the computation system over terms in CTT is Turing complete. We can use these untyped lambda terms to guarantee efficient execution of programs extracted from proofs. Moreover, CTT supports reasoning about these untyped functions in various ways, such as stating a theorem that an untyped term inhabits a certain type without knowing this fact in advance and thus discovering that it is a realizer for a proposition. The fact that the basic objects are untyped makes it possible to annotate terms with types for specific purposes, such as annotations that support good compilation or annotations that approximate the most informative typing and allow fast "annotation checking" as an extensible approximation to type checking; see [Howe, 1998] for an implementation of some of these ideas. These untyped terms not only live outside the type system, they live outside the system of orders, yet even self-referential terms are perfectly safe and extremely useful. Thus for some purposes, the system of orders is not at all restrictive.

The application of a function f to an argument a is the operator $ap(f; a)$. To make the official syntax readable, the Nuprl implementation of CTT uses *display forms* that allow terms to be displayed in a variety of ways, e.g. we could display the identity function as $\lambda x.x$ if we wished, and display function application as $f(a)$ or even as $f a$.

CTT takes the decimal numerals as primitive terms, they are officially written in the uniform syntax and displayed in the standard way. There are also terms for the basic arithmetic operations on integers, and a computable term for the induction principle on integers.

As with ITT, CTT specifies a computation system on terms defined by Plotkin-style *operational semantics* (small step) [Plotkin, 1981; Kahn, 1987]. In most presentations of CTT, this type of all terms is not made explicit, but it is possible to define a type *Base* whose elements are *all closed terms* of the theory, and to define the primitive equality relation on *Base* to be Howe's "squiggle" relation from his fundamental 1989 LICS paper [1989]. We say $a \sim b$ iff either term can be obtained from the other by rewriting any subexpression by any number of steps of symbolic computation.

Definition 6.1. *Base* is the type of all closed terms where $x = y$ in *Base* is $s \sim y$.

[15] In CIC and HOL, all terms are *explicitly typed* as in programming languages such as Java, and unlike the terms of ML for which types are inferred or for Lisp and Scheme, which are considered essentially typeless.

A key feature of CTT and ITT is that all types come with an equality relation; for type A it is usually written $a = b\ in\ A$. Results from Howe establish the following theorem:

Theorem 6.2. If $a = b\ in\ Base$, a and b are in type A, then $a = b\ in\ A$.

For some types such as the integers, Int (displayed as $\mathbb{Z}$), we have $a = b\ in\ Int$ implies $a \sim b\ in\ Base$. We call these types "squiggle types" – see the book *Nuprl Basics* by Stuart Allen in the Math Library at www.nuprl.org. Reasoning about equality in these types is especially easy because it can be done by computation which does not introduce any new proof subgoals.

Another type for which it is easy to reason about equality is Top.

Definition 6.3. *Top is the type of all closed terms with the trivial equality relation that all terms are equal. Thus Top has only one element.*

CTT has a type constructor that produces a separation type or *refinement type* $\{x : A|B(x)\}$ which is the collection of all elements of A that satisfy the proposition $B(x)$. For any type A we can define the set of elements of $Base$ on which it rests, namely $\{x : Base \mid x\ in\ A\}$. This is a subtype of $Base$, and a subtype of A because for C to be a subtype of A it is sufficient that for all a and a', if $a = a'\ in\ C$, then $a = a'\ in\ A$. We write this *subtyping relation* as $C \sqsubseteq A$.

The following theorems hold:

Theorem 6.4. For all $A : \mathbb{U}_i$ we have $A \sqsubseteq Top$.

Theorem 6.5. For all types A, we have that $\{x : Base \mid x\ in\ A\} \sqsubseteq A$.

Defining a type

To define a type, we specify a unique *canonical name* for it, an irreducible term. The type is denoted by terms that reduce to this canonical name, e.g. the type of integers has canonical name Int. To define the elements of a type we specify terms that are irreducible in the computation system which are the *canonical elements of the type*, and we define an equality relation declaring when two canonical terms denote the same abstract object. The equality relation creates abstract objects out of terms. Moreover, any term which reduces to a canonical term of the type is itself a member of the type. For example, the type of integers includes the decimal numbers with the normal equality, and it also includes terms such as $ap(\lambda(x.x); 0)$ which reduces to the canonical integer 0. Likewise, $1 + 2$ is an integer whose canonical form is 3.

A short version of this definition is to say that the extension of a type is a collection of *equivalence classes of terms*. These equivalence classes are understood in a computational way, not as sets of sets, but as types distinguished by their equality relations. This concept is formalized in CTT as the *quotient type*, a type that overcomes some of the computational limitations of the standard definition of equivalence classes to which Bishop objected. Instead of defining the elements of the quotient type as equivalence classes, they are simply elements of the type on which a new equality is imposed. The role of equality leads to the so-called PER *semantics*, for *partial equivalence relation* semantics. Stuart F. Allen [1987] gave a very elegant and precise semantics for CTT and ITT using this approach, and Robert W. Harper [1992] reformulated the method.

Truth and reflected proof

In order to compare how KTT and CTT handle notions of truth and use orders in defining truth, we look at the results in the 1990 article "The semantics of reflected proof" [Allen *et al.*, 1990] which builds a reflection of the type of proofs into the object language by taking a fixed point between the object language L, which is CTT, and its meta-language, which is the ClassicML [Gordon *et al.*, 1979] formalization of CTT described in the Nuprl book, *Implementing Mathematics* [Constable *et al.*, 1986]. The proof mechanism for Nuprl is essentially Milner's LCF mechanism [Gordon *et al.*, 1979], but using *tactic-tree proofs* [Griffin, 1988]. We see immediately a similarity with our presentation of Kripke's KTT, namely, we must start with a reflection of the syntax into the object logic In the case of CTT we must also reflect the notion of *sequent* and *proof.*

Definition 6.6. In all LCF-like proof systems [Gordon *et al.*, 1979], proofs are built using tactics. Indeed, in CTT proofs are tactic-trees [Griffin, 1988], and they use two languages, one for the proof object and one for the tactic program. We can say that justifications involve primitive rules, coded in the object language, and tactics, written in ML, the meta logic.

$$justification = rule + tactic.$$

But if programs can be constructed from proofs, then perhaps one language will suffice, and the tactics can be programs in the object language extracted from proofs. The question we consider here is whether there can be a single notion of tactic-tree proofs, in which the tactics are the programs coming from the tactic-trees as *program principle*, i.e.

$$justification = rule + tactic - from(proof).$$

We will show how this is done in [Allen *et al.*, 1990].

A basic concept for all that follows is the notion that a type of the language L represents a mathematical class S defined on the terms of L.

Definition 6.7. Given a relation t *reps* s between terms t and members of a class of objects S, we say that a type T of the language L *represents* class S (via *reps*) iff there are computable maps, *rep* from objects of S to terms such that *rep*(t) *reps* t and *unrep* a partial function from objects of S to terms such that *unrep*(*rep*$(t) = t$.

In the case of the class of terms itself from the meta-language, there is a *natural representation* relation that can be designed into the object language. To do this, let the class of terms be the least class containing variables and operation names, and such that if t_i are terms and *op* an *operator name*, then $op(\overline{v}_1.t_1;\ldots;\overline{v}_n.t_n)$ is a term where the $\overline{v}_i$ are lists of variables. We say that t_i is a subterm; the $\overline{v}_i$s are binding variables with scope t_i and they bind the variables occurring in t_i. Operator names are included among the terms in anticipation of the need for a term to represent each operator name. This is the approach taken by Howe [1987].

Let us suppose that there are types in L that represent variables and operator names; call them Var and Op. Then the *natural representation* of a term $op(\overline{v}_1.t_1;\ldots;\overline{v}_n.t_n)$ is a pair $\langle OP, (\langle\overline{V}_1.T_1\rangle,\ldots,\langle\overline{V}_1.T_1\rangle)\rangle$ where OP represents *op*, $\overline{V}_i$ represents $\overline{v}_i$ and T_i represents t_i. Using this we can show that if L admits recursive types, as CTT does, then there is a type, call it $Term$, which represents $term$.

Definition 6.8. $Term \stackrel{def}{=} rec\ t.Op \mid Var \ \mid (Op \times ((Var\ list \times t)list))$.

The informal concept of proof that we start with is called a *primitive proof*; it is essentially a finite tree of sequents justified by rules or taken as premises. A *primitive rule* is basically a function from a sequent s_o to a list of premises $s_1,\ldots,s_n$ such that if $s_1,\ldots,s_n$ are true, then s_o is true. We say that a primitive rule justifies s from l provided the rule applied to s yields the sequents at the roots of the subproofs in the list l. A premise of a proof is an unproved sequent at a leaf.

It is easy to see that in a language L with recursive types, the class of primitive proofs can be represented by a type, namely:

Definition 6.9.

$$\begin{array}{rcl}
PrimProofTree & \stackrel{def}{=} & rec\ p.Sequent \times (\{1\} \mid PrimRule \times p\ list \\
IsPrimProof & \stackrel{def}{=} & rec\text{-}ind(P; h, \langle s,j,l\rangle.map\text{-}list(h;l) \times case; j\ of \\
 & & \quad inl(l) \rightarrow l = nil \in Sequent\ list; \\
 & & \quad inr(l) \rightarrow PrimJustifies(s;p;roots(l)) \\
PrimProof & \stackrel{def}{=} & \{p : PrimProofTree \mid IsPrimProof(p)\},
\end{array}$$

where $PrimRule$ is a type containing representatives of the primitive rules, and $PrimJustifies(g;j;s)$ is a predicate that represents the primitive justification relation, that is, it is inhabited exactly when the sequent represented by g is justified by the list of sequents represented by s using the primitive rule represented by j.

In building a *tactic_tree*, there are different ways to check that a tactic produces the right thing. We ask whether such proofs of tactic correctness could be part of this proof type. That is, can we extend the justification to include

Definition 6.10.

$$\begin{array}{rcl}
justification & = & leaf + prim_rule + \{t : term \mid unreps(t) \in proof\} \\
 & & + reflection_rule
\end{array}$$

The reflection rule should require that we build a representation of the class of proofs whose goal is a representation of the sequent we are trying to prove, say s. That is, we want to require:

Definition 6.11.

$$\exists p : Proof.goal(p) = reps(s)$$

where $Proof$ is intended to be a type of the object logic (L_type) which represents the class of proofs which we are trying to define from outside of L. $Proof$ is an internal version of the very type we are now attempting to design. We already know a great deal about this type, and we could write down a formal definition of most of it except for the clause we are examining now.

Before writing the clause, we note that one understanding of this reflection proceeds in levels, as in Kripke. We first reflect those proofs which do not use any reflection in them, say $proof_0$. Then having grasped that type, and defined it internally, we understand a new level of proof, say $proof_1$. We can now reflect these proofs to obtain $proof_2$ and so forth. The reflection rule will be indexed by an integer i which provides this level. We will give a non-hierarchical definition below.

Definition 6.12. Notice that the reflected proof may be incomplete, so it will have subgoals. These subgoals are the goals of the subproofs in the list l that prove them. So the requirement on the subgoals of a proof step to prove s by reflection is that the first subgoal has the special form:

$$\begin{array}{lrcl} \exists p : Proof. & goal(p) & = & rep(s) \\ & & \& & \\ & premises(p) & = & \text{"roots of the remaining subproofs of } l\text{" \&} \\ & & & \text{"all uses of reflection in } p \text{ are of level } j < i.\text{"} \end{array}$$

The name of the reflection rule will have an integer to denote the level and a term t to represent the subgoals generated by the rule application.

Definition 6.13.

$$\begin{array}{rcl} justification & = & leaf + prim\_rule + \{t : term \mid unreps(t) \in proof\} \\ & & +\ (N \times term) \end{array}$$

where if j is a reflection rule, that is $\langle i, t \rangle$, then the goal of the first subproof of l, call it P_o, has the form:

$$\begin{array}{rl} \exists p : & Proofroot(p) = rep(s)\ \& \\ & premises(p) = goals(rest(l))\ \& \\ & reflection\ level(p) < i. \end{array}$$

We can see this more clearly if we depict the relationship between s, j and l in this fourth clause of j.

$$\begin{array}{ll} & s \quad \langle i,t\rangle \\ \exists p:PF.\ \ root(p)=rep(s) & s_1 \quad \ldots \quad s_n \\ \quad \& \ premise(p)= & P_1 \quad \ldots \quad P_n \\ \quad (s_1,\ldots,s_n) & \\ \quad \& \ reflection\_level(p)<i & \\ P_0 & \end{array}$$

Proof is a type in L which we can express except for the clause that mentions Proof itself. To deal with this *self-reference* we will take a fixed point and to facilitate that we replace the entire first subgoal of the reflection rule, the one mentioning Proof, by a piece of syntax, the expression PF.

Definition 6.14.

$$\begin{array}{lll} ProofTree & \stackrel{def}{=} & rec\ p.Sequent \\ & & \times \\ & & (\{1\} \mid PrimRule \mid \{t:Term \mid unreps(t)\in p\} \mid (int\times Term)) \\ & & \times \\ & & p\ list \end{array}$$

$$\begin{array}{lll} IsProof & \stackrel{def}{=} & rec\text{-}ind(P;h,\langle s,j,l\rangle.map\text{-}list(h;l)\times case;j\ of \\ & & \quad in1(l)\to l=nil\in Sequent\ list; \\ & & \quad in2(l)\to PrimJustifies(s;p;roots(l)) \\ & & \quad in3(l)\to s=root(unreps(r))\in Sequent\times \\ & & \qquad\qquad l=frontier(unreps(r))\in Sequent\ list; \\ & & \quad in4(l)\to cons(\langle nil,PFof(s;t;i;PF)\rangle;unreps(t))=l \\ & & \qquad\qquad\qquad \in Sequent\ list \end{array}$$

$$Proof \stackrel{def}{=} \{p:PrimProofTree \mid IsPrimProof(p)\}.$$

To be more precise about proofs and proof representation, notice that in the reflection rule used in proofs as described earlier, a term $Proof$ was used in the reflection subgoal. The intention was that it should be a term which is a type representing this very class of proofs. Here we shall construct such a term directly as a *syntactic fixed point.*

We parameterize the notion of proof by a term to be used where $proof$ was used above; call this family of classes $proof_R$. Our aim is to find a term R which is a type representing the class $proof_R$ (thereafter using $proof_R$ as our proof class). Now, consider the definition of the term $Proof$ given above. It can be parameterized by a term analogously to the parameterisation of the class of proofs, namely, replace PF by another term r; let us call this family of terms $Proof_r$. $Proof_r$ represents the class $Proof_R$ whenever r represents the term R.

Consider the family of terms $Proof_r$; it should be clear that we can find a term q such that whenever a represents term r, then the application $q(a)$ represents the term $Proof_r$. There is a term AP such that $AP(a)(b)$ represents the application $c(d)$ whenever a and b represent the terms c and d. Also the standard representation function on terms can be represented by a term Rep such that $Rep(a)$ represents the standard representation of b if a represents b.

Now for a familiar construction. Let B be the term $\lambda u.q(AP(u)(Rep(u)))$. Let b be the standard representation of term B, then $B(b)$ has the same value as $q(AP(u)(Rep(u)))$. Let b be the standard representation of term B. Then $B(b)$ has the same value as $q(AP(u)(Rep(u))$, which represents $Proof_{B(b)}$, hence, so does $B(b)$. Letting the term $Proof$ be $B(b)$, and letting the class $proof$ be $Proof_{Proof}$, we can now state that:

Theorem 6.15. *$Proof$ represents the class $proof$.*

Another less mysterious approach to defining a proof type which contains a representation of itself is to exploit the library. The library can contain specifications for how to expand defined terms to primitive terms. The truth of a sequent is then parameterized by libraries to mean that the sequent is simply true after all terms of the sequent have been expanded using that library. Representation is parameterized by libraries so that a term represents whatever the term it expands to represents. Because representation will be relevant to some of the rules, the class of proofs will become parameterized by libraries. Let $proof_\Lambda$ be this family of classes.

Now, what we shall do is let the term PF stand for a fixed definable non-primitive in the definition of the term $Proof$ above. Hence we have the following theorem:

Theorem 6.16. With respect to any library Λ:

1. $Proof$ represents $proof_\Lambda$.

2. If PF expands to the standard representation of the term $Proof$, PF represents $proof_\Lambda$.

The following definition and theorems are to be considered under both the above interpretations of $proof$ and $truth$; viz., $proof$ as defined for theorem 6.15 with simple truth and $proof$ and $truth$ implicitly parameterized by libraries as in theorem 6.16.

Definition 6.17. A proof is valid iff the goal sequent is true whenever the premise sequents are.

Now we have the following basic results:

Theorem 6.18.

1. All proofs are valid.

2. Without the level restriction, the class proof is not valid.

Here we can adapt Löb's theorem to prove a contradiction.

Theorem 6.19. Every proof can be reduced to a primitive proof.

This metatheorem could prove quite useful because it allows us to keep all proofs at level o if we wish.

One might want to take Theorem 6.19 as a reducibility axiom for proofs. But the fact is that adding it as an axiom makes it false!

We could use this reflection mechanism to increase the power of the logic, e.g., by adding rules asserting the consistency of the logic. We have chosen to provide a logically conservative extension with considerable practical benefits which we discuss next.

In Nuprl tactics are used to build significant parts of a proof tree. Sometimes they are costly to run because they are building a large piece of data; this often happens in circumstances in which it is easy to discover beforehand whether or not the tactic will succeed. For example, there is a type checking tactic, called *get_type*, which will prove all sequents of the form $H \vdash T$ in $Type$, for example $A : Type \vdash A \times B$ in $Type$, as long as the type T is of restricted form, say first-order. In the reflected system we can express this as a metatheorem, assuming that the relation First-order and Well-formedness are defined.

Theorem 6.20. $\forall s : sequent \,.\; First - order(S) \Rightarrow Well - formedness(s) \Rightarrow$
$\exists p : Proof.goal(p) = s \;\&\; complete(p)$

Now in the course of any proof in which the well-formedness goal $H \vdash T$ in $Type$ appears, we can prove it by reflection. The cost of checking that a type is first-order and recognizing a well-formedness goal will often be vastly simpler than proving the theorem; checking will be linear in size of T whereas building the proof is quadratic. The reflection mechanism will enable users of tactic-based proof assistants with reflection to provide efficient type checking for those subclasses of terms for which a good algorithm is discovered, e.g., for ordinary logic.

These results use a syntactic fixed point to avoid a tower of levels while Kripke used a semantic fixed point to avoid them. We have not considered in detail a formalization of the notion of truth used in this discussion, but it is possible to formalize truth in universe $\mathbb{U}_i$ at the next level, $\mathbb{U}_{i+1}$.

It is less clear whether we need to *reflect* the structure of terms and proofs from the meta logic to accomplish semantic results. It might be possible to take the approach of Lisp and include terms directly in the object logic along with mechanisms like quoting to prevent operators from reducing terms when they are viewed as syntax. Our use of the type *Base* shows that such an idea is sensible in computational theories. For CTT we can introduce the object type Term with syntactic equality and then define *Base* as a quotient type over Term or over Term quotiented by alpha-equality.

7 CONCLUSIONS

At the beginning of the 20th century, the paradoxes led to many new formulations of logical systems and an amazing variety of ideas and approaches. Later on, some of these ideas were abandoned when they should not have been. Even more, some of the ideas proposed were found later to contribute nothing to the solution of the paradoxes. For example, even though ZF set theory uses the foundation axiom, it is clear now that it is the separation rather than the foundation axiom which was responsible for the avoidance of the paradoxes.

Our standpoint in this chapter is not to defend one line against another. Rather, we aim to clarify the different notions and philosophies assumed in the foundation of logic, mathematics, and computing. In this chapter, our chosen notion is that of Russell's orders as found in the famous Ramified Theory of Types RTT. This idea appears also in the different foundational approaches to informatics and computer science. Russell, whose contribution to modern logic is historical, avoided the self-reference paradox (that he himself discovered) by adopting two layers: types and orders.

Later it was found that orders contributed nothing to the avoidance of the paradox and Ramsey's work led to the abandonment of Russell's orders. Later still with CTT it was realized that self-reference in the computation system could be safely allowed in the logic using implicit typing. It is not clear to us whether Russell did actually know at the time that orders do not contribute to the avoidance of the paradox. We believe however that his intuition of using orders (as well as types) is a solid one and we have seen this intuition being repeated in many predicative style logics. In this chapter, we show that Russell's orders come back in Kripke's account of levels of truths and are also present in Martin-Löf's type theory and in Computational Type Theory implemented by the Nuprl proof assistant. Of course the word "orders" is not used by Kripke, Martin-Löf, and Constable. Our study however shows that formally representing (with orders) the informal hierarchies of these systems is informative about these hierarchies, about the systems themselves and about the philosophies behind them, especially about notions of truth and proof.

We conclude this chapter, as so often has been concluded in logic, mathematics, and natural language semantics that Russell's theory of types is too restrictive. However, the usual objections against RTT in logic and mathematics is the use of orders since Ramsey [1926] and Hilbert and Ackerman [1928] had given motivations for leaving out these orders (though Weyl [1918] tried to build-up mathematics within RTT). Here we show that the situation is completely different. Orders in RTT and truth levels in KTT go hand in hand; while the types do not appear in KTT. This establishes KTT as the dual to Church's Simple Type Theory STT which removes the orders from RTT. In particular we show that

1. A proposition of RTT of order n is true if and only if it is true at level n in Kripke's truth hierarchy (Theorem 4.9).

2. The truth of some propositions of order n of RTT cannot be established in KTT at a level of truth hierarchy smaller than n (Theorem 4.11). Yet for some other propositions, it can be established at an earlier level (Example 4.12).

We also saw that Russell's theory has many restrictions. On the one hand, all propositional functions of RTT can be coded in Kripke's theory; on the other hand there are formulas of Kripke's theory that cannot be expressed in RTT, respecting both hierarchies. Our conclusion is that, contra Ramsey, it is the restriction of the *mixture of orders and types* on predication rather than *order* restriction on quantification *alone* that accounts for the very restrictive nature of RTT.[16]

Kripke's theory has a transfinite hierarchy of orders whereas Russell did not investigate such transfiniteness. It would be interesting hence to see how far one can build types in Russell's theory and what properties would hold at such level.

We concluded that some order-n properties of RTT get their truth-value only at level n of KTT whilst others get it already at an earlier level. This divides propositions into two classes, and an accurate description of these classes may be interesting.

As to the question of Kripke being more liberal in that any well-formed sentence can be expressed but its truth value may not be calculated (think of the paradoxical sentences), one may compare this approach to the implicit typing of Curry's Type Theory [Curry and Feys, 1958; Curry *et al.*, 1972] where self-referential sentences may be expressed but are not typeable and to CTT where implicit typing allows controlled self-reference. Hence, even though we said that KTT is the dual of STT, it may be the twin brother of CTT where only truth or falsehood of *typeable* terms can be determined.

BIBLIOGRAPHY

[Abramsky *et al.*, 1992] S. Abramsky, D. M. Gabbay, and T. S. E. Maibaum, eds. *Handbook of Logic in Computer Science, Volume 2: Background: Computational Structures*, Oxford University Press, 1992.

[Allen *et al.*, 1990] S. F. Allen, R. L. Constable, D. J. Howe, and W. E. Aitken. The semantics of reflected proof. In *LICS90*, pp. 95–105. IEEE Computer Society, 1990.

[Allen *et al.*, 2006] S. Allen, M. Bickford, R. L. Constable, R. Eaton, C. Kreitz, L. Lorigo, and E. Moran. Innovations in computational type theory using Nuprl. *Journal of Applied Logic*, 4:428–469, 2006.

[Allen, 1987] S. F. Allen. A non-type-theoretic definition of Martin-Löf's types. In Gries [1987], pp. 215–224.

[Barendregt, 1992] H. P. Barendregt. Lambda calculi with types. In [Abramsky *et al.*, 1992], pp. 117–309. Oxford University Press, 1992.

[Basin and Howe, 1991] D. A. Basin and D. J. Howe. Some normalization properties of Martin-Löf's type theory, and applications. In *Theoretical Aspects of Computer Software, International Conference TACS '91*, Sendai, Japan, pp. 475–494. LNCS Vol 526, Springer-Verlag, 1991.

[16]This is important and takes on an added significance when seen in the context of the logicisation of second order arithmetic in a type free first-order logic utilizing Kripke-Gilmore models which realises the hope of Russell's earlier type free substitutional theory.

[Bates and Constable, 1985] J. L. Bates and R. L. Constable. Proofs as programs. *ACM Transactions on Programming Languages and Systems*, 7:53–71, 1985.
[Beeson, 1985] M. J. Beeson. *Foundations of Constructive Mathematics.* Springer-Verlag, 1985.
[Bertot and Castéran, 2004] Y. Bertot and P. Castéran. *Interactive Theorem Proving and Program Development; Coq'Art: The Calculus of Inductive Constructions.* Texts in Theoretical Computer Science. Springer-Verlag, 2004.
[Bickford and Constable, 2008] M. Bickford and R. L. Constable. Formal foundations of computer security. In *Formal Logical Methods for System Security and Correctness*, volume 14, pp. 29–52, 2008.
[Bickford, 2010] M. Bickford. Automated proof of authentication protocols in a logic of events. *VERIFY-2010, 6th International Workshop, Edinburgh*, 2010.
[Bishop, 1967] E. Bishop. *Foundations of Constructive Analysis.* McGraw Hill, New York, 1967.
[Chirimar and Howe, 1991] J. Chirimar and D. J. Howe. Implementing constructive real analysis: a preliminary report. In *Symposium on Constructivity in Computer Science*, pp. 165–178. Springer-Verlag, 1991.
[Church, 1940] A. Church. A formulation of the simple theory of types. *The Journal of Symbolic Logic*, 5:56–68, 1940.
[Church, 1941] A. Church. *The Calculi of Lambda Conversion.* Princeton University Press, 1941.
[Church, 1976] A. Church. Comparison of Russell's resolution of the semantic antinomies with that of Tarski. *The Journal of Symbolic Logic*, 41:747–760, 1976.
[Constable and Crary, 2001] R. L. Constable and K. Crary. Computational complexity and induction for partial computable functions in type theory. In W. Sieg, R. Sommer, and C. Talcott, eds., *Reflections on the Foundations of Mathematics: Essays in Honor of Solomon Feferman*, Lecture Notes in Logic, pp. 166–183. Association for Symbolic Logic, 2001.
[Constable and Smith, 1993] R. L. Constable and S. F. Smith. Computational foundations of basic recursive function theory. *Theoretical Computer Science*, 121:89–112, December 1993.
[Constable *et al.*, 1986] R. L. Constable, S. F. Allen, H. M. Bromley, W. R. Cleaveland, J. F. Cremer, R. W. Harper, D. J. Howe, T. B. Knoblock, N. P. Mendler, P. Panangaden, J. T. Sasaki, and S. F. Smith. *Implementing Mathematics with the Nuprl Proof Development System.* Prentice-Hall, New Jersey, 1986.
[Coquand and Huet, 1988] T. Coquand and G. P. Huet. The calculus of constructions. *Information and Computation*, 76:95–120, 1988.
[Curry and Feys, 1958] H. B. Curry and R. Feys. *Combinatory Logic I.* Studies in Logic and the Foundations of Mathematics. North-Holland, Amsterdam, 1958.
[Curry *et al.*, 1972] H. B. Curry, J. R. Hindley, and J. P. Seldin. *Combinatory Logic II.* Studies in Logic and the Foundations of Mathematics. North-Holland, Amsterdam, 1972.
[Dummett, 2000] M. Dummett. *Elements of Intuitionism.* Oxford logic guides 39. Clarendon Press, 2nd edition, 2000.
[Feferman, 1964] S. Feferman. Systems of predicative analysis. *The Journal of Symbolic Logic*, 29:1–30, 1964.
[Feferman, 1984] S. Feferman. Toward useful type-free theories I. *The Journal of Symbolic Logic*, 49:75–111, 1984.
[Frege, 1892] G. Frege. *Grundgesetze der Arithmetik, begriffschriftlich abgeleitet*, volume I. Pohle, Jena, 1892. Reprinted 1962 (Olms, Hildesheim).
[Gordon and Melham, 1993] M.J.C. Gordon and T.F. Melham, eds. *Introduction to HOL: A Theorem Proving Environment for Higher Order Logic.* Cambridge University Press, London/New York, 1993.
[Gordon *et al.*, 1979] M. Gordon, R. Milner, and C. Wadsworth. *Edinburgh LCF: a mechanized logic of computation*, volume 78 of *Lecture Notes in Computer Science.* Springer-Verlag, NY, 1979.
[Gries, 1987] D. Gries, ed. *Proceedings of the 2nd IEEE Symposium on Logic in Computer Science.* IEEE Computer Society Press, June 1987.
[Griffin, 1988] T. Griffin. Notational definition - a formal account. In *LICS88*, pp. 372–383, Edinburgh, UK, 1988. IEEE Computer Society Press.
[Harper and Pollack, 1991] R. Harper and R. Pollack. Type checking with universes. *Theoretical Computer Science*, 89:107–136, 1991.
[Harper, 1992] R. Harper. Constructing type systems over an operational semantics. *Journal of Symbolic Computation*, 14:71–84, 1992.

[Heijenoort, 1967] J. van Heijenoort, ed. *From Frege to Gödel: A Source Book in Mathematical Logic, 1879–1931*. Harvard University Press, Cambridge, Massachusetts, 1967.

[Heyting, 1975] A. Heyting, ed. *Brouwer: Collected Works*, volume 1. North-Holland, Amsterdam, 1975.

[Hickey *et al.*, 2003] J. Hickey, A. Nogin, R. L. Constable, B. E. Aydemir, E. Barzilay, Y. Bryukhov, R. Eaton, A. Granicz, A. Kopylov, C. Kreitz, V. Krupski, L. Lorigo, S. Schmitt, C. Witty, and X. Yu. Metaprl - a modular logical environment. In D. A. Basin and B. Wolff, eds., *TPHOLs*, volume 2758 of *Lecture Notes in Computer Science*, pp. 287–303. Springer, 2003.

[Hilbert and Ackermann, 1928] D. Hilbert and W. Ackermann. *Grundzüge der Theoretischen Logik.* Die Grundlehren der Mathematischen Wissenschaften in Einzeldarstellungen, Band XXVII. Springer Verlag, Berlin, first edition, 1928.

[Howe, 1987] D. J. Howe. The computational behaviour of Girard's paradox. In Gries [1987], pp. 205–214.

[Howe, 1989] D. J. Howe. Equality in lazy computation systems. In *LICS89*, pp. 198–203, Asilomar Conference Center, Pacific Grove, California, June 1989. IEEE Computer Society Press.

[Howe, 1996] D. J. Howe. Semantic foundations for embedding HOL in Nuprl. In M. Wirsing and M. Nivat, eds., *Algebraic Methodology and Software Technology*, volume 1101 of *Lecture Notes in Computer Science*, pp. 85–101. Springer-Verlag, Berlin, 1996.

[Howe, 1998] D. J. Howe. A type annotation scheme for Nuprl. In *Theorem Proving in Higher-Order Logics*. Springer, 1998.

[Jackson, 1994] P. B. Jackson. *The Nuprl Proof Development System, Version 4.1 Reference Manual and User's Guide*, Cornell University, Ithaca, NY, 1994.

[Jackson, 1995] P. B. Jackson. *Enhancing the Nuprl Proof Development System and Applying it to Computational Abstract Algebra.* PhD thesis, Cornell University, Ithaca, New York, 1995.

[Kahn, 1987] G. Kahn. Natural semantics. In G. Vidal-Naquet F. Brandenburg and M. Wirsing, eds. *STACS '87*, volume 247 of *Lecture Notes in Computer Science*, pp. 22–39. Springer-Verlag, Berlin, 1987.

[Kamareddine and Laan, 1996] F. Kamareddine and T. Laan. A reflection on Russell's ramified types and Kripke's hierarchy of truths. *Journal of the Interest Group in Pure and Applied Logic*, 4:195–213, 1996.

[Kamareddine and Laan, 2001] F. Kamareddine and T. Laan. A correspondence between Martin-Löf type theory, the ramified theory of types and pure type systems. *Journal of Logic, Language and Information*, 10:375–402, 2001.

[Kamareddine *et al.*, 2004] F. Kamareddine, T. Laan, and R. Nederpelt. *A Modern Perspective on Type Theory.* Kluwer Academic Publishers, 2004.

[Kopylov, 2003] A. Kopylov. Dependent intersection: A new way of defining records in type theory. In *Proceedings of 18th IEEE Symposium on Logic in Computer Science*, pp. 86–95, 2003.

[Kreitz, 2003] C. Kreitz. *The Nuprl Proof Development System, Version 5.1 Reference Manual and User's Guide.* Cornell University, Ithaca, NY, February 2003.

[Kreitz, 2004] C. Kreitz. Building reliable, high-performance networks with the nuprl proof development system. *Journal of Functional Programming*, 14:21–68, 2004.

[Kripke, 1975] S. Kripke. Outline of a theory of truth. *Journal of Philosophy*, 72:690–716, 1975.

[Laan and Nederpelt, 1996] T. Laan and R. P. Nederpelt. A modern elaboration of the Ramified Theory of Types. *Studia Logica*, 57:243–278, 1996.

[Laan, 1997] T. Laan. *The Evolution of Type Theory in Logic and Mathematics.* PhD thesis, Eindhoven University of Technology, 1997.

[Liu *et al.*, 1999] X. Liu, C. Kreitz, R. van Renesse, J. J. Hickey, M. Hayden, K. Birman, and R. L. Constable. Building reliable, high-performance communication systems from components. In D. Kotz and J. Wilkes, eds., *17th ACM Symposium on Operating Systems Principles (SOSP'99)*, volume 33(5) of *Operating Systems Review*, pp. 80–92. ACM Press, December 1999.

[Martin-Löf, 1975] P. Martin-Löf. An intuitionistic theory of types: predicative part. In H. E. Rose and J. C. Shepherdson, eds., *Logic Colloquium '73*, volume 80 of *Studies in Logic and the Foundations of Mathematics*, pp. 73–118, Amsterdam, 1975. North-Holland.

[Martin-Löf, 1982] P. Martin-Löf. Constructive mathematics and computer programming. In *Proceedings of the Sixth International Congress for Logic, Methodology, and Philosophy of Science*, pp. 153–175, Amsterdam, 1982. North Holland.

[Martin-Löf, 1984] P. Martin-Löf. *Intuitionistic Type Theory.* Studies in Proof Theory. Bibliopolis, Napoli, 1984.

[Mendler, 1987] P. F. Mendler. Recursive types and type constraints in second-order Lambda calculus. In D. Gries, ed., *Proceedings of the 2nd IEEE Symposium on Logic in Computer Science*, pp. 30–36, IEEE Computer Society Press, 1987.

[Montague, 1973] R. Montague. The proper treatment of quantification in ordinary English. In J. Hintikka, J. M. E. Moravcsik, and P. Suppes, eds., *Approaches to Natural Language.* Dordrecht, 1973.

[Murthy, 1991] C. Murthy. An evaluation semantics for classical proofs. In *LICS91*, pp. 96–109, Vrije University, Amsterdam, The Netherlands, July 1991. IEEE Computer Society Press.

[Nordström *et al.*, 1990] B. Nordström, K. Petersson, and J. M. Smith. *Programming in Martin-Löf's Type Theory.* Oxford Sciences Publication, Oxford, 1990.

[Plotkin, 1981] G. D. Plotkin. A structural approach to operational semantics. Technical Report DAIMI-FN-19, Aarhus University, Aarhus University, Computer Science Department, Denmark, 1981.

[Ramsey, 1926] F. P. Ramsey. The foundations of mathematics. *Proceedings of the London Mathematical Society,* 2nd series, 25:338–384, 1926.

[Ranta, 1994] A. Ranta. *Type-theoretical grammar.* Oxford Science Publications. Clarendon Press, Oxford, England, 1994.

[Russell, 1903] B. Russell. *The Principles of Mathematics.* Allen & Unwin, London, 1903.

[Russell, 1908] B. Russell. Mathematical logic as based on the theory of types. *American Journal of Mathematics*, 30:222–262, 1908. Also in [Heijenoort, 1967], pp. 150–182.

[Scott, 1976] D. S. Scott. Data types as lattices. *SIAM J. Comput.*, 5:522–587, 1976.

[Tarski, 1936] A. Tarski. Der Wahrheitsbegriff in den formalisierten Sprachen. *Studia Philosophica*, 1:261–405, 1936. German translation by L. Blauwstein from the Polish original (1933) with a postscript added.

[Terlouw, 1989] J. Terlouw. Een nadere bewijstheoretische analyse van GSTT's. Technical report, Department of Computer Science, University of Nijmegen, 1989.

[The Coq Development Team, 2004] The Coq Development Team. *The Coq Proof Assistant Reference Manual – Version V8.0*, April 2004.

[Weyl, 1918] H. Weyl. *Das Kontinuum.* Veit, Leipzig, 1918. German; also in: Das Kontinuum und andere Monographien, Chelsea Publishing Company, New York, 1960.

[Whitehead and Russell, 1910, 1927] A. N. Whitehead and B. Russell. *Principia Mathematica, volumes I,II,III.* Cambridge University Press, 1910, 1927. All references are to the first volume, unless otherwise stated.

[Zermelo, 1908] E. Zermelo. Untersuchungen über die Grundlagen der Mengenlehre. *Mathematische Annalen*, 65:261–281, 1908.

INDEX

CPSIA information can be obtained at www.ICGtesting.com
Printed in the USA
BVOW040548270212

283694BV00006B/1/P

9 780444 516213